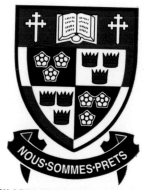

MODERN PHOSPHONATE CHEMISTRY

MODERN PHOSPHONATE CHEMISTRY

PHILIPPE SAVIGNAC
BOGDAN IORGA

CRC PRESS

Boca Raton   London   New York   Washington, D.C.

Cover Art: The front cover shows the X-ray structure of 1-cyanomethylene*bis*(5,5-dimethyl-2-oxo-1,3,2-λ^5-dioxaphosphorinane), representation obtained using ORTEP-3 for Windows (Farrugia, L.G., *J. Appl. Cryst.*, 30, 565, 1997).

Library of Congress Cataloging-in-Publication Data

Savignac, Philippe, 1939-
  Modern phosphonate chemistry / by Philippe Savignac and Bogdan Iorga.
      p. cm.
Includes bibliographical references and index.
  ISBN 0-8493-1099-7
  1. Phosphonates. I. Iorga, Bogdan, 1975- II. Title.
  QD305.P46S38 2003
  547'.07--dc21   2003046162

Visit the CRC Press Web site at www.crcpress.com

Acknowledgments

Many people and organizations, the Centre National de la Recherche Scientifique (CNRS), and the Ecole Polytechnique (Palaiseau) have contributed to the completion of this book, and their efforts are wholeheartedly acknowledged. We are truly grateful for the financial support provided by Jean-Claude Bernier, Directeur du Département des Sciences Chimiques du CNRS, Maurice Robin, Directeur Général Adjoint Chargé de la Recherche à l'Ecole Polytechnique, and Roland Sénéor, Directeur des Relations Extérieures de l'Ecole Polytechnique. The staff of the Bibliothèque Centrale de l'Ecole Polytechnique (BCX) kindly provided continuous access to their collections during the writing and were extremely helpful in obtaining materials through interlibrary loan. Thanks are also due to Duncan Carmichael, Chargé de Recherche au CNRS (DCPH), for a number of comments and criticisms.

Philippe Savignac, Bogdan Iorga
March 2003

The Authors

Philippe Savignac, Ph.D., is Directeur de Recherche (CNRS) at the Ecole Polytechnique (Palaiseau). Dr. Savignac graduated as an Ingénieur of the ENSCT in 1963 and obtained his Ph.D. from the Université de Paris (Sorbonne) in 1968. He became an Attaché de Recherche (CNRS) in 1970 in the laboratory of Professor Henri Normant and Directeur de Recherche in 1976. In 1977 he joined the newly formed CNRS phosphorus chemistry center in Thiais. Since 1987 he has been at the Ecole Polytechnique. Dr. Savignac has worked in several areas of organic phosphorus chemistry. The majority of his work has concerned the organometallic chemistry of phosphorus with particular emphasis on the synthesis of new phosphonylated reagents, phosphoramidates, phosphonates, and α-halogenated phosphonates.

Bogdan Iorga, Ph.D., was born in 1975 in Ploiesti, Romania. He received his B.Sc. in chemistry in 1997 at the University of Bucharest working on the synthesis of carbonic anhydrase inhibitors with Professor Claudiu T. Supuran. In 1997 he joined the European Program at the Ecole Polytechnique, where he obtained M.Sc. (1998) and Ph.D. (2001) degrees on the electrophilic halogenation of phosphonates under the supervision of Dr. Savignac. He currently holds a postdoctoral position in Dr. Jean-Marc Campagne's group at the Institut de Chimie des Substances Naturelles (Gif sur Yvette), where he is investigating catalytic asymmetric reactions.

Preface

A century after their discovery, phosphonates have become very important compounds. They are recognized for their use and efficiency as reagents in organic synthesis and for their biological and industrial importance. The vast literature devoted to phosphonate chemistry reflects an exciting field with many opportunities for research and development. The value of these reagents stems from a combination of easy access and the reactions of anions on carbon adjacent to the phosphoryl group.

The electron-withdrawing properties of the phosphoryl group were explored and exploited in the very earliest studies. In 1927, A.E. Arbuzov and Dunin reported a study of the reaction of diethyl 1-(ethoxycarbonyl)methylphosphonate with sodium or potassium metal and showed that α-metallated phosphonate products react with alkyl halides to form α-substituted phosphonoacetates. This approach not only complemented the Michaelis–Arbuzov reaction but also provided access to a more diversified range of phosphonates. The development of phosphonate carbanions became more sophisticated with the introduction of powerful bases such as NaH and $NaNH_2$ and lithium reagents. Subsequently, two major synthetic advances revolutionized phosphonate chemistry.

The first of these was the discovery, in 1958, that phosphonate carbanions generated from functionalized phosphonates can effect olefination of aldehydes and ketones in a reaction similar to that described by Wittig. The efficiency of this reaction immediately made phosphonate carbanions competitive with Wittig reagents for the preparation of olefins. Except for the cases of cyanophosphonates and formylphosphonates, we do not discuss this Horner–Wadsworth–Emmons reaction, which has been covered admirably in several papers and books. For more information in this area, Johnson's 1993 text describing the preparation of phosphono ylides, their general nucleophilic behavior, reactions with carbonyl compounds, and a comparison with phosphonium ylides in the Wittig reaction is highly recommended.

The second major breakthrough was the report by Corey and Kwiatkowski, in 1966, that simple dialkyl alkylphosphonates react with butyllithium to give phosphonate carbanions. These may be converted into functionalized phosphonates by simple treatment with electrophiles and have great value for the synthesis of complex phosphonates and phosphonate analogues of natural phosphates. This methodology has been enlarged by the use of halogen–metal exchange reactions between butyllithium and halogenophosphonates to generate the phosphonate carbanions. Most of the recent advances in phosphonate chemistry are now based on this concept.

The purpose of this volume is to present a concise summary of the state of the art in phosphonate chemistry, with particular emphasis being given to carbanionic methodologies. It is divided into eight chapters, the 1-alkynylphosphonates, the silylphosphonates, the halogenophosphonates, the epoxyphosphonates, the formylphosphonates, the cyanophosphonates, the ketophosphonates, and the carboxyphosphonates, which treat the most important phosphonates and their applications in organic synthesis as a function of the nature of the groups present in the side chain. Overall, we attempt to give examples of fundamental early studies and comprehensive coverage of the more modern aspects in a way that will be useful to organic chemists and biochemists working in synthetic chemistry, rather than providing coverage for organophosphorus specialists alone. It is quite clear that carbanionic procedures occupy an important place in phosphonate chemistry and that they have evolved to the point where they are now probably the best approach to elaborated phosphonates; an example of this is the elegant carbanionic syntheses of β-ketophosphonates. It is also clear that functionalized phosphonates have an ever-increasing importance in organic synthesis as precursors of elaborated organic compounds, for the synthesis of complex biologically important phosphonates, and in the synthesis of aminophosphonic acids and derivatives. We trust that this volume will serve as a useful guide for the synthetic organic chemist in the design and execution of phosphonate chemistry.

Philippe Savignac, Bogdan Iorga
Ecole Polytechnique, Palaiseau
March 2003

Abbreviations

acac	acetylacetonate
Ad	adamantyl
AEP	2-aminoethylphosphonate
AEPA	2-aminoethylphosphonic acid
AIBN	2,2′-azobisisobutyronitrile
Alk	alkyl
AMPA	DL-α-amino-3-hydroxy-5-methylisoxazole-4-propionic acid
An	4-methoxyphenyl (anisyl)
APB	L-2-amino-4-phosphonobutanoic acid
Ar	aryl
Bn	benzyl
BOC-F	*tert*-butoxycarbonyl fluoride
BSA	*N,O-bis*(trimethylsilyl)acetamide
Bz	benzoyl
CAN	ceric ammonium nitrate
COD	1,4- or 1,5-cyclooctadiene
Cy	cyclohexyl
DAST	(diethylamino)sulfur trifluoride
DBU	1,8-diazabicyclo[5.4.0]undec-7-ene
DCC	dicyclohexylcarbodiimide
DEAD	diethyl azodicarboxylate
DET	diethyl tartrate
DIBAL-H	diisobutylaluminum hydride
DIBOC	di-*tert*-butyl dicarbonate
DIEA	diisopropylethylamine
DMA	*N,N*-dimethyl acetamide
DMAP	dimethylaminopyridine
DMD	dimethyldioxirane
DME	1,2-dimethoxyethane
DMF	*N,N*-dimethyl formamide
DMPU	*N,N′*-dimethylpropyleneurea
dppe	1,2-*bis*(diphenylphosphino)ethane
dppf	1,1′-*bis*(diphenylphosphino)ferrocene
EAA	excitatory amino acids
EDCI	1-ethyl-3-(3′-dimethylaminopropyl)carbodiimide
EWG	electron withdrawing group
HetAr	heteroaryl
HMDS	1,1,1,3,3,3-hexamethyldisilazane
HMPA	hexamethylphosphoramide
HMPT	hexamethylphosphorous triamide
HPP	(*S*)-2-hydroxypropylphosphonic acid

KDA	potassium diisopropylamide
KDO	3-deoxy-D-manno-2-octulosonic acid
LAH	lithium aluminum hydride
LDA	lithium diisopropylamide
LiTMP	lithium 2,2,6,6-tetramethylpiperidide
MCA	monochloroacetic acid
MCPBA	*meta*-chloroperbenzoic acid
MeCN	acetonitrile
MOM	methoxymethyl
MPTP	1-methyl-4-phenyl-1,2,3,6-tetrahydropyridine
Ms	methanesulfonyl (mesyl)
MS	molecular sieves
MTBD	7-methyl-1,5,7-triazabicyclo[4.4.0]dec-5-ene
MTO	methyl trioxorhenium
NADP	nicotinamide adenine dinucleotide phosphate
NBA	*N*-bromoacetamide
NBS	*N*-bromosuccinimide
NCS	*N*-chlorosuccinimide
Nf	nonaflyl
NFBS	*N*-fluorobenzenesulfonimide
NMDA	*N*-methyl-D-aspartic acid
PAAl	phosphonoacetaldehyde
PAla	β-phosphonoalanine
PCC	pyridinium chlorochromate
PDC	pyridinium dichromate
PDP	pyrrolidinopyridine
Pent	pentyl
PEP	phosphoenolpyruvate
PFA	phosphonoformic acid
PhF	9-(9-phenylfluorenyl)
Pht	phthalimido
PhTRAP	2,2″-*bis*[1-(diphenylphosphino)ethyl]-1,1″-biferrocene
PnPyr	phosphonopyruvate
PPA	polyphosphoric acid
PTOC	pyridine-2-thione-*N*-oxycarbonyl
Py	pyridine
Red-Al	sodium *bis*(2-methoxyethoxy)aluminum hydride
SAMP	(S)-(−)-1-amino-2-(methoxymethyl)pyrrolidine
SEM	2-(trimethylsilyloxy)ethoxymethyl
TASF	*tris*(dimethylamino)sulfonium difluorotrimethylsiliconate
TBAB	tetrabutylammonium bromide
TBAC	tetrabutylammonium chloride
TBAF	tetrabutylammonium fluoride
TBAS	tetrabutylammonium hydrogensulfate
TBAT	tetrabutylammonium triphenylsilyldifluoride
TBD	1,5,7-triazabicyclo[4.4.0]dec-5-ene
TBDMS	*tert*-butyldimethylsilyl

TBDPS	*tert*-butyldiphenylsilyl
TEBAC	triethylbenzylammonium chloride
Tf	trifluoromethanesulfonyl (triflyl)
TFA	trifluoroacetic acid
TFAA	trifluoroacetic acid anhydride
TFD	methyl(trifluoromethyl)dioxirane
THF	tetrahydrofurane
TMG	tetramethylguanidine
TMU	tetramethyl urea
Ts	4-toluenesulfonyl (tosyl)
TTMS	*tris*(trimethylsilyl)silane
UHP	urea hydrogen peroxide adduct

Contents

1   The 1-Alkynylphosphonates

The methods developed for the synthesis of 1-alkynylphosphonates use synthetic concepts such as charge affinity inversion and positive halogen abstraction. Four main reaction categories allow the preparation of 1-alkynylphosphonates: the "apparent" Michaelis–Arbuzov and Michaelis–Becker reactions, the propargyl phosphite–allenylphosphonate rearrangement, carbanionic displacements at quinquevalent phosphorus centers SNP(V),[1] and conversions of vinyl- to alkynylphosphonates by addition–elimination reactions. 1-Alkynylphosphonates possess remarkable potential, and cycloaddition reactions appear to offer very attractive synthetic procedures.[2]

1.1 SYNTHESIS OF 1-ALKYNYLPHOSPHONATES

1.1.1 MICHAELIS–ARBUZOV REACTIONS

The 1-bromoalkynes whose triple bond is conjugated to a benzene ring or another multiple bond react surprisingly easily with trialkyl phosphites. This Michaelis–Arbuzov-type rearrangement gives dialkyl 1-alkynylphosphonates in fair to good yields (48–67%, Scheme 1.1).[3] Chloro-, bromo-, and iodoalkylacetylenes not conjugated to an unsaturated system do not react in these conditions.

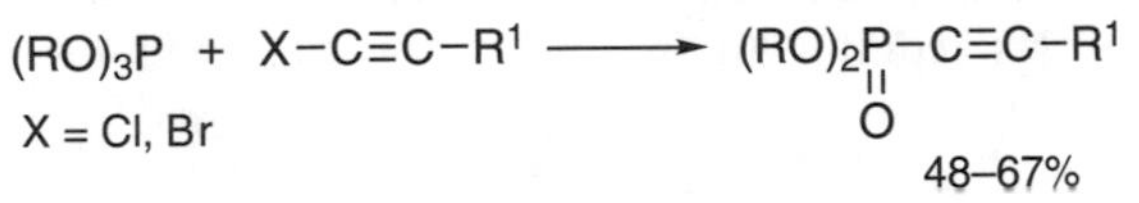

$$(RO)_3P \ + \ X-C{\equiv}C-R^1 \ \longrightarrow \ (RO)_2\overset{\text{II}}{\underset{O}{P}}-C{\equiv}C-R^1$$

X = Cl, Br

48–67%   (1.1)

The presence of electronegative atoms or groups increases the reactivity of the haloacetylene functionality, so that substituted haloacetylenes (X = Cl, Br) undergo an "apparent" Michaelis–Arbuzov reaction with trialkyl phosphites (Scheme 1.1) when R^1 is an electron-withdrawing or accommodating group [Ph,[4] C_6F_5,[5] R-C≡C,[6,7] CH_2=CH,[8,9] Cl,[4,10–13] Br,[12,14] $(EtO)_2$P(O),[10,15,16] SR,[17–19] $SiMe_3$,[20–22] $SnEt_3$[23,24]].

The Michaelis–Arbuzov rearrangement of dichloroacetylene proceeds under surprisingly mild conditions even in the cold (0°C) in Et_2O to give mainly the monosubstitution product. For example, the reaction of triethyl phosphite (1 eq) with a large excess of dichloroacetylene (5 eq) furnishes diethyl 2-chloroethynylphosphonate in 50% yield.[4] A better yield (90%) is achieved at –20°C (Scheme 1.2).[25] The chlorine atom in the resulting phosphonate, which is rather labile because of the polarization induced by the highly electron-withdrawing phosphoryl group, can undergo a further Michaelis–Arbuzov rearrangement with formation of tetraethyl acetylenediphosphonate in 52% yield.[4]

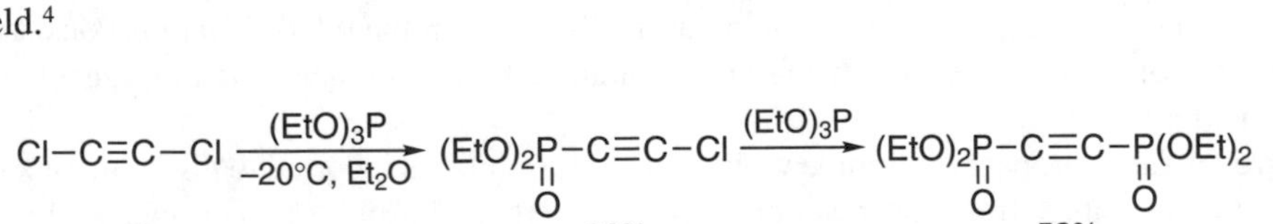

$$Cl-C{\equiv}C-Cl \xrightarrow[\text{–20°C, Et}_2O]{(EtO)_3P} (EtO)_2\overset{\text{II}}{\underset{O}{P}}-C{\equiv}C-Cl \xrightarrow{(EtO)_3P} (EtO)_2\overset{\text{II}}{\underset{O}{P}}-C{\equiv}C-\overset{\text{II}}{\underset{O}{P}}(OEt)_2$$

90%   52%   (1.2)

Dialkyl ethynylphosphonates are the simplest but not the easiest of these compounds to prepare. Several approaches had been described, but no convenient high-yield synthesis appeared until the introduction of the trimethylsilyl functionality as a protecting group for the acetylenic C–H linkage.

1

The synthesis of dialkyl ethynylphosphonates then involves an "apparent" Michaelis–Arbuzov reaction between chloroethynyltrimethylsilane and trialkyl phosphites under heating at reflux. The trimethylsilyl group is sufficiently electron accommodating to promote the Michaelis–Arbuzov reaction. The resulting dialkyl (trimethylsilyl)ethynylphosphonates, obtained in high yields (70–82%), are readily desilylated by 10% aqueous Na_2CO_3 to give the required dialkyl ethynylphosphonates in overall yields up to 77% (Scheme 1.3).[20,26]

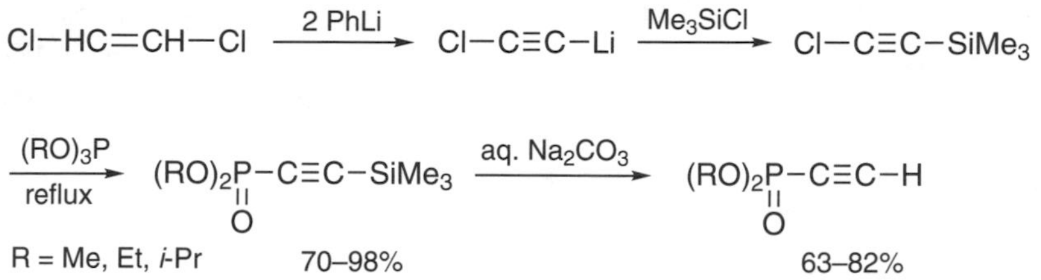

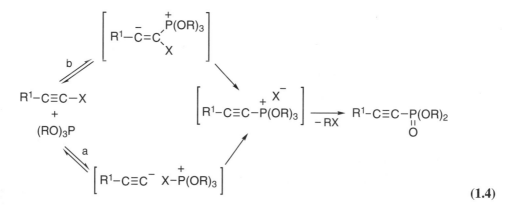

(1.3)

In view of the supposed inertness of halo-unsaturated substrates in the Michaelis–Arbuzov reaction, it is remarkable that arylalkynyl halides are more reactive than comparable alkyl, aryl, or vinyl halides toward trialkyl phosphites. The success of the Michaelis-Arbuzov reaction in the synthesis of dialkyl 1-alkynylphosphonates from 1-haloalkynes has provoked a number of mechanistic studies. Three sites, the halogen, α-carbon, and β-carbon have been investigated as the zone of the initial nucleophilic attack by trialkyl phosphites. Studies to date suggest that the reaction proceeds by at least two different mechanisms, and that probably the most important of these involves positive halogen-abstraction (path a) as outlined in Scheme 1.4.[21,27–29]

(1.4)

The second mechanism operating may well be the addition–elimination mechanism (path b, Scheme 1.4). The negative charge developing at the β-carbon in the intermediate is consistent with the requirement for electron-withdrawing or -accommodating R^1 groups. The first mechanism also satisfies this requirement.

The treatment of propargyl bromide with trialkyl phosphites is reported to give complex reaction mixtures. For example, from the reaction between triethyl phosphite and propargyl bromide at 90°C, diethyl 1-propynylphosphonate is isolated only in low yields (15%).[30–32]

Early transformation of trialkyl phosphites and haloalkynes into dialkyl 1-alkynylphosphonates was limited to haloalkynes in which R^1 was an electron-withdrawing group.[29,33] In 1990, two innovations worthy of merit were reported.[34,35] The first featured alkynylphenyliodonium salts[34] as alkynylating agents, and the second featured $NiCl_2$ as a catalyst.[35]

Alkynylphenyliodonium tosylates react with trialkyl phosphites in a formal Michaelis–Arbuzov reaction to give dialkyl 1-alkynylphosphonates (Scheme 1.5). It seems plausible that the first step in this reaction is a Michael addition of trialkyl phosphite to the electron-deficient β-carbon to form an ylide, which undergoes loss of iodobenzene and formation of vinylidene. The rearrangement of this carbene yields the alkynylphosphonium salt, which is transformed into dialkyl 1-alkynylphosphonate and ROTs according to the second step of the Michaelis–Arbuzov reaction.[36] For example, when an excess of neat trimethyl phosphite is added to solid (*tert*-butylethynyl)phenyliodonium tosylate at room temperature, the iodonium salt rapidly disappears to give, on workup, a 90% yield of dimethyl *tert*-butylethynylphosphonate. Similar treatment with triethyl and triisopropyl phosphites gives the corresponding diethyl and diisopropyl *tert*-butylethynylphosphonates in high yields.[34]

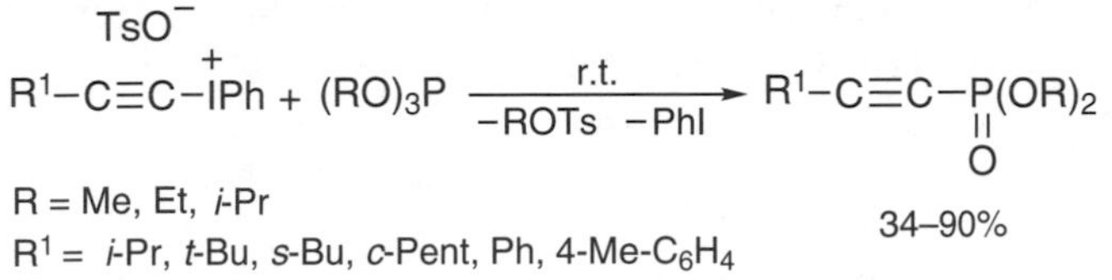

$$\text{R}^1\text{-C}\equiv\text{C-IPh} + (\text{RO})_3\text{P} \xrightarrow[-\text{ROTs} \ -\text{PhI}]{\text{r.t.}} \text{R}^1\text{-C}\equiv\text{C-P(OR)}_2$$

with TsO⁻ counterion and the product bearing a P=O group.

R = Me, Et, *i*-Pr

R¹ = *i*-Pr, *t*-Bu, *s*-Bu, *c*-Pent, Ph, 4-Me-C₆H₄

34–90%

(1.5)

The reactions of a range of alkynylphenyliodonium tosylates with trialkyl phosphites have been investigated. In general, the reactions are conducted with an excess of the neat trialkyl phosphite and the solid iodonium salt. It is clear that the trialkyl phosphite–induced cleavage of the alkynylphenyliodonium ion is remarkably selective in favor of the alkynyl group. This preparation of dialkyl 1-alkynylphosphonates from alkynylphenyliodonium tosylates is a useful complement to the traditional Michaelis–Arbuzov synthesis.[34] Simultaneously, it was reported that *tert*-butylchloroacetylene reacts with trialkyl phosphites at 110–160°C in the presence of a catalytic amount of anhydrous NiCl₂ to give the corresponding alkynyl esters in good yields: 73% for the ethyl ester and 75% for the isopropyl ester.[35]

1.1.2 MICHAELIS–BECKER REACTIONS

The first attempt to prepare dialkyl 1-alkynylphosphonates by a Michaelis–Becker reaction was reported in 1965.[37] Only octyne is produced when 1-bromooctyne is allowed to react with sodium diethyl phosphite in liquid ammonia, the coproduct being diethyl phosphoramidate. It seems likely that reduction of the bromo compound occurs by a halogen–metal exchange, with the resulting diethyl bromophosphate being immediately converted into diethyl phosphoramidate by the liquid ammonia.[37]

The difficulties associated with this approach, namely the conversion of the bromoalkynes into sodium acetylides and the subsequent reaction of the latter with the bromophosphate byproducts, are effectively eliminated by working at low temperature. Thus, sodium diethyl phosphite reacts with 1-bromoalkynes in THF at low temperature to give diethyl 1-alkynylphosphonates in fair to good yields (37–75%, Scheme 1.6).[38–41] Usually, yields are improved by further lowering the temperature and adding the bromoalkyne slowly.[39,41]

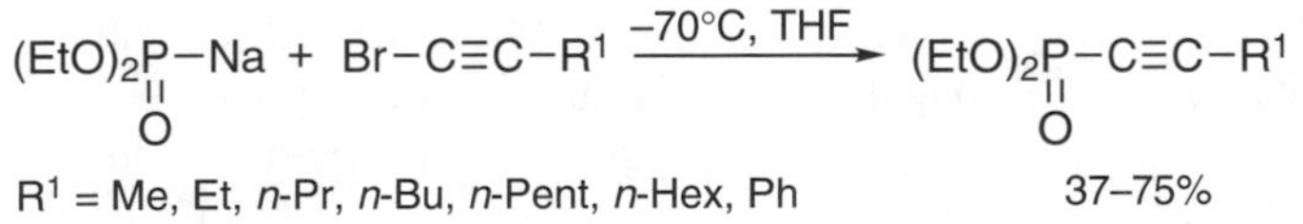

$$(\text{EtO})_2\text{P-Na} + \text{Br-C}\equiv\text{C-R}^1 \xrightarrow{-70°C, \text{THF}} (\text{EtO})_2\text{P-C}\equiv\text{C-R}^1$$

with P=O groups on both phosphorus atoms.

R¹ = Me, Et, *n*-Pr, *n*-Bu, *n*-Pent, *n*-Hex, Ph

37–75%

(1.6)

However, such precautions are not invariably necessary, and fair yields (20–63%) are observed in the synthesis of a number of heterocycle-containing ethynylphosphonic esters, even in C₆H₆ at 80°C on reaction of the corresponding bromoacetylenic alcohols.[42] Likewise, the reaction between diethyl phosphite and 4-chloro-2-methyl-3-butyn-2-ol in dry Et₂O at room temperature in the presence of a catalytic amount of CuCl and diethylamine gives good yields (60–72%) of dialkyl 3-hydroxy-3-methyl-1-butynylphosphonate.[43]

Treatment of alkynyl iodides with copper(I) dimethyl phosphite in a THF suspension under ultrasonic irradiation resulting in the formation of dimethyl phenylethynyl phosphonate in 68% yield has been recently described.[44] A promising method describes the preparation of dialkyl 1-alkynylphosphonates in good yields (65–93%) from sodium dialkyl phosphites and alkynylphenyliodonium tosylates in DMF at 70–80°C (Scheme 1.5).[45]

Nevertheless, the Michaelis–Becker route does not always proceed cleanly, and dialkyl 1-alkynylphosphonates frequently contain undesired side products that are not easily removed. Because of the difficulties associated with the use of 1-bromoalkynes, it is unsurprising that this reaction is used infrequently, and the general synthetic utility of the procedure remains to be proven. Haloalkynes are triphilic, and the approach of anionic and neutral nucleophiles at the haloalkyne has been discussed and evaluated in terms of the three sites.[46]

Propargyl bromide reacts with sodium diethyl phosphite to give diethyl 1-propynylphosphonate. Unfortunately, irrespective of the reaction conditions (sodium in refluxing THF[47] or liquid ammonia,[37] K_2CO_3 in C_6H_6 at 70°C[48] or KF at 60°C without solvent[49]), complex mixtures containing diethyl 1-propynylphosphonate, 2-propynylphosphonate, and diethyl allenylphosphonate are obtained.

1.1.3 SNP(V) Reactions

1.1.3.1 From Alkynylmagnesiums

The parent dialkyl ethynylphosphonate was first prepared in 1960 by the addition of stoichiometric ethynylmagnesium bromide to the appropriate dialkyl chlorophosphates in THF.[50] The product was isolated in low yield (12–25%), presumably because of side reactions involving the relatively acidic alkynyl proton. The yields can be increased slightly (25–35%) by the use of (toxic) dialkyl fluorophosphates.[51,52] Despite advances in developing methodologies for the elaboration of dialkyl 1-alkynylphosphonates, the preparation of the parent ethynylphosphonates remained difficult for a prolonged period. Eventually, diethyl ethynylphosphonate became easily accessible through reaction of a protected synthetic equivalent, trimethylsilylethyne, with methylmagnesium bromide in Et_2O, followed by addition to a solution of diethyl chlorophosphate [SNP(V) reaction]. The resulting diethyl (trimethylsilyl)ethynylphosphonate is deprotected by hydrolysis with 10% aqueous Na_2CO_3 to give the parent diethyl ethynylphosphonate in good overall yield (74%, Scheme 1.7).[26]

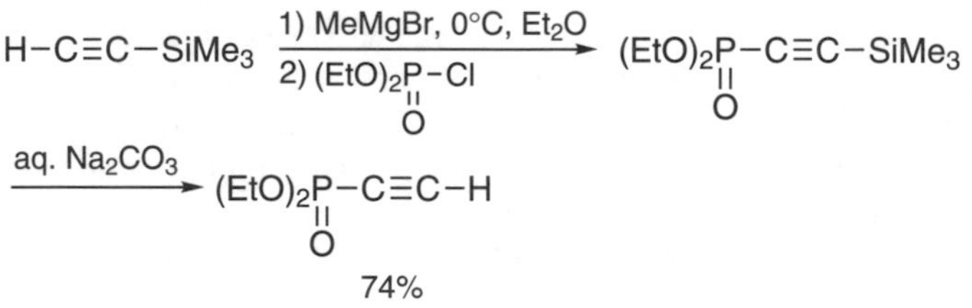

$$H-C\equiv C-SiMe_3 \xrightarrow[\text{2) (EtO)}_2\text{P-Cl}]{\text{1) MeMgBr, 0°C, Et}_2\text{O}} (EtO)_2\overset{\underset{\|}{O}}{P}-C\equiv C-SiMe_3$$

$$\xrightarrow{\text{aq. Na}_2\text{CO}_3} (EtO)_2\overset{\underset{\|}{O}}{P}-C\equiv C-H$$

$$74\% \tag{1.7}$$

The extension of this principle to higher homologues of acetylene gives superior results. These compounds are readily obtained from dialkyl or diphenyl chlorophosphates and the appropriate terminal alkynylmagnesium bromide, prepared in turn from alkynes and ethylmagnesium bromide in Et_2O or THF (Scheme 1.8).[53] The most widely employed phosphorus reagent is diethyl chlorophosphate, which reacts with the alkynylmagnesium bromides through direct[40,41,54,55] or inverse[53] addition at 0°C in Et_2O or at –30°C in THF. This procedure has been successfully employed for the preparation of a variety of dialkyl 1-alkynylphosphonates in fair to good yields (50–76%), with the best results generally obtained using addition of the alkyne salt to the chlorophosphate.

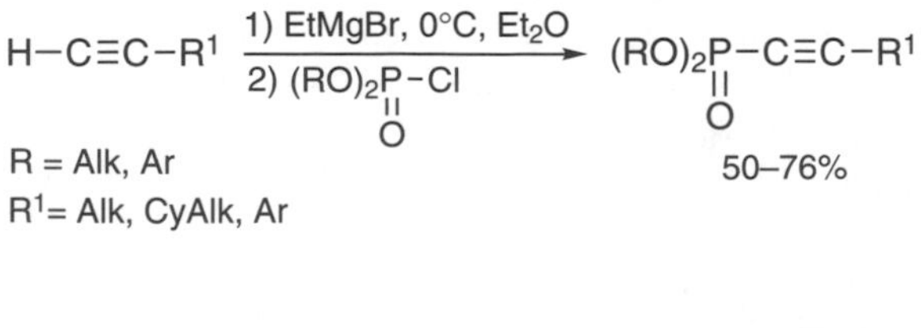

$$H-C\equiv C-R^1 \xrightarrow[\text{2) (RO)}_2\text{P-Cl}]{\text{1) EtMgBr, 0°C, Et}_2\text{O}} (RO)_2\overset{\underset{\|}{O}}{P}-C\equiv C-R^1$$

R = Alk, Ar
R^1 = Alk, CyAlk, Ar

$$50\text{–}76\% \tag{1.8}$$

1.1.3.2 From Alkynyllithiums

Because lithium reagents are generally more reactive carbanion equivalents than their Grignard counterparts, it might be expected that they would be able to react more cleanly with chlorophosphates. Furthermore, the facile and quantitative generation of alkynyllithiums under low-temperature conditions makes the use of lithium reagents for the generation of dialkyl 1-alkynylphosphonates a useful and especially attractive methodology. This approach was developed with simple and functionalized alkynes. Alkynes are metallated with n-BuLi in THF at low temperature, and the resultant lithium acetylides react with diethyl chlorophosphate at the same temperature. This procedure minimizes side reactions and provides high and reproducible yields of diethyl 1-alkynylphosphonates (Scheme 1.9).[56–70]

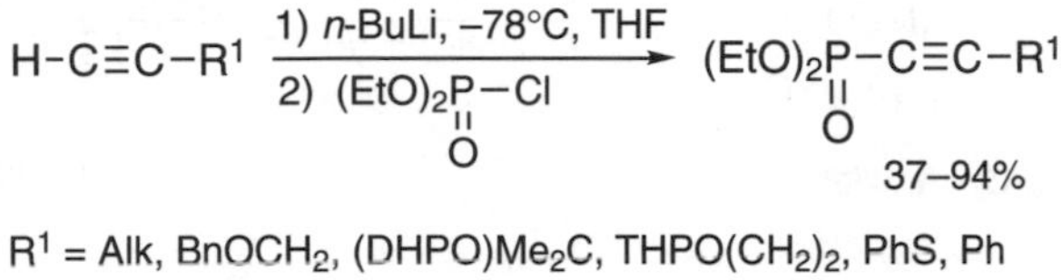

$$H-C\equiv C-R^1 \xrightarrow[\text{2) }(EtO)_2\overset{O}{\underset{}{P}}-Cl]{\text{1) }n\text{-BuLi, }-78°C,\text{ THF}} (EtO)_2\overset{O}{\underset{}{P}}-C\equiv C-R^1$$

37–94%

R^1 = Alk, BnOCH$_2$, (DHPO)Me$_2$C, THPO(CH$_2$)$_2$, PhS, Ph   **(1.9)**

The synthesis of diethyl ethynylphosphonate has recently been described using lithium *bis*(diisopropylamino)boracetylide as a synthetic equivalent for lithium acetylide. After reaction with diethyl chlorophosphate in THF at –78°C, the protected diethyl ethynylphosphonate is hydrolyzed with a 3 M HCl solution to produce diethyl ethynylphosphonate in good overall yield (72%),[71] comparable with those previously obtained (Scheme 1.10).[20,26] Synthesis of diethyl ethynylphosphonate has also been reported with an overall yield of 68% by a three-step synthesis including the reaction of lithium trimethylsilylacetylide with diethyl chlorophosphite (83%) followed by oxidation with MCPBA and deprotection with KF in EtOH (82%).[72,73]

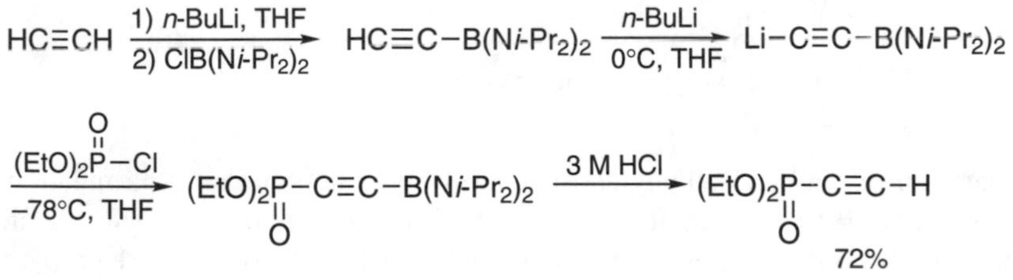

$$HC\equiv CH \xrightarrow[\text{2) ClB(N}i\text{-Pr}_2)_2]{\text{1) }n\text{-BuLi, THF}} HC\equiv C-B(Ni\text{-Pr}_2)_2 \xrightarrow[\text{0°C, THF}]{n\text{-BuLi}} Li-C\equiv C-B(Ni\text{-Pr}_2)_2$$

$$\xrightarrow[\text{–78°C, THF}]{(EtO)_2\overset{O}{\overset{\|}{P}}-Cl} (EtO)_2\overset{O}{\underset{}{P}}-C\equiv C-B(Ni\text{-Pr}_2)_2 \xrightarrow{\text{3 M HCl}} (EtO)_2\overset{O}{\underset{}{P}}-C\equiv C-H$$

72%   **(1.10)**

The advantages of lithium acetylides over haloalkynes (in Michaelis–Arbuzov or Michaelis–Becker routes) have been clearly demonstrated in several comparable syntheses. For example, from the condensation of 3,3-diethoxy-1-lithio-1-propyne with diethyl chlorophosphate at –65°C, diethyl 3,3-diethoxy-1-propynylphosphonate is obtained in 70–80% yield and easily purified.[56] By way of contrast, the reaction of 1-bromo-3,3-diethoxy-1-propyne with potassium or sodium diethyl phosphite (Michaelis–Becker reaction) gives only 30–40% yields of product from which impurities cannot easily be removed by distillation. A comparison of the reaction sequences between magnesium and lithium acetylides, with respect to diethyl chlorophosphate, has also been undertaken under the same experimental conditions. It reveals that in all cases the use of lithium acetylides in place of the Grignard counterparts results in significantly higher yields.[41]

Dialkyl 1-alkynylphosphonates can also be synthesized by the reactions of lithium tetraorganoaluminates with dialkyl chlorophosphates. The reactions of these lithium tetraorganoaluminates, prepared from LiAlH$_4$ and the corresponding substituted alkynes in Py medium, are often sluggish and require high temperatures (105°C) and lengthy reaction times (5 h). However, they produce dialkyl 1-alkynylphosphonates in good yields (60–80%).[74]

1.1.4 Coupling Reactions

1,1-Dibromoalkenes undergo coupling reactions with dimethyl phosphite in DMF at 80°C in the presence of catalytic $Pd(OAc)_2$ and 1,1′-*bis*(diphenylphosphino)ferrocene (dppf) to give the corresponding dimethyl 1-alkynylphosphonates in moderate to good yields (16–76%, Scheme 1.11). The solvent effects are quite marked. When DMF is replaced by toluene or THF, alkynylphosphonates are formed in reduced yields, accompanied by dimethyl 1-bromo-1-alkenylphosphonates. The reaction probably proceeds by the oxidative insertion of Pd(0) into the C-Br bond, ligand–solvent exchange, and HBr elimination to give an alkynyl–palladium complex, subsequently coupled with dimethyl phosphite.[75]

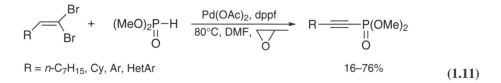

Br
Br
R
+
(MeO)2P–H
O
Pd(OAc)2, dppf
80°C, DMF,
O
R
P(OMe)2
O
R = n-C7H15, Cy, Ar, HetAr
16–76%

(1.11)

1.1.5 β-Elimination Reactions

Because heteroatom-substituted vinylphosphonates such as enol phosphates and vinyl halides are masked acetylenic compounds, a variety of methods have been developed for their elaboration and subsequent use in the synthesis of dialkyl 1-alkynylphosphonates. These eliminations usually have the leaving group β to the phosphoryl group, although cases in which the leaving group is α to the phosphorus center are also known (Scheme 1.12).

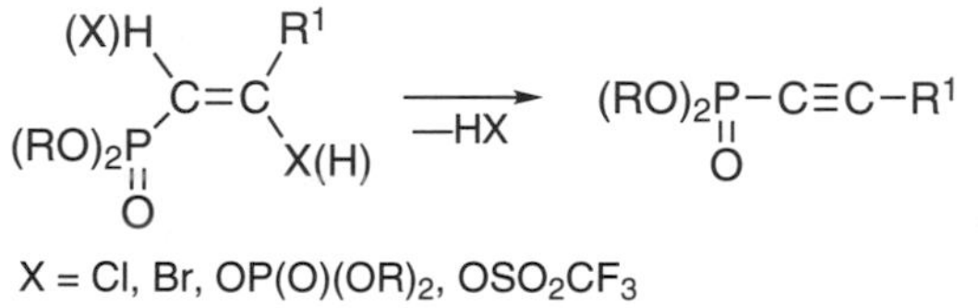

(X)H
R1
C=C
(RO)2P
O
X(H)
—HX
(RO)2P–C≡C–R1
O
X = Cl, Br, OP(O)(OR)2, OSO2CF3

(1.12)

The preparation of dialkyl 1-alkynylphosphonates by an appropriate β-elimination procedure was first recorded in 1957,[76] when it was reported that the action of EtONa on diethyl 2-[3-(diethoxyphosphinyl)propenyl] phosphate in refluxing EtOH leads to diethyl 1-propynylphosphonate in 69% yield (Scheme 1.13).[76,77] The experimental conditions are crucial, and it has been shown that successful elimination of phosphate takes place to the exclusion of the ethanolysis reaction only at elevated temperature. At room temperature, diethyl 2-[3-(diethoxyphosphinyl)propenyl] phosphate is reported to undergo competing ethanolysis to triethyl phosphate and diethyl 2-oxopropylphosphonate. Because of the drastic reaction conditions required for the subsequent conversion to 1-alkynylphosphonates, this method was rarely used and remained underdeveloped for a significant period of time. Fortunately, many efforts have been made to discover bases that might cause elimination of the phosphate from the enol without also bringing about isomerization or hydrolysis. The development of a variety of milder alternative methods has led to the much more widespread adoption of this elimination sequence.[78]

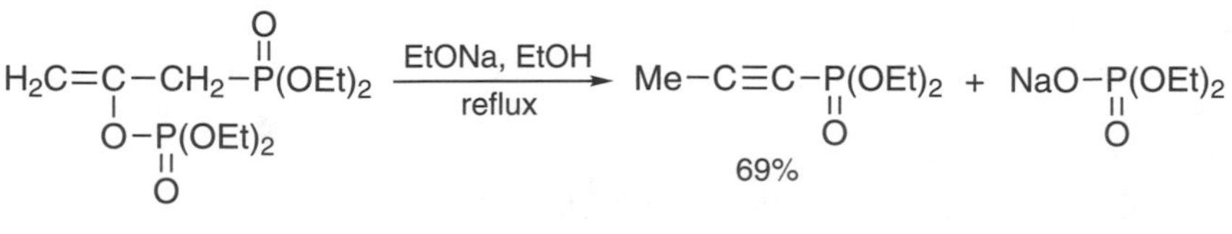

O
H2C=C–CH2–P(OEt)2
O–P(OEt)2
O
EtONa, EtOH
reflux
Me–C≡C–P(OEt)2
O
69%
+ NaO–P(OEt)2
O

(1.13)

Recently, the conversion of diethyl 2-oxoalkylphosphonates into diethyl 1-alkynylphosphonates via transient enol phosphates has been shown to occur in good yields. Thus, treatment of diethyl 2-oxoalkylphosphonates with NaH in THF followed by addition of diethyl chlorophosphate gives the enol phosphate. The β-elimination reaction from the enol phosphate is carried out at low temperature using t-BuOK. The reaction proceeds satisfactorily (43–95% overall yields) with 2-oxoalkylphosphonates except for R^1 = Et. In this case, the product undergoes a prototropic isomerization to give a mixture of diethyl 1-butynyl- and 2-butynylphosphonates (Scheme 1.14).[79]

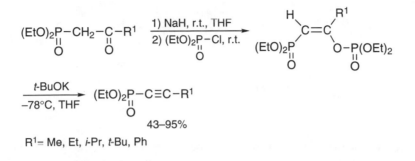

R¹= Me, Et, i-Pr, t-Bu, Ph

(1.14)

This approach has been adapted to the preparation of diethyl 3,3,3-trifluoro-1-propynylphosphonates in 80% yields through trifluoromethanesulfonation of a mixture of diethyl 3,3,3-trifluoro-2-oxopropylphosphonate and its enolic form using trifluoromethanesulfonic anhydride in CH_2Cl_2 at 0°C in the presence of DIEA (3 eq).[80]

The procedures for the conversion of enol phosphates into the corresponding alkynes involve strongly basic conditions, but several milder methods are reported in the related case of vinyl halides, which are ideal precursors for triple-bond generation by a β-elimination reaction. Two complementary approaches for the synthesis of dialkyl 1-alkynylphosphonates have been developed, one using simple dehydrochlorination of dialkyl 2-chloro-1-alkenylphosphonates and the other by dehydrochlorination of 2-haloallylphosphonates with subsequent double-bond isomerisation. For example, dialkyl 2-chloro-1-alkenylphosphonates undergo a β-elimination on heating with an equimolecular amount of KOH in absolute EtOH to give the corresponding dialkyl 1-alkynylphosphonates in 73–85% yields.[81–83] Similarly, by the judicious selection of base, diethyl 2-bromo-2-propenylphosphonate, obtained by a Michaelis–Arbuzov reaction from 2,3-dibromo-1-propene and triethyl phosphite, may be converted into allenyl- or 1-propynylphosphonates. Thus, treatment of diethyl 2-bromopropenylphosphonate with ethylmagnesium bromide in Et_2O at room temperature produces diethyl allenylphosphonate in 45% yield, whereas treatment with the more basic NaH in the same conditions gives the diethyl 1-propynylphosphonate in 90% yield (Scheme 1.15).[84,85]

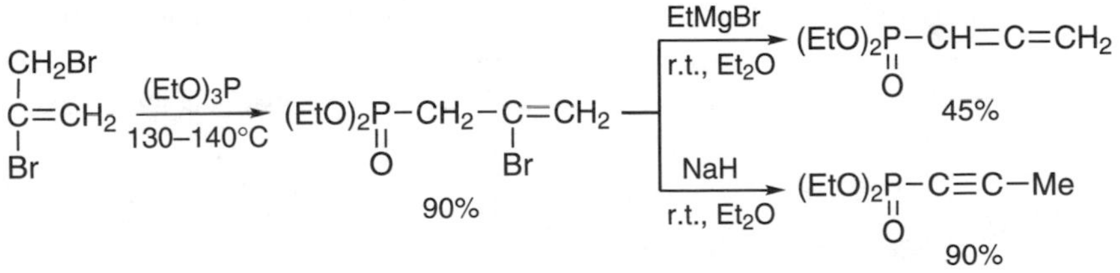

Similarly, treatment of diisopropyl 2-chloro-2-propenylphosphonate with 10% aqueous methanolic Na_2CO_3 at room temperature induces facile dehydrochlorination with subsequent double-bond isomerization to produce diisopropyl 1-propynylphosphonate in essentially quantitative yields.[86]

Another procedure for the synthesis of dialkyl 1-propynylphosphonates is based on a simple dehydrochlorination of dialkyl 1-chloro-cis-propenylphosphonate or a double elimination (dehydrochlorination and detosylation) of dialkyl 1-chloro-2-tosyloxypropylphosphonates. The reaction conditions for the elimination reaction, which involves alkali metal hydroxides or amides in MeOH at room temperature, are relatively mild.[87]

An even gentler method for the dehydrobromination of diethyl (Z)-2-(ethoxycarbonyl)-2-bromovinylphosphonate to diethyl (ethoxycarbonyl)ethynylphosphonate employs Et_3N in refluxing Et_2O[88] or 1,8-diazabicyclo[5.4.0]undec-7-one (DBU) in Et_2O at low temperature.[89] The results are better with the former reagent (Scheme 1.16).

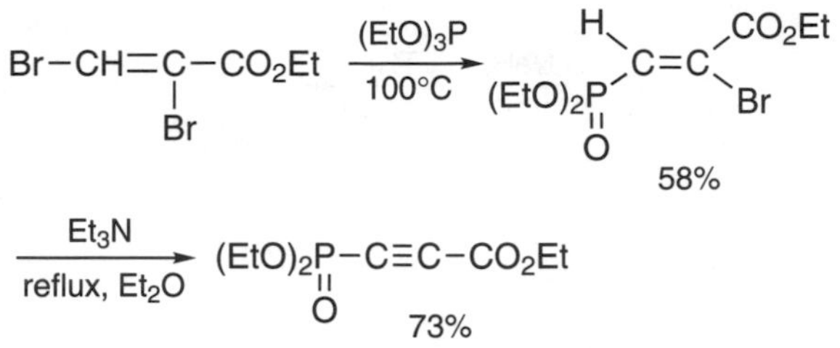

$$\text{Br-CH=C-CO}_2\text{Et} \xrightarrow[100°C]{(EtO)_3P} \begin{array}{c} \text{H} \\ \end{array}\text{C=C}\begin{array}{c}\text{CO}_2\text{Et}\\ \text{Br}\end{array}$$

58%

$$\xrightarrow[\text{reflux, Et}_2\text{O}]{Et_3N} (EtO)_2\overset{\text{O}}{\underset{\|}{P}}\text{-C≡C-CO}_2\text{Et}$$

73%

(1.16)

Diethyl trichloromethylphosphonate appears to be a very useful reagent for the generation of a triple bond from an α-chlorovinylphosphonate. The lithiated β-hydroxyphosphonates, which are readily obtained by the addition of diethyl 1-lithio-1-chloro-1-(trimethylsilyl)methylphosphonate to aromatic or heteroaromatic aldehydes, undergo a Peterson reaction to give diethyl 1-chloro-1-alkenylphosphonates as a mixture of (Z)- and (E)-isomers. They readily undergo dehydrochlorination at low temperature with LiHMDS to produce high overall yields (87–96%) of a wide range of diethyl 1-alkynylphosphonates bearing an aromatic or heteroaromatic ring in the β position (Scheme 1.17).[90]

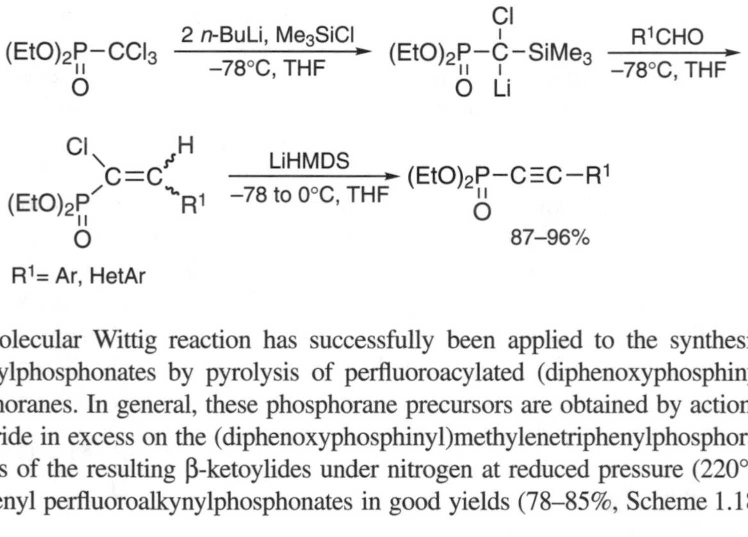

$$(EtO)_2\overset{\text{O}}{\underset{\|}{P}}\text{-CCl}_3 \xrightarrow[-78°C, \text{THF}]{2\ n\text{-BuLi, Me}_3\text{SiCl}} (EtO)_2\overset{\text{O}}{\underset{\|}{P}}\overset{\text{Cl}}{\underset{\text{Li}}{C}}\text{-SiMe}_3 \xrightarrow[-78°C, \text{THF}]{R^1\text{CHO}}$$

$$\begin{array}{c}\text{Cl}\\(EtO)_2\overset{\text{O}}{\underset{\|}{P}}\end{array}\text{C=C}\begin{array}{c}\text{H}\\R^1\end{array} \xrightarrow[-78\ \text{to}\ 0°C, \text{THF}]{LiHMDS} (EtO)_2\overset{\text{O}}{\underset{\|}{P}}\text{-C≡C-R}^1$$

87–96%

R^1= Ar, HetAr

(1.17)

An intramolecular Wittig reaction has successfully been applied to the synthesis of diphenyl perfluoroalkynylphosphonates by pyrolysis of perfluoroacylated (diphenoxyphosphinyl)methylenetriphenylphosphoranes. In general, these phosphorane precursors are obtained by action of the perfluoroacetyl chloride in excess on the (diphenoxyphosphinyl)methylenetriphenylphosphorane at 50°C in C_6H_6. Pyrolysis of the resulting β-ketoylides under nitrogen at reduced pressure (220°C at 10^{-5} torr) produces diphenyl perfluoroalkynylphosphonates in good yields (78–85%, Scheme 1.18).[91]

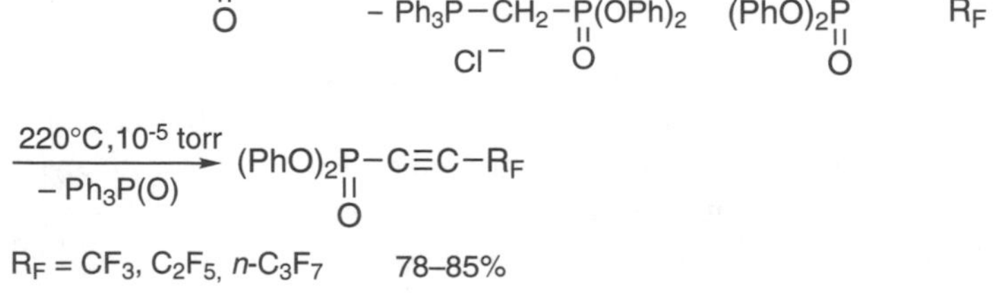

$$2\ Ph_3P\text{=CH-P(OPh)}_2 \xrightarrow[-\ Ph_3\overset{+}{P}\text{-CH}_2\text{-P(OPh)}_2]{R_FCOCl} \begin{array}{c}Ph_3\overset{+}{P}\\(PhO)_2\overset{\text{O}}{\underset{\|}{P}}\end{array}\text{C=C}\begin{array}{c}\text{O}^-\\R_F\end{array}$$

$$\xrightarrow[-\ Ph_3P(O)]{220°C, 10^{-5}\ \text{torr}} (PhO)_2\overset{\text{O}}{\underset{\|}{P}}\text{-C≡C-R}_F$$

R_F = CF_3, C_2F_5, n-C_3F_7 78–85%

(1.18)

1.1.6 REARRANGEMENT AND ISOMERIZATION REACTIONS

In 1962, three laboratories in the Soviet Union[92–94] and United States[95,96] independently reported the rearrangement of dialkyl propargyl phosphites to dialkyl allenylphosphonates. Propargyl phosphites, readily obtained from propargyl alcohols and dialkyl chlorophosphites in Et_2O at 0°C in the presence of Et_3N or Py, rearrange slowly on standing at room temperature to produce excellent overall yields of readily isolable dialkyl allenylphosphonates (Scheme 1.19). In Et_2O at room temperature, the rearrangement stops at the stage of allenylphosphonate.[97,98] The particular ease with which the acetylenic–allenic rearrangement takes place is evidently the result of a combination of the nucleophilic properties of the trivalent phosphorus atom and the rather electrophilic character of the triple bond. This rearrangement has been widely explored because the allenylphosphonates are particularly well suited for elaboration into enaminophosphonates by addition reactions at the allenic carbon atom.

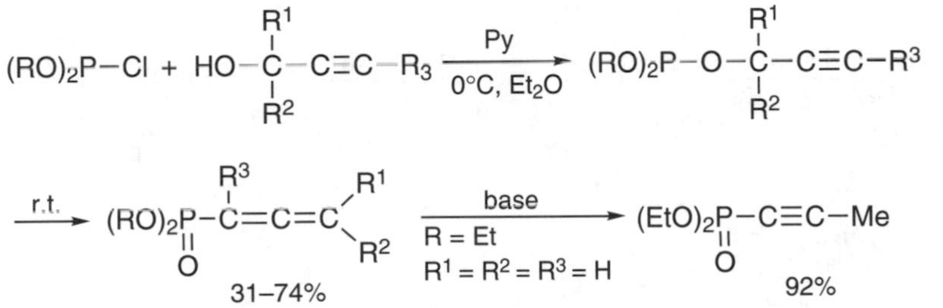

$$(RO)_2P-Cl + HO-\underset{\underset{R^2}{|}}{\overset{\overset{R^1}{|}}{C}}-C\equiv C-R_3 \xrightarrow[0°C, Et_2O]{Py} (RO)_2P-O-\underset{\underset{R^2}{|}}{\overset{\overset{R^1}{|}}{C}}-C\equiv C-R^3$$

$$\xrightarrow{r.t.} \underset{\underset{O}{||}}{(RO)_2P}-\overset{\overset{R^3}{|}}{C}=C=C\overset{R^1}{\underset{R^2}{}} \xrightarrow[\substack{R = Et \\ R^1 = R^2 = R^3 = H}]{base} \underset{\underset{O}{||}}{(EtO)_2P}-C\equiv C-Me$$

$$31–74\% \qquad\qquad 92\%$$

$$(1.19)$$

The prototropic isomerization of allenylphosphonate into 1-propynylphosphonate proceeds partially or completely under the influence of an organic base that is generally present in the reaction mixture in a catalytic amount.[99–101] When pure dialkyl allenylphosphonates are heated to 200°C, they remain unchanged (apart from partial polymerization). In the presence of basic catalysts (EtONa,[102,103] NaOH,[104] and NaH[84,85] at room temperature or Et_3N[105–107] and triethyl phosphite[105] at high temperature), they undergo more or less readily prototropic isomerization into dialkyl 1-propynylphosphonates. Thus, diethyl allenylphosphonate is straightforwardly and almost completely isomerized into diethyl 1-propynylphosphonate in excellent yield (92%, Scheme 1.19). This method has allowed the preparation of a variety of dialkyl 1-propynylphosphonates bearing different R groups at the phosphorus atom. The rearrangement of γ-substituted propargyl phosphites ($R^3 \neq$ H, Scheme 1.19) is accompanied by inversion of the radicals, but the rearrangement stops at the allenic isomers through the lack of a labile hydrogen atom at the carbon bound to the phosphonyl group. Similarly, dialkyl γ,γ-dimethylallenylphosphonate (R^1, $R^2 \neq$ H, Scheme 1.19), which is thermodynamically more stable than the acetylenic isomer, does not undergo prototropic transformation in the presence of base. Even heating dialkyl γ,γ-dimethylallenylphosphonates in the presence of EtONa or Et_3N does not cause prototropic transformation but instead provokes an intramolecular dimerization of the product, according to a kinetic study.

Two variants of this rearrangement have been reported. One uses triethyl phosphite instead of diethyl chlorophosphite.[108] Thus, in the presence of a catalytic amount of *para*-toluenesulfonic acid, triethyl phosphite reacts with propargyl alcohol in DMF at room temperature to give a mixture of diethyl allenylphosphonate (51%) and diethyl 1-propynylphosphonate (14%). The second variant is based on the conversion of dialkyl allenylphosphonates (R^1, $R^2 \neq$ H, $R^3 =$ H, Scheme 1.19) into dialkyl 1-alkynylphosphonates by a photochemically allowed [1,3s]-sigmatropic shift in C_6H_6 with 46–50% yields.[109]

1.2 SYNTHETIC APPLICATIONS OF 1-ALKYNYLPHOSPHONATES

1.2.1 REDUCTION OF THE TRIPLE BOND

Partial reduction of the triple bond is used for the selective conversion of dialkyl 1-alkynylphos-phonates into *cis*-1-alkenylphosphonates. The first racemic synthesis of the antibiotic fosfomycin in 1969 serves as an illustration. The methodology is based on the stereospecific reduction of dibutyl 1-propynylphosphonate into dibutyl (*Z*)-1-propenylphosphonate using the Lindlar catalyst (Pd/CaCO$_3$) poisoned with Pb(OAc)$_2$ in MeOH[110–113] or with quinoline.[57] Dibutyl 1-propynylphosphonate is obtained in reasonable yield (49%, Scheme 1.20) by reaction of propynylmagnesium bromide with di-*n*-butyl chlorophosphate in a C$_6$H$_6$–THF solution (see Section 1.1.3.1).[113]

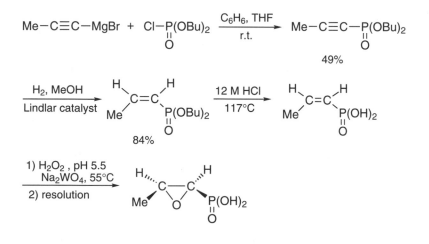

$$(1.20)$$

Unfortunately, the subsequent step using the Lindlar catalyst met with little success, and nonstereoselective partial reduction of the triple bond is observed.[114] It was reported that most catalytic hydrogenations of dialkyl 1-alkynylphosphonates in EtOH using 5% Pd/CaCO$_3$ poisoned with quinoline gave mixtures comprising *cis*- and *trans*-1-alkenylphosphonate and starting material, from which the predominant *cis* isomer was isolated.[114–117]

On the basis of these results, a systematic analysis of the relative formation of these three products from the reduction of tetraethyl acetylenediphosphonate was undertaken using a range of catalysts (Lindlar, Pd/CaCO$_3$, Pd/BaSO$_4$/Pb, Pd/BaSO$_4$, Pd/BaCO$_3$, Pd/SrCO$_3$).[118] It appears that quinoline-poisoned Pd/BaSO$_4$ is superior to the other catalysts for the selective reduction of acetylenediphosphonate to *cis*-ethenediphosphonate.[13] Hydrogenation of di-*n*-butyl 3-hydroxy-1-propynylphosphonate using Pd/BaSO$_4$ and quinoline in MeOH affords the desired *cis*-olefin in 95% yield.[119,120] Similarly, diethyl 4-methanesulfonyl-1-butenylphosphonate is partially hydrogenated in 63% yield using Pd/BaSO$_4$ in THF/Py.[121,122] The use of Red-Al in THF at low temperature provides (*E*)-vinylphosphonates in modest yield (33%).[123]

1.2.2 HYDRATION OF THE TRIPLE BOND

The effectiveness of dialkyl 1-alkynylphosphonates as acetonyl equivalents for the preparation of dialkyl 2-oxoalkylphosphonates has long been established. Because 2-oxoalkylphosphonates them-selves are versatile synthetic intermediates, especially as the reagents of choice for promoting a number of Horner–Wadsworth–Emmons cyclization reactions, procedures that effect the direct conversion of dialkyl 1-alkynylphosphonates into dialkyl 2-oxoalkylphosphonates are of special importance. The procedure for the hydration of dialkyl 1-alkynylphosphonates has remained unchanged since the first report in 1966 (Scheme 1.21).[39] Thus, treatment of diethyl 1-alkynylphos-phonates with aqueous H$_2$SO$_4$ in MeOH in the presence of HgSO$_4$ gives, after reflux for 15 h[39,124] or at room temperature for 48 h,[59,125] a quantitative yield of pure diethyl 2-oxoalkylphosphonates. In all cases studied, the acetylenic phosphonates are transformed without formation of the isomeric α-ketophosphonates. When the phosphorus substrate contains an acid-sensitive group, the hydration reaction can be performed in good yield at room temperature in aqueous THF in the presence of $HgCl_2$ (1 eq) and Py (1.5 eq).[126,127]

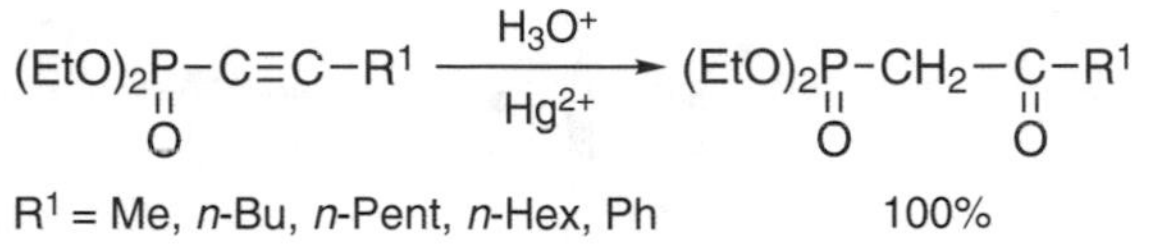

$$(EtO)_2\underset{\underset{O}{\|}}{P}-C\equiv C-R^1 \xrightarrow[Hg^{2+}]{H_3O^+} (EtO)_2\underset{\underset{O}{\|}}{P}-CH_2-\underset{\underset{O}{\|}}{C}-R^1$$

R^1 = Me, n-Bu, n-Pent, n-Hex, Ph   100%   (1.21)

Sensitive enyne phosphonates have been converted into γ,δ-unsaturated β-ketophosphonates in reasonable yields (25–70%) by treatment with a mixture of $HgO/ClCH_2CO_2H/BF_3 \cdot Et_2O$ in refluxing EtOH.[128]

1.2.3 CYCLOADDITION REACTIONS

1.2.3.1 [2+2] Reactions

The cycloaddition of 1-pyrrolidylcyclopentene with activated alkynes, such as dimethyl acetylene-dicarboxylate, is reported to occur at room temperature. In contrast, the less reactive diethyl 1-alkynylphosphonates require reaction temperatures of at least 85°C to permit cycloaddition. Under such conditions, spontaneous ring opening of the thermally unstable cyclobutene intermediate yields the ring-enlarged product. Acid hydrolysis of the product enamine gives the unsaturated β-keto-phosphonate (Scheme 1.22).[62] The best results for the cycloaddition are obtained with rigorous exclusion of moisture and with temperatures held below 100°C. Reaction times varied from 24 h (R^1 = H) to 8 days (R^1 = MOM).[62]

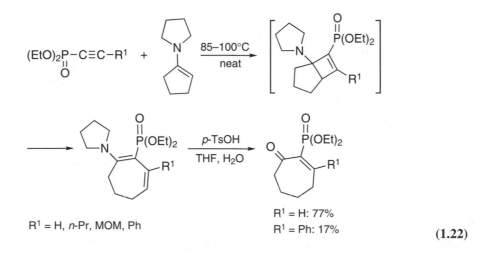

R^1 = H, n-Pr, MOM, Ph   R^1 = H: 77%
                                                 R^1 = Ph: 17%   (1.22)

The reaction must be carefully controlled in order to optimize the yield. Cycloadditions with substituted diethyl 1-alkynylphosphonates are less satisfactory.[62]

1.2.3.2 [3+2] Reactions

Cycloaddition of diazo and azido compounds with acetylenes constitutes a well-established method for the synthesis of pyrazoles and triazoles. Addition of diazomethane to ethynylphosphonates provides a convenient method for synthesizing phosphonopyrazoles.[51,129] Thus, diisopropyl ethynylphosphonate reacts smoothly with an excess of diazomethane to give the diisopropyl 1-methyl-5-phosphonopyrazole **A** in 28% yield, with the diazomethane acting as an *N*-methylation reagent for the first-formed phosphonopyrazole.[51] The tetramethyl acetylenediphosphonate reacts spontaneously with diazomethane in cooled Et$_2$O to give the 4,5-diphosphonopyrazole **B** in 95% yield (Scheme 1.23).[15]

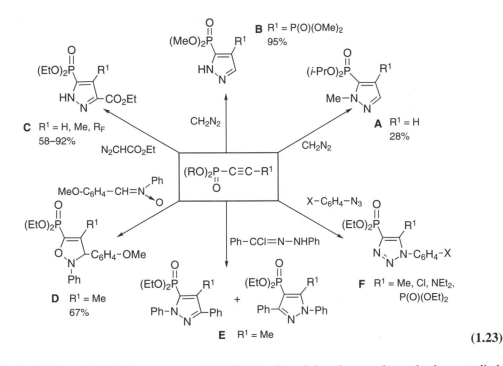

$$(1.23)$$

Cycloadditions of ethyl diazoacetate with dialkyl 1-alkynylphosphonates have also been studied. The data indicate that the cycloaddition of ethyl diazoacetate with diethyl ethynyl-, 1-propynyl-, and perfluoroalkynylphosphonates, wherein the imino group participates in a strong intermolecular hydrogen bond, gives tautomeric pyrazolylphosphonates (**C**) in 58–92% yields as a mixture of regioisomers. Because of its instability, the intermediate 3*H*-pyrazole is rapidly isomerized into the 1*H*-pyrazole. The electron-withdrawing ethoxycarbonyl group, by conferring acidic character on the C$_3$–H bond, facilitates this aromatization to give the 1*H*-pyrazole. A study of the influence of the phosphorus substituent (Cl, EtO, iPrO, *n*-BuO, Et, Ph) on the cycloaddition reaction has been done. Ethyl diazoacetate reacts most easily with dialkyl 1-alkynylphosphonates bearing dichloro substituents at phosphorus, whereas the reaction with diethyl substituents requires more severe conditions.[130]

Other 1,3-dipoles such as *C*-aryl-*N*-phenylnitrones,[131] *para*-substituted phenylazides,[132–135] α-azidoalkylcarboxylates,[136,137] α-azidoalkylphosphonates,[138,139] *C*-substituted-*N*-arylnitrilimines,[134] *para*-substituted benzonitrile oxide,[54,136,140] and *N*-phenylsydnone[141] react with diethyl 1-alkynylphosphonates. They give, respectively, 5-isoxazolinephosphonates (**D**)[131] in 42–72% yield, 1-aryl-4-triazolylphosphonates (**F**),[130,133,134,136,142] 1-(alkoxycarbonyl)-[136–138] and 1-[(diethoxyphosphinyl)methyl]-4-triazolylphosphonates,[139] 4- or 5-pyrazolylphosphonates (**E**),[134] 4-isoxazolinephosphonates[54,136] in 80–92% yields, and 3-pyrazolylphosphonates[141] in 53% yield (Scheme 1.23). The reactivity of phosphonylated propynes and the chemo- and regioselectivity of these reactions is controlled both by donor–acceptor interactions between the reactants and the steric requirements of their substituents.

1.2.3.3 [4+2] Reactions

The facile Diels–Alder reaction of dialkyl 1-alkynylphosphonates means that organophosphorus-substituted acetylenes are potentially useful precursors for introducing organophosphorus substituents into diverse organic structures. This chemistry has been developed mainly with dienophiles such as ethynyl-,[60] haloethynyl- (Scheme 1.24),[14,15,143] formylethynyl-,[114] sulfonylethynyl-,[60] sulfoxylethynyl-,[60] phenylethynylphosphonates,[144] and acetylenediphosphonates.[15,25,143]

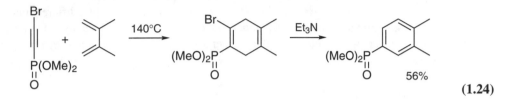

(1.24)

Only one activating group on the alkyne is necessary for the cycloaddition to occur, and the monophosphorylated acetylene reacts as readily as the diphosphorylated one. Dienes such as isoprene,[114] 2,3-dimethyl-1,3-butadiene,[14] cyclopentadiene,[15] 1,3-cyclohexadiene,[14] anthracene,[60] 9-methylanthracene,[60] 4-methyl-5-propoxyoxazole,[145] 1-phenyl-3,4-dimethylphosphole (Scheme 1.25),[144] and α-pyrone[25] have been employed.

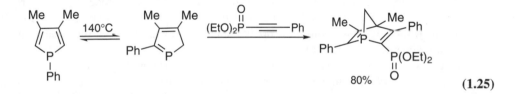

(1.25)

The reactions may be conducted with, or more rarely without, a solvent. Benzene, toluene, or xylene at reflux is generally preferred, and yields are usually good to excellent (65–93%). The strong points of this synthetic procedure are clearly illustrated by the preparation of tetramethyl *ortho*-phenylenediphosphonate in 93% yield from the reaction of tetramethyl acetylenediphosphonate with 1,3-cyclohexadiene at 150°C (Scheme 1.26).[15] With 9-methylanthracene, the reaction is highly selective, and the cycloadduct contains only one isomer.[60] The Diels–Alder reaction has been investigated in particular detail for arylsulfonyl- and arylsulfoxylethynylphosphonates.[60] The 2*H*-pyran-2-ones (α-pyrones) react as dienes in Diels–Alder reaction with tetramethyl acetylenediphosphonate at 200°C to give dimethyl 1,2-phenylenediphosphonates in good yields (62–94%).[146]

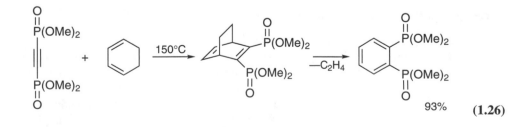

(1.26)

1.2.3.4 Intramolecular Cyclization Reactions

Depending on the position of unsaturation, two different examples of intramolecular ene-type reactions of enynes have been described. The first involves an intramolecular ene reaction of 1,6-enynes whose triple and double bonds form part of the same phosphorus substituent. Catalytic amounts of Pd(OAc)$_2$ in the presence of Ph$_3$P are used for the cyclization. In toluene, slight heating (60°C) for 4 h allows complete reaction at a reasonable rate. Comparative tests showed that Pd(OAc)$_2$ and Ph$_3$P give the best results among all the ligands employed. The yield of isolated methylcyclopentylidenephosphonate from a gram-scale reaction was 68%.[61]

In the second reaction type, the ene and yne functionalities are found in two different phosphorus substituents. These 1-alkynylphosphonates undergo facile Pd(OAc)$_2$–LiCl–catalyzed cyclization to give oxaphospholanes in good yields (77–89%, Scheme 1.27) but with low stereoselectivity, (Z)/(E) ratios at the exo double bonds being approximately 60/40.[147]

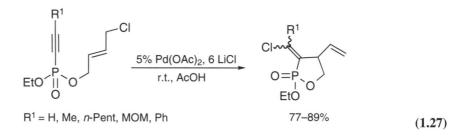

R^1 = H, Me, n-Pent, MOM, Ph   77–89%

(1.27)

AcOH appears to be the best solvent, and the reaction does not proceed particularly well in MeCN, C$_6$H$_6$, EtOH, THF, or MeNO$_2$. The mechanism may involve cis- and trans-halogenopalladation of the triple bond, insertion of the double bond into the C–Pd bond, followed by dehalopalladation to regenerate the catalyst.[147]

The regioselective ring-closing metathesis of the symmetric or unsymmetric bis(alkenyl) ethynylphosphonates under the influence of Grubbs' ruthenium catalyst affords fused bicyclic [m.n.0] phosphonates as well as the monocyclic derivatives (Scheme 1.28). Diallyl ethynylphosphonate (m = n = 1) leads to the exclusive formation of the monocyclic product, whereas 3-butenyl-4-pentenyl ethynylphosphonate (m = 2, n = 3) gives only the bicyclic phosphonate. Moreover, the exclusive formation of the bicyclo [m.n.0] rings depends not only on the length of the alkenyl tethers but also on the quantity of catalyst used.[148]

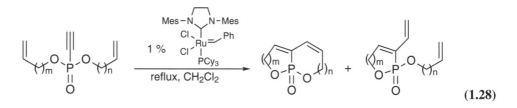

(1.28)

1.2.4 REACTIONS INVOLVING ZIRCONACYCLES

Diethyl 1-alkynylphosphonates react with Negishi's reagent, Cp$_2$ZrCl$_2$/2 n-BuLi, to give three-membered zirconacycles, which appear to be useful intermediates in phosphonate chemistry. On reaction with a large variety of reagents such as aldehydes,[149] ketones,[150] acyl chlorides,[151] nitriles,[151] and chloroformates,[151] they give insertion reactions to produce substituted vinylphosphonates in high yields in a highly stereo- and regioselective manner (Scheme 1.29). The reaction takes place in THF at room temperature for 24 h to produce mainly one isomer, with coupling taking place on C-2. This regioselectivity is apparently a result of steric factors. Consequently, these zirconacycles can be considered as the synthetic equivalents of β-vinylphosphonate carbanions.[149–151] On hydrolysis, the zirconacycle phosphonates are converted into the corresponding *cis*-vinylphosphonates, isolated in good yields (63–79%, Scheme 1.29).[152]

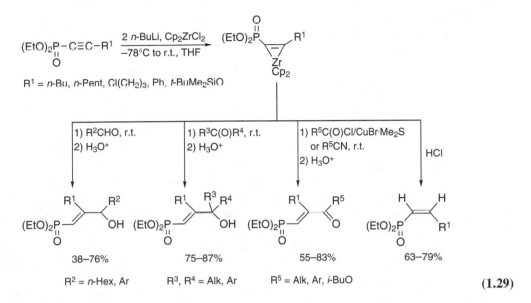

$$(1.29)$$

When three-membered zirconacycle phosphonates are treated with terminal alkynes, and the reaction mixture is hydrolyzed, two isomeric dienylphosphonates are obtained. In all cases the alkyne coupling occurs mainly on C-1, probably because of the steric interactions. On hydrolysis, the dienylphosphonate having the (1*Z*,3*E*) configuration is isolated in 57–73% yields as the major product, whereas the minor product (1*E*,3*E* configuration) is obtained in 11–20% yields (Scheme 1.30).[152]

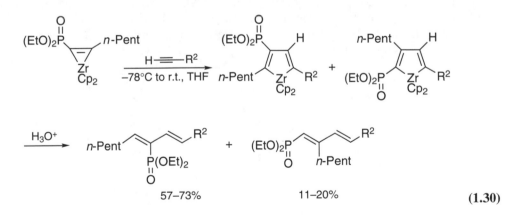

$$(1.30)$$

1.2.5 Addition Reactions of the Triple Bond

Three types of compounds, dialkyl ethynyl-, 1-alkynylphosphonates and chloroethynylphosphonates, predominate within this category. Treatment of diethyl ethynylphosphonates with chlorine in CCl$_4$[12] or bromine in Et$_2$O[51] leads to the formation of isomeric (*Z*)- and (*E*)-1,2-dihalovinylphosphonates in good yields. By selecting the correct conditions, either one or two molecules of EtOH can be added across the triple bond in the presence of EtONa to give the 2-ethoxyvinylphosphonates and 2,2-diethoxyethylphosphonates, respectively.[51] The related addition of EtSH gives 2-ethylthiovinylphosphonates, but the product is most conveniently obtained as a mixture of both isomers by adding the ethynyl compound to an excess of the thiol containing dissolved sodium.[51]

Recently, an efficient stereoselective hydrohalogenation reaction of dialkyl ethynylphosphonates affording the (Z)-2-halovinylphosphonates has been reported.[153] Thus, heating diethyl ethynylphosphonate with LiI in AcOH at 70°C affords diethyl 2-iodovinylphosphonate as the sole product in 85% yield (Scheme 1.31). The corresponding bromo and chloro analogues are prepared with high regio- and stereoselectivity, although in lower yields (50% for the bromo and 30% for the chloro derivatives).[153]

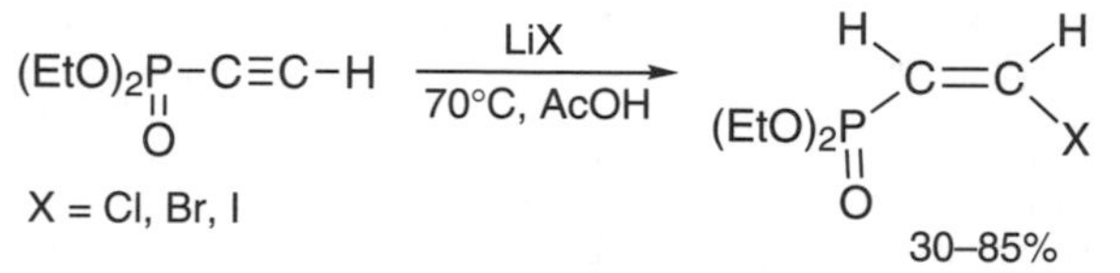

$$(EtO)_2\underset{\underset{O}{\|}}{P}-C\equiv C-H \xrightarrow[\text{70°C, AcOH}]{\text{LiX}} \underset{\underset{\underset{O}{\|}}{(EtO)_2P}}{\overset{H}{\diagdown}}C=C\overset{H}{\underset{X}{\diagup}}$$

X = Cl, Br, I

30–85%

(1.31)

Michael addition to the triple bond is also used to prepare alkenylphosphonic acid analogues of nucleosides. Thus, the conjugate addition of heterocyclic bases (adenine, uracil, cytosine, thymine) to diethyl ethynylphosphonate has been investigated in depth. With t-BuOK used as base, 18-crown-6 as catalyst, and MeCN or DMF as solvent, the alkylation of heterocyclic bases leads to a mixture of (Z)- and (E)-isomers in moderate yields (32–41%), the (Z)-isomer being slightly predominant. The yields are greater (70–80%) with K_2CO_3 as base in DMF at room temperature.[22,154–156] The addition of $para$-methylthiophenol to diethyl ethynylphosphonate results exclusively in the formation of (Z)-alkene, which isomerizes on distillation.[26]

A relatively general procedure for the conversion of diethyl 1-alkynylphosphonates to diethyl 2-oxoalkylphosphonates using enaminophosphonates has been developed. When diethyl 1-alkynylphosphonates are heated under reflux with a 10 M excess of primary or secondary amine, enaminophosphonates are produced in fair to good yields.[157–159] The addition to the triple bond is complete in 20 h for Et_2NH[160] and in 3–5 days for n-BuNH$_2$.[159] Recently, it has been found that the addition of primary and secondary amines proceeds much better in alcohols in the presence of catalytic quantities of CuCl to give the (E)-enaminophosphonates resulting from a cis-addition in 36–80% yields.[161,162] The subsequent hydrolysis of the resulting enaminophosphonates with oxalic acid in a two-phase system at room temperature yields diethyl 2-oxoalkylphosphonates in excellent yields (76–94%, Scheme 1.32).[159] Similarly, amines add across tetraethyl acetylenediphosphonate to give enamine derivatives.[163]

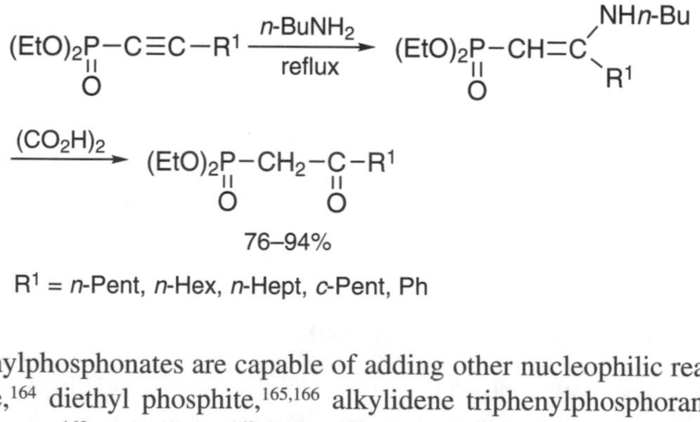

$$(EtO)_2\underset{\underset{O}{\|}}{P}-C\equiv C-R^1 \xrightarrow[\text{reflux}]{n\text{-BuNH}_2} (EtO)_2\underset{\underset{O}{\|}}{P}-CH=C\overset{NHn\text{-Bu}}{\underset{R^1}{\diagup}}$$

$$\xrightarrow{(CO_2H)_2} (EtO)_2\underset{\underset{O}{\|}}{P}-CH_2-\underset{\underset{O}{\|}}{C}-R^1$$

76–94%

$R^1 = n$-Pent, n-Hex, n-Hept, c-Pent, Ph

(1.32)

Dialkyl 1-alkynylphosphonates are capable of adding other nucleophilic reagents, in particular methylbenzylamine,[164] diethyl phosphite,[165,166] alkylidene triphenylphosphoranes,[167] sodium phenyltellurate,[168] malonate,[169] malonitrile,[169] SeBr$_4$,[170] $(EtO)_2P(S)SH$,[171] and EtSH,[165,172] forming products of the addition of one or two molecules of the reagent at the triple bond.[165,166] For example, $tert$-butylethynylphosphonate is subject to a nucleophilic attack by sodium diethyl phosphite in THF at room temperature for 24 h to give a triphosphonate in good yield.[35] A decrease in the activity of sodium diethyl phosphite is observed in ethanolic solutions. The triple bond is attacked by the sodium diethyl phosphite in the β position with respect to the bulky group and not β to the phosphoryl group. Steric hindrance predominates over the electronic directing influence of the phosphoryl group.[35]

The addition of mercaptans proceeds more readily, and in the presence of excess of mercaptans, addition results in the exclusive formation of 2,2-dithioalkoxypropanes.[172] By way of contrast, heating equimolar amounts of dialkyl 1-propynylphosphonates with alcohols in the presence of sodium alkoxide at 60–70°C leads to only one addition product, 2-alkoxypropene.[172] In the reaction of benzenesulfenyl chloride with diethyl 1-propynylphosphonate, the two regioisomeric adducts showing an (*E*) arrangement of the phenylthio group and the chlorine atom relative to the double bond are obtained.[173] Alkyl or aryl chalcogenate (S, Se, Te) anions add to diethyl 1-alkynylphosphonates to give diethyl 2-chalcogenylvinylphosphonates in satisfactory yields (26–70%). The reaction is stereoselective, producing the (*Z*)-isomer predominantly or exclusively.[174]

Diethyl 1-alkynylphosphonates react readily with an excess of alkyl- or arylmagnesium halides in the presence of CuCl at –30°C in Et_2O to provide high yields (75–95%) of β-alkylated or arylated diethyl 1-alkenylphosphonates (Scheme 1.33). The addition is highly stereoselective, producing almost exclusively the product of *cis* addition. Stoichiometric lithium dialkyl- or diarylcuprates can also react with diethyl 1-alkynylphosphonate in Et_2O at low temperature to produce diethyl 1-alkenylphosphonates in high regio- and stereoselectivity. The intermediate alkenylcuprates are particularly suited to provide trisubstituted diethyl 1-alkenylphosphonates by reaction with a wide range of electrophiles.[64,115,116,175]

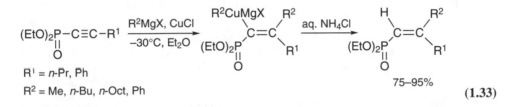

(EtO)2P–C≡C–R1
O
R2MgX, CuCl
–30°C, Et2O
R2CuMgX
R2
(EtO)2P
O
C=C
R1
aq. NH4Cl
H
R2
(EtO)2P
O
C=C
R1
R1 = n-Pr, Ph
R2 = Me, n-Bu, n-Oct, Ph
75–95%

(1.33)

Recently, the hydroboration of 1-alkynylphosphonates with pinacolborane has been reported. However, the vinylphosphonoboronates so obtained are difficult to isolate, and they are immediately subjected to Suzuki coupling reactions with aryl iodides.[69,176] Similarly, hydrophenylation of diethyl 1-octinylphosphonate with phenylboronic acid takes place in dioxane–water (10:1) at 100°C in the presence of a diphosphine-rhodium catalyst to afford diethyl 2-phenyl-1-octenylphosphonate in 87% yield (Scheme 1.34).[177]

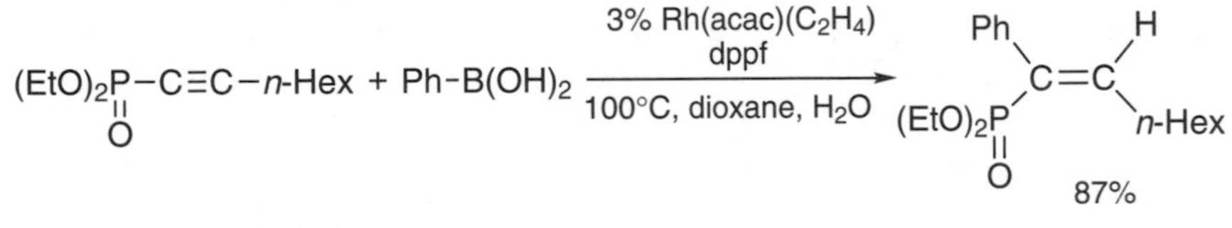

(EtO)2P–C≡C–n-Hex + Ph–B(OH)2
O
3% Rh(acac)(C2H4)
dppf
100°C, dioxane, H2O
Ph
H
C=C
(EtO)2P
n-Hex
O
87%

(1.34)

The phosphorus-containing ynamines have been prepared through the reaction of diethyl 2-chloroethynylphosphonate with dialkylamines (Scheme 1.35).[178] The ynaminophosphonates readily undergo characteristic reactions of ynamines: hydration by water with the formation of acetamides[178,179] and the addition of HCN (generated *in situ* from acetone cyanohydrin) with formation of diethyl 2-cyano-2-(diethylamino)vinylphosphonate (Scheme 1.35).[180] The use of *tert*-butylamine in the reaction with diethyl 2-chloroethynylphosphonate also induces the replacement of halogen, but the reaction is accompanied by isomerization, giving almost quantitative yields of stable ketenimines.[181,182]

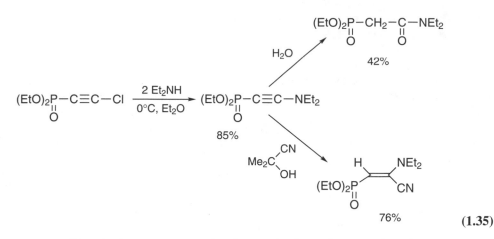

(1.35)

The presence of two hydrogen atoms at the nitrogen in the amines permits subsequent protoropic isomerization with the formation of ketenimines. By way of contrast, reactions with primary phosphines lead to halogen-substitution products with retention of the C≡C bond and P–H bonds.[182] The reaction of ketenimines with water in the presence of a catalytic amount of HCl proceeds smoothly and leads to a high yield of the corresponding amide.[183]

Reactions of diethyl 2-chloroethynylphosphonate have also been studied with a number of charged (Et⁻, t-BuO⁻, PhO⁻)[184–186] and neutral[187] nucleophiles containing one or two nucleophilic centers. Treatment of diethyl 2-chloroethynylphosphonate with binucleophilic reagents such as *ortho*-phenylenediamine, *ortho*-aminophenol, or 2-aminoethanol leads to phosphonylated benzimidazoles, benzoxazoles, and 4,5-dihydroxazoles by a mechanism involving initial substitution of the halogen followed by addition of the second nucleophilic center at the same carbon atom (Scheme 1.36).[187]

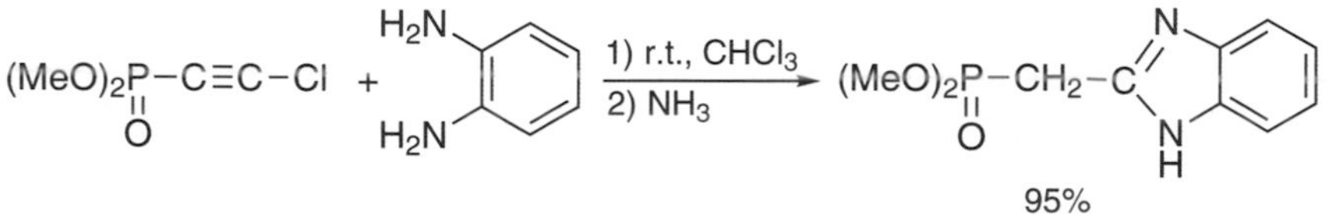

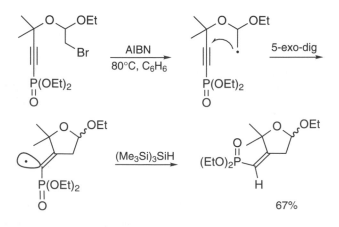

(1.36)

One intramolecular radical addition to the C≡C bond has recently been described. Oxacyclopentane ring systems with an attached exocyclic phosphonomethylene functionality are formed in excellent yields (67%) under the thermal AIBN/(Me₃Si)₃SiH initiation conditions. The very bulky (Me₃Si)₃SiH selectively transfers hydrogen to the (Z)-isomer of vinyl radical intermediate to avoid the easily envisioned steric repulsions encountered with the (E)-isomer to produce exclusively the (Z)-vinylphosphonate (Scheme 1.37).[68]

(1.37)

1.2.6 Isomerization of Alkynylphosphonates

Because of the ready accessibility of dialkyl 1-alkynylphosphonates, their isomerization to conjugated dienes represents a useful synthetic transformation. In the presence of $Pd_2(dba)_3CHCl_3$, the thermal isomerization of diethyl 1-alkynylphosphonates into diethyl (1E,3E)-alkadienylphosphonates occurs at 30°C in toluene, whereas at 110°C isomeric (2E,3E)-alkadienylphosphonates are formed. The transformation is effected with good yields (79–92%) but requires lengthy reaction times (24–69 h).[188] The same isomerization reaction can be effected with Bu_3P.[189]

The diethyl 2-alkynylphosphonate reacts with a stoichiometric amount of Cp_2ZrHCl to give, after hydrolysis, diethyl 1-propynylphosphonate in good yield (82%) and only traces of the expected allylphosphonate. Thus, the Schwartz reagent promotes the isomerization of propargylphosphonate into the 1-propynyl isomer (Scheme 1.38).[190]

(1.38)

1.2.7 Reactions of 3-Halogenopropynylphosphonates

3-Halogenopropynylphosphonates have been used as phosphonate-containing acetonyl equivalents. 3-Hydroxypropynylphosphonate intermediates, readily obtained through the reaction of lithium derivatives of protected propargyl alcohol on diethyl chlorophosphate (see Section 1.1.3.2), are converted into mesylates then into iodides by displacement with sodium iodide[59] or bromides using Ph_3P/CBr_4 in CH_2Cl_2.[191] Subsequent treatment of diethyl 3-iodopropynylphosphonate with ketone enolates (formed using KHMDS/Et$_3$B in THF) gave the corresponding alkylation products which are hydrolyzed to the expected β-ketophosphonates (Scheme 1.39).[59] The reaction has been successfully extended to the preparation of diketophosphonates[126,127] as well as amino derivatives of phosphonopentynoic acid.[191]

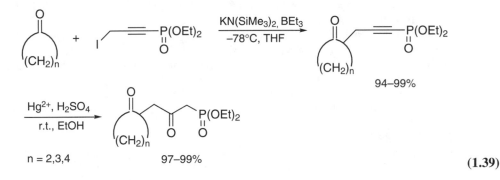

(1.39)

1.2.8 Reactions of the Phosphonyl Group

Bromo- and iodotrimethylsilane, reagents compatible with alkyne and other functionalities, are suitable for the mild P–O dealkylation of dialkyl 1-alkynylphosphonates to give the corresponding phosphonic acids.[192–196] The greater reactivity of iodotrimethylsilane will probably prove advantageous for the low-temperature dealkylation of phosphonates having triple bonds. However, transesterification of diethyl 2-propynylphosphonate with iodotrimethylsilane at –30°C followed by solvolysis with MeOH led quantitatively to the 1-propynylphosphonic acid.[194]

Two procedures for the transformation of diesters of 1-alkynylphosphonic acids into monoesters have been described. They involve the selective replacement of one ethoxy group by chlorine without addition to the triple bond. Treatment of diethyl 1-propynylphosphonate with $POCl_3$ neat (1.2 eq) at 60°C for 3 h gives ethyl 1-propynylphosphonochloridate in 79% yield (Scheme 1.40).[197] Similarly, treatment of diethyl 1-propynylphosphonate with trichloro(*ortho*-phenylenedioxy)phosphorane for 24 h gives the ethyl 1-propynylphosphonochloridate in comparable yield (74%).[198] The transformation into phosphonic dichlorides is achieved by the reaction of diethyl 1-propynylphosphonate with PCl_5 on heating at 110–135°C.[199]

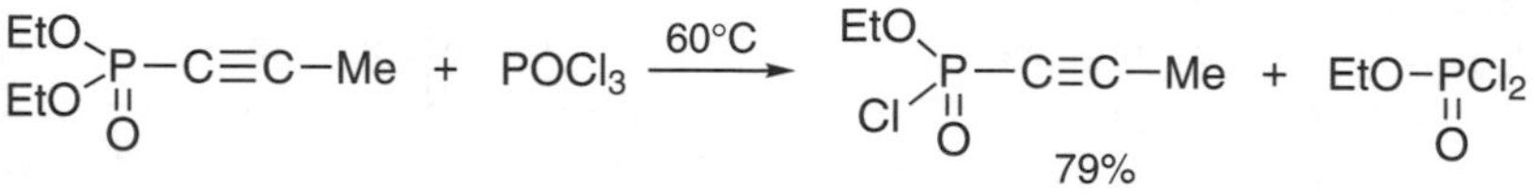

$$\underset{\underset{O}{\overset{\|}{}}}{\overset{EtO}{\underset{EtO}{\diagup}}}P-C\equiv C-Me \;+\; POCl_3 \;\;\xrightarrow{60°C}\;\; \underset{\underset{O}{\overset{\|}{}}}{\overset{EtO}{\underset{Cl}{\diagup}}}P-C\equiv C-Me \;+\; EtO-\underset{\underset{O}{\overset{\|}{}}}{PCl_2}$$

$$\text{79\%}$$

$$(1.40)$$

REFERENCES

1. Eymery, F., Iorga, B., and Savignac, P., Synthesis of phosphonates by nucleophilic substitution at phosphorus. The SNP(V) reaction, *Tetrahedron*, 55, 13109, 1999.
2. Iorga, B., Eymery, F., Carmichael, D., and Savignac, P., Dialkyl 1-alkynylphosphonates. A range of promising reagents, *Eur. J. Org. Chem.*, 3103, 2000.
3. Ionin, B.I., and Petrov, A.A., Arbuzov rearrangement involving acetylenic halides with the halogen atom at the triple bond, *Zh. Obshch. Khim.*, 32, 2387, 1962; *J. Gen. Chem. USSR (Engl. Transl.)*, 32, 2355, 1962.
4. Ionin, B.I., and Petrov, A.A., Arbuzov rearrangement with the participation of fluoro-, chloro-, bromo- and iodo-acetylenes, *Zh. Obshch. Khim.*, 35, 1917, 1965; *J. Gen. Chem. USSR (Engl. Transl.)*, 35, 1910, 1965.
5. Bogoradovskii, E.T., Zavgorodnii, V.S., Mingaleva, K.S., Maksimov, V.L., and Petrov, A.A., Synthesis and properties of α-halo-substituted pentafluorophenylacetylenes, *Zh. Obshch. Khim.*, 48, 1754, 1978; *J. Gen. Chem. USSR (Engl. Transl.)*, 48, 1601, 1978.
6. Ionin, B.I., Lebedev, V.B., and Petrov, A.A., Esters of phosphonic acids with diacetylenic radicals, *Dokl. Akad. Nauk SSSR, Ser. Khim.*, 152, 1354, 1963; *Dokl. Phys. Chem. (Engl. Transl.)*, 152, 831, 1963.
7. Ionin, B.I., Mingaleva, K.S., and Petrov, A.A., Dipole moments of phosphonic esters with an unsaturated *P*-group, *Zh. Obshch. Khim.*, 34, 2630, 1964; *J. Gen. Chem. USSR (Engl. Transl.)*, 34, 2651, 1964.
8. Ionin, B.I., and Petrov, A.A., Investigations in the field of conjugation systems. Part 173. Synthesis and properties of the ethyl esters of phosphoric acids with enyne radicals, *Zh. Obshch. Khim.*, 33, 2863, 1963; *J. Gen. Chem. USSR (Engl. Transl.)*, 33, 2791, 1963.
9. Peiffer, G., Guillemonat, A., Traynard, J.C., and Faure, M., Synthesis of phosphonates containing conjugated enyne groups, *C.R. Acad. Sci., Ser. C*, 268, 358, 1969.
10. Kruglov, S.V., Ignat'ev, V.M., Ionin, B.I., and Petrov, A.A., Synthesis of symmetrical and mixed esters of diphosphonic acids, *Zh. Obshch. Khim.*, 43, 1480, 1973; *J. Gen. Chem. USSR (Engl. Transl.)*, 43, 1470, 1973.
11. Maier, L., Organic phosphorus compounds. Part 58. Preparation and properties of acetylene diphosphonates, haloethene-1,2-diphosphonates, and the corresponding acids, *Phosphorus*, 2, 229, 1973.
12. Garibina, V.A., Dogadina, A.V., Zakharov, V.I., Ionin, B.I., and Petrov, A.A., (Haloethynyl)phosphonates. Synthesis and electrophilic reactions of (chloroethynyl)phosphonic esters, *Zh. Obshch. Khim.*, 49, 1964, 1979; *J. Gen. Chem. USSR (Engl. Transl.)*, 49, 1728, 1979.
13. Epoxyalkylphosphonic acids, *Italchemi*, Belgian Patent Appl. BE 834591, 1976; *Chem. Abstr.*, 85, 192888, 1976.
14. Senderikhin, A.I., Dogadina, A.V., Ionin, B.I., and Petrov, A.A., Bromoacetylene phosphonate as a dienophile in the Diels–Alder reaction, *Zh. Obshch. Khim.*, 58, 1662, 1988; *J. Gen. Chem. USSR (Engl. Transl.)*, 58, 1483, 1988.
15. Seyferth, D., and Paetsch, J.D.H., Diels–Alder reaction in organometallic chemistry. Part 5. Tetramethyl acetylenediphosphonate and dimethyl chloroacetylenephosphonate and their reactions with cyclopentadiene, 1,3-cyclohexadiene, and diazomethane, *J. Org. Chem.*, 34, 1483, 1969.

16. Fluck, E., and Kazenwadel, W., On the reaction of phosphoryl acetylenes with phosphanes, *Z. Naturforsch., Ser. B*, 31B, 172, 1976.

17. Atavin, A.S., Lutskaya, N.V., Zorina, E.F., and Mirskova, A.N., Acetylenic phosphorus containing thio esters, *Irkutsk Institute of Organic Chemistry*, Soviet Patent Appl. SU 449061, 1976; *Chem. Abstr.,* 82, 72527, 1975.

18. Mirskova, A.N., Lutskaya, N.V., and Voronkov, M.G., Reaction of alkyl chloroethynyl sulfides with trialkyl phosphites, *Zh. Obshch. Khim.*, 49, 2668, 1979; *J. Gen. Chem. USSR (Engl. Transl.)*, 49, 2365, 1979.

19. Mirskova, A.N., Seredkina, S.G., Kalikhman, I.D., and Voronkov, M.G., Synthesis and reactions of organyl chloroethynyl sulfides with phosphorus-containing nucleophiles, *Izv. Akad. Nauk SSSR, Ser. Khim.*, 2818, 1985; *Bull. Acad. Sci. USSR, Div. Chem. Sci. (Engl. Transl.)*, 2614, 1985.

20. Burt, D.W., and Simpson, P., Synthesis of ethynylphosphonate esters. A novel organosilicon re-arrangement, *J. Chem. Soc. (C)*, 2273, 1969.

21. Burt, D.W., and Simpson, P., The mechanism of the reaction between trialkyl phosphites and halo-genoacetylenes, *J. Chem. Soc. (C)*, 2872, 1971.

22. Lazrek, H.B., Khaider, H., Rochdi, A., Barascut, J.L., and Imbach, J.L., Synthesis of the nucleotide analogue (R,S)-9-[1-(2-hydroxyethylthio)-2-phosphonylethyl]adenine, *Nucleosides Nucleotides*, 13, 811, 1994.

23. Petrov, A.A., Rogozev, B.I., Krizhanskii, L.M., and Zavgorodnii, V.S., Quadrupole splitting in spectra of nuclear gamma resonance of tin acetylenes, *Zh. Obshch. Khim.*, 38, 1196, 1968; *J. Gen. Chem. USSR (Engl. Transl.)*, 38, 1151, 1968.

24. Zavgorodnii, V.S., Ionin, B.I., and Petrov, A.A., Unsaturated stannyl hydrocarbons. Part 7. Arbuzov rearrangement of tin-containing α-haloacetylenes, *Zh. Obshch. Khim.*, 37, 949, 1967; *J. Gen. Chem. USSR (Engl. Transl.)*, 37, 898, 1967.

25. Kyba, E.P., Rines, S.P., Owens, P.W., and Chou, S.-S.P., A novel synthesis of 1,2-diphosphorylben-zenes, *Tetrahedron Lett.*, 22, 1875, 1981.

26. Acheson, R.M., Ansell, P.J., and Murray, J.R., Addition reactions of heterocyclic compounds. Part 82. The synthesis and reactions of diethyl ethynylphosphonate and tetraethyl ethynyldiphosphonate, *J. Chem. Res. (M)*, 3001, 1986.

27. Fujii, A., Dickstein, J.I., and Miller, S.I., Nucleophilic substitution at an acetylenic carbon. Carbon *vs.* halogen attack by phosphorus nucleophiles, *Tetrahedron Lett.*, 11, 3435, 1970.

28. Simpson, P., and Burt, D.W., The mechanism of the reaction between trialkyl phosphites and halo-genoacetylenes, *Tetrahedron Lett.*, 11, 4799, 1970.

29. Fujii, A., and Miller, S.I., Nucleophilic substitution at acetylenic carbon. Kinetics and mechanism of the Arbuzov reaction of substituted phenylbromo- and phenylchloroacetylenes with triethyl phosphite, *J. Am. Chem. Soc.*, 93, 3694, 1971.

30. Gordon, M., and Griffin, C.E., Phosphonic acids and esters. Part 13. Nylen and Arbuzov reactions with propargyl bromide, *J. Org. Chem.*, 31, 333, 1966.

31. Kondrat'ev, Y.A., Knobel, Y.K., and Ivin, S.Z., Study of the reaction of propargyl bromide with triethyl phosphite by infrared spectroscopy, *Zh. Obshch. Khim.*, 37, 1094, 1967; *J. Gen. Chem. USSR (Engl. Transl.)*, 37, 1037, 1967.

32. Rudinskas, A.J., Hullar, T.L., and Salvador, R.L., Phosphonic acid chemistry. Part 2. Studies on the Arbuzov reaction of 1-bromo-4,4-diethoxy-2-butyne and Rabinowitch method of dealkylation of phosphonate diesters using chloro- and bromotrimethylsilane, *J. Org. Chem.*, 42, 2771, 1977.

33. Jennings, L.J., and Parratt, M.J., Synthesis and antiviral activity of 9-(phosphonoalkynyloxy)purines. Novel acyclonucleotides, *Bioorg. Med. Chem. Lett.*, 3, 2611, 1993.

34. Lodaya, J.S., and Koser, G.F., Alkynyliodonium salts as alkynylating reagents. Direct conversion of alkynylphenyliodonium tosylates to dialkyl alkynylphosphonates with trialkyl phosphites, *J. Org. Chem.*, 55, 1513, 1990.

35. Hägele, G., Goudetsidis, S., Wilke, E., Seega, J., Blum, H., and Murray, M., Synthesis and properties of compounds related to 1-*tert*-butylacetylene-2-phosphonic acid and 1-*tert*-butylethane-1,2,2-triph-osphonic acid. Sterically overcrowded phosphorus compounds. Part 1, *Phosphorus, Sulfur Silicon Relat. Elem.*, 48, 131, 1990.

36. Stang, P.J., and Diederich, F., *Modern Acetylene Chemistry*, VCH, Weinheim, 1995.

37. Meisters, A., and Swan, J. M., Organophosphorus compounds. Part 5. Dialkyl alkylphosphonates from alkyl halides and sodium dialkyl phosphonates in liquid ammonia, *Aust. J. Chem.*, 18, 163, 1965.

38. Sturtz, G., and Charrier, C., Nucleophilic substitution of bromoalkynes by sodium diethyl phosphite. Preparation of long-chain β-ketophosphonates, *C.R. Acad. Sci.*, 261, 1019, 1965.

39. Sturtz, G., Charrier, C., and Normant, H., Effect of 1-bromoacetylene derivatives on sodium dialkyl phosphites. Preparation of long-chain β-oxo phosphonates, *Bull. Soc. Chim. Fr.*, 1707, 1966.

40. Brestkin, A.P., Vikhreva, L.A., Godovikov, N.N., Garbatyuk, V.S., Moralev, S.N., and Kabachnik, M.I., Substituted ethynylphosphonates and their anticholinesterase activity, *Izv. Akad. Nauk SSSR, Ser. Khim.*, 2118, 1988; *Bull. Acad. Sci. USSR, Div. Chem. Sci. (Engl. Transl.)*, 1900, 1988.

41. Tronchet, J.M.J., and Bonenfant, A.P., Utilization of phosphorus ylides in sugar chemistry. Part 45. Acetylenesugars heterosubstituted on the triple bond, *Carbohydr. Res.*, 93, 205, 1981.

42. Azerbaev, I.N., Godovikov, N.N., Abdullaev, N.B., and Abiyurov, B.D., Heterocyclic unsaturated organophosphorus compounds. Part 1. Synthesis of 2,6-diphenyl-3-methyl-4-(dialkoxyphosphinyl-ethynyl)piperidin-4-ols, *Zh. Obshch. Khim.*, 48, 1271, 1978; *J. Gen. Chem. USSR (Engl. Transl.)*, 48, 1163, 1978.

43. Gafurov, E.K., Sal'keeva, L.K., and Shostakovskii, M.F., New organophosphorus compounds from 2-methyl-3-butyn-2-ol, *Zh. Obshch. Khim.*, 52, 2730, 1982; *J. Gen. Chem. USSR (Engl. Transl.)*, 52, 2408, 1982.

44. Suzuki, H., and Abe, H., A new straightforward synthesis of alkynyl sulfones via the sonochemical coupling between alkynyl halides and copper sulfinates, *Tetrahedron Lett.*, 37, 3717, 1996.

45. Zhang, J.-L., and Chen, Z.-C., Hypervalent iodine in synthesis. Part 25. Alkynylphenyliodonium tosylates as alkynylating reagents. Direct conversion of alkynylphenyliodonium tosylates to dialkyl alkynylphosphonates with sodium dialkylphosphonates, *Synth. Commun.*, 28, 175, 1998.

46. Miller, S.I., and Dickstein, J.I., Nucleophilic substitution at acetylenic carbon. The last holdout, *Acc. Chem. Res.*, 9, 358, 1976.

47. Welch, C.M., Gonzales, E.J., and Guthrie, J.D., Derivatives of unsaturated phosphonic acids, *J. Org. Chem.*, 26, 3270, 1961.

48. Khachatryan, R.A., Ovsepyan, S.A., and Indzhikyan, M.G., Synthesis of 2-propenyl- and 2-propynyl-phosphonates from diethyl hydrogen phosphite and potassium carbonate, *Zh. Obshch. Khim.*, 57, 1709, 1987; *J. Gen. Chem. USSR (Engl. Transl.)*, 57, 1524, 1987.

49. Opaleva, E.N., Dogadina, A.V., and Ionin, B.I., Reaction of dialkyl hydrogen phosphites with halogen-containing alkenes and alkynes in the presence of potassium fluoride, *Zh. Obshch. Khim.*, 65, 1467, 1995; *Russ. J. Gen. Chem. (Engl. Transl.)*, 65, 1344, 1995.

50. Hunt, B.B., Saunders, B.C., and Simpson, P., Esters of ethynylphosphonic acid, *Chem. Ind. (London)*, 47, 1960.

51. Saunders, B.C., and Simpson, P., Esters containing phosphorus. Part 18. Esters of ethynylphosphonic acid, *J. Chem. Soc.*, 3351, 1963.

52. Tarasov, V.V., Arbisman, Y.S., Kondrat'ev, Y.A., and Ivin, S.Z., Investigations in the series of phosphorus-containing compounds with acetylene and allene groupings. Part 3. Infrared spectra of some derivatives of substituted alkynylphosphonic acids, *Zh. Obshch. Khim.*, 38, 130, 1968; *J. Gen. Chem. USSR (Engl. Transl.)*, 38, 129, 1968.

53. Chattha, M.S., and Aguiar, A.M., A convenient synthesis of 1-alkynylphosphonate, *J. Org. Chem.*, 36, 2719, 1971.

54. Tronchet, J.M.J., Bonenfant, A.P., Pallie, K.D., and Habashi, F., Phosphorus-containing sugars. Part 3. Derivatives of enose- and ynosephosphonates and related compounds, *Helv. Chim. Acta*, 62, 1622, 1979.

55. Guo, M., Synthesis and application of *O,O*-dialkyl alkynylphosphonates, *Huaxue Shiji*, 23, 51, 2001; *Chem. Abstr.*, 134, 366959, 2001.

56. Rudinskas, A.J., and Hullar, T.L., Pyridoxal phosphate. Part 5. 2-Formylethynylphosphonic acid and 2-formylethylphosphonic acid, potent inhibitors of pyridoxal phosphate binding probes of enzyme topography, *J. Med. Chem.*, 19, 1367, 1976.

57. L'vova, S.D., Kozlov, Y.P., and Gunar, V.I., Synthesis of diethyl esters of *cis*- and *trans*-2-(β-pyridyl)vinylphosphonic acids, *Zh. Obshch. Khim.*, 47, 1251, 1977; *J. Gen. Chem. USSR (Engl. Transl.)*, 47, 1153, 1977.

58. Stepanova, S.V., L'Vova, S.D., and Gunar, V.I., Synthesis of phosphonic acid analogs of pyridoxal 5'-phosphate modified at the 5'-position, *Bioorg. Khim.*, 4, 682, 1978; *Sov. J. Bioorg. Chem. (Engl. Transl.)*, 4, 498, 1978.

59. Poss, A.J., and Belter, R.K., Diethyl 3-iodopropynylphosphonate. An alkylative β-keto phosphonate equivalent, *J. Org. Chem.*, 52, 4810, 1987.

60. Acheson, R.M., and Ansell, P.J., The synthesis of diethyl (*p*-tolylsulfonyl)ethynylphosphonate and related acetylenes, and their reactions with nucleophiles. Pyridinium-1-dicyanomethylides, and dienes, *J. Chem. Soc., Perkin Trans. 1*, 1275, 1987.

61. Knierzinger, A., Grieder, A., and Schönholzer, P., Palladium-catalyzed ene-type cyclizations of terpenoid 1,6 enynes, *Helv. Chim. Acta*, 74, 517, 1991.

62. Ruder, S.M., and Norwood, B.K., Cycloaddition of enamines with alkynylphosphonates. A route to functionalized medium sized rings, *Tetrahedron Lett.*, 35, 3473, 1994.

63. Saalfrank, R.W., Welch, A., Haubner, M. and Bauer, U., 1-Halo-1-acceptor-/1,1-diacceptor-substituted allenes. Part 9. Functionalized allenes, haloallenes, and *bis*allenes via [2,3]/[3,3]-sigmatropic rearrangements and their reactivity, *Liebigs Ann.*, 171, 1996.

64. Gil, J.M., Sung, J.W., Park, C.P., and Oh, D.Y., One-pot synthesis of 1-alkynylphosphonates, *Synth. Commun.*, 27, 3171, 1997.

65. Dikusar, E.A., Beresnevich, L.B., Moiseichuk, K.L., Zalesskaya, E.G., and Yuvchenko, A.P., Synthesis of peroxy-containing alkynylphosphonates, *Zh. Obshch. Khim.*, 68, 576, 1998; *Russ. J. Gen. Chem.*, 68, 539, 1998.

66. Slowinski, F., Aubert, C., and Malacria, M., Highly stereoselective induction in the cobalt-mediated [2 + 2 + 2] cycloaddition of chiral phosphine oxides substituted linear enediynes, *Tetrahedron Lett.*, 40, 5849, 1999.

67. Baxter, R.J., Knox, G.R., McLaughlin, M., Pauson, P.L., and Spicer, M.D., The preparation and reactions of alkynylphosphonate hexacarbonyldicobalt complexes, *J. Organomet. Chem.*, 579, 83, 1999.

68. Jiao, X.-Y., and Bentrude, W.G., Vinylphosphonate formation via a novel cyclization-vinyl radical trapping sequence, *J. Am. Chem. Soc.*, 121, 6088, 1999.

69. Pergament, I., and Srebnik, M., Hydroboration of unsaturated phosphonic esters. Synthesis of boronophosphonates and trisubstituted vinylphosphonates, *Org. Lett.*, 3, 217, 2001.

70. Bookser, B.C., Dang, Q., and Reddy, K.R., Preparation of arylheterocycle phosphates as antidiabetics and aryl fructose-1,6-bisphosphatase inhibitors, *Metabasis Therapeutics*, Int. Patent Appl. WO 2001066553, 2001; *Chem. Abstr.*, 135, 211059, 2001.

71. Blanchard, C., Vaultier, M., and Mortier, J., Lithium *bis*(diisopropylamino)boracetylide [LiC≡C≡B(N*i*-Pr₂)₂]. A new reagent for the preparation of terminal alkynes, *Tetrahedron Lett.*, 38, 8863, 1997.

72. Cox, J.M., Hawkes, T.R., Bellini, P., Ellis, R.M., Barrett, R., Swanborough, J.J., Russell, S.E., Walker, P.A., Barnes, N.J., Knee, A.J., Lewis, T., and Davies, P.R., The design and synthesis of inhibitors of imidazoleglycerol phosphate dehydratase as potential herbicides, *Pestic. Sci.*, 50, 297, 1997.

73. Cox, J.M., Bellini, P., Barrett, R., Ellis, R.M., and Hawkes, T.R., Preparation of triazolylalkylphosphonic acids as herbicides, *Zeneca*, U.S. Patent Appl. WO 9315610, 1993; *Chem. Abstr.*, 120, 134813, 1994.

74. Yagudeev, T.A., Kushembaev, R.K., Nurgalieva, A.N., Zhumagaliev, S., Dzhakiyaev, G.M., and Godovikov, N.N., Synthesis of dialkyl [(1-chlorohexen-1-yl)ethynyl]-, [(3,6-dihydro-2,2-dimethyl-2*H*-pyran-4-yl)ethynyl]-, [(3,6-dihydro-2,2-dimethyl-2*H*-thiopyran-4-yl)ethynyl]-, and [(1,2,3,6-tetrahydro-1,2,5-trimethyl-4-pyridyl)ethynyl]-phosphonates, *Zh. Obshch. Khim.*, 50, 2236, 1980; *J. Gen. Chem. USSR (Engl. Transl.)*, 50, 1804, 1980.

75. Lera, M., and Hayes, C.J., A new one-pot synthesis of alkynylphosphonates, *Org. Lett.*, 2, 3873, 2000.

76. Jacobson, H.I., Griffin, M.J., Preis, S., and Jensen, E.V., Phosphonic acids. Part 4. Preparation and reactions of β-ketophosphonate and enol phosphate esters, *J. Am. Chem. Soc.*, 79, 2608, 1957.

77. Cymerman Craig, J., Bergenthal, M.D., Fleming, I., and Harley-Mason, J., Synthesis of alkynes from enol esters, *Angew. Chem., Int. Ed. Engl.*, 8, 429, 1969.

78. Negishi, E., King, A.O., and Klima, W.L., Conversion of methyl ketones into terminal acetylenes and (*E*)-trisubstituted olefins of terpenoid origin, *J. Org. Chem.*, 45, 2526, 1980.

79. Hong, J.E., Lee, C.-W., Kwon, Y., and Oh, D.Y., Facile synthesis of 1-alkynylphosphonates, *Synth. Commun.*, 26, 1563, 1996.

80. Shen, Y., and Qi, M., New synthesis of dialkyl fluoroalkynylphosphonates, *J. Chem. Soc., Perkin Trans. 1*, 2153, 1993.

81. Anisimov, K.N., and Nesmeyanov, A.N., Derivatives of unsaturated phosphonic acids. Part 15. Neutral esters of β,β-phenylchlorovinylphosphonic and β-phenylethynylphosphonic acids, *Izv. Akad. Nauk SSSR, Ser. Khim.*, 1006, 1955; *Chem. Abstr.*, 50, 11267h, 1956.

82. Anisimov, K.N., and Kopylova, B.V., An investigation in the field of the derivatives of unsaturated phosphinic acids. Part 24. Reaction of phosphorus pentachloride with alkoxyacetylenes, *Izv. Akad. Nauk SSSR, Ser. Khim.*, 277, 1961; *Bull. Acad. Sci. USSR, Div. Chem. Sci. (Engl. Transl.)*, 253, 1961.

83. Petrov, K.A., Raksha, M.A., and Le Dong, H., Synthesis and study of the properties of alkenephosphonic derivatives. Part 6. [2-Chloro- and 2-alkyl-2-(alkylthio)vinyl]phosphonic and -phosphonothioic diesters, *Zh. Obshch. Khim.*, 46, 1991, 1976; *J. Gen. Chem. USSR (Engl. Transl.)*, 46, 1918, 1976.

84. Normant, H., and Sturtz, G., Preparation of 1,3-diphosphonoacetone and its synthetic applications, *C.R. Acad. Sci.*, 260, 1984, 1965.

85. Sturtz, G., Elimination-addition reactions of β-halo-β-ethylenic phosphonates. Obtaining of β-oxo compounds, *Bull. Soc. Chim. Fr.*, 1345, 1967.

86. Slates, H.L., and Wendler, N.L., Novel and efficient syntheses of 1-propynyl- and 1-propenylphosphonates. Precursors of fosfomycin, *Chem. Ind. (London)*, 430, 1978.

87. Murayama, M., Matsumura, S., Etsure, Y., and Ozaki, M., Propynylphosphonate, *Nippon Shinyaku*, Japanese Patent Appl. JP 50010571, 1975; *Chem. Abstr.*, 83, 164368, 1975.

88. Hall, R.G., and Trippett, S., The preparation and Diels–Alder reactivity of ethyl (diethoxyphosphinyl)propynoate, *Tetrahedron Lett.*, 23, 2603, 1982.

89. Jungheim, L.N., and Sigmund, S.K., 1,3-Dipolar cycloaddition reactions of pyrazolidinium ylides with acetylenes. Synthesis of a new class of antibacterial agents, *J. Org. Chem.*, 52, 4007, 1987.

90. Dizière, R., and Savignac, P., A new simple method for the synthesis of 1-alkynylphosphonates using $(EtO)_2P(O)CCl_3$ as precursor, *Tetrahedron Lett.*, 37, 1783, 1996.

91. Shen, Y., Lin, Y., and Xin, Y., Application of elemento-organic compounds of the fifth and sixth groups in organic synthesis. Part 39. Synthesis of 1-perfluoroalkynyl phosphonates, *Tetrahedron Lett.*, 26, 5137, 1985.

92. Pudovik, A.N., and Aladzheva, I.M., Acetylene-allene-acetylenic rearrangements of phosphites with a β,γ-acetylene linkage in the ester radical, *Zh. Obshch. Khim.*, 33, 707, 1963; *J. Gen. Chem. USSR (Engl. Transl.)*, 33, 700, 1963.

93. Pudovik, A.N., and Aladzheva, I.M., Thermal or "pseudo-Claisen" rearrangement of allyl and propargyl esters of phosphorous acid, *Dokl. Akad. Nauk SSSR*, 151, 1110, 1963; *Dokl. Chem. (Engl. Transl.)*, 151, 634, 1963.

94. Pudovik, A.N., Aladzheva, I.M., and Yakovenko, L.M., Synthesis and rearrangement of diethyl propargyl phosphite, *Zh. Obshch. Khim.*, 33, 3444, 1963; *J. Gen. Chem. USSR (Engl. Transl.)*, 33, 3373, 1963.

95. Boisselle, A.P., and Meinhardt, N.A., Acetylene-allene rearrangements. Reactions of trivalent phosphorus chlorides with α-acetylenic alcohols and glycols, *J. Org. Chem.*, 27, 1828, 1962.

96. Mark, V., A facile $S_{N}i'$ rearrangement. The formation of 1,2-alkadienylphosphonates from 2-alkynyl phosphites, *Tetrahedron Lett.*, 3, 281, 1962.

97. Cherbuliez, E., Jaccard, S., Prince, R., and Rabinowitz, J., Formation and transformation of esters. Part 57. Acids and acid chlorides of P(III) and α-acetylenic alcohols. Esterification with or without rearrangement, *Helv. Chim. Acta*, 48, 632, 1965.

98. Guillemin, J.C., Savignac, P., and Denis, J.M., Primary alkynylphosphines and allenylphosphines, *Inorg. Chem.*, 30, 2170, 1991.

99. Khusainova, N.G., Bredikhina, Z.A., Sinitsa, A.D., Kal'chenko, V.I., and Pudovik, A.N., Reaction of dialkyl [α-(diethylamino)benzylidene]phosphoramidites with phosphorylated allenes and acetylenes, *Zh. Obshch. Khim.*, 52, 789, 1982; *J. Gen. Chem. USSR (Engl. Transl.)*, 52, 684, 1982.

100. Kise, M., Morita, I., and Tsuda, M., Preparation of cyclic phosphates as intermediates for circulation disease agents, *Nippon Shinyaku*, Japanese Patent Appl. JP 62226992, 1987; *Chem. Abstr.*, 111, 39581, 1989.

101. Morita, I., Tsuda, M., Kise, M., and Sugiyama, M., Improved synthesis of methyl 2,6-dimethyl-4-(2-nitrophenyl)-5-(2-oxo-1,3,2-dioxaphosphorinan-2-yl)-1,4-dihydropyridine-3-carboxylate (DHP-218), *Chem. Pharm. Bull.*, 36, 1139, 1988.

102. Ionin, B.I., and Petrov, A.A., Prototropic isomerization of phosphonic esters containing acetylenic, dienic, and enynic groups, *Zh. Obshch. Khim.*, 34, 1174, 1964; *J. Gen. Chem. USSR (Engl. Transl.)*, 34, 1165, 1964.

103. Abramov, V.S., and Il'ina, N.A., Rearrangement of mixed dialkylamidophosphites, *Zh. Obshch. Khim.*, 38, 677, 1968; *J. Gen. Chem. USSR (Engl. Transl.)*, 38, 656, 1968.

104. Pines, S.H., and Karady, S., Preparation of 1-propynylphosphoric acid and salts thereof, *Merck*, German Patent Appl. DE 1924148, 1970; *Chem. Abstr.*, 72, 132965, 1970.

105. Pudovik, A.N., Aladzheva, I.M., and Yakovenko, L.M., Synthesis and rearrangements of dialkyl 2-propynyl phosphites and dialkyl propadienylphosphonates, *Zh. Obshch. Khim.*, 35, 1210, 1965; *J. Gen. Chem. USSR (Engl. Transl.)*, 35, 1214, 1965.

106. Ivakina, N.M., Kondrat'ev, Y.A., and Ivin, S.Z., Propargylpyrocatechinic ester of phosphorous acid, *Zh. Obshch. Khim.*, 37, 1691, 1967; *J. Gen. Chem. USSR (Engl. Transl.)*, 37, 1612, 1967.

107. Kondrat'ev, Y.A., Tarasov, V.V., Vasil'ev, A.S., Ivakina, N.M., and Ivin, S.Z., Organophosphorus-containing compounds with acetylenic and allenic groups. Part 4. Synthesis and thermal rearrangement of cyclic *o*-phenylene 2-propynyl phosphite, *Zh. Obshch. Khim.*, 38, 1791, 1968; *J. Gen. Chem. USSR (Engl. Transl.)*, 38, 1745, 1968.

108. Iuchi, K., and Iwashiro, S., *cis*-Propenylphosphonic acid derivatives, *Kanebo*, Japanese Patent Appl. JP 52131532, 1977; *Chem. Abstr.*, 88, 105562, 1978.

109. Welter, W., Hartmann, A., and Regitz, M., Carbenes. Part 18. Isomerization reactions of phosphoryl-vinyl-carbenes to phosphorylated cyclopropenes, allenes, acetylenes, indenes, and 1,3-butadienes, *Chem. Ber.*, 111, 3068, 1978.

110. Christensen, B.G., Beattie, T.R., and Leanza, W.J., *cis*-Propenylphosphonic acid and derivatives, *Merck*, German Patent Appl. DE 1805676, 1968; *Chem. Abstr.*, 72, 67108, 1970.

111. Christensen, B.G., Leanza, W.J., Beattie, T.R., Patchett, A.A., Arison, B.H., Ormond, R.E., Kuehl, F.A., Jr., Albers-Schonberg, G., and Jardetzky, O., Phosphonomycin. Structure and synthesis, *Science*, 166, 123, 1969.

112. Christensen, B.G., Albers-Schonberg, G., and Leanza, W.J., Antibiotic (−)-*cis*-(1,2-epoxypropyl)phosphonates, *Merck*, German Patent Appl. DE 1805682, 1973; *Chem. Abstr.*, 72, 67109, 1970.

113. Christensen, B.G., Leanza, W.J., and Albers-Schonberg, G., Labile esters of (−)-*cis*-(1,2-epoxypropyl)phosphonic acid, *Merck*, U.S. Patent Appl. US 3929840, 1975; *Chem. Abstr.*, 84, 90302, 1976.

114. Rudinskas, A.J., and Hullar, T.L., Phosphonic acid chemistry. Part 1. Synthesis and dienophilic properties of diethyl 2-formylvinylphosphonate and diethyl 2-formylethynylphosphonate, *J. Org. Chem.*, 41, 2411, 1976.

115. Cristau, H.-J., Gasc, M.-B., and Mbianda, X.Y., A convenient stereoselective synthesis of disubstituted alk-1-enyl phosphonates, *J. Organomet. Chem.*, 474, C14, 1994.

116. Cristau, H.-J., Mbianda, X.Y., Beziat, Y., and Gasc, M.-B., Facile and stereoselective synthesis of vinylphosphonates, *J. Organomet. Chem.*, 529, 301, 1997.

117. Cristau, H.J., Pirat, J.L., Drag, M., and Kafarski, P., Regio- and stereoselective synthesis of 2-amino-1-hydroxy-2-aryl ethylphosphonic esters, *Tetrahedron Lett.*, 41, 9781, 2000.

118. Blackburn, G.M., Forster, A.R., Guo, M.-J., and Taylor, G.E., Stereochemical studies on some *vic*-bisphosphonates, *J. Chem. Soc. Perkin Trans. 1*, 2867, 1991.

119. Machida, Y., and Saito, I., Facile synthesis of 1,2-oxaphosphol-3-ene derivatives, *J. Org. Chem.*, 44, 865, 1979.

120. Machida, Y., and Saito, I., 2-Hydroxy(or alkoxy)-2-oxo-1,2-oxaphosphol-3-enes, *Eisai*, Japanese Patent Appl. JP 55035043, 1980; *Chem. Abstr.*, 93, 114706, 1980.

121. Bigge, C.F., Johnson, G., Ortwine, D.F., Drummond, J.T., Retz, D.M., Brahce, L.J., Coughenour, L.L., Marcoux, F.W., and Probert, A.W., Jr., Exploration of *N*-phosphonoalkyl-, *N*-phosphonoalkenyl-, and *N*-(phosphonoalkyl)phenyl-spaced α-amino acids as competitive *N*-methyl-D-aspartic acid antagonists, *J. Med. Chem.*, 35, 1371, 1992.

122. Bigge, C.F., and Johnson, G., *N*-Substituted α-amino acids and derivatives thereof having pharmaceutical activity, *Warner-Lambert*, U.S. Patent Appl. US 5179085, 1993; *Chem. Abstr.*, 119, 96171, 1993.

123. Hayakawa, K., Mori, I., Iwasaki, G., and Matsunaga, S., Preparation of triazolylalkylphosphonates as herbicides, *Japat*, Eur. Patent Appl. EP 528760, 1993; *Chem. Abstr.*, 119, 72837, 1993.

124. Christov, V.C., Aladinova, V.M., and Prodanov, B., Hydration reactions of phosphorylated 1,3-enynes, *Phosphorus, Sulfur Silicon Relat. Elem.*, 155, 67, 1999.

125. Todd, R.S., Reeve, M., and Davidson, A.H., Preparation of phosphorus containing alkynyl derivatives useful as intermediates in the preparation of keto phosphonates and mevinolinic acid derivatives, *British Bio-Technology*, Int. Patent Appl. WO 9322321, 1992; *Chem. Abstr.,* 120, 245504, 1994.

126. Corey, E.J., and Virgil, S.C., Enantioselective total synthesis of a protosterol, 3β,20-dihydroxyprotost-24-ene, *J. Am. Chem. Soc.*, 112, 6429, 1990.

127. Guile, S.D., Saxton, J.E., and Thornton-Pett, M., Synthetic studies towards paspalicine. Part 2. An alternative approach to the synthesis of the C/D ring system, *J. Chem. Soc., Perkin Trans. 1*, 1763, 1992.

128. Peiffer, G., and Courbis, P., New access route to γ,δ-ethylenic β-keto phosphonates, *Can. J. Chem.*, 52, 2894, 1974.

129. Öhler, E. and Zbiral, E., Synthesis, reactions and NMR spectra of dialkyl 2-bromo-3-oxo-1-alke-nylphosphonates and dialkyl 3-oxo-1-alkynylphosphonates, *Monatsh. Chem.*, 115, 493, 1984.

130. Pudovik, A.N., Khusainova, N.G., and Timoshina, T.V., Cycloaddition of ethyl diazoacetate to 1-propynylphosphonates, 1-propynylphosphine oxides, and (3-methyl-1,2-butadienyl)phosphonates, *Zh. Obshch. Khim.*, 44, 272, 1974; *J. Gen. Chem. USSR (Engl. Transl.)*, 44, 257, 1974.

131. Khusainova, N.G., Irtuganova, E.A., and Cherkasov, R.A., Reactions of nitrones with allenyl- and propynylphosphonates, *Zh. Obshch. Khim.*, 65, 1115, 1995; *Russ. J. Gen. Chem. (Engl. Transl.)*, 65, 1017, 1995.

132. Pudovik, A.N., Khusainova, N.G., Berdnikov, E.A., and Nasybullina, Z.A., 1-Propynylphosphonates and 1-propynylphosphine oxides in cycloaddition reactions with phenyl azide, *Zh. Obshch. Khim.*, 44, 222, 1974; *J. Gen. Chem. USSR (Engl. Transl.)*, 44, 213, 1974.

133. Pudovik, A.N., Khusainova, N.G., Bredikhina, Z.A., and Berdnikov, E.A., Effects of substituents in the cycloaddition of azides to phosphinylacetylenes, *Dokl. Akad. Nauk SSSR, Ser. Khim.*, 226, 364, 1976; *Dokl. Chem. (Engl. Transl.)*, 226, 52, 1976.

134. Khusainova, N.G., Trishin, Y.G., Irtuganova, E.A., Tamm, L.A., Chistokletov, V.N., and Pudovik, A.N., Cycloaddition of nitrilimines to allenyl- and propynylphosphonates, *Zh. Obshch. Khim.*, 61, 601, 1991; *J. Gen. Chem. USSR (Engl. Transl.)*, 61, 545, 1991.

135. Khusainova, N.G., Galkin, V.I., and Cherkasov, R.A., Steric effect of substituents in reactions of 1,3-dipolar cycloaddition to phosphorylated allenes and acetylenes, *Zh. Obshch. Khim.*, 60, 995, 1990; *J. Gen. Chem. USSR (Engl. Transl.)*, 60, 876, 1990.

136. Shen, Y., Zheng, J., Xin, Y., Lin, Y., and Qi, M., Synthesis of perfluoroalkylated heterocyclic phosphonates, *J. Chem. Soc., Perkin Trans. 1*, 997, 1995.

137. Palacios, F., Ochoa de Retana, A.M., Pagalday, J., and Sanchez, J.M., Cycloadditions of azidoalkyl-carboxylates to acetylenes and enamines. Regioselective synthesis of substituted triazoles, *Org. Prep. Proc. Int.*, 27, 603, 1995.

138. Palacios, F., Ochoa de Retana, A.M., and Pagalday, J., Synthesis of diethyl 1,2,3-triazolealkylphos-phonates through 1,3-dipolar cycloaddition of azides with acetylenes, *Heterocycles*, 38, 95, 1994.

139. Louërat, F., Bougrin, K., Loupy, A., Ochoa de Retana, A.M., Pagalday, J., and Palacios, F., Cycload-dition reactions of azidomethyl phosphonate with acetylenes and enamines. Synthesis of triazoles, *Heterocycles*, 48, 161, 1998.

140. Tronchet, J.M.J., and Bonenfant, A.P., *C*-Glycosyl derivatives. Part 42. Synthesis of novel types of *C*-glycosyl-derivatives from acetylenic sugars or their partial synthetic equivalents. Preliminary communication, *Helv. Chim. Acta*, 64, 2322, 1981.

141. Pudovik, A.N., Khusainova, N.G., and Frolova, T.I., 1,3-Dipolar addition of *N*-phenylsydnone to 1-propynylphosphonic esters, *O,O*-diethyl 1-propynylphosphonothioate, and diphenyl 1-propynylphos-phine oxide, *Zh. Obshch. Khim.*, 41, 2420, 1971; *J. Gen. Chem. USSR (Engl. Transl.)*, 41, 2446, 1971.

142. Matoba, K., Yonemoto, H., Fukui, M., and Yamazaki, T., Structural modification of bioactive compounds. Part 2. Syntheses of aminophosphonic acids, *Chem. Pharm. Bull.*, 32, 3918, 1984.

143. Tverdomed, S.N., Dogadina, A.V., and Ionin, B.I., Substituted phosphonates and diphosphonates. The synthesis strategy, *Russ. J. Gen. Chem. (Engl. Transl.)*, 71, 1821, 2001; *Chem. Abstr.*, 137, 20416, 2002.

144. Lelièvre, S., Mercier, F., and Mathey, F., Phosphanorbornadienephosphonates as a new type of water-soluble phosphines for biphasic catalysis, *J. Org. Chem.*, 61, 3531, 1996.

145. Stepanova, S.V., Reaction of 4-methyl-5-propoxyoxazole with substituted vinylethynylcarbinols, *Zh. Org. Khim.*, 12, 1568, 1976; *J. Org. Chem. USSR (Engl. Transl.)*, 12, 1544, 1976.

146. Ziegler, T., Layh, M., and Effenberger, F., Syntheses of highly substituted benzenes via Diels–Alder reactions with 2*H*-pyran-2-ones, *Chem. Ber.*, 120, 1347, 1987.

147. Ma, C., Lu, X., and Ma, Y., Novel synthesis of 3-chloromethylidene-2-ethoxy-1,2λ^5-oxaphospholan-2-ones catalysed by palladium(II), *J. Chem. Soc., Perkin Trans. 1*, 2683, 1995.

148. Timmer, M.S.M., Ovaa, H., Filippov, D.V., van der Marel, G.A., and van Boom, J.H., Synthesis of phosphorus mono- and bicycles by catalytic ring-closing metathesis, *Tetrahedron Lett.*, 42, 8231, 2001.

149. Quntar, A.A.A., and Srebnik, M., Carbon–carbon bond formation of alkenylphosphonates by aldehyde insertion into zirconacycle phosphonates, *J. Org. Chem.*, 66, 6650, 2001.

150. Quntar, A.A.A., Melman, A., and Srebnik, M., Highly selective preparation of 2 (hydroxymethyl)vinylphosphonates by insertion of ketones into zirconacycle phosphonates, *Synlett*, 61, 2002.

151. Quntar, A.A.A., Melman, A., and Srebnik, M., Selective preparation of (*E*)-3-oxo-1-alkenylphosphonates by insertion of acyl chlorides and nitriles into zirconacycles, *J. Org. Chem.*, 67, 3769, 2002.

152. Quntar, A.A.A., and Srebnik, M., *cis*-Vinylphosphonates and 1,3-butadienylphosphonates by zirconation of 1-alkynylphosphonates, *Org. Lett.*, 3, 1379, 2001.

153. Huang, X., Zhang, C., and Lu, X., A convenient stereoselective synthesis of 1,3-dienylphosphonates and 1-en-2-ynylphosphonates and their phosphine oxide analogs, *Synthesis*, 769, 1995.

154. Lazrek, H.B., Redwane, N., Rochdi, A., Barascut, J.-L., Imbach, J.-L., and De Clercq, E., Synthesis of acycloalkenyl derivatives of pyrimidines and purines, *Nucleosides Nucleotides*, 14, 353, 1995.

155. Lazrek, H.B., Khaïder, H., Rochdi, A., Barascut, J.-L., and Imbach, J.-L., Synthesis of new acyclic nucleoside phosphonic acids by Michael addition, *Tetrahedron Lett.*, 37, 4701, 1996.

156. Lazrek, H.B., Rochdi, A., Khaïder, H., Barascut, J.-L., Imbach, J.-L., Balzarini, J., Witvrouw, M., Pannecouque, C., and De Clercq, E., Synthesis of (*Z*) and (*E*) α-alkenyl phosphonic acid derivatives of purines and pyrimidines, *Tetrahedron*, 54, 3807, 1998.

157. Chattha, M.S., and Aguiar, A.M., Enamine phosphonates. Their use in the synthesis of α,β-ethylenic ketimines and the corresponding ketones, *Tetrahedron Lett.*, 12, 1419, 1971.

158. Chattha, M.S., and Aguiar, A.M., Organophosphorus enamines. Part 7. Synthesis and stereochemistry of enamine phosphonates, *J. Org. Chem.*, 38, 820, 1973.

159. Chattha, M.S., and Aguiar, A.M., Organophosphorus enamines. Part 8. Convenient preparation of diethyl β-ketophosphonates, *J. Org. Chem.*, 38, 2908, 1973.

160. Chattha, M.S., Synthesis and NMR spectrum of diethyl 3-buten-3-methyl-2-oxo-1-phosphonate, *Chem. Ind. (London)*, 1031, 1976.

161. Panarina, A.E., Dogadina, A.V., and Ionin, B.I., Addition of *tert*-butylamine to diethyl alkynephosphonates catalyzed by CuCl, *Russ. J. Gen. Chem. (Engl. Transl.)*, 71, 147, 2001; *Chem. Abstr.*, 135, 331485, 2001.

162. Panarina, A.E., Dogadina, A.V., Zakharov, V.I., and Ionin, B.I., Addition of secondary amines to alkynylphosphonates, *Tetrahedron Lett.*, 42, 4365, 2001.

163. Whitesell, M.A., and Kyba, E.P., Addition of amine nucleophiles to diphosphorylalkynes. The chemistry of the derived enamines, *Tetrahedron Lett.*, 24, 1679, 1983.

164. Sauveur, F., Collignon, N., Guy, A., and Savignac, P., Access to optically active 2-aminopropylphosphonic acid, *Phosphorus Sulfur*, 14, 341, 1983.

165. Pudovik, A.N., Khusainova, N.G., and Aladzheva, I.M., Nucleophilic addition reactions of alkynylphosphonic esters, *Zh. Obshch. Khim.*, 33, 1045, 1963; *J. Gen. Chem. USSR (Engl. Transl.)*, 33, 1034, 1963.

166. Khusainova, N.G., Romanov, G.V., Nazmutdinov, R.Y., and Pudovik, A.N., Reactinos of *bis*(trimethylsilyl) hypophosphite with phosphorylated allenes and acetylenes, *Zh. Obshch. Khim.*, 51, 2202, 1981; *J. Gen. Chem. USSR (Engl. Transl.)*, 51, 1893, 1981.

167. Jiang, G.F., Sun, J., and Shen, Y., Synthesis, reactivity and crystal structure of perfluoroalkylated diethoxyphosphinyl triphenylphosphorane, *J. Fluorine Chem.*, 108, 207, 2001.

168. Huang, X., Liang, C.-G., Xu, Q., and He, Q.-W., Alkyne-based, highly stereo- and regioselective synthesis of stereodefined functionalized vinyl tellurides, *J. Org. Chem.*, 66, 74, 2001.

169. Pudovik, A.N., Khusainova, N.G., and Galeeva, R.G., Addition of compounds with an active hydrogen atom in a methylene group to esters of propynylphosphonic acid, *Zh. Obshch. Khim.*, 36, 69, 1966; *J. Gen. Chem. USSR (Engl. Transl.)*, 36, 73, 1966.

170. Zborovskii, Y.L., Levon, V.F., and Staninets, V.I., Heterocyclization of phenylethynylphosphonic acid under the action of selenium dioxide or hydrogen bromide, *Zh. Obshch. Khim.*, 64, 1567, 1994; *Russ. J. Gen. Chem. (Engl. Transl.)*, 64, 1401, 1994.

171. Pudovik, A.N., and Khusainova, N.G., Addition of diethyl phosphorothioate to esters of unsaturated phosphonic acids, *Zh. Obshch. Khim.*, 36, 1345, 1966; *J. Gen. Chem. USSR (Engl. Transl.)*, 36, 1359, 1966.

172. Pudovik, A.N., Khusainova, N.G., and Ageeva, A.B., Reactions of nucleophilic reagents with esters of propynylphosphinic acid, *Zh. Obshch. Khim.*, 34, 3938, 1964; *J. Gen. Chem. USSR (Engl. Transl.)*, 34, 3998, 1964.

173. Khusainova, N.G., Naumova, L.V., Berdnikov, E.A., and Pudovik, A.N., Addition of arenesulfenyl chlorides to diethyl 1-propynylphosphonate, *Zh. Obshch. Khim.*, 54, 1971, 1984; *J. Gen. Chem. USSR (Engl. Transl.)*, 54, 1758, 1984.

174. Braga, A.L., Alves, E.F., Silveira, C.C., and de Andrade, L.H., Stereoselective addition of sodium organyl chalcogenolates to alkynylphosphonates. Synthesis of diethyl 2-(organyl)-2-(organochalcogenyl)vinylphosphonates, *Tetrahedron Lett.*, 41, 161, 2000.

175. Gil, J.M., and Oh, D.Y., Carbocupration of diethyl 1-alkynylphosphonates. Stereo- and regioselective synthesis of 1,2,2-trisubstituted vinylphosphonates, *J. Org. Chem.*, 64, 2950, 1999.

176. Pergament, I., and Srebnik, M., Control of hydroboration of 1-alkynylphosphonates, followed by Suzuki coupling provides regio- and stereospecific synthesis of di-substituted 1-alkenylphosphonates, *Tetrahedron Lett.*, 42, 8059, 2001.

177. Hayashi, T., Inoue, K., Taniguchi, N., and Ogasawara, M., Rhodium-catalyzed hydroarylation of alkynes with arylboronic acids. 1,4-Shift of rhodium from 2-aryl-1-alkenylrhodium to 2-alkenylarylrhodium intermediate, *J. Am. Chem. Soc.*, 123, 9918, 2001.

178. Garibina, V.A., Dogadina, A.V., Ionin, B.I., and Petrov, A.A., Phosphorus-containing ynamines, *Zh. Obshch. Khim.*, 49, 2385, 1979; *J. Gen. Chem. USSR (Engl. Transl.)*, 49, 2104, 1979.

179. Ionin, B.I., and Petrov, A.A., Diethyl (diethylamino)ethynylphosphonate, *Zh. Obshch. Khim.*, 35, 2255, 1965; *J. Gen. Chem. USSR (Engl. Transl.)*, 35, 2247, 1965.

180. Lukashev, N.V., Kazantsev, A.V., Borisenko, A.A., and Beletskaya, I.P., Cyanation of nucleophilic alkynes. Easy approach to substituted α-cyanoenamines, *Tetrahedron*, 57, 10309, 2001.

181. Leonov, A.A., Dogadina, A.V., Ionin, B.I., and Petrov, A.A., Phosphorylated aldoketenimines, *Zh. Obshch. Khim.*, 53, 233, 1983; *J. Gen. Chem. USSR (Engl. Transl.)*, 53, 205, 1983.

182. Leonov, A.A., Tuzhikov, O.I., Lomakin, V.Y., Komarov, V.Y., Dogadina, A.V., Ionin, B.I., and Petrov, A.A., Reactions of (chloroethynyl)phosphonates with primary phosphines, *Zh. Obshch. Khim.*, 54, 1422, 1984; *J. Gen. Chem. USSR (Engl. Transl.)*, 54, 1269, 1984.

183. Leonov, A.A., Komarov, V.Y., Dogadina, A.V., Ionin, B.I., and Petrov, A.A., Phosphorylated aldoketenimines. Synthesis and properties, *Zh. Obshch. Khim.*, 55, 32, 1985; *J. Gen. Chem. USSR (Engl. Transl.)*, 55, 26, 1985.

184. Garibina, V.A., Dogadina, A.V., Ionin, B.I., and Petrov, A.A., Phosphorus-containing acetylenic ethers and thio ethers, *Zh. Obshch. Khim.*, 49, 2152, 1979; *J. Gen. Chem. USSR (Engl. Transl.)*, 49, 1888, 1979.

185. Garibina, V.A., Leonov, A.A., Dogadina, A.V., Ionin, B.I., and Petrov, A.A., Interaction of haloacetylenephosphonates with anionic nucleophiles, *Zh. Obshch. Khim.*, 55, 1994, 1985; *J. Gen. Chem. USSR (Engl. Transl.)*, 55, 1771, 1985.

186. Drozd, V.N., Komarova, E.N., and Garibina, V.A., Reaction of (chloroethynyl)phosphonates with salts of vinylidenedithiols, *Zh. Org. Khim.*, 23, 2467, 1987; *J. Org. Chem. USSR (Engl. Transl.)*, 23, 2180, 1987.

187. Garibina, V.A., Leonov, A.A., Dogadina, A.V., Ionin, B.I. and Petrov, A.A., Reaction of (chloroethynyl)phosphonates with neutral nucleophiles, *Zh. Obshch. Khim.*, 57, 1481, 1987; *J. Gen. Chem. USSR (Engl. Transl.)*, 57, 1321, 1987.

188. Ma, C., Lu, X., and Ma, Y., Palladium(0) catalyzed isomerization of alkynyldiethyl phosphonate and alkynyldiphenyl phosphine oxide, *Main Group Met. Chem.*, 18, 391, 1995.

189. Ma, C.L., Lu, X.Y., and Ma, Y.X., A convenient stereoselective synthesis of conjugated diethyl 1*E*,3*E*-dienylphosphonates and 1*E*,3*E*-dienyldiphenylphosphine oxides, *Chin. Chem. Lett.*, 6, 747, 1995; *Chem. Abstr.*, 124, 56093, 1995.

190. Orain, D., and Guillemin, J.-C., Synthesis of functionalized deuterioallylic compounds, *J. Org. Chem.*, 64, 3563, 1999.

191. Harde, C., Neff, K.-H., Nordhoff, E., Gerbling, K.-P., Laber, B., and Pohlenz, H.-D., Syntheses of homoserine phosphate analogs as potential inhibitors of bacterial threonine synthase, *Bioorg. Med. Chem. Lett.*, 4, 273, 1994.

192. Blackburn, G.M., and Ingleson, D., Specific dealkylation of phosphonate esters using iodotrimethyl-silane, *J. Chem. Soc., Chem. Commun.*, 870, 1978.

193. McKenna, C.E., and Schmidhauser, J., Functional selectivity in phosphonate ester dealkylation with bromotrimethylsilane, *J. Chem. Soc., Chem. Commun.*, 739, 1979.

194. Blackburn, G.M., and Ingleson, D., The dealkylation of phosphate and phosphonate esters by iodo-trimethylsilane. A mild and selective procedure, *J. Chem. Soc., Perkin Trans. 1*, 1150, 1980.

195. Blackburn, G.M., Kent, D.E., and Kolkmann, F., The synthesis and metal binding characteristics of novel, isopolar phosphonate analogs of nucleotides, *J. Chem. Soc., Perkin Trans. 1*, 1119, 1984.

196. Blackburn, G.M., Eckstein, F., Kent, D.E., and Perree, T.D., Isopolar vs. isosteric phosphonate analogs of nucleotides, *Nucleosides Nucleotides*, 4, 165, 1985.

197. Morise, X., Savignac, P., Guillemin, J.C., and Denis, J.M., A convenient method for the synthesis of α-functionalized chlorophosphonic esters, *Synth. Commun.*, 21, 793, 1991.

198. Khusainova, N.G., Mironov, V.F., and Cherkasov, R.A., Synthesis of ethyl 1-propynylphosphonochlo-ridate by reaction of trichloro(*o*-phenylenedioxy)phosphorane with diethyl 1-propynylphosphonate, *Zh. Obshch. Khim.*, 65, 1578, 1995; *Russ. J. Gen. Chem. (Engl. Transl.)*, 65, 1445, 1995.

199. Mashlyakovskii, L.N., and Ionin, B.I., Unsaturated phosphonic acids and their derivatives. Part 1. Synthesis of alkadienyl-, alkenynyl-, and alkynyl-phosphonic dichlorides, *Zh. Obshch. Khim.*, 35, 1577, 1965; *J. Gen. Chem. USSR (Engl. Transl.)*, 35, 1582, 1965.

2 The Silylphosphonates

Silylated phosphonates have been known since 1956,[1] but their use as reagents in phosphorus chemistry has become important only in recent years.[2–8] The first participation of the trimethylsilyl group in phosphonate chemistry was reported in 1972,[9] as an extension of the Peterson reaction, and trimethylsilyl was first used as a protecting group in 1988.[10] Undoubtedly, the combination of phosphonate and trimethylsilyl groups has given rise to new synthetic developments and applications for phosphonate chemistry.

2.1 SYNTHESIS OF SILYLPHOSPHONATES

2.1.1 DIALKYL 1-SILYLALKYLPHOSPHONATES

2.1.1.1 Michaelis–Arbuzov Reactions

Diethyl 1-(trimethylsilyl)methylphosphonate, the most useful reagent, was first prepared in 57% yield by refluxing a mixture of triethyl phosphite and (chloromethyl)trimethylsilane for 68.5 h at 100–185°C (Scheme 2.1).[1] Reduction of reaction time significantly reduces the yield.[11] Replacement of triethyl phosphite by trimethyl phosphite has little effect on the yield, which is comparable to the previous one (65%).[12] By contrast, the use of (bromomethyl)trimethylsilane in reaction with triethyl phosphite dramatically decreases the yield to 25%.[13] In further examples, chloromethylsilanes bearing different groups (Et, Ph) on the silicon atom have been used for the preparation of the corresponding silylmethylphosphonates in yields ranging from 44% to 75%.[14–16] One example of conversion of chloromethyl- to iodomethylsilanes followed by the Michaelis–Arbuzov reaction has been reported.[17] Secondary α-silylated bromides are reactive, but the reactions require severe conditions (3–12 h at 170–200°C). Thus, by the Michaelis–Arbuzov rearrangement of triethyl phosphite with α-(trimethylsilyl)- and α-(triethylsilyl)benzyl bromides, the corresponding α-silylated benzylphosphonates have been obtained in 56% and 24% yields, respectively.[18] Similarly, diethyl 1-(methylthio)-1-(trimethylsilyl)methylphosphonate is prepared in 20% yield by heating to reflux a mixture of triethyl phosphite and 1-bromo-1-(methylthio)tetramethylsilane.[19,20] Because of the formation of side products, the yield cannot be increased. Triethyl phosphite reacts with (trimethylsilyl)formaldehyde O,S-dimethyl acetal in CH_2Cl_2 at low temperature in the presence of $TiCl_4$ or $SnCl_4$ to give diethyl 1-(phenylthio)-1-(trimethylsilyl)methylphosphonate in 47% yield in a highly selective Michaelis–Arbuzov-type reaction.[21,22]

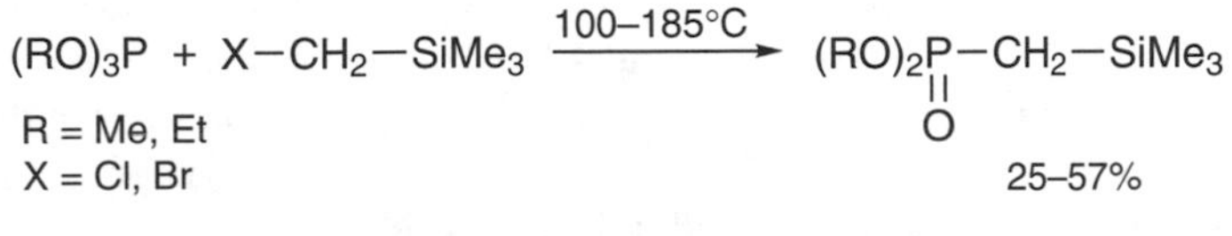

$$(RO)_3P \ + \ X{-}CH_2{-}SiMe_3 \xrightarrow{100{-}185°C} (RO)_2\overset{\displaystyle \|}{\underset{\displaystyle O}{P}}{-}CH_2{-}SiMe_3$$

R = Me, Et
X = Cl, Br   25–57%   (2.1)

2.1.1.2 Michaelis–Becker Reactions

(Halogenomethyl)trimethylsilanes (X = Cl, Br) undergo the typical Michaelis–Becker reaction with a variety of dialkyl phosphites to give dialkyl 1-(trimethylsilyl)methylphosphonates in modest yields (23–38%, Scheme 2.2).[12,13,23,24] This method, which is not economical in terms of yields, has only limited utility and has never been developed.

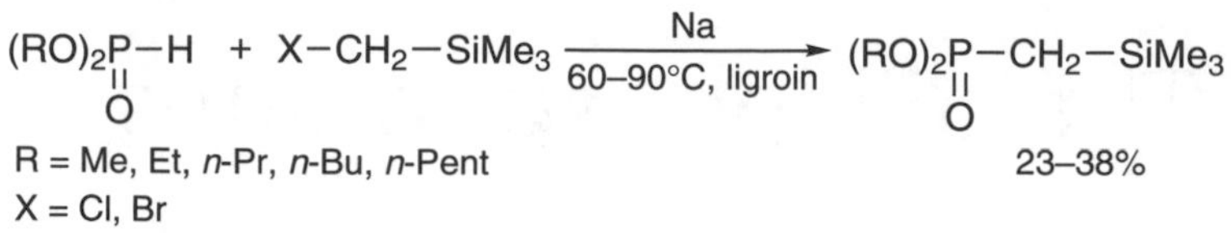

$$(RO)_2\underset{\underset{O}{\|}}{P}-H \;+\; X-CH_2-SiMe_3 \xrightarrow[60-90^\circ C,\ ligroin]{Na} (RO)_2\underset{\underset{O}{\|}}{P}-CH_2-SiMe_3$$

R = Me, Et, n-Pr, n-Bu, n-Pent   23–38%
X = Cl, Br

$$(2.2)$$

2.1.1.3 Kinnear–Perren Reactions

The 1-(trimethylsilyl)methylphosphonic dichloride is prepared in large scale, but in modest yield (31%), by reaction among (chloromethyl)trimethylsilane, phosphorus trichloride, and aluminum trichloride followed by hydrolysis with a limited amount of water. Although this procedure gives moderate yields, the large number of phosphonic diesters subsequently prepared in fair to good yields (40–71%) by alcoholysis of 1-(trimethylsilyl)methylphosphonic dichloride is compensatory (Scheme 2.3).[25,26]

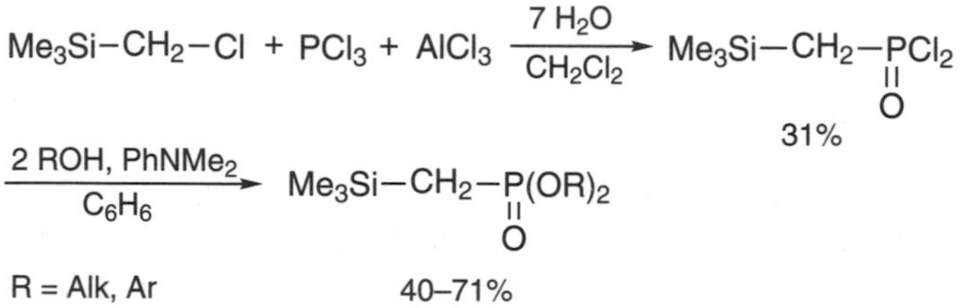

$$Me_3Si-CH_2-Cl \;+\; PCl_3 \;+\; AlCl_3 \xrightarrow[CH_2Cl_2]{7\ H_2O} Me_3Si-CH_2-\underset{\underset{O}{\|}}{P}Cl_2$$

31%

$$\xrightarrow[C_6H_6]{2\ ROH,\ PhNMe_2} Me_3Si-CH_2-\underset{\underset{O}{\|}}{P}(OR)_2$$

R = Alk, Ar   40–71%

$$(2.3)$$

1-(Trimethylsilyl)methylphosphonic dichloride can also be prepared in a yield roughly comparable to the previous one (23%) by reaction of tetramethylsilane with a large excess of phosphorus trichloride in the presence of oxygen.[27]

2.1.1.4 Carbanionic Silylations

Unquestionably, the most attractive procedure for the preparation of dialkyl 1-(trimethylsilyl)methylphosphonates is the carbanionic route. The obvious transmetallation difficulties occurring when carbanionic reagents are employed in conjunction with activated methylene groups have led to the development of procedures based on the trapping of α-metallated phosphonates with chlorosilanes in the presence of lithium diisopropylamide (LDA) in excess.[28-30] For example, the addition at −70°C of chlorotrimethylsilane (1 eq) to a solution of dialkyl 1-lithioalkylphosphonates prepared from dialkyl alkylphosphonates (1 eq) and LDA (2 eq) produces, after workup, dialkyl 1-(trimethylsilyl)alkylphosphonates in 75–90% yields of isolated product, via the quantitative and clean generation of stable dialkyl 1-lithio-1-(trimethylsilyl)alkylphosphonates (Scheme 2.4).[28-30] Varying the phosphorus reactants among dialkyl methyl-, alkyl-, halogenomethyl-, or benzylphosphonates readily produces a number of dialkyl 1-(trimethylsilyl)phosphonates, thus significantly extending the scope and synthetic utility of the carbanionic route.[30-32] This expeditious and versatile technique has largely supplanted the previously reported carbanionic approaches. For example, addition of an excess of dimethyl 1-lithiomethylphosphonate, generated with n-BuLi (1 eq), to a cooled THF solution of Me$_3$SiCl or addition of Me$_3$SiCl in excess to a cooled solution of dimethyl 1-lithiomethylphosphonate resulted in the two cases in low yields (27–31%) of dimethyl 1-(trimethylsilyl)methylphosphonate.[33,34]

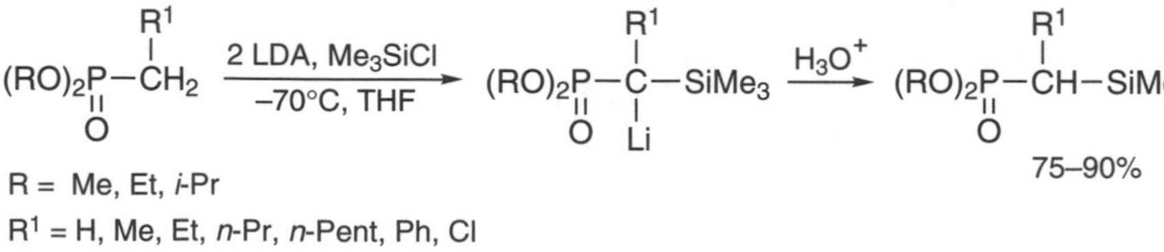

$$(RO)_2\underset{\underset{O}{\|}}{P}-\overset{\overset{R^1}{|}}{C}H_2 \xrightarrow[-70^\circ C,\ THF]{2\ LDA,\ Me_3SiCl} (RO)_2\underset{\underset{O}{\|}}{P}-\overset{\overset{R^1}{|}}{\underset{\underset{Li}{|}}{C}}-SiMe_3 \xrightarrow{H_3O^+} (RO)_2\underset{\underset{O}{\|}}{P}-\overset{\overset{R^1}{|}}{C}H-SiMe_3$$

R = Me, Et, i-Pr   75–90%
R^1 = H, Me, Et, n-Pr, n-Pent, Ph, Cl

$$(2.4)$$

The synthetic advantages of the carbanionic route are evident. Compared with the Michaelis–Arbuzov or Michaelis–Becker routes involving (chloromethyl)trimethylsilane, a more expensive reagent, the reaction sequence uses Me_3SiCl, an inexpensive and readily accessible starting material. All the phosphonates are commercially available or readily accessible on laboratory scale. In addition, the quantitative generation of the intermediate dialkyl 1-lithio-1-(trimethylsilyl)alkylphosphonates, which may be directly used for further transformations, is of considerable importance. This method of generating α-metallated phosphonates by proton–metal exchange reaction from dialkyl alkylphosphonates is complemented by the halogen–metal exchange reaction from dialkyl trichloromethylphosphonates.[10,35,36] Compared with the carbanionic route using α-lithiophosphonates, the use of α-cuprophosphonates provides another, albeit less attractive, synthetic access to dialkyl 1-(trimethylsilyl)alkylphosphonates.[37,38] Recently, the X-ray structure of the α,α-dilithiated 1-(trimethylsilyl)methylphosphonate has been reported.[39,40]

Chloromethylsilanes are highly useful reagents for the trapping of phosphonate carbanions and Me_3SiCl is the most currently used. It has been effectively employed with a great variety of secondary lithiated[10,41–45] tertiary lithiated,[46–54] vinylic lithiated,[55] allylic lithiated,[56] copper(I),[57] magnesium,[58] cadmium,[51,59] and electrochemically generated carbanions.[60,61] Trimethylsilyl trifluoromethansulfonate has also been used.[62]

Abstraction of positive halogen by trivalent phosphorus compounds represents an expeditious route for the synthesis of dialkyl 1-(trimethylsilyl)methylphosphonates. Thus, treatment of diethyl trichloromethylphosphonate with hexamethylphosphorous triamide (HMPT) in the presence of Me_3SiCl in C_6H_6 at room temperature affords diethyl 1-(trimethylsilyl)-1,1-dichloromethylphosphonate in 90% yield.[63] Similarly, CCl_4 and $ClCl_3$ have been reacted with trimethylsilylmethyl dialkyl phosphites.[64,65]

2.1.2 Dialkyl 2-Silylalkylphosphonates

2.1.2.1 Michaelis–Arbuzov Reactions

Because 2-haloethylsilanes undergo decomposition to halosilanes and ethylene on reaction with trialkyl phosphites at high temperature, the preparation of dialkyl 2-(trimethylsilyl)ethylphosphonates by the Michaelis–Arbuzov reaction results in low yields.[16] Although heating of a mixture of (2-bromovinyl)trimethylsilane with triethyl phosphite at 160°C did not lead to the formation of diethyl 2-(trimethylsilyl)vinylphosphonate, it has been found that the reaction can be effected at 150°C in the presence of catalytic amount of $NiCl_2$ to give *trans*-diethyl 2-(trimethylsilyl)vinylphosphonate in 73% yield (Scheme 2.5).[66]

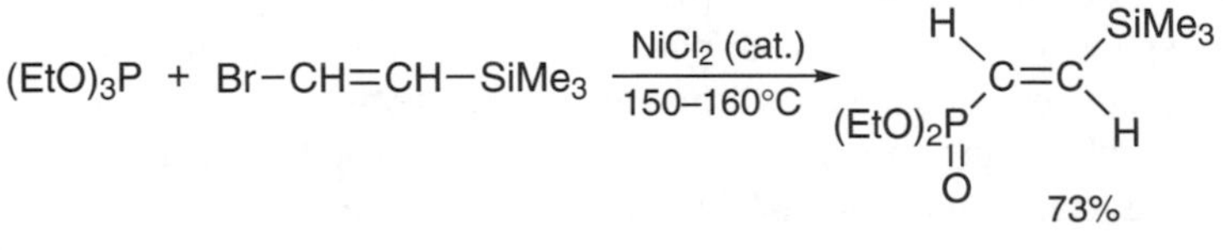

$$(EtO)_3P \ + \ Br-CH{=}CH-SiMe_3 \ \xrightarrow[150-160°C]{NiCl_2 \ (cat.)} \ \begin{array}{c} H \\ (EtO)_2P \end{array} C{=}C \begin{array}{c} SiMe_3 \\ H \end{array}$$

$$\underset{O}{\overset{\|}{}} \qquad 73\%$$

$$(2.5)$$

In contrast to the behavior of (chloroacyl)- and (bromoacyl)silanes, which react according to the Perkow rearrangement, the corresponding iodoacylsilane reacts almost completely with trimethyl phosphite in a Michael–Arbuzov reaction to afford dimethyl 2-(*tert*-butyldimethylsilyl)-2-oxoethylphosphonate in 96% yield.[67,68]

2.1.2.2 Addition Reactions

Although the formation of 2-(trimethylsilyl)ethylphosphonates by addition of dialkyl phosphites to vinylsilanes in the presence of di-*tert*-butylperoxide[12,69–71] or 2,2'-azobisisobutyronitrile (AIBN)[72] generally needs severe conditions, milder methods have been reported that employ the 1,2-addition of dialkyl trimethylsilyl phosphite or *tris*(trimethylsilyl) phosphite to the carbon–carbon double bond of acrylonitrile,[73–76] acrylates,[73,77] or maleimide[78,79] (Scheme 2.6). A complementary procedure exploits the reaction of Michael acceptors with the mixed reagent trialkyl phosphites-Me₃SiCl. However, the yields of functionalized 2-(trimethylsilyl)ethylphosphonates are slightly lower (30–67%).[75]

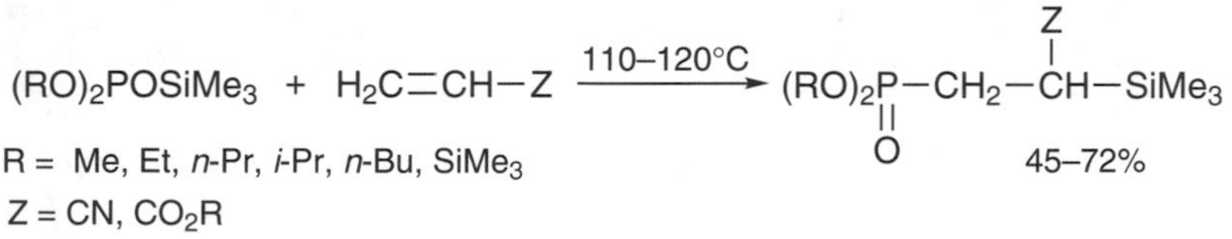

$$(RO)_2POSiMe_3 \ + \ H_2C{=}CH{-}Z \ \xrightarrow{110–120°C} \ (RO)_2\underset{\underset{O}{\|}}{P}{-}CH_2{-}\underset{\overset{Z}{|}}{C}H{-}SiMe_3$$

R = Me, Et, *n*-Pr, *i*-Pr, *n*-Bu, SiMe₃   45–72%

Z = CN, CO₂R

(2.6)

Similarly, a convenient procedure for the synthesis of diethyl 1-methyl-2-nitroethylphosphonate in 71% yield involves the Michael addition of dialkyl trimethylsilyl phosphite to 1-nitropropene.[80] The Michael addition of 1-lithio-1-chloromethylphosphonamide to *tert*-butyl β-silylacrylate proceeds at low temperature. The intermediate γ-chlorophosphonamide ester undergoes intramolecular displacement by the incipient enolate to give the corresponding cyclopropane in 75% yield.[81]

Recently, the preparation of diethyl 1-(ethoxycarbonyl)-2-(trimethylsilyl)-2-arylethylphosphonates by treatment of diethyl 1-(ethoxycarbonyl)-2-arylvinylphosphonates with Mg turnings (6 eq) in DMF at room temperature in the presence of Me₃SiCl has been described (Scheme 2.7). The reaction takes place smoothly to give regioselectively the corresponding β-silylated phosphonate in excellent yields (84–97%) as a mixture of diastereomers. Because Me₃SiCl plays a critical role of continuous activation of Mg metal surface, a large excess is required for smooth and complete reaction.[82]

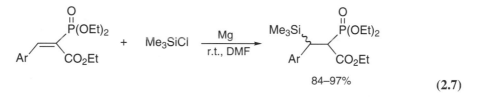

84–97%

(2.7)

2.1.2.3 Carbanionic Silylations

The most widely employed methods for the synthesis of dialkyl 2-(trimethylsilyl)ethylphosphonate involve the reaction of either (halomethyl)trimethylsilane with phosphonylated carbanions in the α-position to phosphorus or Me₃SiCl with phosphonylated carbanions in the β-position to phosphorus. Thus, α-metallated phosphonates are often alkylated with (iodomethyl)trimethylsilane[83–93] or (triethylsilyl)methyl triflate[94] in yields ranging from 65% to 96% (Scheme 2.8). This alkylation reaction is best conducted in dioxane or DME at 70–90°C but fails in THF.[90]

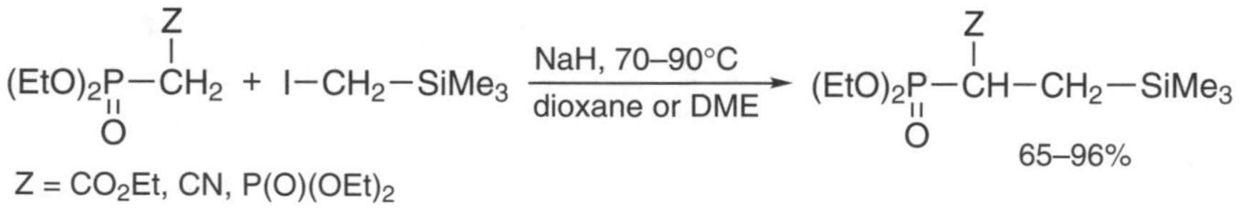

$$(EtO)_2\underset{\underset{O}{\|}}{P}{-}\underset{\overset{Z}{|}}{C}H_2 \ + \ I{-}CH_2{-}SiMe_3 \ \xrightarrow[\text{dioxane or DME}]{NaH, 70–90°C} \ (EtO)_2\underset{\underset{O}{\|}}{P}{-}\underset{\overset{Z}{|}}{C}H{-}CH_2{-}SiMe_3$$

Z = CO₂Et, CN, P(O)(OEt)₂   65–96%

(2.8)

Diethyl 1-(ethoxycarbonyl)methylphosphonate can be selectively *C*-alkylated in comparable yield (65%) using the less expensive (chloromethyl)trimethylsilane in heterogeneous conditions with dry K₂CO₃ at 60°C in the presence of a small amount of NaI.[95] The *ortho*-(trimethylsilyl)arylphosphonates are readily obtained via the *ortho*-lithiation with *n*-BuLi or *sec*-BuLi/TMEDA at low temperature followed by trapping with Me₃SiCl (70–80% yields).[96,97]

2.1.2.4 Miscellaneous

The (2-bromovinyl)trimethylsilane reacts with dialkyl phosphites in toluene at 50°C in the presence of catalytic amounts of Pd(PPh₃)₄/Et₃N to give *trans*-dialkyl 2-(trimethylsilyl)vinylphosphonates in 76–80% yields. This coupling reaction is very sensitive to the steric hindrance of the substrate. The dialkyl 2-(trimethylsilyl)vinylphosphonates so obtained have been used in a Diels–Alder reaction with cyclopentadiene.[98]

When dimethoxyphosphinyl phenyl ketene, prepared by an Rh-catalyzed Wolff rearrangement of dimethyl 1-diazo-2-oxo-2-phenylethylphosphonate, is submitted to the action of (trimethylsilyl)diazomethane, it gives rise to the corresponding dimethyl 2-(trimethylsilyl)-1-phenylvinylphosphonate in reasonable yield (44%, Scheme 2.9).[99]

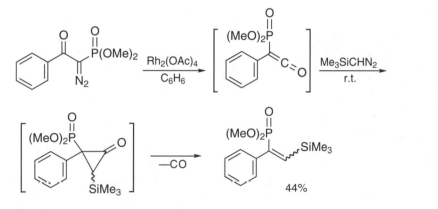

(2.9)

2.1.3 Dialkyl 3- and 4-Silylalkylphosphonates

2.1.3.1 Michaelis–Arbuzov Reactions

Diethyl 3-(triallylsilyl)propylphosphonate has been prepared by refluxing a mixture of triethyl phosphite and (3-bromopropyl)triallylsilane at 165°C and then further converted into the phosphonic acid by hydrolysis in acidic aqueous solution.[100]

2.1.3.2 Carbanionic Silylations

Generally, the preparation of silylphosphonates containing a trimethylsilyl group at the γ-position to phosphorus is efficiently achieved in 80–85% yields using the regioselective silylation of 2-propenylphosphonate carbanions generated at low temperature in THF with LiHMDS (2 eq),[101] *n*-BuLi (1 eq),[56,102,103] or LDA (1 eq).[56,102,103] The diethyl 3-(trimethylsilyl)-1-propenylphosphonate is formed exclusively as the (*E*)-isomer. On heating, it partially undergoes a thermal 1,3-sigmatropic shift to give the (*E*)- and (*Z*)-isomers of diethyl 3-(trimethylsilyl)-2-propenylphosphonate (Scheme 2.10).[103]

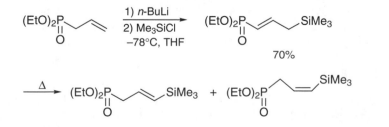

(2.10)

1-Phosphorylenolphosphates[104] and 1-phosphorylenamines,[105] after deprotonation with *t*-BuLi in THF at low temperature, react regio- and stereoselectively with Me₃SiCl to give the corresponding γ-silylated (*E*)-enolphosphate and (*E*)-enamine in good yields (67–79%). The γ-silylated (*E*)-1-phosphorylenolphosphate on reaction with aldehydes or ketones affords the silylated butadienes in fair yields (26–61%) via a Horner–Wadsworth–Emmons reaction.[104]

Similarly, diethyl *ortho*- and *para*-tolylphosphonates are regioselectively lithiated at the benzylic positions using LDA at low temperature and then reacted with Me₃SiCl to give the corresponding silylated derivatives in good yields.[106,107]

Diisopropyl 4-(trimethylsilyl)-2-oxobutylphosphonate is prepared by the regioselective silylation of the δ carbon of the β-ketophosphonate 1,4-dianion via a low-temperature tin–lithium exchange reaction (Scheme 2.11).[108]

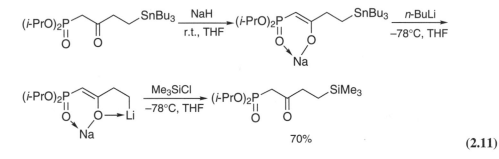

(2.11)

2.1.3.3 Addition Reactions

Addition of diethyl iododifluoromethylphosphonate to vinylsilanes is catalyzed by Pd(PPh₃)₄ under mild conditions.[109–111] The reaction proceeds readily at room temperature in the absence of solvent and requires only a few minutes to be complete. The diethyl 1,1-difluoro-3-iodo-3-(trimethylsilyl)propylphosphonate is further converted into diethyl 1,1-difluoro-3-(trimethylsilyl)propylphosphonate in 65% overall yield by a reductive process using the combination NiCl₂·6H₂O/Zn in moist THF (Scheme 2.12).[109,110]

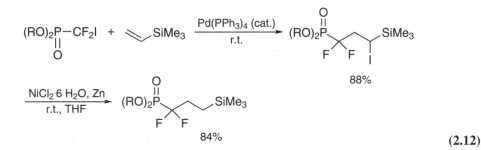

(2.12)

Heating diethyl 2-propenylphosphonate with Et₃SiH in the presence of di-*tert*-butylperoxide at 140°C affords diethyl 3-(triethylsilyl)propylphosphonate in low yield (21%).[112] Under the same conditions, diethyl 4-(triethylsilyl)butylphosphonate is prepared in 63% yield from diethyl phosphite and 4-(triethylsilyl)-1-butene.[15] Addition of (trimethylsilyl)methylmagnesium chloride to diethyl 1,1-difluoro-2-oxo-2-(*tert*-butoxycarbonyl)ethylphosphonate in THF at low temperature results in the formation of γ-silylated β-hydroxyphosphonate in 65% yield.[113]

2.1.3.4 SNP(V) Reactions

On treatment with *n*-BuLi, 2-(trimethylsilyl)-4-[(2,2-dibromo)ethenyl]furan undergoes a Fritsch–Buttenberg–Wiechell rearrangement to provide the lithium acetylide, which reacts with diethyl chlorophosphate according to an SNP(V) reaction[114] to give diethyl [2-(trimethylsilyl)-4-furyl]ethynylphosphonate (Scheme 2.13).[115]

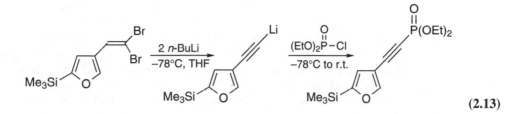

(2.13)

2.1.3.5 Cyclopropanation

When dimethyl diazomethylphosphonate is reacted at 0°C with a large excess of olefin in the presence of copper powder, the carbenoid addition to the double bond of the allyltrimethylsilane provides dimethyl 2-[(trimethylsilyl)methyl]cyclopropylphosphonate in 67% yield (Scheme 2.14).[116,117]

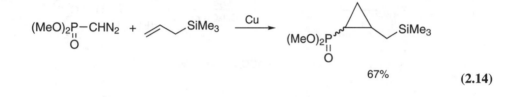

67%

(2.14)

2.2 SYNTHETIC APPLICATIONS OF SILYLPHOSPHONATES

2.2.1 DIALKYL 1-(TRIMETHYLSILYL)METHYLPHOSPHONATES

2.2.1.1 (Trimethylsilyl)methyl as a Masked Carbanion

Several groups have demonstrated that desilylation of dialkyl 1-(trimethylsilyl)methylphosphonates initiated by means of fluoride ion (CsF, KF, or TBAF) is an effective process for the transfer of carbanions to electrophilic centers. Some examples involving the fluoride-induced formation of α-phosphonylated carbanions from dialkyl 1-(trimethylsilyl)methylphosphonates containing a fragile C–Si bond have been described. Cleavage of the carbon–silicon bond under these conditions offers the advantages of neutral conditions and contributes to obtain better yields from sensitive substrates than those obtained under basic conditions. This procedure, which appears operationally simpler and cleaner than traditional protocols, has been applied to the preparation of alkenes[34,118] and 1-alkenylphosphonates[119] by Horner–Wadsworth–Emmons and Peterson reactions, respectively.

When dimethyl 1-(trimethylsilyl)benzylphosphonate is treated with benzaldehyde in the presence of a fluoride ion source (CsF, KF, or TBAF), stilbene and dimethyl benzylphosphonate, a protodesilylation product, are produced. The best yield of stilbene (85%) is obtained on heating in THF for 1 day with freshly dried CsF. Use of MeCN gives a similar result, whereas in toluene the reaction becomes very slow. KF is less effective even in the presence of 18-crown-6 ether. TBAF is efficient at room temperature; however, the yield seems to be modest because of difficulty in drying[120] or the existence of acidic hydrogen, easily transmetallated by the resulting carbanion. When CsF or KF is used, only (E)-stilbene is obtained, whereas TBAF afforded a mixture of (E)- and (Z)-stilbene in a 90/10 ratio. The formation of dimethyl benzylphosphonate seems to result from protonation by water that still remained in the system, and the yield of the olefin becomes water dependent. In the case of the reaction with isobutyraldehyde, the corresponding olefin is obtained in low yield (35%) as an (E)/(Z) mixture (70/30). By contrast, cinnamaldehyde gives stereochemically pure (E,E)-1,4-diphenyl-1,3-butadiene in 67% yield.[34,118]

Several variations on the reaction between diethyl 1-(trimethylsilyl)methylphosphonate and aromatic aldehydes have been produced by solvent-free techniques in a domestic oven. At room temperature in the presence of dried CsF, unsupported or supported on magnesium oxide, diethyl 2-(trimethylsilyloxy)-2-arylethylphosphonates are produced selectively in 60–80% yields. The hygroscopic nature of CsF can complicate the reaction, and under nonanhydrous conditions, with nondried CsF for example, the formation of diethyl arylvinylphosphonates is observed.[121]

Generation of α-silylated carbanions by fluoride ion from diethyl *bis*(trimethylsilyl)methylphosphonate constitutes a promising extension of the Peterson reaction. Diethyl *bis*(trimethylsilyl)methylphosphonate has been reacted with aromatic aldehydes in THF at room temperature and with pivalaldehyde in refluxing THF under *tris*(dimethylamino)sulfonium difluorotrimethylsiliconate (TASF) catalysis. The reaction gives diethyl 1-alkenylphosphonates in 60–70% yield as a mixture of (*E*)- and (*Z*)-isomers in an 85/15 ratio for aromatic aldehydes and 50/50 for pivalaldehyde (Scheme 2.15).[119] This chemistry, proceeding under mild and almost neutral conditions via metal-free carbanionic species, appears to be the method of choice for preparing compounds possessing base- or acid-sensitive functionalities.

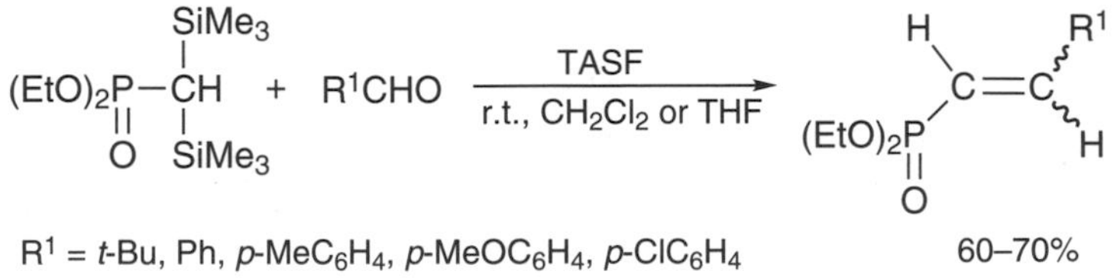

$$
\begin{array}{c}
\underset{\substack{| \\ O \;\; SiMe_3}}{(EtO)_2P-CH} + R^1CHO \xrightarrow[\text{r.t., } CH_2Cl_2 \text{ or THF}]{TASF} \text{product}
\end{array}
$$

R^1 = *t*-Bu, Ph, *p*-MeC_6H_4, *p*-MeOC_6H_4, *p*-ClC_6H_4   60–70%   (2.15)

The readily available diethyl 1,1-difluoro-1-(trimethylsilyl)methylphosphonate[122] is particularly suited to fluoride-ion-catalyzed reactions. In the presence of CsF, it reacts with aromatic or heteroaromatic aldehydes in THF at room temperature to produce silylated adducts, which are readily hydrolyzed to produce diethyl 1,1-difluoro-2-hydroxy-2-arylethylphosphonates in 57–87% yields.[123] Further applications of diethyl 1,1-difluoro-1-(trimethylsilyl)methylphosphonate have been demonstrated in the synthesis of [1–^{14}C]2,2-difluoroethene. The fluoride-ion-induced desilylation of diethyl 1,1-difluoro-1-(trimethylsilyl)methylphosphonate with catalytic amount of anhydrous CsF has been accomplished in the presence of commercial [^{14}C]formaldehyde. The key intermediate undergoes a Horner–Wadsworth–Emmons reaction to give the highly volatile [1–^{14}C]2,2-difluoroethene, which is collected in 10–15% yields with a purity exceeding 97% (Scheme 2.16).[124]

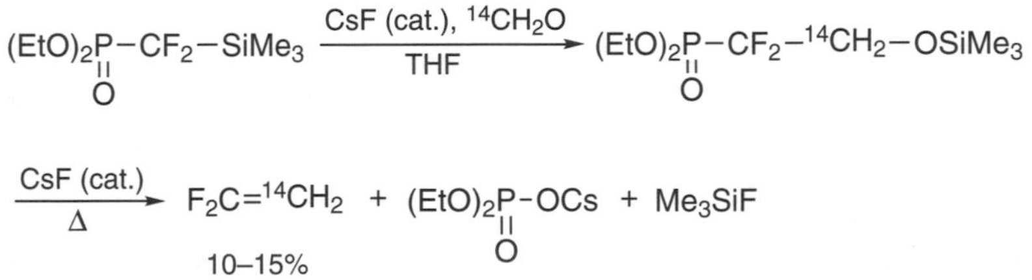

$$
\underset{O}{(EtO)_2P-CF_2-SiMe_3} \xrightarrow[THF]{CsF\ (cat.),\ ^{14}CH_2O} \underset{O}{(EtO)_2P-CF_2-^{14}CH_2-OSiMe_3}
$$

$$
\xrightarrow[\Delta]{CsF\ (cat.)} F_2C=^{14}CH_2 + \underset{O}{(EtO)_2P-OCs} + Me_3SiF
$$

10–15%   (2.16)

2.2.1.2 Trimethylsilyl as a Directing Group

In 1968, Peterson made the important discovery that the anions resulting from lithiation of [(methylthio)methyl]trimethylsilane and [(trimethylsilyl)methyl]diphenylphosphine sulfide reacted with benzophenone to produce lithiated β-hydroxysilanes, which decompose to give olefins by loss of Me_3SiOLi.[125] This olefination reaction resulting in functionally substituted alkenes has been extended to phosphonates and can be considered as an alternative to the Horner–Wadsworth–Emmons reaction.

The reaction has been investigated with dialkyl 1-(trimethylsilyl)alkylphosphonates, unsubstituted or substituted at the α-carbon, and aldehydes or ketones, although aldehydes are more frequently employed in this transformation. The phosphonate version of the Peterson reaction has given the opportunity to develop a one-pot reaction sequence including the successive formation of dialkyl 1-lithio-1-(trimethylsilyl)alkylphosphonates, lithiated β-hydroxysilanes, and 1-alkenylphosphonates (Scheme 2.17). Unfortunately, although the Peterson reaction is under kinetic control and irreversible, its phosphonate version shows little stereoselectivity, and, compared to the Horner–Wadsworth–Emmons reaction, mixtures of (*E*)- and (*Z*)-isomers are frequently obtained. Fortunately, the two isomers are chromatographically separable, thereby proving the usefulness of the route for both isomers.[126]

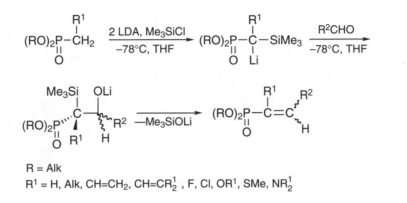

R = Alk
R[1] = H, Alk, CH=CH$_2$, CH=CR$_2^1$, F, Cl, OR[1], SMe, NR$_2^1$

(2.17)

Despite a large number of investigations, a systematic study of the Peterson reaction including solvent, temperature, additives, and nature and size of the α-substituents is lacking. Although the reported reactions have not all been effected under the same experimental conditions, it is useful to consider the nature of the carbonyl compounds (aldehydes or ketones) and of the group attached to the α-carbon atom (H,[9,28,30,127,128] alkyl,[9,28,29,129] OR,[130] SR,[20,131] NR$_2$,[132] F,[133–137] Cl[10]) as experimental variables that affect the stereoselectivity. However, because aldehydes have been more systematically used than ketones, the comparison is best limited to this class of carbonyl compounds. Thus, the reaction of dialkyl 1-(trimethylsilyl)methylphosphonates with aromatic aldehydes favors the (*E*)-isomer [*cis*-P(O)/H relationship; Figure 2.1], whereas aliphatic aldehydes give a mixture

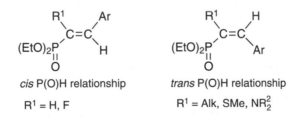

cis P(O)H relationship

R[1] = H, F

trans P(O)H relationship

R[1] = Alk, SMe, NR$_2^2$

FIGURE 2.1

of (*E*)- and (*Z*)-isomers in approximately equal amounts.[9,28,30,128] Similarly, the presence of a fluorine atom at the α-carbon atom promotes the (*E*)-isomer with aromatic aldehydes and a mixture of (*E*)- and (*Z*)-isomers with aliphatic aldehydes.[133–137] With a chlorine substituent, the two isomers are formed in approximately equal amounts.[10] Introduction of an alkyl group (Me, Et, *n*-Pr, *n*-Pent, *neo*-Pent) at the α-carbon atom leads predominantly to the (*Z*)-isomer [*trans*-P(O)/H relationship; Figure 2.1] with aromatic aldehydes whereas a mixture of (*E*)- and (*Z*)-isomers is produced with aliphatic aldehydes.[9,28,29,129] The presence of an NR$_2$ group at the α-carbon atom strongly favors

(Z)-configured products with both aromatic and aliphatic aldehydes.[132] Introduction of an SMe group at the α-carbon atom leads exclusively to the (E)-isomer [trans-P(O)/H relationship; Figure 2.1] with aromatic aldehydes and predominantly to the same isomer with aliphatic aldehydes.[20]

For lithiated allylic phosphonates, the nature of the carbon substituent affects the regioselectivity of the initial silylation as well as the reactivity of the resulting silylated anion.[138] Whereas diethyl 2-propenylphosphonate and 2-methyl-2-propenylphosphonate undergo silylation exclusively at the γ-carbon,[138] the 3-methyl derivative gives exclusively the α-silylated product (Scheme 2.18).[139]

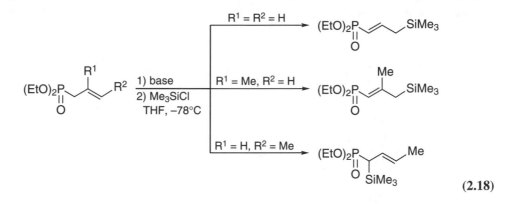

(2.18)

Thus, treatment of diethyl crotylphosphonate at low temperature with LDA (3 eq) followed by addition of Me₃SiCl provides exclusively the α-silylated crotylphosphonate carbanion. By virtue of the trimethylsilyl group, this carbanion reacts with ethyl formate in the γ-position. On acidic hydrolysis, diethyl (E)-3-formyl-2-butenylphosphonate is isolated in 78% yield (Scheme 2.19).[139]

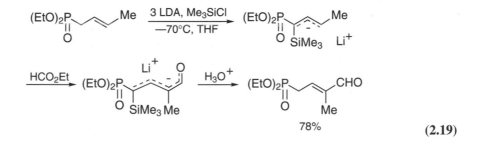

(2.19)

The metallation of diethyl 3-(trimethylsilyl)-1-propenylphosphonate with LDA at low temperature gives the delocalized allylphosphonate carbanion, which reacts regioselectively in the α-position with aliphatic or aromatic aldehydes to afford the corresponding β-alcohol as a mixture of diastereomers.[140] Subsequent dehydration with DCC in CH₂Cl₂ at reflux in the presence of catalytic CuCl₂ stereospecifically produces the desired 2-phosphonylated 1,3-dienes.[140,141] A one-pot approach is based on the β-elimination reaction of an intermediate enol phosphate using t-BuOK.[140]

The α-silylated carbanion generated from diethyl prenylphosphonate, on reaction with alkyl formates, behaves as a Peterson reagent leading to 1-(Z)-2-(diethoxyphosphinyl)-1-alkoxy-4-methyl-1,3-pentadienes in high yields (72–92%).[142] The presence of the two methyl groups at the C-4 determines the α-regioselectivity of this anion toward the alkyl formates and favors the fast decomposition of the hindered kinetic erythro-adduct, leading to the (Z)-enol ether after a Peterson syn-elimination of trimethylsilanolate.[142]

The reaction of lithiated carbanions of 1-(trimethylsilyl)cinnamyl-, prenyl-, and crotylphosphonates with aldehydes has been studied. The cinnamyl derivative undergoes a Peterson reaction with aromatic or aliphatic aldehydes to give the phosphonodienes in high yields and high stereoselectivity. Thus, with aromatic aldehydes, almost stereomerically pure (*E,E*)-phosphonodienes are obtained, whereas predominantly (*Z,E*)-phosphonodienes resulted from the reaction with aliphatic aldehydes. In contrast, the prenyl derivative shows a strict γ-regioselectivity in the reaction with aromatic aldehydes leading to phosphonolactones. A mixture of phosphonolactones (major product) and (*Z,E*)-phosphonodienes (minor product) is formed on reaction with aliphatic aldehydes.[143]

The decomposition of lithiated β-hydroxysilanes to 1-alkenylphosphonates at low temperature can be avoided by trapping the adduct *in situ* at −90°C with Me₃SiCl. For example, the treatment of diethyl 1-lithio-1-(trimethylsilyl)alkylphosphonates with ethyl formate followed by addition of Me₃SiCl provides a mixed acetal as the major product. Although some difficulties are encountered in the hydrolysis of enol ethers, the mixed acetals are easily and completely hydrolyzed in a few minutes at room temperature with diluted HCl, via the sensitive silylated enol ethers, to give α-substituted diethyl formylphosphonates as a mixture of aldehyde and enol forms. Silylated enol ethers are used advantageously instead of the traditional enol ethers, thus offering an efficient approach to the synthesis of a large variety of diethyl 1-formylalkylphosphonates in fair to good yields (Scheme 2.20).[35]

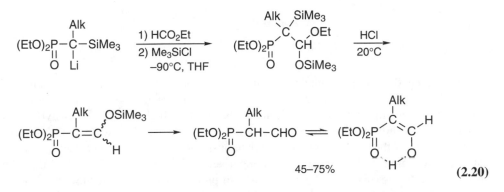

(2.20)

An interesting feature of the Peterson olefination reaction concerns the subsequent transformation of dialkyl 1-alkenylphosphonates into new functionalized phosphonates. For example, diethyl 1-lithio-1-chloro-1-(trimethylsilyl)methylphosphonate, resulting from the treatment of diethyl trichloromethylphosphonate with *n*-BuLi (2 eq) in the presence of Me₃SiCl, reacts at low temperature with aldehydes to produce a mixture of diethyl (*E*)- and (*Z*)-1-chloro-1-alkenylphosphonates. When the carbonyl compound is an aromatic or heteroaromatic aldehyde, the diethyl 1-chloro-1-alkenylphosphonates are readily converted into diethyl 1-alkynylphosphonates in high yields (87–96%) in a one-pot process by treatment with LiHMDS at low temperature (Scheme 2.21).[36] This class of diethyl 1-alkynylphosphonates is difficult to prepare by other methods, which, generally, are closely dependent on the availability of the terminal alkyne.

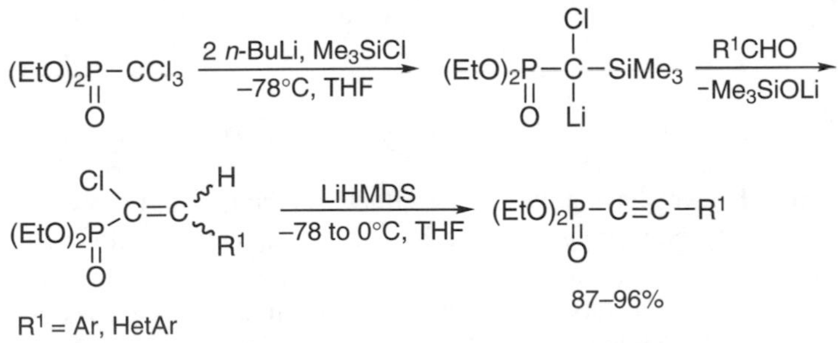

(2.21)

In further examples, diethyl 1-(trimethylsilyl)alkylphosphonates are efficiently converted into sulfines after deprotonation with *n*-BuLi at low temperature and subsequent treatment with SO₂.

Two conditions are crucial to circumvent side reactions: (a) the reaction must be performed at −78°C, and (b) the anion must be added to an excess of SO_2. This reaction appears to be fairly general, and decomposition of the adducts led to good yields of a large number of sulfines bearing various substituents R^1 at the α-carbon atom. In these compounds the S = O function is positioned *syn* with respect to R^1. The sulfines are trapped as [4 + 2] cycloadducts by Diels–Alder reactions with 2,3-dimethyl-1,3-butadiene (Scheme 2.22).[86]

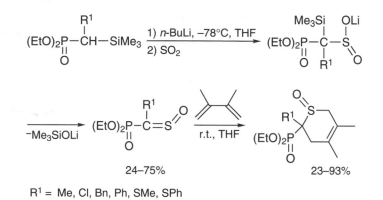

R¹ = Me, Cl, Bn, Ph, SMe, SPh

(2.22)

More recently, another example involving the intermediacy of phosphonylated sulfines has been reported.[144] Treatment of diethyl 3-methyl-1,2-butadienylphosphonate with LDA at low temperature followed by reaction with Me_3SiCl leads to the desired diethyl 1-(trimethylsilyl)-3-methyl-1,2-butadienylphosphonate. The addition of alkyllithiums to the α,β-double bond of this very unstable compound gives the α-phosphonylated carbanions, which are allowed to react with an excess of SO_2. The resulting sulfines are stirred overnight, and only low yields of diethyl 2-thienylphosphonates (12–25%) are obtained, contaminated with protonated anion (Scheme 2.23). The separation of these two compounds proved to be problematic. The major disadvantage of the method is the incomplete reaction of the anion with SO_2.[144]

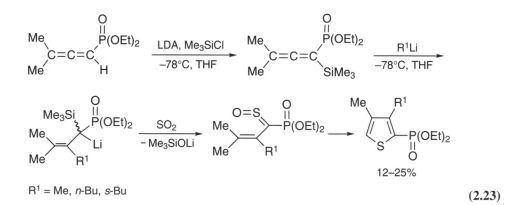

R¹ = Me, n-Bu, s-Bu

(2.23)

α-Silylvinylphosphonates have proved to be effective precursors of more elaborated phosphonates by Friedel–Crafts-type reactions (Scheme 2.24).[145] Thus, diethyl α-silylated phosphonoketene dithioacetals react with acyl chlorides to produce α-acetylated phosphonoketene dithioacetals. In the case of acyclic dithioacetals, the α-silylated intermediates were prepared in fair yields (64%) by addition of LiTMP to a mixture of diethyl phosphonoketene dithioacetals and Me_3SiCl at −78°C, in order to prevent the elimination of the thiolate anion. Replacement of the acyclic dithioacetal by a cyclic dithioacetal provides, in the same experimental conditions, a better yield of α-silylated phosphonoketene cyclic dithioacetals (81%). The reaction of cyclic dithioacetals with a mixture of acyl chlorides (2 eq) and AlCl₃ (2 eq) in CH₂Cl₂ at 0°C gives α-acylated phosphonoketene dithioacetals in satisfactory yields (68–89%, Scheme 2.24). A comparative study between silylated and unsilylated phosphonoketene dithioacetals has demonstrated that the α-trimethylsilyl substituent is essential to increase the reaction rate and yields.[145–147]

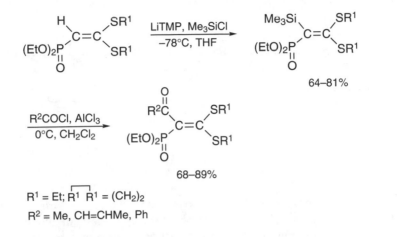

(2.24)

The α- and γ-silylated allylphosphonates, prepared by the carbanionic silylation of allyl- and vinylphosphonates and a subsequent double bond migration, undergo the Friedel–Crafts reaction with acyl chlorides in CH₂Cl₂ at room temperature in the presence of AlCl₃ to give high yields of 4-oxo-2-alkenylphosphonate and 4-oxo-1-alkenylphosphonate, respectively. The latter is converted into the former by a double-bond migration promoted by Et₃N.[148] Thus, diethyl 4-oxo-2-alkenylphosphonates can be obtained in 67–98% yields from allylic and vinylic phosphonates without any effort to separate the mixture of α- and γ-silylated phosphonates (Scheme 2.25).

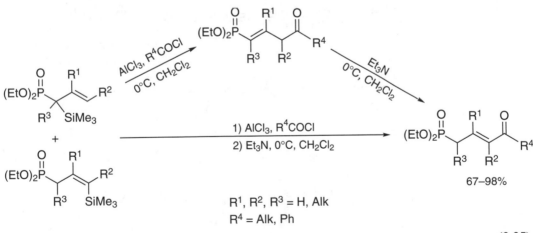

(2.25)

2.2.1.3 Trimethylsilyl as a Protecting Group

In a carbanionic process, dialkyl 1-lithioalkylphosphonates, on reaction with electrophiles bearing electron-withdrawing groups, produce new species containing a more acidic methylene group. Consequently, the initially formed carbanion undergoes a facile transprotonation by the new species to regenerate the starting material and to produce α-substituted 1-lithioalkylphosphonates, which can in turn undergo further reactions. The resulting crude product is unavoidably a complex mixture of expected, starting, and undesired phosphonates that cannot be usefully separated (Scheme 2.26).[134] These drawbacks are overcome when a proton of the methylene group is replaced with a trimethylsilyl group, which stabilizes the adjacent carbanion and controls its reaction with electrophiles. Removal of the trimethylsilyl group as trimethylsilanolate or fluorotrimethylsilane results in the formation of functionally substituted phosphonates. Each of the steps of this reaction sequence may be conveniently executed in one pot without isolation of intermediates.

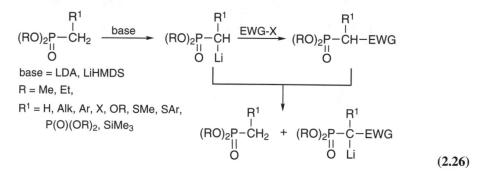

$$(2.26)$$

Together with these innovations, the use of α-silylated phosphonate carbanions has become one of the most useful and general methods available for the synthesis of pure diethyl α-monohalogenoalkylphosphonates in high yields. Two different procedures may be employed to generate α-monohalogenoalkylphosphonates, one by nucleophilic phosphonohalogenomethylation of various substrates (Scheme 2.27),[10,41,149–153] and the other by electrophilic halogenation of substituted dialkyl 1-lithio-1-(trimethylsilyl)methylphosphonates (Scheme 2.28).[154–156] The former, using diethyl trihalogenomethylphosphonates as starting materials, has been developed for fluorine and chlorine substituents, whereas the latter, which appears to be fairly general, has a greater synthetic utility and has been employed for the four halogens.

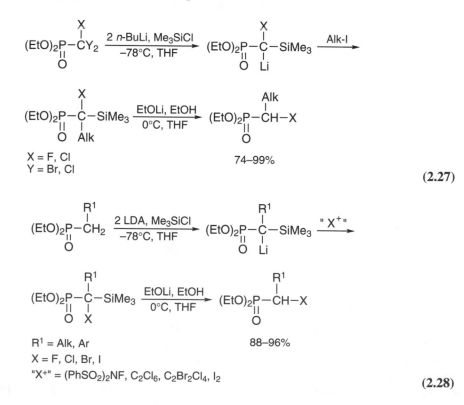

$$(2.27)$$

$$(2.28)$$

In the first approach, diethyl trihalogenomethylphosphonates are submitted to a double halogen-metal exchange reaction with n-BuLi in the presence of Me$_3$SiCl to give diethyl 1-lithio-1-halogeno-1-(trimethylsilyl)methylphosphonates. Under these conditions, these stable carbanions safely undergo alkylation reactions leading to the corresponding trisubstituted phosphonates. Because of the presence of the halogen atoms (F, Cl), the C–Si bond is very sensitive, and the trimethylsilyl group is easily eliminated with EtOLi in EtOH to produce the substituted α-monofluoro- or α-monochloroalkylphosphonates in high yields and free of byproducts (Scheme 2.27).[10,41,150,151,153]

In the second approach, the substituted diethyl 1-lithio-1-(trimethylsilyl)methylphosphonates result from a double proton–metal exchange with LDA from diethyl alkylphosphonates in the presence of Me$_3$SiCl. At low temperature, these carbanions undergo halogenation with commercial electrophilic halogenating reagents: N-fluorobenzenesulfonimide (NFBS), hexachloroethane, dibromotetrachloroethane, or dibromotetrafluoroethane and iodine to give cleanly the monohalogenated derivatives. After removal of the trimethylsilyl group with EtOLi in EtOH, diethyl α-monohalogenoalkylphosphonates are isolated in pure form and almost quantitative yields (Scheme 2.28).[154–156]

Electrophilic halogenation can be coupled with the phosphate–phosphonate transformation[114,157–160] to provide a convenient one-pot procedure applicable for the construction of α-monohalogenoalkylphosphonates bearing widely varied alkyl appendages R^1 (Scheme 2.29).[156] Thus, the reaction of alkyllithiums with triethyl phosphate proceeds as expected to give the intermediate diethyl 1-lithioalkylphosphonates, which are subjected to sequential silylation and halogenation to generate the monohalogenated derivatives in 85–97% yields after desilylation. By this approach, diethyl α-monohalogenoalkylphosphonates, which are relatively elaborate compounds, are obtained from starting materials as simple as triethyl phosphate and alkyllithiums.[156]

$$(EtO)_2\underset{\underset{O}{\|}}{P}\!-\!OEt \xrightarrow[\text{–78 to 0°C, THF}]{3\ \text{Alk-CH}_2\text{Li}} (EtO)_2\underset{\underset{O}{\|}}{P}\!-\!\underset{\underset{Li}{|}}{\overset{\overset{Alk}{|}}{C}}H \xrightarrow[\text{3) "X}^+\text{"}\ \text{–78°C, THF}]{\begin{array}{l}\text{1) 1.5 }i\text{-Pr}_2\text{NH}\\ \text{2) 2 Me}_3\text{SiCl}\end{array}}$$

$$(EtO)_2\underset{\underset{O}{\|}}{P}\!-\!\underset{\underset{X}{|}}{\overset{\overset{Alk}{|}}{C}}\!-\!SiMe_3 \xrightarrow[\text{0°C, THF}]{\text{EtOLi, EtOH}} (EtO)_2\underset{\underset{O}{\|}}{P}\!-\!\overset{\overset{Alk}{|}}{C}H\!-\!X$$

$$X = F, Cl, Br, I \qquad\qquad 85\text{–}92\%$$

$$(2.29)$$

Of great synthetic importance is the preparation of diethyl α-fluorophosphonocarboxylates, which are valuable precursors of fluoroacrylates. They are readily obtained in high yields from diethyl dibromofluoromethylphosphonate via diethyl 1-lithio-1-fluoro-1-(trimethylsilyl)methylphosphonate as outlined in Scheme 2.30. Removal of the trimethylsilyl group is highly dependent on the nature of the substituents of the carbon atom. In the case of diethyl 1-fluoro-1-(trimethylsilyl)-1-(alkoxycarbonyl-methyl)phosphonate, anhydrous EtOH is the preferred reagent to cleave the C–Si bond, whereas the use of EtOLi in EtOH results in cleavage of the sensitive C–P bond (Scheme 2.30).[149,152]

$$(EtO)_2\underset{\underset{O}{\|}}{P}\!-\!\overset{\overset{F}{|}}{C}Br_2 \xrightarrow[\text{–78°C,THF}]{2\ n\text{-BuLi, Me}_3\text{SiCl}} (EtO)_2\underset{\underset{O}{\|}}{P}\!-\!\underset{\underset{Li}{|}}{\overset{\overset{F}{|}}{C}}\!-\!SiMe_3 \xrightarrow{\text{ClCO}_2\text{R}^1}$$

$$(EtO)_2\underset{\underset{O}{\|}}{P}\!-\!\underset{\underset{SiMe_3}{|}}{\overset{\overset{F}{|}}{C}}\!-\!CO_2R^1 \xrightarrow[\text{0°C, THF}]{\text{EtOH}} (EtO)_2\underset{\underset{O}{\|}}{P}\!-\!\overset{\overset{F}{|}}{C}H\!-\!CO_2R^1$$

$$80\text{–}91\%$$

$$R^1 = \text{Me, CH}_2\text{Cl, Et, (CH}_2)_2\text{Cl, }i\text{-Pr, }i\text{-Bu, CH=CH}_2\text{, Ph}$$

$$(2.30)$$

The principle of masking a methylene proton by the trimethylsilyl group has been applied to the preparation of functionalized tetraethyl methylenediphosphonates.[161] The extension of this work provides an efficient method for the incorporation of deuterium into phosphonates.[162,163] Deuteration of functionalized dialkyl 1-lithio-1-(trimethylsilyl)methylphosphonates with heavy water followed by desilylation with *in situ*–generated LiOD has been effected with a number of compounds containing a variety of groups R^1 on the α-carbon (alkyl, aryl, heteroaryl, thioalkyl, thioaryl, halogen, phosphoryl). Incorporation of deuterium and chemical yields are good to excellent, and the reaction conditions are mild enough to preserve the ester groups at phosphorus (Scheme 2.31).[162,163]

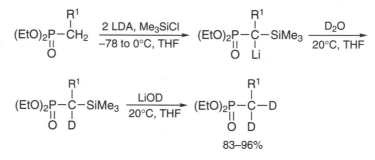

2 LDA, Me3SiCl
−78 to 0°C, THF
D2O
20°C, THF
LiOD
20°C, THF
83–96%
R1 = Me, F, Cl, CH=CHMe, SEt, SPh, P(O)(OR)2, Ar, HetAr

(2.31)

Another illustration of the advantages of 1-(trimethylsilyl)phosphonate reagents is provided by the transformations involving the readily available diethyl 1-(trimethylsilyl)vinylphosphonate.[129,164,165] This reagent has been found to be valuable for the conversion of vinylphosphonates into alkyl-,[129] alkenyl-,[129] β-keto-,[165] and β-amidophosphonates[165] in high yields. Because of the trimethylsilyl group, which activates the vinylphosphonates toward nucleophilic addition of organometallic reagents by virtue of its polarizing effect, Grignard reagents and alkyllithiums add to the double bond at low temperature to give stable anions. They undergo a variety of reactions with electrophiles as alkyl halides,[129] acyl chlorides,[165] carbonyl compounds,[129] and isocyanates[165] to form functionalized diethyl 1-(trimethylsilyl)alkylphosphonates. All the intermediates are successfully protodesilylated by TBAF in moist THF to give functionalized phosphonates in near quantitative yields (Scheme 2.32).

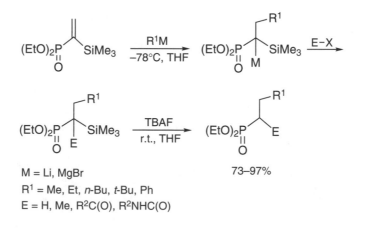

R1M
−78°C, THF
E−X
TBAF
r.t., THF
73–97%
M = Li, MgBr
R1 = Me, Et, n-Bu, t-Bu, Ph
E = H, Me, R2C(O), R2NHC(O)

(2.32)

Other applications of dialkyl 1-(trimethylsilyl)methylphosphonates include the reaction with dielectrophilic reagents for the preparation of three-, four-, five-, and six-membered cycloalkylphosphonates.[166] Diethyl trichloromethylphosphonate is particularly well suited to this approach, and the exchange of the three halogen atoms with *n*-BuLi followed by sequential addition of Me_3SiCl and ω-dibromoalkanes results in the formation of a series of α-silylated cycloalkylphosphonates

(Scheme 2.33). Because the C–Si bond is not activated by the presence of electron-withdrawing groups, the trimethylsilyl group is removed by the use of moist TBAF in THF.[166] The same principle is applied to the synthesis of phosphonates bearing two symmetric or disymmetric, saturated or unsaturated, alkyl groups on the α-carbon atom.[167]

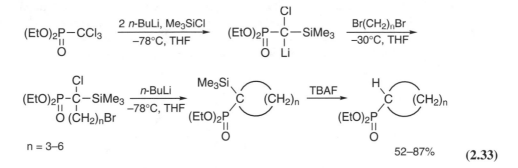

n = 3–6

52–87%   (2.33)

An interesting autodesilylation reaction is observed with diethyl 1-(methylthio)-1-(trimethyl-silyl)methylphosphonate, which appears to be a useful precursor of symmetric or disymmetric diethyl 1,1-bis(alkylthio)methylphosphonates (Scheme 2.34). The diethyl 1-lithio-1-(methylthio)-1-(trimethylsilyl)methylphosphonate, because of its high stability within a wide range of temper-atures, reacts smoothly with dimethyl or diethyl disulfide to give, after desilylation by the lithium alkanethiolates generated in the reaction medium, the 1,1-bis(alkylthio)methylphosphonates in 62–75% and 65% yields, respectively.[20]

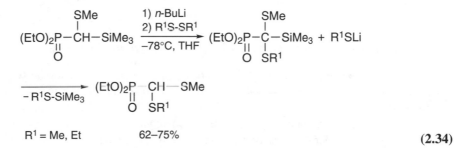

R^1 = Me, Et   62–75%   (2.34)

The dimethyl (tert-butyldimethylsilyl)methylphosphonate, an intermediate in the synthesis of dimethyl selenoformylphosphonate, is prepared by a carbanionic process from dimethyl meth-ylphosphonate and chloro-tert-butyldimethylsilane in 85–95% yields. The reactivity of dimethyl 1-lithio-1-(tert-butyldimethylsilyl)methylphosphonate is attenuated by conversion into lower-order cyanocuprates by treatment with copper(I) cyanide in THF. Addition of the cyanocuprates to selenocyanogen at low temperature results in efficient and rapid generation of dimethyl 1-seleno-cyano-1-(tert-butyldimethylsilyl)methylphosphonate in 73% yield. On treatment with TBAF in THF at room temperature, dimethyl selenoformylphosphonate is generated by desilylation and subse-quently reacted in situ with suitable [4 + 2] cycloaddition trapping reagents (Scheme 2.35).[168–171]

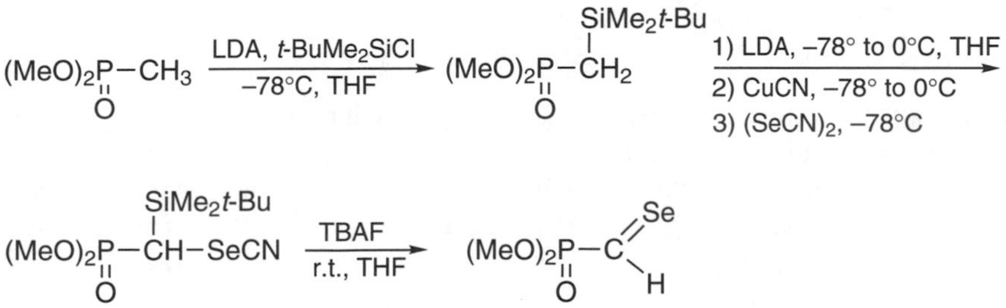

$$(MeO)_2\overset{O}{\underset{||}{P}}-CH_3 \xrightarrow[-78°C, THF]{LDA, \textit{t-}BuMe_2SiCl} (MeO)_2\overset{SiMe_2\textit{t-}Bu}{\underset{\underset{O}{||}}{P-CH_2}} \xrightarrow[\substack{2)\ CuCN, -78°\ to\ 0°C \\ 3)\ (SeCN)_2, -78°C}]{1)\ LDA, -78°\ to\ 0°C, THF}$$

$$(MeO)_2\overset{SiMe_2\textit{t-}Bu}{\underset{\underset{O}{||}}{P-CH-SeCN}} \xrightarrow[r.t., THF]{TBAF} (MeO)_2\overset{O}{\underset{||}{P}}-C\overset{Se}{\underset{H}{\diagdown}}$$

(2.35)

2.2.2 DIETHYL 2-(TRIMETHYLSILYL)-1-(ETHOXYCARBONYL)ETHYLPHOSPHONATE

The allylsilanes are effective intermediates in the construction of various carbocyclic systems, and the Horner–Wadsworth–Emmons reaction appears to be well suited to prepare this class of compounds. Thus, diethyl 2-(trimethylsilyl)-1-(ethoxycarbonyl)ethylphosphonate reacts with aldehydes[84,85,90,91,94,172–190] and ketones[83,191,192] under standard Horner–Wadsworth–Emmons reaction conditions (NaH, THF, or DME, room temperature) to give α-(trimethylsilyl)methyl-α,β-unsaturated esters in good yields (61–87%) as a mixture of (E)- and (Z)-isomers separable by silica gel chromatography. These reaction conditions introduce the (Z)-configured double bond with good (Z)/(E) selectivity (Scheme 2.36).

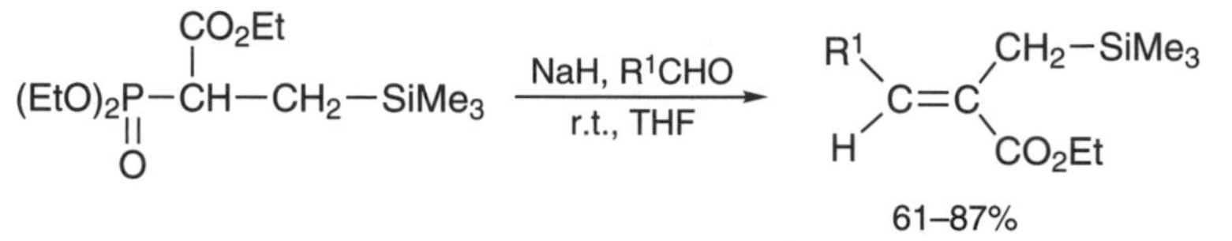

$$\text{(2.36)}$$

Complementary selectivity is observed when the reaction is carried out at low temperature with KH in THF in the presence of 18-crown-6 ether.[176] With the use of n-BuLi in DME at low temperature, the yields of α-(trimethylsilyl)methyl-α,β-unsaturated esters are roughly comparable to those obtained in standard conditions (NaH, THF, room temperature).[94,172] Other modified Horner–Wadsworth–Emmons reaction procedures proved to be applicable. For example, addition of two equivalents of an aqueous solution of K_2CO_3 and formaldehyde to diethyl 2-(trimethylsilyl)-1-(ethoxycarbonyl)ethylphosphonate gives ethyl α-(trimethylsilyl)methyl acrylate in 48% yield.[95]

2.2.3 DIMETHYL 2-(TERT-BUTYLDIMETHYLSILYL)-2-OXOETHYLPHOSPHONATE

The dimethyl 2-(tert-butyldimethylsilyl)-2-oxoethylphosphonate reacts smoothly with a variety of aldehydes under standard Horner–Wadsworth–Emmons reaction conditions (NaH, THF, room temperature) or mild nonbasic conditions (LiCl, DBU, MeCN, room temperature) to give the corresponding trans-α,β-unsaturated acylsilanes in excellent yields (54–97%) with high stereoselectivity (Scheme 2.37). However, this reaction is considerably slower under these mild conditions (24 h) compared to standard conditions (90 min). The efforts to extend the reaction to ketones were disappointing.[67,68,193]

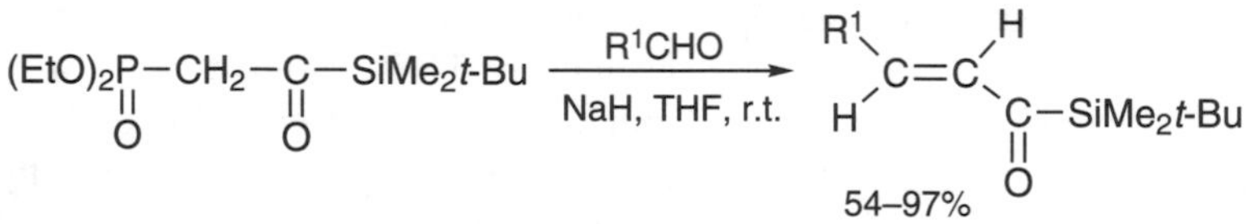

$$\text{(2.37)}$$

2.2.4 DIETHYL 2-CYANO-2-(TRIMETHYLSILYL)ETHYLPHOSPHONATE

The readily available diethyl 2-cyano-2-(trimethylsilyl)ethylphosphonate[75] appears to be an attractive reagent for the synthesis of 2-cyano-1,3-butadienes. Their formation combines the Peterson and Horner–Wadsworth–Emmons reactions in a one-pot process. The reaction has been developed with a variety of aldehydes, and the yields are generally high (34–89%, Scheme 2.38).[76]

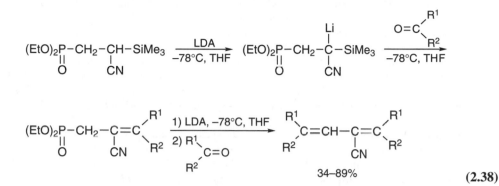

$$(2.38)$$

REFERENCES

1. Gilbert, A.R., Organosilicon compounds containing phosphorus, *General Electric,* U.S. Patent Appl. US 2768193, 1956; *Chem. Abstr.,* 51, 5816h, 1957.
2. Chan, T.-H., Alkene synthesis via β-functionalized organosilicon compounds, *Acc. Chem. Res.,* 10, 442, 1977.
3. Ager, D.J., The Peterson reaction, *Synthesis,* 384, 1984.
4. Lalonde, M., and Chan, T.H., Use of organosilicon reagents as protective groups in organic synthesis, *Synthesis,* 817, 1985.
5. Colvin, E.W., *Silicon Reagents in Organic Synthesis,* Academic Press, New York, 1988.
6. Ager, D.J., The Peterson olefination reaction, *Org. React.,* 38, 1, 1990.
7. Barrett, A.G.M., Hill, J.M., Wallace, E.M., and Flygare, J.A., Recent studies on the Peterson olefination reaction, *Synlett,* 764, 1991.
8. van Staden, L.F., Gravestock, D., and Ager, D.J., New developments in the Peterson olefination reaction, *Chem. Soc. Rev.,* 31, 195, 2002.
9. Carey, F.A., and Court, A.S., Silicon-containing carbanions. Part 1. Synthesis of vinyl thioethers and vinylphosphonates via silicon-modified organolithium reagents, *J. Org. Chem.,* 37, 939, 1972.
10. Teulade, M.-P., and Savignac, P., α-Lithiated *O,O*-diethyl chloro(trimethylsilyl)methylphosphonate. Part 1. Preparation and properties, *J. Organomet. Chem.,* 338, 295, 1988.
11. Ginsburg, V.A., and Yakubovich, A.Y., Synthesis of heteroorganic compounds of aliphatic series by the diazo method. Part 8. Synthesis of compounds of elements of group V. Organophosphorus compounds. Syntheses of *bis-* and *tris*(haloalkyl)phosphines and some transformations of chloroalkyl derivatives of phosphorus, *Zh. Obshch. Khim.,* 28, 728, 1958; *Chem. Abstr.,* 52, 17091g, 1958.
12. Canavan, A.E., and Eaborn, C., Organosilicon compounds. Part 22. The thermal decomposition of some silicon-substituted organophosphorus compounds, *J. Chem. Soc.,* 592, 1962.
13. Canavan, A.E., and Eaborn, C., Organosilicon compounds. Part 21. Some compounds containing phosphorus, *J. Chem. Soc.,* 3751, 1959.
14. Bugerenko, E.F., Chernyshev, E.A., and Petrov, A.D., Synthesis and some organosilicon monomers with a phosphorus-bearing functional group, *Dokl. Akad. Nauk. SSSR,* 143, 840, 1962; *Dokl. Chem. (Engl. Transl.),* 143, 254, 1962.
15. Chernyshev, E.A., Bugerenko, E.F., and Petrov, A.D., Synthesis of some triethylsilyl-substituted alkylphosphonic acids and their esters, *Dokl. Akad. Nauk SSSR,* 148, 875, 1963; *Dokl. Chem. (Engl. Transl.),* 148, 130, 1963.
16. Bugerenko, E.F., Chernyshev, E.A., and Petrov, A.D., Synthesis of compounds containing phosphorus and silicon, *Izv. Akad. Nauk SSSR, Ser. Khim.,* 286, 1965; *Bull. Acad. Sci. USSR, Div. Chem. Sci. (Engl. Transl.),* 268, 1965.
17. Géhanne, S., Giammaruco, M., Taddei, M., and Ulivi, P., Approaches to the synthesis of 2-sila-1-carba-cephalosporins, *Tetrahedron Lett.,* 35, 2047, 1994.
18. Novikova, Z.S., Zdorova, S.N., and Lutsenko, I.F., Esters of silicon-substituted benzylphosphonic acids, *Zh. Obshch. Khim.,* 42, 112, 1972; *J. Gen. Chem. USSR (Engl. Transl.),* 42, 108, 1972.

19. Mikolajczyk, M., Balczewski, P., and Graczyk, P., Phosphonates containing sulfur and silicon. New stereochemical and mechanistic aspects, *Phosphorus Sulfur*, 30, 225, 1987.

20. Mikolajczyk, M., and Balczewski, P., Organosulfur compounds. Part 48. Synthesis and reactivity of diethyl (methylthio)(trimethylsilyl)methylphosphonate, *Synthesis*, 101, 1989.

21. Han, D.I., Kim, D.Y., and Oh, D.Y., Reaction of *O,S*-acetals and phosphites in the presence of Lewis acids. Chemoselectivity in the cleavage of the acetals, *Bull. Korean Chem. Soc.*, 10, 329, 1989.

22. Kim, D.Y., Han, D.I., and Oh, D.Y., Reaction of phosphite with acetal derivatives. Syntheses of 1-alkoxymethylphosphonates and 1-alkylthiomethylphosphonates, *Bull. Korean Chem. Soc.*, 19, 226, 1998.

23. Keeber, W.H., and Post, H.W., Studies in silico-organic compounds. Part 31. The chemistry of sila compounds containing phosporus. Preliminary paper, *J. Org. Chem.*, 21, 509, 1956.

24. Sekiguchi, A., and Ando, W., Formation of phosphorus ylide via 1,3-silyl migration of dimethyl trimethylsilylmethylphosphonate. Application of the phosphonate to a Wittig reaction, *Chem. Lett.*, 1385, 1978.

25. Hartle, R.J., Structural stabilities of the trimethylsilylmethyl and neopentyl groups in the preparation of phosphonyl dichlorides, *J. Org. Chem.*, 31, 4288, 1966.

26. Liptuga, N.I., Vasil'ev, V.V., and Derkach, G.I., Trimethylsilylmethylphosphonic acid derivatives, *Zh. Obshch. Khim.*, 42, 293, 1972; *J. Gen. Chem. USSR (Engl. Transl.)*, 42, 284, 1972.

27. Gilbert, A.R., and Precopio, F., Organosilicon compositions containing phosphorus and their preparations, *General Electric*, U.S. Patent Appl. US 2835651, 1958; *Chem. Abstr.*, 52, 15125b, 1958.

28. Aboujaoude, E.E., Liétjé, S., Collignon, N., Teulade, M.P., and Savignac, P., A one step generation of α-lithio(trimethylsilyl)alkanephosphonates. A new preparative route to dialkyl 1-(trimethylsilyl)alkanephosphonates and vinylphosphonates, *Synthesis*, 934, 1986.

29. Savignac, P., Aboujaoude, E.E., Collignon, N., and Teulade, M.P., α-Substituted silyl and phosphonyl phosphonates. An extensive study of their synthesis and properties, *Phosphorus Sulfur*, 30, 491, 1987.

30. Savignac, P., Teulade, M.-P., and Collignon, N., Preparation and properties of α-silylphosphonates, $(RO)_2P(O)CR^1R^2SiR^3R^4R^5$, and α,α-disilylphosphonates, $(RO)_2P(O)CR^1(SiMe_3)_2$, *J. Organomet. Chem.*, 323, 135, 1987.

31. Patois, C., Ricard, L., and Savignac, P., 2-Alkyl-5,5-dimethyl-1,3,2-dioxaphosphorinan-2-ones α-lithiated carbanions. Synthesis, stability, and conformation, *J. Chem. Soc., Perkin Trans. 1*, 1577, 1990.

32. Aisa, A.M.A., Enke, S., and Richter, H., Synthesis and characterization of long chain alkylsilylphosphonates and alkylsilylphosphites, *J. Organomet. Chem.*, 575, 126, 1999.

33. Paulsen, H., and Bartsch, W., Phosphorus-containing carbohydrates. Part 12. Preparation of olefinic sugar phosphonates and allenic sugar phosphonates, *Chem. Ber.*, 108, 1732, 1975.

34. Kawashima, T., Ishii, T., and Inamoto, N., Fluoride ion-induced Horner–Emmons reaction of α-silylalkylphosphonic derivatives with carbonyl compounds, *Bull. Chem. Soc. Jpn.*, 60, 1831, 1987.

35. Zanella, Y., Berté-Verrando, S., Dizière, R., and Savignac, P., New route to 1-formylalkylphosphonates using diethyl trichloromethylphosphonate as a precursor, *J. Chem. Soc., Perkin Trans. 1*, 2835, 1995.

36. Dizière, R., and Savignac, P., A new simple method for the synthesis of 1-alkynylphosphonates using $(EtO)_2P(O)CCl_3$ as precursor, *Tetrahedron Lett.*, 37, 1783, 1996.

37. Savignac, P., and Mathey, F., A new route to dialkyl 1-(trimethylsilyl)- and 1-(trimethylstannyl)alkanephosphonates, *Synthesis*, 725, 1982.

38. Kolb, U., Dräger, M., Fischer, E., and Jurkschat, K., Preparation and structure of $(EtO)_2P(O)CH_2Si(Me)_2CH_2SnMe_2Cl$, a six-membered ring chelate with chair conformation and $P=O\cdots Sn(Cl)Me_2CH_2$ trigonal bipyramid about Lewis acid tin, *J. Organomet. Chem.*, 423, 339, 1992.

39. Müller, J.F.K., Neuburger, M., and Spingler, B., Structural investigation of a dilithiated phosphonate in the solid state, *Angew. Chem.*, 111, 97, 1999; *Angew. Chem. Int. Ed. Engl.*, 38, 92, 1999.

40. Müller, J.F.K., Kulicke, K.J., Neuburger, M., and Spichty, M., Carbanions substituted by transition metals. Synthesis, structure, and configurational restrictions of a lithium titanium phosphonate, *Angew. Chem. Int. Ed. Engl.*, 40, 2890, 2001.

41. Patois, C., and Savignac, P., A new route to α-fluoromethyl- and α-fluoroalkyl-phosphonates, *J. Chem. Soc., Chem. Commun.*, 1711, 1993.

42. Couture, A., Deniau, E., Woisel, P., and Grandclaudon, P., A convenient synthetic route to *N*-aryl and *N*-alkylamino(alkyl)phosphonates and phosphine oxides, *Tetrahedron Lett.*, 36, 2483, 1995.

43. Makomo, H., Saquet, M., Simeon, F., Masson, S., About-Jaudet, E., Collignon, N., and Gulea-Purcarescu, M., [2.3]-Wittig sigmatropic rearrangement of α-phosphorylated sulfonium and ammonium ylides, *Phosphorus, Sulfur Silicon Relat. Elem.*, 109, 445, 1996.

44. Gulea-Purcarescu, M., About-Jaudet, E., Collignon, N., Saquet, M., and Masson, S., Sigmatropic [2,3]-Wittig rearrangement of α-allylic-heterosubstituted methylphosphonates. Part 2. Rearrangement in the nitrogen series, *Tetrahedron*, 52, 2075, 1996.

45. Castelot-Deliencourt, G., Roger, E., Pannecoucke, X., and Quirion, J.-C., Diastereoselective synthesis of chiral amidophosphonates by 1,5-asymmetric induction, *Eur. J. Org. Chem.*, 3031, 2001.

46. Seyferth, D., and Marmor, R.S., Halomethyl-metal compounds. Part 63. Diethyl lithiodichloromethylphosphonate and tetraethyl lithiochloromethylenediphosphonate, *J. Organomet. Chem.*, 59, 237, 1973.

47. Regitz, M., Weber, B., and Eckstein, U., Investigations on diazo compounds and azides. Part 32. Substitution reactions at the diazo carbon of diazomethylphosphoryl compounds, *Liebigs Ann. Chem.*, 1002, 1979.

48. Eisch, J.J., and Galle, J.E., Organosilicon compounds with functional groups proximate to silicon. Part 17. Synthetic and mechanistic aspects of the lithiation of α,β-epoxyalkylsilanes and related α-heterosubstituted epoxides, *J. Organomet. Chem.*, 341, 293, 1988.

49. Guillemin, J.C., Janati, T., Guenot, P., Savignac, P., and Denis, J.M., Synthesis of non-stabilized phosphaalkynes by hydrogen chloride elimination in a vacuum-gas-solid reaction, *Angew. Chem.*, 103, 191, 1991; *Angew. Chem. Int. Ed. Engl.*, 30, 196, 1991.

50. Grandin, C., About-Jaudet, E., Collignon, N., Denis, J.M., and Savignac, P., 1,1-Dichloroalkylphosphonates. Convenient starting materials for low-coordinates trivalent phosphorus compounds, *Heteroatom Chem.*, 3, 337, 1992.

51. Burton, D.J., and Yang, Z.-Y., Fluorinated organometallics. Perfluoroalkyl and functionalized perfluoroalkyl organometallic reagents in organic synthesis, *Tetrahedron*, 48, 189, 1992.

52. Patois, C., and Savignac, P., Easy access to α-fluorinated phosphonic acid esters, *Phosphorus, Sulfur Silicon Relat. Elem.*, 77, 163, 1993.

53. Carran, J., Waschbüsch, R., Marinetti, A., and Savignac, P., An improved synthesis of dichloroalkylphosphonates, chloroalkynes and terminal alkynes via diethyl dichloromethylphosphonate, *Synthesis*, 1494, 1996.

54. Gueguen, C., About-Jaudet, E., Collignon, N., and Savignac, P., A new and efficient synthesis of α-amino cycloalkylphosphonic acids via electrophilic azidation of cycloalkylphosphonates, *Synth. Commun.*, 26, 4131, 1996.

55. Atta, F.M., Betz, R., Schmid, B., and Schmidt, R.R., Functionally substituted vinyl carbanions. Part 25. Heteroatom influence on vinylic deprotonation, *Chem. Ber.*, 119, 472, 1986.

56. Phillips, A.M.M.M., and Modro, T.A., Phosphonic systems. Part 2. Functional group and skeleton modifications in diethyl esters of 2-propenyl (and 2-pentenyl) phosphonic acids, *Phosphorus, Sulfur Silicon Relat. Elem.*, 55, 41, 1991.

57. Gil, J.M., and Oh, D.Y., Carbocupration of diethyl 1-alkynylphosphonates. Stereo- and regioselective synthesis of 1,2,2-trisubstituted vinylphosphonates, *J. Org. Chem.*, 64, 2950, 1999.

58. Waschbüsch, R., Samadi, M., and Savignac, P., A useful magnesium reagent for the preparation of 1,1-difluoro-2-hydroxyphosphonates from diethyl bromodifluoromethylphosphonate via a metal-halogen exchange reaction, *J. Organomet. Chem.*, 529, 267, 1997.

59. Burton, D.J., Takei, R., and Shin-Ya, S., Preparation, stability, reactivity and synthetic utility of a cadmium stabilized complex of difluoromethylene phosphonic acid ester, *J. Fluorine Chem.*, 18, 197, 1981.

60. Jubault, P., Feasson, C., and Collignon, N., Electrochemical activation of magnesium. A new, rapid and efficient method for the electrosynthesis of 1,1-dichloroalkylphosphonates from diisopropyl trichloromethylphosphonate, *Tetrahedron Lett.*, 36, 7073, 1995.

61. Martynov, B.I., and Stepanov, A.A., Electrochemical synthesis of fluoro-organosilanes, *J. Fluorine Chem.*, 85, 127, 1997.

62. Allspach, T., Gümbel, H., and Regitz, M., Studies on diazo compounds and azides. Part 63. Silylation of α-diazo phosphonates and carboxylates with silyl triflates, *J. Organomet. Chem.*, 290, 33, 1985.

63. Filonenko, L.P., Bespal'ko, G.K., Marchenko, A.P., and Pinchuk, A.M., Silylation of compounds containing trichloromethyl groups, *Zh. Obshch. Khim.*, 57, 2320, 1987; *J. Gen. Chem. USSR (Engl. Transl.)*, 57, 2074, 1987.

64. Kolodyazhnyi, O.I., and Golokhov, D.B., *p*-Chloro ylides with alkoxyl groups on phosphorus, *Zh. Obshch. Khim.*, 57, 2640, 1987; *J. Gen. Chem. USSR (Engl. Transl.)*, 57, 2353, 1987.

65. Kolodyazhnyi, O.I., and Ustenko, S.N., *p*-Iodo ylides, *Zh. Obshch. Khim.*, 62, 2144, 1992; *J. Gen. Chem. USSR (Engl. Transl.)*, 62, 1764, 1992.

66. Chernyshev, E.A., Bugerenko, E.F., Matveicheva, G.P., Bochkarev, V.N., and Kisin, A.V., Silicon- and phosphorus-containing ethylene derivatives, *Zh. Obshch. Khim.*, 45, 1768, 1975; *J. Gen. Chem. USSR (Engl. Transl.)*, 45, 1733, 1975.

67. Nowick, J.S., and Danheiser, R.L., Application of (α-phosphonoacyl)silane reagents to the synthesis of α,β-unsaturated acylsilanes, *J. Org. Chem.*, 54, 2798, 1989.

68. Thomas, S.E., Tustin, G.J., and Ibbotson, A., Synthesis and reactivity of iron carbonyl complexes of α,β-unsaturated acyl silanes, *Tetrahedron*, 48, 7629, 1992.

69. Barnes, G.H., Jr., and David, M.P., Synthesis and hydrolytic stability of some organosilicon phosphonate esters, *J. Org. Chem.*, 25, 1191, 1960.

70. Dolgov, O.N., and Voronkov, M.G., Telomerization of dialkyl hydrogen phosphites and *bis*(trialkylsilyl) hydrogen phosphites with vinylsilane derivatives, *Zh. Obshch. Khim.*, 40, 1668, 1970; *J. Gen. Chem. USSR (Engl. Transl.)*, 40, 1660, 1970.

71. Moskalenko, M.A., and Terent'ev, A.B., Radical telomerization of vinyltrimethylsilane with diethyl phosphite, *Izv. Akad. Nauk SSSR, Ser. Khim.*, 1883, 1990; *Bull. Acad. Sci. USSR, Div. Chem. Sci. (Engl. Transl.)*, 1711, 1990.

72. Hägele, G., Bönigk, W., Dickopp, H., and Wendisch, D., Silylated phosphonic acids. Radical addition of diethyl phosphite to trimethylsilylethylene, *Phosphorus Sulfur*, 26, 253, 1986.

73. Novikova, Z.S., Mashoshina, S.N., Sapozhnikova, T.A., and Lutsenko, I.F., Reaction of trimethylsilyl diethyl phosphite with carbonyl compounds, *Zh. Obshch. Khim.*, 41, 2622, 1971; *J. Gen. Chem. USSR (Engl. Transl.)*, 41, 2655, 1971.

74. Lebedev, E.P., Pudovik, A.N., Tsyganov, B.N., Nazmutdinov, R.Y., and Romanov, G.V., Silyl esters of phosphorous and arylphosphonous acids, *Zh. Obshch. Khim.*, 47, 765, 1977; *J. Gen. Chem. USSR (Engl. Transl.)*, 47, 698, 1977.

75. Nakano, M., Okamoto, Y., and Sakurai, H., Preparation of dialkyl 2-cyano-2-(trimethylsilyl)ethanephosphonates, *Synthesis*, 915, 1982.

76. Nakano, M., and Okamoto, Y., A convenient synthesis of 2-cyano-1,3-butadienes by the reaction of diethyl 2-cyano-2-trimethylsilylethanephosphonate with carbonyl compounds, *Synthesis*, 917, 1983.

77. Novikova, Z.S., and Lutsenko, I.F., Reaction of trialkylsilyl dialkyl phosphites with unsaturated compounds, *Zh. Obshch. Khim.*, 40, 2129, 1970; *J. Gen. Chem. USSR (Engl. Transl.)*, 40, 2110, 1970.

78. Pudovik, A.N., Batyeva, E.S., and Zamaletdinova, G.U., Reaction of trimethylsilyl diethyl phosphite with maleimides, *Zh. Obshch. Khim.*, 45, 940, 1975; *J. Gen. Chem. USSR (Engl. Transl.)*, 45, 922, 1975.

79. Pudovik, A.N., Batyeva, E.S., Zamaletdinova, G.U., Anoshina, N.P., and Kondranina, V.Z., Reaction of trimethylsilyl diethyl phosphite with maleimides, *Zh. Obshch. Khim.*, 46, 953, 1976; *Chem. Abstr.*, 85, 108709g, 1976.

80. Gareev, R.D., Borisova, E.E., and Shermergorn, I.M., New method for the synthesis of (2-nitroalkyl)phosphonic esters, *Zh. Obshch. Khim.*, 45, 944, 1975; *J. Gen. Chem. USSR (Engl. Transl.)*, 45, 928, 1975.

81. Hanessian, S., Cantin, L.-D., Roy, S., Andreotti, D., and Gomtsyan, A., The synthesis of enantiomerically pure, symmetrically substituted cyclopropane phosphonic acids. A constrained analog of the GABA antagonist phaclophen, *Tetrahedron Lett.*, 38, 1103, 1997.

82. Kyoda, M., Yokoyama, T., Kuwahara, T., Maekawa, H., and Nishiguchi, I., Mg-promoted regioselective carbon-silylation of α-phosphorylacrylate derivatives, *Chem. Lett.*, 228, 2002.

83. Gopalan, A., Moerck, R., and Magnus, P., Regiospecific control of the ene reaction of *N*-phenyl-1,2,4-triazoline-3,5-dione, *J. Chem. Soc., Chem. Commun.*, 548, 1979.

84. Henning, R., and Hoffmann, H.M.R., A novel approach to complex terpenoid methylenecyclohexanes, *Tetrahedron Lett.*, 23, 2305, 1982.

85. Hoffmann, H.M.R., and Henning, R., Synthesis of 2-norzizaene and 9,10-dehydro-2-norzizaene (7,7-dimethyl-6-methylidenetricyclo[6.2.1.0^{1,5}]undec-9-ene) via intramolecular allyl cation induced cycloaddition, *Helv. Chim. Acta*, 66, 828, 1983.

86. Porskamp, P.A.T.W., Lammerink, B.H.M., and Zwanenburg, B., Synthesis and reactions of phosphoryl-substituted sulfines, *J. Org. Chem.*, 49, 263, 1984.

87. Binder, J., and Zbiral, E., Diethyl (trimethylsilylethoxymethyl)phosphonates as new reagents for variable strategies for synthesis of carbonyl compounds, α-hydroxycarbonyl compounds and vinylphosphonates, *Tetrahedron Lett.*, 25, 4213, 1984.

88. Tsuge, O., Kanemasa, S., and Suga, H., Synthesis of a new phosphorous-functionalized nitrile oxide, α-(diethylphosphono)acetonitrile oxide, and cycloaddition leading to 3-(diethylphosphonomethyl)-Δ²-isoxazolines, *Chem. Lett.*, 183, 1986.

89. Tsuge, O., Kanemasa, S., Suga, H., and Nakagawa, N., Synthesis of (diethoxyphosphoryl)acetonitrile oxide and its cycloaddition to olefins. Synthetic applications to 3,5-disubstituted 2-isoxazolines, *Bull. Chem. Soc. Jpn.*, 60, 2463, 1987.

90. Giguere, R.J., Duncan, S.M., Bean, J.M., and Purvis, L., Intramolecular [3 + 4] allyl cation cycloaddition. Novel route to hydroazulenes, *Tetrahedron Lett.*, 29, 6071, 1988.

91. Nishitani, K., Fukuda, H., and Yamakawa, K., Studies on the terpenoids and related alicyclic compounds. Part 42. Diastereoselective cyclization of ω-formylated allylsilanes into bicyclic α-methylene-γ-butyrolactones. A facile synthesis of *p*-menthanolides, *Heterocycles*, 33, 97, 1992.

92. Loreto, M.A., Pompili, C., and Tardella, P.A., α-Methylene β-amino phosphonic ester derivatives by amination of (1-trimethylsilanylmethyl-vinyl) phosphonic esters, *Tetrahedron*, 57, 4423, 2001.

93. Hirose, T., Sunazuka, T., Shirahata, T., Yamamoto, D., Harigaya, Y., Kuwajima, I., and Omura, S., Short total synthesis of (+)-madindolines A and B, *Org. Lett.*, 4, 501, 2002.

94. Johnson, W.S., Newton, C., and Lindell, S.D., The carboalkoxyallylsilane terminator for biomimetic polyene cyclizations. A route to 21-hydroxyprogesterone types, *Tetrahedron Lett.*, 27, 6027, 1986.

95. Kirschleger, B., and Queignec, R., Heterogeneous mediated alkylation of ethyl diethyl phosphonoacetate. A "one pot" access to α-alkylated acrylic esters, *Synthesis*, 926, 1986.

96. Dashan, L., and Trippett, S., The *ortho*-lithiation of *N,N,N′,N′*-tetramethylphenylphosphonic diamide, *Tetrahedron Lett.*, 24, 2039, 1983.

97. Vedejs, E., Daugulis, O., Diver, S.T., and Powell, D.R., Generation of the 1,3-phosphasilolene skeleton from *ortho* silylated biarylphosphonates, *J. Org. Chem.*, 63, 2338, 1998.

98. Maffei, M., and Buono, G., Stereoselective synthesis of dimethyl- and diethyl-*E*-2-(trimethylsilyl)vinyl phosphonates and their use in the Diels–Alder reaction, *Phosphorus, Sulfur Silicon Relat. Elem.*, 79, 297, 1993.

99. Léost, F., and Doutheau, A., Trimethylsilylated 1,3-dienes or styrenes from the reaction between trimethylsilyldiazomethane and vinyl or aryl ketenes, *Tetrahedron Lett.*, 40, 847, 1999.

100. Hanggi, D.A., Babu, G.N., and Davis, T.L., Adsorbent, its preparation and use in chromatography, *3M*, Int. Patent Appl. WO 9725140, 1997; *Chem. Abstr.*, 127, 177473, 1997.

101. Kolodyazhnyi, O.I., and Yastenko, S.N., Silylation of an allylphosphonate anion, *Zh. Obshch. Khim.*, 60, 695, 1990; *J. Gen. Chem. USSR (Engl. Transl.)*, 60, 610, 1990.

102. Gerber, J.P., and Modro, T.A., Phosphonic systems. Part 13. Alkylation of diethyl cyclohexenylphosphonates, *Phosphorus, Sulfur Silicon Relat. Elem.*, 84, 107, 1993.

103. Probst, M.F., and Modro, T.A., 3-Silanyl-propenylphosphonates. Properties and alkylation potential, *Heteroatom Chem.*, 6, 579, 1995.

104. Ahlbrecht, H., König, B., and Simon, H., Deprotonation of 1-phosphonato enol phosphates, *Tetrahedron Lett.*, 19, 1191, 1978.

105. Ahlbrecht, H., and Farnung, W., 3-Metalated enamines. Part 10. Formation and reactivity of 1-(diethoxyphosphoryl)-1-(dimethylamino)allyl anions, *Chem. Ber.*, 117, 1, 1984.

106. Boumekouez, A., About-Jaudet, E., and Collignon, N., Synthesis, metalation and functionalisation of diethyl tolylphosphonates, *Phosphorus, Sulfur Silicon Relat. Elem.*, 77, 177, 1993.

107. Boumekouez, A., About-Jaudet, E., and Collignon, N., Side-chain metallation of diethyl phosphonotoluenes, *J. Organomet. Chem.*, 466, 89, 1994.

108. Goswami, R., A novel β-ketophosphonate 1,4-dianion. Tin/lithium exchange, *J. Am. Chem. Soc.*, 102, 5973, 1980.

109. Yang, Z.-Y., and Burton, D.J., A novel, general method for the preparation of α,α-difluoro functionalized phosphonates, *Tetrahedron Lett.*, 32, 1019, 1991.

110. Yang, Z.-Y., and Burton, D.J., A novel and practical preparation of α,α-difluoro functionalized phosphonates from iododifluoromethylphosphonate, *J. Org. Chem.*, 57, 4676, 1992.

111. Zhang, X., Qiu, W., and Burton, D.J., Preparation of α-fluorophosphonates, *J. Fluorine Chem.*, 89, 39, 1998.

112. Chernyshev, E.A., Bugerenko, E.F., Lubuzh, E.D., and Petrov, A.D., Synthesis of γ-organosilyl derivatives of propylphosphonic dichloride and diethyl ester, *Izv. Akad. Nauk SSSR, Ser. Khim.*, 1001, 1962; *Bull. Acad. Sci. USSR, Div. Chem. Sci. (Engl. Transl.)*, 937, 1962.

113. Phillion, D.P., and Cleary, D.G., Disodium salt of 2-[(dihydroxyphosphinyl)difluoromethyl] propenoic acid. An isopolar and isosteric analogue of phosphoenolpyruvate, *J. Org. Chem.*, 57, 2763, 1992.

114. Eymery, F., Iorga, B., and Savignac, P., Synthesis of phosphonates by nucleophilic substitution at phosphorus. The SNP(V) reaction, *Tetrahedron*, 55, 13109, 1999.

115. Lee, G.C.M., 4-Ethyl and 4-ethenyl-5-hydroxy-2(5*H*)-furanones substituted on alpha carbon of the ethyl or ethenyl side chain with a long chain alkyl group and on the beta carbon with a polar group, as anti-inflammatory agents, *Allergan*, U.S. Patent Appl. US 5013850, 1991; *Chem. Abstr.*, 115, 135904, 1991.

116. Seyferth, D., and Marmor, R.S., Dimethyl diazomethylphosphonate. Its preparation and reactions, *Tetrahedron Lett.*, 11, 2493, 1970.

117. Seyferth, D., Marmor, R.S., and Hilbert, P., Some reactions of dimethylphosphono-substituted diazoalkanes. $(MeO)_2P(O)CR$ transfer to olefins and 1,3-dipolar additions of $(MeO)_2P(O)C(N_2)R^1$, *J. Org. Chem.*, 36, 1379, 1971.

118. Kawashima, T., Ishii, T., and Inamoto, N., Fluoride ion induced Horner–Emmons reaction of α-silylalkylphosphonates with carbonyl compounds, *Tetrahedron Lett.*, 24, 739, 1983.

119. Palomo, C., Aizpurua, J.M., Garcia, J.M., Ganboa, I., Cossio, F.P., Lecea, B., and Lopez, C., A new version of the Peterson olefination using *bis*(trimethylsilyl)methyl derivatives and fluoride ion as catalyst, *J. Org. Chem.*, 55, 2498, 1990.

120. Sharma, R.K., and Fry, J.L., Instability of anhydrous tetra-*n*-alkylammonium fluorides, *J. Org. Chem.*, 48, 2112, 1983.

121. Latouche, R., Texier-Boullet, F., and Hamelin, J., Reactivity of silylated compounds with active methylene under microwave irradiation in heterogeneous dry media. Application to the alkali metal fluoride-mediated silyl-Reformatsky reaction, *Bull. Soc. Chim. Fr.*, 130, 535, 1993.

122. Obayashi, M., Ito, E., Matsui, K., and Kondo, K., (Diethylphosphinyl)difluoromethyllithium. Preparation and synthetic application, *Tetrahedron Lett.*, 23, 2323, 1982.

123. Obayashi, M., and Kondo, K., An improved procedure for the synthesis of 1,1-difluoro-2-hydroxyalkylphosphonates, *Tetrahedron Lett.*, 23, 2327, 1982.

124. Ruzicka, J.A., Qiu, W., Baker, M.T., and Burton, D.J., Synthesis of [1–^{14}C]-2,2-difluoroethene from [^{14}C]-formaldehyde, *J. Labelled Compd. Radiopharm.*, 34, 59, 1994.

125. Peterson, D.J., Carbonyl olefination reaction using silyl-substituted organometallic compounds, *J. Org. Chem.*, 33, 780, 1968.

126. Harnden, M.R., Parkin, A., Parratt, M.J., and Perkins, R.M., Novel acyclonucleotides. Synthesis and antiviral activity of alkenylphosphonic acid derivatives of purines and a pyrimidine, *J. Med. Chem.*, 36, 1343, 1993.

127. L'vova, S.D., Mamontova, T.A., and Gunar, V.I., Synthesis of 5′-unsaturated analogs of pyridoxal 5′-phosphate, *Zh. Org. Khim.*, 12, 448, 1976; *J. Org. Chem. USSR (Engl. Transl.)*, 12, 440, 1976.

128. Tian, F., Migaud, M.E., and Frost, J. W., *myo*-Inositol 1-phosphate synthase. Does a single active-site amino acid catalyze multiple proton transfers? *J. Am. Chem. Soc.*, 121, 5795, 1999.

129. Chang, K., Ku, B., and Oh, D.Y., New synthesis of diethyl alkylphosphonates and diethyl alkenylphosphonates, *Synth. Commun.*, 19, 1891, 1989.

130. Binder, J., and Zbiral, E., A new procedure for homologation of carbonyl compounds to α-hydroxycarboxylic esters by means of diethyl (trimethylsilylethoxymethyl)phosphonate, *Tetrahedron Lett.*, 27, 5829, 1986.

131. Ahlbrecht, H., Farnung, W., and Simon, H., Formation and reactivity of 1-phosphoryl-1-(phosphoryloxy)allyl anions and related compounds, *Chem. Ber.*, 117, 2622, 1984.

132. Dufrechou, S., Combret, J.C., Malhiac, C., and Collignon, N., Efficient synthesis of α-enaminophosphonates in the series of piperidine and morpholine. Examples of synthetic applications, *Phosphorus, Sulfur Silicon Relat. Elem*, 127, 1, 1997.

133. Blackburn, G.M., and Parratt, M.J., The synthesis of α-fluoroalkylphosphonates, *J. Chem. Soc., Chem. Commun.*, 886, 1983.

134. Blackburn, G.M., and Parratt, M.J., Synthesis of α-fluoroalkylphosphonates, *J. Chem. Soc., Perkin Trans. 1*, 1425, 1986.

135. Keeney, A., Nieschalk, J., and O'Hagan, D., The synthesis of α-monofluorovinylphosphonates by a Peterson type olefination reaction, *J. Fluorine Chem.*, 80, 59, 1996.

136. Schmitt, L., Cavusoglu, N., Spiess, B., and Schlewer, G., Synthesis of arylalkylmonofluorophosphonates as *myo*-inositol monophosphatase ligands, *Tetrahedron Lett.*, 39, 4009, 1998.

137. Waschbüsch, R., Carran, J., and Savignac, P., A new route to α-fluorovinylphosphonates utilizing Peterson olefination methodology, *Tetrahedron*, 52, 14199, 1996.

138. Probst, M.F., Modro, A.M., and Modro, T.A., 3-Phosphorylated 1,5-hexadienes as precursors for vinylic or 3 silanylvinylic phosphonates, *Can. J. Chem.*, 75, 1131, 1997.

139. Al-Badri, H., About-Jaudet, E., and Collignon, N., New and efficient synthesis of (*E*)-4-diethoxyphosphonyl-2-methyl-2-butenal and of ethyl (*E*)-4-diethoxyphosphonyl-2-methyl-2-butenoate, important building blocks in retinoid chemistry, *Tetrahedron Lett.*, 36, 393, 1995.

140. Chevalier, F., Al-Badri, H., and Collignon, N., Stereoselective synthesis of substituted 2-(diethylphosphonyl)-4-(trimethylsilyl)buta-1,3-dienes, *Bull. Soc. Chim. Fr.*, 134, 801, 1997.

141. Chevalier, F., Al-Badri, H., and Collignon, N., Reactivity of substituted 3-(diethylphosphonyl)-1-(trialkylsilyl)alka-1,3-dienes. Regioselective epoxidation and cyclopropanation reactions, *Heteroatom Chem.*, 10, 231, 1999.

142. Al-Badri, H., About-Jaudet, E., and Collignon, N., Unusual and efficient (*Z*)-stereoselective Peterson synthesis of 2-diethoxyphosphonyl-1-alkoxy-3-methylpenta-1,3-dienes. Their use in the Diels–Alder reaction, *Tetrahedron Lett.*, 37, 2951, 1996.

143. Al-Badri, H., About-Jaudet, E., and Collignon, N., Reaction of *in situ* generated α-silylated allylic phosphonate carbanions with aldehydes. An unexpected cyclization reaction, *J. Chem. Soc., Perkin Trans. 1*, 931, 1996.

144. van der Linden, J.B., Lucassen, A.C.B., and Zwanenburg, B., Synthesis of thiophene-2-phosphonates via α,β-unsaturated sulfines as intermediates, *Recl. Trav. Chim. Pays-Bas*, 113, 547, 1994.

145. Kouno, R., Okauchi, T., Nakamura, M., Ichikawa, J., and Minami, T., Synthesis and synthetic utilization of α-functionalized vinylphosphonates bearing β-oxy or β-thio substituents, *J. Org. Chem.*, 63, 6239, 1998.

146. Okauchi, T., Fukamachi, T., Nakamura, F., Ichikawa, J., and Minami, T., 2-Phosphono-1,1,4,4-tetrathio-1,3-butadienes. Synthesis and synthetic application to highly functionalized dienes and heterocycles, *Bull. Soc. Chim. Belg.*, 106, 525, 1997.

147. Minami, T., Kouno, R., Okauchi, T., Nakamura, M., and Ichikawa, J., Synthetic utilization of α-phosphonovinyl anions, *Phosphorus, Sulfur Silicon Relat. Elem.*, 144–146, 689, 1999.

148. Lee, B.S., Lee, S.Y., and Oh, D.Y., Efficient approach to 4-oxo-2-alkenylphosphonates via regiospecific Friedel–Crafts acylation, *J. Org. Chem.*, 65, 4175, 2000.

149. Patois, C., and Savignac, P., Preparation of triethyl 2-fluoro-2-phosphonoacetate, *Synth. Commun.*, 24, 1317, 1994.

150. Nieschalk, J., Batsanov, A.S., O'Hagan, D., and Howard, J.A.K., Synthesis of monofluoro- and difluoromethylenephosphonate analogs of *sn*-glycerol-3-phosphate as substrates for glycerol-3-phosphate dehydrogenase and the X-ray structure of the fluoromethylenephosphonate moiety, *Tetrahedron*, 52, 165, 1996.

151. Nieschalk, J., and O'Hagan, D., Monofluorophosphonates as phosphate mimics in bioorganic chemistry. A comparative study of CH_2-, CHF- and CF_2-phosphonate analogs of *sn*-glycerol-3-phosphate as substrates for *sn*-glycerol-3-phosphate dehydrogenase, *J. Chem. Soc., Chem. Commun.*, 719, 1995.

152. Waschbüsch, R., Carran, J., and Savignac, P., New routes to diethyl 1-fluoromethylphosphonocarboxylates and diethyl 1-fluoromethylphosphonocarboxylic acid, *Tetrahedron*, 53, 6391, 1997.

153. Waschbüsch, R., Carran, J., and Savignac, P., A new route to α-fluoroalkylphosphonates, *J. Chem. Soc., Perkin Trans. 1*, 1135, 1997.

154. Iorga, B., Eymery, F., and Savignac, P., A highly selective synthesis of α-monofluoro- and α-monochlorobenzylphosphonates using electrophilic halogenation of benzylphosphonates carbanions, *Tetrahedron Lett.*, 39, 3693, 1998.

155. Iorga, B., Eymery, F., and Savignac, P., Controlled monohalogenation of phosphonates. A new route to pure α-monohalogenated diethyl benzylphosphonates, *Tetrahedron*, 55, 2671, 1999.

156. Iorga, B., Eymery, F., and Savignac, P., Controlled monohalogenation of phosphonates. Part 2. Preparation of pure diethyl α-monohalogenated alkylphosphonates, *Synthesis*, 576, 2000.

157. Teulade, M.-P., and Savignac, P., Direct conversion of phosphates in α-lithiated phosphonates using alkyllithiums, *Tetrahedron Lett.*, 28, 405, 1987.

158. Savignac, P., Teulade, M.-P., and Patois, C., Phosphate–phosphonate conversion. Nucleophilic displacement reactions involving phosphoric esters and methyllithium, *Heteroatom Chem.*, 1, 211, 1990.

159. Patois, C., and Savignac, P., Phosphate–phosphonate conversion. A versatile route to linear or branched alkylphosphonates, *Bull. Soc. Chim. Fr.*, 130, 630, 1993.

160. Savignac, P., and Patois, C., Diethyl 1-propyl-2-oxoethylphosphonate (diethyl 1-formylbutylphosphonate), *Org. Synth.*, 72, 241, 1995.

161. Liu, X., Zhang, X.-R., and Blackburn, G.M., Synthesis of three novel supercharged β,γ-methylene analogs of adenosine triphosphate, *J. Chem. Soc., Chem. Commun.*, 87, 1997.

162. Berté-Verrando, S., Nief, F., Patois, C., and Savignac, P., General synthesis of α,α-dideuteriated phosphonic esters, *J. Chem. Soc., Perkin Trans. 1*, 821, 1994.

163. Berté-Verrando, S., Nief, F., Patois, C., and Savignac, P., Preparation of α-dideuteriated aminomethylphosphonic acid, *Phosphorus, Sulfur Silicon Relat. Elem.*, 103, 91, 1995.

164. Chang, K., Ku, B., and Oh, D.Y., Palladium(0) catalyzed stereoselective synthesis of dialkyl 1-(trimethylsilyl)alkenylphosphonates, *Bull. Korean Chem. Soc.*, 10, 320, 1989.

165. Hong, S., Chang, K., Ku, B., and Oh, D.Y., New synthesis of β-keto phosphonates, *Tetrahedron Lett.*, 30, 3307, 1989.

166. Grandin, C., Collignon, N., and Savignac, P., A practical synthesis of cycloalkylphosphonates from trichloromethylphosphonates, *Synthesis*, 239, 1995.

167. Savignac, P., and Iorga, B., unpublished data.

168. Krafft, G.A., and Meinke, P.T., Selenoaldehydes. Preparation and dienophilic reactivity, *J. Am. Chem. Soc.*, 108, 1314, 1986.

169. Meinke, P.T., and Krafft, G.A., Regiochemical preferences in selenoaldehyde cycloadditions, *Tetrahedron Lett.*, 28, 5121, 1987.

170. Meinke, P.T., and Krafft, G.A., Synthesis and cycloaddition reactivity of selenoaldehydes, *J. Am. Chem. Soc.*, 110, 8671, 1988.

171. Meinke, P.T., Krafft, G.A., and Guram, A., Synthesis of selenocyanates via cyanoselenation of organocopper reagents, *J. Org. Chem.*, 53, 3632, 1988.

172. Johnson, W.S., Lindell, S.D., and Steele, J., Rate enhancement of biomimetic polyene cyclizations by a cation-stabilizing auxiliary, *J. Am. Chem. Soc.*, 109, 5852, 1987.

173. Kuroda, C., Shimizu, S., and Satoh, J.Y., A simple synthesis of *cis*- and *trans*-fused 14,15-dinoreudesmanolides, *J. Chem. Soc., Chem. Commun.*, 286, 1987.

174. Giguere, R.J., Tassely, S.M., Rose, M.I., and Krishnamurthy, V.V., Diastereoselectivity and regiocontrol in intramolecular allyl cation cycloadditions. Selective formation of [3 + 2] or [3 + 4] cycloadducts, *Tetrahedron Lett.*, 31, 4577, 1990.

175. Kuroda, C., Shimizu, S., and Satoh, J.Y., A short-step synthesis of 14,15-dinoreudesmanolides using intramolecular cyclization of an allylsilane, *J. Chem. Soc., Perkin Trans. 1*, 519, 1990.

176. Yee, N.K.N., and Coates, R.M., Total synthesis of (+)-9,10-*syn*- and (+)-9,10-*anti*-copalol via epoxy trienylsilane cyclizations, *J. Org. Chem.*, 57, 4598, 1992.

177. Kuroda, C., Shimizu, S., Haishima, T., and Satoh, J.Y., Synthesis of a stereoisomer of frullanolide utilizing the intramolecular cyclization of ω-formyl-2-alkenylsilane, *Bull. Chem. Soc. Jpn.*, 66, 2298, 1993.

178. Kuroda, C., Inoue, S., Kato, S., and Satoh, J.Y., Synthesis of α-methylene-γ-lactones fused to a perhydroazulene carbon framework through intramolecular cyclization of allylsilanes, *J. Chem. Res. (M)*, 458, 1993.

179. Nishitani, K., Nakamura, Y., Orii, R., Arai, C., and Yamakawa, K., Studies on the terpenoids and related alicyclic compounds. Part 43. Stereoselective intramolecular cyclization of β-alkoxycarbonyl-ω-formylallylsilanes into bicyclic α-methylene-γ-lactones, *Chem. Pharm. Bull.*, 41, 822, 1993.

180. Kuroda, C., Ohnishi, Y., and Satoh, J.Y., Intramolecular cyclization of β-(alkoxycarbonyl)allylsilane with conjugated ketone. A new entry to bicyclo[4.3.0]nonane, *Tetrahedron Lett.*, 34, 2613, 1993.

181. Kuroda, C., Inoue, S., Takemura, R., and Satoh, J.Y., Intramolecular cyclization of allylsilanes in the synthesis of guaian-8,12-olide. Stereoselective formation of *trans*- and *cis*-fused methylenelactones, *J. Chem. Soc., Perkin Trans. 1*, 521, 1994.

182. Hashimoto, K., Ohfune, Y., and Shirahama, H., Synthesis of conformationally restricted analogs of kainic acid. Is the conformation of the C4-substituent of kainoid important to its neuroexcitatory activity? *Tetrahedron Lett.*, 36, 6235, 1995.

183. Kuja, E., and Giguere, R.J., On the effect of diene geometry in intramolecular allyl cation cyclo-additions, *Synth. Commun.*, 25, 2105, 1995.

184. Kuroda, C., Mitsumata, N., and Tang, C.Y., An acylative C-C single-bond cleavage and a self-cyclization of ethyl 2-(trimethylsilylmethyl)penta-2,4-dienoate of its free acid under Ritter conditions, *Bull. Chem. Soc. Jpn.*, 69, 1409, 1996.

185. Kuroda, C., and Ito, K., Synthesis of cadinanolide type of tricyclic α-methylene-γ-lactone using intramolecular cyclization of α-trimethylsilylmethyl-α,β-unsaturated ester with cyclic ketone, *Bull. Chem. Soc. Jpn.*, 69, 2297, 1996.

186. Kuroda, C., Nogami, H., Ohnishi, Y., Kimura, Y., and Satoh, J.Y., Stereochemistry of Lewis acid and fluoride promoted intramolecular cyclization of β-(alkoxycarbonyl)allylsilane with enones. Synthesis of bicyclo[4.3.0]nonanes, *Tetrahedron*, 53, 839, 1997.

187. Kuroda, C., and Anzai, S., Synthesis of α-methylene-γ-lactone fused to seven, eight, and fourteen-membered carbocycle through intramolecular cyclization of functionalized allylsilane with acid chloride, *Chem. Lett.*, 875, 1998.

188. Kuroda, C., Kimura, Y., and Nogami, H., Intramolecular cyclization of 2-(alkoxycarbonyl)allylsilanes with ynones. Nucleophilic and electrophilic aspects of the 2-(alkoxycarbonyl)allylsilane moiety, *J. Chem. Res. (M)*, 822, 1998.

189. Kuroda, C., and Koshio, H., New cyclization reaction of 2-(trimethylsilylmethyl)pentadienal. Synthesis of spiro[4.5]decane ring system, *Chem. Lett.*, 962, 2000.

190. Suzuki, H., Monda, A., and Kuroda, C., Synthesis of eleven-membered carbocycles by a new five-carbon ring-expansion reaction, *Tetrahedron Lett.*, 42, 1915, 2001.

191. Kuroda, C., and Hirono, Y., Synthesis of spiro[4.5]decane ring system through allylsilane promoted spiroannulation, *Tetrahedron Lett.*, 35, 6895, 1994.

192. Kuroda, C., Sumiya, H., Murase, A., and Koito, A., Iron(III)-induced tandem Nazarov cyclization-rearrangement of α-(trimethylsilylmethyl)divinyl ketone. Synthesis of the bicyclo[4.3.0]nonane ring system via spiro[4.4]nonane, *J. Chem. Soc., Chem. Commun.*, 1177, 1997.

193. DiFranco, E., Ravikumar, V.T., and Salomon, R.G., Total synthesis of halichondrins. Enantioselective construction of a homochiral tetracyclic KLMN-ring intermediate from D-mannitol, *Tetrahedron Lett.*, 34, 3247, 1993.

3   The α-Halogenophosphonates

The chemistry of α-halogenophosphonates is an area that has been the focus of much recent interest. Syntheses employing electrophilic halogenation of phosphonate carbanions have provided a new flexibility in this area. α-Halogenophosphonates have been used principally as precursors for alkynes and more elaborate phosphonates and for the preparation of fluorophosphonate analogues of biologically active phosphates. This chapter describes their preparation and reactions, including those involving both retention and removal of halogen atoms.

3.1 SYNTHESIS OF α-HALOGENOPHOSPHONATES

3.1.1 α-MONOHALOGENOPHOSPHONATES

3.1.1.1 Michaelis–Arbuzov Reactions

Diethyl iodomethylphosphonate was obtained for the first time in 1936 by A. E. Arbuzov and Kushkova by the reaction of diiodomethane with triethyl phosphite in 60% yield (Scheme 3.1).[1] Similar results were obtained in later studies (30–59%),[2–7] even after experimental modifications aimed at limiting the formation of the major byproduct, tetraethyl methylenediphosphonate.[8] Under the optimized conditions, dimethyl iodomethylphosphonate is obtained in 27% yield. This low yield can be explained by a second side reaction, the Michaelis–Arbuzov rearrangement between trimethyl phosphite and the methyl iodide coproduct, which gives dimethyl methylphosphonate.[9]

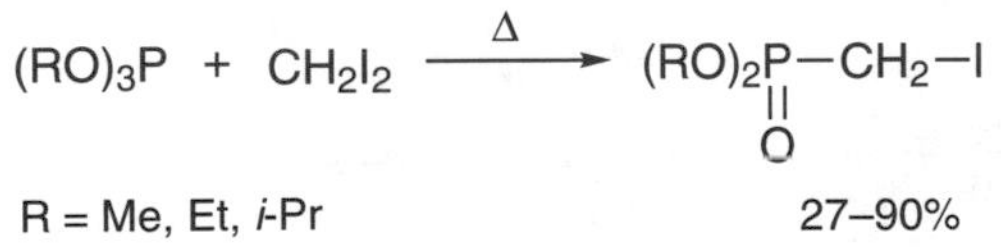

$$(RO)_3P \ + \ CH_2I_2 \ \xrightarrow{\Delta} \ (RO)_2\underset{O}{\overset{\parallel}{P}}-CH_2-I$$

$$R = Me, Et, \textit{i-}Pr \qquad\qquad 27\text{–}90\%$$

$$(3.1)$$

These yields have been greatly improved by the very slow addition of triethyl or trimethyl phosphite to diiodomethane at reflux to allow the elimination of alkyl iodide as the reaction proceeds (90% and 41%, respectively).[10]

The reaction of triethyl phosphite with dibromomethane affords diethyl bromomethylphosphonate in only 10–15% yield.[3,7,11] The use of a fourfold excess of triethyl phosphite[12] or triisopropyl phosphite[13] gives better results (40–48%).

Trialkyl phosphites can react with secondary bromides bearing electron-withdrawing (carboxyl) or stabilizing (vinyl, phenyl) groups. Thus, the reaction of alkyl bromofluoroacetates with trialkyl phosphites at 140–150°C produces dialkyl 1-(alkoxycarbonyl)-1-fluoromethylphosphonates in good yields (Scheme 3.2, Table 3.1), which vary according to the stability of the secondary bromide at the reaction temperature. Recently, diethyl 1-(ethoxycarbonyl)-1-fluoromethylphosphonate has been obtained in 74% yield from triethyl phosphite and ethyl iodofluoroacetate.[14]

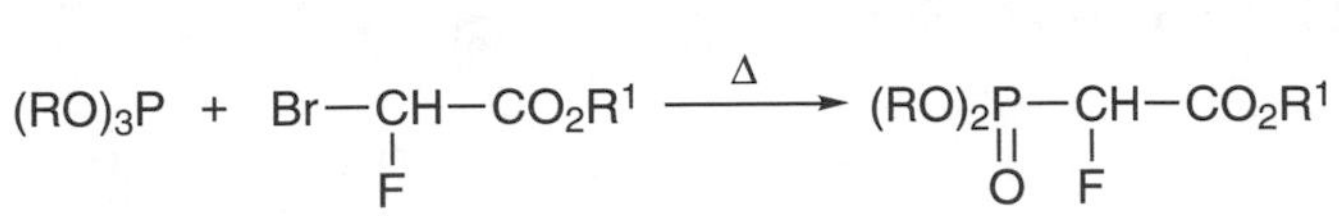

$$(RO)_3P \ + \ Br-\underset{F}{\overset{|}{C}H}-CO_2R^1 \ \xrightarrow{\Delta} \ (RO)_2\underset{O}{\overset{\parallel}{P}}-\underset{F}{\overset{|}{C}H}-CO_2R^1$$

$$(3.2)$$

TABLE 3.1

	R	R¹	Temp. (°C)	Yield (%)[Ref.]
a	Et	Me	reflux	65[15]
b	Et	Et	140 145–148	69,[16] 85[17] 64,[18] 71,[19] 75,[20] 77,[21] 83,[22,23] 98[24]
c	i-Pr	Me	140 150	—[17] 77[25]
d	i-Pr	Et	145–148	—,[20] 74[26]

Analogously, ethyl 4-fluoro-3-methylcrotonate is brominated with NBS in the presence of AIBN and then reacted with triethyl phosphite to produce good yields (72–78%) of α-fluorophosphonate, a key intermediate in the synthesis of fluorinated vitamin A esters (Scheme 3.3).[27] Following a similar procedure (radical bromination and Michaelis–Arbuzov rearrangement), (*E*)-1-fluoro-3-(1-trityl-1,2,4-triazol-3-yl)-2-propenylphosphonate, a good inhibitor of imidazoleglycerol phosphate dehydratase, has been prepared in 25% overall yield.[28,29]

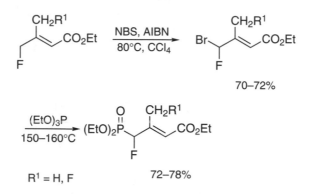

$$(3.3)$$

The reaction of ethyl 4-(bromofluoromethyl)benzoate with triethyl phosphite at 200°C produces the corresponding α-fluorobenzylphosphonate in 68% yield. After Horner–Wadsworth–Emmons olefination, this gives fluorinated aromatic retinoic acid analogues.[30,31]

3.1.1.2 Michaelis–Becker Reactions

The first synthesis of dialkyl fluoromethylphosphonates by a Michaelis–Becker reaction was described in 1967.[32] The halogenofluoromethane, which is no longer commercially available because of its mutagenic properties, is added to a refluxing toluene solution of sodium dialkyl phosphite. The corresponding fluoromethylphosphonate is obtained in modest yields (36–40%, Scheme 3.4).[32]

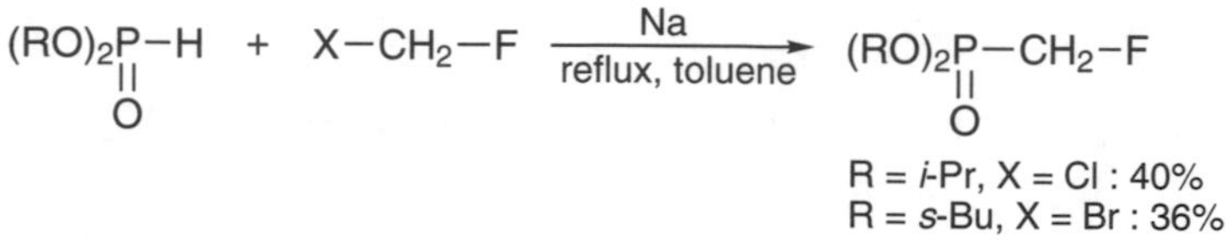

$$(RO)_2\overset{\text{O}}{\underset{\|}{P}}-H \;+\; X-CH_2-F \xrightarrow[\text{reflux, toluene}]{\text{Na}} (RO)_2\overset{\text{O}}{\underset{\|}{P}}-CH_2-F$$

R = i-Pr, X = Cl : 40%
R = s-Bu, X = Br : 36%

$$(3.4)$$

Similarly, the reaction of ethyl chlorofluoroacetate with sodium diethyl phosphite in C_6H_6 at room temperature affords diethyl 1-(ethoxycarbonyl)-1-fluoromethylphosphonate in 40% yield.[33]

The reaction of sodium diethyl phosphite with a large excess of dichloromethane at room temperature generates mainly tetraethyl methylenediphosphonate (51% yield) and diethyl chloromethylphosphonate (14% yield). These results clearly demonstrate that diethyl chloromethylphosphonate is a more reactive substrate in a nucleophilic substitution reaction than dichloromethane.[34]

The competing halogen–metal and proton–metal exchange reactions mean that the Michaelis–Becker reaction has only limited use for the synthesis of α-halogenomethylphosphonates.

3.1.1.3 Kinnear–Perren Reactions

This powerful method of formation of a P–C bond is based on the formal insertion of phosphorus trihalide into a C–X bond. An equimolecular mixture of PCl_3, $AlCl_3$, and CH_2Cl_2 is progressively heated until it becomes solid, then dissolved in CH_2Cl_2 and treated at –20°C with H_2O (11 eq) with vigorous stirring. Filtration affords chloromethylphosphonic dichloride in 85% yield.[35] Dialkyl chloromethylphosphonates are obtained by simple alcoholysis under anhydrous conditions (Scheme 3.5).[36,37] The use of CH_2ClBr or CH_2Br_2 instead of CH_2Cl_2 also gives chloromethylphosphonic dichloride in low yield (35%) by a halogen exchange during the complex formation.[35] Diethyl and diallyl bromomethylphosphonates are obtained by the same procedure in 49% and 43% yields, respectively, using PBr_3, $AlBr_3$, and CH_2Br_2 (Scheme 3.5).[10]

3.1.1.4 Kabachnik Reactions

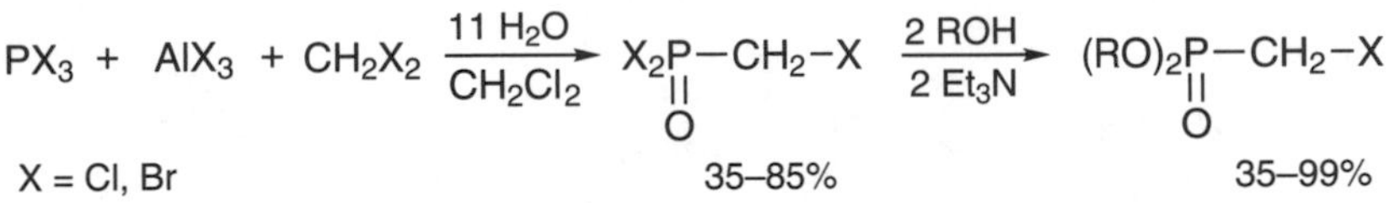

$$PX_3 \ + \ AlX_3 \ + \ CH_2X_2 \ \xrightarrow[CH_2Cl_2]{11\ H_2O} \ \underset{\underset{O}{\|}}{X_2P}-CH_2-X \ \xrightarrow[2\ Et_3N]{2\ ROH} \ \underset{\underset{O}{\|}}{(RO)_2P}-CH_2-X$$

X = Cl, Br   35–85%   35–99%   **(3.5)**

These are particularly simple reactions involving the heating of a mixture of PCl_3 and dry formaldehyde in a sealed tube at 190–260°C.[2,3,38–43] Distillation of the crude product produces the chloromethylphosphonic dichloride in 50–67% yield. Subsequent alcoholysis in the presence of tertiary bases gives dialkyl chloromethylphosphonates (Scheme 3.6).[36,37] The reaction can be extended to other aldehydes, but the yields are somewhat lower.[44]

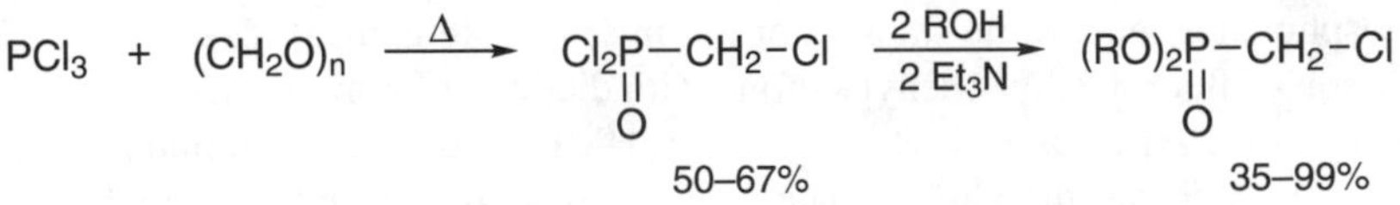

$$PCl_3 \ + \ (CH_2O)_n \ \xrightarrow{\Delta} \ \underset{\underset{O}{\|}}{Cl_2P}-CH_2-Cl \ \xrightarrow[2\ Et_3N]{2\ ROH} \ \underset{\underset{O}{\|}}{(RO)_2P}-CH_2-Cl$$

50–67%   35–99%   **(3.6)**

3.1.1.5 Addition–Elimination Reactions of Phosphites with Difluoroalkenes

The reaction of trialkyl phosphites with perfluoroalkenes in an autoclave at 100–140°C results in the formation of dialkyl perfluoroalkenylphosphonates in low to good yields (21–81%, Scheme 3.7).[45–47] The mechanism involves the attack of the phosphite nucleophile at the polarized terminal CF_2 position to form a perfluoroalkenyltrialkoxyfluorophosphorane, which is stable under the usual conditions. This fluorophosphorane, which may be regarded as an isolable Michaelis–Arbuzov intermediate, decomposes on heating with elimination of alkyl fluoride to give dialkyl perfluoroalkenylphosphonates (Scheme 3.7).[45,48]

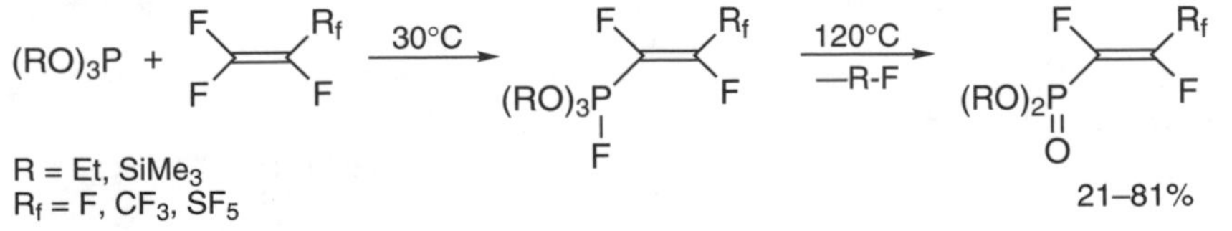

R = Et, SiMe$_3$
R$_f$ = F, CF$_3$, SF$_5$

21–81%   **(3.7)**

With perfluoroethylene, where the polarization is absent, the main product is diethyl ethylphosphonate. When a tertiary base is present, the isomerization rate is lower, and diethyl trifluorovinylphosphonate can be isolated in 10–15% yields.[45] The reactivity of perfluoroalkenes in the reaction with trialkyl phosphites falls as showed in Figure 3.1.[49]

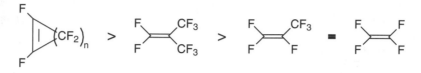

FIGURE 3.1

When a fluorine atom in C_2F_4 is replaced by another halogen, a fluorine atom rather than a chlorine[47] or iodine[50] atom is eliminated as alkyl fluoride from the reaction with trialkyl phosphites. Trimethylsilyl phosphites react with perfluoroalkenes to give *bis*(trimethylsilyl) perfluoroalkenylphosphonates in very good yields (81–95%),[51–55] probably because of the greater nucleophilicity of the trimethylsilyl phosphites and the stability of fluorotrimethylsilane.[56]

Sodium dialkyl phosphites readily react with 1,1-difluoroalkenes at low temperature to provide the corresponding dialkyl α-fluorovinylphosphonates in excellent yields (88–93%, Scheme 3.8).[57–59] Some of them (R^2 = OBn) can be converted into β-fluoro-α-keto esters, which are useful precursors of β-fluoro-α-amino acids.[58]

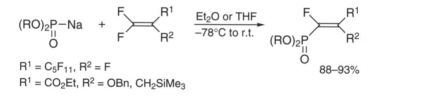

$$R^1 = C_5F_{11}, R^2 = F$$
$$R^1 = CO_2Et, R^2 = OBn, CH_2SiMe_3$$

88–93%

(3.8)

3.1.1.6 SNP(V) Reactions

The electrophilic phosphonylation of a carbanion [SNP(V) reaction] represents an efficient method for the synthesis of phosphonates.[60] The usefulness of this reaction for the preparation of α-halogenophosphonates reflects the stability of the intermediate carbenoid.

The reaction of fluoromethyl phenyl sulfone with diethyl chlorophosphate at room temperature requires the use of 2 eq of base (*n*-BuLi, LDA, LiHMDS) to give the lithium salt of diethyl 1-(phenylsulfonyl)-1-fluoromethylphosphonate, which can be hydrolyzed and isolated in 73% yield[61] or reacted *in situ* with a large variety of carbonyl compounds to form the corresponding α-fluorovinyl phenyl sulfones in moderate to good yields (44–95%, Scheme 3.9).[62–69] These sulfones are readily converted into 1-fluoroalkenes via the corresponding (fluorovinyl)stannanes.[63,64,66,67]

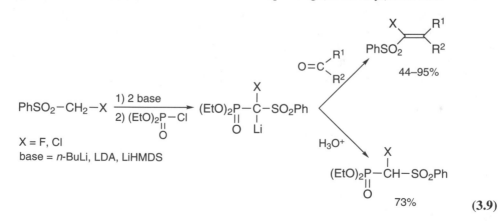

(3.9)

TABLE 3.2

Base	X¹	X²	Product
LDA	Cl	Cl	**A**
LDA	Br	Br	**B**
LiHMDS	Cl	Cl	—[a]
LiHMDS	Cl	Br	**A** (70%) + **B** (30%)
LiHMDS	Cl	I	**A**
LiHMDS	Br	Br	**B**
LiHMDS	I	I	—[b]

[a] No reaction.
[b] Complex mixture.

The electrophilic phosphonylation of chloromethyl phenyl sulfone in the presence of *n*-BuLi (2 eq) proceeds smoothly at low temperature, probably via the sulfone dianion, to give diethyl 1-lithio-1-chloro-1-(phenylsulfonyl)methylphosphonate. This is treated *in situ* with aliphatic or aromatic aldehydes and ketones to obtain α-chloro-α,β-unsaturated ketones by the Horner–Wadsworth–Emmons reaction in good overall yields (76–85%).[70]

A useful method for the preparation of tetraethyl chloro- and bromomethylenediphosphonates has recently been described.[71] It involves the slow addition at low temperature of a mixure of dihalogenomethane and diethyl chlorophosphate to a THF solution of base (LDA or LiHMDS), warming to room temperature, and hydrolysis of the resulting tetraethyl 1-lithio-1-halogenomethylenediphosphonates with 3 M HCl to give tetraethyl halogenomethylenediphosphonates in very good yields (85–95%, Scheme 3.10).[71] The choice and stoichiometry of the base are very important and depend on the dihalogenomethane used. LDA can be used for both dichloro- and dibromomethanes, whereas LiHMDS is able to deprotonate only dibromomethane. A mixture of chloro and bromo derivatives is obtained from chlorobromomethane because of the competition between chlorine–lithium and bromine–lithium exchange reactions (Table 3.2).

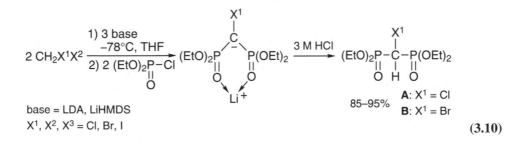

$$
\text{base = LDA, LiHMDS} \qquad \text{X}^1, \text{X}^2, \text{X}^3 = \text{Cl, Br, I}
$$

(3.10)

The proposed mechanism involves two competitive pathways (Scheme 3.11), presenting two common steps, the deprotonation of dihalomethane and the subsequent phosphonylation. The resulting dihalogenomethylphosphonate may undergo a deprotonation–phosphonylation–halogen–metal exchange sequence (path a) or, alternatively, a halogen–metal exchange followed by phosphonylation and deprotonation (path b) to give the lithiated derivative of halogenomethylenediphosphonate (Scheme 3.11). The second mechanism seems to be more probable.[71]

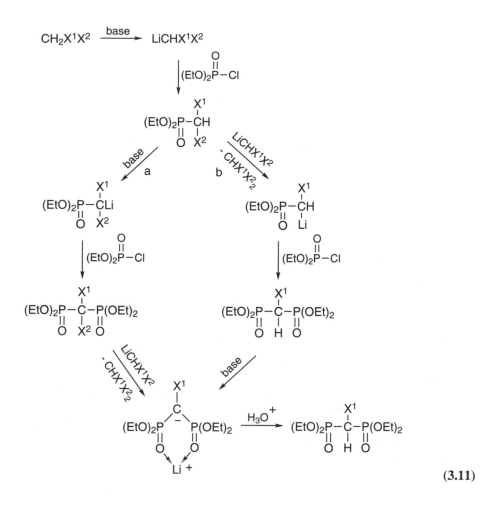

$$(3.11)$$

3.1.1.7 Nucleophilic Halogenation Reactions

The synthesis of α-halogenophosphonates by nucleophilic halogenation usually involves the reaction of α-hydroxyphosphonates with different nucleophilic halogenating agents. The α-hydroxyphosphonates are generally obtained by the Pudovik reaction between a dialkyl phosphite and a carbonyl compound in the presence of a catalytic amount of base (see also Section 7.1.1.2). Because of the reversibility of this reaction, better yields are obtained by trapping the resulting alcoholate with Me$_3$SiCl or diethyl chlorophosphate.[72]

3.1.1.7.1 Nucleophilic Fluorination

The first conversion of a diethyl α-hydroxybenzylphosphonate into the corresponding fluoro derivative was described in 1968 and used (2-chloro-1,1,2-trifluoroethyl)diethylamine.[73] The low yield (25%) was subsequently improved by using (diethylamino)sulfur trifluoride (DAST), a reagent developed by DuPont.[74] The substitution of the hydroxy group by fluorine proceeds smoothly with a small excess of DAST in CH$_2$Cl$_2$ at 0°C or at –78°C (Scheme 3.12), and diethyl α-fluorobenzylphosphonates bearing various substituents on the aromatic ring are isolated in moderate to good yields (40–87%, Table 3.3). An exception is the 4-methoxy derivative, which gives a complex mixture of products.[75,76] Similar nucleophilic fluorination of dibenzyl and di-*tert*-butyl benzylphosphonates has also been described.[77]

TABLE 3.3

	R¹	Yield (%)[Ref.]
a	H	,[78] 45,[75,76] 52[79–81]
b	3-Cl	85[75,76]
c	3-(EtO)₂P(O)	93[82]
d	4-Cl	70[75,76]
e	4-Me	40,[75,76] 81[72]
f	4-(Cbz)Ala	55,[a83] 69,[84] 87[72]
g	2-SCy-5-NO₂	66[85]
h	2,4,6-Me₃	60[75,76]

[a] Reaction carried out in CHCl₃ at −78°C.

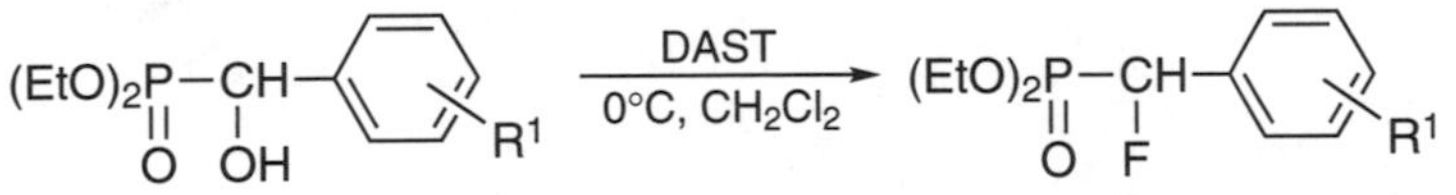

$$(EtO)_2P\text{—}CH\text{—}\overset{}{\bigcirc}\text{—}R^1 \xrightarrow[0°C,\ CH_2Cl_2]{DAST} (EtO)_2P\text{—}CH\text{—}\overset{}{\bigcirc}\text{—}R^1$$

$$\underset{O\ \ OH}{} \qquad\qquad \underset{O\ \ F}{}$$

(3.12)

The reaction of α-hydroxyallylphosphonates with DAST gives exclusively the (E)-γ-fluoro-α,β-unsaturated phosphonates and not the expected α-fluoroallylphosphonates.[75,76,86] In contrast, the regiospecific fluorination of the hydroxy group of α-hydroxypropargylphosphonates produces α-fluoropropargylphosphonates in moderate to good yields (41–94%).[87–91] The subsequent partial catalytic hydrogenation (Pd/Al₂O₃) occurs without any apparent loss of fluorine and gives access to (Z)-α-fluoroallylphosphonates, a series of compounds that are difficult to obtain by other methods (Scheme 3.13).[87,92]

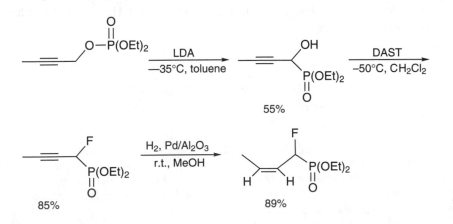

(3.13)

Very low (3–8%) enantiomeric excesses are obtained in the fluorination of optically active α-hydroxybenzylphosphonates by DAST or (1,1,2,3,3,3-hexafluoropropyl)diethylamine (PPDA) at 0°C or −78°C.[72] Therefore, despite a previous report that described a stereoselective reaction with complete inversion of configuration,[84] it seems that this class of substrates generally react through an S_N1 pathway. On the other hand, glucose-derived α-hydroxyphosphonates react stereoselectively with DAST (S_N2 mechanism) to provide the corresponding α-fluorophosphonates in yields not exceeding 40%.[93] The modest yields can be explained by a competing dehydration reaction which in some cases becomes predominant.[75]

Attempts to prepare diethyl fluoromethylphosphonate by DAST fluorination of diethyl hydroxymethylphosphonate gave very low yields (<2%). A different methodology, involving the conversion of the alcohol into the corresponding triflate[94] and a subsequent nucleophilic displacement with TBAF, provided a safe, cheap, and simple route to diethyl fluoromethylphosphonate, which was isolated in 67% yield (Scheme 3.14).[95]

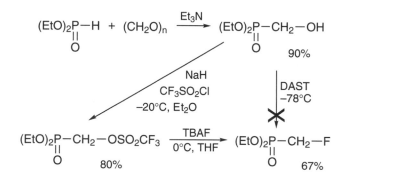

$$(3.14)$$

3.1.1.7.2 Nucleophilic Chlorination

The first synthesis of an α-halogenophosphonic acid by nucleophilic halogenation, described in 1922, involved treating α-phenyl-α-hydroxyethylphosphonic acid with concentrated HCl. It gave α-phenyl-α-chloroethylphosphonic acid in 96% yield.[96]

Subsequently, the nucleophilic chlorination of α-hydroxyphosphonates has become one of the most important routes for the synthesis of α-chlorophosphonates (Scheme 3.15). A wide variety of chlorinating agents has been employed, some of which are presented in Table 3.4. Two couples, CCl_4/PPh_3 and $POCl_3/PhNEt_2$, seem to give the best results, with the latter being more easily removed from the reaction mixture. The α-chlorophosphonates are important reagents for the preparation of chloroalkenes and alkynes, which are obtained after Horner–Wadsworth–Emmons reactions with carbonyl compounds.

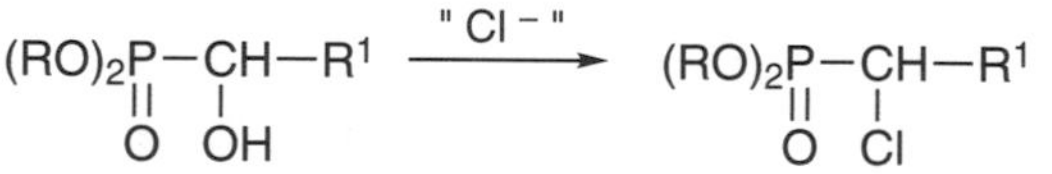

$$(RO)_2\underset{\underset{O}{\|}}{P}-\underset{\underset{OH}{|}}{CH}-R^1 \xrightarrow{\text{"Cl}^-\text{"}} (RO)_2\underset{\underset{O}{\|}}{P}-\underset{\underset{Cl}{|}}{CH}-R^1$$

$$(3.15)$$

The methodology has been successfully applied to the synthesis of diphenyl α-chloro-4-pyridylmethylphosphonate, which was obtained in 80% yield from diphenyl α-hydroxy-4-pyridylmethylphosphonate with $POCl_3$ and catalytic quantities of diethylaniline. The α-chlorophosphonate products react with various 4-substituted benzaldehydes in the presence of t-BuOK (2 eq) to produce, by the Horner–Wadsworth–Emmons reaction, 4-pyridylacetylenes showing interesting nonlinear optical properties (Scheme 3.16).[110]

TABLE 3.4

	R	R^1	" Cl$^-$ "	Yield (%)[Ref.]
a	Et	H	CCl$_4$/PPh$_3$ SOCl$_2$	86,[97] 42,[98] 48[99]
b	Me	Me	SOCl$_2$	40[98]
c	Et	Me	SOCl$_2$ CCl$_4$/PPh$_3$	44[98] 56,[7] 90[97]
d	Et	Et	CCl$_4$/PPh$_3$	84[97]
e	Ph	Et	SOCl$_2$/Py	91[100]
f	Et	n-Pr	CCl$_4$/PPh$_3$	83[97]
g	Et	i-Bu	SOCl$_2$	40[98]
h	Et	Ph	CCl$_4$/PPh$_3$ NCS/PPh$_3$	81[97] —[78]
i	1/2 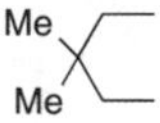 Me$_2$C<	Ph	SOCl$_2$	81[101]
j	1/2 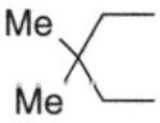 Me$_2$C<	3-Me-C$_6$H$_4$	SOCl$_2$	81[101]
k	Et	3-(EtO)$_2$P(O)-C$_6$H$_4$	SOCl$_2$	61[82]
l	1/2 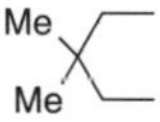 Me$_2$C<	4-Me-C$_6$H$_4$	SOCl$_2$	75[101]
m	Et	4-Cl-C$_6$H$_4$	SOCl$_2$/Py SOCl$_2$	98[102,103] 99[104]
n	Ph	4-Cl-C$_6$H$_4$	POCl$_3$/PhNEt$_2$	56[105]
o	1/2 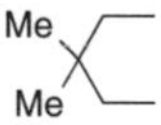 Me$_2$C<	4-Cl-C$_6$H$_4$	SOCl$_2$	75[101]
p	1/2 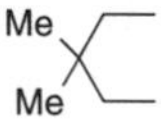 Me$_2$C<	2,4-Cl$_2$-C$_6$H$_3$	SOCl$_2$	64[101]
q	Ph	4-Br-C$_6$H$_4$	POCl$_3$/PhNEt$_2$	29[105]
r	Et	4-MeO-C$_6$H$_4$	SOCl$_2$	98[106]
s	Ph	4-MeO-C$_6$H$_4$	POCl$_3$/PhNEt$_2$	27[105]
t	Ph	4-O$_2$N-C$_6$H$_4$	POCl$_3$ POCl$_3$/Py POCl$_3$/PhNEt$_2$	—[107,108] 77[109] 84[105]
u	Ph	4-CN-C$_6$H$_4$	POCl$_3$/PhNEt$_2$	68[105]
v	Ph	4-C$_5$H$_4$N	POCl$_3$	80[110]

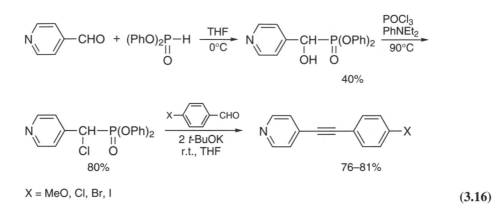

X = MeO, Cl, Br, I

(3.16)

3.1.1.7.3 Nucleophilic Bromination

Like the α-chlorophosphonates, the α-bromophosphonates are valuable precursors of bromoalkenes and alkynes. They are obtained by reacting α-hydroxyphosphonates with nucleophilic brominating agents (Scheme 3.17, Table 3.5). The systems CBr_4/PPh_3 and Ph_3PBr_2/Py, which are generally used for alkyl derivatives, give broadly similar results. $SOBr_2$ and allyl bromide/N,N'-carbonyldiimidazole give good results with aromatic derivatives. Unlike the α-chlorophosphonates, which are obtained in good yields irrespective of the α-substituent, only primary α-hydroxyphosphonates (R^1 = H, Scheme 3.17) are efficiently converted into the corresponding bromo derivatives. Secondary α-hydroxyphosphonates ($R^1 \neq$ H, Scheme 3.17) give lower yields.[111]

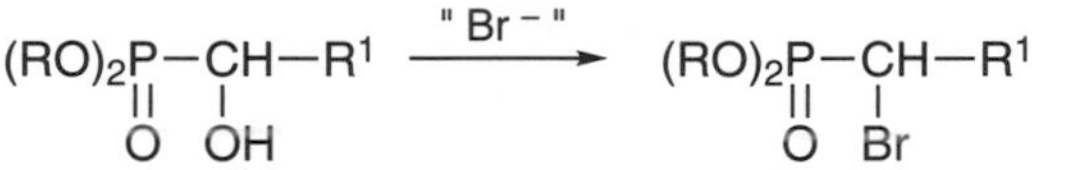

$$(RO)_2\overset{\|}{\underset{O}{P}}-\overset{|}{\underset{OH}{C}}H-R^1 \xrightarrow{\text{" Br}^-\text{"}} (RO)_2\overset{\|}{\underset{O}{P}}-\overset{|}{\underset{Br}{C}}H-R^1$$

(3.17)

3.1.1.7.4 Nucleophilic Iodination

Only a few examples of the preparation of α-iodophosphonates by nucleophilic iodination have been described (Scheme 3.18). Two halogenating systems have been used, PI_3/Et_3N and MeI/N,N'-carbonyldiimidazole, but the instability of the products and the very few data available make the evaluation of their potential difficult (Table 3.6).[101]

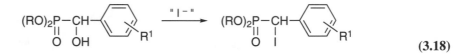

(3.18)

3.1.1.8 Electrophilic Halogenation Reactions

This is one of the most important and attractive methods for preparing α-monohalogenated phosphonates. The main advantages of this approach, which often succeeds when others methods fail, are the ready availability of starting materials and halogenating reagents and the mild reaction conditions. However, the degree of halogenation may be difficult to control because of the existence of acid–base equilibria. The α-protons in the halogenated product are more acidic than those in the starting material, so that phosphonate transmetallation reactions usually give mixtures of mono-, di-, and nonhalogenated phosphonates (Scheme 3.19).

TABLE 3.5

	R	R¹	" Br – "	Yield (%)[Ref.]
a	Et	H	CBr$_4$/PPh$_3$ Ph$_3$PBr$_2$/Py Ph$_3$PBr$_2$	65[111] 67[111] 72[103]
b	Bn	H	CBr$_4$/PPh$_3$	73[112]
c	Et	Me	CBr$_4$/PPh$_3$ Ph$_3$PBr$_2$/Py	57[111] 49[111]
d	Et	Et	CBr$_4$/PPh$_3$ Ph$_3$PBr$_2$/Py	43[111] 46[111]
e	Et	n-Pr	CBr$_4$/PPh$_3$ Ph$_3$PBr$_2$/Py	39[111] 22[111]
f	Et	Ph	CBr$_4$/PPh$_3$ Ph$_3$PBr$_2$/Py NBS/PPh$_3$	42[111] 42[111] —[78]
g	1/2 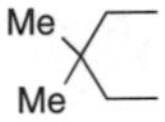 Me$_2$C<	Ph	SOBr$_2$	79[101]
h	1/2 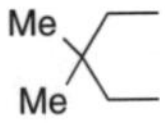 Me$_2$C<	3-Me-C$_6$H$_4$	SOBr$_2$	97[101]
i	1/2 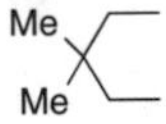 Me$_2$C<	4-Me-C$_6$H$_4$	SOBr$_2$	74[101]
j	1/2 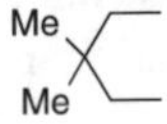 Me$_2$C<	4-Cl-C$_6$H$_4$	SOBr$_2$	76[101]
k	1/2 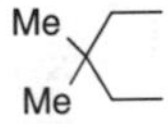 Me$_2$C<	2,4-Cl$_2$-C$_6$H$_3$	SOBr$_2$	91[101]
l	i-Pr	4-MeO-C$_6$H$_4$	CH$_2$=CHCH$_2$Br/CDI	quant.[113]
m	Et	4-EtO-C$_6$H$_4$	CH$_2$=CHCH$_2$Br/CDI	quant.[113]
n	i-Pr	4-EtO-C$_6$H$_4$	CH$_2$=CHCH$_2$Br/CDI	quant.[113]
o	Et	2,3,4-(MeO)$_3$-C$_6$H$_2$	CH$_2$=CHCH$_2$Br/CDI	quant.[113]
p	i-Pr	3-EtO-4-MeO-C$_6$H$_3$	CH$_2$=CHCH$_2$Br/CDI	quant.[113]

TABLE 3.6

	R	R[I]	" I - "	Yield (%)[Ref.]
a	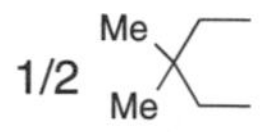 1/2 (Me)(Me)C<	H	PI$_3$/NEt$_3$	9 (dec.)[101]
b	 1/2 (Me)(Me)C<	4-Me	PI$_3$/NEt$_3$	10 (dec.)[101]
c	i-Pr	4-MeO	MeI/CDI	quant.[113]
d	Et	4-EtO	MeI/CDI	quant.[113]
e	i-Pr	4-EtO	MeI/CDI	quant.[113]
f	Et	2,3,4-(MeO)$_3$	MeI/CDI	quant.[113]
g	i-Pr	3-EtO-4-MeO	MeI/CDI	quant.[113]

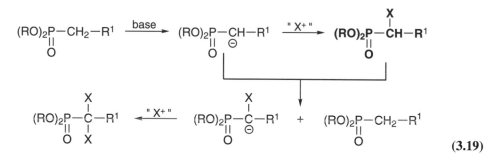

$$(3.19)$$

As a general rule, and particularly where X = F, it is necessary to match the reactivity of the phosphonylated carbanion to an appropriate halogenating reagent. Thus, stabilized carbanions react with strong halogenating reagents, whereas less powerful halogenating reagents are suited to nonstabilized carbanions.

3.1.1.8.1 *Electrophilic Fluorination*

The recent interest in electrophilic fluorination of phosphonates reflects the importance of α-fluorophosphonates as isosteres of biologically important phosphates. However, the use of electrophilic fluorination methods is increasing with the availability of safe and easy to handle fluorine sources. Several reviews of the synthesis and applications of electrophilic fluorination reagents are now available.[114–117] Here only the most important of these reagents and their use in the electrophilic fluorination of phosphonates are presented.

F$_2$ is the simplest reagent that can be used as source of electrophilic fluorine. Special equipment is required, and fluorination is not selective. When a 10% mixture of fluorine in nitrogen is passed through a solution of stabilized phosphonate carbanions in MeCN, a mixture of the non-, mono-, and difluorinated derivatives is obtained. After purification, the monofluorophosphonate can be isolated in yields not exceeding 50%.[118,119]

On the other hand, when the α-position is protected with a formyl group, monofluorination is observed exclusively, and fluoromethylphosphonates having electron-withdrawing groups in the α position are obtained in good yields (62–88%, 14–62% after purification, Scheme 3.20).[120] The use of a preformed sodium or potassium salt of the phosphonate instead of a mixture of the phosphonate and aqueous $NaHCO_3$ or $KHCO_3$ limits decomposition during purification.[120]

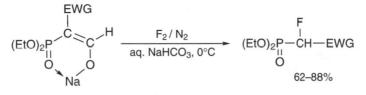

EWG = CN, CO_2Et, C(O)Me, C(O)t-Bu, C(O)Ph

(3.20)

Perchloryl fluoride ($FClO_3$)[121] was the first electrophilic reagent used for the fluorination of phosphonates[122] when it was found to react smoothly with the potassium salt of diethyl 1,2-bis(ethoxycarbonyl)-2-oxoethylphosphonate at −10°C. The fluorinated product was deprotected *in situ*, and diethyl 1-(ethoxycarbonyl)-1-fluoromethylphosphonate was isolated in 77% yield.[122] $FClO_3$ was the only electrophilic fluorinating reagent available until the early 1980s and was used extensively, despite its toxicity and explosive properties, until safer reagents were developed. An interesting study on the reactivity of $FClO_3$ with phenylsulfonyl, phenylsulfoxyl, and phenylsulfenyl phosphonylated carbanions has appeared.[123] The first class gives the expected monofluorinated 1-(phenylsulfonyl)methylphosphonates in 50–60% isolated yields along with a small amount of the starting material. The fluorination of 1-(phenylsulfoxyl)methylphosphonates produces 1-(phenyl-sulfoxyl)-1-fluoromethylphosphonates in 30% yield, and 1-(phenylsulfenyl)methylphosphonates do not yield the expected monofluorination product.[123]

The reaction of the unprotected sodium enolate of dimethyl 2-oxoheptylphosphonate with $FClO_3$ in toluene at −35°C gives the monofluorinated phosphonate in 29% yield.[124,125] The fluorination of sodium or potassium methylenediphosphonate anions with $FClO_3$ cannot be controlled easily, and mixtures of mono- and difluorinated derivatives, which are difficult to separate, are always obtained. The precise mole ratio is governed by the experimental conditions.[126–129] Analogously, when ethyl methyl (+)-(R)-methylphosphonate is treated with *n*-BuLi, and the resulting anion is quenched with perchloryl fluoride, a mixture of mono- (28%) and difluorinated (7%) methylphosphonates is obtained.[130,131] However, treatment of a tertiary carbanion, sodium diethyl tetraisopropyl methinyltriphosphonate, with $FClO_3$ in THF at −75°C gives the monofluorinated derivative in 77% yield. Under the same conditions, NFBS is unreactive even at room temperature.[132]

Acetyl hypofluorite (CH_3CO_2F)[133,134] has been used to fluorinate tetraisopropyl methylene-diphosphonate in two steps. In the first step, the sodium salt of tetraisopropyl methylenediphosphonate is added to a solution of AcOF in $CFCl_3$ at −78°C. After 5 min, the reaction is quenched, and tetraisopropyl fluoromethylenediphosphonate is isolated in 66% yield. This compound is metallated with NaH and submitted to a second electrophilic fluorination as above to obtain tetraisopropyl difluoromethylenediphosphonate in 40% yield.[135] Even if acetyl hypofluorite seems to be a more selective reagent for electrophilic monofluorination than $FClO_3$, its relative instability and the special equipment required to manipulate molecular fluorine greatly limit its use.

N–F electrophilic fluorinating agents [NFBS,[136] NFoBS,[137,138] $(CF_3SO_2)_2NF$,[139] F-TEDA-BF_4[140]] largely overcome the explosive, hygroscopic, gaseous, and toxic properties of the first generation of fluorinating agents. These new reagents are often more stable and selective, and many are solids, which are easy to handle. At present, two of them, *N*-fluorobenzenesulfonimide (NFBS) and 1-(chloromethyl)-4-fluoro-1,4-diazoniabicyclo[2.2.2]octane *bis*(tetrafluoroborate) (F-TEDA-BF_4, Selectfluor™), are commercially available. The others, *N*-fluoro-*o*-benzenedisulfonimide (NFoBS) and *N*-fluorobis(trifluoromethanesulfonyl)imide, are seldom used because of their limited availability (Figure 3.2). An electrochemical study on the reactivity of several N–F electrophilic fluorinating agents has recently been reported.[141]

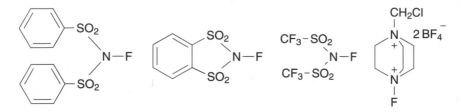

FIGURE 3.2

NFoBS reacts with diethyl 1-(ethoxycarbonyl)methylphosphonate in THF at −78°C in the presence of NaHMDS (1 eq) to afford the expected monofluorination product in 78% yield. The cyclic disulfonimide byproduct is highly water soluble and easy to remove by aqueous workup.[138]

N-Fluoro*bis*(trifluoromethanesulfonyl)imide is one of the most powerful electrophilic fluorinating agents. It was successfully used to fluorinate diethyl cyanomethylphosphonate in the presence of *n*-BuLi (1 eq), and the resulting diethyl 1-fluoro-1-cyanomethylphosphonate can be isolated (51% yield) or metallated *in situ* and reacted with carbonyl compounds in a Horner–Wadsworth–Emmons reaction to give α-fluoroacrylonitriles in 30–58% overall yields (Scheme 3.21).[142] These results are especially noteworthy because the fluorination of diethyl cyanomethylphosphonate with NFBS in the presence of LiHMDS (2 eq) produces the *N*-fluoro and not the expected *C*-fluoro phosphonate.[143]

$$(EtO)_2P(O)-CH_2-CN \xrightarrow[\text{2) } (CF_3SO_2)_2NF]{\text{1) } n\text{-BuLi, }-78°C,\text{ THF}} (EtO)_2P(O)-CHF-CN \quad 51\%$$

$$\xrightarrow[\text{2) } O=CR^1R^2]{\text{1) } n\text{-BuLi, }-78°C,\text{ THF}} \begin{array}{c} F \diagup R^1 \\ NC \diagdown R^2 \end{array} \quad 30\text{–}58\%$$

$$(3.21)$$

The use of F-TEDA-BF$_4$ for the fluorination of stabilized phosphonate carbanions is limited by its poor solubility in THF. However, potassium salts of sulfonyl-substituted phosphonates react with F-TEDA-BF$_4$ in a THF–DMF mixture to afford the desired monofluorosulfonylphosphonates in 47–61% yield.[144–147] The best result on fluorination of the sodium enolate of diethyl 1-(ethoxycarbonyl)methylphosphonate was obtained using F-TEDA-BF$_4$ in a THF–DMF solvent mixture at room temperature, though product yields were quite poor (17%).[95]

NFBS has been one of the most utilized electrophilic fluorination agents in recent years. The fluorination of diethyl alkylphosphonates in the presence of KDA (1 eq) produces diethyl α-fluoroalkylphosphonates in moderate yields (45–54%) except for the parent case, where diethyl fluoromethylphosphonate is obtained impure in only 11% yield.[148] Similarly, several diethyl alkoxymethylphosphonates have been metallated using *n*-BuLi or *sec*-BuLi (1 eq) and then fluorinated with NFBS at −78°C to produce the monofluorinated derivatives (27–29% yields) along with variable amounts of dimeric phosphonates (5–30%).[149] Tetraethyl fluoromethylenediphosphonate is an important starting material for the synthesis of the P$_\beta$,P$_\gamma$-fluoromethylenediphosphonate analogue of *nor*-carbovir triphosphate, a potent inhibitor of HIV reverse transcriptase. In this case, the potassium salt of tetraethyl methylenediphosphonate is treated with NFBS in THF–toluene at −78°C to generate tetraethyl fluoromethylenediphosphonate in 26% yield.[150]

To date, there is a single paper reporting the diastereoselective electrophilic fluorination of α-carbanions of asymmetric phosphonamidates. The reagent system used was NFBS (1 eq) in THF at −78°C.[151] The first attempts were made using *trans*-(*R,R*)-1,2-*bis*(*N*-methylamino)cyclohexane as a chiral auxiliary, but the separation of the diastereomeric α-fluorophosphonates proved to be exceedingly difficult. The corresponding diastereomeric phosphonates bearing (−)-ephedrine as chiral auxiliary were separated, and each isomer was submitted to the electrophilic fluorination under the same conditions as above. The results for each isomer are roughly comparable in terms of yield and diastereomeric excess and show no dependence on the nature of the counterion (Scheme 3.22).[151]

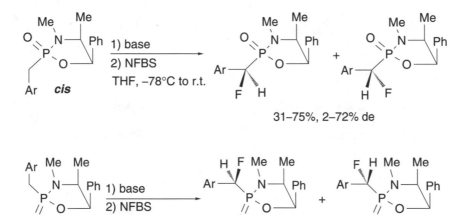

$$Ar = Ph, C_{10}H_7, Ph\text{-}C_6H_4 \tag{3.22}$$

The fluorinated diastereomers are separated by silica gel chromatography. Removal of the ephedrine auxiliary by TMSBr and TFA provides a racemization-free procedure for liberating the desired α-fluorophosphonic acids.[151]

Representative results summarizing electrophilic monofluoration of phosphonate carbanions (Scheme 3.23) are collected in Table 3.7.

$$(RO)_2\underset{\underset{O}{\|}}{P}-CH_2-R^1 \xrightarrow[\text{2) " } F^+ \text{ "}]{\text{1) base}} (RO)_2\underset{\underset{O}{\|}}{P}-\underset{\underset{F}{|}}{C}H-R^1 \tag{3.23}$$

The main drawbacks of the electrophilic fluorination procedures presented above are the lack of selectivity and the inseparable mixtures of mono- and difluorinated phosphonates that are often obtained. To overcome these problems, two methodologies involving the temporary protection of the α-position have been developed.

In the first case, the α-position of phosphonates bearing electron-withdrawing groups [CN, CO_2R, C(O)R] is protected with an oxalyl[122] or formyl[120] group, and the resulting potassium enolate is isolated and subjected to electrophilic fluorination using $FClO_3$[122] or F_2/N_2.[120] The protecting group is removed *in situ,* and the desired α-monofluorophosphonates are obtained pure in good yields (66–88%, Scheme 3.24).[120,122] The disadvantage implicit in the two-step procedure is more than compensated by the high yields and purity of the final product. The isolation of the intermediate protected phosphonate enolate rather than its protonated form minimizes decomposition during purification.[120]

TABLE 3.7

	R	R¹	"F⁺"	Base	Yield (%)$^{Ref.}$
a	Et	H	NFBS	KDA	11^{a148}
			FClO$_3$	n-BuLi	46^{152}
b	Et	Me	NFBS	KDA	45^{148}
c	Et	n-Bu	NFBS	KDA	48^{148}
d	Et	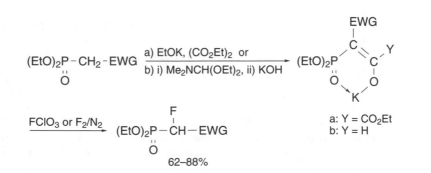	NFBS	KDA	54^{148}
e	Et		NFBS	LDA	—^{29}
f	Et	2,3,6-Me$_3$-4-MeO-C$_6$H	FClO$_3$	n-BuLi	—^{153}
g	Et	TBDMSO(CH$_2$)$_2$O	NFBS	s-BuLi	27^{149}
h	Et	PhS(O)	FClO$_3$	KH	29^{123}
i	Et	PhSO$_2$	FClO$_3$	KH	50^{123}
			NFBS	NaH	38^{154}
j	Me	n-C$_5$H$_{11}$C(O)	FClO$_3$	Na	29^{125}
k	Et	CO$_2$Et	F-TEDA-BF$_4$	NaH	17^{95}
			F$_2$/N$_2$		35^{b118}
l	Et	CN	(CF$_3$SO$_2$)$_2$NF	n-BuLi	51^{142}
m	Et	(EtO)$_2$P(O)	NFBS	KHMDS	26^{150}
n	i-Pr	(i-PrO)$_2$P(O)	F$_2$/N$_2$	NaH	50^{b118}

^{a} Not obtained pure.
^{b} Mixture with the difluorinated derivative.

$$(3.24)$$

A second methodology has recently been developed that is best suited to nonstabilized and semistabilized phosphonate carbanions. It involves an initial silylation in the presence of an appropriate base (LDA or LiHMDS) followed by fluorination of the protected carbanions. The resulting fluorinated phosphonates are easily desilylated to afford pure α-monofluorophosphonates in good to excellent yields (68–97%, Scheme 3.25).[155-157]

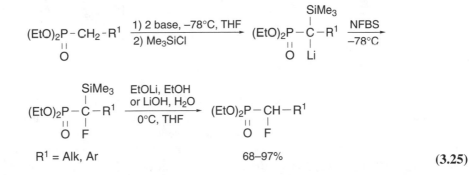

$$R^1 = Alk, Ar \qquad\qquad 68–97\% \qquad\qquad (3.25)$$

The principal parameters governing a clean and quantitative process are (a) the nature and the quantity of the base (LDA or LiHMDS), (b) the temperature for the addition of Me₃SiCl, and (c) the reagent used in the deprotection step (EtOLi, LiOH, or TBAF). This methodology has been successfully combined with the phosphate–phosphonate conversion to obtain α-monofluoroalkyl-phosphonates in good yields (85%) in a one-pot synthesis starting from triethyl phosphate and alkyllithiums (Scheme 3.26). In this case, 3 eq of base are needed, one to react with triethyl phosphate, the second to deprotonate the resulting phosphonate, and the third, after conversion to LDA, to metallate the silylated phosphonate. The first equivalent of Me₃SiCl neutralizes the lithium ethoxide formed in the first step, and the second reacts with the phosphonate carbanion (Scheme 3.26).[157]

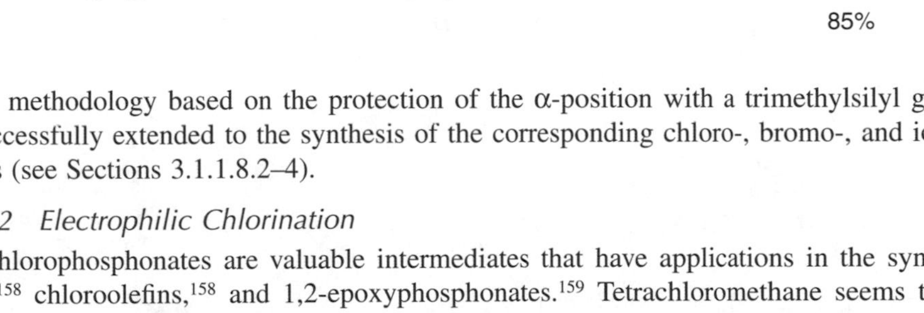

(EtO)₂P–OEt →(3 n-PrCH₂Li, −78°C, THF)→ (EtO)₂P–CH–n-Pr (O, Li) →(1) 1.5 i-Pr₂NH 2) 2 Me₃SiCl −78°C, THF)→

(EtO)₂P–C–n-Pr (SiMe₃, O, Li) →(NFBS, −78°C)→ (EtO)₂P–C–n-Pr (SiMe₃, O, F) →(EtOLi, EtOH, 0°C, THF)→ (EtO)₂P–CH–n-Pr (O, F)

85%   (3.26)

This methodology based on the protection of the α-position with a trimethylsilyl group has been successfully extended to the synthesis of the corresponding chloro-, bromo-, and iodophosphonates (see Sections 3.1.1.8.2–4).

3.1.1.8.2 Electrophilic Chlorination

The α-chlorophosphonates are valuable intermediates that have applications in the synthesis of alkynes,[158] chloroolefins,[158] and 1,2-epoxyphosphonates.[159] Tetrachloromethane seems to be the most widely used reagent in the electrophilic monochlorination of phosphonates, presumably because it gives very clean chlorine transfer and produces an easily removed byproduct (chloroform) (Scheme 3.27). Only 1eq of base (n-BuLi) is needed to achieve the two successive deprotonations. A large number of chloroalkenes, some of them showing interesting insecticidal properties,[160,161] have been obtained in 27–87% yields by this method (Scheme 3.27).[160-165]

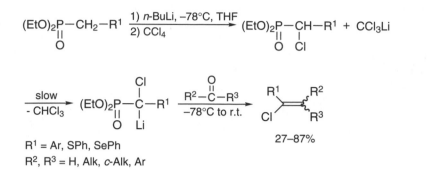

R^1 = Ar, SPh, SePh
R^2, R^3 = H, Alk, c-Alk, Ar

(3.27)

N-Chlorosuccinimide is the most widely used reagent for the chlorination of β-ketophosphonates and 1-(alkoxycarbonyl)methylphosphonates to give structures that are often incorporated into complex carbacyclins,[166,167] prostaglandins,[168] or bicyclic sesqui- and diterpenes.[169] The reaction takes place in the presence of NaH (2 eq) in DME or THF at room temperature, and the resulting chlorinated carbanions are allowed to react *in situ* with the appropriate carbonyl compounds to give the desired chloroolefins in 38–80% yields.[166,167,169]

The reaction of diethyl alkylphosphonates with phenylsulfonyl chloride in the presence of *n*-BuLi (2 eq) in THF at low temperature followed by the addition of carbonyl compounds produces diethyl 1-alkyl-1,2-epoxyalkylphosphonates in 30–80% yields,[170] whereas under the same conditions, diethyl benzylphosphonates give comparable quantities of 1-phenyl-1-chloroalkenes (35–55%, Scheme 3.28).[170] Similarly, β-iminophosphonate carbanions, obtained from diethyl 1-lithiomethylphosphonate and nitriles, react with phenylsulfonyl chloride in THF at –78°C. Hydrolysis (3 M H$_2$SO$_4$) of the resulting α-chloro-β-enaminophosphonates produces diethyl α-chloro-β-ketoalkylphosphonates in satisfactory yields (53–67%).[171]

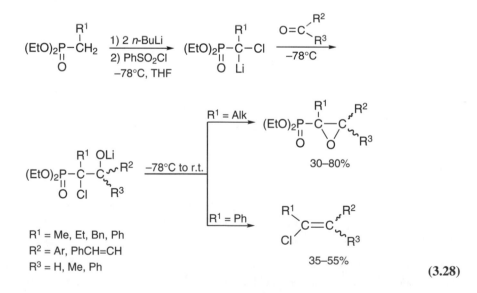

R^1 = Me, Et, Bn, Ph
R^2 = Ar, PhCH=CH
R^3 = H, Me, Ph

(3.28)

The electrophilic chlorination of tertiary phosphonate carbanions proceeds smoothly to give exclusively the monochlorinated derivatives. Thus, tetramethyl α-alkylmethylenediphosphonates, diethyl 1-(ethoxycarbonyl)ethylphosphonate, and diethyl tetraisopropyl methinyltriphosphonate react with sodium hypochlorite at 0°C[132,172,173] or with NaH/CuCl$_2$ in DMSO at 40°C[174] to afford the expected α-chlorophosphonates in moderate to good yields (31–92%).

Similarly, electrophilic chlorination of alkyl- or benzylphosphonates temporarily protected in the α-position by a trimethylsilyl group affords diethyl α-chloroalkyl- or benzylphosphonates in excellent yields (90–99%) after deprotection. In these cases hexachloroethane, a commercially available nonhygroscopic solid reagent producing the volatile tetrachloroethylene as sole byproduct, has been used (see also Section 3.1.1.8.1).[155–157] In the case of phosphonates bearing electron-withdrawing groups such as cyano, sulfonyl, or phosphoryl in the α-position, no protection is needed, and quantitative and selective monochlorination is observed on treatment with hexachloroethane in THF at room temperature in the presence of LiHMDS or LDA.[71,175,176]

3.1.1.8.3 Electrophilic Bromination

The first reported electrophilic bromination reaction was the reaction between diethyl 2,2-diethoxyethylphosphonate with bromine in Et$_2$O in the presence of calcium carbonate. The monobromination product, obtained in 70% yield, was too unstable to be characterized.[177] The reaction of sodium dialkyl 1-(alkoxycarbonyl)methylphosphonates[169,178,179] and sodium dialkyl 2-oxoalkylphosphonates[166,180] with N-bromosuccinimide or bromine in THF produces the corresponding sodium derivatives of α-bromophosphonates, which are converted into α-bromoalkenes by a Horner–Wadsworth–Emmons reaction with carbonyl compounds under standard conditions (Scheme 3.29).[169,178–181] The bromination of the magnesium salt of diethyl methylphosphonate with molecular bromine in THF at –70°C generates diethyl bromomethylphosphonate, a compound rather difficult to obtain by other routes, in 64% yield. In this case, the Grignard reagent gives much better yields than the lithium or copper(I) derivatives.[182] Dialkyl methylphosphonates bearing electron-withdrawing substituents [C(O)R, CO$_2$R, (RO)$_2$P(O)] in the α-position can be metallated with Na and then reacted with molecular bromine to produce the α-bromo derivatives in moderate to good yields (43–95%).[183,184]

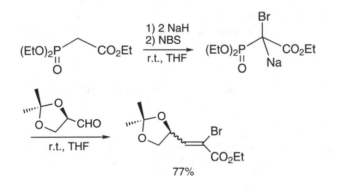

(3.29)

Diethyl 1-(ethoxycarbonyl)ethylphosphonate, on treatment with sodium hypobromite at 10°C, gave the expected bromo derivative in 98% yield.[172] Analogously, when the sodium enolate of a 2-oxoalkylphosphonate is treated with molecular bromine at 0°C, the α-bromo derivative can be isolated in 85% yield.[185]

α-(Trimethylsilyl)-substituted carbanions of diethyl alkyl- or benzylphosphonates can be brominated at low temperature using 1,2-dibromoethane,[186] 1,1,2,2-tetrafluoro-1,2-dibromoethane,[156,157] or 1,1,2,2-tetrachloro-1,2-dibromoethane[156,157] to give diethyl α-bromoalkyl- or benzylphosphonates, which are isolated in very good yields (76–98%) after deprotection with EtOLi in EtOH.[156,157,186]

3.1.1.8.4 Electrophilic Iodination

The reaction of sodium diethyl 1-(ethoxycarbonyl)methylphosphonate with iodine in DME at 0–25°C leads to the α-halogenated carbanion, which can subsequently be alkylated[187] or treated with aromatic[178] or aliphatic[179] aldehydes. After an elimination reaction to complete the one-pot process, substituted ethyl propiolates are obtained in good overall yields (Scheme 3.30).

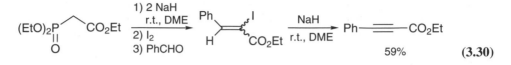

(3.30)

Dimethyl 1-(*tert*-butoxycarbonyl)methylphosphonate reacts analogously with *N*-iodosuccinimide in the presence of NaH in THF at 25°C to give an iodinated carbanion which reacted with (*R*)-2,2-dimethyl-1,3-dioxolan-4-carboxaldehyde in a Horner–Wadsworth–Emmons reaction to produce the corresponding iodoalkene in 64% overall yield.[169]

Diethyl 1-(ethoxycarbonyl)methylphosphonate can be treated with iodine and K_2CO_3 under phase-transfer catalysis conditions to give the crude monoiodinated derivative in good yield (88%).[188,189]

In the case of alkyl- or benzylphosphonates, the protection of the α-position by a trimethylsilyl group is necessary to avoid a second halogenation. Under these conditions, the reaction of silylated carbanions with iodine in THF at low temperature leads quantitatively, after deprotection, to the desired monoiodophosphonates.[156,157]

In general, the stability of α-iodophosphonates decreases with the electron-withdrawing character of the α-substituent, but, in most cases, it seems to be advantageous to use such reagents *in situ* to avoid partial decomposition during isolation.

3.1.1.9 Radical Halogenation Reactions

The reaction of phosphonates with radical halogen sources produces the corresponding α-halogenophosphonates only in the case of chloro and bromo derivatives (Scheme 3.31, Table 3.8). The reaction proceeds smoothly with NCS or NBS on heating in CCl_4 to afford good yields of the desired α-chloro- or α-bromophosphonates, which are key intermediates in the synthesis of acyliminophosphonates[190,191] and α-aminophosphonates.[192] The bromination requires an initiator (benzoyl peroxide or UV light).

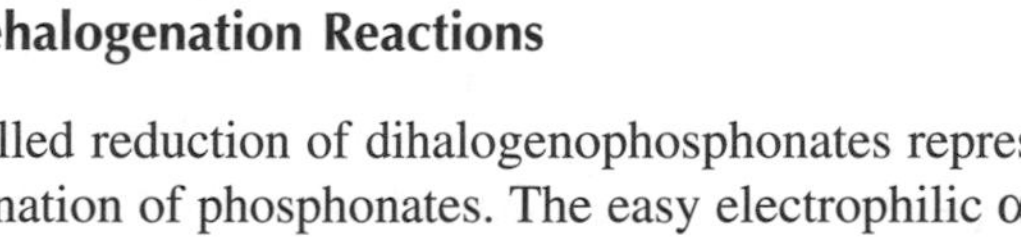

$$(RO)_2\overset{\|}{\underset{O}{P}}-CH_2-R^1 \xrightarrow[\text{initiator}]{"X\cdot"} (RO)_2\overset{\|}{\underset{O}{P}}-\underset{X}{\overset{|}{C}}H-R^1$$

(3.31)

Tertiary bromides derived from carbohydrate phosphonates are also obtained in moderate to good yields (25–64%) by treatment with NBS in CCl_4 under irradiation.[202,203]

3.1.1.10 Reductive Dehalogenation Reactions

In some cases, the controlled reduction of dihalogenophosphonates represents an interesting alternative to the monohalogenation of phosphonates. The easy electrophilic α-dihalogenation of phosphonates bearing electron-withdrawing groups using such cheap reagents as NaOCl,[204–209] NaOBr[204,205,208,209] or KOBr[207] affords the symmetric dihalogenated phosphonates ($X^1 = X^2 = Cl$ or Br, Scheme 3.32) in good yields on large scale (see Section 3.1.2.7). The corresponding diiodophosphonates, which are somewhat unstable, are obtained with lower yields.[205]

$$(RO)_2\overset{\|}{\underset{O}{P}}-\overset{X}{\underset{X}{\overset{|}{C}}}-R^1 \xrightarrow{\text{Reducing agent}} (RO)_2\overset{\|}{\underset{O}{P}}-\underset{X}{\overset{|}{C}}H-R^1$$

(3.32)

TABLE 3.8

	R	R¹	X	" X· "	Initiator	Yield (%)[Ref.]
a	Et	McS	Cl	NCS/CCl$_4$	—	—[193–195]
b	Et	PhS	Cl	NCS/CCl$_4$ NCS/PhH	— —	—,[193,196] 96[197] 90[198]
c	Et	4-Me-C$_6$H$_4$	Cl	NCS/CCl$_4$	—	—[196]
d	Et	Ph$_2$P(O)NHN=C(Me)	Cl	NCS/CHCl$_3$	—	96[199]
e	i-Pr	H	Br	NBS/CCl$_4$	Bz$_2$O$_2$	42[192]
f	i-Pr	Me	Br	NBS/CCl$_4$	Bz$_2$O$_2$	58[192]
g	i-Pr	i-Pr	Br	NBS/CCl$_4$	Bz$_2$O$_2$	74[192]
h	i-Pr	i-Bu	Br	NBS/CCl$_4$	Bz$_2$O$_2$	76[192]
i	Me	Ph	Br	NBS/CCl$_4$	Bz$_2$O$_2$	41[200]
j	Et	Ph	Br	NBS/CCl$_4$	Bz$_2$O$_2$	47[184]
k	i-Pr	Ph	Br	NBS/CCl$_4$	Bz$_2$O$_2$	82[192]
l	Et	4-Br-C$_6$H$_4$	Br	NBS/CCl$_4$	Bz$_2$O$_2$	53[184]
m	Et	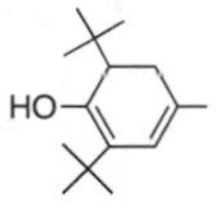	Br	NBS/CCl$_4$	hv	55[201]
n	Et	PhC(O)NH	Br	NBS/CCl$_4$	hv	95[190,191]
o	Et	4-Cl-C$_6$H$_4$C(O)NH	Br	NBS/CCl$_4$	hv	70[190]
p	Et	PhSO$_2$NH	Br	NBS/CCl$_4$	hv	95[191]
q	Et	Cl$_3$C-CH$_2$-OC(O)NH	Br	NBS/CCl$_4$	hv	95[190]

A variety of reducing agents have been described for the conversion of dihalogenophosphonates into monohalogenophosphonates (Scheme 3.32, Table 3.9). However, most of them produce complex mixtures from which the desired product is isolated by distillation or by chromatography. An exhaustive study shows that the reagent of choice is sodium hydrosulfide, which allows the isolation of monohalogenated compounds in almost quantitative yields.[205] Higher temperatures (25°C) are appropriate for isopropyl esters and chlorinated compounds, whereas lower temperatures (−25 to 0°C) are better for ethyl esters and brominated derivatives.[205]

The reduction of diethyl chlorofluoromethylphosphonate with H$_2$/Raney Ni in the presence of Et$_3$N affords diethyl fluoromethylphosphonate in 43%, contaminated with the C–F bond cleavage product, diethyl methylphosphonate (32%).[131]

TABLE 3.9

	R	R^1	X	Reducing agent	Yield (%)$^{Ref.}$
a	Et	EtO$_2$C	Cl	NaSH Na$_2$SO$_3$	—[204,205] 95[172]
b	Et	NO$_2$	Cl	Na$_2$SO$_3$	40[210]
c	Et	(EtO)$_2$P(O)	Cl	n-BuLi Na$_2$SO$_3$ NaSH NaCN+NaOH	79[206] 84,[129] 90,[209] 94[208] 91[204,205] 53[204,205]
d	i-Pr	(i-PrO)$_2$P(O)	Cl	n-BuLi Na$_2$SO$_3$ NaSH KF/18-Crown-6	71[207,211] 54,[204,205] 96[208] 94[204,205] 55[212]
e	Et	EtO$_2$C	Br	SnCl$_2$	85[172]
f	CF$_3$CH$_2$O	MeO$_2$C	Br	SnCl$_2$	70[213,214]
g	Et	(EtO)$_2$P(O)	Br	SnCl$_2$ NaSH NaCN+NaOH	70,[209] 75,[129] 84,[204,205] 90[208] 82[204,205] 54[204,205]
h	i-Pr	(i-PrO)$_2$P(O)	Br	n-BuLi SnCl$_2$ NaSH KF/18-Crown-6	64[207,211] 93[208] 95[204,205] 50[212]

3.1.1.11 Miscellaneous

Regiospecific α-monofluorination of diethyl 1-(phenylsulfenyl)methylphosphonate can be carried out by anodic oxidation in the presence of excess Et$_3$N·3HF (10 eq) to give diethyl 1-(phenylsulfenyl)fluoromethylphosphonate in 78% yield (Scheme 3.33).[215] The extension of the procedure to 2-pyridyl sulfides affords the monofluorinated compound in only 20% yield.[216]

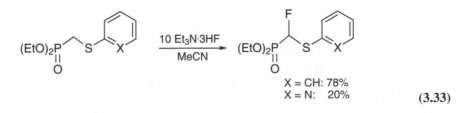

X = CH: 78%
X = N: 20%

(3.33)

The addition of copper(I) reagents to diethyl 1-alkynylphosphonates and trapping of the resulting 1-copper(I)vinylphosphonates with molecular iodine at low temperature afford diethyl α-iodovinylphosphonates in excellent overall yields (97–98%).[217,218] The α-deprotonation of diethyl vinylphosphonates with strong bases and subsequent treatment with iodine has also been reported (Scheme 3.34).[219,220]

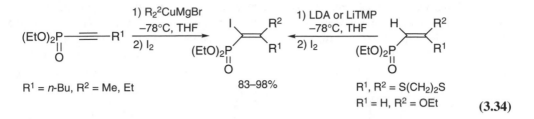

(3.34)

Dialkyl 1-formylmethylphosphonates and their enol ethers react with molecular chlorine or bromine at 0°C by an addition–elimination mechanism to give the α-monohalogenated 1-formyl-methylphosphonates in good yields (66–92%). The reaction temperature has to be carefully controlled in order to avoid a second halogenation.[221–224]

3.1.2 α,α-DIHALOGENOPHOSPHONATES

3.1.2.1 Michaelis–Becker Reactions

The reaction of sodium dialkyl phosphites with difluorochloromethane in petroleum ether at 30–35°C affords the desired dialkyl difluoromethylphosphonates in 48–67% yield.[225] Improved yields (54–77%) are obtained when the reaction is conducted in THF (Scheme 3.35).[226–230] This procedure allows the synthesis of difluoromethylphosphonates bearing functionalized substituents (allyl, β-silylethyl) on the phosphorus.[231] The reaction presumably proceeds via the generation and subsequent trapping of difluorocarbene.[232]

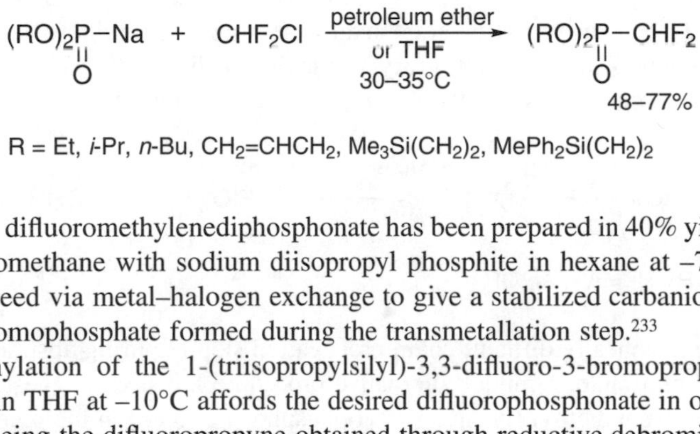

$$(RO)_2P-Na \ + \ CHF_2Cl \xrightarrow[\substack{30-35°C}]{\substack{\text{petroleum ether} \\ \text{or THF}}} (RO)_2P-CHF_2$$

with O double bonds on both phosphorus atoms.

48–77%

R = Et, i-Pr, n-Bu, CH$_2$=CHCH$_2$, Me$_3$Si(CH$_2$)$_2$, MePh$_2$Si(CH$_2$)$_2$

(3.35)

Tetraisopropyl difluoromethylenediphosphonate has been prepared in 40% yield by the reaction of dibromodifluoromethane with sodium diisopropyl phosphite in hexane at –78°C. The reaction is believed to proceed via metal–halogen exchange to give a stabilized carbanion, which reattacks the diisopropyl bromophosphate formed during the transmetallation step.[233]

The phosphonylation of the 1-(triisopropylsilyl)-3,3-difluoro-3-bromopropyne with sodium diethyl phosphite in THF at –10°C affords the desired difluorophosphonate in only 21% yield, the major byproduct being the difluoropropyne obtained through reductive debromination. The mechanism presumably involves an initial halogen–metal exchange reaction, followed by the electrophilic attack of the bromophosphate on the resulting difluoropropargyl anion.[234]

3.1.2.2 Kinnear–Perren Reactions

Dichloromethylphosphonic dichloride can be readily obtained by the reaction of CHCl$_3$, PCl$_3$, and AlCl$_3$ and subsequent hydrolysis (63–70% yields).[35,235,236] Esterification with alcohols[236] or substituted phenols[235] in the presence of bases gives the corresponding dichloromethylphosphonates in 70–81% yields (Scheme 3.36).

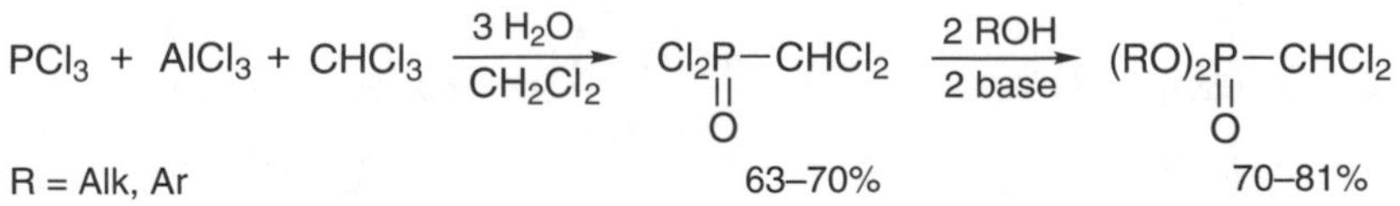

$$PCl_3 \ + \ AlCl_3 \ + \ CHCl_3 \xrightarrow[\substack{CH_2Cl_2}]{\substack{3\ H_2O}} Cl_2P-CHCl_2 \xrightarrow[\substack{2\ base}]{\substack{2\ ROH}} (RO)_2P-CHCl_2$$

R = Alk, Ar   63–70%   70–81%

(3.36)

The intermediate tetrachloroaluminate can be reacted directly with a suitable alcohol to isolate dialkyl dichloromethylphosphonates in low (23%)[237] to good overall yields (71–92%).[238,239] Similarly, dibromomethylphosphonic dibromide can be isolated in 88% yield from the reaction of $CHBr_3$, PBr_3, and $AlBr_3$ or the intermediate tetrabromoaluminate can be reacted directly with various alcohols to give dialkyl dibromomethylphosphonates in very good overall yields (93–95%).[239,240]

3.1.2.3 SNP(V) Reactions

Perfluorohexyl magnesium iodide, generated *in situ* from perfluorohexyl iodide and phenyl magnesium iodide, reacts with diethyl chlorophosphate in Et_2O at –50°C to give diethyl perfluorohexylphosphonate in 56% yield (Scheme 3.37).[241]

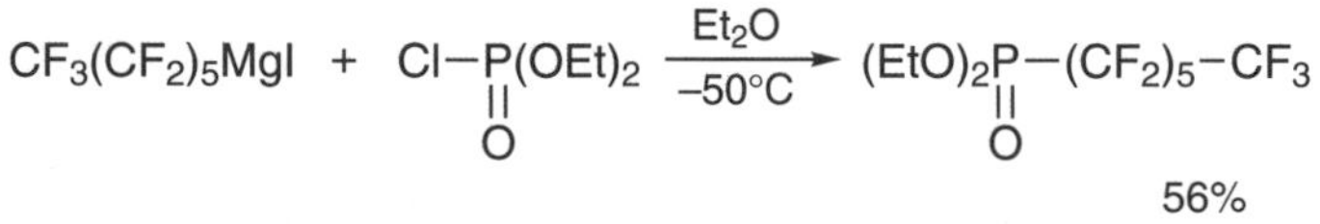

$$CF_3(CF_2)_5MgI \ + \ Cl-\underset{\underset{O}{\|}}{P}(OEt)_2 \ \xrightarrow[-50°C]{Et_2O} \ (EtO)_2\underset{\underset{O}{\|}}{P}-(CF_2)_5-CF_3$$

$$56\% \qquad\qquad (3.37)$$

3.1.2.4 Addition Reactions of Phosphites to Dihalogenoalkenes

The reaction of dimethyl phosphite with tetrafluoroethylene in a sealed tube in the presence of a free radical initiator (*t*-Bu_2O_2, Bz_2O, AIBN) leads to a mixture[242–244] from which the major product, dimethyl 1,1,2,2-tetrafluoroethylphosphonate, is isolated in 48% yield.[243] In 1996, the radical addition of a phosphoryl fragment to difluoroalkenes to produce the corresponding difluoroalkylphosphonates was described independently by two groups (Scheme 3.38).[245,246] The phosphoryl radical was generated by heating diethyl phosphite (3 eq) and a catalytic amount (0.5 eq) of peroxide in C_6H_6[245] or octane.[246,247] The yields are good in the case of aliphatic difluoroalkenes (56–72%)[246,247] but disappointingly low with less reactive (aryl or vinyloxy) alkenes (0–47%), presumably because of the relatively harsh conditions and the possibility of competing reactions.

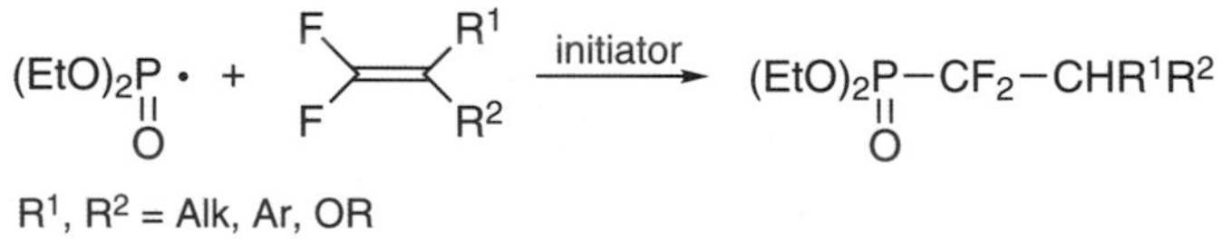

$$(EtO)_2\underset{\underset{O}{\|}}{P}\cdot \ + \ \underset{F}{\overset{F}{>}}\!\!=\!\!\underset{R^2}{\overset{R^1}{<}} \ \xrightarrow{initiator} \ (EtO)_2\underset{\underset{O}{\|}}{P}-CF_2-CHR^1R^2$$

$$R^1, R^2 = Alk, Ar, OR \qquad\qquad (3.38)$$

To avoid the energetically difficult chain propagation step requiring abstraction of a hydrogen atom from diethyl phosphite, a milder alternative procedure has been suggested. The phosphoryl radical is generated from diethyl (phenylselenyl)phosphonate (3 eq) in the presence of Bu_3SnH (4 eq) and catalytic amount of AIBN (0.5 eq). This method, which tolerates free hydroxyl functionalities and benzyl protecting groups, has allowed a number of sugar difluorophosphonates to be prepared in moderate to good yields (14–73%).[245,248]

A peculiar case of phosphoryl radical addition to difluoroalkenes involves the reaction of trialkyl phosphites with 1-bromo-2-iodo-1,1,2,2-tetrafluoroethane under ultraviolet irradiation (254 nm). Surprisingly, the corresponding 2-iodo-1,1,2,2-tetrafluoroethylphosphonates are formed in 42–48% yields with no detectable amount of the bromo derivative (Scheme 3.39). The proposed mechanism involves a halide-induced dealkylation of the trialkyl phosphite radical cation followed by addition of the product phosphoryl radical to tetrafluoroethene (generated by halide anion elimination) and iodide radical abstraction from the starting haloalkane.[249]

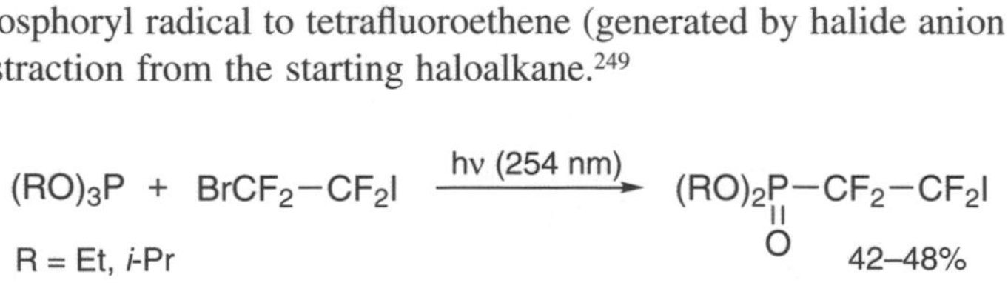

$$(RO)_3P \ + \ BrCF_2-CF_2I \ \xrightarrow{h\nu\ (254\ nm)} \ (RO)_2\underset{\underset{O}{\|}}{P}-CF_2-CF_2I$$

$$R = Et, \textit{i}\text{-}Pr \qquad\qquad\qquad 42\text{–}48\% \qquad\qquad (3.39)$$

3.1.2.5   Reactions of Tetraethyl Pyrophosphite with Perfluoroalkyl Iodides and Subsequent Oxidation

The reaction of tetraethyl pyrophosphite with perfluoroalkyl iodides in the presence of di-*tert*-butyl peroxide in 1,1,2-trichloro-1,2,2-trifluoroethane (F-113) was described for the first time in 1981.[250] Thermal decomposition of di-*tert*-butyl peroxide leads to the abstraction of an iodine atom and gives the reactive perfluoroalkyl radical, which reacts with tetraethyl pyrophosphite to produce the perfluoroalkyl phosphonite. Subsequent oxidation with *tert*-butyl hydroperoxide provides the desired perfluoroalkylphosphonates in 40–71% yields (Scheme 3.40).[249–252] A photochemical variant, which avoids heating the reaction mixture with a peroxide, was reported later. This milder method allows the preparation of functionalized perfluoroalkylphosphonates in good yields (45–77%).[232,253,254]

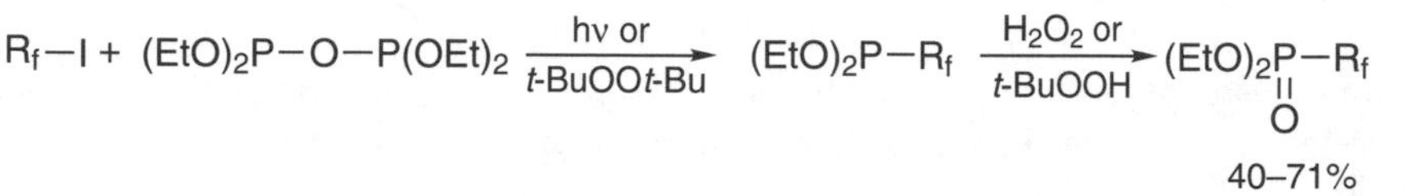

$$R_f{-}I\ +\ (EtO)_2P{-}O{-}P(OEt)_2 \xrightarrow[t\text{-BuOO}t\text{-Bu}]{hv\ or} (EtO)_2P{-}R_f \xrightarrow[t\text{-BuOOH}]{H_2O_2\ or} (EtO)_2\underset{O}{\overset{\parallel}{P}}{-}R_f$$

$$40\text{--}71\% \qquad \textbf{(3.40)}$$

3.1.2.6   Nucleophilic Halogenation Reactions

The synthesis of difluorophosphonates by the nucleophilic fluorination of α-ketophosphonates with DAST has been mainly used to obtain (phosphonodifluoromethyl)phenylalanine derivatives for use as nonhydrolyzable phosphotyrosine mimics. The reaction is carried out neat, at room temperature, to afford the desired α,α-difluorobenzylphosphonates in 43–79% yields (Scheme 3.41).[83,255–261]

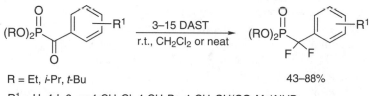

R = Et, *i*-Pr, *t*-Bu   43–88%

R¹ = H, 4-I, 3- or 4-CH₂Cl, 4-CH₂Br, 4-CH₂CH(CO₂Me)NHBoc

$$\textbf{(3.41)}$$

The same procedure has been applied to the synthesis of di-*tert*-butyl 2-naphthyldifluoromethylphosphonate derivatives in 51–59% yields; these show interesting protein-tyrosine phosphatase (PTP) inhibitor properties.[262,263] It has also been used to prepare diethyl 1,1-difluoro-2-alkynylphosphonates, where the yields are slightly lower (30–59%).[264,265] The main drawbacks of this method are the large excess of DAST (up to 15 eq) needed to complete the reaction[255,256] and difficulties associated with scale up: the DAST fluorination becomes uncontrollably exothermic when run neat in large quantities.[259] However, a recent report describing the fluorination of substituted benzoylphosphonates to give the corresponding α,α-difluorobenzylphosphonates in 73–76% yields uses only 3 eq of DAST in CH₂Cl₂ at room temperature.[82] This approach greatly improves the synthetic utility of the method, even if the scale is still rather restricted.

Dialkyl 1-(methylsulfenyl)-1,1-difluoromethylphosphonates can be prepared in 42–63% yields from the corresponding dichloro derivatives on treatment with Et₃N·3HF (6 eq) and ZnBr₂ (1 eq) in MeCN.[266]

3.1.2.7   Electrophilic Halogenation Reactions

As noted previously, the electrophilic halogenation of phosphonate carbanions often suffers from competing transmetallation reactions, which generate product mixtures containing mono-, di-, and nonhalogenated phosphonates (see also Section 3.1.1.8).[267,268] For example, when FClO$_3$ is passed into a solution of diethyl 1-(ethoxyphenylphosphinyl)methylphosphonate in toluene in the presence of *t*-BuOK (1 eq), an inseparable mixture of the corresponding di-, mono-, and nonfluorinated derivatives in a 52/32/16 ratio is obtained.[129] This is in part a result of difficulties in accurately controlling the quantity of the gaseous fluorinating agent introduced into the mixture. With the appearance of N–F fluorinating reagents, a few reliable procedures have been described for the synthesis of difluorophosphonates. Thus, dialkyl 1,1-difluoroalkylphosphonates,[148] α,α-difluorobenzylphosphonates,[82,269–271] 1-(2-pyridyl)-1,1-difluoromethylphosphonates,[82] and 1-(1-naphthyl)-1,1-difluoromethylphosphonates[269] have been prepared in 23–90% yields from the corresponding phosphonates by the electrophilic fluorination with NFBS in the presence of KDA[148] or NaHMDS[82,269–271] (Scheme 3.42). Di-*tert*-butyl phosphonates are too bulky to be fluorinated cleanly by this procedure, and side reactions such as elimination of phosphate or formation of isobutylene are observed.[270] The electrophilic fluorination of a dimethyl benzylphosphonate having a (benzyloxycarbonyl)methyl moiety in the *para* position was unsuccessful, with only unidentified nonfluorinated products being isolated. This may be a consequence of the elimination of fluorine from the first formed monofluorinated products, yielding quinodimethane derivatives, which may undergo further reaction.[270]

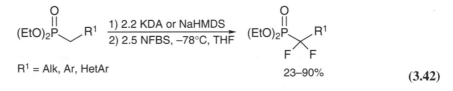

$$R^1 = \text{Alk, Ar, HetAr} \qquad\qquad\qquad\qquad 23\text{–}90\% \qquad\qquad\qquad\qquad \textbf{(3.42)}$$

Another approach employs the electrophilic fluorination of a preformed monofluorinated phosphonate. Diethyl 1-(ethoxycarbonyl)-1,1-difluoromethylphosphonate has been prepared in 60% yield from the commercially available monofluorinated derivative using Selectfluor™ and NaH in THF at room temperature.[272] However, only partial fluorination of sodium tetraisopropyl fluoromethylenediphosphonate occurs with AcOF in CFCl$_3$ at –78°C, and the desired difluoromethylenediphosphonate is isolated in 40% yield.[135]

Diethyl methylphosphonate reacts with phenylsulfonyl chloride (2 eq) in the presence of *n*-BuLi (3 eq) to produce diethyl 1-lithio-1,1-dichloromethylphosphonate, which can be quenched in quantitative yield or reacted directly with a large variety of carbonyl compounds (Horner–Wadsworth–Emmons reaction), thus providing a useful one-pot process for the synthesis of dichloroalkenes.[273]

Dichloro-, dibromo-, and diiodophosphonates bearing electron-withdrawing groups in the α-position are generally obtained from the corresponding phosphonates and sodium hypohalite (Scheme 3.43). The yields and the purity of the product depend on the experimental conditions (Table 3.10). The pH must be maintained between 7.0 and 7.5[129,172] and the temperature kept below 10°C[209,274] to provide optimum yields. In some cases, the use of a phase-transfer reagent also improves the yield.[209]

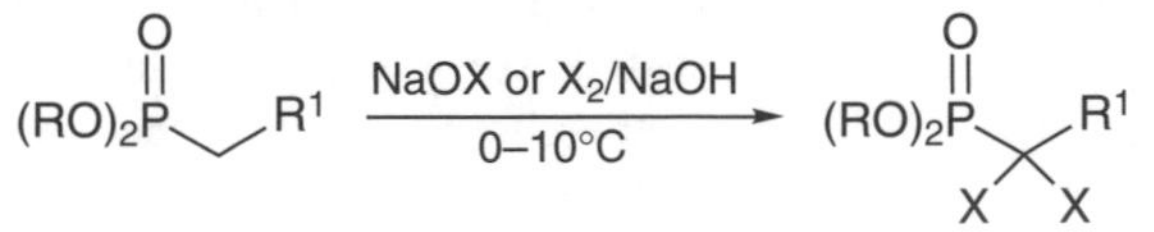

$$\begin{array}{ccc} \underset{(RO)_2P}{\overset{O}{\|}}\diagdown R^1 & \xrightarrow[\text{0–10°C}]{\text{NaOX or } X_2/\text{NaOH}} & \underset{(RO)_2P}{\overset{O}{\|}}\diagup[X][X] R^1 \end{array} \qquad\qquad \textbf{(3.43)}$$

Mixed α,α-dihalogenophosphonates can be obtained by electrophilic halogenation of the corresponding α-halogenophosphonates. Thus, tetraalkyl (α-chloro and α-bromo)-α-fluoromethylenediphosphonates are obtained in 81–94% yields by the reaction of tetraalkyl fluoromethylenediphosphonates with sodium hypochlorite or hypobromite respectively at 0°C.[208] By the same method,

TABLE 3.10

	R	R¹	X	"X⁺"	Yield (%)[Ref.]
a	Me	$P(O)(OMe)_2$	Cl	NaOCl	72,[208] 85[209]
			Br	NaOBr	80,[208] 92[209]
b	Et	$P(O)(OEt)_2$	Cl	NaOCl	62,[129] 81,[209] 82[208]
			Br	NaOBr	68,[129] 85[208]
c	i-Pr	$P(O)(Oi\text{-}Pr)_2$	Cl	NaOCl	82[211]
				Cl_2/Na	94[a][183]
			Br	KOBr	78[211]
				NaOBr	89[208]
d	$n\text{-}C_4H_9$	$P(O)(On\text{-}C_4H_9)_2$	Cl	NaOCl	83[209]
e	$n\text{-}C_5H_{11}$	$P(O)(On\text{-}C_5H_{11})_2$	Cl	NaOCl	90[209]
f	$n\text{-}C_6H_{13}$	$P(O)(On\text{-}C_6H_{13})_2$	Cl	NaOCl	91[209]
			Br	NaOBr	91[209]
g	$n\text{-}C_7H_{15}$	$P(O)(On\text{-}C_7H_{15})_2$	Cl	NaOCl	80–95[274]
h	$n\text{-}C_{10}H_{21}$	$P(O)(On\text{-}C_{10}H_{21})_2$	Br	NaOBr	75–90[274]
i	$n\text{-}C_{22}H_{45}$	$P(O)(On\text{-}C_{22}H_{45})_2$	I	NaOI	70–90[274]
j	$t\text{-}BuOCH_2$	$P(O)(OCH_2Ot\text{-}Bu)_2$	Cl	NaOCl	61[275]
k	CF_3CH_2O	CO_2Me	Br	NaOBr	85[213,214]
l	Et	CO_2Et	Cl	NaOCl	95[172]
			Br	NaOBr	93[172]
m	Et	$CO_2t\text{-}Bu$	Cl	NaOCl	96[276]
n	$n\text{-}C_8H_{17}$	$CO_2n\text{-}C_8H_{17}$	Cl	NaOCl	90–99[274]
o	$n\text{-}C_{15}H_{31}$	$CO_2n\text{-}C_{15}H_{31}$	Br	NaOBr	80–90[274]
p	$n\text{-}C_{20}H_{41}$	$CO_2n\text{-}C_{20}H_{41}$	I	NaOI	70–90[274]

[a] Crude product.

diethyl α-chloro-α-fluoro-, α-bromo-α-fluoro-, and α-bromo-α-chloro-(ethoxycarbonyl)methylphosphonates are obtained in 90–96% yields.[172]

3.1.2.8 Reduction of Dialkyl Trihalogenomethylphosphonates

The most reliable method to convert trihalogenomethylphosphonates into dihalogenomethylphosphonates is the halogen-exchange reaction with i-PrMgCl followed by acid workup. The reaction has been described for the synthesis of diethyl difluoro-,[277] fluorobromo-,[36] and dichloromethylphosphonates[278,279] in 71–94% yields (Scheme 3.44). The intermediate Grignard derivatives are more stable towards α-elimination than the corresponding lithium salts[206,280] and the reaction is cleaner than when phosphines or phosphites are used as reducing agents.[281–283]

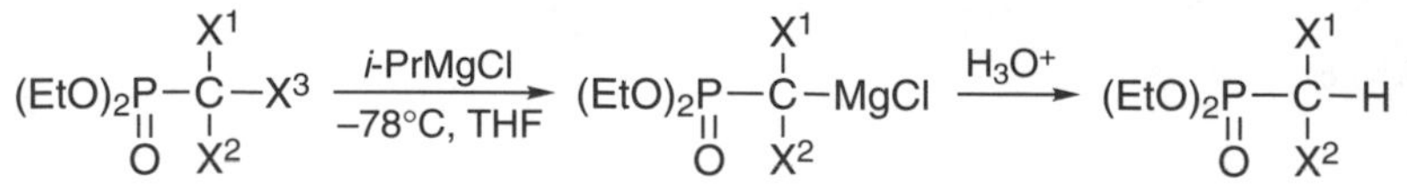

$$(EtO)_2\overset{\displaystyle X^1}{\underset{\displaystyle O}{\overset{|}{\underset{||}{P}}}}\overset{|}{\underset{|}{C}}-X^3 \quad \xrightarrow[-78°C,\ THF]{i\text{-}PrMgCl} \quad (EtO)_2\overset{\displaystyle X^1}{\underset{\displaystyle O}{\overset{|}{\underset{||}{P}}}}\overset{|}{\underset{|}{C}}-MgCl \quad \xrightarrow{H_3O^+} \quad (EtO)_2\overset{\displaystyle X^1}{\underset{\displaystyle O}{\overset{|}{\underset{||}{P}}}}\overset{|}{\underset{|}{C}}-H$$

$$X^1 = X^2 = F\ ;\ X^3 = Br\ : 85\%$$
$$X^1 = F\ ;\ X^2 = X^3 = Br\ : 71\%$$
$$X^1 = X^2 = X^3 = Cl\qquad : 83\text{–}94\%$$

$$\text{(3.44)}$$

3.1.2.9 Miscellaneous

The electrochemical difluorination of diethyl 1-(phenylsulfenyl)methylphosphonate using the powerful fluorine source $Et_4NF\cdot4HF$ in MeCN takes place smoothly without passivation of the anode to give diethyl 1-(phenylsulfenyl)-1,1-difluoromethylphosphonate selectively in 50% yield.[284]

The addition of diethyl 1-fluoro- or 1,1-difluoro-1-iodomethylphosphonates to nonactivated alkenes at room temperature in the presence of catalytic amounts of $Pd(PPh_3)_4$ affords good to excellent yields (63–91%) of mono- or difluorinated diethyl 3-iodo-1,1-difluoroalkylphosphonates.[285–287] The CuCl-mediated addition of diethyl trichloromethylphosphonate to activated and non-activated alkenes has also been described.[288]

Diethyl 2-oxopropylphosphonate may be treated with excess SO_2Cl_2 in CH_2Cl_2[289] or molecular chlorine in CCl_4[290] to obtain diethyl 1,1-dichloro-2-oxopropylphosphonate in 93% yield. However, the extent of chlorination cannot be easily controlled, and the reaction of substituted methylphosphonates with SO_2Cl_2 generally gives a mixture of mono- and dichloromethylphosphonates.[291–298]

3.1.3 α,α,α-TRIHALOGENOPHOSPHONATES

3.1.3.1 Michaelis–Arbuzov Reactions

The most general method for the synthesis of dialkyl α,α,α-trihalogenophosphonates is undoubtedly the Michaelis–Arbuzov reaction between a trialkyl phosphite and a tetrahalomethane in a suitable solvent.

Because of the high volatility of the trifluoromethyl halides and their inertness in the thermal Michaelis–Arbuzov reaction,[299] the synthesis of diethyl trifluoromethylphosphonate under photolytic conditions has been investigated. The modest yield initially obtained (51%)[300] can be improved using vacuum line techniques, and the desired trifluoromethylphosphonate is isolated in almost quantitative yield (Scheme 3.45).[301]

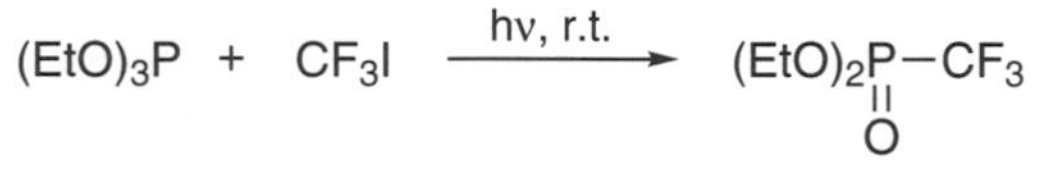

$$(EtO)_3P \ + \ CF_3I \quad \xrightarrow{h\nu,\ r.t.} \quad (EtO)_2\underset{\displaystyle O}{\overset{||}{P}}-CF_3$$

$$51\text{–}100\%$$

$$\text{(3.45)}$$

Diethyl 1-bromo-1,1-difluoromethylphosphonate is usually obtained from the Michaelis–Arbuzov reaction between triethyl phosphite and dibromodifluoromethane in Et_2O at reflux for 24 h (Scheme 3.46).[302] Other dialkyl 1-bromo-1,1-difluoromethylphosphonates were obtained in the same way (Table 3.11). When the reaction is performed in triglyme, the yields are significantly lower.[302] Recently, it has been shown that the reaction time can be shortened to 1.5 h by using refluxing THF.[277]

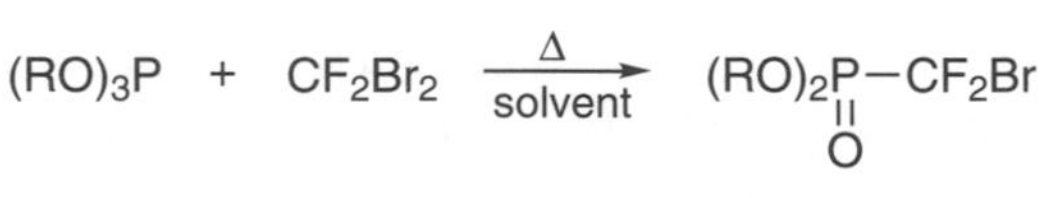

$$(RO)_3P \ + \ CF_2Br_2 \quad \xrightarrow[\text{solvent}]{\Delta} \quad (RO)_2\underset{\displaystyle O}{\overset{||}{P}}-CF_2Br$$

$$\text{(3.46)}$$

TABLE 3.11

	R	Temp. (°C)	Solvent	Reaction time (h)	Yield (%)[Ref.]
a	Me	50	Triglyme	24	55[302]
b	Et	50	Triglyme	24	55[302]
c	Et	35	Et$_2$O	24	90,[303,304] 95[302]
d	Et	60	THF	1.5	96[277]
e	i-Pr	50	Triglyme	24	42[302]
f	n-Bu	35	Et$_2$O	24	65[302]
g	n-Bu	25	—	36	quant.[301]

TABLE 3.12

	R	Temp. (°C)	Solvent	Reaction time (h)	Yield (%)[Ref.]
a	Et	0	—	4	77[251]
b	Et	25	Hexane	11–12 d	97[308]
c	Et	50	Hexane	48	79–85[307]
d	Et	35	Et$_2$O	24	78[302]
e	Et	50	Triglyme	24	60[302]
f	i-Pr	35	Et$_2$O	24	22,[302] 78[233]

Diethyl 1-bromo-1,1-difluoromethylphosphonate has also been obtained in 92% yield by treating (bromodifluoromethyl)triphenylphosphonium bromide with excess triethyl phosphite in CH$_2$Cl$_2$ at room temperature for 10 min. The proposed mechanism involves an insertion of difluorocarbene into bromotriethoxyphosphonium bromide.[305]

A single Michaelis–Arbuzov synthesis of diethyl 1,1-difluoro-1-iodomethylphosphonate has been described (Et$_2$O at reflux for 5 h, 96% yield). However, the usefulness of this reaction is limited because diiododifluoromethane is not readily available.[306]

The simplest and most frequently used method for the preparation of dialkyl 1,1-dibromo-1-fluoromethylphosphonates involves the reaction between dibromodifluoromethane and a trialkyl phosphite (Scheme 3.47). The yield and purity of the product vary widely as a function of the reaction conditions (temperature, solvent) and the alkyl group (Table 3.12). The reaction in refluxing Et$_2$O affords low to good yields of diethyl 1,1-dibromo-1-fluoromethylphosphonate,[302] but the synthesis has recently been improved by performing the reaction in sunlight in hexane at 50°C.[307] This is a good compromise between the long reaction time required under milder conditions[308] and the formation of byproducts at higher temperatures. The cleanest reaction is obtained in a nonpolar solvent such as hexane.[307,308]

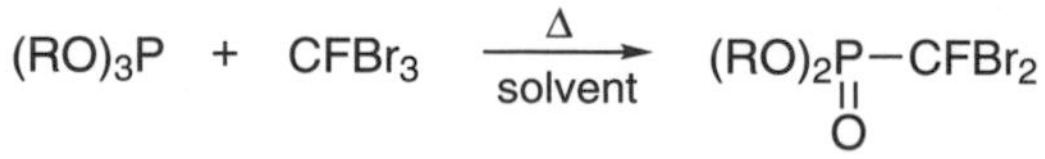

$$(RO)_3P \quad + \quad CFBr_3 \xrightarrow[\text{solvent}]{\Delta} \quad (RO)_2\underset{\underset{O}{\|}}{P}-CFBr_2$$

$$(3.47)$$

The first synthesis of dialkyl trichloromethylphosphonates, which involved the reaction of trialkyl phosphites and excess CCl_4 at reflux, was described in 1946 (Scheme 3.48).[309] This method is now widely used, and trichloromethylphosphonates bearing various alkyl groups are available in yields over 50%.[309-318] Trimethyl phosphite gives relatively low yields of dimethyl trichloromethylphosphonate,[315] but very good yields are obtained with triethyl phosphite (85–95%, Scheme 3.48). Phosphites having alkyl,[313] adamantyl,[319,320] or silyl[321] groups and halides such as bromotrichloromethane[314,322] can be successfully used to prepare dialkyl trichloromethylphosphonates.

$$(EtO)_3P \quad + \quad CCl_4 \xrightarrow{\text{reflux}} \quad (EtO)_2\underset{\underset{O}{\|}}{P}-CCl_3$$

$$85\text{–}95\%$$

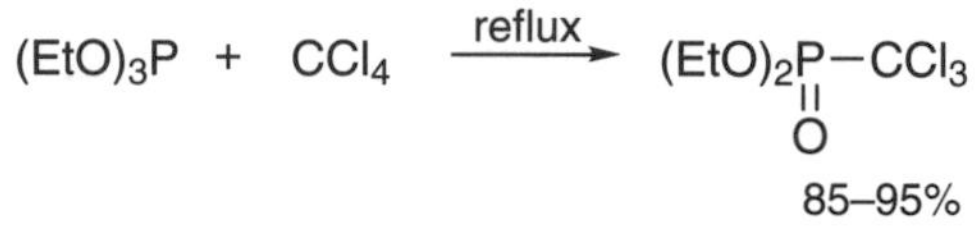

$$(3.48)$$

The reaction can be conducted under "thermal" conditions (reflux), "radical" conditions (the presence of traces of benzoyl peroxide induces a fourfold increase in the thermal reaction rate and a slightly better yield), or "photochemical" conditions (where the reaction proceeds under UV irradiation at room temperature to give the same yield as above; no reaction is observed in the dark at room temperature).[314] The mechanism of the reaction has been studied extensively,[237,314,323-328] and it has been concluded[327] that the thermal reaction of triethyl phosphite with CCl_4 involves an S_NCl^+ substitution. In the presence of UV light or free-radical chain initiators, the radical mechanism generally dominates. The ability of the trichloromethyl radical to initiate a radical chain reaction depends on the relative concentrations of the reagents. The final product mixture is the same as in the ionic case.[327]

The only preparation of diethyl tribromomethylphosphonate described to date involves the reaction of triethyl phosphite with carbon tetrabromide at room temperature. Triethyl phosphite reacts more quickly with diethyl tribromomethylphosphonate than with carbon tetrabromide, so tetraethyl dibromomethylenediphosphonate is the major product (63%). The yield of the desired diethyl tribromomethylphosphonate after distillation is only 15%.[329]

3.1.3.2 Kinnear–Perren Reactions

Trichloromethylphosphonic dichloride can be conveniently prepared on large scale by the reaction of PCl_3 with CCl_4 or CCl_3Br in the presence of $AlCl_3$ and subsequent hydrolysis of the resulting complex with 3–4.5 eq of water (81–90% yields).[35,330,331] When the hydrolysis is replaced by alcoholysis, dialkyl trichloromethylphosphonates are obtained in moderate to good yields (40–95%).[36] The trichloromethyl trichlorophosphonium chloroaluminate complex can also be treated with a perfluoroalcohol to give perfluorinated trichloromethylphosphonates in moderate yields (41–52%).[332]

3.1.3.3 SNP(V) Reactions

The trifluoromethyl anion, generated *in situ* from (trimethylsilyl)trifluoromethane and catalytic quantities of dry KF, reacts with (toxic) dibutyl fluorophosphate to give dibutyl trifluoromethylphosphonate in 93% yield.[333]

3.1.3.4 Oxidation Reactions

α-Halogenated dialkyl methylphosphonates have been prepared from methylphosphonous dihalides via the corresponding phosphonite. In the oxidation step, air[334] or H_2O_2[335] is used to obtain symmetric

TABLE 3.13

	R_X	X	R	R^1	Yield (%)[Ref.]
a	CF_3	I	H	—	83[335,338]
b	CF_3	I	i-Bu	—	58[334]
c	CF_3	Cl	Me	Me	38[336]
d	CF_3	Cl	Et	Et	32[336]
e	CF_3	Cl	Et	n-Pr	35[336]
f	CF_3	Cl	Et	n-Bu	32[336]
g	CF_3	Cl	Et	4-O_2N-C_6H_4	45[337]
h	C_3F_7	Cl	H	—	87[339]
i	CCl_3	Cl	Et	4-O_2N-C_6H_4	31[337]
j	$CHCl_2$	Cl	Et	4-O_2N-C_6H_4	44[337]
k	CH_2Cl	Cl	Et	4-O_2N-C_6H_4	39[337]

diesters, whereas chlorine and alcohols/Et_3N[336] or phenoxides[337] produce disymmetric diesters (Scheme 3.49, Table 3.13). However, the overall yields are modest, and the synthesis is relatively complicated, so it is often better to prepare these products by the corresponding Michaelis–Arbuzov reaction.

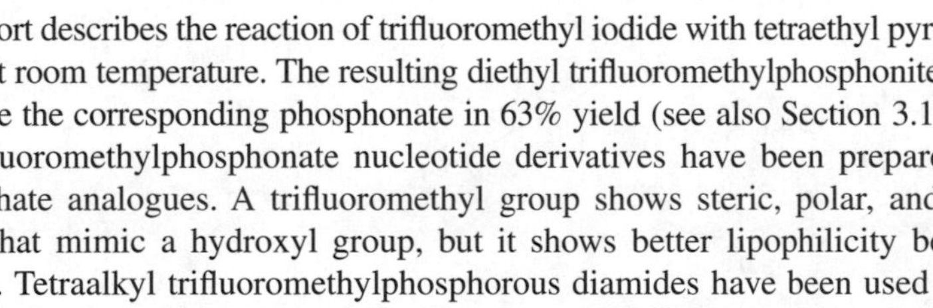

(3.49)

A recent report describes the reaction of trifluoromethyl iodide with tetraethyl pyrophosphite under UV irradiation at room temperature. The resulting diethyl trifluoromethylphosphonite is oxidized with t-BuOOH to give the corresponding phosphonate in 63% yield (see also Section 3.1.2.5).[254]

Several trifluoromethylphosphonate nucleotide derivatives have been prepared and tested as nonionic phosphate analogues. A trifluoromethyl group shows steric, polar, and electronegative characteristics that mimic a hydroxyl group, but it shows better lipophilicity because it has no negative charge. Tetraalkyl trifluoromethylphosphorous diamides have been used as precursors of nucleotide trifluoromethylphosphonate mono-[340] or diesters.[341,342] Thus, trifluoromethylphosphorous bistriazolide has been prepared in two steps (bromination with PBr_3, followed by reaction with $1H$-triazole in the presence of Et_3N) from tetraethyl trifluoromethylphosphorous diamide and then reacted with 5'-O-dimethoxytrityl thymidine. The resulting monotriazolide intermediate has been hydrolyzed, and subsequent oxidation with t-BuOOH afforded the desired nucleotide trifluoromethylphosphonate monoester (Scheme 3.50).[340] Nucleotide trifluoromethylphosphonate diesters[341,342] and a trifluoromethylphosphonate hapten[343] have been prepared by a similar procedure, using 2,4-dichlorophenyl-*N*-tosyloxaziridine and MCPBA for the oxidation step, respectively.

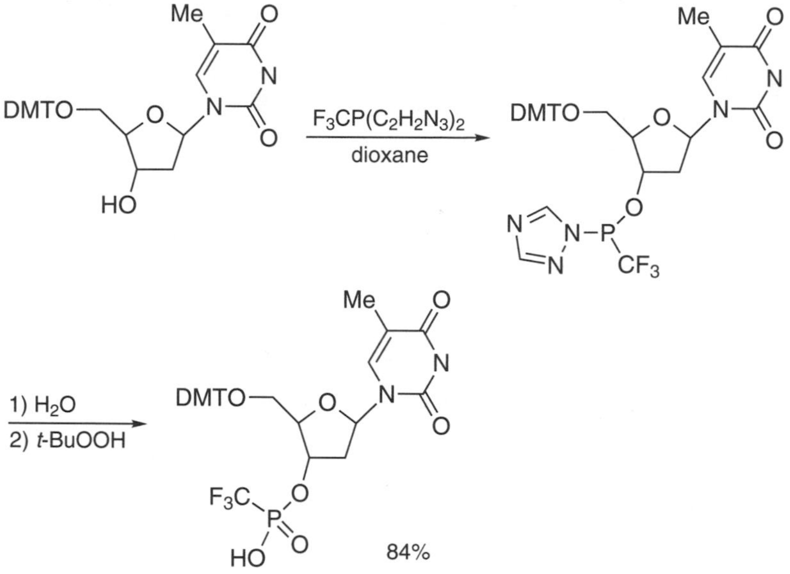

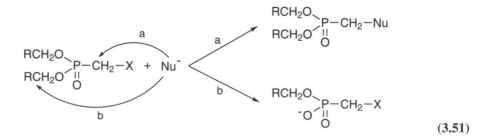

(3.50)

3.2 REACTIONS OF α-HALOGENOPHOSPHONATES

3.2.1 α-MONOHALOGENOPHOSPHONATES

3.2.1.1 Nucleophilic Substitution Reactions

Because of the inertness of the α-halogen in halogenomethylphosphonates, nucleophilic substitution reactions are difficult, and relatively harsh conditions are often needed. The nucleophilic substitution of the halide atom (path a) competes with the direct attack of the nucleophile at the carbon atom of the ester groups (path b), which gives cleavage of the C–O bond (Scheme 3.51). The undesirable dealkylation reaction can be inhibited by use of phosphonate esters bearing electron-withdrawing groups ($R = CCl_3$ or CF_3, Scheme 3.51) or by using phosphonamides instead of phosphonate esters.[37]

(3.51)

3.2.1.1.1 Formation of a C–N Bond

The analogy between aminocarboxylic acids and aminophosphonic acids and the applications of the latter in the inhibition of various metabolic processes have led to ongoing interest in the synthesis of aminophosphonic acids, especially by the nucleophilic amination of halogenoalkylphosphonates. The reaction of diethyl chloromethylphosphonate with a 25% aqueous solution of NH_3 in a sealed tube at 150°C exclusively produces the aminated monodealkylation product (45% yield), which can be quantitatively converted with 6 M HCl into aminomethylphosphonic acid (Scheme 3.52).[2,344,345]

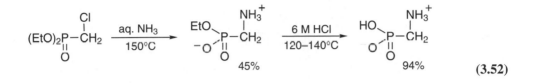

$$(3.52)$$

Analogously, various α-aminophosphonic acids have been prepared by heating chloromethylphosphonic acid with aqueous NH_3,[2,344,345] an aqueous solution of methylamine,[346] aniline,[347] hydrazine hydrate,[348] iminodiacetic acid,[349] and α,ω-diaminoalkanes.[350,351]

The reaction of *bis*(2,2,2-trifluoroethyl) 1,1-dideutero-1-chloromethylphosphonate with NaN_3 in DMSO at 90°C affords an almost quantitative yield of azidomethylphosphonate, which, on subsequent reduction and acidic hydrolysis, gives the 1,1-dideuteriated aminomethylphosphonic acid (Scheme 3.53). This reaction allows milder conditions and higher yields than with chloromethylphosphonates bearing groups that are less electron withdrawing than trifluoromethyl.[352,353] An asymmetric synthesis of α-aminophosphonic acids from chiral bicyclic phosphonamides derived from (*R,R*)- and (*S,S*)-1,2-diaminocyclohexane by a similar protocol has also been described.[354]

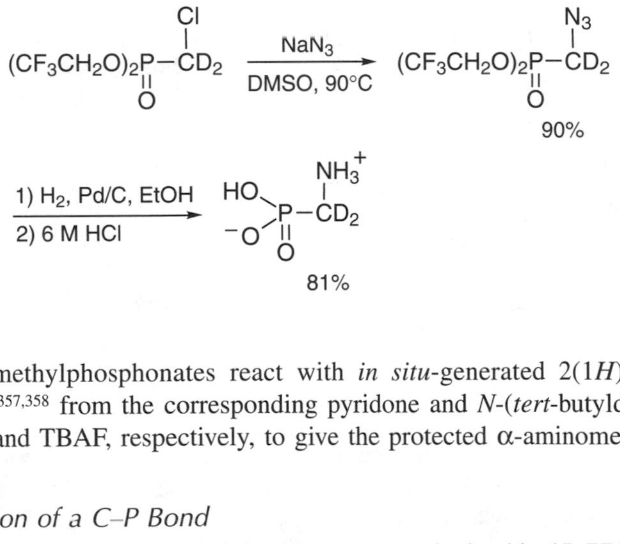

$$(3.53)$$

Dialkyl bromomethylphosphonates react with *in situ*-generated 2(1*H*)-pyridone[355,356] and phthalimide anions,[357,358] from the corresponding pyridone and *N*-(*tert*-butyldimethylsilyl)phthalimide with K_2CO_3 and TBAF, respectively, to give the protected α-aminomethylphosphonates in 70–98% yields.

3.2.1.1.2 Formation of a C–P Bond

Diaryl triphenylphosphoranylidenemethylphosphonates are obtained in 45–77% yields by the quaternization of triphenylphosphine with diaryl chloromethylphosphonate and subsequent treatment with NaOH.[359–362] The use of diethyl iodomethylphosphonate leads only to intractable gums and small quantities of ethyltriphenylphosphonium iodide, which are presumably formed by the dealkylation of an ethyl ester by iodide ion.[359] The ylide can react with a large variety of aliphatic[359–362] and aromatic[359] aldehydes to give *trans*-configured substituted diaryl 1-alkenylphosphonates in 63–95% yields. This ylide has been advantageously replaced later by the more nucleophilic tetraethyl methylenediphosphonate carbanion (Horner–Wadsworth–Emmons reaction, *vide infra*).

Tetraethyl methylenediphosphonate can be prepared from diethyl chloromethylphosphonate by reaction either with triethyl phosphite under drastic conditions (neat, 190°C, 27%, Michaelis–Arbuzov reaction)[363] or with sodium diethyl phosphite in refluxing C_6H_6 (47%, Michaelis–Becker reaction).[364–366] However, the yields are significantly lower than those obtained by SNP(V) reaction[60] of diethyl 1-lithiomethylphosphonate with diethyl chlorophosphate (Scheme 3.54).

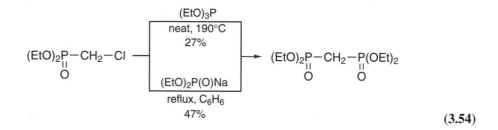

(3.54)

Diethyl chloromethylphosphonate can also be reacted with tetraethyl pentane-1,5-diphosphonite on heating to produce pentane-1,5-*bis*{[(diethoxyphosphonyl)methyl]phosphinate} in 88% yield.[367] Treatment of diphenyl chloromethylphosphonate with *tris*(trimethylsilyl) phosphite at 220–230°C gives, after hydrolysis of the silyl groups, (diphenoxyphosphinyl)methylphosphonic acid in 49% yield.[368] Heating dibenzyl chloromethylphosphonate with methyl dibenzyl phosphite and tribenzyl phosphite under vacuum has been reported to give the corresponding diphosphonates in 65% and 92% yields, respectively.[369,370]

3.2.1.1.3   Formation of a C–O Bond

Several aryloxymethylphosphonates have been prepared in low to moderate yields by the reaction of a variety of sodium aryloxides with diethyl iodomethylphosphonate.[371–373] The first step is a dealkylation, and the second is the nucleophilic substitution of the iodine atom (Scheme 3.55).[372]

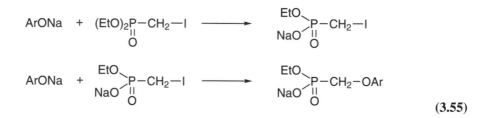

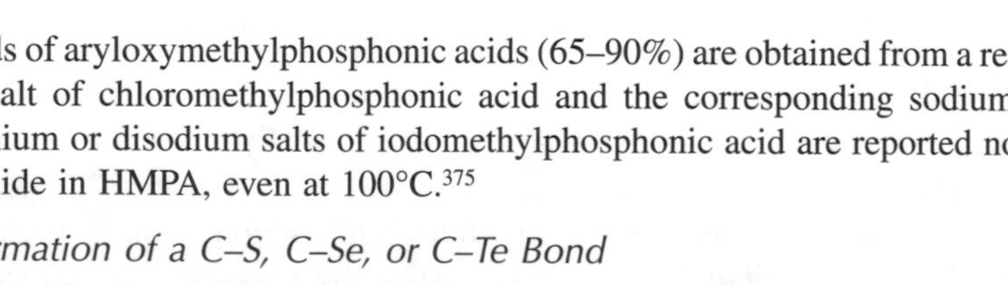

(3.55)

Better yields of aryloxymethylphosphonic acids (65–90%) are obtained from a reaction between the disodium salt of chloromethylphosphonic acid and the corresponding sodium aryloxides.[374] However, dilithium or disodium salts of iodomethylphosphonic acid are reported not to react with sodium phenoxide in HMPA, even at 100°C.[375]

3.2.1.1.4   Formation of a C–S, C–Se, or C–Te Bond

Chloromethylphosphonic acid[347] and diethyl chloromethylphosphonate[147,376–378] react with sodium salts of thioalcohols, thiophenols, and sulfinates to give the corresponding sulfides and sulfones in moderate to good yields (40–86%). The reaction of diethyl iodomethylphosphonate with (1*S*)-10-mercaptoisoborneol under phase-transfer conditions has also been described (89% yield), and the resulting sulfide has been oxidized using MCPBA in a diastereoselective manner to the corresponding sulfoxides (Scheme 3.56).[379] However, use of this thioalkylation is often handicapped by the formation of alkyl sulfides through a dealkylation process.[380]

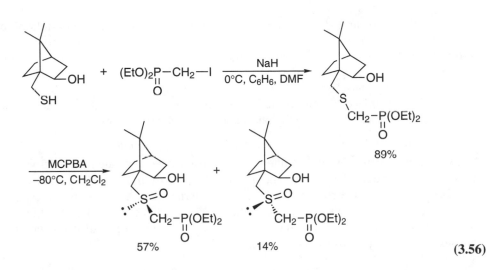

(3.56)

Tetra-*n*-butylammonium or potassium thioacetates react with dialkyl chloro-, bromo-, or iodo-methylphosphonate in THF or DME at room temperature to give dialkyl 1-(acetylthio)methylphos-phonate in 59–95% yields.[381–384] Ethyl hydrogen [(aminoiminomethyl)thio]methylphosphonate can be obtained in 20–38% yields from diethyl chloro-[385] or bromomethylphosphonates[386] and thiourea on heating at 125–130°C. Subsequent treatment with NaOH gives ethyl hydrogen mercaptomethyl-phosphonate.[386]

Diethyl 1-(phenylseleno)-[387,388] and 1-(phenyltelluro)methylphosphonates[389] have been prepared in 77–92% yields by the reaction of diethyl iodomethylphosphonate with sodium phenylselenide and lithium phenyltelluride, respectively.

3.2.1.1.5   Formation of a C–C Bond

The coupling reaction of various aryl organocopper(I) reagents, prepared *in situ* from aryl bromides, with diethyl chloromethylphosphonate in THF at reflux affords the desired benzylic phosphonates in 52–69% yields (Scheme 3.57). When chloromethylphosphonate is replaced by iodomethylphos-phonate the yield drops to 31%.[390]

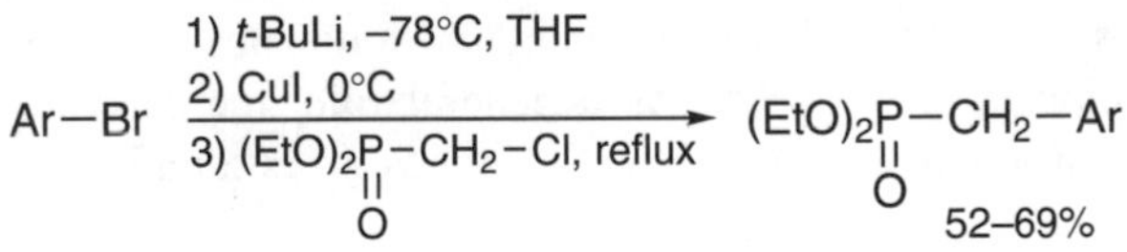

$$Ar-Br \xrightarrow[\text{3) (EtO)}_2\text{P}-\text{CH}_2-\text{Cl, reflux}]{\substack{\text{1) }t\text{-BuLi, }-78°C, THF \\ \text{2) CuI, }0°C}} (EtO)_2P-CH_2-Ar$$

1) *t*-BuLi, −78°C, THF
2) CuI, 0°C
3) (EtO)₂P–CH₂–Cl, reflux   →   (EtO)₂P–CH₂–Ar   52–69%

(3.57)

In contrast with these results, α-lithiated thiophenes give better results with iodomethylphos-phonate in the CuI-mediated coupling reaction, and 2-thienylmethylphosphonates can be isolated in yields up to 64%.[391,392]

3.2.1.2   Carbanionic Reactions

The relative instability of α-halogenomethylphosphonate carbanions[393] means that they require reactive electrophiles at low temperature to obtain satisfactory results. The incorporation of a trimethylsilyl group at the α-position provides a useful method for increasing the stability of the carbanion and preventing undesirable reactions.

3.2.1.2.1   Dialkyl Fluoromethylphosphonates

Dialkyl fluoromethylphosphonates can be metallated with strong bases such as *n*-BuLi or LDA, and the resulting fluorinated carbanion reacts with dimethyl sulfate,[394,395] allyl bromide,[394,395] carbon dioxide,[396] ethyl formate,[152] diethyl chlorophosphate,[152] or (S)-(−)-menthyl *para*-toluenesulfinate[397] to give the corresponding alkyl-, carboxyl-, phosphonyl-, and sulfoxylfluorophosphonates in 37–60% yields. The reaction with chloro- or bromotrimethylsilane under the same conditions gives a mixture of mono-, di-, and nonsilylated phosphonates.[394,395] Zinc(II) and copper(I) derivatives of diethyl fluoromethylphosphonate readily react with electrophiles such as allyl halides, ethyl chloroformate, and diethyl chlorophosphate as well as vinyl, alkynyl, and aryl halides to give the corresponding substituted fluoromethylphosphonates in 34–80% yields.[398]

3.2.1.2.2 Diethyl 1-(Trimethylsilyl)-1-fluoromethylphosphonate

Diethyl 1-lithio-1-(trimethylsilyl)-1-fluoromethylphosphonate, generated *in situ* from diethyl 1,1-dibromo-1-fluoromethylphosphonate and *n*-BuLi (2 eq) or from diethyl fluoromethylphosphonate and LDA (2 eq) in the presence of Me$_3$SiCl under internal quench conditions, is readily alkylated at low temperature with alkyl and allyl iodides[399,400] or triflates.[401,402] The resulting fluorinated alkylphosphonates can be desilylated with EtOLi in EtOH at 0°C to give diethyl α-fluoroalkylphosphonates in excellent yields (83–96%, Scheme 3.58).

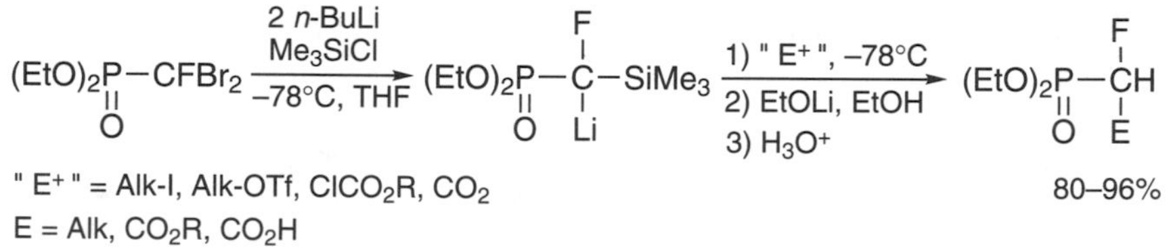

$$(EtO)_2P-CFBr_2 \xrightarrow[\substack{-78°C, \ THF}]{\substack{2 \ n\text{-BuLi} \\ Me_3SiCl}} (EtO)_2\underset{\underset{Li}{O}}{\overset{F}{P}}-\overset{F}{\underset{}{C}}-SiMe_3 \xrightarrow[\substack{2) \ EtOLi, \ EtOH \\ 3) \ H_3O^+}]{1) \ "E^+", \ -78°C} (EtO)_2\underset{O}{\overset{F}{P}}-\overset{F}{\underset{E}{CH}}$$

" E$^+$ " = Alk-I, Alk-OTf, ClCO$_2$R, CO$_2$
E = Alk, CO$_2$R, CO$_2$H

                                                                                          80–96%

                                                                                                    (3.58)

The reaction of silylated fluoromethylphosphonate carbanions with a large variety of alkyl, vinyl, or aryl chloroformates or with carbon dioxide proceeds smoothly at low temperature to give silylated 1-(alkoxycarbonyl)-1-fluoromethylphosphonates and 1-(hydroxycarbonyl)-1-fluoromethylphosphonate, respectively (Scheme 3.58). For the electron-withdrawing carboxyl group, the traces of base already present in the reaction medium are sufficient to effect a desilylation if EtOH is added to the mixture. The desilylated products are isolated in 80–91% yields.[403,404] The addition of EtOLi leads to decomposition with loss of phosphate.

The Peterson reaction of diethyl 1-lithio-1-(trimethylsilyl)-1-fluoromethylphosphonate with a large number of aliphatic, aromatic, and heteroaromatic aldehydes as well as with enolizable and nonenolizable ketones produces 1-fluoro-1-alkenylphosphonates in 39–93% yields with good (E)/(Z) selectivity (Scheme 3.59).[308,405,406] As a general rule, the reaction with aldehydes gives mainly the (E)-isomer, whereas the reaction with ketones gives mainly the (Z)-isomer.[308]

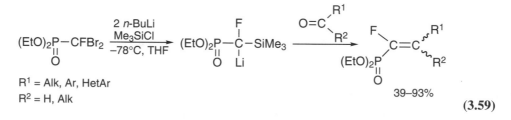

R^1 = Alk, Ar, HetAr
R^2 = H, Alk

                                                                                          39–93%

                                                                                                    (3.59)

3.2.1.2.3 Dialkyl α-Substituted Fluoromethylphosphonates

The potassium salt of diethyl 1-(phenylsulfonyl)-1-fluoromethylphosphonate is alkylated with alkyl triflates or iodides at low temperature in THF in the presence of HMPA to generate diethyl 1-(phenylsulfonyl)-1-fluoroalkylphosphonates in 50–78% yields. These can be cleanly desulfonated with Na(Hg) in MeOH/THF/NaH$_2$PO$_4$ to the resulting diethyl α-fluoroalkylphosphonates, which are isolated in 71–89% yields.[407]

Lithium diethyl 1-(ethoxycarbonyl)-1-fluoromethylphosphonate undergoes acylation with per-fluorinated acyl chlorides at low temperature to form the corresponding acylated phosphonates, which, under mild basic hydrolysis (5% aq. NaHCO$_3$), give the monofluorinated ketoesters in good yields (60–77%) as their hydrates.[408,409] Acylation with ethyl oxalyl chloride (Scheme 3.60),[410–412] perfluoroacyl chlorides,[413] propanoyl or benzoyl chlorides,[19] followed by subsequent reaction of the acylated phosphonates with Grignard reagents, gives α-fluoro-α,β-unsaturated diesters in moderate to good yields (48–68%, Scheme 3.60).

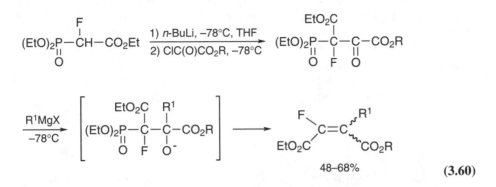

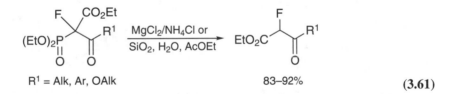

$$(3.60)$$

Substituted diethyl 1-(ethoxycarbonyl)-1-fluoromethylphosphonates bearing acyl[414,415] or alkoxycarbonyl[416] groups in α-position, on room temperature treatment with MgCl$_2$/NH$_4$Cl[414,416] or wet silica gel/AcOEt,[415] produce the corresponding α-fluoro-β-keto esters and fluoromalonates in very good yields (83–92%, Scheme 3.61).[414–416]

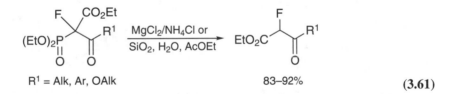

$$(3.61)$$

Diethyl 1-(ethoxycarbonyl)-1-fluoromethylphosphonate undergoes a double acylation reaction with aroyl chlorides in the presence of MgCl$_2$/Et$_3$N with concomitant loss of the phosphoryl group. The resulting diacylfluoroacetate can be selectively monodeacylated with SiO$_2$ in aqueous AcOEt at 40°C to afford α-fluoro-β-ketoesters in 78–94% yields.[417,418] A wide range of diethyl α-fluoro-β-ketophosphonates can be prepared in good yields (66–78%) from dilithium diethyl 1-(hydroxycarbonyl)-1-fluoromethylphosphonate by acylation with acyl chlorides and subsequent decarboxylation.[419]

3.2.1.2.4 *Dialkyl Chloromethylphosphonates*

The α-metallation of dialkyl chloromethylphosphonates with strong bases such as *n*-BuLi, *s*-BuLi, LDA, or LiTMP quantitatively gives dialkyl 1-lithio-1-chloromethylphosphonates. In contrast, the use of *t*-BuLi at low temperature leads to a mixture of H/Li and Cl/Li exchange products in a 9:1 ratio. In these conditions, the 1-lithio-1-chloromethylphosphonate is stable only at low temperature. The same carbanion prepared using 2 eq of hindered amides such as LDA or LiTMP is stable at 0°C for 30 min without apparent degradation. The first equivalent of base effects the deprotonation, and the second probably stabilizes the resulting carbanion.[37]

The asymmetric alkylation of chloromethylphosphonamides derived from (*R,R*)-1,2-*bis*(*N*-methylamino)cyclohexane,[420] (*S*)-2-anilinomethylpyrolidine,[421] or camphor[422] with alkyl halides in THF at low temperature in the presence of *n*-BuLi or LDA followed by hydrolysis with diluted

HCl at room temperature produces optically active 1-chloroalkylphosphonic acids in 65–86% overall yields.[420–422] The Michael addition of the same chloromethylphosphonamide to β-substituted *tert*-butyl acrylates gives, after cyclization, the corresponding phosphonylated cyclopropanes in 46–80% yields. The high selectivities observed in some cases seem to result from a combination of steric and electronic factors.[423]

The reaction of diethyl chloromethylphosphonate with CCl₄ in the presence of LDA at low temperature quantitatively gives diethyl 1-lithio-1,1-dichloromethylphosphonate, which can be hydrolyzed to give diethyl dichloromethylphosphonate[424] or reacted *in situ* with various electrophiles to give the expected products of alkylation or olefination (see Section 3.2.2.2).[425] Similarly, the reaction with CBr₄ transiently generates diethyl 1-lithio-1-bromo-1-chloromethylphosphonate, which, in the presence of an excess of LiBr, undergoes a formal chlorine–bromine exchange reaction to give the dibromo carbanion. This can be hydrolyzed to obtain diethyl dibromomethylphosphonate in 80–90% yield[425] or reacted with a large number of carbonyl compounds to give in 45–70% yields the corresponding dibromoolefins (Scheme 3.62),[425,426] valuable precursors of alkynes (Fritsch–Buttenberg–Wiechell rearrangement).[158]

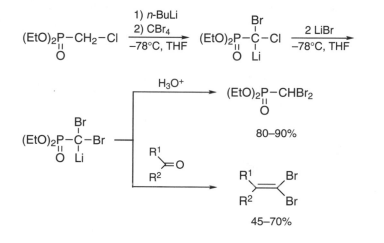

$$(3.62)$$

When diethyl 1-lithio-1-chloromethylphosphonate is treated with carbonyl compounds (aliphatic and aromatic aldehydes or ketones) in THF at low temperature, the intermediate chlorohydrin can eliminate upon warming to room temperature to give diethyl 1,2-epoxyphosphonates in good yields (51–90%)[427,428] or, in the case of aldehydes, can be treated at low temperature with LDA to give, after acidic work-up, diethyl 2-oxoalkylphosphonates in 65–92% yields (Scheme 3.63).[429,430]

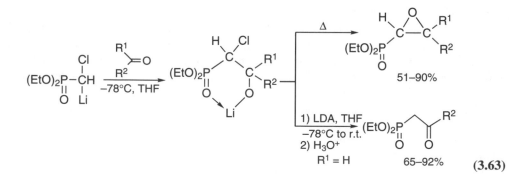

$$(3.63)$$

The reaction of diethyl chloromethylphosphonate with imines has also been described. The reaction proceeds in THF at low temperature in the presence of *n*-BuLi to give diethyl 1,2-aziridinylphosphonates in moderate to good yields (30–95%).[431] When chiral groups are present on the imine nitrogen[432–435] or phosphorus[436] atoms (Scheme 3.64), the corresponding phosphonylated aziridines are obtained in good yields with high enantiomeric purity.

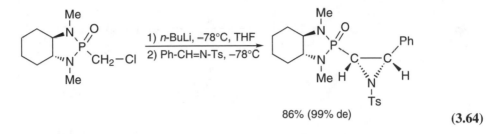

86% (99% de)   **(3.64)**

Diethyl 1-formyl-1-chloromethylphosphonate is generally obtained in high yields by the reaction of diethyl chloromethylphosphonate with a formylating agent in the presence of a base: HC(OEt)$_3$/Na,[437] HCO$_2$Et/*n*-BuLi,[438,439] or DMF/*n*-BuLi.[438] In the same manner, acylation of dialkyl 1-lithio-1-chloromethylphosphonate with aliphatic and aromatic esters[189,439] or diethyl oxalate[439] affords the corresponding 1-chloro-2-oxoalkylphosphonates in good yields. The reaction of diethyl 1-lithio-1-chloromethylphosphonate with CO$_2$[440,441] or diethyl carbonate in THF at low temperature[439,442] gives, respectively, the hydroxycarbonyl and ethoxycarbonyl derivatives, valuable precursors of α-chloroacrylates (Scheme 3.65).[442]

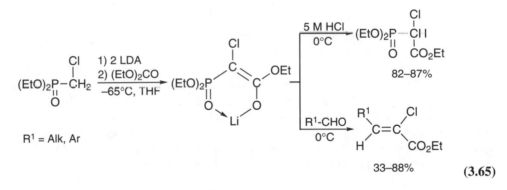

82–87%

33–88%   **(3.65)**

Diethyl 1-lithio-1-chloromethylphosphonate also reacts with dialkyl chlorophosphates in THF at low temperature to give the corresponding symmetric and disymmetric chloromethylenediphosphonates in moderate to good yields (51–81%).[443–445]

3.2.1.2.5 *Diethyl 1-(Trimethylsilyl)-1-chloromethylphosphonate*

Diethyl 1-lithio-1-(trimethylsilyl)-1-chloromethylphosphonate is readily generated in THF at low temperature either from diethyl chloromethylphosphonate or diethyl trichloromethylphosphonate and *n*-BuLi (2 eq) in the presence of Me$_3$SiCl (Scheme 3.66).

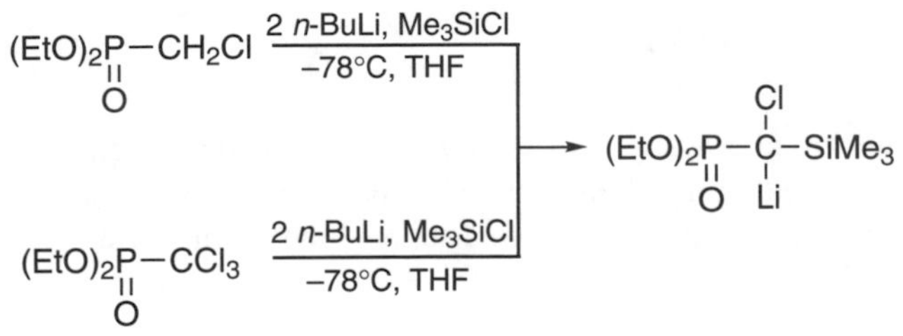

$(EtO)_2P-CH_2Cl$ $\xrightarrow[-78°C, THF]{2\ n\text{-BuLi, Me}_3\text{SiCl}}$

$(EtO)_2P-CCl_3$ $\xrightarrow[-78°C, THF]{2\ n\text{-BuLi, Me}_3\text{SiCl}}$

$\longrightarrow (EtO)_2P-\underset{Li}{\overset{Cl}{C}}-SiMe_3$

(3.66)

As the trimethylsilyl group acts as a stabilizing and protecting group, this carbanion is stable at room temperature and reacts with a variety of electrophiles. Thus, when the carbanion is treated with D_2O at room temperature, 1,1-dideuterated chloromethylphosphonate is obtained in 88–91% yield.[186,446] Alternatively, treatment with HCO_2H then with LiOD in D_2O gives the 1-monodeuterated chloromethylphosphonate in 89% yield (Scheme 3.67).[186]

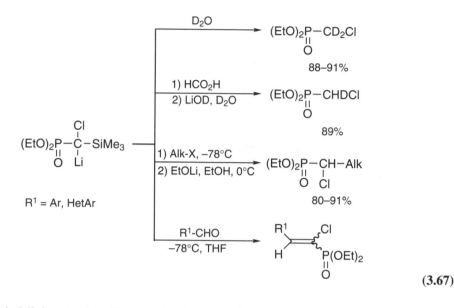

(3.67)

The trimethylsilyl protection allows a selective monoalkylation of diethyl 1-lithio-1-(trimethylsilyl)-1-chloromethylphosphonate on treatment with alkyl, allyl, and benzyl halides in THF at low temperature.[186,447,448] The subsequent mild deprotection with EtOLi in EtOH at 0°C affords diethyl 1-chloroalkylphosphonates in very good yields (80–91%, Scheme 3.67).[186,448] A double alkylation can be realized with 1,ω-dibromoalkanes, by successive removal of the three chlorine atoms from dialkyl trichloromethylphosphonates, to produce dialkyl 1-(trimethylsilyl)cycloalkylphosphonates in 18–75% yields. Although the α-position is not activated by other electron-withdrawing groups than phosphoryl, the removal of the silyl group is conveniently achieved using TBAF in moist THF, and dialkyl cycloalkylphosphonates are isolated in 52–87% yields (Scheme 3.68).[449]

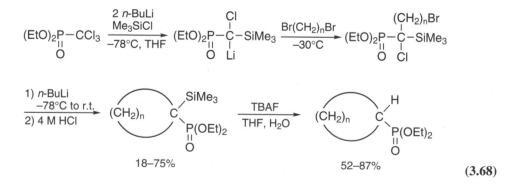

(3.68)

The trimethylsilyl group participates directly in the Peterson reaction of diethyl 1-lithio-1-(trimethylsilyl)-1-chloromethylphosphonate with aromatic or heteroaromatic aldehydes in THF at low temperature to produce diethyl 1-chloro-1-alkenylphosphonates as a mixture of (Z)- and

(*E*)-isomers (Scheme 3.67). The mixture is then treated *in situ* with LiHMDS to give the desired diethyl 1-alkynylphosphonates in excellent yields (87–96%, Scheme 3.69).[450]

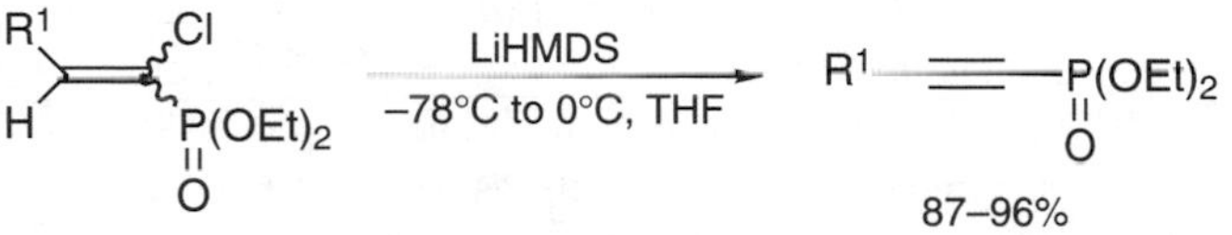

$$ \textbf{(3.69)} $$

3.2.1.2.6   Dialkyl 1-Substituted Chloromethylphosphonates

Dialkyl 1-substituted chloromethylphosphonates, prepared either by electrophilic or nucleophilic chlorination (see Sections 3.1.1.8.2 and 3.1.7.2), readily react with aromatic or heteroaromatic aldehydes in the presence of a base (*n*-BuLi, NaH, or *t*-BuOK, Horner–Wadsworth–Emmons reaction), and the resulting chloroalkenes are directly dehydrohalogenated to give the corresponding alkynes in moderate to good overall yields.[105,109,110,451–455] The best suited base for the last step seems to be *t*-BuOK (Scheme 3.70).[105,110,451]

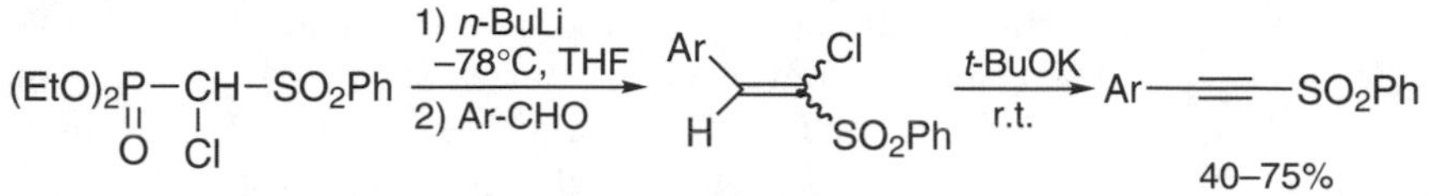

$$ \textbf{(3.70)} $$

The reaction of diethyl 1-lithio-1-chloroalkylphosphonates, prepared from diethyl 1,1-dichloroalkylphosphonates and *n*-BuLi in the presence of LiBr, with aldehydes and aliphatic ketones gives diethyl 1,2-epoxyalkylphosphonates in 61–81% overall yields and not the Horner–Wadsworth–Emmons reaction product.[456] These carbanions also react with CO_2 at –100°C to afford diethyl 1-(hydroxycarbonyl)-1-chloroalkylphosphonates in 59–79% yields.[456]

3.2.1.2.7   Miscellaneous

The iodine–lithium exchange reaction of *bis*(2,2,2-trifluoroethyl) iodomethylphosphonate with *t*-BuLi followed by a CO_2 quench gives *bis*(2,2,2-trifluoroethyl) 1-(hydroxycarbonyl)methylphosphonate, which cannot be obtained by hydrolysis of the carboxylic methyl ester.[457]

Tetraethyl iodomethylenediphosphonate, generated *in situ* by electrophilic iodination of tetraethyl methylenediphosphonate, tetramethyl bromomethylenediphosphonate, and dimethyl 1-(alkoxycarbonyl)-1-bromomethylphosphonate, undergoes Michael addition to C_{60} in toluene at room temperature in the presence of NaH or DBU, and the resulting methano[60]fullerenediphosphonates are isolated in 38–41% yields.[458–460] The reaction of tetraalkyl bromomethylenediphosphonates with electron-deficient alkenes as Michael acceptors in refluxing THF in the presence of EtOTl leads to 2-substituted 1,1-cyclopropanediyldiphosphonates in 37–63% yields.[461]

3.2.2   α,α-DIHALOGENOPHOSPHONATES

3.2.2.1   Reactions of Difluoromethylphosphonate Carbanions

3.2.2.1.1   Alkylation

Dialkyl 1-lithio-1,1-difluoromethylphosphonates are usually generated *in situ* at low temperature from difluoromethylphosphonates in the presence of LDA or LiHMDS. Because of their instability (rapid dissociation to give difluorocarbene and lithium diethyl phosphite) and poor nucleophilicity,[152,462–464] these highly deactivated carbanions require very powerful electrophiles in order to obtain satisfactory results.

Primary alkyl bromides[226,228,465,466] and iodides,[462,467] as well as allyl bromide[226] and benzylic chlorides[468] or bromides,[230,469] have been reported to react with dialkyl 1-lithio-1,1-difluoromethylphosphonates at low temperature to give alkylated products (Scheme 3.71, M = Li, X = Br, I) in low to moderate yields (23–66%). The more reactive primary alkyl triflates (Scheme 3.71, M = Li, X = OTf) give better results, and alkyl-,[470] carbohydrate-,[231,402,471–473] amino acid-,[231,474,475] or deoxypurine-derived[476,477] difluoromethylphosphonates are isolated reproducibly in yields of 35–96%. Secondary triflates do not react under the same conditions, apparently because they are too hindered to undergo nucleophilic displacement.[478] In some cases, the presence of HMPA in the reaction medium is crucial to induce the reaction.[462,471,474,475,477]

$$
\begin{array}{c}
(RO)_2P-CF_2Br \\
\quad \parallel \\
\quad O
\end{array}
\searrow
\begin{array}{c}
(RO)_2P-CF_2M \\
\quad \parallel \\
\quad O
\end{array}
\xrightarrow{R^1CH_2-X}
\begin{array}{c}
(RO)_2P-CF_2-CH_2R^1 \\
\quad \parallel \\
\quad O
\end{array}
$$

$$
\begin{array}{c}
(RO)_2P-CF_2H \\
\quad \parallel \\
\quad O
\end{array}
\nearrow
$$

M = Li, MgBr, CdBr, ZnBr
X = Cl, Br, I, OTf, OTs, OAc

(3.71)

The reaction of diethyl 1-lithio-1,1-difluoromethylphosphonate with diphenyl disulfide, diphenyl diselenide, or phenylselenyl chloride affords diethyl 1-(phenylsulfenyl)- and 1-(phenylselenyl)-1,1-difluoromethylphosphonates in 52% and 71–83% yields, respectively. These compounds are good precursors of the difluoromethylphosphonate radical, which can react with various alkenes carrying alkyl, electron-rich and electron-withdrawing groups. This gives the desired 1,1-difluoroalkylphosphonates in moderate to good yields (22–75%).[266]

In an attempt to alleviate the problems associated with the instability of diethyl 1-lithio-1,1-difluoromethylphosphonate, a study of the reactivity of the corresponding organocadmium compound was performed in 1981.[479] It reacts with a variety of allyl, propargyl, and benzyl bromides or iodides in THF at room temperature to afford the alkylated products in 7–64% yields (Scheme 3.71, M = CdBr, X = Br, I), but these compounds are contaminated with variable quantities of protonated carbanion.[479–482]

Treatment of dialkyl 1-bromo-1,1-difluoromethylphosphonate with acid-washed zinc dust or powder in ethereal solvents such as dioxane, THF, monoglyme, or triglyme at temperatures of 20–60°C gives the corresponding organozinc bromide. This compound is significantly more stable than the lithium salt but also less reactive. The addition of catalytic quantities of CuBr greatly increases the reactivity of the zinc reagent, which then reacts with allylic chlorides and bromides to give difluorinated homoallylic phosphonates in 47–83% yields (Scheme 3.71, M = ZnBr, X = Cl, Br).[483,484] These can subsequently be converted into the corresponding aldehydes by ozonolysis, to carboxylic acids by $RuCl_3/KIO_4$ oxidation and 3,4-epoxyphosphonates in the presence of $Hg(OAc)_2/Br_2/KOH$.[463] The reaction with N-(bromomethyl)phthalimide affords 30–54% yields of the corresponding dialkyl 1,1-difluoro-2-(N-phthalimido)ethylphosphonates, which are precursors of 2-amino-1,1-difluoroethylphosphonic acid, a nonhydrolysable analogue of glycerol-3-phosphate.[466,485,486] S_N2' Alkylation with propargylic tosylates and acetates gives allenic difluoromethylphosphonates in good to excellent yields (72–98%, Scheme 3.71, M = ZnBr, X = Cl, Br).[487]

3.2.2.1.2 Coupling Reactions with Aryl, Vinyl, and Alkynyl Halides

The reaction of the cadmium derivative of diethyl 1-bromo-1,1-difluoromethylphosphonate with aryl iodides proceeds smoothly in DMF at room temperature in the presence of CuCl to give diethyl α,α-difluorobenzylphosphonates in excellent yields (70–88%, Scheme 3.72). A variety of functional groups, such as nitro, ether, ester, halide, azo, urethane, as well as free carboxylic acid are tolerated by the reaction.[488–490]

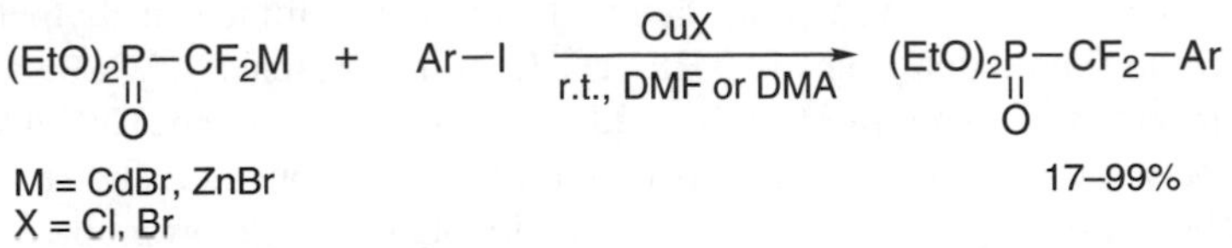

$$(EtO)_2P-CF_2M \quad + \quad Ar-I \xrightarrow[\text{r.t., DMF or DMA}]{CuX} (EtO)_2P-CF_2-Ar$$

$$M = CdBr, ZnBr$$
$$X = Cl, Br$$

$$17\text{–}99\%$$

(3.72)

Variable yields of diethyl α,α-difluorobenzylphosphonates (17–99%) are obtained from the corresponding organozinc reagent and aryl iodides in DMF or DMA at room temperature in the presence of CuBr (Scheme 3.72).[491–497]

It is noteworthy that, under these conditions, the Cu-activated zinc phosphonate and 1,4-diiodobenzene produce diethyl α,α-difluoro-4-iodobenzylphosphonate (37–52%),[491,493,495] a valuable intermediate for the synthesis of (phosphonodifluoromethyl)phenylalanine (F_2Pmp) analogues of phosphotyrosine (pTyr).[258] In contrast, the cadmium derivative afforded exclusively the *bis*-coupling product under similar conditions.[488] As a general rule, the coupling reactions with iodobenzenes possessing an electron-withdrawing substituent give higher yields than with those bearing donor groups. In some cases, the reaction must be conducted under ultrasound irradiation.[491–493,495,498] A stoichiometric amount of CuBr is necessary to induce a good yield. Because the coupling reaction with alkenyl halides proceeds with catalytic amounts of CuBr under similar conditions (*vide infra*),[499] the present mechanism seems to differ from that for the coupling with alkenyl halides. A single electron transfer process has been proposed on the basis of the slow decomposition of the copper(I) derivative in DMF to generate a difluoromethylphosphonate radical.[491,499]

The reaction of the zinc derivative of diethyl 1-bromo-1,1-difluoromethylphosphonate with alkenyl iodides in DMF at room temperature in the presence of CuBr affords diethyl alkenyldifluoromethylphosphonates in good to excellent yields (62–99%) with complete retention of configuration (Scheme 3.73).[498–505] A large variety of (*E*)- and (*Z*)-alkenyl iodides possessing aliphatic substituents in the β-position, nonterminal alkenyl iodides, and β-styryl bromides have been tested with success, although the bromides give somewhat lower yields.[499]

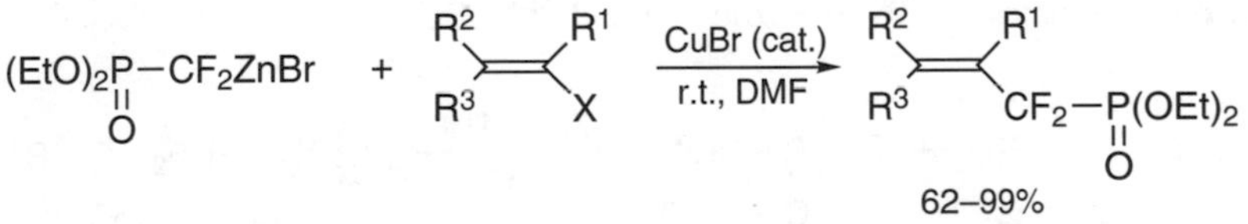

$$(EtO)_2P-CF_2ZnBr \quad + \quad \underset{R^3}{\overset{R^2}{\diagup}}\!\!=\!\!\underset{X}{\overset{R^1}{\diagdown}} \xrightarrow[\text{r.t., DMF}]{CuBr\ (cat.)} \underset{R^3}{\overset{R^2}{\diagup}}\!\!=\!\!\underset{CF_2-P(OEt)_2}{\overset{R^1}{\diagdown}}$$

$$62\text{–}99\%$$

(3.73)

In the presence of 10% of CuBr as catalyst, β-bromostyrene gives the desired β,γ-unsaturated difluoromethylphosphonate in 41% yield. The following mechanism has been proposed (Scheme 3.74). Smooth transmetallation of the zinc derivative of diethyl 1-bromo-1,1-difluoromethylphosphonate with CuBr followed by oxidative addition to β-bromostyrene presumably gives a complex that reductively eliminates CuBr to produce the coupling compound with retention of stereochemistry.[499]

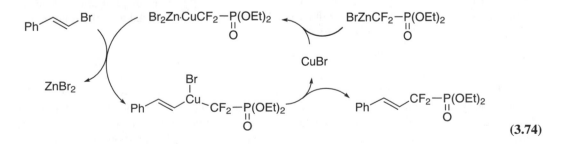

(3.74)

The organozinc reagent derived from diethyl 1-bromo-1,1-difluoromethylphosphonate reacts with various bromo- and iodoalkynes in THF at 5°C in the presence of CuBr to give the corresponding propargylphosphonates in 31–61% yields.[506,507] Analogously, the CuCl/Cd-promoted coupling reaction between diethyl 1-bromo-1,1-difluoromethylphosphonate and (trimethylsilyl)iodoacetylene in DMF at room temperature affords the silylated 1,1-difluoropropargylphosphonate in 67% yield.[265]

3.2.2.1.3   Acylation

The acylation of organozinc reagents derived from dialkyl 1-bromo-1,1-difluoromethylphosphonates with acyl chlorides proceeds smoothly at room temperature to afford the desired products in moderate to good yields (60–77%, Scheme 3.75). Esters, lactones, and triflates did not react at all, even at 60°C.[508–512] Although these zinc and cadmium reagents appear to have similar reactivities, the organozincs provide better yields of dialkyl 2-oxoalkylphosphonates (41%).[479]

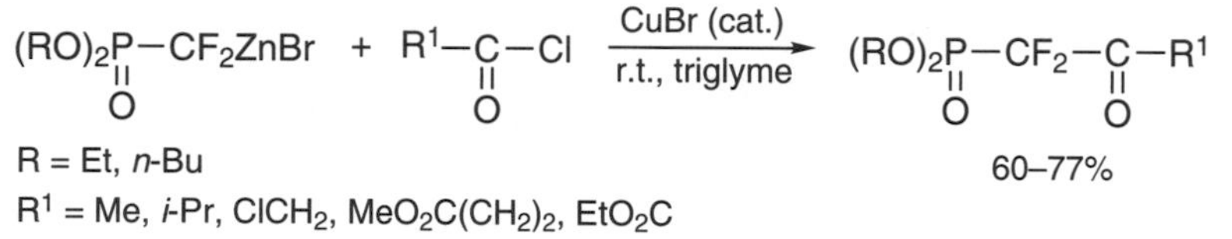

$$(RO)_2\overset{\underset{\|}{O}}{P}-CF_2ZnBr \ + \ R^1-\overset{\underset{\|}{O}}{C}-Cl \ \xrightarrow[\text{r.t., triglyme}]{\text{CuBr (cat.)}} \ (RO)_2\overset{\underset{\|}{O}}{P}-CF_2-\overset{\underset{\|}{O}}{C}-R^1$$

R = Et, n-Bu

R^1 = Me, i-Pr, ClCH$_2$, MeO$_2$C(CH$_2$)$_2$, EtO$_2$C

$\qquad\qquad\qquad\qquad\qquad\qquad\qquad\qquad\qquad\qquad$ 60–77%

$\hfill$ (3.75)

Diethyl 1-lithio-1,1-difluoromethylphosphonate reacts in THF at low temperature with functionalized unactivated esters to give the corresponding diethyl 2-oxo-1,1-difluoroalkylphosphonates in good to excellent yields (60–99%, Scheme 3.76).[478,513–516] The bulkiness of the carboxylic ester substituents seems not to be important because both methyl[478,514,515] and tert-butyl[513] groups can be successfully used. In some cases, the reaction can be performed in the presence of LiAlH$_4$ or MeMgBr, which react with the intermediate β-ketophosphonate to give the corresponding β-hydroxyphosphonates. A subsequent reaction with phenyl chlorothionoformate and methyl oxalyl chloride, followed by radical deoxygenation with n-Bu$_3$SnH, provides a pathway to diethyl 1,1-difluoroalkylphosphonates in 40–50% overall yields.[478]

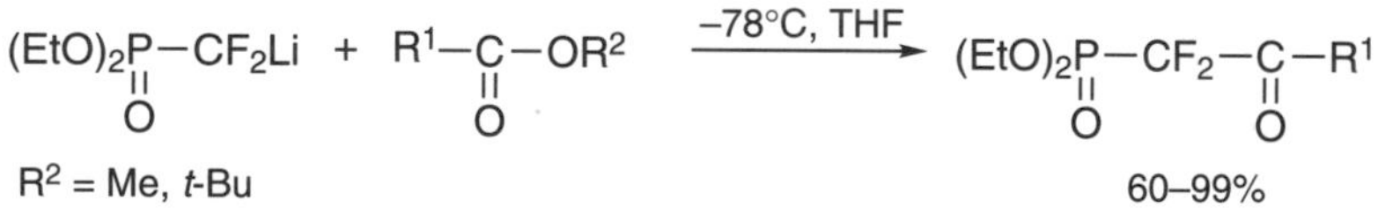

$$(EtO)_2\overset{\underset{\|}{O}}{P}-CF_2Li \ + \ R^1-\overset{\underset{\|}{O}}{C}-OR^2 \ \xrightarrow{-78°C,\ THF} \ (EtO)_2\overset{\underset{\|}{O}}{P}-CF_2-\overset{\underset{\|}{O}}{C}-R^1$$

R^2 = Me, t-Bu

$\qquad\qquad\qquad\qquad\qquad\qquad\qquad\qquad\qquad\qquad$ 60–99%

$\hfill$ (3.76)

A useful variation involves the cerium(III)-mediated reaction of diethyl 1-lithio-1,1-difluoromethylphosphonate with various carboxylic esters. A wide range of substrates, including aliphatic and aromatic compounds, α,β-unsaturated esters, and lactones, can be used to obtain reproducibly high yields of diethyl 2-oxo-1,1-difluoroalkylphosphonates (57–88%).[517,518]

DMF fails to give useful products when exposed to diethyl 1-lithio-1,1-difluoromethylphosphonate. However, in the presence of dry CeCl$_3$ the reaction proceeds smoothly in THF at low temperature and diethyl 1-formyl-1,1-difluoromethylphosphonate is isolated as its hydrate in good yields (77–86%).[517–520]

3.2.2.1.4   Hydroxy- and Alkoxycarbonylation

Bubbling dry carbon dioxide gas at low temperature through a THF solution of diethyl 1-lithio-1,1-difluoromethylphosphonate followed by acidification gives diethyl 1-(hydroxycarbonyl)-1,1-difluoromethylphosphonate in 60–69% yields.[95,396] The reaction of the organozinc reagent derived from diethyl 1-bromo-1,1-difluoromethylphosphonate with ethyl chloroformate at room temperature in the presence of catalytic quantities of CuBr gives diethyl 1-(ethoxycarbonyl)-1,1-difluoromethylphosphonate in 50% yield.[509,521]

3.2.2.1.5 Phosphonylation

The reaction of diethyl 1-lithio-1,1-difluoromethylphosphonate in THF at low temperature with diethyl chlorophosphate[150,226,522] or methyl alkylphosphochloridates[522–524] gives the corresponding difluorinated derivatives in 30–74% yields (Scheme 3.77).

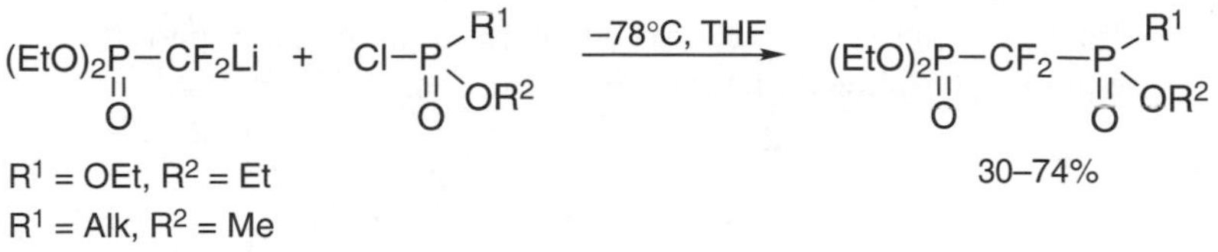

$$(EtO)_2\underset{\underset{O}{\|}}{P}-CF_2Li \;\; + \;\; Cl-\underset{\underset{O}{\|}}{P}\overset{R^1}{\underset{OR^2}{<}} \;\;\xrightarrow{-78°C,\;THF}\;\; (EtO)_2\underset{\underset{O}{\|}}{P}-CF_2-\underset{\underset{O}{\|}}{P}\overset{R^1}{\underset{OR^2}{<}}$$

R^1 = OEt, R^2 = Et
R^1 = Alk, R^2 = Me   30–74%

(3.77)

Tetraalkyl difluoromethylenediphosphonates can be also obtained in modest yields (34–47%) by the reaction of dialkyl 1-bromo-1,1-difluoromethylphosphonates with sodium dialkyl phosphites at 0°C in a formal Michaelis–Becker reaction.[126,282,303,525] The mechanism probably involves an intermediate halogen–metal exchange scrambling reaction to give sodium dialkyl difluoromethylphosphonates and dialkyl bromophosphates, which subsequently react through an SNP(V) mechanism (Scheme 3.78).[282,526,527]

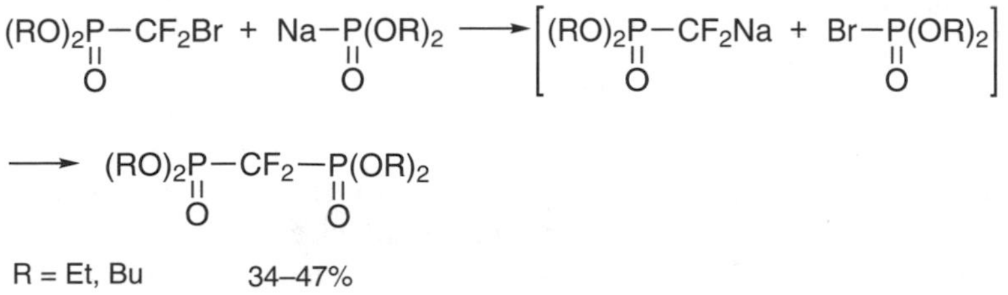

$$(RO)_2\underset{\underset{O}{\|}}{P}-CF_2Br \; + \; Na-\underset{\underset{O}{\|}}{P}(OR)_2 \;\longrightarrow\; \left[(RO)_2\underset{\underset{O}{\|}}{P}-CF_2Na \; + \; Br-\underset{\underset{O}{\|}}{P}(OR)_2\right]$$

$$\longrightarrow \; (RO)_2\underset{\underset{O}{\|}}{P}-CF_2-\underset{\underset{O}{\|}}{P}(OR)_2$$

R = Et, Bu   34–47%

(3.78)

When the phosphonate and the phosphite carry different alkyl ester groups, a mixture of the expected disymmetric and two symmetric diphosphonates is obtained.[126] Two mechanisms have been proposed, one suggesting the intermediacy of difluorocarbene (for the reaction conducted at 0°C)[126,282,526,527] and the second involving the nucleophilic attack of the starting difluoromethylphosphonate carbanion on the difluoromethylenediphosphonate product (for the reaction performed at low temperature).[522]

3.2.2.1.6 Reactions with Carbonyl Compounds

Lithium, magnesium, cerium, and cobalt derivatives of diethyl difluoromethylphosphonate readily react with a large variety of carbonyl compounds to give 2-hydroxy-1,1-difluoroalkylphosphonates in good yields (46–99%, Scheme 3.79) after acidic workup. Aliphatic,[28,277,401,528–531] aromatic,[277,530,532–534] heteroaromatic,[77,277,530] and α,β-unsaturated[277,535] aldehydes as well as cyclic,[277] α,β-unsaturated,[535,536] and ethoxycarbonyl-substituted[484] ketones have been used successfully. The presence of lithium salts in the solution increases the stability of the difluorinated carbanion.[277]

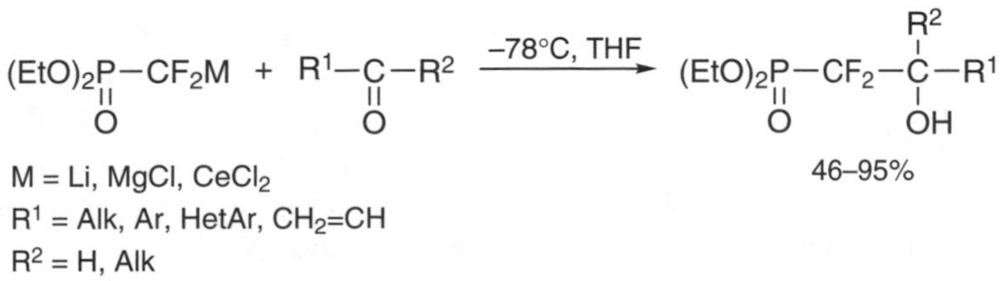

$$(EtO)_2\underset{\underset{O}{\|}}{P}-CF_2M \; + \; R^1-\underset{\underset{O}{\|}}{C}-R^2 \;\xrightarrow{-78°C,\;THF}\; (EtO)_2\underset{\underset{O}{\|}}{P}-CF_2-\underset{\underset{OH}{|}}{\overset{\overset{R^2}{|}}{C}}-R^1$$

M = Li, MgCl, CeCl$_2$
R^1 = Alk, Ar, HetAr, CH$_2$=CH   46–95%
R^2 = H, Alk

(3.79)

The resulting 2-hydroxy-1,1-difluoroalkylphosphonates can be submitted to PDC or Swern oxidation to give 2-oxo-1,1-difluoroalkylphosphonates in 56–94% yields[530] or treated with phenyl chlorothionoformate and then deoxygenated with n-Bu$_3$SnH to produce the reduced 1,1-difluoroalkylphosphonates in good yields (68–89%).[537–539]

Polyhydroxylated pyrrolidines bearing a difluoromethylphosphonate group in the pseudoanomeric position have been prepared in 31–33% overall yields by nucleophilic opening of arabino-, ribo-, and xylofuranosylamines with diethyl 1-lithio-1,1-difluoromethylphosphonate followed by cyclization of the amino phosphonate intermediates (Scheme 3.80).[540,541]

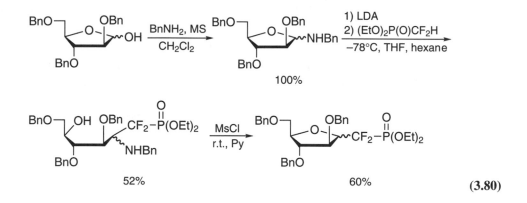

$$(3.80)$$

The diethyl difluoromethylphosphonate carbanion can be also generated *in situ* from diethyl 1-trimethylsilyl-1,1-difluoromethylphosphonate in the presence of catalytic quantities of dry CsF, TBAF, or TBAT and then reacted with aliphatic, aromatic, or heteroaromatic aldehydes and ketones to form 2-hydroxy-1,1-difluoroalkylphosphonates (38–87% yields) after acidic workup of the O-silylated intermediate.[542,543] Under these conditions, the reaction of diethyl 1-trimethylsilyl-1,1-difluoromethylphosphonate with [^{14}C]formaldehyde generates the corresponding silylated 2-hydroxyphosphonate, which, on heating with CsF, gives [^{14}C]-labeled difluoroethylene in a formal Horner–Wadsworth–Emmons reaction (Scheme 3.81).[544] The low reported yields (not exceeding 15%) probably reflect the volatility of the product.

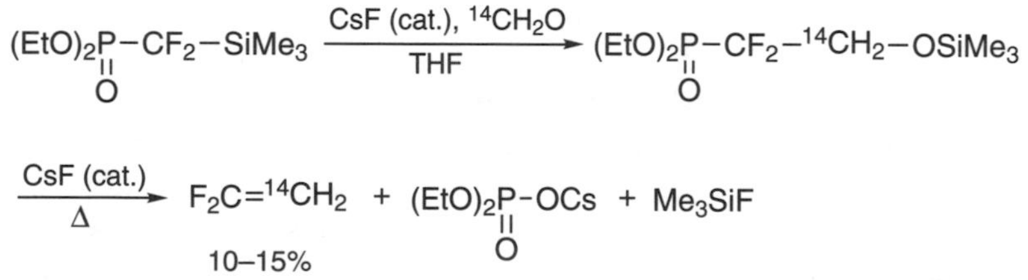

$$(EtO)_2\underset{O}{\overset{||}{P}}-CF_2-SiMe_3 \xrightarrow[THF]{CsF\ (cat.),\ ^{14}CH_2O} (EtO)_2\underset{O}{\overset{||}{P}}-CF_2-{}^{14}CH_2-OSiMe_3$$

$$\xrightarrow[\Delta]{CsF\ (cat.)} \underset{10-15\%}{F_2C={}^{14}CH_2} + (EtO)_2\underset{O}{\overset{||}{P}}-OCs + Me_3SiF$$

$$(3.81)$$

Diethyl difluoromethylphosphonate has been reported to react at −78°C with a large variety of aldehydes and ketones in the presence of LDA. On heating, the corresponding difluoroalkenes are obtained in 30–75% yields along with small amounts of difluorophosphates.[226,545] This report has been questioned in a later publication, where it has been found that difluoromethylphosphates are almost always the major products (48–52% isolated yields) of the reaction, with only traces (less than 3%) of the desired difluoroalkenes being formed (Scheme 3.82).[546]

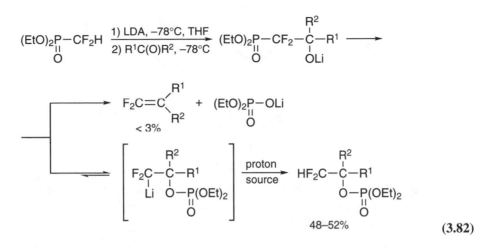

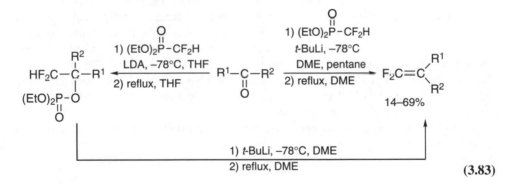

(3.82)

As shown in Scheme 3.83, difluoroalkenes and phosphates are the only products, the latter probably being obtained through the abstraction of a proton from diisopropylamine. In strictly aprotic conditions (use of *t*-BuLi at –78°C instead of LDA), the protonation pathway is suppressed. Thus, diethyl difluoromethylphosphonate can be treated with *t*-BuLi at –78°C in a DME–pentane mixture followed by the carbonyl compound. The pentane is distilled off, and the remaining DME solution is refluxed for 6–13 h. Aqueous workup and a short chromatography on silica gel afford reproducibly the desired difluoroalkenes in 14–69% yield (Scheme 3.83).[546]

(3.83)

In some cases, better yields are obtained by using a two-step process, where the intermediate phosphates are isolated, treated with *t*-BuLi at –65°C, and then refluxed in DME (Scheme 3.83).[546]

3.2.2.1.7 Addition Reactions to Alkenes and Alkynes

The reaction of diethyl 1-lithio-1,1-difluoromethylphosphonate with 4-chloro-β-nitrostyrene to give a 78% yield of the 1,4-addition product, diethyl 3-nitro-2-(4-chlorophenyl)-1,1-difluoropropylphosphonate, was described for the first time in 1991. The product phosphonate can be subsequently reduced using a Raney Ni catalyst in an atmosphere of hydrogen to give the amino derivative, a weak GABA$_B$ agonist.[547]

A series of addition reactions use the more nucleophilic cerium(III)[548–550] or zinc(II)[304,484] derivatives prepared *in situ*. Thus, the organozinc reagent derived from diethyl 1-bromo-1,1-difluoromethylphosphonate reacts smoothly with α-bromoacrylic acid in THF at room temperature in the presence of catalytic quantities of CuI to give the 1,4-addition product in 33% yield. After several transformations, this is converted into diethyl 3-(hydroxycarbonyl)-3-amino-1,1-difluoropropylphosphonate, an analogue of phosphoserine (Scheme 3.84).[304,484]

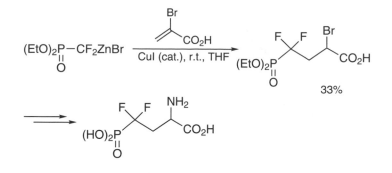

(3.84)

The cerium(III)-mediated conjugate addition of a difluoromethylphosphonate carbanion to a wide range of aliphatic nitroalkenes,[548] acyclic and cyclic vinyl sulfones,[549] and vinyl sulfoxides[550] has been reported recently (Scheme 3.85).

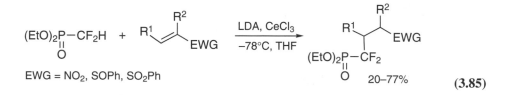

(3.85)

β-Alkyl-substituted nitroalkenes react conveniently with diethyl 1-lithio-1,1-difluoromethylphosphonate in the presence of CeCl₃ to give the corresponding diethyl 3-nitro-1,1-difluoroalkylphosphonates in 25–62% yields, but nitroalkenes derived from aromatic aldehydes react poorly. They require a threefold excess of nucleophile, which is very difficult to remove from the product mixture.[548] Good yields of the desired adducts (43–75%) are obtained from cyclic vinyl sulfones and sulfonates, but the yields drop (5–55%) in the case of aromatic or bulky β-substituents.[549,550] The resulting γ-sulfonyl derivatives are reduced smoothly to diethyl 1,1-difluoroalkylphosphonates with 5% Na(Hg) in MeOH.[549] The corresponding sulfoxides eliminate PhSOH in toluene at reflux to give 1,1-difluoro-2-alkenylphosphonates.[550]

Diethyl 1,1-difluoro-2-alkenylphosphonates have recently been synthesized in moderate yields (23–62%) through a copper(I)-catalyzed addition of zinc[499] or cadmium[551] diethyl difluoromethylphosphonate to terminal[499] or perfluorinated internal[551] acetylenes in DMF.

3.2.2.2 Reactions of Dichloromethylphosphonate Carbanions

Dichloromethylphosphonate carbanions are generally prepared *in situ* from dialkyl trichloromethylphosphonates by a halogen–metal exchange reaction under Barbier conditions, by electrophilic chlorination of a chloromethylphosphonate carbanion, or by a double electrophilic chlorination of methylphosphonate with phenylsulfonyl chloride (2 eq) in the presence of *n*-BuLi (3 eq).[36]

3.2.2.2.1 Alkylation

Dialkyl 1-lithio-1,1-dichloromethylphosphonates, generated *in situ* from dialkyl trichloromethylphosphonate by halogen–metal exchange, from dialkyl dichloromethylphosphonates by deprotonation, or from dialkyl chloromethylphosphonates by electrophilic chlorination (Scheme 3.86), react at low temperature with various alkyl iodides, benzyl, allyl, crotyl bromides, or Me₃SiCl to give dialkyl 1,1-dichloroalkylphosphonates in excellent yields (67–97%).[206,278,448,552–554] When HMPA (1 eq) is added to the reaction mixture, alkylation can be effected with the corresponding alkyl bromides to obtain slightly better yields.[553]

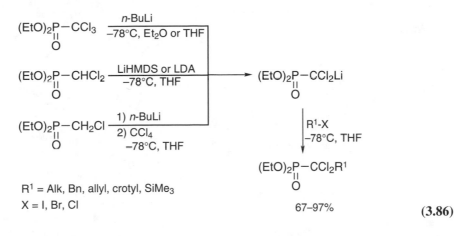

(3.86)

A different approach involving the abstraction of a chlorine atom from diethyl trichloromethylphosphonate by means of $P(NMe_2)_3$ in C_6H_6 at room temperature in the presence of Me_3SiCl gives diethyl 1-(trimethylsilyl)-1,1-dichloromethylphosphonate in 90% yield.[280]

Electrochemical reduction of diethyl trichloromethylphosphonate has been carried out in aprotic medium in the presence of alkyl halides to afford diethyl 1,1-dichloroalkylphosphonates in moderate yields (25–63%).[555] Improved yields (65–78%) have been obtained using a carbon felt cathode and a sacrificial magnesium anode.[556] The electrogenerated dichloromethylphosphonate carbanion exhibits a higher nucleophilicity toward alkyl halides than the corresponding lithiated reagent, which cannot be alkylated by alkyl chlorides. Two factors are responsible for this enhanced reactivity: (a) the temperature is higher in the electrochemical procedure because electrolyses are performed at room temperature; and (b) the absence of a metallic cation provides a more reactive "naked" carbanion.[555]

3.2.2.2.2 Reactions with Carbonyl Compounds
(Horner–Wadsworth–Emmons Reaction)

Diethyl 1-lithio-1,1-dichloromethylphosphonate readily reacts with various aldehydes at low temperature (−100°C) in a THF–Et₂O mixture, then allowed to warm to room temperature. The resulting dichloroalkenes are dehydrohalogenated on treatment with bases such as t-BuOK, LDA, or LiHMDS to give chloroalkynes in good yields (67–92%, Scheme 3.87).[278,557–559] The sequence can be performed in two steps from diethyl trichloromethylphosphonate[557–559] or in one pot from diethyl dichloromethylphosphonate.[278] LiHMDS is well adapted to the conversion of aromatic dichloroolefins into the corresponding chloroalkynes,[278] but it is not strong enough in the case of the aliphatic derivatives, which require the use of LDA or Et₂NLi.[278,557]

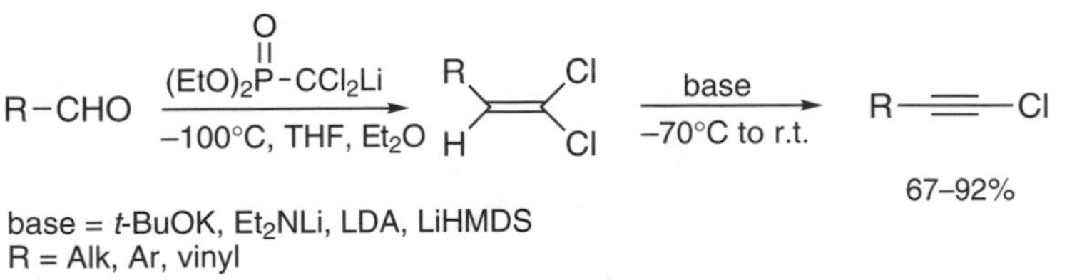

$$R-CHO \xrightarrow[-100°C, \; THF, \; Et_2O]{\underset{}{(EtO)_2\overset{O}{\overset{\|}{P}}-CCl_2Li}} \underset{H}{\overset{R}{>}}C=C\underset{Cl}{\overset{Cl}{<}} \xrightarrow[-70°C \; to \; r.t.]{base} R\equiv\!\!\equiv\!\!-Cl$$

base = t-BuOK, Et₂NLi, LDA, LiHMDS
R = Alk, Ar, vinyl

67–92%

(3.87)

When geminal dichloroalkenes obtained from aldehydes are treated with n-BuLi (2 eq) at low temperature, a Fritsch–Buttenberg–Wiechell rearrangement is observed to give lithiated terminal alkynes,[158] which can be readily hydrolyzed[279,557,560–562] or directly reacted with electrophiles such as alkyl halides,[557,563] trialkylsilyl chlorides,[545,564–567] or aldehydes[568,569] (Scheme 3.88, Table 3.14).

TABLE 3.14

	R	"E+"	E	Yield (%)[Ref.]
a	C_9H_{19}	H_2SO_4 HCl	H	94[557] 73[278]
b	Et_2CH	H_2SO_4	H	83[557]
c	$CH_3CH(Ph)CH_2$	HCl	H	69[278]
d	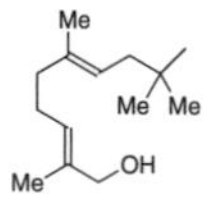 (Me/Me Me / Me / OH structure)	H_2SO_4	H	83[561]
e	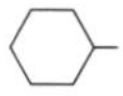 (cyclohexyl-methyl structure)	CH_3OCH_2Cl	CH_3OCH_2	88[557]
f	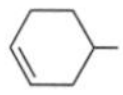 (cyclohexenyl-methyl structure)	CH_3OCH_2Cl	CH_3OCH_2	88[557]
g	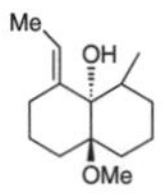 (Me/OH/Me/OMe decalin structure)	$ClCO_2Me$	CO_2Me	67[569]
h	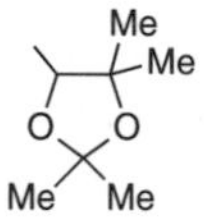 (dioxolane Me Me / Me Me structure)	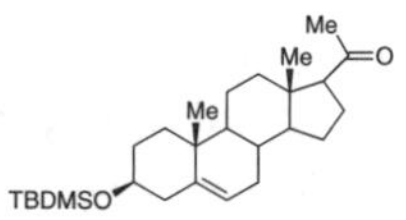 (steroid ketone, TBDMSO structure)	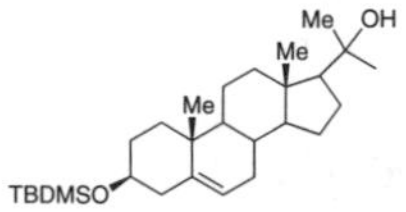 (steroid OH, TBDMSO structure)	53[568]
i	 (iodo-phenyl structure)	Me_3SiCl	Me_3Si	87[545]
j	$CH_3-CH=CH$	CH_3OCH_2Cl	CH_3OCH_2	76[557]
k	$EtS-CH=CH$	H_2SO_4	H	78[557]
l	$Ph-CH=CH$	H_2SO_4	H	57[557]
m	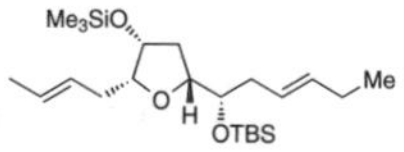 (Me_3SiO / OTBS tetrahydrofuran structure)	NH_4Cl	H	98[562]
n	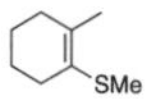 (SMe cyclohexenyl structure)	MeI	Me	67[563]
o	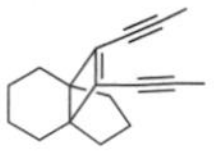 (diyne bicyclic structure)	Me_3SiCl TIPSCl	Me_3Si TIPS	78[566,567] 47[566,567]
p	Ph	HCl	H	57[278]
q	$2\text{-MeO-}C_6H_4$	H_2SO_4 HCl	H	91[557] 72[278]
r	$4\text{-MeO-}C_6H_4$	H_2SO_4 HCl	H	86[557] 63,[a][279] 84[278]
s	$2\text{-Cl-}C_6H_4$	H_2SO_4	H	73[557]
t	$4\text{-Cl-}C_6H_4$	H_2SO_4	H	83[557]
u	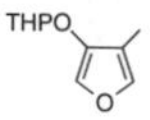 (THPO furan structure)	Me_3SiCl	Me_3Si	71[564,565]

[a] Overall yield (2 steps).

The migrating ability decreases from hydrogen to aryl, then to alkyl functions. In the last case, only low yields (<10%) of aliphatic alkynes are obtained.[570]

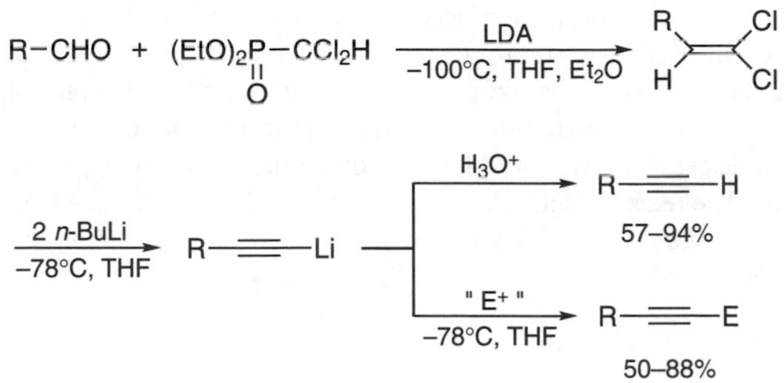

$$(3.88)$$

The reaction of 1-aryl-2,2-dichloroalkenes with n-BuLi at low temperature proceeds by a similar mechanism. A wide range of aromatic alkynes can be synthesized in excellent yields (86–97%) by this method (Scheme 3.89).[571]

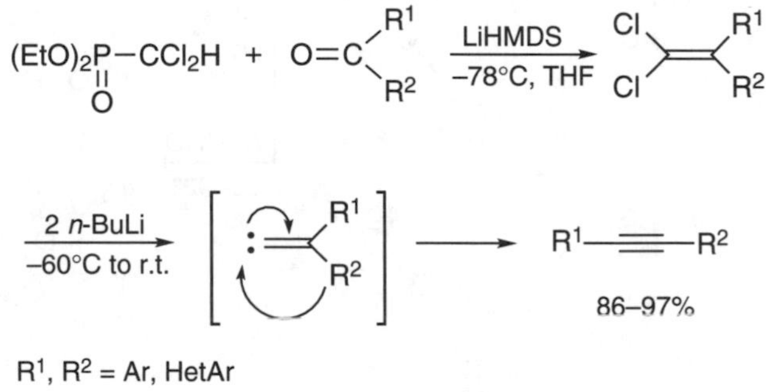

$R^1, R^2 = Ar, HetAr$

$$(3.89)$$

3.2.2.3 Miscellaneous

The addition of diethyl 1-bromo-1,1-difluoromethylphosphonate to various electron-deficient alkenes initiated by a cobaloxime(III)/Zn bimetallic system gives the corresponding diethyl difluoroalkylphosphonates (1:1 Michael-type adducts) in moderate to good yields (34–72%). Cyclic α,β-unsaturated ketones and electron-rich alkenes produce somewhat lower yields of addition products (18–58%), along with small quantities of diethyl difluoromethylphosphonate. A radical mechanism has been proposed for this process.[572]

Benzaldehyde and hexanal react with diethyl iodomethylphosphonate (2 eq) in the presence of 2 eq of KHMDS to give the corresponding iodoacetylenes in 85% and 58% overall yields, respectively, in a one-pot process via the diiodoalkenes.[573]

3.2.3 α,α,α-TRIHALOGENOPHOSPHONATES

3.2.3.1 Monodehalogenation Reactions

Dialkyl trihalogenomethylphosphonates are the most used precursors of dialkyl dihalogenomethylphosphonate anions, which can be hydrolyzed to isolate the corresponding reduced phosphonates (see Section 3.1.2.8) or reacted *in situ* with a large variety of electrophiles (see Sections 3.2.2.1–2).

3.2.3.2 Autocondensation (Self-Trapping Reactions)

Diethyl 1,1-dibromo-1-fluoromethylphosphonate reacts with *n*-BuLi at low temperature in the presence of Me_3SiCl to give quantitatively dietyl 1-lithio-1-(trimethylsilyl)-1-fluoromethylphospho-nate, which is further reacted with a variety of electrophiles (see Section 3.2.1.2.2). In the absence of Me_3SiCl, the addition of diethyl 1,1-dibromo-1-fluoromethylphosphonate to a THF solution of *n*-BuLi at –78°C affords, after acidic workup, tetraethyl fluoromethylenediphosphonate in 95% yield (Scheme 3.90).[574] The intermediate fluoromethylenediphosphonate anion can also be treated with powerful electrophiles such as MeI or C_2Cl_6 to provide the corresponding substituted deriva-tives in very high yields. The reaction with aromatic aldehydes (2 eq) allows the isolation of diethyl 1-fluoro-1-alkenylphosphonates in 83–92% yields with greater selectivities than those obtained by the Peterson reaction (Scheme 3.90).[574]

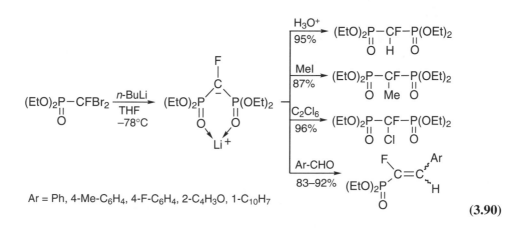

Ar = Ph, 4-Me-C$_6$H$_4$, 4-F-C$_6$H$_4$, 2-C$_4$H$_3$O, 1-C$_{10}$H$_7$

(3.90)

Similarly, tetraalkyl fluoromethylenediphosphonates have been obtained in yields up to 40% from dialkyl 1-chloro-1-fluoromethylphosphonates in the presence of LDA[152] or *n*-BuLi[575] at low temperature. The lithiation of diethyl trichloromethylphosphonate with *n*-BuLi at –100°C gives a stable carbanion, but at temperatures higher than –80°C the lithiation is accompanied by sponta-neous reactions leading exclusively to tetraethyl 1-lithio-1-chloromethylenediphosphonate, which can be protonated (95% yield) or reacted with aldehydes and ketones to generate diethyl 1-chloro-1-alkenylphosphonates in high yields (59–92%) in a single step.[576–578]

Two mechanisms have been proposed for this transformation (Scheme 3.91).[577] Both begin with a halogen–metal exchange reaction of trihalogenomethylphosphonate with *n*-BuLi. In the first mechanism, the resulting carbanion undergoes fragmentation to lithium diethyl phosphite and dihalocarbene (path a), followed by the reaction between the new nucleophile and the starting trihalogenomethylphosphonate (path b). The resulting dihalogenomethylenediphosphonate is next dehalogenated by *n*-BuLi (path c). The second mechanism involves the displacement of CX_3^- from trihalogenomethylphosphonate by its carbanionic derivative (path d), then the dehalogenation (path c). The nucleophilic displacement (path d) favored by the presence of the good leaving group CX_3 seems to be more plausible than the reaction of lithium diethyl phosphite with trihalogenom-ethylphosphonate (path b). Additionally, when the reaction is realized in the presence of an olefin, the corresponding isomeric 1,1-dihalogenocyclopropanes are formed in a 2.4:1 ratio, indicating the intermediacy of a carbenoid rather than a true carbene (Scheme 3.91).[176,579]

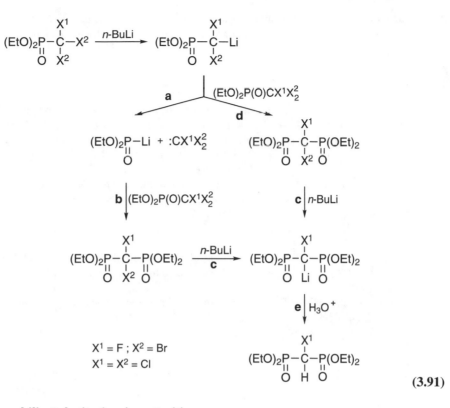

$$X^1 = F \; ; \; X^2 = Br$$
$$X^1 = X^2 = Cl$$

(3.91)

3.2.3.3 Nucleophilic Substitution in α-Position

Diethyl 1,1-dibromo-1-fluoromethylphosphonate reacts with $Na_2S_2O_4$ in the presence of $NaHCO_3$ in a $H_2O/MeCN$ system to give the expected sulfinate in 85% yield.[251,580]

3.2.4 REACTIONS OF THE PHOSPHONYL GROUP

Most of the procedures describing the conversion of halogenoalkylphosphonates into the corresponding phosphonic acids use Me_3SiBr neat[581–583] or Me_3SiCl/NaI in MeCN,[584,585] followed by treatment with MeOH or H_2O. The reaction tolerates substituents such as CCl_3, CH_2Br, or CH_2I and proceeds smoothly at room temperature, and the yields obtained are quantitative (Scheme 3.92).

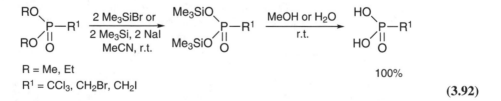

R = Me, Et
$R^1 = CCl_3, CH_2Br, CH_2I$

100%

(3.92)

Labile *tert*-butyl ester groups of phosphonates are stable to electrophilic fluorination but require an alkaline pH during quenching and workup. The main advantage of this methodology is facile deprotection of the phosphonate under exposure to mild acid conditions.[255,262] A complementary approach involves the $Pd(OH)_2/C$-catalyzed hydrogenation of dibenzyl 1,1-difluoroalkylphosphonates to give the corresponding phosphonic acids in quantitative yields.[475]

Treatment of dialkyl 1-halogenoalkylphosphonates with PCl_5[586] or PCl_5/PCl_3[587] at 90–110°C affords the corresponding phosphonic dichlorides in 56% and 78–89% yields, respectively, whereas alkyl phosphonochloridates are obtained on treatment with $POCl_3$ at 70°C (70–85% yields).[588–591] Chlorinated and brominated substituents are tolerated, and no halogen exchange is observed.

REFERENCES

1. Arbuzov, A.E., and Kushkova, N.P., Action of dihalo hydrocarbons on ethyl phosphite and the salts of diethyl phosphite, *Zh. Obshch. Khim.*, 6, 283, 1936; *Chem. Abstr.*, 30, 4813h, 1936.
2. Kabachnik, M.I., and Medved, T.Y., Organophosphorus compounds. Part 14. Synthesis of aminophosphonic acids. Part 1, *Izv. Akad. Nauk SSSR, Ser. Khim.*, 635, 1950; *Chem. Abstr.*, 45, 8444b, 1951.
3. Crofts, P.C., and Kosolapoff, G.M., The preparation and the determination of apparent dissociation constants of some substituted aliphatic phosphonic acids, *J. Am. Chem. Soc.*, 75, 5738, 1953.
4. Kosolapoff, G.M., The chemistry of aliphatic phosphonic acids. Part 1. Alkylation of methanediphosphonic acid, *J. Am. Chem. Soc.*, 75, 1500, 1953.
5. Gold, A.M., Synthesis of a series of organophosphorus esters containing alkylating groups, *J. Org. Chem.*, 26, 3991, 1961.
6. Harger, M.J.P., and Williams, A., Reactions of *N*-phenyl α-halogenophosphonamidates with alkoxide. Migration of the anilino group from phosphorus to the α carbon atom, *J. Chem. Soc., Perkin Trans. 1,* 1681, 1986.
7. Balczewski, P., and Mikolajczyk, M., Free radical reaction of α-haloalkylphosphonates with alkenes and alkynes. A new approach to modified phosphonates, *Synthesis*, 392, 1995.
8. Ford-Moore, A.H., and Williams, J.H., Reaction between trialkyl phosphites and alkyl halides, *J. Chem. Soc.*, 1465, 1947.
9. Rowley, G.L., Greenleaf, A.L., and Lenyon, G.L., Specificity of creatine kinase. New glycocyamines and glycocyamine analogs related to creatine, *J. Am. Chem. Soc.*, 93, 5542, 1971.
10. Cade, J.A., Methylenediphosphonates and related compounds. Part 1. Synthesis from methylene halides, *J. Chem. Soc.*, 2266, 1959.
11. Etemad-Moghadam, G., and Seyden-Penne, J., Stereoselective synthesis of *E* or *Z* α-methyl α,β-ethylenic esters from phosphonates or phosphine oxides by the Wittig–Horner reaction, *Tetrahedron*, 40, 5153, 1984.
12. Ezquerra, J., Yruretagoyena, B., Moreno-Manas, M., and Roglans, A., Improved preparation of diethyl bromomethylphosphonate and diiodomethane-catalyzed triethylphosphite Michaelis–Arbuzov isomerization, *Synth. Commun.*, 25, 191, 1995.
13. Göbel, R., Richter, F., and Weichmann, H., Synthesis and reactivity of methylene bridged diphosphonyl compounds, *Phosphorus, Sulfur Silicon Relat. Elem.*, 73, 67, 1992.
14. Kvicala, J., Vlaskova, R., Plocar, J., Paleta, O., and Pelter, A., Preparation of intermediates for fluorinated ligands by conjugated and tandem additions on 3-fluorofuran-2(5*H*)-one, *Collect. Czech. Chem. Commun.*, 65, 772, 2000.
15. Etemad-Moghadam, G., and Seyden-Penne, J., Stereoselective synthesis of *E*-α-fluoro α,β-unsaturated esters by Wittig–Horner reaction from methyl α-(*O*,*O*-diethylphosphono)-α-fluoroacetate. Comparison with methyl α-(diphenylphosphinyl)-α-fluoroacetate, *Bull. Soc. Chim. Fr.*, 448, 1985.
16. Blade, R.J., Peek, R.J., and Cockerill, G.S., Preparation of (quinolinyloxy)dodecadienamides and related compounds as pesticides, U.S. Patent Appl. US 5114940, 1992; *Chem. Abstr.,* 118, 101822, 1993.
17. Grison, C., Boulliung, N., and Coutrot, P., Preparation of 2-fluoroacrylate esters from fluorophosphonoacetate esters and formaldehyde, *Elf Atochem*, Eur. Patent Appl. EP 521752, 1993; *Chem. Abstr.,* 119, 160985, 1993.
18. Thenappan, A., and Burton, D.J., Reduction-olefination of esters. A new and efficient synthesis of α-fluoro α,β-unsaturated esters, *J. Org. Chem.*, 55, 4639, 1990.
19. Tsai, H.-J., Thenappan, A., and Burton, D.J., A one-pot synthesis of unsymmetrical and symmetrical tetrasubstituted α-fluoro-α,β-unsaturated esters, *Phosphorus, Sulfur Silicon Relat. Elem.*, 105, 205, 1995.
20. Tsai, H.-J., Application of fluorocarbethoxy-substituted phosphonate. A facile entry to substituted 2-fluoro-3-oxoesters, *Phosphorus, Sulfur Silicon Relat. Elem.*, 126, 1, 1997.

21. Biller, S.A., and Magnin, D.R., Preparation of α-phosphonoalkanoates as squalene synthetase inhibitors, *Bristol-Myers Squibb*, U.S. Patent Appl. US 5312814, 1994; *Chem. Abstr.,* 122, 10300, 1995.

22. Machleidt, H., and Wessendorf, R., Organic fluorine compounds. Part 4. Carbonyl fluoroolefination, *Justus Liebigs Ann. Chem.,* 674, 1, 1964.

23. Machleidt, H., Wessendorf, R., and Strehlke, G., Lower alkyl esters of β-cyclocitrylidenefluoroacetic acid and β-ionylidenefluoroacetic acid, *Olin Mathieson Chem.,* U.S. Patent Appl. US 3277147, 1966; *Chem. Abstr.,* 66, 55609, 1967.

24. Elkik, E., and Imbeaux, M., A convenient synthesis of ethyl (diethoxyphosphoryl)fluoroacetate from ethyl fluoroacetate, *Synthesis,* 861, 1989.

25. Milner, D.J., Spinney, M.A., and Robson, M.J., Preparation of hydroxymethylbenzyl cyclopropane-carboxylate esters as insecticides, *Imperial Chemical Industries*, Eur. Patent Appl. EP 381317, 1990; *Chem. Abstr.,* 114, 82191, 1991.

26. Thenappan, A., and Burton, D.J., Alkylation of (fluorocarbethoxymethylene)-tri-*n*-butylphosphorane. A facile entry to α-fluoroalkanoates, *J. Org. Chem.,* 55, 2311, 1990.

27. Machleidt, H., and Strehlke, G., Organic fluorine derivatives. Part 8. Syntheses and reactions of new fluoro vitamin A acid esters, *Justus Liebigs Ann. Chem.,* 681, 21, 1965.

28. Cox, J.M., Bellini, P., Barrett, R., Ellis, R.M., and Hawkes, T.R., Preparation of triazolylalkylphosphonic acids as herbicides, *Zeneca*, U.S. Patent Appl. WO 9315610, 1993; *Chem. Abstr.,* 120, 134813, 1994.

29. Cox, J.M., Hawkes, T.R., Bellini, P., Ellis, R.M., Barrett, R., Swanborough, J.J., Russell, S.E., Walker, P.A., Barnes, N.J., Knee, A.J., Lewis, T., and Davies, P.R., The design and synthesis of inhibitors of imidazoleglycerol phosphate dehydratase as potential herbicides, *Pestic. Sci.,* 50, 297, 1997.

30. Dawson, M.I., Chan, R., Hobbs, P.D., Chao, W.-R., and Schiff, L.J., Aromatic retinoic acid analogues. Part 2. Synthesis and pharmacological activity, *J. Med. Chem.,* 26, 1282, 1983.

31. Dawson, M.I., and Hobbs, P.D., Aromatic retinoic acid analogues, *SRI International*, U.S. Patent Appl. US 4532343, 1985; *Chem. Abstr.,* 101, 192249, 1984.

32. Gryszkiewicz-Trochimowski, E., Fluoromethanephosphonic acid and some of its derivatives, *Bull. Soc. Chim. Fr.,* 4289, 1967.

33. Fokin, A.V., Zimin, V.I., Studnev, Y.N., and Rapkin, A.I., Reactions of fluorochloroacetic acid derivatives with nucleophilic reagents, *Zh. Org. Khim.,* 7, 249, 1971; *Chem. Abstr.,* 74, 124736, 1971.

34. Hormi, O.E.O., Pajunen, E.O., Avall, A.-K.C., and Pennanen, P., A cheap one-pot approach to tetraethyl methylenediphosphonate, *Synth. Commun.,* 20, 1865, 1990.

35. Kinnear, A.M., and Perren, E.A., Formation of organophosphorus compounds by the reaction of alkyl chlorides with phosphorus trichloride in the presence of aluminum chloride, *J. Chem. Soc.,* 3437, 1952.

36. Waschbüsch, R., Carran, J., Marinetti, A., and Savignac, P., The synthesis of dialkyl α-halogenated methylphosphonates, *Synthesis,* 727, 1997.

37. Waschbüsch, R., Carran, J., Marinetti, A., and Savignac, P., Synthetic applications of dialkyl (chloromethyl)phosphonates and *N,N,N',N'*-tetraalkyl(chloromethyl)phosphonic diamides, *Chem. Rev.,* 97, 3401, 1997.

38. Kabachnik, M.I., and Shepeleva, E.S., Reaction of aldehydes with halogen derivatives of trivalent phosphorus, *Dokl. Akad. Nauk SSSR,* 75, 219, 1950; *Chem. Abstr.,* 45, 6569, 1951.

39. Kabachnik, M.I., and Shepeleva, E.S., Organophosphorus compounds. Part 15. Reaction of formaldehyde with phosphorus trichloride, *Izv. Akad. Nauk SSSR, Ser. Khim.,* 185, 1951; *Chem. Abstr.,* 45, 10191a, 1951.

40. Korshak, V.V., Gribova, I.A., and Andreeva, M.A., Organophosphorus polymers. Part 1. Polymerization of cyclic esters of alkyl- and arylphosphonic acids, *Izv. Akad. Nauk SSSR, Ser. Khim.,* 631, 1957; *Chem. Abstr.,* 51, 14621g, 1957.

41. McConnell, R.L., McCall, M.A., and Coover, H.W., Jr., Chloroalkyl and chloroaryl (chloromethyl)phosphonates, *J. Org. Chem.,* 22, 462, 1957.

42. Hudson, R.F., and Moss, G.E., Mechanism of hydrolysis of phosphonochloridates and related compounds. Part 5. Inductive effect, *J. Chem. Soc.,* 1040, 1964.

43. Kabachnik, M.I., Medved, T.Y., Polikarpov, Y.M., and Yudina, K.S., Synthesis of polyphosphine oxides, *Izv. Akad. Nauk SSSR, Ser. Khim.,* 591, 1967; *Chem. Abstr.,* 68, 39743, 1968.

44. Kabachnik, M.I., and Shepeleva, E.S., Reaction of benzaldehyde with phosphorus trichloride, *Izv. Akad. Nauk S.S.S.R., Otdel Khim. Nauk,* 39, 1950; *Chem. Abstr.,* 44, 7257f, 1950.

45. Knunyants, I.L., Pervova, E.Y., and Tyuleneva, V.V., Esters of perfluoroalkenylphosphonic acids, *Dokl. Akad. Nauk SSSR*, 129, 576, 1959; *Chem. Abstr.*, 54, 7536e, 1960.

46. Knunyants, I.L., and Pervova, E.Y., Reaction of triethyl phosphite with α,β-dichloro-ω-iodoperfluoroalkanes, *Izv. Akad. Nauk SSSR, Otd. Khim. Nauk*, 1409, 1962; *Chem. Abstr.*, 58, 2468h, 1963.

47. Knunyants, I.L., Sterlin, R.N., Tyuleneva, V.V., and Pinkina, L.N., Pseudohaloid properties of perfluoroalkenyl radicals in esters of perfluoroalkenylphosphonic acids, *Izv. Akad. Nauk SSSR, Otd. Khim. Nauk*, 1123, 1963; *Chem. Abstr.*, 59, 8784d, 1963.

48. Knunyants, I.L., Tyuleneva, V.V., Pervova, E.Y., and Sterlin, R.N., Quasiphosphonium compounds from triethyl phosphite and perfluorolefins, *Izv. Akad. Nauk SSSR, Ser. Khim.*, 1797, 1964; *Chem. Abstr.*, 62, 2791c, 1965.

49. Furin, G.G., Phosphorus-containing nucleophiles in reactions with polyfluorinated organic compounds, *Usp. Khim.*, 62, 267, 1993; *Russ. Chem. Rev.*, 62, 243, 1993.

50. Dittrich, R., and Hägele, G., Michaelis–Arbuzov perhalogenation reaction of olefins. Part 3. The trifluorovinyl halide, $CF_2{:}CFX$, *Phosphorus Sulfur*, 10, 127, 1981.

51. von Allwörden, U., and Röschenthaler, G.V., New phosphorus derivatives of perfluoropropene, *Chem.-Ztg.*, 109, 81, 1985.

52. von Allwörden, U., and Röschenthaler, G.V., Phosphorus derivatives of perfluoropropene, *Chem.-Ztg.*, 112, 69, 1988.

53. Wessolowski, H., Röschenthaler, G.-V., Winter, R., and Gard, G.L., [(*E*)-1,2-Difluoro-2-(pentafluoro-λ^6-sulfanyl)ethenyl]phosphonates, *Z. Naturforsch. (B)*, 46, 126, 1991.

54. Wessolowski, H., Röschenthaler, G.-V., Winter, R., and Gard, G.L., [2,2-Difluoro-1-(pentafluoro-λ^6-sulfanyl)]ethylene and 1,1,3,3,3-pentafluoropropene. Reactions with diethyl(trimethylsilyl) and *tris*(trimethylsilyl)phosphite, *Phosphorus, Sulfur Silicon Relat. Elem.*, 60, 201, 1991.

55. Wessolowski, H., Gard, G.L., and Röschenthaler, G.-V., Novel perfluoroalkenylphosphonates and iodoperfluoroalkenes from 3,3-*bis*(trifluoromethyl)-1,1,2,4,4,4-hexafluoro-1-butylene and nonafluoro-*n*-butoxy-1,1,2-trifluoroethylene, *J. Fluorine Chem.*, 80, 149, 1996.

56. Wozniak, L., and Chojnowski, J., Silyl esters of phophorus — common intermediates in synthesis, *Tetrahedron*, 45, 2465, 1989.

57. Dietrich, P., Engler, G., Gross, U., Lunkwitz, K., Prescher, D., and Schulze, J., Straight-chain perfluorinated olefin derivatives, *Akademie der Wissenschaften der DDR*, German Patent Appl. DE 2808754, 1978; *Chem. Abstr.*, 90, 40551, 1979.

58. Shi, G.-Q., and Cao, Z.-Y., Ethyl 2-benzyloxy-3,3-difluoropropenoate as a novel synthon of β-fluoro-α-keto acid derivatives. Preparation and reactions with nucleophiles, *J. Chem. Soc., Chem. Commun.*, 1969, 1995.

59. Huang, X.-H., He, P.-Y. and Shi, G.-Q., Highly stereoselective addition-elimination reaction of nucleophiles with ethyl 3,3-difluoro-2-[(trimethylsilyl)methyl]propenoate, *J. Org. Chem.*, 65, 627, 2000.

60. Eymery, F., Iorga, B., and Savignac, P., Synthesis of phosphonates by nucleophilic substitution at phosphorus. The SNP(V) reaction, *Tetrahedron*, 55, 13109, 1999.

61. Appell, R.B., Straightforward and scalable synthesis of diethyl fluoro(phenylsulfonyl)methylphosphonate, *Synth. Commun.*, 25, 3583, 1995.

62. McCarthy, J.R., Matthews, D.P., Edwards, M.L., Stemerick, D.M., and Jarvi, E.T., A new route to vinyl fluorides, *Tetrahedron Lett.*, 31, 5449, 1990.

63. McCarthy, J.R., Matthews, D.P., Stemerick, D.M., Huber, E.W., Bey, P., Lippert, B.J., Snyder, R.D., and Sunkara, P.S., Stereospecific method to *E* and *Z* terminal fluoro olefins and its application to the synthesis of 2′-deoxy-2′-fluoromethylene nucleosides as potential inhibitors of ribonucleoside diphosphate reductase, *J. Am. Chem. Soc.*, 113, 7439, 1991.

64. McCarthy, J.R., Matthews, D.P., and Paolini, J.P., Stereoselective synthesis of 2,2-disubstituted 1-fluoroalkenes. (*E*)-{[Fluoro(2-phenylcyclohexylidene)methyl]sulfonyl}benzene and (*Z*)-[2-(fluoromethylene)cyclohexyl]benzene, *Org. Synth.*, 72, 216, 1993.

65. Gross, R.S., Mehdi, S., and McCarthy, J.R., A stereoselective method to (*E*)- and (*Z*)-fluorovinyl phosphonates utilizing palladium(0) coupling methodology, *Tetrahedron Lett.*, 34, 7197, 1993.

66. Matthews, D.P., Gross, R.S., and McCarthy, J.R., A new route to 2-fluoro-1-olefins utilizing a synthetic equivalent for the 1-fluoroethene anion, *Tetrahedron Lett.*, 35, 1027, 1994.

67. McCarthy, J.R., Huber, E.W., Le, T.-B., Laskovics, M.F., and Matthews, D.P., Stereospecific synthesis of 1-fluoro olefins via (fluorovinyl)stannanes and an unequivocal NMR method for the assignment of fluoro olefin geometry, *Tetrahedron*, 52, 45, 1996.

68. Matthews, D.P., Bitonti, A.J., Edwards, M.L., and McCarthy, J.R., Treatment of carcinoma by administration of 2'-halomethylidenyl-2'-deoxynucleosides, *Merrell Pharmaceuticals*, U.S. Patent Appl. US 5607925, 1997; *Chem. Abstr.*, 126, 251359, 1997.

69. van der Donk, W.A., Gerfen, G.J., and Stubbe, J., Direct EPR spectroscopic evidence for an allylic radical generated from (E)-2'-fluoromethylene-2'-deoxycytidine 5'-diphosphate by *E. coli* ribonucleotide reductase, *J. Am. Chem. Soc.*, 120, 4252, 1998.

70. Lee, J.W., and Oh, D.Y., A convenient one-pot synthesis of α-functionalized α,β-unsaturated sulfones, *Synth. Commun.*, 20, 273, 1990.

71. Iorga, B., and Savignac, P., Controlled monohalogenation of phosphonates. Part 4. Selective synthesis of monohalogenomethylenediphosphonates, *J. Organomet. Chem.*, 624, 203, 2001.

72. Yokomatsu, T., Yamagishi, T., Matsumoto, K., and Shibuya, S., Stereocontrolled synthesis of hydroxymethylene phosphonate analogues of phosphorylated tyrosine and their conversion to monofluoromethylene phosphonate analogues, *Tetrahedron*, 52, 11725, 1996.

73. Bergmann, E.D., Shahak, I., and Appelbaum, J., Synthesis and reactions of diethyl fluorobenzyl phosphonate, *Isr. J. Chem.*, 6, 73, 1968.

74. Middleton, W.J., New fluorinating reagents. Dialkylaminosulfur fluorides, *J. Org. Chem.*, 54, 574, 1975.

75. Blackburn, G.M., and Kent, D.E., A novel synthesis of α- and γ-fluoroalkylphosphonates, *J. Chem. Soc., Chem. Commun.*, 511, 1981.

76. Blackburn, G.M., and Kent, D.E., Synthesis of α- and γ-fluoroalkylphosphonates, *J. Chem. Soc., Perkin Trans. 1*, 913, 1986.

77. Sethi, R., and Haque, W., Pyridoxal-5'-phosphate, pyridoxal, pyridoxine, pyridoxamine, and related analogs for the treatment of cerebrovascular disease, and preparation of compounds, *Medicure*, Int. Patent Appl. WO 2001072309, 2001; *Chem. Abstr.*, 135, 267243, 2001.

78. Taylor, W.P., Zhang, Z.-Y., and Widlanski, T.S., Quiescent affinity inactivators of protein tyrosine phosphatases, *Bioorg. Med. Chem.*, 4, 1515, 1996.

79. Tsai, H.-J., Synthesis of phenyl substituted fluoro-olefins, *Tetrahedron Lett.*, 37, 629, 1996.

80. Tsai, H.-J., and Burton, D.J., Synthesis of phenyl and ester substituted vinyl fluorides via reduction and olefination of esters, *Phosphorus, Sulfur Silicon Relat. Elem.*, 140, 135, 1998.

81. Tsai, H.-J., Lin, K.-W., Ting, T.-H., and Burton, D.J., A general and efficient route for the preparation of phenyl-substituted vinyl fluorides, *Helv. Chim. Acta*, 82, 2231, 1999.

82. Caplan, N.A., Pogson, C.I., Hayes, D.J., and Blackburn, G.M., The synthesis of novel biphosphonates as inhibitors of phosphoglycerate kinase (3-PGK), *J. Chem. Soc., Perkin Trans. 1*, 421, 2000.

83. Burke, T.R.S., Jr., Smyth, M.S., and Lim, B.B., Phosphonoalkyl phenylalanine compounds suitably protected for use in peptide synthesis, *United States Dept. of Health and Human Services*, U.S. Patent Appl. US 5475129, 1995; *Chem. Abstr.*, 124, 203096, 1996.

84. Burke, T.R., Jr., Smyth, M.S., Nomizu, M., Otaka, A., and Roller, P.P., Preparation of fluoro- and hydroxy-4-(phosphonomethyl)-D,L-phenylalanine suitably protected for solid-phase synthesis of peptides containing hydrolytically stable analogues of *O*-phosphotyrosine, *J. Org. Chem.*, 58, 1336, 1993.

85. Beers, S.A., Malloy, E.A., Wei, W., Wachter, M.P., Gunnia, U., Cavender, D., Harris, C., Davis, J., Brosius, R., Pellegrino-Gensey, J.L., and Siekierka, J., Nitroarylhydroxymethylphosphonic acids as inhibitors of CD45, *Bioorg. Med. Chem.*, 5, 2203, 1997.

86. Hammond, G.B., and deMendonca, D.J., An efficient synthesis of (Z)-γ-fluoroallylphosphonates using a base-promoted deconjugation of (E)-γ-fluorovinylphosphonates, and its utility as fluoroolefin-containing building block, *J. Fluorine Chem.*, 102, 189, 2000.

87. Sanders, T.C., and Hammond, G.B., A regiospecific fluorination strategy. Synthesis of (α-fluoropropargyl)- and (α-fluoroallyl)phosphonate esters, *J. Org. Chem.*, 58, 5598, 1993.

88. Benayoud, F., deMendonca, D.J., Digits, C.A., Moniz, G.A., Sanders, T.C., and Hammond, G.B., Efficient syntheses of (α-fluoropropargyl)phosphonate esters, *J. Org. Chem.*, 61, 5159, 1996.

89. Sanders, T.C., Golen, J.A., Williard, P.G., and Hammond, G.B., The crystal structure of 2-fluoro-1-(*p*-methoxyphenyl)-1-penten-3-yne, a fluorinated vinylacetylene prepared via Horner–Wadsworth–Emmons condensation, *J. Fluorine Chem.*, 85, 173, 1997.

90. Hammond, G.B., and Zapata, A.J., Preparation and applications of fluorinated propargyl phosphonate reagents, *University of Massachusetts*, Int. Patent Appl. WO 9918138, 1999; *Chem. Abstr.*, 130, 296825, 1999.

91. Zapata, A.J., Gu, Y., and Hammond, G.B., The first α-fluoroallenylphosphonate. The synthesis of conjugated fluoroenynes, and the stereoselective synthesis of vinylfluorophosphonates using a new multifunctional fluorine-containing building block, *J. Org. Chem.*, 65, 227, 2000.

92. Benayoud, F., Chen, L., Moniz, G.A., Zapata, A.J., and Hammond, G.B., Synthesis of α-fluoro-β,γ-alkenylphosphonates and conjugated fluoroenynes from a common intermediate (α-fluoropropargyl)phosphonate, *Tetrahedron*, 54, 15541, 1998.

93. Berkowitz, D.B., Bose, M., Pfannenstiel, T.J., and Doukov, T., α-Fluorinated phosphonates as substrate mimics for glucose 6-phosphate dehydrogenase. The CHF stereochemistry matters, *J. Org. Chem.*, 65, 4498, 2000.

94. Phillion, D.P., and Andrew, S.S., Synthesis and reactivity of diethyl phosphonomethyltriflate, *Tetrahedron Lett.*, 27, 1477, 1986.

95. Hamilton, C.J., and Roberts, S.M., Synthesis of fluorinated phosphonoacetate derivatives of carbocyclic nucleoside monophosphonates and activity as inhibitors of HIV reverse transcriptase, *J. Chem. Soc., Perkin Trans. 1*, 1051, 1999.

96. Conant, J.B., and Coyne, B.B., Addition reactions of the phosphorus halides. Part 5. The formation of an unsaturated phosphonic acid, *J. Am. Chem. Soc.*, 44, 2530, 1922; *Chem. Abstr.*, 17, 273a, 1923.

97. Gajda, T., A convenient synthesis of diethyl 1-chloroalkylphosphonates, *Synthesis*, 717, 1990.

98. Pudovik, A.N., Zimin, M.G., and Sobanov, A.A., Reactions of dialkyl phosphites with ketones activated by electronegative groups, *Zh. Obshch. Khim.*, 42, 2174, 1972; *Chem. Abstr.*, 78, 58543, 1973.

99. Johnson, R.A., Arylmethylphosphonates and phosphonic acids useful as anti-inflammatory agents, their preparation, and their activity, *Upjohn*, U.S. Patent Appl. US 5500417, 1996; *Chem. Abstr.*, 124, 307587, 1996.

100. Cabioch, J.-L., Pellerin, B., and Denis, J.-M., Synthesis of primary α-chlorophosphines by a chemoselective reduction of α-chlorophosphonates, *Phosphorus, Sulfur Silicon Relat. Elem.*, 44, 27, 1989.

101. Kumaraswamy, S., Selvi, R.S., and Swamy, K.C.K., Synthesis of new α-hydroxy-, α-halogeno- and vinylphosphonates derived from 5,5-dimethyl-1,3,2-dioxaphosphinan-2-one, *Synthesis*, 207, 1997.

102. Diehr, H.-J., and Fuchs, R.A., Chlorostyrylcyclopropanecarboxylic acid derivatives, *Bayer*, Eur. Patent Appl. EP 0006205, 1980; *Chem. Abstr.*, 92, 215078, 1980.

103. Hofmann, H., Haloalkenes, *Bayer*, German Patent Appl. DE 3000065, 1981; *Chem. Abstr.*, 95, 150041, 1981.

104. Roush, D.M., Davis, S.G., Lutomski, K.A., Meier, G.A., Phillips, R.B., and Burkart, S.E., Preparation of 2-(2-thienylethynyl)benzothiophenes as acaricides and insecticides, *FMC*, U.S. Patent Appl. US 5073564, 1991; *Chem. Abstr.*, 117, 90127, 1992.

105. Kondo, K., Fujitani, T., and Ohnishi, N., Synthesis and non-linear properties of disubstituted diphenylacetylene and related compounds, *J. Mater. Chem.*, 7, 429, 1997.

106. Jautelat, M., Arlt, D., Lantzsch, R., Fuchs, R.A., Riebel, H., Schroeder, R., and Harnisch, H., Styrylcyclopropanecarboxylic acid esters and intermediates, *Bayer*, German Patent Appl. DE 2916321, 1980; *Chem. Abstr.*, 94, 121121, 1981.

107. Zimmer, H., Bercz, P.J., Maltenieks, O.J., and Moore, M.W., Acetylenes and α-chlorostilbenes via phosphonate anions, *J. Am. Chem. Soc.*, 87, 2777, 1965.

108. Zimmer, H., Hickey, K.R., and Schumacher, R., Syntheses with α-heterosubstituted phosphonate carbanions. Part 3. Diacetylenes and substituted vinyl chlorides, *Chimia*, 28, 656, 1974.

109. Gallagher, M.J., and Noerdin, H., The synthesis and photochemistry of *p*-dimethylamino-*p*′-nitrodiphenylethyne. A push-pull acetylene, *Aust. J. Chem.*, 38, 997, 1985.

110. Kondo, K., Ohnishi, N., Takemoto, K., Yoshida, H., and Yoshida, K., Synthesis of optically quadratic nonlinear phenylpyridylacetylenes, *J. Org. Chem.*, 57, 1622, 1992.

111. Gajda, T., Preparation of diethyl 1-bromoalkylphosphonates, *Phosphorus, Sulfur Silicon Relat. Elem.*, 53, 327, 1990.

112. Ikeda, H., Abushanab, E., and Marquez, V.E., The assembly of β-methylene-TAD, a metabolically stable analogue of the antitumor agent TAD, by the stepwise esterification of the monodeprotected methylene*bis*(phosphonate)benzyl esters under Mitsunobu conditions, *Bioorg. Med. Chem. Lett.*, 9, 3069, 1999.

113. Green, D., Elgendy, S., Patel, G., Baban, J.A., Skordalakes, E., Husman, W., Kakkar, V.V., and Deadman, J., The facile synthesis of *O,O*-dialkyl α-halobenzylphosphonates from *O,O*-dialkyl α-hydroxybenzylphosphonates, *Tetrahedron*, 52, 10215, 1996.

114. Sharts, C.M., and Sheppard, W.A., Modern methods to prepare monofluoroaliphatic compounds, *Org. React.*, 21, 125, 1974.

115. Wilkinson, J.A., Recent advances in the selective formation of the C–F bond, *Chem. Rev.*, 92, 505, 1992.

116. Burton, D.J., Yang, Z.-Y., and Qiu, W., Fluorinated ylides and related compounds, *Chem. Rev.*, 96, 1641, 1996.

117. Taylor, S.D., Kotoris, C.C., and Hum, G., Recent advances in electrophilic fluorination, *Tetrahedron*, 55, 12431, 1999.

118. Chambers, R.D., and Hutchinson, J., Elemental fluorine. Part 9. Catalysis of the direct fluorination of 2-substituted carbonyl compounds, *J. Fluorine Chem.*, 92, 45, 1998.

119. Chambers, R.D., and Hutchinson, J., Preparation of fluorinated phosphonate compounds, *F2 Chemicals*, Int. Patent Appl. WO 9905080, 1999; *Chem. Abstr.*, 130, 139451, 1999.

120. Tessier, J., Demoute, J.-P., and Truong, V.T., Process for preparing fluorinated derivatives of phosphonic acid, and products obtained using this process, *Roussel Uclaf*, Eur. Patent Appl. EP 224417, 1987; *Chem. Abstr.*, 107, 96889y, 1987.

121. Bode, H., and Klesper, E., Action of fluorine on chlorates (preliminary report), *Z. Anorg. Allg. Chem.*, 266, 275, 1951.

122. Grell, W., and Machleidt, H., Syntheses with organophosphorus compounds. Part 1. Ester olefinations, *Justus Liebigs Ann. Chem.*, 693, 134, 1966.

123. Koizumi, T., Hagi, T., Horie, Y., and Takeuchi, Y., Diethyl 1-fluoro-1-phenylsulfonylmethanephosphonate, a versatile agent for the preparation of monofluorinated building blocks, *Chem. Pharm. Bull.*, 35, 3959, 1987.

124. Grieco, P.A., Yokoyama, Y., Nicolaou, K.C., Barnette, W.E., Smith, J.B., Ogletree, M., and Lefer, A.M., Total synthesis of 14-fluoroprostaglandin $F_{2\alpha}$ and 14-fluoroprostacyclin, *Chem. Lett.*, 1001, 1978.

125. Grieco, P.A., Schillinger, W.J., and Yokoyama, Y., Carbon-14 fluorinated prostaglandins. Synthesis and biological evaluation of the methyl esters of (+)-14-fluoro-, (+)-15-*epi*-14-fluoro-, (+)-13(*E*)-14-fluoro-, and (+)-13(*E*)-15-*epi*-14-fluoroprostaglandin $F_{2\alpha}$, *J. Med. Chem.*, 23, 1077, 1980.

126. Blackburn, G.M., England, D.A., and Kolkmann, F., Monofluoro- and difluoro-methylenebisphosphonic acids. Isopolar analogues of pyrophosphoric acid, *J. Chem. Soc., Chem. Commun.*, 930, 1981.

127. McKenna, C.E., and Shen, P.-D., Fluorination of methanediphosphonate esters by perchloryl fluoride. Synthesis of fluoromethanediphosphonic acid and difluoromethanediphosphonic acid, *J. Org. Chem.*, 46, 4573, 1981.

128. McKenna, C.E., α-Fluorinated alkanediphosphonates, *University of Southern California*, U.S. Patent Appl. US 4478763, 1984; *Chem. Abstr.*, 102, 95821, 1985.

129. McKenna, C.E., Pham, P.-T.T., Rassier, M.E., and Dousa, T.P., α-Halo [(phenylphosphinyl)methyl]phosphonates as specific inhibitors of Na^+-gradient-dependent Na^+-phosphate cotransport across renal brush border membrane, *J. Med. Chem.*, 35, 4885, 1992.

130. Hall, C.R., Inch, T.D., and Williams, N.E., The stereochemical course of substitution reactions at halomethylphosphonates, *Phosphorus Sulfur*, 18, 213, 1983.

131. Hall, C.R., Inch, T.D., and Williams, N.E., The preparation and properties of some chiral fluoromethylphosphonates, phosphonothioates, and phosphonamidothioates, *J. Chem. Soc., Perkin Trans. 1*, 233, 1985.

132. Liu, X., Adams, H., and Blackburn, G.M., Synthesis of novel "supercharged" analogues of pyrophosphoric acid, *J. Chem. Soc., Chem. Commun.*, 2619, 1998.

133. Rozen, S., Lerman, O., and Kol, M., Acetyl hypofluorite, the first member of a new family of organic compounds, *J. Chem. Soc., Chem. Commun.*, 443, 1981.

134. Lerman, O., Tor, Y., and Rozen, S., Acetyl hypofluorite as a taming carrier of elemental fluorine for novel electrophilic fluorination of activated aromatic rings, *J. Org. Chem.*, 46, 4629, 1981.

135. Hebel, D., Kirk, K.L., Kinjo, J., Kovacs, T., Lesiak, K., Balzarini, J., De Clercq, E., and Torrence, P.F., Synthesis of a difluoromethylenephosphonate analog of AZT 5′-triphosphate and its inhibition of HIV-1 reverse transcriptase, *Bioorg. Med. Chem. Lett.*, 1, 357, 1991.

136. Differding, E., and Ofner, H., *N*-Fluorobenzenesulfonimide. A practical reagent for electrophilic fluorinations, *Synthesis*, 187, 1991.

137. Davis, F.A., and Han, W., *N*-Fluoro-*o*-benzenedisulfonimide. A useful new fluorinating reagent, *Tetrahedron Lett.*, 32, 1631, 1991.

138. Davis, F.A., Han, W., and Murphy, C.K., Selective electrophilic fluorinations using *N*-fluoro-*o*-benzenedisulfonimide, *J. Org. Chem.*, 60, 4730, 1995.

139. Singh, S., DesMarteau, D.D., Zuberi, S.S., Witz, M., and Huang, H.-N., *N*-Fluoroperfluoroalkylsulfonimides. Remarkable new fluorination reagents, *J. Am. Chem. Soc.*, 109, 7194, 1987.

140. Banks, R.E., Mohialdin-Khaffaf, S.N., Lal, G.S., Sharif, I., and Syvret, R.G., 1-Alkyl-4-fluoro-1,4-diazoniabicyclo[2.2.2]octane salts. A novel family of electrophilic fluorinating agents, *J. Chem. Soc., Chem. Commun.*, 595, 1992.

141. Oliver, E.W., and Evans, D.H., Electrochemical studies of six N-F electrophilic fluorinating reagents, *J. Electroanal. Chem.*, 474, 1, 1999.

142. Xu, Z.-Q., and DesMarteau, D.D., A convenient one-pot synthesis of α-fluoro-α,β-unsaturated nitriles from diethyl cyanofluoromethanephosphonate, *J. Chem. Soc., Perkin Trans. 1*, 313, 1992.

143. Iorga, B., Synthesis of α-monohalogenated phosphonates by electrophilic halogenation, Ph.D. thesis, Ecole Polytechnique, Palaiseau, 2001.

144. Lal, G.S., Site-selective fluorination of organic compounds using 1-alkyl-4-fluoro-1,4-diazabicyclo[2.2.2]octane salts (Selectfluor reagents), *J. Org. Chem.*, 58, 2791, 1993.

145. Lal, G.S., Method of selective fluorination, *Air Products and Chemicals*, U.S. Patent Appl. US 5442084, 1995; *Chem. Abstr.*, 123, 340399, 1995.

146. Wnuk, S.F., and Robins, M.J., Stannyl radical-mediated cleavage of π-deficient heterocyclic sulfones. Synthesis of α-fluoro esters and first homonucleoside α-fluoromethylene phosphonate, *J. Am. Chem. Soc.*, 118, 2519, 1996.

147. Wnuk, S.F., Bergolla, L.A., and Garcia, P.I., Jr., Studies toward the synthesis of α-fluorinated phosphonates via tin-mediated cleavage of α-fluoro-α-(pyrimidin-2-ylsulfonyl)alkylphosphonates. Intramolecular cyclization of the α-phosphonyl radicals, *J. Org. Chem.*, 67, 3065, 2002.

148. Differding, E., Duthaler, R.O., Krieger, A., Rüegg, G.M., and Schmit, C., Electrophilic fluorinations with *N*-fluorobenzenesulfonimide. Convenient access to α-fluoro- and α,α-difluorophosphonates, *Synlett*, 395, 1991.

149. Chen, W., Flavin, M.T., Filler, R., and Xu, Z.-Q., Synthesis and antiviral activities of fluorinated acyclic nucleoside phosphonates, *J. Chem. Soc., Perkin Trans. 1*, 3979, 1998.

150. Hamilton, C.J., Roberts, S.M., and Shipitsin, A., Synthesis of a potent inhibitor of HIV reverse transcriptase, *J. Chem. Soc., Chem. Commun.*, 1087, 1998.

151. Kotoris, C.C., Wen, W., Lough, A., and Taylor, S.D., Preparation of chiral α-monofluoroalkylphosphonic acids and their evaluation as inhibitors of protein tyrosine phosphatase 1B, *J. Chem. Soc., Perkin Trans. 1*, 1271, 2000.

152. Blackburn, G.M., Brown, D., Martin, S.J., and Parratt, M.J., Studies on selected transformations of some fluoromethanephosphonate esters, *J. Chem. Soc., Perkin Trans. 1*, 181, 1987.

153. Pawson, B.A., Chan, K.-K., DeNoble, J., Han, R.J.L., Piermattie, V., Specian, A.C., Srisethnil, S., Trown, P.W., Bohoslawec, O., Machlin, L.J., and Gabriel, E., Fluorinated retinoic acids and their analogs. Part 1. Synthesis and biological activity of (4-methoxy-2,3,6-trimethylphenyl)nonatetraenoic acid analogs, *J. Med. Chem.*, 22, 1059, 1979.

154. Palmer, J.T., Rasnick, D., and Klaus, J.L., Irreversible cysteine protease inhibitors containing vinyl groups conjugated to electron withdrawing groups, *Khepri Pharmaceuticals*, Int. Patent Appl. WO 9523222, 1995; *Chem. Abstr.*, 123, 309464, 1995.

155. Iorga, B., Eymery, F., and Savignac, P., A highly selective synthesis of α-monofluoro- and α-monochlorobenzylphosphonates using electrophilic halogenation of benzylphosphonates carbanions, *Tetrahedron Lett.*, 39, 3693, 1998.

156. Iorga, B., Eymery, F., and Savignac, P., Controlled monohalogenation of phosphonates. A new route to pure α-monohalogenated diethyl benzylphosphonates, *Tetrahedron*, 55, 2671, 1999.

157. Iorga, B., Eymery, F., and Savignac, P., Controlled monohalogenation of phosphonates. Part 2. Preparation of pure diethyl α-monohalogenated alkylphosphonates, *Synthesis*, 576, 2000.

158. Eymery, F., Iorga, B., and Savignac, P., The usefulness of phosphorus compounds in alkyne synthesis, *Synthesis*, 185, 2000.

159. Iorga, B., Eymery, F., and Savignac, P., The syntheses and properties of 1,2-epoxyphosphonates, *Synthesis*, 207, 1999.

160. Engel, J.F., Insecticidal 3-styryl-2,2-dimethylcyclopropane-1-carboxylic acid esters, *FMC*, German Patent Appl. DE 2738150, 1978; *Chem. Abstr.*, 89, 42821, 1978.

161. Engel, J.F., Insecticidal (β-phenyl-β-substituted-vinyl)cyclopropanecarboxylates, *FMC*, U.S. Patent Appl. US 4183942, 1980; *Chem. Abstr.*, 92, 215077, 1980.

162. Petrova, J., Coutrot, P., Dreux, M., and Savignac, P., The α-chlorination and carbonyl olefination. Arylchloromethanephosphonic acid esters. Preparation, alkylation, and Wittig–Horner reactions, *Synthesis*, 658, 1975.

163. Coutrot, P., Laurenço, C., Petrova, J., and Savignac, P., α-Chlorination and carbonyl olefination. Synthesis of phenyl 1-chloro-1-alken-1-yl sulfides, *Synthesis*, 107, 1976.

164. Crenshaw, M.D., and Zimmer, H., Synthesis of trisubstituted vinyl chlorides, *J. Org. Chem.*, 48, 2782, 1983.

165. Coutrot, P., Grison, C., and Youssefi-Tabrizi, M., α-Chlorination and carbonylolefination. Synthesis of phenyl 1-chloro-1-alken-1-yl selenides (chlorovinyl phenyl selenides), *Synthesis*, 169, 1987.

166. Iseki, K., Shinoda, M., Ishiyama, C., Hayasi, Y., Yamada, S.-I., and Shibasaki, M., Synthesis of (Z)-4,5,13,14-tetradehydro-9(0)-methano-$\Delta^{6(9\alpha)}$-PGI$_1$, *Chem. Lett.*, 559, 1986.

167. Tomiyama, T., Wakabayashi, S., and Yokota, M., Synthesis and biological activity of novel carbacyclins having bicyclic substituents on the ω-chain, *J. Med. Chem.*, 32, 1988, 1989.

168. Gandolfi, C., Pellegata, R., Ceserani, R., and Usardi, M.M., ω-*Nor*-cycloalkyl-13,14-dehydroprostaglandins, *Farmitalia Carlo Erba*, German Patent Appl. DE 2539116, 1976; *Chem. Abstr.*, 85, 77748, 1976.

169. Braun, N.A., Klein, I., Spitzner, D., Vogler, B., Braun, S., Borrmann, H., and Simon, A., Cascade reactions with chiral Michael acceptors. Synthesis of enantiomerically pure tricyclo[3.2.1.0^{2,7}]- and bicyclo[3.2.1]octanes, *Liebigs Ann. Org. Bioorg. Chem.*, 2165, 1995.

170. Lee, K., Shin, W.S., and Oh, D.Y., A one-pot generation of α-chloro-α-lithioalkanephosphonates. A new preparative route to 1,2-epoxyalkanephosphonates, *Synth. Commun.*, 22, 649, 1992.

171. Lee, K., and Oh, D.Y., Synthesis of α-hetero atom substituted β-keto and enamine phosphonates, *Synth. Commun.*, 21, 279, 1991.

172. McKenna, C.E., and Khawli, L.A., Synthesis of halogenated phosphonoacetate esters, *J. Org. Chem.*, 51, 5467, 1986.

173. Nguyen, L.M., Niesor, E., and Bentzen, C.L., *gem*-Diphosphonate and *gem*-phosphonate-phosphate compounds with specific high density lipoprotein inducing activity, *J. Med. Chem.*, 30, 1426, 1987.

174. Shi, X.-X., and Dai, L.X., Mild halogenation of stabilized ester enolates by cupric halides, *J. Org. Chem.*, 58, 4596, 1993.

175. Iorga, B., Ricard, L., and Savignac, P., Carbanionic displacement reactions at phosphorus. Part 3. Cyanomethylphosphonate *vs.* cyanomethylenediphosphonate. Synthesis and solid state structure, *J. Chem. Soc., Perkin Trans. 1,* 3311, 2000.

176. Savignac, P., and Iorga, B., unpublished results.

177. Dawson, N.D., and Burger, A., Some alkyl thiazolephosphonates, *J. Am. Chem. Soc.*, 74, 5312, 1952.

178. Wadsworth, W.S., Jr., and Emmons, W.D., The utility of phosphonate carbanions in olefin synthesis, *J. Am. Chem. Soc.*, 83, 1733, 1961.

179. Miryan, N.I., Isaev, S.D., Kovaleva, S.A., Petukh, N.V., Dvornikova, E.V., Kardakova, E.V., and Yurchenko, A.G., Horner–Emmons reaction in the synthesis of esters of unsaturated acids from adamantane series and related carcass compounds, *Zh. Org. Khim.*, 35, 882, 1999; *Russ. J. Org. Chem. (Engl. Transl.)*, 35, 857, 1999.

180. Skuballa, W., Schillinger, E., Stürzebecher, C.-S., and Vorrbrüggen, H., Synthesis of a new chemically and metabolically stable prostacyclin analogue with high and long-lasting oral activity, *J. Med. Chem.*, 29, 313, 1986.

181. Qing, F.-L., and Zhang, X., A one-pot synthesis of (E)-α-bromo-α,β-unsaturated esters and their trifluoromethylation. A general and stereoselective route to (E)-α-trifluoromethyl-α,β-unsaturated esters, *Tetrahedron Lett.*, 42, 5929, 2001.

182. Coutrot, P., Youssefi-Tabrizi, M., and Grison, C., Diethyl-1-magnesium chloride methanephosphonate, a novel Grignard reagent and its use in organic synthesis, *J. Organomet. Chem.*, 316, 13, 1986.

183. Roy, C.H., Substituted methylenediphosphonic acid compounds and detergent compositions containing them, *Procter & Gamble*, U.K. Patent Appl. GB 1026366, 1966; *Chem. Abstr.*, 65, 30501, 1966.

184. Grinev, G.V., Chervenyuk, G.I., and Dombrovskii, A.V., Synthesis of α-bromophosphonates, *Zh. Obshch. Khim.*, 39, 1253, 1969; *Chem. Abstr.*, 71, 101932, 1969.

185. Balczewski, P., and Mikolajczyk, M., An expeditious synthesis of (±)-desepoxy-4,5-didehydromethylenomycin A methyl ester, *Org. Lett.*, 2, 1153, 2000.

186. Teulade, M.-P., and Savignac, P., α-Lithiated *O,O*-diethyl chloro(trimethylsilyl)methylphosphonate. Part 1. Preparation and properties, *J. Organomet. Chem.*, 338, 295, 1988.

187. Ide, J., Endo, R., and Muramatsu, S., Synthesis of ethyl α-(diethylphosphono)acrylate and its homologs. Versatile synthetic reagents, *Chem. Lett.*, 401, 1978.

188. Töke, L., Jaszay, Z.M., Petnehazy, I., Clementis, G., Vereczkey, G., Kövesdi, I., Rockenbauer, A., and Kovats, K., A versatile building block for the synthesis of substituted cyclopropanephosphonic acid esters, *Tetrahedron*, 51, 9167, 1995.

189. Balczewski, P., and Pietrzykowski, W.M., A new, effective approach for the C-C bond formation utilizing 1-, 2- and 3-phosphonyl substituted radicals derived from iodoalkylphosphonates and *n*-Bu₃SnH/Et₃B/O₂ system, *Tetrahedron*, 53, 7291, 1997.

190. Schrader, T., Kober, R., and Steglich, W., Synthesis of 1-aminophosphonic acid derivatives via (acylimino)phosphonic esters, *Synthesis*, 372, 1986.

191. Schrader, T., and Steglich, W., Phosphorus analogs of amino acids. Part 4. Syntheses of unusual 1-aminophosphonic acids via Diels–Alder reactions of diethyl (*N*-acyliminomethyl)phosphonates, *Synthesis*, 1153, 1990.

192. Chakraborty, S.K., and Engel, R., A novel synthesis of 1-aminoalkylphosphonathes, *Synth. Commun.*, 21, 1039, 1991.

193. Kim, T.H., and Oh, D.Y., New synthetic method of *O,S*-thioacetals of formylphosphonates, *Tetrahedron Lett.*, 26, 3479, 1985.

194. Kim, T.H., and Oh, D.Y., Synthesis of 1-(substituted-aryl) methylthiomethanephosphonates by Friedel–Crafts reaction of aromatic compounds with chloro(methylthio)methanephosphonate, *Tetrahedron Lett.*, 27, 1165, 1986.

195. Ishibashi, H., Sato, T., Irie, M., Ito, M., and Ikeda, M., Carbon–carbon bond forming reactions via α-phosphoryl-α-thio carbocations, *J. Chem. Soc., Perkin Trans. 1*, 1095, 1987.

196. Kim, T.H., and Oh, D.Y., Synthesis of *S,S*-thioacetals of formylphosphonate from chloro(arylthio)metanephosphonate, *Synth. Commun.*, 18, 1611, 1988.

197. Yamamoto, I., Sakai, T., Yamamoto, S., Ohta, K., and Matsuzaki, K., Synthesis of chlorovinyl phenyl sulfides and sulfones from diethyl chloro-(phenylthio)-methanephosphonate, *Synthesis*, 676, 1985.

198. Balczewski, P., Free radical desulfenylation and deselenylation of α-sulfur and α-seleno substituted phosphonates with the *n*-Bu₃SnH/AIBN reagents system, *Phosphorus, Sulfur Silicon Relat. Elem.*, 104, 113, 1995.

199. Palacios, F., Aparicio, D., and de los Santos, J.M., An efficient strategy for the regioselective synthesis of 3-phosphorylated-1-aminopyrroles from β-hydrazono phosphine oxides and phosphonates, *Tetrahedron*, 55, 13767, 1999.

200. Collins, D.J., Drygala, P.F., and Swan, J.M., Organophosphorus compounds. Part 20. Approaches to the synthesis of 2,3-dihydro-1*H*-1,2-benzazaphospholes involving carbon–carbon and carbon–phosphorus ring closure, *Aust. J. Chem.*, 37, 1009, 1984.

201. Gross, H., and Ozegowski, S., α-Substituted phosphonates. Part 54. Synthesis of 4-hydroxyphenylmethanebisphosphonic acid, *Phosphorus, Sulfur Silicon Relat. Elem.*, 47, 1, 1990.

202. Vasella, A., and Wyler, R., Synthesis of a phosphonic acid analogue of *N*-acetyl-2,3-didehydro-2-deoxyneuraminic acid, an inhibitor of *Vibrio cholerae* sialidase, *Helv. Chim. Acta*, 74, 451, 1991.

203. Barnes, N.J., Probert, M.A., and Wightman, R.H., Synthesis of 2-deoxy-α- and -β-D-arabino-hexopyranosyl phosphonic acids and related compounds. Analogues of early intermediates in the shikimate pathway, *J. Chem. Soc., Perkin Trans. 1*, 431, 1996.

204. Nicholson, D.A., Process for the preparation of monohalogenated methylenediphosphonate esters, and phosphonoacetate esters, *Procter & Gamble*, U.S. Patent Appl. US 3627842, 1971; *Chem. Abstr.*, 76, 99826, 1972.

205. Nicholson, D.A., and Vaughn, H., New approaches to the preparation of halogenated methylenediphosphonates, phosphonoacetates, and malonates, *J. Org. Chem.*, 36, 1835, 1971.

206. Seyferth, D., and Marmor, R.S., Halomethyl-metal compounds. Part 63. Diethyl lithiodichloromethylphosphonate and tetraethyl lithiochloromethylenediphosphonate, *J. Organomet. Chem.*, 59, 237, 1973.

207. Hutchinson, D.W., and Semple, G., Relative reactivities of tetraalkyl esters of methylene bisphosphonic acid, *J. Organomet. Chem.*, 309, C7, 1986.

208. McKenna, C.E., Khawli, L.A., Ahmad, W.-Y., Pham, P., and Bongartz, J.-P., Synthesis of α-halogenated methanediphosphonates, *Phosphorus Sulfur*, 37, 1, 1988.

209. Vepsäläinen, J., Nupponen, H., Pohjala, E., Ahlgren, M., and Vainiotalo, P., Bisphosphonic compounds. Part 3. Preparation and identification of tetraalkyl methylene- and (α-halomethylene)bisphosphonates by mass spectrometry, NMR spectroscopy and X-ray crystallography, *J. Chem. Soc., Perkin Trans. 2*, 835, 1992.

210. Petrov, K.A., Chauzov, V.A., and Bogdanov, N.N., Nitroalkyl organophosphorus compounds. Part 5. Chloro derivatives of α-nitroalkylphosphonic acids, *Zh. Obshch. Khim.*, 46, 1499, 1976; *Chem. Abstr.*, 85, 160257, 1976.

211. Hutchinson, D.W., and Semple, G., Synthesis of alkylated methylene bisphosphonates via organothallium intermediates, *J. Organomet. Chem.*, 291, 145, 1985.

212. Hutchinson, D.W., and Semple, G., The dehalogenation of dihalogenomethylenebisphosphonates, *Phosphorus Sulfur*, 21, 1, 1984.

213. Tago, K., and Kogen, H., *Bis*(2,2,2-trifluoroethyl)bromophosphonoacetate, a novel HWE reagent for the preparation of (*E*)-α-bromoacrylates. A general and stereoselective method for the synthesis of trisubstituted alkenes, *Org. Lett.*, 2, 1975, 2000.

214. Tago, K., and Kogen, H., A highly stereoselective synthesis of (*E*)-α-bromoacrylates, *Tetrahedron*, 56, 8825, 2000.

215. Fuchigami, T., Shimojo, M., and Konno, A., Electrolytic partial fluorination of organic compounds. Part 17. Regiospecific anodic fluorination of sulfides bearing electron-withdrawing substituents at the position α to the sulfur atom, *J. Org. Chem.*, 60, 3459, 1995.

216. Erian, A.W., Konno, A., and Fuchigami, T., Electrolytic partial fluorination of organic compounds. Part 19. A novel synthesis of fluorothieno[2,3-*b*]pyridines using anodic fluorination of heterocyclic sulfides as a key step, *J. Org. Chem.*, 60, 7654, 1995.

217. Gil, J.M., and Oh, D.Y., Carbocupration of diethyl 1-alkynylphosphonates. Stereo- and regioselective synthesis of 1,2,2-trisubstituted vinylphosphonates, *J. Org. Chem.*, 64, 2950, 1999.

218. Minami, T., Okauchi, T., and Kouno, R., α-Phosphonovinyl carbanions in organic synthesis, *Synthesis*, 349, 2001.

219. Kouno, R., Okauchi, T., Nakamura, M., Ichikawa, J., and Minami, T., Synthesis and synthetic utilization of α-functionalized vinylphosphonates bearing β-oxy or β-thio substituents, *J. Org. Chem.*, 63, 6239, 1998.

220. Minami, T., Kouno, R., Okauchi, T., Nakamura, M., and Ichikawa, J., Synthetic utilization of α-phosphonovinyl anions, *Phosphorus, Sulfur Silicon Relat. Elem.*, 144, 689, 1999.

221. Ismailov, V.M., Moskva, V.V., Dadasheva, L.A., Zykova, T.V., and Guseinov, F.I., Dichloro(dialkoxyphosphinyl)acetaldehydes, *Zh. Obshch. Khim.*, 52, 2140, 1982; *J. Gen. Chem. USSR (Engl. Transl.)*, 52, 1906, 1982.

222. Ismailov, V.M., Moskva, V.V., and Zykova, T.V., (Chloroformylmethyl)phosphonic esters, *Zh. Obshch. Khim.*, 53, 2793, 1983; *J. Gen. Chem. USSR (Engl. Transl.)*, 53, 2518, 1983.

223. Shagidullin, R.R., Pavlov, V.A., Buzykin, B.I., Aristova, N.V., Chertanova, L.F., Vandyukova, I.I., Plyamovatyi, A.K., Enikeev, K.M., Sokolov, M.P., and Moskva, V.V., Structures of (dialkoxyphosphoryl)haloacetaldehydes and their nitrosation products, *Zh. Obshch. Khim.*, 61, 1590, 1991; *J. Gen. Chem. USSR (Engl. Transl.)*, 61, 1459, 1991.

224. Guseinov, F.I., Moskva, V.V., and Ismailov, V.M., Preparative methods of the synthesis of (dialkoxyphosphinyl)-mono- and -di-chloroacetaldehydes, *Zh. Obshch. Khim.*, 63, 93, 1993; *Russ. J. Gen. Chem. (Engl. Transl.)*, 63, 66, 1993.

225. Soborovskii, L.Z., and Baina, N.F., Difluorochloromethane as a difluoromethylating agent. Part 2. Reaction of difluorochloromethane with sodium dialkyl phosphites, *Zh. Obshch. Khim.*, 29, 1144, 1959; *J. Gen. Chem. USSR (Engl. Transl.)*, 29, 1115, 1959.

226. Obayashi, M., Ito, E., Matsui, K., and Kondo, K., (Diethylphosphinyl)difluoromethyllithium. Preparation and synthetic application, *Tetrahedron Lett.*, 23, 2323, 1982.

227. Bergstrom, D.E., and Shum, P.W., Synthesis and characterization of a new fluorine substituted nonionic dinucleoside phosphonate analogue, *P*-deoxy-*P*-(difluoromethyl)thymidylyl(3′→5′)thymidine, *J. Org. Chem.*, 53, 3953, 1988.

228. Bigge, C.F., Drummond, J.T., and Johnson, G., Synthesis and NMDA receptor binding of 2-amino-7,7-difluoro-7-phosphonoheptanoic acid, *Tetrahedron Lett.*, 30, 7013, 1989.

229. Biller, S.A., Preparation of alkenylphosphinylmethylphosphonates as squalene synthetase inhibitors, *Squibb*, Eur. Patent Appl. EP 324421, 1989; *Chem. Abstr.,* 112, 77633, 1990.

230. Halazy, S., and Danzin, C., Phosphonoalkylpurine derivatives as purine nucleoside phosphorylase inhibitors, *Merrell Dow Pharmaceuticals*, Eur. Patent Appl. EP 338168, 1989; *Chem. Abstr.,* 112, 158267, 1990.

231. Berkowitz, D.B., and Sloss, D.G., Diallyl (lithiodifluoromethyl)phosphonate. A new reagent for the introduction of the (difluoromethylene)phosphonate functionality, *J. Org. Chem.*, 60, 7047, 1995.

232. Pedersen, S.D., Qiu, W., Qiu, Z.-M., Kotov, S.V., and Burton, D.J., The synthesis of phosphonate ester containing fluorinated vinyl ethers, *J. Org. Chem.*, 61, 8024, 1996.

233. Hutchinson, D.W., and Thornton, D.M., A simple synthesis of monofluoromethylene bisphosphonic acid, *J. Organomet. Chem.*, 340, 93, 1988.

234. Wang, Z., and Hammond, G.B., A highly efficient synthesis of triisopropylsilyldifluorobromopropyne yields a versatile *gem*-difluoromethylene building block, *J. Chem. Soc., Chem. Commun.*, 2545, 1999.

235. Roy, N.K., and Mukerjee, S.K., Synthesis of potential organophosphorus insecticides. Part 1. Synthesis of diaryldichloromethylphosphonates, *Indian J. Chem.*, 10, 1159, 1972.

236. Hoffmann, H., Maurer, F., Priesnitz, U., and Riebel, H.J., 3-(2,2-Dichlorovinyl)-2,2-dimethyl-cyclopropane-1-carboxylic acid derivatives, *Bayer*, Eur. Patent Appl. EP 22971, 1981; *Chem. Abstr.,* 95, 6617, 1981.

237. Bunyan, P.J., and Cadogan, J.I.G., Reactivity of organophosphorus compounds. Part 13. Radical-chain transfer reactions of triethyl phosphite. A new phosphorothiolate synthesis, *J. Chem. Soc.*, 2953, 1962.

238. Elkaim, J.C., and Riess, J.G., Direct syntheses of (dichloromethyl)phosphonyl esters and amides, *Tetrahedron Lett.*, 16, 4409, 1975.

239. Elkaim, J.C., and Riess, J., Haloalkylphosphonic acid derivatives as fireproofing agents, *ANVAR*, French Patent Appl. FR 2364223, 1978; *Chem. Abstr.,* 90, 39585, 1979.

240. Elkaim, J.C., Casabianca, F., and Riess, J.G., The direct synthesis of dibromomethylphosphonic dibromide and some of its derivatives, *Synth. React. Inorg. Met.-Org. Chem.*, 9, 479, 1979.

241. Cen, W., and Shen, Y., A new synthesis of perfluoroalkanephosphonates, *J. Fluorine Chem.*, 52, 369, 1991.

242. Bittles, J.A., Jr., and Joyce, R.M., Jr., Fluoroalkanephosphonic compounds, *E.I. du Pont de Nemours*, U.S. Patent Appl. US 2559754, 1951; *Chem. Abstr.,* 46, 5653, 1952.

243. Brace, N.O., ω-Hydroperfluoroalkylphosphonic acids, esters, and dichlorides, *J. Org. Chem.*, 26, 3197, 1961.

244. Haszeldine, R.N., Hobson, D.L., and Taylor, D.R., Organophosphorus chemistry. Part 19. Free-radical addition of dialkyl phosphites to polyfluoroolefins, *J. Fluorine Chem.*, 8, 115, 1976.

245. Herpin, T.F., Houlton, J.S., Motherwell, W.B., Roberts, B.P., and Weibel, J.-M., Preparation of some new anomeric carbohydrate difluoromethylenephosphonates via phosphonyl radical addition to *gem*-difluoroenol ethers, *J. Chem. Soc., Chem. Commun.*, 613, 1996.

246. Piettre, S.R., Simple and efficient synthesis of 2,2-disubstituted-1,1-difluorophosphonates and phosphonothioates, *Tetrahedron Lett.*, 37, 2233, 1996.

247. Kovensky, J., McNeil, M., and Sinay, P., D-Galactofuranosylphosphonates. First synthesis of UDP-*C*-D-galactofuranose, *J. Org. Chem.*, 64, 6202, 1999.

248. Herpin, T.F., Motherwell, W.B., Roberts, B.P., Roland, S., and Weibel, J.-M., Free radical chain reactions for the preparation of novel anomeric carbohydrate difluoromethylene -phosphonates and -phosphonothioates, *Tetrahedron*, 53, 15085, 1997.

249. Nair, H.K., and Burton, D.J., Novel dialkyl (β-halotetrafluoroethyl)phosphonates. Facile synthesis via thermally and photochemically induced radical reactions. A unique photochemical transformation of BrCF₂CF₂I, *J. Am. Chem. Soc.*, 116, 6041, 1994.

250. Kato, M., and Yamabe, M., A new synthetic route to perfluoroalkylphosphonates involving facile formation of the CF-P linkage, *J. Chem. Soc., Chem. Commun.*, 1173, 1981.

251. Su, D., Cen, W., Kirchmeier, R.L., and Shreeve, J.M., Synthesis of fluorinated phosphonic, sulfonic, and mixed phosphonic/sulfonic acids, *Can. J. Chem.*, 67, 1795, 1989.

252. Nair, H.K., Guneratne, R.D., Modak, A.S., and Burton, D.J., Synthesis of novel fluorinated bisphosphonates and bisphosphonic acids, *J. Org. Chem.*, 59, 2393, 1994.

253. Nair, H.K., and Burton, D.J., A new synthetic route to perfluoroalkylidene-α,ω-bisphosphonates, *Tetrahedron Lett.*, 36, 347, 1995.

254. Nair, H.K., and Burton, D.J., Facile synthesis of fluorinated phosphonates via photochemical and thermal reactions, *J. Am. Chem. Soc.*, 119, 9137, 1997.

255. Smyth, M.S., Ford, H., and Burke, T.R., Jr., A general method for the preparation of benzylic α,α-difluorophosphonic acids. Non-hydrolyzable mimetics of phosphotyrosine, *Tetrahedron Lett.*, 33, 4137, 1992.

256. Wrobel, J., and Dietrich, A., Preparation of L-(phosphonodifluoromethyl)phenylalanine derivatives as non-hydrolyzable mimetics of O-phosphotyrosine, *Tetrahedron Lett.*, 34, 3543, 1993.

257. Burke, T.R., Jr., Smyth, M.S., Otaka, A., and Roller, P.P., Synthesis of 4-phosphono(difluoromethyl)-D,L-phenylalanine and N-Boc and N-Fmoc derivatives suitably protected for solid-phase synthesis of nonhydrolysable phosphotyrosyl peptide analogues, *Tetrahedron Lett.*, 34, 4125, 1993.

258. Smyth, M.S., and Burke, T.R., Jr., Enantioselective synthesis of N-Boc and N-Fmoc protected diethyl 4-phosphono(difluoromethyl)-L-phenylalanine. Agents suitable for the solid-phase synthesis of peptides containing nonhydrolyzable analogues of O-phosphotyrosine, *Tetrahedron Lett.*, 35, 551, 1994.

259. Smyth, M.S., and Burke, T.R., Jr., Enantioselective synthesis of N-Boc and N-Fmoc protected diethyl 4-phosphono(difluoromethyl)-L-phenylalanine (F$_2$Pmp), *Org. Prep. Proced. Int.*, 28, 77, 1996.

260. Solas, D., Hale, R.L., and Patel, D.V., An efficient synthesis of N-α-Fmoc-4-(phosphonodifluoromethyl)-L-phenylalanine, *J. Org. Chem.*, 61, 1537, 1996.

261. Liu, W.-Q., Roques, B.P., and Garbay, C., Synthesis of L-2,3,5,6-tetrafluoro-4-(phosphonomethyl)phenylalanine, a novel non-hydrolyzable phosphotyrosine mimetic and L-4-(phosphonodifluoromethyl)phenylalanine, *Tetrahedron Lett.*, 38, 1389, 1997.

262. Ye, B., and Burke, T.R., Jr., Synthesis of a difluorophosphonomethyl-containing phosphatase inhibitor designed from the X-ray structure of a PTP1B-bound ligand, *Tetrahedron*, 52, 9963, 1996.

263. Yao, Z.-J., Ye, B., Wu, X.-W., Wang, S., Wu, L., Zhang, Z.-Y., and Burke, T.R., Jr., Structure-based design and synthesis of small molecule protein-tyrosine phosphatase 1B inhibitors, *Bioorg. Med. Chem.*, 6, 1799, 1998.

264. Benayoud, F., and Hammond, G.B., An expedient synthesis of (α,α-difluoroprop-2-ynyl)phosphonate esters, *J. Chem. Soc., Chem. Commun.*, 1447, 1996.

265. Wang, Z., Gu, Y., Zapata, A.J., and Hammond, G.B., An improved preparation of α-fluorinated propargylphosphonates and the solid phase synthesis of α-hydroxy-γ-TIPS propargylphosphonate ester, *J. Fluorine Chem.*, 107, 127, 2001.

266. Lequeux, T., Lebouc, F., Lopin, C., Yang, H., Gouhier, G., and Piettre, S.R., Sulfanyl- and selanyldifluoromethylphosphonates as a source of phosphonodifluoromethyl radicals and their addition onto alkenes, *Org. Lett.*, 3, 185, 2001.

267. Quimby, O.T., Curry, J.D., Allan Nicholson, D., Prentice, J.B., and Roy, C.H., Metalated methylenediphosphonate esters. Preparation, characterization and synthetic applications, *J. Organomet. Chem.*, 13, 199, 1968.

268. Quimby, O.T., and Prentice, J.B., Hypohalogenation of *gem*-diphosphonate esters and phosphonoacetate esters, *Procter and Gamble*, U.S. Patent Appl. US 3772412, 1973; *Chem. Abstr.*, 80, 37278, 1974.

269. Taylor, S.D., Dinaut, A.N., Thadani, A.N., and Huang, Z., Synthesis of benzylic mono(α,α-difluoromethylphosphonates) and benzylic *bis*(α,α-difluoromethylphosphonates) via electrophilic fluorination, *Tetrahedron Lett.*, 37, 8089, 1996.

270. Taylor, S.D., Kotoris, C.C., Dinaut, A.N., and Chen, M.-J., Synthesis of aryl(difluoromethylenephosphonates) via electrophilic fluorination of α-carbanions of benzylic phosphonates with N-fluorobenzenesulfonimide, *Tetrahedron*, 54, 1691, 1998.

271. Taylor, S.D., Kotoris, C.C., Dinaut, A.N., Wang, Q., Ramachandran, C., and Huang, Z., Potent non-peptidyl inhibitors of protein tyrosine phosphatase 1B, *Bioorg. Med. Chem.*, 6, 1457, 1998.

272. Arnone, A., Bravo, P., Frigerio, M., Viani, F., and Zappala, C., Synthesis of 3'-arylsulfonyl-4'-[(diethoxyphosphoryl)difluoromethyl]thymidine analogs, *Synthesis*, 1511, 1998.

273. Lee, K., Shin, W.S., and Oh, D.Y., A facile method for the preparation of 1,1-dichloroolefins using benzenesulfonyl chloride as a chlorenium ion source, *Synth. Commun.*, 21, 1657, 1991.

274. Nicholson, D.A., Halogenated methylenediphosphonates, malonates, and phosphonoacetates, *Procter & Gamble*, German Patent Appl. DE 1948475, 1970; *Chem. Abstr.*, 73, 25656, 1970.

275. Niemi, R., Vepsäläinen, J., Taipale, H., and Järvinen, T., Bisphosphonate prodrugs. Synthesis and *in vitro* evaluation of novel acyloxyalkyl esters of clodronic acid, *J. Med. Chem.*, 42, 5053, 1999.

276. Brown, R.J., Daniel, D.J., Frasier, D.A., Howard, M.H., Jr., Koether, G.M., and Rorer, M.P., Preparation of heterocyclic dihydrazole compounds as agrochemical fungicides, *E.I. Du Pont De Nemours*, U.S. Patent Appl. US 6096895, 2000; *Chem. Abstr.*, 133, 131165, 2000.

277. Waschbüsch, R., Samadi, M., and Savignac, P., A useful magnesium reagent for the preparation of 1,1-difluoro-2-hydroxyphosphonates from diethyl bromodifluoromethylphosphonate via a metal-halogen exchange reaction, *J. Organomet. Chem.*, 529, 267, 1997.

278. Carran, J., Waschbüsch, R., Marinetti, A., and Savignac, P., An improved synthesis of dichloroalkylphosphonates, chloroalkynes and terminal alkynes via diethyl dichloromethylphosphonate, *Synthesis*, 1494, 1996.

279. Marinetti, A., and Savignac, P., Diethyl (dichloromethyl)phosphonate. Preparation and use in the synthesis of alkynes. (4-Methoxyphenyl)ethyne, *Org. Synth.*, 74, 108, 1995.

280. Filonenko, L.P., Bespal'ko, G.K., Marchenko, A.P., and Pinchuk, A.M., Silylation of compounds containing trichloromethyl groups, *Zh. Obshch. Khim.*, 57, 2320, 1987; *J. Gen. Chem. USSR (Engl. Transl.)*, 57, 2074, 1987.

281. Atkinson, R.E., Cadogan, J.I.G., and Dyson, J., The reactivity of organophosphorus compounds. Part 22. Preparation of diethyl trichloromethylphosphonite and diethyl dichloromethylphosphonate, *J. Chem. Soc. (C)*, 2542, 1967.

282. Burton, D.J., and Flynn, R.M., Preparation of *F*-methylene *bis* phosphonates, *J. Fluorine Chem.*, 15, 263, 1980.

283. Patois, C., and Savignac, P., Easy access to α-fluorinated phosphonic acid esters, *Phosphorus, Sulfur Silicon Relat. Elem.*, 77, 163, 1993.

284. Konno, A., and Fuchigami, T., Electrolytic partial fluorination of organic compounds. Part 23. Regioselective anodic difluorination of sulfides using novel fluorine source $Et_4NF \cdot 4HF$, *J. Org. Chem.*, 62, 8579, 1997.

285. Yang, Z.-Y., and Burton, D.J., A novel, general method for the preparation of α,α-difluoro functionalized phosphonates, *Tetrahedron Lett.*, 32, 1019, 1991.

286. Yang, Z.-Y., and Burton, D.J., A novel and practical preparation of α,α-difluoro functionalized phosphonates from iododifluoromethylphosphonate, *J. Org. Chem.*, 57, 4676, 1992.

287. Zhang, X., Qiu, W., and Burton, D.J., Preparation of α-fluorophosphonates, *J. Fluorine Chem.*, 89, 39, 1998.

288. Villemin, D., Sauvaget, F., and Hajek, M., Addition of diethyl trichloromethylphosphonate to olefins catalysed by copper complexes, *Tetrahedron Lett.*, 35, 3537, 1994.

289. Haelters, J.P., Corbel, B., and Sturtz, G., Synthesis of [1- or 3-(arylamino)-2-oxopropyl]phosphonates. Bischler cyclization to indolylphosphonates, *Phosphorus Sulfur*, 37, 65, 1988.

290. Moskva, V.V., Guseinov, F.I., Ismailov, V.M., and Gallyamov, M.R., α,α-Dichloro-α-(diethoxyphosphoryl)acetone, *Zh. Obshch. Khim.*, 57, 234, 1987; *Chem. Abstr.*, 107, 217736, 1987.

291. Pudovik, A.N., Anomalous reaction of α-halo ketones with esters of phosphorous acid. Part 4. Reactions of esters of phosphorus acid with mono- and dichloroacetylacetone, phosphonoacetone, and acetoacetic ester, *Zh. Obshch. Khim.*, 26, 2238, 1956; *Chem. Abstr.*, 51, 1827a, 1957.

292. Shevchenko, V.I., Bondarchuk, P.D., and Kirsanov, A.V., Phosphorylation of esters of malonic acid, *Zh. Obshch. Khim.*, 32, 2994, 1962; *Chem. Abstr.*, 58, 9130f, 1963.

293. Gross, H., and Seibt, H., α-Substituted phosphonic acid esters. Part 10. Synthesis of chloro(methoxy)methylphosphonic esters and chloro(methylthio)methylphosphonic esters, *J. Prakt. Chem.*, 312, 475, 1970.

294. Gross, H., and Seibt, H., Chlorosubstituted methanephosphonic acid esters or thiophosphonic acid esters, German (East) Patent Appl. DD 76972, 1970; *Chem. Abstr.*, 75, 140982, 1971.

295. Mikolajczyk, M., Zatorski, A., Grzejszczak, S., Costisella, B., and Midura, W., α-Phosphoryl sulfoxides. Part 4. Pummerer rearrangements of α-phosphoryl sulfoxides and asymmetric induction in the transfer of chirality from sulfur to carbon, *J. Org. Chem.*, 43, 2518, 1978.

296. Motoyoshiya, J., and Hirata, K., Cycloaddition reactions of (diethylphosphono)ketenes, *Chem. Lett.*, 211, 1988.

297. Mikolajczyk, M., Midura, W.H., Grzejszczak, S., Montanari, F., Cinquini, M., Wieczorek, M.W., and Karolak-Wojciechowska, J., α-Phosphoryl sulfoxides. Part 8. Stereochemistry of α-chlorination of α-phosphoryl sulfoxides, *Tetrahedron*, 50, 8053, 1994.

298. Balczewski, P., Novel 1-phosphonyl radicals derived from 1-mono and 1,1-diheterosubstituted 2-oxoalkylphosphonates as useful phosphoroorganic intermediates in organic synthesis, *Tetrahedron*, 53, 2199, 1997.

299. Isbell, A.F., Synthesis of organophosphorus compounds, *U.S. Dept. Com., Office Tech. Serv.*, 266675, 1961; *Chem. Abstr.*, 58, 11394f, 1963.

300. Burton, D.J., and Flynn, R.M., A facile synthesis of trifluoromethane- and pentafluorobenzenephosphonates, *Synthesis*, 615, 1979.

301. Mahmood, T., and Shreeve, J.M., Simple preparation of dialkyl polyfluoroalkyl phosphonates, *Synth. Commun.*, 17, 71, 1987.

302. Burton, D.J., and Flynn, R.M., Michaelis–Arbuzov preparation of halo-*F*-methylphosphonates, *J. Fluorine Chem.*, 10, 329, 1977.

303. Davisson, V.J., Woodside, A.B., Neal, T.R., Stremler, K.E., Muehlbacher, M., and Poulter, C.D., Phosphorylation of isoprenoid alcohols, *J. Org. Chem.*, 51, 4768, 1986.

304. Kawamoto, A.M., and Campbell, M.M., A new method for the synthesis of a phosphonic acid analogue of phosphoserine via a novel 1,1-difluorophosphonate intermediate, *J. Fluorine Chem.*, 81, 181, 1997.

305. Burton, D.J., Naae, D.G., Flynn, R.M., Smart, B.E., and Brittelli, D.R., Phosphine- and phosphite-mediated difluorocarbene exchange reactions of (bromodifluoromethyl)phosphonium salts. Evidence for facile dissociation of (difluoromethylene)triphenylphosphorane, *J. Org. Chem.*, 48, 3616, 1983.

306. Li, A.-R., and Chen, Q.-Y., Diethyl iododifluoromethylphosphonate. A new synthetic method and its reaction with alkynes, *Synthesis*, 606, 1996.

307. Waschbüsch, R., Carran, J., and Savignac, P., A high yielding synthesis of diethyl 1-fluoromethylphosphonate in pure form, *C.R. Acad. Sci., Ser. IIc*, 49, 1998.

308. Waschbüsch, R., Carran, J., and Savignac, P., A new route to α-fluorovinylphosphonates utilizing Peterson olefination methodology, *Tetrahedron*, 52, 14199, 1996.

309. Kamai, G., and Egorova, L.P., Action of carbon tetrachloride on alkyl esters of phosphorous acid, *Zh. Obshch. Khim.*, 16, 1521, 1946; *Chem. Abstr.*, 41, 5439g, 1947.

310. Kamai, G., Action of carbon tetrachloride on the esters of phosphorous acid and benzenephosphonous acid, *Dokl. Akad. Nauk SSSR*, 55, 219, 1947; *Chem. Abstr.*, 41, 5863g, 1947.

311. Kosolapoff, G.M., Isomerization of alkyl phosphites. Part 6. Reactions with chlorides of singular structure, *J. Am. Chem. Soc.*, 69, 1002, 1947.

312. Bengelsdorf, I.S., and Barron, L.B., Trichloromethylphosphonic acid, *J. Am. Chem. Soc.*, 77, 2869, 1955.

313. Kamai, G., and Kharrasova, F.M., Action of carbon tetrachloride on mixed alkyl esters of phosphorous acid, *Zh. Obshch. Khim.*, 27, 953, 1957; *Chem. Abstr.*, 52, 3666a, 1958.

314. Griffin, C.E., Reaction of trichloromethyl radicals with triethyl phosphite, *Chem. Ind. (London)*, 415, 1958.

315. Kharrasova, F.M., Zykova, T.V., Salakhutdinov, R.A., Efimova, V.D., and Shafigullina, R.D., Reaction of phosphorus(III) acid esters with carbon tetrachloride, *Zh. Obshch. Khim.*, 44, 2419, 1974; *Chem. Abstr.*, 82, 73097, 1975.

316. Downie, I.M., Wynne, N., and Harrison, S., A high yield route to ethyl esters of carboxylic acids, *Tetrahedron*, 38, 1457, 1982.

317. Steinbach, J., Herrmann, E., and Riesel, L., Reaction of the two-component system trialkyl phosphite/carbon tetrachloride with nucleophiles containing hydrogen. Part 2. Reaction with ammonia and amines, *Z. Anorg. Allg. Chem.*, 523, 180, 1985.

318. Le Menn, J.C., Tallec, A., and Sarrazin, J., Organic electrosynthesis without a potentiostat, *J. Chem. Ed.*, 68, 513, 1991.

319. Yurchenko, R.I., Klepa, T.I., Bobrova, O.B., Yurchenko, A.G., and Pinchuk, A.M., Phosphorylated adamantanes. Part 2. Adamantyl esters of phosphorous acid in an Arbuzov reaction, *Zh. Obshch. Khim.*, 51, 786, 1981; *Chem. Abstr.*, 95, 133031, 1981.

320. Yurchenko, R.I., and Klepa, T.I., Leaving group in the Arbuzov reaction of diethyl 1-adamantyl phosphite, *Zh. Obshch. Khim.*, 54, 714, 1984; *Chem. Abstr.*, 101, 91060, 1984.

321. Gazizov, T.K., Ustanova, L.N., Ryzhikov, D.V., and Pudovik, A.N., Interaction of dialkyl trimethylsilyl phosphites with carbon tetrachloride, *Izv. Akad. Nauk SSSR, Ser. Khim.*, 1830, 1984; *Chem. Abstr.*, 101, 230664, 1984.

322. Kamai, G., Action of haloforms on alkyl esters of phosphorous acid, *Dokl. Akad. Nauk SSSR*, 79, 795, 1951; *Chem. Abstr.*, 46, 6081f, 1952.

323. Walling, C., and Rabinowitz, R., The reaction of trialkyl phosphites with thiyl and alkoxy radicals, *J. Am. Chem. Soc.*, 81, 1243, 1959.

324. Kamai, G., and Kharrasova, F.M., Reaction rate of some esters of phosphorous acid with carbon tetrachloride, *Izv. Vysshikh Ucheb. Zaved. Khim. Khim. Tekhnol.*, 4, 229, 1961; *Chem. Abstr.*, 55, 21762e, 1961.

325. Cadogan, J.I.G., and Foster, W.R., Reactivity of organophosphorus compounds. Part 5. Reaction of trichloromethyl radicals with trialkyl phosphites, *J. Chem. Soc.*, 3071, 1961.

326. Cadogan, J. I.G., and Sharp, J.T., Duality of mechanism in the reaction of triethyl phosphite and carbon tetrachloride, *Tetrahedron Lett.*, 7, 2733, 1966.

327. Bakkas, S., Julliard, M., and Chanon, M., Reactivity of triethyl phosphite with tetrachloromethane. Electron transfer versus ionic substitution on "positive" halogen, *Tetrahedron*, 43, 501, 1987.

328. Bakkas, S., Mouzdahir, A., Khamliche, L., Julliard, M., Peralez, E., and Chanon, M., Difference in behavior of the reactive electrophiles diethyl trichloromethylphosphonate and carbon tetrachloride towards triethyl phosphite, *Phosphorus, Sulfur Silicon Relat. Elem.*, 157, 211, 2000.

329. Kukhar, V.P., and Sagina, E.I., Reactions of trialkyl phosphites with polyhalomethanes, *Zh. Obshch. Khim.*, 49, 1470, 1979; *Chem. Abstr.*, 91, 140921, 1979.

330. Kennard, K.C., and Hamilton, C.S., The synthesis of certain organophosphorus compounds containing the trichloromethyl group, *J. Am. Chem. Soc.*, 77, 1156, 1955.

331. Kennard, K.C., and Hamilton, C.S., Trichloromethylphosphonyl dichloride [phosphonic dichloride, (trichloromethyl)-], *Org. Synth. Coll.*, IV, 950, 1963.

332. Makarov, A.M., and Gabov, N.I., *O,O-Bis*(polyfluoroalkyl) (trichloromethyl)phosphonates, *Zh. Obshch. Khim.*, 51, 963, 1981; *Chem. Abstr.*, 95, 81130, 1981.

333. Semchenko, F.M., Eremin, O.G., and Martynov, B.I., New method for trifluoromethylation of 4-coordinate phosphorus derivatives, *Zh. Obshch. Khim.*, 62, 473, 1992; *Chem. Abstr.*, 118, 22316, 1993.

334. Shibaev, V.I., Garabadzhiu, A.V., and Rodin, A.A., Reaction of perfluoroiodoalkanes with phosphorus(III) acid esters. Part 6. Synthesis of trifluoromethyl- and heptafluoropropyl-phosphonates, *Zh. Obshch. Khim.*, 53, 1743, 1983; *Chem. Abstr.*, 100, 6679, 1984.

335. Bennett, F.W., Emeleus, H.J., and Haszeldine, R.N., Organometallic and organometalloidal fluorine compounds. Part 10. Trifluoromethyl-phosphonous and -phosphonic acids, *J. Chem. Soc*, 3598, 1954.

336. Maslennikov, I.G., Lavrent'ev, A.N., Lyubimova, M.V., Shvedova, Y.I., and Lebedev, V.B., Dialkyl (trifluoromethyl)phosphonates, *Zh. Obshch. Khim.*, 53, 2681, 1983; *Chem. Abstr.*, 100, 121230, 1984.

337. Blinova, G.G., Burova, O.N., Lavrent'ev, A.N., and Mel'nikova, L.N., Ethyl *p*-nitrophenyl (haloalkyl)phosphonates, *Zh. Obshch. Khim.*, 56, 1277, 1986; *Chem. Abstr.*, 106, 50314, 1987.

338. Emeléus, H.J., Haszeldine, R.N., and Paul, R.C., Organometallic and organometalloidal fluorine compounds. Part 12. Bistrifluoromethylphosphinic acid and related phosphorus oxyacids, *J. Chem. Soc*, 563, 1955.

339. Emeléus, H.J., and Smith, J.D., The heptafluoropropyliodophosphines and their derivatives, *J. Chem. Soc.*, 375, 1959.

340. Blackburn, G.M., and Guo, M.-J., Trifluoromethylphosphinyl *bis*-triazolides in the synthesis of trifluoromethylphosphonate analogues of nucleotides, *Tetrahedron Lett.*, 34, 149, 1993.

341. Mayer, M., Ugi, I., and Richter, W., Studies on trifluoromethylphosphonamidite analogues as building blocks in oligonucleotide synthesis, *Tetrahedron Lett.*, 36, 2047, 1995.

342. Karl, R.M., Richter, W., Klösel, R., Mayer, M., and Ugi, I., The 1,1-dianisyl-2,2,2-trichloroethyl moiety as a new protective group for the synthesis of dinucleoside trifluoromethylphosphonates, *Nucleosides Nucleotides*, 15, 379, 1996.

343. Moriarty, R.M., Liu, K., Tuladhar, S.M., Guo, L., Condeiu, C., Tao, A., Xu, W., Lenz, D., and Brimfield, A., Synthesis of pentacoordinate phosphorus haptens for catalytic antibody production. Bait and switch concept, *Phosphorus, Sulfur Silicon Relat. Elem.*, 109, 237, 1996.

344. Kabachnik, M.I., and Medved, T.Y., Synthesis of aminomethanephosphonic acid, *Izv. Akad. Nauk S.S.S.R., Otdel. Khim. Nauk*, 95, 1951; *Chem. Abstr.*, 46, 421c, 1952.

345. Kabachnik, M.I., and Medved, T.Y., Aminomethylphosphonic acid, *Akad. Nauk S.S.S.R., Inst. Org. Khim., Sintezy Org. Soedinenii, Sbornik*, 2, 12, 1952; *Chem. Abstr.*, 48, 564c, 1954.

346. Maier, L., Organic phosphorus compounds. Part 98. Synthesis and properties of *N*-methylaminomethylphosphonic acid and derivatives, *Phosphorus, Sulfur Silicon Relat. Elem.*, 62, 29, 1991.

347. Kreutzkamp, N., and Mengel, W., Carbonyl- and cyanophosphonic acid esters. Part 8. Diethylphospho-noalkylmalonates, *Arch. Pharm. Ber. Dtsch. Pharm. Ges.*, 295, 773, 1962.

348. Tsirul'nikova, N.V., Temkina, V.Y., Sushitskaya, T.M., and Rykov, S.V., Iminobis(methylenephospho-nic acid) and its reactivity, *Zh. Obshch. Khim.*, 51, 1028, 1981; *Chem. Abstr.*, 95, 204054, 1981.

349. Schwarzenbach, G., Ackermann, H., and Ruckstuhl, P., Complex ions. Part 15. New derivatives of iminodiacetic acid and their alkaline earth complexes. Connection between acidity and complex formation, *Helv. Chim. Acta*, 32, 1175, 1949.

350. Cameron, D.G., Hudson, H.R., Ojo, I.A.O., and Pianka, M., Organophosphorus compounds as poten-tial fungicides. Part 1. *N*-(ω-guanidinoalkyl)aminoalkanephosphonic acids and their aminophosphonic precursors. Preparation, NMR spectroscopy, and fast atom bombardment mass spectrometry, *Phos-phorus Sulfur*, 40, 183, 1988.

351. Hudson, H.R., and Pianka, M., An approach to the development of organophosphorus fungicides, *Phosphorus, Sulfur Silicon Relat. Elem.*, 109, 345, 1996.

352. Berté-Verrando, S., Nief, F., Patois, C., and Savignac, P., Preparation of α-dideuterated aminometh-ylphosphonic acid, *Phosphorus, Sulfur Silicon Relat. Elem.*, 103, 91, 1995.

353. Berté-Verrando, S., Dizière, R., Samadi, M., and Savignac, P., New route to amino[²H₂]methylphosphonic acid via *bis*(trifluoroethyl) phosphonate transesterification, *J. Chem. Soc., Perkin Trans. 1*, 3125, 1995.

354. Hanessian, S., and Bennani, Y.L., A versatile asymmetric synthesis of α-amino α-alkylphosphonic acids of high enantiomeric purity, *Tetrahedron Lett.*, 31, 6465, 1990.

355. Fu, J.-M., and Castelhano, A.L., Design and synthesis of a pyridone-based phosphotyrosine mimetic, *Bioorg. Med. Chem. Lett.*, 8, 2813, 1998.

356. Fu, J.-M., Chen, Y., and Castelhano, A.L., Synthesis of functionalized pyridones via palladium catalyzed cross coupling reactions, *Synlett*, 1408, 1998.

357. Tanaka, J., Kuwano, E., and Eto, M., Synthesis and pesticidal activities of phosphonate analogs of amino acids, *J. Fac. Agric., Kyushu Univ.*, 30, 209, 1986; *Chem. Abstr.*, 106, 133666, 1987.

358. Chun, Y.-J., Park, J.-H., Oh, G.-M., Hong, S.-I., and Kim, Y.-J., Synthesis of ω-phthalimidoalkylphos-phonates, *Synthesis*, 909, 1994.

359. Jones, G.H., Hamamura, E.K., and Moffatt, J.G., A new stable Wittig reagent suitable for the synthesis of α,β-unsaturated phosphonates, *Tetrahedron Lett.*, 9, 5731, 1968.

360. Böhringer, M.P., Graff, D., and Caruthers, M.H., Synthesis of 5′-deoxy-5′-methylphosphonate linked thymidine oligonucleotides, *Tetrahedron Lett.*, 34, 2723, 1993.

361. Szabo, T., and Stawinski, J., Synthesis and some conformational features of the 5′-deoxy-5′-meth-ylphosphonate linked dimer, 5′-deoxy-5′-*C*-(phosphonomethyl)thymidin-3′-yl (thymidin-5′-yl)meth-ylphosphonate [*p*(CH₂)T*p*(CH₂)T], *Tetrahedron*, 51, 4145, 1995.

362. Zhao, Z., and Caruthers, M.H., Synthesis and preliminary biochemical studies with 5′-deoxy-5′-methylidyne phosphonate linked thymidine oligonucleotides, *Tetrahedron Lett.*, 37, 6239, 1996.

363. Ginsburg, V.A., and Yakubovich, A.Y., Synthesis of heteroorganic compounds of aliphatic series by the diazo method. Part 8. Synthesis of compounds of elements of group V. Organophosphorus compounds. Syntheses of *bis*- and *tris*(haloalkyl)phosphines and some transformations of chloroalkyl derivatives of phosphorus, *Zh. Obshch. Khim.*, 28, 728, 1958; *Chem. Abstr.*, 52, 17091g, 1958.

364. Schwarzenbach, G., and Zurc, J., The pyro- and hypophosphorus acids in comparison with organic diphosphonic acids, *Monatssch. Chem.*, 81, 202, 1950.

365. Petrov, K.A., Maklyaev, F.L., and Bliznyuk, N.K., Diphosphonates. Part 1. Esters of methylenediphos-phonic acid, *Zh. Obshch. Khim.*, 30, 1602, 1960; *J. Gen. Chem. USSR (Engl. Transl.)*, 30, 1604, 1960.

366. Richard, J.J., Burke, K.E., O'Laughlin, J.W., and Banks, C.V., *gem-Bis*(disubstituted-phosphi-nyl)alkanes. Part 1. Synthesis and properties of *bis*(di-*n*-hexylphosphinyl)methane and related com-pounds, *J. Am. Chem. Soc.*, 83, 1722, 1961.

367. Maier, L., Organic phosphorus compounds. Part 40. Preparation of pentamethylene-1,5-*bis*(dihydrox-yphosphonylmethylphosphinic acid), *Helv. Chim. Acta*, 53, 1940, 1970.

368. Vaghefi, M.M., Bernacki, R.J., Hennen, W.J., and Robins, R.K., Synthesis of certain nucleoside methylenediphosphonate sugars as potential inhibitors of glycosyltransferases, *J. Med. Chem.*, 30, 1391, 1987.

369. Saady, M., Lebeau, L., and Mioskowski, C., First use of benzyl phosphites in the Michaelis–Arbuzov reaction synthesis of mono-, di-, and triphosphate analogs, *Helv. Chim. Acta*, 78, 670, 1995.

370. Saady, M., Lebeau, L., and Mioskowski, C., Synthesis of di- and triphosphate ester analogs via a modified Michaelis–Arbuzov reaction, *Tetrahedron Lett.*, 36, 5183, 1995.

371. Craniades, P., and Rumpf, P., Phosphonic acids, a new series comparable to cytoactive carboxylic acids, *Bull. Soc. Chim. Fr.*, 1194, 1955.

372. Maguire, M.H., and Shaw, G., Synthetic plant hormones. Part 3. Aryloxymethylphosphonates, *J. Chem. Soc.*, 1756, 1955.

373. Dillard, R.D., Bach, N.J., Draheim, S.E., Berry, D.R., Carlson, D.G., Chirgadze, N.Y., Clawson, D.K., Hartley, L.W., Johnson, L.M., Jones, N.D., McKinney, E.R., Mihelich, E.D., Olkowski, J.L., Schevitz, R.W., Smith, A.C., Snyder, D.W., Sommers, C.D., and Wery, J.-P., Indole inhibitors of human non-pancreatic secretory phospholipase A_2. Part 2. Indole-3-acetamides with additional functionality, *J. Med. Chem.*, 39, 5137, 1996.

374. Walsh, E.N., Beck, T.M., and Toy, A.D.F., Phenoxymethylphosphonic acids and phosphonic acid ion-exchange resins, *J. Am. Chem. Soc.*, 78, 4455, 1956.

375. Cornforth, J., and Wilson, J.R.H., A general reagent for *O*-phosphonomethylation of phenols, *J. Chem. Soc., Perkin Trans. 1*, 1897, 1994.

376. Arbuzov, B.A., and Bogonostseva, N.P., Synthesis of some phosphonosulfides and phosphonosulfones, *Zh. Obshch. Khim.*, 27, 2360, 1957; *Chem. Abstr.*, 52, 7122g, 1958.

377. Wegener, W., and Courault, K., Synthesis of β-substituted vinyl methyl sulfones, *Z. Chem.*, 20, 337, 1980.

378. Villemin, D., and Thibault-Starzyk, F., Synthesis of a new sulfur analog of PMEA, *Synth. Commun.*, 23, 1053, 1993.

379. Arai, Y., Matsui, M., and Koizumi, T., Diastereoselective synthesis of 10-(alkylsulfinyl)- and 10-(alkenylsulfinyl)isoborneols by oxidation of the corresponding sulfides with 3-chloroperoxybenzoic acid, *Synthesis*, 320, 1990.

380. Savignac, P., and Lavielle, G., Monodealkylation of phosphoric and phosphonic esters by thiolates and thiophenates, *Bull. Soc. Chim. Fr.*, 1506, 1974.

381. Mikolajczyk, M., Grzejszczak, S., Chefczynska, A., and Zatorski, A., Addition of elemental sulfur to phosphonate carbanions and its application for the synthesis of α-phosphoryl organosulfur compounds. Synthesis of aromatic ketones, *J. Org. Chem.*, 44, 2967, 1979.

382. Logusch, E.W., A simple preparation of *S*-alkyl homocysteine derivatives. *S*-Phosphonomethyl homocysteines as inhibitors of glutamine synthetase, *Tetrahedron Lett.*, 29, 6055, 1988.

383. Bigge, C.F., Johnson, G., Ortwine, D.F., Drummond, J.T., Retz, D.M., Brahce, L.J., Coughenour, L.L., Marcoux, F.W., and Probert, A.W., Jr., Exploration of *N*-phosphonoalkyl-, *N*-phosphonoalkenyl-, and *N*-(phosphonoalkyl)phenyl-spaced α-amino acids as competitive *N*-methyl-D-aspartic acid antagonists, *J. Med. Chem.*, 35, 1371, 1992.

384. Hadd, M.J., Smith, M.A., and Gervay-Hague, J., A novel reagent for the synthesis of geminal di-sulfones, *Tetrahedron Lett.*, 42, 5137, 2001.

385. Mizrakh, L.I., Yakovlev, V.G., Yukhno, E.M., Mamonov, V.I., and Svergun, V.I., Phosphorus-containing isothiuronium derivatives, *Zh. Obshch. Khim.*, 41, 2654, 1971; *J. Gen. Chem. USSR (Engl. Transl.)*, 41, 2687, 1971.

386. Allan, R.D., and Tran, H.W., Facile synthesis of β-phenylethylamine derivatives related to baclofen via aziridine ring opening, *Aust. J. Chem.*, 43, 1123, 1990.

387. Comasseto, J.V., and Petragnani, N., The reaction of selenophosphonates with carbonyl compounds. Vinylic selenides, *J. Organomet. Chem.*, 152, 295, 1978.

388. Ley, S.V., O'Neil, I.A., and Low, C.M.R., Ultrasonic formation and reactions of sodium phenylse-lenide, *Tetrahedron*, 42, 5363, 1986.

389. Lee, C.-W., Koh, Y.J., and Oh, D.Y., Preparation of 1-telluroalkylphosphonates. New synthetic route to vinyl tellurides, *J. Chem. Soc., Perkin Trans. 1*, 717, 1994.

390. Poindexter, M.K., and Katz, T.J., Simple conversion of aryl bromides to arylmethylphosphonates, *Tetrahedron Lett.*, 29, 1513, 1988.

391. Wang, C., and Dalton, L.R., A facile synthesis of thienylmethylphosphonates. Direct conversion from thiophenes, *Tetrahedron Lett.*, 41, 617, 2000.

392. Turbiez, M., Frere, P., Blanchard, P., and Roncali, J., Mixed π-conjugated oligomers of thiophene and 3,4-ethylenedioxythiophene (EDOT), *Tetrahedron Lett.*, 41, 5521, 2000.

393. Teulade, M.-P., Savignac, P., Aboujaoude, E.E., and Collignon, N., α-Lithiated phosphonate carbanions. Synthesis, basicity and stability to self-condensation, *J. Organomet. Chem.*, 312, 283, 1986.

394. Blackburn, G.M., and Parratt, M.J., The synthesis of α-fluoroalkylphosphonates, *J. Chem. Soc., Chem. Commun.*, 886, 1983.

395. Blackburn, G.M., and Parratt, M.J., Synthesis of α-fluoroalkylphosphonates, *J. Chem. Soc., Perkin Trans. 1,* 1425, 1986.

396. Blackburn, G.M., Brown, D., and Martin, S.J., A novel synthesis of fluorinated phosphonoacetic acids, *J. Chem. Res. (S)*, 92, 1985.

397. Van Steenis, J. H., Boer, P.W. S., Van der Hoeven, H.A., and Van der Gen, A., Formation of enantiomerically pure 1-fluorovinyl and 1-fluoromethyl sulfoxides, *Eur. J. Org. Chem.*, 911, 2001.

398. Zhang, X., Qiu, W., and Burton, D.J., The preparation of $(EtO)_2P(O)CFHZnBr$ and $(EtO)_2P(O)CFHCu$ and their utility in the preparation of functionalized α-fluorophosponates, *Tetrahedron Lett.*, 40, 2681, 1999.

399. Patois, C., and Savignac, P., A new route to α-fluoromethyl- and α-fluoroalkyl-phosphonates, *J. Chem. Soc., Chem. Commun.*, 1711, 1993.

400. Waschbüsch, R., Carran, J., and Savignac, P., A new route to α-fluoroalkylphosphonates, *J. Chem. Soc., Perkin Trans. 1,* 1135, 1997.

401. Nieschalk, J., and O'Hagan, D., Monofluorophosphonates as phosphate mimics in bioorganic chemistry. A comparative study of CH_2-, CHF- and CF_2-phosphonate analogs of *sn*-glycerol-3-phosphate as substrates for *sn*-glycerol-3-phosphate dehydrogenase, *J. Chem. Soc., Chem. Commun.*, 719, 1995.

402. Nieschalk, J., Batsanov, A.S., O'Hagan, D., and Howard, J.A.K., Synthesis of monofluoro- and difluoromethylenephosphonate analogs of *sn*-glycerol-3-phosphate as substrates for glycerol-3-phosphate dehydrogenase and the X-ray structure of the fluoromethylenephosphonate moiety, *Tetrahedron*, 52, 165, 1996.

403. Patois, C., and Savignac, P., Preparation of triethyl 2-fluoro-2-phosphonoacetate, *Synth. Commun.*, 24, 1317, 1994.

404. Waschbüsch, R., Carran, J., and Savignac, P., New routes to diethyl 1-fluoromethylphosphonocarboxylates and diethyl 1-fluoromethylphosphonocarboxylic acid, *Tetrahedron*, 53, 6391, 1997.

405. Keeney, A., Nieschalk, J., and O'Hagan, D., The synthesis of α-monofluorovinylphosphonates by a Peterson type olefination reaction, *J. Fluorine Chem.*, 80, 59, 1996.

406. Schmitt, L., Cavusoglu, N., Spiess, B., and Schlewer, G., Synthesis of arylalkylmonofluorophosphonates as *myo*-inositol monophosphatase ligands, *Tetrahedron Lett.*, 39, 4009, 1998.

407. Berkowitz, D.B., Bose, M., and Asher, N.G., A convergent triflate displacement approach to (α-monofluoroalkyl)phosphonates, *Org. Lett.*, 3, 2009, 2001.

408. Thenappan, A., and Burton, D.J., An expedient synthesis of α-fluoro-β-keto esters, *Tetrahedron Lett.*, 30, 6113, 1989.

409. Thenappan, A., and Burton, D.J., Acylation of fluorocarbethoxy-substituted ylids. A simple and general route to α-fluoro β-keto esters, *J. Org. Chem.*, 56, 273, 1991.

410. Tsai, H.-J., Thenappan, A., and Burton, D.J., An expedient synthesis of α-fluoro-α,β-unsaturated diesters, *Tetrahedron Lett.*, 33, 6579, 1992.

411. Tsai, H.-J., Thenappan, A., and Burton, D.J., A novel intramolecular Horner–Wadsworth–Emmons reaction. A simple and general route to α-fluoro-α,β-unsaturated diesters, *J. Org. Chem.*, 59, 7085, 1994.

412. Tsai, H.-J., Isolation and characterization of intermediate in the synthesis of α-fluorodiesters, *J. Chin. Chem. Soc. (Taipei)*, 45, 543, 1998; *Chem. Abstr.*, 129, 244849, 1998.

413. Shen, Y., and Ni, J., A novel sequential transformation of phosphonate. Highly stereoselective synthesis of perfluoroalkylated α-fluoro-α,β-unsaturated esters, *J. Org. Chem.*, 62, 7260, 1997.

414. Kim, D.Y., Choi, J.S., and Rhie, D.Y., P–C bond cleavage of triethyl 2-fluoro-3-oxo-2-phosphonoacetates with magnesium chloride. A synthesis of α-fluoro-β-keto esters, *Synth. Commun.*, 27, 1097, 1997.

415. Kim, D.Y., A facile P–C bond cleavage of 2-fluoro-2-phosphonyl-1,3-dicarbonyl compounds on silica gel, *Synth. Commun.*, 30, 1205, 2000.

416. Kim, D.Y., and Kim, J.Y., A new synthesis of α-fluoromalonates from α-fluoro-α-phosphonylmalonates using P-C bond cleavage, *Synth. Commun.*, 28, 2483, 1998.

417. Kim, D.Y., Rhie, D.Y., and Oh, D.Y., Acylation of diethyl (ethoxycarbonyl)fluoromethylphosphonate using magnesium chloride-triethylamine. A facile synthesis of α-fluoro β-keto esters, *Tetrahedron Lett.*, 37, 653, 1996.

418. Kim, D.Y., Lee, Y.M., and Choi, Y.J., Acylation of α-fluorophosphonoacetate derivatives using magnesium chloride-triethylamine, *Tetrahedron*, 55, 12983, 1999.

419. Kim, D.Y., and Choi, Y.J., Synthesis of α-fluoro-β-keto phosphonates from α-fluoro phosphonoacetic acid, *Synth. Commun.*, 28, 1491, 1998.

420. Hanessian, S., Bennani, Y.L., and Delorme, D., The asymmetric synthesis of α-chloro-α-alkyl- and α-methyl-α-alkylphosphonic acids of high enantiomeric purity, *Tetrahedron Lett.*, 31, 6461, 1990.

421. Yuan, C., Li, S., and Wang, G., A new and efficient asymmetric synthesis of 1-amino-1-alkylphosphonic acids, *Heteroatom Chem.*, 11, 528, 2000.

422. Giovenzana, G.B., Pagliarin, R., Palmisano, G., Pilati, T., and Sisti, M., Camphor-based oxazaphospholanes as chiral templates for the enantioselective synthesis of α-chlorophosphonic acids, *Tetrahedron: Asymmetry*, 10, 4277, 1999.

423. Hanessian, S., Cantin, L.-D., Roy, S., Andreotti, D., and Gomtsyan, A., The synthesis of enantiomerically pure, symmetrically substituted cyclopropane phosphonic acids. A constrained analog of the GABA antagonist phaclophen, *Tetrahedron Lett.*, 38, 1103, 1997.

424. Savignac, P., Dreux, M., and Coutrot, P., Halogen–metal exchange. Synthesis of dichloromethylphosphonates, *Tetrahedron Lett.*, 16, 609, 1975.

425. Savignac, P., and Coutrot, P., Preparation of 1,1-dibromoalkenes by halogen exchange, *Synthesis*, 197, 1976.

426. Sato, H., Isono, N., Miyoshi, I., and Mori, M., 1,1-, 1,2-, and 1,4-eliminations from the corresponding dihalogenated compounds using Bu$_3$SnSiMe$_3$-F-, *Tetrahedron*, 52, 8143, 1996.

427. Coutrot, P., and Savignac, P., A one-step synthesis of diethyl 1,2-epoxyalkanephosphonates, *Synthesis*, 34, 1978.

428. Teulade, M.-P., and Savignac, P., Synthesis of 1-formyl 1,1-dialkylmethylphosphonates, *Synth. Commun.*, 17, 125, 1987.

429. Lavielle, G., Carpentier, M., and Savignac, P., Stereochemical orientation in Darzens reaction applied to (chloromethyl)phosphonates and phosphonamides, *Tetrahedron Lett.*, 14, 173, 1973.

430. Savignac, P., and Coutrot, P., A direct conversion of aldehydes into 2-oxoalkanephosphonates via the diethyl α-lithiochloromethanephosphonate anion, *Synthesis*, 682, 1978.

431. Coutrot, P., Elgadi, A., and Grison, C., Highly functionalized aziridines. Part 2. A facile synthesis of diethyl aziridinylphosphonates and 2-chloroaziridinylphosphonates, *Heterocycles*, 28, 1179, 1989.

432. Davis, F.A., and McCoull, W., Asymmetric synthesis of aziridine 2-phosphonates and azirinyl phosphonates from enantiopure sulfinimines, *Tetrahedron Lett.*, 40, 249, 1999.

433. Davis, F.A., McCoull, W., and Titus, D.D., Asymmetric synthesis of α-methylphosphophenylalanine derivatives using sulfinimine-derived enantiopure aziridine-2-phosphonates, *Org. Lett.*, 1, 1053, 1999.

434. Kim, D.Y., Suh, K.H., Choi, J.S., Mang, J.Y., and Chang, S.K., Asymmetric synthesis of aziridinyl phosphonates using Darzens-type reaction of chloromethyl phosphonate to chiral sulfinimines, *Synth. Commun.*, 30, 87, 2000.

435. Davis, F.A., Wu, Y., Yan, H., Prasad, K.R., and McCoull, W., 2*H*-Azirine 3-phosphonates. A new class of chiral iminodienophiles. Asymmetric synthesis of quaternary piperidine phosphonates, *Org. Lett.*, 4, 655, 2002.

436. Hanessian, S., Bennani, Y.L., and Hervé, Y., A novel asymmetric synthesis of α- and β-amino aryl phosphonic acids, *Synlett*, 35, 1993.

437. Yoffe, S.T., Vatsuro, K.V., Petrovskii, P.V., and Kabachnik, M.I., The influence of structural factors and the solvent on the nature of the enolization of the formyl group, *Izv. Akad. Nauk SSSR, Ser. Khim.*, 731, 1971; *Bull. Acad. Sci. USSR, Div. Chem. Sci. (Engl. Transl.)*, 655, 1971.

438. Aboujaoude, E.E., Collignon, N., and Savignac, P., Synthesis of β-carbonylated phosphonates. Part 1. Carbanionic route, *J. Organomet. Chem.*, 264, 9, 1984.

439. Teulade, M.P., Savignac, P., Aboujaoude, E.E., and Collignon, N., Acylation of diethyl α-chloromethylphosphonates by carbanionic route. Preparation of α-clorinated β-carbonylated phosphonates, *J. Organomet. Chem.*, 287, 145, 1985.

440. Savignac, P., Snoussi, M., and Coutrot, P., Carboxychloro olefination. A convenient synthesis of α-chloro-α,β-ethylenic carboxylic acids, *Synth. Commun.*, 8, 19, 1978.

441. Coutrot, P., and Ghribi, A., A facile and general, one-pot synthesis of 2-oxoalkane phosphonates from diethylphosphonocarboxylic acid chlorides and organometallic reagents, *Synthesis*, 661, 1986.

442. Tay, M.K., About-Jaudet, E., Collignon, N., Teulade, M.P., and Savignac, P., α-Lithioalkylphosphonates as functional group carriers. An *in situ* acrylic ester synthesis, *Synth. Commun.*, 18, 1349, 1988.

443. Aboujaoude, E.E., Lietjé, S., Collignon, N., Teulade, M.P., and Savignac, P., Direct conversion of alkylphosphonates to vinylphosphonates, *Tetrahedron Lett.*, 26, 4435, 1985.

444. Teulade, M.-P., Savignac, P., Aboujaoude, E.E., Lietgé, S., and Collignon, N., Alkylidenediphosphonates and vinylphosphonates. A selective synthetic course via carbanions, *J. Organomet. Chem.*, 304, 283, 1986.

445. Teulade, M.-P., and Savignac, P., A general one-pot synthesis of 1,3-butadienyl phosphanes, *Tetrahedron Lett.*, 30, 6327, 1989.

446. Berté-Verrando, S., Nief, F., Patois, C., and Savignac, P., General synthesis of α,α-dideuteriated phosphonic esters, *J. Chem. Soc., Perkin Trans. 1*, 821, 1994.

447. Zanella, Y., Berté-Verrando, S., Dizière, R., and Savignac, P., New route to 1-formylalkylphosphonates using diethyl trichloromethylphosphonate as a precursor, *J. Chem. Soc., Perkin Trans. 1*, 2835, 1995.

448. Pilard, J.-F., Gaumont, A.-C., Friot, C., and Denis, J.-M., Intramolecular [4 + 2] cycloadditions involving transient phosphaalkene intermediates as dienophiles. A useful entry to phosphabicyclo[4.3.0]non-4-ene derivatives, *J. Chem. Soc., Chem. Commun.*, 457, 1998.

449. Grandin, C., Collingnon, N., and Savignac, P., A practical synthesis of cycloalkylphosphonates from trichloromethylphosphonates, *Synthesis*, 239, 1995.

450. Dizière, R., and Savignac, P., A new simple method for the synthesis of 1-alkynylphosphonates using $(EtO)_2P(O)CCl_3$ as precursor, *Tetrahedron Lett.*, 37, 1783, 1996.

451. Lee, J.W., Kim, T.H., and Oh, D.Y., One-pot synthesis of acetylenic sulfones from chloro(phenylsulfonyl)methanephosphonate, *Synth. Commun.*, 19, 2633, 1989.

452. Kumaraswamy, S., and Swamy, K.C.K., A convenient route to aryl substituted chloro and bromo olefins, *Tetrahedron Lett.*, 38, 2183, 1997.

453. Muthiah, C., Kumar, K.P., Kumaraswamy, S., and Swamy, K.C.K., An easy access to trisubstituted vinyl chlorides and improved synthesis of chloro/bromostilbenes, *Tetrahedron*, 54, 14315, 1998.

454. Hewkin, C.T., Di Fabio, R., Conti, N., Cugola, A., Gastaldi, P., Micheli, F., and Quaglia, A.M., New synthesis of substituted 2-carboxyindole derivatives. Versatile introduction of a carbamoylethynyl moiety at the C-3 position, *Arch. Pharm. (Weinheim Ger.)*, 332, 55, 1999.

455. Muthiah, C., Kumar, K.P., Mani, C.A., and Swamy, K.C.K., Chlorophosphonates. Inexpensive precursors for stereodefined chloro-substituted olefins and unsymmetrical disubstituted acetylenes, *J. Org. Chem.*, 65, 3733, 2000.

456. Perriot, P., Villieras, J., and Normant, J.F., Diethyl 1-chloro-1-lithioalkanephosphonates. A general synthesis of diethyl 1,2-epoxyalkanephosphonates and 1-alkoxycarbonyl-1-chloroalkanephosphonates, *Synthesis*, 33, 1978.

457. Kim, Y., Singer, R.A., and Carreira, E.M., Total synthesis of macrolactin A with versatile catalytic, enantioselective dienolate aldol addition reactions, *Angew. Chem.*, 110, 1321, 1998; *Angew. Chem. Int. Ed. Engl.*, 37, 1261, 1998.

458. Cheng, F., Yang, X., Zhu, H., and Song, Y., Synthesis and optical properties of tetraethyl methano[60]fullerenediphosphonate, *Tetrahedron Lett.*, 41, 3947, 2000.

459. Pellicciari, R., Natalini, B., Amori, L., Marinozzi, M., and Seraglia, R., Synthesis of methano[60]fullerenephosphonic- and methano[60]fullerenediphosphonic acids, *Synlett*, 1816, 2000.

460. Nuretdinov, I.A., Gubskaya, V.P., Berezhnaya, L.S., Il'yasov, A.V., and Azancheev, N.M., Synthesis of phosphorylated methanofullerenes, *Russ. Chem. Bull. (Engl. Transl.)*, 49, 2048, 2000; *Chem. Abstr.*, 135, 46238, 2001.

461. Yuan, C., Li, C., and Ding, Y., Studies on organophosphorus compounds. Part 51. A new and facile route to 2-substituted 1,1-cyclopropanediyl*bis*(phosphonic acids), *Synthesis*, 854, 1991.

462. Kim, C.U., Luh, B.Y., Misco, P.F., Bronson, J.J., Hitchcock, M.J.M., Ghazzouli, I., and Martin, J.C., Acyclic purine phosphonate analogues as antiviral agents. Synthesis and structure–activity relationships, *J. Med. Chem.*, 33, 1207, 1990.

463. Burton, D.J., and Yang, Z.-Y., Fluorinated organometallics. Perfluoroalkyl and functionalized perfluoroalkyl organometallic reagents in organic synthesis, *Tetrahedron*, 48, 189, 1992.

464. Piettre, S.R., and Raboisson, P., Easy and general access to α,α-difluoromethylene phosphonothioic acids. A new class of compounds, *Tetrahedron Lett.*, 37, 2229, 1996.

465. Halazy, S., Ehrhard, A., Eggenspiller, A., Berges-Gross, V., and Danzin, C., Fluorophosphonate derivatives of N^9-benzylguanine as potent, slow-binding multisubstrate analogue inhibitors of purine nucleoside phosphorylase, *Tetrahedron*, 52, 177, 1996.

466. Jakeman, D.L., Ivory, A.J., Williamson, M.P., and Blackburn, G.M., Highly potent bisphosphonate ligands for phosphoglycerate kinase, *J. Med. Chem.*, 41, 4439, 1998.

467. Halazy, S., Ehrhard, A., and Danzin, C., 9-(Difluorophosphonoalkyl)guanines as a new class of multisubstrate analogue inhibitors of purine nucleoside phosphorylase, *J. Am. Chem. Soc.*, 113, 315, 1991.

468. Stirtan, W.G., and Withers, S.G., Phosphonate and α-fluorophosphonate analogue of the ionization state of pyridoxal 5′-phosphate (PLP) in glycogen phosphorylase, *Biochemistry*, 35, 15057, 1996.

469. Halazy, S., Eggenspiller, A., Ehrhard, A., and Danzin, C., Phosphonate derivatives of N^9-benzylguanine. A new class of potent purine nucleoside phosphorylase inhibitors, *Bioorg. Med. Chem. Lett.*, 2, 407, 1992.

470. Chen, R., Schlossman, A., Breuer, E., Hagele, G., Tillmann, C., Van Gelder, J.M., and Golomb, G., Long-chain functional bisphosphonates. Synthesis, anticalcification, and antiresorption activity, *Heteroatom Chem.*, 11, 470, 2000.

471. Berkowitz, D.B., Eggen, M., Shen, Q., and Sloss, D.G., Synthesis of (α,α-difluoroalkyl)phosphonates by displacement of primary triflates, *J. Org. Chem.*, 58, 6174, 1993.

472. Vinod, T.K., Griffith, O.H., and Keana, J.F.W., Synthesis of isosteric and isopolar phosphonate substrate analogues designed as inhibitors for phosphatidylinositol-specific phospholipase C from *Bacillus cereus*, *Tetrahedron Lett.*, 35, 7193, 1994.

473. Matulic-Adamic, J., Haeberli, P., and Usman, N., Synthesis of 5′-deoxy-5′-difluoromethyl phosphonate nucleotide analogs, *J. Org. Chem.*, 60, 2563, 1995.

474. Berkowitz, D.B., Shen, Q., and Maeng, J.-H., Synthesis of the (α,α-difluoroalkyl)phosphonate analogue of phosphoserine, *Tetrahedron Lett.*, 35, 6445, 1994.

475. Berkowitz, D.B., Bhuniya, D., and Peris, G., Facile installation of the phosphonate and (α,α-difluoromethyl)phosphonate functionalities equipped with benzyl protection, *Tetrahedron Lett.*, 40, 1869, 1999.

476. Wolff-Kugel, D., and Halazy, S., Synthesis of new carbocyclic phosphonate analogs of dideoxypurine nucleotides, *Tetrahedron Lett.*, 32, 6341, 1991.

477. Yokomatsu, T., Sato, M., and Shibuya, S., Lipase-catalyzed enantioselective acylation of prochiral 2-(ω-phosphono)alkyl-1,3-propanediols. Application to the enantioselective synthesis of ω-phosphono-α-amino acids, *Tetrahedron: Asymmetry*, 7, 2743, 1996.

478. Berkowitz, D.B., Eggen, M., Shen, Q., and Shoemaker, R.K., Ready access to fluorinated phosphonate mimics of secondary phosphates. Synthesis of the (α,α-difluoroalkyl)phosphonate analogues of L-phosphoserine, L-phosphoallothreonine, and L-phosphothreonine, *J. Org. Chem.*, 61, 4666, 1996.

479. Burton, D.J., Takei, R., and Shin-Ya, S., Preparation, stability, reactivity and synthetic utility of a cadmium stabilized complex of difluoromethylene phosphonic acid ester, *J. Fluorine Chem.*, 18, 197, 1981.

480. Chambers, R.D., Jaouhari, R., and O'Hagan, D., Synthesis of a difluoromethylenephosphonate analogue of glycerol-3-phosphate. A substrate for NADH linked glycerol-3-phosphate dehydrogenase, *J. Chem. Soc., Chem. Commun.*, 1169, 1988.

481. Chambers, R.D., Jaouhari, R., and O'Hagan, D., Fluorine in enzyme chemistry. Part 2. The preparation of difluoromethylenephosphonate analogues of glycolytic phosphates. Approaching an isosteric and isoelectronic phosphate mimic, *Tetrahedron*, 45, 5101, 1989.

482. Chambers, R.D., Jaouhari, R., and O'Hagan, D., Fluorine in enzyme chemistry. Part 1. Synthesis of difluoromethylenephosphonate derivatives as phosphate mimics, *J. Fluorine Chem.*, 44, 275, 1989.

483. Burton, D.J., and Sprague, L.G., Allylations of [(diethoxyphosphinyl)difluoromethyl]zinc bromide as a convenient route to 1,1-difluoro-3-alkenephosphonates, *J. Org. Chem.*, 54, 613, 1989.

484. Kawamoto, A.M., and Campbell, M.M., Novel class of difluorovinylphosphonate analogues of PEP, *J. Chem. Soc., Perkin Trans. 1*, 1249, 1997.

485. Chambers, R.D., O'Hagan, D., Lamont, R.B., and Jain, S.C., The difluoromethylenephosphonate moiety as a phosphate mimic. X-ray structure of 2-amino-1,1-difluoroethylphosphonic acid, *J. Chem. Soc., Chem. Commun.*, 1053, 1990.

486. Blackburn, G.M., Jakeman, D.L., Ivory, A.J., and Willamson, M.P., Synthesis of phosphonate analogues of 1,3-*bis*phosphoglyceric acid and their binding to yeast phosphoglycerate kinase, *Bioorg. Med. Chem. Lett.*, 4, 2573, 1994.

487. Yokomatsu, T., Ichimura, A., Kato, J., and Shibuya, S., Synthesis of allenic (α,α-difluoromethylene)phosphonates from propargylic tosylates and acetates, *Synlett*, 287, 2001.

488. Qiu, W., and Burton, D.J., A facile and general preparation of α,α-difluoro benzylic phosphonates by the CuCl promoted coupling reaction of the (diethylphosphonyl)difluoromethylcadmium reagent with aryl iodides, *Tetrahedron Lett.*, 37, 2745, 1996.

489. Qabar, M.N., Urban, J., and Kahn, M., A facile solution and solid phase synthesis of phosphotyrosine mimetic L-4-[diethylphosphono(difluoromethyl)]-phenylalanine (F$_2$Pmp(EtO)$_2$) derivatives, *Tetrahedron*, 53, 11171, 1997.

490. Park, S.B., and Standaert, R.F., α,α-Difluorophosphonomethyl azobenzene derivatives as photoregulated phosphoamino acid analogs. Part 1. Design and synthesis, *Tetrahedron Lett.*, 40, 6557, 1999.

491. Yokomatsu, T., Murano, T., Suemune, K., and Shibuya, S., Facile synthesis of aryl(difluoromethyl)phosphonates through CuBr-mediated cross coupling reactions of [(diethoxyphosphinyl)difluoromethyl]zinc bromide with aryl iodides, *Tetrahedron*, 53, 815, 1997.

492. Yokomatsu, T., Minowa, T., Murano, T., and Shibuya, S., Enzymatic desymmetrization of prochiral 2-benzyl-1,3-propanediol derivatives. A practical chemoenzymatic synthesis of novel phosphorylated tyrosine analogues, *Tetrahedron*, 54, 9341, 1998.

493. Yokomatsu, T., Murano, T., Umesue, I., Soeda, S., Shimeno, H., and Shibuya, S., Synthesis and biological evaluation of α,α-difluorobenzylphosphonic acid derivatives as small molecular inhibitors of protein-tyrosine phosphatase 1B, *Bioorg. Med. Chem. Lett.*, 9, 529, 1999.

494. Shakespeare, W.C., Bohacek, R.S., Narula, S.S., Azimioara, M.D., Yuan, R.W., Dalgarno, D.C., Madden, L., Botfield, M.C., and Holt, D.A., An efficient synthesis of a 4′-phosphonodifluoromethyl-3′-formyl-phenylalanine containing Src SH2 ligand, *Bioorg. Med. Chem. Lett.*, 9, 3109, 1999.

495. Cockerill, G.S., Easterfield, H.J., and Percy, J.M., Facile syntheses of aryldifluorophosphonate building blocks, *Tetrahedron Lett.*, 40, 2601, 1999.

496. Cockerill, G.S., Easterfield, H.J., Percy, J.M., and Pintat, S., Facile syntheses of building blocks for the construction of phosphotyrosine mimetics, *J. Chem. Soc., Perkin Trans. 1*, 2591, 2000.

497. Jia, Z., Ye, Q., Dinaut, A.N., Wang, Q., Waddleton, D., Payette, P., Ramachandran, C., Kennedy, B., Hum, G., and Taylor, S.D., Structure of protein tyrosine phosphatase 1B in complex with inhibitors bearing two phosphotyrosine mimetics, *J. Med. Chem.*, 26, 4584, 2001.

498. Yokomatsu, T., Hayakawa, Y., Suemune, K., Kihara, T., Soeda, S., Shimeno, H., and Shibuya, S., Synthesis and biological evaluation of 1,1-difluoro-2-(tetrahydro-3-furanyl)ethylphosphonic acids possessing a N^9-purinylmethyl functional group at the ring. A new class of inhibitors for purine nucleoside phosphorylases, *Bioorg. Med. Chem. Lett.*, 9, 2833, 1999.

499. Yokomatsu, T., Suemune, K., Murano, T., and Shibuya, S., Synthesis of (α,α-difluoroallyl)phosphonates from alkenyl halides or acetylenes, *J. Org. Chem.*, 61, 7207, 1996.

500. Yokomatsu, T., Abe, H., Sato, M., Suemune, K., Kihara, T., Soeda, S., Shimeno, H., and Shibuya, S., Synthesis of 1,1-difluoro-5-(1H-9-purinyl)-2-pentenylphosphonic acids and the related methano analogues. Remarkable effect of the nucleobases and the cyclopropane rings on inhibitory activity toward purine nucleoside phosphorylase, *Bioorg. Med. Chem.*, 6, 2495, 1998.

501. Yokomatsu, T., Abe, H., Yamagishi, T., Suemune, K., and Shibuya, S., Convenient synthesis of cyclopropylalkanol derivatives possessing a difluoromethylenephosphonate group at the ring, *J. Org. Chem.*, 64, 8413, 1999.

502. Otaka, A., Mitsuyama, E., Kinoshita, T., Tamamura, H., and Fujii, N., Stereoselective synthesis of CF$_2$-substituted phosphothreonine mimetics and their incorporation into peptides using newly developed deprotection procedures, *J. Org. Chem.*, 65, 4888, 2000.

503. Butt, A.H., Percy, J.M., and Spencer, N.S., A rearrangement-based approach to secondary difluorophosphonates, *J. Chem. Soc., Chem. Commun.*, 1691, 2000.

504. Yokomatsu, T., Hayakawa, Y., Kihara, T., Koyanagi, S., Soeda, S., Shimeno, H., and Shibuya, S., Synthesis and evaluation of multisubstrate analogue inhibitors of purine nucleoside phosphorylases, *Bioorg. Med. Chem.*, 8, 2571, 2000.

505. Yokomatsu, T., Katayama, S., and Shibuya, S., Diels–Alder cycloaddition of novel buta-1,3-diene derivatives possessing a (diethoxyphosphinoyl)difluoromethyl unit, *J. Chem. Soc., Chem. Commun.*, 1878, 2001.

506. Zhang, X., and Burton, D.J., An alternative route for the preparation of α,α-difluoropropargylphosphonates, *Tetrahedron Lett.*, 41, 7791, 2000.

507. Zhang, X., and Burton, D.J., The preparation of α,α-difluoropropargylphosphonates, *J. Fluorine Chem.*, 116, 15, 2002.

508. Burton, D.J., Ishihara, T., and Maruta, M., A useful zinc reagent for the preparation of 2-oxo-1,1-difluoroalkylphosphonates, *Chem. Lett.*, 755, 1982.

509. Burton, D.J., and Sprague, L.G., Preparation of difluorophosphonoacetic acid and its derivatives, *J. Org. Chem.*, 53, 1523, 1988.

510. Lindell, S.D., and Turner, R.M., Synthesis of potential inhibitors of the enzyme aspartate transcarbamoylase, *Tetrahedron Lett.*, 31, 5381, 1990.

511. Chen, S., and Yuan, C., Studies on organophosphorus compounds. Part 82. Synthesis of some functionalized 1,1-difluoromethylphosphonates, *Phosphorus, Sulfur Silicon Relat. Elem.*, 82, 73, 1993.

512. Tsai, H.-J., Preparation and synthetic application of diethyl 2-oxo-1,1-difluorophosphonates, *Phosphorus, Sulfur Silicon Relat. Elem.*, 122, 247, 1997.

513. Phillion, D.P., and Cleary, D.G., Disodium salt of 2-[(dihydroxyphosphinyl)difluoromethyl]-propenoic acid. An isopolar and isosteric analogue of phosphoenolpyruvate, *J. Org. Chem.*, 57, 2763, 1992.

514. Bouvet, D., and O'Hagan, D., The synthesis of 1-fluoro- and 1,1-difluoro- analogues of 1-deoxy-D-xylulose, *Tetrahedron*, 55, 10481, 1999.

515. Inoue, M., Hiratake, J., and Sakata, K., Synthesis and characterization of intermediate and transition-state analogue inhibitors of γ-glutamyl peptide ligases, *Biosci. Biotechnol. Biochem.*, 63, 2248, 1999.

516. Ladame, S., Bardet, M., Perié, J., and Willson, M., Selective inhibition of *Trypanosoma brucei* GAPDH by 1,3-bisphospho-D-glyceric acid (1,3-diPG) analogues, *Bioorg. Med. Chem.*, 9, 773, 2001.

517. Lequeux, T.P., and Percy, J.M., Facile syntheses of α,α-difluoro-β-ketophosphonates, *J. Chem. Soc., Chem. Commun.*, 2111, 1995.

518. Blades, K., Lequeux, T.P., and Percy, J.M., A reproducible and high-yielding cerium-mediated route to α,α-difluoro-β-ketophosphonates, *Tetrahedron*, 53, 10623, 1997.

519. Blades, K., Lequeux, T.P., and Percy, J. M., Reactive dienophiles containing a difluoromethylenephosphonato group, *J. Chem. Soc., Chem. Commun.*, 1457, 1996.

520. Blades, K., Butt, A.H., Cockerill, G.S., Easterfield, H.J., Lequeux, T.P., and Percy, J.M., Synthesis of activated alkenes bearing the difluoromethylenephosphonate group. A range of building blocks for the synthesis of secondary difluorophosphonates, *J. Chem. Soc., Perkin Trans. 1*, 3609, 1999.

521. Burton, D.J., Sprague, L.G., Pietrzyk, D.J., and Edelmuth, S.H., A safe facile synthesis of difluorophosphonoacetic acid, *J. Org. Chem.*, 49, 3437, 1984.

522. Shipitsin, A.V., Victorova, L.S., Shirokova, E.A., Dyatkina, N.B., Goryunova, L.E., Beabealashvilli, R.S., Hamilton, C.J., Roberts, S.M., and Krayevsky, A., New modified nucleoside 5'-triphosphates. Synthesis, properties towards DNA polymerases, stability in blood serum and antiviral activity, *J. Chem. Soc., Perkin Trans. 1*, 1039, 1999.

523. Biller, S.A., Forster, C., Gordon, E.M., Harrity, T., Scott, W.A., and Ciosek Carl, P., Isoprenoid (phosphinylmethyl)phosphonates as inhibitors of squalene synthetase, *J. Med. Chem.*, 31, 1869, 1988.

524. Biller, S.A., and Forster, C., The synthesis of isoprenoid (phosphinylmethyl)phosphonates, *Tetrahedron*, 46, 6645, 1990.

525. Bystrom, C.E., Pettigrew, D.W., Remington, S.J., and Branchaud, B.P., ATP analogs with non-transferable groups in the γ position as inhibitors of glycerol kinase, *Bioorg. Med. Chem. Lett.*, 7, 2613, 1997.

526. Burton, D.J., Ishihara, T., and Flynn, R.M., Difluoromethylene exchange in the preparation of fluorinated *bis*-phosphonates, *J. Fluorine Chem.*, 20, 121, 1982.

527. Burton, D.J., and Flynn, R.M., *Bis*(phosphonic acid)difluoromethane, *Univ. Iowa Res. Found.*, U.S. Patent Appl. US 4330486, 1982; *Chem. Abstr.*, 97, 145014, 1982.

528. Stemerick, D.M., Farnesyl protein transferase inhibitors as anticancer agents, *Merrell Dow Pharmaceuticals*, Int. Patent Appl. WO 9419357, 1994; *Chem. Abstr.*, 121, 301073, 1994.

529. Barnes, N.J., Cox, J.M., Hawkes, T.R., and Knee, A.J., Preparation and herbicidal properties of triazole substituted phosphonic acid derivatives, *Zeneca*, U.K. Patent Appl. GB 2280676, 1995; *Chem. Abstr.*, 123, 83729, 1995.

530. Piettre, S.R., Girol, C., and Schelcher, C.G., A new strategy for the conversion of aldehydes into difluoromethyl ketones, *Tetrahedron Lett.*, 37, 4711, 1996.

531. Yokomatsu, T., Takechi, H., Akiyama, T., Shibuya, S., Kominato, T., Soeda, S., and Shimeno, H., Synthesis and evaluation of a difluoromethylene analogue of sphingomyelin as an inhibitor of sphingomyelinase, *Bioorg. Med. Chem. Lett.*, 11, 1277, 2001.

532. Ganzhorn, A.J., Hoflack, J., Pelton, P.D., Strasser, F., Chanal, M.-C., and Piettre, S.R., Inhibition of *myo*-inositol monophosphatase isoforms by aromatic phosphonates, *Bioorg. Med. Chem.*, 6, 1865, 1998.

533. Orsini, F., Reformatsky-type co-mediated synthesis of β-hydroxyphosphonates, *Tetrahedron Lett.*, 39, 1425, 1998.

534. Chetyrkina, S., Estieu-Gionnet, K., Lain, G., Bayle, M., and Deleris, G., Synthesis of *N*-Fmoc-4-[(diethylphosphono)-2′,2′-difluoro-1′-hydroxyethyl]phenylalanine, a novel phosphotyrosyl mimic for the preparation of signal transduction inhibitory peptides, *Tetrahedron Lett.*, 41, 1923, 2000.

535. Halazy, S., and Gross-Berges, V., Short and stereoselective synthesis of δ-functionalized *E*-α,α-difluoroallylphosphonates, *J. Chem. Soc., Chem. Commun.*, 743, 1992.

536. Blades, K., Cockerill, G.S., Easterfield, H.J., Lequeux, T.P., and Percy, J.M., A DAST-free route to aryl(difluoromethyl)phosphonic esters, *J. Chem. Soc., Chem. Commun.*, 1615, 1996.

537. Martin, S.F., Dean, D.W., and Wagman, A.S., A general method for the synthesis of 1,1-difluoroalkylphosphonates, *Tetrahedron Lett.*, 33, 1839, 1992.

538. Otaka, A., Miyoshi, K., Burke, T.R., Jr., Roller, P.P., Kubota, H., Tamamura, H., and Fujii, N., Synthesis and application of *N*-Boc-L-2-amino-4-(diethylphosphono)-4,4-difluorobutanoic acid for solid-phase synthesis of nonhydrolyzable phosphoserine peptide analogues, *Tetrahedron Lett.*, 36, 927, 1995.

539. Levy, S.G., Wasson, D.B., Carson, D.A., and Cottam, H.B., Synthesis of 2-chloro-2′,5′-dideoxy-5′-difluoromethylphosphinyladenosine. A nonhydrolyzable isosteric, isopolar analog of 2-chlorodeoxyadenosine monophosphate, *Synthesis*, 843, 1996.

540. Behr, J.-B., Evina, C.M., Phung, N., and Guillerm, G., Synthesis of (difluoromethyl)phosphonate azasugars designed as inhibitors for glycosyl transferases, *J. Chem. Soc., Perkin Trans. 1*, 1597, 1997.

541. Gautier-Lefebvre, I., Behr, J.-B., Guillerm, G., and Ryder, N.S., Synthesis of new (difluoromethylphosphono)azadisaccharides designed as bisubstrate analogue inhibitors for GlcNAc:β-1,4 glycosyltransferases, *Bioorg. Med. Chem. Lett.*, 10, 1483, 2000.

542. Obayashi, M., and Kondo, K., An improved procedure for the synthesis of 1,1-difluoro-2-hydroxyalkylphosphonates, *Tetrahedron Lett.*, 23, 2327, 1982.

543. Cox, R.J., Hadfield, A.T., and Mayo-Martin, M.B., Difluoromethylene analogues of aspartyl phosphate. The first synthetic inhibitors of aspartate semi-aldehyde dehydrogenase, *J. Chem. Soc., Chem. Commun.*, 1710, 2001.

544. Ruzicka, J.A., Qiu, W., Baker, M.T., and Burton, D.J., Synthesis of [1–[14]C]-2,2-difluoroethene from [^{14}C]-formaldehyde, *J. Labelled Compd. Radiopharm.*, 34, 59, 1994.

545. Adcock, W., and Kok, G.B., Polar substituent effects on ^{19}F chemical shifts of aryl and vinyl fluorides. A fluorine-19 nuclear magnetic resonance study of some 1,1-difluoro-2-(4-substituted-bicyclo[2.2.2]oct-1-yl)ethenes, *J. Org. Chem.*, 50, 1079, 1985.

546. Piettre, S.R., and Cabanas, L., Reinvestigation of the Wadsworth–Emmons reaction involving lithium difluoromethylenephosphonate, *Tetrahedron Lett.*, 37, 5881, 1996.

547. Howson, W., Hills, J.M., Blackburn, G.M., and Broekman, M., Synthesis and biological evaluation of [3-amino-2-(4-chlorophenyl)-1,1-difluoropropyl]phosphonic acid, *Bioorg. Med. Chem. Lett.*, 1, 501, 1991.

548. Lequeux, T.P., and Percy, J.M., Cerium-mediated conjugate additions of a difluorophosphonate carbanion to nitroalkenes, *Synlett*, 361, 1995.

549. Blades, K., Lapôtre, D., and Percy, J.M., Conjugate addition reactions of a (diethoxyphosphinoyl)difluoromethyl anion equivalent to acyclic and cyclyc vinyl sulfones, *Tetrahedron Lett.*, 38, 5895, 1997.

550. Blades, K., and Percy, J.M., A conjugate addition/sulfoxide elimination route to allylic difluorophosphonates, *Tetrahedron Lett.*, 39, 9085, 1998.

551. Guneratne, R.D., and Burton, D.J., Preparation and functionalization of an alkenylcopper reagent formed by the *syn* addition of (dialkoxyphosphinyl)difluoromethylcopper to hexafluoro-2-butyne, *J. Fluorine Chem.*, 98, 11, 1999.

552. Savignac, P., Petrova, J., Dreux, M., and Coutrot, P., Halogen-metal exchange. Preparation of dialkylchloromethylphosphonates. Effect of salt, *J. Organomet. Chem.*, 91, C45, 1975.

553. Coutrot, P., Laurenço, C., Normant, J.F., Perriot, P., Savignac, P., and Villieras, J., Synthesis of diethyl 1,1-dichloroalkane- and 1-chloroalkanephosphonates, *Synthesis*, 615, 1977.

554. Denis, J.M., Guillemin, J.C., and Le Guennec, M., Synthesis of primary α,α'-dichlorophosphines, precursors of unhindered *C*-chlorophospha-alkenes and synthetic equivalents of λ^3-phosphaalkynes, *Phosphorus, Sulfur Silicon Relat. Elem.*, 49, 317, 1990.

555. Le Menn, J.C., and Sarrazin, J., Electrochemical synthesis of diethyl 1,1-dichloroalkylphosphonates, *J. Chem. Res. (S)*, 26, 1989.

556. Jubault, P., Feasson, C., and Collignon, N., Electrochemical activation of magnesium. A new, rapid and efficient method for the electrosynthesis of 1,1-dichloroalkylphosphonates from diisopropyl trichloromethylphosphonate, *Tetrahedron Lett.*, 36, 7073, 1995.

557. Villieras, J., Perriot, P., and Normant, J.F., Simple route from aldehydes to alkynes and 1-chloro-1-alkynes, *Synthesis*, 458, 1975.

558. Roedig, A., Ganns, E.M., Henrich, C., and Schnutenhaus, H., Polyhalogenated bicyclo[4.2.0]octa-1,5,7-trienes. Part 8. Synthesis and cyclodimerization of polyhalogenated 2-phenyl- and 2-pentachlorophenyl substituted butenynes, *Liebigs Ann. Chem.*, 1674, 1981.

559. Roedig, A., and Ganns, E.M., Polyhalogenated bicyclo[4.2.0]octa-1,5,7-trienes. Part 9. Synthesis and cyclodimerization of 1,2,4-trichloro-1-phenyl-1-buten-3-yne, *Liebigs Ann. Chem.*, 1685, 1981.

560. Marcacci, F., Giacomelli, G., and Menicagli, R., A synthetic approach to chiral acetylene derivatives from optically active aldehydes, *Gazz. Chim. Ital.*, 110, 195, 1980.

561. Miyaura, N., and Suginome, H., New stereo- and regiospecific synthesis of humulene by means of the palladium catalyzed cyclization of haloalkenylboranes, *Tetrahedron Lett.*, 25, 761, 1984.

562. Brown, M.J., Harrison, T., and Overman, L.E., General approach to halogenated tetrahydrofuran natural products from red algae of the genus *Laurencia*. Total synthesis of (±)-*trans*-kumausyne and demonstration of an asymmetric synthesis strategy, *J. Am. Chem. Soc.*, 113, 5378, 1991.

563. Frejd, T., Karlsson, J.O., and Gronowitz, S., Ring-opening reactions. Part 18. Synthesis of cyclic thioenols ethers, *J. Org. Chem.*, 46, 3132, 1981.

564. Ebe, H., Nakagawa, T., Iyoda, M., and Nakagawa, M., Syntheses of an isoannulated annulene, a *bis*dehydro[14]annuleno[c]furan and *bis*dehydro[14]annulene derivatives, *Tetrahedron Lett.*, 22, 4441, 1981.

565. Iyoda, M., Nakagawa, T., Ebe, H., Oda, M., Nakagawa, M., Yamamoto, K., Higuchi, H., and Ojima, J., Synthesis and properties of a bisdehydro[14]annuleno[*c*]furan and an *ortho*-annelated tetrakisdehydro[14]annuleno[14]annulene, *Bull. Chem. Soc. Jpn.*, 67, 778, 1994.

566. Tobe, Y., Fujii, T., and Naemura, K., Photochemical method for generation of linear polyynes. [2 + 2] Cycloreversion of [4.3.2]propellatrienes extruding indan, *J. Org. Chem.*, 59, 1236, 1994.

567. Tobe, Y., Fujii, T., Matsumoto, H., Tsumuraya, K., Noguchi, D., Nakagawa, N., Sonodo, M., Naemura, K., Achiba, Y., and Wakabayashi, T., [2+2] Cycloreversion of [4.3.2]propella-1,3,11-trienes. An approach to cyclo[n]carbons from propellane-annelated dehydro[n]annulenes, *J. Am. Chem. Soc.*, 122, 1762, 2000.

568. Burger, A., Hetru, C., and Luu, B., Stereoselective synthesis of cholesteryl derivatives bearing a chiral allenic group in the side chain, *Synthesis*, 93, 1989.

569. Nakamura, T., Matsui, T., Tanino, K., and Kuwajima, I., A new approach for ingenol synthesis, *J. Org. Chem.*, 62, 3032, 1997.

570. Marek, I., Synthesis and reactivity of sp^2 geminated organobismetallic derivatives, *Chem. Rev.*, 100, 2887, 2000.

571. Mouriès, V., Waschbüsch, R., Carran, J., and Savignac, P., A facile and high yielding synthesis of symmetrical and unsymmetrical diarylalkynes using diethyl dichloromethylphosphonate as precursor, *Synthesis*, 271, 1998.

572. Hu, C.-M., and Chen, J., Addition of diethyl bromodifluoromethylphosphonate to various alkenes initiated by Co(III)/Zn bimetal redox system, *J. Chem. Soc., Perkin Trans. 1*, 327, 1993.

573. Bonnet, B., Le Gallic, Y., Plé, G., and Duhamel, L., Expedient synthesis of 1,1-diiodoalkenes, *Synthesis*, 1071, 1993.

574. Iorga, B., Eymery, F., and Savignac, P., An efficient synthesis of tetraethyl fluoromethylenediphosphonate and derivatives from diethyl dibromofluoromethylphosphonate, *Tetrahedron Lett.*, 39, 4477, 1998.

575. Martynov, B.I., Sokolov, V.B., Aksinenko, A.Y., Goreva, T.V., Epishina, T.A., and Pushin, A.N., Convenient method for the synthesis and some transformations of the lithium salt of *bis*(diethoxy-phosphoryl)fluoromethane, *Izv. Akad. Nauk, Ser. Khim.*, 2039, 1998; *Russ. Chem. Bull. (Engl. Transl.)*, 1983, 1998.

576. Lowen, G.T., and Almond, M.R., A novel synthesis of phosphonates from diethyl (trichloro-methyl)phosphonate, *J. Org. Chem.*, 59, 4548, 1994.

577. Perlikowska, W., Modro, A.M., Modro, T.A., and Mphahlele, M.J., Lithiation of diethyl trichloro-methylphosphonate and the transformations of the α-lithiated derivative, *J. Chem. Soc., Perkin Trans. 2*, 2611, 1996.

578. Perlikowska, W., Mphahlele, M.J., and Modro, T.A., One-pot stereoselective synthesis of (Z)-diethyl α-chlorovinylphosphonates, *J. Chem. Soc., Perkin Trans. 2*, 967, 1997.

579. Burton, D.J., and Hahnfeld, J.L., Fluorochloro-, fluorobromo-, and monofluorocarbene generation via organolithium reagents, *J. Org. Chem.*, 42, 828, 1977.

580. Burton, D.J., Modak, A.S., Guneratne, R., Su, D., Cen, W., Kirchmeier, R.L., and Shreeve, J.M., Synthesis of (sulfodifluoromethyl)phosphonic acid, *J. Am. Chem. Soc.*, 111, 1773, 1989.

581. McKenna, C.E., Higa, M.T., Cheung, N.H., and McKenna, M.C., The facile dealkylation of phosphonic acid dialkyl esters by bromotrimethylsilane, *Tetrahedron Lett.*, 18, 155, 1977.

582. Gross, H., Böck, C., Costisella, B., and Glöde, J., α-Substituted phosphonates. Part 30. Dealkylation of phosphonates with labile functional groups using trimethylsilyl bromide, *J. Prakt. Chem.*, 320, 344, 1978.

583. McKenna, C.E., and Schmidhauser, J., Functional selectivity in phosphonate ester dealkylation with bromotrimethylsilane, *J. Chem. Soc., Chem. Commun.*, 739, 1979.

584. Morita, T., Okamoto, Y., and Sakurai, H., A convenient dealkylation of dialkyl phosphonates by chlorotrimethylsilane in the presence of sodium iodide, *Tetrahedron Lett.*, 19, 2523, 1978.

585. Morita, T., Okamoto, Y., and Sakurai, H., Dealkylation reaction of acetals, phosphonate, and phosphate esters with chlorotrimethylsilane/metal halide reagent in acetonitrile, and its application to the synthesis of phosphonic acids and vinyl phosphates, *Bull. Chem. Soc. Jpn.*, 54, 267, 1981.

586. Harger, M.J.P., and Sreedharan-Menon, R., Alkoxide induced rearrangement of alkyl bromometh-ylphosphonamidates. Steric influences on the direction of ring opening of the azaphosphiridine oxide intermediate, *J. Chem. Soc., Perkin Trans. 1*, 211, 1998.

587. Patois, C., Berté-Verrando, S., and Savignac, P., Easy preparation of alkylphosphonyl dichlorides, *Bull. Soc. Chim. Fr.*, 130, 485, 1993.

588. Morise, X., Savignac, P., Guillemin, J.C., and Denis, J.M., A convenient method for the synthesis of α-functionalized chlorophosphonic esters, *Synth. Commun.*, 21, 793, 1991.

589. Berté-Verrando, S., Nief, F., Patois, C., and Savignac, P., Preparation of alkyl- and aryl-amino[²H₂]methylphosphinic acids, *J. Chem. Soc., Perkin Trans. 1*, 2045, 1995.

590. Morise, X., Savignac, P., and Denis, J.-M., New syntheses of 1-chloroalkylphosphinates, *J. Chem. Soc., Perkin Trans. 1*, 2179, 1996.

591. Iorga, B., Carmichael, D., and Savignac, P., Phosphonate-phosphonochloridate conversion, *C.R. Acad. Sci., Ser. IIc*, 3, 821, 2000.

4 The Epoxyphosphonates

The popularity of epoxyphosphonates emanates from their simplicity, ease of access, and reactivity. The discovery of the antibiotic fosfomycin[1,2] [(1R,2S) (−)-1,2-epoxypropylphosphonic acid] in 1969 generated interest in the chemistry of 1,2-epoxyalkylphosphonates and higher homologues and gave rise to a large number of synthetic preparations. It has also given a biochemical significance to 1,2-epoxyalkylphosphonates and their derivatives. The fosfomycin phenomenon has largely contributed to the interest in and development of this topic.[3]

4.1 SYNTHESIS OF EPOXYPHOSPHONATES

4.1.1 DIALKYL 1,2-EPOXYALKYLPHOSPHONATES

4.1.1.1 Darzens Reactions

The Darzens synthesis of glycidic esters by condensation of carbonyl compounds with α-haloesters is an important and useful method[4] that has been extended in phosphorus chemistry. Unquestionably, the most general and perhaps most widely employed method for the synthesis of dialkyl 1,2-epoxyalkylphosphonates involves the reaction of dialkyl chloromethylphosphonates with carbonyl compounds. These synthetically useful phosphonates are readily obtained by standard alcoholysis of chloromethylphosphonic dichloride under anhydrous conditions.[5,6] The latter is obtained in up to 67% yield from phosphorus trichloride and paraformaldehyde at 250°C.[7–12] Several variations on the preparation of dialkyl 1,2-epoxyalkylphosphonates from dialkyl chloromethylphosphonates and different carbonyl partners have been reported.[13–16] The conditions for generating the α-metallated dialkyl chloromethylphosphonates were found to be critical because of their notorious instability.

The addition of alkoxide bases (EtONa, t-BuOK) to an equimolecular amount of dialkyl chloromethylphosphonate and a carbonyl compound (aromatic aldehydes or ketones) in Et$_2$O or t-BuOH at a temperature between −10°C and +10°C leads to dialkyl 1,2-epoxyalkylphosphonates in low to good yields (10–68%).[13,14]

Best yields are obtained when dialkyl chloromethylphosphonates are metallated with n-BuLi at low temperature in THF.[15,16] The resulting carbanions undergo facile addition to carbonyl compounds to induce the almost immediate formation of chlorohydrins, which on warming conduce quantitatively to dialkyl 1,2-epoxyalkylphosphonates without trace or side products (Scheme 4.1). These conditions are preferable for the preparation of dialkyl 1,2-epoxyalkylphosphonates in pure form and in good to excellent yields (50–90%). The method appears to be applicable to a wide range of carbonyl compounds so that aliphatic and aromatic aldehydes as well as aliphatic, cyclic, and aromatic ketones may be introduced in the reaction (Table 4.1). The Darzens process is exempt from side reactions but is not stereoselective, and mixtures of (E)- and (Z)-isomers are frequently obtained, the (E)-form being preferred.

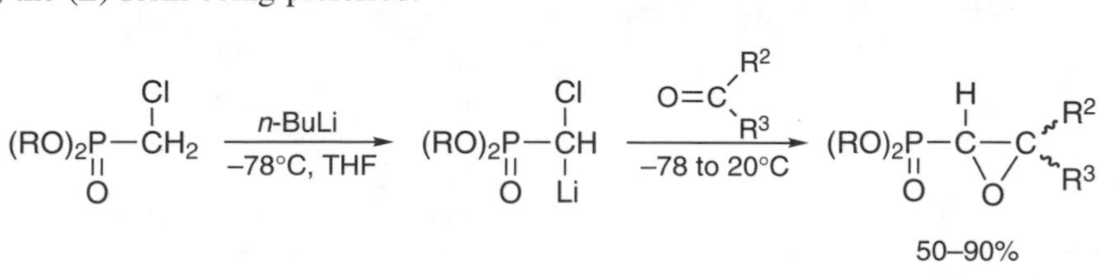

50–90%

(4.1)

TABLE 4.1

	R	R^2	R^3	Base	Solvent	Yield (%)[Ref.]
a	Et	Me	Me	n-BuLi	THF	80[16]
				NaH	DMSO	61[17]
b	Et	Et	Et	n-BuLi	THF	89[16]
c	Et	i-Pr	i-Pr	n-BuLi	THF	88[16]
d	Et	Me	Et	n-BuLi	THF	85[16]
e	Et	Me	i-Pr	n-BuLi	THF	86[16]
f	Et	Me	t-Bu	n-BuLi	THF	75[16]
g	Et	Me	i-Bu	n-BuLi	THF	81[16]
h	Et	Me	-CH=CH$_2$	n-BuLi	THF	36[16]
i	Et	Me	-C(Me)=CH$_2$	n-BuLi	THF	75[16]
j	Et	Me	-CH(OMe)$_2$	n-BuLi	THF	76[16]
k	Et	Me		n-BuLi	THF	90[16]
l	Et	Me	Ph	n-BuLi	THF	51[16]
				NaH	DMSO	—[18]
m	Et	H	Me	n-BuLi	THF	65[16]
n	Et	H	i-Pr	n-BuLi	THF	66[16]
o	Et	H	i-Bu	n-BuLi	THF	57[16]
p	Et		-(CH$_2$)$_4$-	n-BuLi	THF	76[16]
q	Me	Me	-(CH$_2$)$_2$CH(CH$_3$)$_2$	EtONa	Et$_2$O	37[13,14]
r	Et	H		EtONa	Et$_2$O	31[13,14]
				t-BuOK	t-BuOH	43[13,14]
s	Me	Me	Ph	EtONa	Et$_2$O	21[13,14]
				t-BuOK	t-BuOH	36[13,14]
t	Me	H	Ph	t-BuOK	t-BuOH	65[13,14]
u	Et	H	Ph	EtONa	Et$_2$O	14[13,14]
				t-BuOK	t-BuOH	61[13,14]
				NaH	DMSO	—[18]

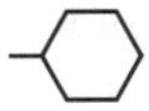

TABLE 4.1 (continued)

	R	R^2	R^3	Base	Solvent	Yield (%)[Ref.]
v	Mc	Me	Me	NaH	DMSO	—[18]
w	Me	Ph	Ph	NaH	DMSO	—[18]
x	Et	Ph	Ph	NaH	DMSO	—[18]
y	Et		–(CH$_2$)$_5$–	NaH	DMSO	—[18]
				NaNH$_2$	C$_6$H$_6$	—[19]
z	Me	H	4-Me-C$_6$H$_4$	t-BuOK	t-BuOH	68[13,14]
aa	Me	H	4-MeO-C$_6$H$_4$	t-BuOK	t-BuOH	10[13,14]
ab	i-Pr	H	4-Cl-C$_6$H$_4$	n-BuLi	THF	—[15]
ac	Et	H	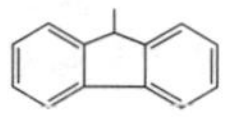 (fluorenyl)	NaH	C$_6$H$_6$	—[20]

Several innovations have significantly extended the scope and synthetic utility of the classical Darzens reaction. For example, a useful supplement has been described to convert diethyl alkyl-phosphonates into diethyl 1-substituted 1,2-epoxyalkylphosphonates by a one-pot process.[21] Thus, metallation of diethyl alkylphosphonates with n-BuLi (2 eq) at low temperature in THF followed by electrophilic chlorination using benzenesulfonyl chloride yields diethyl 1-lithio-1-chloroalkyl-phosphonates, which can be converted in moderate to good overall yields (30–80%) into diethyl 1-substituted 1,2-epoxyalkylphosphonates by reaction with carbonyl compounds (Scheme 4.2).[21]

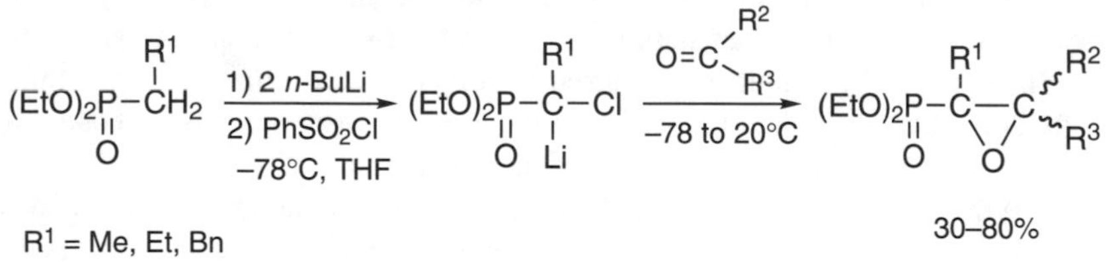

R^1 = Me, Et, Bn
R^2 = Ar, PhCH=CH
R^3 = H, Me, Ph

30–80%

(4.2)

The halogen–metal exchange reaction of diethyl 1,1-dichloroalkylphosphonates with n-BuLi provides a convenient preparation of diethyl 1-lithio-1-chloroalkylphosphonates in a yield somewhat higher than those obtained by the chlorination procedure.[22] For example, trapping of diethyl 1-lithio-1-chloroethylphosphonate with aliphatic or aromatic aldehydes and aliphatic ketones provides a suitable route to diethyl 1-methyl-1,2-epoxyalkylphosphonates in good yields (61–81%, Scheme 4.3) as a mixture of (Z)- and (E)-isomers in approximately equal amounts.[23]

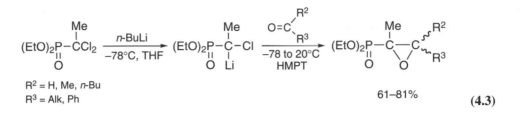

$$R^2 = H, Me, n\text{-Bu}$$
$$R^3 = Alk, Ph$$

61–81%   (4.3)

Another technique for the formation of diethyl 1,2-epoxyalkylphosphonates is based on the reactivity of an electrochemically generated carbanion of diethyl chloromethylphosphonate in DMF toward aromatic aldehydes. The yield of 72% obtained with *para*-methoxybenzaldehyde is promising.[24] However, the major disadvantage that attends the use of the electrochemical route is the difficulty that may be encountered in the electrogeneration of species sufficiently basic to deprotonate chloromethylphosphonates.

A useful variant of the Darzens reaction has been reported for the preparation of fosfomycin from diethyl chloromethylphosphonate.[25] Nucleophilic substitution of chlorine in diethyl chloromethylphosphonate with dimethyl sulfide gives the sulfonium salt, further converted into its stable ylide by the use of NaH in DMSO. The addition of the ylide to acetaldehyde produces diethyl 1,2-epoxypropylphosphonate, thus avoiding the problems associated with the stability of chloromethylphosphonate carbanions (Scheme 4.4).

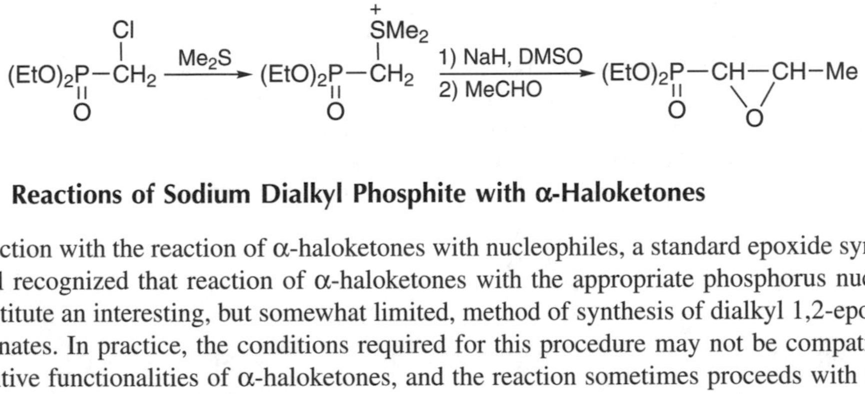

$$(EtO)_2\underset{\underset{O}{\|}}{P}-CH_2 \xrightarrow{Me_2S} (EtO)_2\underset{\underset{O}{\|}}{P}-CH_2 \xrightarrow[\text{2) MeCHO}]{\text{1) NaH, DMSO}} (EtO)_2\underset{\underset{O}{\|}}{P}-CH-CH-Me$$

(4.4)

4.1.1.2 Reactions of Sodium Dialkyl Phosphite with α-Haloketones

In connection with the reaction of α-haloketones with nucleophiles, a standard epoxide synthesis,[26] it is well recognized that reaction of α-haloketones with the appropriate phosphorus nucleophile can constitute an interesting, but somewhat limited, method of synthesis of dialkyl 1,2-epoxyalkylphosphonates. In practice, the conditions required for this procedure may not be compatible with the sensitive functionalities of α-haloketones, and the reaction sometimes proceeds with a lack of regiospecificity to produce a mixture of phosphorus compounds. The reaction can produce the epoxide alone,[27,28] the vinyl phosphate alone,[29,30] a mixture of epoxide and vinyl phosphate,[31–34] or a mixture of epoxide and β-ketophosphonate.[35–38] A large number of investigations have led to the proposal of four types of mechanisms involving an initial attack of the phosphorus on the halogen,[39–42] the carbonyl oxygen,[43–45] the α-carbon bound to the halogen,[46–49] or the carbonyl carbon.[50–56] Formation of the vinyl phosphate, epoxyphosphonate, and phosphate halohydrin is believed to result from attack by the phosphorus at the carbonyl carbon. Formation of the β-ketophosphonate results from attack at the α-halo carbon. Alkali metal derivatives of dialkyl phosphites attack α-haloketones primarily at the carbonyl carbon atom, forming an alkoxide anion, which displaces the β-halogen to give dialkyl 1,2-epoxyalkylphosphonates in reasonable yields (55–71%, Scheme 4.5, Table 4.2). This procedure appears to be limited to the use of α-halo ketones having only one nonhydrogen substituent on the α-halo carbon. In some cases, however, formation of vinyl phosphate can be a competing reaction (Scheme 4.5).[31–34,57] If two substituents are present, the vinyl phosphate becomes the main product.[30–33,38,58,59]

TABLE 4.2

	R	R¹	R²	R³	X	Yield (%)[Ref.]
a	Me	H	H	Me	Cl	74[60]
b	Me	Ph	H	Ph	Cl	—[19]
c	Me	Ph	H	4-MeO-C_6H_4	Cl	—[19]
d	Et	H	H	Ph	Cl	56[34]
e	Et	H	H	Me	Cl	27,[38] 65,[28] 77[60]
f	Et	H	H	Et	Cl	72[28]
g	Et	H	H	Ph	Br	—[33]
h	Et	H	$-(CH_2)_3-$		Cl	—[32]
i	Et	H	$-(CH_2)_4-$		Cl	—,[32] 53[61]
					Br	—[31]
j	Et	Me	H	Me	Cl	53[60]
					Br	—[38]
k	Et	Me	H	Ph	Br	—[33]
l	Et	Ph	H	Me	Cl	—[19]
m	Et	Ph	H	t-Bu	Cl	—[19]
n	i-Pr	H	H	Me	Cl	—,[38] 69[60]
o	n-Bu	H	H	Me	Cl	69[28,60]

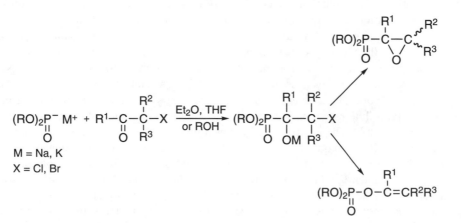

$$(4.5)$$

TABLE 4.3

	R	R¹	R²	Yield (%)[Ref.]
a	Me	Me	H	84[59]
b	Et	Me	H	83,[59] 81[62]
c	Bn	Me	H	48[59]
d	Me	t-Bu	H	87[59]
e	Me	t-Bu	Me	70[59]
f	Me	Me	Me	53[59]
g	Me	–(CH₂)₄–		33[59]

An alternative procedure for the preparation of dialkyl 1,2-epoxyalkylphosphonates from α-haloketones is the action of MeONa (1 eq) on a mixture of α-chloroketones and dimethyl or diethyl phosphite (1 eq) in MeOH at room temperature.[59] The yields of 53–87% obtained by this technique[59] are reproducible and represent a significant improvement over the yields of 27%[38] and 65%[28] reported for the one-step reaction of α-chloroketones with sodium diethyl phosphite. This procedure appears to be limited to aliphatic α-chloroketones (Scheme 4.6). The reaction of α-chloroacetophenone with dimethyl phosphite resulted in the formation of dimethyl 1-phenylvinyl phosphate.[59] However, in spite of the difficulties frequently encountered in the choice of α-chloroketones, this facile and convenient procedure is a useful variant of the classical approach as illustrated by the results presented in Table 4.3.

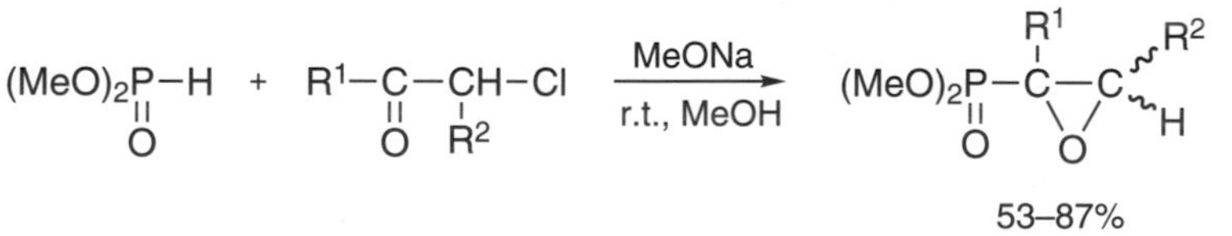

$$\text{(MeO)}_2\overset{\text{O}}{\underset{\|}{\text{P}}}-\text{H} \quad + \quad R^1-\overset{\text{O}}{\underset{\|}{\text{C}}}-\underset{R^2}{\text{CH}}-\text{Cl} \quad \xrightarrow[\text{r.t., MeOH}]{\text{MeONa}} \quad (\text{MeO})_2\overset{\text{O}}{\underset{\|}{\text{P}}}-\overset{R^1}{\underset{\text{O}}{\text{C}}}-\overset{R^2}{\underset{H}{\text{C}}}$$

$$53–87\%$$ **(4.6)**

The reaction between diethyl phosphite and α-chloroketones can also be accomplished by fluoride ion-mediated deprotonation reaction in DMF (dry KF or KF·2H₂O as fluoride ion source), which constitutes a nonbasic route to diethyl 1,2-epoxyalkylphosphonates. Unfortunately, the application of this technique provides a mixture of 1,2-epoxyalkylphosphonate (about 50%) and vinyl phosphate (about 30%) whatever the α-chloroketones.[63] The reaction between 3-(ω-bromoacetyl)coumarin and diethyl or di-n-butyl phosphite has been performed with 50% NaOH in C₆H₆ under phase-transfer catalysis conditions with TEBA as catalyst.[64] The only products of the reaction are the epoxyphosphonates isolated in 70% and 34% yields, respectively.[64] In the presence of Et₃N, diethyl phosphite reacts with 2-chloroacetoacetate to give diethyl 2-(ethoxycarbonyl)-1-methyl-1,2-epoxyethylphosphonate in 72% yield as a mixture of cis and trans isomers.[65]

The reaction conditions required for the preparation of halogenoketones frequently may not be compatible with the sensitive functionalities present in complex polyfunctional molecules. Recent innovations in the standard synthetic procedure using α-haloketones have featured the use of α-tosyloxyketones.[66–72] Thus, a relatively general procedure for the quantitative generation of dialkyl 1,2-epoxyalkylphosphonates possessing a sugar or a steroid appendage using α-tosyloxyketones has been developed. The generation of epoxides may be conveniently executed under a variety of reaction conditions. They include sodium diethyl phosphite in EtOH at reflux (69%),[66] diethyl phosphite in the two-phase system CH_2Cl_2/NaOH 50%/TBAC at room temperature[71] or diethyl phosphite in MeOH in the presence of DBU at 0°C (62–90%, Scheme 4.7).[67,68,72] This attractive procedure appears as a modification having a great synthetic utility because the leaving group is introduced in mild conditions by merely treating the sensitive substrate containing a primary alcohol with TsCl in the presence of Py.[72]

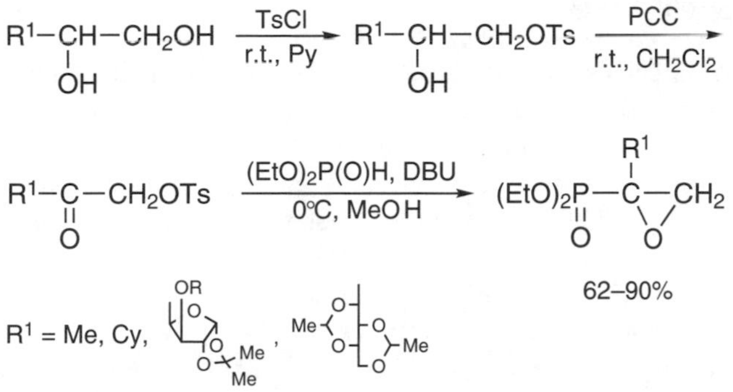

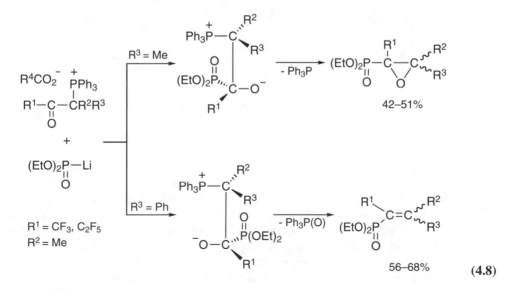

(4.7)

Another variation of the classic Darzens reaction is a one-pot synthesis of diethyl 1-perfluoro-alkyl-1,2-epoxyalkylphosphonates using the nucleophilic attack of lithium diethyl phosphite on the carbonyl carbon of perfluorinated β-oxophosphonium salts (Scheme 4.8).[73] The resulting interme-diate can eliminate in two directions. When the oxygen anion in the least sterically hindered position ($R^2 = R^3 = Me$) attacks the neighboring carbon atom, diethyl 1-perfluoroalkyl-1,2-epoxyalkylphos-phonates are obtained in moderate yields (42–51%) after elimination of Ph_3P (*anti* fashion). The formation of α-(perfluoroalkyl)vinylphosphonates by attack at phosphorus (*syn* fashion) can be a competing reaction. The results indicate that the selectivity can be controlled to produce exclusively either epoxyphosphonates or vinylphosphonates.[73]

(4.8)

The most recent development in the field has featured the use of phase-transfer catalysis conditions in the preparation of dialkyl 1,2-epoxyalkylphosphonates from dialkyl phosphites and α-haloketones (X = Cl).[60] A first procedure uses K_2CO_3 in excess as base and a second a mixture of K_2CO_3 and $NaHCO_3$, TBAB being introduced as phase-transfert catalyst. Each reaction is carried out in the absence of solvent, the liquid phase being formed by the starting reagents. This method does not improve the yields of dialkyl 1,2-epoxyalkylphosphonates, which are roughly comparable to those previously reported. However, the absence of side products, which usually accompany this reaction, renders this methodology a viable alternative to the methods that require the use of a solvent.[60]

4.1.1.3 Reactions of Dialkyl 1,2-Halohydrinphosphonates and Related Structures with Bases

In addition to the two previous methodologies, there is a technique that employs the conversion of 1,2-halohydrinphosphonates into 1,2-epoxyphosphonates by deprotonation. This two-step proce-dure appears to be limited to the synthesis of unsubstituted or monosubstituted dialkyl 1,2-epoxy-ethylphosphonates. The 1,2-halohydrinphosphonates are prepared by two procedures, which exploit either the condensation of a dialkyl phosphite with an α-chlorocarbonyl compound or the halohy-droxylation of a vinylphosphonate.

The reaction proceeds on heating dialkyl phosphites with α-chloroketones for a long time at 100–165°C with a catalyst (MeONa/MeOH)[74] or at 100–120°C without catalyst,[75–79] and with chloroacetaldehyde at 100°C without catalyst[80] to give, in 50–100% yields, the chlorohydrinphos-phonates having the hydroxyl group in the α-position. The reaction between diethyl phosphite and chloroacetaldehyde can also proceed at room temperature in the presence of a catalytic amount of Et_3N.[81]

The halohydroxylation method has been developed for exploiting the reactivity of the double bond present in dialkyl vinyl- and propenylphosphonates. Preparation of the halohydrins of diethyl vinylphosphonate failed using NBS or NBA.[82] Treatment of diethyl vinylphosphonate on large scale with NaOCl or NaOBr in aqueous medium at pH < 3 gives the halohydrins with the hydroxyl group in the β-position (anti-Markovnikoff addition) with 95% or 65–85% purity, respectively (Scheme 4.9).[82] As a consequence of the acid conditions, the preparation of halohydrins is generally accompanied by the formation of undesired diethyl 1,2-dihalogenophosphonates.[82]

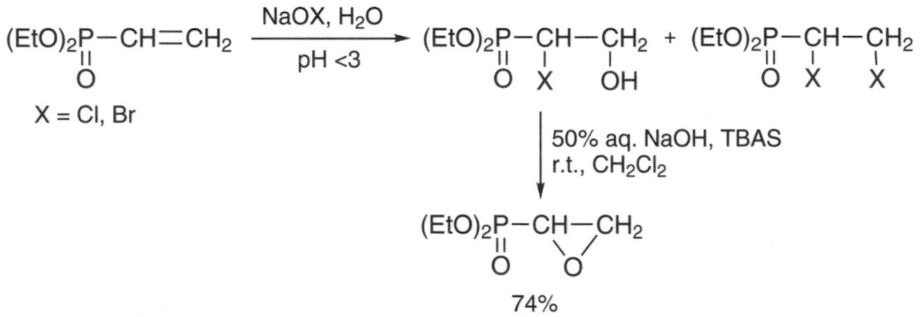

(4.9)

Synthetic reactions that result in the bromohydroxylation of the monoester of 1-propenylphos-phonate with NBA in H_2O at 15°C and in the chloro- or bromohydroxylation of the diacid with NCS or NBS were particularly useful in providing, in good conditions, the two isomeric *threo*-halohydrins, which can be separated before epoxidation.[83] For example, crystallization from acetone of the mixture of diastereomeric bromohydrins prepared by bromohydroxylation of the monoester of propenylphosphonate provided the preferred *threo*-bromohydrin in diastereomerically pure form.[83]

A variety of procedures have been developed to effect the conversion of halohydrins with the hydroxy group in α- or β-positions into 1,2-epoxyethylphosphonates (NaH/THF,[82] NaH/dioxane,[84] KOH/THF,[82] EtO$^-$/EtOH,[82] KOH/EtOH,[74,75,77,79,80,82] K_2CO_3/MeOH or EtOH,[82] or NaOH/CH_2Cl_2/TBAS[82]). Generally, bromohydrinphosphonates give better results than chlorohy-drinphosphonates. Phase-transfer catalysis using the chlorohydrin of diethyl vinylphosphonate and concentrated NaOH (50%) at room temperature has been found to be the method of choice for preparing the diethyl 1,2-epoxyethylphosphonate on large scale (74%, Scheme 4.9).[82] The methods using alkoxide anion or KOH in alcohol tend to be accompanied by important monodealkylation at phosphorus. The preferred *threo*-bromohydrin obtained from propenylphosphonic acid is converted into epoxide by reaction with MeONa in MeOH at 40°C.[83]

Oxidation of tetraethyl ethenylidenediphosphonate with aqueous NaOCl gives tetraethyl oxiranylidene-1,1-diphosphonate directly via Michael β-addition of the hypochlorite anion followed by ring closure (Scheme 4.10). Preparation of the tetrasodium salt of the corresponding epoxide is accomplished in a two-step process through the bromohydrin. The latter is prepared from tetraethyl ethenylidenediphosphonate by treatment with Br_2/H_2O, followed by ring closure with a concentrated solution of NaOH.[85,86]

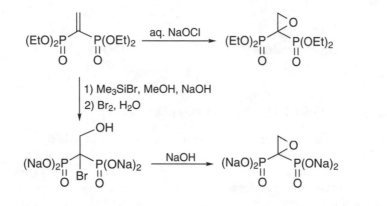

(4.10)

Oxidation of the double bond in diethyl 1-(ethoxycarbonyl)vinylphosphonate has been investigated in detail using O_2, catalytic RuO_4 (generated *in situ* from $RuO_2/NaOCl$ or $RuO_2/NaIO_4$) or stoichiometric RuO_4 (Scheme 4.11).[87] Treatment of diethyl 1-(ethoxycarbonyl)vinylphosphonate on large scale in a two-phase (CCl_4/H_2O) reaction mixture with catalytic RuO_4 using NaOCl as cooxidant or with NaOCl alone at room temperature produces diethyl 1-ethoxycarbonyl-1,2-epoxyethylphosphonate in up to 79% isolated yields. As above, epoxidation of the double bond would begin with nucleophilic addition of the ClO^- anion followed by ring closure. When $NaIO_4$ is used as the RuO_2 reoxidant in a two-phase ($CHCl_3/H_2O$) reaction mixture, the major product is diethyl 1,1-dihydroxy-1-(ethoxycarbonyl)methylphosphonate, the hydrate of diethyl 1-ethoxycarbonyl-1-oxomethylphosphonate.

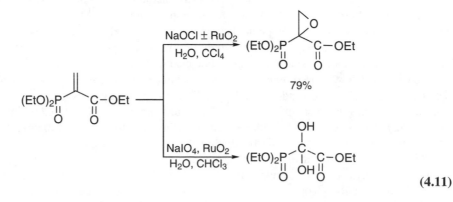

(4.11)

One ingenious procedure for the construction of a nucleoside epoxyphosphonate from a 1,2-dihydroxyalkylphosphonate exploits the interesting properties of triflates, which serve as good leaving groups.[88,89] Treatment of the 1,2-diol with trifluoromethanesulfonyl chloride and 4-DMAP directly produces the nucleoside 1,2-epoxyphosphonate in 31% yield via the displacement of an intermediate 2′-triflate. Unfortunately, this reaction is accompanied by the formation of the methylated uracil derivative. When the 1,2-diol is treated with trifluoromethanesulfonyl chloride and 4-PDP in CH_2Cl_2 at 0°C, the nucleoside 1,2-epoxyphosphonate is obtained in 74% yield, and no alkylated byproducts are detected (Scheme 4.12). The same strategy can also be used to prepare the isomeric 2′,3′-epoxy-2′-phosphonyl derivative.[88,89]

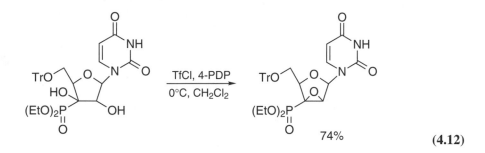

(4.12)

4.1.1.4   Epoxidation of 1,2-Unsaturated Phosphonates

The most attractive and potentially most general route for the synthesis of dialkyl 1,2-epoxyalkylphosphonates appears to be the direct epoxidation of the corresponding dialkyl vinylphosphonates (Scheme 4.13). The syntheses of dialkyl vinylphosphonates with essentially any desired combination of α and β substituents can be achieved readily by a number of procedures.[90] The use of vinylphosphonates offers appreciable advantages. The *cis*- and *trans*-vinylphosphonates (R^2 or R^3 ≠ H) are readily separable, and each one converted to the corresponding 1,2-epoxyphosphonate (Scheme 4.13). Moreover, the use of these unsaturated intermediates allows the acid-catalyzed hydrolysis of the ester functions before epoxidation.

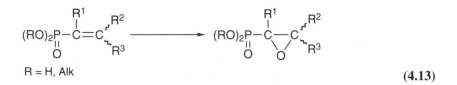

R = H, Alk

(4.13)

   In view of the relatively weak electrophilicity of the double bond of vinylphosphonates and the tendency of these compounds to undergo nucleophilic addition (Michael addition), epoxidations with either strongly electrophilic peracids or nucleophilic oxidants (akaline hydrogen peroxide and *tert*-butylhydroperoxide) were investigated. Full utilization of peracids is often handicapped by the appearance of competing side reactions.[91] For example, when a buffered trifluoroperacetic acid solution in CH_2Cl_2 was used as oxidant, ring opening of the formed epoxide could be observed. The use of di-*tert*-butylperoxide in C_6H_6 with Triton B as catalyst results in the Michael addition product of the butoxy moiety to the olefin.[91–95] In a further example of a $Mo(CO)_6$-catalyzed reaction, *tert*-butylhydroperoxide in 1,2-dichloroethane at reflux fails to react with diethyl 2-methyl- and 1-methylvinylphosphonate.[96] Similarly, diethyl 1-chlorovinylphosphonates do not react with *t*-BuOOLi but are converted to the corresponding epoxides by treatment with H_2O_2 in CH_2Cl_2 at reflux for 10 days. However, under these conditions, the resulting 1-chloro-1,2-epoxyethylphosphonates are partially or completely isomerized to β-chlorinated α-ketophosphonates.[97] The highly electrophilic dimethyl perfluoroisopropenylphosphonate is epoxidized with particular ease using *para*-(methoxycarbonyl)perbenzoic acid in Et_2O to give dimethyl 1,2-pentafluoroepoxyethylphosphonate in 60% yield.[94]

These modest results are overcome when the reagents are replaced by an alkaline H_2O_2 solution (30% solution) in alcohol. Alkaline hydroperoxide is recognized as the most common and perhaps most generally useful reagent for epoxidation of double bonds conjugated with electron-withdrawing groups. Although H_2O_2 has proved to be ineffective at pH 4.7 in the oxidation of diethyl 2-methyl- and 1-methylvinylphosphonate in the presence of Na_2WO_4 at 55–60°C without solvent, H_2O_2 in alcoholic solution at more basic pH appears to have a great synthetic utility in the epoxidation of phosphonate diesters or diacids bearing electron-withdrawing groups. Several oxidizing systems have been developed with success: $H_2O_2/Na_2WO_4/Et_3N/n$-PrOH/pH 5.8–5.9/40–55°C, 1 h[98]; H_2O_2/Na_2CO_3/MeOH/15°C, 3–12 h[99–102]; H_2O_2/NaOH/MeOH/r.t., 1 h[103]; $H_2O_2/NaHCO_3$/EtOH/r.t., 2 h.[104,105] The epoxidation of tetraethyl ethenylidenediphosphonate (Scheme 4.10) is efficiently achieved in 95% EtOH using 30% aqueous H_2O_2 and $NaHCO_3$ at room temperature, instead of NaOCl, to give tetraethyl oxiranylidene-1,1-diphosphonate as a pure product in excellent yield (93%).[104] Similarly, in the route to 2-hydroxy-2-phosphonyl-3-(3-pyridyl)propanoic acid (NE-10864), 2-(3-pyridyl)ethylenediphosphonate has been epoxydized with H_2O_2/Na_2WO_4 in 52% yield.[106]

Similarly, dimethyl 3-oxo-1-cycloalkenylphosphonates ($n = 1,2$) have been converted into the corresponding epoxy derivatives by treatment with H_2O_2 in MeOH in the presence of Na_2CO_3 in 63% ($n = 1$) and 81% ($n = 2$) yields, respectively (Scheme 4.14).[99,101,102,107]

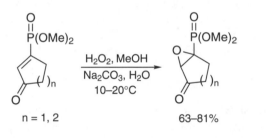

$$n = 1, 2 \qquad\qquad 63\text{–}81\%$$

(4.14)

A variety of diethyl vinylphosphonates have been oxidized in moderate to good yields (19–84%) using dioxirane generated *in situ* from buffered monopersulfate (Oxone®) and a ketone in a two-phase (CH_2Cl_2/H_2O) reaction mixture.[108,109] Because of the pH dependence of dioxirane stability, the process needs an accurate control of the pH at 7.4. However, the reaction is often very sluggish and requires regular addition of fresh portions of oxone and lengthy reaction times.

4.1.2 DIALKYL 2,3-EPOXYALKYLPHOSPHONATES

4.1.2.1 Michaelis–Arbuzov Reactions

Dialkyl 2,3-epoxypropylphosphonates are readily prepared by heating trialkyl phosphites and 1-bromo- or 1-iodo-2,3-epoxypropane at 130–140°C. Between these two precursors, the 1-bromo-2,3-epoxypropane appears to be the best reagent, as demonstrated by the synthesis of dialkyl 2,3-epoxypropylphosphonates in 51–86% yields on molar scale (Scheme 4.15).[91,110–124] When 1-iodo-2,3-epoxypropane is allowed to react with triethyl phosphite, only modest yields of diethyl 2,3-epoxypropylphosphonate are obtained (28–30%), contaminated mainly with diethyl ethylphosphonate (23%).[125–128] Under the conditions required for the Michaelis–Arbuzov rearrangement, the 1-chloro-2,3-epoxypropane does not undergo the expected reaction. Triethyl phosphite reacts with 1-chloro-2,3-epoxypropane mainly at the epoxy group, and preferentially at the CH_2 group of the epoxide ring, to give mixtures of products containing triethyl phosphite, diethyl methylphosphonate, and diethyl vinyl phosphate.[125,127,129] Similarly, on reaction with diethyl trimethylsilyl phosphite, 1-chloro-2,3-epoxypropane undergoes selective opening of the epoxide ring with formation of diethyl 3-chloro-2-(trimethylsiloxy)propylphosphonate, resulting in the transfer of the silyl ester linkage to the oxirane oxygen.[130]

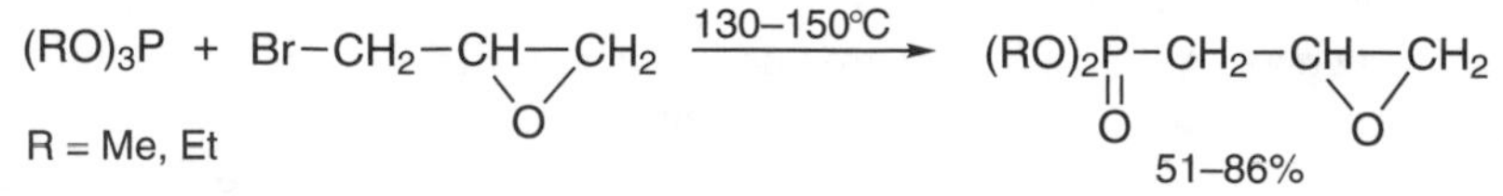

$$(RO)_3P \; + \; Br-CH_2-CH-CH_2 \xrightarrow{130-150°C} (RO)_2P-CH_2-CH-CH_2$$

R = Me, Et

51–86%

(4.15)

The Michaelis–Arbuzov reaction has been extended to substituted bromomethyl epoxides such as *cis*- and *trans*-1,3-diphenyl-2,3-epoxy-4-bromo-1-butanone[131] and *cis*-4-benzyloxy-1-bromo-2,3-epoxybutane[132] to give the substituted diethyl 2,3-epoxypropylphosphonates in high yields.

4.1.2.2 Reactions of Phosphonate Carbanions with α-Haloketones

The reaction between α-halo ketones and sodium dialkyl phosphites has been extended to phosphonate carbanions. For example, diisopropyl 1-lithio-1-fluoromethylphosphonate, generated from diisopropyl fluoromethylphosphonate and LDA, reacts with 3-chloro-2-butanone at low temperature in THF to give diisopropyl 1-fluoro-2-methyl-2,3-epoxybutylphosphonate in 46% yield.[133,134]

4.1.2.3 Reactions of 2,3-Halohydrinphosphonates with Bases

Diethyl 3-chloro-2-oxopropylphosphonate has been conveniently transformed into diethyl 3-chloro-2-hydroxypropylphosphonate in 82% yield and 72% ee with baker's yeast. Although the reduction needs several hours in H_2O, no dechlorinated phosphonate could be detected in the reaction medium. Diethyl 3-chloro-2-hydroxypropylphosphonate is readily converted into diethyl 2,3-epoxypropylphosphonate (92% yield) using K_2CO_3 in THF at reflux.[135]

4.1.2.4 Abramov and Pudovik Reactions

Dialkyl phosphites react slowly with epoxy ketones at room temperature in the presence of a catalytic amount of sodium metal to form dialkyl 1-hydroxy-2,3-epoxypropylphosphonates in 41–99% yields.[136] Similarly, dialkyl phosphites react with *cis*-1-formyl-1-trimethylsilyl-1,2-epoxyalkanes in the presence of DBU in THF to give dialkyl 1-hydroxy-2-trimethylsilyl-2,3-epoxyalkylphosphonate in high yields (84%, Scheme 4.16).[137] By contrast, DBU or Et_3N in C_6H_6 at reflux was found inefficient for the addition of dimethyl phosphite to the epoxide derived from chalcone. The reaction is best carried out in the presence of $KF \cdot 2H_2O$ to produce a diastereomeric mixture of hydroxyphosphonates free from other organophosphorus compounds.[138,139]

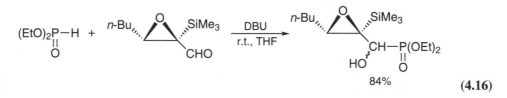

84%

(4.16)

Aminoepoxyphosphonates such as dialkyl 1-amino-2,3-epoxypropylphosphonates and their *N*-substituted derivatives can be prepared by a three-component reaction using dialkyl phosphites, 2,3-epoxypropionaldehyde, and primary amines. However, the reaction is retarded by water, so it is preferable to add dialkyl phosphites to the corresponding aldimines in C_6H_6 at room temperature.[140]

4.1.2.5 Epoxidation of 2,3-Unsaturated Phosphonates

As for dialkyl vinylphosphonates, the combination $H_2O_2/Na_2CO_3/MeOH$ at room temperature is very efficient for the epoxidation of (*E*)-4-oxo-2-alkenylphosphonates in good yields (65–75%).[100,107,141,142] Epoxidation of 2,3-unsaturated glycosylphosphonates proved to be difficult with H_2O_2 (30% in H_2O) and Na_2WO_4. At room temperature, the reaction is sluggish, requiring lengthy reaction times

(5 days), and the yields do not exceed 55%. They are improved (80%) by the addition of AcONa. However, the reaction is accompanied by some epimerization at C-1. Satisfactory results are also achieved by employing H_2O_2 (30%) with dodecatungstophosphoric acid ($H_3PW_{12}O_4{\cdot}21H_2O$) and AcONa.[143] Epoxidation of dialkyl allylphosphonates has also been effected by treatment with H_2O_2 (30%) in acetone in the presence of $V(acac)_3$[144] or by treatment with H_2O_2 (30%) in MeOH at room temperature at pH 9.5 in the presence of Na_2HPO_4 and benzonitrile.[62]

Because the use of a protic solvent (H_2O, ROH) may lead to the destruction of sensitive products or a reduction in the stereoselectivity through competitive hydrogen bonding by the solvent, recent innovations have significantly extended the synthetic utility of the classical direct epoxidation. The introduction of methyl trioxorhenium (MTO) in combination with urea hydrogen peroxide adduct (UHP) as a reoxidant in nonprotic solvents for the catalytic epoxidation of allylic phosphonates is a useful modification.[145–147] Thus, allylic α-hydroxyphosphonates react with MTO/UHP in CH_2Cl_2 at room temperature to give the corresponding 2,3-epoxyphosphonates in 88–93% yields as a mixture of diastereomers (Scheme 4.17).[145]

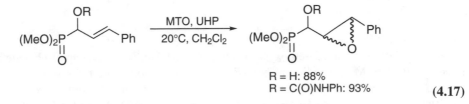

R = H: 88%
R = C(O)NHPh: 93%

(4.17)

Epoxidation of allylic phosphonates is achieved with success at room temperature with $MeCO_3H$ in Et_2O, CF_3CO_3H in $CHCl_3$, MCPBA in CH_2Cl_2, or MoO_5/HMPA complex in CH_2Cl_2 to give the corresponding 2,3-epoxyalkylphosphonates as a mixture of diastereomers.[148–153] Allylic phosphonates may also be converted into 2,3-epoxyphosphonates via the 1,1,1-trifluorodimethyl-dioxirane-mediated oxidation.[154] Dimethyldioxirane (DMD) in acetone at room temperature[122,155] or methyl(trifluoromethyl)dioxirane (TFD) in CH_2Cl_2 at low temperature[156] can be used instead of MCPBA. Because the reaction is quantitative, evaporation of acetone and excess of DMD allow the direct isolation of the pure product.[122,155]

4.1.3 DIALKYL 3,4-EPOXYALKYLPHOSPHONATES

4.1.3.1 Reactions of Dialkyl 3,4-Halohydrinphosphonates and Related Structures with Bases

The treatment of diethyl 1,1-difluoro-3-butenylphosphonate with mercuric acetate and bromine gives rise to a mixture of isomeric masked bromhydrins that, on reaction with KOH in MeOH, produce diethyl 1,1-difluoro-3,4-epoxybutylphosphonate in 30% yield (Scheme 4.18).[157,158] The replacement of the halogen atom by a tosylate group gives better yields of 3,4-epoxybutylphos-phonates. Thus, dialkyl 4-tosyloxy-3-hydroxybutylphosphonates, prepared by tosylation of 3,4-dihydroxybutylphosphonates[159] or by reaction of glycidol tosylate with dialkyl 1-lithiomethylphos-phonate,[160] undergo base cyclization with RONa in ROH to produce quantitatively dialkyl 3,4-epoxybutylphosphonates.[159,160]

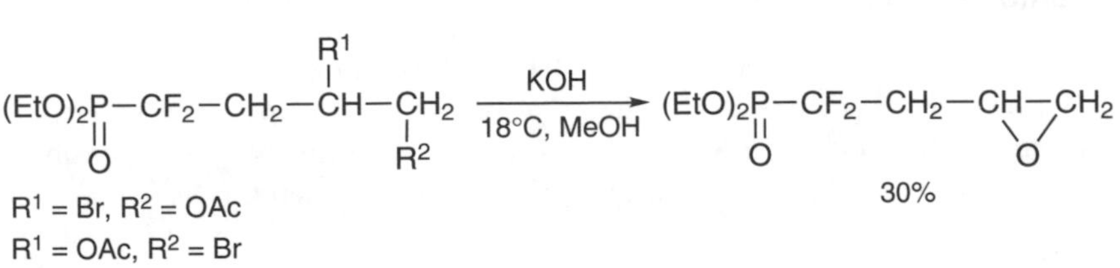

$$(EtO)_2\overset{\displaystyle \underset{O}{\|}}{P}-CF_2-CH_2-\overset{\displaystyle \underset{R^2}{\overset{R^1}{|}}}{C}H-CH_2 \quad \xrightarrow[18°C,\ MeOH]{KOH} \quad (EtO)_2\overset{\displaystyle \underset{O}{\|}}{P}-CF_2-CH_2-CH-CH_2$$

R^1 = Br, R^2 = OAc
R^1 = OAc, R^2 = Br

30%

(4.18)

4.1.3.2 Epoxidation of 3,4-Unsaturated Phosphonates

The most general and convenient procedure for the synthesis of dialkyl 3,4-epoxyalkylphosphonates involves the direct epoxidation of homoallylic phosphonates with MCPBA or t-BuO$_2$H. A great variety of diethyl 3,4-epoxyalkylphosphonates have been prepared in fair to good yields (40–94%) from the corresponding homoallylic phosphonates on treatment with MCPBA in CH$_2$Cl$_2$. The procedure tolerates the substitution of the α, γ, and δ positions (Scheme 4.19).[161–170] This procedure has also been applied to cyclic dienic phosphonates.[171]

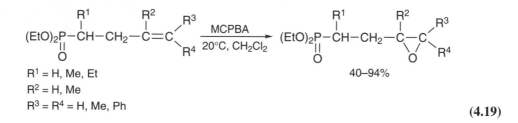

R^1 = H, Me, Et
R^2 = H, Me
R^3 = R^4 = H, Me, Ph
                                                                          40–94%

(4.19)

Epoxidation of diethyl 3-(ethoxycarbonyl)-2-vinyl-1,1-difluoropropylphosphonate is very slow under MCPBA or MTO conditions, but TFD prepared *in situ* proved to be more effective, and the desired 3,4-epoxyphosphonate is obtained in 91% yield as a mixture of diastereomers.[172]

By using the catalytic Sharpless asymmetric oxidation [(−)-DET, Ti(Oi-Pr)$_4$, t-BuO$_2$H, CH$_2$Cl$_2$], dimethyl 5-hydroxy-3,4-epoxypentylphosphonate is obtained in excellent enantiomeric excess (98%) and high chemical yield (73%).[173] Under the same conditions, diisopropyl (E)-2-hydroxy-3-pentenylphosphonate yields a mixture of diastereoisomeric epoxides.[174] Oxidation of homoallylic phosphonates with H$_2$O$_2$/Na$_2$WO$_4$·2H$_2$O has been reported to generate 3,4-epoxybutylphosphonates in comparable yields.[175]

4.1.3.3 Reactions of Phosphonate Carbanions with Epihalohydrines

The reaction of diethyl 1-copper(I)alkylphosphonates with 1-chloro- or 1-bromo-2,3-epoxypropane produces diethyl 3,4-epoxyalkylphosphonates. Although this procedure gives moderate yields (30–47%), the possibility of readily generating α-substituted 3,4-epoxyalkylphosphonates is compensatory (Scheme 4.20).[161]

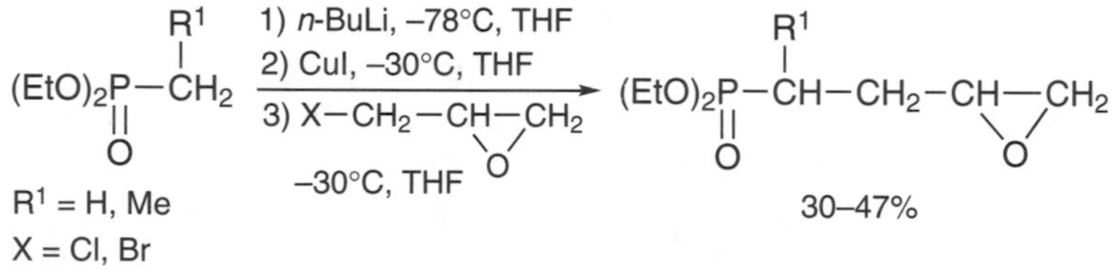

(4.20)

4.1.3.4 Addition Reactions

Cerium-mediated conjugate addition of diethyl 1-lithio-1,1-difluoromethylphosphonate to 1-phenylsulfonyl-3,4-epoxycyclopentene and -cyclohexene proceeds smoothly at low temperature in THF to give the corresponding adduct in 53% and 65% yields, respectively.[176] The possibility of mild reductive removal of the SO$_2$Ph moiety makes these reactions attractive.

4.1.4 Dialkyl 4,5-, 5,6-, 6,7-, 7,8-, and 8,9-Epoxyalkylphosphonates

4.1.4.1 Alkylation of Phosphonate Carbanions

Long-chain phosphonylated aldehydes are generally prepared to achieve the formation of macrocycles via an intramolecular Horner–Wadsworth–Emmons reaction. The phosphonate group is frequently incorporated at one extremity of the chain by a carbanionic approach. Thus, displacement of iodide of the alkyl chain containing epoxide by the sodium enolate of diethyl 1-(ethoxycarbonyl)methylphosphonate at 50°C in DMF leads to diethyl 1-(ethoxycarbonyl)-4,5-epoxyalkylphosphonate in 78% yield (Scheme 4.21).[177]

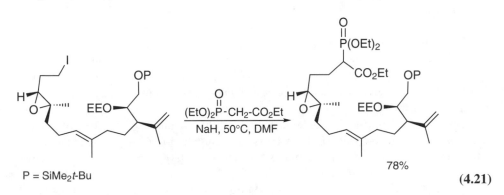

$$\text{(4.21)}$$

It is noteworthy that diethyl 1-(ethoxycarbonyl)methylphosphonate can be first metallated, then alkylated by an homoallyl iodide to give 4,5-unsaturated phosphonate, and further converted into epoxides by treatment with MCPBA in CH_2Cl_2 in 93% yield.[178,179]

4.1.4.2 Addition Reactions

Diethyl 1,1-difluoro-3-iodo-6,7-epoxyheptylphosphonate is prepared in 63% yield by the palladium-catalyzed addition of diethyl iododifluoromethylphosphonate to 1,2-epoxy-5-hexene at room temperature (Scheme 4.22).[180,181]

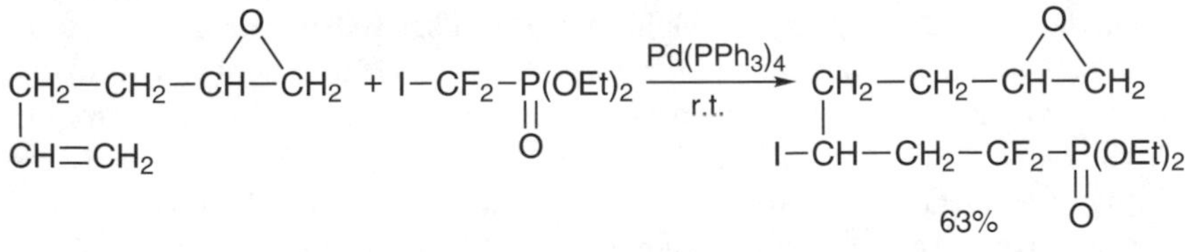

$$\text{(4.22)}$$

Similarly, addition of diethyl fluoroiodomethylphosphonate to 1,2-epoxy-5-hexene and 1,2-epoxy-7-octene on heating at 75°C in the presence of $Pd(PPh_3)_4$ gives the corresponding diethyl 1-fluoro-3-iodo-6,7-epoxyheptylphosphonate and diethyl 1-fluoro-3-iodo-8,9-epoxynonylphosphonate in 69% and 65% yields, respectively.[182]

4.1.4.3 Epoxidation of Unsaturated Phosphonates

Diethyl 4,5- and 5,6-epoxyalkylphosphonates are prepared by treatment of the corresponding 4,5- and 5,6-unsaturated phosphonates with MCPBA in CH_2Cl_2 at 40°C[183] or aqueous solution of monoperoxyphthalate in i-PrOH at room temperature.[184] The epoxidation of the double bond of a 6-oxo-7,8-unsaturated phosphonate involves the use of H_2O_2 in NaOH–MeOH solution to provide the corresponding 7,8-epoxyphosphonate in 90% yield.[185]

4.2 SYNTHESIS OF FOSFOMYCIN

Fosfomycin, the (1*R*,2*S*) (−)-1,2-epoxypropylphosphonic acid, formerly called phosphonomycin, is a low-molecular-weight antibiotic of unusual structure that was originally isolated in 1969 from used culture medium of *Streptomyces fradiae* (Figure 4.1).[1,186–188] The structure was established by

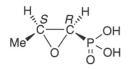

FIGURE 4.1

means of nuclear magnetic resonance (NMR), infrared (IR), and mass spectrometry (MS) studies of its dimethylester and by direct comparison of the monobenzylammonium salt of the natural and synthetic products.[189] Fosfomycin is also produced by various species of *Pseudomonas* such as *Pseudomonas syringae*[190] or *Pseudomonas viridiflava*.[191] Fosfomycin is a broad-spectrum antibiotic effective against both Gram-positive and Gram-negative infections in mammals, and its effectiveness is comparable with that of tetracycline or chloramphenicol.[192] The antibiotic irreversibly inactivates the uridine-5′-diphospho-*N*-acetylglucosamine-3-*O*-enolpyruvyltransferase (MurA), the enzyme that catalyzes the first committed step in cell wall biosynthesis.[193,194] The inactivation occurs through the alkylation of a cysteine residue (Cys-115 of the enzyme from *Escherichia coli*) in the active site of MurA.[193–195] The antibiotic, whose action has been investigated in depth,[196–200] can be administrated orally or intravenously and shows little toxicity to humans.

Early reported works using isotope-labeling techniques have shown that labeling of carbons C-1 and C-2 was provided by [1,6–14C]glucose, which primarily labels C-1, [2–14C]glucose, which labels C-2, and [2–14C]acetate, which labels both C-1 and C-2 in equal amounts.[201] From these data, phosphoenolpyruvate (PEP) was suggested as the most likely precursor of carbons C-1 and C-2 of fosfomycin.[201] 2-Hydroxyethylphosphonic and 2-aminoethylphosphonic acids were equally found to be intermediates in the biosynthesis of fosfomycin.[202] In order to obtain information about the mechanism of oxirane ring formation, several studies using isotope-labeling techniques have been carried out to examine the incorporation of various labeled 2-hydroxyalkylphosphonic acids into fosfomycin in *Streptomyces fradiae*. For this purpose, labeled 2-hydroxyethylphosphonic, 2-hydroxypropylphosphonic, and 1,2-dihydroxypropylphosphonic acids have been prepared and used in feeding experiments.[203–206] From the incorporation of 18O into the alcohol function of (2-hydroxypropyl)phosphonic acid, the oxirane oxygen atom has been unequivocally confirmed to arise from the hydroxy group.[207] On the basis of these results, the biogenesis of fosfomycin begins with the isomerization of phosphoenolpyruvate to give phosphonopyruvic acid, which presumably decarboxylates to yield phosphonoacetaldehyde,[201] which in turn is methylated to provide (*S*)-2-hydroxypropylphosphonic acid, the last intermediate before the formation of fosfomycin.[207] In the transformation of phosphonoacetaldehyde into (*S*)-2-hydroxypropylphosphonic acid, it is assumed that the methyl group of L-methionine is transferred via a corrin to the phosphonoacetaldehyde (Scheme 4.23).[208]

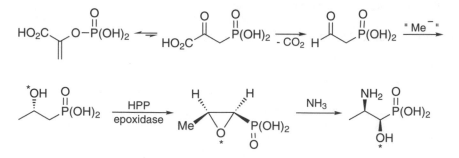

(4.23)

More recently, fosfomycin mono- and diphosphates have been identified as the inactivation products of fosfomycin, formed by the action of gene products of *fomA*[209] and *fomB*[210] expressed in *Escherichia coli*.[211] The enzyme responsible for the conversion of (*S*)-2-hydroxypropylphosphonic acid into fosfomycin (HPP epoxidase, Scheme 4.23) has been characterized, and a mechanism for this unusual epoxidation has been proposed. Both $Fe(NH_4)_2(SO_4)_2$ and NAD(P)H are essential to effect this conversion.[212]

Fosfomycin is present on the pharmaceutical market as the disodium, calcium, and *tris*(hydroxymethyl)ammonium salts.[213] The structures of the diester, monoester, disodium, calcium, and phenethylammonium salts of fosfomycin have been studied by ^{1}H, ^{13}C, and ^{31}P NMR, IR and Raman spectroscopy, and X-ray crystallography.[214-217] A number of methods for the synthesis of fosfomycin are presently available, and only the more attractive have been selected.

The first racemic synthesis of fosfomycin is based on the epoxidation of (±) (*Z*)-1-propenylphosphonic acid (Scheme 4.24).[189,218] The reaction was effectively achieved using an excess of H_2O_2 (30% solution) at 60°C in the presence of Na_2WO_4 as catalyst (2%) at pH 5.5. The (*Z*)-1-propenylphosphonic acid was prepared by stereospecific hydrogenation, with Lindlar catalyst, of dibutyl 1-propynylphosphonate into dibutyl (*Z*)-1-propenylphosphonate, followed by hydrolysis of the ester functions with concentrated HCl. The resolution of the racemic (±) (*Z*)-1,2-epoxypropylphosphonic acid to fosfomycin was accomplished by way of optically active amines, α-phenylethylamine,[218,219] or quinine.[189,220,221]

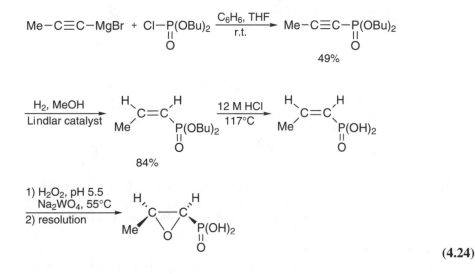

(4.24)

An alternative procedure for the synthesis of 1-propynylphosphonic ester uses the Michaelis–Arbuzov reaction of 2,3-dichloro-1-propene with triisopropyl phosphite to produce almost quantitatively diisopropyl 2-chloroallylphosphonate. On treatment with aqueous methanolic Na_2CO_3 solution at room temperature, it is converted into diisopropyl 1-propynylphosphonate by a dehydrochlorination–isomerization sequence (Scheme 4.25).[222]

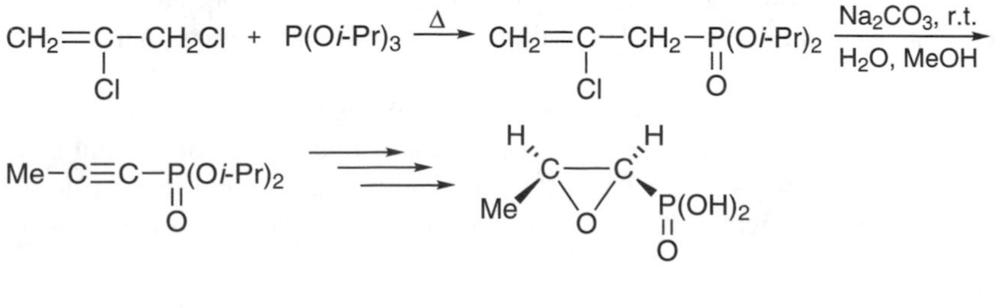

$$CH_2{=}\underset{\underset{Cl}{|}}{C}{-}CH_2Cl \ + \ P(Oi\text{-}Pr)_3 \ \xrightarrow{\Delta} \ CH_2{=}\underset{\underset{Cl}{|}}{C}{-}CH_2{-}\underset{\underset{O}{\|}}{P}(Oi\text{-}Pr)_2 \ \xrightarrow[H_2O,\ MeOH]{Na_2CO_3,\ r.t.}$$

$$Me{-}C{\equiv}C{-}\underset{\underset{O}{\|}}{P}(Oi\text{-}Pr)_2 \ \xrightarrow{\hspace{2cm}} \ \underset{Me}{\overset{H}{\diagdown}}\underset{O}{C}{-}C\underset{\underset{O}{\|}}{\overset{H}{\diagup}}P(OH)_2$$

(4.25)

The original synthesis of fosfomycin has been improved by using the acid-sensitive *t*-BuOH as blocking group instead of *n*-BuOH and propargyl alcohol instead of propyne. A sequence of reactions for the conversion of the propargyl alcohol into (±) (Z)-1-propenylphosphonic acid via the alkadienylphosphonate has been developed (Scheme 4.26).[98,223] The rearrangement of the alkynyl phosphite produces the alkadienylphosphonate,[224–228] which is selectively and stereospecifically hydrogenated with Pd/C in C_6H_6 to give di-*tert*-butyl (Z)-1-propenylphosphonate. The rapid removal of the two *tert*-butyl groups is accomplished by refluxing the C_6H_6 solution with a strong acid catalyst. An elegant improvement was the use of (Z)-1-propenylphosphonic acid as catalyst to avoid the contamination of the product with a foreign acid.[98,223] Removal of the solvent affords (Z)-1-propenylphosphonic acid in 81% overall yield. The conversion of (±) (Z)-1-propenylphosphonic acid into fosfomycin is accomplished by using (+)-α-phenylethylamine as resolving base in propanol, followed by treatment at pH of 5.8–5.9 with H_2O_2 in the presence of Na_2WO_4 at 50–55°C. A single recrystallization from aqueous propanol affords fosfomycin salt with 100% optical purity.[98,223]

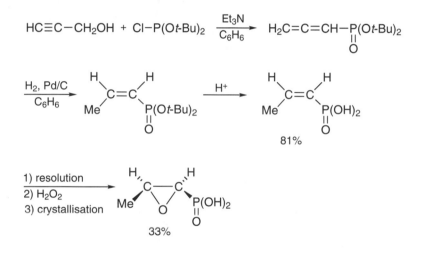

(4.26)

An alternative synthesis of fosfomycin has been effected by the halohydrin route.[229] Thus, treatment of (±) (Z)-1-propenylphosphonic acid in aqueous solution with sodium hypochlorite gave *threo*-1-chloro-2-hydroxypropylphosphonic acid (85%). Resolution is accomplished by means of (−)-α-phenylethylamine to yield (+)-chlorohydrin (80%), which is converted with 10 M aqueous NaOH into fosfomycin (85–90%).[229]

Because of the commercial importance of fosfomycin, it is not surprising that several important and attractive synthetic methods are reported in patents. They include, for example, precursors such as dimethyl hydroxymethylphosphonate,[230,231] dimethyl, dibenzyl, and diallyl formylphosphonate,[232,233] trimethyl phosphite and 2-cyano-1-hydroxypropene,[234] trialkyl phosphite and 2-chloropropionaldehyde or 2-acetoxypropionaldehyde,[235] diethyl chloromethylphosphonate,[25] dibenzyl phosphite, and 1-chloro-1,2-propylene oxide,[236,237] propynylphosphonic acid,[238–240] propenylphosphonic acid,[241–244] 2-chloro-(*cis*-1,2-epoxypropyl)phosphonic acid,[245] and extrusion reactions on thermolysis.[246,247] The resolution of racemic acids has also been reported.[248] In search of new effective antibiotics, a large variety of substituted epoxyethylphosphonic acids have been prepared.[249]

The first nonmicrobial asymmetric synthesis of fosfomycin was based on the use of tartaric acid as chiral auxiliary to induce the appropriate bifunctionalization of prochiral (Z)-1-propenylphosphonic acid (Scheme 4.27).[83] Thus, the reaction of (2S,3S)-tartaric acid derivatives with (Z)-1-propenylphosphonic dichloride in CH_2Cl_2 at −10°C gives cyclic phosphonates that undergo ring opening in aqueous media.[218] The resulting monoesters are crystallized (70% yield) and then subjected to bromohydroxylation with NBA in H$_2$O at 15°C under neutral or acidic conditions. This reaction is highly chemoselective, regiospecific, and stereospecific, providing *threo*-bromohydrins (1*R*,2*S*) and (1*S*,2*R*) in a 7:3 ratio with yields higher than 90%. Crystallization of the mixture of diastereomeric bromohydrins provides the (1*R*,2*S*) derivative in diastereomerically pure form. After the acid-mediated elimination of tartaric acid, the *threo*-(1*R*,2*S*)-bromohydrin is converted into enantiomerically pure fosfomycin as the sodium salt by reaction with MeONa in MeOH at 40°C.[83]

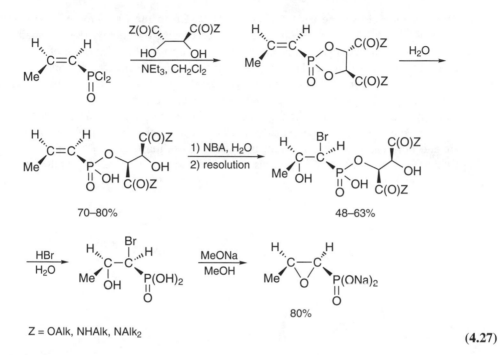

Z = OAlk, NHAlk, NAlk$_2$

(4.27)

An original approach to the large-scale synthesis of fosfomycin is based on the enantioselective hydrogenation of prochiral β-ketophosphonates to obtain *threo*-(1*R*,2*S*)-bromohydrin (Scheme 4.28).[250] The racemic dimethyl 1-bromo-2-oxopropylphosphonate starting material is prepared by the reaction of dimethyl 2-oxopropylphosphonate with HBr (8.8 M) and H$_2$O$_2$ (30%) in THF at 25°C. Although hydrogenation of a racemic α-substituted-β-ketophosphonate usually gives four possible stereoisomers, the lability of the α-stereogenic center confers an *in situ* stereoinversion, which allows the stereoselective synthesis of a single isomer under conditions of dynamic kinetic resolution. A combination of experimentation and computer-aided analysis of the stereochemical and kinetic parameters allowed the determination of suitable experimental conditions. The (*S*)-BINAP-Ru(II) complex formed *in situ* by simple heating of a mixture of [RuCl$_2$(C$_6$H$_6$)]$_2$ and (*S*)-BINAP in DMF acts as an excellent catalyst for the enantioselective hydrogenation of prochiral α-bromo-β-ketophosphonate. The reaction takes place smoothly in MeOH under 4 atm at 25°C over 100 h to give a mixture of dimethyl (1*R*,2*S*)-1-bromo-2-hydroxypropylphosphonate (98% ee), dimethyl (1*S*,2*S*)-1-bromo-2-hydroxypropylphosphonate (94% ee), and dimethyl (*S*)-2-hydroxypropylphosphonate (98% ee), resulting from partial hydrogenolytic debromination, in a 76.5/8.5/15 ratio.[250] Similarly, hydrogenolitic debromination of dimethyl 2-bromo-1-hydroxypropylphosphonate has been achieved with Pd/CaCO$_3$ in 90% MeOH.[229] Fosfomycin, as its sodium salt, can be obtained by acid-catalyzed hydrolysis of the ester groups of the α-bromhydrin, followed by ring closure during treatment with NaOH.[229]

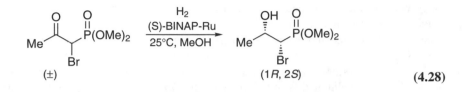

(4.28)

Another asymmetric synthesis of fosfomycin is based on the formation of a protected 1,2-dihydroxyphosphonate by stereoselective addition of dibenzyl trimethylsilyl phosphite to (S)-(triisopropylsilyloxy)lactaldehyde at low temperature (Scheme 4.29).[251] Exposure of the crude reaction mixture to citric acid in MeOH affords the deprotected α-hydroxyphosphonate (80% yield, 80% de), which, on reaction with methanesulfonyl chloride and after separation of the diastereomers by flash chromatography, yields the corresponding methanesulfonate. Treatment with TBAF in THF simultaneously deprotects the β-hydroxy group and achieves the ring closure to give the dibenzyl ester of fosfomycin as a single diastereoisomer (77%). After deprotection at phosphorus by hydrogenolysis and purification, the fosfomycin is isolated as sodium salt in 76% yield.[251] The same strategy using (S)-benzyloxylactaldehyde has also been reported independently at the same time.[252]

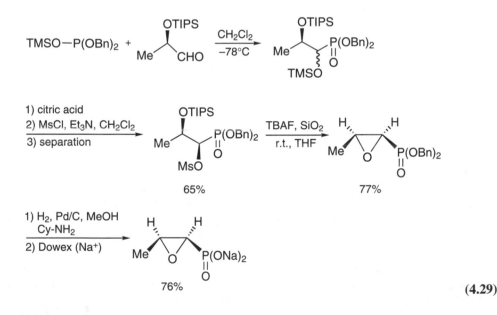

(4.29)

An alternative asymmetric synthesis of fosfomycin based on the same strategy uses (2S)-acetoxypropionyl chloride, which undergoes a Michaelis–Arbuzov reaction with dimethyl trimethylsilyl phosphite to give the corresponding α-ketophosphonate in excellent yield (90%). As illustrated in Scheme 4.30, the α-ketophosphonate is reduced to α-hydroxyphosphonate, then reacted with methanesulfonyl chloride in the presence of Et₃N. Treatment with methanesulfonic acid in MeOH at reflux deprotects the β-hydroxy group, and the ring closure is readily achieved with K₂CO₃ in MeOH.[253] A similar synthesis of fosfomycin using (2S)-benzyloxypropionyl chloride has also been described.[254]

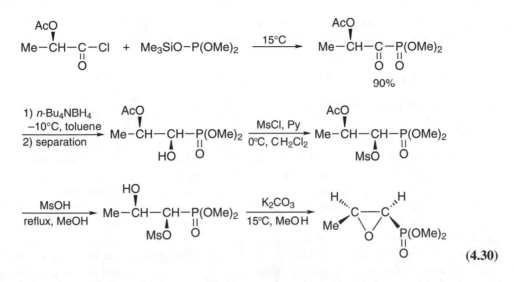

(4.30)

The key step of a recent synthesis of fosfomycin is the Sharpless asymmetric dihydroxylation (AD) reaction.[255] Dibenzyl (*E*)-1-propenylphosphonate is prepared in 94% yield from (*E*)-propenyl bromide and dibenzyl phosphite in THF at 60°C in the presence of Et$_3$N and catalytic amount of Pd(PPh$_3$)$_4$. A modified AD-mix-α is used to speed up the dihydroxylation reaction. The resulting *syn*-α,β-dihydroxyphosphonate is isolated in 65% yield after recrystallization and then submitted to regioselective α-sulfonylation followed by cyclization by treatment with K$_2$CO$_3$ in acetone at room temperature to produce dibenzyl (1*R*,2*S*)-1,2-epoxypropylphosphonate in 67% yield (Scheme 4.31).[255] Fosfomycin can be obtained by hydrogenolysis of this epoxyphosphonate.[251]

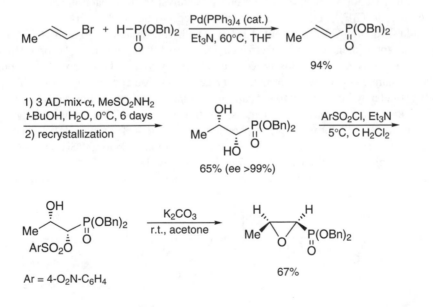

(4.31)

The oxidation of (*Z*)-propenylphosphonic acid to fosfomycin by a fermentation process has been investigated. It has been demonstrated that *Penicillium spinolusum* may effectively be employed to transform the biologically inactive (*Z*)-propenylphosphonic acid into the active fosfomycin. At optimum levels of glucose, used as carbon source, an epoxidation efficiency approaching 90% of olefin charged was obtained after 10 days of incubation at 28°C at pH 5.6. Moreover, the epoxidizing systems show substrate specificity because (E)-propenylphosphonic acid was not epoxidized.[256]

4.3 SYNTHETIC APPLICATIONS OF EPOXYPHOSPHONATES

4.3.1 1,2-EPOXYALKYLPHOSPHONATES

4.3.1.1 Rearrangements

The first example of thermal and acid-catalyzed phosphoryl shifts from carbon to carbon was reported in 1966.[18] It was reported that diethyl 2,2-disubstituted-1,2-epoxyethylphosphonates readily undergo specific rearrangement to yield, as a result of the phosphoryl group migration, diethyl 1,1-disubstituted-1-formylmethylphosphonate (Scheme 4.32). The thermal specificity, and hence applicability, is limited to the 2,2-disubstituted substrates, which can rearrange at a temperature not exceeding 170°C (0.7 mmHg), and evidence for an efficient reaction was obtained in only one instance.[18] In other cases, the rearranged products are unstable at the temperature required for rearrangement (270–300°C, 0.6–0.7 mmHg), and only dephosphorylated aldehydes were isolated accompanied by polymeric materials.

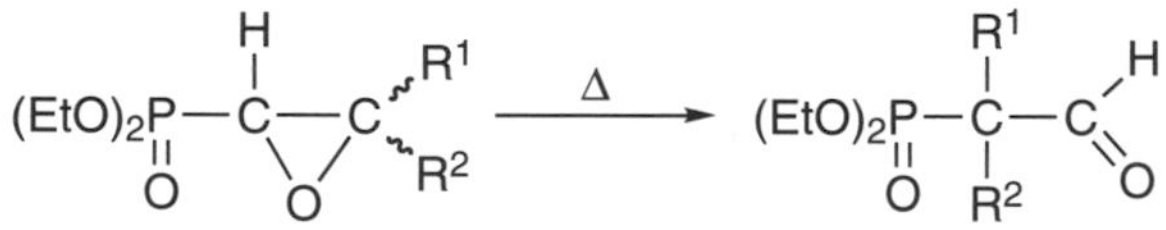

$$(4.32)$$

The thermal technique has largely been supplanted by the use of Lewis acids, which are more efficient in this reaction.[17–20,91,257–261] The preferred catalyst for this reaction was found to be boron trifluoride etherate (BF$_3$·Et$_2$O). Its use increases reaction rates and yields. Rearrangement of the readily available diethyl 2,2-disubstituted 1,2-epoxyethylphosphonates to diethyl 1,1-disubstituted-1-formylmethylphosphonates can be effected on preparative scale in C$_6$H$_6$ or CH$_2$Cl$_2$ at room temperature in a few minutes. This procedure selectively yields the carbonyl compound in pure form. The mechanism involves the formation of a carbonium ion and the migration of the phosphoryl group rather than a hydrogen atom to the carbonium ion. The boron trifluoride–catalyzed rearrangement of 2,2-disubstituted epoxyphosphonates constitutes an expeditious route for the synthesis of diethyl 1,1-disubstituted 1-formylalkylphosphonates, and a variety of compounds have been synthesized on a large scale in excellent yields from symmetric or asymmetric ketones (Scheme 4.33).[261] The use of other Lewis acids such as SnCl$_4$, SnBr$_2$, ZnCl$_2$, ZnBr$_2$, or TaF$_5$ in C$_6$H$_6$ or CH$_2$Cl$_2$ gives mixtures of products in which 1,1-disubstituted 1-formylmethylphosphonates remain the major compounds.

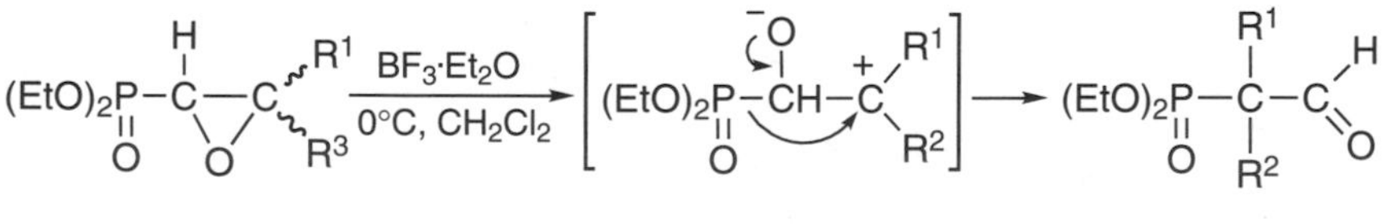

25–78% $$(4.33)$$

This rearrangement is not perfectly general. For example, the 1,2-epoxyphosphonates prepared from cyclic ketones (n = 1–3, Scheme 4.34) undergo a competing proton migration on treatment with BF$_3$·Et$_2$O and consequently give a mixture of diethyl 1,1-disubstituted 1-formylphosphonates (70%) and diethyl 1-hydroxyallylphosphonates (30%).[261]

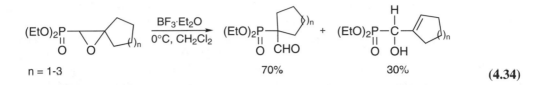

$$(4.34)$$

The occurrence of the rearrangement depends on the stability of the incipient carbonium ion at C-2. 1,2-Epoxyethylphosphonate undergoes ring opening without rearrangement on treatment with BF$_3$·Et$_2$O. In the same conditions, phosphoryl group migration occurs in preference to alkyl group migration in 2,2-disubstituted 1-alkyl-1,2-epoxyalkylphosphonates. By contrast, 1,2,2-triaryl-1,2-epoxyethylphosphonates react with BF$_3$·Et$_2$O to give a mixture of α- and β-ketophosphonates as a consequence of the competing migration of aryl and phosphoryl groups, respectively. The following order of migratory aptitudes has been established: *para*-anisyl > *para*-tolyl > phenyl = (RO)$_2$P(O) > H > alkyl.[260]

Whereas α-halohydrinphosphonates are converted into 1,2-epoxyphosphonates by treatment with strong bases, the corresponding β-halohydrins rearrange instantaneously in the presence of aqueous NaHCO$_3$ and slowly in water alone to give 1-formylethylphosphonic acid (Scheme 4.35).[229]

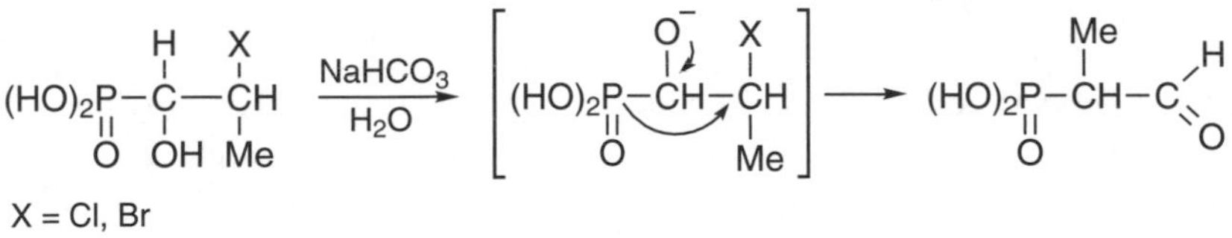

$$(4.35)$$

4.3.1.2 Ring-Opening Reactions

Most of the synthetic applications of dialkyl 1,2-epoxyalkylphosphonates are based on the ring-opening reactions with nucleophiles, which provide a convenient method for incorporating a functional group into the molecule. Generally, the opening of the oxirane ring is directionally specific at the C-2 carbon, and only in one instance was a C-1 attack observed. Nucleophilic substitution in α-position to the phosphoryl group is generally difficult because of both steric and electronic factors. The formation of α,β-difunctionalized alkylphosphonates is readily achieved by both acid-catalyzed and noncatalyzed openings of the oxirane ring. The preferred nucleophiles are water, alcohols, phenols, aqueous ammonia, amines, and phosphites.

Treatment of diethyl 1,2-epoxyethylphosphonate with aqueous H$_2$SO$_4$ at reflux leads to the formation of 1,2-dihydroxyethylphosphonate (55%).[91] The opening of the epoxide ring with MeOH or EtOH containing H$_2$SO$_4$ takes place at the C-2 carbon to give 1-hydroxy-2-methoxy- or 1-hydroxy-2-ethoxyethylphosphonates in fair yields (Scheme 4.36).[91]

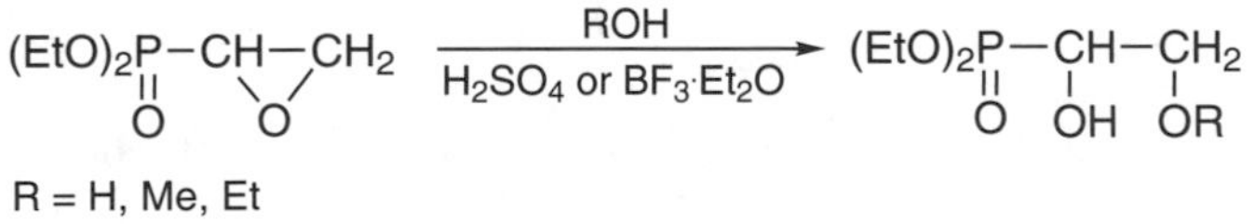

$$(4.36)$$

However, it has been reported, and proved by mass spectrometry, that diethyl 1-methyl-1,2-epoxyethylphosphonate is attacked at the α-position by alcohols under acid-catalysis conditions (BF$_3$·Et$_2$O or H$_2$SO$_4$) to give diethyl 1-alkoxy-1-methyl-2-hydroxyethylphosphonates.[262]

An attractive procedure for the preparation of phosphonylated amino alcohols from 1,2-epoxy-phosphonates involves oxirane ring opening by amines. The reaction follows the general pattern of nucleophilic attack at C-2 established with electronegative substituents at C-1. The structure of reaction products demonstrate that ring opening in diethyl 1-methyl-1,2-epoxyethylphosphonate and diethyl 1,2-epoxyethylphosphonate is directionally specific and results from an attack at C-2 by ammonia or amines. Thus, diethyl 1,2-epoxyethylphosphonate and diethyl 1-methyl-1,2-epoxy-ethylphosphonate undergo ring-opening reactions with aqueous ammonia at 0°C or amines at room temperature, followed by acid hydrolysis to produce the corresponding 1-hydroxy-2-aminophos-phonic acids in 40–70% yields (Scheme 4.37).[263,264] By contrast, ring opening of diethyl 1,2-epoxyethylphosphonate with aniline is realized by heating at 120°C in 90% yield.[91] Treatment of diethyl 1-methyl-1,2-epoxyethylphosphonate with diethanolamine at 80°C produces diethyl 1-hydroxy-1-methyl-2-[*bis*(2-hydroxyethyl)amino]ethylphosphonate.[62]

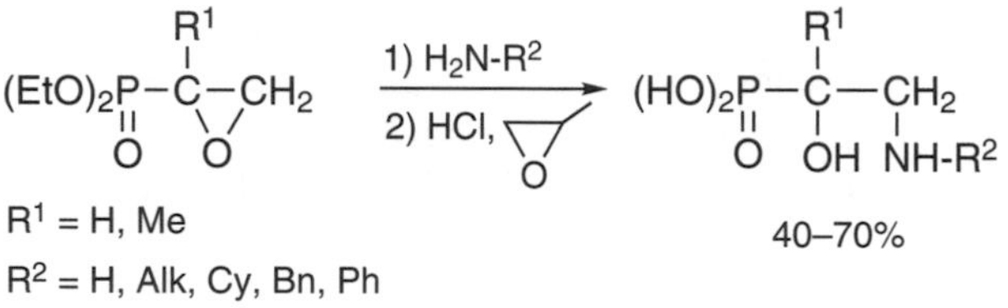

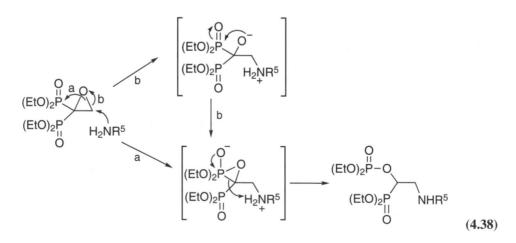

$$R^1 = H, Me$$
$$R^2 = H, Alk, Cy, Bn, Ph$$

40–70%

(4.37)

Exposed to hydrolytic conditions at different pH (2, 6, 10), tetraethyl oxiranylidene-1,1-diphos-phonate reacts very slowly with water.[104] The reaction of tetraethyl oxiranylidene-1,1-diphosphonate with several primary amines, including *n*-propylamine, cyclohexylamine, benzylamine, and ally-lamine, has been examined, and the phosphonyl phosphates were isolated as the major products in 40–63% yields (Scheme 4.38). It seems very likely that the formation of phosphonyl phosphates is the result of a rearrangement either in concert with (path a) or subsequent to (path b) the opening of the epoxide with amine and generation of an intermediate alkoxide ion. When di-*n*-propylamine was used, a second product, ethenylidene-1-phosphonyl-1-phosphate, corresponding to the loss of amine by elimination, is formed (32%) in addition to aminoethyl-1-phosphonyl-1-phosphate (20%).[104]

(4.38)

On treatment with saturated aqueous ammonia for 3 days at 55–60°C, fosfomycin gives as major component the (1*R*,2*R*) (–)-(2-amino-1-hydroxypropyl)phosphonic acid (58%) resulting from preferential attack of nucleophile at the C-2 carbon (Scheme 4.39).[203,206,208,265–269] Similarly, treatment of the fosfomycin with 0.05 M H_2SO_4 delivers the corresponding diol.[205]

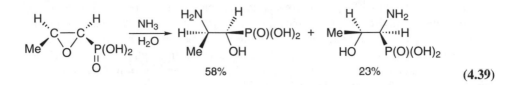

$$(4.39)$$

It is established that under normal physiologic conditions, the opening of the fosfomycin oxirane ring can occur either at C-2 or at C-1. Thus, the inactivation of enzyme MurA by fosfomycin occurs through alkylation at the C-2 of the antibiotic by the thiol of a cysteine residue in the active site of MurA. By contrast, the plasmid-mediated resistance to the antibiotic involves the reaction of fosfomycin at C-1 with the thiol group of glutathione catalyzed by FosA (fosfomycin-specific metalloglutathione transferase).[270–273] The latter addition is somewhat surprising in that nucleophilic substitutions at C-1 are difficult. Very recently it has been shown that L-cysteine can act as an alternative thiol substrate for FosA (Scheme 4.40).[274] The regiochemistry of the reaction is the same as with normal substrate, glutathione, involving addition of the nucleophile at the most hindered carbon, C-1. The stereochemical course of the reaction occurs with inversion of configuration at C-1 of fosfomycin, suggesting that the enzyme catalyzes the reaction via an S_N2 or borderline S_N2 mechanism.

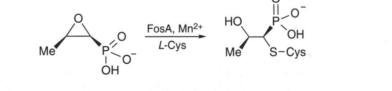

$$(4.40)$$

Treatment of diethyl 1,2-epoxyethylphosphonate with R^1MgBr failed to result in the formation of the expected hydroxyalkylphosphonate. The oxirane ring is opened by reaction with $MgBr_2$ to yield the halohydrin, which is hydrolyzed to the glycol during workup.[91]

The readily available diethyl 1,2-epoxyalkylphosphonates have proved to be valuable synthetic intermediates. On treatment with sodium dialkyl phosphites in EtOH, they undergo a rearrangement via an oxirane ring-opening reaction. Sodium dialkyl phosphite attacks the β carbon atom of the epoxide ring to give unstable sodium α-hydroxyphosphonates, which eliminate sodium diethyl phosphite to produce dialkyl 1-formylethyl- (R^1 = H) or 2-oxopropyl- (R^1 = Me) phosphonates in 50–85% yields (Scheme 4.41).[275]

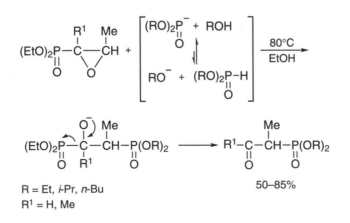

R = Et, *i*-Pr, *n*-Bu
R^1 = H, Me

50–85%

$$(4.41)$$

The diisopropyl (*E*)-1,2-epoxy-3-oxoalkylphosphonates, easily prepared by epoxidation of the corresponding alkenyl compounds, undergo heterocyclization with various 1,3-bidentate nucleophiles such as ethyl 2-pyridylacetate, 2-aminopyridine, 2-aminopyrimidine, thiourea, and thiocarboxamides (Scheme 4.42) to produce α-hydroxyheteroarylmethylphosphonates (indolizyl, imidazo[1,2-*a*]pyridyl, imidazo[1,2-*a*]pyrimidyl, and thiazolyl) arising from the regioselective oxirane ring opening and subsequent dehydration.[99–102,276,277]

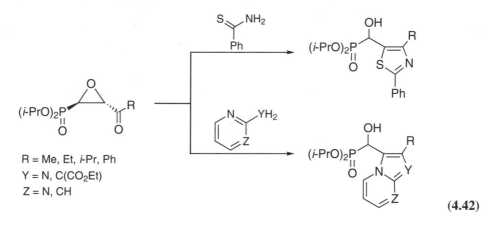

R = Me, Et, *i*-Pr, Ph
Y = N, C(CO$_2$Et)
Z = N, CH

(4.42)

4.3.1.3 Miscellaneous

α-Metallated 1,2-epoxyethylphosphonate can serve as precursor to higher homologues (Scheme 4.43).[278,279] Reaction of diethyl 1,2-epoxyethylphosphonate with LDA in THF at very low temperature results in deprotonation at the α-position to phosphorus, giving quantitatively diethyl 1-lithio-1,2-epoxyethylphosphonate, which can be trapped under internal quench conditions by electrophiles (Me$_3$SiCl, 91% and MeI, 57%), as illustrated in Scheme 4.43.

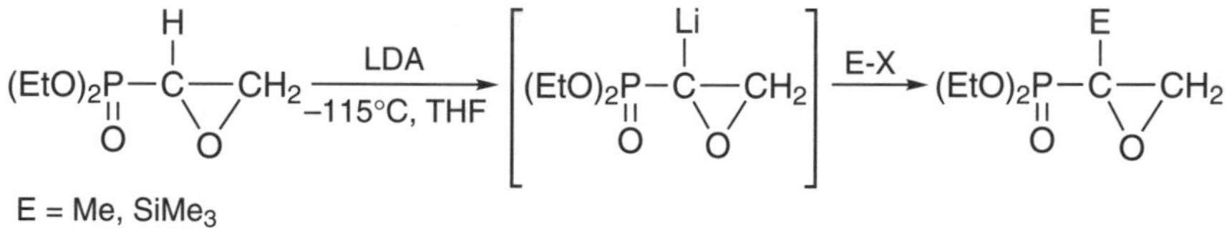

$$(EtO)_2P\overset{H}{\underset{O}{|}}C\text{—}CH_2 \xrightarrow[-115°C, THF]{LDA} \left[(EtO)_2P\overset{Li}{\underset{O}{|}}C\text{—}CH_2 \right] \xrightarrow{E\text{-}X} (EtO)_2P\overset{E}{\underset{O}{|}}C\text{—}CH_2$$

E = Me, SiMe$_3$

(4.43)

Kinetic resolution of (±) diethyl 1,2-epoxyethylphosphonate has been achieved by stirring the epoxide in the presence of catalytic quantities of (*R*,*R*)-*N*,*N*′-*bis*(3,5-di-*tert*-butylsalicylidene)-1,2-cyclohexanediaminocobalt(III) acetate and H$_2$O at 20°C for 4 days.[280] The reaction mixture becomes very viscous within the first few hours, consistent with the rapid conversion of some of the epoxide into the corresponding diol. After 4 days, the unreacted epoxide is isolated by flash chromatography (39%). It consists essentially of a single enantiomer. The (*R*)-configuration of the resolved epoxide has been established by conversion into the natural product (*R*)-2-amino-1-hydroxyethylphosphonic acid (Scheme 4.44).[280]

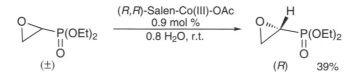

(4.44)

4.3.2 2,3-Epoxyalkylphosphonates

Dialkyl 2,3-epoxyalkylphosphonates undergo a series of directionally specific ring-opening reactions. No evidence was obtained for the formation of isomeric products resulting from attack of nucleophiles at the C-2 position. When diethyl 2,3-epoxypropylphosphonate is warmed in water, hydrolytic cleavage of both epoxide ring and phosphonate ester groups takes place to yield dihydroxypropylphosphonic acid.[126] Diethyl 2,3-epoxypropylphosphonate reacts with aqueous H_2SO_4 and with dry HCl yielding diethyl 2,3-dihydroxypropylphosphonates[91,118,138] and 2-hydroxy-3-chloropropylphosphonates.[118,152] Reaction with aqueous 48% HBr at 0°C gives diethyl 2-hydroxy-3-bromopropylphosphonate in 86% yield.[281,282] Pure diethyl 2-hydroxy-3-bromopropylphosphonate is conveniently prepared in high yields by treating diethyl 2,3-epoxypropylphosphonate with dry $MgBr_2$ followed by aqueous workup.[118] Diethyl 2-hydroxy-3-alkoxypropylphosphonates are prepared by acid-catalyzed alcoholysis ($BF_3 \cdot Et_2O$ in C_6H_6 or toluene,[91,115] H_2SO_4 in alcohol[91]) in good yields (74–90%).

Treatment of diethyl 2,3-epoxypropylphosphonate with a catalytic amount of MeONa in MeOH at 0°C, followed by stirring with Dowex 50W produces *trans*-3-hydroxy-1-propenylphosphonate in excellent yields (95–100%) by a ring-opening–elimination sequence (Scheme 4.45).[113,132,152,283]

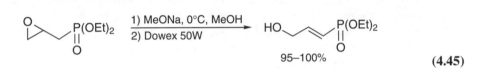

$$(4.45)$$

The reaction of diethyl 4-chloro-2,3-epoxyalkylphosphonates with alcoholates induces the elimination of hydrogen chloride accompanied by the conjugate opening and closing of the oxirane ring with formation of diethyl 3,4-epoxy-1-pentenylphosphonate.[284] Acid-catalyzed treatment of dialkyl 2,3-epoxypropylphosphonates with Ac_2O produces dialkyl 2,3-diacetoxypropylphosphonates in fair yields (18–56%).[285]

On treatment with R^1MgBr, diethyl 2,3-epoxypropylphosphonate does not effect any extension of the carbon skeleton but produces the corresponding 2,3-diols in moderate yields.[91,118] Apparently, the oxirane ring is opened by reaction with $MgBr_2$ to yield the halohydrin, which undergoes hydrolysis to the glycol during workup.[91,118] Formation of a new carbon–carbon bond is achieved in the reactions of diethyl 2,3-epoxypropylphosphonate with Grignard reagents in the presence of catalytic CuI or with organocuprates.[118,124,150,151] In all cases, the attack of the nucleophile takes place at carbon C-3, and all diethyl 2-hydroxyalkylphosphonates are formed in fair to good yields (51–90%, Scheme 4.46). However, it has been reported that the resulting diethyl 2-hydroxyalkylphosphonates are contaminated with 5–15% of diethyl 2-hydroxy-3-bromopropylphosphonate resulting from the reaction of diethyl 2,3-epoxypropylphosphonate with $MgBr_2$ present in equilibrium with the Grignard reagent.[120] The dialkyl 2-hydroxyalkylphosphonates are more conveniently prepared by reduction of the readily available dialkyl 2-oxoalkylphosphonates with $NaBH_4$.[151]

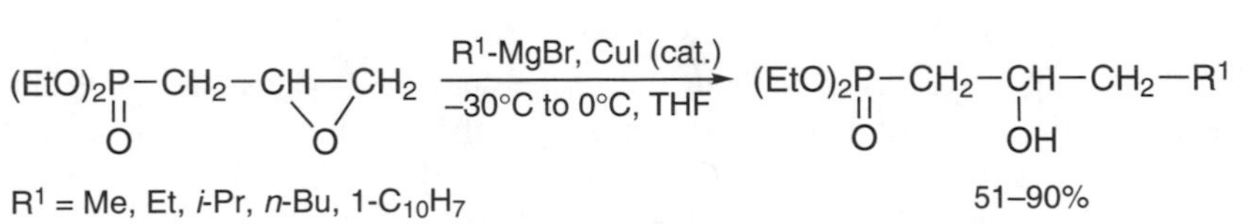

$$(EtO)_2\underset{\underset{O}{\|}}{P}-CH_2-CH\overset{\overset{O}{\diagdown\diagup}}{-}CH_2 \xrightarrow[-30°C \text{ to } 0°C, \text{ THF}]{R^1\text{-MgBr, CuI (cat.)}} (EtO)_2\underset{\underset{O}{\|}}{P}-CH_2-\underset{\underset{OH}{|}}{CH}-CH_2-R^1$$

$$R^1 = Me, Et, \textit{i}-Pr, \textit{n}-Bu, 1-C_{10}H_7 \qquad\qquad 51–90\%$$

$$(4.46)$$

An attractive route to diethyl 2-aminoalkylphosphonates involves the intermediacy of the diethyl 2-azidoalkylphosphonates. The 2-azidoalkylphosphonates are formed by treatment of 2-hydroxyalkylphosphonates with the preformed betaine-type adduct $PPh_3/DEAD/HN_3$ at room temperature in CH_2Cl_2 (Mitsunobu conditions) and converted *in situ* by the Staudinger reaction into iminophos-phoranes that, on hydrolysis, produce directly the corresponding diethyl 2-aminoalkylphosphonates in moderate to good yields (36–81%, Scheme 4.47).[120]

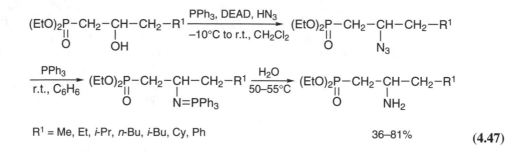

R¹ = Me, Et, *i*-Pr, *n*-Bu, *i*-Bu, Cy, Ph

36–81%   (4.47)

Diethyl 2-hydroxy-3-(phenylamino)propylphosphonate is formed by the reaction of diethyl 2,3-epoxypropylphosphonate with aniline at 120°C, and diethyl 2-hydroxy-3-aminopropylphosphonate is produced by the reaction with aqueous ammonia at room temperature.[91] Ring opening at C-3 is also achieved with secondary amines such as piperidine in MeOH at 25–65°C,[286] diethylamine at 100°C,[287] or diethanolamine[62] or 1,2,4-triazole in methyl ethyl ketone at 50°C in presence of a base (K_2CO_3, $NaHCO_3$, CsF, or R_4NX).[114]

Diethyl 2,3-epoxypropylphosphonate can be converted into diethyl 2,3-diaminopropylphospho-nate by combining epoxide ring opening at C-3 by an amine followed by azidation at C-2.[121] Thus, ring opening with dibenzylamine in refluxing MeOH and subsequent treatment of alcohol with mesyl chloride and NaN_3 under phase-transfer conditions affords the corresponding azide in 60% yield. Reduction of the latter with $NaBH_4$ in THF/MeOH and debenzylation using $H_2/Pd(OH)_2/C$ in MeOH gives diethyl 2,3-diaminopropylphosphonate.[121]

The use of lithium halides for the opening of epoxides to give halohydrins has also been reported.[288] Thus, diethyl 2,3-epoxypropylphosphonate undergoes highly regioselective attack at C-3 by LiI in THF in the presence of AcOH in excess to give the corresponding iodhydrin in 96% yield (Scheme 4.48). The reaction time can be shortened by using a large excess of lithium halide, and the reactivity of lithium halides follows the order LiI > LiBr > LiCl. It is quite clear that AcOH plays an important role because almost all of the epoxide is recovered when it is not present in the reaction.[288]

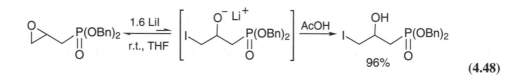

(4.48)

Diethyl 2,3-epoxy-4-oxoalkylphosphonates can be converted on reaction with thioureas and thioamides into diethyl 2-hydroxy-2-(5-thiazolyl)ethylphosphonates or into 2-(5-thiaz-olyl)vinylphosphonates (Scheme 4.49).[107] Cleavage of both systems into phosphorus-free thiazoles can be readily achieved with CsF in DMF. Reaction of diethyl 2,3-epoxy-4-oxoalkylphosphonates with 2-mercaptobenzimidazole yields α-substituted enones arising from regioselective oxirane opening and subsequent dehydration (Scheme 4.49).[100] These derivatives exist, mainly or exclu-sively, as cyclic tautomers. The cyclization step depends on the reaction conditions.[100]

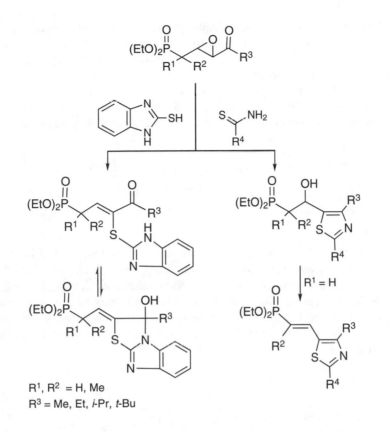

R^1, R^2 = H, Me
R^3 = Me, Et, *i*-Pr, *t*-Bu

(4.49)

Kinetic resolution of (±) diethyl 2,3-epoxypropylphosphonate by enantioselective hydrolysis has recently been described (Scheme 4.50).[289] In the presence of (R,R)-N,N'-*bis*(3,5-di-*tert*-butyl-salicylidene)-1,2-cyclohexanediaminocobalt(III) acetate and H_2O for 19 h, racemic diethyl 2,3-epoxpropylphosphonate is converted into a mixture of (S)-(-) diethyl 2,3-epoxypropylphosphonate (82% ee) and diethyl (R)-(−) 2,3-dihydroxypropylphosphonate (98% ee). An improved enantiomeric excess (93% ee) of (S)-(−) diethyl 2,3-epoxypropylphosphonate has been obtained after a 72-h hydrolytic kinetic resolution experiment.[289]

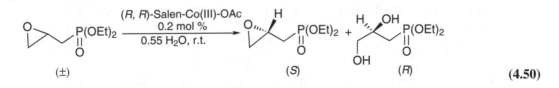

(4.50)

4.3.3 3,4-EPOXYALKYLPHOSPHONATES

Dialkyl 3,4-epoxybutylphosphonates undergo a series of ring-opening reactions. Thus, on treatment with glacial AcOH at room temperature in the presence of $BF_3\cdot Et_2O$, diethyl 3,4-epoxybutylphosphonate undergoes attack at C-4 to give 3-hydroxy-4-acetoxybutylphosphonate in 67% yield.[163] Ring-opening reactions with hexadecanol under Lewis acid catalysis conditions [$BF_3\cdot Et_2O$, $ZnCl_2$, $FeCl_3$, Et_2AlCl, Al_2O_3, or $Ti(Oi\text{-}Pr)_4$] occurs at C-3 and C-4 resulting in the formation in poor yields of two isomeric products separated by column chromatography.[160] Diethyl 1,1-difluoro-3,4-epoxy-butylphosphonate is efficiently converted into diethyl 3,4-dihydroxybutylphosphonate at room temperature in DMSO under catalytic acid conditions (HCl 5%).[157,158]

Diethyl 1-phenylsulfonyl-3,4-epoxybutylphosphonate is smoothly converted into the corresponding iodohydrin with complete regioselectivity by means of n-Bu$_4$NI and BF$_3$·Et$_2$O in CH$_2$Cl$_2$ at 0°C.[166] Replacing the BF$_3$·Et$_2$O with *tert*-butyldimethylsilyl triflate and changing the solvent from CH$_2$Cl$_2$ to THF allowed protected diethyl 1-(phenylsulfonyl)-3-hydroxy-4-iodobutylphosphonate to be prepared directly from the epoxide (Scheme 4.51). This reaction presumably proceeds via the generation *in situ* of *tert*-butyldimethylsilyl iodide.[166] Similarly, diethyl 1-(ethoxycarbonyl)-3,4-epoxybutylphosphonate undergoes regioselective ring opening at C-4 by amines in DMF at 50°C.[167]

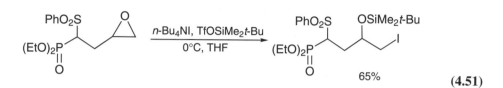

$$(4.51)$$

Diethyl 1-substituted-3,4-epoxybutylphosphonates are treated with n-BuLi at low temperature in THF, and the resulting anions regioselectively rearrange on warming into diethyl 1-substituted-2-(hydroxymethyl)-1,2-cyclopropylphosphonates via an intramolecular epoxide ring-opening reaction (Scheme 4.52).[168] Regiospecific epoxide ring opening occurs to give the cyclopropane ring exclusively. The stereochemistry of the hydroxymethyl and phosphonyl groups are specifically *trans*, and no *cis* products are detected. The ester groups at phosphorus and the α-substituents have no influence on the stereoselectivity.[168]

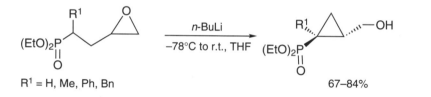

$$(4.52)$$

REFERENCES

1. Hendlin, D., Stapley, E.O., Jackson, M., Wallick, H., Miller, A.K., Wolf, F.J., Miller, T.W., Chaiet, L., Kahan, F.M., Foltz, E.L., Woodruff, H.B., Mata, J.M., Hernandez, S., and Mochales, S., Phosphonomycin, a new antibiotic produced by strains of *Streptomyces*, *Science*, 166, 122, 1969.
2. Chaiet, L., Miller, T.W., Goegelman, R.T., Kempf, A.J., and Wolf, F.J., Phosphonomycin. Isolation from fermentation sources, *J. Antibiot.*, 23, 336, 1970.
3. Iorga, B., Eymery, F., and Savignac, P., The synthesis and properties of 1,2-epoxyalkylphosphonates, *Synthesis*, 207, 1999.
4. Newman, M.S., and Magerlein, B.J., The Darzens glycidic ester condensation, *Org. React.*, 5, 413, 1949.
5. Waschbüsch, R., Carran, J., Marinetti, A., and Savignac, P., The synthesis of dialkyl α-halogenated methylphosphonates, *Synthesis*, 727, 1997.
6. Waschbüsch, R., Carran, J., Marinetti, A., and Savignac, P., Synthetic applications of dialkyl (chloromethyl)phosphonates and *N,N,N′,N′*-tetraalkyl(chloromethyl)phosphonic diamides, *Chem. Rev.*, 97, 3401, 1997.
7. Kabachnik, M.I., and Shepeleva, E.S., Reaction of aldehydes with halogen derivatives of trivalent phosphorus, *Dokl. Akad. Nauk SSSR*, 75, 219, 1950; *Chem. Abstr.*, 45, 6569, 1951.
8. Kabachnik, M.I., and Medved, T.Y., Organophosphorus compounds. Part 14. Synthesis of aminophosphonic acids. Part 1, *Izv. Akad. Nauk SSSR, Ser. Khim.*, 635, 1951; *Chem. Abstr.*, 45, 8444b, 1951.
9. Kabachnik, M.I., and Shepeleva, E.S., Organophosphorus compounds. Part 15. Reaction of formaldehyde with phosphorus trichloride, *Izv. Akad. Nauk SSSR, Ser. Khim.*, 185, 1951; *Chem. Abstr.*, 45, 10191a, 1951.

10. Crofts, P.C., and Kosolapoff, G.M., The preparation and the determination of apparent dissociation constants of some substituted aliphatic phosphonic acids, *J. Am. Chem. Soc.*, 75, 5738, 1953.

11. Korshak, V.V., Gribova, I.A., and Andreeva, M.A., Organophosphorus polymers. Part 1. Polymerization of cyclic esters of alkyl- and arylphosphonic acids, *Izv. Akad. Nauk SSSR, Ser. Khim.*, 631, 1957; *Chem. Abstr.*, 51, 14621g, 1957.

12. McConnell, R.L., McCall, M.A., and Coover, H.W., Jr., Chloroalkyl and chloroaryl (chloromethyl)phosphonates, *J. Org. Chem.*, 22, 462, 1957.

13. Martynov, V.F., and Timofeev, V.E., Darzens reaction with ethyl chloromethylphosphinate, *Zh. Obshch. Khim.*, 32, 3449, 1962; *J. Gen. Chem. USSR (Engl. Transl.)*, 32, 3383, 1962.

14. Martynov, V.F., and Timofeev, V.E., Compounds containing a three-membered epoxide ring. Part 35. Application of the Darzens reaction for the synthesis of epoxyalkylphosphonic esters, *Zh. Obshch. Khim.*, 34, 3890, 1964; *J. Gen. Chem. USSR (Engl. Transl.)*, 34, 3950, 1964.

15. Lavielle, G., Carpentier, M., and Savignac, P., Stereochemical orientation in Darzens reaction applied to (chloromethyl)phosphonates and phosphonamides, *Tetrahedron Lett.*, 14, 173, 1973.

16. Coutrot, P., and Savignac, P., A one-step synthesis of diethyl 1,2-epoxyalkanephosphonates, *Synthesis*, 34, 1978.

17. Cann, P.F., Howells, D., and Warren, S., Rearrangements of a cation of the neopentyl-type containing a diphenylphosphinyl substituent, *J. Chem. Soc., Perkin Trans. 2*, 304, 1972.

18. Churi, R.H., and Griffin, C.E., 1,2-Shifts of dialkoxyphosphono groups in skeletal rearrangements of α,β-epoxyvinylphosphonates, *J. Am. Chem. Soc.*, 88, 1824, 1966.

19. Sprecher, M., and Kost, D., The rearrangement of dialkyl α,β-epoxyphosphonates, *Tetrahedron Lett.*, 9, 703, 1969.

20. Ulman, A., and Sprecher, M., Thermal and photochemical reactions of sodium salts of β-phosphono tosylhydrazones, *J. Org. Chem.*, 44, 3703, 1979.

21. Lee, K., Shin, W.S., and Oh, D.Y., A one-pot generation of α-chloro-α-lithioalkanephosphonates. A new preparative route to 1,2-epoxyalkanephosphonates, *Synth. Commun.*, 22, 649, 1992.

22. Coutrot, P., Laurenço, C., Normant, J.F., Perriot, P., Savignac, P., and Villieras, J., Synthesis of diethyl 1,1-dichloroalkane- and 1-chloroalkanephosphonates, *Synthesis*, 615, 1977.

23. Perriot, P., Villieras, J., and Normant, J.F., Diethyl 1-chloro-1-lithioalkanephosphonates. A general synthesis of diethyl 1,2-epoxyalkanephosphonates and 1-alkoxycarbonyl-1-chloroalkanephosphonates, *Synthesis*, 33, 1978.

24. Le Menn, J.-C., Sarrazin, J., and Tallec, A., Electrochemically generated phosphonate carbanions. Formation and reactivity towards aldehydes, *Can. J. Chem.*, 67, 1332, 1989.

25. Christensen, B.G., and Firestone, R.A., (cis-1,2-Epoxypropyl)phosphonates, *Merck*, German Patent Appl. DE 1924135, 1969; *Chem. Abstr.*, 72, 43870n, 1970.

26. Lewars, E.G., Oxiranes and oxirenes, in *Comprehensive Heterocyclic Chemistry*, Katritzky, A.R., and Rees, C.W., Eds., Pergamon Press, Oxford, 1984, p. 95.

27. Arbuzov, B.A., and Movsesyan, M.E., Esters of β-oxophosphonic acids. Part 4. Infrared spectra of the reaction products of α-halo ketones with triethyl phosphite and diethyl sodiophosphite, *Izv. Akad. Nauk SSSR, Ser. Khim.*, 267, 1959; *Chem. Abstr.*, 53, 19850f, 1959.

28. Sturtz, G., The action of sodium phosphites on the ω-halogenated ketones, *Bull. Soc. Chim. Fr.*, 2333, 1964.

29. Abramov, V.S., Shalman, A.L., and Molodykh, Z.V., Reaction of dialkylphosphites with aldehydes and ketones. Part 33. Products of the dehydrochlorination of dimethyl and *bis*(β-chloroethyl) esters of α-hydroxy-chloroisopropylphosphonic acids, *Zh. Obshch. Khim.*, 38, 541, 1968; *J. Gen. Chem. USSR (Engl. Transl.)*, 38, 529, 1968.

30. Russell, G.A., and Ros, F., Reaction of α-halo ketones with nucleophiles, *J. Am. Chem. Soc.*, 107, 2506, 1985.

31. Arbuzov, B.A., Vinogradova, V.S., and Polezhaeva, N.A., Structure of products of reaction of some halo ketones of the carbocyclic series with triethyl phosphite and diethyl sodiophosphite, *Dokl. Akad. Nauk SSSR*, 121, 641, 1958; *Chem. Abstr.*, 53, 1180i, 1959.

32. Arbuzov, B.A., Vinogradova, V.S., and Polezhaeva, N.A., Esters of β-oxo phosphonic acids. Part 5. Structure of the reaction products of some α-halo ketones of the carbocyclic series with triethyl phosphite and sodium diethyl phosphite, *Izv. Akad. Nauk SSSR, Ser. Khim.*, 832, 1960; *Chem. Abstr.*, 54, 24454e, 1960.

33. Arbuzov, B.A., Vinogradova, V.S., Polezhaeva, N.A., and Shamsutdinova, A.K., Esters of β-oxophosphonic acids. Part 12. Structure of reaction products of some α-halo ketones of the aromatic series with triethyl phosphite and diethyl sodium phosphite, *Izv. Akad. Nauk SSSR, Ser. Khim.*, 1380, 1963; *Bull. Acad. Sci. USSR, Div. Chem. Sci. (Engl. Transl.)*, 1257, 1963.

34. Meisters, A., and Swan, J.M., Organophosphorus compounds. Part 6. Michaelis–Becker reaction. Diethyl 1-phenylepoxyethylphosphonate and diethyl 1-phenylvinyl phosphate from diethyl phosphonate and phenacyl chloride, *Aust. J. Chem.*, 18, 168, 1965.

35. Arbuzov, B.A., Lugovkin, B.P., and Bogonostseva, N.P., Action of α- and γ-bromoacetoacetic ester and 2-chlorocyclohexanone on triethyl phosphite and sodium diethyl phosphite, *Zh. Obshch. Khim.*, 20, 1468, 1950; *Chem. Abstr.*, 45, 1506a, 1951.

36. Pudovik, A.N., Anomalous reaction of α-halo ketones with triethyl ester of phosphorus acid, *Zh. Obshch. Khim.*, 25, 2173, 1955; *Chem. Abstr.*, 50, 8486i, 1956.

37. Bogonostseva, N.P., Reaction of some halogen derivatives with sodium diethyl phosphite, *Uch. Zap. Kazan. Gos. Univ.*, 116, 71, 1956; *Chem. Abstr.*, 51, 6581g, 1957.

38. Arbuzov, B.A., Vinogradova, V.S., and Polezhaeva, N.A., Esters of β-oxophosphonic acids. Part 3. Structure of reaction products of some halogenated ketones with triethyl phosphite and sodium diethyl phosphite, *Izv. Akad. Nauk SSSR, Ser. Khim.*, 41, 1959; *Chem. Abstr.*, 53, 15035e, 1959.

39. Miller, B., Reactions of 4-bromocyclohexadienones with esters of phosphorous acids, *J. Org. Chem.*, 26, 4781, 1961.

40. Hoffmann, H. and Diehr, H.J., Reaction of tertiary phosphines on α-halocarbonyl compounds. Model experiment on the Perkow reaction, *Tetrahedron Lett.*, 3, 583, 1962.

41. Miller, B., Reactions of 4-bromocyclohexadienones with nucleophiles containing phosphorus and sulfur, *J. Org. Chem.*, 28, 345, 1963.

42. Borowitz, I.J., and Virkhaus, R., Mechanism of the reactions of α-bromo ketones with triphenylphosphine, *J. Am. Chem. Soc.*, 85, 2183, 1963.

43. Kamai, G., and Kukhtin, V.A., Addition of trialkyl phosphites to α,β-unsaturated aldehydes, *Dokl. Akad. Nauk SSSR*, 112, 868, 1957; *Chem. Abstr.*, 51, 13742f, 1957.

44. Kukhtin, V.A., and Pudovik, A.N., Some new forms of the Arbuzov rearrangement, *Usp. Khim.*, 28, 96, 1959; *Chem. Abstr.*, 53, 9025c, 1959.

45. Tripett, S., New synthesis of acetylenes. Part 2. Reaction of triphenylphosphine with α-halo carbonyl compounds, *J. Chem. Soc.*, 2337, 1962.

46. Perkow, W., Reactions of alkyl phosphites. Part 1. Conversions with Cl$_3$CCHO and Br$_3$CCHO, *Chem. Ber.*, 87, 755, 1954.

47. Perkow, W., Krockow, E.W., and Knoevenagel, K., Rearrangements with alkyl phosphites. Part 2. The reaction with α-halogenated aldehydes, *Chem. Ber.*, 88, 662, 1955.

48. Spencer, E.Y., Todd, A.R., and Webb, R.F., Studies on phosphorylation. Part 17. Hydrolysis of methyl 3-(*O,O*-dimethylphosphoryloxy)but-2-enoate, *J. Chem. Soc.*, 2968, 1958.

49. Cramer, F., Preparation of esters, amides, and phosphoric acid anhydrides, *Angew. Chem.*, 72, 236, 1960.

50. Allen, J.F., and Johnson, O.H., The synthesis of monovinyl esters of phosphorus(V) acids, *J. Am. Chem. Soc.*, 77, 2871, 1955.

51. Kharasch, M.S., and Bengelsdorf, I.S., The reaction of triethyl phosphite with α-trichloromethyl carbonyl compounds, *J. Org. Chem.*, 20, 1356, 1955.

52. Chopard, P.A., Clark, V.M., Hudson, R.F., and Kirby, A.J., Mechanism of the reaction between trialkyl phosphites and α-halogenated ketones, *Tetrahedron*, 21, 1961, 1965.

53. Sekine, M., Okimoto, K. and Hata, T., Silyl phosphites. Part 6. Reactions of *tris*(trimethylsilyl) phosphite with α-halocarbonyl compounds, *J. Am. Chem. Soc.*, 100, 1001, 1978.

54. Sekine, M., Okimoto, K., Yamada, K., and Hata, T., Silyl phosphites. Part 15. Reactions of silyl phosphites with α-halo carbonyl compounds. Elucidation of the mechanism of the Perkow reaction and related reactions with confirmed experiments, *J. Org. Chem.*, 46, 2097, 1981.

55. Sekine, M., Nakajima, M., and Hata, T., Silyl phosphites. Part 16. Mechanism of the Perkow reaction and the Kukhtin–Ramirez reaction. Elucidation by means of a new type of phosphoryl rearrangements utilizing silyl phosphites, *J. Org. Chem.*, 46, 4030, 1981.

56. Malenko, D.M., Simurova, N.V., and Sinitsa, A.D., Synthesis of phosphorylated oxiranes from silylated dichloroacetone and triethyl phosphite, *Zh. Obshch. Khim.*, 63, 943, 1993; *Russ. J. Gen. Chem. (Engl. Transl.)*, 63, 657, 1993.

57. Herzig, C., and Gasteiger, J., Reaction of 2-chlorooxiranes with phosphites and phosphanes. A new route to β-carbonylphosphonic esters and -phosphonium salts, *Chem. Ber.*, 115, 601, 1982.

58. Kreutzkamp, N., and Kayser, H., Carbonyl- and cyanophosphonic esters. Part 2. The course of reaction of phosphites with bromo- and chloroacetone, *Chem. Ber.*, 89, 1614, 1956.

59. Springs, B., and Haake, P., A one-step synthesis of epoxyphosphonates, *J. Org. Chem.*, 41, 1165, 1976.

60. Kossev, K., Troev, K., and Roundhill, D.M., Synthesis of dialkyl 1,2-epoxyphosphonates under phase-transfer catalyst conditions, *Phosphorus, Sulfur Silicon Relat. Elem.*, 83, 1, 1993.

61. Guseinov, F.I., Burangulova, R.N., and Moskva, V.V., Reactions of α-chloro-β-oxoaldehydes with dialkyl phosphites, *Zh. Obshch. Khim.*, 67, 1663, 1997; *Russ. J. Gen. Chem. (Engl. Transl.)*, 67, 1564, 1997.

62. Boutevin, B., Hervaud, Y., Mouledous, G., and Vera, R., Synthesis of polyols bearing phosphonate groups. Part 2. Epoxides used as starting materials, *Phosphorus, Sulfur Silicon Relat. Elem.*, 161, 9, 2000.

63. Texier-Boullet, F., and Foucaud, A., Reactions in heterogenous liquid-solid medium. Wittig-Horner reactions and diethyl phosphite addition on the carbonyl compounds in the presence of potassium fluoride dihydrate, *Tetrahedron Lett.*, 21, 2161, 1980.

64. Nikolova, R., Bojilova, A., and Rodios, N.A., Reaction of 3-bromobenzyl and 3-bromoacetyl coumarin with phosphites. Synthesis of some new phosphonates and phosphates in the coumarin series, *Tetrahedron*, 54, 14407, 1998.

65. Pudovik, A.N., and Gareev, R.D., *cis*- And *trans*-1-(diethoxyphosphinyl)-2-(ethoxycarbonyl)-1-methyloxiranes [*cis*- and *trans*-2,3-epoxy-3-phosphonobutyric acid triethyl esters], *Zh. Obshch. Khim.*, 42, 1861, 1972; *J. Gen. Chem. USSR (Engl. Transl.)*, 42, 1845, 1972.

66. Hirai, S., Harvey, R.G., and Jensen, E.V., Phosphonic acids. Part 21. Steroidal 21a-phosphonate esters, *Tetrahedron*, 22, 1625, 1966.

67. Inokawa, S., Kawata, Y., Yamamota, K., Kawamoto, H., Yamamoto, H., Takagi, K., and Yamashita, M., Synthesis of 5(*R*)- or 5(*S*)-5,6-anhydro-1,2-*O*-isopropylidene-5-*C*-phosphinyl-α-D-*xylo*-hexofuranose derivatives, *Carbohydr. Res.*, 88, 341, 1981.

68. Kashino, S., Inokawa, S., Haisa, M., Yasuoka, N., and Kakudo, M., The structure and absolute configuration of (5*R*)-5,6-anhydro-3-*O*-benzyl-1,2-*O*-isopropylidene-5-*C*-(dimethoxyphosphinoyl)-α-D-*xylo*-hexofuranose, *Acta Crystallogr., Sect. B*, 37, 1572, 1981.

69. Penz, G., and Zbiral, E., Synthesis of dialkyl (±)-*cis*-1,2-epoxy-3-oxoalkylphsphonates. Phosphonomycin analogues, *Monatsh. Chem.*, 113, 1169, 1982.

70. Inokawa, S., and Yamamoto, H., A new synthetic approach to α,β-unsaturated phosphonates (1-methylenealkanephosphonates), *Phosphorus Sulfur*, 16, 79, 1983.

71. Glebova, Z.I., Eryuzheva, O.V., and Zhdanov, Y.A., Simple method for synthesizing carbohydrate-containing α,β-epoxy phosphonates from 1-*O*-tosyl-3,5:4,6-di-*O*-ethylidene-L-*xylo*-hex-2-ulose under phase-transfer catalysis, *Zh. Obshch. Khim.*, 63, 1677, 1993; *Russ. J. Gen. Chem. (Engl. Transl.)*, 63, 1172, 1993.

72. Hanaya, T., Yasuda, K., and Yamamoto, H., Stereoselectivity in the preparation of 5,6-dideoxy-5-dimethoxyphosphinyl-D- and -L-hexofuranoses, and an efficient synthesis of 5,6-dideoxy-5-hydroxyphosphinyl-L-galactopyranose (a P-in-the-ring L-fucose analogue), *Bull. Chem. Soc. Jpn.*, 66, 2315, 1993.

73. Shen, Y., Liao, Q., and Qiu, W., A novel synthesis of α-fluoroalkylvinyl- or α-fluoroepoxyalkylphosphonates, *J. Chem. Soc., Perkin Trans. 1*, 695, 1990.

74. Arbuzov, B.A., Vinogradova, V.S., and Polezhaeva, N.A., Product of reaction of diethyl sodiophosphite with bromoacetone, *Dokl. Akad. Nauk SSSR*, 111, 107, 1956; *Chem. Abstr.*, 51, 8001g, 1957.

75. Abramov, V.S., and Kapustina, A.S., Esters of α-hydroxy-β-chloro-isopropylphosphonic acid and 1,2-epoxy-2-propylphosphonic acid, *Dokl. Akad. Nauk SSSR*, 111, 1243, 1956; *Chem. Abstr.*, 51, 9473g, 1957.

76. Abramov, V.S., and Kapustina, A.S., Reaction of dialkyl esters of phosphorus acids with aldehydes and ketones. Part 13. Esters of α-hydroxy-β-chloroisopropylphosphonic and 1,2-epoxyisopropylphosphonic acids, *Zh. Obshch. Khim.*, 27, 1012, 1957; *Chem. Abstr.*, 52, 3666i, 1958.

77. Popova, Z.V., Yanovskii, D.M., Kirpichnikov, P.A., Kapustina, A.S., and Davydova, V.M., Stabilization of polyvinyl chloride by alkyl phosphonic acid esters, *Zh. Prikl. Khim. (Leningrad)*, 36, 187, 1963; *J. Appl. Chem. USSR (Engl. Transl.)*, 36, 173, 1963.

78. Kirpichnikov, P.A., Kapustina, A.S., and Tokareva, G.N., Synthesis of esters of α-hydroxy-β-chloro-cyclohexylphosphonic acid and esters of 1,2-epoxycyclohexylphosphonic acid, *Trans. Kazan. Khim.-Tekhnol. Inst.*, 33, 188, 1964; *Chem. Abstr.*, 66, 2619, 1967.

79. Bogatyreva, T.K., Kapustina, A.S., Kirpichnikov, P.A., Tikhova, N.V., and Yanovskii, D.M., Stabilization of polyvinyl chloride by esters of 1,2-epoxy-1-phenylethylphosphonic acids, *Zh. Prikl. Khim. (Leningrad)*, 39, 1572, 1966; *J. Appl. Chem. USSR (Engl. Transl.)*, 39, 1465, 1966.

80. Abramov, V.S., and Savintseva, R.N., Reactions of dialkyl phosphinates with aldehydes and ketones. Part 28. Esters of α-hydroxy-β-chloroethylphosphonic acids and α,β-epoxyethylphosphonic acids, *Khim. Org. Soedin. Fosfora, Akad. Nauk SSSR Otd. Obshch. Tekh. Khim.*, 129, 1967; *Chem. Abstr.*, 69, 67465y, 1968.

81. Agawa, T., Kubo, T., and Ohshiro, Y., Synthesis of dialkyl epoxyphosphonates, *Synthesis*, 27, 1971.

82. Sturtz, G., and Pondaven-Raphalen, A., Cyclization of 1,2-halohydrins of diethyl ethylphosphonate for the synthesis of diethyl 1,2-epoxyethylphosphonate, *Phosphorus Sulfur*, 20, 35, 1984.

83. Giordano, C., and Castaldi, G., First asymmetric synthesis of enantiomerically pure (1*R*,2*S*)-(−)-(1,2-epoxypropyl)phosphonic acid (fosfomycin), *J. Org. Chem.*, 54, 1470, 1989.

84. Coppola, G.M., Novel heterocycles. Part 10. Reactions of 1,3,2-benzodiazaphosphorin-4(1*H*)-one 2-oxides, *J. Heterocycl. Chem.*, 20, 331, 1983.

85. Foucaud, A., and Bakouetila, M., Facile epoxidation of alumina-supported electrophilic alkenes and montmorillonite-supported electrophilic alkenes with sodium hypochlorite, *Synthesis*, 854, 1987.

86. Duncan, G.D., Li, Z.-M., Khare, A.B., and McKenna, C.E., Oxiranylidene-2,2-*bis*(phosphonate). Unambiguous synthesis, hydrolysis to 1,2-dihydroxyethylidene-1,1-*bis*(phosphonate), and identification as the primary product from mild Na$_2$WO$_4$/H$_2$O$_2$ oxidation of ethylidene-1,1-*bis*(phosphonate), *J. Org. Chem.*, 60, 7080, 1995.

87. Levy, J.N., and McKenna, C.E., Oxidations of triethyl α-phosphonoacrylate. Epoxidation to triethyl α-phosphonoacrylate oxide by hypochlorite and formation of triethyl dihydroxyphosphonoacetate with RuO$_4$-periodate, *Phosphorus, Sulfur Silicon Relat. Elem.*, 85, 1, 1993.

88. McEldoon, W.L., and Wiemer, D.F., Synthesis of nucleoside epoxyphosphonates, *Tetrahedron*, 52, 11695, 1996.

89. McEldoon, W.L., and Wiemer, D.F., Synthesis of nucleoside epoxphosphonates (Erratum), *Tetrahedron*, 53, 1546, 1997.

90. Minami, T., and Motoyoshiya, J., Vinylphosphonates in organic synthesis, *Synthesis*, 349, 1992.

91. Griffin, C.E., and Kundu, S.K., Phosphonic acids and esters. Part 20. Preparation and ring-opening reactions of α,β- and β,γ-epoxyalkylphosphonates. The proton magnetic resonance spectra of vicinally substituted ethyl- and propylphosphonates, *J. Org. Chem.*, 34, 1532, 1969.

92. Sobolev, V.G., and Ionin, B.I., Stereospecific epoxidation of 1-alkenephosphonates, *Zh. Obshch. Khim.*, 55, 225, 1985; *J. Gen. Chem. USSR (Engl. Transl.)*, 55, 198, 1985.

93. Hunger, K., Epoxidation of α,β-unsaturated phosphonic acid esters, *Chem. Ber.*, 101, 3530, 1968.

94. Kadyrov, A.A., Rokhlin, E.M., and Knunyants, I.L., Epoxidation of highly electrophilic unsaturated fluorine-containing compounds by peracids, *Izv. Akad. Nauk SSSR, Ser. Khim.*, 2344, 1982; *Bull. Acad. Sci. USSR, Div. Chem. Sci. (Engl. Transl.)*, 2063, 1982.

95. Ryabov, B.V., Ionin, B.I., and Petrov, A.A., Oxirane–carbonyl isomerization of phosphorylated α-halooxiranes, *Zh. Obshch. Khim.*, 59, 272, 1989; *J. Gen. Chem. USSR (Engl. Transl.)*, 59, 233, 1989.

96. Sturtz, G., and Pondaven-Raphalen, A., Catalytic epoxidation of α,β- and β,γ-unsaturated phosphonates with hydroperoxides, *Bull. Soc. Chim. Fr.*, 125, 1983.

97. Coutrot, P., Grison, C., Lecouvey, M., Kribii, A., and El Gadi, A., Highly functionalized new small molecules. Preparation of α-chlorinated epoxyphosphonates. Study of their transformation to β-halogenated α-ketophosphonates, *Phosphorus, Sulfur Silicon Relat. Elem.*, 133, 167, 1998.

98. Glamkowski, E.J., Gal, G., Purick, R., Davidson, A.J., and Sletzinger, M., A new synthesis of the antibiotic phosphonomycin, *J. Org. Chem.*, 35, 3510, 1970.

99. Öhler, E., Zbiral, E., and El-Badawi, M., A novel and versatile synthesis of heterocyclic aldehydes using dialkyl 3-oxo-1-alkenyl-phosphonates, *Tetrahedron Lett.*, 24, 5599, 1983.

100. Öhler, E., Kang, H.-S., and Zbiral, E., Regioselective cyclisation reactions of acyl-substituted epoxy-phosphonates with 2-mercaptoazoles. Syntheses of thiazolo[3,2-*a*]benzimidazole, imidazo[2,1-*b*]thiazole, and thiazolo[3,2-*b*]-[1,2,4]triazole derivatives, *Chem. Ber.*, 121, 977, 1988.

101. Öhler, E., and Zbiral, E., Oxidative rearrangement of phosphorus containing tertiary allylic alcohols. Synthesis of (3-oxo-1-cycloalkenyl)phosphonates, -methylphosphonates, -methyldiphenylphosphine oxides and their epoxy derivatives, *Synthesis*, 357, 1991.

102. Öhler, E., A new heteroannulation method to 2-cyclohexenone, mediated by phosphonate auxiliaries. Synthesis of 4,5,6,7-tetrahydrobenzothiazole derivatives, *Monatsh. Chem.*, 124, 763, 1993.

103. Morel, G., Scux, R., and Foucaud, A., Vinyl α-phenylphosphonates β-substituted by two electron-attracting groups. Preparation and reactivity toward various nucleophiles, *Bull. Soc. Chim. Fr.*, 177, 1976.

104. Burgos-Lepley, C.E., Mizsak, S.A., Nugent, R.A., and Johnson, R.A., Tetraalkyl oxiranylidenebis(phosphonates). Synthesis and reactions with nucleophiles, *J. Org. Chem.*, 58, 4159, 1993.

105. Page, P.C.B., McKenzie, M.J., and Gallagher, J.A., Simple synthesis of oxiranylidene-2,2-*bis*(phosphonic acid). Tetrabenzyl geminal bisphosphonate esters as useful intermediates, *Synth. Commun.*, 32, 211, 2002.

106. Ebetino, F.H., Bayless, A.V., Amburgey, J., Ibbotson, K.J., Dansereau, S., and Ebrahimpour, A., Elucidation of a pharmacophore for the bisphosphonate mechanism of bone antiresorptive activity, *Phosphorus, Sulfur Silicon Relat. Elem.*, 109, 217, 1996.

107. Öhler, E., Kang, H.-S., and Zbiral, E., Dialkyl (2,3-epoxy-4-oxoalkyl)phosphonates as synthons for thiazoles, *Chem. Ber.*, 121, 533, 1988.

108. Cristau, H.-J., Yangkou Mbianda, X., Geze, A., Beziat, Y., and Gasc, M.-B., Dioxirane oxidation of substituted vinylphosphonates. A novel efficient route to 1,2-epoxyalkylphosphonates, *J. Organomet. Chem.*, 571, 189, 1998.

109. Cristau, H.-J., Pirat, J.-L., Drag, M., and Kafarski, P., Regio- and stereoselective synthesis of 2-amino-1-hydroxy-2-aryl ethylphosphonic esters, *Tetrahedron Lett.*, 41, 9781, 2000.

110. Coover, H.W., Jr., Epoxyalkylphosphonates, *Eastman Kodak*, U.S. Patent Appl. US 2627521, 1953; *Chem. Abstr.*, 48, 1417i, 1954.

111. Hardy, E.E., and Reetz, T., Esters of epoxidized phosphonic acids, *Monsanto Chem.*, U.S. Patent Appl. US 2770610, 1956; *Chem. Abstr.*, 51, 7401e, 1957.

112. Nuretdinova, O.N., and Arbuzov, B.A., Reaction of epoxy compounds with thio acid salts, *Izv. Akad. Nauk SSSR, Ser. Khim.*, 353, 1971; *Bull. Acad. Sci. USSR, Div. Chem. Sci. (Engl. Transl.)*, 287, 1971.

113. Just, G., Potvin, P., and Hakimelahi, G.H., 3-Diethylphosphonoacrolein diethylthioacetal anion (6*), a reagent for the conversion of aldehydes to α,β-unsaturated ketene dithioacetals and three-carbon homologated α,β-unsaturated aldehydes, *Can. J. Chem.*, 58, 2780, 1980.

114. Cox, J.M., Substituted propyl phosphonic acid derivatives and their use as herbicides, *Imperial Chemical Industries*, Eur. Patent Appl. EP 78613, 1983; *Chem. Abstr.*, 99, 140148, 1983.

115. Disselnkötter, H., Lieb, F., Oediger, H., and Wendisch, D., Synthesis of phosphono analogs of 2-O-acetyl-1-O-hexadecyl(octadecyl)-sn-3-glycerylphosphorylcholine (platelet-activating factor), *Arch. Pharm. (Weinheim Ger.)*, 318, 695, 1985.

116. Courregelongue, J., Frehel, D., Maffrand, J.P., Paul, R., and Rico, I., Preparation of 3-hydroxy-4-phosphonobutanoic acid derivatives as hypolipidemics, *Sanofi*, French Patent Appl. FR 2596393, 1987; *Chem. Abstr.*, 108, 94780, 1988.

117. Deneke, U., Güthlein, W., Weckerle, W., and Wielinger, H., Substituted oxazoles and thiazoles and their use as redox indicators, *Boehringer Mannheim*, German Patent Appl. DE 3425118, 1986; *Chem. Abstr.*, 106, 67288, 1987.

118. Phillips, A.M.M.M., Mphahlele, M.J., Modro, A.M., Modro, T.A., and Zwierzak, A., Phosphonic systems. Part 7. Reactions of 2,3-epoxyphosphonates with nucleophiles. Preparation of 2,3-disubstituted alkylphosphonic esters and related systems, *Phosphorus, Sulfur Silicon Relat. Elem.*, 71, 165, 1992.

119. Lauth de Viguerie, N., Willson, M., and Perié, J., Synthesis and inhibition studies on glycolytic enzymes of phosphorylated epoxides and α-enones, *New J. Chem.*, 18, 1183, 1994.

120. Gajda, T., Nowalinska, M., Zawadzki, S., and Zwierzak, A., New approach to diethyl 2-aminoalkylphosphonates, *Phosphorus, Sulfur Silicon Relat. Elem.*, 105, 45, 1995.

121. Verbruggen, C., De Craecker, S., Rajan, P., Jiao, X.-Y., Borloo, M., Smith, K., Fairlamb, A.H., and Haemers, A., Phosphonic acid and phosphinic acid tripeptides as inhibitors of glutathionylspermidine synthetase, *Bioorg. Med. Chem. Lett.*, 6, 253, 1996.

122. de Macedo Puyau, P., and Perié, J.J., Synthesis of substrate analogues and inhibitors for the phosphoglycerate mutase enzyme, *Phosphorus, Sulfur Silicon Relat. Elem.*, 129, 13, 1997.

123. Thatcher, G.R., Bennett, B.M., Reynolds, J.N., Boegman, R.J., and Jhamandas, K., Nitrate esters, their preparation and use for treatment of neurological conditions, *Queen's University at Kingston*, U.S. Patent Appl. US 6310052, 2001; *Chem. Abstr.*, 135, 339273, 2001.

124. Cameron, K.O.K., and Lefker, B.A., Preparation of EP4 receptor selective agonists for the treatment of osteoporosis, *Pfizer*, Int. Patent Appl. WO 2002042268, 2002; *Chem. Abstr.*, 137, 6083, 2002.

125. Arbuzov, B.A., and Lugovkin, B.P., Synthesis of esters of phosphonic acids containing heterocyclic radicals. Part 2. Ethyl esters of phosphonic acids with oxygen-bearing heterocyclic radicals, *Zh. Obshch. Khim.*, 22, 1193, 1952; *Chem. Abstr.*, 47, 4871c, 1953.

126. Preis, S., Myers, T.C., and Jensen, E.V., Phosphonic acids. Part 3. Hydroxyl substituted propylphosphonic acids, *J. Am. Chem. Soc.*, 77, 6225, 1955.

127. Abramov, V.S., and Savintseva, R.N., Reaction of epihalohydrins with trialkyl phosphites, *Zh. Obshch. Khim.*, 37, 2784, 1967; *J. Gen. Chem. USSR (Engl. Transl.)*, 37, 2650, 1967.

128. Abramov, V.S., Savintseva, R.N., and Ermakova, V.E., Reaction of epihalohydrins with phosphorus(III) compounds. Part 2. Reactions of epiiodohydrin with esters, esteramides, and amides of phosphorous acid, *Zh. Obshch. Khim.*, 38, 2281, 1968; *J. Gen. Chem. USSR (Engl. Transl.)*, 38, 2207, 1968.

129. Abramov, V.S., and Savintseva, R.N., Reaction of epihalohydrins with trivalent phosphorus derivatives. Part 3. Effect of epichlorohydrin on trialkyl phosphites, *Zh. Obshch. Khim.*, 39, 849, 1969; *J. Gen. Chem. USSR (Engl. Transl.)*, 39, 812, 1969.

130. Prishchenko, A.A., Livantsov, M.V., Livantsova, L.I., Pol'shchikov, D.G., and Grigor'ev, E.V., Addition of trimethylsilyl esters of trivalent phosphorus acids to heterosubstituted derivatives of propylene oxide, *Zh. Obshch. Khim.*, 67, 1917, 1997; *Russ. J. Gen. Chem. (Engl. Transl.)*, 67, 1806, 1997.

131. Padwa, A., and Eastman, D., The reaction of organophosphorus compounds with α- and β-diphenylacyl bromides, *J. Org. Chem.*, 35, 1173, 1970.

132. Lau, W.Y., Zhang, L., Wang, J., Cheng, D., and Zhao, K., Preparation of 3'-phosphonate analogs of 2',3'-dideoxynucleosides, *Tetrahedron Lett.*, 37, 4297, 1996.

133. Blackburn, G.M., and Parratt, M.J., The synthesis of α-fluoroalkylphosphonates, *J. Chem. Soc., Chem. Commun.*, 886, 1983.

134. Blackburn, G.M., and Parratt, M.J., The synthesis of α-fluoroalkylphosphonates, *J. Chem. Soc., Perkin Trans. 1*, 1425, 1986.

135. Yuan, C.-Y., Wang, K., and Li, Z.-Y., Studies on organophosphorus compounds. Part 110. Enantioselective reduction of 2-keto-3-haloalkane phosphonates by baker's yeast, *Heteroatom Chem.*, 12, 551, 2001.

136. Pudovik, A.N., Zimin, M.G., and Sobanov, A.A., Reactions of dialkylphosphites with some epoxy ketones, *Zh. Obshch. Khim.*, 45, 1232, 1975; *J. Gen. Chem. USSR (Engl. Transl.)*, 45, 1212, 1975.

137. Kabat, M.M., A novel method of highly enantioselective synthesis of γ-hydroxy-β-keto phosphonates via allene oxides, *Tetrahedron Lett.*, 34, 8543, 1993.

138. Wroblewski, A.E., and Karolczak, W., Chalcone epoxide derived hydroxyphosphonates. Synthesis, stereochemistry and ring opening reactions, *Pol. J. Chem.*, 72, 1160, 1998.

139. Wroblewski, A.E., and Karolczak, W., Novel α-hydroxyphosphonates–enol phosphates rearrangement, *Pol. J. Chem.*, 73, 1191, 1999.

140. Tesoro, G.C., Amino epoxy phosphonates and compositions containing them, *J.P. Stevens*, U.K. Patent Appl. GB 1172916, 1970; *Chem. Abstr.*, 72, 91448, 1970.

141. Öhler, E., Kang, H.-S., and Zbiral, E., Synthesis of dialkyl (2,3-epoxy-4-oxoalkyl)phosphonates from (1-formylalkyl)phosphonates, *Chem. Ber.*, 121, 299, 1988.

142. Öhler, E., Kang, H.-S., and Zbiral, E., A convenient synthesis of new diethyl (2,4-dioxoalkyl)phosphonates, *Synthesis*, 623, 1988.

143. Liem, D.X., Jenkins, I.D., Skelton, B.W., and White, A.H., Synthesis of 2,3-anhydro glycosyl phosphonates, *Aust. J. Chem.*, 49, 371, 1996.

144. Kötzsch, H.J., Seiler, C.D., and Vahlensieck, H.J., Functional group-containing organophosphoric acid esters as adhesive or coatings for metal, *Dynamit Nobel A.-G.*, German Patent Appl. DE 2344197, 1975; *Chem. Abstr.*, 83, 44864, 1975.

145. Boehlow, T.R., and Spilling, C.D., The regio- and stereo-selective epoxidation of alkenes with methyl trioxorhenium and urea-hydrogen peroxide adduct, *Tetrahedron Lett.*, 37, 2717, 1996.

146. Boehlow, T., de la Cruz, A., Rath, N.P., and Spilling, C.D., Dimethyl (±)-(1S*,2R*,3S*)-[3-Phenyl-1-(N-phenylcarbamoyloxy)-2,3-epoxypropyl]phosphonate, *Acta Crystallogr., Sect. C: Cryst. Struct. Commun.*, 53, 1947, 1997.

147. de la Cruz, M.A., Shabany, H., and Spilling, C.D., Stereoselective reactions of allylic hydroxy phosphonates, *Phosphorus, Sulfur Silicon Relat. Elem.*, 144, 181, 1999.

148. Arbuzov, B.A., and Dianova, E.N., Esters of 2-cyclopentenyl-1-phosphonic acids and some of their derivatives, *Izv. Akad. Nauk SSSR, Ser. Khim.*, 1399, 1960; *Bull. Acad. Sci. USSR, Div. Chem. Sci. (Engl. Transl.)*, 1301, 1960.

149. Arbuzov, B.A., and Dianova, E.N., Esters of cyclohexen-2-yl-1-phosphinic acid and some of their derivatives, *Izv. Akad. Nauk SSSR, Ser. Khim.*, 1288, 1961; *Bull. Acad. Sci. USSR, Div. Chem. Sci. (Engl. Transl.)*, 1197, 1961.

150. Nagase, T., Kawashima, T., and Inamoto, N., A new synthetic route of β-hydroxyalkylphosphonates from β,γ-epoxyalkylphosphonates, *Chem. Lett.*, 1997, 1984.

151. Nagase, T., Kawashima, T., and Inamoto, N., Diastereoselective synthesis of 2,3-epoxyalkylphosphonates and phosphinates by epoxidation, *Chem. Lett.*, 1655, 1985.

152. Ryabov, B.V., Ionin, B.I., and Petrov, A.A., 2,3-Epoxyphosphonates. Synthesis and reactions with nucleophiles, *Zh. Obshch. Khim.*, 58, 969, 1988; *J. Gen. Chem. USSR (Engl. Transl.)*, 58, 859, 1988.

153. Chevalier, F., Al-Badri, H., and Collignon, N., Reactivity of substituted 3-(diethylphosphonyl)-1-(trialkylsilyl)alka-1,3-dienes. Regioselective epoxidation and cyclopropanation reactions, *Heteroatom Chem.*, 10, 231, 1999.

154. Cox, J.M., Hawkes, T.R., Bellini, P., Ellis, R.M., Barrett, R., Swanborough, J.J., Russell, S.E., Walker, P.A., Barnes, N.J., Knee, A.J., Lewis, T., and Davies, P.R., The design and synthesis of inhibitors of imidazoleglycerol phosphate dehydratase as potential herbicides, *Pestic. Sci.*, 50, 297, 1997.

155. de Macedo Puyau, P., and Perié, J.J., Selective epoxidation of polar substrates by dimethyldioxirane, *Synth. Commun.*, 28, 2679, 1998.

156. Stoianova, D.S., and Hanson, P.R., A ring-closing metathesis strategy to phosphonosugars, *Org. Lett.*, 3, 3285, 2001.

157. Chambers, R.D., Jaouhari, R., and O'Hagan, D., Synthesis of a difluoromethylenephosphonate analogue of glycerol-3-phosphate. A substrate for NADH linked glycerol-3-phosphate dehydrogenase, *J. Chem. Soc., Chem. Commun.*, 1169, 1988.

158. Chambers, R.D., Jaouhari, R., and O'Hagan, D., Fluorine in enzyme chemistry. Part 2. The preparation of difluoromethylenephosphonate analogues of glycolytic phosphates. Approaching an isosteric and isoelectronic phosphate mimic, *Tetrahedron*, 45, 5101, 1989.

159. Tang, K.-C., Tropp, B.E., and Engel, R., The synthesis of phosphonic acid and phosphate analogues of glycerol-3-phosphate and related metabolites, *Tetrahedron*, 34, 2873, 1978.

160. Li, Z., Racha, S., and Abushanab, E., Phosphonate isosteres of phospholipids, *Tetrahedron Lett.*, 34, 3539, 1993.

161. Savignac, P., Brèque, A., and Mathey, F., Homoallylic phosphonates. Useful precursors for 3,4-epoxyalkanephosphonates, *Synth. Commun.*, 487, 1979.

162. Minami, T., Yamanouchi, T., Tokumasu, S., and Hirao, I., The reaction of butadienylphosphonates with a oxosulfonium ylide, phosphonium ylides, and ketone enolates, *Bull. Chem. Soc. Jpn.*, 57, 2127, 1984.

163. Reist, E.J., and Sturm, P.A., Preparation and formulation of alkoxyphosphonoguanines as antiviral agents, *SRI International*, Int. Patent Appl. WO 8805438, 1988; *Chem. Abstr.*, 109, 230649, 1988.

164. Casara, P., and Jund, K., Preparation of (acyclo)nucleoside (fluoroalkyl)phosphonates as antivirals and antitumors, *Merrell Dow Pharmaceuticals*, Eur. Patent Appl. EP 335770, 1989; *Chem. Abstr.*, 112, 119356, 1990.

165. Katagiri, N., Yamamoto, M., and Kaneko, C., 3-Acetoxy-2-dimethylphosphonoacrylates. New dienophiles and their use for the synthesis of carbocyclic C-nucleoside precursors by the aid of RRA reaction, *Chem. Lett.*, 1855, 1990.

166. Norcross, R.D., Matt, P.v., Kolb, H.C., and Bellus, D., Synthesis of novel cyclobutylphosphonic acids as inhibitors of imidazole glycerol phosphate dehydratase, *Tetrahedron*, 53, 10289, 1997.

167. Yoshida, I., Ikuta, H., Fukuda, Y., Eguchi, Y., Kaino, M., Tagami, K., Kobayashi, N., Hayashi, K., Hiyoshi, H., Ohtsuka, I., Nakagawa, M., Abe, S., and Souda, S., Preparation of phosphonic acid derivatives useful for medically treating hyperlipemia, *Eisai*, Int. Patent Appl. WO 9420508, 1994; *Chem. Abstr.*, 121, 301075, 1994.

168. Hah, J.H., Gil, J.M., and Oh, D.Y., The stereoselective synthesis of cyclopropylphosphonate analogs of nucleotides, *Tetrahedron Lett.*, 40, 8235, 1999.

169. Afarinkia, K., and Mahmood, F., The anionic route to tricyclanes, *Tetrahedron Lett.*, 41, 1287, 2000.

170. Defacqz, N., Touillaux, R., Cordi, A., and Marchand-Brynaert, J., β-Aminophosphonic compounds derived from methyl 1-dimethoxy-phosphoryl-2-succinimidocyclohex-3-ene-1-carboxylates, *J. Chem. Soc., Perkin Trans. 1*, 2632, 2001.

171. Kashman, Y., and Sprecher, M., Phosphorus containing steroids, *Tetrahedron*, 27, 1331, 1971.

172. Butt, A.H., Percy, J.M., and Spencer, N.S., A rearrangement-based approach to secondary difluoro-phosphonates, *J. Chem. Soc., Chem. Commun.*, 1691, 2000.

173. Zucco, M., Le Bideau, F., and Malacria, M., Palladium-catalyzed intramolecular cyclization of vinyloxirane regioselective formation of cyclobutanol derivative, *Tetrahedron Lett.*, 36, 2487, 1995.

174. Wawrzenczyk, C., Zon, J., and Leja, E., Synthesis of diisopropyl (2-hydroxy-3-alkene-1-yl) and (5-carbethoxy-2-alkene-1-yl)phosphonates, *Phosphorus, Sulfur Silicon Relat. Elem.*, 71, 179, 1992.

175. Dufau, C., and Sturtz, G., Preparation of tetraethyl but-3-enylidene-1,1-bisphosphonate and study of its reactivity as a synthon of tetraethyl 3,4-diaminobutylidene-1,1-bisphosphonate, *Phosphorus, Sulfur Silicon Relat. Elem.*, 69, 93, 1992.

176. Blades, K., Lapôtre, D., and Percy, J.M., Conjugate addition reactions of a (diethoxyphosphinoyl)difluormethyl anion equivalent to acyclic and cyclyc vinyl sulfones, *Tetrahedron Lett.*, 38, 5895, 1997.

177. Tius, M.A., and Fauq, A., Total synthesis of (−)-asperdiol, *J. Am. Chem. Soc.*, 108, 6389, 1986.

178. Li, Y., Liu, Z., Lan, J., Li, J., Peng, L., Li, W.Z., Li, Y., and Chan, A.S.C., Enantioselective total synthesis of natural 11,12-epoxycembrene-C, *Tetrahedron Lett.*, 41, 7465, 2000.

179. Liu, Z., Li, W.Z., Peng, L., Li, Y., and Li, Y., First enantioselective total synthesis of (natural) (+)-11,12-epoxy-11,12-dihydrocembrene-C and (−)-7,8-epoxy-7,8-dihydrocembrene-C, *J. Chem. Soc., Perkin Trans. 1*, 4250, 2000.

180. Yang, Z.-Y., and Burton, D.J., A novel, general method for the preparation of α,α-difluoro functionalized phosphonates, *Tetrahedron Lett.*, 32, 1019, 1991.

181. Yang, Z.-Y., and Burton, D.J., A novel and practical preparation of α,α-difluoro functionalized phosphonates from iododifluoromethylphosphonate, *J. Org. Chem.*, 57, 4676, 1992.

182. Zhang, X., Qiu, W., and Burton, D.J., Preparation of α-fluorophosphonates, *J. Fluorine Chem.*, 89, 39, 1998.

183. Blades, K., Lequeux, T.P., and Percy, J.M., Reactive dienophiles containing a difluoromethylenephosphonato group, *J. Chem. Soc., Chem. Commun.*, 1457, 1996.

184. Halazy, S., Preparation of purinylcyclopentylphosphonates as virucides, *Merrell Dow Pharmaceuticals*, Eur. Patent Appl. EP 468119, 1992; *Chem. Abstr.*, 116, 194803, 1992.

185. Kuo, F., and Fuchs, P.L., Use of 1-penten-3-one-4-phosphonate as a kinetic ethyl vinyl ketone equivalent in the Robinson annulation reaction, *Synth. Commun.*, 16, 1745, 1986.

186. Horiguchi, M., *Biochemistry of natural C–P compounds,* Japanese Association for Research on the Biochemistry of C–P Compounds, Tokyo, 1984, p. 88.

187. Smith, J.D., Metabolism of phosphonates, in *The Role of Phosphonates in Living Systems,* Hilderbrand, R.L., Ed., CRC Press, Boca Raton, Florida, 1983, p. 31.

188. Seto, H., and Kuzuyama, T., Bioactive natural products with carbon-phosphorus bonds and their biosynthesis, *Nat. Prod. Rep.*, 589, 1999.

189. Christensen, B.G., Leanza, W.J., Beattie, T.R., Patchett, A.A., Arison, B.H., Ormond, R.E., Kuehl, F.A., Jr., Albers-Schonberg, G., and Jardetzky, O., Phosphonomycin. Structure and synthesis, *Science*, 166, 123, 1969.

190. Shoji, J., Kato, T., Hinoo, H., Hattori, T., Hirooka, K., Matsumoto, K., Tanimoto, T., and Kondo, E., Production of fosfomycin (phosphonomycin) by *Pseudomonas syringae*, *J. Antibiot.*, 39, 1011, 1986.

191. Katayama, N., Tsubotani, S., Nozaki, Y., Harada, S., and Ono, H., Fosfadecin and fosfocytocin, new nucleotide antibiotics produced by bacteria, *J. Antibiot.*, 43, 238, 1990.

192. Kawakami, Y., Furuwatari, C., Akahane, T., Okimura, Y., Furihata, K.I., Katsuyama, T., and Matsumoto, H., *In vitro* activity of arbekacin against clinical isolates of methicillin-resistant *Staphylococcus aureus* in a hospital, *J. Antibiot.*, 47, 507, 1994.

193. Cassidy, P.J., and Kahan, F.M., A stable enzyme-phosphoenolpyruvate intermediate in the synthesis of uridine-5′-diphospho-*N*-acetyl-2-amino-2-deoxyglucose 3-*O*-enolpyruvyl ether, *Biochemistry*, 12, 1364, 1973.

194. Kahan, F.M., Kahan, J.S., Cassidy, P.J., and Kropp, H., The mechanism of action of fosfomycin (phosphonomycin), *Ann. N.Y. Acad. Sci.*, 235, 364, 1974.

195. Marquardt, J.L., Brown, E.D., Lane, W.S., Haley, T.M., Ichikawa, Y., Wong, C.H., and Walsh, C.T., Kinetics, stoichiometry, and identification of the reactive thiolate in the inactivation of UDP-GlcNAc enolpyruvoyl transferase by the antibiotic fosfomycin, *Biochemistry*, 33, 10646, 1994.

196. Genua, M.I., Giraldez, J., Rocha, E., and Monge, A., Effects of antibiotics on platelet functions in human plasma *in vitro* and dog plasma *in vivo*, *J. Pharm. Sci.*, 69, 1282, 1980.

197. Ullmann, U., and Lindemann, B., *In vitro* investigations on the action of fosfomycin alone and in combination with other antibiotics on *Pseudomonas aeruginosa* and *Serratia marcescens*, *Arzneim. Forsch.*, 30, 1247, 1980.

198. Radda, T.M., Gnad, H.D., and Paroussis, P., Fosfomycin levels in human aqueous humor after intravenous administration, *Arzneim. Forsch.*, 35, 1329, 1985.

199. Fernandez Lastra, C., Marino, E.L., and Dominguez-Gil, A., Linearity of the pharmacokinetics of phosphomycin in serum and interstitial tissue fluid in rabbits, *Arzneim. Forsch.*, 36, 1518, 1986.

200. Vogt, K., Rauhut, A., Trautmann, M., and Hahn, H., Comparison of fosfomycin and teicoplanin in serum bactericidal activity against *Staphylococci*, *Arzneim. Forsch.*, 45, 894, 1995.

201. Rogers, T.O., and Birnbaum, J., Biosynthesis of fosfomycin by *Streptomyces fradiae*, *Antimicrob. Agents Chemother.*, 5, 121, 1974.

202. Imai, S., Seto, H., Ogawa, H., Satoh, A., and Otake, N., Studies on the biosynthesis of fosfomycin. Conversion of 2-hydroxyethylphosphonic acid and 2-aminoethylphosphonic acid to fosfomycin, *Agric. Biol. Chem.*, 49, 873, 1985.

203. Hammerschmidt, F., and Kählig, H., Biosynthesis of natural products with a phosphorus-carbon bond. Part 7. Synthesis of $[1,1-^2H_2]$-, $[2,2-^2H_2]$-, (*R*)- and (*S*)-$[1-^2H_1]$(2-hydroxyethyl)phosphonic acid and (*R,S*)-$[1-^2H_1]$(1,2-dihydroxyethyl)phosphonic acid and incorporation studies into fosfomycin in *Streptomyces fradiae*, *J. Org. Chem.*, 56, 2364, 1991.

204. Hammerschmidt, F., Labelled representatives of a possible intermediate of fosfomycin biosynthesis in *Streptomyces fradiae*. Preparation of (*R,S*)-(2-hydroxypropyl)-, (*R,S*)-, (*R*)- and (*S*)-(2-hydroxy-$[1,1\ ^2H_2]$propyl)- and (*R,S*)-(2-$[^{18}O]$hydroxypropyl)phosphonic acid, *Monatsh. Chem.*, 122, 389, 1991.

205. Hammerschmidt, F., Addition of dialkyl phosphites and dialkyl trimethylsilyl phosphites to 2-(benzyloxy)propanal. Preparation of all four stereoisomeric (1,2-dihydroxy-$[1-^2H_1]$propyl)phosphonic acids from chiral lactates, *Liebigs Ann. Chem.*, 469, 1991.

206. Hammerschmidt, F., Biosynthesis of natural products with a P–C bond. Part 9. Synthesis and incorporation of (*S*)- and (*R*)-2-hydroxy$[2-^2H_1]$ethylphosphonic acid into fosfomycin by *Streptomyces fradiae*, *Liebigs Ann. Chem.*, 553, 1992.

207. Hammerschmidt, F., Biosynthesis of natural products with a P–C bond. Part 8. On the origin of the oxirane oxygen atom of fosformycin in *Streptomyces fradiae*, *J. Chem. Soc., Perkin Trans. 1*, 1993, 1991.

208. Hammerschmidt, F., Biosynthesis of natural substances with a P–C bond. Part 11. Incorporation of L-[methyl-$^2H_3$]methionine and 2-[hydroxy-^{18}O]hydroxyethylphosphonic acid in phosphomycin in *Streptomyces fradiae* — an unusual methyl transfer, *Angew. Chem., Int. Ed. Engl.*, 33, 341, 1994.

209. Suarez, J.E., and Mendoza, M.C., Plasmid-encoded fosfomycin resistance, *Antimicrob. Agents Chemother.*, 35, 791, 1991.

210. Zilhao, R., and Courvalin, P., Nucleotide sequence of the *fosB* gene conferring fosfomycin resistance in *Staphylococcus epidermidis*, *FEMS Microbial. Lett.*, 68, 267, 1990.

211. Kuzuyama, T., Kobayashi, S., O'Hara, K., Hidaka, T., and Seto, H., Fosfomycin monophosphate and fosfomycin diphosphate, two inactivated fosfomycin derivatives formed by gene products of *fom*A and *fom*B from a fosfomycin producing organism *Streptomyces wedmorensis*, *J. Antibiot.*, 49, 502, 1996.

212. Liu, P., Murakami, K., Seki, T., He, X., Yeung, S.-M., Kuzuyama, T., Seto, H., and Liu, H.-w., Protein purification and function assignment of the epoxidase catalyzing the formation of fosfomycin, *J. Am. Chem. Soc.*, 123, 4619, 2001.

213. *Merck Index*, 10th ed., Merck and Co. Inc., Rahway, NJ, 1983, p. 607.

214. Perales, A., and Garcia-Blanco, S., The crystal and molecular structure of the phenethylammonium salt of fosfomycin [phenethylammonium (−)-(1*R*,2*S*)-epoxypropylphosphonate monohydrate: $C_8H_{12}N^+C_3H_6OPO_3^- \cdot H_2O$], *Acta Crystallogr., Sect. B*, 34, 238, 1978.

215. Perales, A., Martinez-Ripoll, M., Fayos, J., von Carstenn-Lichterfelde, C., and Fernandez, M., The absolute configuration of active and inactive fosfomycin, *Acta Crystallogr., Sect. B*, 38, 2763, 1982.

216. von Carstenn-Lichterfelde, C., Fernandez-Ibanez, M., Galvez-Ruano, E., and Bellanato, J., Structural study of fosfomycin [(−)-*cis*-1,2-epoxypropylphosphonic acid] salts and related compounds, *J. Chem. Soc., Perkin Trans. 2*, 943, 1983.

217. Fernandez-Ibanez, M., Prieto, A., Bellanato, E., Galvez-Ruano, E., and Arias-Perez, M.S., Spectroscopic study of (−)-*trans*-1,2-epoxypropylphosphonic acid derivatives, *J. Mol. Struct.*, 142, 391, 1986.

218. Christensen, B.G., Albers-Schonberg, G., and Leanza, W.J., Antibiotic (−)-*cis*-(1,2-epoxypropyl)phosphonates, *Merck*, German Patent Appl. DE 1805682, 1968; *Chem. Abstr.*, 72, 67109, 1970.

219. Christensen, B.G., Leanza, W.J., and Albers-Schonberg, G., Labile esters of (−)(*cis*-1,2-epoxypropyl)phosphonic acid, *Merck*, U.S. Patent Appl. US 3929840, 1975; *Chem. Abstr.*, 84, 90302, 1976.

220. Christensen, B.G., Beattie, T.R. and Leanza, W.J., (±)-*cis*-(1,2-Epoxypropyl)phosphonic acid and derivatives, antibacterials, *Merck*, German Patent Appl. DE 1805678, 1969; *Chem. Abstr.*, 72, 67106, 1970.

221. Christensen, B.G., Beattie, T.R., and Leanza, W.J., *cis*-Propenylphosphonic acid and derivatives, *Merck*, Belgian Patent Appl. BE 723070, 1969; *Chem. Abstr.*, 72, 67089, 1970.

222. Slates, H.L., and Wendler, N.L., Novel and efficient syntheses of 1-propynyl- and 1-propenylphosphonates. Precursors of fosfomycin, *Chem. Ind. (London)*, 430, 1978.

223. Firestone, R.A., Process for the preparation of (±) *cis*-1,2-epoxypropylphosphonic acid and derivatives, *Merck*, German Patent Appl. DE 1924098, 1970; *Chem. Abstr.*, 72, 90629a, 1970.

224. Mark, V., A facile $S_N i'$ rearrangement. The formation of 1,2-alkadienylphosphonates from 2-alkynyl phosphites, *Tetrahedron Lett.*, 3, 281, 1962.

225. Boisselle, A.P., and Meinhardt, N.A., Acetylene–allene rearrangements. Reactions of trivalent phosphorus chlorides with α-acetylenic alcohols and glycols, *J. Org. Chem.*, 27, 1828, 1962.

226. Pudovik, A.N., and Aladzheva, I.M., Acetylene-allene-acetylenic rearrangements of phosphites with a β,γ-acetylene linkage in the ester radical, *Zh. Obshch. Khim.*, 33, 707, 1963; *J. Gen. Chem. USSR (Engl. Transl.)*, 33, 700, 1963.

227. Pudovik, A.N., and Aladzheva, I.M., Thermal or "pseudo-Claisen" rearrangement of allyl and propargyl esters of phosphorous acid, *Dokl. Akad. Nauk SSSR*, 151, 1110, 1963; *Dokl. Chem. (Engl. Transl.)*, 151, 634, 1963.

228. Pudovik, A.N., Aladzheva, I.M., and Yakovenko, L.M., Synthesis and rearrangement of diethyl propargyl phosphite, *Zh. Obshch. Khim.*, 33, 3444, 1963; *J. Gen. Chem. USSR (Engl. Transl.)*, 33, 3373, 1963.

229. Girotra, N.N., and Wendler, N.L., Synthesis and transformations in the phosphomycin series, *Tetrahedron Lett.*, 4647, 1969.

230. Firestone, R.A., and Glamkowski, E.J., *cis*-(1,2-Epoxypropyl)phosphonic acid and derivatives, *Merck*, German Patent Appl. DE 1924105, 1970; *Chem. Abstr.*, 72, 132952h, 1970.

231. Firestone, R.A., and Sletzinger, M., Antimicrobial *cis*-1,2-epoxypropylphosphonic acid and its salt and ester derivatives, *Merck*, U.S. Patent Appl. US 3584014, 1971; *Chem. Abstr.*, 75, 63977t, 1971.

232. Firestone, R.A., Hydroxy- and imino-containing phosphonic acid diesters, *Merck*, U.S. Patent Appl. US 3784590, 1974; *Chem. Abstr.*, 80, 60031, 1974.

233. Firestone, R.A., Esters and amides of (diazomethyl)phosphonic acid, *Merck*, U.S. Patent Appl. US 3668197, 1972; *Chem. Abstr.*, 77, 114560, 1972.

234. Firestone, R.A., Phosphonic acid derivatives useful as preservatives and bactericides, *Merck*, German Patent Appl. DE 1924260, 1970; *Chem. Abstr.*, 72, 90622t, 1970.

235. Pollak, P.I., Christensen, B.G., and Wendler, N.L., (1,2-Epoxypropyl)phosphonic acids, esters, and salts, *Merck*, German Patent Appl. DE 1924169, 1970; *Chem. Abstr.*, 72, 100882u, 1970.

236. Chemerda, J.M., and Glamkowski, E.J., (±)-(*cis*-1,2-Epoxypropyl)phosphonic acid and esters, *Merck*, German Patent Appl. DE 1924118, 1970; *Chem. Abstr.*, 72, 132953j, 1970.

237. Chemerda, J.M., and Glamkowski, E.J., (±)-(*cis*-1,2-Epoxypropyl)phosphonic acid esters and salts, *Merck*, German Patent Appl. DE 1924173, 1969; *Chem. Abstr.*, 72, 43871p, 1970.

238. Chemerda, J.M., and Sletzinger, M., Phosphonic acid esters, *Merck*, German Patent Appl. DE 1924172, 1970; *Chem. Abstr.*, 72, 90631v, 1970.

239. Pines, S.H., and Karady, S., 1-Propynylphosphonic acid, *Merck*, German Patent Appl. DE 1924148, 1970; *Chem. Abstr.*, 72, 132965q, 1970.

240. Sletzinger, M., and Karady, S., Phosphonic acid esters and salts, *Merck*, German Patent Appl. DE 1924149, 1970; *Chem. Abstr.*, 72, 90628z, 1970.

241. Christensen, B.G., and Cama, L.D., Antibacterial optically-active epoxyphosphonic acid and its derivatives, *Merck*, German Patent Appl. DE 2002415, 1970; *Chem. Abstr.*, 75, 77031s, 1971.

242. Christensen, B.G., and Leanza, W.J., (*cis*-1,2-Epoxypropyl)phosphonic acid and its derivatives and intermediates, *Merck*, U.K. Patent Appl. GB 1244910, 1971; *Chem. Abstr.*, 75, 129940j, 1971.

243. Pollak, P.I., and Slates, H.L., Allylphosphonic acid and derivatives, *Merck*, German Patent Appl. DE 1924251, 1970; *Chem. Abstr.*, 72, 111614j, 1970.

244. Glamkowski, E.J., Rosas, C.B., Sletzinger, M., and Wantuck, J.A., *cis*-1-Propenylphosphonic acid, *Merck*, French Patent Appl. FR 2074329, 1971; *Chem. Abstr.*, 77, 62132, 1972.

245. Schoenewaldt, E.F., (±)- And (−)-*cis*-(1,2-epoxypropyl)phosphonic acids and salts, *Merck*, German Patent Appl. DE 1924231, 1970; *Chem. Abstr.*, 72, 132972q, 1970.

246. Firestone, R.A., Production of (*cis*-1,2-epoxypropyl)phosphonic acid and its salts and esters, *Merck*, German Patent Appl. DE 1924093, 1970; *Chem. Abstr.*, 72, 132971p, 1970.

247. Firestone, R.A., Antibiotic phosphonic acid derivatives, *Merck*, German Patent Appl. DE 1924138, 1970; *Chem. Abstr.*, 72, 111613h, 1970.

248. Shuman, R.F., Enantiomer separation of *cis*-(1,2-epoxypropyl)phosphonic acid derivatives, *Merck*, German Patent Appl. DE 1924085, 1970; *Chem. Abstr.*, 72, 90634y, 1970.

249. Christensen, B.G., Beattie, T.R., and Graham, D.W., Antibiotic substituted phosphonic acids and salts, *Merck*, French Patent Appl. FR 2034480, 1971; *Chem. Abstr.*, 75, 88759, 1971.

250. Kitamura, M., Tokunaga, M., and Noyori, R., Asymmetric hydrogenation of β-keto phosphonates. A practical way to fosfomycin, *J. Am. Chem. Soc.*, 117, 2931, 1995.

251. Bandini, E., Martelli, G., Spunta, G., and Panunzio, M., Silicon directed asymmetric synthesis of (1*R*,2*S*)-(-)-(1,2-epoxypropyl)phosphonic acid (fosfomycin) from (*S*)-lactaldehyde, *Tetrahedron: Asymmetry*, 6, 2127, 1995.

252. Oshikawa, T., and Yamashita, K., Preparation of fosfomycin diesters by stereoselective carbon–phosphorus bond formation, *Nippon Soda*, Japanese Patent Appl. JP 07324092, 1995; *Chem. Abstr.*, 124, 232132, 1996.

253. Castaldi, G., and Giordano, C., Preparation of chiral fosfomycin intermediates, *Zambon Group*, Eur. Patent Appl. EP 299484, 1989; *Chem. Abstr.*, 111, 57412, 1989.

254. Christensen, B.G., Dialkyl propylphosphonates, *Merck*, German Patent Appl. DE 1924104, 1970; *Chem. Abstr.*, 72, 90630, 1970.

255. Kobayashi, Y., William, A.D., and Tokoro, Y., Sharpless asymmetric dihydroxylation of *trans*-propenylphosphonate by using a modified AD-mix-α and the synthesis of fosfomycin, *J. Org. Chem.*, 66, 7903, 2001.

256. White, R.F., Birnbaum, J., Meyer, R.T., ten Broeke, J., Chemerda, J.M., and Demain, A.L., Microbial epoxidation of *cis*-propenylphosphonic to (−)-*cis*-1,2-epoxyphosphonic acid, *Appl. Microbiol.*, 22, 55, 1971.

257. Arbuzov, B.A., Polezhaeva, N.A., and Vinogradova, V.S., Isomerization of diethoxyphosphorylalkene oxides, *Izv. Akad. Nauk SSSR, Ser. Khim.*, 1146, 1967; *Bull. Acad. Sci. USSR, Div. Chem. Sci. (Engl. Transl.)*, 1110, 1967.

258. Redmore, D., Heterocyclic systems bearing phosphorus substituents. Synthesis and chemistry, *Chem. Rev.*, 71, 315, 1971.

259. Razumov, A.I., Liorber, B.G., Moskva, V.V., and Sokolov, M.P., Phosphorylated aldehydes, *Russ. Chem. Rev.*, 42, 538, 1973.

260. Griffin, C.E., and Ranieri, R.L., Acid-catalyzed rearrangements of dimethyl α,β,β-triaryl-α,β-epoxyethylphosphonates, *Phosphorus*, 6, 161, 1976.

261. Teulade, M.-P., and Savignac, P., Synthesis of 1-formyl 1,1-dialkylmethylphosphonates, *Synth. Commun.*, 17, 125, 1987.

262. Ryazantsev, E.N., Ponomarev, D.A., and Al'bitskaya, V.M., Reactions of organophosphorus α-epoxides with alcohols and phenols, *Zh. Obshch. Khim.*, 57, 2300, 1987; *J. Gen. Chem. USSR (Engl. Transl.)*, 57, 2056, 1987.

263. Zygmunt, J., and Mastalerz, P., Synthesis of phosphonic analogs of serine and isoserine, *Pol. J. Chem.*, 52, 2271, 1978.

264. Zygmunt, J., Walkowiak, U., and Mastalerz, P., 1-Hydroxy-2-aminoethanephosphonic acid and derivatives, *Pol. J. Chem.*, 54, 233, 1980.

265. Hammerschmidt, F., Bovermann, G., and Bayer, K., Biosynthesis of natural products with a P-C bond. Part 5. The oxirane oxygen atom of phosphomycin is not derived from atmospheric oxygen, *Liebigs Ann. Chem.*, 1055, 1990.

266. Hammerschmidt, F., Kählig, H., and Müller, N., Biosynthesis of natural products with a P–C bond. Part 6. Preparation of deuterium- and carbon-13-labelled L-alanyl- and L-alanyl-L-alanyl-(2-aminoethyl)phosphonic acids and their use in biosynthetic studies of fosfomycin in *Streptomyces fradiae*, *J. Chem. Soc., Perkin Trans. 1*, 365, 1991.

267. Hammerschmidt, F., and Kählig, H., Biosynthesis of natural products with a P–C bond. Part 10. Incorporation of D-[1–^{2}H$_1$]glucose into 2-aminoethylphosphonic acid in *Tetrahymena thermophila* and into fosfomycin in *Streptomyces fradiae*. The stereochemical course of a phosphoenolpyruvate mutase-catalyzed reaction, *Liebigs Ann. Chem.*, 1201, 1992.

268. Hammerschmidt, F., Biosynthesis of natural products with P–C bond. Incorporation of D-[1–^{2}H$_1$]glucose into 2-aminoethylphosphonic acid in *Tetrahymena thermophila* and of D-[1–^{2}H$_1$]glucose and L-[methyl-^{2}H$_3$]methionine into fosfomycin in *Streptomyces fradiae*, *Phosphorus, Sulfur Silicon Relat. Elem.*, 76, 111, 1993.

269. Simov, B.P., Wuggenig, F., Lammerhofer, M., Lindner, W., Zarbl, E., and Hammerschmidt, F., Biosynthesis of natural products with a P–C bond. Part 12. Indirect evidence for the biosynthesis of (1S,2S)-1,2-epoxypropylphosphonic acid as a co-metabolite of fosfomycin [(1R,2S)-1,2-epoxypropylphosphonic acid] by *Streptomyces fradiae*, *Eur. J. Org. Chem.*, 1139, 2002.

270. Arca, P., Rico, M., Brana, A.F., Villar, C.J., Hardisson, C., and Suarez, J.E., Formation of an adduct between fosfomycin and glutathione. A new mechanism of antibiotic resistance in bacteria, *Antimicrob. Agents Chemother.*, 32, 1552, 1988.

271. Arca, P., Hardisson, C., and Suarez, J.E., Purification of a glutathione S-transferase that mediates fosfomycin resistance in bacteria, *Antimicrob. Agents Chemother.*, 34, 844, 1990.

272. Armstrong, R.N., Structure, catalytic mechanism, and evolution of the glutathione transferases, *Chem. Res. Toxicol.*, 10, 2, 1997.

273. Bernat, B.A., Laughlin, L.T., and Armstrong, R.N., Fosfomycin resistance protein (FosA) is a manganese metalloglutathione transferase related to glyoxalase I and the extradiol dioxygenases, *Biochemistry*, 36, 3050, 1997.

274. Bernat, B.A., Laughlin, L.T., and Armstrong, R.N., Regiochemical and stereochemical course of the reaction catalyzed by the fosfomycin resistance protein, *fos*A, *J. Org. Chem.*, 63, 3778, 1998.

275. Baboulene, M., and Sturtz, G., Reactions of diethyl 1,2-epoxyalkanephosphonates. A new synthesis of 2-oxoalkanephosphonates, *Synthesis*, 456, 1978.

276. Drescher, M., Öhler, E., and Zbiral, E., A convenient approach to dialkyl heteroarylmethylphosphonates and [alkylthio(heteroaryl)methyl]phosphonates. Synthesis of imidazo[1,2-*a*]pyridine, imidazo[1,2-*a*]pyrimidine, indolizine and thiazole derivatives, *Synthesis*, 362, 1991.

277. Öhler, E., El-Badawi, M., and Zbiral, E., Dialkyl (1,2-epoxy-3-oxoalkyl)phosphonates as synthons for heterocyclic carbonyl compounds. Synthesis of acyl-substituted thiazoles, indolizines, imidazo[1,2-*a*]pyridines and imidazo[1,2-*a*]pyrimidines, *Chem. Ber.*, 118, 4099, 1985.

278. Eisch, J.J., and Galle, J.E., Stereospecific preparation of 1,2-epoxyalkyllithium reagents via the generalized lithiation of α-heterosubstituted epoxides, *J. Organomet. Chem.*, 121, C10, 1976.

279. Eisch, J.J., and Galle, J.E., Organosilicon compounds with functional groups proximate to silicon. Part 17. Synthetic and mechanistic aspects of the lithiation of α,β-epoxyalkylsilanes and related α-heterosubstituted epoxides, *J. Organomet. Chem.*, 341, 293, 1988.

280. Wyatt, P.B., and Blakskjaer, P., An enantioselective synthesis of (*R*)-2-amino-1-hydroxyethylphosphonic acid by hydrolytic kinetic resolution of (±)-diethyl oxiranephosphonate, *Tetrahedron Lett.*, 40, 6481, 1999.

281. Kuroda, Y., Okuhara, M., Iguchi, E., Aoki, H., Imanaka, H., Kamiya, T., Hashimoto, M., Hemmi, K., and Takeno, H., Hydroxylaminohydrocarbylphosphonic acid derivatives, *Fujisawa Pharm.*, Belgian Patent Appl. BE 857211, 1977; *Chem. Abstr.*, 89, 6413, 1978.

282. Patterson, D.R., Preparation of herbicidal hydroxyaminoalkylphosphonic acid derivatives, *Rohm and Haas*, U.S. Patent Appl. US 4693742, 1987; *Chem. Abstr.*, 108, 6183, 1988.

283. Rakov, A.P., and Alekseev, A.V., Reaction of β,γ-epoxypropylphosphonic acid esters with aliphatic alcohols, *Zh. Obshch. Khim.*, 43, 276, 1973; *J. Gen. Chem. USSR (Engl. Transl.)*, 43, 275, 1973.

284. Ryabov, B.V., Ionin, B.I., and Petrov, A.A., Conjugated opening–closing of the oxirane ring in the reactions of (4-chloro-2,3-epoxyalkyl)phosphonates with nucleophiles, *Zh. Obshch. Khim.*, 56, 2647, 1986; *J. Gen. Chem. USSR (Engl. Transl.)*, 56, 2344, 1986.

285. Rakov, A.P., and Andreev, G.F., Reaction of β,γ-epoxypropylphosphonates with acetic anhydride, *Zh. Obshch. Khim.*, 44, 952, 1974; *J. Gen. Chem. USSR (Engl. Transl.)*, 44, 916, 1974.

286. Biftu, T., Feng, D.D., Liang, G.-B., Ponpipom, M.M., Qian, X., Fisher, M.H., Wyvratt, M.J., and Bugianesi, R.L., Aliphatic hydroxy substituted piperidyl diaryl pyrrole derivatives as antiprotozoal agents, *Merck*, Int. Patent Appl. WO 0134632, 2001; *Chem. Abstr.*, 134, 366805, 2001.

287. Rakov, A.P., Andreev, G.F., and Alekseev, A.V., Reaction of dialkyl β,γ-epoxypropylphosphonates with amines, *Zh. Obshch. Khim.*, 45, 241, 1975; *J. Gen. Chem. USSR (Engl. Transl.)*, 45, 229, 1975.

288. Bajwa, J.S., and Anderson, R.C., A highly regioselective conversion of epoxides to halohydrins by lithium halides, *Tetrahedron Lett.*, 32, 3021, 1991.

289. Wroblewski, A.E., and Halajewska-Wosik, A., Towards enantiomeric 2,3-epoxypropylphosphonates, *Tetrahedron: Asymmetry*, 11, 2053, 2000.

5   The Formylphosphonates

The last 25 years have seen very significant progress in the chemistry of formylphosphonates.[1,2] This chapter describes their preparation, their use in intermolecular and intramolecular Horner–Wadsworth–Emmons reactions, and, given their discovery in nature, their importance as precursors for aminophosphonic acids.

5.1 SYNTHESIS OF FORMYLPHOSPHONATES

5.1.1 DIALKYL FORMYLPHOSPHONATES

5.1.1.1 Dialkyl Formylphosphonates

The stability of compounds containing the formyl group attached directly to the phosphoryl group was much debated, and until recently, information on their synthesis was limited and to some extent unreliable. Preparation of dimethyl formylphosphonate from sodium dimethyl phosphite and formic acid mixed anhydride was first claimed in a patent describing the synthesis of phosphonates having utility as intermediates for the production of fosfomycin.[3] The reaction, when reproduced on a laboratory scale, gave a mixture containing the expected dimethyl formylphosphonate (47%) and what appeared to be oligomers.[4] The ^{1}H-NMR data given in the original publication do not support the proposed structure. In 1987 it was shown that diethyl formylphosphonate decomposes at $-10°C$ into diethyl phosphite and carbon monoxide (Scheme 5.1). As diethyl phosphite accumulates in the reaction mixture, it adds to the carbonyl group of formylphosphonate to give tetraethyl 1-hydroxymethylenediphosphonate, which, in turn, isomerizes into diethyl 1-(diethoxyphosphinyloxy)methylphosphonate (79%).[5]

$$(5.1)$$

Difficulties in isolating free dialkyl formylphosphonates by acid hydrolysis of the corresponding acetal[6–9] or iminium salt[10] have been reported. Attempts to convert diethyl 1,1-diethoxymethylphosphonate into the corresponding aldehyde by acid hydrolysis gave only diethyl phosphite and ethyl formate.[9] However, the smooth conversion of dimethyl 1,1-dimethoxymethylphosphonate into silyl esters with Me_3SiBr followed by hydrolysis at 80°C and salification with 3 M NaOH gives the sodium salt of phosphonoformaldehyde as its hydrate in excellent yield (89%).[11] The adduct formed on addition of HCN to the hydrate is efficiently converted by hydrolysis of the nitrile group with concentrated HCl into 1-hydroxycarbonyl-1-hydroxymethylphosphonic acid.[11,12] This compound has been found to be active against herpes viruses in warm-blooded animals.[11,12]

Recently, diethyl formylphosphonate hydrate has been obtained in quantitative yield by treatment of diethyl diazomethylphosphonate with dimethyldioxirane (DMD) in acetone at room temperature (Scheme 5.2).[13,14] The condensation of diethyl formylphosphonate hydrate with primary amines leads to the corresponding aldimines. With secondary amines it acts as a formylating agent.[14] More recently, diethyl 1-chloro-1-ethoxymethylphosphonate has been prepared by treatment of diethyl 1,1-diethoxymethylphosphonate with $TiCl_4$ in Et_2O. In spite of its chloroether structure, it is stable to cold water and reacts with a wide variety of nucleophiles to give good to excellent yields of α-functionalized phosphonates.[15]

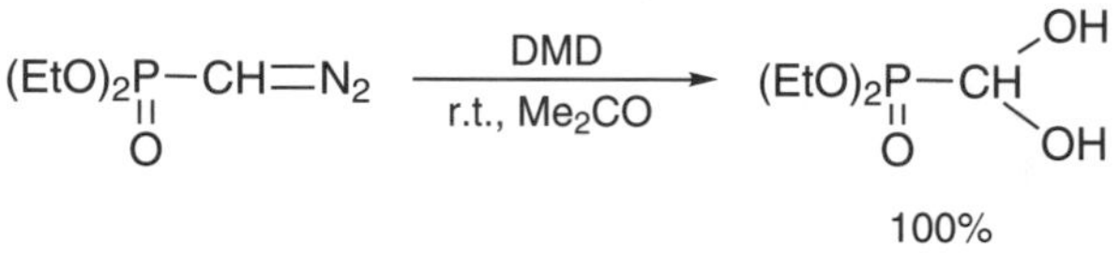

$$(EtO)_2\overset{\displaystyle O}{\underset{\displaystyle \|}{P}}-CH{\equiv}N_2 \quad \xrightarrow[\text{r.t., Me}_2\text{CO}]{\text{DMD}} \quad (EtO)_2\overset{\displaystyle O}{\underset{\displaystyle \|}{P}}-CH\overset{\displaystyle OH}{\underset{\displaystyle OH}{}}$$

$$100\%$$

$$(5.2)$$

5.1.1.2 Masked Forms of Dialkyl Formylphosphonates

Masked forms of diethyl formylphosphonate, such as acetals, hemithioacetals, and dithioacetals, are known and readily available. Unlike the parent formylphosphonate, their acetal and thioacetal derivatives are chemically stable, and a variety of synthetic methods have been developed for their preparation. Condensation of dialkyl phosphites with orthoformic esters provides dialkyl 1,1-dialkoxymethylphosphonates in moderate yields, but the reaction conditions are drastic (182°C for 5 h), and this method is not often used.[6,16] The addition of a catalytic amount of $BF_3{\cdot}Et_2O$ is crucial to the success of the reaction. It makes the reaction conditions considerably milder and gives superior results.[17–19] Variations using other hard (HCl, Me_3SiCl)[20–22] or soft (BnI, I_2)[22] electrophiles as initiators for the direct preparation of dialkyl 1,1-dialkoxymethylphosphonates from trialkyl phosphites and orthoformates have also been reported. An attractive route to dialkyl 1,1-dialkoxy-methylphosphonates involves the reaction of dialkyl chlorophosphites with orthoformic esters.[23–33] One such procedure, which uses moderate conditions, seems generally applicable for the high-yield (70–100%) preparation of 1,1-dialkoxymethylphosphonates bearing a wide range of substituents on both carbon and phosphorus atoms. By this method, diethyl 1,1-diethoxymethylphosphonate can be prepared on large scale (16 mol) from triethyl orthoformate and diethyl chlorophosphite in 89% yield.[30] The procedure can be refined by reacting a mixture of trialkyl phosphite, PCl_3, and orthoformic esters at 0°C in the presence of $ZnCl_2$[34,35] or simply reacting a 1:3 mixture of PCl_3 and orthoformic esters overnight at room temperature with a trace of anhydrous $ZnCl_2$ as catalyst.[34,36–40] Preparation of dialkyl 1,1-dialkoxymethylphosphonates from trialkyl phosphites, N,N-dimethylformamide dialkyl acetals, and methyl iodide is reported to give low to moderate yields.[41]

Hydrolysis of diethyl formylphosphonate acetals with 1 M HCl at 117°C does not lead to the corresponding aldehydes but involves the cleavage of the P–C bond with formation of diethyl phosphite, EtOH, and HCO_2Et.[6,42]

Although the reaction of dialkyl phosphites with orthoformic esters has been extended to triethyl trithioorthoformate,[43] the preferred synthesis of formylphosphonate dithioacetals is usually the high-yield Michaelis–Arbuzov reaction of trivalent phosphorus compounds with the appropriate chlorodithioacetals.[44–49] For the corresponding hemithioacetals, a Pummerer-type reaction of α-phosphoryl sulfoxides with alcohols in the presence of iodine is usually the method of choice (Scheme 5.3).[44,46,50,51] However, hemithioacetal formation is solvent dependent and generally gives a moderate yield of product in a mixture with several other byproducts arising from transesterification reactions.

Michaelis—Arbuzov reaction:

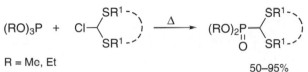

R = Me, Et
R^1 = Me, Et, Ph   50–95%
R^1-R^1 = (CH$_2$)$_3$, CH$_2$SCH$_2$, (CH$_2$)$_2$

Pummerer-type reaction:

$$(EtO)_2\underset{O}{\overset{\parallel}{P}}-CH_2-\underset{O}{\overset{\parallel}{S}}-R^2 \ + \ R^3OH \ \xrightarrow[\text{reflux}]{I_2} \ (EtO)_2\underset{O}{\overset{\parallel}{P}}\overset{SR^2}{\underset{OR^3}{\diagdown}}$$

R^2 = Me, Ph   28–47%
R^3 = Me, Et

(5.3)

A variety of other methods may be employed for the synthesis of dithio- and hemithioacetals. They involve the treatment of diethyl 1-lithio-1-(methylthio)methylphosphonate with dialkyl or diaryl disulfides,[52] with sulfur followed by alkylation,[53] and the treatment of diethyl formylphosphonate dithioacetals[47] or 1-chloro-1-(arylthio)methylphosphonates with thiols in CH$_2$Cl$_2$[54,55] or alcohols at reflux.[56]

A careful investigation of that reaction with diphenyl disulfide has shown that varying the reaction conditions can lead to either mixed or symmetric dithioacetal.[57] When the reaction mixture is treated in acidic medium before evaporation, the mixed dithioacetal is obtained predominantly in 70–90% yield accompanied by 5–20% of symmetric dithioacetal. In contrast, evaporation of the solvents before acidification results mainly in the formation of the symmetric dithioacetal (60–70%) accompanied by mixed dithioacetal (10%).[57]

Recently, the synthesis of dithio- and diselenoacetals of formylphosphonates by the BF$_3$·Et$_2$O-catalyzed reaction between dialkyl diazomethylphosphonate and disulfides or diselenides has been reported.[49] The reaction proceeds by the insertion of the dialkoxyphosphorylcarbene moiety into the S-S or Se-Se bonds of the disulfide or diselenide. Preparation of diethyl 1,1-*bis*(methylthio)methylphosphonate is generally carried out in methyl disulfide as solvent, and it has been found that both BF$_3$·Et$_2$O and disulfide had to be used in excess to produce the expected dithioacetal in a good yield (85%).[49]

Anions obtained by metallation of the dithioacetals and mixed dithioacetals of formylphosphonates undergo Horner–Wadsworth–Emmons reaction with carbonyl compounds to give the corresponding ketene *S,S*- or *O,S*-thioacetals in good to excellent yields.[46,58,59] Formylphosphonate thioacetals are the most useful reagents, being easily metallated, whereas mixed thioacetals form lithium derivatives only on treatment with strong bases (*t*-BuLi).[46] Metallated thioacetals are also available through the thiophilic addition of organolithium reagents to phosphonodithioformates. On treatment with aldehydes or ketones, ketene thioacetals are obtained in yields ranging from 31% to 82% (Scheme 5.4).[60]

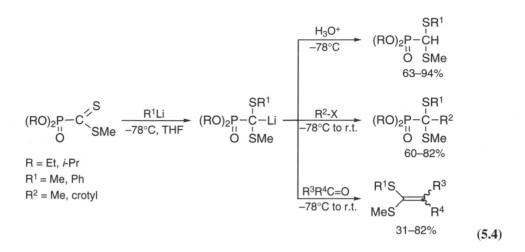

R = Et, *i*-Pr
R^1 = Me, Ph
R^2 = Me, crotyl

                                                                                                        (5.4)

5.1.2 DIALKYL 1-FORMYLALKYLPHOSPHONATES

5.1.2.1 Michaelis–Arbuzov Reactions

Possibly the most frequently used and most widely known phosphonylated aldehyde is diethyl 1-formylmethylphosphonate. A variety of methods for the preparation of this aldehyde are reported in the literature. The oldest employs the Michaelis–Arbuzov reaction[61] between triethyl phosphite and bromoacetaldehyde diethyl acetal, which yields diethyl 2,2-diethoxyethylphosphonate on heating to 160°C (Scheme 5.5).[62] Subsequent acid hydrolysis gives diethyl 1-formylmethylphosphonate. The aldehyde function requires protection because it is known that α-haloaldehydes react with trialkyl phosphites in a Perkow reaction affording dialkyl vinylphosphates isomeric with the expected phosphonates.[63–65]

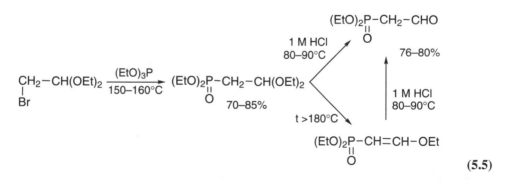

                                                                                                        (5.5)

This Michaelis–Arbuzov methodology, first carried out by Dawson and Burger,[62] used stoichiometric quantities of each reagent and gave, in 53% yield, a compound that was identified later as diethyl 2-ethoxyvinylphosphonate. This resulted from the thermal degradation of diethyl 2,2-diethoxyethylphosphonate.[66] Detailed studies of this reaction were made by Razumov et al.,[67–72] who employed various phosphorus(III) acid esters (phosphites, aliphatic and aromatic phosphonites, and phosphinites) and a wide range of haloacetals. Monitoring the reaction by ^{31}P-NMR spectroscopy has established that the reaction temperature is a crucial parameter. Formation of diethyl 2,2-diethoxyethylphosphonate [δ^{31}P(THF) = +27.2] takes place cleanly at 150–160°C, but decomposition into diethyl 2-ethoxyvinylphosphonate [δ^{31}P(THF) = +21.6][73] occurs at temperatures above 180°C, with the sensitivity of the phosphonylated acetals to thermolysis depending on the nature of the substituents at phosphorus. Fortunately, formation of the phosphonylated vinyl ether pyrolysis product is not a major obstacle to the isolation of diethyl 1-formylmethylphosphonate because hydrolysis of both acetal and the vinyl ether gives the expected phosphonylated aldehyde. However, hydrolysis of the phosphonylated vinyl ether requires more severe conditions, which lowers the yield of the desired 1-formylmethylphosphonate.[74,75]

The use of an appropriate hydrolytic procedure for the conversion of diethyl 2,2-diethoxyeth-ylphosphonate into diethyl 1-formylmethylphosphonate is an important step in the reaction sequence. A high concentration and a large excess of acid should be avoided to prevent hydrolysis of the acid-sensitive ester groups attached at phosphorus. The use of drastic conditions may provoke undesired side reactions, resulting in loss of the expected product. In the case of diethyl 1-formylmethylphosphonate, a variety of acids have been used (HCl,[1,68–72,76–81] H_2SO_4,[82] H_2SO_4–AcOH,[79] HCO_2H,[79,83] $HClO_4$,[76,84] and TFA[79]) with varying degrees of success. It suffices to use the theoretical amount of 1 M HCl at 80–90°C to obtain a complete hydrolysis of the diethyl acetal without degradation of ester groups. For example, when 1 M HCl at 90°C is used, 81% of diethyl 1-formylmethylphosphonate is obtained after 1.5 h, and 45% after 6 h. Clearly, prolonged hydrolysis has deleterious effects.[77,79]

An attractive procedure for hydrolysis of the acetal into the aldehyde involves the use of acid ion-exchange resins. Under these conditions, the phosphonylated diethyl acetal is treated overnight with a strongly acidic ion-exchange resin such as Dowex 50W[85,86] or Amberlyst-15[87,88] to generate phosphonoacetaldehyde diester in good to excellent yields (66–93%). The nature of the procedure means that the phosphonylated aldehydes, isolated by vacuum distillation, are often hydrated.[82] During the distillation of diethyl 1-formylmethylphosphonate, a small amount of viscous product remains in the distillation flask, presumed to be a trimer.[67] Phosphonylated acetaldehydes display an aldo–enol tautomer equilibrium, and spectral data indicate that the presence of water drives the equilibrium toward the enol form.[70,72,82,89–98]

The phosphonylation of haloacetaldehyde diethyl acetals coupled with a smooth hydrolysis of the acetals has been applied with success to the preparation of phosphonylated acetaldehydes[1,66,67,77,79,82,99–102] and is particularly useful on a large scale.[103] Unfortunately, this procedure cannot be extended to the synthesis of α-substituted dialkyl 1-formylmethylphosphonates.[104]

A variation of the Michaelis–Arbuzov reaction uses chloromethyloxazine as a masked formyl group (Scheme 5.6). Chloromethyloxazine, prepared in 55–63% yields by condensation of chloro-acetonitrile and 2-methyl-2,4-pentanediol in cold H_2SO_4, reacts with trialkyl phosphites at reflux to furnish dialkyl oxazinemethylphosphonates in moderate (40%, R = Me) to good yields (80%, R = Et).[105,106]

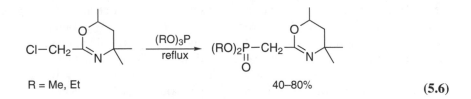

R = Me, Et   40–80%   (5.6)

It is not possible to prepare stable α-metallated derivatives of 2,2-diethoxyethylphosphonates because of the favorable elimination of ethoxide. However, oxazine phosphonates not containing a leaving group β to phosphorus undergo classical metallation at the active methylene group, which makes them reliable reagents for the production of chain-substituted phosphonoacetaldehydes. For instance, diethyl 1,1-dimethyloxazinemethylphosphonate is obtained in high yield (93%) by treating the sodium salt of diethyl oxazinemethylphosphonate in DME with excess methyl iodide (Scheme 5.7).[106] Subsequent unmasking of the formyl group is accomplished in 62% overall yield by a standard reduction–hydrolysis technique using $NaBH_4$ at –45°C followed by heating with oxalic acid.[106]

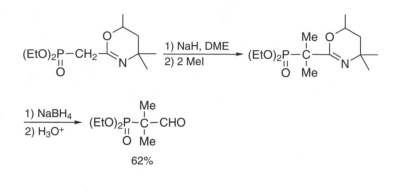

$$62\%$$

(5.7)

5.1.2.2 Anionic Formylation

Despite the progress made in the preparation of carbon-substituted 1-formylmethylphosphonates, techniques involving the phosphonylation of masked formyl groups seem to be synthetically useful only for preparing relatively simple compounds. More decisive progress has been made through the formylation of phosphonate carbanions. This methodology has been found to be the procedure of choice for preparing both unsubstituted and α-substituted phosphonoacetaldehydes with or without functional groups next to the formyl group. In addition to the phosphorus-containing group, aldehydes synthesized include those with alkyl, aryl, and electron-withdrawing groups in the α-position.

Two complementary procedures have been successfully developed. One employs HCO_2Me,[107–110] HCO_2Et,[111–118] DMF,[119–123] N-formylmorpholine,[124] or orthoformic esters[89,90,111] as formylating agents, and the other N,N-dimethylformamide dimethyl acetal as a latent formyl group.[110,125–127]

In the first procedure, alkyllithiums or lithium amides are employed to deprotonate the α-methylene group of diethyl alkylphosphonates to generate diethyl 1-lithioalkylphosphonates. These readily react with DMF or N-formylmorpholine at low temperature to give transient lithium β-aminoalkoxides. After treatment with aqueous acid (3 M HCl), dialkyl 1-formylalkylphosphonates are isolated in near-quantitative yields (Scheme 5.8).[119,124] Diethyl 1-lithioalkylphosphonates react with HCO_2Et in a similar procedure to give diethyl 1-formylalkylphosphonates in comparable yields. This procedure has been applied to the synthesis of a variety of alkylphosphonates including alkyl-,[114] allyl-,[117] benzyl-,[112,128] chloromethyl-,[113] bromomethyl-,[111] 3,3-diethoxybutyl-,[118] and thioethylmethyl-[112] derivatives. The hydrate of diethyl 1-formyl-1,1-difluoromethylphosphonate is prepared via a high-yielding cerium(III)-mediated reaction between diethyl 1,1-difluoromethylphosphonate and LDA in THF at low temperature, followed by quenching with DMF and an acidic workup.[129–132] The yields of substituted dialkyl 1-formylalkylphosphonates are high, with the only contaminants being the starting alkylphosphonates.

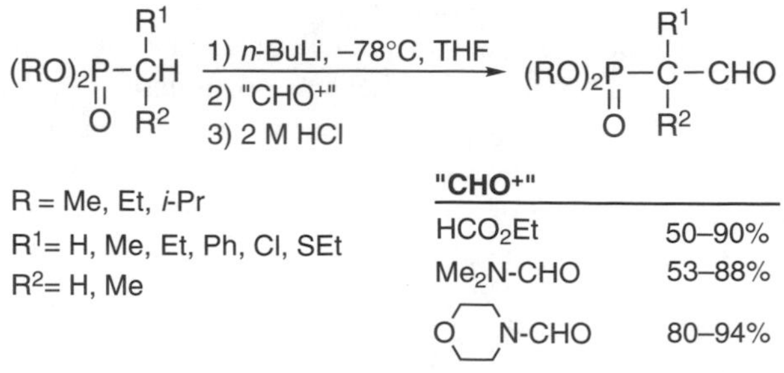

R = Me, Et, *i*-Pr

R¹ = H, Me, Et, Ph, Cl, SEt

R² = H, Me

"CHO⁺"	
HCO_2Et	50–90%
Me_2N-CHO	53–88%
O⌒N-CHO	80–94%

(5.8)

The ethyl formate approach can be combined with the phosphate–phosphonate conversion to provide a useful one-pot transformation of triethyl phosphate into diethyl 1-formylalkylphosphonates in good overall yield (90–94%, Scheme 5.9).[114,116,133,134] This is an attractive synthetic method for the generation of diethyl 1-lithioalkylphosphonates and provides a viable alternative to methods requiring the separate preparation of diethyl alkylphosphonates.

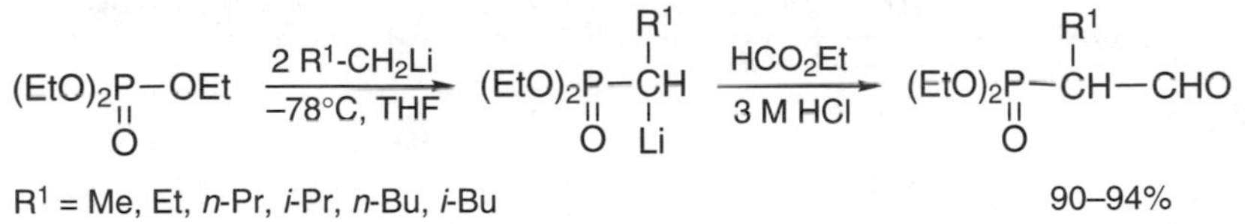

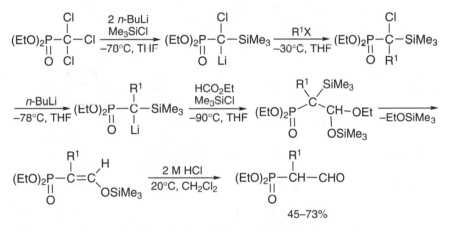

$$R^1 = Me, Et, n\text{-}Pr, i\text{-}Pr, n\text{-}Bu, i\text{-}Bu \qquad\qquad\qquad 90\text{–}94\%$$

$$(5.9)$$

Readily available diethyl trichloromethylphosphonate[135] has proved to be a useful reagent for the one-pot synthesis of diethyl 1-formylalkylphosphonates (Scheme 5.10). Diethyl trichloromethylphosphonate reacts at low temperature with n-BuLi (2 eq) in the presence of Me₃SiCl to give diethyl 1-lithio-1-chloro-1-(trimethylsilyl)methylphosphonate, which, on reaction with a variety of alkyl halides, affords α-substituted chloromethylphosphonates in excellent yields.[115] A third halogen–metal exchange gives diethyl 1-lithio-1-(trimethylsilyl)alkylphosphonates, which, on treatment with HCO₂Et and Me₃SiCl, are converted into the corresponding mixed acetals. Subsequent facile hydrolysis of the mixed acetal (2 M HCl) produces diethyl 1-formylalkylphosphonates in good overall yields (45–73%), usually as a mixture of aldehyde and enol forms.[115]

$$R^1 = n\text{-}Pr, n\text{-}C_5H_{11}, n\text{-}C_7H_{15}, n\text{-}C_{12}H_{25}, Cl(CH_2)_3, Cl(CH_2)_4, \text{allyl, crotyl}$$

$$(5.10)$$

For phosphonates containing an electron-withdrawing group Z, formylation reactions using orthoformic esters have limited scope. Phosphonates having relatively (Z = Cl,[90] Br,[90] Ph,[90]) or very acidic phosphonates (Z = CN,[89,111] CO₂Et,[89] Ac[90,111]) react with orthoformic esters in the presence of sodium in Et₂O at room temperature to afford reasonable yields of the corresponding enol ethers. However, after aqueous acid hydrolysis, yields of formylated phosphonates are modest (32–50%), so this methodology is less competitive than methods using formylating agents at low temperature.

When phosphonates containing electron-withdrawing groups are subjected to formylation with HCO₂Et, the corresponding phosphonylated aldehydes are obtained in low yields (25–30%),[109,111,112] and, in general, the carbanionic route works least well for stabilized anions, which are poorly reactive towards formyl group transfer. In these cases, N,N-dimethylformamide dimethyl acetal (Scheme 5.11) provides a highly useful reagent for the nucleophilic transfer of an aminomethylene fragment to the functionalized phosphonates under mild conditions, thus providing phosphonylated enamines in pure form.[125–127,136] Subsequent unmasking of the formyl group by acid or basic hydrolysis of the phosphonylated enamines affords α-functionalized diethyl 1-formylmethylphosphonates. In marked contrast to the previous route, the effect of the electron-withdrawing Z groups is favorable in promoting the abstraction of a proton by *N,N*-dimethylformamide dimethyl acetal.

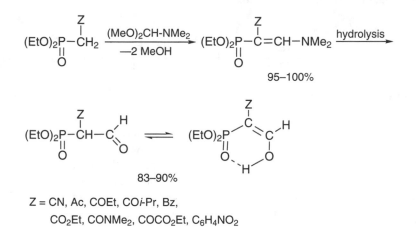

Z = CN, Ac, COEt, CO*i*-Pr, Bz,
CO$_2$Et, CONMe$_2$, COCO$_2$Et, C$_6$H$_4$NO$_2$

(5.11)

N,N-Dimethylformamide dialkyl acetals are better formylating reagents than orthoformate esters. They offer the two advantages of neutral conditions and higher yields than other methods. *N,N*-Dimethylformamide dimethyl acetal and *tert*-butoxy *bis*(dimethylamino)methane have been shown to be useful reagents for phosphonates bearing electron-withdrawing groups Z in α-position. *N,N*-Dimethylformamide dimethyl acetal undergoes spontaneous elimination of MeOH at room temperature when Z = CN or COCO$_2$R, whereas the reaction requires reflux for 2 h with Z = CONR$_2$, CO$_2$R, or COR and for 8 h with Z = *o*-C$_6$H$_4$NO$_2$ or *p*-C$_6$H$_4$NO$_2$. These reactions give phosphonylated enamines in nearly quantitative yields. However, phosphonates with Z = CH(OEt)$_2$, Cl, CH=CH$_2$, CH=CHCl, Ph, *o*-C$_6$H$_4$Cl, *p*-C$_6$H$_4$Cl, and SEt are inert. Deprotection of the formyl group can be performed at room temperature by treatment with 3 M HCl in a biphasic medium (Z = *o*-C$_6$H$_4$NO$_2$, *p*-C$_6$H$_4$NO$_2$, CO$_2$R, CONR$_2$) or with 2 M NaOH. The yields are excellent (83–90%), and the phosphonates being isolated are predominantly in the enol form.[126,127]

5.1.2.3 Rearrangement of 1,2-Epoxyphosphonates

An attractive route to dialkyl 1-formylalkylphosphonates employs the rearrangement of readily available 1,2-epoxyphosphonates to carbonyl compounds, a well-established reaction offering true synthetic utility. Thus, an attractive synthesis of diethyl 1,1-disubstituted 1-formylmethylphosphonates uses Lewis acid-induced catalytic isomerization of diethyl 1,2-epoxyalkylphosphonates (Scheme 5.12). BF$_3$·Et$_2$O is the more effective catalyst and gives the best results. 1,2-Epoxyalkylphosphonates readily undergo [1,2]phosphoryl group migration to produce both symmetric and disymmetric aliphatic and cyclic phosphonylated aldehydes with high selectivity and in high yields.[137–142]

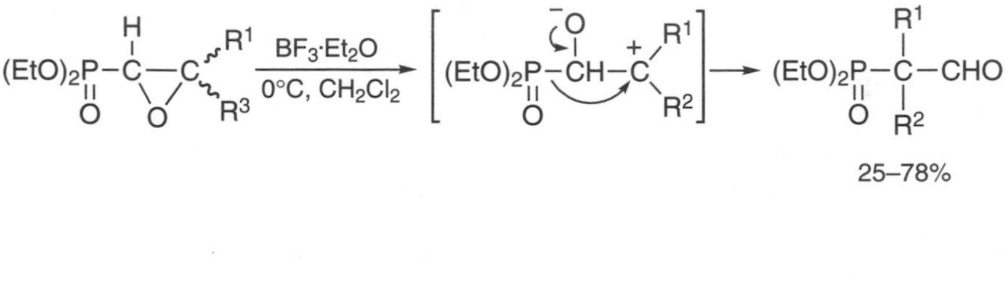

$$(EtO)_2P-\overset{H}{\underset{O}{C}}-\overset{R^1}{\underset{R^3}{C}} \xrightarrow[0°C,\ CH_2Cl_2]{BF_3·Et_2O} \left[(EtO)_2P-CH-\overset{+}{C}\overset{R^1}{\underset{R^2}{}} \right] \longrightarrow (EtO)_2P-\overset{R^1}{\underset{R^2}{C}}-CHO$$

25–78%   (5.12)

However, there are limitations to this rearrangement. For example, 1,2-epoxyalkylphosphonates prepared from cyclic ketones ($n = 1–3$, Scheme 5.13) undergo a competing proton migration on treatment with $BF_3 \cdot Et_2O$ and consequently give a mixture of 1,1-disubstituted-1-formylphosphonates (major product, 25–71%) and 1-hydroxyallylphosphonates (minor product, 10–15%).[142]

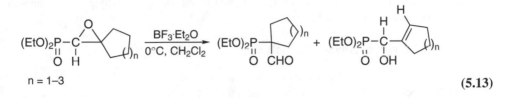

$$ \text{(5.13)} $$

In some cases, diethyl 1,2-epoxyalkylphosphonates undergo thermal phosphoryl group migration at high temperatures (170–300°C) to produce 1-substituted 1-formylmethylphosphonates. However, this method has only limited utility because the rearrangement products themselves can be thermally labile with respect to dephosphorylation.[137,143,144]

Ring opening of diethyl 1,2-epoxyalkylphosphonates can be effected by sodium dialkyl phosphites in refluxing ethanol (Scheme 5.14), which attack the β-carbon atom of the epoxide via the sodium phosphite–alkoxide equilibrium. For example, diethyl 1,2-epoxypropylphosphonate undergoes ring opening by sodium dialkyl phosphites with concomitant elimination of diethyl phosphite anion to afford dialkyl 1-formylethylphosphonates in good yields (50–85%).[145]

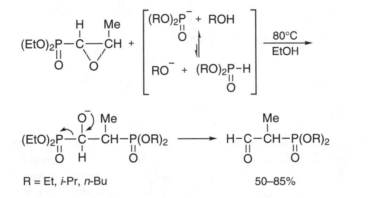

$$ \text{(5.14)} $$

5.1.2.4 Hydroformylation of Vinylphosphonates

Preparation of dialkyl 1-formylethylphosphonates can be achieved in low to good yields (27–71%) by reacting dialkyl vinylphosphonates with carbon monoxide and hydrogen (1/1 CO/H_2 ratio) in the presence of a conventional rhodium-containing hydroformylation catalyst [Rh(COD)Cl]₂ at temperatures ranging from 60°C to 130°C under a pressure of 20–1500 bars (Scheme 5.15). Hydroformylation can be carried out with or without solvent. Loading of rhodium catalyst from 0.5 to 500 ppm have proved suitable.[146,147]

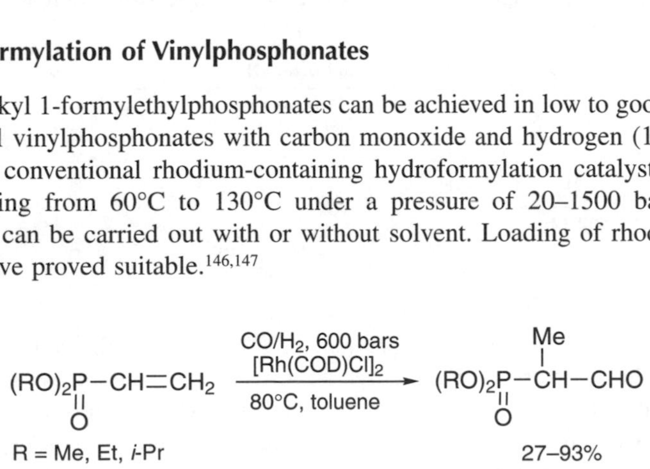

$$ (RO)_2\underset{\underset{O}{\parallel}}{P}-CH=CH_2 \quad \xrightarrow[\text{80°C, toluene}]{\substack{CO/H_2, \text{ 600 bars} \\ [Rh(COD)Cl]_2}} \quad (RO)_2\underset{\underset{O}{\parallel}}{P}-\underset{\underset{}{\overset{\overset{Me}{|}}{}}}{C}H-CHO $$

R = Me, Et, *i*-Pr   27–93%

$$ \text{(5.15)} $$

5.1.2.5 Additions to Dialkyl Ethynylphosphonates

Depending on the conditions employed, one or two molecules of EtOH or Et_2NH can be added to dialkyl ethynylphosphonates to give dialkyl 2-ethoxyvinylphosphonate, 2,2-diethoxyethylphosphonate or 2-diethylaminovinylphosphonate. All of these compounds give dialkyl 1-formylmethylphosphonates on acid hydrolysis. However, this method using dialkyl ethynylphosphonates as starting materials does not offer real synthetic utility.[148]

5.1.2.6 Hydrolysis of Allylic Alcohols

Diethyl 2-ethoxyvinylphosphonate can be used as precursor of diethyl 1-formylvinylphosphonate. The reaction between the β-ethoxy-α-phosphonovinyl anion, generated from diethyl 2-ethoxyvinylphosphonate and LDA, and aldehydes or ketones leads to the expected diethyl 2-ethoxy-1-(hydroxymethyl)vinylphosphonates in good to excellent yields. Subsequent treatment of the allylic alcohol product with TFA in CH_2Cl_2 at 0°C gives diethyl 1-formylvinylphosphonates in almost quantitative yields (95–99%, Scheme 5.16).[149]

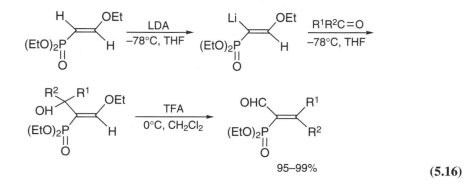

(5.16)

5.1.2.7 Ozonolysis

The palladium-catalyzed Michaelis–Arbuzov reaction of allyl acetates with triethyl phosphite provides reasonable yield (65%) of diethyl allylphosphonates. The reaction has been employed in the synthesis of diethyl 1-formylmethylphosphonate on a preparative scale (52%) by reductive ozonolysis of diethyl allylphosphonate in CH_2Cl_2 at low temperature (Scheme 5.17).[150]

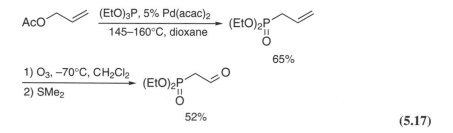

(5.17)

The synthetic potential of singlet oxygen may be used to effect the conversion of dimethyl 1-(nitromethyl)alkylphosphonates into dimethyl 1-formylalkylphosphonates. Thus, treatment of a Rose Bengal-sensitized solution of dimethyl 1-(nitromethyl)ethylphosphonate and MeONa in MeOH with singlet oxygen affords dimethyl 1-formylethylphosphonate in 80% yield. A temperature dependence of the yield has been observed. For the reaction with dimethyl 1-(nitromethyl)ethylphosphonate in MeOH at 10°C, 0°C, and –78°C, the conversions are 50%, 60%, and 100%, respectively. The higher yield at low temperature may reflect increased stability of the intermediate dioxazetidine.[151,152] The procedure seems to be applicable across a wide spectrum of 1-formylalkyl-phosphonates, but the relatively long reaction sequence is not compensated by the originality of formylphosphonate structures and the yields obtained (Scheme 5.18).

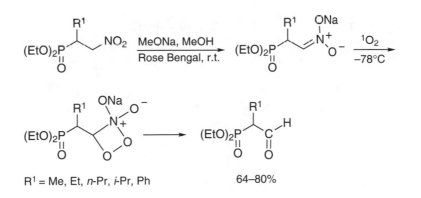

R[1] = Me, Et, *n*-Pr, *i*-Pr, Ph   64–80%

(5.18)

5.1.3 DIALKYL 2-FORMYLALKYLPHOSPHONATES

5.1.3.1 Michaelis–Arbuzov Reactions

The reaction of triethyl phosphite with linear or branched primary 3-halopropionaldehyde diethyl acetals is used for the synthesis of diethyl 3,3-diethoxypropylphosphonates (Scheme 5.19).[67,77,78,101,153–156] 3,3-Diethoxypropylphosphonates may be metallated by treatment with *n*-BuLi in THF at low temperature to generate stable α-lithiated carbanions that react with alkyl halides to give 1-alkyl-3,3-diethoxypropylphosphonates in good yields (62–70%) in a reaction of significant synthetic importance. The acetals are hydrolyzed under the same conditions as above (1 M HCl) to give diethyl 2-formylethylphosphonates and 1-alkyl-2-formylethylphosphonates in yields ranging from 50 to 85%. Distillation of these compounds is frequently accompanied by extensive polymerization.[79]

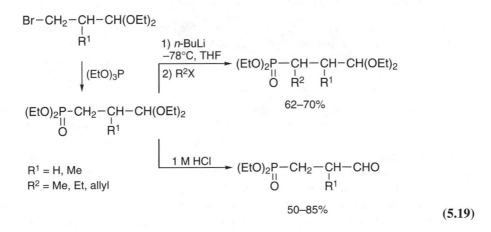

(5.19)

Reaction of triethyl phosphite with 1,3-dihalogenopropenes at 120–140°C provides diethyl 3-halogeno-2-propenylphosphonates (X = Cl, Br) in 56–90% yield. On treatment with EtONa or EtSNa in EtOH at room temperature, these compounds are selectively transferred into 3-ethoxy- or 3-thioethoxy-2-propenylphosphonates (Scheme 5.20). Hydrolysis with 5 M HCl converts these enol and thioenol ethers to the corresponding diethyl 2-alkyl-2-formylethylphosphonates (R[1] ≠ H) and 2-formylethylphosphonate (R^1 = H), thus giving an entry into β-substituted phosphonopropionaldehydes.[157]

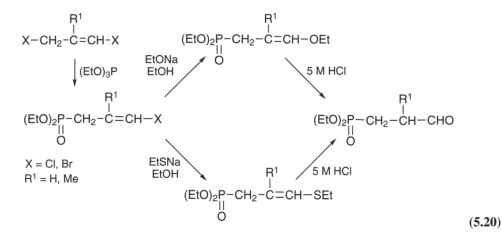

$$(5.20)$$

These two methods are complementary and give access to a variety of carbon-substituted diethyl 2-formylalkylphosphonates.

Diethyl *ortho*-formylphenylphosphonate can be prepared by the photolysis of *ortho*-iodobenzaldehyde in the presence of triethyl phosphite.[158] This reaction, which provides a general access to diethyl phenylphosphonates, is formally a photolytic modification of the Michaelis–Arbuzov reaction. The good yields and the experimental simplicity of the process suggest that the reaction might profitably be applied to the synthesis of many ring-substituted phenylphosphonates. Unfortunately, irradiation of *ortho*-iodobenzaldehyde in a fourfold excess of triethyl phosphite at 0°C gives only a 34% yield of the relatively stable diethyl *ortho*-formylphenylphosphonate. This yield, the lowest found in a series of ring-substituted iodobenzenes, suggests the use of *ortho*-iodobenzaldehyde protected as the corresponding 1,3-dioxolane.[158]

5.1.3.2   Phosphonylation of α,β-Unsaturated Aldehydes

A reaction having significant synthetic importance for the construction of dialkyl 2-formylalkylphosphonates involves the addition of trivalent phosphorus reagents to α,β-unsaturated aldehydes.[159] This Michael–Michaelis–Arbuzov strategy is the method of choice for incorporating, from readily available Michael-type substrates, one or two substituents at any position of the carbon chain between the phosphoryl and formyl groups.

The addition of trialkyl phosphites to acrolein was first investigated in dioxane solution.[160] However, despite a rather effective control of the conditions, isolation of a pure product was difficult. The reaction generally furnishes the conjugated addition product in low yields in the enol ether form. Among a number of factors that may affect the completion of such a conjugate addition reaction, the dealkylation of the zwitterionic adduct is crucial to the success of the reaction to exclude the inter- and intramolecular competition that can lead to production of mixtures of products, including acetals, enol ethers, and simple carbonyl compounds.[160–163]

Further efforts demonstrate that the reaction of triethyl phosphite occurs under quite mild conditions in a conjugate manner with a wide range of α,β–unsaturated aldehydes provided that a proton source is present (Scheme 5.21). The use of protic solvents, such as alcohols or phenols, not only provides a source of proton for the anionic site of the zwitterionic adduct but also furnishes a nucleophile for the required dealkylation step. The simplest hypothesis consistent with all of these facts is that the protonation-valency expansion of the quasiphosphonium ion intermediate using a proton source proceeds at a rate greater than the intermolecular pathway to enol ether.[164]

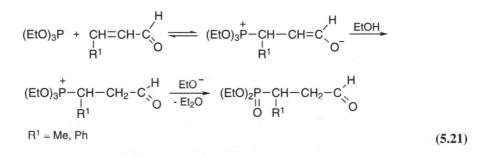

(EtO)3P + CH=CH-C(H)=O ⇌ (EtO)3P+-CH-CH=C(H)-O−
R1
EtOH
(EtO)3P+-CH-CH2-C(H)=O
EtO−
- Et2O
(EtO)2P(=O)-CH-CH2-C(H)=O
R1 = Me, Ph

(5.21)

Standard conditions for these reactions were chosen on the basis of preliminary test reactions from which several generalizations were derived: (a) reactions occur more rapidly in MeOH than EtOH and with triethyl phosphite than with trimethyl phosphite; (b) where the solvent alcohol is not identical with the alcohol from which the P(III) ester is derived, partial ester exchange results in a mixed P(V) ester; and (c) reactions in phenol are cleaner and result in higher yields with only minor or no contamination by products from ester exchange.[164] It has been demonstrated that when phenol is used as the proton source, crotonaldehyde undergoes hydrophosphonylation,[164] even at 0°C, to selectively furnish the phosphonate as its diphenyl acetal. The yield of phosphonylated crotonaldehyde diphenyl acetal in phenol at 100°C was 82% against 59% of diethyl acetal in refluxing EtOH for the same time.[164]

A procedure employing the conjugate addition of $(RO)_3P$ to acrolein in the presence of the corresponding alcohol has been developed on an industrial scale (Scheme 5.22). It allows different dialkyl 3,3-dialkoxyethylphosphonates to be isolated in excellent yields (80–91%).[163] The acetals are hydrolyzed to the aldehyde using a sulfonated resin and water or 0.35 M HCl at 40°C.[163] Another large-scale industrial procedure for the preparation of dialkyl 2-formylethylphosphonates in 86% yields involves the addition of dialkyl phosphites to 3,3-diacetoxypropene at 120–160°C in the presence of *tert*-butyl peroctoate followed by unmasking of the formyl group.[162]

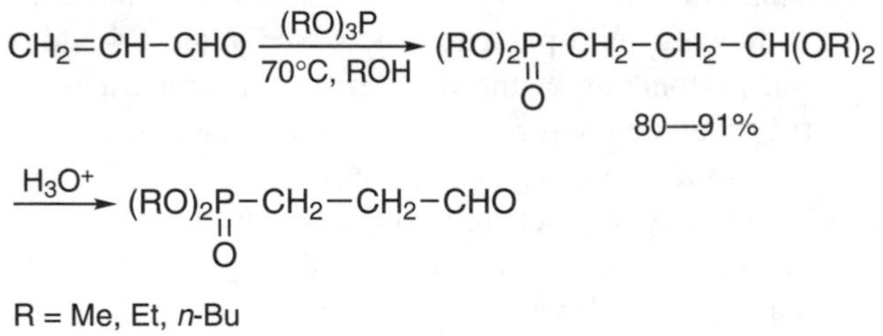

CH2=CH-CHO
(RO)3P
70°C, ROH
(RO)2P(=O)-CH2-CH2-CH(OR)2
80—91%
H3O+
(RO)2P(=O)-CH2-CH2-CHO
R = Me, Et, n-Bu

(5.22)

Besides alcohols or phenols, a variety of related reagents including dialkyl chlorophosphates[165,166] and silyl halides[167,168] have been added to the reaction medium as traps. In recent years, some striking developments using conjugate addition reactions have been accomplished with phosphorus reagents bearing either silyl ester linkages such as $(RO)_2POSiR_3$ or $(RO)_3P$ and R_3SiCl. Reactions with α,β-unsaturated aldehydes are carried out with stoichiometric quantities of each reagent either neat or in a solvent at ambient temperature (Scheme 5.23). A comparison of different reagents reveals that the reaction of acrolein, methacrolein, and cinnamaldehyde proceeds quite readily, but with some complications. Although in certain instances low yields of conjugate addition products can be isolated, the major product arises from addition to the carbonyl carbon (Abramov product).[167,168]

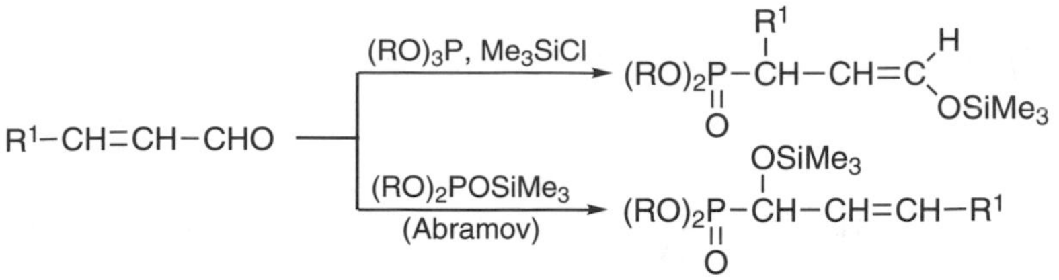

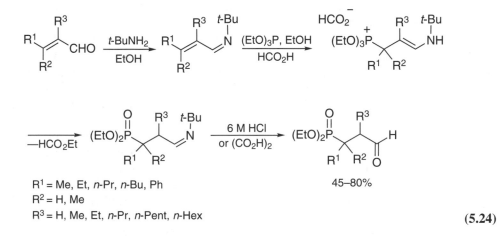

(5.23)

The exploitation of the full potential of α,β-unsaturated aldehydes for synthetic purposes is often handicapped by the complications reported above. These disadvantages can be overcome when the unprotected α,β-unsaturated aldehydes are replaced by α,β-unsaturated aldimines, which are activated as iminium salts (Scheme 5.24).[169]

(5.24)

Triethyl phosphite does not react with aldimines in EtOH. However, the addition of HCO_2H (1 eq) to the $(EtO)_3P$/aldimine mixture in EtOH induces an exothermic reaction resulting in the exclusive formation of the iminoalkylphosphonate. Undoubtedly, the 1,4-addition of triethyl phosphite is promoted by initial protonation at the nitrogen atom, which activates the double bond toward Michael addition and generates the formate anion for the dealkylation step (Michaelis–Arbuzov reaction). The dialkyl-substituted formylphosphonates are obtained by hydrolysis of the imine function with an appropriate hydrolytic procedure using either 6 M HCl or 1 M oxalic acid.[169] The method provides a simple and convenient access to a broad range of substituted diethyl 2-formylalkylphosphonates in satisfactory overall yields (45–80%).[169]

5.1.3.3 Anionic Phosphonylation

Diethyl formylethynylphosphonate is prepared in high yield by the SNP(V) reaction of 3,3-diethoxy-1-lithio-1-propyne with diethyl chlorophosphate followed by deprotection of the acetal using 97% HCO_2H.[170] It is a colorless distillable liquid that proved to be stable when stored for several months at 0°C. Catalytic hydrogenation of diethyl 3,3-diethoxypropynylphosphonate using quinoline-poisoned 5% Pd/CaCO$_3$, gives a mixture of (Z)- and (E)-acetals, from which the (Z)-isomer is isolated in 60% yield by distillation. Formolysis of pure (Z)-3,3-diethoxypropenylphosphonate using 97% HCO_2H gives a mixture of (Z)- and (E)-aldehydes from which diethyl (Z)-2-formylvinylphosphonate is obtained by distillation in 36% yield (Scheme 5.25).[171]

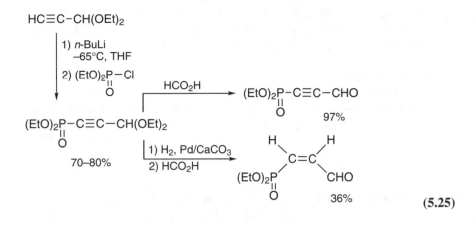

(5.25)

The preparation of diethyl 3,3-diethoxy-2-hydroxypropylphosphonate in 38% yield is possible through the addition of lithium diethyl phosphite to a mixture of D-glycidaldehyde diethyl acetal and BF$_3$Et$_2$O at −80°C in THF (Scheme 5.26).[172] The relatively low yield may be explained by deactivation of the epoxide and the weakly nucleophilic character of the phosphite anion. Deprotection of the formyl group is accomplished under mild conditions with 0.1 M HCl at 40°C.[172,173]

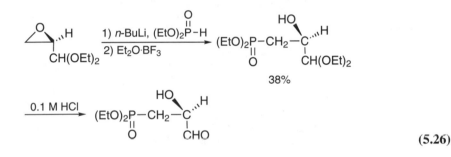

(5.26)

5.1.3.4 Oxidation

The methodology for the conversion of epoxyphosphonates into allylic alcohols has been used in the preparation of diethyl 2-formylvinylphosphonate by basic ring opening of diethyl 2,3-epoxypropylphosphonate. When the latter is allowed to react with MeONa in MeOH at 0°C and then treated with an ion-exchange resin (Dowex 50W), diethyl (E)-3-hydroxy-1-propenylphosphonate is isolated in quantitative yield.[174] It is converted by room temperature oxidation with PCC in CH$_2$Cl$_2$ into pure diethyl (E)-2-formylvinylphosphonate in 52% yield (Scheme 5.27).[156,174]

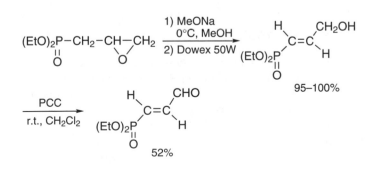

(5.27)

By a similar procedure, 90% yields of dialkyl 1-amino-2-formylethylphosphonates are obtained by oxidation of the corresponding alcohols in CH_2Cl_2 using DMSO/COCl)$_2$/Et$_3$N.[175] Phosphono-cyclopropanes bearing *tert*-butyl and trimethylsilylethyl protected carboxyl groups are oxidized with $NaIO_4$ in a THF-H_2O solution at room temperature to give the 2-formylcyclopropylphospho-nate in 97% yield.[176]

Lactols are used as starting materials for the conversion of phosphonylated 1,2-glycols to the corresponding aldehydes via oxidation with basic $NaIO_4$ (Malaprade reaction). The oxidative ring opening of the lactol provides diethyl 2-formyl-2-(hydroxymethyl)ethylphosphonate, which elim-inates water during silica gel chromatography to produce diethyl 2-formylallylphosphonate.[177]

5.1.3.5 Ozonolysis

The method of converting vinyl to carbonyl groups by ozonolysis has received much attention and has been successfully applied to a variety of phosphonates containing homoallyl groups, including α–functionalized phosphonates.[178–180] The homoallyl group is introduced by the reaction of allyl-bromide with diethyl 1-lithioalkylphosphonates.[179–181] A convenient preparation of diethyl 1,1-difluoro-2-formylethylphosphonate (Scheme 5.28) involves the ozonolysis of diethyl 1-allyl-1,1-difluoromethylphosphonate, which, in turn, is prepared from diethyl bromodifluoromethylphospho-nate in 62% yield by treatment of its cadmium bromide derivative with allyl bromide.[182–184]

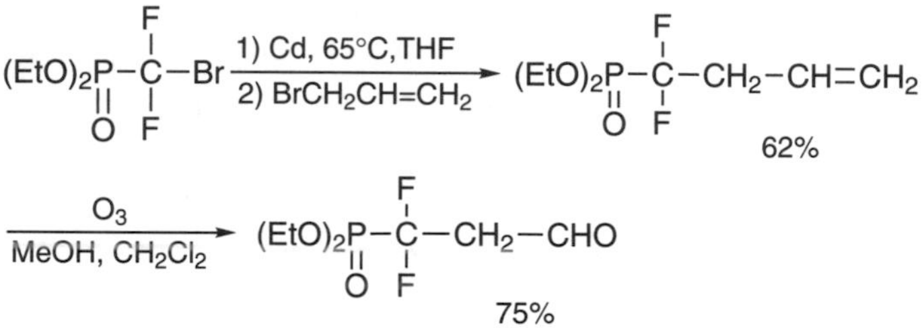

$$(EtO)_2\overset{\overset{F}{|}}{\underset{\underset{F}{|}}{\underset{||}{P}}}-C-Br \xrightarrow[\text{2) BrCH}_2\text{CH=CH}_2]{\text{1) Cd, 65°C, THF}} (EtO)_2\overset{\overset{F}{|}}{\underset{\underset{F}{|}}{\underset{||}{P}}}-C-CH_2-CH=CH_2$$
$$62\%$$

$$\xrightarrow[\text{MeOH, CH}_2\text{Cl}_2]{O_3} (EtO)_2\overset{\overset{F}{|}}{\underset{\underset{F}{|}}{\underset{||}{P}}}-C-CH_2-CHO$$
$$75\%$$

$$(5.28)$$

A Wacker-type PdCl$_2$- and CuCl$_2$-catalyzed oxidation by air of an aqueous solution of diethyl allylphosphonate gives an equimolar mixture of 2-formylethyl- and 2-oxopropylphosphonates.[185] The low-temperature ozonolysis of diethyl 1-cyano-3-butenylphosphonate affords diethyl 1-cyano-2-formylethylphosphonate, a key intermediate in the synthesis of a biologically active cyanocyclitol.[179]

5.1.4 DIALKYL 3-FORMYLALKYLPHOSPHONATES

5.1.4.1 Michaelis–Arbuzov Reactions

Despite the low reactivity of the carbon chain, the Michaelis–Arbuzov reaction is still used with success. For example, diethyl 3-formylpropylphosphonate can be prepared from triethyl phosphite and 4-bromobutyraldehyde diethyl acetal or 2-(3-bromopropyl)-1,3-dioxolane (Scheme 5.29). The latter reacts with triethyl phosphite at 110°C to give 83% of the cyclic acetal, which is hydrolyzed with 0.35 M HCl in dioxane at 100°C to give the diethyl 3-formylpropylphosphonate in nearly quantitative yield.[101] Similarly, diethyl *ortho*-formylbenzylphosphonate is obtained in 81% yield on heating triethyl phosphite and the unprotected *ortho*-(chloromethyl)benzaldehyde at 160°C.[186]

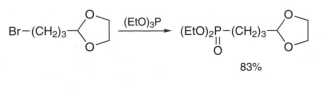

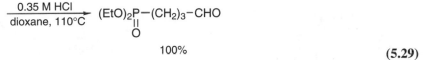

                                                                                    (5.29)

Unsaturated 3-formylphosphonates can be prepared by a Michaelis–Arbuzov reaction that employs 1,1-diethoxy-4-bromo-2-butene[187] as the masked formyl group. Reaction with triethyl phosphite at 135°C gives diethyl 4,4-diethoxy-2-butenylphosphonate in 50% yield. Deprotection is accomplished in 98% yield using a cold-saturated aqueous solution of tartaric acid.[187]

5.1.4.2 Phosphonomethylation

As previously described for $n = 2$, the ring-opening reaction of D-glycidaldehyde diethyl acetal at −80°C in THF by diethyl 1-lithiomethylphosphonate in the presence of $BF_3 \cdot Et_2O$ provides a convenient method for the incorporation of the formyl group at the carbon atom γ to the phosphonate group. With this methodology, diethyl 4,4-diethoxy-3-hydroxybutylphosphonate is isolated in 78% yield (Scheme 5.30).[188] Unmasking of the latent aldehyde functionality to give the diethyl 3-hydroxy-3-formylpropylphosphonate is accomplished by using 0.1 M HCl at 40°C.[172,188,189]

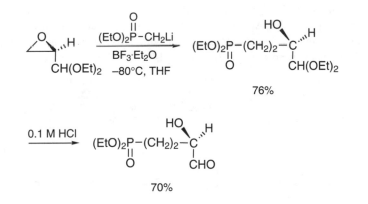

                                                                                    (5.30)

Synthesis of a variety of benzylphosphonates bearing electron-withdrawing groups (CO_2Et, CN, SO_2Me) at the α-carbon and a formyl group on the aromatic ring has been described. It involves the arylation of sodium diethyl 1-(ethoxycarbonyl)methylphosphonate with aryl halides in DMF or HMPA in the presence of CuI at 100°C (Scheme 5.31).[190] In the case of *ortho*-iodobenzaldehyde, after protection of the formyl group as *ortho*-(1,3-dioxolan-2-yl)iodobenzene and a coupling reaction with sodium diethyl 1-(ethoxycarbonyl)methylphosphonate, the expected diethyl α-(ethoxycarbonyl)-*ortho*-formylbenzylphosphonate is isolated in 75% yield after acid hydrolysis.[190]

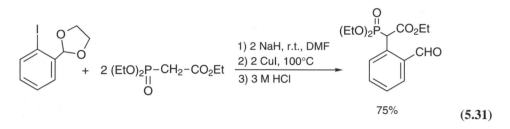

(5.31)

5.1.4.3 Michael Addition

The dicyclohexylammonium salt of diethyl 1-(hydroxycarbonyl)vinylphosphonate appears to be a readily accessible Michael acceptor. It reacts with aldehydes in C_6H_6 at 50°C to provide the desired adducts in satisfactory yields (70–75%). Chromatography using a strongly acidic ion-exchange resin (Dowex 50W) affords the corresponding aldehydes quantitatively (Scheme 5.32).[191]

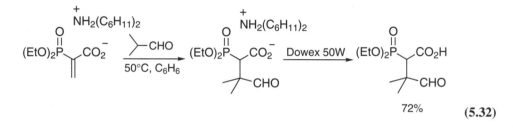

(5.32)

5.1.4.4 Anionic Formylation

Diethyl 3-formyl-3-methylallylphosphonate can be prepared using a carbanionic route from the readily available diethyl crotylphosphonate (Scheme 5.33).[192] Treatment with LDA (3 eq), silylation with Me_3SiCl (1 eq), and subsequent formylation with HCO_2Et give an oxoanion that, on acid hydrolysis, leads exclusively to diethyl (E)-3-formyl-3-methylallylphosphonate in 78% yield.[192]

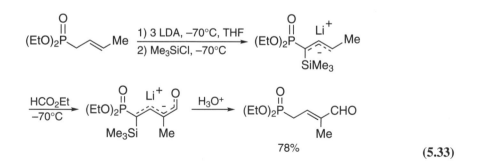

(5.33)

Diethyl 2-oxoalkylphosphonates are synthetically useful compounds that are known to give stable 1,3-dianions at ambient temperature. After generation using NaH and LDA (2 eq) at low temperature, 1,3-dianions react smoothly with HCO_2Et to provide stabilized water-soluble *bis*-enolates. Acidification with 4 M HCl produces diethyl 3-formyl-2-oxoalkylphosphonates in high yields (81–89%).[193] This sequential one-pot procedure offers a short and efficient route to a variety of functionalised diethyl 3-formylphosphonates from readily available starting materials (Scheme 5.34).[193] An alternative preparation of diethyl 3-formyl-2-oxopropylphosphonate from diethyl 2-oxopropylphosphonate involves a multistep route.[194]

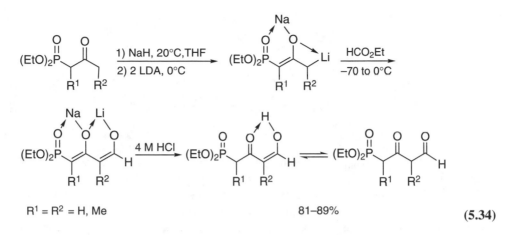

$R^1 = R^2 = H, Me$ 81–89%

$$(5.34)$$

The versatile diethyl 3-oxocyclobutylphosphonate is readily converted in high yield (83%) to the homologous diethyl 3-formylcyclobutylphosphonate by reaction with diethyl 1-lithio-1-isocyanomethylphosphonate in THF at low temperature and subsequent hydrolysis with 6 M HCl.[195]

5.1.4.5 Oxidation

Oxidation provides a general and efficient route to phosphonylated aldehydes. For example, diethyl 3-(*tert*-butyldimethylsilyloxy)-4-hydroxybutylphosphonate is converted by Swern oxidation to diethyl 3-(*tert*-butyldimethylsilyloxy)-3-formylpropylphosphonate in good yield.[196]

1-Acetoxy-4-chloro-2-butene reacts with trialkyl phosphites at 125–140°C to give the corresponding phosphoryl acetates in 54–81% yields (E : Z = 90 : 10).[197,198] By refluxing the acetates in EtOH or MeOH with a catalytic amount of TsOH, the alcohols are obtained smoothly in almost quantitative yield (95–97%). The Jones reagent proved to be the most useful for converting the alcohols into the corresponding aldehydes (65–68% yield).[198] Dialkyl 3-formylallylphosphonates are reasonably stable if stored under argon at −20°C. As an illustration of the above process, the preparation of diethyl 3-formyl-2-methylallylphosphonate is accomplished in 86% yield (E : Z = 55 : 45) from 1-bromo-2-methyl-4-acetoxy-2-butene (Scheme 5.35).[198]

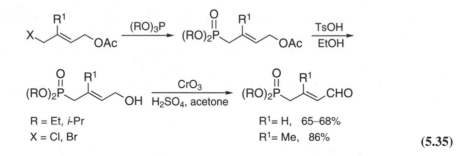

R = Et, *i*-Pr
X = Cl, Br

R^1= H, 65–68%
R^1= Me, 86%

$$(5.35)$$

A procedure involving the addition of ethyl diazoacetate to the double bond of diethyl allylphosphonate followed by conversion of the ester group into aldehyde via the alcohol has been developed (Scheme 5.36).[199] The CuSO₄-induced addition of ethyl diazoacetate to the readily available diethyl allylphosphonate in refluxing cyclohexane produces diethyl cyclopropylmethylphosphonate in moderate yield (30%) as a 1 : 3 mixture of *cis* and *trans* isomers. The ester is hydrolyzed to the acid and then treated with borane to give the alcohol. Subsequent oxidation of the alcohol using PCC (Scheme 5.36)[199] or DMSO/(COCl)₂/Et₃N[200] in CH₂Cl₂ affords the aldehydes in 83–98% yields.

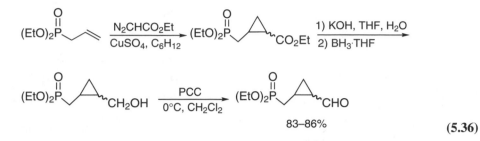

(5.36)

An attractive preparative procedure involves the oxidation of alcohols formed by hydroboration. Metallation of diethyl 1′-methoxy-4-fluorobenzylphosphonate at low temperature in THF and alkylation of the carbanion produced thereby with allyl bromide gives the required homoallylphosphonate. Elaboration of the double bond by a sequence involving, hydroboration, basic hydrolysis, and oxidation of the resulting alcohol provides the desired diethyl 1′-methoxy-1′-(3-oxopropyl)-4-fluorobenzylphosphonate (Scheme 5.37).[201]

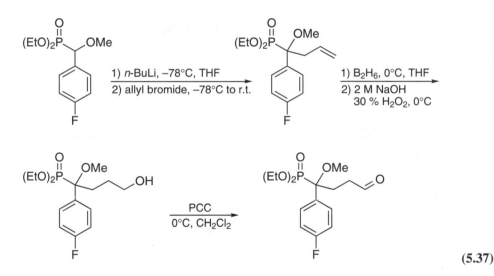

(5.37)

Phosphonylation of α^4-3-O-isopropylidene-α^5-pyridoxyl chloride via a Michaelis–Becker reaction, followed by deprotection with 1 M HCl and oxidation of the resulting primary alcohol with MnO$_2$ produces diethyl (4-formyl-3-hydroxy-2-methyl-5-pyridyl)methylphosphonate[202] (Scheme 5.38), an analogue of pyridoxal 5′-phosphate whose 5-position side chain has been replaced by a phosphonomethyl group. The alcohol oxidation step can be accomplished with a wide range of reagents, such as activated MnO$_2$ in CHCl$_3$ at room temperature (53%),[202] PCC in CH$_2$Cl$_2$ (83–86%),[199] or the Swern reaction (>95%).[203]

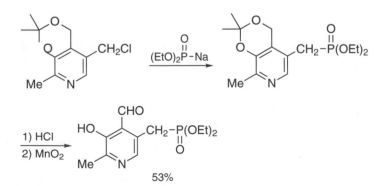

(5.38)

The phosphate–phosphonate rearrangement of phosphonylated 4-hydroxybenzyl alcohol provides the basis of an efficient procedure for the preparation of arylphosphonates bearing a formyl group. On treatment of dimethyl 4-[(*tert*-butyldimethylsilyloxy)methyl]phenyl phosphate with LDA in THF at −78°C, the aromatic ring undergoes metallation *ortho*- to the phosphate with concomitant migration of the phosphoryl group from oxygen to the carbon ring, giving the corresponding phenylphosphonate in 77% yield.[204] Subsequent benzylation of the phenolic OH and desilylation affords the primary alcohol, which is oxidized with activated MnO_2 in THF at room temperature to give the dimethyl 3-formylphenylphosphonate in 96% yield (Scheme 5.39).[204] The aldehyde can also be generated by reduction of the corresponding ester with DIBAL-H at −60°C. Any small amounts of the corresponding overreduced alcohol can be transformed quantitatively into aldehyde with activated MnO_2.[205]

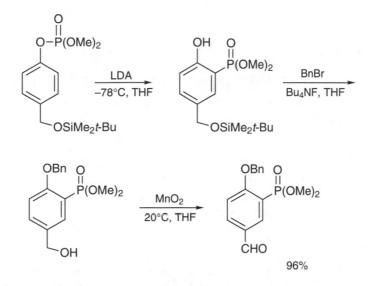

(5.39)

An attractive application of the oxidative cleavage of phosphonylated 1,2-diols with $NaIO_4$ is demonstrated in the synthesis of diethyl 3-formylalkylphosphonates (Scheme 5.40),[206] which is accomplished on a preparative scale using the 1,2:5,6-di-*O*-isopropylidene-α-D-glucofuranose. This mild method is advantageous for complex molecules. The phosphonomethyl group is first introduced with sodium tetraethyl methylenediphosphonate for the one-carbon chain extension followed by catalytic hydrogenation (H_2, Pd/C) and then oxidative cleavage with $NaIO_4$ to give the expected aldehyde in almost quantitative yield.[206]

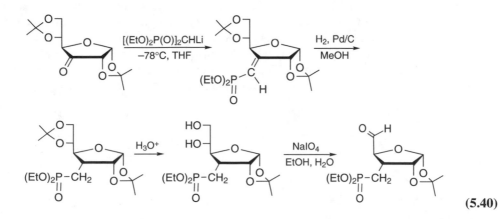

(5.40)

The oxidative cleavage of 1,2-diols with $NaIO_4$ appears to be a powerful method to prepare phosphonylated aldehydes with variable carbon chain lengths. It has been applied successfully to the synthesis of diethyl 1-(ethoxycarbonyl)-4-oxobutylphosphonate, a precursor in the preparation of glycinoeclepin A, in 70% yield (Scheme 5.41).[207]

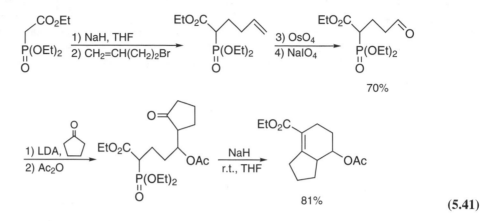

(5.41)

5.1.4.6 Ozonolysis

Diethyl 2-hydroxy-3-formylpropylphosphonate can be prepared in the masked form through the allylation of diethyl 1-formylmethylphosphonate using allyl bromide and zinc in H_2O-THF.[208,209] Allylation in a protic medium gives a cleaner reaction and better yield than the corresponding Grignard addition using allyl magnesium bromide. Ozonolysis of the resulting diethyl 2-hydroxy-4-pentenylphosphonate provides the expected aldehyde.[208,209]

5.1.4.7 Reduction

The configurationally unstable diethyl 3-(S)-tert-butoxycarbonylamino-3-formylphosphonate is prepared in 80% yield by low-temperature reduction of the corresponding methyl ester with DIBAL-H in toluene and subsequent hydrolysis with 1 M HCl.[210,211]

5.1.5 DIALKYL 4-FORMYLALKYLPHOSPHONATES

5.1.5.1 Michaelis–Arbuzov Reactions

The Michaelis–Arbuzov reaction is rarely used for the synthesis of diethyl 4-formylalkylphosphonate. However, it has been shown to provide straightforward access to diethyl 4-formylalkylphosphonates ($n = 2$) from unprotected aldehydes through the reaction of triethyl phosphite and bromoaldehydes. At 110°C, the Michaelis–Arbuzov reaction produces the diethyl 4-formylalkylphosphonate ($n = 2$, Scheme 5.42), but yields are much lower for cis-cyclopropylaldehydes than for the trans isomers (27% versus 62%).[199] This presumably reflects an unwanted reaction with the cis-disposed aldehyde group. The reaction has been extended to the preparation of the higher homologue ($n = 3$, Scheme 5.42, see Section 5.1.6.1).[199]

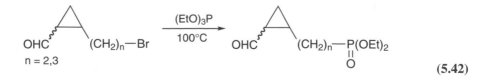

(5.42)

For carbohydrates, 5-phosphonylated-D-ribofuranoside is prepared by a Michaelis–Arbuzov reaction from the 5-iodo derivative and then deprotected to give the corresponding aldehyde.[212,213]

Several other reported preparative procedures include steps involving the phosphonylation of aromatic or heteroaromatic substrates containing a free formyl group. For example, the Michaelis–Arbuzov reaction between triethyl phosphite and the corresponding 4- or 6-methoxy-3-formylbenzyl chlorides at 170–180°C gives diethyl 4- or 6-methoxy-3-formylbenzylphosphonates in 83% or 74% yields, respectively.[214] Similarly, 3,5-*bis*(bromomethyl)benzaldehyde undergoes a double Michaelis–Arbuzov reaction with triethyl phosphite at 140°C to give the corresponding diphosphonate in quantitative yield.[215] Another example of the Michaelis–Arbuzov reaction applied to an unprotected aldehyde-containing halide was employed in the synthesis of 2-formylated 4-phosphonomethylthiophene.[216] Reaction between trimethyl phosphite and 5,5′-dibromo-4,4′-*bis*(bromomethyl)-2,2′-diformyl-3,3′-bithienyl gave the corresponding dimethyl 4-formylthienylphosphonate in 95% yield.[216]

The aromatic version of the Michaelis–Arbuzov reaction between trialkyl phosphites and aryl bromides in the presence of Ni(II) chloride has extensively been studied with a variety of substituted halobenzenes. It has been shown that electron-acceptor substituents, regardless of their position on the aromatic nucleus, facilitate reaction with P(III) compounds, whereas electron-donor substituents render the reaction more difficult. Substituents lying *ortho* to the halogen lower the reactivity of aryl bromides, probably because of steric hindrance.[217] Thus, the NiCl$_2$-catalyzed direct phosphonylation of *para*-bromobenzaldehyde with triisopropyl phosphite at 160°C affords diisopropyl *para*-formylphenylphosphonate in 54% yield (Scheme 5.43).[217]

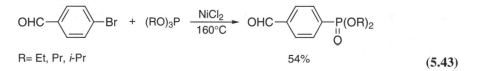

R= Et, Pr, *i*-Pr   54%   (5.43)

5.1.5.2 Anionic Phosphonylation

The preparation of dialkyl phenylphosphonates by the coupling of aryl bromides or iodides with dialkyl phosphites in the presence of Et$_3$N and a catalytic amount of Pd(PPh$_3$)$_4$ has been exploited for the synthesis of diethyl *para*-formylphenylphosphonate. The *para*-bromobenzaldehyde has to be protected as its 1,3-dioxolane to allow an efficient coupling with diethyl phosphite (Scheme 5.44).[218]

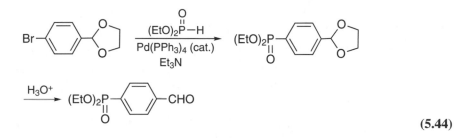

(5.44)

5.1.5.3 Michael Addition

Although seldom used, the conjugate addition of magnesium derivative of the 2-(2-bromoethyl)-1,3-dioxolane to an activated vinylphosphonate has occasionally been found to be a good method for chain-homologation of formylphosphonates (Scheme 5.45).[219–221] Unmasking of the protected carbonyl group with 1 M HCl produces the aldehyde.

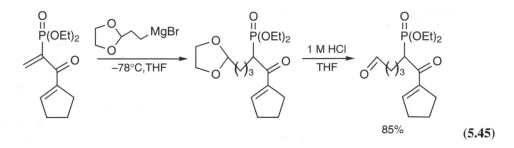

85%   **(5.45)**

Treatment of a mixture of diethyl 2-nitropropylphosphonate and acrolein by Triton B in MeCN at 0°C and subsequent hydrolysis provides diethyl 2-nitro-2-methyl-4-formylbutylphosphonate in 64% yield.[222]

5.1.5.4 Cyclopropanation

An innovative approach to diethyl 4-formylalkylphosphonates employs a metallic copper-induced decomposition of diethyl diazomethylphosphonate to generate a phosphonylated carbenoid that can be used to cyclopropanate functionalized olefins (Scheme 5.46).[223] However, competing carbene dimerization makes the process difficult to control, even in the presence of a large excess of olefin. Subsequent work showed that cuprous trifluoromethanesulfonate is a particularly effective catalyst for this reaction and that, in contrast to the original procedure, use of a two- to threefold excess of olefin is then sufficient to minimize competing carbene dimerization. By this CuOTf-catalyzed procedure, diethyl 2-(3,3-dialkoxy-2,2-dimethylpropyl)cyclopropylphosphonates are obtained in 53–73% yield from the corresponding unsaturated acetals.[223] The tolerance of the acetal group without competing cleavage reactions is noteworthy.

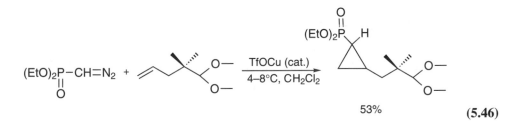

53%   **(5.46)**

5.1.5.5 Heck Coupling

Under modified Heck reaction conditions, coupling diethyl (*Z*)-2-iodovinylphosphonate with acrolein in MeCN at room temperature produces the diethyl (1*Z*,3*E*)-dienylphosphonate in 92% yield with high stereoselectivity (Scheme 5.47).[224]

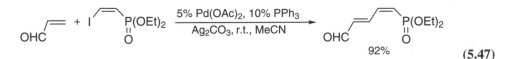

92%   **(5.47)**

5.1.5.6 Oxidation

Lactones have proved to be useful precursors in several synthetic schemes, especially those relating to the synthesis of diethyl 4-formylalkylphosphonates. When treated with diethyl 1-lithioalkylphosphonates, a γ-lactone undergoes ring opening to give a hydroxypentylphosphonate having a carbonyl group β to phosphorus (Scheme 5.48). Subsequent Swern oxidation of the terminal primary alcohol provides diethyl 4-formyl-2-oxobutylphosphonate.[225]

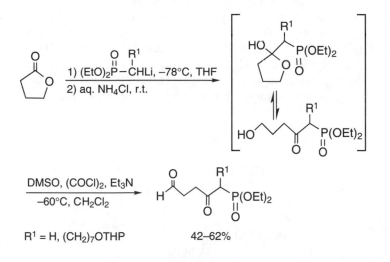

R¹ = H, (CH₂)₇OTHP   42–62%

$$\tag{5.48}$$

Treatment of a lactone with DIBAL-H in toluene at −78°C results in rapid and quantitative reduction to the lactol. The reduction is quite solvent-sensitive, and only toluene gives good results. Reaction of the lactol with the sodium tetraethyl methylenediphosphonate in a Horner–Wadsworth–Emmons reaction gives in good yields the hydroxyphosphonate with one-carbon chain extension of the aldehyde to an α,β-unsaturated phosphonate. Catalytic hydrogenation of the double bond α,β to phosphorus, followed by Swern oxidation of the hydroxyphosphonate, leads to the aldehyde (Scheme 5.49).[226] The primary alcohol can also be oxidized using the SO₃·Py complex in DMSO and Et₃N. This procedure allows the use of higher temperatures than the Swern protocol and provides the aldehyde in good yield. Thus, lactones appear to be valuable synthetic intermediates well suited for the preparation of a variety of phosphonylated aldehydes.[226]

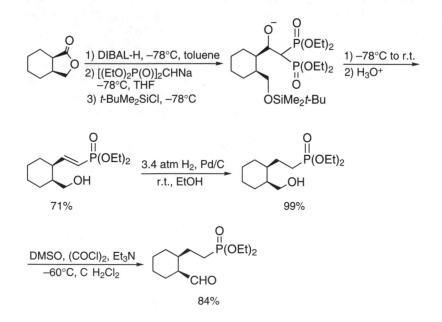

$$\tag{5.49}$$

A quite similar method is employed in the one-carbon chain extension of a 5′-phosphonate isostere of pyridoxal 5′-phosphate (Scheme 5.50).[227] The α^4,3-O-isopropylideneisopyridoxal is converted into an α,β-unsaturated phosphonate in yields up to 65% by condensation with sodium tetraethyl methylenediphosphonate. Selective hydrolysis with 10% HCO_2H at reflux gives the key diethyl 2-(3-hydroxy-4-hydroxymethyl-2-methyl-5-pyridyl)vinylphosphonate intermediate in 98% yield. After catalytic hydrogenation of the double bond using 5% Pd/C in EtOH, a comparison of methods for the oxidation of the alcohol has been made, including use of MnO_2 in water, CrO_3·Py, and DMSO-DCC. Finally, oxidation with activated MnO_2 in $CHCl_3$ was found to proceed smoothly to give the diethyl 2-(4-formyl-3-hydroxy-2-methyl-5-pyridyl)ethylphosphonate.[227]

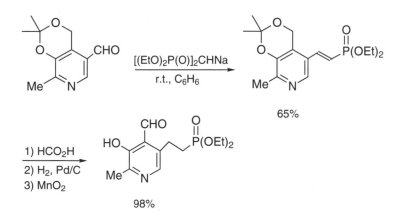

(5.50)

The vinylic and acetylenic analogues of pyridoxal 5′-phosphate are known. The former can be prepared as a *cis–trans* mixture by Peterson reaction between α^4,3-O-isopropylideneisopyridoxal and diethyl 1-lithio-1-(trimethylsilyl)methylphosphonate. Isolation of the *cis* derivative and subsequent deprotection of the alcohol followed by oxidation with MnO_2 in $CHCl_3$ leads to the diethyl 2-(4-formyl-3-hydroxy-2-methyl-5-pyridyl)vinylphosphonate.[228] The latter is prepared by reaction of the lithium ethynyl analogue of pyridoxal with diethyl chlorophosphate in Et_2O at low temperature [SNP(V) reaction]. After deprotection of the alcohol by heating with 10% aqueous TFA and oxidation with MnO_2, the diethyl 2-(4-formyl-3-hydroxy-2-methyl-5-pyridyl)ethynylphosphonate is obtained in 26% yield.[228]

The Swern oxidation is widely used to convert terminal alcohols into the corresponding aldehydes. Reaction of dimethyl 1-(methoxycarbonyl)methylphosphonate with butadienemonoepoxide in the presence of 5% $Pd(PPh_3)_4$ furnishes the dimethyl 5-hydroxy-1-methoxycarbonyl-3-pentenylphosphonate in a highly regio- and stereoselective manner in fair yields. After a Sharpless asymmetric epoxidation, the alcohol is submitted to a Swern oxidation, which produces dimethyl 1-methoxycarbonyl-3,4-epoxy-4-formylbutylphosphonate in good yield (Scheme 5.51).[229] Diethyl 4-hydroxy-2,3-epoxybutylphosphonate undergoes an analogous Swern oxidation to give the α,β-epoxyaldehyde in 65% yield.[230]

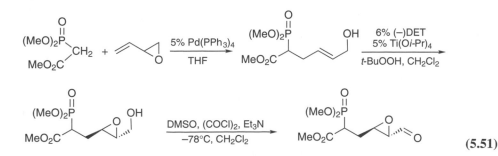

(5.51)

The terminal alcohol group can also be generated by a hydroboration–oxidation sequence. Thus, the reaction of diethyl (3-methylene-4-tetrahydropyranyl)-1,1-difluoroethylphosphonate with borane, followed by treatment with 30% aqueous H_2O_2 in the presence of AcONa, affords the primary alcohol, which can be converted into the corresponding aldehyde with DMSO/(COCl)$_2$/Et$_3$N.[231]

Phosphonylation and formylation reactions can be used successively in multistep procedures. For example, 1,3-*bis*(bromomethyl)benzene undergoes an initial phosphonylation through Michaelis–Becker[232] or Michaelis–Arbuzov[233] reactions, followed by oxidation of the benzylic bromide with 2-nitropropane and EtONa[232] (Scheme 5.52) or Fe/H$_2$O[233] to give the corresponding 3-formylbenzylphosphonate in satisfactory overall yield (30–37%).

$$31\% \qquad\qquad 97\% \qquad (5.52)$$

Diethyl 4-formylalkylphosphonate can be prepared in almost quantitative yield (98%) through the mild oxidation of phosphonylated 1,2-dideoxy-3,4:5,6-di-*O*-isopropyliden-D-arabinose with NaIO$_4$ at 0°C in H_2O or a H_2O/EtOH mixture.[234]

5.1.5.7 Ozonolysis

Diethyl β-keto-ω-alkenylphosphonates, prepared by γ-alkylation of the corresponding diethyl β-ketophosphonates with allyl or homoallyl bromide, undergo ozonolysis at low temperature in CH$_2$Cl$_2$ to give acyclic dicarbonyl phosphonates. Cyclization is accomplished with the aid of TsOH and Et$_3$N as catalysts.[235] Similarly, the diethyl 3-hydroxy-4-formylbutylphosphonate is prepared by allylation of diethyl 2-formylethylphosphonate using allyl bromide and zinc in a protic medium followed by ozonolysis and reductive workup.[208,209]

5.1.6 DIALKYL ω-FORMYLALKYLPHOSPHONATES

5.1.6.1 Thermal and Anionic Phosphonylation

1,2,3,4-Tetra-*O*-acetyl-6-deoxy-6-phosphono-β-D-glucopyranose is obtained in near-quantitative yield by heating 2,3,4-tri-*O*-acetyl-6-bromo-6-deoxy-α-D-glucopyranosyl bromide with a large excess of triethyl phosphite.[236] An extension of the methodology reported for the preparation of diethyl 4-formylalkylphosphonates ($n = 2$, Scheme 5.42) from an unprotected aldehyde functionality provides one of the few known methods for the transformation of bromoaldehydes into the corresponding diethyl 5-formylalkylphosphonates in 12–58% yields ($n = 3$, Scheme 5.42).[199]

The copper(I)-mediated coupling reaction of sodium diethyl 1-(ethoxycarbonyl)methylphosphonate with aryl halides in DMF at 100°C can be applied to *para*-iodobenzaldehyde protected as its 1,3-dioxolane by a similar procedure to those described in Scheme 5.31 to produce diethyl α-(ethoxycarbonyl)-*para*-formylbenzylphosphonate in 68% yield.[190]

Most the diethyl 5- and 6-formylalkylphosphonates incorporating an *ortho*-substituted aromatic ring having a masked aldehyde group are obtained by direct displacement reactions employing Michaelis–Becker attack at alkyl halides or mesylates. For example, *ortho*-methylbenzyl alcohol is elaborated to masked aldehyde in low yield by treatment with *n*-BuLi and alkylation of the resulting carbanion by 2-(2-bromoethyl)-1,3-dioxolane or bromoacetaldehyde dimethylacetal. The benzylic alcohol function is mesylated, then attacked with sodium diethyl phosphite to introduce the phosphoryl group. The aldehydes are unmasked by treatment with 3 M HCl in THF at 25°C (Scheme 5.53).[232]

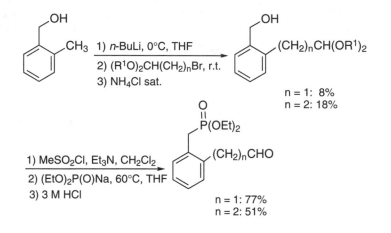

n = 1: 8%
n = 2: 18%

n = 1: 77%
n = 2: 51%

(5.53)

Significant progress has been achieved by using α-hydroxyphosphonates resulting from a Pudovik reaction, as illustrated by the synthesis of di-*tert*-butyl *para*-formylbenzylphosphonate (Scheme 5.54). Thus, a Pudovik reaction of terephthalaldehyde, semiprotected as a diethyl acetal, with acid-labile di-*tert*-butyl phosphite in the presence of basic alumina oxide allows the synthesis of the di-*tert*-butyl α-hydroxy-*para*-(diethoxymethyl)benzylphosphonate. Radical deoxygenation of the secondary alcohol can be achieved in a two-step process wherein a xanthate intermediate is reduced by *n*-Bu$_3$SnH. The diethyl acetal is again converted to the aldehyde (29%) by treatment with a saturated solution of 1 M HCl in CHCl$_3$.[237–241]

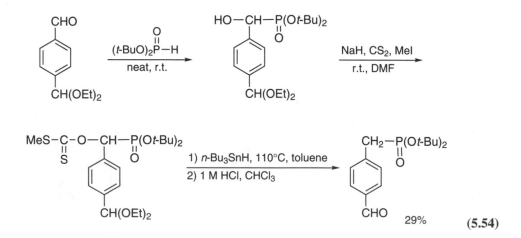

29%

(5.54)

5.1.6.2 Phosphonomethylation

Ring opening of a lactam, using dimethyl 1-lithiomethylphosphonate in THF at −78°C, generates the corresponding *N*-methyl enamine. This is converted on workup into dimethyl 6-formylphosphonate, which, in turn, gives a perhydroazulenone on treatment with *t*-BuOK/*t*-BuOH (Scheme 5.55).[242]

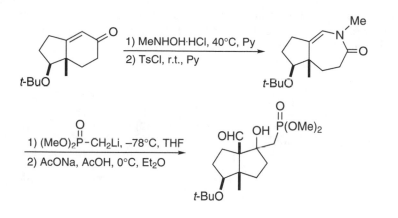

(5.55)

5.1.6.3 Michael Addition

An interesting approach to the central eight-membered ring of ceroplastol I is based on an intramolecular Horner–Wadsworth–Emmons reaction employing a diethyl 7-formylalkylphosphonate. The C-7 aldehyde, protected as a 1,3-dioxolane, is obtained in 20% yield by a Michael addition of an alkenyllithium to 2-(diethoxyphosphinyl)cyclopentenone in THF at −78°C. Hydrolysis of the acetal using H_2O and TsOH in refluxing acetone yields 89% of the phosphonylated ketoaldehyde (Scheme 5.56).[243]

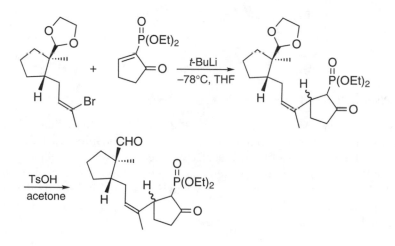

(5.56)

The same strategy has been used to prepare *trans* bicyclic enones.[244] The protected C_5 phosphonylated aldehyde is obtained in 84% yield by a CuBr·SMe$_2$-mediated Michael addition of the Grignard reagent derived from 4-chlorobutyraldehyde diethyl acetal to a 5-phosphonylated 2,3-dihydro-4-pyridone in THF. Subsequent room-temperature hydrolysis of the acetal using aqueous oxalic acid in THF affords a near-quantitative yield of the crude aldehyde, which undergoes an intramolecular Horner–Wadsworth–Emmons reaction under treatment with Et$_3$N/LiCl in THF at room temperature (89%).[244]

5.1.6.4 Electrophilic Substitution

A preparation of 5-formyl-2-[(diethoxyphosphinyl)methyl]furan in 40% yield by electrophilic substitution has been described. It involves the reaction between diethyl 2-furylmethylphosphonate and the complex DMF-POCl$_3$ at 50°C.[245] A related methodology allows the preparation of 5-substituted diethyl 2-furylmethylphosphonates (Scheme 5.57). In this case diethyl 2-furylmethylphosphonate is alkylated under similar conditions in modest yield by acrolein and crotonaldehyde at room temperature in the presence of catalytic H_2SO_4 in AcOH. The yields of the alkylation products increase with increasing stability of the carbenium ion formed by the protonation in the double bond of the unsaturated compound. Depending on the position of the phosphonomethyl group on the furan ring, this unusual methodology provides an access to 5-substituted diethyl 2- or 3-furylmethylphosphonates.[246,247]

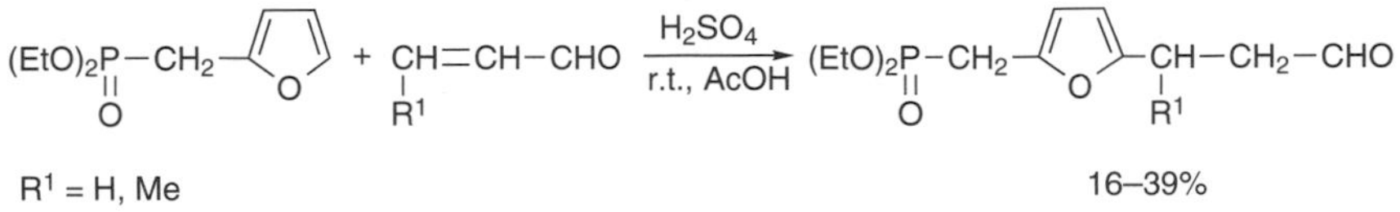

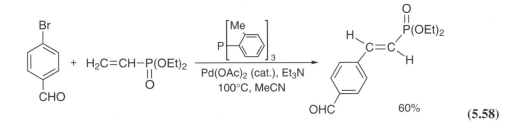

$$R^1 = H, Me \qquad\qquad\qquad\qquad 16\text{--}39\% \qquad\qquad (5.57)$$

5.1.6.5 Heck Coupling

The palladium-catalyzed arylation of diethyl vinylphosphonate using aryl bromides has been reported. The experimental procedure is simple, and the yields are moderate to good (Scheme 5.58). For example, with *para*-bromobenzaldehyde in MeCN at 100°C, diethyl *para*-formyl-*trans*-styrylphosphonate is formed exclusively in 60% yield.[248]

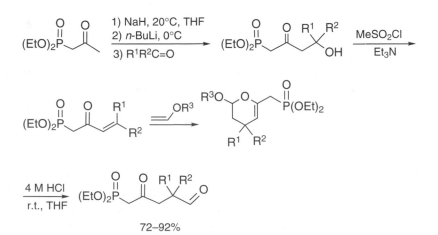

$$(5.58)$$

5.1.6.6 Diels–Alder Reactions

Diethyl 2-oxo-3-alkenylphosphonates, which are readily accessible from diethyl 2-oxopropylphosphonate, are useful heterodienes in Diels–Alder reactions. Cycloaddition reactions with vinyl ethers in C_6H_6 in a sealed tube at 85–130°C give satisfactory yields (57–88%) of dihydro-2*H*-pyrans. The products are isolated as a mixture of 2,4-*trans* and 2,4-*cis* isomers. When treated with 4 M HCl in THF, the dehydropyran hemiacetal moiety is readily hydrolyzed to provide diethyl 5-formyl-2-oxopentylphosphonates in excellent yields (72–92%, Scheme 5.59). When $R^1 = R^2 = H$, the aldehyde undergoes smooth cyclization leading to diethyl 2-oxocyclohexenephosphonate.[249]

$$72\text{--}92\% \qquad\qquad\qquad\qquad (5.59)$$

5.1.6.7 Oxidation

The EtAlCl$_2$-catalyzed ene reaction of dimethyl 1-(methoxycarbonyl)vinylphosphonate with alkenes at 25°C has been investigated in a series of experiments (Scheme 5.60).[250] Lewis acid catalysis offers significant advantages over the corresponding thermal ene reaction, which generally occurs at 200–300°C. These mild conditions have been applied successfully to the synthesis of dimethyl 6-formylalkylphosphonates. EtAlCl$_2$-mediated reaction of a primary alcohol with dimethyl 1-(methoxycarbonyl)vinylphosphonate in CH$_2$Cl$_2$ at 0°C gives the ene adduct in 40% yield. Oxidation of the alcohol with PDC gives a 68% yield of dimethyl 1-methoxycarbonyl-3-(2-propenyl)-6-formylheptylphosphonate, which is a useful candidate for an intramolecular Horner–Wadsworth–Emmons reaction.[250]

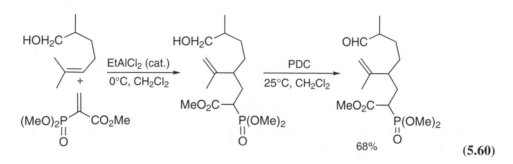

(5.60)

Phosphonates bearing an aldehyde group at C-5 or C-9 are important intermediates in the incorporation of a C-1–C-9 fragment into macrolide antibiotics (pikronolide,[251,252] carbonolide B,[253] erythromycin A[254]). They are prepared by treatment of a C$_7$ dialdehyde masked at one terminus as double bond or *tert*-butyldimethylsilyl ether with dimethyl 1-lithiomethylphosphonate at –80°C in THF to give β–hydroxyphosphonates in 74–79% yields. Subsequent oxidation (PDC/DMF[251,252] or TPAP/NMO/CH$_2$Cl$_2$[253,254]) produces the corresponding β–ketophosphonate in excellent overall yield. Unmasking of the terminal aldehyde at C-5 or C-9 is performed by oxidation of the primary alcohol with PCC/CH$_2$Cl$_2$[255] or TPAP/NMO[253] or of the double bond with OsO$_4$/NMO/acetone followed by cleavage with NaIO$_4$.[251,252]

5.1.6.8 Ozonolysis

An elegant synthetic procedure for the oxygen-mediated ring opening of furans bearing either a diethyl ethyl- or diethyl propylphosphonate moiety in the 2-position has been described (Scheme 5.61).[256] The reaction occurs under irradiation in MeOH at 0°C to provide the corresponding diethyl 5-formyl-3-oxo-4-pentenylphosphonates ($n = 2$) and 6-formyl-4-oxo-5-hexenylphosphonates ($n = 3$). After reduction of the double bond, the two derivatives are converted to diethyl 5-formyl-3-oxo-pentylphosphonates ($n = 2$) and 6-formyl-4-oxo-hexylphosphonates ($n = 3$) in excellent yields (92–96%).[256]

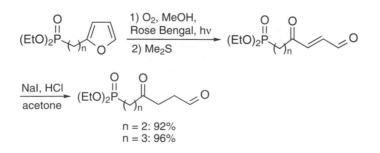

n = 2: 92%
n = 3: 96%

(5.61)

A similar methodology starting from lactones has been successfully developed. A mixed anhydride, obtained by the opening of a δ-lactone, after suitable protection, is treated with diethyl 1-copper(I)methylphosphonate to produce the corresponding β-ketophosphonate in good yield (65%). Ozonolysis of the terminal double bond then gives the diethyl 7-formylalkylphosphonate in 85% yield (Scheme 5.62).[257–259]

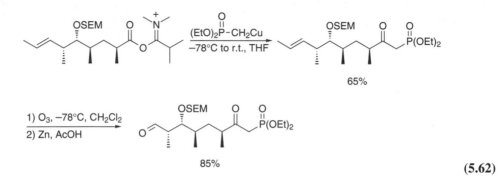

65%

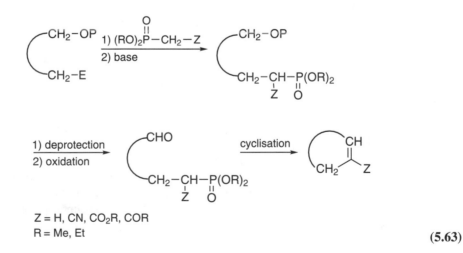

85%

(5.62)

5.1.6.9   General Strategy for the Preparation of Long-Chain Formylphosphonates

To date long-chain phosphonylated aldehydes ($n > 10$) have been prepared exclusively as precursors for the synthesis of 11- to 17-membered macrocycles via intramolecular Horner–Wadsworth–Emmons reactions (Scheme 5.63). This widely used cyclization step provides the best method for preparing macrocycles.

Z = H, CN, CO₂R, COR
R = Me, Et

(5.63)

Generally, the phosphonyl group is installed using carbanion-based chemistry. Methods for introducing the phosphonate group as part of a one-,[260–277] three-,[278,279] or four-carbon[194] fragment have been described.[280,281] In the C-1 case, stabilized (Z = CO₂R)[260–264,271,276,277] or nonstabilized (Z = H)[265–270,272–275] methylphosphonate carbanions have been made to react with chloride-,[276] bromide-,[277] iodide-,[260–266,271] aldehyde-,[272,273,275] or carboxyl-containing[267–270,274] electrophiles. The aldehyde group at the remote terminus is usually generated by oxidation of an alcohol or by hydrolysis of an acetal group.

5.2 SYNTHETIC APPLICATIONS OF FORMYLPHOSPHONATES

5.2.1 REACTIONS OF THE CARBONYL GROUP

5.2.1.1 Protection as Acetals

The reaction of 2,2-dialkoxyethylphosphonates with the corresponding diols produces moderate to good yields of phosphonylated 1,3-dioxolanes,[282,283] 1,3-dioxans,[284] and polymethylene acetals.[285] The acetal transfer is carried out by heating an equimolar mixture of the reactants at 140–150°C with simultaneous distillation of the liberated alcohol.[156] Heating a mixture of diethyl 2,2-diethoxyethylphosphonate and crotyl alcohol at 130°C gives the phosphonylated allyl vinyl ether, whose rearrangement to 1-formyl-2-methyl-3-butenylphosphonate is accompanied by extensive decomposition.[286]

5.2.1.2 Reduction

Reduction of phosphonylated aldehydes may be effected using H_2 over Adams catalyst (PtO_2),[107] $NaBH_4$,[152,287–289] or $NaBH_3CN$[152] to give the corresponding hydroxy derivatives in high yields.

5.2.1.3 Oxidation

Phosphonylated aldehydes are oxidized to the corresponding phosphonylated carboxylic acids by peracids[290] at room temperature or by alkaline $KMnO_4$[291] in quantitative yields. The same compounds can be obtained, even more simply, by combining the hydrolysis and oxidation steps. Thus, diethyl 2,2-diethoxyethylphosphonate can be first hydrolyzed with 2 M HCl and then oxidized with 78% peracetic acid to give diethyl 1-(hydroxycarbonyl)methylphosphonate in almost quantitative yield (93–97%) without isolation of the intermediate aldehyde.[290]

5.2.1.4 Enamines

The most important reaction class involving the formyl group of phosphonoacetaldehydes is the transformation into phosphonylated enamines. On reaction with both primary or secondary amines, dialkyl 1-formylmethylphosphonates generate the corresponding phosphonylated enamines.[292–295] These compounds are highly versatile synthetic intermediates that may be converted into the enamine of a more complex aldehyde or ketone on treatment with a variety of electrophiles under basic conditions. Dialkyl 1-formylmethylphosphonates react with primary amines to give a mixture of 2-iminoethylphosphonates and 2-aminovinylphosphonates. The latter are useful precursors for the two-carbon chain extension of aldehydes and ketones to masked α,β-unsaturated aldehydes (Horner–Wadsworth–Emmons reaction), or the preparation of 2-aminoalkylphosphonates. These aspects are discussed in Sections 5.2.4 and 5.2.7, respectively.

Phosphonylated enamines prepared from secondary amines are used in the preparation of aminoalkylphosphonates (see Section 5.2.7). They are also employed in the conversion of diethyl 1-formylmethylphosphonates into diethyl 2-oxoalkylphosphonates via diethyl 2-pyrrolidinovinylphosphonate intermediates. The process involves treatment of diethyl 1-formylalkylphosphonates and pyrrolidine in the presence of a drying agent, which provides the diethyl 2-pyrrolidinovinylphosphonate in 60–70% yield. On treatment under kinetic control with s-BuLi in THF at low temperature, metallation of these pyrrolidine enamines occurs β to phosphorus. The resulting anions react with typical electrophiles such as primary alkyl halides and chloroformates to give substituted enamine intermediates that generate diethyl β-ketoalkylphosphonates in low to good yields on acid hydrolysis (22–85%, Scheme 5.64). On warming to room temperature, the β-phosphonylated anions undergo isomerization to the thermodynamically favored α-phosphonylated counterparts.[296]

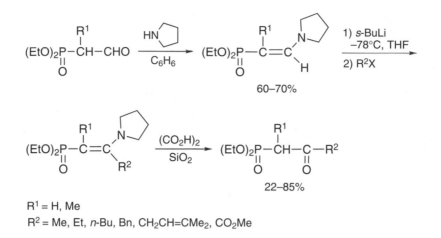

R^1 = H, Me
R^2 = Me, Et, *n*-Bu, Bn, CH$_2$CH=CMe$_2$, CO$_2$Me

(5.64)

β-Enaminophosphonates derived from primary amines react with *n*-BuLi in THF at 0°C to give delocalized anions whose reactions with electrophiles provide α-functionalized β-enaminophosphonates in high yields.[297]

5.2.1.5 Aldol Reactions

The Sn(OTf)$_2$-mediated aldol reaction of diethyl 1-formylalkylphosphonates with α-bromomethylketones provides the corresponding bromohydrins in 30–73% yields. On treatment with Et$_3$N, these compounds undergo cyclization to give 2,3-epoxy-4-oxoalkylphosphonates (81–91% yields, Scheme 5.65). These epoxides are readily isomerized into 2,4-dioxoalkylphosphonates on heating in toluene with catalytic amounts of Pd(PPh$_3$)$_4$ and dppe.[298]

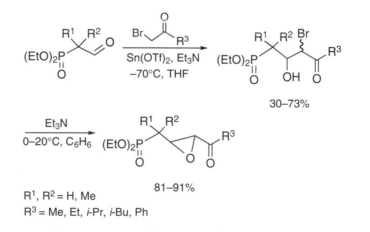

R^1, R^2 = H, Me
R^3 = Me, Et, *i*-Pr, *i*-Bu, Ph

(5.65)

Potassium nitroethanide attacks the formyl group of diethyl 1,1-difluoro-1-formylmethylphosphonate hydrate to give phosphonylated nitroalcohols,[299] but the purification of the crude products, obtained in almost quantitative yield, is difficult because of their low thermal stability. As solutions in nonpolar solvents, they exist as monomers, with the hydrogen of the hydroxy group being intramolecularly hydrogen-bonded to the phosphoryl oxygen.[299] The reaction has been extended to the preparation of diethyl 3-nitro-2-hydroxy-1,1-difluoropropylphosphonate by the KF-promoted reaction of diethyl 1-formyl-1,1-difluoromethylphosphonate hydrate with nitromethane at room temperature. A subsequent mesylation–elimination sequence leads to 3-nitro-1,1-difluoro-2-propenyl-phosphonate, which was isolated in 75% overall yield (Scheme 5.66).[129,130]

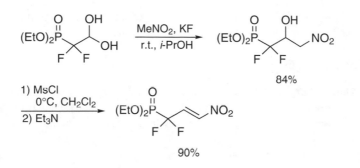

84%

90%   **(5.66)**

5.2.1.6 Wittig and Horner–Wadsworth–Emmons Reactions

Diethyl *trans*-2,3-epoxy-4-oxoalkylphosphonates are synthetically attractive intermediates that can be prepared by the epoxidation of diethyl (*E*)-4-oxo-2-alkenylphosphonates, which are available through the Wittig reaction of diethyl 1-formylalkylphosphonates with 2-oxoalkylidenetriphenylphosphoranes in toluene at reflux (Scheme 5.67).[120]

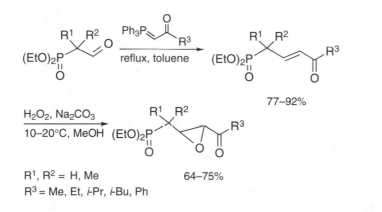

77–92%

R^1, R^2 = H, Me   64–75%
R^3 = Me, Et, *i*-Pr, *i*-Bu, Ph

(5.67)

The reaction of diethyl 1-formylmethylphosphonate with dimethyl *N*-(benzyloxycarbonyl)-1-amino-1-(methoxycarbonyl)methylphosphonate to give the β,γ-unsaturated phosphonate is promoted by a variety of bases (*n*-BuLi, NaH, *t*-BuOK). Under these conditions, *t*-BuOK gives the best results in terms of selectivity (*Z/E* = 33/1) and yield (78%, Scheme 5.68).[300]

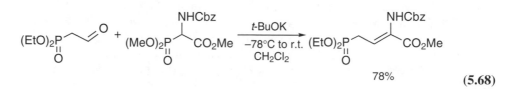

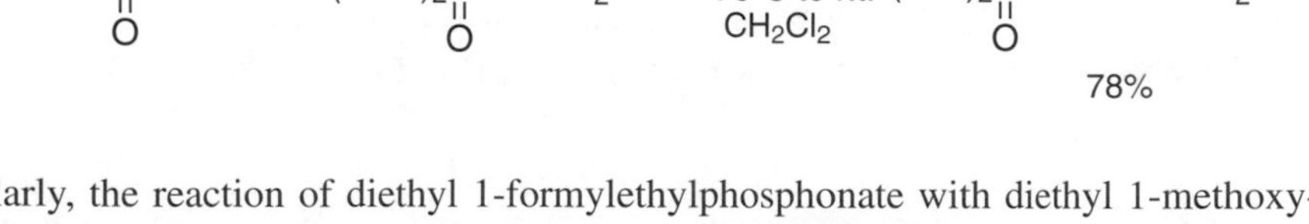

78%   **(5.68)**

Similarly, the reaction of diethyl 1-formylethylphosphonate with diethyl 1-methoxy-1-(methoxycarbonyl)methylphosphonate in the presence of NaHMDS in THF at room temperature affords the expected γ-methoxy-β,γ-unsaturated phosphonate in 39% yield.[301] Diethyl 1,1-difluoro-1-formylmethylphosphonate hydrate undergoes an olefination reaction with stabilized methylphosphonate carbanions under a variety of conditions to give 3-substituted diethyl 1,1-difluoro-2-propenylphosphonates in 45–78% yields.[129,130,132]

A report of the synthesis of thermodynamically stable (*E*)-vinylphosphonates from diethyl 1-formyl-1-phenylalkylphosphonates by a Horner–Wadsworth–Emmons-type reaction has appeared.[302] Sodium β-hydroxyphosphonates, generated by the reaction of sodium dialkyl phosphites with diethyl 1-formyl-1-phenylalkylphosphonates in THF at room temperature, undergo *syn*-elimination to provide the corresponding *trans*-dialkyl vinylphosphonates in excellent yields (82–94%).[302]

5.2.1.7 Addition Reactions

Diethyl 1-formylalkylphosphonates react with $NaCN/NaHCO_3$ in aqueous solution[303] or with trimethylsilyl cyanide in the presence of tributylstannyl cyanide as catalyst[304] to give the corresponding hydroxynitriles. These can be hydrolyzed to the hydroxycarboxylic acids.

5.2.2 REACTIONS OF THE METHYLENE GROUP

Reaction of dialkyl 1-formylmethylphosphonates with Cl_2/CCl_4 results in formation of 1-chloro or 1,1-dichloro derivatives (Scheme 5.69). This reaction occurs selectively as a function of the temperature: (a) in CCl_4 between −10°C and −15°C, the monochlorinated aldehydes are formed in 70–80% (appearance of a yellow color indicates the completion of monochlorination), whereas (b) raising the temperature to 0°C results in the replacement of both hydrogen atoms in the active methylene group (80–90%).[305–308] On warming to room temperature, the HCl coproduct can induce the cleavage of the P–C bond with formation of phosphorochloridates. The tendency towards cleavage of the P–C bond increases with the bulkiness of the ester groups attached to phosphorus.

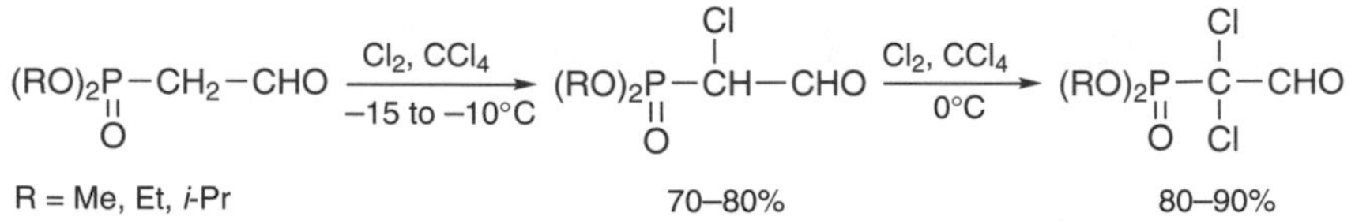

$$(RO)_2\overset{O}{\underset{\parallel}{P}}-CH_2-CHO \xrightarrow[-15 \text{ to } -10°C]{Cl_2,\ CCl_4} (RO)_2\overset{O}{\underset{\parallel}{P}}-\overset{Cl}{\underset{H}{C}}H-CHO \xrightarrow[0°C]{Cl_2,\ CCl_4} (RO)_2\overset{O}{\underset{\parallel}{P}}-\overset{Cl}{\underset{Cl}{C}}-CHO$$

R = Me, Et, *i*-Pr   70–80%   80–90%   **(5.69)**

The chlorination process is extended to the preparation of 1-substituted-1-formylmethylphosphonates with comparable yields (53–92%).[305–308] Contrary to the previously reported results,[309] the reaction of triethyl phosphite with trichloroacetaldehyde does not give diethyl 1,1-dichloro-1-formylmethylphosphonate (Michaelis–Arbuzov product); instead, the isomeric diethyl 2,2-dichlorovinylphosphate (Perkow product) is formed in 98% yield.[63,64,310] However, protection of the formyl group before reaction with triethyl phosphite provides the expected diethyl 1,1-dichloro-1-formylmethylphosphonate selectively in high yield (80%).[311]

Because the exploitation of the chemistry of unprotected phosphonylated acetaldehydes is handicapped by the sensitivity of the P–C bond in acidic media, a reliable alternative procedure for the preparation of dialkyl 1-chloro-1-formylmethylphosphonates from dialkyl 2-ethoxyvinylphosphonates has been developed. Room-temperature chlorination of 2-ethoxyvinylphosphonates with dry chlorine in CCl_4 followed by hydrolysis of the phosphonylated α,β-dichloro ethers with water at 50–60°C gives the expected monochlorinated aldehydes in 73–82% yields (Scheme 5.70).[312–315] A similar treatment with bromine in water at 0°C converts diisopropyl 2-ethoxyvinylphosphonate smoothly into diisopropyl 1-bromo-1-formylmethylphosphonate in 92% yield.[313]

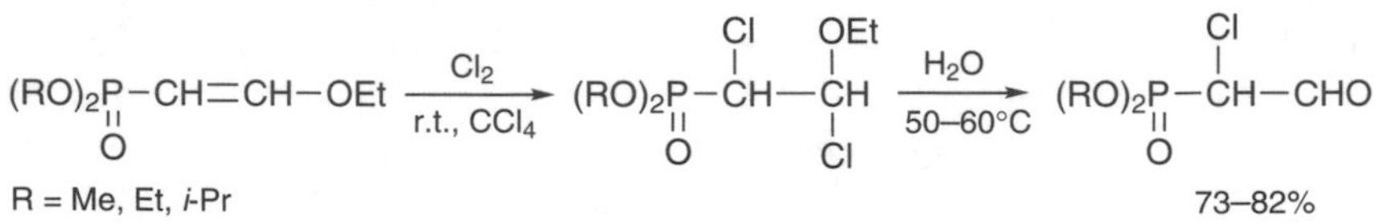

$$(RO)_2\overset{O}{\underset{\parallel}{P}}-CH{=}CH-OEt \xrightarrow[\text{r.t., } CCl_4]{Cl_2} (RO)_2\overset{O}{\underset{\parallel}{P}}-\overset{Cl}{\underset{}{C}}H-\overset{OEt}{\underset{Cl}{C}}H \xrightarrow[50-60°C]{H_2O} (RO)_2\overset{O}{\underset{\parallel}{P}}-\overset{Cl}{\underset{}{C}}H-CHO$$

R = Me, Et, *i*-Pr   73–82%   **(5.70)**

By a similar procedure, dialkyl 2-ethoxyvinylphosphonates are conveniently converted into dialkyl 1,1-dichloro-1-formylmethylphosphonates (phosphonochlorals) in aqueous solution in

80–90% yields by reaction with chlorine (Scheme 5.71).[315] These phosphonochlorals give stable crystalline hydrates in aqueous solution.

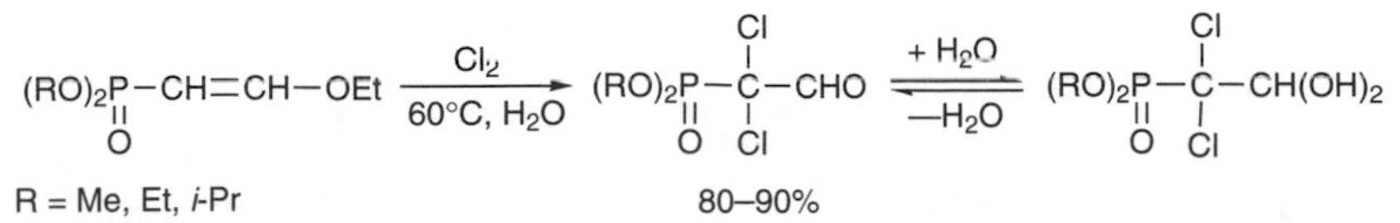

$$(RO)_2P(O)-CH=CH-OEt \xrightarrow[60°C, H_2O]{Cl_2} (RO)_2P(O)-\underset{\underset{Cl}{|}}{\overset{\overset{Cl}{|}}{C}}-CHO \underset{-H_2O}{\overset{+ H_2O}{\rightleftharpoons}} (RO)_2P(O)-\underset{\underset{Cl}{|}}{\overset{\overset{Cl}{|}}{C}}-CH(OH)_2$$

R = Me, Et, i-Pr   80–90%

(5.71)

Under carefully optimized experimental conditions, dialkyl 2,2-diethoxyethylphosphonates react with chlorine in refluxing CCl_4 under UV irradiation to provide dialkyl 1,1-dichloro-2,2-diethoxyethylphosphonates.[315] These products can also be prepared in quantitative yield by reaction of diethyl 1,1-dichloro-1-formylmethylphosphonate with orthoformic esters in the presence of sulfuric acid.[316] However, subsequent hydrolysis of these acetals gives modest yields of the corresponding aldehydes (30–40%).[315]

Systematic studies of the chemistry of the dialkyl 1-chloro- and 1,1-dichloro-1-formylmethylphosphonates show that they have high reactivity and a rich synthetic potential.[308] Treatment of dialkyl 1,1-dichloro-1-formylmethylphosphonate with $LiAlH_4$ in Et_2O gives dialkyl 1,1-dichloro-2-hydroxyethylphosphonates in 65–75% yields.[317]

Nitrosylation of dialkyl 1-chloro-1-formylmethylphosphonate with $NaNO_2/HCl$ below 0°C results in a hydrolytic rupture of the C–C bond in the initially formed 1-nitroso compound. This provokes loss of the formyl group and formation of dialkyl chloroformylphosphonate oximes in high yields (85–95%, Scheme 5.72). The reaction also occurs with the corresponding bromo aldehydes to give bromoformylphosphonate oximes in comparable yields. As the reaction temperature rises, loss of the halogen atom, chlorine or bromine, becomes competitive with the deformylation, and the α–nitroso derivative of 1-formylmethylphosphonate becomes the dominant product.[313,318–322]

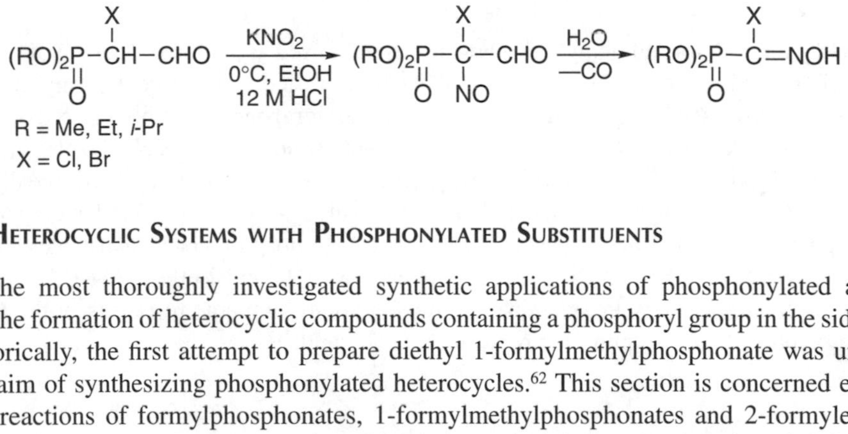

$$(RO)_2P(O)-\underset{\underset{}{}}{\overset{\overset{X}{|}}{C}}H-CHO \xrightarrow[\substack{0°C, EtOH \\ 12\ M\ HCl}]{KNO_2} (RO)_2P(O)-\underset{\underset{NO}{|}}{\overset{\overset{X}{|}}{C}}-CHO \xrightarrow[-CO]{H_2O} (RO)_2P(O)-\underset{\underset{}{}}{\overset{\overset{X}{|}}{C}}=NOH$$

R = Me, Et, i-Pr
X = Cl, Br

(5.72)

5.2.3 HETEROCYCLIC SYSTEMS WITH PHOSPHONYLATED SUBSTITUENTS

One of the most thoroughly investigated synthetic applications of phosphonylated aldehydes involves the formation of heterocyclic compounds containing a phosphoryl group in the side chain[323] and, historically, the first attempt to prepare diethyl 1-formylmethylphosphonate was undertaken with the aim of synthesizing phosphonylated heterocycles.[62] This section is concerned essentially with the reactions of formylphosphonates, 1-formylmethylphosphonates and 2-formylethylphosphonates. In most cases they are employed in their masked forms, as phosphonylated acetals or enamines, rather than in the free form.

Dialkyl 1,1-dialkoxymethylphosphonates can be condensed with amino-, hydroxy-, or mercapto-1,2-disubstituted benzene derivatives to produce a series of benzoheterocyclic phosphonates. The reactions are carried out under drastic conditions that involve heating a mixture of equimolecular amounts of the reactants at 180°C for several hours in a stream of nitrogen. Under these conditions, the reactions of the phosphonylated acetals go exclusively with cleavage of the C–O–C link in the acetal group. The yields of benzoxazole and benzothiazole derivatives are acceptable (68% and 30%),[283,324–326] but 2-benzimidazolylphosphonate is obtained in only 8% yield (Scheme 5.73).[327] All these syntheses of phosphonylated azoles go through intermediates azolines, which are converted onto azoles in a stream of dry oxygen.[326] The same reactions have also been performed with dialkyl 2,2-diethoxyethylphosphonates and 3,3-diethoxypropylphosphonates.[283,324–327]

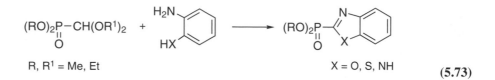

$$R, R^1 = Me, Et \qquad\qquad\qquad X = O, S, NH$$

(5.73)

In the first phosphonylated indole synthesis, diethyl 2,2-diethoxyethyl- and 3,3-diethoxypropylphosphonates were reacted with phenylhydrazine under classical Fischer indole synthesis conditions. The reaction resulted in the formation of 2-indolylphosphonates in low yield.[68,328] The reaction has been reinvestigated by a similar procedure using the unprotected 1-formylmethyl- and 2-formylethylphosphonates. Thus, diethyl 1-formylmethylphosphonate reacts with a wide variety of arylhydrazines in refluxing toluene in the presence of PPA to give the 3-indolyl-phosphonates in moderate yields (35–45%, Scheme 5.74).[329] With diethyl 2-formylethylphosphonate and phenylhydrazine chlorhydrate, the yields are comparable (4–45%). The substitution of the nitrogen atom is also important because the use of N-benzylated arylhydrazines gives superior results (28–83%).[329]

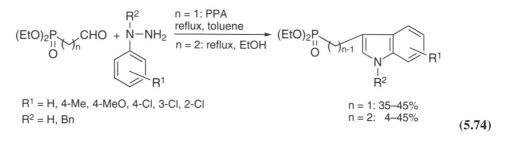

R^1 = H, 4-Me, 4-MeO, 4-Cl, 3-Cl, 2-Cl
R^2 = H, Bn

n = 1: 35–45%
n = 2: 4–45%

(5.74)

Diethyl 1-chloro-1-formylmethylphosphonate has been used as precursor for several heterocycle syntheses via phenylhydrazone intermediates. Treatment of diethyl 1-chloro-1-formylmethylphosphonate hydrazones with benzylamine gives N^3-benzylamidrazones, which can be oxidized with H_2O_2, $KMnO_4$, or Ag_2O in a biphasic medium to 3-phosphonylated-1,2,4-triazoles in 40–80% yields.[330] Similarly, diisopropyl 1-chloro-1-formylmethylphosphonate oxime affords the diisopropyl phosphonoacetonitrile oxide when reacted with Et_3N or $NaHCO_3$. This compound undergoes dipolar cycloaddition with terminal acetylenes at 20°C to give 3-phosphonylated isoxazoles in good to excellent yields (69–94%, Scheme 5.75).[100,331] The reaction of diethyl 2-ethoxy-1-(*ortho*-formylphenyl)vinylphosphonate with hydroxylamine in EtOH at room temperature results in the formation of 4-(diethoxyphosphinyl)isoquinoline N-oxide in 85% yield.[332]

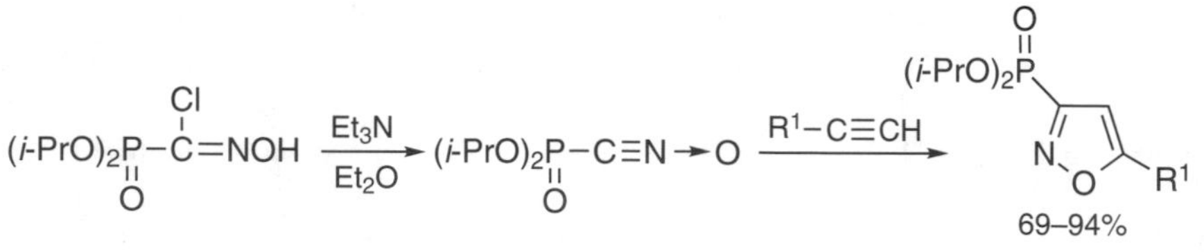

$$(i\text{-PrO})_2\overset{\displaystyle Cl}{\underset{\displaystyle O}{P}}-C=NOH \xrightarrow[Et_2O]{Et_3N} (i\text{-PrO})_2\overset{}{\underset{\displaystyle O}{P}}-C\equiv N\rightarrow O \xrightarrow{R^1-C\equiv CH}$$

69–94%

(5.75)

Phosphonylated enamines having an electron-withdrawing functional group (CHO, COMe, COPh, CO_2Et) in α-position are effective precursors for phosphonylated heterocycles. They undergo nucleophilic cyclization reactions with compounds containing labile hydrogen atoms, and this reactivity provides the basis for a new synthesis of heterocycles. The dimethylamino group of phosphonylated enamines can be selectively and totally transaminated by treatment with symmetric or disymmetric hydrazines, guanidine, acetamidine, or methylisothiourea in refluxing EtOH (Scheme 5.76), with the corresponding phosphonylated pyrazolones (R^1 = OH, R^2 = H), pyrazoles (a,b) and pyrimidines (c,d) being isolated in excellent yield (90–95%).[126,127] Dialkyl 1-formyl-1-(ethoxycarbonyl)methylphosphonates can also be used as precursors for phosphonylated pyrazolones on reaction with hydrazines.[333] Similarly, 1-phenyl-5-(diethoxyphosphinyl)pyrazoles can be readily prepared from diethyl 1-chloro-2-formylvinylphosphonate and phenylhydrazine.[334,335]

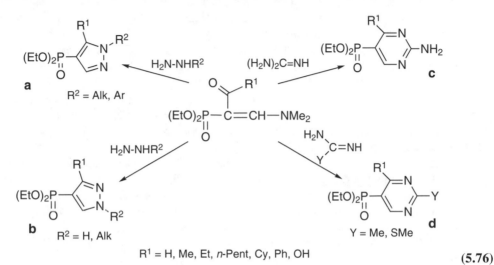

(5.76)

The report of an efficient one-pot procedure for the preparation of 3-indolylphosphonate in 88% yield using diethyl 2-dimethylamino-1-(2-nitrophenyl)vinylphosphonate is of special interest (Scheme 5.77). After reduction, the presence of a nitro group in the *ortho* position provokes a spontaneous transamination reaction. This provides a straightforward approach to 3-indolylphosphonate derivatives from easily available benzylphosphonates.[126]

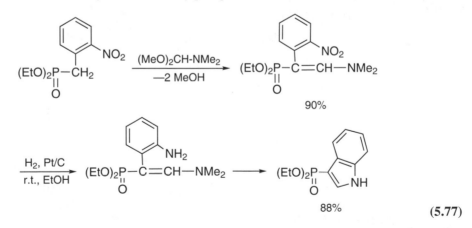

(5.77)

Addition of tosylazide to diethyl 1-formylalkylphosphonate yields the corresponding triazoline which evolves to diethyl diazomethylphosphonate (Scheme 5.78).[336,337]

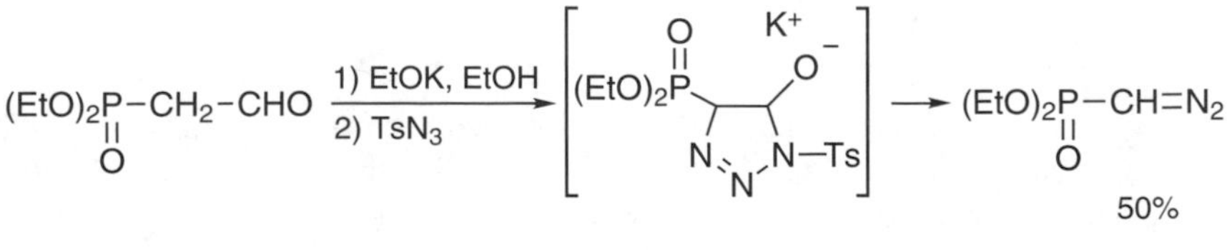

$$(EtO)_2P-CH_2-CHO \xrightarrow[\text{2) TsN}_3]{\text{1) EtOK, EtOH}} \left[(EtO)_2P \right] \longrightarrow (EtO)_2P-CH=N_2$$

50%

(5.78)

5.2.4 Horner–Wadsworth–Emmons Reactions

The Horner–Wadsworth–Emmons reaction is among the most important reactions for effecting a two-carbon chain elongation of a carbonyl functional group.[338–344] It can be described as a general synthetic protocol for the conversion of aldehydes and ketones into α,β-unsaturated carbonyl compounds and their derivatives.[345]

The first example of preparation of an α,β-unsaturated aldehyde from unprotected 1-formyl-methylphosphonate involved the rather reactive substrate hexafluoroacetone. Thus, diethyl 1-formylmethylphosphonate reacts with hexafluoroacetone in *n*-Bu₂O at room temperature in the presence of *t*-BuOK to give the 3,3-*bis*(trifluoromethyl)propenal in 54% yield.[346,347] Another example describes the formation of an α,β-unsaturated aldehyde in 62% yield from unprotected 1-formylmethylphosphonate under Masamune conditions (LiCl, DIEA, MeCN, room temperature).[348]

A number of attractive phosphonate-based reagents have been developed for preparing α,β-unsaturated aldehydes from carbonyl compounds. They involve the formation of the carbon–carbon double bond before unmasking the new carbonyl group. The first reagent used for the formylolefination of aldehydes and ketones, diethyl 2-(cyclohexylimino)ethylphosphonate, was prepared in 60–74% yield by reaction of diethyl 1-formylmethylphosphonate with cyclohexylamine in MeOH and distilled in the presence of K₂CO₃ (Scheme 5.79).[103,349–352] The phosphonate carbanion is generated by treatment of diethyl 2-(cyclohexylimino)ethylphosphonate with NaH in THF at 0°C, under conditions that allow the usual equilibrium between the phosphonylated enamino and imino forms, and then allowing the product to react at room temperature with a variety of cabonyl compounds. Biphasic hydrolysis of the resulting α,β-unsaturated aldimines is effected with dilute aqueous oxalic acid (0.57 M) or with an AcOH-AcONa buffer and SiO₂/weak acid. The use of concentrated oxalic acid solution (1–1.3 M) is responsible for isomerization of the double bond. Satisfactory yields of (*E*)-α,β-unsaturated aldehydes are usually obtained.[102,349–358] It is noteworthy that steroidal 3- and 17-ketones produce the (*E*)-α,β-unsaturated aldehydes in good to excellent yields (65–84%).[351,356] Unfortunately, because this important reagent is not commercially available and requires a multistep preparation, its full potential has not been exploited, and the conversion of α,β-unsaturated nitriles into α,β-unsaturated aldehydes using DIBAL-H is generally preferred (see Chapter 6, Section 6.2.4.1).

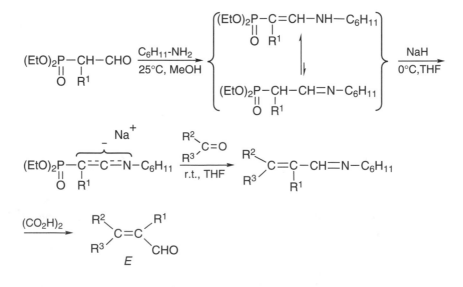

$$(5.79)$$

Several variations on the formylolefination reaction using protecting groups for aldehydes such as oxazine,[105] semicarbazone,[359,360] *tert*-butylimine,[361] 1,1-dimethylhydrazone,[362–364] *para*-toluenesulfonylhydrazone,[365,366] and benzylimine[81] have been reported (Figure 5.1).

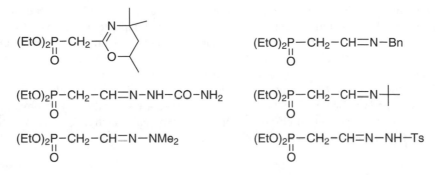

FIGURE 5.1

Diethyl 1-formylmethylphosphonate has recently been used as an effective acetaldehyde synthon for the preparation of a tricyclic quinoline. The reaction proceeds efficiently through the formation of an aromatic enamine, which undergoes ring closure under basic conditions (Scheme 5.80).[367]

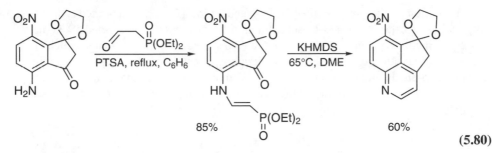

$$(5.80)$$

The semicarbazone[359,360] and *para*-toluenesulfonylhydrazone[365,366] of diethyl 1-formylmeth-ylphosphonate can be metallated with NaH in THF or DME for reaction with carbonyl compounds at room temperature, whereas the oxazine[368] and dimethylhydrazine[362–364] derivatives require met-allation with *n*-BuLi in THF at low temperature. The oxazine protecting group is removed by sequential treatment with NaBH₄ and aqueous oxalic acid at reflux.[368] The *para*-toluenesulfonyl-hydrazone participates in cyclization reactions with the release of sodium *para*-toluenesulfinate to produce pyrazoles (Scheme 5.81).[365,366]

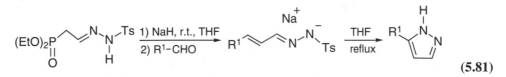

$$(5.81)$$

The dimethylhydrazine derivative of diethyl 1-formylmethylphosphonate, on reaction with *O*-propargyl-2-hydroxybenzaldehyde, gives an azadiene that, on heating in mesitylene at 138–162°C for several hours, participates directly in an intramolecular [4 + 2] cycloaddition (Scheme 5.82).[362]

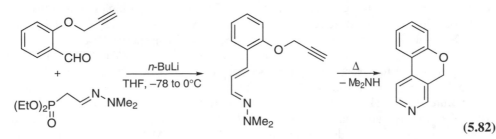

$$(5.82)$$

An alternative route for preparing (*E*)-α,β-unsaturated aldehydes relies on the nucleophilic attack of lithiated *N-tert*-butylacetaldimine on a chlorophosphate in an SNP(V) reaction (Scheme 5.83).[361] In this attractive one-pot procedure, the imine is deprotonated with LDA (2 eq) at low temperature and then allowed to react with diethyl chlorophosphate to generate the lithiated phosphonate coupling partner directly. Addition of the carbonyl reagent and workup with aqueous oxalic acid solution affords (*E*)-α,β-unsaturated aldehydes in good overall yields (53–94%). When a mixture of (*E*)- and (*Z*)-isomers is obtained, the isomers require separation by chromatography.[369] This sequential one-pot procedure eliminates the need to prepare diethyl 2,2-diethoxyethylphosphonate, 1-formylmethylphosphonate, and the phosphonylated aldimine.[361]

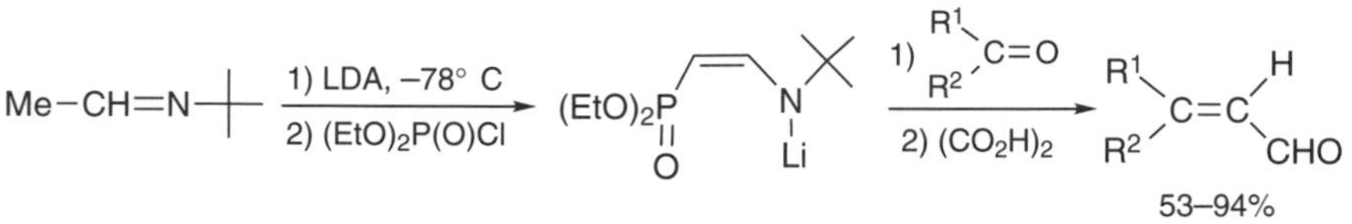

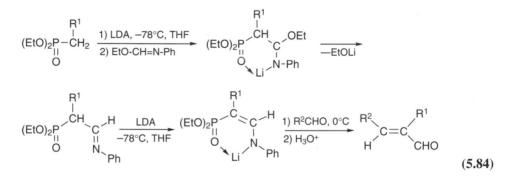

$$\text{(5.83)}$$

In addition to the previous methods for generating (*E*)-α,β-unsaturated aldehydes, a one-pot preparation of α,β-unsaturated-α–substituted aldehydes has been reported (Scheme 5.84).[370] The procedure involves the intermediacy of lithiated α,β–enaminophosphonates, which are prepared by nucleophilic addition of diethyl 1-lithioalkylphosphonates to ethyl *N*-phenylformimidate.[371] The lithiated α,β-enaminophosphonate is reacted with aliphatic or aromatic aldehydes to produce α,β-unsaturated-α-substituted imines, which can be efficiently converted to α,β-unsaturated-α-substituted aldehydes by acid hydrolysis in yields ranging from 45% to 78%.[370]

$$\text{(5.84)}$$

The intramolecular version of the Horner–Wadsworth–Emmons methodology is of significant synthetic importance. It is the method of choice for preparing five-,[219,221] six-[271,274,372,373] and seven-membered[242] ring systems as well as macrocyclic ring systems, including rings containing 11,[270,277] 12,[280,281] 14,[260–267,279] 15,[268,278] 16,[268] and 17[194,275,278] atoms. High-dilution procedures are usually required to obtain satisfactory yields in the preparation of large rings. The use of mild base (LiCl/DBU) or crown ether catalysts has also been shown to be of importance. To demonstrate the utility of this technique, some typical macrocyclizations that provide precursors of biologically active compounds (with conditions in brackets) are listed below: 11-membered ring (–)-bertyadionol (NaH/toluene/42°C, 28–32%),[270] stolonidiol (DBU/LiCl/18-crown-6/MeCN/room temperature, 69%),[277] 12-membered ring (+)-cleomeolide (K$_2$CO$_3$/18-crown-6/toluene/room temperature, 38%),[280,281] 12-membered ring pseudopterane and 14-membered ring furanocembrane systems (LiCl/DBU/MeCN/room temperature, 50% and 42%),[264] 14-membered ring (–)-asperdiol and (+)-desepoxyasperdiol (LiCl/DBU/MeCN/room temperature, 61% and 30%),[260,261] 14-membered ring (±)-methyl ceriferate-I (NaH/DME/80°C, 24%),[265,266] 14-membered ring anisomelic acid (LiCl/DBU/MeCN/room temperature, 71%),[262,263] 14-membered ring cembratrienediols and thunbergols (LiCl/DBU/MeCN/40°C, 63%),[279] 14-membered

ring macrocyclic trienone (LiCl/DBU/MeCN/room temperature, 63%),[267] 15-membered ring mus-
cone precursor (NaH/DME/high dilution, 50%,[268] K_2CO_3/18-crown-6/C_6H_6, 65°C, 35–65%[269]),
17-membered ring civetone precursor (t-BuOH/H_2O/KHCO$_3$/reflux, 56%),[270] and 17-membered
ring lankacidin (K_2CO_3/18-crown-6/toluene/100°C, 37%).[194] The synthesis of (−)-cylindrocyclo-
phane A via a double Horner–Wadsworth–Emmons reaction has been reported in 53% yield using
DBU/LiCl in MeCN at room temperature.[276] The stereochemistry, as well as the ease of the
intramolecular reaction, is largely determined by the nature of the carbon chain that forms the ring.

5.2.5 PHOSPHONOACETALDEHYDE IN NATURE

Much of the work in this area has been performed on *Tetrahymena*, where it can be shown that
phosphoenolpyruvate (PEP) is the immediate precursor of phosphonates and that its decarboxylation
product, phosphonoacetaldehyde (PAAl), is the direct precursor of 2-aminoethylphosphonate (AEP)
(Scheme 5.85).[374–381] β-Phosphonoalanine (PAla) is a side product in the synthesis and apparently
is not directly converted to AEP. The radioactive carbon atom in phosphonoenol [3–^{14}C]pyruvate
is incorporated into the phosphonate carbon atom in 2-aminoethylphosphonic acid (AEPA), con-
firming that, in *Tetrahymena*, an intramolecular rearrangement of PEP takes place during the
biosynthesis of AEP. Because incorporation of label into AEP is inhibited more by PAAl than PAla,
this rearrangement may not be the main biosynthetic pathway to AEP. The phosphonopyruvate
(PnPyr) formed by the intramolecular rearrangement of PEP probably undergoes decarboxylation
to phosphonoacetaldehyde before amination to AEP. The latter is then incorporated into phospho-
nolipids. Since the discovery of PEP phosphonomutases, which govern P–C bond formation in
Tetrahymena or *Streptomyces hygroscopicus*, new systems have been reported.[382] Another enzyme
[phosphonoacetaldehyde hydrolase (phosphonatase) from *Bacillus cereus*] catalyzes the dephosphonylation reaction of PAAl to inorganic phosphate and acetaldehyde. This involves activation
towards P–C bond cleavage by Schiff base formation between PAAl and an active site lysine in
phosphonatase.[383]

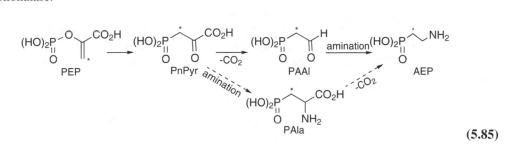

$$(5.85)$$

Fosfomycin, a clinically used antibiotic, is biosynthetically derived from a P–C_2 unit, probably
phosphonoacetaldehyde, and an L-methionine methyl group (Scheme 5.86). Phosphoenolpyruvate
(PEP) is the most likely precursor of carbons 1 and 2 of fosfomycin.[384–386]

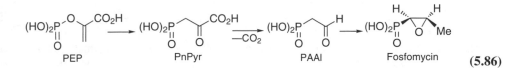

$$(5.86)$$

5.2.6 REDUCTIVE AMINATION

Because phosphonoacetaldehyde is an ideal precursor for the generation of AEP, an attractive
procedure for the preparation of 2-aminoalkylphosphonic acids from diethyl 1-formylalkylphos-
phonates has been developed using reductive amination of the carbonyl in the presence of NaBH$_3$CN

(Scheme 5.87). The method appears to be fairly general and may be used for the preparation of diethyl 2-aminoalkylphosphonates containing primary, secondary, or tertiary amino groups simply by treating the carbonyl compound at room temperature with primary or secondary, aliphatic or aromatic amines[77,387] and protected diamines.[388] When AcONH$_4$, as ammonia source, is subjected to the reaction, it requires lengthy reaction times and produces a mixture of primary aminophosphonate (7%) and imino*bis*(ethylphosphonate) (30%).[389] The primary aminophosphonate is more nucleophilic than ammonia, and the attack of the carbonyl group by the generated aminophosphonate represents a competing side reaction. Hydrolysis of diethyl aminoalkylphosphonates with 8 M HCl followed by purification using ion-exchange resins (Amberlite IRA 410) provides pure 2-aminoalkylphosphonic acids in moderate to good yields (40–81%).

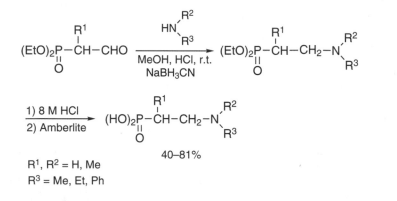

$$R^1, R^2 = H, Me$$
$$R^3 = Me, Et, Ph$$

(5.87)

The reductive amination process has been modified for the preparation of *N*-unsubstituted aminophosphonic acids (Scheme 5.88).[390] Thus, the use of benzylamine in reaction with phosphonylated aldehydes increases chemospecificity, reaction rates, and yields. Aminobenzylphosphonates are isolated in yields of around 85% and then converted to aminoalkylphosphonates. Subsequent catalytic hydrogenation followed by acid hydrolysis and purification gives aminoalkylphosphonic acids.[390] Together with its modifications, reductive amination is one of the most useful and general methods available for the preparation of a wide variety of aminoalkylphosphonic acids from phosphonylated aldehydes.

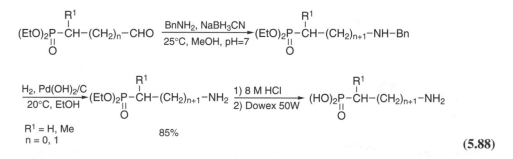

$$R^1 = H, Me$$
$$n = 0, 1 \qquad 85\%$$

(5.88)

The preparation of AEP via the reduction of oxime of diethyl 1-formylmethylphosphonate is of synthetic importance. The oxime is prepared from diethyl 1-formylmethylphosphonate and hydroxylamine hydrochloride in EtOH/Py, treated with (AcO)$_2$O in AcOH, and then hydrogenated using 5% Pd/C. After acid hydrolysis using 6 M HCl and purification with ion-exchange resin (Dowex 50W), AEP is isolated in 43% yield.[291]

The synthesis of 1-aminoalkyldiphosphonic acids from phosphonylated aldehydes employs the thioureidoalkylphosphonate method (Scheme 5.89). Phosphonylated aldehydes (*n* = 1, 2, 3), react with *N*-phenylthiourea and diphenyl phosphite in glacial AcOH to give the corresponding thioure-idoalkylphosphonates in good yields. Hydrolysis in a refluxing mixture of AcOH/HCl provides 1-aminoalkyldiphosphonic acids in satisfactory overall yields (50–70%). These compounds are purified by chromatography using a strongly acidic ion-exchange resin (Dowex 50W).[101,391]

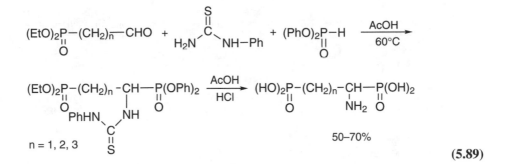

$$(5.89)$$

5.2.7 Aminocarboxylation (Strecker and Related Reactions)

Receptors for glutamic and aspartic acids[392,393] have been implicated in the pathology of several neurologic and neurodegenerative illnesses including epilepsy, cerebral ischemia, hypoglycemia, and Huntington's, Alzheimer's, and Parkinson's diseases. The growing number of these disorders, in which excitatory amino acids (EAA) have been implicated, has resulted in considerable attention focused on the development of excitatory amino acid neurotransmission antagonists. Compounds that act selectively and competitively at the subclass of EAA receptor selectively activated by *N*-methyl-D-aspartate (NMDA) have received particular attention.[394] Several amino phosphonocarboxylic acids have been characterized as being potent and selective competitive NMDA antagonists[394–399] (Figure 5.2) and have been found to be effective against neuronal damage. With these considerations in mind, the Strecker and Bucherer–Bergs reactions for the synthesis of ω–amino-ω-carboxyalkylphosphonic acids have been applied to a variety of phosphonylated aldehydes.

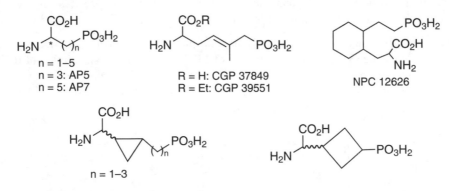

FIGURE 5.2

For their synthesis, phosphonylated aldehydes are smoothly converted to amino nitriles, which are further treated with concentrated HCl to provide the amino acids (Scheme 5.90). Typically, treatment of freshly purified phosphonylated aldehydes in the absence of light at room temperature in water (or MeOH/water or MeCN) with NaCN (or KCN) and NH_4Cl [or NH_4OH, HCO_2NH_4, or $(NH_4)_2CO_3$ at 50–60°C in the Bucherer–Bergs reaction[203,226]] produces the amino nitriles selectively in moderate to high yields (44–91%).[79,154,187,199,226,232] Improved yields have been reported through the use of alumina/ultrasound[199] or a modified Strecker reaction.[400] Hydrolysis of the amino nitriles with 6–8 M HCl at reflux delivers the crude amino acids, which can be purified by ion-exchange chromatography using a strongly acidic ion-exchange resin (Dowex 50W). The Bucherer–Bergs route is less efficient because of difficulties in the hydrolysis of the hydantoin.[79]

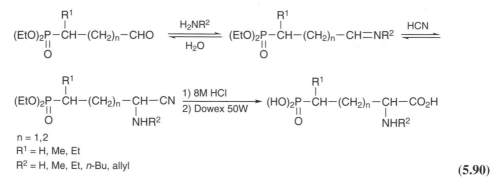

$$n = 1,2$$
$$R^1 = H, Me, Et$$
$$R^2 = H, Me, Et, \textit{n-Bu, allyl}$$

(5.90)

Although initially prepared and evaluated as a racemate, the NMDA antagonist activity was likely to reside primarily in a single enantiomer. The stereoselective nature of the NDMA receptor is well established, albeit not completely understood. Consequently, several attempts have been undertaken to develop synthetic protocols that would allow preparation of optically active compounds. Early reports of preparation of optically active ω-amino-ω-carboxyalkylphosphonic acids describe the preparation of (S)-AP-3 from an optically active amino nitrile prepared by reaction of diethyl 1-formylphosphonate with hydrogen cyanide and (S)-(–)-α-methylbenzylamine.[401] Acid hydrolysis, enrichment of the diastereomers by fractional recrystallization, and debenzylation lead to the isolation of (S)-AP-3 in 86% enantiomeric excess.[401] Recently reported procedures, which use chemoenzymatic processes, offer a more convenient and mild approach for the production of optically pure aminophosphonic acids. Enzymatic hydrolysis of amides using penicillinacylase (EC 3.5.1.11) from *Escherichia coli* provides a mild and efficient, highly enantioselective method of resolution.[195] Enzymatic hydrolysis of esters uses Subtilisin A[226] or Carlsberg esterase[402] in phosphate buffer. For example, hydrolysis of a diastereomeric mixture of hydantoins, epimeric at the C-5 hydantoin carbon atom, with D-hydantoinase enzyme from a strain of *Agrobacterium*, in alkaline buffer results in efficient conversion to D-carbamoyl acid.[226]

Another attractive route to aminocarboxylalkylphosphonic acids involves the intermediacy of the ethyl α–azidoacetate (Scheme 5.91). Addition of MeONa in MeOH to a solution of dialkyl 4-formylbenzylphosphonate and ethyl α-azidoacetate in MeOH at −30°C provides the vinyl azide. Hydrogenation of the vinyl azide with 10% Pd/C in MeOH gives the aminoester,[238,239,403] which at reflux with 3 M HCl delivers the free amino acid.[238,239]

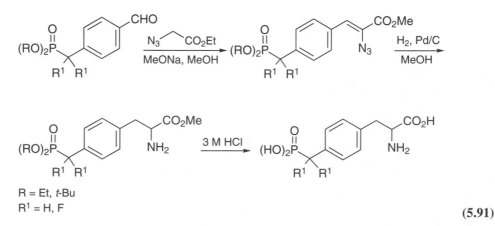

$$R = Et, \textit{t-Bu}$$
$$R^1 = H, F$$

(5.91)

A promising methodology using an Horner–Wadsworth–Emmons reaction between a phosphonylated aldehyde and the sodium *N*-acyl-2-(dialkoxyphosphinyl)glycine provides the dehydro amino acid, albeit in poor yield.[226] Subsequent hydrogenation gives a diastereomeric mixture of aminocarboxylalkylphosphonates resolved by treatment with Subtilisin A in a buffered aqueous solution.[226]

5.2.8 Reactions of the Phosphonyl Group: Dealkylation

The reaction of dialkyl 2,2-alkoxyethylphosphonates or 2-alkoxyvinylphosphonates with Me₃SiCl/KI occurs rapidly at room temperature to give quantitatively the corresponding *bis*(trimethylsilyl)phosphonates. The dealkylation proceeds to completion in few minutes. Treatment of silyl esters with MeOH at room temperature gives the corresponding phosphonic acids in high yields (Scheme 5.92).[404–406]

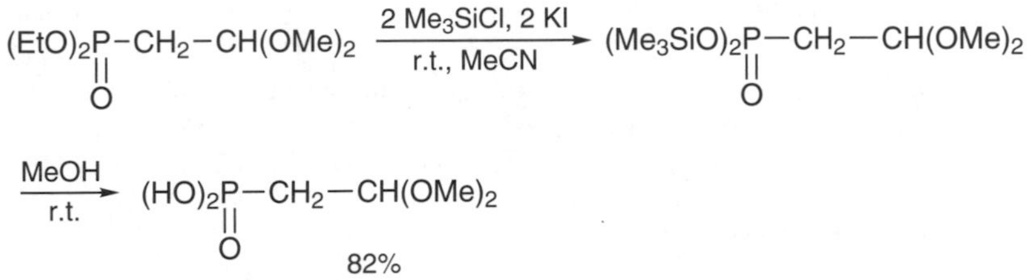

$$(EtO)_2\underset{\underset{O}{\|}}{P}-CH_2-CH(OMe)_2 \xrightarrow[\text{r.t., MeCN}]{2\ Me_3SiCl,\ 2\ KI} (Me_3SiO)_2\underset{\underset{O}{\|}}{P}-CH_2-CH(OMe)_2$$

$$\xrightarrow[\text{r.t.}]{MeOH} (HO)_2\underset{\underset{O}{\|}}{P}-CH_2-CH(OMe)_2$$

82%

(5.92)

REFERENCES

1. Razumov, A.I., Liorber, B.G., Moskva, V.V., and Sokolov, M.P., Phosphorylated aldehydes, *Russ. Chem. Rev.*, 42, 538, 1973.
2. Iorga, B., Eymery, F., Mouriès, V., and Savignac, P., Phosphorylated aldehydes. Preparations and synthetic uses, *Tetrahedron*, 54, 14637, 1998.
3. Firestone, R.A., Hydroxy- and imino-containing phosphonic acid diesters, *Merck*, U.S. Patent Appl. US 3784590, 1974; *Chem. Abstr.*, 80, 60031, 1974.
4. Vasella, A., and Voeffray, R., Asymmetric synthesis of α-aminophosphonic acids by cycloaddition of *N*-glycosyl-*C*-dialkoxyphosphonoylnitrones, *Helv. Chim. Acta*, 65, 1953, 1982.
5. Moskva, V.V., and Mavrin, V.Y., Diethyl formylphosphonate, *Zh. Obshch. Khim.*, 57, 2793, 1987; *J. Gen. Chem. USSR (Engl. Transl.)*, 57, 2492, 1987.
6. Razumov, A.I., and Moskva, V.V., Derivatives of phosphonic and phosphinic acids. Part 32. Reaction of orthoformate esters with partial esters of phosphorous and phosphinic acids, *Zh. Obshch. Khim.*, 35, 1595, 1965; *Chem. Abstr.*, 63, 18144a, 1965.
7. Moskva, V.V., Maikova, A.I., and Razumov, A.I., Phosphonic and phosphonous acid derivatives. Part 64. Reaction of phosphites and arylphosphonates with orthoacetic ester, *Zh. Obshch. Khim.*, 39, 2451, 1969; *J. Gen. Chem. (Engl. Transl.)*, 39, 2391, 1969.
8. Wagenknecht, J., An electrochemical method for the preparation of iminodimethylenediphosphonic acid, *Synth. Inorg. Met. Org. Chem.*, 4, 567, 1974.
9. Dingwall, J.G., Ehrenfreund, J., and Hall, R.G., Diethoxymethylphosphonites and phosphinates. Intermediates for the synthesis of α,β- and γ-aminoalkylphosphonous acids, *Tetrahedron*, 45, 3787, 1989.
10. Möhrle, H., and Vetter, W., Participation of phosphonate neighbour groups with dehydrogenations of amines, *Z. Naturforsch.*, 43b, 1662, 1988.
11. Oediger, H., Lieb, F., and Disselnkotter, H., Phosphonoformaldehyde and its salts, *Bayer A.-G.*, Eur. Patent Appl. EP 50778, 1982; *Chem. Abstr.*, 97, 163251, 1982.
12. Oediger, H., Lieb, F., and Disselnkotter, H., Phosphonohydroxyacetonitrile, *Bayer A.-G.*, Eur. Patent Appl. EP 50792, 1982; *Chem. Abstr.*, 97, 163250, 1982.
13. Hamilton, R., McKervey, M.A., Rafferty, M.D., and Walker, B.J., The reaction of dimethyl dioxirane with diazomethylphosphonates. The first synthesis of a formylphosphonate hydrate, *J. Chem. Soc., Chem. Commun.*, 37, 1994.

14. Hamilton, R., McKervey, M.A., Rafferty, M.D., and Walker, B.J., Reactions of diazomethylphospho-nate. The first synthesis of a formylphosphonate hydrate, *Phosphorus, Sulfur Silicon Relat. Elem.*, 109–110, 441, 1996.

15. Cairns, J., Dunne, C., Franczyk, T.S., Hamilton, R., Hardacre, C., Stern, M.K., Treacy, A., and Walker, B.J., The synthesis and chemistry of formylphosphonate, *Phosphorus, Sulfur Silicon Relat. Elem.*, 144, 385, 1999.

16. Razumov, A.I., and Moskva, V.V., Reaction of dialkylphosphorous acids with orthoformic esters, *Zh. Obshch. Khim.*, 34, 3125, 1964; *J. Gen. Chem. USSR (Engl. Transl.)*, 34, 3167, 1964.

17. Gallagher, M.J., and Honegger, H., Dialkoxymethylation of phosphorus with trialkyl orthoformates. Reactions of phosphonic and phosphinic acids via their trivalent tautomers, *Tetrahedron Lett.*, 34, 2987, 1977.

18. Livantsov, M.V., Boiko, V.I., Proskurnina, M.V., and Lutsenko, I.F., Phosphorylation of orthoformates, *Zh. Obshch. Khim.*, 52, 930, 1982; *J. Gen. Chem. USSR (Engl. Transl.)*, 52, 811, 1982.

19. Livantsov, M.V., Proskurnina, M.V., Prishchenko, A.A., and Lutsenko, I.F., Synthesis and some properties of phosphorylated formals, *Zh. Obshch. Khim.*, 54, 2504, 1984; *J. Gen. Chem. USSR (Engl. Transl.)*, 54, 2237, 1984.

20. Malenko, D.M., and Gololobov, Y.G., New method for the synthesis of phosphorylated formals, *Zh. Obshch. Khim.*, 51, 1432, 1981; *J. Gen. Chem. USSR (Engl. Transl.)*, 51, 1214, 1981.

21. Nesterov, L.V., and Aleksandrova, N.A., Reactions of triethyl phosphite with triethyl orthoformate in the presence of strong electrophilic reagents, *Izv. Akad. Nauk SSSR, Ser. Khim.*, 1919, 1984; *Bull. Acad. Sci. USSR, Div. Chem. Sci. (Engl. Transl.)*, 1754, 1984.

22. Nesterov, L.V., and Aleksandrova, N.A., The Arbuzov reaction for dialkyl(trialkylsilyl)phosphites with esters of orthocarboxylic acids. A new heterolytic chain process in the chemistry of organophosphorus compounds, *Zh. Obshch. Khim.*, 55, 222, 1985; *J. Gen. Chem. USSR (Engl. Transl.)*, 55, 195, 1985.

23. Dietsche, W., Reaction of trivalent phosphorus chlorides with ortho esters and their thio analogs. Preparation of *C*-phosphorylated formaldehyde acetals, *Liebigs Ann. Chem.*, 712, 21, 1968.

24. Moskva, V.V., Maikova, A.I., and Razumov, A.I., Phosphinic and phosphinous acid derivatives. Part 58. Reaction of ortho esters of carboxylic acids and acetals with phosphorus(III) acid chlorides, *Zh. Obshch. Khim.*, 39, 595, 1969; *Chem. Abstr.*, 71, 50076r, 1969.

25. Gazizov, M.V., Razumov, A.I., and Sekerin, E.A., Reactions of diethoxymethyl acetate with dialkyl-chlorophosphites, *Zh. Obshch. Khim.*, 43, 1407, 1973; *J. Gen. Chem. USSR (Engl. Transl.)*, 43, 1394, 1973.

26. Gross, H., and Costisella, B., α-Substituted phosphonates. Part 28. Synthesis of 2-phosphorylated 1,3-benzodioxoles, 1,3-benzoxathioles and 1,3-benzodithioles, *Synthesis*, 622, 1977.

27. Gazizov, M.B., Khairullin, R.A., Kadirova, R.F., Lewis, E.S., and Kook, A.M., Catalysis phenomena and intermediates in the reaction of phosphorous trichloride with aldehydes, *J. Chem. Soc., Chem. Commun.*, 1133, 1990.

28. Ebetino, F.H., and Jamieson, L.A., The design and synthesis of bone-active phosphinic acid analogues. Part 1. The pyridylaminomethane phosphonoalkylphosphinates, *Phosphorus, Sulfur Silicon Relat. Elem.*, 51/52, 493, 1990.

29. Endova, M., Masojidkova, M., Budesinsky, M., and Rosenberg, I., Synthesis of 1-phosphonoalkylidene and -arylidene derivatives of nucleosides, *Tetrahedron Lett.*, 37, 3497, 1996.

30. Ebetino, F.H., and Berk, J.D., A stereoselective process for the preparation of novel phosphonoalkyl-phosphinates, *J. Organomet. Chem.*, 529, 135, 1997.

31. Endova, M., Masojikova, M., Budesinsky, M., and Rosenberg, I., 2′,3′-*O*-Phosphonoalkylidene deriv-atives of ribonucleosides. Synthesis and reactivity, *Tetrahedron*, 54, 11151, 1998.

32. Endova, M., Masojidkova, M., Budesinsky, M., and Rosenberg, I., 3′,5′-*O*-Phosphonoalkylidene derivatives of 1-(2-deoxy-β-D-*threo*-pentofuranosyl)thymine. Synthesis and reactivity, *Tetrahedron*, 54, 11187, 1998.

33. Pressova, M., Endova, M., Tocik, Z., Liboska, R., and Rosenberg, I., UV spectroscopy study on complexes of phosphonate ApA analogs with poly(U). Promising step in prediction of oligonucleotide analog properties? *Bioorg. Med. Chem. Lett.*, 8, 1225, 1998.

34. Gross, H., Freiberg, J., and Costisella, B., α-Haloethers. Part 35. Existence of halodialkoxyalkanes. Simple synthesis of dialkoxymethanephosphonates, *Chem. Ber.*, 101, 1250, 1968.

35. Costisella, B., and Gross, H., α-Substituted phosphonates. Part 14. Phosphoryl derivatives of 2,2-*bis*(hydroxymethyl)-5-norbornene, *J. Prakt. Chem.*, 314, 532, 1972.

36. Krokhina, S.S., Pyrkin, R.I., Levin, Y.A., and Ivanov, B.E., Reaction of triethyl orthoformate with some trivalent phosphorus derivatives, *Izv. Akad. Nauk SSSR, Ser. Khim.*, 1420, 1968; *Bull. Acad. Sci. USSR, Div. Chem. Sci. (Engl. Transl.)*, 1349, 1968.

37. Gross, H., and Costisella, B., α-Substituted phosphonates. Part 7. Reaction of orthoformic acid esters and phosphorous(III) compounds, *J. Prakt. Chem.*, 311, 571, 1969.

38. Gross, H., and Costisella, B., Dialkyl (dialkoxyalkyl)phosphonates, German (East) Patent Appl. DD 77712, 1970; *Chem. Abstr.*, 75, 129939r, 1971.

39. Nguyen-Ba, P., Lee, N., Mitchell, H., Chan, L., and Quimpère, M., Design and synthesis of a novel class of nucleotide analogs with anti-HCMV activity, *Bioorg. Med. Chem. Lett.*, 8, 3555, 1998.

40. Nguyen-Ba, N., Chan, L., Quimpere, M., Turcotte, N., Lee, N., Mitchell, H., and Bedard, J., Design and SAR study of a novel class of nucleotide analogues as potent anti-HCMV agents, *Nucleosides Nucleotides*, 18, 821, 1999.

41. Costisella, B., and Gross, H., α-Substituted phosphonates. Part 22. Synthesis and NMR spectra of open-chain and cyclic acetals of formyl phosphonates and formyl diphenylphosphine oxide, *J. Prakt. Chem.*, 319, 8, 1977.

42. Gross, H., and Costisella, B., α-Substituted phosphonates. Part 12. Reactivity of formylphosphonate acetals, *J. Prakt. Chem.*, 313, 265, 1971.

43. Akhmetkhanova, F.M., Rol'nik, L.Z., and Proskurnina, M.V., Reaction of triethyl trithioorthoformate with dipropyl hydrogen phosphite, *Zh. Obshch. Khim.*, 57, 1918, 1987; *J. Gen. Chem. USSR (Engl. Transl.)*, 57, 1715, 1987.

44. Mikolajczyk, M., Costisella, B., Grzejszczak, S., and Zatorski, A., Synthesis of *O,S*-thioacetals of formylphosphonates, *Tetrahedron Lett.*, 17, 477, 1976.

45. Mlotkowska, B., Gross, H., Costisella, B., Mikolajczyk, M., Grzejszczak, S., and Zatorski, A., α-Substituted phosphonates. Part 23. Synthesis of open-chain and cyclic *S,S*-acetals of formyl phosphonates, *J. Prakt. Chem.*, 319, 17, 1977.

46. Mikolajczyk, M., Grzejszczak, S., Zatorski, A., Mlotkowska, B., Gross, H., and Costisella, B., Organosulphur compounds. Part 18. A new and general synthesis of ketene *S,S*- and *O,S*-thioacetals based on the Horner–Wittig reaction, *Tetrahedron*, 34, 3081, 1978.

47. Gross, H., Keitel, I., Costisella, B., Mikolajczyk, M., and Midura, W., α-Substituted phosphonates. Part 44. Synthesis of *bis*-alkylmercaptomethane phosphonates, *Phosphorus Sulfur*, 16, 257, 1983.

48. Mikolajczyk, M., Graczyk, P., and Wieczorek, M.W., Conformational preference in 1,3-dithianes containing 2-phosphoryl, -(thiophosphoryl), and -(selenophosphoryl) groups. Chemical and crystallographic implications of the nature of the anomeric effect, *J. Org. Chem.*, 59, 1672, 1994.

49. Mikolajczyk, M., Mikina, M., Graczyk, P.P., and Balczewski, P., Insertion of α-phosphorylcarbene moiety into S-S and Se-Se bonds. Synthesis of dithio- and diselenoacetals of formylphosphonates, *Synthesis*, 1232, 1996.

50. Mikolajczyk, M., Zatorski, A., Grzejszczak, S., Costisella, B., and Midura, W., α-Phosphoryl sulfoxides. Part 4. Pummerer rearrangements of α-phosphoryl sulfoxides and asymmetric induction in the transfer of chirality from sulfur to carbon, *J. Org. Chem.*, 43, 2518, 1978.

51. Costisella, B., and Keitel, I., α-Substituted phosphonates. *O,S*-Acetals of diethoxyphosphinylformaldehyde and orthoesters of diethoxyphosphinylformic acid by anodic methoxylation, *Synthesis*, 44, 1987.

52. Mikolajczyk, M., Balczewski, P., and Grzejszczak, S., Sulphenylation of phosphonates. A facile synthesis of α-phosphoryl sulphides and *S,S*-acetals of oxomethanephosphonates, *Synthesis*, 127, 1980.

53. Mikolajczyk, M., Grzejszczak, S., Chefczynska, A., and Zatorski, A., Addition of elemental sulfur to phosphonate carbanions and its application for the synthesis of α-phosphoryl organosulfur compounds. Synthesis of aromatic ketones, *J. Org. Chem.*, 44, 2967, 1979.

54. Kim, T.H., and Oh, D.Y., Synthesis of *S,S*-thioacetals of formylphosphonate from chloro(arylthio)metanephosphonate, *Synth. Commun.*, 18, 1611, 1988.

55. Kim, T.H., and Oh, D.Y., Formation and reactions of α-phosphoryl thiocarbocations. Synthesis of α-sulfenyl phosphonates, *Bull. Korean Chem. Soc.*, 16, 609, 1995.

56. Kim, T.H., and Oh, D.Y., New synthetic method of *O,S*-thioacetals of formylphosphonates, *Tetrahedron Lett.*, 26, 3479, 1985.

57. Balczewski, P., and Mikolajczyk, M., Sulfenylation of α-phosphoryl sulfides. Chemical evidence for intermediate formation of α-phosphoryl trithioorthoformate, *Heteroatom Chem.*, 5, 487, 1994.

58. Mikolajczyk, M., and Balczewski, P., Organosulphur compounds. Part 71. Diverse reactivity of α-carbanions derived from α-phosphoryl dithioacetals and α-phosphoryl sulphides towards α,β-unsaturated carbonyl compounds. A general synthesis of conjugated ketene dithioacetals, *Tetrahedron*, 48, 8697, 1992.

59. Iorga, B., Mouriès, V., and Savignac, P., Carbanionic displacement reaction at phosporus. Synthesis and reactivity of 5,5-dimethyl-2-oxo-2-(1,3-dithian-2-yl)-1,3,2-dioxaphosphorinane, *Bull. Soc. Chim. Fr.*, 134, 891, 1997.

60. Bulpin, A., Masson, S., and Sene, A., The use of phosphonodithioformates for the synthesis of ketene dithioacetals, *Tetrahedron Lett.*, 30, 3415, 1989.

61. Battacharya, A.K., and Thyagarajan, G., The Michaelis–Arbuzov rearrangement, *Chem. Rev.*, 81, 415, 1981.

62. Dawson, N.D., and Burger, A., Some alkyl thiazolephosphonates, *J. Am. Chem. Soc.*, 74, 5312, 1952.

63. Perkow, W., Ullerich, K., and Meyer, F., New phosphoric acid ester with miotic activity, *Naturwissenschaften*, 39, 353, 1952.

64. Perkow, W., Reactions of alkyl phosphites. Part 1. Conversions with Cl_3CCHO and Br_3CCHO, *Chem. Ber.*, 87, 755, 1954.

65. Sekine, M., Okimoto, K., Yamada, K., and Hata, T., Silyl phosphites. Part 15. Reactions of silyl phosphites with α-halo carbonyl compounds. Elucidation of the mechanism of the Perkow reaction and related reactions with confirmed experiments, *J. Org. Chem.*, 46, 2097, 1981.

66. Gryszkiewicz-Trochimowski, E., and Chmelevsky, A., On the synthesis of dimethyl (formylmethyl)phosphonate, *Bull. Soc. Chim. Fr.*, 2043, 1966.

67. Razumov, A.I., and Moskva, V.V., Phosphonic and phosphonous derivatives. Part 23. Synthesis and some properties of aliphatic phosphorylated acetals and aldehydes, *Zh. Obshch. Khim.*, 34, 2589, 1964; *J. Gen. Chem. USSR (Engl. Transl.)*, 34, 2612, 1964.

68. Razumov, A.I., and Gurevich, P.A., Phosphonic and phosphonous acid derivatives. Part 44. Synthesis and some properties of phosphorylated indoles, *Zh. Obshch. Khim.*, 37, 1615, 1967; *J. Gen. Chem. USSR (Engl. Transl.)*, 37, 1532, 1967.

69. Razumov, A.I., Savicheva, G.A., Zykova, T.V., Sokolov, M.P., Smirnova, G.G., Liorber, B.G., and Salakhutdinov, R.A., Reactivity and structure of phosphorylated carbonyl compounds. Part 8. NMR and IR spectra of arylalkoxyphosphinylacetaldehydes, *Zh. Obshch. Khim.*, 41, 2164, 1971; *Chem. Abstr.*, 76, 85036, 1972.

70. Razumov, A.I., Savicheva, G.A., Zykova, T.V., Sokolov, M.P., Liorber, B.G., and Salakhutdinov, R.A., Reactivity and structure of phosphorylated carbonyl compounds. Part 7. NMR and IR spectra of dialkoxyphosphonylacetaldehydes, *Zh. Obshch. Khim.*, 41, 1954, 1971; *J. Gen. Chem. USSR (Engl. Transl.)*, 41, 1970, 1971.

71. Razumov, A.I., Sokolov, M.P., Zykova, T.V., Liorber, B.G., Savicheva, G.A., and Salakhutdinov, R.A., Reactivity and structure of phosphorylated carbonyl compounds. Part 9. Aldo–enol equilibrium of phosphorylated acetaldehydes, *Zh. Obshch. Khim.*, 42, 47, 1972; *J. Gen. Chem. USSR (Engl. Transl.)*, 42, 43, 1972.

72. Razumov, A.I., Liorber, B.G., Sokolov, M.P., Moskva, V.V., Hazvanova, G.F., Zykova, T.V., Chemodanova, L.A., and Salakhutdinov, R.A., Reactivity and structure of phosphorylated carbonyl compounds. Part 11. Temperature dependence of the aldo–enol equilibrium of phosphorylated aldehydes, *Zh. Obshch. Khim.*, 43, 568, 1973; *J. Gen. Chem. USSR (Engl. Transl.)*, 43, 570, 1973.

73. Iorga, B., Eymery, F., Mouriès, V., and Savignac, P., Unpublished results.

74. Petrov, K.A., Raksha, M.A., Korotkova, V.P., and Shmidt, E., Synthesis and properties of alkenylphosphonic derivatives. Part 4. β-Aldo phosphonic esters, *Zh. Obshch. Khim.*, 41, 324, 1971; *J. Gen. Chem. USSR (Engl. Transl.)*, 41, 319, 1971.

75. Moskva, V.V., Nazvanova, G.F., Zykova, T.V., Razumov, A.I., and Chemodanova, L.A., Substituted vinylphosphonic derivatives. Part 10. (2-Alkoxy-1-alkylvinyl)-phosphonic and -phosphonothioic derivatives, *Zh. Obshch. Khim.*, 41, 1680, 1971; *J. Gen. Chem. USSR (Engl. Transl.)*, 41, 1687, 1971.

76. Halmann, M., Vofsi, D., and Yanai, S., Reaction of phosphorylated acetals. Acid-catalysed hydrolysis of dialkyl 2,2-dialkoxyethylphosphonates, *J. Chem. Soc., Perkin Trans. 2*, 1210, 1976.

77. Varlet, J.M., Collignon, N., and Savignac, P., A new route to 2-aminoalkanephosphonic acids, *Synth. Commun.*, 8, 335, 1978.

78. Cates, L.A., Jones, G.S., Jr., Good, D.J., Tsai, H.Y., Li, V.-S., Caron, N., Tu, S.-C., and Kimball, A.P., Cyclophosphamide potentiation and aldehyde oxidase inhibition by phosphorylated aldehydes and acetals, *J. Med. Chem.*, 23, 300, 1980.

79. Varlet, J.M., Fabre, G., Sauveur, F., Collignon, N., and Savignac, P., Preparation and conversion of ω-formylalkylphosphonates in aminocarboxyalkylphosphonic acids, *Tetrahedron*, 37, 1377, 1981.

80. Tsuge, O., Kanemasa, S., and Suga, H., Synthesis of a new phosphorous-functionalized nitrile oxide, α-(diethylphosphono)acetonitrile oxide, and cycloaddition leading to 3-(diethylphosphonomethyl)-Δ^2-isoxazolines, *Chem. Lett.*, 183, 1986.

81. Olson, G.L., Cheung, H.-C., Chiang, E., Madison, V.S., Sepinwall, J., Vincent, G.P., Winokur, A., and Gary, K.A., Peptide mimetics of thyrotropin-releasing hormone based on a cyclohexane framework. Design, synthesis, and cognition-enhancing properties, *J. Med. Chem.*, 38, 2866, 1995.

82. Pfeiffer, F.R., Mier, J.D., and Weisbach, J.A., Synthesis of phosphonic acid isosteres of 2-phospho-, 3-phospho-, and 2,3-diphosphoglyceric acid, *J. Med. Chem.*, 17, 112, 1974.

83. Gorgues, A., Formation of aldehydes and ketones by reaction of formic acid with their acetals. Laboratory note, *Bull. Soc. Chim. Fr.*, 529, 1974.

84. Yanai, S., Vofsi, D., and Halmann, M., Reaction of phosphonated acetals. Part 3. Acid-catalysed hydrolysis of diethyl 3,3-diethoxypropylphosphonate, *J. Chem. Soc., Perkin Trans. 2*, 517, 1978.

85. Tavs, P., Synthesis of *trans*-ethylene-1,2-diphosphonic acid esters and conversion into cyclohexane-1,2-diphosphonic acid, *Chem. Ber.*, 100, 1571, 1967.

86. Clare, B.W., Ferro, V., Skelton, B.W., Stick, R.V., and White, A.H., Approaches to the synthesis of retronecine from some pyrrolidine precursors, *Aust. J. Chem.*, 46, 805, 1993.

87. Coppola, G.M., Amberlyst-15, a superior acid catalyst for the cleavage of acetals, *Synthesis*, 1021, 1984.

88. Laatsch, H., and Pudleiner, A., Synthesis of ω-acetyl-α-methylenpolyene carboxylic esters, *J. Prakt. Chem./Chem.-Ztg.*, 336, 663, 1994.

89. Yoffe, S.T., Vatsuro, K.V., Petrovskii, P.V., Fedin, E.I., and Kabachnik, M.I., Enolization of phosphorylated derivatives of formylacetic acid, *Izv. Akad. Nauk SSSR, Ser. Khim.*, 1504, 1970; *Bull. Acad. Sci. USSR, Div. Chem. Sci. (Engl. Transl.)*, 1420, 1970.

90. Yoffe, S.T., Vatsuro, K.V., Petrovskii, P.V., and Kabachnik, M.I., The influence of structural factors and the solvent on the nature of the enolization of the formyl group, *Izv. Akad. Nauk SSSR, Ser. Khim.*, 731, 1971; *Bull. Acad. Sci. USSR, Div. Chem. Sci. (Engl. Transl.)*, 655, 1971.

91. Matrosov, E.I., Yoffe, S.T., and Kabachnik, M.I., Infrared spectra and hydrogen-bonding of substituted (formylmethyl)phosphonic esters, *Zh. Obshch. Khim.*, 42, 2625, 1972; *J. Gen. Chem. USSR (Engl. Transl.)*, 42, 2617, 1972.

92. Moskva, V.V., Nazvanova, G.F., Zykova, T.V., Razumov, A.I., Remizov, A.B., and Salakhutdinov, R.A., Derivatives of substituted vinylphosphonic acids. Part 13. Acylation of phosphorylated aldehydes, *Zh. Obshch. Khim.*, 42, 498, 1972; *J. Gen. Chem. USSR (Engl. Transl.)*, 42, 497, 1972.

93. Razumov, A.I., Moskva, V.V., Sokolov, M.P., and Sazonova, Z.Y., Tautomerism of phosphorylated carbonyl compounds, *Zh. Obshch. Khim.*, 46, 2011, 1976; *J. Gen. Chem. USSR (Engl. Transl.)*, 46, 1936, 1976.

94. Razumov, A.I., Liorber, B.G., Pavlov, V.A., Sokolov, M.P., Zykova, T.V., and Salakhutdinov, R.A., Reactivity and structure of phosphorylated carbonyl compounds. Part 13. Methylalkoxyphosphinyl-acetaldehydes, *Zh. Obshch. Khim.*, 47, 243, 1977; *J. Gen. Chem. USSR (Engl. Transl.)*, 47, 224, 1977.

95. Razumov, A.I., Sokolov, M.P., Liorber, B.G., Zykova, T.V., Salakhutdinov, R.A., and Urmancheeva, G.V., Reactivity and structure of phosphorylated carbonyl compounds. Part 14. Study of the reaction of phosphorylated acetaldehydes with β-aminocarbonyl compounds, *Zh. Obshch. Khim.*, 47, 2717, 1977; *J. Gen. Chem. USSR (Engl. Transl.)*, 47, 2474, 1977.

96. Razumov, A.I., Sokolov, M.P., Liorber, B.G., Zykova, T.V., and Zhikhareva, N.A., Reactivity and structure of phosphorylated carbonyl compounds. Part 12. The 2,4-diphosphorylated 2-butenals, *Zh. Obshch. Khim.*, 47, 563, 1977; *J. Gen. Chem. USSR (Engl. Transl.)*, 47, 515, 1977.

97. Zykova, T.V., Ismailov, V.M., Moskva, V.V., and Salakhutdinov, R.A., Study of the tautomerism of phosphorylated aldehydes by means of ^{13}C-NMR, *Zh. Obshch. Khim.*, 54, 1288, 1984; *J. Gen. Chem. USSR (Engl. Transl.)*, 54, 1152, 1984.

98. Sokolov, M.P., Gazizov, I.G., and Mavrin, G.V., Ionization of α-phosphorylated carbonyl compounds and hydrazones, *Zh. Obshch. Khim.*, 59, 1043, 1989; *J. Gen. Chem. USSR (Engl. Transl.)*, 59, 921, 1989.

99. Sturz, G., Clement, J. C., Daniel, A., Molinier, J., and Lenzi, M., Extracting power of β-carbonylated phosphonates with respect to copper, cobalt and nickel, *Bull. Soc. Chim. Fr.*, 161, 1981.

100. Nkusi, G., and Neidlein, R., Convenient synthesis of diethyl 3-methylisoxazoline and isoxazolephosphonates, potent synthons to biological active compounds, *J. Prakt. Chem./Chem.-Ztg.*, 334, 278, 1992.

101. Kudzin, Z.H., Kotynski, A., and Andrijewski, G., Aminoalkanediphosphonic acids. Synthesis and acidic properties, *J. Organomet. Chem.*, 479, 199, 1994.

102. Friese, A., Hell-Momeni, K., Zuendorf, I., Winckler, T., Dingermann, T., and Dannhardt, G., Synthesis and biological evaluation of cycloalkylidene carboxylic acids as novel effectors of Ras/Raf interaction, *J. Med. Chem.*, 45, 1535, 2002.

103. Nagata, W., Wakabayashi, T., and Hayase, Y., Diethyl 2-(cyclohexylamino)vinylphosphonate, *Org. Synth. Coll.*, VI, 448, 1988.

104. Reichel, L., and Jahns, H.J., Reactions of α-bromopropionaldehyde and its diethylacetal with triethyl phosphite and triphenylphosphine, *Justus Liebigs Ann. Chem.*, 751, 69, 1971.

105. Malone, G.R., and Meyers, A.I., The chemistry of 2-chloromethyloxazines. Formation of phosphoranes phosphonates. The use of α,β-unsaturated oxazines as a common intermediate for the synthesis of aldehydes, ketones, and acids, *J. Org. Chem.*, 39, 623, 1974.

106. Malone, G.R., and Meyers, A.I., The chemistry of 2-chloromethyl-5,6-dihydro-1,3-oxazines. Grignard coupling metalation studies. A synthesis of α-chloro aldehydes and arylacetic acids, *J. Org. Chem.*, 39, 618, 1974.

107. Tammelin, L.-E., and Fagerlind, L., Organophosphorus analogues of tropic acid esters. Part 1. 2-Phenyl-2-diethoxyphosphoryl ethanol, *Acta Chem. Scand.*, 14, 1353, 1960.

108. Larsson, L., and Tammelin, L.-E., Organophosphorus analogues of tropic acid esters. Part 2. Tautomerism of phenyl-diethoxyphosphoryl acetaldehyde, *Acta Chem. Scand.*, 15, 349, 1961.

109. Kirilov, M., and Petrov, G., Preparation and properties of dialkyl esters of α-acylphosphonoacetonitriles, *Monatsh. Chem.*, 99, 166, 1968.

110. Kantlehner, W., Wagner, F., and Bredereck, H., Orthoamides. Part 32. Reaction of *tert*-butoxy-*N,N,N′,N′*-tetramethylmethanediamine with NH- and CH-acidic compounds, *Liebigs Ann. Chem.*, 344, 1980.

111. Yoffe, S.T., Petrovskii, P.V., Goryunov, Y.I., Yershova, T.V., and Kabachnik, M.I., The aldo-*cis-trans*-enolic equilibrium of substituted alkyl formylacetates and alkyl formylmethylphosphonates, *Tetrahedron*, 28, 2783, 1972.

112. Aboujaoude, E.E., Collignon, N., and Savignac, P., Synthesis of β-carbonylated phosphonates. Part 1. Carbanionic route, *J. Organomet. Chem.*, 264, 9, 1984.

113. Teulade, M.P., Savignac, P., Aboujaoude, E.E., and Collignon, N., Acylation of diethyl α-chloromethylphosphonates by carbanionic route. Preparation of α-clorinated β-carbonylated phosphonates, *J. Organomet. Chem.*, 287, 145, 1985.

114. Teulade, M.-P., and Savignac, P., Direct conversion of phosphates in α-lithiated phosphonates using alkyllithiums, *Tetrahedron Lett.*, 28, 405, 1987.

115. Zanella, Y., Berté-Verrando, S., Dizière, R., and Savignac, P., New route to 1-formylalkylphosphonates using diethyl trichloromethylphosphonate as a precursor, *J. Chem. Soc., Perkin Trans. 1*, 2835, 1995.

116. Savignac, P., and Patois, C., Diethyl 1-propyl-2-oxoethylphosphonate (diethyl 1-formylbutylphosphonate), *Org. Synth.*, 72, 241, 1995.

117. Al-Badri, H., About-Jaudet, E., Combret, J.-C., and Collignon, N., Synthesis, acetylation and silylation of (1*E*)-2-diethoxyphosphonylbuta-1,3-dien-1-ols. A convenient route to new, activated dienylphosphonates, *Synthesis*, 1401, 1995.

118. Truel, I., Mohamed-Hachi, A., About-Jaudet, E., and Collignon, N., A simple and efficient synthesis of 2-substituted-3-diethylphosphono-5-methylfurans, *Synth. Commun.*, 27, 1165, 1997.

119. Aboujaoude, E.E., Collignon, N., and Savignac, P., A simple synthesis of dialkyl 1-formylalkanephosphonates, *Synthesis*, 634, 1983.

120. Öhler, E., Kang, H.-S., and Zbiral, E., Synthesis of dialkyl (2,3-epoxy-4-oxoalkyl)phosphonates from (1-formylalkyl)phosphonates, *Chem. Ber.*, 121, 299, 1988.

121. Bourne, S., Modro, A.M., and Modro, T.A., Phosphonic systems. Part 17. Solution and solid state studies on dimethyl 2-hydroxy-3-benzoylpropylphosphonate, *Phosphorus, Sulfur Silicon Relat. Elem.*, 102, 83, 1995.

122. Bode, J.W., and Carreira, E.M., Stereoselective syntheses of epothilones A and B via directed nitrile oxide cycloaddition, *J. Am. Chem. Soc.*, 123, 3611, 2001.

123. Bode, J.W., and Carreira, E.M., Stereoselective syntheses of epothilones A and B via nitrile oxide cycloadditions and related studies, *J. Org. Chem.*, 66, 6410, 2001.

124. Olah, G.A., Ohannesian, L., and Arvanaghi, M., Synthetic methods and reactions. Part 119. *N*-Formylmorpholine, a new and effective formylating agent for the preparation of aldehydes and dialkyl (1-formylalkyl)phosphonates from Grignard or organolithium reagents, *J. Org. Chem.*, 49, 3856, 1984.

125. Grassberger, M.A., Formylation of diethyl benzylphosphonate and (arylmethyl)phosphonium salts, *Justus Liebigs Ann. Chem.*, 1872, 1974.

126. Aboujaoude, E.E., Collignon, N., and Savignac, P., α-Functional dialkyl formyl-1-methylphosphonates. Part 2. Thermical preparation and conversion into α-phosphonic heterocycles, *Tetrahedron*, 41, 427, 1985.

127. Aboujaoude, E.E., Collignon, N., and Savignac, P., Synthesis of α-phosphonic heterocycles. New developments, *Phosphorus Sulfur*, 31, 231, 1987.

128. Phan, H.T., Nguyen, L.M., Diep, V.V., Azoulay, R., Eschenhof, H., Niesor, E.J., Bentzen, C.L., and Ife, R.J., Preparation of α-substituted β-aminoethyl phosphonates and their use in lowering lipoprotein A and apolipoprotein levels, and for lowering plasma levels of total cholesterol, *Ilex Oncology Research*, Int. Patent Appl. WO 2002026752, 2002; *Chem. Abstr.*, 136, 294946, 2002.

129. Lequeux, T.P., and Percy, J.M., Facile syntheses of α,α-difluoro-β-ketophosphonates, *J. Chem. Soc., Chem. Commun.*, 2111, 1995.

130. Blades, K., Lequeux, T.P., and Percy, J.M., Reactive dienophiles containing a difluoromethylenephosphonato group, *J. Chem. Soc., Chem. Commun.*, 1457, 1996.

131. Blades, K., Lequeux, T.P., and Percy, J.M., A reproducible and high-yielding cerium-mediated route to α,α-difluoro-β-ketophosphonates, *Tetrahedron*, 53, 10623, 1997.

132. Blades, K., Butt, A.H., Cockerill, G.S., Easterfield, H.J., Lequeux, T.P., and Percy, J.M., Synthesis of activated alkenes bearing the difluoromethylenephosphonate group. A range of building blocks for the synthesis of secondary difluorophosphonates, *J. Chem. Soc., Perkin Trans. 1*, 3609, 1999.

133. Savignac, P., Teulade, M.-P., and Patois, C., Phosphate–phosphonate conversion. Nucleophilic displacement reactions involving phosphoric esters and methyllithium, *Heteroatom Chem.*, 1, 211, 1990.

134. Patois, C., and Savignac, P., Phosphate-phosphonate conversion. A versatile route to linear or branched alkylphosphonates, *Bull. Soc. Chim. Fr.*, 130, 630, 1993.

135. Kosolapoff, G.M., Isomerization of alkyl phosphites. Part 6. Reactions with chlorides of singular structure, *J. Am. Chem. Soc.*, 69, 1002, 1947.

136. Mel'nikov, N.N., Kozlov, V.A., Churusova, S.G., Buvashkina, N.I., Ivanchenko, V.I., Negrebetskii, V.V., and Grapov, A.F., β-Phosphorylated enamines, *Zh. Obshch. Khim.*, 53, 1689, 1983; *J. Gen. Chem. USSR (Engl. Transl.)*, 53, 1519, 1983.

137. Churi, R.H., and Griffin, C.E., 1,2-Shifts of dialkoxyphosphono groups in skeletal rearrangements of α,β-epoxyvinylphosphonates, *J. Am. Chem. Soc.*, 88, 1824, 1966.

138. Sprecher, M., and Kost, D., The rearrangement of dialkyl α,β-epoxyphosphonates, *Tetrahedron Lett.*, 10, 703, 1969.

139. Cann, P.F., Howells, D., and Warren, S., Rearrangements of a cation of the neopentyl-type containing a diphenylphosphinyl substituent, *J. Chem. Soc., Perkin Trans. 2*, 304, 1972.

140. Griffin, C.E., and Ranieri, R.L., Acid catalyzed rearrangements of dimethyl α,β,β-triaryl-α,β-epoxyethylphosphonates, *Phosphorus*, 6, 161, 1976.

141. Ulman, A., and Sprecher, M., Thermal and photochemical reactions of sodium salts of β-phosphono tosylhydrazones, *J. Org. Chem.*, 44, 3703, 1979.

142. Teulade, M.-P., and Savignac, P., Synthesis of 1-formyl 1,1-dialkylmethylphosphonates, *Synth. Commun.*, 17, 125, 1987.

143. Griffin, C.E., and Kundu, S.K., Phosphonic acids and esters. Part 20. Preparation and ring-opening reactions of α,β- and β,γ-epoxyalkylphosphonates. The proton magnetic resonance spectra of vicinally substituted ethyl- and propylphosphonates, *J. Org. Chem.*, 34, 1532, 1969.

144. Agawa, T., Kubo, T., and Ohshiro, Y., Synthesis of dialkyl epoxyethylphosphonates, *Synthesis*, 27, 1971.

145. Baboulene, M., and Sturtz, G., Reactions of diethyl 1,2-epoxyalkanephosphonates. A new synthesis of 2-oxoalkanephosphonates, *Synthesis*, 456, 1978.

146. Schrepfer, H.J., Siegel, H., and Theobald, H., Manufacture of α-formylethanephosphonic acid esters, *BASF A.-G.*, German Patent Appl. DE 2715923, 1977; *Chem. Abstr.*, 90, 23253g, 1979.

147. Mizushima, E., Han, L.P., Hayashi, T., and Tanaka, M., Preparation of (formylethyl)phosphine oxides and (formylethyl)phosphonates, *Ministry of Economy, Trade and Industry & National Industrial Research Institute, Japan*, Japanese Patent Appl. JP 2001253890, 2001; *Chem. Abstr.*, 135, 242341, 2001.

148. Saunders, B.C., and Simpson, P., Esters containing phosphorus. Part 18. Esters of ethynylphosphonic acid, *J. Chem. Soc.*, 3351, 1963.

149. Kouno, R., Tsubota, T., Okauchi, T., and Minami, T., Synthesis and synthetic application of α-formylvinylphosphonates. Facile synthesis of phosphono-substituted heterocyclic compounds, *J. Org. Chem.*, 65, 4326, 2000.

150. Malet, R., Moreno-Mañas, M., and Pleixats, R., Palladium-catalyzed preparation of dialkyl allylphosphonates. A new preparation of diethyl 2-oxoethylphosphonate, *Synth. Commun.*, 22, 2219, 1992.

151. Yamashita, M., Nomoto, H., and Imoto, H., Preparation of dimethyl (1-formylalkyl)phosphonates via singlet oxygen adducts, *Synthesis*, 716, 1987.

152. Yamashita, M., Sugiura, M., Oshikawa, T., and Inokawa, S., Synthesis of dimethyl (1-nitromethyl-alkyl)phosphonates and their conversion to dimethyl (1-formylalkyl)phosphonates by oxidation with ozone, *Synthesis*, 62, 1987.

153. Normant, H., and Sturtz, G., Preparation of phosphonates containing a carbonyl function, *C.R. Acad. Sci., Ser. C*, 2366, 1961.

154. Gruszeka, E., Mastalerz, P., and Soroka, M., New synthesis of phosphinothricin and analogs, *Rocz. Chem.*, 49, 2127, 1975.

155. Cox, J. M., Bellini, P., Barrett, R., Ellis, R.M. and Hawkes, T.R., Preparation of triazolylalkylphosphonic acids as herbicides, *Zeneca*, U.S. Patent Appl. US 5393732, 1995; *Chem. Abstr.*, 120, 134813, 1994.

156. de Macedo Puyau, P., and Perié, J.J., Synthesis of substrate analogues and inhibitors for the phosphoglycerate mutase enzyme, *Phosphorus, Sulfur Silicon Relat. Elem.*, 129, 13, 1997.

157. Lavielle, G., Sturtz, G., and Normant, H., The action of nucleophilic reagents on γ-halogeno-β-ethylenephosphonates, *Bull. Soc. Chim. Fr.*, 4186, 1967.

158. Obrycki, R., and Griffin, C.E., Phosphonic acids esters. Part 19. Syntheses of substituted phenyl- and arylphosphonates by the photoinitiated arylation of trialkyl phosphites, *J. Org. Chem.*, 33, 632, 1968.

159. Engel, R., *Synthesis of Carbon–Phosphorus Bonds*, CRC Press, Boca Raton, 1988, p. 137.

160. Kamai, G., and Kukhtin, V.A., Addition of trialkyl phosphites to α,β-unsaturated aldehydes, *Dokl. Akad. Nauk SSSR, Ser. Khim.*, 112, 868, 1957; *Chem. Abstr.*, 51, 13742f, 1957.

161. Kamai, G., and Kukhtin, V.A., Addition of complete esters of phosphorous acid and phosphonous acids to conjugated systems. Part 2. Addition of trialkyl phosphites to acrolein and crotonaldehyde, *Zh. Obshch. Khim.*, 27, 2376, 1956; *Chem. Abstr.*, 52, 7127h, 1958.

162. Block, H.D., Phosphorus-containing aldehydes, *Bayer A.-G.*, German Patent Appl. DE 2516341, 1976; *Chem. Abstr.*, 86, 55580v, 1977.

163. Merger, F., and Fouquet, G., 1,1-Dialkoxy-3-(dialkoxyphosphono)propanes, *BASF A.-G.*, German Patent Appl. DE 2517448, 1976; *Chem. Abstr.*, 86, 72869p, 1977.

164. Harvey, R.G., Reactions of triethyl phosphite with activated olefins, *Tetrahedron*, 22, 2561, 1966.

165. Okamoto, Y., A convenient preparation of dialkyl 3-(dialkoxyphosphinyl)-1-propenyl phosphate derivatives. Addition of the mixed reagent trialkyl phosphite/dialkyl phosphorochloridate to α,β-unsaturated aldehyde, *Chem. Lett.*, 87, 1984.

166. Okamoto, Y., and Azuhata, T., Synthesis of dialkyl 3-(dialkylphosphinyloxy)-2-alkenephosphonate and diphenyl-[3-(diphenylphosphinyloxy)-2-propenyl]phosphine oxide, *Synthesis*, 941, 1986.

167. Evans, D.A., Hurst, K.M., Truesdale, L.K., and Takacs, J.M., The carbonyl insertion reactions of mixed tervalent phosphorus–organosilicon reagents, *Tetrahedron Lett.*, 18, 2495, 1977.

168. Evans, D.A., Hurst, K.M., and Takacs, J.M., New silicon–phosphorus reagents in organic synthesis. Carbonyl conjugate addition reactions of silicon phosphite esters and related systems, *J. Am. Chem. Soc.*, 100, 3467, 1978.

169. Teulade, M.-P., and Savignac, P., A new and easy synthesis of diethyl 2-formylalkylphosphonates, *Synthesis*, 1037, 1987.

170. Rudinskas, A.J., and Hullar, T.L., Pyridoxal phosphate. Part 5. 2-Formylethynylphosphonic acid and 2-formylethylphosphonic acid, potent inhibitors of pyridoxal phosphate binding probes of enzyme topography, *J. Med. Chem.*, 19, 1367, 1976.

171. Rudinskas, A.J., and Hullar, T.L., Phosphonic acid chemistry. Part 1. Synthesis and dienophilic properties of diethyl 2-formylvinylphosphonate and diethyl 2-formylethynylphosphonate, *J. Org. Chem.*, 41, 2411, 1976.

172. Page, P., Blonski, C., and Perié, J., An improved chemical and enzymatic synthesis of new fructose derivatives for import studies by the glucose transporter in parasites, *Tetrahedron*, 52, 1557, 1996.

173. von der Osten, C.H., Sinskey, A.J., Barbas, C.F., III, Pederson, R.L., Wang, Y.F., and Wong, C.H., Use of a recombinant bacterial fructose-1,6-diphosphate aldolase in aldol reactions. Preparative syntheses of 1-deoxynojirimycin, 1-deoxymannojirimycin, 1,4-dideoxy-1,4-imino-D-arabinitol, and fagomine, *J. Am. Chem. Soc.*, 111, 3924, 1989.

174. Just, G., Potvin, P., and Hakimelahi, G.H., 3-Diethylphosphonoacrolein diethylthioacetal anion (6*), a reagent for the conversion of aldehydes to α,β-unsaturated ketene dithioacetals and three-carbon homologated α,β-unsaturated aldehydes, *Can. J. Chem.*, 58, 2780, 1980.

175. Hannour, S., Ryglowski, A., Roumestant, M.L., Viallefont, P., Martinez, J., Ouazzani, F., and El Hallaoui, A., Enantiospecific synthesis of α-amino phoshonic acids, *Phosphorus, Sulfur Silicon Relat. Elem.*, 134/135, 419, 1998.

176. Yamazaki, S., Takada, T., Imanishi, T., Moriguchi, Y., and Yamabe, S., Lewis acid-promoted [2 + 1] cycloaddition reactions of a 1-seleno-2-silylethene to 2-phosphonoacrylates. Stereoselective synthesis of a novel functionalized α-aminocyclopropanephosphonic acid, *J. Org. Chem.*, 63, 5919, 1998.

177. Blackburn, G.M., and Rashid, A., Stereoselective syntheses of the methylene- and α-fluoromethyl-enephosphonate analogues of 2-phospho-D-glyceric acid, *J. Chem. Soc., Chem. Commun.*, 317, 1988.

178. Mikolajczyk, M., Zurawinski, R., and Kielbasinski, P., A new synthesis of (±)-sarkomycin from a β-ketophosphonate, *Tetrahedron Lett.*, 30, 1143, 1989.

179. Gijsen, H.J.M., and Wong, C.-H., Synthesis of a cyclitol via a tandem enzymatic aldol-intramolecular Horner–Wadsworth–Emmons reaction, *Tetrahedron Lett.*, 36, 7057, 1995.

180. Balczewski, P., Pietrzykowski, W.M., and Mikolajczyk, M., Studies on the free radical carbon–carbon bond formation in the reaction of α-phosphoryl sulfides and selenides with alkenes, *Tetrahedron*, 51, 7727, 1995.

181. Mikolajczyk, M., Zurawinski, R., Kielbasinski, P., Wieczorek, M.W., Blaszczyk, J., and Majzner, W.R., Total synthesis of racemic and optically active sarkomycin, *Synthesis*, 356, 1997.

182. Chambers, R.D., Jaouhari, R., and O'Hagan, D., Fluorine in enzyme chemistry. Part 1. Synthesis of difluoromethylenephosphonate derivatives as phosphate mimics, *J. Fluorine Chem.*, 44, 275, 1989.

183. Chambers, R.D., Jaouhari, R., and O'Hagan, D., Fluorine in enzyme chemistry. Part 2. The preparation of difluoromethylenephosphonate analogues of glycolytic phosphates. Approaching an isosteric and isoelectronic phosphate mimic, *Tetrahedron*, 45, 5101, 1989.

184. Austin, R.E., and Cleary, D.G., Synthesis of a difluorophosphonate analog of the oxathiolanyl nucleoside (−)-β-L-(2R,5S)-1,3-oxathianyl-5-fluorocytosine (FTC), *Nucleosides Nucleotides*, 14, 1803, 1995.

185. Sturtz, G., and Pondaven-Raphalen, A., Synthesis of keto- and aldophosphonates by applying the Wacker method to the corresponding ethylenic compounds, *J. Chem. Res. (M)*, 2512, 1980.

186. Peyman, A., Uhlmann, E., Winkler, I., Helsberg, M., and Meichsner, C., 2-Formylbenzylphosphonic acid derivatives useful for the treatment of diseases caused by viruses, *Hoechst A.-G.*, Eur. Patent Appl. EP 433928, 1991; *Chem. Abstr.*, 115, 136389, 1991.

187. Natchev, I.A., Total synthesis, enzyme–substrate interactions and herbicidal activity of plumbemicin A and B (N-1409), *Tetrahedron*, 44, 1511, 1988.

188. Page, P., Blonski, C., and Perié, J., An improved synthesis of phosphonomethyl analogues of glyceraldehyde-3-phosphate and dihydroxy-acetone phosphate, *Tetrahedron Lett.*, 36, 8027, 1995.

189. Page, P., Blonski, C., and Perié, J., Synthesis of phosphono analogues of dihydroxyacetone phosphate and glyceraldehyde 3-phosphate, *Bioorg. Med. Chem.*, 7, 1403, 1999.

190. Minami, T., Isonaka, T., Okada, Y., and Ichikawa, J., Copper(I) salt-mediated arylation of phosphinyl-stabilized carbanions and synthetic application to heterocyclic compounds, *J. Org. Chem.*, 58, 7009, 1993.

191. Krawczyk, H., Michael addition mediated by an internal catalyst. A novel route to 2-diethylphosphonoalkanoic acids, *Synlett*, 1114, 1998.

192. Al-Badri, H., About-Jaudet, E., and Collignon, N., New and efficient synthesis of *E*-4-diethoxyphosphonyl-2-methyl-2-butenal and of ethyl *E*-4-diethoxyphosphonyl-2-methyl-2-butenoate, important building blocks in retinoid chemistry, *Tetrahedron Lett.*, 36, 393, 1995.

193. Fouqué, D., About-Jaudet, E., Collignon, N., and Savignac, P., A convenient synthesis of diethyl 2,4-dioxoalkylphosphonates, *Synth. Commun.*, 22, 219, 1992.

194. Mata, E.G., and Thomas, E.J., Development of a synthesis of lankacidins. An investigation into 17-membered ring formation, *J. Chem. Soc., Perkin Trans. 1*, 785, 1995.

195. Hanrahan, J.R., Taylor, P.C., and Errington, W., The synthesis of 3-phosphonocyclobutyl amino acid analogues of glutamic acid via diethyl 3-oxycyclobutylphosphonate, a versatile synthetic intermediate, *J. Chem. Soc., Perkin Trans. 1*, 493, 1997.

196. Ott, G.R., and Heathcock, C.H., A method for constructing the C44-C51 side chain of altohyrtin C, *Org. Lett.*, 1, 1475, 1999.

197. Rein, T., Åkermark, B., and Helquist, P., Synthesis of polyconjugated aldehydes using a new Horner–Wadsworth–Emmons reagent, *Acta Chem. Scand., Ser. B*, 42, 569, 1988.

198. Kann, N., Rein, T., Åkermark, B., and Helquist, P., New functionalized Horner–Wadsworth–Emmons reagents. Useful building blocks in the synthesis of polyunsaturated aldehydes. A short synthesis of (±)-(*E,E*)-coriolic acid, *J. Org. Chem.*, 55, 5312, 1990.

199. Dappen, M.S., Pellicciari, R., Natalini, B., Monahan, J.B., Chiorri, C., and Cordi, A.A., Synthesis and biological evaluation of cyclopropyl analogues of 2-amino-5-phosphonopentanoic acid, *J. Med. Chem.*, 34, 161, 1991.

200. Yokomatsu, T., Yamagishi, T., Suemune, K., Abe, H., Kihara, T., Soeda, S., Shimeno, H., and Shibuya, S., Stereoselective reduction of cyclopropylalkanones possessing a difluoromethylenephosphonate group at the ring. Application to stereoselective synthesis of novel cyclopropane nucleotide analogues, *Tetrahedron*, 56, 7099, 2000.

201. Ambler, S.J., Dell, C.P., Gilmore, J., Hotten, T.M., Sanchiz Suarez, A., Simmonds, R.G., Timms, G.H., Tupper, D.E., and Urquhart, M.W.J., Novel *N*-substituted α-amino acid amides as calcium channel modulators, *Lilly Industries*, Eur. Patent Appl. EP 805147, 1997; *Chem. Abstr.*, 128, 13437, 1997.

202. Han, C.-N.A., Iwata, C., and Metzler, D.E., Phosphonate analogues of pyridoxal phosphate with shortened side chains, *J. Med. Chem.*, 26, 595, 1983.

203. Yokomatsu, T., Nakabayashi, N., Matsumoto, K., and Shibuya, S., Lipase-catalysed kinetic resolution of *cis*-1-diethylphosphonomethyl-2-hydroxymethylcyclohexane. Application to enantioselective synthesis of 1-diethylphosphonomethyl-2-(5'-hydantoinyl)cyclohexane, *Tetrahedron: Asymmetry*, 6, 3055, 1995.

204. Paladino, J., Guyard, C., Thurieau, C., and Fauchère, J.-L., Enantioselective synthesis of (2*S*)-2-amino-3-(4-hydroxy-3-phosphonophenyl)propionic acid (= 3'-phosphono-L-tyrosine) and its incorporation into peptides, *Helv. Chim. Acta*, 76, 2465, 1993.

205. Shakespeare, W.C., Bohacek, R.S., Narula, S.S., Azimioara, M.D., Yuan, R.W., Dalgarno, D.C., Madden, L., Botfield, M.C., and Holt, D.A., An efficient synthesis of a 4'-phosphonodifluoromethyl-3'-formyl-phenylalanine containing Src SH2 ligand, *Bioorg. Med. Chem. Lett.*, 9, 3109, 1999.

206. Albrecht, H.P., Jones, G.H., and Moffatt, J.G., Homonucleoside phosphonic acids. Part 1. A general synthesis of isosteric phosphonate analogs of nucleoside 3'-phosphates, *Tetrahedron*, 40, 79, 1984.

207. Kraus, G.A., and Jones, C., The reaction of ketone enolates with a δ-oxo phosphonate. A carbanion-based [4 + 2] annulation, *Synlett*, 793, 2001.

208. Guanti, G., Banfi, L., and Zannetti, M.T., Phosphonic derivatives of carbohydrates. Chemoenzymatic synthesis, *Tetrahedron Lett.*, 41, 3181, 2000.

209. Guanti, G., Zannetti, M.T., Banfi, L., and Riva, R., Enzymatic resolution of acetoxyalkenylphospho-nates and their exploitation in the chemoenzymatic synthesis of phosphonic derivatives of carbohy-drates, *Adv. Synth. Catal.*, 343, 682, 2001.

210. Schick, A., Kolter, T., Giannis, A., and Sandhoff, K., Synthesis of phosphonate analogues of sphin-ganine-1-phosphate and sphingosine-1-phosphate, *Tetrahedron*, 51, 11207, 1995.

211. Schick, A., Schwarzmann, G., Kolter, T., and Sandhoff, K., Synthesis of tritium labeled phosphonate analogs of sphinganine-1-phosphate, *J. Labelled Compd. Radiopharm.*, 39, 441, 1997.

212. Parikh, J.R., Analogs of nucleotides. Part 2. Phosphonate esters of ribose and glucopyranosyl purine derivatives, *J. Am. Chem. Soc.*, 79, 2778, 1957.

213. Wolff, M.E., and Burger, A., Analogs of nucleotides. Part 3. Syntheses in the series of adenosine phosphonate derivatives, *J. Am. Pharm. Assoc.*, 48, 56, 1959.

214. Czekanski, T., Witek, S., Costisella, B., and Gross, H., Synthesis of β-nitroalkenylbenzylphosphoryl derivatives, *J. Prakt. Chem.*, 323, 353, 1981.

215. Diez-Barra, E., Garcia-Martinez, J.C., Merino, S., del Rey, R., Rodriguez-Lopez, J., Sanchez-Verdu, P., and Tejeda, J., Synthesis, characterization, and optical response of dipolar and non-dipolar poly(phe-nylenevinylene) dendrimers, *J. Org. Chem.*, 66, 5664, 2001.

216. Takimiya, K., Otsubo, T., and Ogura, F., Naphtho[1,8-*bc*:5,4-*b'c'*]dithiophene. A new heteroarene isoelectronic with pyrene, *J. Chem. Soc., Chem. Commun.*, 1859, 1994.

217. Sentemov, V.V., Krasil'nikova, E.A., and Berdnik, I.V., Mechanism of the catalysis of the Arbuzov reaction by complex compounds of transition metals. Part 3. Influence of the nature of the aryl halide used on the ease of Arbuzov reaction catalyzed by metal complexes, *Zh. Obshch. Khim.*, 59, 2692, 1989; *J. Gen. Chem. USSR (Engl. Transl.)*, 59, 2406, 1989.

218. Issleib, K., Döpfer, K.-P., and Balszuweit, A., Aminodiphosphonic acids and diaminodiphosphonic acids. Synthesis and transamination behavior, *Phosphorus Sulfur*, 14, 171, 1983.

219. Minami, T., Watanabe, K., Chikugo, T., and Kitajima, Y., A new synthesis of γ-lactones with α,β-fused ring systems using α-diethoxyphosphinyl-Δ^{α,β}-butenolides, *Chem. Lett.*, 2369, 1987.

220. Minami, T., Nakayama, M., Fujimoto, K., and Matsuo, S., A new approach to cyclopentane annulated compounds via 1-(cyclopent-1-enylcarbonyl)vinylphosphonates, *J. Chem. Soc., Chem. Commun.*, 190, 1992.

221. Minami, T., Nakayama, M., Fujimoto, K., and Matsuo, S., A new approach to cyclopentane annulated compounds via 1-(cyclopent-1-enylcarbonyl)vinylphosphonates, and synthesis and synthetic applica-tion of α-diethoxyphosphoryl-Δ^{α,β}-butenolides, *Phosphorus, Sulfur Silicon Relat. Elem.*, 75, 135, 1993.

222. Roubaud, V., Mercier, A., Olive, G., Le Moigne, F., and Tordo, P., 5-(Diethoxyphosphorylmethyl)-5-methyl-4,5-dihydro-3*H*-pyrrole *N*-oxide. Synthesis and evaluation of spin trapping properties, *J. Chem. Soc., Perkin Trans. 2*, 1827, 1997.

223. Lewis, R.T., and Motherwell, W.B., An improved preparation of cyclopropyl phosphonates and their application in arylidene cyclopropane formation, *Tetrahedron Lett.*, 29, 5033, 1988.

224. Huang, X., Zhang, C., and Lu, X., A convenient stereoselective synthesis of 1,3-dienylphosphonates and 1-en-3-ynylphosphonates and their phosphine oxide analogs, *Synthesis*, 769, 1995.

225. Altenbach, H.-J., Holzapfel, W., Smerat, G., and Finkler, S.H., A simple, regiospecific synthesis of cycloalkenones and lactones, *Tetrahedron Lett.*, 26, 6329, 1985.

226. Hamilton, G.S., Huang, Z., Yang, X.-J., Patch, R.J., Narayanan, B.A., and Ferkany, J.W., Asymmetric synthesis of a potent selective competitve NMDA antagonist, *J. Org. Chem.*, 58, 7263, 1993.

227. Hullar, T.L., Pyridoxal phosphate. Part 1. Phosphonic acid analogs of pyridoxal phosphate. Synthesis via Wittig reactions and enzymic evaluation, *J. Med. Chem.*, 12, 58, 1969.

228. Stepanova, S.V., L'vova, S.D., and Gunar, V.I., Synthesis of 5'-modified phosphonate analogs of pyridoxal 5'-phosphate, *Bioorg. Khim.*, 4, 682, 1978; *Sov. J. Bioorg. Chem. (Engl. Transl.)*, 4, 498, 1978.

229. Zucco, M., Le Bideau, F., and Malacria, M., Palladium-catalyzed intramolecular cyclization of vinyl-oxirane. Regioselective formation of cyclobutanol derivative, *Tetrahedron Lett.*, 36, 2487, 1995.

230. Mitchell, M., Qaio, L., and Wong, C.-H., Chemical-enzymatic synthesis of iminocyclitol phosphonic acids, *Adv. Synth. Catal.*, 343, 596, 2001.

231. Yokomatsu, I., Hayakawa, Y., Kihara, T., Soeda, S., Urano, K., and Shibuya, A., Purine derivatives and purine nucleoside phosphorylase inhibitors containing them, *Hisamitsu Pharmaceutical*, Japanese Patent Appl. JP 2002080486, 2002; *Chem. Abstr.*, 136, 241663, 2002.

232. Bigge, C.F., Drummond, J.T., Johnson, G., Malone, T., Probert, A.W., Jr., Marcoux, F.W., Coughenour, L.L., and Brahce, L.J., Exploration of phenyl-spaced 2-amino-(5–9)-phosphonoalkanoic acids as competitive N-methyl-D-aspartic acid antagonists, *J. Med. Chem.*, 32, 1580, 1989.

233. Meier, H., Lehmann, M., and Kolb, U., Stilbenoid dendrimers, *Chem. Eur. J.*, 6, 2462, 2000.

234. Paulsen, H., and Bartsch, W., Phosphorus-containing carbohydrates. Part 13. Rearrangement of sugar-1-enophosphonates to sugar-2-enophosphonates, *Chem. Ber.*, 108, 1745, 1975.

235. Gil, J.M., Hah, J.H., Park, K.Y., and Oh, D.Y., One pot synthesis of mono- and spirocyclic α-phosphonato-α,β-unsaturated cycloenones, *Tetrahedron Lett.*, 39, 3205, 1998.

236. Griffin, B.S., and Burger, A., D-Glucopyranose-6-deoxy-6-phosphonic acid, *J. Am. Chem. Soc.*, 78, 2336, 1956.

237. Burke, T.R., Jr., Li, Z.-H., Bolen, J.B., and Marquez, V.E., Phosphonate-containing inhibitors of tyrosine-specific protein kinases, *J. Med. Chem.*, 34, 1577, 1991.

238. Burke, T.R., Jr., Russ, P., and Lim, B., Preparation of 4-[*bis*(*tert*-butoxy)phosphorylmethyl]-N-(fluoren-9-ylmethoxycarbonyl)-D,L-phenylalanine. A hydrolytically stable analogue of O-phosphotyrosine potentially suitable for peptide synthesis, *Synthesis*, 1019, 1991.

239. Burke, T.R., Jr., Smyth, M.S., Nomizu, M., Otaka, A., and Roller, P.P., Preparation of fluoro- and hydroxy-4-(phosphonomethyl)-D,L-phenylalanine suitably protected for solid-phase synthesis of peptides containing hydrolytically stable analogues of O-phosphotyrosine, *J. Org. Chem.*, 58, 1336, 1993.

240. Burke, T.R., Jr., Smyth, M.S., Otaka, A., and Roller, P.P., Synthesis of 4-phosphono(difluoromethyl)-D,L-phenylalanine and N-Boc and N-Fmoc derivatives suitably protected for solid-phase synthesis of nonhydrolysable phosphotyrosyl peptide analogues, *Tetrahedron Lett.*, 34, 4125, 1993.

241. Burke, T.R., Jr., Lim, B.B., and Smyth, M.S., Process of making benzylic α,α-difluorophosphonates from benzylic α-ketophosphonates, *US Health Dept.*, U.S. Patent Appl. US 5264607, 1993; *Chem. Abstr.*, 124, 203096, 1996.

242. Roberts, M.R., and Schlessinger, R.H., Total synthesis of *dl*-helenalin, *J. Am. Chem. Soc.*, 101, 7626, 1979.

243. Snider, B.B., and Yang, K., A short enantiospecific synthesis of the ceroplastin nucleus, *J. Org. Chem.*, 57, 3615, 1992.

244. Comins, D.L., and Ollinger, C.G., Inter- and intramolecular Horner–Wadsworth–Emmons reactions of 5-(diethoxyphosphoryl)-1-acyl-2-alkyl(aryl)-2,3-dihydro-4-pyridones, *Tetrahedron Lett.*, 42, 4115, 2001.

245. Pevzner, L.M., Terekhova, M.I., Ignat'ev, V.M., Petrov, E.S., and Ionin, B.I., Synthesis and CH acidities of some diethyl (furylmethyl)phosphonates, *Zh. Obshch. Khim.*, 54, 1990, 1984; *J. Gen. Chem. USSR (Engl. Transl.)*, 54, 1775, 1984.

246. Pevzner, L.M., Ignat'ev, V.M., and Ionin, B.I., Reaction of diethyl furfurylphosphonate with acrolein, crotonaldehyde, and mesityl oxide, *Zh. Obshch. Khim.*, 54, 1200, 1984; *J. Gen. Chem. USSR (Engl. Transl.)*, 54, 1074, 1984.

247. Pevzner, L.M., Ignat'ev, V.M., and Ionin, B.I., Reaction of 3-(diethoxyphosphorylmethyl)furans with α,β-unsaturated carbonyl compounds, *Zh. Obshch. Khim.*, 67, 241, 1997; *J. Gen. Chem. USSR (Engl. Transl.)*, 67, 224, 1997.

248. Xu, Y., Jin, X., Huang, G., and Huang, Y., A facile synthesis of diethyl 2-arylethenephosphonates, *Synthesis*, 556, 1983.

249. Wada, E., Kanemasa, S., and Tsuge, O., New synthesis of 2-oxo-3-alkenylphosphonates and hetero Diels–Alder reactions with vinyl ethers leading to 5-substituted 2-phosphinyl-2-cyclohexen-1-ones, *Bull. Chem. Soc. Jpn.*, 62, 860, 1989.

250. Snider, B.B., and Phillips, G.B., Ene reactions of 2-phosphonoacrylates, *J. Org. Chem.*, 48, 3685, 1983.

251. Nakajima, N., Hamada, T., Tanaka, T., Oikawa, Y., and Yonemitsu, O., Stereoselective synthesis of pikronolide, the aglicon of the 14-membered ring macrolide pikromycin, from D-glucose. Role of MPM and DMPM protection, *J. Am. Chem. Soc.*, 108, 4645, 1986.

252. Nakajima, N., Tanaka, T., Hamada, T., Oikawa, Y., and Yonemitsu, O., Highly stereoselective total synthesis of pikronolide, the aglicon of the first macrolide antibiotic pikromycin. Crucial role of benzyl-type protecting groups removable by 2,4-dichloro-5,6-dicyanobenzoquinone oxidation, *Chem. Pharm. Bull.*, 35, 2228, 1987.

253. Keck, G.E., Palani, A., and McHardy, S.F., Total synthesis of (+)-carbonolide B, *J. Org. Chem.*, 59, 3113, 1994.

254. Nishida, A., Yagi, K., Kawahara, N., Nishida, M., and Yonemitsu, O., Chemical modification of erythromycin A. Synthesis of the C1–C9 fragment from erythromycin A and reconstruction of the macrolactone ring, *Tetrahedron Lett.*, 36, 3215, 1995.

255. Nicotra, F., Panza, L., Russo, G., Senaldi, A., Burlini, N., and Tortora, P., Stereoselective synthesis of the isosteric bisphosphono analogue of β-D-fructose-2,6-bisphosphate, *J. Chem. Soc., Chem. Commun.*, 1396, 1990.

256. D'Onofrio, F., Piancatelli, G., and Nicolai, M., Photo-oxidation of 2-furylalkylphosphonates. Synthesis of new cyclopentenone derivatives, *Tetrahedron*, 51, 4083, 1995.

257. Ditrich, K., Bube, T., Stürmer, R., and Hoffmann, R.W., Total synthesis of mycinolide V, the aglycone of a macrolide antibiotic of the mycinamycin series, *Angew. Chem., Int. Ed. Engl.*, 25, 1028, 1986.

258. Hoffmann, R.W., and Ditrich, K., Total synthesis of mycinolide V, *Liebigs Ann. Chem.*, 23, 1990.

259. Ditrich, K., Total synthesis of methynolide, *Liebigs Ann. Chem.*, 789, 1990.

260. Tius, M.A., and Fauq, A.H., Total synthesis of (+)-desepoxyasperdiol, *J. Am. Chem. Soc.*, 108, 1035, 1986.

261. Tius, M.A., and Fauq, A.H., Total synthesis of (–)-asperdiol, *J. Am. Chem. Soc.*, 108, 6389, 1986.

262. Marshall, J.A., and DeHoff, B.S., Stereoselective total synthesis of the cembranolide diterpene anisomelic acid, *Tetrahedron Lett.*, 40, 4873, 1986.

263. Marshall, J.A., and DeHoff, B.S., Cembranolide total synthesis. Anisomelic acid, *Tetrahedron*, 43, 4849, 1987.

264. Marshall, J.A., and DuBay, W.J., Synthesis of the pseudopterane and furanocembrane ring systems by intraannular cyclization of β and γ-alkynyl allylic alcohols, *J. Org. Chem.*, 59, 1703, 1994.

265. Kodama, M., Shiobara, Y., Sumitomo, H., Fukuzumi, K., Minami, H., and Miyamoto, Y., Synthesis of macrocyclic terpenoids by intramolecular cyclization. Part 10. Total synthesis of methyl ceriferate-I, *Tetrahedron Lett.*, 27, 2157, 1986.

266. Kodama, M., Shiobara, Y., Sumitomo, H., Fukuzumi, K., Minami, H., and Miyamoto, Y., Synthesis of macrocyclic terpenoids by intramolecular cyclization. Part 12. Total synthesis of methyl ceriferate I, a 14-membered ring sesterterpene from scale insects, *J. Org. Chem.*, 53, 1437, 1988.

267. Roush, W.R., Warmus, J.S., and Works, A.B., Synthesis and transannular Diels–Alder reactions of (E,E,E)-cyclotetradeca-2,8,10-trienones, *Tetrahedron Lett.*, 34, 4427, 1993.

268. Nicolaou, K.C., Seitz, S.P., Pavia, M.R., and Petasis, N.A., Synthesis of macrocycles by intramolecular ketophosphonate reactions. Stereoselective construction of the "left-wing" of carbomycin B and a synthesis of dl-muscone from oleic acid, *J. Org. Chem.*, 44, 4011, 1979.

269. Nicolaou, K.C., Pastor, J., Winssinger, N., and Murphy, F., Solid phase synthesis of macrocycles by an intramolecular ketophosphonate reaction. Synthesis of a (dl)-muscone library, *J. Am. Chem. Soc.*, 120, 5132, 1998.

270. Smith, A.B., III, Dorsey, B.D., Visnick, M., Maeda, T., and Malamas, M.S., Synthesis of (–)-bertyadionol, *J. Am. Chem. Soc.*, 108, 3110, 1986.

271. Kawamura, M., and Ogasawara, K., Stereo- and enantio-controlled synthesis of (+)-juvabione and (+)-epijuvabione from (+)-norcamphor, *J. Chem. Soc., Chem. Commun.*, 2403, 1995.

272. Richardson, T.I., and Rychnovsky, S.D., Total synthesis of filipin III, *J. Am. Chem. Soc.*, 119, 12360, 1997.

273. Shimura, T., Komatsu, C., Matsumura, M., Shimada, Y., Ohta, K., and Mitsunobu, O., Preparation of the C1-C10 fragment of carbonolide B. A relay approach to carbomycin B, *Tetrahedron Lett.*, 38, 8341, 1997.

274. Wenkert, E., Guo, M., and Mancini, P., A formal total synthesis of the longipinenes, *Bull. Soc. Chim. Fr.*, 127, 714, 1990.

275. Tatsuta, K., Masuda, N., and Nishida, H., The first total synthesis of (±)-terpestacin, HIV syncytium formation inhibitor, *Tetrahedron Lett.*, 39, 83, 1998.

276. Hoye, T.R., Humpal, P.E., and Moon, B., Total synthesis of (–)-cylindrocyclophane A via a double Horner–Emmons macrocyclic dimerization event, *J. Am. Chem. Soc.*, 122, 4982, 2000.

277. Miyaoka, H., Baba, T., Mitome, H., and Yamada, Y., Total synthesis of marine diterpenoid stolonidiol, *Tetrahedron Lett.*, 42, 9233, 2001.

278. Büchi, G., and Wüest, H., Macrocycles by olefination of dialdehydes with 1,3-*bis*(dimethylphosphono)-2-propanone, *Helv. Chim. Acta*, 62, 2661, 1979.

279. Astles, P.C., and Thomas, E.J., Syntheses of thunbergols α– and β-cembra-2,7,11-triene-4,6-diols, *J. Chem. Soc., Perkin Trans. 1*, 845, 1997.

280. Paquette, L.A., Wang, T.-Z., Wang, S., and Philippo, C.M.G., Enantioselective synthesis of (+)-cleomeolide, the structurally unique diterpene lactone constituent of *Cleome viscosa*, *Tetrahedron Lett.*, 34, 3523, 1993.

281. Paquette, L.A., Wang, T.-Z., Philippo, C.M.G., and Wang, S., Total synthesis of the cembranoid diterpene lactone (+)-cleomeolide. Some remarkable conformational features of nine-membered belts linked in a 2,6-fashion to a methylenecyclohexane core, *J. Am. Chem. Soc.*, 116, 3367, 1994.

282. Moskva, V.V., and Razumov, A.I., Phosphinic and phosphinous acids. Part 34. Phosphorylated 1,3-dioxolanes, *Trans. Kazan. Khim.-Tekhnol. Inst.*, 34, 273, 1965; *Chem. Abstr.*, 68, 39729y, 1968.

283. Razumov, A.I., and Gurevich, P.A., Synthesis of phosphorylated 1,3-dioxoles, *Zh. Obshch. Khim.*, 38, 944, 1968; *J. Gen. Chem. USSR (Engl. Transl.)*, 38, 909, 1968.

284. Razumov, A.I., Gurevich, P.A., and Moskva, V.V., Synthesis of phosphorylated *m*-dioxanes, *Zh. Obshch. Khim.*, 37, 961, 1967; *J. Gen. Chem. USSR (Engl. Transl.)*, 37, 910, 1967.

285. Gurevich, P.A., Shelepova, N.I., and Razumov, A.I., Synthesis of cyclic alkylene acetals of phosphinyl-acetaldehydes, *Zh. Obshch. Khim.*, 38, 1905, 1968; *J. Gen. Chem. USSR (Engl. Transl.)*, 38, 1854, 1968.

286. Cooper, D., and Trippett, S., Some rearrangements of unsaturated phosphonate esters, *J. Chem. Soc., Perkin Trans. 1*, 2127, 1981.

287. Razumov, A.I., Gurevich, P.A., Baigil'dina, S.Y., Zykova, T.V., and Akhmadullina, A.V., 2-Phospho-rylated ethanols, *Zh. Obshch. Khim.*, 44, 459, 1974; *J. Gen. Chem. USSR (Engl. Transl.)*, 44, 439, 1974.

288. Paulsen, H., and Bartsch, W., Phosphorus-containing carbohydrates. Part 10. Reaction of carbonyl sugars with lithio methanephosphonic esters. Preparation of dimethyl 1-deoxy-D-fructose-1-phospho-nate, *Chem. Ber.*, 108, 1229, 1975.

289. Dondoni, A., Daninos, S., Marra, A., and Formaglio, P., Synthesis of ketosyl and ulosonyl phospho-nates by Arbuzov-type glycosidation of thiazolylketol acetates, *Tetrahedron*, 54, 9859, 1998.

290. Razumov, A.I., and Moskva, V.V., Derivatives of phosphonic and phosphinic acids. Part 26. Synthesis of phosphorylated carboxylic acids from aldehydes and acetals, *Zh. Obshch. Khim.*, 35, 1149, 1965; *J. Gen. Chem. USSR (Engl. Transl.)*, 35, 1151, 1965.

291. Isbell, A.F., Englert, L.F., and Rosenberg, H., Phosphonoacetaldehyde, *J. Org. Chem.*, 34, 755, 1969.

292. Moskva, V.V., Razumov, A.I., Sazonova, Z.Y., and Zykova, T.V., Reaction of phosphonoacetaldehydes with secondary amines, *Zh. Obshch. Khim.*, 41, 1874, 1971; *Chem. Abstr.*, 76, 3974a, 1972.

293. Razumov, A.I., Sokolov, M.P., Liorber, B.G., Moskva, V.V., Sazonova, Z.Y., and Loginova, N.G., Synthesis and properties of phosphorylated imines and enamines, *Zh. Obshch. Khim.*, 43, 1019, 1973; *J. Gen. Chem. USSR (Engl. Transl.)*, 43, 1012, 1973.

294. Alikin, A.Y., Liorber, B.G., Sokolov, M.P., Razumov, A.I., Zykova, T.V., and Salakhutdinov, R.A., Derivatives of organophosphorus acids. Part 105. Phosphorylated enamines and their *E,Z* isomeriza-tion, *Zh. Obshch. Khim.*, 52, 316, 1982; *J. Gen. Chem. USSR (Engl. Transl.)*, 52, 274, 1982.

295. Moskva, V.V., Sitdikova, T.S., Zykova, T.V., Alparova, M.V., and Shagvaleev, F.S., Reactions of α-phosphorylated carbonyl compounds with amino alcohols, *Zh. Obshch. Khim.*, 56, 1051, 1986; *J. Gen. Chem. USSR (Engl. Transl.)*, 56, 923, 1986.

296. Boeckman, R.K., Walters, M.A., and Koyano, H., Regiocontrolled metalation of diethyl β-dialkylami-novinylphosphonates. A new synthesis of substituted β-ketophosphonates, *Tetrahedron Lett.*, 30, 4787, 1989.

297. Palacios, F., Ochoa de Retana, A.M., and Oyarzabal, J., A simple synthesis of 3-phosphonyl-4-aminoquinolines from β-enaminophosphonates, *Tetrahedron*, 55, 5947, 1999.

298. Öhler, E., Kang, H.-S., and Zbiral, E., A convenient synthesis of new diethyl (2,4-dioxoalkyl)phos-phonates, *Synthesis*, 623, 1988.

299. Razumov, A.I., Liorber, B.G., Khammatova, Z.K., Sokolov, M.P., Zykova, T.V., and Alparova, M.V., Phosphonous, phosphinous, phosphonic and phosphinic compounds. Part 98. Phosphorylated β-nitro alcohols, *Zh. Obshch. Khim.*, 47, 567, 1977; *J. Gen. Chem. USSR (Engl. Transl.)*, 47, 519, 1977.

300. Kroona, H.B., Peterson, N.L., Koerner, J.F., and Johnson, R.L., Synthesis of the 2-amino-4-phospho-nobutanoic acid analogues (*E*)- and (*Z*)-2-amino-2,3-methano-4-phosphonobutanoic acid and their evaluation as inhibitors of hippocampal excitatory neurotransmission, *J. Med. Chem.*, 34, 1692, 1991.

301. Toshima, K., Jyojima, T., Yamaguchi, H., Noguchi, Y., Yoshida, T., Murase, H., Nakata, M., and Matsumura, S., Total synthesis of bafilomycin A_1, *J. Org. Chem.*, 62, 3271, 1997.

302. Koh, Y.J., and Oh, D.Y., A new synthesis of vinyl phosphonates from α-phenyl-β-oxo phosphonates and dialkyl phosphite, *Synth. Commun.*, 25, 2587, 1995.

303. Tang, K.C., Tropp, B.E., and Engel, R., The synthesis of phosphonic acid and phosphate analogues of glycerol-3-phosphate and related metabolites, *Tetrahedron*, 34, 2873, 1978.

304. Kazantsev, A.V., Averin, A.D., Lukashev, N.V., and Beletskaya, I.P., Addition of Me₃SiCN to α-, β- and γ-ketophosphonates and phosphinoylacetaldehyde as a route to phosphorylated 4α-hydroxycarboxylic acids and the corresponding nitriles, *Zh. Org. Khim.*, 34, 1495, 1998; *Russ. J. Org. Chem. (Engl. Transl.)*, 34, 1432, 1998.

305. Ismailov, V.M., Moskva, V.V., Dadasheva, L.A., Zykova, T.V., and Guseinov, F.I., Dichloro(dialkoxyphosphinyl)acetaldehydes, *Zh. Obshch. Khim.*, 52, 2140, 1982; *J. Gen. Chem. USSR (Engl. Transl.)*, 52, 1906, 1982.

306. Ismailov, V.M., Moskva, V.V., and Zykova, T.V., (Chloroformylmethyl)phosphonic esters, *Zh. Obshch. Khim.*, 53, 2793, 1983; *J. Gen. Chem. USSR (Engl. Transl.)*, 53, 2518, 1983.

307. Ismailov, V.M., Moskva, V.V., Guseinov, F.I., Zykova, T.V., and Sadyakov, I.S., Synthesis and properties of α-chloro-α-phosphinyl-substituted aldehydes, *Zh. Obshch. Khim.*, 56, 2005, 1986; *J. Gen. Chem. USSR (Engl. Transl.)*, 56, 1768, 1986.

308. Ismailov, V.M., Aydin, A., and Guseinov, F., Derivatives of α-phosphorylated aldehydes, *Tetrahedron*, 55, 8423, 1999.

309. Arbuzov, A.E., and Alimov, P.I., Products of condensation of esters of pyrophosphorous acid with aldehydes, *Izv. Akad. Nauk SSSR, Otdel. Khim. Nauk*, 530, 1951; *Chem. Abstr.*, 47, 696, 1953.

310. Kharasch, M.S., and Bengelsdorf, I.S., The reaction of triethyl phosphite with α-trichloromethyl carbonyl compounds, *J. Org. Chem.*, 20, 1356, 1955.

311. Mao, L.-J., and Chen, R.-Y., Highly selective Arbuzov reaction of α-chlorocarbonyl compounds with P(OEt)₃ and substituted amino urea, *Phosphorus, Sulfur Silicon Relat. Elem.*, 111, 167, 1996.

312. Sokolov, M.P., Buzykin, B.I., and Mavrin, G.V., 2-*Bis*(dialkylamino)phosphoryl-2-chloro(bromo)acetaldehydes, *Zh. Obshch. Khim.*, 60, 1986, 1990; *J. Gen. Chem. USSR (Engl. Transl.)*, 60, 1773, 1990.

313. Shagidullin, R.R., Pavlov, V.A., Buzykin, B.I., Aristova, N.V., Chertanova, L.F., Vandyukova, I.I., Plyamovatyi, A.K., Enikeev, K.M., Sokolov, M.P., and Moskva, V.V., Structures of (dialkoxyphosphoryl)haloacetaldehydes and their nitrosation products, *Zh. Obshch. Khim.*, 61, 1590, 1991; *J. Gen. Chem. USSR (Engl. Transl.)*, 61, 1459, 1991.

314. Buzykin, B.I., Sokolov, M.P., Zyablikova, T.A., and Chertanova, L.F., Halogenation of ethoxyvinylphosphonic acid dimorpholide, *Zh. Obshch. Khim.*, 62, 1489, 1992; *J. Gen. Chem. USSR (Engl. Transl.)*, 62, 1222, 1992.

315. Guseinov, F.I., Moskva, V.V., and Ismailov, V.M., Preparative methods of the synthesis of (dialkoxyphosphinyl)-mono- and -di-chloroacetaldehydes, *Zh. Obshch. Khim.*, 63, 93, 1993; *Russ. J. Gen. Chem. (Engl. Transl.)*, 63, 66, 1993.

316. Moskva, V.V., Guseinov, F.I., and Ismailov, V.M., Phosphorylated acetals, *Zh. Obshch. Khim.*, 57, 1668, 1987; *J. Gen. Chem. USSR (Engl. Transl.)*, 57, 1487, 1987.

317. Guseinov, F.I., Klimentova, G.Y., Kol'tsova, O.L., Egereva, T.N., and Moskva, V.V., Intramolecular nucleophilic reactions of dialkyl 1,1-dichloro-2-hydroxyethylphosphonates, *Zh. Obshch. Khim.*, 66, 455, 1996; *Russ. J. Gen. Chem. (Engl. Transl.)*, 66, 441, 1996.

318. Sokolov, M.P., Buzykin, B.I., and Pavlov, V.A., Oximes of (diisopropoxyphosphoryl)formyl halides, *Zh. Obshch. Khim.*, 60, 223, 1990; *J. Gen. Chem. USSR (Engl. Transl.)*, 60, 195, 1990.

319. Pavlov, V.A., Aristova, N.V., and Moskva, V.V., Novel aspects of the nitrosation of phosphorus-containing carbonyl compounds with a reactive methine group, *Dokl. Akad. Nauk SSSR, Ser. Khim.*, 315, 1137, 1990; *Dokl. Chem. (Engl. Transl.)*, 315, 358, 1990.

320. Pavlov, V.A., Aristova, N.V., Moskva, V.V., Makhaeva, G.F., Yankovskaya, V.L., and Malygin, V.V., Synthesis of phosphorus-containing formhydroxamoyl halides and their anticholine esterase properties, *Khim. Farm. Zh.*, 25, 231, 1991; *Pharm. Chem. J. (Engl. Transl.)*, 25, 255, 1991.

321. Pavlov, V.A., Tkachenko, S.E., Aristova, N.V., Moskva, V.V., Komalov, R.M., and Pushin, A.N., Features of reactions of (diisopropoxyphosphoryl)carbonyl halide oximes with nitrogen-containing nucleophilic reagents, *Zh. Obshch. Khim.*, 62, 1772, 1992; *J. Gen. Chem. USSR (Engl. Transl.)*, 62, 1457, 1992.

322. Zyablikova, T.A., Pavlov, V.A., Smith, J.A.S., and Liorber, B.G., Investigation of the structure of products of the nitrosation of (dialkoxyphosphinyl)acetaldehydes by NMR and ¹⁴N-NQR methods, *Zh. Obshch. Khim.*, 62, 1272, 1992; *J. Gen. Chem. USSR (Engl. Transl.)*, 62, 1046, 1992.

323. Redmore, D., Heterocyclic systems bearing phosphorus substituents. Synthesis and chemistry, *Chem. Rev.*, 71, 315, 1971.

324. Razumov, A.I., Liorber, B.G., and Gurevich, P.A., Synthesis of phosphorylated benzoxazoles, *Zh. Obshch. Khim.*, 37, 2782, 1967; *J. Gen. Chem. USSR (Engl. Transl.)*, 37, 2648, 1967.

325. Razumov, A.I., Liorber, B.G., and Gurevich, P.A., Synthesis of phosphorylated benzothiazolines, *Zh. Obshch. Khim.*, 38, 199, 1968; *J. Gen. Chem. USSR (Engl. Transl.)*, 38, 203, 1968.

326. Razumov, A.I., Gurevich, P.A., Liorber, B.G., and Borisova, T.B., Phosphonic and phosphonous derivatives. Part 47. Synthesis of phosphorylated heterocyclic compounds, *Zh. Obshch. Khim.*, 39, 392, 1969; *J. Gen. Chem. USSR (Engl. Transl.)*, 39, 369, 1969.

327. Razumov, A.I., and Gurevich, P.A., Phosphonic and phosphonous acid derivatives. Part 45. Synthesis some properties of phosphorylated benzimidazoles, *Zh. Obshch. Khim.*, 37, 1620, 1967; *J. Gen. Chem. USSR (Engl. Transl.)*, 37, 1537, 1967.

328. Razumov, A.I., and Gurevich, P.A., Phosphonic and phosphonous acids derivatives. Part 43. Some transformations of phosphorylated acetals, *Tr. Kazan. Khim. Tekhnol. Inst.*, 36, 480, 1967; *Chem. Abstr.*, 70, 20160a, 1969.

329. Haelters, J.P., Corbel, B., and Sturtz, G., Synthesis of indolphosphonates by Fischer cyclisation of arylhydrazones phosphonates, *Phosphorus Sulfur*, 37, 41, 1988.

330. Buzykin, B.I., Bredikhina, Z.A., Sokolov, M.P., and Gazetdinova, N.G., 1-Aryl-3-dialkoxyphosphoryl-1,2,4-triazoles from *C*-phosphorylated amidrazones, *Zh. Obshch. Khim.*, 62, 551, 1992; *J. Gen. Chem. USSR (Engl. Transl.)*, 62, 452, 1992.

331. Aristova, N.V., Gorin, B.I., Kovalenko, S.V., Pavlov, V.A., and Moskva, V.V., Phosphorylated isoxazoles, *Khim. Geterotsikl. Soedin.*, 1287, 1990; *Chem. Heterocycl. Compd. (Engl. Transl.)*, 1076, 1990.

332. Kouno, R., Okauchi, T., Nakamura, M., Ichikawa, J., and Minami, T., Synthesis and synthetic utilisation of α-functionalized vinylphosphonates bearing β-oxy or β-thio substituents, *J. Org. Chem.*, 63, 6239, 1998.

333. Nifant'ev, E.E., Patlina, S.I., and Matrosov, E.I., Synthesis of phosphorylated pyrazolones, *Khim. Geterotsikl. Soedin.*, 513, 1977; *Chem. Heterocycl. Compd. (Engl. Transl.)*, 414, 1977.

334. Chen, H., Qian, D.-Q., Xu, G.-X., Liu, Y.-X., Chen, X.-D., Shi, X.-D., Cao, R.-Z., and Liu, L.-Z., New strategy for the synthesis of phosphonyl pyrazoles, *Synth. Commun.*, 29, 4025, 1999.

335. Chen, H., Qian, D.-Q., Xu, G.-X., Liu, Y.-X., Chen, X.-D., Shi, X.-D., Cao, R.-Z., and Liu, L.-Z., The synthesis of phosphonyl pyrazoles, *Phosphorus, Sulfur Silicon Relat. Elem.*, 144, 85, 1999.

336. Regitz, M., Anschütz, W., and Liedhegener, A., Reactions of CH-active compounds with azides. Part 23. Synthesis of α-diazophosphonic acid esters, *Chem. Ber.*, 101, 3734, 1968.

337. Regitz, M., and Anschuetz, W., Reactions of CH-active compounds with azides. Part 27. Diethylphosphonodiazomethane, *Justus Liebigs Ann. Chem.*, 730, 194, 1969.

338. Johnson, A.W., *Ylid Chemistry*, Academic Press, New York, 1966.

339. Pudovik, A.N., and Yastrebova, G.E., Organophosphorus compounds with an active methylene group, *Usp. Khim.*, 39, 1190, 1970; *Russ. Chem. Rev.*, 39, 562, 1970.

340. Boutagy, J., and Thomas, R., Olefin synthesis with organic phosphonate carbanions, *Chem. Rev.*, 74, 87, 1974.

341. Wadsworth, W.S., Jr., Synthetic applications of phosphoryl-stabilised anions, *Org. React.*, 25, 73, 1977.

342. Walker, B.J., *Transformations via Phosphorus-Stabilised Anions. Part 2. PO-Activated Olefination*, Academic Press, New York, 1979, p. 155.

343. Johnson, A.W., Kaska, W.C., Ostoja Starzewski, K.A., and Dixon, D.A., *Ylides and Imines of Phosphorus*, John Wiley & Sons, New York, 1993.

344. Burton, D.J., Yang, Z.-Y., and Qiu, W., Fluorinated ylides and related compounds, *Chem. Rev.*, 96, 1641, 1996.

345. Martin, S.F., Synthesis of aldehydes, ketones, and carboxylic acids from lower carbonyl compounds by C-C coupling reactions, *Synthesis*, 633, 1979.

346. Abele, H., Haas, A., and Lieb, M., Efforts in synthesizing α,β-unsaturated trifluoromethyl-substituted aldehydes, *Chem. Ber.*, 119, 3502, 1986.

347. Haas, A., Lieb, M., and Zwingenberger, J., Synthesis of *bis*(trifluoromethyl)substituted olefins. Methods of preparation and properties, *Liebigs Ann. Org. Bioorg. Chem.*, 2027, 1995.

348. Meinke, P.T., O'Connor, S.P., Mrozik, H., and Fisher, M.H., Synthesis of ring-contracted, 25-*nor*-6,5-spiroketal-modified avermectin derivatives, *Tetrahedron Lett.*, 33, 1203, 1992.

349. Nagata, W., and Hayase, Y., Formylolefination of carbonyl compounds, *Tetrahedron Lett.*, 9, 4359, 1968.

350. Nagata, W., and Hayase, Y., 2-Hydrocarbyl aminovinyl phosphonates or phosphinates, *Shionogi*, German Patent Appl. DE 1911768, 1968; *Chem. Abstr.*, 72, 32128a, 1970.

351. Nagata, W., and Hayase, Y., Formylolefination of carbonyl compounds, *J. Chem. Soc. (C)*, 460, 1969.

352. Nagata, W., Yoshioka, M., Okumura, T., and Murakami, M., Hydrocyanation. Part 9. Synthesis of β-cyano-aldehydes by conjugate hydrocyanation of allylideneamines followed by hydrolysis, *J. Chem. Soc. (C)* 2355, 1970.

353. Venton, D.L., Enke, S.E., and LeBreton, G.C., Azaprostanoic acid derivatives. Inhibitors of arachidonic acid induced platelet aggregation, *J. Med. Chem.*, 22, 824, 1979.

354. Farnum, D.G., Ghandi, M., Raghu, S., and Reitz, T., A new entry to the $C_{12}H_{12}$ energy surface. Pyrolysis and photolysis of *trans*-β-[*anti*-9-bicyclo[6.1.0]nona-2,4,6-trienyl]acrolein tosylhydrazone salt, *J. Org. Chem.*, 47, 2598, 1982.

355. Nagata, W., Wakabayashi, T., and Hayase, Y., Preparation of α,β-unsaturated aldehydes via the Wittig reaction. Cyclohexylideneacetaldehyde, *Org. Synth. Coll.*, VI, 358, 1988.

356. Drew, J., Letellier, M., Morand, P., and Szabo, A.G., Synthesis from pregnenolone of fluorescent cholesterol analogue probes with conjugated unsaturation in the side chain, *J. Org. Chem.*, 52, 4047, 1987.

357. Tietze, L.F., Geissler, H., Fennen, J., Brumby, T., Brand, S., and Schultz, G., Simple and induced diastereoselectivity in intramolecular hetero-Diels–Alder reactions of 1-oxa-1,3-butadienes. Experimental data and calculations, *J. Org. Chem.*, 59, 182, 1994.

358. Deussen, H.-J., Hendrickx, E., Boutton, C., Krog, D., Clays, K., Bechgaard, K., Persoons, A., and Bjørnholm, T., Novel chiral *bis*-dipolar 6,6'-disubstituted binaphtol derivatives for second-order non-linear optics. Synthesis and linear and nonlinear optical properties, *J. Am. Chem. Soc.*, 118, 6841, 1996.

359. Hermann, K., and Dreiding, A.S., Total synthesis of betalaines, *Helv. Chim. Acta*, 60, 673, 1977.

360. Hilpert, H., and Dreiding, A.S., About the total synthesis of betalaines, *Helv. Chim. Acta*, 67, 1547, 1984.

361. Meyers, A.I., Tomioka, K., and Fleming, M.P., Convenient preparation of α,β-unsaturated aldehydes, *J. Org. Chem.*, 43, 3788, 1978.

362. Dolle, R.E., Armstrong, W.P., Shaw, A.N., and Novelli, R., Intramolecular [4 + 2] cycloaddition of α,β-unsaturated hydrazones as a route to annelated pyridines, *Tetrahedron Lett.*, 29, 6349, 1988.

363. Bushby, N., Moody, C.J., Riddick, D.A., and Waldron, I.R., Double intramolecular Diels–Alder reaction of α,β-unsaturated hydrazones. A new route to 2,2'-bipyridines, *J. Chem. Soc., Chem. Commun.*, 793, 1999.

364. Bushby, N., Moody, C.J., Riddick, D.A., and Waldron, I.R., Double intramolecular hetero Diels–Alder reactions of α,β-unsaturated hydrazones as 1-azadienes. A new route to 2,2'-bipyridines, *J. Chem. Soc., Perkin Trans. 1*, 2183, 2001.

365. Almirante, N., Cerri, A., Fedrizzi, G., Marazzi, G., and Santagostino, M., A general, [1 + 4] approach to the synthesis of 3(5)-substituted pyrazoles from aldehydes, *Tetrahedron Lett.*, 39, 3287, 1998.

366. Almirante, N., Benicchio, A., Cerri, A., Fedrizzi, G., Marazzi, G., and Santagostino, M., β-Tosylhydrazono phosphonates in organic synthesis. An unambiguous entry to polysubstituted pyrazoles, *Synlett*, 299, 1999.

367. Walz, A.J., and Sundberg, R.J., Diethyl formylmethylphosphonate as an acetaldehyde synthon for quinoline synthesis, *Synlett*, 75, 2001.

368. Politzer, I.R., and Meyers, A.I., Aldehydes from 2-benzyl-4,4,6-trimethyl-5,6-dihydro-1,3(4*H*)-oxazine. 1-Phenylcyclopentanecarboxaldehyde, *Org. Synth. Coll.*, VI, 905, 1988.

369. Molin, H., and Pring, B.G., Regio- and stereoselective synthesis of the carbocyclic analogue of 3-deoxy-β-D-mann-2-octulopyranosonic acid (β-KDO) from (−)-quinic acid, *Tetrahedron Lett.*, 26, 677, 1985.

370. Tay, M.K., Aboujaoude, E.E., Collignon, N., and Savignac, P., α-Lithiated phosphonates as functional transfert agents. Preparation of α,β-unsaturated-α-substituted aldehydes, *Tetrahedron Lett.*, 28, 1263, 1987.

371. Roberts, R.M., and Vogt, P.J., Ethyl *N*-phenylformimidate, *Org. Synth. Coll.*, IV, 464, 1963.

372. Darling, S.D., Muralidharan, F.N., and Muralidharan, V.B., Enolate alkylations with diethylbutadiene phosphonate. Part 2, *Tetrahedron Lett.*, 20, 2761, 1979.

373. McClure, C.K., and Jung, K.-Y., Pentacovalent oxaphosphorane chemistry in organic synthesis. Part 2. Total syntheses of (±)-*trans*- and (±)-*cis*-neocnidilides, *J. Org. Chem.*, 56, 2326, 1991.

374. La Nauze, J.M., and Rosenberg, H., The identification of 2-phosphonoacetaldehyde as an intermediate in the degradation of 2-aminoethylphosphonate by *Bacillus cereus*, *Biochim. Biophys. Acta*, 165, 438, 1968.

375. Warren, W.A., Biosynthesis of phosphonic acids in *Tetrahymena*, *Biochim. Biophys. Acta*, 156, 340, 1968.

376. Horiguchi, M., Biosynthesis of 2-aminoethylphosphonic acid in cell-free preparations from *Tetrahymena*, *Biochim. Biophys. Acta*, 261, 102, 1972.

377. Horiguchi, M., *Biochemistry of Natural C-P Compounds*, Japanese Association for Research on the Biochemistry of C–P Compounds, Tokyo, 1984, p. 88.

378. McQueney, M.S., Lee, S.-L., Bowman, E., Mariano, P.S., and Dunaway-Mariano, D., A remarkable pericyclic mechanism for enzyme-catalyzed P–C bond formation, *J. Am. Chem. Soc.*, 111, 6885, 1989.

379. Hammerschmidt, F., Biosynthesis of natural products with a P–C bond. Part 1. Incorporation of D-[6,6-D$_2$]glucose into (2-aminoethyl)phosphonic acid in *Tetrahymena thermophila*, *Liebigs Ann. Chem.*, 531, 1988.

380. Hammerschmidt, F., Biosynthesis of natural products with a P–C bond. Part 2. Hydroxylation of deuterated (2-aminoethyl)phosphonic acids to (2-amino-1-hydroxyethyl)phosphonic acid in *Acanthamoeba castellanii* (Neff), *Liebigs Ann. Chem.*, 537, 1988.

381. Nakashita, H., Watanabe, K., Hara, O., Hidaka, T., and Seto, H., Studies on the biosynthesis of bialaphos. Biochemical mechanism of C–P bond formation. Discovery of phosphonopyruvate decarboxylase which catalyzes the formation of phosphonoacetaldehyde from phosphonopyruvate, *J. Antibiot.*, 50, 212, 1997.

382. Hidaka, T., and Seto, H., Biosynthetic mechanism of C–P bond formation. Isolation of carboxyphosphonoenolpyruvate and its conversion to phosphinopyruvate, *J. Am. Chem. Soc.*, 111, 8012, 1989.

383. Lee, S.-L., Hepburn, T.W., Swartz, W.H., Ammon, H.L., Mariano, P.S., and Dunaway-Mariano, D., Stereochemical probe for the mechanism of P–C bond cleavage catalyzed by the *Bacillus cereus* phosphonoacetaldehyde hydrolase, *J. Am. Chem. Soc.*, 114, 7346, 1992.

384. Hammerschmidt, F., Bovermann, G., and Bayer, K., Biosynthesis of natural products with a P–C bond. Part 5. The oxirane oxygen atom of phosphomycin is not derived from atmospheric oxygen, *Liebigs Ann. Chem.*, 1055, 1990.

385. Hammerschmidt, F., Biosynthesis of natural products with a P–C bond. Part 8. On the origin of the oxirane oxygen atom of fosfomycin in *Streptomyces fradiae*, *J. Chem. Soc., Perkin Trans. 1*, 1993, 1991.

386. Hammerschmidt, F., and Kählig, H., Biosynthesis of natural products with a P–C bond. Part 7. Synthesis of [1,1-^{2}H$_2$]-, [2,2-^{2}H$_2$]-, (*R*)- (*S*)-[1-^{2}H$_1$](2-hydroxyethyl)phosphonic acid and (*R,S*)-[1-^{2}H$_1$](1,2-dihydroxyethyl)phosphonic acid incorporation studies into phosphomycin in *Streptomyces fradiae*, *J. Org. Chem.*, 58, 2364, 1991.

387. Savignac, P., and Collignon, N., Some aspects of aminoalkylphosphonic acids. Synthesis by the reductive amination approach, *ACS Symp. Ser. (Phosphorus Chem.)*, 171, 255, 1981; *Chem. Abstr.*, 96, 35344, 1982.

388. Kinney, W.A., Abou-Gharbia, M., Garrison, D.T., Schmid, J., Kowal, D.M., Bramlett, D.R., Miller, T.L., Tasse, R.P., Zaleska, M.M., and Moyer, J.A., Design and synthesis of [2-(8,9-dioxo-2,6-diaza-bicyclo[5.2.0]non-1(7)-en-2-yl)-ethyl]phosphonic acid (EAA-090), a potent *N*-methyl-D-aspartate antagonist, via the use of 3-cyclobutene-1,2-dione as an achiral α-amino acid bioisostere, *J. Med. Chem.*, 41, 236, 1998.

389. Collignon, N., Fabre, G., Varlet, J.M., and Savignac, P., Reductive amination of phosphonic aldehydes. A reexamination, *Phosphorus Sulfur*, 10, 81, 1981.

390. Fabre, G., Collignon, N., and Savignac, P., Aminobenzylation of phosphonic aldehydes. Preparation of aminoalkylphosphonic acids, *Can. J. Chem.*, 59, 2864, 1981.

391. Kudzin, Z.H., Kotynski, A., and Andrijewski, G., Aminoalkanediphosphonic acids. The synthesis via thioureidoalkanephosphonate method, *Phosphorus, Sulfur Silicon Relat. Elem.*, 77, 208, 1993.

392. Mayer, M.L., and Westbrook, G.L., The physiology of excitatory amino acids in the vertebrate central nervous system, *Prog. Neurobiol. (England)*, 28, 197, 1990.

393. Monaghan, D.T., Bridges, R.J., and Cotman, C.W., The excitatory amino acid receptors: their classes, pharmacology, and distinct properties in the function of the central nervous system, *Annu. Rev. Pharmacol. Toxicol.*, 29, 365, 1989.

394. Lehmann, J., Hutchison, A.J., McPherson, S.E., Mondadori, C., Schmutz, M., Sinton, C.M., Tsai, C., Murphy, D.E., Steel, D.J., and Williams, M., CGS 19755, a selective and competitive *N*-methyl-D-aspartate-type excitatory amino acid receptor antagonist, *J. Pharmacol. Exp. Ther.*, 246, 65, 1988.

395. Perkins, M.N., Stone, T.W., Collins, J.F., and Curry, K., Phosphonate analogues of carboxylic acids as aminoacid antagonists on rat cortical neurones, *Neurosci. Lett.*, 23, 333, 1981.

396. Davies, J., Francis, A.A., Jones, A.W., and Watkins, J.C., 2-Amino-5-phosphonovalerate (2APV), a potent and selective antagonist of amino acid-induced and synaptic excitation, *Neurosci. Lett.*, 21, 77, 1981.

397. Davies, J., Evans, R.H., Herrling, P.L., Jones, A.W., Olverman, H.J., Pook, P., and Watkins, J.C., CPP, a new potent and selective NMDA antagonist. Depression of central neuron responses, affinity for [3H]D-AP5 binding sites on brain membranes and anticonvulsant activity, *Brain. Res.*, 382, 169, 1986.

398. Harris, E.W., Ganong, A.H., Monaghan, D.T., Watkins, J.C., and Cotman, C.W., Action of 3-((±)-2-carboxypiperazin-4-yl)-propyl-1-phosphonic acid (CPP), a new and highly potent antagonist of *N*-methyl-D-aspartate receptors in the hippocampus, *Brain. Res.*, 382, 174, 1986.

399. Boast, C.A., Gerhardt, S.C., Pastor, G., Lehmann, J., Etienne, P.E., and Liebman, J.M., The *N*-methyl-D-aspartate antagonists CGS 19755 and CPP reduce ischemic brain damage in gerbils, *Brain. Res.*, 442, 345, 1988.

400. Crossley, R., and Curran, A.C.W., A novel synthesis of thiols from α-amino-nitriles, *J. Chem. Soc., Perkin Trans. 1*, 2327, 1974.

401. Villanueva, J.M., Collignon, N., Guy, A., and Savignac, P., Preparation of optically active aminocarboxyalkylphosphonic acids, *Tetrahedron*, 39, 1299, 1983.

402. Garbay-Jaureguiberry, C., McCort-Tranchepain, I., Barbe, B., Ficheux, D., and Roques, B.P., Improved synthesis of (*p*-phosphono and *p*-sulfo)methylphenylalanine. Resolution of (*p*-phosphono-, *p*-sulfo, *p*-carboxy- and *p*-*N*-hydroxycarboxamido-)methylphenylalanine, *Tetrahedron: Asymmetry*, 3, 637, 1992.

403. Chetyrkina, S., Estieu-Gionnet, K., Lain, G., Bayle, M., and Deleris, G., Synthesis of *N*-Fmoc-4-[(diethylphosphono)-2′,2′-difluoro-1′-hydroxyethyl]phenylalanine, a novel phosphotyrosyl mimic for the preparation of signal transduction inhibitory peptides, *Tetrahedron Lett.*, 41, 1923, 2000.

404. Morita, T., Okamoto, Y., and Sakurai, H., The preparation of phosphonic acids having labile functional groups, *Bull. Chem. Soc. Jpn.*, 51, 2169, 1978.

405. Morita, T., Okamoto, Y., and Sakurai, H., A convenient dealkylation of dialkyl phosphonates by chlorotrimethylsilane in the presence of sodium iodide, *Tetrahedron Lett.*, 19, 2523, 1978.

406. Morita, T., Okamoto, Y., and Sakurai, H., Dealkylation reaction of acetals, phosphonate, and phosphate esters with chlorotrimethylsilane/metal halide reagent in acetonitrile, and its application to the synthesis of phosphonic acids and vinyl phosphates, *Bull. Chem. Soc. Jpn.*, 54, 267, 1981.

6 The Cyanophosphonates

The cyanomethylphosphonates are readily available reagents, on laboratory scale or commercially. Since their first preparation, they have given rise to a surprisingly large number of synthetic applications. This chapter describes their preparations and their reactions including preservation and removal of the phosphoryl group.

6.1 SYNTHESIS OF CYANOPHOSPHONATES

6.1.1 DIALKYL CYANOPHOSPHONATES

6.1.1.1 Michaelis–Arbuzov Reactions

The first reported synthesis of diethyl cyanophosphonate (DECP), also called diethyl phosphorocyanidate (DEPC), employed a reaction between cyanogen iodide and triethyl phosphite (46% yield).[1] In a later variation, the use of cyanogen bromide provided DECP in 65% yield.[2] Freshly prepared and fractionally distilled, DECP shows sharp bands of medium intensity at 2,200 and 2,085 cm[-1]. The band at the longer wavelength has been attributed[3] to the presence of isomeric isocyanidate rather than free HCN formed by hydrolysis. Redistillation after 10 weeks of standing at room temperature affords pure DECP in 70% yield as a colorless oil having a single absorption band at 2,200 cm[-1].[3]

6.1.1.2 Phosphonylation of Cyanides

The synthesis of DECP by the reaction of diethyl phosphite with an alkali metal cyanide[4] in CCl_4 is reported to be difficult to reproduce.[5] In contrast, when the experiment is done with diethyl chlorophosphate and fresh NaCN containing catalytic amounts of NaOH in refluxing C_6H_6, consistent yields of DECP (up to 46%) are obtained (Scheme 6.1).[5] The addition of a phase-transfer catalyst does not improve the yield, and the use of polar aprotic solvents (MeCN, DMF, DMSO) completely inhibits the reaction.[5] The reaction of diethyl chlorophosphate with trimethylsilyl cyanide in the presence of $TiCl_4$ leads to a mixture of DECP and partially silylated DECP.[6]

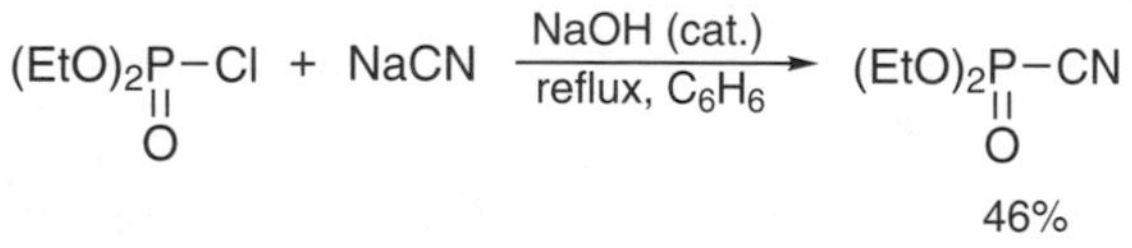

$$(EtO)_2\underset{\underset{O}{\|}}{P}-Cl \ + \ NaCN \ \xrightarrow[\text{reflux, } C_6H_6]{\text{NaOH (cat.)}} \ (EtO)_2\underset{\underset{O}{\|}}{P}-CN$$

46%

(6.1)

6.1.1.3 Other Reactions

2-Cyano-5,5-dimethyl-2-oxo-1,3,-2-dioxaphosphorinane was first prepared by the isomerization of 2-isocyano-5,5-dimethyl-2-oxo-1,3,2-dioxaphosphorinane, a very reactive and unstable compound.[7] An alternative preparation involves the oxidation of the corresponding phosphorocyanidite with N_2O_4 in CH_2Cl_2.[8]

TABLE 6.1

	R^1	Yield (%)[Ref.]
a	Me	45,[26] 55,[18] 70[27]
b	Et	76[29]
c	i-Pr	40[29]
d	n-C$_6$H$_{13}$	70[29]
e	OMe	54[28]
f	Ot-Bu	57[28]

6.1.2 DIALKYL 1-CYANOALKYLPHOSPHONATES

6.1.2.1 Michaelis–Arbuzov Reactions

Dawson and Burger[9] first demonstrated that the Michaelis–Arbuzov reaction of chloroacetonitrile and triethyl phosphite furnishes diethyl cyanomethylphosphonate. In its most general form, it involves the addition of triethyl phosphite to chloroacetonitrile at 150–170°C to give diethyl cyanomethylphosphonate in up to 80% yields (Scheme 6.2).[10–21] This procedure has been developed on an industrial scale.[22,23] The use of bromo- or iodoacetonitrile (X = Br, I, Scheme 6.2) was also reported.[15,16] The reaction may be used to prepare any dialkyl cyanomethylphosphonate, and a variety of trialkyl phosphites have produced the corresponding phosphonates in reasonable to good yields (29–90%, Scheme 6.2).[14,16,23,24] Cyclic phosphites, 4,5-dimethyl-2-methoxy-1,3,2-dioxaphospholane (*meso* and racemic) and 5,5-dimethyl-2-methoxy-1,3,2-dioxaphosphorinane, react with chloroacetonitrile to produce 5- and 6-membered cyclic phosphonates in 50% and 72% yields, respectively.[25]

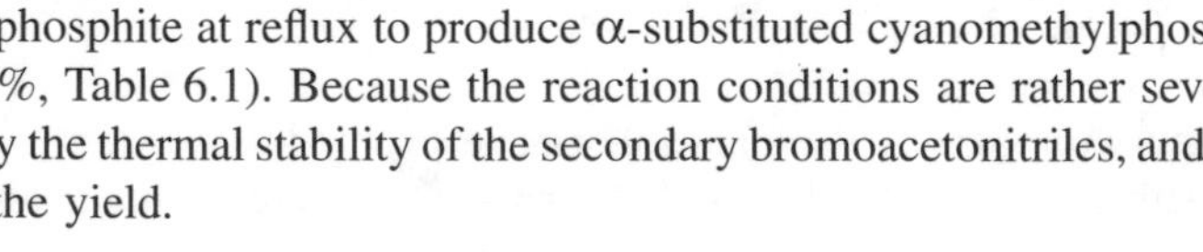

$$(RO)_3P + X-CH_2-CN \xrightarrow{150-170°C} (RO)_2\overset{O}{\underset{\parallel}{P}}-CH_2-CN$$

R = Me, Et, n-Pr, i-Pr, n-Bu
X = Cl, Br, I

29–90%

(6.2)

Substituted α-haloacetonitriles (R^1 ≠ H, Scheme 6.3) are exceptions to the rule that branched-chain halides are unreactive in Michaelis–Arbuzov reactions. A variety of α-bromo nitriles[18,26–30] react with triethyl phosphite at reflux to produce α-substituted cyanomethylphosphonates in fair to good yield (40–76%, Table 6.1). Because the reaction conditions are rather severe, the rearrangement is governed by the thermal stability of the secondary bromoacetonitriles, and prolonged heating does not increase the yield.

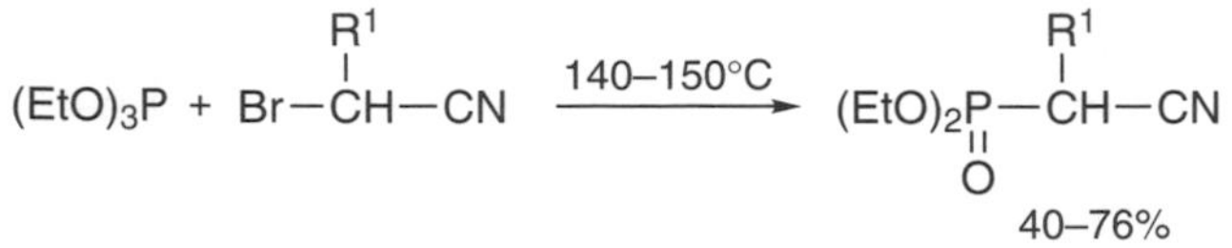

$$(EtO)_3P + Br-\overset{R^1}{\underset{|}{CH}}-CN \xrightarrow{140-150°C} (EtO)_2\overset{O}{\underset{\parallel}{P}}-\overset{R^1}{\underset{|}{CH}}-CN$$

40–76%

(6.3)

Trichloroacetonitrile reacts with trialkyl phosphites (1:2 ratio) in a Michaelis–Arbuzov-type reaction to give tetraalkyl cyanomethylenediphosphonate. The intermediacy of dialkyl cyanodichloromethylphosphonate and tetraalkyl cyanochloromethylenediphosphonate was evidenced in this case.[31] Similarly, dichloromalonitrile reacts with trialkyl phosphites to give tetraalkyl dicyanomethylenediphosphonates.[32]

6.1.2.2 Phosphonylation of α-Lithiated Nitriles [SNP(V) Reactions]

The preparation of dialkyl cyanoalkylphosphonates through the nucleophilic attack of chlorophosphates by metallated acetonitrile and its higher homologues is of significant synthetic utility. Although the condensation of Li, Na, and Ca derivatives of phenylacetonitrile with diethyl chlorophosphate in Et_2O is a low-yield reaction (21%), it was the first example of a general strategy for the introduction of a phosphonyl group through anionic rather than thermal Michaelis–Arbuzov chemistry.[33] Thus, metallation of acetonitrile and aliphatic nitriles with lithiated bases (n-BuLi, LDA, LiTMP, LiHMDS) in THF at low temperature leads quantitatively to α-cyano carbanions, which are converted into cyanoalkylphosphonates in high yields by treatment with chlorophosphates (Scheme 6.4).[34–43] Even in the best case, the method requires the use of 2 eq of weakly nucleophilic base per equivalent of nitrile to guarantee the success of the reaction. The first equivalent metallates the nitrile and gives the cyanoalkylphosphonate. However, this product, being more acidic than the starting nitrile, undergoes a second proton abstraction to give the cyanoalkylphosphonate carbanion. The use of two equivalents of base to generate the phosphonate carbanion prevents the loss of α-cyano carbanion by transmetallation reactions.

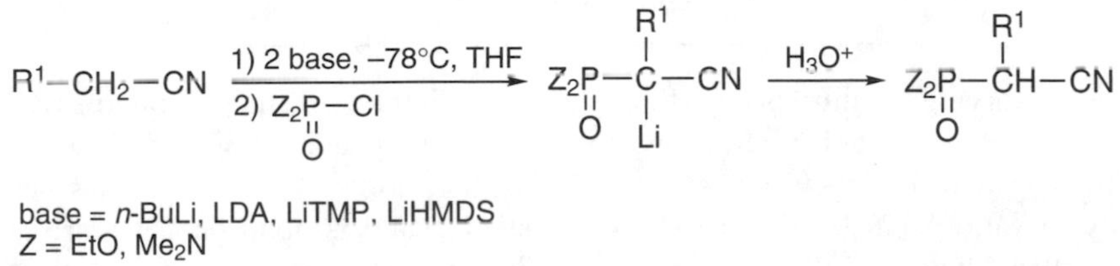

$$R^1-CH_2-CN \xrightarrow[\text{2) } Z_2P-Cl]{\text{1) 2 base, } -78°C, \text{ THF}} \underset{O}{\overset{R^1}{Z_2P-\underset{Li}{\overset{|}{C}}-CN}} \xrightarrow{H_3O^+} \underset{O}{\overset{R^1}{Z_2P-CH-CN}}$$

base = n-BuLi, LDA, LiTMP, LiHMDS
Z = EtO, Me_2N

$$(6.4)$$

The effect of the structure of the lithiated base on the nature of reaction products has been evidenced in the case of acetonitrile.[43] In contrast to previously reported results using LDA (2 eq) for deprotonation of acetonitrile, LiHMDS (2 eq) was found to be the preferred base for best yields of dialkyl cyanomethylphosphonate. LDA induces side reactions resulting in the competing formation of tetraalkyl cyanomethylenediphosphonate (Scheme 6.5). In marked contrast, when the carbanion of higher nitriles is generated with LDA (2 eq), the sterically hindered lithium cyanoalkylphosphonates are obtained in high yields and without traces of byproducts. LiHMDS is not sufficiently basic for the metallation of higher homologues of acetonitrile. Recently, the structure of lithium diethyl cyanomethylphosphonate and tetraethyl cyanomethylenediphosphonate by X-ray diffraction has been reported.[43,44]

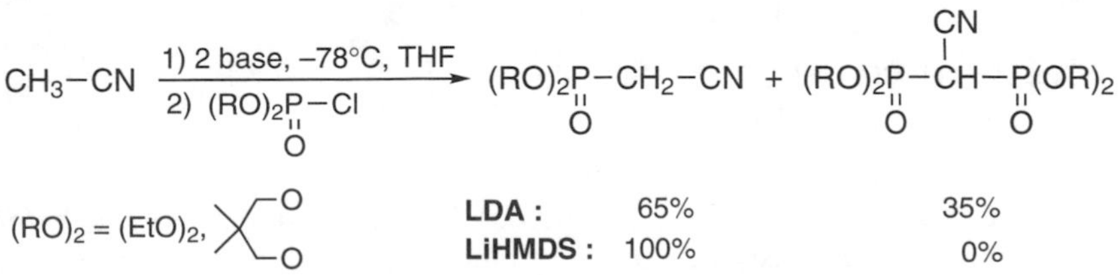

$$CH_3-CN \xrightarrow[\text{2) } (RO)_2P-Cl]{\text{1) 2 base, } -78°C, \text{ THF}} \underset{O}{(RO)_2P-CH_2-CN} + \underset{O \quad O}{\overset{CN}{(RO)_2P-CH-P(OR)_2}}$$

$(RO)_2 = (EtO)_2,$ [structure]

	LDA :	65%	35%
	LiHMDS :	100%	0%

$$(6.5)$$

TABLE 6.2

	R^1	Yield (%)[Ref.]
a	H	47,[36,38] 53,[36] 88,[43] 90[41]
b	Me	52,[36] 63,[36] 73,[38] 89[43]
c	Et	98[43]
d	n-Pr	99[43]
e	CH$_2$CH=CH$_2$	70[38]
f	Ph	47,[36] 99[43]
g	F	—[45]
h	(CH$_2$)$_3$CN	—[46]
i	(CH$_2$)$_4$CN	—[46]
j	CH$_2$CH(OMe)$_2$	—[37]
k	(CH$_2$)$_2$CMe(OMe)$_2$	—[47]

Judiciously varying the nitriles and chlorophosphates influences the phosphorus reactivity and thus provides a method of choice for the one-pot large scale preparation of dialkyl cyanoalkylphosphonates bearing a wide variety of ester groups at phosphorus. The reaction has been applied successfully to various nitriles (Table 6.2). The dialkyl cyanoalkylphosphonates are obtained in good to excellent yields (47–99%) after mild acidic workup. The potential weaknesses of the procedure lie in two competing reactions: the formation of a phosphoroamidate by reaction between chlorophosphate and a nucleophilic base and the rapid self-condensation of the α-cyano carbanion to give α,β-aminoacrylonitrile.[28,38,43]

An added and valuable advantage found in this attractive and mild approach to dialkyl cyanoalkylphosphonates is the possibility of trapping the phosphonate carbanions *in situ* by reaction with an aldehyde or ketone when the desired product is the olefin resulting from the Horner–Wadsworth–Emmons reaction (Scheme 6.6).

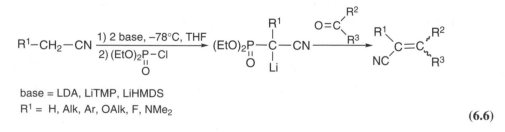

base = LDA, LiTMP, LiHMDS
R^1 = H, Alk, Ar, OAlk, F, NMe$_2$

(6.6)

Although some procedures involve the isolation of the cyanoalkylphosphonate before olefination, most of the reaction sequences are executed in one pot.[36,38,45,48] The feasibility of this methodology is illustrated by a synthetic program developed around chemically modified retinals and spheroidenes.[46,47,49–52] An application of the procedure is reported in the preparation of specifically enriched retinal by using the appropriately labeled [1–¹³C],[50] [2–¹³C],[49–51] [1,2–¹³C],[50] [1–¹⁴C],[41] or [1,1,1–²H₃]acetonitrile.[50]

6.1.2.3 Phosphonylation by Michael Addition

There are relatively few reports involving the Michael addition of cyanomethylphosphonate anions to unsaturated compounds. The reaction provides a methodology for the elaboration of new reagents but more frequently is used for the preparation of phosphonylated heterocycles. Under basic conditions, diethyl cyanomethylphosphonate adds to benzalacetophenone and 2-benzylidene-3-methyl-4-nitro-3-thiolene-1,1-dioxide to give the addition products in modest yields.[53,54]

Offering the advantages of neutral conditions, diethyl 1-cyanoethylphosphonate in the presence of Rh(I) catalyst undergoes Michael addition to vinyl ketones or acrylaldehyde, giving the adduct in high yield (80–98%) and high enantioselectivity (92–93% ee) without formation of the 1,2-addition product (Scheme 6.7).[55]

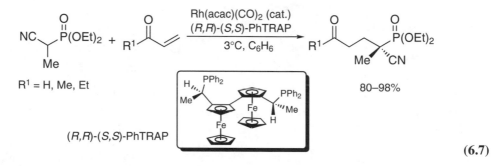

$$(6.7)$$

Michael addition of diethyl cyanomethylphosphonate to α-nitroalkenes in the presence of LDA (1 eq) followed by treatment with Me₃SiCl provides a one-pot synthesis of 2-isoxazoline derivatives. The reaction proceeds through a 1,3-dipolar cycloaddition of trimethylsilyl nitronate with an α-cyanovinylphosphonate.[56] Similarly, the 1,4-addition of the sodium diethyl cyanomethylphosphonate to the azo-ene system of conjugated alkenes results in the formation of 1,2-diamino-3-(diethoxyphosphinyl)pyrroles in moderate to good yields via a hydrazonic intermediate.[57,58] In a route to phosphonylated pyrones, the sodium diethyl cyanomethylphosphonate reacts with 3-anilinomethylene derivatives of 4-hydroxycoumarins in DMF with displacement of aniline.[59]

The 1-cyanovinylphosphonates, readily accessible by the Knoevenagel reaction, undergo reaction with charged nucleophiles such as cyanide,[60–62] lithium dialkylcuprate,[63] nitroalkanes in the presence of KF/Al₂O₃,[64] or with neutral nucleophiles such as pyrazolidinones (Scheme 6.8)[65,66] or pyridazinones[67] to give more elaborated products.

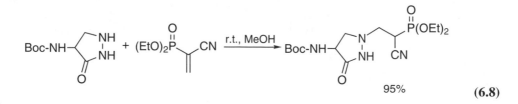

95%

$$(6.8)$$

6.1.2.4 α-Functionalization of Cyanomethylphosphonates

The ready formation of nucleophilic cyanomethylphosphonate carbanions provides a useful route to a variety of new reagents bearing alkyl, aryl, acyl, or formyl groups at the α-carbon. The full scope of these reactions has been explored in depth.

6.1.2.4.1 Knoevenagel Reactions

The condensation–dehydration reaction of carbonyl compounds with phosphonates containing active methylene groups provides α,β-unsaturated phosphonates in which the carbonyl carbon becomes the β carbon of the α,β-unsaturated phosphonates. This method of converting cyanomethylphosphonates to 1-cyanovinylphosphonates was first described in the 1960s.[13,68,69] It has been applied to a variety of substrates and often succeeds where other methods have failed. The Knoevenagel reaction using dialkyl cyanomethylphosphonates and aldehydes has been found to be the method of choice for preparing dialkyl (E)-1-cyanovinylphosphonates in good yields. For instance, heating a mixture of dialkyl cyanomethylphosphonate and benzaldehyde in the presence of piperidine and piperidinium acetate as catalyst in C_6H_6 at 130–145°C for several hours with azeotropic removal of water gives the dialkyl 1-cyano-2-phenylvinylphosphonate in fair to good yields (47–75%, Scheme 6.9).[13,68]

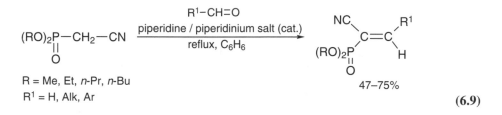

$$R = Me, Et, n\text{-}Pr, n\text{-}Bu$$
$$R^1 = H, Alk, Ar$$

$$(6.9)$$

The reaction can be applied successfully to phosphonates carrying a variety of substituents at phosphorus using simple (paraformaldehyde),[68,70] long-chain aliphatic, aromatic, or heteroaromatic aldehydes (Scheme 6.9).[13,68,69,71–79] The condensation is achieved in homogeneous systems (ROH, $CHCl_3$, C_6H_6/DMF); however, in low-boiling-point solvents ($CHCl_3$), the reaction is particularly slow. Several combinations of amines and ammonium salts (ammonium acetate, cyclohexylammonium acetate, piperidinium acetate or phosphate) are active catalysts, but the piperidine/piperidinium acetate couple appears to be the preferred catalyst. Side reactions are reported with isobutyraldehyde,[80] which is prone to migration of the double bond (vinyl–allyl), and acrolein, which gives the 1-cyano-1,3-butadienylphosphonate contaminated with the acrolein crotonization product.[81] When diethyl cyanomethylphosphonate is subjected to the Knoevenagel reaction with triethyl orthoformate in the presence of $ZnCl_2$, the synthetically useful diethyl 1-cyano-2-ethoxyvinylphosphonate is obtained in excellent yield (95%).[82]

By a similar procedure, the dialkyl cyanomethylphosphonates undergo Knoevenagel reaction with ketones, but the results are rather disappointing. The condensation is sensitive to the nature and structure of the ketone, and the reaction leads to the corresponding 1-cyanovinylphosphonates as a mixture of geometric isomers in low to modest yields (13–51%).[80,83,84]

Other applications include the Knoevenagel reaction under microwave irradiation in a solvent-free reaction. With this technique, benzaldehyde reacts in a few minutes with diethyl cyanomethylphosphonate in the presence of carefully controlled amounts of piperidine to give diethyl 1-cyano-2-phenylvinylphosphonate in 71% yield. The workup involves a simple solvent wash or a short-path distillation.[85]

6.1.2.4.2 Alkylation

Alkylation of dialkyl cyanomethylphosphonates gives somewhat higher yields of α-substituted cyanomethylphosphonates than those obtained by Michaelis–Arbuzov rearrangement from trialkyl phosphites and substituted haloacetonitriles.

The use of magnesium cyanomethylphosphonate, prepared from magnesium in liquid ammonia, was rather disappointing.[86] Although the combination of potassium metal and diethyl cyanomethylphosphonate, on heating in Et_2O–dioxane, gives fair yields of α-substituted cyanomethylphosphonate,[87,88] the best results are obtained with the use of alkali metal hydrides (NaH, KH) in THF or DME. Addition of primary alkyl bromides[89–91] or iodides,[92–95] secondary alkyl bromides,[95–98] or PhSCl[28] to alkali metal salts of diethyl cyanomethylphosphonate at room temperature gives, after workup, α-substituted cyanomethylphosphonates (25–91%, Scheme 6.10). With *bis*-electrophiles, diethyl cycloalkylcyanophosphonates are readily obtained by two successive alkylations in fair to excellent yields (40–90%).[99–102] Diethyl cyanomethylphosphonate undergoes dialkylation with benzylbromide in the presence of DBU in DMF (60% yield).[103]

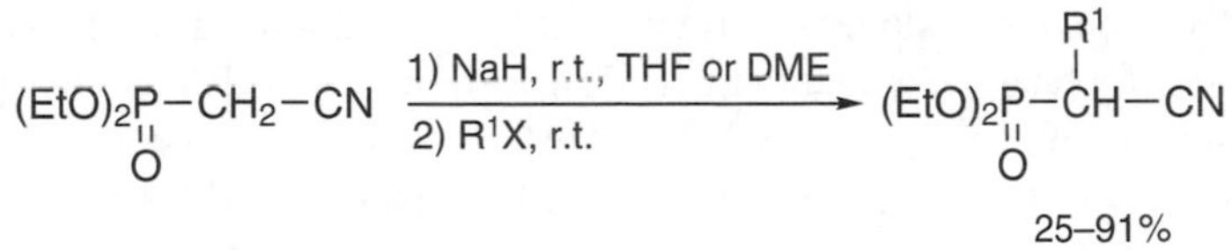

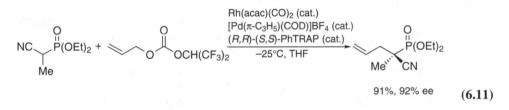

$$ \text{(EtO)}_2\overset{\text{O}}{\underset{\|}{\text{P}}}-\text{CH}_2-\text{CN} \quad \xrightarrow[\text{2) R}^1\text{X, r.t.}]{\text{1) NaH, r.t., THF or DME}} \quad \text{(EtO)}_2\overset{\text{O}}{\underset{\|}{\text{P}}}-\overset{\text{R}^1}{\underset{\,}{\text{CH}}}-\text{CN} $$

25–91%   **(6.10)**

A two-component catalyst system consisting of two different transition metal complexes (Rh–Pd) has been applied to the enantioselective allylation of diethyl 1-cyanoethylphosphonate. The reaction proceeds with high enantioselectivity to give optically active phosphonic acid derivatives (Scheme 6.11).[104]

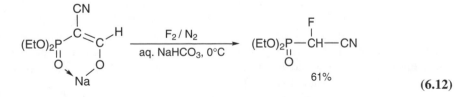

91%, 92% ee **(6.11)**

The ion-pair extraction method (phase-transfer catalysis) can offer improved syntheses of α-substituted cyanomethylphosphonates. It is effectively employed in the alkylation of dialkyl cyanomethylphosphonate to smoothly achieve specific monoalkylation.[34,105,106] This technique is also well suited to the *bis*-alkylation and cycloalkylation of cyanomethylphosphonates.[107,108]

6.1.2.4.3 Halogenation

Diethyl cyanomethylphosphonate anion undergoes halogenation, rendering it a convenient precursor of diethyl 1-cyano-1-fluoromethylphosphonate. Thus, treatment of an aqueous solution of sodium enolate of diethyl 1-cyano-1-formylmethylphosphonate with fluorine gives diethyl 1-cyano-1-fluoromethylphosphonate in 61% yield after concomitant deformylation (Scheme 6.12).[109,110] In contrast, treatment of unprotected lithium diethyl cyanomethylphosphonate with $(CF_3SO_2)_2NF$ in THF at low temperature leads to 1-cyano-1-fluoromethylphosphonate in lower yield (51%).[111]

$$ \xrightarrow[\text{aq. NaHCO}_3, \, 0^\circ\text{C}]{\text{F}_2 / \text{N}_2} \quad \text{(EtO)}_2\overset{\text{O}}{\underset{\|}{\text{P}}}-\overset{\text{F}}{\underset{\,}{\text{CH}}}-\text{CN} $$

61%   **(6.12)**

6.1.2.4.4 Arylation

As anticipated, sodium diethyl cyanomethylphosphonate does not undergo arylation by condensation of halobenzenes on simple heating. In contrast, the Cu(I)-mediated reaction produces diethyl cyanobenzylphosphonate in moderate yield (50%).[112,113] The reaction occurs in DMF or HMPA in the presence of CuI (1 eq) at 100°C (Scheme 6.13). This approach has provided a novel route to 4-phosphonylated 3-aminoisoquinolines.[114]

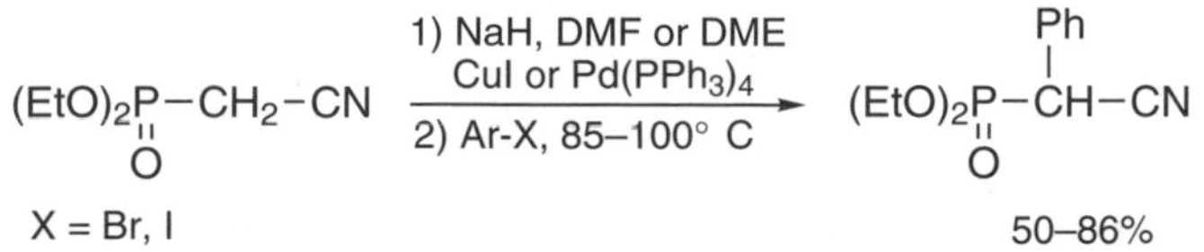

$$
\begin{array}{c}
\text{(EtO)}_2\text{P}-\text{CH}_2-\text{CN} \\
\overset{\|}{\underset{O}{}} \\
X = \text{Br, I}
\end{array}
\quad
\xrightarrow[\text{2) Ar-X, 85–100° C}]{\substack{\text{1) NaH, DMF or DME} \\ \text{CuI or Pd(PPh}_3)_4}}
\quad
\begin{array}{c}
\overset{\text{Ph}}{\underset{|}{}} \\
\text{(EtO)}_2\text{P}-\text{CH}-\text{CN} \\
\overset{\|}{\underset{O}{}} \\
\text{50–86\%}
\end{array}
\qquad \textbf{(6.13)}
$$

Arylation of diethyl cyanomethylphosphonate can also be efficiently catalyzed by Pd(0).[115,116] When 4-substituted bromo- or iodobenzenes are heated with sodium diethyl cyanomethylphosphonate in the presence of a catalytic amount of Pd(PPh$_3$)$_4$ in refluxing DME, diethyl cyanobenzylphosphonates are isolated in 69–86% yields (Scheme 6.13). By a similar procedure, 2- and 3-bromopyridines and 2-bromothiophene are phosphonylated in high yields (59–80%).[116]

Nucleophilic aromatic substitution of activated substrates (pentafluoropyridine, octafluorotoluene, hexafluorobenzene, metal arene π-complexes) by anions of diethyl cyanomethylphosphonate has been achieved. The reaction is carried out with satisfactory yields in DMF, MeCN, or THF at room temperature in the presence of NaH, CsF, K$_2$CO$_3$ or Cs$_2$CO$_3$.[117–119] In a similar manner, 3,6-dihalopyridazines react with sodium diethyl cyanomethylphosphonate in refluxing THF to give phosphonosubstituted pyridazines in 22–68% yields.[120]

6.1.2.4.5 Acylation

Acylation of dialkyl cyanomethylphosphonates was first accomplished using potassium metal in Et$_2$O and acyl chlorides. The resulting α-cyano-β-ketoalkylphosphonates were isolated in modest yields (24–36%), usually as a mixture of the keto and enol forms.[121,122] The recently reported procedure using powdered NaOH or KOH in MeCN offers a more convenient route and better yields (17–90%, Scheme 6.14).[123] 2-Chloroethyl chloroformate reacts with sodium diethyl cyanomethylphosphonate in THF to produce diethyl 1-cyano-(1,3-dioxolan-2-ylidene)methylphosphonate in 83% yield by a C-acylation–O-alkylation sequence.[124]

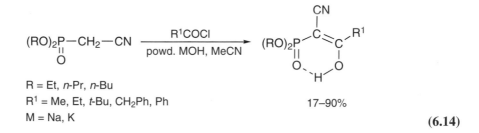

$$
\begin{array}{c}
\text{(RO)}_2\text{P}-\text{CH}_2-\text{CN} \\
\overset{\|}{\underset{O}{}}
\end{array}
\quad
\xrightarrow[\text{powd. MOH, MeCN}]{\text{R}^1\text{COCl}}
$$

R = Et, n-Pr, n-Bu
R^1 = Me, Et, t-Bu, CH$_2$Ph, Ph
M = Na, K

17–90%

$$\qquad \textbf{(6.14)}$$

The yields are not so good when the ester function serves as leaving group. Thus, diethyl 1-cyano-1-oxalylmethylphosphonate is obtained in 55% yield from diethyl oxalate[125] and lithium diethyl cyanomethylphosphonate in THF at low temperature and in 64% yield from the sodium salt in refluxing Et$_2$O.[126] Acylation using TFA esters and TFAA has also been reported.[127–130]

Formylation of dialkyl cyanomethylphosphonates with methyl formate using sodium metal in Et$_2$O gives modest yields (25–30%) of dialkyl 1-cyano-1-formylmethylphosphonate.[122] In contrast, when the reaction is realized in MeOH in the presence of MeONa, the corresponding sodium enolate is isolated in quantitative yield.[131]

An alternative procedure is the reaction of N,N-dimethylformamide dimethyl acetal with diethyl cyanomethylphosphonate.[132–136] In this case, the reaction proceeds spontaneously at room temperature to give the corresponding enaminophosphonate in high yield (95%, Scheme 6.15).[134] Hydrolysis under basic conditions (2 M NaOH) at room temperature leads to diethyl 1-cyano-1-formylmethylphosphonate in 83% yield in the enolic form exclusively.[134]

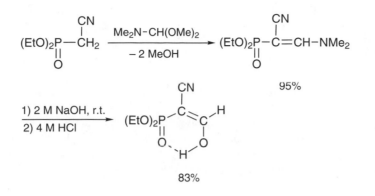

$$(6.15)$$

6.1.2.4.6 Miscellaneous

Most of the information available on this topic is derived from works on the reactions of alkali metal salts (sodium or potassium) of dialkyl cyanomethylphosphonates with sp^2 or sp electrophilic center. Thus, in the presence of a further equivalent of sodium hydride, sodium dialkyl cyanomethylphosphonates add to carbon disulfide in Et_2O at room temperature to give alkenedithiolates, which are further characterized by alkylation or oxidation. For example, reaction with methyl iodide provides dialkyl 1-cyano-2,2-*bis*(methylthio)vinylphosphonates in 21–40% yields (Scheme 6.16),[20,137–139] whereas treatment with acetyl chloride in refluxing THF leads to 1,3-dithietane-2,4-diylidene*bis*(cyanomethylphosphonate) in low to fair yields (10–54%, Scheme 6.16).[140–142]

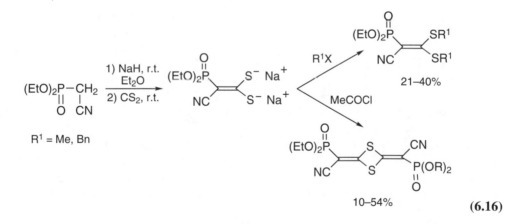

$$(6.16)$$

Similarly, the alkali metal salts of dialkyl cyanomethylphosphonates react with thionocarboxylic, dithiocarboxylic, and xanthogenic esters to give alkali metal salts of 2-substituted 2-mercapto-1-cyanovinylphosphonates. These salts are protonated or alkylated to the thioenol ethers, isolated as mixtures of (*E*)- and (*Z*)-isomers.[143,144]

Alkali metal salts of dialkyl cyanomethylphosphonates react with phenyl isocyanate or isothiocyanate in THF at room temperature to form the corresponding phosphonylated acetanilides or thioacetanilides, which undergo further alkylation reactions.[145–151]

Sodium dialkyl cyanomethylphosphonate reacts smoothly at room temperature in THF with isoamyl nitrite to give (hydroxyimino)cyanomethylphosphonate in excellent yield (80%), and this is further converted to 1-amino-1-cyanomethylphosphonate (64%) by aluminium amalgam reduction in refluxing Et_2O (Scheme 6.17).[152–154]

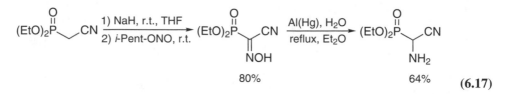

$$ 80\% \qquad\qquad\qquad\qquad 64\% \qquad (6.17) $$

Dialkyl cyanomethylphosphonates, possessing both an active methylene group and an electrophilic cyano group, react under basic conditions with arylazides,[155] arylnitrile oxides,[155] diazonium salts of indazoles, pyrazoles, and triazoles[156] and isatoic anhydride[157] to produce intermediates that spontaneously cyclize, giving heterocyclic systems.

6.1.2.5 Other Reactions

The expected product from the reaction of diethyl acetylphosphonate with sodium bisulfite and sodium cyanide is diethyl 1-cyano-1-hydroxyethylphosphonate.[158,159] However, a rearrangement takes place during the preparation under alkaline conditions (pH > 8) that leads to diethyl 1-cyanoethyl phosphate (Scheme 6.18).[81,159] By a similar procedure, long-chain diethyl α-cyanohydroxyphosphonates are prepared in high yields (85–94%) from the corresponding acylphosphonates by successive treatment with aqueous $NaHSO_3$ and KCN at pH < 8.[160,161] The same type of 1-cyano-1-hydroxyalkylphosphonates is obtained by addition of diethyl acetyl,[162] diethyl trimethylsilyl,[163] or diethyl[164] phosphite to RCOCN.

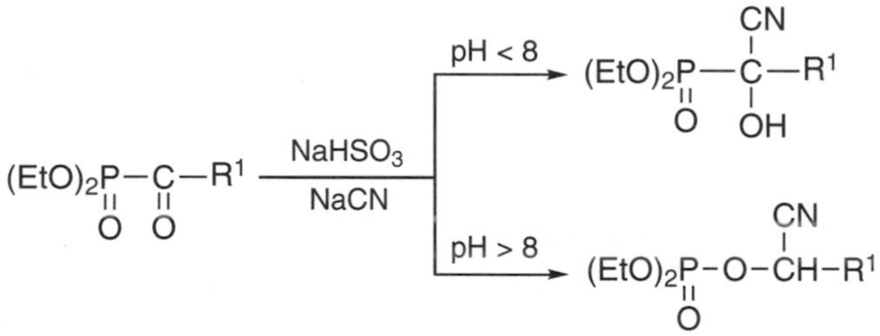

$$ (6.18) $$

An alternative procedure for the preparation of diethyl cyanobenzylphosphonates uses the addition of diethyl trimethylsilyl phosphite to β-nitrostyrenes in the presence of $TiCl_4$. The silyl functionality is transferred to the oxygen to form nitronates, which, in the presence of low-valence titanium generated *in situ*, are converted to diethyl cyanobenzylphosphonates in 74–91% yields (Scheme 6.19).[165]

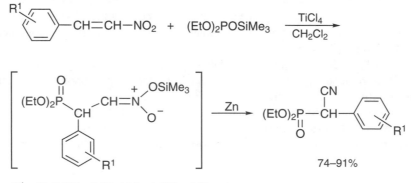

R^1 = H, 2-OMe, 2-Cl, 4-Me, 4-OMe, 4-Cl

$$ (6.19) $$

As evidence of the intermediacy of 2-aminoethylphosphonate (AEP) in the metabolism of fosfomicyn, treatment of diisopropyl tosyloxymethylphosphonate with Na[13]CN in DMF in the presence of NaI at 110°C gives labeled cyanomethylphosphonate.[166] The umpolung reaction of sodium diethyl 1-(ethoxycarbonyl)methylphosphonate with ArOCN in THF gives a low yield of diethyl 1 (ethoxycarbonyl)-1-cyanomethylphosphonate (24%).[167]

6.1.3 DIALKYL 2-CYANOALKYLPHOSPHONATES

6.1.3.1 Michaelis–Arbuzov Reactions

Triethyl phosphite, on heating at 150–160°C with 2-(diethylamino)- or 2-piperidino propionitrile methiodides, undergoes a Michaelis–Arbuzov rearrangement leading to diethyl 2-cyanoethylphospho-nate in 37% and 50% yields, respectively (Scheme 6.20).[168] The rearrangement can be initiated with a comparable yield (39%) by adding AcOH to a mixture of triethyl phosphite and 2-(diethylamino)pro-pionitrile heated at 180°C.[169] An alternative procedure for the preparation of diethyl 2-cyanoethylphos-phonate is the reaction of 2-acetoxypropionitrile with triethyl phosphite at 140–150°C (62% yield).[170] More traditionally, dialkyl 2-cyano-2-propenylphosphonates are prepared in excellent yields by heating 2-chloro- or 2-bromomethacrylonitrile with the corresponding trialkyl phosphites.[171–174]

$$(EtO)_3P \ + \ \underset{CH_2-CH_2-CN}{\overset{Me}{N^+}} \ I^- \quad \xrightarrow{150-160°C} \quad \underset{O}{\overset{\|}{(EtO)_2P}}-CH_2-CH_2-CN$$

50%   **(6.20)**

6.1.3.2 Michaelis–Becker Reactions

A Michaelis–Becker reaction between sodium diethyl phosphite and 3-chloropropionitrile in reflux-ing Et₂O has been reported (70% yield, Scheme 6.21). However, this approach does not seem to have any advantage as a synthetic method and has not been used as such.[175]

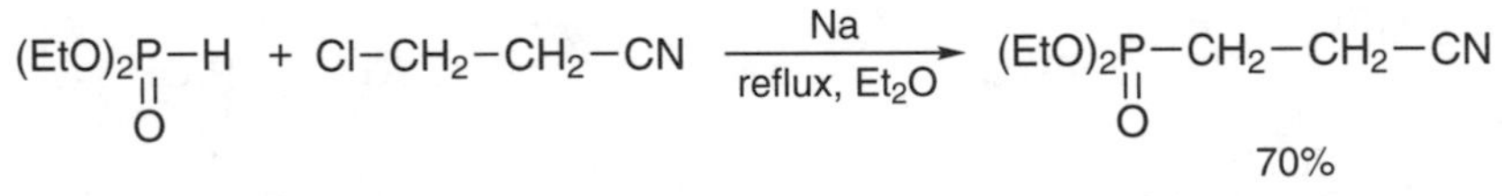

$$\underset{O}{\overset{\|}{(EtO)_2P}}-H \ + \ Cl-CH_2-CH_2-CN \quad \xrightarrow[\text{reflux, Et}_2O]{Na} \quad \underset{O}{\overset{\|}{(EtO)_2P}}-CH_2-CH_2-CN$$

70%   **(6.21)**

6.1.3.3 Phosphonylation by Michael Addition (Pudovik Reaction)

Possibly the most frequently used route to dialkyl 2-cyanoethylphosphonate is the conjugate addition of dialkyl phosphites to α,β-unsaturated nitriles (Pudovik reaction). There have been systematic attempts to rationalize the process according to the phosphorus reagents and the nitrile substrates.[176,177] In the presence of MeONa, diethyl phosphite reacts vigorously with acrylonitrile in MeOH to give the Michael addition product in 83% yield (Scheme 6.22).[178,179] This hydrophos-phonylation is applied with success to the synthesis of dialkyl 2-cyanoethylphosphonates bearing a large variety of acyclic (symmetric[178–194] or disymmetric[197]), cyclic (symmetric[198,199] or disymmetric[198]), or silyl[200,201] substituents at phosphorus.

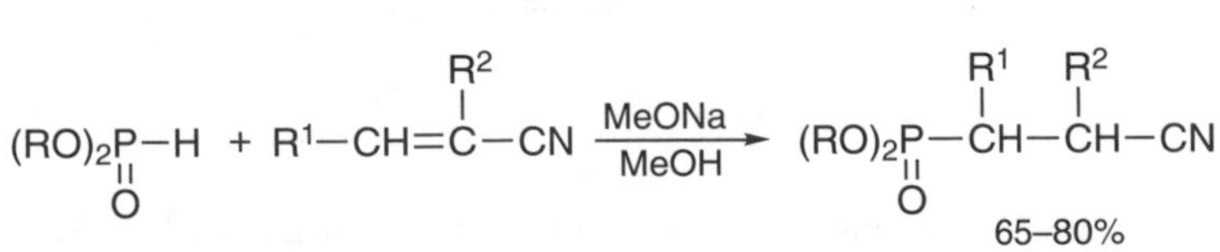

$$\underset{O}{\overset{\|}{(RO)_2P}}-H \ + \ R^1-CH=\overset{R^2}{\underset{}{C}}-CN \quad \xrightarrow[\text{MeOH}]{MeONa} \quad \underset{O}{\overset{\|}{(RO)_2P}}-\overset{R^1}{\underset{}{C}}H-\overset{R^2}{\underset{}{C}}H-CN$$

65–80%

R = Me, Et, *i*-Pr, *n*-Bu, *i*-Bu

R¹, R² = H, Me

(6.22)

The reaction has been extended to methacrylonitrile, α-chloromethacrylonitrile, 2-furylacrylonitrile, crotonitrile, allyl cyanide, and cyclohexenecarbonitrile to give the corresponding 2-cyanoethylphosphonates.[171–173,202–205] The yields are variable, 65% to 80%, depending on the substituents attached at phosphorus and the structure of the nitrile. Under basic conditions, allyl cyanide undergoes an isomerisation–Michael addition leading to the same product as for crotonitrile.[202,203] Recently, progress has been achieved with the use of tetramethylguanidine (TMG) as catalyst. In these conditions, addition of diethyl phosphite to acrylonitrile takes place smoothly at 0°C, giving rise to diethyl 2-cyanoethylphosphonate in 47% yield after an easy workup (Scheme 6.23).[206]

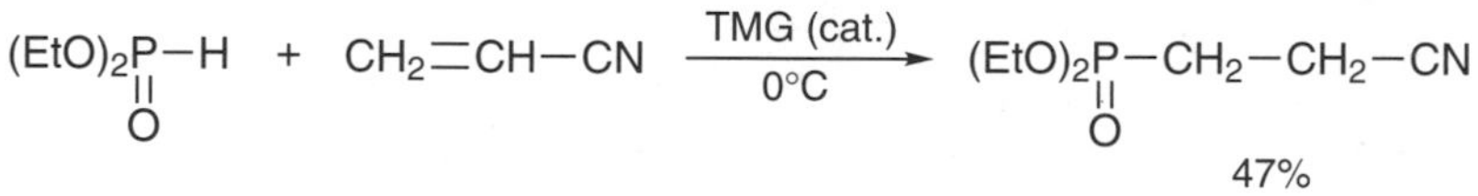

$$(EtO)_2\underset{\underset{O}{\|}}{P}{-}H \quad + \quad CH_2{=}CH{-}CN \quad \xrightarrow[0°C]{TMG\ (cat.)} \quad (EtO)_2\underset{\underset{O}{\|}}{P}{-}CH_2{-}CH_2{-}CN$$

47%   **(6.23)**

Addition of diethyl phosphite to acrylonitrile can also be accomplished in a solid–liquid two-phase system at 80°C in the presence of a catalytic amount of anhydrous K_2CO_3 and TBAB, providing diethyl 2-cyanoethylphosphonate in 79% yield.[207] Similarly, the K_2CO_3/EtOH catalytic system at room temperature appears as a promising combination, superior to the traditional sodium alcoholate catalyst, for hydrophosphonylation of unsaturated nitriles. It makes it possible to almost completely avoid side processes that result in tarring of the starting and final compounds, and, consequently, the yields are increased (66–90%).[208,209]

Other applications of the hydrophosphonylation reaction include the addition of dialkyl phosphite to 2-(alkoxycarbonyl)- or 2-cyanoacrylonitriles. The reaction is carried out in the absence of catalyst[210–216] or in the presence of NaNH$_2$,[182,183] RONa,[217,218] or TFAA.[219] The Michael addition can be combined with an elimination reaction. For example, because of the presence of a methylthio leaving group, dialkyl 2,2-dicyano- or 2-cyano-2-ethoxycarbonyl-1-methylthio-(E)-vinylphosphonates are the products of the conjugate addition of dialkyl phosphite to the corresponding ketene dithioacetals in the presence of NaH (2 eq) in THF (Scheme 6.24).[220]

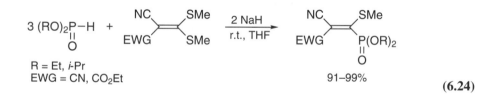

R = Et, i-Pr
EWG = CN, CO$_2$Et

91–99%

(6.24)

The Michael reaction can also be applied to the formation of cyclopropanic systems. Thus, Michael addition of phosphonate carbanions, generated by electrochemical technique or by thallium(I) ethoxide in refluxing THF, to α,β-unsaturated nitriles provides a convenient preparation of substituted 2-cyanocyclopropylphosphonates via a tandem Michael addition–cycloalkylation sequence (Scheme 6.25).[221–224]

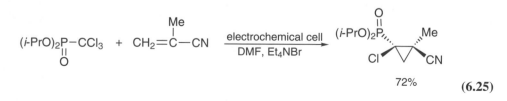

72%   **(6.25)**

In a Michael-type reaction, unsaturated phosphorus substrates such as diethyl styrylphosphonate[225] and tetraethyl ethenylidenediphosphonate[226] (Scheme 6.26) undergo smooth addition of cyanide ion to give β-cyanophosphonates in fair yields (40–75%). Allenylphosphonates can also be used as substrate for the synthesis of 2-cyanoalkylphosphonates by addition of cyanide.[227]

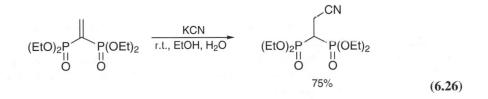

$$(6.26)$$

The reaction between trialkyl phosphites and α,β-unsaturated nitriles has been less thoroughly investigated. Whereas the hydrophosphonylation of unsaturated nitriles with diakyl phosphites proceeds smoothly, the addition of trialkyl phosphites requires more severe conditions. It involves the nucleophilic attack by the trivalent phosphorus reagent at the terminal carbon atom of the conjugated system followed by valency expansion of phosphorus in agreement with the Michaelis–Arbuzov mechanism (Scheme 6.27).[228–233]

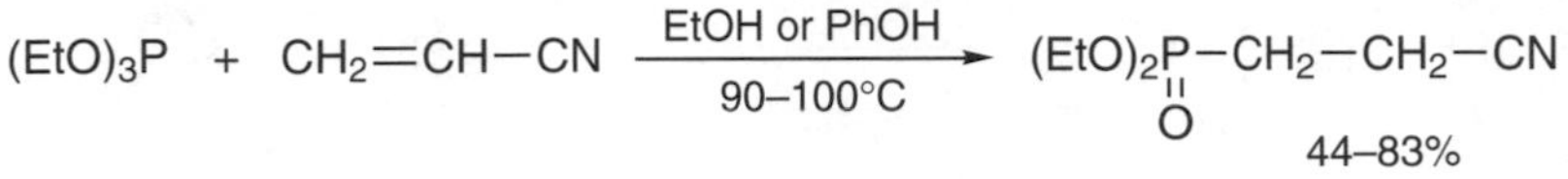

$$(EtO)_3P \ + \ CH_2{=}CH{-}CN \xrightarrow[\text{90–100°C}]{\text{EtOH or PhOH}} (EtO)_2\underset{\underset{O}{\|}}{P}{-}CH_2{-}CH_2{-}CN$$

44–83%

$$(6.27)$$

The reaction occurs in protic solvents (alcohol or phenol) at 100°C to form the β-cyanophosphonate as the predominant product in fair to good yields (44–83%, Scheme 6.27).[229] The use of a protic solvent provides a source of a proton for the anionic site of the zwitterionic adduct and a nucleophile for the dealkylation step. In the presence of ammonium iodide as proton donor, acrylonitrile is conveniently hydrophosphonylated in good yield (60%) by heating with triethyl phosphite at 70°C for only 2 h.[230]

Both diethyl trimethylsilyl phosphite and *tris*(trimethylsilyl) phosphite participate in conjugate addition reactions with α,β-unsaturated nitriles.[234,235] With acrylonitrile, addition occurs readily at 120°C at the β position, with transfer of the silyl ester linkage to the α position of the nitrile, to give 2-silylated 2-cyanoethylphosphonates in 46–72% yields.[236] Protodesilylation with a proton donor leads to the simple 2-cyanoethylphosphonates (Scheme 6.28).[234]

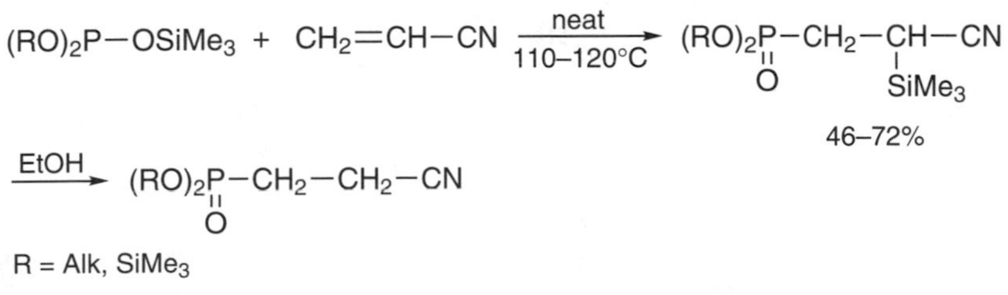

$$(RO)_2P{-}OSiMe_3 \ + \ CH_2{=}CH{-}CN \xrightarrow[\text{110–120°C}]{\text{neat}} (RO)_2\underset{\underset{O}{\|}}{P}{-}CH_2{-}\underset{\underset{SiMe_3}{|}}{CH}{-}CN$$

46–72%

$$\xrightarrow{\text{EtOH}} (RO)_2\underset{\underset{O}{\|}}{P}{-}CH_2{-}CH_2{-}CN$$

R = Alk, SiMe₃

$$(6.28)$$

As for the reaction between dialkyl phosphites and α,β-unsaturated systems, α-cyano-[237–239] and α-carboxy acrylonitriles[237,238,240] react with triethyl phosphite in alcohol to give the conjugate addition products in good yields (85–90%).

Triethyl phosphite reacts with 1-(1-acetoxyalkyl)acrylonitriles to give diethyl 2-cyano-2-alkenylphosphonates in 85–94% yields as a mixture of (*Z*)- and (*E*)-isomers. The reaction proceeds by the conjugate addition of triethyl phosphite to acrylonitrile followed by the β-elimination of acetate, which represents the nucleophile in the dealkylation step.[241]

6.1.3.4 Reactions of 2-Cyanoethylphosphonates

Diethyl 2-cyanoethylphosphonate reacts with bases (NaH or LDA) to give the corresponding anions at the β-position to phosphorus. Their reactions with ethyl formate[242] and diethyl oxalate[243] give diethyl 3-oxo-2-cyanoalkylphosphonates in high yields.

Diethyl 2-cyano-2-(trimethylsilyl)ethylphosphonate smoothly reacts with carbonyl compounds under basic conditions (LDA) to provide 2-cyano-2-alkenylphosphonates by a Peterson reaction. Subsequent Horner–Wadsworth–Emmons reaction affords substituted 2-cyano-1,3-butadienes in moderate to good yields (34–89%) by treatment with LDA and then with a second equivalent of carbonyl compound (Scheme 6.29).[244]

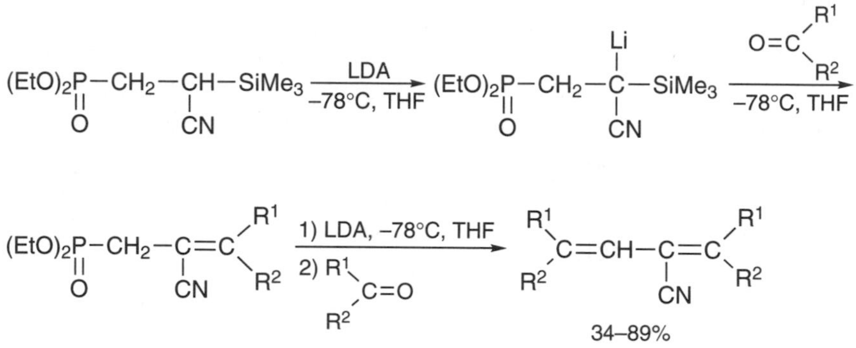

$$(6.29)$$

6.1.3.5 Other Reactions

Several promising processes are applied to the synthesis of 2-cyanoethylphosphonates, including, for example, the thermally induced Michaelis–Arbuzov-type rearrangement of diethyl 2-cyanoallylphosphite to diethyl 2-cyanoallylphosphonate (Scheme 6.30).[245,246]

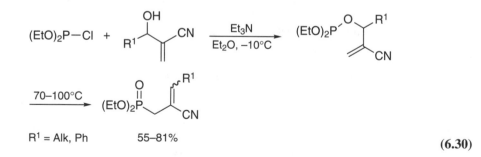

$$(6.30)$$

The addition of trimethylsilylcyanide to diethyl 2-oxopropylphosphonate, catalyzed by KCN/18-crown-6[247] or n-Bu$_3$SnCN,[248] gives the corresponding cyanohydrin trimethylsilyl ethers in high yields.

Alkylation of sodium diethyl α-(ethoxycarbonyl)benzylphosphonates with bromoacetonitrile gives diethyl α-ethoxycarbonyl-α-(cyanomethyl)benzylphosphonates in 68–70% yields.[249] The 1-acetamido-2-cyanoethylphosphonate has been prepared in 53% yield by dehydration of the corresponding amide with Ac$_2$O in the presence of a catalytic amount of Ni(OAc)$_2$.[250]

6.1.4 Dialkyl 3-, 4-, 5-, and 6-Cyanoalkylphosphonates

6.1.4.1 Michaelis–Arbuzov Reactions

The reaction is mainly used to prepare dialkyl 3-cyano-2-propenylphosphonates, which are used as precursors for retinal and derivatives. Thus, diethyl and *bis*(trifluoroethyl) 3-cyano-2-methyl-2-propenylphosphonates are conveniently prepared in good to excellent yields (75–90%) by heating at 180–200°C 4-chloro-[251,252] or 4-bromo-3-methylcrotonitriles[23] and the corresponding phosphite. Similarly, diethyl 3-cyano-2-butenylphosphonate is obtained in 95% yield from triethyl phosphite and 4-chloro-2-methylcrotonitrile (Scheme 6.31).[49] In a further example, 4-bromo-3-methyl-2-hexenenitrile has been reacted with triethyl phosphite at 170–180°C to prepare diethyl 3-cyano-2-methyl-1-ethyl-2-propenylphosphonate in fair yield (44%).[253] The reaction of a polyfunctional iodide, 3-amino-2-(iodoacetyl)crotonitrile, with triethyl phosphite affords the expected 3-cyanoalkylphosphonate in 78% yield.[254]

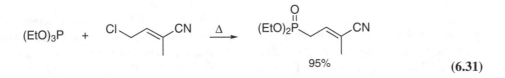

$$(6.31)$$

Diethyl 2-, 3-, and 4-cyanobenzylphosphonates are generally obtained in high yields by the Michaelis–Arbuzov reaction of triethyl phosphite with the corresponding cyanobenzyl chloride or bromide.[255–259] Diethyl 3-(9-cyano-9-fluorenyl)propylphosphonate is analogously obtained in 75% yield from triethyl phosphite and the corresponding bromide.[260]

The use of microwave irradiation in the Michaelis–Arbuzov reaction increases both reaction rates and yields. Under these conditions, a high yield of diethyl 3-cyanopropylphosphonate (75%) is obtained from triethyl phosphite and 4-bromobutyronitrile.[261]

6.1.4.2 Michaelis–Becker Reactions

This reaction can efficiently replace the Michaelis–Arbuzov reaction. Thus, diethyl 3-cyanopropylphosphonate is prepared from sodium diethyl phosphite and 4-chloro-[262] or 4-bromobutyronitrile[263] in 40% and 80% yields, respectively. Similarly, diethyl 3-furylmethylphosphonates bearing a cyano group in the 2- or 4-position are prepared in 23–39% yields from sodium diethyl phosphite and the corresponding chloromethylfurans in boiling C_6H_6.[264]

Dimethyl 2-cyanobenzylphosphonate is prepared by treatment of a solution of 2-cyanobenzyl chloride and dimethyl phosphite with MeONa in boiling C_6H_6 to give a yield (58%) comparable with those obtained by the Michaelis–Arbuzov reaction (Scheme 6.32).[265]

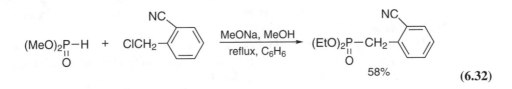

$$(6.32)$$

6.1.4.3 Michael Additions

The synthesis of dialkyl 3-cyanopropylphosphonates by Michael reaction can be effected by two different routes, the phosphorus reagent being the nucleophile or the Michael acceptor. The Michael addition of compounds containing active methylene groups such as cyanomethylphosphonate,[10,266]

1-(alkoxycarbonyl)methylphosphonates,[10,11,88,267–271] β-ketophosphonates,[11,269,272,273] 2-pyridylmethyl-phosphonate,[274] carbamoylmethylphosphonate,[275] enaminophosphonates,[276] methylenediphosphonate[277,278] and 1-(ethoxycarbonyl)-1,3-propylenediphosphonate[279,280] to α,β-unsaturated nitriles, generally in EtOH in the presence of EtONa, leads to 1:1 or 1:2 adducts resulting from addition to one or two molecules of the unsaturated nitriles, respectively (Scheme 6.33).

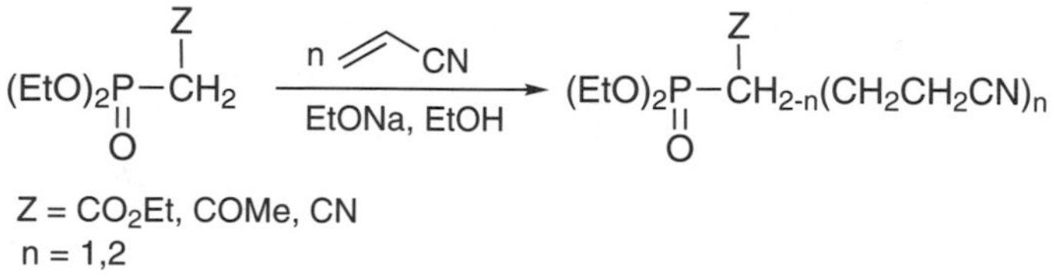

$$
\begin{array}{c}
\text{Z} \\
| \\
(EtO)_2\overset{\displaystyle O}{\underset{\displaystyle ||}{P}}-CH_2 \quad \xrightarrow[\text{EtONa, EtOH}]{n\; \diagup\!\diagup^{CN}} \quad (EtO)_2\overset{\displaystyle O}{\underset{\displaystyle ||}{P}}-CH_{2-n}(CH_2CH_2CN)_n
\end{array}
$$

Z = CO$_2$Et, COMe, CN
n = 1,2

(6.33)

Formation of the 1:1 or 1:2 adducts is determined by the structure of the Michael acceptor. The presence of substituents on the α-carbon at the double bond favors almost exclusive formation of the 1:1 adduct. To a lesser extent, the product ratio depends both on the reactant ratio and on the amount of base. By a careful choice of conditions, either compound can be favored. The addition of α-substituted compounds containing active methylene groups to acrylonitrile or methacrylonitrile gives 1:1 Michael-type adducts. The yields are fair to good in the first case and rather modest in the second.

The stabilized α-substituted cyanomethylphosphonate carbanion resulting from the addition of potassium diethyl cyanomethylphosphonate to one equivalent of acrylonitrile in THF/HMPA appears to be a useful reagent in Horner–Wadsworth–Emmons olefination reaction. Thus, the reaction with aromatic aldehydes is completely stereoselective and produces (E)-1-aryl-2,4-dicyano-1-butenes in high yields (71–83%, Scheme 6.34).[266]

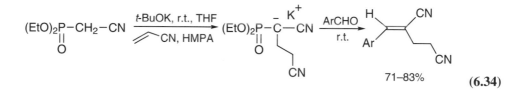

(6.34)

Addition of di-*tert*-butyl 1-(diphenylmethylimino)methylphosphonate to acrylonitrile under phase-transfer catalysis conditions at room temperature (Aliquat 336/KOH/CH$_2$Cl$_2$) leads to the 1:1 Michael adduct in 96% yield.[281]

The addition of trihalomethylphosphonates to Michael acceptors has also been reported. Thus, in the presence of the cobaloxime(III)/Zn redox system, diethyl bromodifluoromethylphosphonate adds smoothly to acrylonitrile in EtOH to give the 1:1 Michael adduct in 67% yield.[282] Similarly, the copper(I)-catalyzed addition of diethyl trichloromethylphosphonate to methacrylonitrile gives the Michael adduct in modest yield (31%).[283]

Other Michael acceptors than acrylonitrile have been investigated. For example, the Michael reaction of β-ketophosphonates with arylmethylenemalonitriles in Et$_2$O in the presence of piperidine produces 5-phosphonylated pyrans in good yields.[284]

Disappointing results have been obtained from the Michael addition of compounds containing active methylene groups to vinylphosphonates. For example, cyanoacetate reacts with the diethyl vinylphosphonate under the conditions of base catalysis to give a mixture of 1:1 and 1:2 adduct resulting of one or two additions to the vinylphosphonate.[285] When 2-pyridylacetonitrile[286] and cyanomethylphosphonate[279,280] are subjected to this reaction, diethyl 3-(2-pyridyl)- or 3-diethoxy-phosphinyl-3-cyanopropylphosphonates are obtained in 61% and 98% yields, respectively (Scheme 6.35).

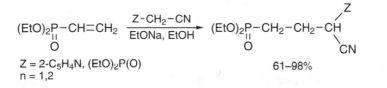

$$Z = 2\text{-}C_5H_4N, (EtO)_2P(O)$$
$$n = 1,2$$

(6.35)

Substituted vinylphosphonates also behave as Michael acceptors. Thus, phenyl-, 2-thienyl-, 2-pyridyl, and 1-naphthylacetonitriles add to tetraethyl ethenylidenediphosphonate under DBU conditions in THF,[287] and ethyl cyanoacetate adds to diethyl 2-(ethoxycarbonyl)vinylphosphonate under basic conditions.[288] In both cases, the corresponding addition products are isolated in low to moderate yields.

Unsuccessful attempts to form 1:1 adducts selectively from the Michael addition of diethyl 1-(ethoxycarbonyl)methylphosphonate or tetraethyl methylenediphosphonate to acrylonitrile under phase-transfer catalysis conditions (K_2CO_3/TEBAC/C_6H_6) have been reported. Despite variation of the reaction conditions, mixtures of 1:1 and 1:2 adducts are obtained.[289]

6.1.4.4 Other Reactions

The addition of diethyl phosphite to 3-cyanopropionaldehyde yields 1-hydroxy-3-cyanopropylphosphonate.[290] Triethyl phosphite reacts with 2,2-dimethyl-4-cyanobutyraldehyde in the presence of Me_3SiCl to give the protected α-hydroxyphosphonate in 67% yield.[291,292]

The applicability of cyanation has been successfully demonstrated in several instances. Thus, a 2′-phosphonylated 3′-trifluoromethylsulfonylnucleoside reacts smoothly with n-Bu$_4$NCN in MeCN to give the cyanation product in 45% yield (Scheme 6.36).[293] With halo derivatives, assistance is needed, and 3-bromo-3-(1-trityl-1,2,4-triazol-3-yl)propylphosphonate reacts with NaCN and 15-crown-5 in DMF at room temperature,[294,295] whereas diethyl 4-(bromomethyl)benzylphosphonate reacts with KCN and NaI in Ac_2O/H_2O.[296] The preparation of (3-cyanaziridin-2-yl)methylphosphonates by addition of Me_3SiCN to (2H-azirin-2-yl)methylphosphonates has also been reported.[297]

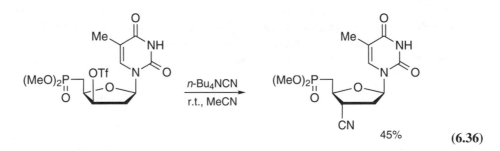

45% (6.36)

The Pd-catalyzed hydrophosphonylation of terminal alkynes with dialkyl phosphites seems destined to become the method of choice for the synthesis of α-substituted vinylphosphonates. The reaction proceeds by regioselective attack of dialkyl phosphite at the internal carbon of the triple bond (Scheme 6.37). Thus, 5-cyano-1-pentyne reacts efficiently with dimethyl phosphite in the presence of cis-PdMe$_2$(PPh$_2$Me)$_2$ affording dimethyl 1-(3-cyanopropyl)vinylphosphonate in 94% yield. In the absence of catalyst, the hydrophosphonylation product is not detected.[298]

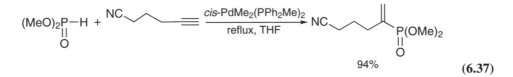

94% (6.37)

An α-sulfonylated diethyl cyanohexylphosphonate has been obtained by *C*-phosphorylation of the lithium cyanohexylsulfone with diethyl chlorophosphate [SNP(V) reaction] in the presence of LDA, followed by reaction *in situ* with formaldehyde to give the Horner–Wadsworth–Emmons reaction product in 74% overall yield.[299] The olefination reaction of 2-cyano-5-formyl-1-methoxycarbonylpyrrolidine with lithium tetraethyl methylenediphosphonate followed by Pd-catalyzed hydrogenation of the resulting double bond affords the corresponding cyanophosphonate in 69% yield.[300]

6.2 SYNTHETIC APPLICATIONS OF CYANOPHOSPHONATES

6.2.1 REACTIONS OF DIETHYL CYANOPHOSPHONATE

Diethyl cyanophosphonate (DECP) is an efficient cyanide source,[301] but as a cyanation reagent DECP is less reactive than TMSCN.[302] DECP reacts sluggishly with pyrrolidine enamines at room temperature in MeCN and rapidly in refluxing THF to give the α-aminonitriles in good yields.[303] Further investigations have revealed that α-aminonitriles can be prepared directly from carbonyl compounds with DECP and amines under mild reaction conditions.[303–308] In contrast to the aqueous conditions of the classical Strecker synthesis, this hydrocyanation route constitutes a convenient, high-yield method for the preparation of α-aminonitriles under nonaqueous conditions. The superiority of this method is demonstrated in the reaction of 4-cholesten-3-one with DECP and pyrrolidine, which affords the α-aminonitrile in 80% yield (Scheme 6.38).[303]

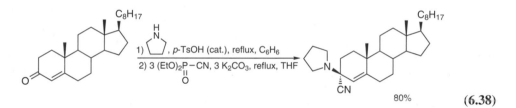

(6.38)

The advantages of DECP have been used in the successful cyanation of aromatic amine oxides (modified Reissert–Henze reaction). Aromatic *N*-oxides react with excess of DECP in the presence of Et₃N in refluxing MeCN to give the deoxygenated α-cyano compounds in fair to good yields (Scheme 6.39).[302,309–314] Although the cyanation with DECP generally results in the attack at the electron-deficient carbon α to the activated *N*-oxide, there is a single example where DECP can be used to α-cyanate quinoxaline without prior activation.[315]

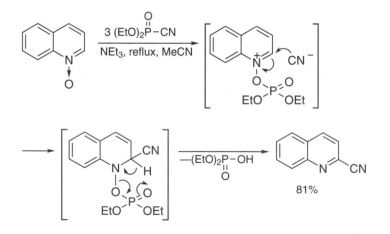

(6.39)

Cyanophosphorylation of aldehydes and ketones is successfully carried out in almost quantitative yield by treatment with DECP in the presence of LiCN in THF at room temperature (Scheme 6.40).[316–321]

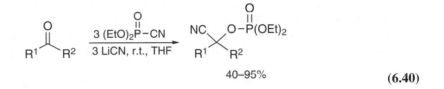

$$40–95\%$$

(6.40)

The resulting cyanophosphates are converted to nitriles by reductive cleavage with freshly prepared SmI$_2$ in THF in the presence of t-BuOH at room temperature.[322–332] This method has been applied with success to a large variety of carbonyl compounds: simple aldehydes and ketones, steroidal and glycosyl ketones, α,β-unsaturated ketones, aliphatic ketones bearing amine, ester, sulfamoyl, vinyl carbamate, and aromatic methoxy substituents (Scheme 6.41).[324] The cyanophosphate–nitrile conversion can also been effected by hydrogenolysis (3 atm H$_2$, 10% Pd/C, AcOEt).[333]

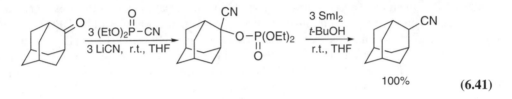

$$100\%$$

(6.41)

The cyanophosphates, prepared by the reaction of ketones with DECP in the presence of LiCN, are readily converted into α,β-unsaturated nitriles by treatment with BF$_3$·Et$_2$O in C$_6$H$_6$ at room temperature (Scheme 6.42).[334–336] This methodology has been used in the synthesis of lysergic acid.[337,338] The same conversion can be realized on treatment with a strong organic acid such as MsOH or TsOH in refluxing toluene.[339–341]

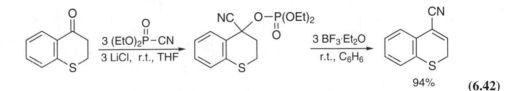

$$94\%$$

(6.42)

The cyanophosphates resulting from α,β-unsaturated ketones are readily transformed into conjugated allylic phosphate via a BF$_3$·Et$_2$O-catalyzed allylic rearrangement (Scheme 6.43).[342–345] Similarly, the cyanophosphates derived from 1,4-benzoquinones react with aromatic and heteroaromatic compounds to give 3-aryl-4-hydroxybenzonitriles.[346,347]

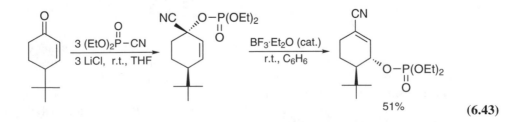

$$51\%$$

(6.43)

Reaction of carboxylic[3,348,349] or phosphinic[350] acids with DECP at room temperature in the presence of Et$_3$N gives transient acyl cyanides. When the reaction is carried out in the presence of nucleophiles, alcohols,[3] or thiols,[350–352] the corresponding esters and thioesters are obtained in fair yields (Scheme 6.44). The main advantage of the transformation may be that the reaction proceeds under mild conditions (room temperature, almost neutral conditions). In the presence of DECP (2 eq), the reaction with carboxylic acids takes a different course to give dicyanophosphates in good yields.[353] The direct C-acylation of active methylene compounds with carboxylic acids using DECP in the presence of Et$_3$N occurs easily in a single operation under exceptionally mild conditions.[354–358]

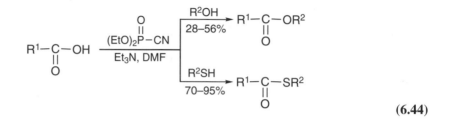

$$(6.44)$$

DECP is an efficient and valuable reagent for amide formation, and both aliphatic and aromatic acids react with aliphatic and aromatic amines.[359] In combination with Et$_3$N, DECP is a coupling reagent for the racemization-free peptide synthesis in DMF, CH$_2$Cl$_2$, or THF with excellent yields.[359–378] It is also useful for solid-phase peptide synthesis in both the stepwise and fragment-condensation approaches.[379] With DECP, the construction of porcine motilin or peptides of aquatic origin has been successfully achieved.[380]

The cyanophosphates prepared from α,β-unsaturated or aromatic ketones are reductively cleaved with lithium in liquid ammonia at low temperature. When the reduction is carried out under refluxing conditions (at −33°C) in the presence of t-BuOH, the methylene compound is obtained in quantitative yield by loss of the cyanophosphate group (Scheme 6.45).[381] In contrast, quenching the reaction mixture at low temperature with isoprene leads to the unsaturated nitrile, resulting from dephosphorylation, as an epimeric mixture.[382]

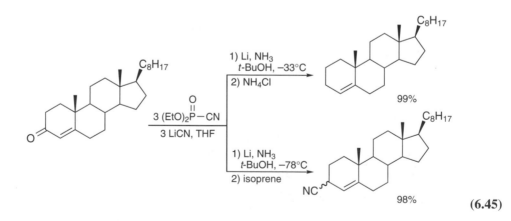

$$(6.45)$$

Similarly, treatment of ynones with DECP and LiCN leads to the corresponding cyanophosphate, which, on reaction with cuprates in THF at low temperature, give the allenic nitrile by elimination of phosphate.[383] Only a few reports describe the properties of DECP as a phosphorylation reagent for imidazoline nitrogen atoms,[384] phenols,[385] and Grignard reagents.[386]

The sodium salts of aryl sulfinic acids react rapidly with DECP in refluxing THF to give aryl thiocyanates in good yields, and even hindered substrates give the corresponding thiocyanates.[387]

Some reports describe the nucleophilic attack at the cyano carbon atom of DECP without P–CN bond splitting. For example, the cycloaddition of lithium trimethylsilyldiazomethane and nitrile oxides[5] to the cyano group of DECP produces 1,2,3-triazole and 1,2,4-oxadiazole-4-phosphonates, respectively. In the presence of a Lewis acid, which increases the electrophilicity of the cyano carbon atom, compounds containing active methylene groups add to the cyano group to give α,β-unsaturated α-aminophosphonates (Scheme 6.46).[388]

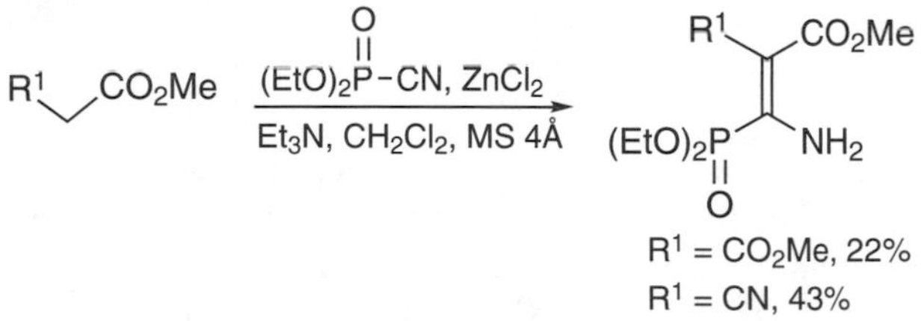

$$R^1 \diagup CO_2Me \quad \xrightarrow[\text{Et}_3\text{N, CH}_2\text{Cl}_2\text{, MS 4Å}]{(EtO)_2\overset{\displaystyle O}{\overset{\|}{P}}-CN,\ ZnCl_2} \quad \begin{array}{c} R^1 \diagdown \diagup CO_2Me \\ (EtO)_2\underset{\underset{O}{\|}}{P} \diagdown NH_2 \end{array}$$

$$R^1 = CO_2Me,\ 22\%$$
$$R^1 = CN,\ 43\%$$

(6.46)

The mild conditions used to dealkylate DECP attest to the strongly activating nature of the cyano group, which renders the phosphoryl group even more electrophilic. Thus, conversion of DECP to *bis*(trimethylsilyl) ester is effected by brief treatment of neat DECP with iodotrimethylsilane.[389] Protodesilylation with isopropyl alcohol gives the cyanophosphonic acid isolated as the inorganic and organic salts. The dealkylation of a single ethyl ester of DECP can also be effected.[389]

6.2.2 HORNER–WADSWORTH–EMMONS REACTIONS OF DIALKYL CYANOMETHYLPHOSPHONATES: PREPARATION OF α,β-UNSATURATED NITRILES

Dialkyl cyanomethylphosphonates are routine reagents readily accessible on laboratory scale and also available from a number of chemical suppliers. Since the review by Pudovik and Yastrebova published in 1970,[390] the use of dialkyl cyanomethylphosphonates in Horner–Wadsworth–Emmons reactions has been covered in several comprehensive and excellent reviews.[391–395] All the factors governing the reaction (nature of the carbanions and carbonyl group, reaction conditions, mechanism, and stereochemistry) have been studied in depth. We invite the reader to refer to these papers. In contrast, we discuss here the synthetic applications resulting directly from the use of dialkyl cyanomethylphosphonate, which is the pathway of choice for the preparation, via α,β-unsaturated nitriles, of α,β-unsaturated aldehydes, cyanoethyl compounds, allylamines, and saturated amines.

6.2.2.1 Reaction Conditions

The starting dialkyl cyanomethylphosphonate is generally dissolved in an aprotic solvent and treated at 0°C under inert conditions with the appropriate base to generate the carbanion. When the resulting mixture becomes clear, the carbonyl compound is added at room temperature, and the reaction is allowed to continue 1 h for aldehydes and several hours for ketones, occasionally with heating. Workup with water to dissolve the phosphate and subsequent extraction provide the crude α,β-unsaturated nitrile as a mixture of (*E*)- and (*Z*)-isomers in which the (*E*)-isomer is predominant. In the great majority of reported examples the two isomers are separated by column chromatography.

Traditionally, NaH is the most commonly used base, followed by *n*-BuLi and *t*-BuOK.[395] Sodium amide,[396–400] LDA,[401–405] LiHMDS,[406–412] NaHMDS,[406,413,414] KHMDS,[90,413,415–418] sodium and magnesium methoxide[419,420] or ethoxide,[421,422] and DBU[423] are less frequently used. The nature of the cation present in the Horner–Wadsworth–Emmons reaction depends on the chosen base and greatly influences the stereoselectivity of the reaction. The chiral lithium 2-aminoalkoxides (1*R*,2*S*) have been used as chiral bases for the enantioselective reaction between diethyl cyanomethylphosphonate and 4-*tert*-butylcyclohexanone.[424]

The solvents usually are aprotic and in the most cases ethers (THF, DME, dioxane, Et_2O), but toluene, C_6H_6, DMF, and DMSO, frequently in combination with NaH,[401,425–431] are also used (Scheme 6.47).

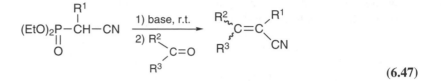

$$\tag{6.47}$$

The formation of α,β-unsaturated nitriles has also been developed in less traditional methods. For example, the search for neutral and mild reaction conditions for generating cyanomethylphosphonate carbanions has promoted the use of simple systems based on the association of an amine and a lithium salt. Thus, LiCl–DBU, LiCl–DEPA, and LiCl–Et_3N in dry MeCN are very effective combinations to generate active species in the presence of base-sensitive substrate or reagent,[423,432] permitting the expected olefination in high yields.

The liquid–liquid two-phase system using aqueous NaOH and C_6H_6, CH_2Cl_2, $CHCl_3$, or CCl_4 in the presence of tetrabutylammonium salts leads to variable yields of α,β-unsaturated nitriles.[433–435] By this technique, (E)- and (Z)-2-methoxycinnamonitrile and 3-cyclohexyl-2-propenenitrile have been prepared on large scale with 93% and 94% yields, respectively.[436,437]

Several solid–liquid two-phase systems have been described: powdered KOH in THF[438] or DMF,[439] $Ba(OH)_2$ in moist dioxane,[440] K_2CO_3 in water or D_2O[441–450] or anhydrous solvents,[39,40,441,451] Na_2CO_3 in moist THF[452] and $NaHCO_3$ in water.[453] The system using aqueous 6–9 M K_2CO_3 solution appears as the most promising. This new technique provides routinely good yields of functional olefins from aldehydes (Scheme 6.48). It allows the use of functional aldehydes without protection or aqueous stabilized solutions of unstable pure aldehydes without previous isolation (formaldehyde, crotonaldehyde).

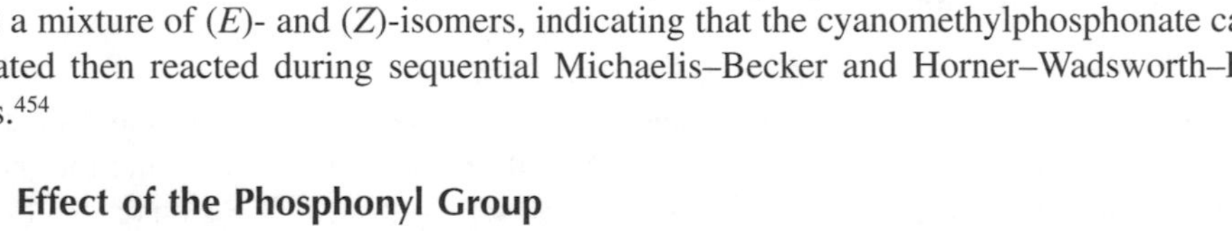

$$(EtO)_2\underset{\underset{O}{\|}}{P}-CH_2-CN \xrightarrow[D_2O]{6\ M\ K_2CO_3} (EtO)_2\underset{\underset{O}{\|}}{P}-CD_2-CN \xrightarrow{R^1-CHO} R^1-CH{=}CD-CN$$

R^1 = Alk, Ar

$$90\text{–}97\% \tag{6.48}$$

Electrolysis of diethyl phosphite in MeCN containing tetraethylammonium salts in the presence of chloroacetonitrile and aldehyde gives the corresponding α,β-unsaturated nitriles in satisfactory yields as a mixture of (E)- and (Z)-isomers, indicating that the cyanomethylphosphonate carbanion is generated then reacted during sequential Michaelis–Becker and Horner–Wadsworth–Emmons reactions.[454]

6.2.2.2 Effect of the Phosphonyl Group

The Horner–Wadsworth–Emmons reaction is generally conducted with diethyl cyanomethylphosphonate, which is easily accessible by the Michaelis–Arbuzov reaction or prepared *in situ* from α-cyano carbanions. On reaction with aldehydes, it leads to a mixture of α,β-unsaturated nitriles in which the (E)-isomer is predominant. Simple replacement of the ethoxy groups attached to phosphorus by isopropoxy groups enhances the (E)/(Z) ratio.[24,442,455–457] However, the effect on the stereoselectivity is too modest to justify the systematic replacement of the ethoxy groups by isopropoxy groups. The *bis*(trifluoroethoxy) group has been reported to be efficient in the preparation of all-*cis* retinal.[252,458] In contrast, replacement of the ethoxy groups by phenoxy groups reverses the selectivity, and the (Z)-isomer becomes the dominant product.[42] The use of five-membered cyclic phosphonate derived from commercially available 2,3-butanediol results in dominant (*E*)-isomer formation with aliphatic and aromatic aldehydes. The effect is similar with six-membered ring-containing phosphonates.[25]

6.2.2.3 Effect of α-Substitution

Several diethyl α-substituted cyanomethylphosphonates participate to the Horner–Wadsworth–Emmons reaction. The most common substituents attached to the α-carbon atom are alkyl groups and only occasionally aromatic moieties: R^1 = Me,[29,30,95,128–130,459–466] Et,[29,91,462,467] *i*-Pr,[29,407,409,412,413,415,462] *n*-C_6H_{13},[29,462] allyl,[128] benzyl,[462] 2,2-dimethoxyethyl,[468] 2-methyl-2-penten-5-yl,[90,416–418] 4-hydroxybutyl,[94] 5-ribofuranosyl,[469] OMe,[28,470–472] O*t*-Bu,[28,462,470] PhS,[28] Ph,[116,473] 3-pyridyl[115,474] (trimethylsilyl)methyl,[93] CH_2CO_2Me,[95] $CH(Me)CO_2Me$,[96] $CH(Me)CO_2Et$,[95] $CH(Et)CO_2t$-Bu,[97] $CH(CH_2Ph)CO_2Et$,[95] F,[110,111] and NMe_2.[475–478] The steric demand of R^1 determines the (*E*)/(*Z*) ratio in the unsaturated nitrile product. The (*E*)-isomer dominates when R^1 = H and (*E*)/(*Z*) unsaturated nitrile ratios near of 85/15 are routinely observed. The proportion of the (*Z*)-isomer rises as the size of the R^1 group increases (R^1 = Me, Et, *n*-C_6H_{13}) with values of 15/85 being recorded when R^1 = *i*-Pr and O*t*-Bu.[413,462] The conditions to provide high (*Z*)-selectivity have been examined in depth in the reaction between diethyl 1-cyanoisobutylphosphonate and farnesal.[413]

6.2.2.4 The Carbonyl Reagents

A great variety of carbonyl compounds, aldehydes or ketones, can participate to the reaction: aliphatic,[479–510] aromatic,[486,511–521] heteroaromatic,[522–530] or α,β-unsaturated aldehydes,[515,531–539] dialdehydes,[540–547] aliphatic ketones,[496,499,519,548–559] four-,[560,561] five-,[561–573] six-,[515,561,574–585] seven-,[29,463] eight-,[586,587] and 12-membered[29,588,589] cyclic ketones, aromatic,[513,519,590–594] mixed,[515,519,555,595–614] and α,β-unsaturated[615–621] ketones, diketones,[622,623] and lactones.[624] Aldehydes are more reactive and require mild conditions (room temperature, short reaction times), whereas ketones often need longer reaction times and sometimes heating to go to completion.

6.2.2.5 Tandem Horner–Wadsworth–Emmons and Michael Reactions

Epimeric ribosylacetonitriles have been prepared by tandem Horner–Wadsworth–Emmons and Michael reactions of protected riboses with sodium diethyl cyanomethylphosphonate in THF or DME.[406,408,625–631] The reaction proceeds through the ring-open intermediate olefins resulting from the Horner–Wadsworth–Emmons reaction, which spontaneously undergo cyclization under the reaction conditions to produce ribosylacetonitriles in good yields as a mixture of α- and β-epimers (Scheme 6.49).[628] Further reduction of the cyano group to the corresponding amines was found to be effective with a borane–dimethyl sulfide complex. The reaction proceeds smoothly, giving the aminoethylated glycosides in high yields.[627,628]

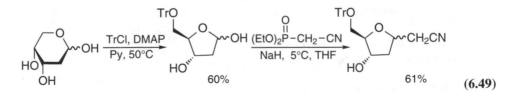

(6.49)

The tandem Horner–Wadsworth–Emmons and Michael reactions have been applied to lactols, providing tetrahydropyranyl or tetrahydrofuranyl derivatives;[632–634] to the hemiaminal derived from *N*-Boc-protected pyroglutamic esters, providing 5-substituted prolinates;[635] and to imidate anion, providing oxazolidine.[636] On reaction with lactols, replacement of the cyanomethylphosphonate by the 1-cyano-1-dimethylaminomethylphosphonate leads to the formation of α-cyano enamines as a mixture of (*E*)- and (*Z*)-isomers.[637] Hydrolysis of the enamines (AcOH/THF/H$_2$O, 60°C) produces the δ-lactones in good overall yield (40–55%). A similar strategy is used for the construction of a benzofuranitrile[638] and for the synthesis of analogues of chuangxinmycin, which contains a heterocyclic skeleton bearing an indole ring.[639] The intramolecular Michael addition is promoted by reaction of methylthioacetate[639] or methoxyacetate carbanions[638] with ethylenic nitriles.

6.2.2.6 Tandem Michael and Horner–Wadsworth–Emmons Reactions

Addition of charged nucleophiles to diethyl 1-cyanovinylphosphonate[640] leads to the generation of α-substituted cyanomethylphosphonate carbanions. In the presence of carbonyl compounds, they undergo Horner–Wadsworth–Emmons reaction to produce the α,β-unsaturated nitriles in fair to good yields as a mixture of (*E*)- and (*Z*)-isomers.[640]

6.2.3 HORNER–WADSWORTH–EMMONS REACTIONS OF PHOSPHONOCROTONONITRILES: PREPARATION OF α,β,γ,δ-UNSATURATED NITRILES

In the synthesis of retinal from β-ionone, diethyl 3-cyano-2-methyl-2-propenylphosphonate is routinely used in association with diethyl cyanomethylphosphonate to generate the four-carbon homologue of β-ionylideneacetaldehyde which gives retinal in high yields (Scheme 6.50).[455,456,466,641–658]

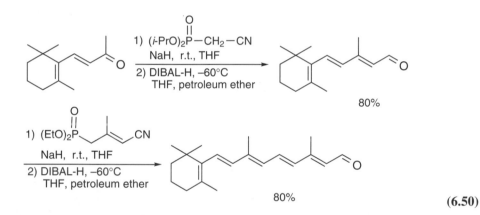

$$(6.50)$$

Diethyl 3-cyano-2-butenylphosphonate[49,659–661] and 3-cyano-2-methyl-1-ethyl-2-propenylphosphonate[253] appear as synthetically useful reagents in the synthesis of modified retinals, whereas *bis*(trifluoroethyl) 3-cyano-2-methyl-2-propenylphosphonate[252,458] is used to obtain *cis*-retinals. Diethyl 3-cyano-3-dimethylamino-2-propenylphosphonate has been evaluated in reaction with aldehydes,[662] and dimethyl 2-oxo-5-cyanopentyl- or 2-oxo-6-cyanohexylphosphonates have found use in prostaglandin synthesis.[663–665]

6.2.4 ELABORATION OF HORNER–WADSWORTH–EMMONS-PRODUCED α,β-UNSATURATED NITRILES

6.2.4.1 Into α,β-Unsaturated Aldehydes

The most frequently used and widely known application of α,β-unsaturated nitriles is their conversion into α,β-unsaturated aldehydes. This is also the most extensively studied reaction for the formation of α,β-unsaturated nitriles. In 1957 it was reported that simple aliphatic or aromatic nitriles are effectively converted to aldehydes in high yields via reduction to intermediate imines with DIBAL-H followed by hydrolysis of the imines.[666,667] This transformation has been introduced for the synthesis of β-ionylideneacetaldehyde, an important intermediate in the synthesis of retinoids and carotenoids, in combination with the preparation of β-ionylideneacetonitrile from diethyl cyanomethylphosphonate and β-ionone (Scheme 6.51). DIBAL-H appears to be the reagent of first choice, and, since β-ionylideneacetaldehyde[24] was synthesized, it is well recommended for the preparation of α,β-unsaturated aldehydes and regularly used with success. Diethyl cyanomethylphosphonate is preferable in most instances to protected diethyl 1-formylmethylphosphonate that is specially elaborated for the formylolefination of carbonyl compounds. The nitrile reagent has the advantage of being more accessible than protected formylphosphonates and commercially available.

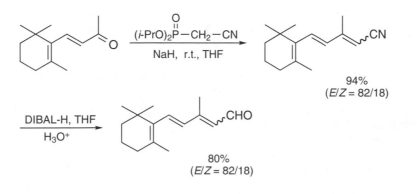

$$94\%$$
$$(E/Z = 82/18)$$

$$80\%$$
$$(E/Z = 82/18)$$

$$(6.51)$$

The mixture of (E)- and (Z)-isomers of α,β-unsaturated nitriles is treated with an excess of DIBAL-H in a large choice of solvents: THF,[24,457,458,668–678] toluene,[39,40,405,641,646,679–685] C_6H_6,[686,687] hexane,[402,410,642,643,645,652,658,688–695] pentane,[696] Et_2O,[648,655,656,697–701] CH_2Cl_2,[465,466,657,702–707] or petroleum ether.[41,50,253,630,651,653,654,659,661,708]

Most of the reactions are carried out by adding DIBAL-H (in hexane or toluene) to the nitrile at low temperature. After a slow return to room temperature to complete the reaction, the resulting imine is hydrolyzed with a dilute acid (tartaric acid, oxalic acid, AcOH, H_2SO_4, HCl), with a saturated aqueous solution of a salt (NH_4Cl or NaH_2PO_4) or with wet silica gel. The reduction can also be carried out at 0°C.[24,668,670,674,677,686,691,704] The reported yields are high (60–92%), and in most cases pure (E)- and (Z)-isomers are isolated by chromatographty with silica gel. Conversion of α,β-unsaturated nitriles to α,β-unsaturated aldehydes is also achieved in fair yields (61–66%) with Raney Ni on heating in HCO_2H or aqueous NaH_2PO_2/Py.[98,425,709–711]

There are many synthetically useful conversion reactions of α,β-unsaturated nitriles to α,β-unsaturated aldehydes using DIBAL-H. To illustrate the value of the method, we have brought together all the examples concerned with chemistry of retinal. Thus, the applicability of the sequence combining the use of diethyl cyanomethylphosphonate with diethyl 3-cyano-2-methyl-2-propenylphosphonate, diethyl 3-cyano-2-butenylphosphonate, or diethyl 3-cyano-2-methyl-1-ethyl-2-propenylphosphonate has been studied intensively in retinal synthesis to prepare retinal analogues: sila-retinal,[698] modified retinal,[50,253,465,466,642,646–648,652,653,655,656,659,668,672,689,695,699–701,703,712–714] fluororetinal,[402,691,708] cis-retinal,[687,694] bicyclic retinal,[457,643,644,650,654,657,688,693] acetylenic retinal,[641,645] and labeled retinal.[455,456,649,651,705]

6.2.4.2 Into Saturated Nitriles

Selective reduction of the olefin unit of the α,β-unsaturated nitrile is accomplished essentially with two systems: one by catalytic hydrogenation using Pd/C, the other using magnesium in MeOH, a reagent combination known for its ability to reduce α,β-unsaturated nitriles.

In the catalytic hydrogenation, the (E)- and (Z)-mixture of olefins is hydrogenated with 5–10% Pd/C in alcoholic medium, generally at room temperature and atmospheric pressure but occasionally under pressure, to afford nitriles in good to excellent yields.[37,437,453,589,715–734] With ethylenic nitriles coming from cyclic ketones, hydrogenation of the double bond gives a mixture of nitrile epimers separable by chromatography.[724] Other catalytic systems have been found to be efficient for the hydrogenation of α,β-unsaturated nitriles, such as the borohydride-reduced palladium catalyst, prepared from PdCl$_2$ and NaBH$_4$ in MeOH,[735,736] the Adams catalyst (PtO$_2$) in EtOH,[737] and RhCl(Ph$_3$P)$_3$ in EtOH/toluene[738] under hydrogen pressure, convert α,β-unsaturated to saturated nitriles in high yields (85–90%).

Magnesium in MeOH is recognized as a performant system offering several advantages.[459] The reaction is carried out with excess of magnesium turnings in dry MeOH, usually at 0°C, and gives rapid conversion with high yields (75–95%).[27,419,426,429,431,739–744] The use of magnesium in MeOH avoids the problem of overreduction to primary amine,[741] and with ethylenic nitriles coming from cyclic ketones, the reduction proceeds in a highly stereoselective manner with delivery of only one epimer (Scheme 6.52).[745–748] An elegant use of the magnesium-MeOH system is reported in the synthesis of sinefungin analogues. The Horner–Wadsworth–Emmons reaction between a suitably functionalized 1-(5-ribosyl)cyanomethylphosphonate and an aliphatic aldehyde is carried out in MeOH containing magnesium methanolate. Without isolation of the α,β-unsaturated nitrile, the double bond is reduced by simple addition of an excess of magnesium to the reaction medium to give the nitrile in 90% overall yield.[419,420]

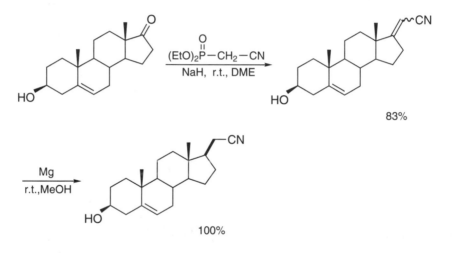

(6.52)

In alcoholic medium, NaBH$_4$ converts α,β-unsaturated nitriles into cyanoethyl compounds under more severe conditions (refluxing alcohol for several hours).[749–751] The enantioselective reduction of the double bond of each (E)- and (Z)-isomer of α,β-unsaturated nitriles derived from ketones has been evaluated using NaBH$_4$ in EtOH/diglyme in the presence of "semicorrin" CoCl$_2$ catalyst; the yields are good (about 75%), but the enantioselectivity remains modest (53–69%).[752] In contrast, it has been found that copper(I) hydride, a reagent that presumably operates by a mechanism akin to the Michael reaction,[753,754] cleanly and stereoselectively effects the reduction of α,β-unsaturated to saturated nitriles in the presence of 2-butanol, as proton donnor, in THF. The use of LAH in THF at room temperature has also been reported in the stereoselective reduction of α,β-unsaturated nitriles.[755,756]

6.2.4.3 Into Allylamines

The choice of metal hydride reducing agent is important, and the selection of the reagents is dictated by the need to prevent concomitant reduction of the double bond. With LAH it is recommended that the reaction be carried out in Et$_2$O at room temperature with a molar ratio nitrile/LAH = 1/0.9 to favor the nitrile reduction.[448,467,757,758] However, even at low temperature, it is not possible to prevent the formation of fully saturated products together with substantial amounts of oligomeric products. To avoid these complications, AlH₃ (prepared *in situ* from one equivalent of LAH and AlCl₃ or from LAH and H₂SO₄) in Et₂O at 0°C appears to be the reagent of choice, giving regular yields of allylamines in the range of 78–94%.[448,759–761] When W2-Raney Ni in methanolic solution saturated with ammonia is used, only reduction of the nitrile to primary amine is observed after prolonged reaction times (Scheme 6.53).[425] Sequential addition of DIBAL-H and NaBH₄ at low temperature has also been reported.[762]

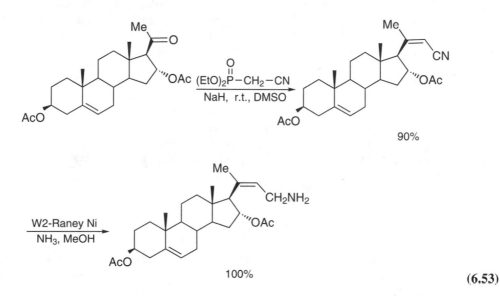

90%

100%

(6.53)

6.2.4.4 Into Saturated Amines

Conversion of α,β-unsaturated nitriles to saturated amines proceeds in one or two steps. Two catalytic systems have been used in the one-step process. Ethylenic nitriles are hydrogenated to propylamine with either Raney Ni in alcohol under basic conditions[396,449,450,763–766] or 10% Pd/C in HCl/EtOH solution with variable hydrogen pressure and temperature.[436,767] In the absence of an exhaustive study, it must be merely recorded that the two systems lead to the fully saturated compounds in fair to good yields (55–75%). Hydrogenation under pressure in EtOH at 50°C in the presence of PtO₂ produces the saturated amines in good yields (Scheme 6.54).[768–770] In contrast, the borohydride-reduced cobalt catalyst, prepared from CoCl₂ and NaBH₄ in MeOH, gives lower yields (32%).[404] In some cases, exocyclic α,β-unsaturated nitriles prepared from cyclic ketones undergo double bond migration to give the thermodynamically favored endocyclic β,γ-unsaturated nitriles. Reduction of the nitrile with LAH[771–773] or AlH₃[774] in Et₂O affords the amine with an endocyclic double bond.

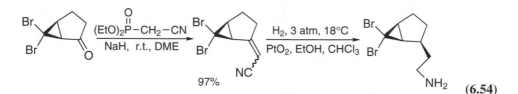

97%

(6.54)

In the two-step process, the catalytic hydrogenation of the double bond is frequently carried out with 10% Pd/C in alcoholic medium[775–780] to produce good yields (85–90%) of cyanoethyl compounds. When hydrides are used, NaBH₄ in refluxing alcohol gives good yields (79–96%),[781,782] whereas LAH in THF at room temperature gives only fair yields (63%).[783] Magnesium in MeOH at room temperature is less often used but gives good results (74–95% yield).[451,784]

In turn, the resulting saturated nitrile is reduced to amine under hydrogen with a Rh/Al$_2$O$_3$ catalytic system[776,777] and more frequently with hydrides. Thus, LAH in Et$_2$O at reflux,[780,781] in THF at 0°C,[451,779] and NaBH$_4$ with CoCl$_2$ in MeOH[778] are preferred to LAH in refluxing THF.[783] The use of diborane in THF has also been reported.[775,782] In all these reactions, the yields of the first step are generally higher than those of the second.

The two operations can be reversed, the nitrile group being reduced first with Raney Ni in MeOH saturated with ammonia and the double bond further hydrogenated with 10% Pd/C in MeOH.[785]

6.2.4.5   Into α,β-Unsaturated Alcohols, Ketones and Acids

The α,β-unsaturated nitriles can be converted to the corresponding α,β-unsaturated alcohols by two successive treatments with DIBAL-H. This reaction has been used in the synthesis of sarco-phytols A and T to produce the allylic alcohol in 71% yield.[409]

Treatment of α,β-unsaturated nitriles with an ethereal solution of alkylmagnesiums in refluxing C$_6$H$_6$ (MeMgI, *i*-BuMgBr),[19,786,787] or MeLi in Et$_2$O at 0°C[788–791] gives imines that are hydrolyzed *in situ* with aqueous NH$_4$Cl to the corresponding α,β-unsaturated ketones (54–88%, Scheme 6.55).

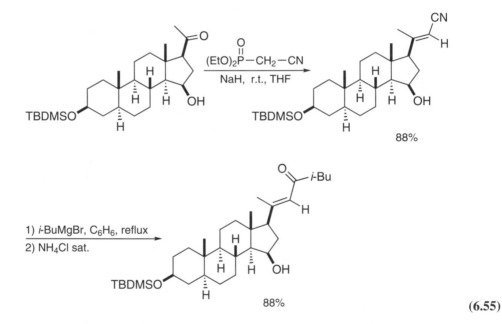

$$(6.55)$$

α,β-Unsaturated acids are obtained by hydrolysis of α,β-unsaturated nitriles with KOH in 95% ethanol solution at reflux (65–70%),[401,792–794] with 40% H$_2$O-MeOH at reflux,[795] or with aqueous H$_2$SO$_4$ (59%).[796,797]

6.2.4.6   Into Alkenes, Allyl Alcohols, and α,β-Unsaturated Ketones
(Decyanation Reaction)

The procedure developed for the decyanation of secondary nitriles[798] has been extended to α,β-unsaturated nitriles.[29] The reaction of α-unsubstituted α,β-unsaturated nitriles with LDA in THF/HMPA results in the abstraction of a γ-hydrogen from a methylene site to afford a delocalized anion. Trapping this anion at the α-position by dimethylation[799,800] followed by reduction–elimination with dissolving metals leads to tetrasubstituted alkenes (Scheme 6.56).[799] Epoxidation of the β,γ-unsaturated nitrile[800,801] before reductive decyanation with dissolving metals leads to allylic alcohols (Scheme 6.56).[801]

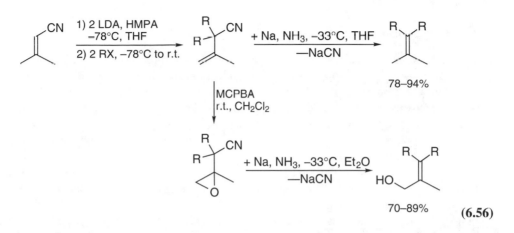

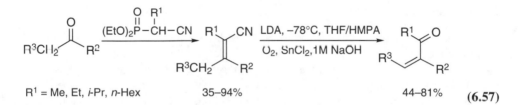

(6.56)

Metallation of α-substituted α,β-unsaturated nitriles followed by introduction of dry oxygen gas results in the regioselective trapping of the delocalized anion at the α carbon to produce the hydroperoxide. Reduction with aqueous Na_2SO_3 and exposure of the cyanohydrin to NaOH affords the β,γ-unsaturated ketones in good yields (Scheme 6.57).[29,463,588]

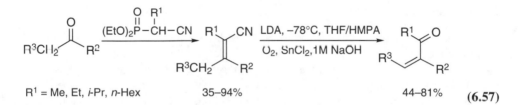

(6.57)

6.2.4.7 Into Carboxylic Acids, Esters, and Amides

An efficient sequence for the one-carbon homologation of aldehydes and ketones to carboxylic acids, esters, and amides involves a Horner–Wadsworth–Emmons reaction of diethyl 1-cyano-1-*tert*-butoxymethylphosphonate. The resulting 1-*tert*-butoxyacrylonitriles undergo the cleavage of the *tert*-butyl ether using $ZnCl_2$ in refluxing Ac_2O to afford acetoxyacrylonitriles. Finally, treatment of the acetoxyacrylonitrile with aqueous KOH provides, after acidification, the carboxylic acids in excellent yields. The reactions with sodium alkoxides or amines furnish the carboxylic esters and amides, respectively (Scheme 6.58).[470]

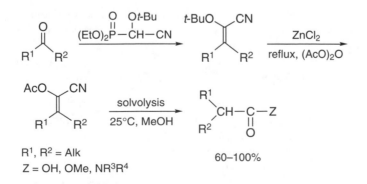

(6.58)

Diethyl 1-cyano-1-(dimethylamino)methylphosphonate has been used as the basis of one-carbon homologation of carbonyl compounds through the conversion of cyanoenamines into carboxylic acids.[475–478]

6.2.5 CYCLOADDITION REACTIONS

Dialkyl α-diazophosphonates readily undergo 1,3-dipolar additions with acrylonitrile to give dialkyl 3-cyano-Δ^2-pyrazolyl-5-phosphonates[802–804] or cyclization reactions in the presence of HCl or H_2S to form the corresponding phosphonylated chloro-1,2,3-triazoles[805] and 5-amino-1,2,3-thiadiazoles,[806] respectively. Similarly, diethyl phosphonoacetonitrile oxide reacts with acrylonitrile, leading to diethyl 3-phosphonomethyl-5-cyano-Δ^2-isoxazoline.[807–809] Diethyl 4-phosphono-5-cyanoisoxazolidine is selectively obtained from the condensation of diethyl 2-cyanovinylphosphonate with aromatic nitrones.[810,811]

One of the key steps in the total synthesis of phorboxazole A and B is the Rh(II)-catalyzed cycloaddition reaction of diethyl cyanomethylphosphonate with dimethyl diazomalonate to form the desired phosphonylated oxazole.[812]

Phosphonylated butadienes are reacted with acrylonitrile[813–816] or tetracyanoethylene[817] in a [4 + 2] cycloaddition reaction to form the corresponding phosphonylated cyanohexenes. Similarly, the isomeric diethyl 4-phosphono-4-cyano-1-cyclohexenes are obtained from the reaction of diethyl 1-cyanovinylphosphonate with the adequate butadiene.[818]

6.2.6 REACTIONS OF THE CYANO GROUP

6.2.6.1 Addition Reactions

With heating in the presence of aqueous HCl, alcohols add to the triple bond of phosphonylated nitriles to produce the corresponding amides.[819,820] Several variations on the preparation of diethyl 1-(thioacetamido)methylphosphonate by mercaptolysis of diethyl cyanomethylphosphonate have been reported. Best yields, up to 90%, are obtained by addition of H_2S to a suspension of diethyl cyanomethylphosphonate, Et_3N, and tetrabutylphosphonium bromide in toluene (Scheme 6.59).[821] The early reported procedure using mercaptolysis at room temperature of a mixture of diethyl cyanomethylphosphonate, Et_3N, and Py led to diethyl 1-(thioacetamido)methylphosphonate in low to good yields (23–85%).[822–825] The addition of cysteamine to the cyano group has also been reported.[826]

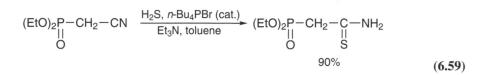

$$(6.59)$$

Mild and selective hydrolysis of diethyl cyanomethylphosphonate leading to carboxylic acid and amide can be achieved under neutral conditions by an immobilized enzyme preparation from *Rhodococus* sp., and no ester cleavage is detected.[827]

6.2.6.2 Reduction Reactions: Synthesis of Aminoalkylphosphonic Acids

Cyanoalkylphosphonates are valuable precursors of aminoalkylphosphonates by hydrogenation of the cyano group. The dialkyl cyanomethylphosphonates are efficiently reduced to aminoethylphosphonates on large scale in almost quantitative yield using Raney Ni[17,35,828] or Adams catalyst (PtO$_2$).[829,830] The reaction is carried out under hydrogen atmosphere, at ordinary or elevated pressure, at room temperature, in EtOH. In each case, it is necessary to adjust the conditions to obtain a full hydrogenation. Reduction with diborane in THF has been less frequently used.[166] Starting from labeled cyanomethylphosphonate, prepared from Na^{13}CN, the reduction with diborane in THF leads to [2–^{13}C]aminoethylphosphonate.[166] The reduction has also been applied to 2-cyanoethylphosphonates. Thus, the cyano group is hydrogenated under a variety of conditions: in alcohol over Raney Ni at ordinary pressure and room temperature (97%),[831] under pressure at room temperature[225] or at 50°C (95%),[832] over Raney Co under pressure at 60°C (80%),[833] or over PtO$_2$ under pressure at room temperature (76%)[193] or at 65°C (94%).[249] However, because hydrogenation can be accompanied by intramolecular and intermolecular side reactions, it is recommended that, before hydrogenation, the diester be converted into a monosalt by saponification with aqueous sodium hydroxide. The aminoalkylphosphonates are further converted into phosphonic acids by hydrolysis in acidic medium (Scheme 6.60). Ion-exchange chromatography represents the best way to obtain pure acids.[832]

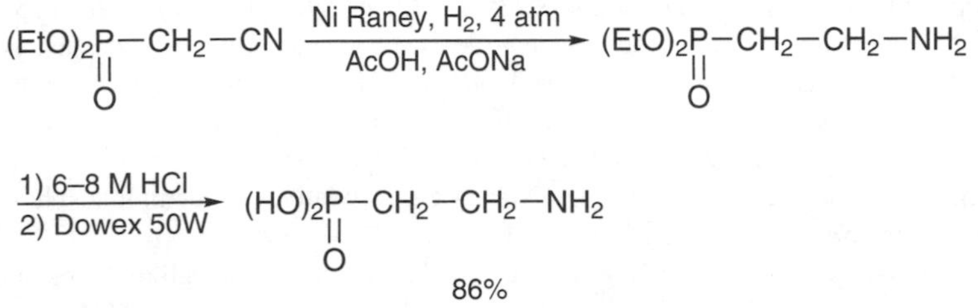

$$(EtO)_2\overset{\underset{\|}{O}}{P}-CH_2-CN \xrightarrow[\text{AcOH, AcONa}]{\text{Ni Raney, H}_2\text{, 4 atm}} (EtO)_2\overset{\underset{\|}{O}}{P}-CH_2-CH_2-NH_2$$

$$\xrightarrow[\text{2) Dowex 50W}]{\text{1) 6–8 M HCl}} (HO)_2\overset{\underset{\|}{O}}{P}-CH_2-CH_2-NH_2$$

$$86\%$$

(6.60)

6.2.7 REACTIONS OF THE PHOSPHONYL GROUP: DEALKYLATION

The reaction of dialkyl cyanomethylphosohonates with Me$_3$SiCl/NaI occurs rapidly at room temperature to give the corresponding *bis*(trimethylsilyl) phosphonates quantitatively. The dealkylation proceeds to completion in a few minutes. The rate of dealkylation decreases according to the bulkiness of the alcoxy groups attached to phosphorus. Treatment of silyl esters with MeOH at room temperature gives the corresponding phosphonic acids in high yields (Scheme 6.61).[834–837] Sodium thiolates and thiophenates in EtOH at reflux[838] and lithium chloride or bromide in methyl *n*-butyl ketone at 80°C[174] induce selective monodealkylation of cyanophosphonates.

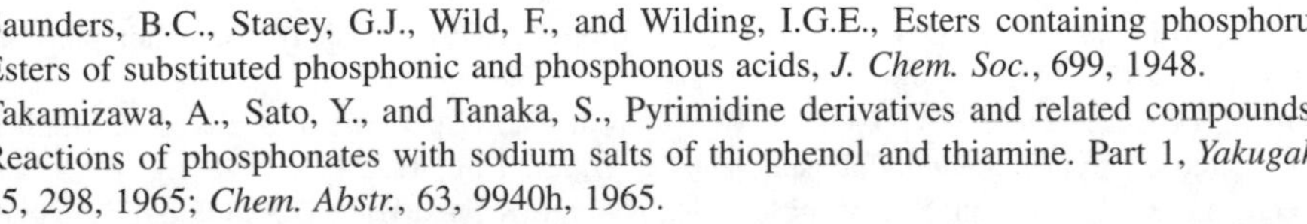

$$(EtO)_2\overset{\underset{\|}{O}}{P}-CH_2-CN \xrightarrow[\text{MeCN, r.t.}]{\text{2 Me}_3\text{SiCl, 2 NaI}} (Me_3SiO)_2\overset{\underset{\|}{O}}{P}-CH_2-CN \xrightarrow[\text{r.t.}]{\text{MeOH}} (HO)_2\overset{\underset{\|}{O}}{P}-CH_2-CN$$

$$100\% \qquad\qquad 85\%$$

(6.61)

REFERENCES

1. Saunders, B.C., Stacey, G.J., Wild, F., and Wilding, I.G.E., Esters containing phosphorus. Part 5. Esters of substituted phosphonic and phosphonous acids, *J. Chem. Soc.*, 699, 1948.
2. Takamizawa, A., Sato, Y., and Tanaka, S., Pyrimidine derivatives and related compounds. Part 33. Reactions of phosphonates with sodium salts of thiophenol and thiamine. Part 1, *Yakugaku Zasshi*, 85, 298, 1965; *Chem. Abstr.*, 63, 9940h, 1965.
3. Shioiri, T., Yokoyama, Y., Kasai, Y., and Yamada, S., Phosphorus in organic synthesis. Part 11. Amino acids and peptides. Part 21. Reaction of diethyl phosphorocyanidate (DEPC) with carboxylic acids. A new synthesis of carboxylic esters and amides, *Tetrahedron*, 32, 2211, 1976.
4. T'ung, T.S., and Tai, C.S., A new method for the preparation of *O,O′*-dialkylphosphoryl cyanides, *Hua Hsueh Hsueh Pao*, 31, 199, 1965; *Chem. Abstr.*, 63, 16380a, 1965.
5. Das, S.K., and Balasubrahmanyam, S.N., Dialkyl (3-aryl-1,2,4-oxadiazol-5-yl)phosphonates. Synthesis and thermal behavior. Evidence for monomeric alkyl metaphosphate, *J. Org. Chem.*, 48, 4232, 1983.
6. Lazukina, L.A., Kukhar, V.P., Romanov, G.V., Khaskin, G.I., Dubinina, T.N., Ofitserov, E.N., Volkova, V.N., and Pudovik, A.N., Reactions of cyanotrimethylsilane with phosphorus and sulfur acid chlorides, *Zh. Obshch. Khim.*, 50, 985, 1980; *J. Gen. Chem. USSR (Engl. Transl.)*, 50, 783, 1980.

7. Stec, W.J., Konopka, A., and Uznanski, B., Synthesis of a dialkyl phosphoroisocyanidate, *J. Chem. Soc., Chem. Commun.*, 923, 1974.

8. Uznanski, B., and Stec, W.J., Synthesis of dialkyl phosphorocyanidites. Axial preference of the *P*-cyano group in the 1,3,2-dioxaphosphorinane ring system, *Synthesis*, 735, 1975.

9. Dawson, N.D., and Burger, A., Some alkyl thiazolephosphonates, *J. Am. Chem. Soc.*, 74, 5312, 1952.

10. Fiszer, B., and Michalski, J., Synthesis of organophosphorus compounds based on phosphonoacetic ester and its analogs. Addition of phosphonoacetic ester, alkylated phosphonoacetic esters, and phosphonoacetic nitrile to α,β-unsaturated esters and nitriles, *Rocz. Chem.*, 28, 185, 1954.

11. Pudovik, A.N., and Lebedeva, N.M., New method of synthesis of esters of phosphonic acids and thiophosphonic acids. Part 23. Addition of phosphonoacetic ester, phosphonoacetone, and its homologs to unsaturated compounds, *Zh. Obshch. Khim.*, 25, 1920, 1955; *Chem. Abstr.*, 50, 8442c, 1956.

12. Arbuzov, B.A., and Yarmukhametova, D.K., Some ethylideneglyceryl esters of phosphoric, phosphorous, and thiophosphoric acids, *Izv. Akad. Nauk SSSR, Ser. Khim.*, 292, 1957; *Chem. Abstr.*, 51, 14542b, 1957.

13. Kirilov, M., and Petrova, J., Phosphonoacetonitrile esters. Condensation of phosphonoacetonitrile esters with benzaldehyde and furfural, *Dokl. Bolg. Akad. Nauk*, 17, 45, 1964; *Chem. Abstr.*, 61, 8335d, 1964.

14. Pudovik, A.N., Yastrebova, G.E., and Nikitina, V.I., Esters of α-cyanovinylphosphonic acid, *Zh. Obshch. Khim.*, 36, 1232, 1966; *Chem. Abstr.*, 65, 15418g, 1966.

15. Derkach, G.I., Malovik, V.V., and Bodnarchuk, N.D., Derivatives of methylphosphonic acid diesters, *Z. Naturforsch., Ser. B*, 24, 139, 1969.

16. Bodnarchuk, N.D., Malovik, V.V., and Derkach, G.I., Derivatives of dialkoxyphosphonoacetonitriles, *Zh. Obshch. Khim.*, 39, 168, 1969; *J. Gen. Chem. USSR (Engl. Transl.)*, 39, 154, 1969.

17. Isbell, A.F., Berry, J.P., and Tansey, L.W., Amino phosphonic acids. Part 3. The synthesis and properties of 2-aminoethylphosphonic and 3-aminopropylphosphonic acids, *J. Org. Chem.*, 37, 4399, 1972.

18. Deschamps, B., Lefebvre, G., and Seyden-Penne, J., Mechanism of Horner–Emmons reaction. Part 1. Reaction of benzaldehyde and phosphononitriles in tetrahydrofuran, *Tetrahedron*, 28, 4209, 1972.

19. Buchecker, R., Hamm, P., and Eugster, C.H., Absolute configuration of xanthphyll (lutein), *Helv. Chim. Acta*, 57, 631, 1974.

20. Schaumann, E., and Grabley, F.F., Thioketene synthesis. Part 3. Reactions of phosphonate carbanions with carbon disulfide, *Liebigs Ann. Chem.*, 1715, 1979.

21. Stache, U., Fritsch, W., and Fehlhaber, H.-W., Synthesis of the highly cardioselective β-sympathicolytic pacrinolol, *Arzneim. Forsch. (Drug Res.)*, 37, 1217, 1987.

22. Ladd, E.C., Alkyl esters of phosphono *bis* β-cyanoethylacetic acid and acetic acid nitrile, *U.S. Rubber*, U.S. Patent Appl. US 2632019, 1953; *Chem. Abstr.*, 48, 1418b, 1954.

23. Stilz, W., and Pommer, H., Process for the production of α,β-unsaturated nitriles, *BASF A.-G.*, German Patent Appl. DE 1108208, 1959; *Chem. Abstr.*, 56, 11422e, 1962.

24. Dugger, R.W., and Heathcock, C.H., An efficient preparation of *trans,trans*-β-ionylideneacetaldehyde, *Synth. Commun.*, 10, 509, 1980.

25. Breuer, E., and Bannet, D.M., The preparation of some cyclic phosphonates and their use in olefin synthesis, *Tetrahedron*, 34, 997, 1978.

26. Ueda, K., and Matsui, M., Studies on chrysantemic acid. Part 20. Synthesis of four geometrical isomers of (±)-pyrethric acid, *Agric. Biol. Chem.*, 34, 1119, 1970.

27. Raggio, M.L., and Watt, D.S., A synthesis of progesterone from dehydroepiandrosterone, *J. Org. Chem.*, 41, 1873, 1976.

28. Dinizo, S.E., Freerksen, R.W., Pabst, W.E., and Watt, D.S., Synthesis of α-alkoxyacrylonitriles using substituted diethyl cyanomethylphosphonates, *J. Org. Chem.*, 41, 2846, 1976.

29. Freerksen, R.W., Selikson, S.J., Wroble, R.R., Kyler, S.K., and Watt, D.S., Oxidative decyanation of secondary nitriles to ketones, *J. Org. Chem.*, 48, 4087, 1983.

30. Armesto, D., Horspool, W.M., Gallego, M.G., and Agarrabeitia, A.R., Steric and electronic effects on the photochemical reactivity of oxime acetates of β,γ-unsaturated aldehydes, *J. Chem. Soc., Perkin Trans. 1*, 163, 1992.

31. Kukhar, V.P., and Sagina, E.I., Reaction of trialkylphosphites with trichloroacetic acid derivatives, *Zh. Obshch. Khim.*, 49, 60, 1979; *J. Gen. Chem. USSR (Engl. Transl.)*, 49, 50, 1979.

32. Kukhar, V.P., Sagina, E.I., and Pavlenko, N.G., Reactions of halocyanomethanes and halophenylmethanes with triethylphosphite and triphenylphosphine, *Zh. Obshch. Khim.*, 49, 2217, 1979; *J. Gen. Chem. USSR (Engl. Transl.)*, 49, 1947, 1979.

33. Kirilov, M., Ivanov, D., Petrov, G., and Golemchinski, G., Phosphonation of polyfunctional organo-metallic reagents. Part 2. Metallic derivative of phenylacetonitrile, *Bull. Soc. Chim. Fr.*, 3053, 1973.

34. Blanchard, J., Collignon, N., Savignac, P., and Normant, H., *C*-Alkylation of cyanomethanephosphonic tetramethyldiamides using phase-transfer catalysis, *Synthesis*, 655, 1975.

35. Blanchard, J., Collignon, N., Savignac, P., and Normant, H., Preparation of β-aminoethylphosphonic acids, *Tetrahedron*, 32, 455, 1976.

36. Comins, D.L., Jacobine, A.F., Marshall, J.A., and Turnbull, M.M., Phosphonate ester preparation from active methylene compounds and phosphorochloridate esters, *Synthesis*, 309, 1978.

37. Kandil, A.A., and Slessor, K.N., A chiral synthesis of (+)-lineatin, the aggregation pheromone of *Trypodendron lineatum* (olivier), from D-ribonolactone, *J. Org. Chem.*, 50, 5649, 1985.

38. Kandil, A.A., Porter, T.M., and Slessor, K.N., One-step synthesis of stabilized phosphonates, *Synthesis*, 411, 1987.

39. Paquette, L.A., Wang, T.-Z., Wang, S., and Philippo, C.M.G., Enantioselective synthesis of (+)-cleomeolide, the structurally unique diterpene lactone constituent of *Cleome viscosa*, *Tetrahedron Lett.*, 34, 3523, 1993.

40. Paquette, L.A., Wang, T.-Z., Philippo, C.M. G., and Wang, S., Total synthesis of the cembranoid diterpene lactone (+)-cleomeolide. Some remarkable conformational features of nine-membered belts linked in a 2,6-fashion to a methylenecyclohexane core, *J. Am. Chem. Soc.*, 116, 3367, 1994.

41. Azim, E.-m., Auzeloux, P., Maurizis, J.-C., Braesco, V., Grolier, P., Veyre, A., and Madelmont, J.-C., Synthesis of all-*trans*-β-carotene retinoids and derivatives labelled with ^{14}C, *J. Labelled Compd. Radiopharm.*, 38, 441, 1996.

42. Zhang, T.Y., O'Toole, J.C., and Dunigan, J.M., An efficient and practical synthesis of diphenyl cyanomethylenephosphonate. Applications to the stereoselective synthesis of *cis*-α,β-unsaturated nitriles, *Tetrahedron Lett.*, 39, 1461, 1998.

43. Iorga, B., Ricard, L., and Savignac, P., Carbanionic displacement reactions at phosphorus. Part 3. Cyanomethylphosphonate *vs.* cyanomethylenediphosphonate. Synthesis and solid state structure, *J. Chem. Soc., Perkin Trans. 1*, 3311, 2000.

44. Henderson, K.W., Kennedy, A.R., McKeown, A.E., and Strachan, D., Lithiated α-cyanophosphonates. Self-assembly of two-dimensional molecular sheets composed of interconnected twenty-four membered rings, *J. Chem. Soc., Dalton Trans.*, 4348, 2000.

45. Patrick, T.B., and Nadji, S., Preparation of fluoroalkenes from fluoroacetonitrile, *J. Fluorine Chem.*, 49, 147, 1990.

46. Groesbeek, M., Kirillova, Y.G., Boeff, R., and Lugtenburg, J., Synthesis of six novel retinals and their interaction with bacterioopsin, *Recl. Trav. Chim. Pays-Bas*, 113, 45, 1994.

47. Groesbeek, M., Robijn, G.W., and Lugtenburg, J., Synthesis and spectroscopic characterization of the doubly locked 9*E*,11*Z* retinal model systems 7*E*,13*E*-11,19–10,20-dimethanoretinal and its 13*Z* isomer, *Recl. Trav. Chim. Pays-Bas*, 111, 92, 1992.

48. Gulbrandsen, T., and Kolsaker, P., The reaction between diazoalkanes and allylic halides carrying electronegative γ-substituents. Part 4. Conformational equilibria of highly substituted 1-pyrazolines, *Acta Chem. Scand., Ser. B*, 37, 219, 1983.

49. Gebhard, R., van der Hoef, K., Lefeber, A.W.M., Erkelens, C., and Lugtenburg, J., Synthesis and spectroscopy of (14′-^{13}C)- and (15′-^{13}C)spheroidene, *Recl. Trav. Chim. Pays-Bas*, 109, 378, 1990.

50. Cromwell, M.E.M., Gebhard, R., Li, X.-Y., Batenburg, E.S., Hopman, J.C.P., Lugtenburg, J., and Mathies, R.A., Synthesis and vibrational analysis of a locked 14-*S-cis* conformer of retinal, *J. Am. Chem. Soc.*, 114, 10860, 1992.

51. Groesbeek, M., Rood, G.A., and Lugtenburg, J., Synthesis of (12,13–^{13}C$_2$)retinal and (13,14–^{13}C$_2$)retinal. A strategy to prepare multiple-^{13}C-labeled conjugated systems, *Recl. Trav. Chim. Pays-Bas*, 111, 149, 1992.

52. Cappon, J.J., Witters, K.D., Baart, J., Verdegem, P.J.E., Hoek, A.C., Luiten, R.J.H., Raap, J., and Lugtenburg, J., Synthesis of L-histidine specifically labelled with stable isotopes, *Recl. Trav. Chim. Pays-Bas*, 113, 318, 1994.

53. Fiszer, B., and Michalski, J., Organophosphorus compounds with active methylene group. Part 3. Addition of phosphinicacetic esters and their analogs to α,β-unsaturated ethylene derivatives, *Rocz. Chem.*, 34, 1461, 1960.

54. Nikolaenko, T.Y., Trukhin, E.V., Tebbi, J., and Berestovitskaya, V.M., Reaction of phosphorylated CH acids with 2-benzylidene-3-methyl-4-nitro-3-thiolene 1,1-dioxide, *Zh. Obshch. Khim.*, 66, 1043, 1996; *Russ. J. Gen. Chem. (Engl. Transl.)*, 66, 1018, 1996.

55. Sawamura, M., Hamashima, H., and Ito, Y., Rhodium-catalyzed enantioselective Michael addition of (1-cyanoethyl)phosphonate. Synthesis of optically active phosphonic acid derivatives with phosphorus-substituted quaternary asymmetric carbon center, *Bull. Chem. Soc. Jpn.*, 73, 2559, 2000.

56. Yuan, C., and Li, C., Studies on organophosphorus compounds. Part 67. Reactions of α-nitroalkenes with compounds bearing active methylene groups. A novel and convenient synthesis of 2-isoxazoline derivatives, *Phosphorus, Sulfur Silicon Relat. Elem.*, 78, 47, 1993.

57. Attanasi, O.A., Filippone, P., Giovagnoli, D., and Mei, A., Conjugated azoalkenes. Part 15. Facile and direct synthesis of new 3-phosphonopyrrole and 3a-phosphonopyrrolo[2,3-*b*]pyrrole derivatives, *Synthesis*, 181, 1994.

58. Attanasi, O.A., De Crescentini, L., Foresti, E., Gatti, G., Giorgi, R., Perrulli, F.R., and Santeusanio, S., Cleavage and reactions of some NH-Boc protected 1-aminopyrroles. A new one-pot route to pyrrolo[1,2-*b*] [1,2,4]triazines together with spectroscopic and X-ray studies, *J. Chem. Soc., Perkin Trans. 1*, 1829, 1997.

59. Knierzinger, A., and Wolfbeis, O.S., Syntheses of fluorescent dyes. Part 9. New 4-hydroxycoumarins, 4-hydroxy-2-quinolones, 2*H*,5*H*-pyrano[3,2-*c*]benzopyran-2,5-diones and 2*H*,5*H*-pyrano[3,2-*c*]quinoline-2,5-diones, *J. Heterocycl. Chem.*, 17, 225, 1980.

60. Danion, D., and Carrié, R., Comparative study of intramolecular *O*-phosphorylations during the condensation of aldehydes with phosphonate carbanions having several nucleophilic sites, *Tetrahedron Lett.*, 12, 3219, 1971.

61. Morel, G., Seux, R., and Foucaud, A., Action of nucleophiles on $(R'O)_2P(O)$-$C(C_6H_5) = C(CN)CO_2R$, *C.R. Acad. Sci., Ser. C*, 275, 629, 1972.

62. Danion, D., and Carrié, R., Nucleophilic addition of cyanide ion to α-cyanovinylphosphonic esters. Synthesis and physicochemical properties of 1,2-dicyanoethylphosphonic esters, *Bull. Soc. Chim. Fr.*, 1538, 1974.

63. Barbot, F., Paraiso, E., and Miginac, P., A novel route to substituted phosphonates via conjugate addition of organometallics to 1-(functionally) substituted alkene phosphonates, *Tetrahedron Lett.*, 25, 4369, 1984.

64. Melot, J.-M., Texier-Boullet, F., and Foucaud, A., Cyclopropanation of electrophilic alkenes with nitroalkanes in the presence of alumina-supported potassium fluoride, *Synthesis*, 364, 1987.

65. Ternansky, R.J., and Draheim, S.E., [3.3.0]Pyrazolidinones. An efficient synthesis of a new class of synthetic antibacterial agents, *Tetrahedron Lett.*, 31, 2805, 1990.

66. Ternansky, R.J., and Draheim, S.E., The chemistry of substituted pyrazolidinones. Applications to the synthesis of bicyclic derivatives, *Tetrahedron*, 48, 777, 1992.

67. Jungheim, L.N., Boyd, D.B., Indelicato, J.M., Pasini, C.E., Preston, D.A., and Alborn, W.E., Jr., Synthesis, hydrolysis rates, supercomputer modeling, and antibacterial activity of bicyclic tetrahydropyridazinones, *J. Med. Chem.*, 34, 1732, 1991.

68. Wadsworth, W.S., Jr., and Park, E., Dialkyl 1-cyanovinylphosphonates, their preparation, and polymers and copolymers thereof, *Rohm and Haas*, U.S. Patent Appl. US 3047606, 1962; *Chem. Abstr.*, 58, 4599b, 1963.

69. Pudovik, A.N., Yastrebova, G.E., and Nikitina, V.I., Cyanophosphonomethane condensation reactions, *Zh. Obshch. Khim.*, 38, 300, 1968; *J. Gen. Chem. USSR (Engl. Transl.)*, 38, 301, 1968.

70. Boyer, N.E., Substituted norbornene compounds and their applications in thermoplastic polymer compositions, *Borg-Warner*, U.S. Patent Appl. US 3928507, 1975; *Chem. Abstr.*, 84, 90938, 1976.

71. Andreae, S., and Seeboth, H., Furan derivatives. Part 7. *O,O*-Diethyl 2-(2-furyl)- and 2-(5-nitro-2-furyl)vinylphosphonates, *J. Prakt. Chem.*, 321, 353, 1979.

72. Zhdanov, Y.A., Uzlova, L.A., and Maksimushkina, L.M., Condensation of aldehydo-pentoses with phosphorus analogs of carbonyl compounds, *Zh. Obshch. Khim.*, 52, 937, 1982; *J. Gen. Chem. USSR (Engl. Transl.)*, 52, 818, 1982.

73. Minami, T., Yamanouchi, T., Takenaka, S., and Hirao, I., Synthesis of butadienylphosphonates containing electronegative substituents and their synthetic applications to functionalized cyclopentenylphosphonates, *Tetrahedron Lett.*, 24, 767, 1983.

74. Minami, T., Yamanouchi, T., Tokumasu, S., and Hirao, I., The reaction of butadienylphosphonates with a oxosulfonium ylide, phosphonium ylides, and ketone enolates, *Bull. Chem. Soc. Jpn.*, 57, 2127, 1984.

75. Minami, T., Tokumasu, S., Mimasu, R., and Hirao, I., Cycloaddition of diazomethane to butadienylphosphonates. A new approach to functionalized pentadienylphosphonates and pyrazoles, *Chem. Lett.*, 1099, 1985.

76. Mignani, G., Leising, F., Meyrueix, R., and Samson, H., Synthesis of new thiophene compounds with large second order optical non-linearities, *Tetrahedron Lett.*, 31, 4743, 1990.

77. Cho, H., Ueda, M., Tamaoka, M., Hamaguchi, M., Aisaka, K., Kiso, Y., Inoue, T., Ogino, R., Tatsuoka, T., Ishihara, T., Noguchi, T., Morita, I., and Murota, S.-i., Novel caffeic acid derivatives. Extremely potent inhibitors of 12-lipoxygenase, *J. Med. Chem.*, 34, 1503, 1991.

78. Bojilova, A., Nikolova, R., Ivanov, C., Rodios, N.A., Terzis, A., and Raptopoulou, C.P., A comparative study of the interaction of salicylaldehydes with phosphonoacetates under Knoevenagel reaction conditions. Synthesis of 1,2-benzoxaphosphorines and their dimers, *Tetrahedron*, 52, 12597, 1996.

79. Kiesel, M., Haug, E., and Kantlehner, W., Orthoamides. Part 50. Contribution to the chemistry of propionaldehyde aminals. Synthesis and transformations to push-pull-substituted buta-1,3-dienes, cyclobutanes, vinylogous amidinium salts and 1,2,3-triazoles, *J. Prakt. Chem./Chem.-Ztg.*, 339, 159, 1997.

80. Danion, D., and Carrié, R., Synthesis and structure of α-cyanovinylphosphonic esters, *Bull. Soc. Chim. Fr.*, 1130, 1972.

81. Pudovik, A.N., Yastrebova, G.E., Nikitina, V.I., and Mukhametzyanova, E.K., Synthesis and reactions of some unsaturated α-cyanophosphonic acids, *Khim. Org. Soedin. Fosfora, Akad. Nauk SSSR*, 18, 1967; *Chem. Abstr.*, 69, 59331n, 1968.

82. Breuer, H., and Treuner, U.D., Vinylaminoacetylpenicillins and cephalosporins, *Chemische Fabrik von Heyden*, German Patent Appl. DE 2362978, 1974; *Chem. Abstr.*, 81, 105546, 1974.

83. Kirilov, M., and Petrova, J., Reaction of esters of (cyanomethyl)phosphonic acid with aliphatic, alicyclic, and aromatic ketones, *Dokl. Bolg. Akad. Nauk*, 21, 339, 1968; *Chem. Abstr.*, 69, 77366f, 1968.

84. Danion, D., and Carrié, R., Synthesis and structure of diethyl 1-cyanovinylphosphonates, *Tetrahedron Lett.*, 9, 4537, 1968.

85. Ayoubi, S.A.-E., Texier-Boullet, F., and Hamelin, J., Minute synthesis of electrophilic alkenes under microwave irradiation, *Synthesis*, 258, 1994.

86. Kirilov, M., and Petrov, G., Synthesis, structure, and reactivity of some metal derivatives of cyanomethylphosphonates (dialkoxyphosphinylacetonitriles), *Chem. Ber.*, 104, 3073, 1971.

87. Pudovik, A.N., and Lebedeva, N.M., Synthesis of esters of phosphonic and thiophosphonic acids. Part 24. Addition of phosphonoacetonitrile and its homologs to esters and nitriles of unsaturated carboxylic acids, *Zh. Obshch. Khim.*, 25, 2235, 1955; *Chem. Abstr.*, 50, 9280d, 1956.

88. Pudovik, A.N., and Khusainova, N.G., Esters and nitriles of propynylphosphonoacetic acid, *Zh. Obshch. Khim.*, 39, 2426, 1969; *J. Gen. Chem. USSR (Engl. Transl.)*, 39, 2366, 1969.

89. Flitsch, W., Rosche, J., and Lubisch, W., Synthetic routes to 1-substituted butadienylphosphonates, *Liebigs Ann. Chem.*, 661, 1987.

90. Zoretic, P.A., Zhang, Y., and Ribeiro, A.A., An intramolecular radical approach to angular electrophores in polycyclic systems, *Tetrahedron Lett.*, 37, 1751, 1996.

91. Bailey, P.D., and Morgan, K.M., A total asymmetric synthesis of (−)-suaveoline, *J. Chem. Soc., Chem. Commun.*, 1479, 1996.

92. Hampton, A., Sasaki, T., and Paul, B., Synthesis of 6′-cyano-6′-deoxyhomoadenosine-6′-phosphonic acid and its phosphoryl and pyrophosphoryl anhydrides and studies of their interactions with adenine nucleotide utilizing enzymes, *J. Am. Chem. Soc.*, 95, 4404, 1973.

93. Gopalan, A., Moerck, R., and Magnus, P., Regiospecific control of the ene reaction of *N*-phenyl-1,2,4-triazoline-3,5-dione, *J. Chem. Soc., Chem. Commun.*, 548, 1979.

94. Kigoshi, H., Sawada, A., Niwa, H., and Yamada, K., Synthesis of a functionalized carbocyclic skeleton of ptaquilosin, the aglycone of a bracken carcinogen ptaquiloside based on an intramolecular Diels–Alder reaction, *Bull. Chem. Soc. Jpn.*, 62, 1639, 1989.

95. Compagnone, R.S., Suarez, A.I., Zambrano, J.L., Pina, I.C., and Dominguez, J.N., A short and versatile synthesis of 3-substituted 2-aminoquinolines, *Synth. Commun.*, 27, 1631, 1997.

96. Gossauer, A., and Hinze, R.P., An improved synthesis of racemic phycocyanobilin dimethyl ester, *J. Org. Chem.*, 43, 283, 1978.

97. Compagnone, R.S., and Rapoport, H., Chirospecific synthesis of (+)-pilocarpine, *J. Org. Chem.*, 51, 1713, 1986.

98. Suarez, A., Lopez, F., and Compagnone, R.S., Stereospecific synthesis of the lignans. 2-*S*-(3,4-Dimethoxybenzyl)-3-*R*-(3,4,5-trimethoxybenzyl)butyrolactone, and its positional isomeric lactone, *Synth. Commun.*, 23, 1991, 1993.

99. Nasser, J., About-Jaudet, E., and Collignon, N., α-Functional cycloalkylphosphonates. Part 1. Synthesis, *Phosphorus, Sulfur Silicon Relat. Elem.*, 54, 171, 1990.

100. Fatima, A., Zaman, F., and Voelter, W., Synthesis of cyclopropylphosphonate pyranosides, *Z. Naturforsch., Ser. B*, 46, 1227, 1991.

101. Neidlein, R., and Eichinger, T., Substituted methylphosphonates as synthons for alicyclic α-functionalised phosphonates, *Monatsh. Chem.*, 123, 1037, 1992.

102. Uneme, H., Mitsudera, H., Yamada, J., Kamikado, T., Kono, Y., Manabe, Y., and Numata, M., Synthesis and biological activity of the functional derivatives of 3- and 4-(dimethylaminomethyl)-1,2-dithiolanes, *Biosci. Biotech. Biochem.*, 56, 1623, 1992.

103. Oediger, H., Dialkylmethylene compounds, *Bayer A.-G.*, German Patent Appl. DE 2206778, 1973; *Chem. Abstr.*, 79, 115163, 1973.

104. Sawamura, M., Sudoh, M., and Ito, Y., An enantioselective two-component catalyst system. Rh-Pd-catalyzed allylic alkylation of activated nitriles, *J. Am. Chem. Soc.*, 118, 3309, 1996.

105. D'Incan, E., and Seyden-Penne, J., Ion pair extraction. Alkylation of phosphonates and the Wittig/Horner reaction, *Synthesis*, 516, 1975.

106. Moshkina, T.M., and Pudovik, A.N., Phosphorus-containing azo and hydrazo compounds, *Zh. Obshch. Khim.*, 35, 2042, 1965; *J. Gen. Chem. USSR (Engl. Transl.)*, 35, 2034, 1965.

107. Singh, R.K., Alkylation studies. Part 2. *Bis*-alkylation of diethyl cyanomethanephosphonate, *Synthesis*, 762, 1986.

108. Vinogradova, N.M., Lysenko, K.A., Odinets, I.L., Petrovskii, P.V., Mastryukova, T.A., and Kabachnik, M.I., Synthesis and X-ray structure of 6-cyano-2,10-dioxa-1-phosphabicyclo[4.4.0]decane-1-oxide, *Phosphorus, Sulfur Silicon Relat. Elem.*, 132, 265, 1998.

109. Tessier, J., Demoute, J.-P., and Truong, V.T., Process for preparing fluorinated derivatives of phosphonic acid, and products obtained using this process, *Roussel Uclaf*, Eur. Patent Appl. EP 224417, 1987; *Chem. Abstr.*, 107, 96889y, 1987.

110. Baader, E., Bartmann, W., Beck, G., Below, P., Bergmann, A., Jendralla, H., Kesseler, K., and Wess, G., Enantioselective synthesis of a new fluoro-substituted HMG-CoA reductase inhibitor, *Tetrahedron Lett.*, 30, 5115, 1989.

111. Xu, Z.-Q., and DesMarteau, D.D., A convenient one-pot synthesis of α-fluoro-α,β-unsaturated nitriles from diethyl cyanofluoromethanephosphonate, *J. Chem. Soc., Perkin Trans. 1*, 313, 1992.

112. Suzuki, H., Watanabe, K. and Qui, Y., Copper(I) iodide-mediated arylation of dialkyl cyanomethanephosphonates. A novel route to α-arylated alkanenitriles, *Chem. Lett.*, 1779, 1985.

113. Minami, T., Isonaka, T., Okada, Y., and Ichikawa, J., Copper(I) salt-mediated arylation of phosphinyl-stabilized carbanions and synthetic application to heterocyclic compounds, *J. Org. Chem.*, 58, 7009, 1993.

114. Suzuki, H., and Abe, H., A simple cyclization route to some 4-substituted 3-aminoisoquinolines, *Synthesis*, 763, 1995.

115. Sakamoto, T., Katoh, E., Kondo, Y., and Yamanaka, H., Palladium(0)-catalyzed condensation of bromopyridines with α-substituted acetonitriles, *Heterocycles*, 27, 1353, 1988.

116. Sakamoto, T., Katoh, E., Kondo, Y., and Yamanaka, H., Palladium-catalyzed condensation of aryl halides with phenylsulfonylacetonitrile and diethyl cyanomethylphosphonate, *Chem. Pharm. Bull.*, 38, 1513, 1990.

117. Tarasenko, E.A., Artamkina, G.A., Voevodskaya, T.I., Lukashev, N.V., and Beletskaya, I.P., Nucleophilic substitution in perhalogenated aromatic compounds by carbanions derived from substituted dialkyl methylphosphonates, *Zh. Org. Khim.*, 34, 1523, 1998; *Russ. J. Org. Chem. (Engl. Transl.)*, 34, 1459, 1998.

118. Artamkina, G.A., Tarasenko, E.A., Lukashev, N.V., and Beletskaya, I.P., Synthesis of perhaloaromatic diethyl methylphosphonates containing α-electron-withdrawing group, *Tetrahedron Lett.*, 39, 901, 1998.

119. Artamkina, G.A., Sazonov, P.K., and Beletskaya, I.P., Arylation of phosphoryl-stabilized carbanions with metal π-complexes of aryl chlorides and fluorides, *Tetrahedron Lett.*, 42, 4385, 2001.

120. Krug, H.G., Neidlein, R., Krieger, C., and Kramer, W., Synthesis of phosphono- and phosphino-substituted six-membered heterocycles, *Heterocycles*, 38, 2695, 1994.

121. Kirilov, M., and Petrov, G., Acylation of phosphonoacetonitrile diethyl ester, *Dokl. Bolg. Akad. Nauk*, 18, 331, 1965; *Chem. Abstr.*, 63, 11606h, 1965.

122. Kirilov, M., and Petrov, G., Preparation properties of dialkyl esters of α-acylphosphonoacetonitriles, *Monatsh. Chem.*, 99, 166, 1968.

123. Odinets, I.L., Artyushin, O.I., Kalyanova, R.M., Antipin, M.Y., Lysenko, K.A., Struchkov, Y.T., Petrovskii, P.V., Masteryukova, T.A., and Kabachnik, M.I., Phase-transfer acylation of phosphorus-containing CH acids. Part 2. Acylation of phosphinoylacetonitriles, *Zh. Obshch. Khim.*, 64, 1957, 1994; *Russ. J. Gen. Chem. (Engl. Transl.)*, 64, 1738, 1994.

124. Neidlein, R., and Eichinger, T., [(1,3-Dioxolan-2-yliden)methyl]phosphonates and -phosphinates as (simple) synthons in heterocyclic synthesis, *Helv. Chim. Acta*, 75, 124, 1992.

125. Neidlein, R., and Feistauer, H., Syntheses of 2,3-dioxoalkylphosphonates and other novel β-ketophosphonates as well as of a phosphinopyruvamide, *Helv. Chim. Acta*, 79, 895, 1996.

126. Grell, W., and Machleidt, H., Syntheses with organophosphorus compounds. Part 1. Ester olefinations, *Justus Liebigs Ann. Chem.*, 693, 134, 1966.

127. Bal'on, Y.G., Kozhushko, B.N., Paliichuk, Y.A., and Shokol, V.A., Derivatives of substituted vinylphosphonic acid, *Zh. Obshch. Khim.*, 62, 2530, 1992; *J. Gen. Chem. USSR (Engl. Transl.)*, 62, 2089, 1992.

128. Shen, Y., and Ni, J., Synthesis of perfluoroalkylated 4-cyanoalka-1,4-dienes, *J. Chem. Res. (S)*, 358, 1997.

129. Shen, Y., and Ni, J., A convenient synthesis of perfluoroalkylated α,β-unsaturated nitriles, *J. Fluorine Chem.*, 86, 173, 1997.

130. Shen, Y., Ni, J., Li, P., and Sun, J., One-pot stereoselective synthesis of trifluoromethylated penta-(2Z,4E)-dienenitriles via double olefination, *J. Chem. Soc., Perkin Trans. 1*, 509, 1999.

131. Katagiri, N., Yamamoto, M., and Kaneko, C., 3-Acetoxy-2-dimethylphosphonoacrylates. New dienophiles and their use for the synthesis of carbocyclic *C*-nucleoside precursors by the aid of RRA reaction, *Chem. Lett.*, 1855, 1990.

132. Kantlehner, W., Wagner, F., and Bredereck, H., Orthoamides. Part 32. Reactions of *tert*-butoxy-*N*,*N*,*N'*,*N'*-tetramethylmethanediamine with NH- and CH-acidic compounds, *Liebigs Ann. Chem.*, 344, 1980.

133. Mel'nikov, N.N., Kozlov, V.A., Churusova, S.G., Buvashkina, N.I., Ivanchenko, V.I., Negrebetskii, V.V., and Grapov, A.F., β-Phosphorylated enamines, *Zh. Obshch. Khim.*, 53, 1689, 1983; *J. Gen. Chem. USSR (Engl. Transl.)*, 53, 1519, 1983.

134. Aboujaoude, E.E., Collignon, N., and Savignac, P., α-Functional dialkyl formyl-1-methylphosphonates. Part 2. Thermical preparation and conversion in α-phosphonic heterocycles, *Tetrahedron*, 41, 427, 1985.

135. Kozlov, V.A., Churusova, S.G., Ivanchenko, V.I., Negrebetskii, V.V., Grapov, A.F., and Mel'nikov, N.N., β-Phosphorylated enamines, *Zh. Obshch. Khim.*, 56, 2013, 1986; *J. Gen. Chem. USSR (Engl. Transl.)*, 56, 1775, 1986.

136. Aboujaoude, E.E., Collignon, N., and Savignac, P., Synthesis of α-phosphonic heterocycles. New developments, *Phosphorus Sulfur*, 31, 231, 1987.

137. Schaumann, E., and Fittkau, S., Synthesis and Wittig–Horner reactions of 1-(functionally)substituted 2,2-dimethylpropanephosphonic esters (1-*tert*-butyl-substituted phosphonic esters), *Synthesis*, 449, 1983.

138. Kozlov, V.A., Zheltova, E.V., Pokrovskaya, L.A., Grapov, A.F., and Mel'nikov, N.N., Phosphorus-containing ethylene dithiolates in the synthesis of heterocyclic compounds, *Zh. Obshch. Khim.*, 62, 1065, 1992; *J. Gen. Chem. USSR (Engl. Transl.)*, 62, 871, 1992.

139. Krug, H.G., Neidlein, R., Boese, R., and Kramer, W., Synthesis and reactions of dialkyl (1-*R*-5-amino-3-methylsulfanyl-1*H*-pyrazol-4-yl)phosphonates, *Heterocycles*, 41, 721, 1995.

140. Neidlein, R., Jochheim, M., Krieger, C., and Kramer, W., Synthesis of 1,3-dithietane-2,4-diylidene*bis*(cyanomethylphosphonates) and -phenylphosphinates and their reaction with carboxylic acid hydrazides, *Heterocycles*, 39, 185, 1994.

141. Neidlein, R., Hahn, D.U., Kramer, W., and Krieger, C., Synthesis and chemical reactions of new phosphono-phosphino substituted γ-thiapyrones, *Heterocycles*, 47, 221, 1998.

142. Abdou, W.M., and El Khoshnieh, Y.O., Synthesis of fused thioxo-pyran systems from α-methylene carbonyl compounds and *bis*methylene-1,3-dithietane, *Synth. Commun.*, 29, 2657, 1999.

143. Hartke, K., and Günther, O., Thioacylation of diethyl cyanomethane-phosphonates, *Arch. Pharm. (Weinheim)*, 307, 144, 1974.

144. Günther, O., and Hartke, K., Heterocyclic *o*-amino-phosphonic esters, *Arch. Pharm. (Weinheim)*, 308, 693, 1975.

145. Hamlet, Z., and Mychajlowskij, W., Reactions of phenyl isocyanate and phenyl isothiocyanate with phosphonate carbanions, *Chem. Ind. (London)*, 829, 1974.

146. Barnikow, G., and Säling, G., Isothiocyanates. Part 40. Reactions of phosphono-substituted CH-acid thiocarboxamides, *J. Prakt. Chem.*, 316, 534, 1974.

147. Dixon, W.D., and Becker, M.E., α-Cyano-α-(dialkylphosphono)acetanilides, *Monsanto*, U.S. Patent Appl. US 3907937, 1975; *Chem. Abstr.*, 84, 58675, 1976.

148. Dixon, W.D., and Becker, M.E., α-Cyano-α-dialkylphosphonato acetanilde herbicides, *Monsanto*, U.S. Patent Appl. US 3955958, 1976; *Chem. Abstr.*, 85, 108768, 1976.

149. Kozlov, V.A., Dol'nikova, T.Y., Ivanchenko, V.I., Negrebetskii, V.V., Grapov, A.F., and Mel'nikov, N.N., β-Phosphorylated ketene aminomercaptals, *Zh. Obshch. Khim.*, 53, 2229, 1983; *J. Gen. Chem. USSR (Engl. Transl.)*, 53, 2008, 1983.

150. Dol'nikova, T.Y., Kozlov, V.A., Grapov, A.F., and Mel'nikov, N.N., Reactions of phosphorylated thioacetamides with mercuric oxide and nucleophilic reagents, *Zh. Obshch. Khim.*, 54, 88, 1984; *J. Gen. Chem. USSR (Engl. Transl.)*, 54, 76, 1984.

151. Kozlov, V.A., Zheltova, E.V., Negrebetskii, V.V., Grapov, A.F., and Mel'nikov, N.N., Phosphorus-containing carboxylic acid thioamides in the synthesis of heterocyclic compounds, *Zh. Obshch. Khim.*, 60, 49, 1990; *J. Gen. Chem. USSR (Engl. Transl.)*, 60, 42, 1990.

152. Bartlett, P.A., Hunt, J.T., Adams, J. L., and Gehret, J.-C.E., Phosphorus-containing purines and pyrimidines. A new class of transition state analogs, *Bioorg. Chem.*, 7, 421, 1978.

153. Strepikheev, Y.A., Khokhlov, P.S., and Kashemirov, B.A., Synthesis of oximinocyanomethylphosphonates, *Zh. Obshch. Khim.*, 51, 1206, 1981; *J. Gen. Chem. USSR (Engl. Transl.)*, 51, 1020, 1981.

154. Buchanan, J.G., McCaig, A.E., and Wightman, R.H., The synthesis of 4-alkylsulphonyl-5-amino- and 5-amino-4-phosphono-imidazole nucleosides as potential inhibitors of purine biosynthesis, *J. Chem. Soc., Perkin Trans. 1*, 955, 1990.

155. Heep, U., On the synthesis of heterocyclic phosphonic acid esters, *Justus Liebigs Ann. Chem.*, 578, 1973.

156. Ankenbrand, T., and Neidlein, R., Syntheses of phosphonato-substituted azolo[1,2,4]triazines with potential biomedical application, *Heterocycles*, 51, 513, 1999.

157. Coppola, G.M., Hardtmann, G.E., and Pfister, O.R., Chemistry of 2*H*-3,1-benzoxazine-2,4(1*H*)-dione (isatoic anhydride). Part 2. Reactions with thiopseudoureas and carbanions, *J. Org. Chem.*, 41, 825, 1976.

158. Kabachnik, M.I., Rossiiskaya, P.A., and Shepeleva, E.S., Esters of α-keto phosphonic acids. Part 3. Two types of carboxylic acid derivatives, *Izv. Akad. Nauk SSSR, Ser. Khim.*, 163, 1947; *Chem. Abstr.*, 42, 4132i, 1948.

159. Hall, L.A.R., Stephens, C.W., and Drysdale, J.J., A rearrangement to form diethyl 1-cyanoethyl phosphate, *J. Am. Chem. Soc.*, 79, 1768, 1957.

160. Okamoto, Y., Nitta, T., and Sakurai, H., The synthesis of long-chain α-hydroxycarboxylic acids from acyl phosphonates, *Kogyo Kagaku Zasshi*, 71, 187, 1968; *Chem. Abstr.*, 69, 35342, 1968.

161. Okamoto, Y., Nitta, T., and Sakurai, H., Preparation of 2-alkenonitrile by means of the pyrolysis of diethyl 1-cyanoalkyl phosphate, *Bull. Chem. Soc. Jpn.*, 42, 543, 1969.

162. Konovalova, I.V., Ofitserova, E.K., Stepanova, T.Y., and Pudovik, A.N., Reaction of alkyl phosphites with α-oxocarboxylic acid nitriles, *Zh. Obshch. Khim.*, 48, 1237, 1978; *J. Gen. Chem. USSR (Engl. Transl.)*, 48, 1132, 1978.

163. Konovalova, I.V., Burnaeva, L.A., Saifullina, N.S., and Pudovik, A.N., On the reaction of trimethyl silyl phosphite with esters and nitriles of α-oxocarboxylic acids, *Zh. Obshch. Khim.*, 46, 18, 1976; *J. Gen. Chem. USSR (Engl. Transl.)*, 46, 17–19, 1976.

164. Pudovik, A.N., Samitov, Y.Y., Gur'yanova, I.V., and Banderova, L.V., Reaction of phosphonic and phosphinic acid esters with the nitriles of α-keto carboxylic acids, *Khim. Org. Soedin. Fosfora., Akad. Nauk SSSR*, 45, 1967; *Chem. Abstr.*, 69, 67489j, 1968.

165. Kim, D.Y., and Oh, D.Y., A new facile synthesis of diethyl 1-aryl-1-cyanomethanephosphonates, *Synth. Commun.*, 17, 953, 1987.

166. Hammerschmidt, F., Kählig, H., and Müller, N., Biosynthesis of naturally products with a P-C bond. Part 6. Preparation of deuterium- and carbon-13-labelled L-alanyl- and L-alanyl-L-alanyl-(2-aminoethyl)phosphonic acids and their use in biosynthetic studies of fosfomycin in *Streptomyces fradiae*, *J. Chem. Soc., Perkin Trans. 1*, 365, 1991.

167. Martin, D., and Niclas, H.-J., Cyanic acid esters. Part 11. Reactions of bromocyanogen and cyanic acid aryl esters with carbanion-active phosphorus compounds, *Chem. Ber.*, 100, 187, 1967.

168. Shono, T., and Oda, R., Some reactions of triethyl phosphite and diethylbenzylphosphonate, *Kogyo Kagaku Zasshi*, 60, 21, 1957; *Chem. Abstr.*, 60, 4176i, 1959.

169. Ivanov, B.E., Zheltukin, V.F., and Vavilov, T.G., Reaction of Mannich bases with triethyl phosphite. Part 1. Reaction in the presence of acetic acid, *Izv. Akad. Nauk SSSR, Ser. Khim.*, 1285, 1967; *Bull. Acad. Sci. USSR, Div. Chem. Sci. (Engl. Transl.)*, 1239, 1967.

170. Ivanov, B.E., and Zheltukhin, V.F., Reaction between triethyl phosphite and β-substituted nitriles, *Izv. Akad. Nauk SSSR, Ser. Khim.*, 2089, 1967; *Bull. Acad. Sci. USSR, Div. Chem. Sci. (Engl. Transl.)*, 2008, 1967.

171. Coover, H.W., Jr., and Dickey, J.B., Cyanopropene phosphonamides and polymers thereof, *Eastman Kodak*, U.S. Patent Appl. US 2725371, 1955; *Chem. Abstr.*, 50, 11054a, 1956.

172. Dickey, J.B., and Coover, H.W., Jr., Dialkyl 2-cyanopropene-3-phosphonates, *Eastman Kodak*, U.S. Patent Appl. US 2721876, 1955; *Chem. Abstr.*, 50, 10123i, 1956.

173. Dickey, J.B., and Coover, H.W., Jr., Polymers of dialkyl 2-cyanopropene-3-phosphonates, *Eastman Kodak*, U.S. Patent Appl. US 2780616, 1957; *Chem. Abstr.*, 51, 7765b, 1957.

174. Krawczyk, H., A convenient route for monodealkylation of diethyl phosphonates, *Synth. Commun.*, 27, 3151, 1997.

175. Arbuzov, B.A., and Lugovkin, B.P., Action of triethyl phosphite and diethyl sodium phosphite on some dihalo derivatives, *Zh. Obshch. Khim.*, 21, 99, 1951; *Chem. Abstr.*, 45, 7002e, 1951.

176. Pudovik, A.N., and Konovalova, I.V., Addition reactions of esters of phosphorus(III) acids with unsaturated systems, *Synthesis*, 81, 1979.

177. Nesterov, L.V., Krepysheva, N.E., and Aleksandova, N.A., Mechanisms of Abramov and Pudovik reactions, *Zh. Obshch. Khim.*, 54, 54, 1984; *J. Gen. Chem. USSR (Engl. Transl.)*, 54, 47, 1984.

178. Pudovik, A.N., and Arbuzov, B.A., Addition of dialkyl phosphites to unsaturated ketones, nitriles and esters, *Dokl. Akad. Nauk SSSR, Ser. Khim.*, 73, 327, 1950; *Chem. Abstr.*, 45, 2853a, 1951.

179. Pudovik, A.N., and Arbuzov, B.A., Addition of dialkylphosphorous acids (dialkyl phosphites) to unsaturated compounds. Part 2. Addition of dialkylphosphorous acids to acrylonitrile and to methyl methacrylate, *Zh. Obshch. Khim.*, 21, 1837, 1951; *Chem. Abstr.*, 46, 6082e, 1952.

180. Bochwic, B., and Michalski, J., Formation of C-P bonds by addition of di-alkyl hydrogen phosphonates and alkyl hydrogen phosphinates to activated ethylenic derivatives, *Nature*, 167, 1035, 1951.

181. Bochwic, B., and Michalski, J., Organic phosphorus compounds. Part 1. Addition of di-alkyl hydrogen phosphonates to ethylenic derivatives, *Rocz. Chem.*, 25, 338, 1951; *Chem. Abstr.*, 48, 12013a, 1954.

182. Johnston, F., Plastic composition comprising a vinyl halide polymer and a phosphorus containing compound as a plasticizer therefor, *Union Carbide and Carbon*, U.S. Patent Appl. US 2668800, 1954; *Chem. Abstr.*, 48, 8586h, 1954.

183. Johnston, F., Production of diesterified phosphono derivatives of polyfunctional organic compounds, *Union Carbide and Carbon*, U.S. Patent Appl. US 2754319, 1956; *Chem. Abstr.*, 51, 466b, 1957.

184. Johnston, F., Production of diesterified phosphono derivatives of monofunctional compounds, *Union Carbide and Carbon*, U.S. Patent Appl. US 2754320, 1956; *Chem. Abstr.*, 51, 466i, 1957.

185. Ladd, E.C., and Harvey, M.P., Organo-phosphorus compounds, *U.S. Rubber*, U.S. Patent Appl. US 2971019, 1961; *Chem. Abstr.*, 55, 16427, 1961.

186. Pudovik, A.N., and Krupnov, G.P., New method of synthesis of esters of phosphonic and thiophosphonic acids. Part 36. Synthesis of derivatives of phosphonic acids with cyclic radicals in the ester groups, *Zh. Obshch. Khim.*, 31, 4053, 1961; *J. Gen. Chem. USSR (Engl. Transl.)*, 31, 3782, 1961.

187. Moshkina, T.M., and Pudovik, A.N., Synthesis of glycol diphosphates and some derivatives of esters of phosphonic acids, *Zh. Obshch. Khim.*, 32, 1671, 1962; *J. Gen. Chem. USSR (Engl. Transl.)*, 32, 1654, 1962.

188. Kreutzkamp, N., and Schindler, H., Carbonyl and cyanophosphonic acid esters. Part 5. Conversion of cyano and carboxylic acid ester groups to substituted phosphonic acid esters, *Arch. Pharm. Ber. Dtsch. Pharm. Ges.*, 295, 28, 1962.

189. Koziara, A., and Zwierzak, A., Organophosphorous compounds. Part 143. Phosphorous acid amides. Part 4. Chemical properties of monoalkyl phosphoroamidites $(RO)(R^1_2N)P(O)H$, *Bull. Acad. Pol. Sci., Ser. Sci. Chim.*, 17, 507, 1969.

190. Alizade, Z.A., and Velieva, R.K., Synthesis and properties of di(α-chloromethyl-β-alkoxyethyl)phosphites, *Zh. Obshch. Khim.*, 39, 599, 1969; *J. Gen. Chem. USSR (Engl. Transl.)*, 39, 567, 1969.

191. Shilov, I.V., and Nifant'ev, E.E., Proton mobility of tetraalkylphosphorodiamidous acids, *Zh. Obshch. Khim.*, 43, 581, 1973; *J. Gen. Chem. USSR (Engl. Transl.)*, 43, 583, 1973.

192. Nifant'ev, E.E., and Shilov, I.V., β-Substituted alkylphosphonic diamides, *Moscow State University*, Soviet Patent Appl. SU 371246, 1973; *Chem. Abstr.,* 79, 53561, 1973.

193. Collins, D.J., Hetherington, J.W., and Swan, J.M., Organophosphorus compounds. Part 15. Synthesis of 2-ethoxy-1-methyl-1,2-azaphospholidine 2-oxide, *Aust. J. Chem.*, 27, 1759, 1974.

194. Yurchenko, R.I., Klepa, T.I., Lozovitskaya, O.V., and Tikhonov, V.P., Phosphorylated adamantanes. Part 5. Some reactions of adamantane-containing dialkyl hydrogen phosphites, *Zh. Obshch. Khim.*, 53, 2445, 1983; *J. Gen. Chem. USSR (Engl. Transl.)*, 53, 2206, 1983.

195. Pudovik, A.N., and Krupnov, V.K., Synthesis of the ethyl ester of diethylamidophosphorous acid and its addition to unsaturated compounds, *Zh. Obshch. Khim.*, 38, 1406, 1968; *J. Gen. Chem. USSR (Engl. Transl.)*, 38, 1359, 1968.

196. Maklyaev, F.L., Kirillov, N.V., Fokin, A.V., and Rudnitskaya, L.S., Synthesis of esters of phosphonocarboxylic acids with different kinds radicals, *Zh. Obshch. Khim.*, 40, 1014, 1970; *J. Gen. Chem. USSR (Engl. Transl.)*, 40, 999, 1970.

197. Laskorin, B.N., Yakshin, V.V., and Bulgakova, V.B., Addition of salts of alkyl dihydrogen phosphites to unsaturated compounds, *Zh. Obshch. Khim.*, 46, 2477, 1976; *J. Gen. Chem. USSR (Engl. Transl.)*, 46, 2372, 1976.

198. Pudovik, A.N., and Golitsyna, G.A., Addition reactions of alkylene phosphites and phosphorothioites, *Zh. Obshch. Khim.*, 34, 876, 1964; *J. Gen. Chem. USSR (Engl. Transl.)*, 34, 870, 1964.

199. Pudovik, A.N., and Pudovik, M.A., Addition of cyclic *bis*(ethylene glycol) diphosphite at double bonds, *Zh. Obshch. Khim.*, 36, 565, 1966; *J. Gen. Chem. USSR (Engl. Transl.)*, 36, 585, 1966.

200. Volodina, L.N., Reactivity of *bis*(trialkylsilyl)phosphites, *Zh. Obshch. Khim.*, 37, 1842, 1967; *J. Gen. Chem. USSR (Engl. Transl.)*, 37, 1755, 1967.

201. Pudovik, A.N., and Khlyupina, N.I., New method of synthesis of esters of phosphonic and thiophosphonic acids. Part 25. Addition of mixed dialkyl phosphites, dialkyl thiophosphites, and diallyl phosphite to unsaturated compounds, *Zh. Obshch. Khim.*, 26, 1672, 1956; *Chem. Abstr.*, 51, 3439f, 1957.

202. Pudovik, A.N., New method of synthesis of esters of phosphonocarboxylic acids and their derivatives, *Dokl. Akad. Nauk SSSR, Ser. Khim.*, 85, 349, 1952; *Chem. Abstr.*, 47, 5351i, 1953.

203. Pudovik, A.N., and Plakatina, N.I., Addition of dialkyl phosphorous acids to unsaturated compounds. Part 10. Addition of dialkylphosphorous acids to unsaturated nitriles, *Sb. Statei Obshch. Khim.*, 2, 831, 1953; *Chem. Abstr.*, 49, 6821b, 1955.

204. Pudovik, A.N., and Sudakova, T.M., Reactivity of incomplete esters of acids of phosphorus in addition reactions with acrylonitrile, *Zh. Obshch. Khim.*, 36, 1113, 1966; *J. Gen. Chem. USSR (Engl. Transl.)*, 36, 1126, 1966.

205. Pevzner, L.M., Ignat'ev, V.M., and Ionin, B.I., Addition of diethyl hydrogen phosphite to furylalkanes, *Zh. Obshch. Khim.*, 54, 69, 1984; *J. Gen. Chem. USSR (Engl. Transl.)*, 54, 59, 1984.

206. Simoni, D., Invidiata, F.P., Manferdini, M., Lampronti, I., Rondanin, R., Roberti, M., and Pollini, G.P., Tetramethylguanidine (TMG)-catalyzed addition of dialkyl phosphites to α,β-unsaturated carbonyl compounds, alkenenitriles, aldehydes, ketones and imines, *Tetrahedron Lett.*, 39, 7615, 1998.

207. Makosza, M., and Wojciechowski, K., Synthesis of phosphonic acid esters in solid–liquid catalytic two-phase system, *Bull. Acad. Pol. Sci., Ser. Sci. Chim.*, 32, 175, 1984.

208. Platonov, A.Y., Sivakov, A.A., Chistokletov, V.N., and Maiorova, E.D., Transformations of electrophilic reagents in a diethyl phosphite–potassium carbonate–ethanol system, *Izv. Akad. Nauk, Ser. Khim.*, 369, 1999; *Russ. Chem. Bull. (Engl. Transl.)*, 367, 1999.

209. Platonov, A.Y., Sivakov, A.A., Chistokletov, V.N., and Maiorova, E.D., Reaction of diethyl hydrogen phosphite with activated alkenes in heterophase K_2CO_3/ethanol system, *Zh. Obshch. Khim.*, 69, 514, 1999; *Russ. J. Gen. Chem. (Engl. Transl.)*, 69, 493, 1999.

210. Gaudemar-Bardone, F., Mladenova, M., and Gaudemar, M., Synthesis of 3-substituted ethyl 4,4,4-trichloro-2-cyanobutanoates via Michael addition to ethyl 4,4,4-trichloro-2-cyano-2-butenoate, *Synthesis*, 611, 1988.

211. Kolomnikova, G.D., Prikhodchenko, D.Y., Petrovskii, P.V., and Gololobov, Y.A., Reaction of α-cyanoacrylic acid and cyanoacrylates with dialkyl and diaryl phosphites, *Izv. Akad. Nauk, Ser. Khim.*, 1913, 1992; *Bull. Russ. Acad. Sci., Div. Chem. Sci. (Engl. Transl.)*, 1497, 1992.

212. Khaskin, B.A., Rymareva, T.C., and Goloshov, S.N., Reaction of dialkyl phosphites with 1,1-dicyano-3,3,3-trichloropropylene, *Zh. Obshch. Khim.*, 57, 2138, 1987; *J. Gen. Chem. USSR (Engl. Transl.)*, 57, 1914, 1987.

213. Abdou, W.M., Khidre, M.D., and Mahran, M.R., Organophosphorus chemistry. Part 16. The reaction of furfurylidenemalononitrile with alkyl phosphites, *J. Prakt. Chem.*, 332, 1029, 1990.

214. Fahmy, A.A., Ismail, N.A., and Hafez, T.S., Reaction of alkyl phosphites on some derivatives of malononitriles, *Phosphorus, Sulfur Silicon Relat. Elem.*, 66, 201, 1992.

215. El Hashash, M., and Mohamed, M.M., Some reactions with arylidene malonitrile, malonic acid and malonic ester, *Pak. J. Sci. Ind. Res.*, 20, 325, 1977; *Chem. Abstr.*, 91, 56722, 1979.

216. Rymareva, T.G., Sandakov, V.B., Khaskin, B.A., Promonenkov, V.K., and Koroleva, T.I., Synthesis of sodium derivatives of α-phosphorylated benzylmalononitriles, *Zh. Obshch. Khim.*, 52, 220, 1982; *J. Gen. Chem. USSR (Engl. Transl.)*, 52, 201, 1982.

217. Kreutzkamp, N., and Cordes, G., Carbonyl- and cyanophosphonic acid esters. Part 6. α-Substituted benzylphosphonic acid and its derivatives, *Arch. Pharm. Ber. Dtsch. Pharm. Ges.*, 295, 276, 1962.

218. Schulze, W., Willitzer, H., and Fritzsche, H., Nitrogen mustard derivatives by the reaction of *N,N*-bis(β-chloroethyl)-*p*-(tricyanovinyl)aniline with the sodium salts of OH, NH, PH, and SH acidic compounds, *Chem. Ber.*, 100, 2640, 1967.

219. Kolomnikova, G.D., Krylova, T.O., Chernoglazova, I.V., Petrovskii, P.V., and Gololobov, Y.G., 2-Cyanoacrylates in the conjugate addition of trifluoroacetic acid and nucleophilic reagents, *Izv. Akad. Nauk SSSR, Ser. Khim.*, 1245, 1993; *Russ. Chem. Bull. (Engl. Transl.)*, 1188, 1993.

220. Lu, R., and Yang, H., A novel approach to phosphonyl-substituted heterocyclic system. Part 1, *Tetrahedron Lett.*, 38, 5201, 1997.

221. Yuan, C., Li, C., and Ding, Y., Studies on organophosphorus compounds. Part 51. A new and facile route to 2-substituted 1,1-cyclopropanediyl*bis*(phosphonic acids), *Synthesis*, 854, 1991.

222. Yuan, C.Y., Li, C.Z., Wang, G.H., and Huang, W.S., New synthetic methods for carbocyclic and heterocyclic compounds bearing phosphonate moiety with biological significances, *Phosphorus, Sulfur Silicon Relat. Elem.*, 75, 147, 1993.

223. Jubault, P., Goumain, S., Feasson, C., and Collignon, N., A new and efficient electrosynthesis of polysubstituted cyclopropylphosphonates, using electrochemically activated magnesium, *Tetrahedron*, 54, 14767, 1998.

224. Goumain, S., Jubault, P., Feasson, C., and Collignon, N., A new and efficient electrosynthesis of 2-substituted 1,1-cyclopropanediyl*bis*(phosphonates), *Synthesis*, 1903, 1999.

225. Chiefari, J., Galanopoulos, S., Janowski, W.K., Kerr, D.I.B., and Prager, R.H., The synthesis of phosphonobaclofen, an antagonist of baclofen, *Aust. J. Chem.*, 40, 1511, 1987.

226. Sturtz, G., and Guervenou, J., Synthesis of novel functionalized *gem*-biphosphonates, *Synthesis*, 661, 1991.

227. Brel, V.K., and Abramkin, E.V., Functionally substituted allenylphosphanates and products of their transformations. Part 4. Nucleophilic substitution of tosyl group in 1,2-alkadienylphosphonic acid derivatives, *Zh. Obshch. Khim.*, 64, 1989, 1994; *Russ. J. Gen. Chem. (Engl. Transl.)*, 64, 1764, 1994.

228. Ginsburg, V.A., and Yakubovich, A.Y., Addition of trialkyl phosphites to acrylic systems, *Zh. Obshch. Khim.*, 30, 3987, 1960; *J. Gen. Chem. USSR (Engl. Transl.)*, 30, 3944, 1960.

229. Harvey, R.G., Reactions of triethyl phosphite with activated olefins, *Tetrahedron*, 22, 2561, 1966.

230. Pande, K.C., Hydrophosphinylation of activated alkenes or alkynes, *Dow Chemical*, U.S. Patent Appl. US 3622654, 1971; *Chem. Abstr.*, 76, 46303, 1972.

231. Tronchet, J.M.J., Neeser, J.-R., and Charollais, E.J., Some sugar phosphates, phosphonates and phosphine oxides. Preliminary communication, *Helv. Chim. Acta*, 61, 1942, 1978.

232. Tronchet, J.M.J., Neeser, J.-R., Gonzales, L., and Charollais, E.J., Preparation of unsaturated sugars phosphonates using nucleophilic conjugate addition, *Helv. Chim. Acta*, 62, 2022, 1979.

233. Neeser, J.-R., Tronchet, J.M.J., and Charollais, E.J., Structural analysis of 3-*C*-phosphonates, -phosphinates, and -phosphine oxides of branched-chain sugars, *Can. J. Chem.*, 61, 2112, 1983.

234. Novikova, Z.S., Mashoshina, S.N., Sapozhnikova, T.A., and Lutsenko, I.F., Reaction of trimethylsilyldiethylphosphite with carbonyl compounds, *Zh. Obshch. Khim.*, 41, 2622, 1971; *J. Gen. Chem. USSR (Engl. Transl.)*, 41, 2655, 1971.

235. Lebedev, E.P., Pudovik, A.N., Tsyganov, B.N., Nazmutdinov, R.Y., and Romanov, G.V., Silyl esters of phosphorous and arylphosphonous acids, *Zh. Obshch. Khim.*, 47, 765, 1977; *J. Gen. Chem. USSR (Engl. Transl.)*, 47, 698, 1977.

236. Nakano, M., Okamoto, Y., and Sakurai, H., Preparation of dialkyl 2-cyano-2-(trimethylsilyl)ethanephosphonates, *Synthesis*, 915, 1982.

237. Seux, R., and Foucaud, A., Participation of a carboxylic ester group in the displacement of a phosphoryl group in a phosphonate, *C.R. Acad. Sci., Ser. C*, 273, 842, 1971.

238. Morel, G., Seux, R., and Foucaud, A., Vinyl α-phenylphosphonates β-substituted by two electron-attracting groups. Preparation and reactivity toward various nucleophiles, *Bull. Soc. Chim. Fr.*, 177, 1976.

239. Malenko, D.M., Randina, L.V., Simurova, N.V., and Sinitsa, A.D., Reactions of substituted 4,4,4-trichloro-2-butenoates with triethyl phosphite, *Zh. Obshch. Khim.*, 59, 1909, 1989; *J. Gen. Chem. USSR (Engl. Transl.)*, 59, 1707, 1989.

240. Shin, C.G., Yonezawa, Y., Sekine, Y., and Yoshimura, J., α,β-Unsaturated carboxylic acid derivatives. Part 7. Reaction of ethyl α,β-unsaturated α-cyanocarboxylates with triethyl or diethyl phosphonate, *Bull. Chem. Soc. Jpn.*, 48, 1321, 1975.

241. Basavaiah, D., and Pandiaraju, S., Nucleophilic addition of triethyl phosphite to acetates of the Baylis–Hillman adducts. Stereoselective synthesis of (*E*)- and (*Z*)-allylphosphonates, *Tetrahedron*, 52, 2261, 1996.

242. Fouqué, D., About-Jaudet, E., and Collignon, N., α-Pyrazolyl-alkylphosphonates. Part 2. A simple and efficient synthesis of diethyl 1-(pyrazol-4-yl)-alkylphosphonates, *Synth. Commun.*, 25, 3443, 1995.

243. Kazakov, P.V., Odinets, I.L., Antipin, M.Y., Petrovskii, P.V., Kovalenko, L.V., Struchkov, Y.T., and Mastryukova, T.A., Synthesis of 2-keto-4-phosphorylbutyric acids and their derivatives, *Izv. Akad. Nauk SSSR, Ser. Khim.*, 2120, 1990; *Bull. Acad. Sci. USSR, Div. Chem. Sci. (Engl. Transl.)*, 1931, 1990.

244. Nakano, M., and Okamoto, Y., A convenient synthesis of 2-cyano-1,3-butadienes by the reaction of diethyl 2-cyano-2-trimethylsilylethanephosphonate with carbonyl compounds, *Synthesis*, 917, 1983.

245. Morita, K., and Miyaji, H., Allylphosphonic acid esters, *Toyo Rayon*, Japanese Patent Appl. JP 45009526, 1970; *Chem. Abstr.*, 72, 132968, 1970.

246. Janecki, T., and Bodalski, R., A convenient method for the synthesis of substituted 2-methoxycarbonyl- and 2-cyanoallylphosphonates. The allyl phosphite-allyl phosphonate rearrangement, *Synthesis*, 799, 1990.

247. Greenlee, W.J., and Hangauer, D.G., Addition of trimethylsilyl cyanide to α-substituted ketones. Catalyst efficiency, *Tetrahedron Lett.*, 24, 4559, 1983.

248. Kazantsev, A.V., Averin, A.D., Lukashev, N.V., and Beletskaya, I.P., Addition of Me$_3$SiCN to α-, β- and γ-ketophosphonates and phosphinoylacetaldehyde as a route to phosphorylated 4α-hydroxycarboxylic acids and the corresponding nitriles, *Zh. Org. Khim.*, 34, 1495, 1998; *Russ. J. Org. Chem. (Engl. Transl.)*, 34, 1432, 1998.

249. Prager, R.H., and Schafer, K., Potential GABA$_B$ receptor antagonists. Part 10. The synthesis of further analogues of baclofen, phaclofen and saclofen, *Aust. J. Chem.*, 50, 813, 1997.

250. Merrett, J.H., Spurden, W.C., Thomas, W.A., Tong, B.P., and Whitcombe, I.W.A., The synthesis and rotational isomerism of 1-amino-2-imidazol-4-ylethylphosphonic acid [phosphonohistidine, His(P)] and 1-amino-2-imidazol-2-ylethylphosphonic acid [phosphonoisohistidine, Isohis(P)], *J. Chem. Soc., Perkin Trans. 1*, 61, 1988.

251. Fujiwara, K., Takahashi, H., and Ohta, M., 4-Chloro-3-methylcrotonic acid derivatives and phosphonates, *Bull. Chem. Soc. Jpn.*, 35, 1743, 1962.

252. Trehan, A., Mirzadegan, T., and Liu, R.S.H., The doubly hindered 7,11-di*cis*, 7,9,11-tri*cis*, 7,11,13-tri*cis* and all-*cis* isomers of retinonitrile and retinal, *Tetrahedron*, 46, 3769, 1990.

253. Groesbeek, M., van der Steen, R., van Vliet, J.C., Vertegaal, L.B.J., and Lugtenburg, J., Synthesis of three retinal models, including the 10-*s-cis*-locked retinal, all-*E*-12,19-methanoretinal, *Recl. Trav. Chim. Pays-Bas*, 108, 427, 1989.

254. Gewald, K., Rehwald, M., Eckert, K., Schäfer, H., and Gruner, M., Syntheses with pyridinium and sulfonium salts of acylated β-enaminonitriles, *Monatsh. Chem.*, 126, 711, 1995.

255. Fleck, F., Schmid, H., Mercer, A.V., and Paver, R., Phenylstilbenyltriazole compounds, *Sandoz*, Swiss Patent Appl. CH 561709, 1975; *Chem. Abstr.*, 83, 81213, 1975.

256. Kormany, G., Benzoxazole compounds, *Ciba-Geigy A.-G.*, German Patent Appl. DE 2756883, 1978; *Chem. Abstr.*, 89, 131049, 1978.

257. Hermant, R.M., Bakker, N.A.C., Scherer, T., Krijnen, B., and Verhoeven, J.W., Systematic study of a series of highly fluorescent rod-shaped donor-acceptor systems, *J. Am. Chem. Soc.*, 112, 1214, 1990.

258. Bergmann, R., and Gericke, R., Synthesis and antihypertensive activity of 4-(1,2-dihydro-2-oxo-1-pyridyl)-2*H*-1-benzopyrans and related compounds. New potassium channel activators, *J. Med. Chem.*, 33, 492, 1990.

259. Scherer, T., Hielkema, W., Krijnen, B., Hermant, R.M., Eijckelhoff, C., Kerkhof, F., Ng, A.K.F., Verleg, R., van der Tol, E.B., Brouwer, A.M., and Verhoeven, J.W., Synthesis and exploratory photophysical investigation of donor-bridge-acceptor systems derived from *N*-substituted 4-piperidones, *Recl. Trav. Chim. Pays-Bas*, 112, 535, 1993.

260. Bellucci, C., Teodori, E., Gualtieri, F., and Piacenza, G., Search for new Ca⁺⁺ antagonists. Lipophilic oximes and phosphonates, *Il Farmaco, Ed. Sci.*, 40, 730, 1985.

261. Villemin, D., Simeon, F., Decreus, H., and Jaffres, P.-A., Rapid and efficient Arbuzov reaction under microwave irradiation, *Phosphorus, Sulfur Silicon Relat. Elem.*, 133, 209, 1998.

262. Nylen, P., Organic phosphorus compounds. Part 2. β-Phosphonopropionic acid and γ-phosphonobutyric acid, *Chem. Ber.*, 59, 1119, 1926.

263. De Lombaert, S., Stamford, L.B., Blanchard, L., Tan, J., Hoyer, D., Diefenbacher, C.G., Wei, D., Wallace, E.M., Moskal, M.A., Savage, P., and Jeng, A.Y., Potent non-peptidic dual inhibitors of endothelin-converting enzyme and neutral endopeptidase 24.11, *Bioorg. Med. Chem. Lett.*, 7, 1059, 1997.

264. Pevzner, L.M., Ignat'ev, V.M., and Ionin, B.I., Reactions of halomethyl derivatives of acylfurans and furylcarbonitriles with sodium diethyl phosphite, *Zh. Obshch. Khim.*, 64, 135, 1994; *Russ. J. Gen. Chem. (Engl. Transl.)*, 64, 125, 1994.

265. Arient, J., Symmetrical and unsymmetrical derivatives of distyrylbenzene, *Collect. Czech. Chem. Commun.*, 46, 101, 1981.

266. Shen, Y., and Zhang, Z., Stereoselective synthesis of (*E*)-4-cyano-γ,δ-unsaturated nitriles, *Synth. Commun.*, 30, 445, 2000.

267. Pudovik, A.N., and Lebedeva, N.M., Addition of phosphonoacetic ester and its homologs to unsaturated electrophilic reagents, *Zh. Obshch. Khim.*, 22, 2128, 1952; *Chem. Abstr.*, 48, 564h, 1954.

268. Fiszer, B., and Michalski, J., Addition of phosphonoacetic esters and their analogs to α,β-unsaturated ethylenic derivatives, *Rocz. Chem.*, 26, 293, 1952.

269. Pudovik, A.N., and Lebedeva, N.M., Reaction of addition and condensation of phosphonoacetone and phosphonoacetic ester, *Dokl. Akad. Nauk SSSR, Ser. Khim.*, 90, 799, 1953; *Chem. Abstr.*, 49, 2429d, 1955.

270. Fiszer, B., Organophosphorus compounds with an active methylene group. Part 5. Thermal decomposition of diethoxyphosphinylacetic acid, *Rocz. Chem.*, 37, 949, 1963.

271. Auel, T., and Ulrich, H., (Carboxyalkyl)phosphonic acids, *Hoechst A.-G.*, German Patent Appl. DE 2333151, 1975; *Chem. Abstr.*, 82, 171201, 1975.

272. Pudovik, A.N., and Khusainova, N.G., Oxidative condensation and addition reactions of 1-dialkoxyphosphinyl-1-acetyl-3-butynes, *Zh. Obshch. Khim.*, 40, 1419, 1970; *J. Gen. Chem. USSR (Engl. Transl.)*, 40, 1403, 1970.

273. Ruder, S.M., and Kulkarni, V.R., Michael-type additions of 2-(diethoxyphosphinyl) cyclohexanone to activated alkenes and alkynes, *J. Chem. Soc., Chem. Commun.*, 2119, 1994.

274. Maruszewska-Wieczorkowska, E., and Michalski, J., Alkyl and alkenyl pyridines. Part 9. Phosphonates and tertiary phosphine oxide containing 2- and 4-pyridylmethyl groups at the phosphorus atom, and some of their reactions, *Rocz. Chem.*, 38, 625, 1964.

275. Pudovik, A.N., Yastrebova, G.E., and Cherkasova, O.A., Condensation and addition reactions of diethyl (carbamoylmethyl)phosphonate, *Zh. Obshch. Khim.*, 42, 88, 1972; *J. Gen. Chem. USSR (Engl. Transl.)*, 42, 83, 1972.

276. Alikin, A.Y., Sokolov, M.P., Liorber, B.G., Razumov, A.I., Zykova, T.V., and Zykova, V.V., Phosphinic, phosphonic, phosphinous, and phosphonous derivatives. Part 106. Reactions of enamines of phosphorylated ketones with α,β-unsaturated carbonyl compounds, *Zh. Obshch. Khim.*, 51, 547, 1981; *J. Gen. Chem. USSR (Engl. Transl.)*, 51, 428, 1981.

277. Pudovik, A.N., Yastrebova, G.E., and Pudovik, O.A., Condensation and addition to unsaturated compounds of tetraethoxydiphosphono methane, *Zh. Obshch. Khim.*, 40, 499, 1970; *J. Gen. Chem. USSR (Engl. Transl.)*, 40, 462, 1970.

278. Biere, H., Rufer, C., and Boettcher, I., Diphosphonic acid derivatives and pharmaceutical preparations containing them, *Schering A.-G.*, German Patent Appl. DE 3225469, 1982; *Chem. Abstr.*, 100, 210141, 1984.

279. Block, H.D., and Kallfass, H., Substituted propane-1,3-diphosphonic acids and -phosphinic acids, *Bayer A.-G.*, German Patent Appl. DE 2621604, 1977; *Chem. Abstr.*, 88, 89844, 1978.

280. Block, H.D., and Hendricks, U.W., Organic derivatives of 1-substituted propane-1,3-diphosphonic acids and -phosphinic acids, *Bayer A.-G.*, German Patent Appl. DE 2621605, 1977; *Chem. Abstr.*, 88, 89843, 1978.

281. Gênet, J.P., Uziel, J., Port, M., Touzin, A.M., Roland, S., Thorimbert, S., and Tanier, S., A practical synthesis of α-aminophosphonic acids, *Tetrahedron Lett.*, 33, 77, 1992.

282. Hu, C.-M., and Chen, J., Addition of diethyl bromodifluoromethylphosphonate to various alkenes initiated by Co(III)/Zn bimetal redox system, *J. Chem. Soc., Perkin Trans. 1*, 327, 1993.

283. Villemin, D., Sauvaget, F., and Hajek, M., Addition of diethyl trichloromethylphosphonate to olefins catalysed by copper complexes, *Tetrahedron Lett.*, 35, 3537, 1994.

284. Marco, J.L., and Chinchon, P.M., Michael reactions of β-keto phosphonates with arylmethylenemalononitriles. The first synthesis of densely functionalized 5-diethylphosphinyl-2-amino-4*H*-pyrans, *Heterocycles*, 51, 1137, 1999.

285. Pudovik, A.N., and Grishina, O.N., Synthesis and properties of vinylphosphonic esters. Part 2. Phosphonoethylation reaction. Addition of malonic, cyanoacetic, acetoacetic ester, and their homologs to vinylphosphonic ester, *Zh. Obshch. Khim.*, 23, 267, 1953; *Chem. Abstr.*, 48, 2573c, 1954.

286. Maruszewska-Wieczorkowska, E., and Michalski, J., Alkyl- and alkenylpyridines. Part 7. 3-(2-Pyridyl)propylphosphonic acid, *Rocz. Chem.*, 37, 1315, 1963.

287. Nugent, R.A., Schlachter, S.T., Murphy, M., Dunn, C.J., Staite, N.D., Galinet, L.A., Shields, S.K., Wu, H., Aspar, D.G., and Richard, K.A., Carbonyl-containing bisphosphonate esters as novel antiinflammatory and antiarthritic agents, *J. Med. Chem.*, 37, 4449, 1994.

288. Pudovik, A.N., and Kuzovleva, R.G., Diethyl esters of α- and β-carbalkoxyvinylphosphonic acid and addition reaction of these with nucleophilic agents, *Zh. Obshch. Khim.*, 35, 354, 1965; *J. Gen. Chem. USSR (Engl. Transl.)*, 35, 354, 1965.

289. Tarasenko, E.A., Mukhaiimana, P., Tsvetkov, A.V., Lukashev, N.V., and Beletskaya, I.P., Michael addition of phosphorylated CH acids under conditions of phase-transfer catalysis, *Zh. Org. Khim.*, 34, 64, 1998; *Russ. J. Org. Chem. (Engl. Transl.)*, 34, 52, 1998.

290. Khaskin, B.A., Rymareva, T.G., and Drozdova, O.N., Synthesis and properties of phosphorylated 3-cyanopropionaldehyde, *Zh. Obshch. Khim.*, 52, 2020, 1982; *J. Gen. Chem. USSR (Engl. Transl.)*, 52, 1795, 1982.

291. Birum, G.H., Process of preparing organic phosphorus compounds, *Monsanto Chemical*, U.S. Patent Appl. US 3014944, 1961; *Chem. Abstr.*, 56, 11622, 1962.

292. Birum, G.H., and Richardson, G.A., Silicon-phosphorus compounds, *Monsanto Chemical*, U.S. Patent Appl. US 3113139, 1963; *Chem. Abstr.*, 60, 5551g, 1964.

293. Hakimelahi, G.H., Moosavi-Movahedi, A.A., Sadeghi, M.M., Tsay, S.-C., and Hwu, J.R., Design, synthesis, and structure–activity relationship of novel dinucleotide analogs as agents against herpes and human immunodeficiency viruses, *J. Med. Chem.*, 38, 4648, 1995.

294. Cox, J.M., Bellini, P., Barrett, R., Ellis, R.M., and Hawkes, T.R., Preparation of triazolylalkylphosphonic acids as herbicides, *Zeneca*, U.S. Patent Appl. US 5393732, 1995; *Chem. Abstr.*, 120, 134813, 1994.

295. Cox, J.M., Hawkes, T.R., Bellini, P., Ellis, R.M., Barrett, R., Swanborough, J.J., Russell, S.E., Walker, P.A., Barnes, N.J., Knee, A.J., Lewis, T., and Davies, P.R., The design and synthesis of inhibitors of imidazoleglycerol phosphate dehydratase as potential herbicides, *Pestic. Sci.*, 50, 297, 1997.

296. Ortwine, D.F., Malone, T.C., Bigge, C.F., Drummond, J.T., Humblet, C., Johnson, G., and Pinter, G.W., Generation of N-methyl-D-aspartate agonist and competitive antagonist pharmacophore models. Design and synthesis of phosphonoalkyl-substituted tetrahydroisoquinolines as novel antagonists, *J. Med. Chem.*, 35, 1345, 1992.

297. Öhler, E., and Kanzler, S., [(2H-Azirin-2-yl)methyl]phosphonates. Synthesis from allylic α- and γ-hydroxyphosphonates and application to diastereoselective formation of substituted [(azirin-2-yl)methyl]phosphonates, *Liebigs Ann. Chem.*, 867, 1994.

298. Han, L.-B., and Tanaka, M., Palladium-catalyzed hydrophosphorylation of alkynes via oxidative addition of HP(O)(OR)$_2$, *J. Am. Chem. Soc.*, 118, 1571, 1996.

299. Bush, E.J., and Jones, D.W., Control of stereochemistry in an intramolecular Diels–Alder reaction by the phenylsulfonyl group. An improved synthesis of pisiferol, *J. Chem. Soc., Perkin Trans. 1*, 3531, 1997.

300. Langlois, N., Rojas-Rousseau, A., and Decavallas, O., Synthesis of conformationally constrained analogues of (R)-2-amino-7-phosphonoheptanoic acid, *Tetrahedron: Asymmetry*, 7, 1095, 1996.

301. Hudson, R.F., and Greenhalgh, R., Influence of the leaving group on the rate of phosphorylation, *J. Chem. Soc. (B)*, 325, 1969.

302. Sato, N., Shimomura, Y., Ohwaki, Y., and Takeuchi, R., Studies on pyrazines. Part 22. Lewis acid-mediated cyanation of pyrazine N-oxides with trimethylsilyl cyanide. New route to 2-substituted 3-cyanopyrazines, *J. Chem. Soc., Perkin Trans. 1*, 2877, 1991.

303. Harusawa, S., Hamada, Y., and Shioiri, T., New methods and reagents in organic synthesis. Part 4. Reaction of enamines with diethyl phosphorocyanidate, *Synthesis*, 716, 1979.

304. Harusawa, S., Hamada, Y., and Shioiri, T., Diethyl phosphorocyanidate (DEPC). A novel reagent for the classical Strecker's α-amino nitrile synthesis, *Tetrahedron Lett.*, 20, 4663, 1979.

305. Bravo, P., Capelli, S., Meille, S.V., Seresini, P., Volonterio, A., and Zanda, M., Enantiomerically pure α-fluoroalkyl-α-amino acids. Synthesis of (R)-α-difluoromethyl-alanine and (S)-α-difluoromethyl-serine, *Tetrahedron: Asymmetry*, 7, 2321, 1996.

306. Kim, Y.J., and Kitahara, T., Novel synthesis of nectrisine and 4-*epi*-nectrisine, *Tetrahedron Lett.*, 38, 3423, 1997.

307. Kim, Y.J., Kido, M., Bando, M., and Kitahara, T., Novel and versatile synthesis of pyrrolidine type azasugars, DAB-1 and LAB-1, potent glucosidase inhibitors, *Tetrahedron*, 53, 7501, 1997.

308. Kutterer, K.M.K., and Just, G., Cyano/diallylamine additions to glycoside and nucleoside aldehydes and its application to the synthesis of polyoxin L and uracil polyoxin C, *Heterocycles*, 51, 1409, 1999.

309. Harusawa, S., Hamada, Y., and Shioiri, T., New methods and reagents in organic synthesis. Part 12. Reaction of diethyl phosphorocyanidate (DEPC) with aromatic amine oxides. A modified Reissert–Henze reaction, *Heterocycles*, 15, 981, 1981.

310. Antonini, I., Cristalli, G., Franchetti, P., Grifantini, M., Martelli, S., and Filippeschi, S., 2,2'-Bipyridyl-6-carbothioamide derivatives as potential antitumor agents, *Il Farmaco, Ed. Sci.*, 41, 346, 1986.

311. Suzuki, H., Yokoyama, Y., Miyagi, C., and Murakami, Y., A new general synthetic route for 1-substituted 4-oxygenated β-carbolines, *Chem. Pharm. Bull.*, 39, 2170, 1991.

312. Rivalle, C., and Bisagni, E., Synthesis of 9-methoxyellipticine N-oxide and its transformation into 1-functionalized ellipticine derivatives, *Heterocycles*, 38, 391, 1994.

313. Suzuki, H., Iwata, C., Sakurai, K., Tokumoto, K., Takahashi, H., Hanada, M., Yokoyama, Y., and Murakami, Y., Synthetic studies on indoles and related compounds. Part 41. A general synthetic route for 1-substituted 4-oxygenated β-carbolines, *Tetrahedron*, 53, 1593, 1997.

314. Suzuki, H., Unemoto, M., Hagiwara, M., Ohyama, T., Yokoyama, Y., and Murakami, Y., Synthetic studies on indoles and related compounds. Part 46. First total syntheses of 4,8-dioxygenated β-carboline alkaloids, *J. Chem. Soc., Perkin Trans. 1*, 1717, 1999.

315. Dawson, M.I., O'Krongly, D., Hobbs, P.D., Barrueco, J.R., and Sirotnak, F.M., Synthesis of the 7-hydroxy metabolites of methotrexate and 10-ethyl-10-deazaaminopterin, *J. Pharm. Sci.*, 76, 635, 1987.

316. Harusawa, S., Yoneda, R., Kurihara, T., Hamada, Y., and Shioiri, T., Cyanophosphorylation of ketones and aldehydes using diethyl phosphorocyanidate (DEPC), *Chem. Pharm. Bull.*, 31, 2932, 1983.

317. Kurihara, T., Hanakawa, M., Wakita, T., and Harusawa, S., Reaction of 3-acylindoles with diethyl phosphorocyanidate. A facile synthesis of 2-cyano-3-indoleacetonitriles, *Heterocycles*, 23, 2221, 1985.

318. Kurihara, T., Hanakawa, M., Harusawa, S., and Yoneda, R., Synthesis and cycloaddition reaction of 2-cyano-3-indoleacetonitriles, *Chem. Pharm. Bull.*, 34, 4545, 1986.

319. Kurihara, T., Santo, K., Harusawa, S., and Yoneda, R., Application of cyanophosphates in organic synthesis. Reactivity of α-cyano-α-diethylphosphonooxy anions, *Chem. Pharm. Bull.*, 35, 4777, 1987.

320. Pudovik, A.N., and Nazmutdinova, V.N., Reactions of three- and four-coordinate phosphorus cyanides with α,β-unsaturated aldehydes and ketones, *Zh. Obshch. Khim.*, 58, 2215, 1988; *J. Gen. Chem. USSR (Engl. Transl.)*, 58, 1973, 1988.

321. Takemoto, Y., Yoshikawa, N., and Iwata, C., Utility of diene-tricarbonyliron complexes as mobile chiral auxiliaries. Highly diastereoselective 1,5-nucleophilic substitution with 1,2-migration of the Fe(CO)$_3$ moiety, *J. Chem. Soc., Chem. Commun.*, 631, 1995.

322. Yoneda, R., Harusawa, S., and Kurihara, T., Cyanophosphate, an efficient intermediate for conversion of carbonyl compounds to nitriles, *Tetrahedron Lett.*, 30, 3681, 1989.

323. Keitel, J., Fischer-Lui, I., Boland, W., and Müller, D.G., Novel C$_9$ and C$_{11}$ hydrocarbons from the brown alga *Cutleria multifida*. Part 6. Sigmatropic and electrocyclic reactions in nature, *Helv. Chim. Acta*, 73, 2101, 1990.

324. Yoneda, R., Harusawa, S., and Kurihara, T., Cyano phosphate, an efficient intermediate for the chemoselective conversion of carbonyl compounds to nitriles, *J. Org. Chem.*, 56, 1827, 1991.

325. Abad, A., Arno, M., Marin, M.L., and Zaragoza, R.J., Spongian pentacyclic diterpenes. Stereoselective synthesis of aplyroseol-1, aplyroseol-2 and deacetylaplyroseol-2, *J. Chem. Soc., Perkin Trans. 1*, 1861, 1993.

326. Konieczny, M.T., Toma, P.H., and Cushman, M., Synthesis of hydroxyethylene isosteres of the transition state of the HIV protease-catalyzed Phe-Pro hydrolysis. Reaction of 2-[(Boc)amino]-1-(2′-oxocyclopentyl)-3-phenylpropanols with diethyl phosphorocyanidate and lithium cyanide followed by samarium iodide, *J. Org. Chem.*, 58, 4619, 1993.

327. Bigge, C.F., Malone, T.C., Hays, S.J., Johnson, G., Novak, P.M., Lescosky, L.J., Retz, D.M., Ortwine, D.F., Probert, A.W., Jr., Coughenour, L.L., Boxer, P.A., Robichaud, L.J., Brahce, L.J., and Shillis, J.L., Synthesis and pharmacological evaluation of 4a-phenanthrenamine derivatives acting at the phencyclidine binding site of the N-methyl-D-aspartate receptor complex, *J. Med. Chem.*, 36, 1977, 1993.

328. Ennis, M.D., Stjernlöf, P., Hoffman, R.L., Ghazal, N.B., Smith, M.W., Svensson, K., Wikström, H., Haadsma-Svensson, S.R., and Lin, C.-H., Structure–activity relationships in the 8-amino-6,7,8,9-tetrahydro-3H-benz[e]indole ring system. Part 2. Effects of 8-amino nitrogen substitution on serotonin receptor binding and pharmacology, *J. Med. Chem.*, 38, 2217, 1995.

329. Han, M., Zorumski, C.F., and Covey, D.F., Neurosteroid analogues. Part 4. The effect of methyl substitution at the C-5 and C-10 positions of neurosteroids on electrophysiological activity at GABA$_A$ receptors, *J. Med. Chem.*, 39, 4218, 1996.

330. Guillaumet, G., Viaud, M.-C., Mamai, A., Charton, I., Renard, P., Bennejean, C., Guardiola, B., and Daubos, P., Chromenes and benzodioxins as melatonin receptor agonists, *Adir*, Int. Patent Appl. WO 9852935, 1998; *Chem. Abstr.*, 130, 25076, 1998.

331. Tung, R.D., Murcko, M.A., and Bhisetti, G.R., Preparation of heterocyclic sulfonamide inhibitors of aspartyl protease, *Vertex Pharmaceuticals*, U.S. Patent Appl. US 5783701, 1998; *Chem. Abstr.*, 129, 136097, 1998.

332. Arno, M., Gonzalez, M.A., Marin, M.L., and Zaragoza, R.J., Synthesis of spongian diterpenes. (−)-Spongian-16-oxo-17-al and (−)-acetyldendrillol-1, *Tetrahedron Lett.*, 42, 1669, 2001.

333. Harusawa, S., Nakamura, S., Yagi, S., Kurihara, T., Hamada, Y. and Shioiri, T., A new synthesis of some non-steroidal anti-inflammatory agents via cyanophosphates, *Synth. Commun.*, 14, 1365, 1984.

334. Harusawa, S., Yoneda, R., Kurihara, T., Hamada, Y., and Shioiri, T., A novel synthesis of α,β-unsaturated nitriles from aromatic ketones via cyanophosphates, *Tetrahedron Lett.*, 25, 427, 1984.

335. Yoneda, R., Terada, T., Harusawa, S., and Kurihara, T., Studies on indenopyridine derivatives and related compounds. Part 4. Synthesis and stereochemistry of ethyl 9,9-dimethyl-1,2,3,9a-tetrahydro-9H-indeno[2,1-b]pyridine-3-carboxylate and its derivatives, *Heterocycles*, 23, 557, 1985.

336. Tagat, J.R., McCombie, S.W., Nazareno, D.V., Boyle, C.D., Kozlowski, J.A., Chackalamannil, S., Josien, H., Wang, Y., and Zhou, G., Synthesis of mono- and difluoronaphthoic acids, *J. Org. Chem.*, 67, 1171, 2002.

337. Kurihara, T., Terada, T., Satoda, S., and Yoneda, R., Studies on indenopyridine derivatives and related compounds. Part 6. Synthesis and stereochemistry of ethyl 9,9-dimethyl-1,2,3,9a-tetrahydro-9H-indeno[2,1-b]pyridine-3-carboxylate as a possible intermediate for the total synthesis of lysergic acid, *Chem. Pharm. Bull.*, 34, 2786, 1986.

338. Kurihara, T., Terada, T., Harusawa, S., and Yoneda, R., Synthetic studies of (±)-lysergic acid and related compounds, *Chem. Pharm. Bull.*, 35, 4793, 1987.

339. Meyer, M.D., Altenbach, R.J., Hancock, A.A., Buckner, S.A., Knepper, S.M., and Kerwin, J.F., Jr., Synthesis and *in vitro* characterization of *N*-[5-(4,5-dihydro-1*H*-imidazol-2-yl)-2-hydroxy-5,6,7,8-tetrahydronaphthalen-1-yl]methanesulfonamide and its enantiomers. A novel selective α_{1A} receptor agonist, *J. Med. Chem.*, 39, 4116, 1996.

340. Meyer, M.D., Hancock, A.A., Tietje, K., Sippy, K.B., Prasad, R., Stout, D.M., Arendsen, D.L., Donner, B.G. and Carroll, W.A., Structure-activity studies for a novel series of *N*-(arylethyl)-*N*-(1,2,3,4-tetrahydronaphthalen-1-ylmethyl)-*N*-methylamines possessing dual 5-HT uptake inhibiting and α_2-antagonistic activities, *J. Med. Chem.*, 40, 1049, 1997.

341. Kloubert, S., Mathé-Allainmat, M., Andrieux, J., Sicsic, S., and Langlois, M., Synthesis of benzocycloalkane derivatives as new conformationally restricted ligands for melatonin receptors, *Bioorg. Med. Chem. Lett.*, 8, 3325, 1998.

342. Harusawa, S., Miki, M., Yoneda, R., and Kurihara, T., Allylic rearrangement of α,β-unsaturated ketone-diethyl phosphorocyanidate adducts, *Chem. Pharm. Bull.*, 33, 2164, 1985.

343. Miki, M., Wakita, T., Harusawa, S., and Kurihara, T., Reaction of methyl vinyl ketone cyanohydrin phosphate with aromatic compounds, *Chem. Pharm. Bull.*, 33, 3558, 1985.

344. Kurihara, T., Miki, M., Yoneda, R., and Harusawa, S., Allylic rearrangement of cyanophosphates. Part 2. Synthesis of β-cyano-α,β-unsaturated ketones, *Chem. Pharm. Bull.*, 34, 2747, 1986.

345. Kurihara, T., Miki, M., Santo, K., Harusawa, S., and Yoneda, R., Allylic rearrangement of cyanophosphates. Part 3. Reaction of acyclic enone cyanophosphates, *Chem. Pharm. Bull.*, 34, 4620, 1986.

346. Harusawa, S., Miki, M., Hirai, J.-I., and Kurihara, T., A new synthesis of biaryls, *Chem. Pharm. Bull.*, 33, 899, 1985.

347. Kurihara, T., Harusawa, S., Hirai, J.-I., and Yoneda, R., Regiospecific arylation of 1,4-benzoquinone cyanohydrin phosphate. Synthesis of 3-aryl-4-hydroxybenzonitriles, *J. Chem. Soc., Perkin Trans. 1*, 1771, 1987.

348. Nagra, B.S., Shaw, G., and Robinson, D.H., A novel synthesis of acyl cyanides from diethyl phosphorocyanidate and some 1-substituted imidazole carboxylic acids including a D-ribofuranoside, *J. Chem. Soc., Chem. Commun.*, 459, 1985.

349. Jones, D.W., Motevalli, M., Shaw, G., and Shaw, J.D., The cyanocarbonyl group. Synthesis and crystal structure of an imidazole carbonyl cyanide, $C_{12}N_5OH_{17}$, *J. Chem. Crystallogr.*, 28, 561, 1998.

350. Manthey, M.K., Huang, D.T.C., Bubb, W.A., and Christopherson, R.I., Synthesis and enzymic evaluation of 4-mercapto-6-oxo-1,4-azaphosphinane-2-carboxylic acid 4-oxide as an inhibitor of mammalian dihydroorotase, *J. Med. Chem.*, 41, 4550, 1998.

351. Yamada, S., Yokohama, Y., and Shioiri, T., A new synthesis of thiol esters, *J. Org. Chem.*, 39, 3302, 1974.

352. Yokoyama, Y., Shioiri, T., and Yamada, S., Phosphorus in organic synthesis. Part 16. Diphenyl phosphorazidate (DPPA) and diethyl phosphorocyanidate (DEPC). Two new reagents for the preparation of thiol esters from carboxylic acids and thiols, *Chem. Pharm. Bull.*, 25, 2423, 1977.

353. Mizuno, M., and Shioiri, T., Reaction of carboxylic acids with diethyl phosphorocyanidate. A novel synthesis of homologated α-hydroxycarboxylic acids from carboxylic acids, *Tetrahedron Lett.*, 39, 9209, 1998.

354. Shioiri, T., and Hamada, Y., New methods and reagents in organic synthesis. Part 3. Diethyl phosphorocyanidate. A new reagent for *C*-acylation, *J. Org. Chem.*, 43, 3631, 1978.

355. Broom, N.J.P., Elder, J.S., Hannan, P.C.T., Pons, J.E., O'Hanlon, P.J., Walker, G., Wilson, J., and Woodall, P., The chemistry of pseudomonic acid. Part 14. Synthesis and *in vivo* biological activity of heterocyclyl substituted oxazole derivatives, *J. Antibiot.*, 48, 1336, 1995.

356. Sim, M.M., Lee, C.L., and Ganesan, A., Solid-phase *C*-acylation of active methylene compounds, *Tetrahedron Lett.*, 39, 2195, 1998.

357. Lardy, C., Barbanton, J., Dumas, H., Collonges, F., and Durbin, P., New nitromethyl ketones for use as aldose reductase inhibitors, *Merck*, Int. Patent Appl. WO 9852906, 1998; *Chem. Abstr.*, 130, 38195, 1998.

358. Barbas, C.F., Lerner, R.A., and Zhong, G., Antibody catalysis of enantio- and diastereo-selective aldol reactions, *Scripps Research Institute*, U.S. Patent Appl. US 6210938, 2001; *Chem. Abstr.*, 134, 265244, 2001.

359. Yamada, S.I., Kasai, Y., and Shioiri, T., Diethylphosphoryl cyanide. A new reagent for the synthesis of amides, *Tetrahedron Lett.*, 14, 1595, 1973.

360. Rosowsky, A., Wright, J.E., Ginty, C., and Uren, J., Methotrexate analogues. Part 15. A methotrexate analogue designed for active-site-directed irreversible inactivation of dihydrofolate reductase, *J. Med. Chem.*, 25, 960, 1982.

361. Rosowsky, A., Freisheim, J.H., Bader, H., Forsch, R.A., Susten, S.S., Cucchi, C.A., and Frei, E., III, Methotrexate analogues. Part 25. Chemical and biological studies on the γ-*tert*-butyl esters of methotrexate and aminopterin, *J. Med. Chem.*, 28, 660, 1985.

362. Rosowsky, A., Forsch, R.A., Moran, R.G., Kohler, W., and Freisheim, J.H., Methotrexate analogues. Part 32. Chain extension, α-carboxyl deletion, and γ-carboxyl replacement by sulfonate and phosphonate. Effect on enzyme binding and cell-growth inhibition, *J. Med. Chem.*, 31, 1326, 1988.

363. Bock, M.G., DiPardo, R.M., Evans, B.E., Freidinger, R.M., Rittle, K.E., Payne, L.S., Boger, J., Whitter, W.L., LaMont, B.I., Ulm, E.H., Blaine, E.H., Schorn, T.W., and Veber, D.F., Renin inhibitors containing hydrophilic groups. Tetrapeptides with enhanced aqueous solubility and nanomolar potency, *J. Med. Chem.*, 31, 1918, 1988.

364. Mori, S., Ohno, T., Harada, H., Aoyama, T., and Shioiri, T., New methods and reagents in organic synthesis. Part 92. A stereoselective synthesis of tilivalline and its analogs, *Tetrahedron*, 47, 5051, 1991.

365. He, G.-X., Browne, K.A., Groppe, J.C., Blasko, A., Mei, H.-Y., and Bruice, T.C., Microgonotropens and their interactions with DNA. Part 1. Synthesis of the tripyrrole peptides dien-microgonotropen-A, -B, and -C and characterization of their interactions with dsDNA, *J. Am. Chem. Soc.*, 115, 7061, 1993.

366. Kantoci, D., Wechter, W.J., Murray, E.D., Jr., Dewind, S.A., Borchardt, D., and Khan, S.I., Endogenous natriuretic factors. Part 6. The stereochemistry of a natriuretic γ-tocopherol metabolite LLU-α[1], *J. Pharm. Exp. Ther.*, 282, 648, 1997.

367. Fässler, A., Bold, G., Capraro, H.-G., and Lang, M., Preparation of azahexane derivatives as substrate isosteres of retroviral aspartate proteases for the treatment of AIDS, *Novartis*, Int. Patent Appl. WO 9719055, 1997; *Chem. Abstr.*, 127, 66217, 1997.

368. Amschler, H., Bar, T., Flockerzi, D., Gutterer, B., Thibaut, U., Hatzelmann, A., Boss, H., Beume, R., Kley, H.-P., Hafner, D., Martin, T., and Ulrich, W.-R., Preparation of dihydrobenzofurans as phosphodiesterase IV inhibitors, *BYK Gulden Lomberg Chemische Fabrik*, Int. Patent Appl. WO 9822453, 1998; *Chem. Abstr.*, 129, 41073, 1998.

369. Fässler, A., Bold, G., Lang, M., Bhagwat, S., and Schneider, P., Antiretroviral hydrazine derivatives, *Novartis*, U.S. Patent Appl. US 5753652, 1998; *Chem. Abstr.*, 129, 28216, 1998.

370. Mabuchi, H., Suzuki, N., and Miki, T., Preparation of 4,1-benzoxazepines as somatostatin agonists, *Takeda Chemical Ind.*, Int. Patent Appl. WO 9847882, 1998; *Chem. Abstr.*, 129, 330745, 1998.

371. Ariga, K., Kamino, A., Cha, X., and Kunitake, T., Multisite recognition of aqueous dipeptides by oligoglycine arrays mixed with guanidinium and other receptor units at the air–water interface, *Langmuir*, 15, 3875, 1999.

372. Oda, T., Notoya, K., Gotoh, M., Taketomi, S., Fujisawa, Y., Makino, H., and Sohda, T., Synthesis of novel 2-benzothiopyran and 3-benzothiepin derivatives and their stimulatory effect on bone formation, *J. Med. Chem.*, 42, 751, 1999.

373. Fujita, T., Wada, K., and Fujiwara, T., Preparation of substituted fused heterocyclic compounds as pharmaceuticals, *Sankyo*, Int. Patent Appl. WO 9918081, 1999; *Chem. Abstr.*, 130, 296681, 1999.

374. Thompson, S.K., Veber, D.F., Tomaszek, T.A., and Tew, D.G., Preparation of amino acid derivatives for treatment of parasitic diseases by inhibition of cysteine proteases of the papain superfamily, *Smithkline Beecham*, Int. Patent Appl. WO 9953039, 1999; *Chem. Abstr.*, 131, 299694, 1999.

375. MacDonald, M., Vander Velde, D., and Aube, J., Effect of progressive benzyl substitution on the conformations of aminocaproic acid-cyclized dipeptides, *J. Org. Chem.*, 66, 2636, 2001.

376. Willemen, H.M., Vermonden, T., Marcelis, A.T.M., and Sudhölter, E.J.R., *N*-Cholyl amino acid alkyl esters. A novel class of organogelators, *Eur. J. Org. Chem.*, 2329, 2001.

377. Pflücker, F., Driller, H., Kirschbaum, M., Scholz, V., and Neunhoeffer, H., Preparation containing quinoxaline derivatives, *Merck*, Int. Patent Appl. WO 2001068047, 2001; *Chem. Abstr.*, 135, 277730, 2001.

378. Shioiri, T., and Irako, N., An efficient synthesis of the piperazinone fragment of pseudotheonamide A₁ via a stereoselective intramolecular Michael ring closure, *Chem. Lett.*, 130, 2002.

379. Yamada, S.I., Ikota, N., Shioiri, T., and Tachibana, S., Diphenyl phosphorazidate (DPPA) and diethyl phosphorocyanidate (DEPC). Two new reagents for solid-phase peptide synthesis and their application to the synthesis of porcine Motilin, *J. Am. Chem. Soc.*, 97, 7174, 1975.

380. Shioiri, T., and Hamada, Y., Efficient synthesis of biologically active peptides of aquatic origin involving unusual amino acids, *Synlett*, 184, 2001.

381. Yoneda, R., Osaki, H., Harusawa, S., and Kurihara, T., Reductive deoxygenation of α,β-unsatured ketones via cyanophosphates by lithium in liquid ammonia, *Chem. Pharm. Bull.*, 37, 2817, 1989.

382. Yoneda, R., Osaki, T., Harusawa, S., and Kurihara, T., Dephosphorylation of cyano diethyl phosphates by reduction with lithium-liquid ammonia. An efficient method for conversion of carbonyl compounds into nitriles, *J. Chem. Soc., Perkin Trans. 1,* 607, 1990.

383. Yoneda, R., Inagaki, N., Harusawa, S., and Kurihara, T., A new synthesis of allenic nitriles from ynones via cyano phosphates, *Chem. Pharm. Bull.*, 40, 21, 1992.

384. Murdock, K.C., Lee, V.J., Citarella, R.V., Durr, F.E., Nicolau, G., and Kohlbrenner, M., *N*-Phosphoryl derivatives of bisantrene. Antitumor prodrugs with enhanced solubility and reduced potential for toxicity, *J. Med. Chem.*, 36, 2098, 1993.

385. Guzman, A., and Diaz, E., A convenient method for the phosphorylation of phenols with diethyl cyanophosphonate, *Synth. Commun.*, 27, 3035, 1997.

386. Guzman, A., Alfaro, R., and Diaz, E., Synthesis of aryl phosphonates by reaction of Grignard reagents with diethyl cyanophosphonate, *Synth. Commun.*, 29, 3021, 1999.

387. Harusawa, S., and Shioiri, T., New methods and reagents in organic synthesis. Part 23. Diethyl phosphorocyanidate (DEPC), a useful reagent for an unprecedented transformation of sulfinic acids to thiocyanates, *Tetrahedron Lett.*, 23, 447, 1982.

388. Sakamoto, M., Fukuda, Y., Kamiyama, T., and Kawasaki, T., Studies on conjugated nitriles. Part 7. Lewis acid-promoted reaction of active methylene compounds with diethyl phosphorocyanidate. Preparation of α,β-unsaturated α-aminophosphonates, *Chem. Pharm. Bull.*, 42, 1919, 1994.

389. Lennon, P.J., Vulfson, S.G., and Civade, E., New preparation of cyanophosphonate salts, *J. Org. Chem.*, 64, 2958, 1999.

390. Pudovik, A.N., and Yastrebova, G.E., Organophosphorus compounds with an active methylene group, *Usp. Khim.*, 39, 1190, 1970; *Russ. Chem. Rev.*, 39, 562, 1970.

391. Boutagy, J., and Thomas, R., Olefin synthesis with organic phosphonate carbanions, *Chem. Rev.*, 74, 87, 1974.

392. Wadsworth, W.S., Jr., Synthetic applications of phosphoryl-stabilised anions, *Org. React.*, 25, 73, 1977.

393. Walker, B.J., *Transformations via Phosphorus-Stabilised Anions. Part 2. PO-Activated Olefination,* Academic Press, New York, 1979, p. 155.

394. Maryanoff, B.E., and Reitz, A.B., The Wittig olefination reaction and modifications involving phosphoryl-stabilized carbanions. Stereochemistry, mechanism, and selected synthetic aspects, *Chem. Rev.*, 89, 863, 1989.

395. Johnson, A.W., *Ylides and Imines of Phosphorus*, John Wiley & Sons, New York, 1993.

396. Blank, B., Zuccarello, W.A., Cohen, S.R., Frishmuth, G.J., and Scaricaciottoli, D., Synthesis and adrenocortical inhibiting activity of substituted diphenylalkylamines, *J. Med. Chem.*, 12, 271, 1969.

397. Ferreira, A.B.B., and Salisbury, K., The photochemistry of 1-cyano-2-methyl-3-phenylpropene and ring substituted derivatives, *J. Chem. Soc., Perkin Trans. 2,* 25, 1982.

398. Metzger, J.O., and Blumenstein, M., Stereoselectivity of the thermally initiated free-radical chain addition of cyclohexane to 1-alkynes, *Chem. Ber.*, 126, 2493, 1993.

399. Grosa, G., Viola, F., Ceruti, M., Brusa, P., Delprino, L., Dosio, F., and Cattel, L., Synthesis and biological activity of a squalenoid maleimide and other classes of squalene derivatives as irreversible inhibitors of 2,3-oxidosqualene cyclase, *Eur. J. Med. Chem.*, 29, 17, 1994.

400. Bierer, D.E., Dener, J.M., Dubenko, L.G., Gerber, R.E., Litvak, J., Peterli, S., Peterli-Roth, P., Truong, T.V., Mao, G., and Bauer, B.E., Novel 1,2-dithiins. Synthesis, molecular modeling studies, and antifungal activity, *J. Med. Chem.*, 38, 2628, 1995.

401. Piers, E., De Waal, W., and Britton, R.W., Total synthesis, stereochemistry, and lithium-ammonia reduction of (±)-demethylaristolone and (±)-5-epi-4-demethylaristolone, *Can. J. Chem.*, 47, 4299, 1969.

402. Liu, R.S.H., Matsumoto, H., Asato, A.E., Denny, M., Yoshizawa, T., and Dahlquist, F.W., Synthesis and properties of 12-fluororetinal and 12-fluororhodopsin. A model system for ^{19}F-NMR studies of visual pigments, *J. Am. Chem. Soc.*, 103, 7195, 1981.

403. Janecki, T., A convenient synthesis of substituted 2-cyano-1,3-butadienes, *Synthesis*, 167, 1991.

404. Ting, P.C., and Solomon, D.M., Synthesis of spiro[4′,5′,10,11-tetrahydro-5*H*-dibenzo[*a,d*]cyclohepten-5-yl-2′ (3′*H*)-furans] as potential cytokine inhibitors, *J. Heterocycl. Chem.*, 32, 1027, 1995.

405. Armesto, D., Gallego, M.G., Horspool, W.M., and Agarrabeitia, A.R., A new photochemical synthesis of cyclopropanecarboxylic acids present in pyrethroids by the aza-di-π-methane rearrangement, *Tetrahedron*, 51, 9223, 1995.

406. Reitz, A.B., Jordan A.D., Jr., and Maryanoff, B.E., Formation of chiral alkoxy dienes in Wittig/Michael reactions of 2,3,5-tri-*O*-benzyl-D-arabinose, *J. Org. Chem.*, 52, 4800, 1987.

407. Takayanagi, H., Kitano, Y., and Morinaka, Y., Stereo- and enantioselective total synthesis of sarcophytol A, *Tetrahedron Lett.*, 31, 3317, 1990.

408. Anastasia, M., Allevi, P., Ciuffreda, P., Fiecchi, A., and Scala, A., Action of alkali on *O*-benzylated aldoses. A simple rationalization of some reactions occurring in alkaline media, *J. Org. Chem.*, 56, 3054, 1991.

409. Kodama, M., Yoshio, S., Yamaguchi, S., Fukuyama, Y., Takayanagi, H., Morinaka, Y., Usui, S., and Fukazawa, Y., Total syntheses of both enantiomers of sarcophytols A and T based on stereospecific [2,3]Wittig rearrangement, *Tetrahedron Lett.*, 34, 8453, 1993.

410. Takayanagi, H., Kitano, Y., and Morinaka, Y., Total synthesis of sarcophytol A, an anticarcinogenic marine cembranoid, *J. Org. Chem.*, 59, 2700, 1994.

411. Auguste, S.P., and Young, D.W., Synthesis of uridine derivatives containing strategically placed radical traps as potential inhibitors of ribonucleotide reductase, *J. Chem. Soc., Perkin Trans. 1*, 395, 1995.

412. Yue, X., Li, Y., and Sun, Y., Studies on the stereo- and enantioselective synthesis of 11,12-epoxysarcophytol A, 11,12-epoxysarcophytol A acetate and 11,12-epoxycembrene C. Synthesis of 5,9,13-trimethyl-2-isopropyl-12(*S*),12(*S*)-epoxy-14-bromo-2(*Z*),4(*E*),8(*E*)-tetradecatrienenitrile, *Bull. Soc. Chim. Belg.*, 104, 509, 1995.

413. Takayanagi, H., Z-Selective formation of trisubstituted α,β-unsaturated nitrile by the Horner–Emmons reaction, *Tetrahedron Lett.*, 35, 1581, 1994.

414. Bleisch, T.J., Ornstein, P.L., Allen, N.K., Wright, R.A., Lodge, D., and Schoepp, D.D., Structure–activity studies of aryl-spaced decahydroisoquinoline-3-carboxylic acid AMPA receptor antagonists, *Bioorg. Med. Chem. Lett.*, 7, 1161, 1997.

415. Takayanagi, H., Sugiyama, S., and Morinaka, Y., A ketal Claisen rearrangement for α-ketol isoprene unit elongation. Application to a practical synthesis of sarcophytol A intermediate, *J. Chem. Soc., Perkin Trans. 1*, 751, 1995.

416. Zoretic, P.A., Chen, Z., Zhang, Y., and Ribeiro, A.A., A radical prototype to steroids. Synthesis of *d,l*-5α-D-homoandrostane-4α-methyl-3,17*a*-dione, *Tetrahedron Lett.*, 37, 7909, 1996.

417. Zoretic, P.A., Zhang, Y., Fang, H., Ribeiro, A.A., and Dubay, G., Advanced tetracycles in a stereoselective approach to *d,l*-spongiatriol and related metabolites. The use of radicals in the synthesis of angular electrophores, *J. Org. Chem.*, 63, 1162, 1998.

418. Zoretic, P.A., Fang, H., and Ribeiro, A.A., Application of a radical methodology toward the synthesis of *d,l*-5α-pregnanes and related steroids. A stereoselective radical cascade approach, *J. Org. Chem.*, 63, 7213, 1998.

419. Geze, M., Blanchard, P., Fourrey, J.L., and Robert-Gero, M., Synthesis of sinefungin and its C-6′ epimer, *J. Am. Chem. Soc.*, 105, 7638, 1983.

420. Blanchard, P., Dodic, N., Fourrey, J.-L., Lawrence, F., Mouna, A.M., and Robert-Gero, M., Synthesis and biological activity of sinefungin analogues, *J. Med. Chem.*, 34, 2798, 1991.

421. Jackmann, L.M., Rüegg, R., Ryser, G., von Planta, C., Gloor, U., Mayer, H., Schudel, P., Kofler, M., and Isler, O., On the chemistry of the vitamin K. Part 2. Total synthesis of *trans*- and *cis*-(7′*R*,11′*R*)phylloquinone and related compounds, *Helv. Chim. Acta*, 48, 1332, 1965.

422. Seitz, G., and Köhler, K., Cyanomethylene and nitromethylene derivatives of squaric acid, *Synthesis*, 216, 1986.

423. Blanchette, M.A., Choy, W., Davis, J.T., Essenfeld, A.P., Masamune, S., Roush, W.R., and Sakai, T., Horner–Wadsworth–Emmons reaction. Use of lithium chloride and an amine for base sensitive compounds, *Tetrahedron Lett.*, 25, 2183, 1984.

424. Kumamoto, T., and Koga, K., Enantioselective Horner–Wadsworth–Emmons reaction using chiral lithium 2-aminoalkoxides, *Chem. Pharm. Bull.*, 45, 753, 1997.

425. Piancatelli, G., and Scettri, A., Synthesis of a C_{23} spiroketalic steroidal sapogenin, *Gazz. Chim. Ital.*, 104, 343, 1974.

426. Heissler, D., and Riehl, J.-J., Synthesis with benzenesulfenyl chloride. On the structure of a $C_{12}H_{18}$ hydrocarbon from East Indian sandalwood oil, *Tetrahedron Lett.*, 21, 4711, 1980.

427. Carnmalm, B., De Paulis, T., Högberg, T., Johansson, L., Persson, M.-L., Thorberg, S.-O., and Ulff, B., Synthesis of the antidepressant zimelidine and related 3-(4-bromophenyl)-3-(3-pyridyl)ally-lamines. Correlation of their configurations, *Acta Chem. Scand., Ser. B*, 36, 91, 1982.

428. Heissler, D., and Ladenburger, C., Synthesis of (+)-tricyclohexaprenol, a possible precursor of a family of tricyclic geoterpanes, and synthesis of an isomer, *Tetrahedron*, 44, 2513, 1988.

429. Heissler, D., Jenn, T., and Nagano, H., A radical-based synthesis of (Z,Z)-tricyclohexaprenol, *Tetrahedron Lett.*, 32, 7587, 1991.

430. Geribaldi, S., and Rouillard, M., Wittig–Horner reaction between phosphonates and β-substituted cyclohexenones, *Tetrahedron*, 47, 993, 1991.

431. Jenn, T., and Heissler, D., Synthesis of tricyclopolyprenols via a radical addition and a stereoselective elimination. Part 1. Methodology, *Tetrahedron*, 54, 97, 1998.

432. Guillaume, M., Janousek, Z., and Viehe, H.G., New trifluoromethylated pyridazines by reductive cyclization of vinyldiazomethanes bearing a carbonyl group, *Synthesis*, 920, 1995.

433. D'Incan, E., Phase transfer catalysis and extraction by ion pairs. Stereoselectivity of the Horner–Emmons reaction, *Tetrahedron*, 33, 951, 1977.

434. Zhdanov, Y.A., Uzlova, L.A., and Maksimushkina, L.M., Aldehydo-L-xylose in a Horner reaction with phase-transfer catalysis, *Zh. Obshch. Khim.*, 53, 484, 1983; *J. Gen. Chem. USSR (Engl. Transl.)*, 53, 426, 1983.

435. Sarmah, C.S., and Kataky, J. C.S., A novel synthesis of substituted 1,6-methano[10]annulenes using a phase transfer catalyst, *Indian J. Chem., Sect. B*, 32, 1149, 1993.

436. Leeson, P.D., Ellis, D., Emmett, J.C., Shah, V.P., Showell, G.A., and Underwood, A.H., Thyroid hormone analogues. Synthesis of 3'-substituted 3,5-diiodo-L-thyronines and quantitative structure-activity studies of *in vivo* thyromimetic activities in rat liver and heart, *J. Med. Chem.*, 31, 37, 1988.

437. Yokohama, S., Miwa, T., Aibara, S., Fujiwara, H., Matsumoto, H., Nakayama, K., Iwamoto, T., Mori, M., Moroi, R., Tsukada, W., and Isoda, S., Synthesis and antiallergy activity of [1,3,4]thiadiazolo[3,2-*a*]-1,2,3-triazolo[4,5-*d*]pyrimidin-9(3*H*)-one derivatives. Part 2. 6-Alkyl- and 6-cycloalkylalkyl derivatives, *Chem. Pharm. Bull.*, 40, 2391, 1992.

438. Griffith, R.K., and DiPietro, R.A., Improved syntheses of vinyl imidazoles, *Synth. Commun.*, 16, 1761, 1986.

439. Weber, K., Liechti, P., Meyer, H.R., and Siegrist, A.E., Use of *bis*-stilbene compounds as fluorescent whitener for organic textile materials, *Ciba-Geigy A.-G.*, Swiss Patent Appl. CH 566420, 1975; *Chem. Abstr.*, 83, 195246, 1975.

440. Sinisterra, J.V., Mouloungui, Z., Delmas, M., and Gaset, A., Barium hydroxide as catalyst in organic reactions. Part 5. Application in the Horner reaction under solid–liquid phase-transfer conditions, *Synthesis*, 1097, 1985.

441. Villieras, J., Rambaud, M., and Kirschleger, B., Wittig–Horner reaction in heterogeneous media. Part 4. Comparated performances of weak bases as K_2CO_3 and $KHCO_3$ in water or anhydrous solvents, *Phosphorus Sulfur*, 14, 385, 1983.

442. Etemad-Moghadam, G., and Seyden-Penne, J., Highly stereoselective synthesis of *E*-α,β-unsaturated nitriles from diphenyl cyanomethyl phosphine oxide, *Synth. Commun.*, 14, 565, 1984.

443. Graff, M., Dilaimi, A.A., Seguineau, P., Rambaud, M., and Villieras, J., The Wittig–Horner reaction in heterogeneous media. Part 9. *Bis*-aldolization of phosphonates from aliphatic aldehydes weakly basic aqueous media. Synthesis of functionnalized cyclenols, *Tetrahedron Lett.*, 27, 1577, 1986.

444. Seguineau, P., and Villieras, J., The Wittig–Horner reaction in heterogeneous media. Part 10. Synthesis of α-deuterated functional olefins using potassium carbonate with deuterium oxide, *Tetrahedron Lett.*, 29, 477, 1988.

445. Baldwin, J.S., Adlington, R.M., Lowe, C., O'Neil, I.A., Sanders, G.L., Schoefield, C.J., and Sweeney, G.B., Reactions of a glycidyl radical equivalent with 2-functionalised allyl stannanes, *J. Chem. Soc., Chem. Commun.*, 1030, 1988.

446. Petit, S., Nallet, J.P., Guillard, M., Dreux, J., Chermat, R., Poncelet, M., Bulach, C., and Simon, P., Synthesis and psychotropic activity of 3,4-diarylpiperidin-2-ones. Structure–activity relationship, *Eur. J. Med. Chem.*, 25, 641, 1990.

447. Csuk, R., Höring, U., and Schaade, M., Chain extension of aldonolactones by samarium iodide mediated Dreiding–Schmidt reactions and samarium assisted Imamoto reactions, *Tetrahedron*, 52, 9759, 1996.

448. Jordis, U., Grohmann, F., and Küenburg, B., Optimized synthesis of some γ,γ-disubstituted allylamines, *Org. Prep. Proced. Int.*, 29, 549, 1997.

449. Furet, P., Gay, B., Caravatti, G., Garcia-Echeverria, C., Rahuel, J., Schoepfer, J., and Fretz, H., Structure-based design and synthesis of high affinity tripeptide ligands of the Grb2-SH2 domain, *J. Med. Chem.*, 41, 3442, 1998.

450. Schoepfer, J., Fretz, H., Gay, B., Furet, P., Garcia-Echeverria, C., End, N., and Caravatti, G., Highly potent inhibitors of the Grb2-SH2 Domain, *Bioorg. Med. Chem. Lett.*, 9, 221, 1999.

451. Contreras, J.-M., Rival, Y.M., Chayer, S., Bourguignon, J.-J., and Wermuth, C.G., Aminopyridazines as acetylcholinesterase inhibitors, *J. Med. Chem.*, 42, 730, 1999.

452. Lassalle, G., Gree, R., and Toupet, L., Stereoselective syntheses of *cis*- and *trans*-divinyl pyrrolidines, *Bull. Soc. Chim. Fr.*, 453, 1990.

453. Buchanan, J.G., Craven, D.A., Wightman, R.H., and Harnden, M.R., *C*-Nucleoside studies. Part 21. Synthesis of some hydroxyalkylated pyrrolo- and thieno-[3,2-d]pyrimidines related to known antiviral acyclonucleosides, *J. Chem. Soc., Perkin Trans. 1*, 195, 1991.

454. Petrosyan, V.A., and Niyazymbetov, M.E., A new electrochemical method of preparation of olefins, *Izv. Akad. Nauk SSSR, Ser. Khim.*, 1151, 1991; *Bull. Acad. Sci. USSR, Div. Chem. Sci. (Engl. Transl.)*, 1032, 1991.

455. Pardoen, J.A., Mulder, P.P.J., van der Berg, E.M.M. and Lugtenburg, J., Synthesis of 8-, 9-, 12-, and 13-mono-^{13}C-retinal, *Can. J. Chem.*, 63, 1431, 1985.

456. Courtin, J.M.L., 't Lam, G.K., Peters, A.J.M., and Lugtenburg, J., Synthesis of 5-, 6-, 7- and 18-mono-^{13}C-labelled retinals, *Recl. Trav. Chim. Pays-Bas*, 104, 281, 1985.

457. Wada, A., Sakai, M., Kinumi, T., Tsujimoto, K., Yamauchi, M., and Ito, M., Conformational study of retinochrome chromophore. Synthesis of 8,18-ethanoretinal and a new retinochrome analog, *J. Org. Chem.*, 59, 6922, 1994.

458. Trehan, A., and Liu, R.S.H., All-*cis*-retinal and 7-*cis*,9-*cis*,11-*cis*-retinal, *Tetrahedron Lett.*, 29, 419, 1988.

459. Corey, E.J., and Watt, D.S., A total synthesis of (±)-α- and (±)-β-copaenes and ylangenes, *J. Am. Chem. Soc.*, 95, 2303, 1973.

460. Freerksen, R.W., Raggio, M.L., Thoms, C.A., and Watt, D.S., Hydroxylation of α,β-unsaturated nitriles and esters in steroid systems, *J. Org. Chem.*, 44, 702, 1979.

461. Nagaoka, H., Rutsch, W., Schmid, G., Iio, H., Johnson, M.R., and Kishi, Y., Total synthesis of rifamycins. Part 1. Stereocontrolled synthesis of the aliphatic building block, *J. Am. Chem. Soc.*, 102, 7962, 1980.

462. Tomizawa, K., Watt, D.S., and Lenz, G.R., Use of β-chloropropionaldehyde as an acrolein equivalent in the Wadsworth–Emmons modification of the Wittig reaction, *Synthesis*, 887, 1985.

463. Wiseman, J.R., and Lee, S.Y., Synthesis of anatoxin A, *J. Org. Chem.*, 51, 2485, 1986.

464. Tyvorskii, V.I., Savchenko, A.I., and Kukharev, A.S., Synthesis of 5-hydroxymethyl-3-methylfuran-2(5*H*)-one, *Zh. Org. Khim.*, 31, 920, 1995; *Russ. J. Org. Chem. (Engl. Transl.)*, 31, 853, 1995.

465. Tan, Q., Lou, J., Borhan, B., Karnaukhova, E., Berova, N., and Nakanishi, K., Absolute sense of twist of the C12–C13 bond of the retinal chromophore in bovine rhodopsin based on exciton-coupled CD spectra of 11,12-dihydroretinal analogues, *Angew. Chem., Int. Ed. Engl.*, 36, 2089, 1997.

466. Tan, Q., Nakanishi, K., and Crouch, R.K., Mechanism of transient dark activity of 13-desmethylretinal/rodopsin complex, *J. Am. Chem. Soc.*, 120, 12357, 1998.

467. Buschauer, A., Friese-Kimmel, A., Baumann, G., and Schunack, W., Synthesis and histamine H$_2$ agonistic activity of arpromidine analogues. Replacement of the pheniramine-like moiety by non-heterocyclic groups, *Eur. J. Med. Chem.*, 27, 321, 1992.

468. Tokoroyama, T., Tsukamoto, M., Asada, T., and Iio, H., One-pot stereospecific construction of *cis*-clerodane skeleton by means of doubly stereocontrolled cyclization. Total synthesis of linaridial, *Tetrahedron Lett.*, 28, 6645, 1987.

469. Mouna, A.M., Blanchard, P., Fourrey, J.-L., and Robert-Gero, M., Synthesis of adenine nucleosides related to sinefungin, *Tetrahedron Lett.*, 31, 7003, 1990.

470. Dinizo, S.E., Freerksen, R.W., Pabst, W.E., and Watt, D.S., A one-carbon homologation of carbonyl compounds to carboxylic acids, esters, and amides, *J. Am. Chem. Soc.*, 99, 182, 1977.

471. Stévenart-De Mesmaeker, N., Merenyi, R., and Viehe, H.G., Synthesis of dienes with 1,1-captodative substitution, *Tetrahedron Lett.*, 28, 2591, 1987.

472. Clive, D.L.J., and Etkin, N., Synthesis of α-amino acids by addition of putative azido radicals to α-methoxy acrylonitriles derived from aldehydes and ketones, *Tetrahedron Lett.*, 35, 2459, 1994.

473. Eicher, T., and Stapperfenne, U., Reaction of triafulvenes with isonitriles. A simple synthesis of diphenyl-substituted functionalized cyclobutene derivatives and related products, *Synthesis*, 619, 1987.

474. Shen, W., Fakhoury, S., Donner, G., Henry, K., Lee, J., Zhang, H., Cohen, J., Warner, R., Saeed, B., Cherian, S., Tahir, S., Kovar, P., Bauch, J., Ng, S.-C., Marsh, K., Sham, H., and Rosenberg, S., Potent inhibitors of protein farnesyltransferase. Heteroarenes as cysteine replacements, *Bioorg. Med. Chem. Lett.*, 9, 703, 1999.

475. Costisella, B., and Gross, H., α-Substituted phosphonates. Part 38. 1-Dimethylamino-1-cyanomethanephosphonic acid diethyl ester, a new approach to the preparation of carboxylic acids, 1-cyanoenamines, and homoenolates, *Tetrahedron*, 38, 139, 1982.

476. Costisella, B., and Gross, H., α-Substituted phosphonates. Part 40. 1-Trimethylammonium-1-diethylphosphono-1-cyanomethylid, a stable *N*-ylid, *J. Prakt. Chem.*, 324, 545, 1982.

477. Theil, F., Costisella, B., Mahrwald, R., Gross, H., Schick, H., and Schwarz, S., Prostaglandins and prostaglandin intermediates. Part 17. Synthesis of the main metabolite of the $PGF_{2\alpha}$ analogue cloprostenol, *J. Prakt. Chem.*, 328, 435, 1986.

478. Costisella, B., and Gross, H., α-Substituted phosphonates. Part 51. Diethyl 1-cyano-1-dimethylaminomethylphosphonate lithium — an instant-Horner-reagent, *Z. Chem.*, 27, 143, 1987.

479. Kovalev, B.G., Vaskan, R.N., and Shamshurin, A.A., Keto aldehydes. Part 6. Nitriles of unsaturated aldehydo and keto acids, *Zh. Org. Khim.*, 5, 450, 1969; *J. Org. Chem. USSR (Engl. Transl.)*, 5, 437, 1969.

480. Zimmermann, H.E., and Klun, R.T., Mechanistic and exploratory organic photochemistry. Part 112. The di-π-methane rearrangement of systems with simple vinyl moieties, *Tetrahedron*, 34, 1775, 1978.

481. Field, D.J., and Jones, D.W., *o*-Quinonoid compounds. Part 16. 1,5-Shift of vinyl groups in 1,3-dimethylindenes. Product studies and migratory aptitudes of substituted vinyl groups, *J. Chem. Soc., Perkin Trans. 1*, 714, 1980.

482. Courtheyn, D., Verhé, R., De Kimpe, N., De Buyck, L., and Schamp, N., Synthesis of electrophilic allyl dichlorides, *J. Org. Chem.*, 46, 3226, 1981.

483. Boeckman, R.K., Jr., and Ko, S.S., Stereochemical control in the intramolecular Diels–Alder reaction. Part 2. Structural and electronic effects on reactivity and selectivity, *J. Am. Chem. Soc.*, 104, 1033, 1982.

484. Smith, P., Brown, L., Boutagy, J., and Thomas, R., Cardenolide analogues. Part 14. Synthesis and biological activity of glucosides of C17β-modified derivatives of digitoxigenin, *J. Med. Chem.*, 25, 1222, 1982.

485. Tyvorskii, V.I., Tishchenko, I.G., Nakhar, P., and Benitsevich, M.S., PO-olefination (Wittig–Horner reaction) of formyloxiranes. Synthesis and transformation of 4,5-epoxy-2-pentenoic acid derivatives, *Zh. Org. Khim.*, 18, 540, 1982; *J. Org. Chem. USSR (Engl. Transl.)*, 18, 472, 1982.

486. Larsen, R.O., and Aksnes, G., Kinetic study of the Horner reaction. Part 1. The effect of the phosphoryl substituents on the stereochemistry of the Horner reaction, *Phosphorus Sulfur*, 15, 219, 1983.

487. Ratier, M., Pereyre, M., Davies, A.G., and Sutcliffe, R., Regioselectivity in the ring opening of 2-alkylcyclopropylmethyl radicals. The effect of electronegative substituents, *J. Chem. Soc., Perkin Trans. 2*, 1907, 1984.

488. Hensel, M.J., and Fuchs, P.L., A highly selective synthesis of a chiral γ-amino trisubstituted Z-vinyl ester, *Synth. Commun.*, 16, 1285, 1986.

489. Majetich, G., Desmond, R.W., Jr., and Soria, J.J., Allylsilane-initiated cyclopentane annulations, *J. Org. Chem.*, 51, 1753, 1986.

490. Park, S.-U., Chung, S.-K., and Newcomb, M., Acceptor, donor, and captodative stabilization in transition states of 5-hexen-1-yl radical cyclizations, *J. Am. Chem. Soc.*, 108, 240, 1986.

491. Schubert, G., Schönecker, B., Wunderwald, M., and Ponsold, K., Synthesis of 17β-configurated acrylic acid derivatives of 19-nor steroids, *Pharmazie*, 41, 469, 1986.

492. Ishii, K., Abe, M., and Sakamoto, M., Photochemical reactions of nitrile compounds. Part 2. Photochemistry of an α,β-unsaturated γ,δ-epoxy nitrile and γ,δ-cyclopropyl nitrile, *J. Chem. Soc., Perkin Trans. 1,* 1937, 1987.

493. Sato, T., Okazaki, H., Otera, J., and Nozaki, H., A new strategy for 1,4- and 1,4,7-polycarbonyl compounds, *J. Am. Chem. Soc.,* 110, 5209, 1988.

494. Rodriguez, J., Brun, P., and Waegell, B., Isomerisation of 1,5-functionnalised dienes by pentacarbonyliron, *J. Organomet. Chem.,* 359, 343, 1989.

495. Park, S.-U., Varick, T.R., and Newcomb, M., Acceleration of the 4-exo radical cyclization to a synthetically useful rate. Cyclization of the 2,2-dimethyl-5-cyano-4-pentenyl radical, *Tetrahedron Lett.,* 31, 2975, 1990.

496. Chou, W.-N., and White, J.B., The use of *trans*-2,3-divinyl epoxides as precursors to carbonyl ylides, *Tetrahedron Lett.,* 32, 7637, 1991.

497. Pouzar, V., Chodounska, H., Cerny, I., and Drasar, P., Derivatives of 5α-androstan-3α- and 3β-ol with acrylonitrile side chain, *Collect. Czech. Chem. Commun.,* 56, 2906, 1991.

498. Angle, S.R., and Rainier, J.D., Reductive cyclization of quinone methides, *J. Org. Chem.,* 57, 6883, 1992.

499. Chou, W.-N., White, J.B., and Smith, W.B., Control over the relative stereochemistry at C4 and C5 of 4,5-dihydrooxepins through the Cope rearrangement of 2,3-divinyl epoxides and a conformational analysis of this ring system, *J. Am. Chem. Soc.,* 114, 4658, 1992.

500. Newcomb, M., Varick, T.R., Ha, C., Manek, M.B., and Yue, X., Picosecond radical kinetics. Rate constants for reaction of benzeneselenol with primary alkyl radicals and calibration of the 6-cyano-5-hexenyl radical cyclization, *J. Am. Chem. Soc.,* 114, 8158, 1992.

501. Lee, G.H., Choi, E.B., Lee, E., and Pak, C.S., Reductive cyclization of ketones tethered to activated olefins mediated by magnesium in methanol, *J. Org. Chem.,* 59, 1428, 1994.

502. Takemoto, Y., Ohra, T., Yonetoku, Y., Nishimine, K., and Iwata, C., Novel intramolecular Michael addition of organomercury halides mediated by a Lewis acid and halide anion, *J. Chem. Soc., Chem. Commun.,* 81, 1994.

503. Batsanov, A.S., Begley, M.J., Fletcher, R.J., Murphy, J.A., and Sherburn, M.S., Stereocontrol in cyclisation of dioxolanyl radicals, *J. Chem. Soc., Perkin Trans. 1,* 1281, 1995.

504. Grissom, J.W., and Klingberg, D., Tandem enyne allene-radical cyclization via base-catalyzed isomerization of enediyne sulfones, *Tetrahedron Lett.,* 36, 6607, 1995.

505. Lanier, M., and Pastor, R., Stereoselective synthesis of Z and E 3-F-alkyl 2-propenoates and derivatives, *J. Fluorine Chem.,* 75, 35, 1995.

506. Lautens, M., Edwards, L.G., Tam, W., and Lough, A.J., Nickel-catalyzed [2π+2π+2π] (homo-Diels–Alder) and [2π+2π] cycloadditions of bicyclo[2.2.1]hepta-2,5-dienes, *J. Am. Chem. Soc.,* 117, 10276, 1995.

507. Nuss, J.M., Chinn, J.P., and Murphy, M.M., Substituent control of excited state reactivity. The intramolecular *ortho* arene-olefin photocycloaddition, *J. Am. Chem. Soc.,* 117, 6801, 1995.

508. Grissom, J.W., Klingberg, D., Huang, D., and Slattery, B.J., Tandem enyne allene-radical cyclization. Low-temperature approaches to benz[*e*]indene and indene compounds, *J. Org. Chem.,* 62, 603, 1997.

509. Boger, D.L., Kochanny, M.J., Cai, H., Wyatt, D., Kitos, P.A., Warren, M.S., Ramcharan, J., Gooljarsingh, L.T., and Benkovic, S.J., Design, synthesis, and evaluation of potential GAR and AICAR transformylase inhibitors, *Bioorg. Med. Chem.,* 6, 643, 1998.

510. Itadani, S., Hashimoto, K., and Shirahama, H., Construction of 3,4-disubstituted tetrahydrofuran ring by radical cyclization. A model study for kainoid synthesis, *Heterocycles,* 49, 105, 1998.

511. Wadsworth, W.S., Jr., and Emmons, W.D., The utility of phosphonate carbanions in olefin synthesis, *J. Am. Chem. Soc.,* 83, 1733, 1961.

512. Harayama, T., Ohtani, M., Oki, M., and Inubushi, Y., Substituent effect on the addition course of the Diels–Alder reaction 2-methyl-5-substituted-1,4-benzoquinones with butadiene, *Chem. Pharm. Bull.,* 21, 25, 1973.

513. Garanti, L., and Zecchi, G., Thermochemical behavior of *o*-azidocinnamonitriles, *J. Org. Chem.,* 45, 4767, 1980.

514. Chen, S.C., and MacTaggart, J.M., Abscisic acid analogs with a geometrically rigid conjugated acid side-chain, *Agric. Biol. Chem.,* 50, 1097, 1986.

515. Angeletti, E., Tundo, P., and Venturello, P., Gas–liquid phase-transfer catalysis. Wittig–Horner reaction in heterogeneous conditions, *J. Chem. Soc., Perkin Trans. 1,* 713, 1987.

516. Parker, K.A., and Fokas, D., Stereochemistry of radical cyclizations to side-chain olefinic bonds. An approach to control of the C-9 center of morphine, *J. Org. Chem.*, 59, 3927, 1994.

517. Parker, K.A., and Fokas, D., The radical cyclization approach to morphine. Models for highly oxygenated ring-III synthons, *J. Org. Chem.*, 59, 3933, 1994.

518. Gopal, V.R., Reddy, A.M., and Rao, V.J., Wavelength dependent *trans* to *cis* and quantum chain isomerizations of anthrylethylene derivatives, *J. Org. Chem.*, 60, 7966, 1995.

519. Stamm, H., and Baumann, T., Synthesis of the first α-ylidene-γ-amidobutyronitriles, *Pharmazie*, 52, 441, 1997.

520. Tsuji, K., Nakamura, K., Konishi, N., Tojo, T., Ochi, T., Senoh, H., and Matsuo, M., Studies on anti-inflammatory agents. Part 4. Synthesis and pharmacological properties of 1,5-diarylpyrazoles and related derivatives, *Chem. Pharm. Bull.*, 45, 987, 1997.

521. Lyngsø, L.O., and Nielsen, J., Solid-phase synthesis of 3-amino-2-pyrazolines, *Tetrahedron Lett.*, 39, 5845, 1998.

522. Brooks, S., Sainsbury, M., and Weerasinge, D.K., The synthesis of 3-vinylindoles and 11*H*-5-cyanobenzo[*a*]carbazole, *Tetrahedron*, 38, 3019, 1982.

523. Glazer, E.A., and Chappel, L.R., Pyridoquinoxaline *N*-oxides. Part 1. A new class of antitrichomonal agents, *J. Med. Chem.*, 25, 766, 1982.

524. Kavadias, G., Luh, B., and Saintonge, R., Synthesis of 4,5-disubstituted imidazoles, *Can. J. Chem.*, 60, 723, 1982.

525. Singh, R., Jain, P.C., and Anand, N., Potential anticancer agents. Synthesis of some substituted pyrrolo[2,1-*c*][1,4]benzodiazepines, *Indian J. Chem., Sect. B*, 21B, 225, 1982.

526. Ahmed, F.R., and Toube, T.P., Pyrrolylpolyenes. Part 7. Synthesis and spectra of pyrrolylpolyenes and 3-chloropyrrolylpolyenes, *J. Chem. Res. (M)*, 3601, 1986.

527. Mercey, J.M., and Toube, T.P., Pyrrolylpolyenes. Part 8. *N*-Alkylation of pyrroles by phosphorus esters, *J. Chem. Res. (M)*, 680, 1987.

528. Chan, W.K., Huang, F.-C., Morrissette, M.M., Warus, J.D., Moriarty, K.J., Galemmo, R.A., Dankulich, W.D., Poli, G., and Sutherland, C.A., Structure–activity relationships study of two series of leukotriene B$_4$ antagonists. Novel indolyl and naphthyl compounds substituted with a 2-[methyl(2-phenethyl)amino]-2-oxoethyl side chain, *J. Med. Chem.*, 39, 3756, 1996.

529. Mercey, J.M., and Toube, T.P., Pyrrolylpolyenes. Part 11. The synthesis of 3-substituted 2-vinylpyrroles, *J. Chem. Res. (M)*, 582, 1996.

530. Sundberg, R.J., Biswas, S., Murthi, K.K., Rowe, D., McCall, J.W., and Dzimianski, M.T., *Bis*-cationic heteroaromatics as macrofilaricides. Synthesis of *bis*-amidine and *bis*-guanylhydrazone derivatives of substituted imidazo[1,2-α]pyridines, *J. Med. Chem.*, 41, 4317, 1998.

531. Yanovskaya, L.A., Stepanova, R.N., and Kucherov, V.F., Chemistry of polyene and polyacetylenic compounds. Part 23. Synthesis of some unsymmetrically α,ω-disubstituted polyenes, *Izv. Akad. Nauk SSSR, Ser. Khim.*, 1334, 1967; *Bull. Acad. Sci. USSR, Div. Chem. Sci. (Engl. Transl.)*, 1282, 1967.

532. Harayama, T., Takatani, M., and Inubushi, Y., Synthetic studies on 8-deoxyserratinine type alkaloids. Selective cyclization of the 1,2-cyclohexanediacetaldehyde derivative by intramolecular aldol condensation, *Chem. Pharm. Bull.*, 28, 1276, 1980.

533. Nesterov, N.I., Belyaev, N.N., Stadnichuk, M.D., Mingaleva, K.S., and Sigolaev, Y.F., Stereoselective synthesis and some properties of functional derivatives of 1,3-dienic hydrocarbons and sila hydrocarbons, *Zh. Obshch. Khim.*, 50, 76, 1980; *J. Gen. Chem. USSR (Engl. Transl.)*, 50, 63, 1980.

534. Gawronski, J.K., and Walborsky, H.M., Circular dichroism of linearly conjugated chromophores, *J. Org. Chem.*, 51, 2863, 1986.

535. Baumeler, A., and Eugster, C.H., Synthesis of enantiomerically pure mimulaxanthin and of its (9Z,9′Z)- and (15Z)-isomers, *Helv. Chim. Acta*, 74, 469, 1991.

536. de Raadt, A., Griengl, H., Klempier, N., and Stütz, A.E., A mild and simple enzymatic conversion of aldono- and aldurononitriles into the corresponding amides and/or carboxylic acids, *J. Org. Chem.*, 58, 3179, 1993.

537. Nuretdinov, I.A., Karaseva, I.P., Gubskaya, V.P., and Shakirov, I.C., Functionally substituted dienes based on α-pinene, *Izv. Akad. Nauk, Ser. Khim.*, 1684, 1994; *Russ. Chem. Bull. (Engl. Transl.)*, 1779, 1994.

538. Voigt, K., von Zezschwitz, P., Rosauer, K., Lansky, A., Adams, A., and Reiser, O., The twofold Heck reaction on 1,2-dihalocycloalkenes and subsequent 6π-electrocyclization of the resulting (E,Z,E)-1,3,5-hexatrienes. A new formal [2 + 2 + 2]-assembly of six-membered rings, *Eur. J. Org. Chem.*, 1521, 1998.

539. Raap, J., Nieuwenhuis, S., Creemers, A., Hexspoor, S., Kragl, U., and Lugtenburg, J., Synthesis of isotopically labelled L-phenylalanine and L-tyrosine, *Eur. J. Org. Chem.*, 2609, 1999.

540. Stilz, W., and Pommer, H., 1,4-Distyrylbenzenes, *BASF A.-G.*, German Patent Appl. DE 1112072, 1961; *Chem. Abstr.*, 56, 2378d, 1962.

541. Grinev, G.V., Dombrovskii, V.A., and Yanovskaya, L.A., Synthesis and some properties of vinylogs of *p-bis*(alkoxycarbonyl)-, *p*-dicyano-, and *p*-diaroylbenzenes, *Izv. Akad. Nauk SSSR, Ser. Khim.*, 635, 1972; *Bull. Acad. Sci. USSR, Div. Chem. Sci. (Engl. Transl.)*, 598, 1972.

542. Weber, K., Liechti, P., Meyer, H.R., and Siegrist, A.E., *Bis*-stilbene fluorescent whiteners, *Ciba-Geigy A.-G.*, Swiss Patent Appl. CH 560736, 1975; *Chem. Abstr.*, 83, 61745, 1975.

543. Fleck, F., and Heller, J., Divinyl stilbenes as fluorescent whiteners, *Sandoz A.-G.*, U.S. Patent Appl. US 4108887, 1978; *Chem. Abstr.*, 90, 88756, 1979.

544. Villieras, J., Rambaud, M., and Graff, M., The Wittig–Horner reaction in heterogeneous media. Part 8. Cyclization during the aldolisation step from aqueous glutaraldehyde, *Synth. Commun.*, 16, 149, 1986.

545. Wunz, T.P., Dorr, R.T., Alberts, D.S., Tunget, C.L., Einspahr, J., Milton, S., and Remers, W.A., New antitumor agents containing the anthracene nucleus, *J. Med. Chem.*, 30, 1313, 1987.

546. Magnus, P., Danikiewicz, W., Katoh, T., Huffman, J.C., and Folting, K., Synthesis of helical poly-β-pyrroles. Multiple atropisomerism resulting in helical enantiomorphic conformations, *J. Am. Chem. Soc.*, 112, 2465, 1990.

547. Deussen, H.J., Hendrickx, E., Boutton, C., Krog, D., Clays, K., Bechgaard, K., Persoons, A., and Bjørnholm, T., Novel chiral *bis*-dipolar 6,6′-disubstituted binaphthol derivatives for second-order nonlinear optics. Synthesis and linear and nonlinear optical properties, *J. Am. Chem. Soc.*, 118, 6841, 1996.

548. Martin, S.F., Assercq, J.M., Austin, R.E., Dantanarayana, A.P., Fishpaugh, J.R., Gluchowski, C., Guinn, D.E., Hartmann, M., Tanaka, T., Wagner, R., and White, J.B., Facile access to the ABC ring system of the taxane diterpenes via anionic oxy-Cope rearrangements, *Tetrahedron*, 51, 3455, 1995.

549. Gijsen, H.J.M., and Wong, C.-H., Synthesis of a cyclitol via a tandem enzymatic aldol-intramolecular Horner–Wadsworth–Emmons reaction, *Tetrahedron Lett.*, 36, 7057, 1995.

550. Kovalev, B.G., Al'tmark, E.M., and Lavrinenko, E.S., Unsymmetrical chain lengthening of dimethyl diketones in the Wittig reaction. Synthesis of esters of α,β-unsaturated keto acids and ketonitriles, *Zh. Org. Khim.*, 6, 2187, 1970; *J. Org. Chem. USSR (Engl. Transl.)*, 6, 2196, 1970.

551. Franke, A., Mattern, G., and Traber, W., Synthetical juvenile hormone. Part 4. *para*-Substituted 2-methyl-5-phenyl-penten(1)-carboxylic acid derivatives, *Helv. Chim. Acta*, 58, 293, 1975.

552. Hejno, K., and Sorm, F., Synthesis of 8-oxa analogues of acyclic juvenoidal substances, *Collect. Czech. Chem. Commun.*, 41, 151, 1976.

553. McFadden, H.G., Harris, R.L.N., and Jenkins, C.L.D., Potential inhibitors of phosphoenolpyruvate carboxylase. Part 2. Phosphonic acid substrate analogues derived from reactions of trialkyl phosphites with halomethacrylates, *Aust. J. Chem.*, 42, 301, 1989.

554. Yang, C.-C., and Fang, J.-M., Free-radical cyclisations of 2-aminoalka-2,5-dienenitriles, *J. Chem. Soc., Perkin Trans. 1*, 879, 1995.

555. Hori, M., Kataoka, T., Shimizu, H., Imai, E., Iwata, N., Kawamura, N., Kurono, M., Nakano, K., and Kido, M., Synthesis and analgesic activity of novel heterocycles, [1]benzothiopyrano[3,4-*b*]pyrrole derivatives, *Chem. Pharm. Bull.*, 37, 1282, 1989.

556. Ohno, M., Ito, Y., Arita, M., Shibata, T., Adachi, K., and Sawai, H., Synthetic studies on biologically active natural products by a chemicoenzymatic approach. Enantioselective synthesis of *C*- and *N*-nucleosides, showdomycin, 6-azapseudouridine and cordycepin, *Tetrahedron*, 40, 145, 1984.

557. Ito, Y., Shibata, T., Arita, M., Sawai, H., and Ohno, M., Chirally selective synthesis of sugar moiety of nucleosides by chemicoenzymatic approach: L- and D-riboses, showdomycin, and cordycepin, *J. Am. Chem. Soc.*, 103, 6739, 1981.

558. Shekhter, O.V., Tsizin, Y.S., and Pridantseva, E.A., Compounds with juvenile hormone activity. Part 8. Synthesis of 3,11-dimethyl-11-alkoxy(chloro)-2-dodecenoic acid derivatives, *Zh. Obshch. Khim.*, 45, 1186, 1975; *J. Gen. Chem. USSR (Engl. Transl.)*, 45, 1166, 1975.

559. Shekhter, O.V., and Tsizin, Y.S., Compounds with juvenile hormone activity. Part 7. Synthesis of some derivatives of 3,11-dimethyl-10,11-epoxy-2-dodecenoic acid, *Zh. Obshch. Khim.*, 45, 1180, 1975; *J. Gen. Chem. USSR (Engl. Transl.)*, 45, 1161, 1975.

560. Erickson, K.L., Markstein, J., and Kim, K., Base-induced reactions methylenecyclobutane derivatives, *J. Org. Chem.*, 36, 1024, 1971.

561. Eberbach, W., Seiler, W., and Fritz, H., Thermolysis of compounds with a geometrically fixed vinyl-oxirane unit. Stereospecific synthesis of 4,5-fused *cis*- and *trans*-2,3-dihydrofurans, *Chem. Ber.*, 113, 875, 1980.

562. Kalai, T., Szabo, Z., Jekö, J., and Hideg, K., Synthesis of new allylic nitroxides via the Wadsworth–Emmons reaction, *Org. Prep. Proced. Int.*, 28, 443, 1996.

563. Pratap, R., Gupta, R.C., and Anand, N., Comparative reactivity of carbonyl groups in 6β-methyl-*cis*-bicyclo[4.3.0]nona-3,7-dione, *Indian J. Chem., Sect. B*, 22, 731, 1983.

564. Hirano, T., Kumagai, T., Miyashi, T., Akiyama, K., and Ikegami, Y., Inversion of the ground-state spin multiplicity by electron-withdrawing groups in trimethylenemethane derivatives generated photochemically from methylenequadricyclane derivatives, *J. Org. Chem.*, 57, 876, 1992.

565. Scopes, D.I.C., Hayes, N.F., Bays, D.E., Belton, D., Brain, J., Brown, D.S., Judd, D.B., McElroy, A.B., Meerholz, C.A., Naylor, A., Hayes, A.G., Sheehan, M.J., Birch, P.J., and Tyers, M.B., New *k*-receptor agonists based upon a 2-[(alkylamino)methyl]piperidine nucleus, *J. Med. Chem.*, 35, 490, 1992.

566. Sowell, C.G., Wolin, R.L., and Little, R.D., Electroreductive cyclization reactions. Stereoselection, creation of quaternary centers in bicyclic frameworks, and a formal total synthesis of quadrone, *Tetrahedron Lett.*, 31, 485, 1990.

567. Kotera, M., Ishii, K., Tamura, O., and Sakamoto, M., 1,3-Dipolar cycloadditions of photoinduced carbonyl ylides from α,β-unsaturated β,γ-epoxy dinitriles, *J. Chem. Soc., Perkin Trans. 1*, 2353, 1994.

568. Kotera, M., Ishii, K., and Sakamoto, M., Photocyclization reactions of epoxy nitriles via carbonyl ylides. Formation of spiroketals, spiroethers, and a spirolactone, *Chem. Pharm. Bull.*, 43, 1621, 1995.

569. Kotera, M., Ishii, K., Tamura, O., and Sakamoto, M., 1,3-Dipolar cycloadditions of photoinduced carbonyl ylides. Part 2. Photoreactions of α,β-unsaturated γ,δ-epoxy dinitriles and ethyl vinyl ether, *J. Chem. Soc., Perkin Trans. 1*, 313, 1998.

570. Ishii, K., Kotera, M., Nakano, T., Zenko, T., Sakamoto, M., Iida, I., and Nishio, T., Photochemistry of δ-hydroxybutyl α,β-unsaturated γ,δ-epoxy nitriles. Formation of spiro ketals, *Liebigs Ann. Org. Bioorg. Chem.*, 19, 1995.

571. Bode, H.E., Sowell, C.G., and Little, R.D., Electrolyte-assisted stereoselection and control of cyclization *vs.* saturation in electroreductive cyclizations, *Tetrahedron Lett.*, 31, 2525, 1990.

572. Swamy, N., Addo, J.K., and Ray, R., Development of an affinity-driven cross-linker. Isolation of a vitamin D receptor associated factor, *Bioorg. Med. Chem. Lett.*, 10, 361, 2000.

573. Phillipps, G.H., and Ewan, G., Steroid derivatives, *Glaxo*, German Patent Appl. DE 2715078, 1977; *Chem. Abstr.*, 88, 38078, 1978.

574. Szabo, L., Honty, K., Töke, L., Toth, I., and Szantay, C., Investigations on the chemistry of berbans. Part 1. Synthesis of dimethoxy-despyrrolo-*b*-yohimbin, *Chem. Ber.*, 105, 3215, 1972.

575. Kubica, Z., Burski, Z., and Piatkowski, K., Stereochemistry of the carane system. Part 5. (−)-*cis*-Caran-4-one in the Wittig–Horner reaction, *Pol. J. Chem.*, 59, 827, 1985.

576. Schönholzer, P., Süss, D., Wan, T.S., and Fischli, A., Cob(I)alamin differentiating alkenes during saturation, *Helv. Chim. Acta*, 67, 669, 1984.

577. Campiani, G., Sun, L.-Q., Kozikowski, A.P., Aagaard, P., and McKinney, M., A palladium-catalyzed route to huperzine A and its analogues and their anticholinesterase activity, *J. Org. Chem.*, 58, 7660, 1993.

578. Gras, J.-L., Soto, T., and Viala, J., Chiral conformationally restricted arachidonic acid analogs based on a 1,3-dioxane core, *Tetrahedron: Asymmetry*, 10, 139, 1999.

579. Alonso, R.A., Burgey, C.S., Rao, B.V., Vite, G.D., Vollerthun, R., Zottola, M.A., and Fraser-Reid, B., Carbohydrates to carbocycles. Synthesis of the densely functionalized carbocyclic core of tetrodotoxin by radical cyclization of an anhydro sugar precursor, *J. Am. Chem. Soc.*, 115, 6666, 1993.

580. Magnus, P., and Fairhurst, R.A., Relative rates of cycloaromatization of dynemicin azabicyclo[7.3.1]enediyne core structures. An unusual change in ΔS(excit.), *J. Chem. Soc., Chem. Commun.*, 1541, 1994.

581. Magnus, P., Eisenbeis, S.A., Fairhurst, R.A., Iliadis, T., Magnus, N.A., and Parry, D., Synthetic and mechanistic studies on the azabicyclo[7.3.1]enediyne core and naphtho[2,3-*h*]quinoline portions of dynemicin A, *J. Am. Chem. Soc.*, 119, 5591, 1997.

582. Chida, N., Yamada, K., and Ogawa, S., Synthesis and absolute configuration of the naturally occurring cyano glucoside simmondsin, *J. Chem. Soc., Perkin Trans. 1*, 1131, 1992.

583. Chida, N., Yamada, K., and Ogawa, S., Total synthesis of simmondsin, *J. Chem. Soc., Chem. Commun.*, 588, 1991.

584. Ornstein, P.L., Schoepp, D.D., Arnold, M.B., Augenstein, N.K., Lodge, D., Millar, J.D., Chambers, J., Campbell, J., Paschal, J.W., Zimmerman, D.M., and Leander, J.D., 6-Substituted decahydroiso-quinoline-3-carboxylic acids as potent and selective conformationally constrained NMDA receptor antagonists, *J. Med. Chem.*, 35, 3547, 1992.

585. Bose, A.K., and Dahill, R.T., Jr., Steroids. Part 3. Transformations of steroid ketones using phosphonate carbanions, *J. Org. Chem.*, 30, 505, 1965.

586. Jamison, T.F., Shambayati, S., Crowe, W.E., and Schreiber, S.L., Tandem use of cobalt-mediated reactions to synthesize (+)-epoxydictymene, a diterpene containing a *trans*-fused 5–5 ring system, *J. Am. Chem. Soc.*, 119, 4353, 1997.

587. Jamison, T.F., Shambayati, S., Crowe, W.E., and Schreiber, S.L., Cobalt-mediated total synthesis of (+)-epoxydictymene, *J. Am. Chem. Soc.*, 116, 5505, 1994.

588. Wroble, R.R., and Watt, D.S., A synthesis of α,β-unsaturated ketones from α,β-unsaturated nitriles, *J. Org. Chem.*, 41, 2939, 1976.

589. Fischli, A., Cob(I)alamin as catalyst. Part 1. Communication. Reduction of saturated nitriles in aqueous solution, *Helv. Chim. Acta*, 61, 2560, 1978.

590. Popp, F.D., Dubois, R.G., and Casey, A.C., Diazepines. Part 6. The chemistry of 3,8-dihalo-11*H*-dibenzo[*c,f*][1,2]diazepin-11-ones, *J. Heterocycl. Chem.*, 6, 285, 1969.

591. Steiner, G., Franke, A., Hädicke, E., Lenke, D., Teschendorf, H.-J., Hofmann, H.P., Kreiskott, H., and Worstmann, W., Tricyclic epines. Novel (*E*)- and (*Z*)-11*H*-dibenz[*b,e*]azepines as potential central nervous system agents. Variation of the basic side chain, *J. Med. Chem.*, 29, 1877, 1986.

592. Heinisch, G., Holzer, W., and Huber, T., Pyridazines. Part 58. 1-Phenyl-1-pyridazinyl-2-substituted ethenes. Synthesis and configuration, *Monatsh. Chem.*, 122, 1055, 1991.

593. Take, K., Okumura, K., Tsubaki, K., Terai, T., and Shiokawa, Y., Agents for the treatment of overactive detrusor. Part 4. Synthesis and structure-activity relationships of cyclic analogues of terodiline, *Chem. Pharm. Bull.*, 41, 507, 1993.

594. Werner, L.H., Ricca, S., Rossi, A., and DeStevens, G., Imidazoline derivatives with antiarrhythmic activity, *J. Med. Chem.*, 10, 575, 1967.

595. Matsumoto, M., and Watanabe, N., Synthesis of 4-(cyanomethylidene)- and 4-(ethoxycarbonylmeth-ylidene)-4,5,6,7-tetrahydroindoles and their dehydrogenation to 4-(cyanomethyl)- and 4-(ethoxycar-bonylmethyl)indoles, *Heterocycles*, 24, 2611, 1986.

596. Matsumoto, M., Watanabe, N., and Ishida, Y., A facile synthesis of 4-(cyanomethyl)indoles and 4-(ethoxycarbonylmethyl)indoles from 5-halo-4-oxo-4,5,6,7-tetrahydroindoles, *Heterocycles*, 24, 3157, 1986.

597. Murakami, Y., Tani, M., Ariyasu, T., Nishiyama, C., Watanabe, T., and Yokoyama, Y., The Friedel–Crafts acylation of ethyl pyrrole-2-carboxylate. Scope, limitations, and application to synthesis of 7-substituted indoles, *Heterocycles*, 27, 1855, 1988.

598. Mosti, L., Menozzi, G., Schenone, P., Molinario, L., Conte, F., Montanario, C., and Marmo, E., Acetic acids bearing the 1-phenyl-1*H*-indazole nucleus with analgesic and antiinflammatory activity, *Il Farmaco, Ed. Sci.*, 43, 763, 1988.

599. Tani, M., Ariyasu, T., Ohtsuka, M., Koga, T., Ogawa, Y., Yokoyama, Y., and Murakami, Y., Synthetic studies on indoles and related compounds. Part 39. New strategy for indole synthesis from ethyl pyrrole-2-carboxylate, *Chem. Pharm. Bull.*, 44, 55, 1996.

600. Mosti, L., Menozzi, G., Fossa, P., Schenone, P., Lampa, E., Parrillo, C., D'Amico, M., and Rossi, F., 4-Substituted 1-methyl-1*H*-indazoles with analgesic, antiinflammatory and antipyretic activities, *Il Farmaco*, 47, 567, 1992.

601. Borthwick, A.D., Biggadike, K., Rocherolle, V., Cox, D.M., and Chung, G.A.C., 5-(Acetamidomethyl)-3-aryldihydrofuran-2-ones, and 5-(acetamidomethyl)-3-aryltetrahydrofuran-2-ones, two new classes antibacterial agents, *Med. Chem. Res.*, 6, 22, 1996.

602. Franke, A., Mattern, G., and Traber, W., Synthetical juvenile hormone. Part 1. *para*-Substituted 2-methyl-cinnamic acid derivatives, *Helv. Chim. Acta*, 58, 268, 1975.

603. Ames, D.E., and Bull, D., Preparation of cinnoline-3,4-dicarbonitrile and -dicarboxylic acid, *Tetrahedron*, 37, 2489, 1981.

604. McKay, R., Proctor, G.R., Scopes, D.I.C., and Sneddon, A.H., Bridged-ring nitrogen compounds. Part 9. Use of a ring-expansion reaction for the synthesis of 1,5-methano-4-benzazonin-12-ones and derivatives of benzo[3,4]cyclohepta[1,2-*b*]pyrrole, *J. Chem. Res. (M)*, 2024, 1989.

605. Ellingboe, J.W., Alessi, T.R., Dolak, T.M., Nguyen, T.T., Tomer, J.D., Guzzo, F., Bagli, J.F., and McCaleb, M.L., Antihyperglycemic activity of novel substituted 3*H*-1,2,3,5-oxathiadiazole 2-oxides, *J. Med. Chem.*, 35, 1176, 1992.

606. Sanchez, J.P., Mich, T.F., and Huang, G.G., An efficient synthesis of 6-formyl-1,2-dihydro-2-oxo-3-pyridinecarboxylic acid and some carbonyl derivatives of it and its 6-acetyl homologue, *J. Heterocycl. Chem.*, 31, 297, 1994.

607. Matsuhisa, A., Kikuchi, K., Sakamoto, K., Yatsu, T., and Tanaka, A., Nonpeptide arginine vasopressin antagonists for both V_{1A} and V_2 receptors. Synthesis and pharmacological properties of 4'-[5-(substituted methylidene)-2,3,4,5-tetrahydro-1*H*-1-benzoazepine-1-carbonyl]benzanilide and 4'-[5-(substituted methyl)-2,3-dihydro-1*H*-1-benzoazepine-1-carbonyl]benzanilide derivatives, *Chem. Pharm. Bull.*, 47, 329, 1999.

608. Geier, M., and Hesse, M., Benzannulated lactones by ring enlargement, *Synthesis*, 56, 1990.

609. El-Hossini, M.S., McCullough, K.J., McKay, R., and Proctor, G.R., Ring-expansion by a Wittig–Prevost sequence, *Tetrahedron Lett.*, 27, 3783, 1986.

610. Kawasaki, T., Terashima, R., Sakaguchi, K.-E., Sekiguchi, H., and Sakamoto, M., A short route to "reverse-prenylated" pyrrolo[2,3-*b*]indoles via tandem olefination and Claisen rearrangement of 2-(3,3-dimethyl-allyloxy)indol-3-ones. First total synthesis of flustramine C, *Tetrahedron Lett.*, 37, 7525, 1996.

611. Nasutavicus, W.A., Tobey, S.W., and Johnson, F., The cyclization of nitriles by halogen acids. Part 2. A new synthesis of substituted 3*H*-azepines, *J. Org. Chem.*, 32, 3325, 1967.

612. Harris, R.L.N., and McFadden, H.G., Acylphosphonates as substrates for Wittig and Horner–Wittig reactions. Unusual stereoselectivity in the synthesis of β-phosphinoylacrylates, *Aust. J. Chem.*, 37, 417, 1984.

613. Leclerc, V., Depreux, P., Lesieur, D., Caignard, D.H., Renard, P., Delagrange, P., Guardiola-Lemaître, B., and Morgan, P., Synthesis and biological activity of conformationally restricted tricyclic analogs of the hormone melatonin, *Bioorg. Med. Chem. Lett.*, 6, 1071, 1996.

614. Wentland, M.P., Bailey, D.M., Alexander, E.J., Castaldi, M.J., Ferrari, R.A., Haubrich, D.R., Luttinger, D.A., and Perrone, M.H., Synthesis and antidepressant properties of novel 2-substituted 4,5-dihydro-1*H*-imidazole derivatives, *J. Med. Chem.*, 30, 1482, 1987.

615. Gerecke, M., and Brossi, A., A novel synthesis of aporphines via 3-phenylphenethylamines, *Helv. Chim. Acta*, 62, 1549, 1979.

616. Lochynski, S., and Walkowicz, M., Stereochemistry of bicyclo[3.1.0]hexane derivatives. Part 14. Wittig reactions. Rearrangmenent of bicyclo[3.1.0]hexane system to cyclohexane system, *Pol. J. Chem.*, 56, 1333, 1982.

617. Beck, A., Hunkler, D., and Prinzbach, H., Synthesis and electrocyclisation of hendecafulvadienes "azulenoid" 14π-annulenes, *Tetrahedron Lett.*, 24, 2151, 1983.

618. McMorris, T.C., Le, P.H., Preus, M.W., Schow, S.R., and Weihe, G.R., Synthesis of dehydrooogoniol, a female-activating hormone of *Achlya*, *J. Org. Chem.*, 48, 3370, 1983.

619. McMorris, T.C., Le, P.H., Preus, M.W., Schow, S.R., and Weihe, G.R., Synthesis of dehydrooogoniol, a female-activating hormone of *Achlya*. The progesterone route, *Steroids*, 53, 345, 1989.

620. Hopf, H., and Kreutzer, M., Novel planar π-systems, *Angew. Chem. Int. Ed. Engl.*, 29, 393, 1990.

621. Cardin, C.J., Kavanagh, P.V., McMurry, T.B.H., and Wilcock, D.J., Synthesis of the tricyclo[6.2.1.0^{3,8}]undecan-9-one skeleton, *J. Chem. Res. (M)*, 701, 1994.

622. Taylor, R.J.K., *Bis*-Wittig reactions of 1,2-diketones. The preparation of 3,4-disubstituted mucononitriles, *Synthesis*, 566, 1977.

623. Hayashi, K.-I., Shinada, T., Sakaguchi, K., Horikawa, M., and Ohfune, Y., Olefination of dialkyl squarates by Wittig and Horner–Emmons reactions. A facile synthesis of 3,4-dioxo-1-cyclobutene-1-acetic acid esters, *Tetrahedron Lett.*, 38, 7091, 1997.

624. Kreiser, W., Bartels, G., Bathe-Burmeister, S., Ernst, L., and Stache, U., Structure elucidation of the 2:1 adduct from diethyl (cyanomethyl)phosphonate and strophanthidin K, *Liebigs Ann. Chem.*, 315, 1989.

625. De Bernardo, S., and Weigele, M., Synthesis of oxazinomycin (minimycin), *J. Org. Chem.*, 42, 109, 1977.

626. Monti, D., Gramatica, P., Speranza, G., and Manitto, P., A convenient synthesis of both the anomers of ethyl (2,3,4,6-tetra-*O*-benzyl-D-glucopyranosyl)acetate, *Tetrahedron Lett.*, 28, 5047, 1987.

627. Scremin, C.L., Boal, J.H., Wilk, A., Phillips, L.R., Zhou, L., and Beaucage, S.L., 1-(2-Deoxy-α- and β-D-*erythro*-pentofuranosyl)-2-(thymin-1-yl)ethane derivatives as conformational probes for *alt*DNA oligonucleotides, *Tetrahedron Lett.*, 36, 8953, 1995.

628. Boal, J.H., Wilk, A., Scremin, C.L., Gray, G.N., Phillips, L.R., and Beaucage, S.L., Synthesis of (2-deoxy-α- and -β-D-*erythro*-pentofuranosyl)(thymin-1-yl)alkanes and their incorporation into oligode-oxyribonucleotides. Effect of nucleobase–sugar linker flexibility on the formation of DNA–DNA and DNA–RNA hybrids, *J. Org. Chem.*, 61, 8617, 1996.

629. Lee, C.S., Du, J., and Chu, C.K., Syntheses of 2′,3′-dideoxy-L-glyceropentofuranosyl-*C*-nucleosides, *Nucleosides Nucleotides*, 15, 1223, 1996.

630. Liang, C., Ma, T., Cooperwood, J.S., Du, J., and Chu, C.K., Synthesis of L-ribofuranosyl *C*-nucleosides, *Carbohydr. Res.*, 303, 33, 1997.

631. Matsuura, N., Yashiki, Y., Nakashima, S., Maeda, M., and Sasaki, S., A shortcut and stereoselective synthesis of 1-β-alkyl-2-deoxy-D-ribose derivatives via Wittig–Horner–Emmons reaction, *Heterocycles*, 51, 975, 1999.

632. Canevet, J.C., and Sharrard, F., Synthesis of cyclic enones and dienic acids by Wittig–Horner–Emmons reaction, *Tetrahedron Lett.*, 23, 181, 1982.

633. Bloch, R., and Seck, M., Optically active five-membered oxygen-containing rings. A synthesis of (+)-eldanolide, *Tetrahedron*, 45, 3731, 1989.

634. Passarotti, C.M., Valenti, M., Ceriani, R., and Grianti, M., Synthesis of some 2-cyanomethyltetrahy-drofuran and 2-cyanomethyltetrahydropyran derivatives, *Boll. Chim. Farm.*, 132, 150, 1993.

635. Collado, I., Ezquerra, J., Jose-Vaquero, J., and Pedregal, C., Diastereoselective functionalization of 5-hydroxy prolinates by tandem Horner–Emmons–Michael reaction, *Tetrahedron Lett.*, 35, 8037, 1994.

636. Fraser-Reid, B., Alonso, R.A., McDevitt, R.E., Rao, B.V., Vite, G.D., and Zottola, M.A., Novel reactions of carbohydrates discovered en route to natural products, *Bull. Soc. Chim. Belg.*, 101, 617, 1992.

637. Theil, F., Costisella, B., Gross, H., Schick, H., and Schwarz, S., A three-step procedure for the conversion of γ-lactones into δ-lactones, *J. Chem. Soc., Perkin Trans. 1*, 2469, 1987.

638. Hutchison, A.J., de Jesus, R., Williams, M., Simke, J.P., Neale, R.F., Jackson, R.H., Ambrose, F., Barbaz, B.J., and Sills, M.A., Benzofuro[2,3-*c*]pyridin-6-ols. Synthesis, affinity for opioid-receptor subtypes, and antinociceptive activity, *J. Med. Chem.*, 32, 2221, 1989.

639. Matsumoto, M., and Watanabe, N., Synthesis of chuangxinmycin analogues, *Heterocycles*, 26, 1743, 1987.

640. Minami, T., Suganuma, H., and Agawa, T., Synthesis and reactions of vinylphosphonates bearing electronegative substituents, *Chem. Lett.*, 285, 1978.

641. Gärtner, W., Oesterhelt, D., Seifert-Schiller, E., Towner, P., Hopf, H., and Böhm, I., Acetylenic retinals form functional bacteriorhodopsins but do not form bovine rhodopsins, *J. Am. Chem. Soc.*, 106, 5654, 1984.

642. Baasov, T., and Sheves, M., C=C stretching vibrational frequencies in the model compounds of protonated Schiff bases of retinal, *Angew. Chem.*, 96, 786, 1984.

643. Baasov, T., and Sheves, M., Model compounds for the study of spectroscopic properties of visual pigments and bacteriorhodopsin, *J. Am. Chem. Soc.*, 107, 7524, 1985.

644. van der Steen, R., Biesheuvel, P.L., Mathies, R.A., and Lugtenburg, J., Retinal analogues with locked 6–7 conformations show that bacteriorhodopsin requires the 6-*s-trans* conformation of the chro-mophore, *J. Am. Chem. Soc.*, 108, 6410, 1986.

645. Krause, N., Hopf, H., and Ernst, L., Retinoids. Part 7. Synthesis of the acetylenic retinoids 9,10-didehydro-19-norretinal and 9,10,11,12-tetradehydro-19-norretinal, *Liebigs Ann. Chem.*, 1398, 1986.

646. Kölling, E., Oesterhelt, D., Hopf, H., and Krause, N., Retinoids. Part 9. Regulation of the 6-*S*-equilibrium conformation of retinal in bacteriorhodopsins by substitution at C-5. 5-Methoxy- and 5-ethylretinalbacteriorhodopsin, *Angew. Chem., Int. Ed. Engl.*, 26, 580, 1987.

647. Ok, H., Caldwell, C., Schroeder, D.R., Singh, A.K., and Nakanishi, K., Synthesis of optically active 3-diazoacetylretinals with triisopropylphenylsulfonylhydrazone, *Tetrahedron Lett.*, 29, 2275, 1988.

648. Park, M.H., Yamamoto, T., and Nakanishi, K., Probes which reflect the distance between the retinal chromophore and membrane surface in Bacteriorhodopsin (bR). Direction of retinal 9-methyl in bR, *J. Am. Chem. Soc.*, 111, 4997, 1989.

649. Gebhard, R., Courtin, J.M.L., Shadid, J.B., van Haveren, J., van Haeringen, C.J., and Lugtenburg, J., Synthesis of retinals labelled with ^{13}C in the cyclohexene ring, *Recl. Trav. Chim. Pays-Bas*, 108, 207, 1989.

650. van der Steen, R., Biesheuvel, P.L., Erkelens, C., Mathies, R.A., and Lugtenburg, J., 8,16- and 8,18-methanobacteriorhodopsin. Synthesis and spectroscopy of 8,16- and 8,18-methanoretinal and their interaction with bacterioopsin, *Recl. Trav. Chim. Pays-Bas*, 108, 83, 1989.

651. van der Berg, E.M.M., van der Bent, A., and Lugtenburg, J., Synthesis of specifically deuteriated 9- and 13-demethylretinals, *Recl. Trav. Chim. Pays-Bas*, 109, 160, 1990.

652. Colmenares, L.U., and Liu, R.S.H., 11-Methyl-9-demethylretinal and 11-methyl-9,13-didemethylretinal. Effect of altered methyl substitution pattern on polyene conformation, photoisomerization and formation of visual pigment analogs, *Tetrahedron*, 47, 3711, 1991.

653. Spijker-Assink, M.B., Robijn, G.W., Ippel, J.H., Lugtenburg, J., Groen, B.H., and van Dam, K., (1R)- and (1S)-5-demethyl-8,16-methanobacteriorhodopsin and its properties. The synthesis and spectroscopy of 5-demethyl-8,16-methanoretinal in optically active and isotopic forms, *Recl. Trav. Chim. Pays-Bas*, 111, 29, 1992.

654. Groesbeek, M., van Galen, A.J.J., Ippel, J.H., Berden, J.A., and Lugtenburg, J., Three bacteriorhodopsins with ring-didemethylated 6-*s*-locked chromophores and their properties, *Recl. Trav. Chim. Pays-Bas*, 112, 237, 1993.

655. Katsuta, Y., Ito, M., Yoshihara, K., Nakanishi, K., Kikkawa, T., and Fujiwara, T., Synthesis of (+)-(4S)- and (−)-(4R)-(11Z)-4-hydroxyretinals and determination of the absolute stereochemistry of a visual pigment chromophore in the firefly squid, *Watasenia scintillans*, *J. Org. Chem.*, 59, 6917, 1994.

656. Katsuta, Y., Yoshihara, K., Nakanishi, K., and Ito, M., Synthesis of (+)-(4S)- and (−)-(4R)-11Z-4-hydroxyretinals and determination of the absolute stereochemistry of a visual pigment chromophore in the bioluminescent squid, *Watasenia scintillans*, *Tetrahedron Lett.*, 35, 905, 1994.

657. Hoischen, D., Steinmüller, S., Gärtner, W., Buss, V., and Martin, H.-D., Merocyanines as extremely bathochromically absorbing chromophores in the halobacterial membrane protein bacteriorhodopsin, *Angew. Chem., Int. Ed. Engl.*, 36, 1630, 1997.

658. Imai, H., Hirano, T., Terakita, A., Shichida, Y., Muthyala, R.S., Chen, R.-L., Colmenares, L.U., and Liu, R.S.H., Probing for the threshold energy for visual transduction. Red-shifted visual pigment analogs from 3-methoxy-3-dehydroretinal and related compounds, *Photochem. Photobiol.*, 70, 111, 1999.

659. Gebhard, R., van Dijk, J.T.M., van Ouwerkerk, E., Boza, M.V.T.J., and Lugtenburg, J., Synthesis and spectroscopy of chemically modified spheroidenes, *Recl. Trav. Chim. Pays-Bas*, 110, 459, 1991.

660. Albeck, A., Livnah, N., Gottlieb, H., and Sheves, M., ^{13}C-NMR studies of model compounds for bacteriorhodopsin. Factors affecting the retinal chromophore chemical shifts and absorption maximum, *J. Am. Chem. Soc.*, 114, 2400, 1992.

661. Verdegem, P.J.E., Monnee, M.C.F., and Lugtenburg, J., Simple and efficient preparation of [10,20-^{13}C$_2$]- and [10-CH3,13-^{13}C$_2$]-10-methylretinal. Introduction of substituents at the 2-position of 2,3-unsaturated nitriles, *J. Org. Chem.*, 66, 1269, 2001.

662. Costisella, B., and Gross, H., α-Substituted phosphonates. Part 46. 1-Cyanodiene-1-amines and 1-cyanotriene-1-amines via the Horner reaction, *Z. Chem.*, 24, 383, 1984.

663. Hess, H.J.E., and Schaaf, T.K., Cyanoprostaglandins, *Pfizer*, U.S. Patent Appl. US 4045465, 1977; *Chem. Abstr.*, 88, 6417, 1978.

664. Lieb, F., 2-Oxoalkanephosphonic acid dialkylesters useful in the synthesis of prostaglandin analogs, *Bayer A.-G.*, German Patent Appl. DE 2711009, 1978; *Chem. Abstr.*, 89, 215553, 1978.

665. Disselnkötter, H., Lieb, F., Oediger, H., and Wendisch, D., Synthesis of prostaglandin analogues, *Liebigs Ann. Chem.*, 150, 1982.

666. Zakharkin, L.I., and Khorlina, I.M., Preparation of aldehydes by reduction of nitriles with diisobutylaluminium hydride, *Dokl. Akad. Nauk SSSR, Ser. Khim.*, 116, 422, 1957; *Chem. Abstr.*, 52, 8040f, 1958.

667. Marshall, J.A., Andersen, N.H., and Schlicher, J.W., The synthesis of bicyclo[5.4.0]undecanones via olefin cyclization, *J. Org. Chem.*, 35, 850, 1970.

668. Ito, M., Kodama, A., Tsukida, K., Fukada, Y., Shichida, Y., and Yoshizawa, T., A novel rhodopsin analogue possessing the cyclopentatrienylidene structure as the 11-*cis*-locked and the full planar chromophore, *Chem. Pharm. Bull.*, 30, 1913, 1982.

669. Baader, E., Bartmann, W., Beck, G., Bergmann, A., Jendralla, H., Kesseler, K., Wess, G., Schubert, W., Granzer, E., Kerekjarto, B.V., and Krause, R., Synthesis of a novel HMG-CoA reductase inhibitor, *Tetrahedron Lett.*, 29, 929, 1988.

670. Jendralla, H., Baader, E., Bartmann, W., Beck, G., Bergmann, A., Granzer, E., Kerekjarto, B.V., Kesseler, K., Krause, R., Schubert, W., and Wess, G., Synthesis and biological activity of new HMG-CoA reductase inhibitors. Part 2. Derivatives of 7-(1*H*-pyrrol-3-yl)-substituted-3,5-dihydroxyhept-6(*E*)-enoic (-heptanoic) acids, *J. Med. Chem.*, 33, 61, 1990.

671. Paul, G.C., and Gajewski, J.J., Thermal isomerization of a vinylcyclobutene to a cyclohexadiene, *J. Org. Chem.*, 57, 1970, 1992.

672. Asato, A.E., Peng, A., Hossain, M.Z., Mirzadegan, T., and Bertram, J.S., Azulenic retinoids. Novel nonbenzenoid aromatic retinoids with anticancer activity, *J. Med. Chem.*, 36, 3137, 1993.

673. Hashizume, H., Ito, H., Kanaya, N., Nagashima, H., Usui, H., Oshima, R., Kanao, M., Tomoda, H., Sunazuka, T., Nagamitsu, T., Kumagai, H., and Omura, S., Synthesis and biological activities of new HMG-CoA synthase inhibitors. 2-Oxetanones with a side chain containing biphenyl, terphenyl or phenylpyridine, *Heterocycles*, 38, 1551, 1994.

674. Heirtzler, F.R., Hopf, H., and Lehne, V., Cyclophanes. Part 40. On the preparation of conjugated polyenes with [2.2.2]paracyclophanyl end groups, *Liebigs Ann. Org. Bioorg. Chem.*, 1521, 1995.

675. von Riesen, C., and Hoffmann, H.M.R., A tricyclic dehydrorubanone and new isomers of the major quinidine metabolite, *Chem. Eur. J.*, 2, 680, 1996.

676. Aurrecoechea, J.M., Lopez, B., Fernandez, A., Arrieta, A., and Cossio, F.P., Diastereoselective synthesis of cycloalkylamines by samarium diiodide-promoted cyclization of α-amino radicals derived from α-benzotriazolylalkenylamines, *J. Org. Chem.*, 62, 1125, 1997.

677. Molander, G.A., and del Pozo Losada, C., Sequenced reactions with samarium(II) iodide. Domino epoxide ring-opening/ketyl olefin coupling reactions, *J. Org. Chem.*, 62, 2935, 1997.

678. Häberli, A., and Pfander, H., Synthesis of bixin and three minor carotenoids from Annatto (*Bixa orellana*), *Helv. Chim. Acta*, 82, 696, 1999.

679. Liu, D., Stuhmiller, L.M., and McMorris, T.C., Synthesis of 15β-hydroxy-24-oxocholesterol and 15β,29-dihydroxy-7-oxofucosterol, *J. Chem. Soc., Perkin Trans. 1*, 2161, 1988.

680. Magnus, P., Lewis, R.T., and Bennett, F., Synthesis of the esperamicin A$_1$/calicheamicin γ-trisulphide functionality. Thermal stability and reduction, *J. Chem. Soc., Chem. Commun.*, 916, 1989.

681. Magnus, P., Lewis, R., and Bennett, F., Synthetic and mechanistic studies on the antitumor antibiotics esperamicin A$_1$ and calicheamicin γ$_1$. Oxidative functionalization of the 13-ketobicyclo[7.3.1]tridecenediyne core structure. Construction of the allylic trisulfide trigger, *J. Am. Chem. Soc.*, 114, 2560, 1992.

682. Jendrzejewski, S., and Ermann, P., Total synthesis of restricticin, *Tetrahedron Lett.*, 34, 615, 1993.

683. Weyerstahl, P., Schwieger, R., Schwope, I., and Hashem, M.A., Synthesis of some naturally occuring drimane and dinorlabdane derivatives, *Liebigs Ann. Org. Bioorg. Chem.*, 1389, 1995.

684. Addo, J.K., and Ray, R., Synthesis and binding-analysis of 5*E*-[19-(2-bromoacetoxy)methyl]-25-hydroxyvitamin D$_3$ and 5*E*-25-hydroxyvitamin D$_3$-19-methyl[(4-azido-2-nitro)phenyl]glycinate. Novel C$_{19}$-modified affinity and photoaffinity analogs of 25-hydroxyvitamin D$_3$, *Steroids*, 63, 218, 1998.

685. Chakraborty, T.K., and Suresh, V.R., Synthetic studies toward potent cytotoxic agents amphidinolides. Synthesis of the C$_1$–C$_{18}$ moiety of amphidinolides G, H and L, *Tetrahedron Lett.*, 39, 9109, 1998.

686. Le, P.H., Preus, M.W., and McMorris, T.C., Synthesis of 3β,29-dihydroxystigmasta-5,24(28)(*E*)-dien-7-one, *J. Org. Chem.*, 47, 2163, 1982.

687. Gärtner, W., Hopf, H., Hull, W.E., Oesterhelt, D., Scheutzow, D., and Towner, P., On the photoisomerisation of 13-desmethyl-retinal, *Tetrahedron Lett.*, 21, 347, 1980.

688. Kodama, A., Ito, M., and Tsukida, K., Retinoids and related compounds. Part 4. Synthesis of bicyclic retinoates, *Chem. Pharm. Bull.*, 30, 4205, 1982.

689. Ito, M., Kodama, A., Hiroshima, T., and Tsukida, K., Retinoids and related compounds. Part 9. Synthesis and spectral characteristics of retinal analogues involving the 11-*cis*-locked-cyclopentatrienylidene structure, *J. Chem. Soc., Perkin Trans. 1*, 905, 1986.

690. Byers, J., Isolation and identification of the polyenes formed during the thermal degradation of β,β-carotene, *J. Org. Chem.*, 48, 1515, 1983.

691. Hanzawa, Y., Suzuki, M., Kobayashi, Y., and Taguchi, T., Synthesis of fluorinated retinal. A C6–C7 *s-trans* fixed retinal containing a trifluoromethyl group at the terminal position of the conjugated system, *Chem. Pharm. Bull.*, 39, 1035, 1991.

692. Tsukuda, T., Umeda, I., Masubuchi, K., Shirai, M., and Shimma, N., Synthesis of restrictinol and 9,10,11,12-tetrahydro-7-desmethylrestricticin, *Chem. Pharm. Bull.*, 41, 1191, 1993.

693. Ito, M., Hiroshima, T., Tsukida, K., Shichida, Y., and Yoshizawa, T., A novel rhodopsin analogue possessing the conformationally 6-*s-cis*-fixed retinylidene chromophore, *J. Chem. Soc., Chem. Commun.*, 1443, 1985.

694. Vogt, P., Schlageter, M., and Widmer, E., Preparation of (9Z,11Z)-vitamin A, *Tetrahedron Lett.*, 32, 4115, 1991.

695. Zhang, H., Lerro, K.A., Yamamoto, T., Lien, T.H., Sastry, L., Gawinowicz, M.A., and Nakanishi, K., The location of the chromophore in rhodopsin. A photoaffinity study, *J. Am. Chem. Soc.*, 116, 10165, 1994.

696. Wu, K.-M., Midland, M.M., and Okamura, W.H., Structural effects on [1,5]-sigmatropic hydrogen shifts of vinylallenes, *J. Org. Chem.*, 55, 4381, 1990.

697. Wess, G., Kramer, W., Han, X.B., Bock, K., Enhsen, A., Glombik, H., Baringhaus, K.-H., Böger, G., Urmann, M., Hoffmann, A., and Falk, E., Synthesis and biological activity of bile acid-derived HMG-CoA reductase inhibitors. The role of 21-methyl in recognition of HMG-CoA reductase and the ileal bile acid transport system, *J. Med. Chem.*, 37, 3240, 1994.

698. Münstedt, R., and Wannagat, U., Silaretinol, the first sila-substituted vitamin, *J. Organomet. Chem.*, 322, 11, 1987.

699. Caldwell, C.G., Derguini, F., Bigge, C.F., Chen, A.-H., Hu, S., Wang, J., Sastry, L., and Nakanishi, K., Synthesis of retinals with eight- and nine-membered rings in the side chain. Models for rhodopsin photobleaching intermediates, *J. Org. Chem.*, 58, 3533, 1993.

700. Zhang, H., Lerro, K.A., Takekuma, S.-I., Baek, D.-J., Moquin-Pattey, C., Boehm, M.F., and Nakanishi, K., Orientation of the retinal 9-methyl group in bacteriorhodopsin as studied by photoaffinity labeling, *J. Am. Chem. Soc.*, 116, 6823, 1994.

701. Borhan, B., Kunz, R., Wang, A.Y., Nakanishi, K., Bojkova, N., and Yoshihara, K., Chemoenzymatic synthesis of 11-*cis*-retinal photoaffinity analog by use of squid retinochrome, *J. Am. Chem. Soc.*, 119, 5758, 1997.

702. Moon, S., Stuhmiller, L.M., Chadha, R.K., and McMorris, T.C., Synthesis of dehydro-oogoniol and oogoniol. The adrenosterone route, *Tetrahedron*, 46, 2287, 1990.

703. Zhang, L., Nadzan, A.M., Heyman, R.A., Love, D.L., Mais, D.E., Croston, G., Lamph, W.W., and Boehm, M.F., Discovery of novel retinoic acid receptor agonists having potent antiproliferative activity in cervical cancer cells, *J. Med. Chem.*, 39, 2659, 1996.

704. Frigerio, M., Santagostino, M., and Sputore, S., 3aβ-Hydroxy-7aβ-methylperhydroinden-4-one derivatives en route to analogues of cardenolides, *Synlett*, 833, 1997.

705. Bennani, Y.L., Marron, K.S., Mais, D.E., Flatten, K., Nadzan, A.M., and Boehm, M.F., Synthesis and characterization of a highly potent and selective isotopically labeled retinoic acid receptor ligand, ALRT1550, *J. Org. Chem.*, 63, 543, 1998.

706. Abe, H., Aoyagi, S., and Kibayashi, C., First total synthesis of the marine alkaloids (±)-fasicularine and (±)-lepadiformine based on stereocontrolled intramolecular acylnitroso-Diels–Alder reaction, *J. Am. Chem. Soc.*, 122, 4583, 2000.

707. Wada, A., Tsutsumi, M., Inatomi, Y., Imai, H., Shichida, Y., and Ito, M., Retinoids and related compounds. Part 26. Synthesis of (11Z)-8,18-propano- and methano-retinals and conformational study of the rhodopsin chromophore, *J. Chem. Soc., Perkin Trans. 1*, 2430, 2001.

708. Groesbeek, M., and Smith, S.O., Synthesis of 19-fluororetinal and 20-fluororetinal, *J. Org. Chem.*, 62, 3638, 1997.

709. Garner, G.V., Mobbs, D.B., Suschitzky, H., and Millership, J.S., Syntheses of heterocyclic compounds. Part 24. Cyclisation studies with *ortho*-substituted arylcarbene arylnitrene precursors, *J. Chem. Soc. (C)*, 3693, 1971.

710. Sliskovic, D.R., Blankley, C.J., Krause, B.R., Newton, R.S., Picard, J.A., Roark, W.H., Roth, B.D., Sekerke, C., Shaw, M.K., and Stanfield, R.L., Inhibitors of cholesterol biosynthesis. Part 6. *trans*-6-[2-(2-*N*-Heteroaryl-3,5-disubstituted-pyrazol-4-yl)ethyl/ethenyl]tetrahydro-4-hydroxy-2*H*-pyran-2-ones, *J. Med. Chem.*, 35, 2095, 1992.

711. Cerri, A., Almirante, N., Barassi, P., Benicchio, A., Fedrizzi, G., Ferrari, P., Micheletti, R., Quadri, L., Ragg, E., Rossi, R., Santagostino, M., Schiavone, A., Serra, F., Zappavigna, M.P., and Melloni, P., 17β-*O*-Aminoalkyloximes of 5β-androstane-3β,14β-diol with digitalis-like activity. Synthesis, cardiotonic activity, structure-activity relationships, and molecular modeling of the Na⁺,K⁺-ATPase receptor, *J. Med. Chem.*, 43, 2332, 2000.

712. Haeck, H.H., Kralt, T., and van Leeuwen, P.H., Syntheses of carotenoidal compounds. Part 1. Preparation of some substituted polyenes with a cross-conjugation, *Recl. Trav. Chim. Pays-Bas*, 85, 334, 1966.

713. Bernard, M., Ford, W.T., and Nelson, E.C., Syntheses of ethyl retinoate with polymer-supported Wittig reagents, *J. Org. Chem.*, 48, 3164, 1983.

714. Valla, A., Prat, V., Laurent, A., Andriamialisoa, Z., Giraud, M., Labia, R., and Potier, P., Synthesis of 9-methylene analogs of retinol, retinal, retinonitrile and retinoic acid, *Eur. J. Org. Chem.*, 1731, 2001.

715. Rosenthal, A., and Baker, D.A., New route to branched-chain amino sugars by application of modified Wittig reaction to ketoses, *Tetrahedron Lett.*, 10, 397, 1969.

716. Pettit, G.R., Fessler, D.C., Paull, K.D., Hofer, P., and Knight, J.C., Bufadienolides. Part 7. Synthesis of 3β-acetoxy-5α,14α-bufa-20,22-dienolide, *J. Org. Chem.*, 35, 1398, 1970.

717. Mukherjee, D., Mukhopadhyay, S.K., Mahalanabis, K.K., Gupta, A.D., and Dutta, P.C., Synthetic studies on terpenoids. Part 16. Synthesis of 3β,17-diacetoxyphyllocladen-15-one, *J. Chem. Soc., Perkin Trans. 1*, 2083, 1973.

718. Reimann, E., and Dammertz, W., Bicyclic α-amino acids. Part 4. Synthesis of 3-(1,2,3,4-tetrahydro-1-naphthalenyl)- and 3-(5,6,7,8-tetrahydro-5-quinolinyl)alanine, *Arch. Pharm. (Weinheim)*, 316, 297, 1983.

719. Neidlein, R., and Hofmann, G., Biotransformation and pharmacokinetics of β-methyl(1,1′-biphenyl)-4-propanenitrile (LU 20884) in rats. Investigations on biotransformation, *Arzneim. Forsch. (Drug Res.)*, 33, 920, 1983.

720. Neidlein, R., and Hofmann, G., Biotransformation and pharmacokinetics of β-methyl(1,1′-biphenyl)-4-propanenitrile (LU 20884) in rats. Synthesis of reference compounds, *Arzneim. Forsch. (Drug Res.)*, 33, 691, 1983.

721. Ananthanarayan, T.P., Magnus, P., and Norman, A.W., Reductive cleavage of the 9,10-bond in 11-oxygenated steroids. A new method for the partial synthesis of the vitamin D skeleton, *J. Chem. Soc., Chem. Commun.*, 1096, 1983.

722. Teague, S.J., and Roth, G.P., The synthesis of highly functionalized naphthalene derivatives, *Synthesis*, 427, 1986.

723. Galemmo, R.A., Jr., Johnson, W.H., Jr., Learn, K.S., Lee, T.D.Y., Huang, F.-C., Campbell, H.F., Youssefyeh, R., O'Rourke, S.V., Schuessler, G., Sweeney, D.M., Travis, J.J., Sutherland, C.A., Nuss, G.W., Carnathan, G.W., and van Ingwegen, R.G., The development of a novel series of (quinolin-2-ylmethoxy)phenyl-containing compounds as high-affinity leukotriene receptor antagonists. Part 3. Structural variation of the acidic side chain to give antagonists of enhanced potency, *J. Med. Chem.*, 33, 2828, 1990.

724. Ornstein, P.L., Schoepp, D.D., Arnold, M.B., Leander, J.D., Lodge, D., Paschal, J.W., and Elzey, T., 4-(Tetrazolylalkyl)piperidine-2-carboxylic acids. Potent and selective *N*-methyl-D-aspartic acid receptor antagonists with a short duration of action, *J. Med. Chem.*, 34, 90, 1991.

725. Lankin, D.C., Nugent, S.T., and Rao, S.N., NMR spectroscopy and conformational analysis of substituted 1,2:5,6-di-*O*-isopropylidene-α-D-allofuranose derivatives, *Carbohydr. Res.*, 229, 245, 1992.

726. Sauers, R.R., and Stevenson, T.A., Further studies in pentacycloundecan-8-one photochemistry, *J. Org. Chem.*, 57, 671, 1992.

727. Yamamoto, T., Eki, T., Nagumo, S., Suemune, H., and Sakai, K., Drastic ring transformation reactions of fused bicyclic rings to bridged bicyclic rings, *Tetrahedron*, 48, 4517, 1992.

728. Mérour, J.Y., and Cossais, F., Regioselective *N*-alkylation of methyl indole-2-carboxylate, *Synth. Commun.*, 23, 1813, 1993.

729. Ellingboe, J.W., Lombardo, L.J., Alessi, T.R., Nguyen, T.T., Guzzo, F., Guinosso, C.J., Bullington, J., Browne, E.N.C., Bagli, J.F., Wrenn, J., Steiner, K., and McCaleb, M.L., Antihyperglycemic activity of novel naphthalenyl 3*H*-1,2,3,5-oxathiadiazole 2-oxides, *J. Med. Chem.*, 36, 2485, 1993.

730. Perrone, R., Berardi, F., Colabufo, N.A., Tortorella, V., Lograno, M.D., Daniele, E., and Govoni, S., Conformationally restricted thiazole derivatives as novel class of 5-HT$_3$ receptor ligands, *Il Farmaco*, 50, 77, 1995.

731. Alonso, F., Mico, I., Najera, C., Sansano, J.M., Yus, M., Èzquerra, J., Yrurctagoyena, B., and Gracia, I., Synthesis of 3- and 4-substituted cyclic α-amino acids structurally related to ACPD, *Tetrahedron*, 51, 10259, 1995.

732. Godard, A., Lamour, P., Ribereau, P., and Quéguiner, G., Synthesis of new substituted quinolizidines as potential inhibitors of ergosterol biosynthesis, *Tetrahedron*, 51, 3247, 1995.

733. Ornstein, P.L., Arnold, M.B., Allen, N.K., Bleisch, T., Borromeo, P.S., Lugar, C.W., Leander, J.D., Lodge, D., and Schoepp, D.D., Structure–activity studies of 6-(tetrazolylalkyl)-substituted decahydroisoquinoline-3-carboxylic acid AMPA receptor antagonists. Part 1. Effects of stereochemistry, chain length, and chain substitution, *J. Med. Chem.*, 39, 2219, 1996.

734. Moe, S.T., Shimizu, S.M., Smith, D.L., van Wagenen, B.C., DelMar, E.G., Balandrin, M.F., Chien, Y., Raszkiewicz, J.L., Artman, L.D., Mueller, A.L., Lobkovsky, E., and Clardy, J., Synthesis, biological activity, and absolute stereochemical assignment of NPS 1392, a potent and stereoselective NMDA receptor antagonist, *Bioorg. Med. Chem. Lett.*, 9, 1915, 1999.

735. Sanchez, I.H., and Mendoza, M.T., Improved formal total synthesis of tetrahydrometinoxocrinine, *Tetrahedron Lett.*, 21, 3651, 1980.

736. Jellimann, C., Mathé-Allainmat, M., Andrieux, J., Kloubert, S., Boutin, J.A., Nicolas, J.-P., Bennejean, C., Delagrange, P., and Langlois, M., Synthesis of phenalene and acenaphthene derivatives as new conformationally restricted ligands for melatonin receptors, *J. Med. Chem.*, 43, 4051, 2000.

737. Bien, S., and Michael, U., Steric aspects of the intramolecular cyclisation of 2-arylcyclohexylacetic acids. Part 5, *J. Chem. Soc. (C)*, 2151, 1968.

738. Neidlein, R., and Schröder, G., Reactions of 4-bromo-1,6-methano[10]annulene-3-carbaldehyde. Syntheses of 3,4-heteroanellated 1,6-methano[10]annulenes, *Helv. Chim. Acta*, 75, 825, 1992.

739. Freerksen, R.W., Pabst, W.E., Raggio, M.L., Sherman, S.A., Wroble, R.R., and Watt, D.S., Photolysis of α-paracetoxynitriles. Part 2. A comparison of two synthetic approaches to 18-cyano-20-ketosteroids, *J. Am. Chem. Soc.*, 99, 1536, 1977.

740. Ornstein, P.L., Arnold, M.B., Augenstein, N.K., Lodge, D., Leander, J.D., and Schoepp, D.D., (3*SR*,4a*RS*,6*RS*,8a*RS*)-6[2-(1*H*-Tetrazol-5-yl)ethyl]decahydroisoquinoline-3-carboxylic acid. A structurally novel, systemically active, competitive AMPA receptor antagonist, *J. Med. Chem.*, 36, 2046, 1993.

741. Ornstein, P.L., Augenstein, N.K., and Arnold, M.B., Stereoselective synthesis of 6-substituted decahydroisoquinoline-3-carboxylates. Intermediates for the preparation of conformationally constrained acidic amino acids, *J. Org. Chem.*, 59, 7862, 1994.

742. Ornstein, P.L., Arnold, M.B., Allen, N.K., and Schoepp, D.D., Synthesis and characterization of phosphonic acid-substituted amino acids as excitatory amino acid receptor antagonists, *Phosphorus, Sulfur Silicon Relat. Elem.*, 109–110, 309, 1996.

743. Efremov, I., and Paquette, L.A., First synthesis of a rearranged *neo*-clerodane diterpenoid. Development of totally regioselective trisubstituted furan ring assembly and medium-ring alkylation tactics for efficient access to (−)-teubrevin G, *J. Am. Chem. Soc.*, 122, 9324, 2000.

744. Hartmann, R.W., Hector, M., Haidar, S., Ehmer, P.B., Reichert, W., and Jose, J., Synthesis and evaluation of novel steroidal oxime inhibitors of P450 17 (17α-hydroxylase/C17–20-lyase) and 5α-reductase types 1 and 2, *J. Med. Chem.*, 43, 4266, 2000.

745. Takano, S., Yamada, S.I., Numata, H., and Ogasawara, K., A new synthesis of a steroid side chain via stereocontrolled protonation. Synthesis of (−)-desmosterol, *J. Chem. Soc., Chem. Commun.*, 760, 1983.

746. Garratt, P.J., Doecke, C.W., Weber, J.C., and Paquette, L.A., Intramolecular anionic cyclization route to capped [3]peristylanes, *J. Org. Chem.*, 51, 449, 1986.

747. Lansbury, P.T., Galbo, J. P., and Springer, J.P., An enantioselective approach to chiral pseudoguaianolide intermediates, *Tetrahedron Lett.*, 29, 147, 1988.

748. Chen, Y.-J., De Clercq, P., and Vandewalle, M., Synthesis a new vitamin D$_3$ analogues with a decalin-type CD-ring, *Tetrahedron Lett.*, 37, 9361, 1996.

749. Ishikawa, F., Cyclic guanidines. Part 10. Synthesis of 2-(2,2-disubstituted ethenyl- and ethyl)-2-imidazolines as potent hypoglycemics, *Chem. Pharm. Bull.*, 28, 1394, 1980.

750. Grissom, J.W., Klingberg, D., Meyenburg, S., and Stallman, B.L., Enediyne- and tributyltin hydride-mediated aryl radical additions onto various radical acceptors, *J. Org. Chem.*, 59, 7876, 1994.

751. Aslanian, R., Brown, J.E., Shih, N.Y., Mutahi, M.W., Green, M.J., She, S., Del Prado, M., West, R., and Hey, J., 4-[(1*H*-Imidazol-4-yl)methyl]benzamidines and benzylamidines. Novel antagonists of the histamine H$_3$ receptor, *Bioorg. Med. Chem. Lett.*, 8, 2263, 1998.

752. Misun, M., and Pfaltz, A., Enantioselective reduction of electrophilic C=C bonds with sodium tetrahydroborate and "semicorrin" cobalt catalysts, *Helv. Chim. Acta*, 79, 961, 1996.

753. Osborn, M.E., Kuroda, S., Muthard, J.L., Kramer, J.D., Engel, P., and Paquette, L.A., Functionalization reactions of C$_{16}$-hexaquinacene and related hemispherical molecules, *J. Org. Chem.*, 46, 3379, 1981.

754. Golebiewski, W.M., Keyes, R.F., and Cushman, M., Exploration of the effects of linker chain modifications on anti-HIV activities in a series of cosalane analogues, *Bioorg. Med. Chem.*, 4, 1637, 1996.

755. Lansbury, P.T., and Vacca, J.P., Superior methodology for γ-lactone annulation. Intramolecular alkoxyhydride reduction of conjugated nitriles, *Tetrahedron Lett.*, 23, 2623, 1982.

756. Lansbury, P.T., and Mojica, C.A., Total synthesis of (±)-arteannuin B, *Tetrahedron Lett.*, 27, 3967, 1986.

757. Trehan, I.R., Vig, M., and Bala, K., Synthesis of 13-aza-3-methoxy-18-nor-5,6-seco-estrone and its 7-methyl analogue, *Indian J. Chem., Sect. B*, 21, 200, 1982.

758. Groundwater, P.W., and Sharp, J.T., Electrocyclic aromatic substitution by nitrile ylides to give 3*H*-2-benzazepines. Substituent effects and mechanism, *Tetrahedron*, 48, 7951, 1992.

759. van der Bent, A., Blommaert, A.G.S., Melman, C.T.M., Ijzerman, A.P., van Wijngaarden, I., and Soudijn, W., Hybrid cholecystokinin A antagonists based on molecular modeling of lorglumide and L-364,718, *J. Med. Chem.*, 35, 1042, 1992.

760. Schlessinger, R.H., and Li, Y.-J., Total synthesis of (−)-virginiamycin M$_2$ using second-generation vinylogous urethane chemistry, *J. Am. Chem. Soc.*, 118, 3301, 1996.

761. Breuilles, P., and Uguen, D., Toward a total synthesis of pristinamycin II$_B$. A chiron approach to a C-9/C-16 fragment, *Tetrahedron Lett.*, 39, 3145, 1998.

762. Dolle, R.E., and Nicolaou, K.C., Total synthesis of elfamycins. Aurodox and efrotomycin. Part 1. Strategy and construction of key intermediates, *J. Am. Chem. Soc.*, 107, 1691, 1985.

763. Buzas, A., Herisson, C., and Lavielle, G., Application of the Wittig–Horner reaction to indolinones. A convenient synthesis of tryptamines, *Synthesis*, 129, 1977.

764. Buzas, A., and Merour, J.Y., Synthesis and reactions of 1-acetyl-2-benzylidene-3-oxo-2,3-dihydroindoles, *Synthesis*, 458, 1989.

765. Bellemin, R., Decerprit, J., and Festal, D., New indole derivatives as ACAT inhibitors. Synthesis and structure–activity relationships, *Eur. J. Med. Chem. Chim. Ther.*, 31, 123, 1996.

766. Mueller, A., Moe, S., and Balandrin, M., Preparation of diarylalkylamines and related compounds active at both the serotonin reuptake site and the *N*-methyl-D-aspartate receptor for treatment depression and other disorders, *NPS Pharmaceuticals*, Int. Patent Appl. WO 2000002551, 2000; *Chem. Abstr.*, 132, 93096, 2000.

767. Hoffman, J.M., Wai, J.S., Thomas, C.M., Levin, R.B., O'Brien, J.A., and Goldman, M.E., Synthesis and evaluation of 2-pyridinone derivatives as HIV-1 specific reverse transcriptase inhibitors. Part 1. Phthalimidoalkyl and -alkylamino analogues, *J. Med. Chem.*, 35, 3784, 1992.

768. Banwell, M.G., and Wu, A.W., A stereoselective total synthesis of (±)-γ-lycorane, *J. Chem. Soc., Perkin Trans. 1*, 2671, 1994.

769. Banwell, M.G., Hockless, D.C.R., and Wu, A.W., (*E*)-3-(2,2-Dibromospiro[2.4]heptan-1-yl)propenenitrile, *Acta Crystallogr., Sect. C: Cryst. Struct. Commun.*, C53, 504, 1997.

770. Suzuki, T., Imanishi, N., Itahana, H., Watanuki, S., Miyata, K., Ohta, M., Nakahara, H., Yamagiwa, Y., and Mase, T., Novel 5-hydroxytryptamine 4 (5-HT$_4$) receptor agonists. Synthesis and gastroprokinetic activity of 4-amino-*N*-[2-(1-aminocycloalkan-1-yl)ethyl]-5-chloro-2-methoxybenzamides., *Chem. Pharm. Bull.*, 46, 1116, 1998.

771. Hattersley, P.J., Lockhart, I.M., and Wright, M., Some alkylation and Grignard reactions with 1-tetralones and related compounds, *J. Chem. Soc. (C)*, 217, 1969.

772. Huff, J.R., Anderson, P.S., Baldwin, J.J., Clineschmidt, B.V., Guare, J.P., Lotti, V.J., Pettibone, D.J., Randall, W.C., and Vacca, J.P., N-(1,3,4,6,7,12b-Hexahydro-2H-benzo[b]furo[2,3-a]quinolizin-2-yl)-N-methyl-2-hydroxyethanesulfonamide, a potent and selective α_2-adrenoceptor antagonist, J. Med. Chem., 28, 1756, 1985.

773. Shiotani, S., Tsuno, M., Tanaka, N., Tsuiki, M., and Itoh, M., Furopyridines. Part 15. Synthesis and properties of ethyl 2-(3-furo[2,3-b]-, -[3,2-b]-, -[2,3-c]- and -[3,2-c]pyridyl)acetate, J. Heterocycl. Chem., 32, 129, 1995.

774. Cesati, R.R., III, and Katzenellenbogen, J.A., Preparation of hexahydrobenzo[f]isoquinolines using a vinylogous Pictet–Spengler cyclization, Org. Lett., 2, 3635, 2000.

775. Takadate, A., and Fishman, J., 6-Aminoalkyl catechol estrogens. Models of steroidal biogenic amines, J. Org. Chem., 44, 67, 1979.

776. Kahn, M., and Devens, B., The design and synthesis of a nonpeptide mimic of an immunosuppressing peptide, Tetrahedron Lett., 27, 4841, 1986.

777. Kahn, M., Chen, B., and Zieske, P., The design and synthesis of a nonpeptide mimic of erabutoxin, Heterocycles, 25, 29, 1987.

778. Tseng, C.C., Handa, I., Abdel-Sayed, A.N., and Bauer, L., N-[(Aryl substituted adamantane)alkyl]-2-mercaptoacetamidines, their corresponding disulfides and S-phosphorothioates, Tetrahedron, 44, 1893, 1988.

779. Papageorgiou, C., Petcher, T.J., and Waldvogel, E., Synthesis of hydroxy- and methoxy-substituted octahydrobenzo[g]isoquinolines as potential ligands for serotonin receptors, Helv. Chim. Acta, 72, 1463, 1989.

780. Sellier, C., Buschauer, A., Elz, S., and Schunack, W., Synthesis of (Z)- and (E)-3-(1H-imidazol-4-yl)-2-propenamine and some 3-(1H-imidazol-4-yl)propanamines, Liebigs Ann. Chem., 317, 1992.

781. Buschauer, A., Synthesis of primary ω-phenyl-ω-pyridylalkylamines, Arch. Pharm. (Weinheim), 322, 165, 1989.

782. Grunewald, G.L., Paradkar, V.M., Stillions, D.M., and Ching, F., A new procedure for regioselective synthesis of 8,9-dichloro-2,3,4,5-tetrahydro-1H-2-benzazepine (LY 134046) and its 3-methyl analogue as inhibitors of phenylethanolamine N-methyltransferase (PNMT), J. Heterocycl. Chem., 28, 1587, 1991.

783. Brown, T.J., Chapman, R.F., Mason, J.S., Palfreyman, M.N., Vicker, N., and Walsh, R.J.A., Syntheses and biological activities of potent potassium channel openers derived from (±)-2-oxo-1-pyridin-3-yl-cyclohexanecarbothioic acid methylamide. New potassium channel openers, J. Med. Chem., 36, 1604, 1993.

784. Hori, M., Kataoka, T., Shimizu, H., Imai, E., Iwata, N., Kawamura, N., and Kurono, M., Formation of novel heterocycles. [1]Benzothiopyrano[3,4-b]pyrrole derivatives by unusual cyclization reaction of 1-benzothiopyran-1-oxide derivatives, Heterocycles, 23, 1381, 1985.

785. Main, A.J., Bhagwat, S.S., Boswell, C., Goldstein, R., Gude, C., Cohen, D.S., Furness, P., Lee, W., and Louzan, M., Thromboxane receptor antagonism combined with thromboxane synthase inhibition. Part 3. Pyridinylalkyl-substituted 8-[(arylsulfonyl)amino]octanoic acids, J. Med. Chem., 35, 4366, 1992.

786. Tori, M., Tsuyuri, T., and Takahashi, T., The reaction of dammarane derivatives. Reactions involving C-20 carbonium ions, Bull. Chem. Soc. Jpn., 50, 3349, 1977.

787. Krafft, M.E., Dasse, O.A., and Fu, Z., Synthesis of the C/D/E and A/B rings of xestobergsterol A, J. Org. Chem., 64, 2475, 1999.

788. Mayer, H., and Rüttimann, A., Synthesis of optically active, natural carotenoids and structurally related compounds. Part 4. Synthesis of (3R,3'R,6'R)-lutein, Helv. Chim. Acta, 63, 1451, 1980.

789. Sierra, M.G., Cravero, R.M., de Los Angeles Laborde, M., and Ruveda, E.A., Stereoselective synthesis of (±)-18,19-dinor-13βH,14αH-cheilanthane, the most abundant tricyclic compound from petroleums and sediments, J. Chem. Soc., Chem. Commun., 417, 1984.

790. Sierra, M.G., Cravero, R.M., de Los Angeles Laborde, M., and Ruveda, E.A., Synthesis of the key intermediate (±)-18,19-dinor-14αH-cheilantha-12,15-dien-17-one and its transformation into the geochemical marker 18,19-dinor-13βH,14αH-cheilanthane and the marine-type sesterterpene methyl scalar-17-en-25-oate, J. Chem. Soc., Perkin Trans. 1, 1227, 1985.

791. Gerspacher, M., and Pfander, H., C_{45} And C_{50}-carotenoids. Synthesis of an optically active cyclic C_{20}-building block and of decaprenoxanthine [(2R,6R,2'R,6'R)-2,2'-bis(4-hydroxy-3-methylbut-2-enyl)-ε,ε-carotene], Helv. Chim. Acta, 72, 151, 1989.

792. Valenti, P., Recanatini, M., Magistretti, M., and Da Re, P., Some basic derivatives of Pummerer's ketone, *Arch. Pharm. (Weinheim)*, 314, 740, 1981.

793. Trehan, I.R., Bala, R., Trikha, D.K., and Singh, J., Synthesis of nuclear diazasteroidal system. Synthesis of 2,13-diaza-3-*O*-methyl-18-nor-17-oxo-estra-1,3,5(10),8(9)-tetraene and its 7,7-dimethyl derivative, *Indian J. Chem., Sect. B*, 20, 1022, 1981.

794. Yu, K.-L., Spinazze, P., Ostrowski, J., Currier, S.J., Pack, E.J., Hammer, L., Roalsvig, T., Honeyman, J.A., Tortolani, D.R., Reczek, P.R., Mansuri, M.M., and Starrett, J.E., Jr., Retinoic acid receptor β,γ-selective ligands. Synthesis and biological activity of 6-substituted 2-naphthoic acid retinoids, *J. Med. Chem.*, 39, 2411, 1996.

795. Popplestone, C.R., and Unrau, A.M., Studies on the biosynthesis of antheridiol, *Can. J. Chem.*, 52, 462, 1974.

796. Mérour, J.Y., Coadou, J.Y., and Tatibouët, F., Syntheses of 2(5)-substituted 1-acetyl-3-oxo-2,3-dihydroindoles, 3-acetoxy-1-acetylindoles, and of 2-methyl-5-methoxyindole-3-acetic acid, *Synthesis*, 1053, 1982.

797. Gupta, K.A., Saxena, A.K., Jain, P.C., and Anand, N., Synthesis and biological activities of 1,3,4-trisubstituted-3,4-dehydropiperidines, *Indian J. Chem., Sect. B*, 26, 344, 1987.

798. Kharasch, M.S., and Sosnovsky, G., Oxidative reactions of nitriles. Part 1. Autoxidation, *Tetrahedron*, 3, 97, 1958.

799. Marshall, J.A., and Bierenbaum, R., Synthesis of olefins via reduction–decyanation of β,γ-unsaturated nitriles, *J. Org. Chem.*, 42, 3309, 1977.

800. Morzycki, J.W., Gryszkiewicz, A., and Jastrzebska, I., Neighboring group participation in epoxide ring cleavage in reactions of some 16α,17α-oxidosteroids with lithium hydroperoxide, *Tetrahedron*, 57, 2185, 2001.

801. Marshall, J.A., Hagan, C.P., and Flynn, G.A., Reductive decyanation of β,γ-epoxy nitriles. A new synthesis of β-isopropylidene alcohols, *J. Org. Chem.*, 40, 1162, 1975.

802. Seyferth, D., Hilbert, P., and Marmor, R.S., Novel diazoalkanes and the first carbene containing the $(MeO)_2P(O)$ group, *J. Am. Chem. Soc.*, 89, 4811, 1967.

803. Seyferth, D., Marmor, R.S., and Hilbert, P., Some reactions of dimethylphosphono-substituted diazoalkanes. $(MeO)_2P(O)CR$ transfer to olefins and 1,3-dipolar additions of $(MeO)_2P(O)C(N_2)R^1$, *J. Org. Chem.*, 36, 1379, 1971.

804. Khokhlov, P.S., Kashemirov, B.A., Mikityuk, A.D., Strepikheev, Y.A., and Chimiskyan, A.L., Diazotization of amino(dialkoxyphosphinyl)acetic esters, *Zh. Obshch. Khim.*, 54, 2785, 1984; *J. Gen. Chem. USSR (Engl. Transl.)*, 54, 2495, 1984.

805. Khokhlov, P.S., Kashemirov, B.A., Mikityuk, A.D., Strepikheev, Y.A., and Chimishkyan, A.L., Diazotization of amino(dialkoxyphosphinyl)acetonitriles, *Zh. Obshch. Khim.*, 54, 2641, 1984; *J. Gen. Chem. USSR (Engl. Transl.)*, 54, 2359, 1984.

806. Mikityuk, A.D., Strepikheev, Y.A., and Khokhlov, P.S., Cyclization of α-diazo-α-(dialkoxyphosphinyl)acetonitriles into 5-chloro-4-(dialkoxyphosphinyl)-1*H*-1,2,3-triazoles, *Zh. Obshch. Khim.*, 56, 1911, 1986; *J. Gen. Chem. USSR (Engl. Transl.)*, 56, 1689, 1986.

807. Tsuge, O., Kanemasa, S., and Suga, H., Synthesis of a new phosphorus-functionalized nitrile oxide, α-(diethylphosphono)acetonitrile oxide, and cycloaddition leading to 3-(diethylphosphonomethyl)-Δ^2-isoxazolines, *Chem. Lett.*, 183, 1986.

808. Tsuge, O., Kanemasa, S., Suga, H., and Nakagawa, N., Synthesis of (diethoxyphosphoryl)acetonitrile oxide and its cycloaddition to olefins. Synthetic applications to 3,5-disubstituted 2-isoxazolines, *Bull. Chem. Soc. Jpn.*, 60, 2463, 1987.

809. Nkusi, G., and Neidlein, R., Convenient synthesis of diethyl 3-methylisoxazoline and isoxazolephosphonates, potent synthons to biological active compounds, *J. Prakt. Chem./Chem.-Ztg.*, 334, 278, 1992.

810. Arbuzov, B.A., Samitov, Y.Y., Dianova, E.N., and Lisin, A.F., Stereochemistry of organophosphorus compounds. Part 10. Structure of 2,3-diphenyl-4-diethoxyphosphono-5-cyanoisoxazolidine and 2,3-diphenyl-4-diethoxyphosphono-5-carbomethoxyisoxazolidine, *Izv. Akad. Nauk SSSR, Ser. Khim.*, 2559, 1975; *Bull. Acad. Sci. USSR, Div. Chem. Sci. (Engl. Transl.)*, 2446, 1975.

811. Arbuzov, B.A., Samitov, Y.Y., Dianova, E.N., and Lisin, A.F., 1,3-Dipolar cycloaddition of *C*-benzoyl-*N*-phenylnitrone to vinyl-, β-cyanovinyl-, and allylphosphonates and *C,N*-diphenylnitrone to allylphosphonate, *Izv. Akad. Nauk SSSR, Ser. Khim.*, 2779, 1976; *Bull. Acad. Sci. USSR, Div. Chem. Sci. (Engl. Transl.)*, 2589, 1976.

812. Wolbers, P., Misske, A.M., and Hoffmann, H.M.R., Synthesis of the enantiopure C15–C26 segment of phorboxazole A and B, *Tetrahedron Lett.*, 40, 4527, 1999.

813. Pudovik, S.I., and Aladzheva, I.M., Acetylene-allene-diene rearrangements of diphosphites with β,γ-acetylenic bond in the entire ester radical, *Zh. Obshch. Khim.*, 33, 708, 1963; *J. Gen. Chem. USSR (Engl. Transl.)*, 33, 702, 1963.

814. Pudovik, A.N., Konovalova, I.V., and Ishmaeva, E.A., Diels–Alder reactions and addition reactions of butadienylphosphonic and butadienylphosphonothioic esters, *Zh. Obshch. Khim.*, 33, 2509, 1963; *J. Gen. Chem. USSR (Engl. Transl.)*, 33, 2446, 1963.

815. Dangyan, Y.M., Voskanyan, M.G., Zurabyan, N.Z., and Badanyan, S.O., Reactions of unsaturated compounds. Part 57. Vinylallenylphosphonates as diene fragments in the Diels–Alder reaction, *Arm. Khim. Zh.*, 32, 460, 1979; *Chem. Abstr.*, 92, 76604v, 1980.

816. Dangyan, Y.M., Panosyan, G.A., Voskanyan, M.G., and Badanyan, S.O., Reactions of unsaturated compounds. Part 88. Vinylallenic phosphonates in cycloaddition reactions with unsymmetric dienophiles, *Zh. Obshch. Khim.*, 53, 61, 1983; *J. Gen. Chem. USSR (Engl. Transl.)*, 53, 47, 1983.

817. Brel, V.K., Abramkin, E.V., and Martynov, I.V., [2 + 4]-Cycloaddition of tetracyanoethylene with dialkyl esters of 3-dialkylamino-1,3-alkadiene-2-phosphonic acid, *Izv. Akad. Nauk SSSR, Ser. Khim.*, 2843, 1989; *Bull. Acad. Sci. USSR, Div. Chem. Sci. (Engl. Transl.)*, 2607, 1989.

818. Defacqz, N., Touillaux, R., and Marchand-Brynaert, J., [4 + 2]Cycloaddition of *N*-buta-1,3-dienyl-succinimide to *gem*-substituted vinyl phosphonates, *J. Chem. Res. (M)*, 2273, 1998.

819. Shishkin, V.E., Elfimova, S.N., and No, B.I., Phosphorus-containing hydrochlorides of imino esters and syntheses related to them, *Zh. Obshch. Khim.*, 44, 526, 1974; *J. Gen. Chem. USSR (Engl. Transl.)*, 44, 504, 1974.

820. Shishkin, V.E., Mednikov, E.V., Isakova, E.V., and No, B.I., First representatives of dialkoxyphosphorylalkyl esters of *C*-phosphorylated imidic acids, *Zh. Obshch. Khim.*, 67, 1031, 1997; *Russ. J. Gen. Chem. (Engl. Transl.)*, 67, 969, 1997.

821. Weferling, N., and Kleiner, H.-J., Process for preparation of thiocarbonylmethanephosphonic acid diesters, *Hoechst A.-G.*, Eur. Patent Appl. EP 676406, 1995; *Chem. Abstr.*, 124, 994719, 1995.

822. Pudovik, A.N., Cherkasov, R.A., Sudakova, T.M., and Evstaf'ev, G.I., Addition of phosphorus dithio acids at the carbon–nitrogen triple bond, *Dokl. Akad. Nauk SSSR, Ser. Khim.*, 211, 113, 1973; *Dokl. Chem. (Engl. Transl.)*, 211, 542, 1973.

823. Wharton, C.J., and Wrigglesworth, R., Inhibitors of pyrimidine biosynthesis. Part 2. The synthesis of amidine phosphonates as potential inhibitors of carbamoyl phosphate synthase, *J. Chem. Soc., Perkin Trans. 1*, 433, 1981.

824. Baimashev, B.A., Polezhaeva, N.A., and Arbuzov, B.A., Preparation of dialkyl thiocarbamoylphosphonates, *Kazan State University*, USSR Patent Appl. SU 1583425, 1990; *Chem. Abstr.*, 114, 122710, 1991.

825. Maehr, H., and Yang, R., A convergent synthesis of Ro24–5913, a novel leukotriene D_4 antagonist, *Tetrahedron Lett.*, 37, 5445, 1996.

826. Caujolle, R., Baziard-Mouysset, G., Favrot, J.D., Payard, M., Loiseau, P.R., Amarouch, H., Linas, M.D., Seguela, J.P., Loiseau, P.M., Bories, C., and Gayral, P., Synthesis, anti-parasitic and anti-fungal activities of arylalkyl- and arylvinylthiazolines, *Eur. J. Med. Chem.*, 28, 29, 1993.

827. de Raadt, A., Klempier, N., Faber, K., and Griengl, H., Chemoselective enzymatic hydrolysis of aliphatic and alicyclic nitriles, *J. Chem. Soc., Perkin Trans. 1*, 137, 1992.

828. Kafarski, P., and Mastalerz, P., Synthesis of dipeptides containing *P*-terminal 2-aminoethylphosphonic acid, *Rocz. Chem.*, 51, 433, 1977.

829. Blackburn, G.M., Jakeman, D.L., Ivory, A.J., and Williamson, M.P., Synthesis of phosphonate analogues of 1,3-*bis*phosphoglyceric acid and their binding to yeast phosphoglycerate kinase, *Bioorg. Med. Chem. Lett.*, 4, 2573, 1994.

830. Jakeman, D.L., Ivory, A.J., Williamson, M.P., and Blackburn, G.M., Highly potent bisphosphonate ligands for phosphoglycerate kinase, *J. Med. Chem.*, 41, 4439, 1998.

831. Nasser, J., About-Jaudet, E., and Collignon, N., α-Functional cycloalkylphosphonates. Part 2. Synthesis of α-aminomethyl and α-(*N*-substituted)-aminomethylcycloalkylphosphonates, *Phosphorus, Sulfur Silicon Relat. Elem.*, 55, 137, 1991.

832. Braden, R., Hendricks, U., Oertel, G., and Schliebs, R., ω-Aminoalkylphosphonates and -phosphinates, *Bayer A.-G.*, German Patent Appl. DE 2032712, 1970; *Chem. Abstr.*, 76, 72656, 1972.

833. Merger, F., Miksovsky, F., Winderl, S., and Fouquet, G., Diaminoalkylphosphonic acid dialkyl ester, *BASF A.-G.*, German Patent Appl. DE 2525508, 1975; *Chem. Abstr.*, 86, 121519, 1977.

834. Morita, T., Okamoto, Y., and Sakurai, H., A convenient dealkylation of dialkyl phosphonates by chlorotrimethylsilane in the presence of sodium iodide, *Tetrahedron Lett.*, 19, 2523, 1978.

835. Morita, T., Okamoto, Y., and Sakurai, H., The preparation of phosphonic acids having labile functional groups, *Bull. Chem. Soc. Jpn.*, 51, 2169, 1978.

836. Andreae, S., and Grimm, A., Furan derivatives. Part 8. Preparation of fur-2-yl- and 5-nitrofur-2-ylphosphonates, *Z. Chem.*, 20, 338, 1980.

837. Morita, T., Okamoto, Y., and Sakurai, H., Dealkylation reaction of acetals, phosphonate, and phosphate esters with chlorotrimethylsilane/metal halide reagent in acetonitrile, and its application to the synthesis of phosphonic acids and vinyl phosphates, *Bull. Chem. Soc. Jpn.*, 54, 267, 1981.

838. Savignac, P., and Lavielle, G., Monodealkylation of phosphoric and phosphonic esters by thiolates and thiophenates, *Bull. Soc. Chim. Fr.*, 1506, 1974.

7 The Ketophosphonates

Good general synthetic routes to dialkyl 1-oxoalkylphosphonates have been available for many years. Although good general methods now exist, dialkyl 2-oxoalkylphosphonates were traditionally less accessible, and many relatively specific protocols have been developed. These ketophosphonates are important reagents for synthesis and useful intermediates for the preparation of elaborate phosphonates. This chapter describes their synthesis and applications.

7.1 SYNTHESIS OF KETOPHOSPHONATES

7.1.1 DIALKYL 1-OXOALKYLPHOSPHONATES

7.1.1.1 Michaelis–Arbuzov Reactions

Dialkyl acylphosphonates were obtained for the first time by Kabachnik and Rossiiskaya in 1945 by the reaction of acyl chlorides with trialkyl phosphites.[1] The reaction proceeds readily at room temperature and even on cooling by slowly adding trialkyl phosphite to acyl chlorides. Improvements have been made, especially by controlling the temperature of the reaction.[2] The different versions of this reaction, which are distinguished by the nature of the acylating agent and the structure of the phosphite, have been covered by the reviews of Zhdanov et al. in 1980,[3] Breuer in 1996,[4] and McKenna and Kashemirov in 2002.[5]

This Michaelis–Arbuzov-type reaction is generally applicable to the preparation of dialkyl 1-oxoalkyl- or 1-oxoarylphosphonates from a great variety of trialkyl phosphites and aliphatic or aromatic acyl halides in good to excellent yields (48–98%, Scheme 7.1, Table 7.1). The reaction has been applied with success to long-chain acyl chlorides,[6,7] unsaturated acyl chlorides,[8–17] cycloalkyl carboxylic acid chlorides,[9,18,19] amido acetyl chlorides,[20] carbohydrate-derived acyl chlorides,[21] coumarin-3-carboxylic acid chloride,[22] ω-azidoacid chlorides,[23] adamantyl acid chlorides,[24,25] and acyl bromides.[26] A large variety of phosphites such as tribenzyl,[27,28] tris(2-chloroethyl),[29] tris(2-cyanoethyl),[30] tris(1-adamantyl),[31] tris(trialkylsilyl),[32–34] and diethyl vinyl[35] phosphites have been successfully used. Usually the reaction is conducted neat, but in some cases solvents as C_6H_6,[22,36,37] toluene,[13,38–41] Et_2O,[42,43] or CH_2Cl_2[23,26] can be used.

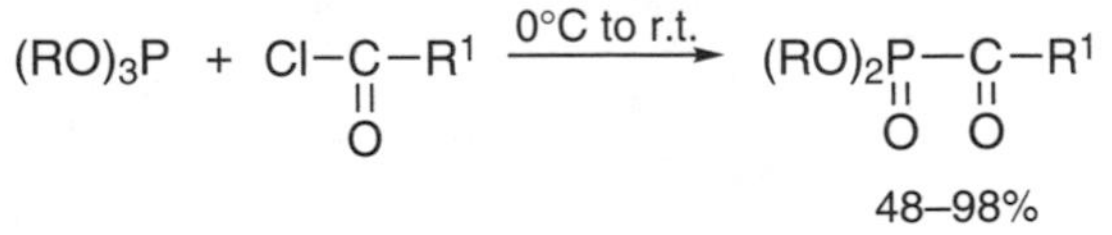

$$(RO)_3P \ + \ Cl-\underset{\underset{O}{\|}}{C}-R^1 \ \xrightarrow{0°C \text{ to r.t.}} \ (RO)_2\underset{\underset{O}{\|}}{P}-\underset{\underset{O}{\|}}{C}-R^1$$

$$48-98\%$$

(7.1)

Although the preparation of dialkyl acetylphosphonates is relatively straightforward, the same method applied to the formation of β,γ-unsaturated α-ketophosphonates leads to a rather large variety of products including the desired one.[8,10,14] For example, dimethyl (E)-1-oxo-2-butenoylphosphonate can be prepared in 53% yield by the slow addition of trimethyl phosphite to a threefold excess of (E)-2-butenoyl chloride.[10] When the reaction is carried out with equimolar quantities of trimethyl phosphite and (E)-2-butenoyl chloride, the main product is a diphosphonylated trans-2-butenoyl ester (Scheme 7.2). This ester is formed through the intermediacy of a pentacovalent oxaphospholene, resulting from the reaction of dimethyl (E)-1-oxo-2-butenoylphosphonate with trimethyl phosphite, which in turn is attacked by (E)-2-butenoyl chloride (Scheme 7.2).[10]

TABLE 7.1

	R	R^1	Temp. (°C)	Yield (%)[Ref.]
a	Me	Me	5,[1] 0[2,9]	59,[1] 80,[2] 90[9]
b	Et	Me	r.t.,[1,44] 15[2,8]	12,[1] 50,[2] 76,[8] 83[44]
c	n-Bu	Me	0,[2] 5[45]	50,[2] 92[45]
d	Cl(CH$_2$)$_2$	Me	D	52[29]
e	Bn	Me, i-Pr, i-Bu, Bn, MeO$_2$C(CH$_2$)$_2$	—	50[27]
f	Et	Me, Et, i-Pr, t-Bu	< 50	65–86[18,44]
g	Et	Et	—[46]	49[46]
h	Me, Et	CH$_3$(CH$_2$)$_5$	5[45]	82–85[45]
i	Et, i-Pr	n-Bu(Et)CH	0	77–78[47]
j	Me	i-Pr, t-Bu, n-Pent, Cl(CH$_2$)$_2$	0	86–97[9]
k	Me	Et, Ph(CH$_2$)$_2$, Cy	r.t.	48–64[19]
l	Et	EtOCH$_2$, EtSCH$_2$, n-BuSCH$_2$	r.t.	52–68[48]
m	Me	Ph	< 35,[49] r.t.,[1] 5[45]	72,[1] 81,[49] 85[45]
n	Et	Ph	r.t.,[1,44] 10–15,[50] 5[45]	67,[1] 79,[44] 87,[50] 88[45]
o	Me	R^{2}C$_6$H$_4$	r.t.	58–98[36,38,40,41,50]
p	Et	2-Cl-C$_6$H$_4$	10–15	70[50]
q	i-Pr	R^{2}C$_6$H$_4$	80	67–92[50]
r	Me, Et, i-Pr, allyl	4-R^{2}C$_6$H$_4$	< 35	44–92[44,49,51]

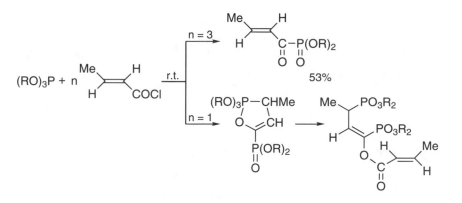

$$(7.2)$$

Analogously, treatment of triethyl phosphite with 3-fluoropropanoyl chloride gives a mixture of products containing the desired diethyl 1-oxo-3-fluoropropylphosphonate (20%) together with two major products (25% of each).[52] One results from attack by a second molecule of acyl halide on the enol form of the first-formed acylphosphonate, whereas the other results from the attack of a second molecule of phosphite on the carbonyl group of the acylphosphonate.[52]

Trimethyl phosphite does not react with chloroacetyl chloride through a Michaelis–Arbuzov/Perkow reaction sequence. It gives an acylpseudophosphonium salt, which, in turn, reacts readily with trimethyl phosphite to give a new adduct. Subsequent warming affords dealkylation to give the final dimethyl 1-(dimethoxyphosphinyloxy)vinylphosphonate (Scheme 7.3). More interesting from the mechanistic standpoint is the use of diethyl trimethylsilyl phosphite. In reaction with chloroacetyl chloride at 30°C, the initial adduct loses Me$_3$SiCl rapidly to give diethyl 2-chloroacetylphosphonate in 37% yield.[53]

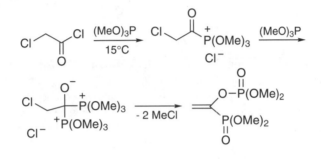

$$(7.3)$$

Reaction between equimolar amounts of dialkyl chlorophosphites and acyclic acid anhydrides at 50°C has been reported to produce the corresponding dialkyl acylphosphonates in satisfactory yields (70–75%), but the general utility of this procedure remains to be demonstrated.[54] On pyrolysis, diethyl 1-oxoethylphosphonate produces ketene and diethyl phosphite.[55]

The previous reactions suggest that trialkyl phosphites are able to react with acylphosphonates. Effectively, it has been shown that the reaction of trimethyl phosphite with dimethyl aroylphosphonates under appropriate conditions leads to the formation of anionic adducts that decompose to give phosphonyl carbenes and trimethyl phosphate.[56–60] These carbene intermediates can react with further trimethyl phosphite to give ylidic phosphonates, or, when suitable *ortho*-substituents are present on the aromatic ring, intramolecular carbene insertion reactions can occur to give cyclic systems (Scheme 7.4).[38,40,56,58–65]

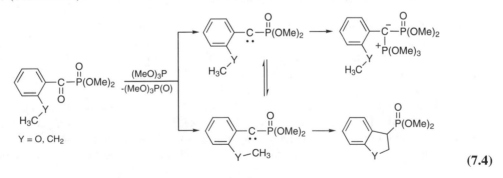

$$(7.4)$$

Evidence for carbene intermediates has been obtained from the reactions of *ortho*-substituted aroylphosphonates such as 2-methoxybenzoyl-, 2-ethylbenzoyl-, and 2-phenylbenzoylphosphonates. They undergo an insertion of the carbene into a C–H bond with the formation of dihydrobenzofuran or indan derivatives. The reaction is accompanied by the formation of ylide in a temperature-dependent manner.

When no *ortho* substituents are present on the benzene ring, the carbenes are readily trapped by trimethyl phosphite present in the reaction mixture to give ylidic phosphonates, which have a strong tendency for a subsequent rearrangement under the reaction conditions to give diphosphonates (Scheme 7.5).[41]

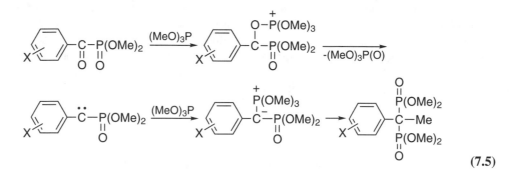

$$(7.5)$$

The Michaelis–Arbuzov reaction has been applied to *tris*(trialkylsilyl) phosphites, which, on reaction with aliphatic or aromatic acyl halides in Et_2O or C_6H_6 at room temperature, undergo the rearrangement in mild conditions to produce the valuable *bis*(trialkylsilyl) 1-oxoalkyl- or 1-oxoarylphosphonates in good yields (50–77%, Scheme 7.6).[32–34,66] Free acylphosphonic acids are obtained by treatment of the *bis*(trimethylsilyl) esters with EtOH or by exposure to air for several hours.[34]

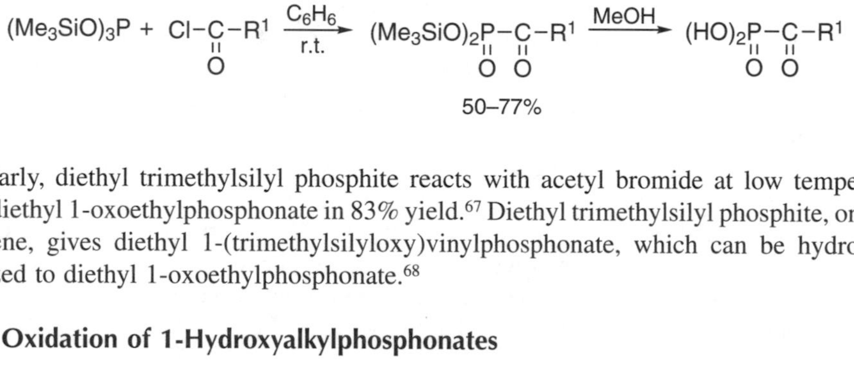

$$(Me_3SiO)_3P + Cl-\overset{\displaystyle O}{\underset{\displaystyle \|}{C}}-R^1 \xrightarrow[\text{r.t.}]{C_6H_6} (Me_3SiO)_2\overset{\displaystyle O}{\underset{\displaystyle \|}{P}}-\overset{\displaystyle O}{\underset{\displaystyle \|}{C}}-R^1 \xrightarrow{MeOH} (HO)_2\overset{\displaystyle O}{\underset{\displaystyle \|}{P}}-\overset{\displaystyle O}{\underset{\displaystyle \|}{C}}-R^1$$

50–77%

$$(7.6)$$

Similarly, diethyl trimethylsilyl phosphite reacts with acetyl bromide at low temperature to produce diethyl 1-oxoethylphosphonate in 83% yield.[67] Diethyl trimethylsilyl phosphite, on reaction with ketene, gives diethyl 1-(trimethylsilyloxy)vinylphosphonate, which can be hydrolyzed or alcoholyzed to diethyl 1-oxoethylphosphonate.[68]

7.1.1.2 Oxidation of 1-Hydroxyalkylphosphonates

The Michaelis–Arbuzov reaction works well for the less complex aroyl and alkanoyl chlorides, where purification by distillation is possible. Moreover, there has been less success in the preparation of α,β-unsaturated acylphosphonates, where multiple addition products are often observed. Recourse has been found in an attractive route that entails oxidation of dialkyl 1-hydroxyalkyl- or 1-hydroxyarylphosphonates produced by nucleophilic addition of dialkyl phosphites to carbonyl compounds under basic conditions (Pudovik reaction). Alkali metal salts of dialkyl phosphites are currently used in the Pudovik reaction, and the more common procedure of generating the anion involves the addition of a small amount of alcoholic alkoxide ion to the reaction mixture.[15,16,69,70] Neutral amines represent an alternative to the use of anionic bases. In recent years, the use of solid-phase materials as basic catalysts has been successfully developed (Scheme 7.7). One system involves the addition of basic alumina to the carbonyl compound and dialkyl phosphite; the other involves the addition of KF or CsF to the mixture of carbonyl compound and dialkyl phosphite.[71–73] Such a process for generating dialkyl 1-hydroxyalkylphosphonates is very flexible and accommodates a large variety of carbonyl compounds.

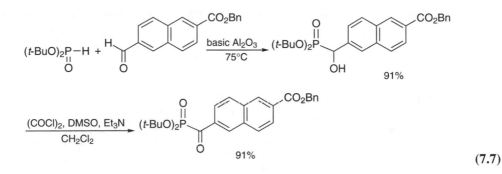

$$(7.7)$$

The first attempt to prepare dialkyl 1-oxophosphonates from 1-hydroxyphosphonates was reported in 1985,[74] but it is only quite recently that the transformation has been elaborated with success.[75] It has been found that a variety of reagents and conditions are commonly used to oxidize 1-hydroxyalkylphosphonates to give stable 1-oxoalkylphosphonates. However, the reaction conditions tend to work well with isolated examples but are not general for a range of substrates. For example, it is well known that MnO_2 in CH_2Cl_2 is a useful reagent, but oxidations are often difficult to reproduce because of the inconsistency between batches of oxidant and a gradual degradation of performance on standing. Benzylic hydroxyphosphonates are cleanly oxidized when treated with MnO_2,[75] whereas in the case of allylic hydroxyphosphonates, sonication in MeCN is required to obtain a highly reproducible conversion.[76]

Treatment of benzylic hydroxyphosphonates with powdered $KMnO_4$ in CH_2Cl_2/toluene results in a clean conversion to the corresponding benzoylphosphonates. Unfortunately, the allylic hydroxyphosphonates decompose with $KMnO_4$, and aliphatic hydroxyphosphonates are unreactive.[76] Allylic and benzylic hydroxyphosphonates are cleanly converted into the corresponding 1-keto-phosphonates using pyridinium dichromate (PDC) in CH_2Cl_2 at room temperature[50,77,78] or magtrieve (CrO_3) in refluxing MeCN.[76] In the same conditions, aliphatic phosphonates proved to be more resistant to oxidation than other phosphonates. Recently, it has been reported that under solvent-free conditions CrO_3/Al_2O_3 and $ZnCr_2O_7 \cdot 3H_2O$ are able to produce high yields of 1-ketophosphonates from 1-hydroxyalkyl-, benzyl-, 1-furylmethyl-, naphthylmethyl-, crotyl-, and cinnamylphosphonates. The reactions occur immediatly and are relatively clean.[79–81]

Swern oxidation [$DMSO/(COCl)_2/Et_3N/CH_2Cl_2$] at low temperature, which accomodates any ester groups at phosphorus, is used with success for the conversion of 1-hydroxybenzyl- and 1-naphthylmethylphosphonates into the corresponding 1-ketophosphonates in excellent yields (Scheme 7.7).[82–84]

Successful application of the Dess–Martin conditions (periodinane/CH_2Cl_2/r.t.) to the synthesis of 1-oxoalkylphosphonates has also been achieved. For example, room-temperature treatment of 1-hydroxy-2-aminoalkylphosphonates with Dess–Martin periodinane in CH_2Cl_2 generates the expected 1-oxo-2-aminoalkylphosphonates in good yields (84–88%).[72] Similarly, 1-hydroxyprop-argylphosphonate is oxidized in satisfactory yields (59%) under Dess–Martin conditions[73] or Pfitzner–Moffat conditions ($DMSO/DCC/Cl_2CHCO_2H$/toluene/0°C).[71] Another related method for the oxidation of dimethyl 1-hydroxycinnamylphosphonate in good yield (87%) has been recently reported under Parikh–Doering conditions (SO_3/Py complex/DMSO).[16]

7.1.1.3 Deprotection of 1,3-Dithianes

Diethyl 1,3-dithianylphosphonate is particularly well suited for the elaboration of complex molecules by carbanionic procedure.[85] For example, metallation of diethyl 1,3-dithianylphosphonate with n-BuLi at low temperature and alkylation of the carbanion produced thereby with 2,3:5,6-di-O-isopropylidene-4-acetyl-D-mannitol triflate followed by deacetylation and unmasking of the latent carbonyl group by treatment with a large excess of NBS in wet acetone gives the phosphonic analogue of KDO (3-deoxy-D-manno-2-octulosonic acid, Scheme 7.8). The resultant 1-ketophosphonate immediately cyclizes *in situ*.[86]

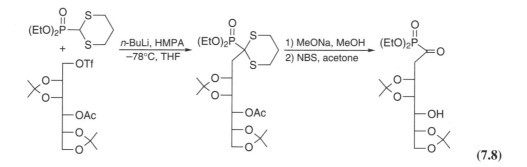

(7.8)

7.1.2 DIALKYL 2-OXOALKYLPHOSPHONATES

7.1.2.1 Michaelis–Arbuzov Reactions

A. E. Arbuzov and Razumov obtained diethyl 2-oxo-2-phenylethylphosphonate in 1934 by the reaction of triethyl phosphite with 2-bromoacetophenone (Scheme 7.9). However, the authors found that the compound formed had an unusually wide boiling range.[87]

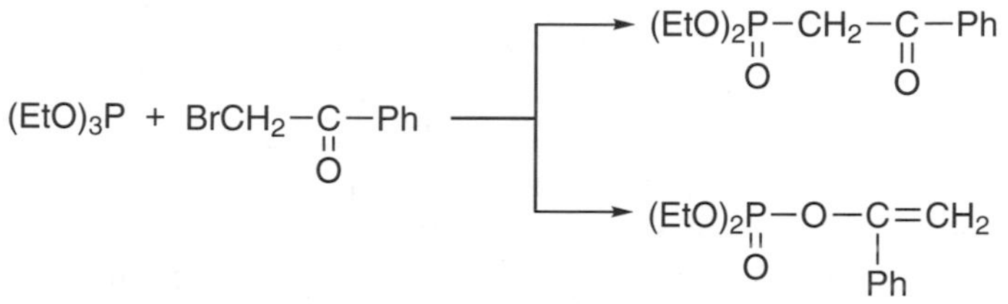

$$(EtO)_3P + BrCH_2{-}\overset{\text{O}}{\underset{\|}{C}}{-}Ph \longrightarrow \begin{cases} (EtO)_2\overset{\|}{\underset{O}{P}}{-}CH_2{-}\overset{\|}{\underset{O}{C}}{-}Ph \\ \\ (EtO)_2\overset{\|}{\underset{O}{P}}{-}O{-}\overset{}{\underset{Ph}{C}}{=}CH_2 \end{cases}$$

(7.9)

In a similar reaction of triethyl phosphite with bromoacetone, Razumov and Petrov obtained two isomeric products determined by analysis to be diethyl 2-oxopropylphosphonate but that differed considerably in their physical properties.[88] They conclude that these compounds were tautomeric forms of the phosphonate.

In all cases mentioned at that time, the products were assumed to be phosphonates of the type expected from a normal Michaelis–Arbuzov reaction.[87,89–92] However, in light of the Perkow reaction, all the β-ketophosphonates described in numerous papers and patents have to be formulated as the isomeric vinyl phosphates.

In an attempt to investigate the assumed keto–enol tautomerism of these compounds, a series of 2-oxoalkylphosphonates were synthesized from triethyl phosphite and α-bromoketones, and the amount of enol was determined by the bromine titration method.[92–94] These results, however, did not support the presence of the enol form because diethyl 1,1-dimethyl-2-oxopropylphosphonate, in which enolization is unlikely, had an enol content of 23% according to the consumption of bromine.

These contradictory results were elucidated later when, in 1952, Perkow[95,96] showed for α-halogenoaldehydes and Pudovik[97,98] and Allen and Johnson[99] for α-halogenoketones that both the Michaelis–Arbuzov reaction, with formation of 2-oxoalkylphosphonates, and the Perkow reaction, with formation of vinyl phosphates, can take place independently or simultaneously. It is important to mention that the formation of vinyl phosphates by an "anomalous Michaelis–Arbuzov reaction" was published at the same time by Whetstone and Harman in a patent issued in 1960.[100] The Michaelis–Arbuzov and Perkow pathways have been covered in several papers and reviews.[88,101–117]

The nature and position of the halogen atom in the carbonyl compounds and the conditions in which the experiments are carried out have a substantial effect on the course of the reaction. This has been well demonstrated by the results of the reactions of different trialkyl phosphites with chloro-, bromo-, and iodoacetone under varying conditions, yielding mixtures of dialkyl 2-oxopropylphosphonates and dialkyl 1-methylvinyl phosphates.[118–120] Thus, it has been shown that the relative yield of 2-oxoalkylphosphonates increases appreciably on going from chloro- to bromo- and iodoacetone.[118,121] 2-Oxoalkylphosphonates are prepared in fair to good yields from primary monobromoketones[91,122–128] and in high yields from primary iodoketones (Scheme 7.10).[118,119,129–134]

The reaction of trimethyl phosphite with an allenic chloromethyl ketone proceeds with the formation of an oxaphospholene and not through the Michaelis–Arbuzov or Perkow rearrangements.[135]

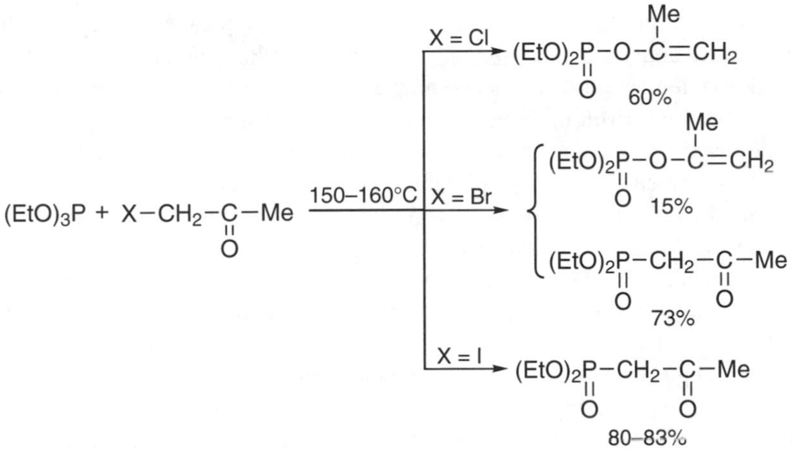

$$(EtO)_3P + X-CH_2-\overset{\overset{\displaystyle O}{\|}}{C}-Me \xrightarrow[150-160°C]{}$$

X = Cl → $(EtO)_2\overset{\overset{\displaystyle O}{\|}}{P}-O-\overset{\overset{\displaystyle Me}{|}}{C}=CH_2$ 60%

X = Br → $\begin{cases} (EtO)_2\overset{\overset{\displaystyle O}{\|}}{P}-O-\overset{\overset{\displaystyle Me}{|}}{C}=CH_2 \quad 15\% \\ (EtO)_2\overset{\overset{\displaystyle O}{\|}}{P}-CH_2-\overset{\overset{\displaystyle O}{\|}}{C}-Me \quad 73\% \end{cases}$

X = I → $(EtO)_2\overset{\overset{\displaystyle O}{\|}}{P}-CH_2-\overset{\overset{\displaystyle O}{\|}}{C}-Me$ 80–83%

(7.10)

With the presence of more than one halogen atom in the α-position of the halo ketone, the formation of 2-oxoalkylphosphonates becomes more difficult, and the reactions proceed exclusively according to the Perkow reaction. Thus, dichloromethyl, dibromomethyl, and tribromomethyl methyl ketones, in reaction with trialkyl phosphites, lead to the corresponding vinyl phosphates (Scheme 7.11).[91,97,122,136,137]

$$(EtO)_3P + Cl-\overset{\overset{\displaystyle |}{\underset{\underset{\displaystyle Cl}{|}}{CH}}}-\overset{\overset{\displaystyle O}{\|}}{C}-Me \longrightarrow (EtO)_2\overset{\overset{\displaystyle O}{\|}}{P}-O-\overset{\overset{\displaystyle |}{\underset{\underset{\displaystyle Me}{|}}{C}}}=CH-Cl$$

(7.11)

α-Haloketones, which have the halogen on a secondary carbon atom, react according to the Perkow reaction as demonstrated by the reaction of triethyl phosphite with 3-chloro-2,4-pentanedione, 3-chloro- and 3-bromo-2-butanones (Scheme 7.12).[122,126,138,139]

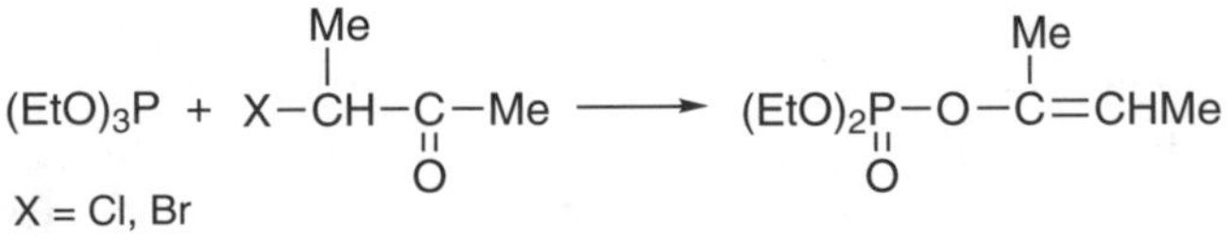

$$(EtO)_3P + X-\overset{\overset{\displaystyle Me}{|}}{CH}-\overset{\overset{\displaystyle O}{\|}}{C}-Me \longrightarrow (EtO)_2\overset{\overset{\displaystyle O}{\|}}{P}-O-\overset{\overset{\displaystyle Me}{|}}{C}=CHMe$$

X = Cl, Br

(7.12)

1,3-dichloro- and 1,3-dibromo-2-propanones react according to the Perkow reaction at low temperature, forming dialkyl 1-(halomethyl)vinyl phosphates.[97,99,136,140] These are further used as the halo substrate in a Michaelis–Arbuzov reaction (Scheme 7.13).[141]

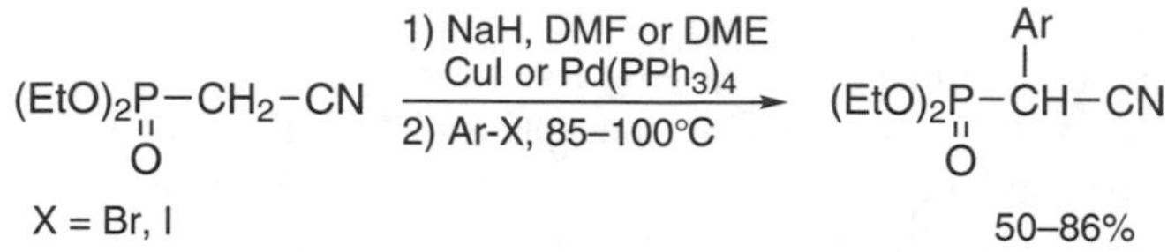

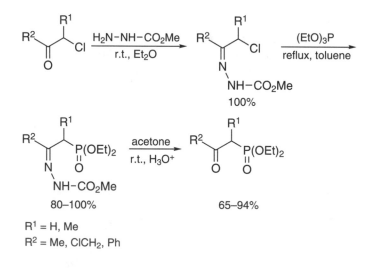

$$(EtO)_2\overset{O}{\underset{||}{P}}-CH_2-CN \xrightarrow[\text{2) Ar-X, 85–100°C}]{\substack{\text{1) NaH, DMF or DME} \\ \text{CuI or Pd(PPh}_3)_4}} (EtO)_2\overset{O}{\underset{||}{P}}-\overset{Ar}{\underset{|}{C}}H-CN$$

X = Br, I   50–86%

(7.13)

The yields of 2-oxoalkylphosphonates are also affected by the solvent and the reaction temperature. Trialkyl phosphites give the highest yields of vinyl phosphate when treated with chloroacetone at the lowest possible temperature at which the reaction would take place. Higher reaction temperatures increase the amount of phosphonate, thus favoring the Michaelis–Arbuzov reaction.[97,118] When bromoacetone is used as the halo component in the reaction, the amount of vinyl phosphate in the resulting mixture is greater the lower the temperature is kept. With iodoacetone the reaction products consisted predominantly of 2-oxopropylphosphonate, even when the reaction is carried out at low temperature. The same marked influence of the nature of the halogen and the temperature has been observed in the reaction of triethyl phosphite with 2-chloro-, 2-bromo-, and 2-iodoacetophenone.[121] When iodoacetone reacts with triethyl phosphite in Et$_2$O, diethyl 2-oxopropylphosphonate is formed in higher yield than in aqueous alcohol.[119]

To overcome the difficulties associated with the Perkow reaction, the carbonyl group has been protected before the Michaelis–Arbuzov reaction with suitable inexpensive, easy to remove, and nondeactivating groups.[142–145] For example, the (methoxycarbonyl)hydrazono group introduced by the reaction of 1-chloroalkyl ketones with methylhydrazinocarboxylate is well suited to this chemistry. The hydrazono derivative is first prepared in Et$_2$O or EtOH/Et$_2$O and then reacted with triethyl phosphite in refluxing toluene to give 2-hydrazonoalkylphosphonates in 80–100% yields. Hydrolysis of the latent carbonyl group with aqueous 3 M HCl/acetone solution at room temperature produces dialkyl 2-oxoalkylphosphonates in good overall yields (Scheme 7.14).[142–145]

R^1 = H, Me
R^2 = Me, ClCH$_2$, Ph

(7.14)

The same srategy has been developed with (bromomethyl)aryl ketones, which, before reacting with triethyl phosphite, are converted into oxime acetates or oxime ethers. A subsequent Michaelis–Arbuzov reaction leads to protected diethyl 2-aryl-2-oxoethylphosphonates in excellent yields (67–99%).[146,147]

7.1.2.2 Michaelis–Becker Reactions

The reaction of sodium dialkyl phosphite with α-halogenoketones leads to a mixture of products consisting of β-ketophosphonates, 1,2-epoxyphosphonates, and vinyl phosphates (Scheme 7.15).[148–151] The subject has been covered in the Section 4.1.1.2.

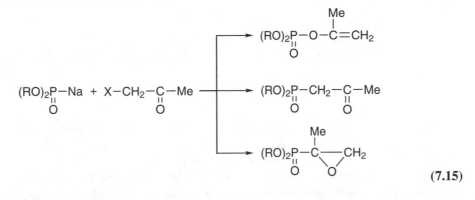

$$(7.15)$$

When the reaction is allowed to proceed with protected carbonyl groups, the expected protected β-ketophosphonates resulting from a nucleophilic substitution at the carbon atom of α-chloroketones are isolated in reasonable yields (Scheme 7.16).[152–154] However, unprotected γ-bromo-β-oxobutanoic ester[155] or amides[156,157] undergo phosphonylation in satisfactory yields with sodium or potassium dialkyl phosphites in THF at 0°C.

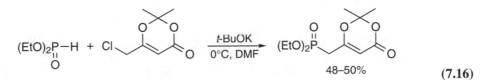

48–50% $$(7.16)$$

7.1.2.3 Hydration of 1-Alkynylphosphonates

The procedure for the hydration of dialkyl 1-alkynylphosphonates remained unchanged since the first report in 1966 (Scheme 7.17).[158] Thus, treatment of diethyl 1-alkynylphosphonates with aqueous H_2SO_4 in MeOH in the presence of $HgSO_4$ gives, after reflux for 15 h[158,159] or 48 h at room temperature,[160,161] a quantitative yield of pure diethyl 2-oxoalkylphosphonates. In all cases studied, the acetylenic phosphonates are transformed without formation of the isomeric 1-oxoalkylphosphonates. When the phosphorus substrate contains an acid-sensitive group, the hydration reaction can be performed in good yield at room temperature in aqueous THF in the presence of $HgCl_2$ (1 eq) and Py (1.5 eq).[162,163] The addition of EtOH to allenylphosphonates followed by acid hydrolysis represents an alternative route.[164]

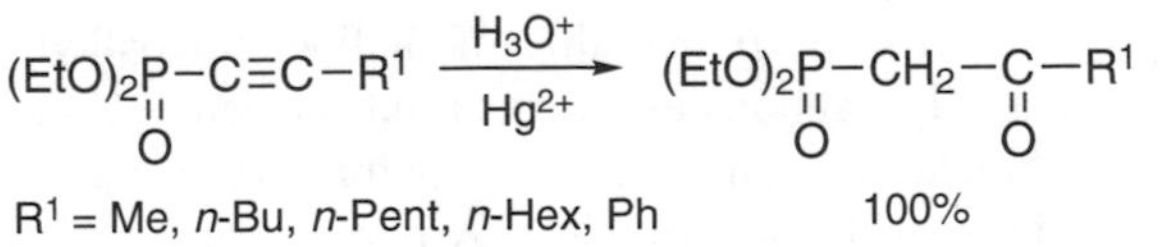

$$(EtO)_2\underset{\underset{O}{\|}}{P}-C{\equiv}C-R^1 \quad \xrightarrow[Hg^{2+}]{H_3O^+} \quad (EtO)_2\underset{\underset{O}{\|}}{P}-CH_2-\underset{\underset{O}{\|}}{C}-R^1$$

R^1 = Me, n-Bu, n-Pent, n-Hex, Ph   100% $$(7.17)$$

7.1.2.4 Hydrolysis of Enaminophosphonates

A relatively general procedure for the conversion of diethyl 1-alkynylphosphonates to diethyl 2-oxoalkylphosphonates via enaminophosphonates has been developed. When diethyl 1-alkynylphosphonates are heated under reflux with a 10 M excess of primary or secondary amines, diethyl

2-(alkylamino)-1-alkenylphosphonates are produced in fair to good yields.[165–167] The addition to the triple bond is complete in 20 h for Et_2NH[168] and in 3–5 days for n-$BuNH_2$.[167] The subsequent hydrolysis of the resulting enaminophosphonates with oxalic acid[168] or TsOH[169] in a two-phase system at room temperature affords diethyl 2-oxoalkylphosphonates in excellent yields (76–94%, Scheme 7.18).

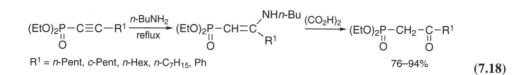

$$R^1 = n\text{-Pent, } c\text{-Pent, } n\text{-Hex, } n\text{-}C_7H_{15}, \text{ Ph} \qquad\qquad 76\text{–}94\%$$

(7.18)

Although some difficulties are encountered in the addition of primary or secondary amines to the triple bond of dialkyl 1-alkynylphosphonates, the related diethyl 2-bromopropenyl-phosphonate[170,171] and diethyl 2,3-dibromopropenylphosphonate[172,173] appear as efficient precursors of β-ketophosphonates. For example, diethyl 2,3-dibromopropenylphosphonate, on treatment at room temperature with diethylamine, readily undergoes an elimination–addition reaction in Et_2O to give diethyl 3-bromo-2-(diethylamino)-2-propenylphosphonate in excellent yield (90%).[173] The bromine atom at C-3 can be easily replaced by phenoxy, phenylthio, alkoxy, or alkylthio groups to give 3-alkoxy-, 3-phenoxy-, and 3-phenylthio-2-(diethylamino)-2-propenylphosphonates in 70–90% yields. Unmasking of the latent carbonyl group by acid hydrolysis of the crude compounds at room temperature produces the corresponding diethyl 2-oxoalkylphosphonates in 52–85% yields (Scheme 7.19).[173]

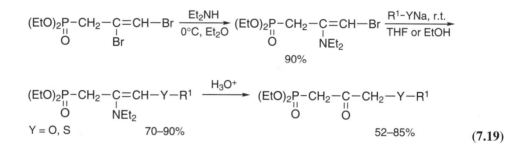

(7.19)

Dialkyl allenylphosphonates, readily obtained from propargyl alcohols and dialkyl chlorophosphites, are converted into dialkyl 2-oxoalkylphosphonates in 60–80% yields by nucleophilic addition of diethylamine and subsequent hydrolysis of the corresponding enamines.[174,175]

7.1.2.5 Acylation of Alkylphosphonates

7.1.2.5.1 With Esters

A relatively general procedure for the preparation of dialkyl 2-oxoalkylphosphonates by direct acylation of dialkyl 1-lithioalkylphosphonates has been introduced by Corey and Kwiatkowski in 1966.[176] The use of phosphonate carbanions as nucleophiles in reaction with carboxylic esters avoids the problems associated with the Michaelis–Arbuzov reaction. The reaction sequence is initiated by the addition at low temperature of dimethyl 1-lithiomethylphosphonate (2 eq and frequently more) to a carboxylic ester (1 eq) to give the transient lithium phosphonoenolate. The dimethyl methylphosphonate, being readily available and easy to eliminate, is the most frequently used phosphonate, but other phosphonates such as diethyl and diisopropyl methylphosphonates can be used. When the resulting enolate is treated with acid, dimethyl 2-oxoalkylphosphonate is produced in moderate to good yields (45–95%, Scheme 7.20). The reaction has been achieved with primary,[176–235] secondary,[236–238] and tertiary carbanions.[239] The reaction has also been applied to the synthesis of tetraalkyl *bis*(β-ketophosphonates) by treatment of dicarboxylic esters with dimethyl 1-lithiomethylphosphonate in excess (4 eq) at low temperature.[240]

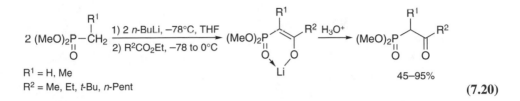

$$R^1 = H, Me$$
$$R^2 = Me, Et, t\text{-}Bu, n\text{-}Pent$$

$$45–95\%$$

(7.20)

In contrast to the results reported with alkylphosphonates, significantly better results are obtained with the more nucleophilic lithium salts derived from dialkyl chloromethylphosphonates. Dialkyl 1-lithio-1-chloromethylphosphonates, prepared by metallation of dialkyl chloromethylphosphonates with *n*-BuLi at low temperature, react with stoichiometric amounts of carboxylic esters to give transient lithium phosphonoenolates. On treatment with diluted HCl, dialkyl 1-chloro-2-oxoalkylphosphonates are isolated in 72–82% yields (Scheme 7.21).[241]

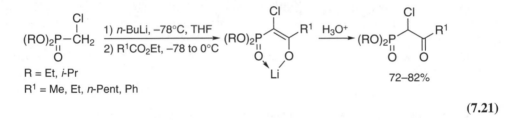

$$R = Et, i\text{-}Pr$$
$$R^1 = Me, Et, n\text{-}Pent, Ph$$

$$72–82\%$$

(7.21)

A drawback in this important procedure is the use of dimethyl 1-lithiomethylphosphonate in excess (2 eq) because this reagent acts as a phosphonomethylation agent and as a base. Moreover, the reaction is not atom efficient because one equivalent of reagent is lost. Effectively, in a carbanionic process, treatment of dialkyl 1-lithioalkylphosphonates with electrophiles bearing electron-withdrawing groups (EWG) generates new species containing a more acidic methylene group. The carbanion initially formed undergoes a facile transprotonation by the new species to regenerate the starting phosphonate and to produce the α-functionalized 1-lithioalkylphosphonate, which can in turn undergo further reactions (Scheme 7.22).

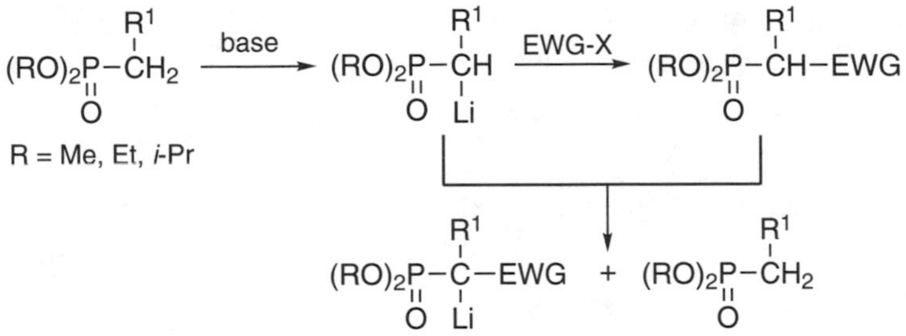

$$R = Me, Et, i\text{-}Pr$$

(7.22)

These drawbacks are overcome when the standard procedure is developed in the presence of a metallation agent in excess. Thus, when the reaction between dialkyl alkylphosphonates and carboxylic acid esters is allowed to proceed in the presence of LDA in excess (2 eq) in THF at low temperature, the resulting lithium phosphonoenolate is obtained in quantitative yield. Monitoring of the reaction by ^{31}P-NMR shows the appearance of one signal from the lithium phosphonoenolate and the complete disappearance of the signal of the starting phosphonate. The dialkyl 2-oxoalkyl-phosphonates are isolated in good to excellent yields on quenching in acidic medium (Scheme 7.23).[242]

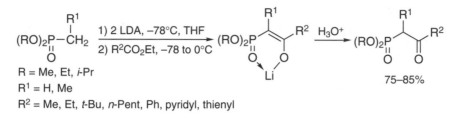

$$(7.23)$$

The synthesis of diethyl 1,1-difluoro-2-oxoalkylphosphonates has been reported by a cerium-mediated reaction between diethyl 1-lithio-1,1-difluoromethylphosphonate and ethyl esters at low temperature in THF. After acidic quench at low temperature, diethyl 1,1-difluoro-2-oxoalkylphos-phonates are isolated in good yield.[243,244] Addition of diethyl 1-lithio-1,1-difluoromethylphosphonate to di-*tert*-butyl oxalate at low temperature in THF affords exclusively the addition product, an unusually stable *tert*-butyl hemiketal.[245] When a solution of this material is heated at reflux in C_6H_6 or chromatographed on silica gel, only a few percent is converted to the hydrate. This conversion is best effected by dissolving the hemiketal in MeCN followed by reaction with aqueous $NaHCO_3$. Subsequent azeotropic dehydration with C_6H_6 affords diethyl 1,1-difluoro-2-oxo-2-(*tert*-butoxycar-bonyl)ethylphosphonate.[245]

7.1.2.5.2   With Acyl Chlorides

Another attractive and general route to dialkyl 2-oxoalkylphosphonares involves the chemoselective acylation of α-metallated dialkyl alkylphosphonates with carboxylic acid chlorides. However, the major drawback that attends the use of dialkyl 1-lithioalkylphosphonates in reaction with acyl chlorides is the difficulty of controlling their high basicity. To attenuate the reactivity of dialkyl 1-lithioalkylphosphonates, they are converted quantitatively by transmetallation into dialkyl 1-cop-per(I)alkylphosphonates by treatment with CuI (1 eq) in THF at low temperature. On coupling with acyl chlorides (1 eq) at low temperature, these copper(I) intermediates are converted into dialkyl 2-oxoalkylphosphonates in good to excellent yields without undesired side reactions (Scheme 7.24).[246–272] Acyl chlorides can advantageously be replaced by anhydrides.[273] The reaction has been applied with success to primary or secondary phosphonate carbanions and to long-chain acyl chlorides, unsaturated acyl chlorides, halogenated acyl halides, aromatic and heteroaromatic acyl chlorides, and functionalized acyl chlorides. A large variety of substituents at phosphorus such as methyl, ethyl, isopropyl, and butyl are tolerated.[246–249]

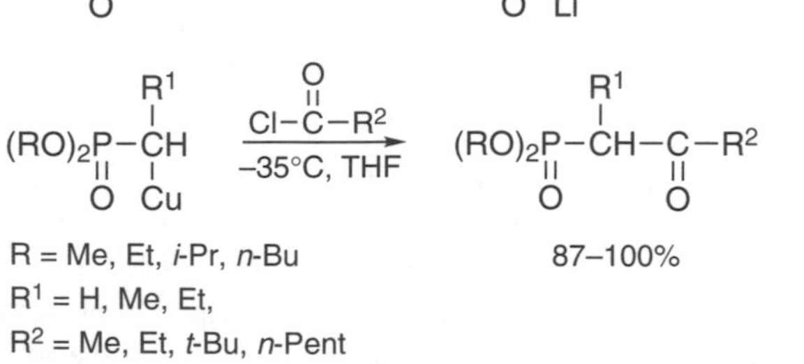

$$(7.24)$$

Acylation of lithiated phosphonate carbanions is observed at low temperature with tertiary carbanions and acyl chlorides[274–281] or anhydride.[282–285] By contrast, acylation of the secondary carbanion is less efficient.[286,287]

Another procedure for the preparation of dialkyl 2-oxoalkylphosphonates features the trapping of acyl chlorides by phosphonate carbanions produced by the conjugate addition of organolithiums[288] or organocuprates[289] to vinylphosphonates, as illustrated by the conversion of diethyl 1-(trimethylsilyl)vinylphosphonate into diethyl 1-ethyl-2-oxobutylphosphonate in 80% yield (Scheme 7.25).[288]

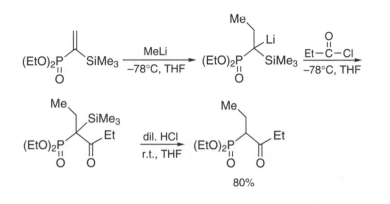

$$ (7.25) $$

Readily available dialkyl 1-bromo-1,1-difluoromethylphosphonates can also participate in the synthesis of 2-oxoalkylphosphonates. Thus, dialkyl 1-bromo-1,1-difluoromethylphosphonates, on reaction with activated Zn metal, are converted into difluoromethylphosphonylzinc bromides, which are coupled with carboxylic acid chlorides at room temperature to give dialkyl 1,1-difluoro-2-oxoalkylphosphonates in good yields.[290–293] Dimethyl 1-lithiomethylphosphonate undergoes transmetallation with MnCl$_2$ in THF to give dimethyl 2-oxobutylphosphonate in 80% yield on reaction with propionyl anhydride.[294]

It has been shown that acylated phosphonoacetates prepared from diethyl 1-(ethoxycarbonyl)methylphosphonate and aromatic or aliphatic carboxylic acid chlorides in the presence of Mg(OEt)$_2$ in THF[295] or MgCl$_2$ and Et$_3$N in CH$_2$Cl$_2$ or toluene[296,297] undergo an acid-catalyzed decarboxylation to give diethyl 2-oxoalkylphosphonates in fair to good yields (26–98%, Scheme 7.26). Although it is apparently limited to aroyl chlorides, the reaction can be used for the preparation of 2-oxoalkylphosphonates in fair yields from acyl chlorides (26–70%). The same protocol has been applied to the preparation of diethyl 2-oxoalkylphosphonates from diethyl 1-(hydroxycarbonyl)methylphosphonate and carboxylic acid chlorides in the presence of MgCl$_2$ and Et$_3$N in MeCN.[298,299] The solid support version of this reaction has also been reported.[300]

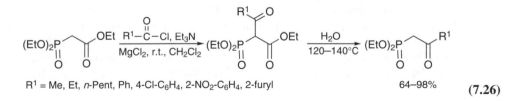

R^1 = Me, Et, n-Pent, Ph, 4-Cl-C$_6$H$_4$, 2-NO$_2$-C$_6$H$_4$, 2-furyl   64–98%

$$ (7.26) $$

The process has been extended to the synthesis of diethyl 1-fluoro-2-oxoalkylphosphonates. Thus, low-temperature acylation of diethyl 1-(hydroxycarbonyl)-1-fluoromethylphosphonate with carboxylic acid chlorides in the presence of n-BuLi in THF followed by decarboxylation provides an expeditious route for the synthesis of diethyl 1-fluoro-2-oxoalkylphosphonates in good yields (66–78%, Scheme 7.27).[301]

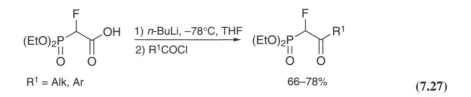

(7.27)

A clear illustration of the advantages of this synthetic procedure is provided by the conversion of diethyl 1-fluoro-1-(ethoxycarbonyl)-2-aryl-2-oxoethylphosphonates into the corresponding α-fluoro-β-keto esters. Because of the sensitive nature of the P–C bond, the reaction proceeds smoothly at room temperature by the MgCl$_2$-induced P–C bond cleavage in diethyl 1-fluoro-1-(ethoxycarbonyl)-2-aryl-2-oxoethylphosphonates, readily obtained by the reaction of diethyl 1-fluoro-1-(ethoxycarbonyl)methylphosphonate with aromatic carboxylic acid chlorides (Scheme 7.28).[302–306]

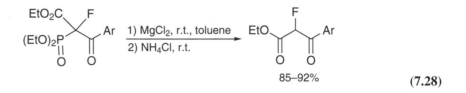

(7.28)

7.1.2.5.3   With Amides

It has been shown that ethyl *N*-Boc-L- or D-pyroglutamate reacts regioselectively at the amide function with diethyl 1-lithiomethylphosphonate in THF at low temperature to give diethyl *N*-(*tert*-butoxycarbonyl)-5-amino-5-(ethoxycarbonyl)-2-oxopentylphosphonate in 60% yield without racemization at the chiral center (Scheme 7.29).[307]

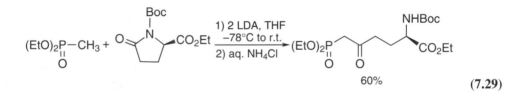

(7.29)

7.1.2.5.4   With Nahm–Weinreb Amides

An alternative route to dialkyl 2-oxoalkylphosphonates is based on the use of Nahm–Weinreb amides. Condensation of Nahm–Weinreb amides with dimethyl 1-lithiomethylphosphonate proceeds readily at low temperature. Compared with the ester route (Section 7.1.2.5.1), the Nahm–Weinreb amide route possesses significant synthetic advantages because only a slight excess of phosphonate carbanion is necessary to afford the β-ketophosphonates in good yields (66–85%).[270,308–310] This approach has been developed in the production of the optically active dimethyl 2-oxo-4-(*tert*-butyldimethylsilyloxy)-5-(methoxycarbonyl)pentylphosphonate (Scheme 7.30).[309]

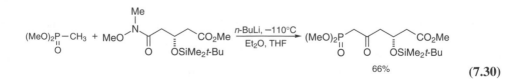

(7.30)

Several attempts have been undertaken to improve the procedure. For example, it has been found that addition of dimethyl 1-lithiomethylphosphonate to the corresponding amide-acid gives a clean reaction. Treatment of the crude product with CH_2N_2 affords the β-keto ester in better yield (77%, Scheme 7.31).[310]

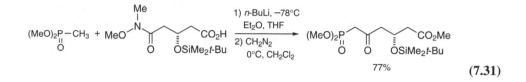

$$(7.31)$$

7.1.2.5.5 With Lactones

Lactones appear as valuable precursors for the synthesis of dialkyl 2-oxoalkylphosphonates. Thus, treatment of γ- or δ-lactones with dimethyl 1-lithiomethylphosphonate in excess (2 eq) at low temperature in THF proceeds via a hemiketal intermediate that, in the presence of phosphonate anion in excess, collapses to generate a dianion. Addition of Me_3SiCl gives the open-chain disilylated β-ketophosphonate derivatives quantitatively, presumably because of a faster rate of silylation of the hydroxyl group compared to the lactol formation.[311,312] After hydrolysis, diethyl β-keto-ω-(trimethylsilyloxy)alkylphosphonates are isolated in 82–96% yields (Scheme 7.32).[311,312] When only one equivalent of phosphonate carbanion is used, the opening of lactones remains incomplete because of the fast proton exchange between the dimethyl 1-lithiomethylphosphonate and the generated β-ketophosphonate. It is better to perform the reaction with 1 eq of dimethyl 1-lithiomethylphosphonate and 1 eq of LDA[311,312] instead of 2 eq of dimethyl 1-lithiomethylphosphonate.[313–321]

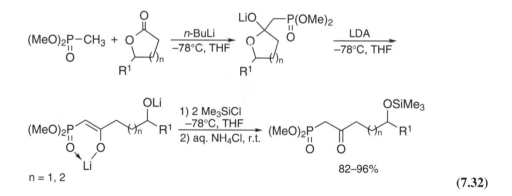

$$(7.32)$$

After quenching in an acidic medium, the hemiketals obtained from dimethyl 1-lithiomethylphosphonate and lactones undergo a reductive ring opening to give the open-chain tautomer (which, under acidic conditions, can recyclize back to the hemiketal). Oxidation of the open-chain tautomer produces dimethyl β-keto-ω-oxoalkylphosphonates in moderate to good yields (42–84%) using Jones reagent (acetone/aqueous 4 M H_2CrO_4),[313,314] DMSO/(COCl)$_2$/Et$_3$N/ CH_2Cl_2,[316,320,322] CrO_3/Py/CH_2Cl_2,[313,315,318,323] or Dess-Martin periodinane in CH_2Cl_2[324,325] (Scheme 7.33). Reduction of the mixture of hemiketal and hydroxyketone isomers with Et_3SiH in the presence of $BF_3 \cdot Et_2O$ traps the cyclic form to provide the corresponding pyran.[321] Alternatively, the hydroxyketone can be trapped from the mixture of the open and cyclic forms by esterification using $BnO(CH_2)_6CO_2H$/DCC/DMAP/DMF.[319]

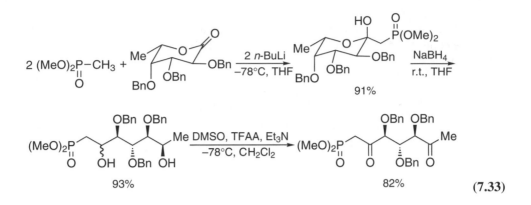

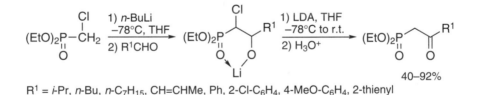

(7.33)

7.1.2.6 Dehydrochlorination of Phosphonochlorohydrins

An unusual application of the reaction between diethyl chloromethylphosphonate and carbonyl compounds is the formation of diethyl 2-oxoalkylphosphonates resulting from the treatment of the intermediate chlorohydrins at low temperature with an excess of LDA (2.2 eq) (Scheme 7.34). On treatment with LDA, the chlorohydrins undergo deprotonation and subsequent elimination of LiCl to provide selectively the stable lithium phosphonoenolates, which, under acidic conditions, give diethyl 2-oxoalkylphosphonates. Thus, under advantageous experimental conditions, diethyl 1-lithio-1-chloromethylphosphonate allows the direct phosphonomethylation of a large variety of aliphatic, aromatic, and heteroaromatic aldehydes in good overall yields (40–92%).[326]

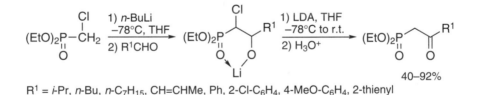

R^1 = i-Pr, n-Bu, n-C$_7$H$_{15}$, CH=CHMe, Ph, 2-Cl-C$_6$H$_4$, 4-MeO-C$_6$H$_4$, 2-thienyl

(7.34)

7.1.2.7 Reactions of 1-(Chlorocarbonyl)alkylphosphonates with Organometallic Reagents

Acylation of organocuprates or organomagnesium reagents with diakyl 1-(chlorocarbonyl)alkyl-phosphonates or 1-(chlorocarbonyl)-1-fluoromethylphosphonate has been reported to be an efficient route to dialkyl 2-oxoalkylphosphonates in moderate to good yields (50–94%, Scheme 7.35).[327–329] The reaction is based on the avaibility of dialkyl 1-(hydroxycarbonyl)alkylphosphonates and may be used with a large variety of α-substituted and unsubstituted phosphonates. In general, the conversion of diethyl 1-(chlorocarbonyl)alkylphosphonates into diethyl 2-oxoalkylphosphonates can be more effectively achieved using the less basic organocuprates.

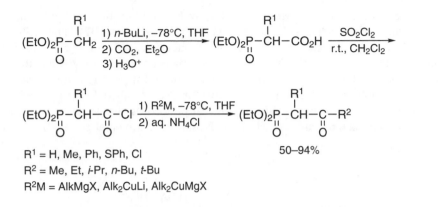

(7.35)

In a similar manner, the lithium derivative of 2-methyl-5-(4-methylphenylsulfinyl)pentene, obtained by treatment with LDA in THF at low temperature, is acylated with diethyl 1-(ethoxycarbonyl)-1,1-difluoromethylphosphonate to give a mixture of diastereomeric β-ketophosphonates.[330,331]

7.1.2.8 Oxidation of 2-Hydroxyalkylphosphonates

Dialkyl 2-oxoalkylphosphonates can also be derived from oxidation of the corresponding β-hydroxyphosphonates, which in turn are obtained by attack of the dimethyl 1-lithiomethylphosphonate on aldehydes. The first reported example was that of active MnO_2 in refluxing $CHCl_3$.[176,332] However, oxidations with MnO_2 are often difficult to reproduce because of inconsistencies among batches of oxidant and gradual degradation of performance on standing.

Several classical reagents including Swern's reagent [DMSO/$(CF_3CO)_2O$/Et_3N/CH_2Cl_2],[333,334] Dess–Martin periodinanc (Scheme 7.36),[335,336] PDC,[337,338] PCC,[187,339] CrO_3/H_2SO_4/H_2O,[340,341] and RuO_2/$NaIO_4$[342] have been successfully used.

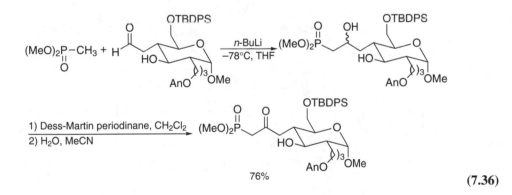

(7.36)

The synthesis of 2-oxoalkylphosphonates may also be accomplished in good yields under palladium catalysis conditions (Wacker process).[343,344]

7.1.2.9 SNP(V) Reactions

Several routes to β-ketophosphonates that rely on electrophilic phosphorus reagents have been developed. One of the most important involves the rearrangement of vinyl phosphates to β-ketophosphonates.[345–348] Thus, when camphor is treated sequentially with LDA and diethyl chlorophosphate at low temperature, the vinyl phosphate is produced quantitatively. When this product is treated at the same temperature with a second equivalent of LDA, the vinyl phosphate rearranges smoothly to the corresponding β-ketophosphonate (72%, Scheme 7.37). This 1,3-migration of phosphorus from oxygen to carbon, which is an intramolecular process, is also observed in other ring systems (cyclopentanone, cyclohexanone, 2-norbornanone, tetralone), yielding the β-ketophosphonates in good yields (70–88%). This reaction is the most useful method available for the synthesis of β-ketophosphonates derived from five- and six-membered-ring ketones. The starting materials are readily available, and the experimental protocol is simple. However, when enolates derived from acyclic ketones are treated under the same experimental conditions, the results are complex.

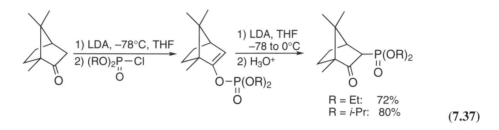

(7.37)

The vinyl phosphate–β-ketophosphonate rearrangement has also been explored with substituted cyclohexenones. The treatment of enones with LDA under conditions of kinetic control and subsequent reaction with diethyl chlorophosphate gives a single dienyl phosphate. On treatment with an additional equivalent of LDA, a single phosphonate is formed, the regioisomer with a C–P bond at the α'-position of the resulting α,β-unsaturated ketone. The respective phosphonates are isolated in an average yield of 70% (Scheme 7.38).[347,349]

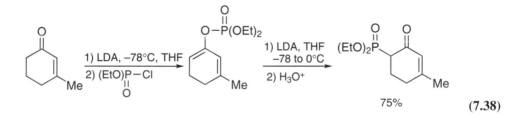

(7.38)

The vinyl phosphate–β-ketophosphonate rearrangement can be used to prepare α-phosphonolactones in high yields (68–78%, Scheme 7.39).[350] By a variable-temperature ^{31}P-NMR study it has been shown that abstraction of a proton from the vinyl phosphate is the slow step in the rearrangement and that, once this anion is formed, the rearrangement to the phosphonolactone is facile.[350]

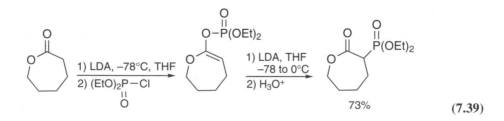

(7.39)

It has been demonstrated that it is possible to initiate a vinyl phosphate–β-ketophosphonate rearrangement via a halogen–metal exchange without significant competition from nucleophilic attack at phosphorus. The rearrangement is regiospecific, with the phosphoryl group of the product attached to the carbon formerly bearing the halogen (Scheme 7.40). This strategy has been employed to prepare cyclic β-ketophosphonates derived from L-(−)-verbenone and (−)-thujone in 90% and 88% isolated yields, respectively.[351]

Br
O
1) L-Selectride
−78°C, THF
2) (i-BuO)2P−Cl
O
O−P(Oi-Bu)2
60%
1) n-BuLi
−78°C, THF
2) NH4Cl
(i-BuO)2P
90%

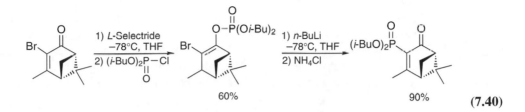

(7.40)

To explore the viability of chiral auxiliaries in the vinyl phosphate–β-ketophosphonate rearrangement, a series of vinyl phosphates derived from ephedrine, pseudoephedrine, (1*R*,2*S*,5*R*)-(−)-menthol, (1*S*,2*S*,3*S*,5*R*)-(+)-isopinocampheol, (*S*)-(−)-2-methylbutanol, binaphtol, and (2*R*,4*R*)- or (2*S*,4*S*)-pentane-2,4-diol have been examined.[352,353] (*S*)-(−)-2-methylbutanol and 2,4-pentane-2,4-diol derivatives appear as the most attractive chiral auxiliaries.

The reaction of diethyl chlorophosphite with ketone enolates has been explored as a direct route to β-ketophosphonates (Scheme 7.41). Instead of attempting isolation of phosphite intermediates, an *in situ* oxidation of the intermediate products is accomplished by leaving the products open to air. The use of these standard conditions results in the formation of a mixture of C–P and O–P products. The effect of solvents on the C–P/O–P ratio has been explored, and in many cases the use of Et_2O/HMPA improves the ratio. By application of the chlorophosphite/oxidation approach a variety of ketones have been converted into their respective β-ketophophonates in attractive yields.[354]

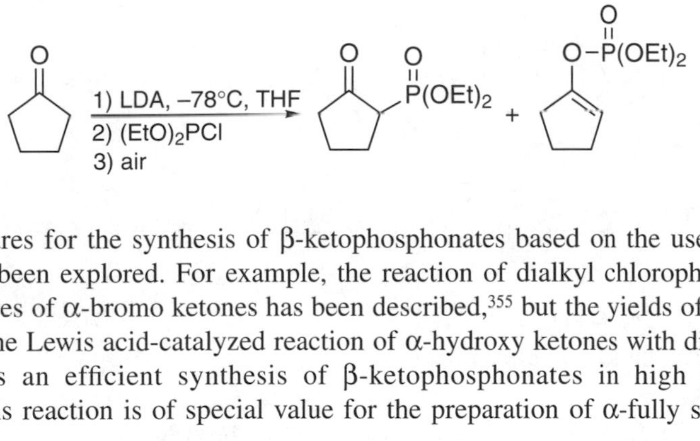

O
1) LDA, −78°C, THF
2) (EtO)2PCl
3) air
O
O
P(OEt)2
+
O
O−P(OEt)2

(7.41)

Other procedures for the synthesis of β-ketophosphonates based on the use of a phosphorus electrophile have been explored. For example, the reaction of dialkyl chlorophosphates with the dilithium derivatives of α-bromo ketones has been described,[355] but the yields of isolated products remain modest. The Lewis acid-catalyzed reaction of α-hydroxy ketones with dialkyl chlorophosphites constitutes an efficient synthesis of β-ketophosphonates in high yields (92–96%, Scheme 7.42). This reaction is of special value for the preparation of α-fully substituted β-ketophosphonates.[356]

(EtO)2P−Cl + HO
n-Pr Me
Me
O
FeCl3
r.t., hexane
(EtO)2P
n-Pr Me
Me
O O
96%

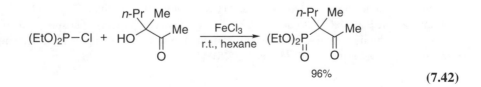

(7.42)

7.1.2.10 Other Preparations

Several general procedures for the synthesis of dialkyl 2-oxoalkylphosphonates have been discussed previously, but there are a number of other important methods worthy of merit. However, these methods are not always perfectly general and often require the synthesis of relatively sophisticated starting materials. Despite their interest, these procedures remain limited by the availability of the starting materials.

7.1.2.10.1 From Diazo Compounds

Dialkyl phosphites react with a variety of diazo compounds in refluxing C_6H_6 in the presence of a homogeneous catalyst, $Cu(acac)_2$, to give the corresponding dialkyl 2-oxoalkylphosphonates (Scheme 7.43). The optimum molecular ratio dialkyl phosphite/diazo compound is 5/1. The reaction proceeds via the carbenoid insertion into the P–H bond of the phosphite.[357–360] The procedure may be useful for the preparation of symmetric dialkyl 2,2′-diketophosphonates.

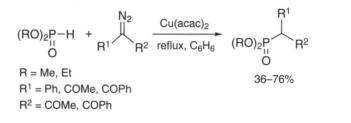

Cu(acac)2
reflux, C6H6
R = Me, Et
R1 = Ph, COMe, COPh
R2 = COMe, COPh
36–76%

(7.43)

7.1.2.10.2 From Nitriles

Since the enamine group is a precursor for the generation of a carbonyl group, a procedure for the preparation of β-ketophosphonates by the addition of diethyl 1-lithiomethylphosphonate to nitriles has been developed.[361] Because the generated anion is more reactive at carbon than at nitrogen, it is able to react with various electrophiles. Usual workup with aqueous NH_4Cl gives enaminophosphonates, which are converted into β-ketophosphonates by room temperature hydrolysis with 3 M H_2SO_4 (Scheme 7.44).[361,362] The reaction has found developements in the synthesis of 3-furylphosphonates.

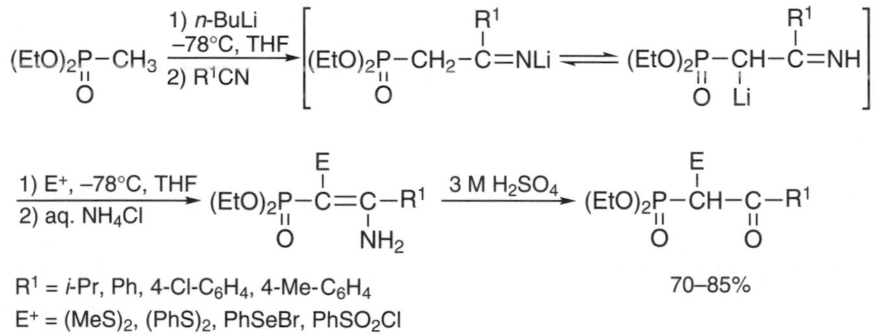

1) n-BuLi
−78°C, THF
2) R1CN
1) E+, −78°C, THF
2) aq. NH4Cl
3 M H2SO4
R1 = i-Pr, Ph, 4-Cl-C6H4, 4-Me-C6H4
E+ = (MeS)2, (PhS)2, PhSeBr, PhSO2Cl
70–85%

(7.44)

7.1.2.10.3 From 1-(Trimethylsilyl)vinylphosphonates by Acylation

The Friedel–Crafts reaction has been used to introduce an acyl group at the α-position of phosphonoketene dithioacetals.[363,364] Under classical conditions using diethyl (1,3-dithiolan-2-ylidene)methylphosphonate and $MeCOCl/AlCl_3$ in CH_2Cl_2, the expected α-(acetyl)phosphonoketene dithioacetal is isolated in modest yield (38%). By contrast, the use of α-silylated phosphonoketene dithioacetals improves the yield up to 89% (Scheme 7.45).[363,364]

MeCH=CHCOCl, AlCl3
0°C, CH2Cl2
R1 = Me, MeCH=CH, Ph
89%

(7.45)

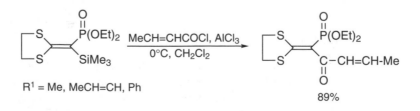

7.1.2.10.4 From 1,2-Epoxysulfones

α-Substituted epoxysulfones on treatment at room temperature with sodium diethyl phosphite in THF undergo ring opening in smooth conditions to produce the corresponding diethyl 2-oxoalkyl-phosphonates in generally good to excellent yields (Scheme 7.46).[365]

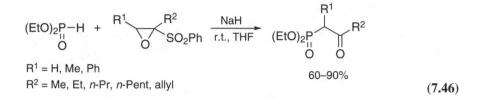

R¹ = H, Me, Ph
R² = Me, Et, n-Pr, n-Pent, allyl

60–90%

(7.46)

7.1.2.10.5 From 1,2-Epoxyphosphonates

Ring opening of diethyl 1,2-epoxyalkylphosphonates by refluxing ethanolic solutions of sodium dialkyl phosphites proceeds by attack at the β-carbon atom of the epoxide via the sodium phosphite–alcoholate equilibrium. For example, diethyl 1-methyl-1,2-epoxypropylphosphonate undergoes ring opening by sodium dialkyl phosphites with concomitant elimination of diethyl phosphite to afford dialkyl 1-methyl-2-oxopropylphosphonates in reasonable yields (Scheme 7.47).[366]

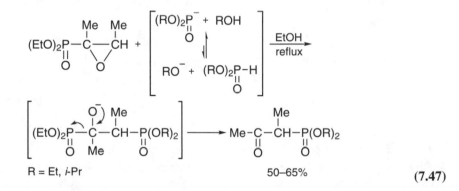

R = Et, i-Pr

50–65%

(7.47)

7.1.2.10.6 From 2,3-Epoxy-4-oxoalkylphosphonates

Diethyl 2,3-epoxy-4-oxoalkylphosphonates isomerize readily to diethyl 2,4-dioxoalkylphosphonates on heating in refluxing toluene in the presence of catalytic amounts of Pd(PPh₃)₄/dppe (Scheme 7.48).[367] The reaction conditions (4–20 mol% of the catalyst, 2–5 days at 120°C) depend on the substituents of the oxirane. Prolonged reaction time induces the formation of substantial amounts of deoxygenated compounds. The reaction is assumed to proceed through regioselective cleavage of the oxirane ring α to the carbonyl group by the Pd(0) reagent.[367]

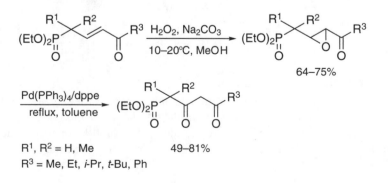

R¹, R² = H, Me
R³ = Me, Et, i-Pr, t-Bu, Ph

64–75%

49–81%

(7.48)

7.1.2.10.7   From 1-Chloroepoxydes

At room temperature trialkyl phosphites in large excess react with α-chloroepoxides to give the corresponding β-ketophosphonates in good yields (87–92%, Scheme 7.49).[368] However, this reaction is very sluggish and requires 3 to 4 days to proceed to completion.

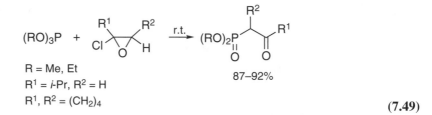

R = Me, Et
R¹ = *i*-Pr, R² = H
R¹, R² = (CH₂)₄

$$R = Me, Et$$
$$R^1 = \textit{i-}Pr, R^2 = H$$
$$R^1, R^2 = (CH_2)_4$$

87–92%

(7.49)

7.1.2.10.8   From 1-Nitroepoxides

A convenient synthesis of cyclic β-ketophosphonates has been reported from sodium dialkyl phosphites and α-nitroepoxides.[369] The phosphite anions attack regiospecifically the carbon β to nitro group with epoxide ring opening and elimination of sodium nitrite. All the cyclic dialkyl β-ketophosphonates are isolated in good yields (56–96%, Scheme 7.50).

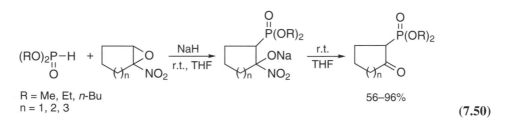

R = Me, Et, *n*-Bu
n = 1, 2, 3

56–96%

(7.50)

7.1.2.10.9   From 2-Oxoalkyliodoniums

Diethyl 2-aryl-2-oxoalkylphosphonates can be prepared from triethyl phosphite and α-ketomethyl aryliodonium intermediates. The precursors are generated from iodosobenzene–BF₃·Et₂O and silyl enol ethers and reacted directly *in situ* with triethyl phosphite to give diethyl 2-aryl-2-oxoalkyl-phosphonates in 62–83% yield (Scheme 7.51).[370]

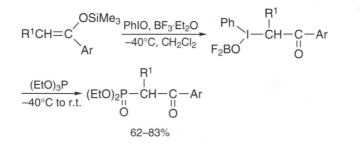

62–83%

(7.51)

7.1.2.10.10   From Diazomethylphosphonates

Addition at room temperature of diethyl diazomethylphosphonate to aldehydes in CH₂Cl₂ in the presence of catalytic amounts of SnCl₂ results in the formation of the corresponding diethyl β-ketoalkylphosphonates (Scheme 7.52). However, the yields of β-ketophosphonates are highly dependent on the nature of the aldehydes, and the reaction works well only with unencumbered aldehydes.[371]

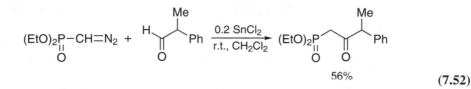

$$(7.52)$$

7.1.2.10.11 From Nitroalkenes

In the presence of DBU and Me₃SiCl in CH₂Cl₂, the addition of diethyl phosphite to the readily available nitroalkenes gives transient α-phosphonyl nitronates, which are smoothly converted into β-ketophosphonates on treatment with MCPBA. The procedure is useful to prepare a variety of diethyl 1-aryl-2-oxoalkylphosphonates (Scheme 7.53).[372]

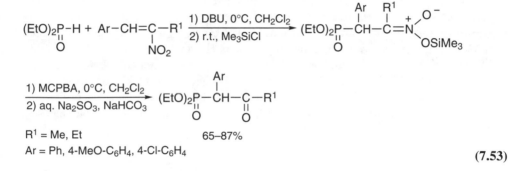

$$(7.53)$$

7.1.2.10.12 From Phosphonoketenes

Phosphonoketenes, generated from the corresponding diethyl 1-(chlorocarbonyl)alkylphosphonates and Et₃N, react with cyclopentadiene in excess to give phosphonocyclobutanones in satisfactory yields (50–64%, Scheme 7.54).[373]

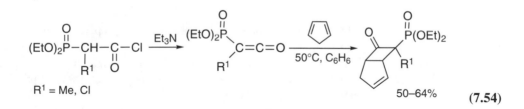

$$(7.54)$$

7.1.3 DIALKYL 3-OXOALKYLPHOSPHONATES

Dialkyl esters of steroidal 21*a*-phosphonic acids, which may be regarded as phosphonate analogues of 21-phosphates with the oxygen link replaced by a methylene group, have been prepared by the Michaelis–Arbuzov reaction of a trialkyl phosphite with a quaternary salt of a Mannich base derivative of a 20-ketosteroid.[374] The methiodides of the various steroidal Mannich bases have been converted into the corresponding phosphonates by heating in boiling trialkyl phosphite. The reaction involves a direct displacement of the nitrogen function to provide the phosphonium intermediate, which undergoes a valency expansion leading to the phosphonate.[374]

Most of the preparations of dialkyl 3-ketoalkylphosphonates employ the phosphonosilylation of α,β-unsaturated ketones. The first reported method involved the addition of dialkyl trimethylsilyl phosphite to α,β-unsaturated ketones.[375] On heating without solvent, methyl vinyl ketone reacts with mixed silicon–phosphorus reagents by exclusive 1,4-addition to give the enolsilane possessing the (Z)-geometry in good yields.[376–380] As long as the β-carbon and the carbonyl carbon atoms are at least comparably hindered, 1,4-adduct formation clearly represents the dominant reaction pathway. However, as the degree of hindrance around the β-carbon increases, the relative amount of 1,4-adducts formed correspondingly decreases.[381] The driving force for the production of trimethylsilyl enol ethers is rationalized by the principles of hard–soft acid–base theory. The soft phosphorus nucleophile preferentially attacks the softer β-carbon atom rather than the harder carbonyl carbon.[381]

The utility of this general strategy has been enhanced when it was reported that the addition of catalytic or stoichiometric Lewis acids such as $AlMe_3$ or Me_3SiOTf (Scheme 7.55) facilitates the conjugate addition of dialkyl phosphites.[382,383] Other Lewis acids (CF_3SO_3H, $TiCl_4$, $BF_3 \cdot Et_2O$) are less effective, and the adducts are obtained in much lower yields. Tetramethylguanidine (TMG) also appears to be an effective catalyst. In the presence of TMG, addition of dimethyl phosphite to α,β-unsaturated ketones proceeds smoothly at 0°C, giving rise essentially to conjugate 1,4-addition products in high yields after an easy workup. However, methyl vinyl ketone failed to give the Michael adduct, the 1,2-addition product being formed preferentially.[384] Recently, it has been found that 1,5,7-triazabicyclo[4.4.0]dec-5-ene (TBD) and its 7-methyl derivative (MTBD) are superior catalysts.[385] Methyl vinyl ketone reacts with dimethyl phosphite according to classical Michael-addition conditions (MeONa, MeOH, 65°C) to give dimethyl 3-oxobutylphosphonate in only 42% yield.[386]

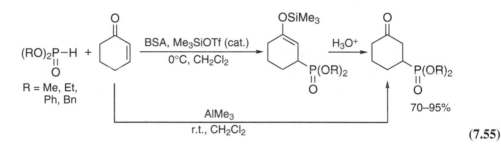

(7.55)

It has been shown recently that 3-oxocycloalkylphosphonates can be prepared by photochemical phosphonosilylation of cyclic α,β-unsaturated ketones. Thus, irradiation of a mixture of dialkyl trimetylsilyl phosphite and cyclic α,β-unsaturated ketones in MeCN provides essentially conjugate 1,4-addition products. Hydrolytic workup of the 1,4-phosphonosilyl adduct affords high isolated yields (78–91%) of cyclic dialkyl 3-ketophosphonates.[387]

A related route to 3-oxoalkylphosphonates involves the conjugate addition of trialkyl phosphites to α,β-unsaturated ketones and hydrolysis of the intermediate oxaphospholenes. The utility of this method is enhanced by the nucleophilic character of the intermediate oxaphosphorane, which facilitates the stereoselective aldol reaction resulting in the formation of β-substituted γ-ketophosphonates.[388–392] The pentacovalent oxaphospholene reacts with dialkyl azidocarboxylates to give β-hydrazido-γ-ketophosphonates in excellent yields.[393]

A large-scale synthesis of dialkyl 3-oxoalkylphosphonates proceeds by treatment of dialkyl 1-copper(I)alkylphosphonates with 2,3-dihalopropenes. Low-temperature hydrolysis of the resulting vinyl halide with H_2SO_4 in biphasic medium releases the γ-keto functionality (Scheme 7.56).[394] Thus, alkylation of 1-copper(I)alkylphosphonates with 1,2-dichloro- or 1,2-dibromo-2-propene is equivalent to the introduction of masked 3- or 4-oxoalkyl groups. In this way, 1-substituted or unsubstituted 3-oxoalkylphosphonates are prepared in high yields (90%).[394]

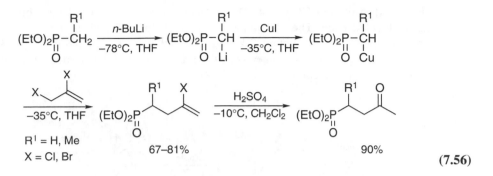

(7.56)

A related preparation of diethyl 3-oxoalkylphosphonates uses a zinc-mediated approach from diethyl 2-bromoethylphosphonate. Thus, treatment of diethyl 2-bromoethylphosphonate with zinc dust in THF at 30°C affords the corresponding alkylzinc bromide in 90% yield. The addition of the soluble CuCN·2LiCl at 0°C transmetallates the intermediate zinc compound to the copper compound. This copper–zinc reagent reacts with acyl chlorides in THF at 0°C to provide the corresponding diethyl 3-oxoalkylphosphonates in high yields (84–96%, Scheme 7.57).[395]

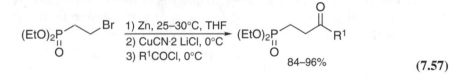

(7.57)

A recent route to diethyl 3-oxoalkylphosphonates employs the conjugate addition of acylcyanocuprate reagents to diethyl vinylphosphonate.[396] Acylcyanocuprate reagents, generated by the carbonylation of dialkylcyanocuprates with CO at −110°C in a solvent system consisting of THF, Et$_2$O, and pentane, react with diethyl vinylphosphonate to give on hydrolysis the corresponding diethyl 3-oxoalkylphosphonates in 69–82% yields (Scheme 7.58). Additionally, the anionic reaction intermediate can be trapped by reactive electrophiles such as allyl bromide.[396]

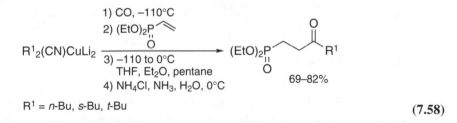

(7.58)

The more recently reported route to diethyl 3-oxoalkylhosphonates uses a zinc carbenoid-mediated approach, which is believed to proceed through the intermediacy of a cyclopropylzinc alkoxide. Thus, treatment of simple diethyl 2-oxoalkylphosphonates with the Furukawa-modified Simmons–Smith reagent provides a rapid and efficient preparation of 3-oxoalkylphosphonates.[397] The chain extension of simple, unfunctionalized β-ketophosphonates requires an excess (6 eq) of both Et$_2$Zn and CH$_2$I$_2$ at room temperature. The presence of α-substitution on the 2-oxoalkylphosphonate does not diminish the efficiency of the reaction (see Section 7.2.3.7).

Exposure of diethyl 2-oxocyclopentylphosphonate to ethyl(iodomethyl)zinc results in efficient ring expansion, providing diethyl 3-oxocyclohexylphosphonate in 83% yield (see Section 7.2.3.7).[397]

The use of diethyl 3-oxo-1-butenylphosphonate as dienophile in Diels–Alder reactions has been investigated. On reaction with cyclopentadiene with and without Lewis acid assistance [ZnCl$_2$,

LiClO$_4$, Eu(fod)$_3$] in Et$_2$O or CH$_2$Cl$_2$ at 0°C, a mixture of endo/exo bicycles is obtained with the acetyl group directed endo. The lanthanide Lewis acid, Eu(fod)$_3$, gives the best result.[398] The Diels–Alder reaction between diethyl 3-ketovinylphosphonate and (E)-1-acetoxy-1,3-butadiene produces only one regio- and stereomer with the acetyl group directed endo in the presence of ZnCl$_2$ in Et$_2$O at room temperature (Scheme 7.59).[399]

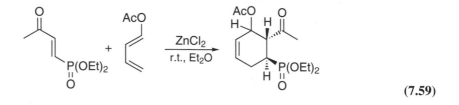

$$(7.59)$$

The reductive amination method can be used for the preparation in mild conditions of diethyl 3-aminoalkylphosphonates containing primary amino groups by simply treating diethyl 3-oxoalkyl-phosphonates with AcONH$_4$ and NaBH$_3$CN in MeOH at room temperature.[251,400] It appears that the reaction is subject to steric influence. Hydrolysis of diethyl 3-aminoalkylphosphonates with 8 M HCl followed by purification using ion-exchange resins (Amberlite IRA 410) provides pure 3-aminoalkylphosphonic acids in good yields (71–75%, Scheme 7.60).[251,400]

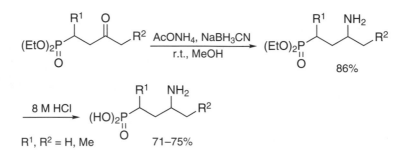

$$(7.60)$$

7.1.4   Dialkyl ω-Oxoalkylphosphonates

Reaction of *cis*- and *trans*-1,3-diphenyl-2,3-epoxy-4-bromo-1-butanones with triethyl phosphite at 120°C goes to completion to give diethyl *cis*- or *trans*-1,3-diphenyl-4-oxo-2,3-epoxybutylphospho-nates as the only isolated products in high yields.[401] Several preparations of 4-oxoalkylphosphonates are based on the use of masked ketones. For example, the Michaelis–Arbuzov reaction of 5-chloro-2-pentanone ethylene ketal with triethyl phosphite followed by unmasking of the latent carbonyl group gives diethyl 4-oxopentylphosphonate in modest yield (38%).[402]

The synthesis of dialkyl 3-oxoalkylphosphonates proceeding by treatment of dialkyl 1-cop-per(I)alkylphosphonates with 2,3-dihalopropenes has been extended to dialkyl 4-oxoalkylphospho-nates. Thus, 1,3-dichloro-2-butene is readily coupled with dialkyl 1-copper(I)alkylphosphonates at –35°C, and then the hydrolysis of the resulting vinyl halide with H$_2$SO$_4$ in biphasic medium at low temperature releases the δ-keto functionality (Scheme 7.61).[394] In this way 1-substituted or unsub-stituted diethyl 4-oxoalkylphosphonates are readily prepared in high yields (79–85%).[394]

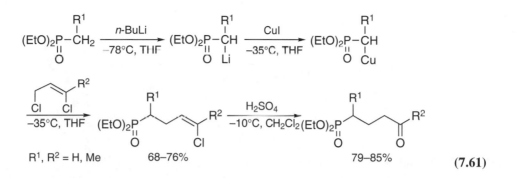

R¹, R² = H, Me   68–76%   79–85%   (7.61)

Preparation of diethyl 3-oxoalkylphosphonates using the copper–zinc reagents derived from diethyl 2-bromoethylphosphonate can be extended to diethyl 5-oxoalkylphosphonates. Effectively, these copper–zinc reagents have been found to react with cyclohexenone and benzylideneacetone in the presence of Me₃SiCl to give the 1,4-addition products in satisfactory yields (71–88%). 3-Iodo-2-cyclohexenone reacts with the copper–zinc reagents to give the addition–elimination product in 86% yield.[395]

The phosphonylation, via a Michaelis–Becker-type reaction,[403] and the phosphonomethylation, with lithiated phosphonate carbanions,[404,405] of protected ketones have been applied to the preparation of 4-oxoalkylphosphonates after acidic hydrolysis.

A number of addition reactions have been reported. For example, diethyl vinylphosphonate reacts with the enolates of cyclohexanone and 2-methylcyclohexanone, generated *in situ* by the treatment of silyl enol ethers with benzyltrimethylammonium fluoride, to give the corresponding Michael adducts in 31% and 44% yields, respectively.[406]

Michael addition has been applied to the formation of cyclopropanic systems. Thus, the addition of phosphonate carbanions generated from tetraethyl dichloromethylenediphosphonate by electrochemical techniques to methyl vinyl ketone provides a preparation of tetraethyl 2-acetylcyclopropane-1,1-diphosphonate in good yield (75%).[407]

The asymmetric Michael addition of metallated SAMP hydrazones to a variety of diethyl 1-alkenylphosphonates followed by oxidative cleavage of the 1,4-adducts produces 2,3-disubstituted 4-oxoalkylphosphonates with good yields (58–80%), but low to moderate de's (6–74%) and excellent ee's (> 93%).[408,409]

Several methods have been developed for incorporating a difluoromethylenephosphonate group. Thus, in the presence of a cobaloxime(III)/Zn redox system, the room-temperature addition of diethyl 1-bromo-1,1-difluoromethylphosphonate to methyl vinyl ketone in EtOH proceeds readily to give diethyl 1,1-difluoro-4-oxopentylphosphonate in 64% yield.[410] The copper-catalyzed cross-coupling reaction of difluoromethylphosphonylzinc bromide, generated from diethyl 1-bromo-1,1-difluoromethylphosphonate and Zn in DMF or DMA, has been explored with iodoalkenes. Thus, the reaction with (*E*)-4-iodobut-3-en-2-one at room temperature gives diethyl 1,1-difluoro-4-oxo-2-pentenylphosphonate in 91% yield.[411]

The addition of diethyl 1-iodo-1,1-difluoromethylphosphonate to 5-hexen-2-one catalyzed by Pd(PPh₃)₄ or copper metal constitutes an approach to 6-oxoalkylphosphonates. The reaction proceeds readily at room temperature in the absence of solvent and requires only few minutes to be complete. Treatment of the adduct with Zn in the presence of NiCl₂ in moist THF at room temperature provides diethyl 6-oxo-3-iodo-1,1-difluoroheptylphosphonate in 55% yield (Scheme 7.62).[412,413]

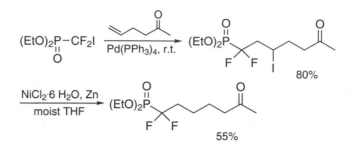

$$(7.62)$$

Radical additions have also been explored for the preparation of dialkyl ω-oxoalkylphospho-nates. Thus, addition of the highly electrophilic trifluoroacetonyl radical to the electrophilic diethyl vinylphosphonate, although not a favorable combination, gives diethyl 4-(trifluoromethyl)-4-oxobutylphosphonate in 42% yield.[414] The reaction of the xanthate derived from 2,4-difluoro-5-(chloroacetyl)toluene with diethyl 1,1-difluoro-3-butenylphosphonate takes place smoothly, on heating in refluxing cyclohexane in the presence of lauroyl peroxide as initiator, to give the corresponding 6-oxoalkylphosphonate in 72% yield.[415,416] Similarly, heating 2-allyl-2-methylcyclohexanone with the xanthate derived from diethyl 1-(ethoxycarbonyl)methylphosphonate in refluxing 1,2-dichloroethane delivers the expected 6-oxoalkylphosphonate (Scheme 7.63).[417] In most cases, the reductive removal of the xanthate group is readily achieved by treatment with Bu$_3$SnH in the presence of AIBN.

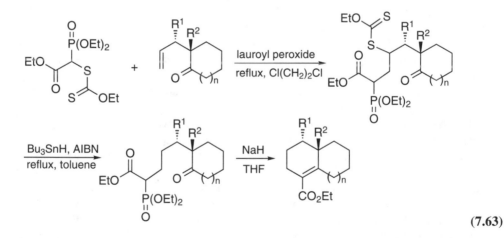

$$(7.63)$$

Radical reaction of bromomethylphosphonate and methyl vinyl ketone (Bu$_3$SnH/AIBN in C$_6$H$_6$ or toluene at reflux) proceeds via the addition of phosphonomethyl radical to give diethyl 4-oxopentylphosphonate in modest yield together with the reduced phosphonate precursor.[418] Another more efficient procedure utilizes 1-, 2-, and 3-iodoalkylphosphonates as radical precursors. The reaction products are regularly obtained in fair to good yields accompanied by small quantities of the reduced phosphonate precursors. Thus, from methyl vinyl ketone, the following ketophosphonates have been isolated: diethyl 4-oxopentylphosphonate in 88% yield from iodomethylphosphonate, diethyl 5-oxohexylphosphonate in 89% yield from iodoethylphosphonate, and diethyl 6-oxoheptylphosphonate in 82% yield from iodopropylphosphonate.[419,420]

Alkylation of a ketone lithium enolate with dialkyl 2-ethoxy-2-(bromo or iodo)-1-propenylphosphonates gives the corresponding 2,5-dioxophosphonates after hydrolysis.[421–423]

Syntheses of diketophosphonates by hydrolysis of α-hydroxy furylphosphonate[424] or by photooxidation of furylalkylphosphonates[425] have also been reported. Thus, diethyl 1-(5-methyl-2-furyl)-1-hydroxymethylphosphonate, prepared from 5-methyl-2-furaldehyde and diethyl phosphite by a base-catalyzed reaction, is treated with 1.25 M HCl in acetone at 58°C to give diethyl (E)-2,5-dioxo-3-hexenylphosphonate in 86% yield. This is transformed into its saturated analogue, diethyl 2,5-dioxohexylphosphonate, by treatment with 2 M HCl and NaI in acetone (Scheme 7.64).[424]

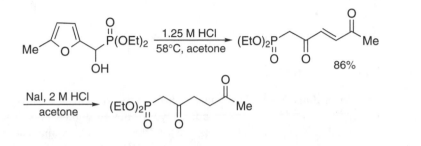

(7.64)

Several preparations of dialkyl ω-oxoalkylphosphonates have been described, with the keto group being generated by oxidation in various conditions: $O_3/CH_2Cl_2/-78°C$,[426] Jones reagent,[427] and PDC/CH_2Cl_2/room temperature.[428]

The reductive amination method has been applied to the preparation of 4-aminoalkylphosphonic acids from 4-oxoalkylphosphonates (Scheme 7.65).[394]

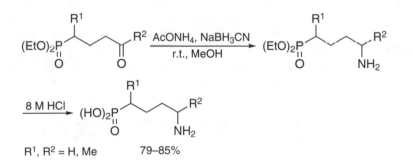

R^1, R^2 = H, Me   79–85%

(7.65)

7.2 SYNTHETIC APPLICATIONS OF KETOPHOSPHONATES

7.2.1 REACTIONS OF DIALKYL 1-OXOALKYLPHOSPHONATES WITH CONSERVATION OF THE P–C BOND

7.2.1.1 Reactions with Benzhydrylamine

One example of reductive amination of acylphosphonates has been reported.[429] Thus, reaction of dimethyl 1-oxoalkylphosphonates with benzhydrylamine results in the formation of an imine reduced *in situ* with $NaBH(OAc)_3$. The reaction is independent of the solvent, and THF, CH_2Cl_2, and $CHCl_3$ give similar results.[429] After selective removal of the benzhydrylic group by catalytic hydrogenation and hydrolysis with concentrated HCl, the 1-aminoalkylphosphonic acids are isolated in satisfactory yields (30–60%, Scheme 7.66).[429]

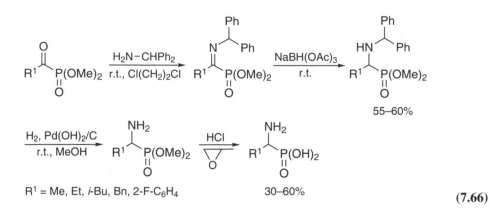

Ph
N
Ph
HN
O
R1
P(OMe)2
H2N–CHPh2
r.t., Cl(CH2)2Cl
NaBH(OAc)3
r.t.
55–60%
H2, Pd(OH)2/C
r.t., MeOH
NH2
HCl
P(OH)2
R1 = Me, Et, i-Bu, Bn, 2-F-C6H4
30–60%

(7.66)

7.2.1.2 Reactions with Hydrazines

Generally, the reaction of dialkyl acylphosphonates proceeds well with phenylhydrazine or 2,4-dinitrophenylhydrazine to provide the corresponding hydrazones in good yields. The reaction is conducted in EtOH at reflux[430] or at room temperature.[18,49,431,432] By contrast, the reaction of dialkyl acylphosphonates with hydrazine,[433,434] methylhydrazine,[433] and disymmetric *N,N*-dimethylhydrazine[435] proceeds only in AcOH to give the corresponding hydrazones in almost quantitative yields. In the absence of AcOH there is formation of acylhydrazides resulting from C–P bond cleavage. The hydrazones are smoothly converted into the corresponding hydrazines by treatment with excess $NaBH_3CN$ in EtOH.[434] The hydrogenolysis of hydrazones over Pd/C in EtOH[430] or AcOH[435] medium at room temperature and atmospheric pressure leads to α-aminoalkyl- or benzylphosphonates in moderate to good yields. However, the reduction of hydrazones with Zn in $AcOH/CF_3CO_2H$ appears as a more attractive procedure. Effectively, a convenient one-pot procedure has been elaborated that seems generally applicable for the preparation of diethyl α-aminoalkyl- or benzylphosphonates directly from diethyl acylphosphonates without isolation of the intermediate hydrazones (Scheme 7.67). The diethyl 1-aminophosphonates are hydrolyzed in boiling AcOH/HCl in a 1:3 ratio to give, after treatment with propylene oxide, the corresponding 1-aminophosphonic acids in good yields (52–75%).[435]

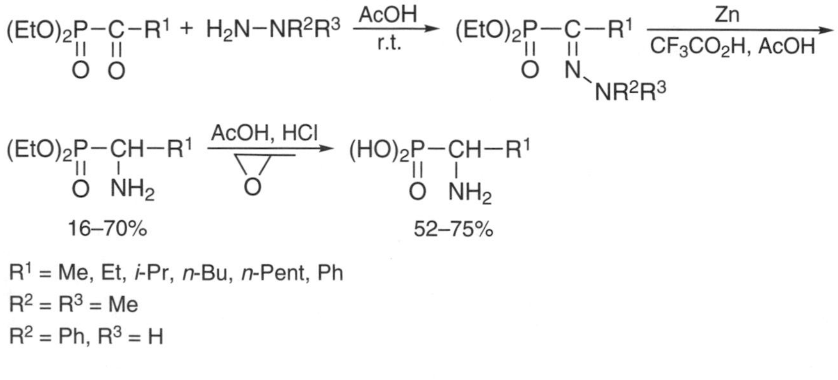

(EtO)2P–C–R1 + H2N–NR2R3
AcOH
r.t.
(EtO)2P–C–R1
NR2R3
Zn
CF3CO2H, AcOH
(EtO)2P–CH–R1
NH2
AcOH, HCl
(HO)2P–CH–R1
16–70%
52–75%
R1 = Me, Et, i-Pr, n-Bu, n-Pent, Ph
R2 = R3 = Me
R2 = Ph, R3 = H

(7.67)

When the phenylhydrazone of diethyl 1-oxopropylphosphonate in glacial AcOH is heated under reflux for 15 h, diethyl 3-methyl-2-indolylphosphonate resulting from a Fischer reaction is isolated in 60% yield (Scheme 7.68).[432] Similarly, reaction of the phenylhydrazones of diethyl 1-oxoalkylphosphonates with the Vilsmaier reagent ($POCl_3$/DMF) at 20–80°C for 5–10 h gives diethyl 4-substituted 1-phenyl-3-pyrazolylphosphonates.[436]

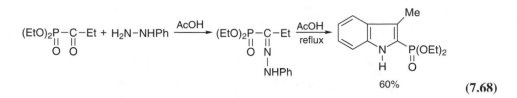

(7.68)

7.2.1.3 Reactions with Hydroxylamine

Hydroxylamine reacts with acylphosphonates to give the corresponding oximes. The reactions are usually carried out by reacting hydroxylamine hydrochloride with dialkyl acylphosphonates in the presence of Py in absolute EtOH (Scheme 7.69).[437–448] Yields are generally near quantitative, and the reaction almost always leads to mixtures of (E)- and (Z)-isomers, which can be observed by ^{31}P-NMR. The method can be applied successfully to the synthesis of 1-hydroxyiminophosphonates derived from protected amino acids.[449] Oxime ethers are prepared, as mixtures of (E)- and (Z)-isomers, similarly to oximes by reacting acylphosphonates with the respective O-methyl[450] or O-benzylhydroxylamine.[451–453] Another useful technique for the preparation of (E)-O-benzylhydroxyiminophosphonates involves the treatment of diethyl 1-hydroxyiminophosphonates with benzyl bromide in the presence of MeONa in boiling MeOH (53–70%).[452] The Michaelis–Arbuzov reaction of triethyl phosphite with O-benzylformylhydroxamyl chloride at 160°C gives diethyl benzyloxyiminomethylphosphonate in moderate yield (40%).[452]

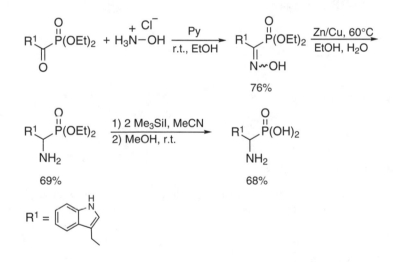

(7.69)

Reduction of diethyl 1-hydroxyiminophosphonates to aminophosphonates occupies a prominent place, and for that reason it has been investigated under a large variety of conditions. These include freshly prepared aluminum amalgam in EtOH/H$_2$O at room temperature (R^1 = aryl, 38–85%),[437,442] B$_2$H$_6$ in THF at 0°C (R^1 = alkyl, 56–58%),[438] H$_2$ with Raney-Ni in EtOH at 100°C under 80 atm (R^1 = Bn, 37–99%),[439,443] Zn dust in 99% HCO$_2$H at 65°C (R^1 = alkyl or aryl, 42–68%),[441,448] Zn/Cu in aqueous EtOH at 60°C (R^1 = 3-indolylmethyl, 69%, Scheme 7.69),[444] sodium triacetoxyborohydride (TABH)/TiCl$_4$ in methanolic solution of acetate buffer (R^1 = alkyl, benzyl, 50–60%),[446] BH$_3$·Py or BH$_3$·Et$_3$N under strongly acidic conditions (R^1 = Me or benzyl, 30–55%),[453] Et$_3$SiH in CF$_3$CO$_2$H at 40°C (R^1 = alkyl, aryl, benzyl, 54–79%),[452] and Me$_3$SiCl/LiBH$_4$ in THF at room temperature (R^1 = alkyl, 43–95%).[454]

Reduction of 1-hydroxyiminoalkylphosphonates to hydroxyaminoalkylphosphonates in yields of 35–90% has been reported using BH$_3$·Py complex in ethanolic HCl solution.[445]

Diisopropyl 1-hydroxyiminophosphonates can be oxidized by MCPBA in CH$_2$Cl$_2$ at room temperature to give 1-nitroalkylphosphonates in 50–65% yields (Scheme 7.70).[455]

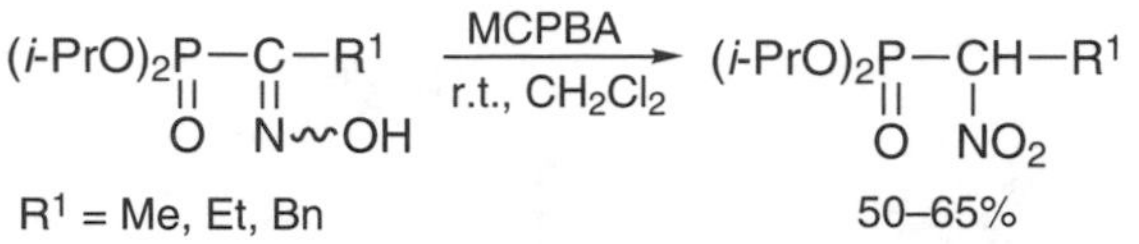

$$(i\text{-PrO})_2\underset{\substack{\| \\ O}}{\overset{}{P}}-\underset{\substack{\| \\ N\sim OH}}{\overset{}{C}}-R^1 \xrightarrow[\text{r.t., CH}_2\text{Cl}_2]{\text{MCPBA}} (i\text{-PrO})_2\underset{\substack{\| \\ O}}{\overset{}{P}}-\underset{\substack{| \\ NO_2}}{\overset{}{C}}H-R^1$$

$$R^1 = \text{Me, Et, Bn} \qquad\qquad 50\text{–}65\%$$

(7.70)

When oxime benzoates, prepared by benzoylation of dialkyl 1-hydroxyiminophosphonates in the presence of Py, are submitted to the action of tin hydrides under radical conditions, the weakness of the N–O bond is responsible for the fast formation of the reactive iminyl radical. In the example displayed in Scheme 7.71, the intermediate iminyl radical undergoes cyclization faster than the β-scission to nitrile to give a cyclic aminophosphonate.[456,457] The reaction has also been explored with dialkyl 1-benzoyloxyimino-2,2-dimethyl-3,4-pentadienylphosphonate and Bu$_3$SnH/AIBN in refluxing cyclohexane. In this case, the iminyl radical generated by stannyl radical addition on the benzoyl moiety leads to the sole formation of phosphonylated dihydropyridine in quantitative yield.[17]

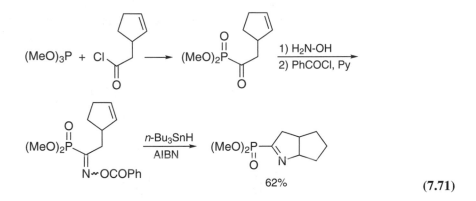

(7.71)

7.2.1.4 Reactions with Halogens

Halogenation is usually carried out by reacting halogenating agents with acylphosphonates in anhydrous conditions to avoid the concurrent C–P bond cleavage reaction. Thus, diethyl 1-oxo-2-bromoalkylphosphonates are prepared from Br$_2$/CCl$_4$ at 5°C.[458] The preferred conditions for chlorination consist in addition of SO$_2$Cl$_2$ to acylphosphonates at room temperature (Scheme 7.72). The acidic medium together with the presence of the electron-withdrawing phosphoryl group facilitates the enolization of the keto function so that the chlorination occurs at room temperature. However, it is quite impossible to completely suppress the dichlorination of the acylphosphonates.[459,460]

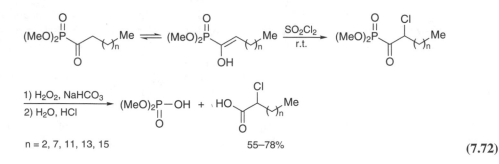

$$n = 2, 7, 11, 13, 15 \qquad\qquad 55\text{–}78\%$$

(7.72)

7.2.1.5 Reactions with Activated Methylene Compounds (Knoevenagel Reactions)

Activated methylene compounds such as malononitrile[461] and isocyanoacetates[462] react with acylphosphonates in the presence of bases, pipcridinc, or NaII to give β-functionalized-α-hydroxyphosphonates. These adducts undergo successive dehydration to the vinyl analogue followed by hydration of either the cyano (Scheme 7.73)[461] or the isocyano groups.[462] In the case of isocyanoacetate, the same result is obtained by conducting the reaction between acylphosphonates and isocyanoacetate in refluxing THF in the presence of Cu_2O.[462]

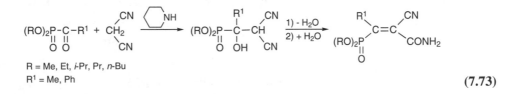

R = Me, Et, *i*-Pr, Pr, *n*-Bu
R¹ = Me, Ph

(7.73)

7.2.1.6 Preparation of 1-Diazoalkylphosphonates

Acylphosphonates, on treatment at room temperature with *p*-toluenesulfonylhydrazide hydrochloride in THF, are converted into *para*-toluenesulfonylhydrazones. In the presence of aqueous Na_2CO_3 at room temperature, these *para*-toluenesulfonylhydrazones undergo a remarkably facile Bamford–Stevens type of elimination to give 1-diazoalkylphosphonates in good to excellent yields (44–87%, Scheme 7.74).[9,11,30,36,37,463–468]

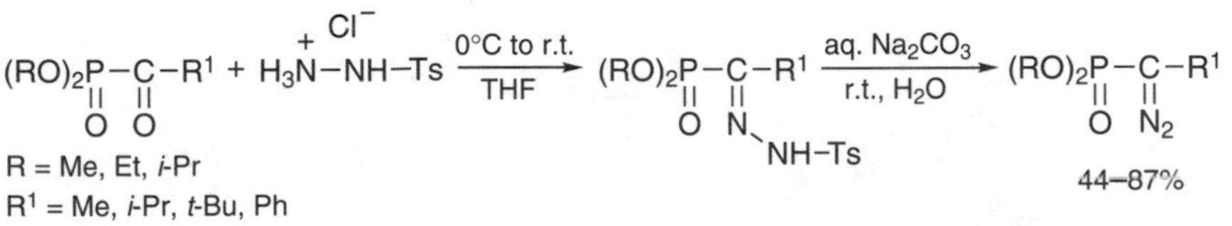

$$(RO)_2P-C-R^1 + H_3N^+-NH-Ts \xrightarrow[THF]{0°C \text{ to r.t.}} (RO)_2P-C-R^1 \xrightarrow[r.t., H_2O]{aq. Na_2CO_3} (RO)_2P-C-R^1$$

R = Me, Et, *i*-Pr
R¹ = Me, *i*-Pr, *t*-Bu, Ph

44–87%

(7.74)

With few exceptions, the dialkyl diazoalkylphosphonates are unusually stable thermally, and in this respect they are far superior to the respective carbonyl analogues.[469] Some can be distilled at reduce pressure and isolated in analytic purity, e.g., dimethyl 1-diazohexylphosphonate and dimethyl 1-cyclohexyl-1-diazomethylphosphonate. Others, such as dimethyl 1-cyclopropyl-1-diazomethylphosphonate, undergo spontaneous decomposition, whereas dimethyl 1-cyclobutyl-1-diazomethylphosphonate decomposes partially under the reaction conditions.[9] All these 1-diazoalkylphosphonates are decomposed photolytically or under Cu-powder catalysis to give carbenoids.

An alternative route to dimethyl 1-diazoalkylphosphonates involves the reduction of tosylhydrazones with $NaBH_4$ in MeOH under reflux. An α-elimination of *para*-toluenesulfinic acid occurs to give dimethyl 1-diazoalkylphosphonates in 80–95% yield after isolation.[470]

7.2.1.7 Preparation of Dialkyl 1-Hydroxy-2-nitroalkylphosphonates

The course of the reaction between acylphosphonates and 1-nitroalkane carbanions depends on the nature of the acyl group. Thus, nucleophilic addition of nitromethane to acetylphosphonates in the presence of di- or triethylamine results in the formation of dialkyl 1-hydroxy-1-methyl-2-nitroethylphosphonates (R¹ = Me, R² = H, Scheme 7.75).[471,472] However, the procedure is not applicable to the synthesis of dialkyl 1-hydroxy-1-phenyl-2-nitroethylphosphonate because the electron-withdrawing ability of the benzene ring destabilizes the molecule toward alkali and results in C–P or C–C bond cleavage. The fact that diisopropyl 1-hydroxy-1-phenyl-2-nitroethylphosphonate was obtained using Et$_3$N as a catalyst may be explained by the steric effect of the isopropyl groups, which hinder the nucleophilic attack of anionic species on the phosphorus atom and subsequently prevent the fragmentation process.[473] An effective and convenient procedure for the preparation of dialkyl 1-alkyl- or 1-aryl-1-hydroxy-2-nitroethylphosphonates is based on the addition of nitromethane to acylphosphonates under phase-transfer catalysis conditions using K$_2$CO$_3$/TBAF at room temperature.[474] The reaction proceeds smoothly and provides the expected nitroethylphosphonates in moderate to good yields.

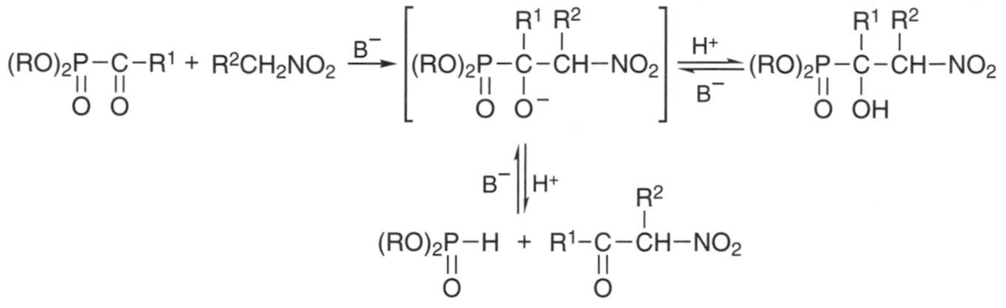

$$(RO)_2\underset{\underset{O}{\|}}{\overset{\overset{O}{\|}}{P}}-C-R^1 + R^2CH_2NO_2 \xrightarrow{B^-} \left[(RO)_2\underset{\underset{O}{\|}}{\overset{\overset{O}{\|}}{P}}-\underset{R^1}{\overset{R^1}{C}}-\underset{O^-}{\overset{R^2}{C}}H-NO_2\right] \underset{B^-}{\overset{H^+}{\rightleftharpoons}} (RO)_2\underset{\underset{O}{\|}}{\overset{\overset{O}{\|}}{P}}-\underset{R^1}{\overset{R^1}{C}}-\underset{OH}{\overset{R^2}{C}}H-NO_2$$

$$B^- {\big\Updownarrow} H^+$$

$$(RO)_2\underset{\underset{O}{\|}}{\overset{\overset{O}{\|}}{P}}-H + R^1-\underset{\underset{O}{\|}}{\overset{\overset{O}{\|}}{C}}-\overset{R^2}{C}H-NO_2$$

(7.75)

These compounds are converted to dialkyl (E)-1-alkyl or 1-aryl-2-nitrovinylphosphonates under mild conditions by using SOCl$_2$ (1 eq)/Py (2 eq) in boiling CHCl$_3$. Under these conditions the dehydration proceeds smoothly to provide a series of (E)-1-alkyl or 1-aryl-2-nitrovinylphosphonates in satisfactory yields (51–66%, Scheme 7.76).[475,476] The elimination may also be effected by using acetate instead of chloride as the leaving group.[471]

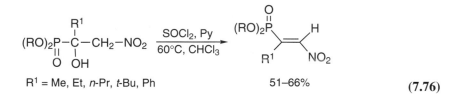

$$R^1 = Me, Et, n\text{-}Pr, t\text{-}Bu, Ph \qquad\qquad\qquad 51\text{–}66\%$$

(7.76)

Reaction between nitromethane carbanion and dialkyl (trichloroacetyl)phosphonate (R^1 = CCl$_3$) results in rearrangement to dialkyl 1-(nitromethyl)-2,2-dichlorovinyl phosphate with additional dehydrochlorination.[477]

7.2.1.8 Preparation of Tetraesters of 1-(Hydroxyalkylidene)-1,1-diphosphonates

Dialkyl acylphosphonates, on treatment with dialkylphosphites at 0°C in Et$_2$O in the presence of a catalytic amount of a relatively weak base (Et$_2$NH, n-Bu$_2$NH or Et$_3$N), produce symmetric or disymmetric tetraesters of 1-hydroxyalkylidenediphosphonates.[478–480] The yields for symmetric tetraesters are rather high (81–92%, Scheme 7.77), and sometimes they may be isolated by crystallization, especially in the case of tetramethyl esters. When the tetraesters are heated at high temperature (>120°C) or treated with a trace of alcoholate in alcohol instead of dialkylamine, they undergo a complete rearrangement to the isomeric phosphates.[481–485] A catalytic amount (about 5 mol%) of diethylamine or di-n-butylamine gives only hydroxy diphosphonate compounds, whereas the use of about 80–100 mol% of the same base yields the corresponding phosphonate–phosphates.[486,487] The presence of electron-withdrawing groups on the HO-bearing carbon atom favors the formation of the phosphate over the cleavage of the ester into diakyl phosphite and a carbonyl compound, which is favored by electron donor groups.[488]

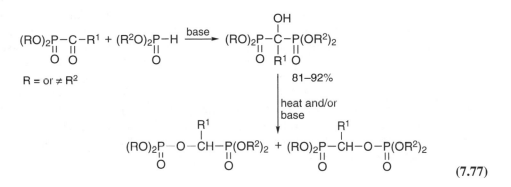

(7.77)

Hydrolysis of hydroxymethylenediphosphonates in aqueous HCl or with Me_3SiBr, followed by treatment with water, gives diphosphonic acids.[487,489] They are stable in solution at pH 1.6 up to 125°C, and in alkaline solution at pH 8.5–11.5 up to 195°C.

Chemically and enzymatically stable diphosphonates are commonly used in treatment of various diseases of bone and calcium metabolism. They bind strongly to calcium phosphate and inhibit its formation, aggregation, and dissolution. The affinity for the bone mineral is the basis for their use as skeletal markers and inhibitors of bone resorption.[490–494]

Diethyl acylphosphonates, on treatment with diethyl trimethylsilyl phosphite at room temperature, produce high yields of symmetric diethyl (1-trimethylsilyloxy)alkylidenediphosphonates (Scheme 7.78).[495,496] The same reaction has been applied to diphenyl(trimethylsilyl)phosphine.[497] By contrast, diethyl 2,2,2-trichloro-1-oxoethylphosphonate reacts with diethyl trimethylsilyl phosphite by abstraction of chlorine and phosphorylation of the generated enol form.[498]

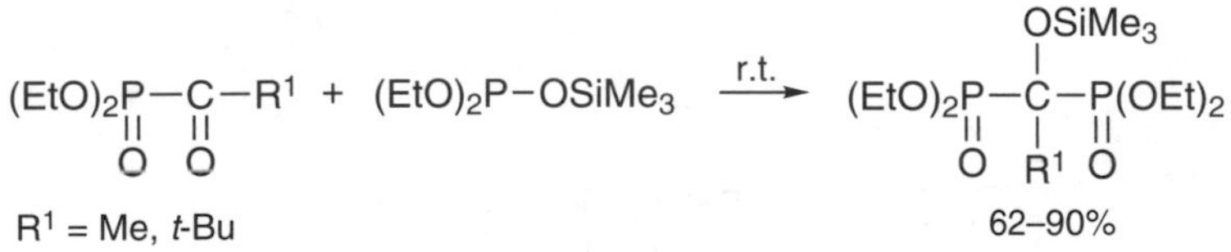

$$(EtO)_2\underset{O}{\underset{\|}{P}}-\underset{O}{\underset{\|}{C}}-R^1 \ + \ (EtO)_2P-OSiMe_3 \ \xrightarrow{\text{r.t.}} \ (EtO)_2\underset{O}{\underset{\|}{P}}-\underset{\underset{R^1}{|}}{\overset{\overset{OSiMe_3}{|}}{C}}-\underset{O}{\underset{\|}{P}}(OEt)_2$$

$R^1 = Me, t\text{-Bu}$ 62–90%

(7.78)

7.2.1.9 Preparation of 1,1-Difluoroalkylphosphonates

Conversion of 1-oxoalkylphosphonates into 1,1-difluoroalkylphosphonates by reaction with (diethylamino)sulfur trifluoride (DAST) has been explored using a variety of solvents and temperatures. The most satisfactory results are obtained by running the reaction neat at room temperature,[39,71,73,75,77,82–84,499–502] but the reagent is also used in large excess in CH_2Cl_2 solution (Scheme 7.79).

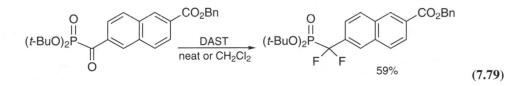

(7.79)

7.2.1.10 Wittig and Horner–Wadsworth–Emmons Reactions

Acylphosphonates are suitable carbonyl substrates for Wittig and Horner–Wadsworth–Emmons reactions, but with certain limitations.[19,503] These limitations are the result of competing side reactions involving proton transfer or acylation. Horner–Wadsworth–Emmons reactions fail with diethyl 1-oxoethylphosphonate and its 2-phenyl derivative but are successful with longer-chain acylphosphonates. Phenyl-stabilized phosphoranes give olefins by the Wittig reaction, but the corresponding Horner–Wadsworth–Emmons reaction is unsuccessful.[504] Ester- and nitrile-stabilized carbanions give olefins by both methods in most reactions studied,[504–506] and 2-oxoalkylidenetriphenylphosphoranes give the expected dialkyl 3-oxo-1-alkenylphosphonates on reaction with aliphatic and aromatic acylphosphonates.[506,507] In contrast to saturated acylphosphonates, dialkyl 1-oxo-2-alkenylphosphonates react with 2-oxoalkylidenetriphenylphosphoranes in C_6H_6 at 80°C to give stereoselectively the (E)-dienones with cis-arrangement of C=O and γ,δ-double bond. Usually, the isomeric dialkyl 2H-pyranylphosphonates are isolated, occasionally together with the nonconjugated isomeric dienones (Scheme 7.80).[508]

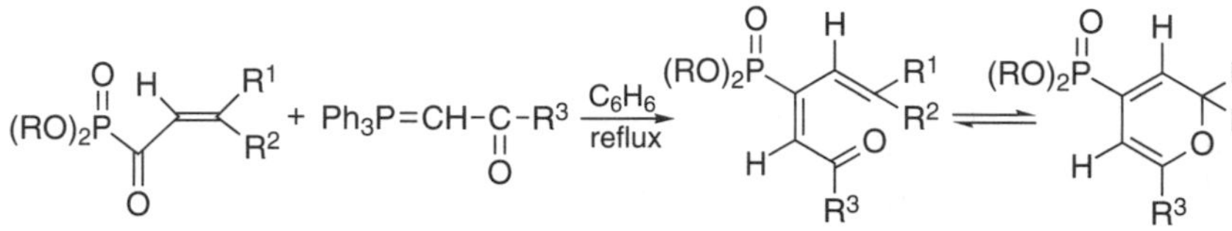

$$(7.80)$$

The Wittig reaction with methylenetriphenylphosphorane, generated by the reaction of $NaNH_2$ with methylenetriphenylphosphonium salts, gives olefins in fair yields (Scheme 7.81).[19,503] The use of n-BuLi is generally not successful. All these reactions give quite satisfying results by working in C_6H_6 or toluene at reflux.[19,504,506,507]

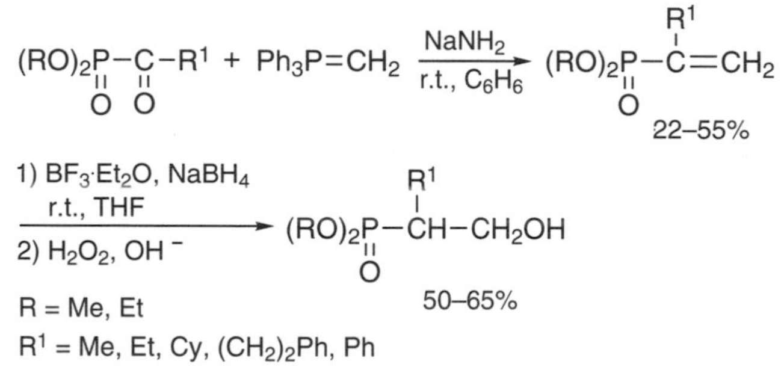

$$(7.81)$$

Whereas the Wittig reaction gives exclusively (E)-isomers, the Horner–Wadsworth–Emmons procedure gives predominantly (Z)-isomers. This result is contrary to expectation because phosphonate carbanions usually give better (E)-specificity than phosphoranes.[504]

7.2.1.11 Enolization of 1-Oxoalkylphosphonates

Dialkyl 2-phenyl-1-oxoethylphosphonates and a number of other β-aryl-α-ketophosphonates, varying in both the aryl group and the phosphorus substituents, are fully enolized in solution and presumably in the solid state. In all cases, the enol has exclusively the (E)-configuration, as evident from small P–H coupling values of 10–12 Hz. However, in spite of unexpected ease of enolization of these compounds, their enolization properties have only recently been investigated.[43] The first investigations were conducted with diethyl 1-oxoethyl- and 1-oxopropylphosphonates under mild basic conditions (Et₃N, CH_2Cl_2, room temperature). It has been shown that the enol form can be trapped on treatment with diethyl chlorophosphite, AcCl and Me₃SiCl,[509,510] Ac₂O,[43] silyltriflates,[43] or nonaflyl fluoride[511] to give the corresponding enol derivatives. In all cases, these derivatives are exclusively formed in the (E)-geometry with the alkyl and phosphonyl groups in trans position.

The palladium-catalyzed cross-coupling reaction of diethyl 1-nonaflatevinylphosphonates with alkynes proceeds well in refluxing toluene to afford diethyl 1-alkynylvinylphosphonates in good to excellent yields (71–100%, Scheme 7.82).[511]

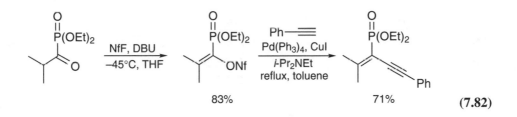

(7.82)

Low-temperature reaction of diethyl 1-oxopropylphosphonate with LiHMDS affords exclusively the (Z)-enolate, which, on condensation with benzaldehyde in excess (2.2 eq), gives a highly diastereoselective reaction leading to the formation of *syn*- and *anti*-3-hydroxy-2-methyl-3-phenylpropionic acids in a 97% yield, together with diethyl 1-hydroxybenzylphosphonate.[512]

Dimethyl 1-oxoalkylphosphonates react with benzoic anhydride and DBU in THF at room temperature to form enolbenzoates in moderate to good yields (43–86%). Only the (E)-isomer is observed. Catalytic asymmetric hydrogenation of enolbenzoate phosphonates has been performed in MeOH at room temperature under hydrogen pressure (4 atm) with DuPHOS-Rh and BPE-Rh catalysts (Scheme 7.83). Ligand optimization studies have been performed with each enolbenzoate to provide optimum ee's.[513]

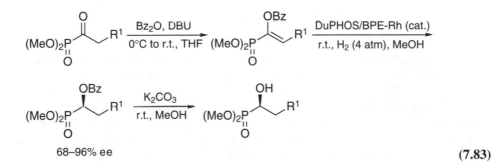

(7.83)

7.2.1.12 Deoxygenative Coupling of Ketones and Aldehydes by Low-Valent Lanthanides

Acyloxyphosphonates are formed in satisfactory yields from diethyl acylphosphonates and aliphatic or aromatic aldehydes in excess (4 eq) using a catalytic amount of Sm metal or SmI$_2$. No reaction takes place with aliphatic acylphosphonates, and the reaction of acylphosphonates with ketones does not afford the corresponding acyloxyphosphonates. On addition of a second carbonyl compound, aldehyde or ketone, the acyloxyphosphonates undergo a reductive elimination of carboxylic acids in the presence of SmI$_2$ (2 eq) to give the corresponding diethyl β-hydroxyphosphonates in reasonable yields (40–64%, Scheme 7.84).[514] The β-hydroxyphosphonates can further be converted into olefins on treatment with bases (Horner–Wadsworth–Emmons reaction), but all attempts to produce the olefins directly from acyloxyphosphonates and carbonyl compounds with an excess of SmI$_2$, elevated temperature, and polar solvents were unsuccessful.[514] Room-temperature treatment of diethyl acylphosphonates with SmI$_2$ (2 eq) in THF gives the reduced diethyl alkylphosphonates in high yield.[514]

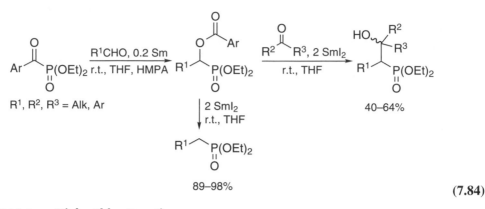

$$(7.84)$$

7.2.1.13 Hetero-Diels–Alder Reactions

It has been shown quite recently that α,β-unsaturated acylphosphonates participate in hetero-Diels–Alder reactions with electron-rich olefins. The heterocycloaddition reactions between dimethyl 1-oxo-2,4-hexadienephosphonate (10:1 mixture of *E,E*- and *E,Z*-dienes) and 2-alkylidene-1,3-dithiane proceeds in CHCl₃ at room temperature with the selective formation of the "head-to-head" cycloadducts in good yields (65–70%, Scheme 7.85).[13]

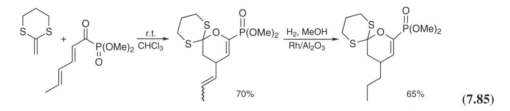

$$(7.85)$$

The enantioselective hetero-Diels–Alder reactions of α,β-unsaturated acylphosphonates with enol ethers catalyzed by Cu(II)*bis*(oxazoline) complexes have been investigated in depth.[15,16,515] It was found that Cu(II)*bis*(oxazoline) complexes activate α,β-unsaturated acylphosphonates to the extent that they undergo facile cycloaddition reactions at low temperature with electron-rich alkenes. For example, dimethyl (*E*)-1-oxo-2-butenylphosphonate reacts with ethyl vinyl ether in the presence of {Cu[(*S,S*)-*t*-Bu-box]}(OTf)₂ complex (10 mol% catalyst) to generate the cycloadduct in 89% yield (Scheme 7.86) with exceptional stereoselectivity (endo/exo = 99/1, 99% ee). Cyclic enol ethers also undergo stereoselective reactions with dimethyl (*E*)-1-oxo-2-butenylphosphonate. It specifically reacts with 2,3-dihydrofuran in the presence of {Cu[(*S,S*)-*t*-Bu-box]}(OTf)₂ to deliver the bicyclic enolphosphonate. Of particular merit is the fact that a large variety of β,γ-unsaturated acylphosphonates may be tolerated with no loss in selectivity for the derived cycloadducts.[15,16,515]

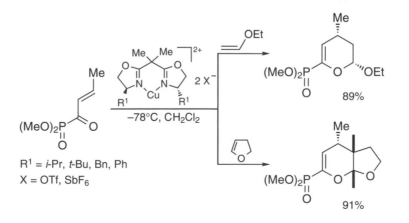

$$(7.86)$$

The enantiomeric excess and absolute stereochemistry for the cycloadducts vary with the pendant oxazoline substituent (R^1 = i-Pr, t-Bu, Bn, Ph) and counterion (X = OTf, SbF$_6$). The cycloaddition may be conducted with as little as 0.2 mol% of the chiral catalyst, and the reaction exhibits a favorable temperature–enantioselectivity profile, with selectivities exceeding 90% even at room temperature (Scheme 7.86).[15,16,515]

7.2.2 REACTIONS OF DIALKYL 1-OXOALKYLPHOSPHONATES WITH CLEAVAGE OF THE P–C BOND

7.2.2.1 Cleavage by Nucleophiles

Two exhaustive studies of the reaction have been published in 1980 by Zhdanov et al.[3] and in 1996 by Breuer.[4] Acylphosphonates behave as *O*- and *N*-acylating agents. The carbon–phosphorus bond is cleaved by nucleophiles such as water, alcohols, ammonia, aliphatic amines, aniline, and hydrazine by preferential attack at the carbonyl carbon atom (Scheme 7.87). Thus, the P–C bond of dialkyl acylphosphonates is rapidly cleaved in neutral and alkaline conditions to give the corresponding carboxylic acids and dialkyl phosphites.[1,459,460,516,517] The reaction of dialkyl acylphosphonates with alcohols proceeds slowly at room temperature to give carboxylate esters and dialkyl phosphite.[518–523] Diethyl benzoylphosphonate has been introduced as an efficient tool for benzoylation of hydroxyl groups of complex molecules in the presence of DBU.[523] Similarly, dialkyl benzoylphosphonates have been developed as benzoylating reagent for amines.[505,521,524–527]

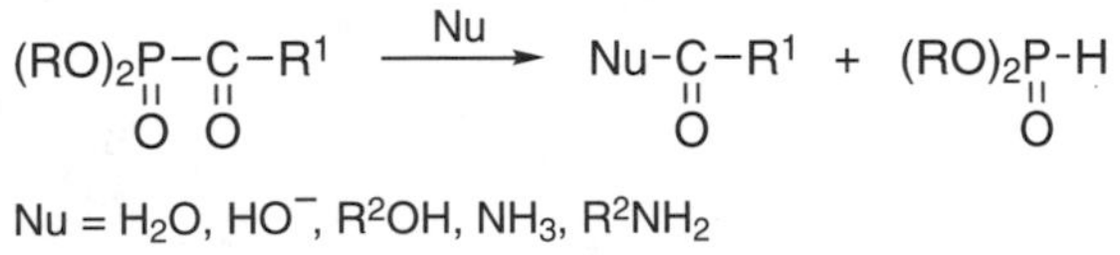

$$(RO)_2\underset{\underset{O}{\|}}{P}-\underset{\underset{O}{\|}}{C}-R^1 \xrightarrow{\text{Nu}} Nu-\underset{\underset{O}{\|}}{C}-R^1 + (RO)_2\underset{\underset{O}{\|}}{P}-H$$

Nu = H$_2$O, HO$^-$, R^{2}OH, NH$_3$, R^{2}NH$_2$

(7.87)

7.2.2.2 Cleavage by Grignard and Reformatsky Reagents

Dephosphorylation of acylphosphonates by Grignard reagents with the ultimate formation of ketones has been described.[505,528,529] Thus, addition of Grignard reagents to a toluene solution of dialkyl acylphosphonates at −78°C gives α-hydroxyalkylphosphonate intermediates, which are directly dephosphonylated by treatment in aqueous NaOH–DME to give the corresponding ketones in low to good yields (12–81%, Scheme 7.88). Various acylphosphonates smoothly enter the reaction course with essentialy methyl- and phenylmagnesium bromide. However, the reaction has a severe limitation in that Grignard reagents with reducing properties act as hydride donors, resulting in the reduction of the keto group. For example, *n*-BuMgBr on reaction with diethyl α-benzoylphosphonate leads to the expected valerophenone in low yield (10–12%), the main product being the diethyl α-hydroxybenzylphosphonate.[505,528,529]

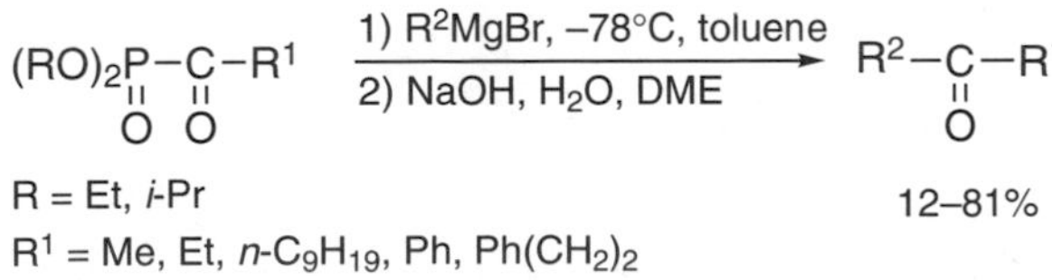

$$(RO)_2\underset{\underset{O}{\|}}{P}-\underset{\underset{O}{\|}}{C}-R^1 \xrightarrow[\text{2) NaOH, H}_2\text{O, DME}]{\text{1) R}^2\text{MgBr, −78°C, toluene}} R^2-\underset{\underset{O}{\|}}{C}-R^1$$

R = Et, *i*-Pr

R^1 = Me, Et, *n*-C$_9$H$_{19}$, Ph, Ph(CH$_2$)$_2$

12–81%

(7.88)

The Reformatsky reaction of ethyl bromoacetate with dimethyl benzoylphosphonate takes place readily, but the yield of the expected final product, ethyl benzoylacetate, is low (30%).[528]

7.2.2.3 Baeyer–Villiger Rearrangement

In a two-phase medium, diethyl 1-oxopropylphosphonate is oxidized with H_2O_2 to give diethyl propionyl phosphate in 90% yield. This Baeyer–Villiger rearrangement involves a migrating phosphoryl moiety that promotes the elimination of water from an intermediate α-hydroxyl hydroperoxide resulting from an attack by H_2O_2 on the α-carbonyl carbon (Scheme 7.89).[530] This transformation provides an efficient synthetic route to acyl phosphates.

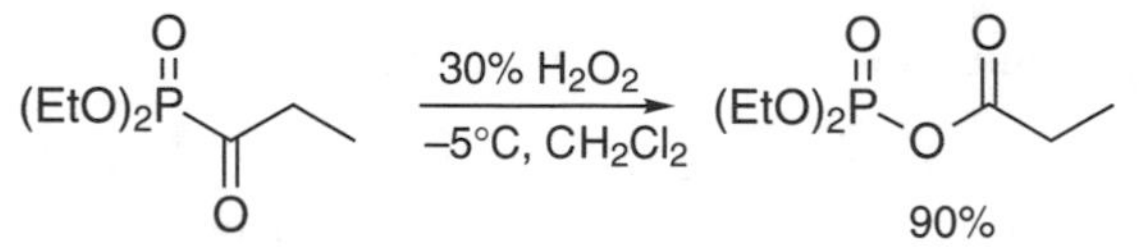

$$\text{(7.89)}$$

7.2.2.4 Decarbonylation

Aroyl- and alkanoylphosphonates are decarbonylated to give aryl- and alkylphosphonates in refluxing toluene in the presence of a catalytic amount of palladium complexes such as $Pd(PPh_3)_4$, *trans*-$PdMe_2(PMe_2Ph)_2$, *trans*-$PdEt_2(PMePh_2)_2$, and *cis*-$PdMe_2(PMePh_2)_2$ (Scheme 7.90). However, aroylphosphonates are more readily decarbonylated than alkanoylphosphonates, and higher yields are obtained for the former (35–100%) than the latter (5–55%).[531–533]

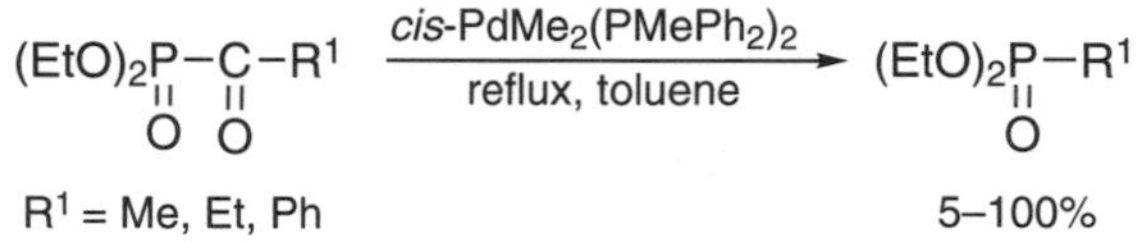

$$\text{(7.90)}$$

Refluxing a toluene solution containing diethyl 1-oxo-1-phenylmethylphosphonate and *trans*-$PdEt_2(PMe_3)_2$ for a few minutes yields a crystalline complex (Scheme 7.91) formed by the oxidative addition of diethyl 1-oxo-1-phenylmethylphosphonate at the C–P bond to a "$Pd(PMe_3)_2$" species. This complex is fairly stable in the solid state and in organic solvents. Its catalytic activity toward decarbonylation has been demonstrated by treating the complex with a large excess of diethyl 1-oxo-1-phenylmethylphosphonate to produce diethyl phenylphosphonate in high yield.[531–533]

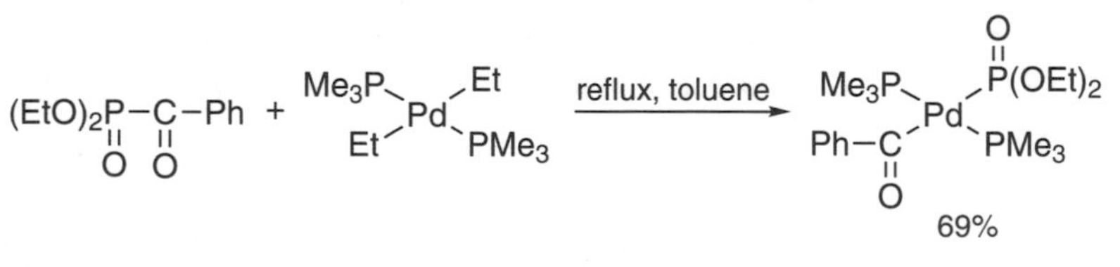

$$\text{(7.91)}$$

7.2.2.5 Reduction

Dialkyl acylphosphonates have been reduced to 1-hydroxyphosphonates using $NaBH_4$,[528,534–539] $NaBH_3CN$,[540] $BH_3 \cdot Me_2S$,[541] catecholborane,[541] $Al(Oi\text{-}Pr)_3$,[505] or activated Zn in AcOH.[505] Reduction of acylphosphonates using $NaBH_4$ needs careful control of the reaction conditions (H_2O or MeOH, KH_2PO_4, 0–5°C, pH = 6–7) to avoid hydrolysis of both starting acylphosphonates and reaction product. Similarly, aqueous solutions of monosodium benzoylphosphonate or 2-phenyl-1-oxoeth-ylphosphonate are readily reduced to the corresponding α-hydroxyphosphonic acids using NaBH₄.[542] Treatment of diethyl benzoylphosphonate with lithium triethylborodeuteride in THF gives the corresponding 1-deuterio-1-hydroxyphosphonate.[543] Formation of the 1-hydroxyphospho-nates may be followed by unmasking of the latent aldehyde by treatment in basic medium.[536] An illustration of this synthetic procedure, which represents a modification of the Rosenmund method, is provided by the conversion of lauroyl chloride into dodecyl aldehyde in 30% overall yield (Scheme 7.92).

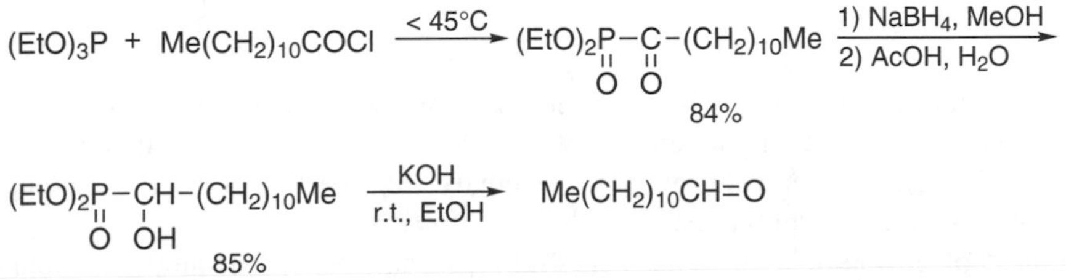

$$(EtO)_3P \ + \ Me(CH_2)_{10}COCl \ \xrightarrow{< 45°C} \ (EtO)_2\underset{O}{\underset{\|}{P}}-\underset{O}{\underset{\|}{C}}-(CH_2)_{10}Me \ \xrightarrow[\text{2) AcOH, H}_2\text{O}]{\text{1) NaBH}_4,\text{ MeOH}}$$

$$84\%$$

$$(EtO)_2\underset{O}{\underset{\|}{P}}-\underset{OH}{\underset{|}{CH}}-(CH_2)_{10}Me \ \xrightarrow[\text{r.t., EtOH}]{\text{KOH}} \ Me(CH_2)_{10}CH=O$$

$$85\%$$

$$(7.92)$$

When submitted to LiAlH₄ in excess (3 eq) in boiling Et₂O, dialkyl acylphosphonates are converted into primary phosphines in good yields (62–84%).[24]

The catalytic enantioselective reduction of 1-ketophosphonates has recently been developed. This approach takes advantage of a development in the enantioselective reduction of prochiral ketones to chiral alcohols by means of catalytic amounts of oxazaborolidines with borane as reducing agent. Thus, the enantioselective reduction of 1-ketophosphonates is accomplished by treatment with different boranes, BH₃·THF (0.9 eq),[544] BH₃·Me₂S (0.66 eq),[50,545] or catecholborane (1.1 eq)[50,546] in different solvent systems in the presence of a catalytic amount of freshly prepared B-n-butyloxazaborolidine, (S) or (R) (Scheme 7.93).[547] The reaction is complete in about 5 h and produces the expected dialkyl 1-hydroxyalkylphosphonates in satisfactory yields (53–98%).

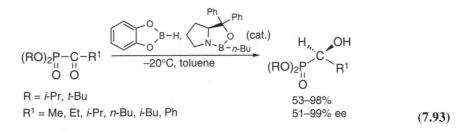

R = i-Pr, t-Bu
R¹ = Me, Et, i-Pr, n-Bu, i-Bu, Ph

53–98%
51–99% ee

$$(7.93)$$

Moderate to good enantioselectivities are obtained (up to 83%) using BH₃·THF or BH₃·Me₂S as the reducing reagent. The highest ee (up to 99%) as well as high chemical yields are obtained using catecholborane in toluene at −20°C. Only small differences between the α-aryl- and α-alkylketophosphonates are observed in the stereochemical outcome of the chiral reduction. Thus, the reduction proceeds with predictable stereochemistry. So the (S)-configurated oxazaborolidine catalyst leads in all cases to (S)-configuration at the new stereogenic center.

A clear illustration of the advantages of this synthetic procedure is provided by the conversion of optically active dialkyl 1-hydroxyalkylphosphonates into dialkyl 1-aminoalkylphosphonates by the Mitsunobu azidation. For example, a sequence of reactions for the conversion of optically active diethyl 1-hydroxyalkylphosphonates into diethyl 1-aminoalkylphosphonates proceeding in 50–88% overall yields and 48–82% ee's has been developed (Scheme 7.94).[544]

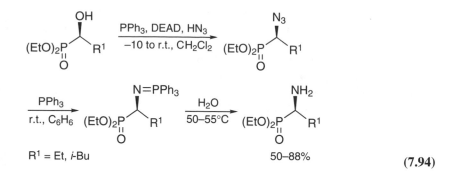

$$R^1 = Et, i\text{-}Bu \qquad\qquad\qquad 50\text{--}88\% \tag{7.94}$$

Low-temperature reduction of β-phthalimido-α-ketophosphonates with catecholborane in toluene in the presence of a catalytic amount of oxazaborolidine affords β-phthalimido-α-hydroxyphosphonates in high yields and high diastereoselectivity. Deprotection of the amino group produces the β-amino-α-hydroxyphosphonates.[541]

Treatment of acylphosphonic acids with $NaBH_4$ in aqueous or alcoholic ammonia solution produces 1-aminoalkylphosphonic acids in good yields (50–70%).[548,549]

7.2.3 REACTIONS OF DIALKYL 2-OXOALKYLPHOSPHONATES

7.2.3.1 Knoevenagel Reactions

Condensation of diethyl 2-oxopropylphosphonate with benzaldehyde on heating in C_6H_6 in the presence of piperidine was reported by Pudovik et al. in 1967 to give a 65% yield of diethyl 1-acetyl-2-phenylvinylphosphonate (Scheme 7.95).[550] The reaction was further repeated in toluene with diethyl 2-phenyl-2-oxoethylphosphonate, piperidine, and AcOH under azeotropical removal of water.[551,552]

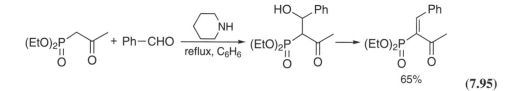

$$65\% \tag{7.95}$$

However, further synthetic investigations have shown that with piperidine/AcOH as catalyst, the reaction proceeds by several competitive processes to give complex mixtures containing the desired phosphonate, the Horner–Wadsworth–Emmons olefination product, the straight-condensation product, and other byproducts.[553]

A new method has been developed for the effective synthesis of diethyl 1-acetyl-2-phenylvinylphosphonate from α,α-dimorpholinotoluene in the presence of α-halo acids (MCA or TFA) (Scheme 7.96). The aminal method is advantageous for large-scale preparation because of the exclusive formation of diethyl 1-acetyl-2-phenylvinylphosphonate. The reaction proceeds under mild conditions through the formation of 4-benzylidenemorpholinium carboxylates as intermediates. The Knoevenagel reaction using such iminium salts provides efficient syntheses of 1-substituted 2-arylvinylphosphonates in good to excellent yields. The reaction products are obtained as mixture of (E)- and (Z)-isomers, the (E)/(Z) ratios being high in all examples.[553,554]

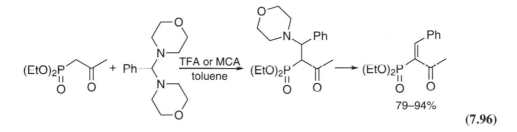

(7.96)

7.2.3.2 Alkylation

7.2.3.2.1 α-Alkylation

Although the combination sodium or potassium metal and diethyl 2-oxoethylphosphonate, on heating in Et$_2$O, gave fair yields of α-substituted β-ketophosphonates on reaction with reactive alkyl halides (MeI, EtI, alkynyl halides, bromoacetates),[555–558] the best results are obtained with the use of alkali metal hydrides (NaH, KH) in THF or DME. Thus, room-temperature alkylation of sodium phosphonoenolates in THF with reactive alkyl halides (allyl halides, alkynyl halides, bromo- and iodoacetates) gives good yields of α-monoalkylated 2-oxoalkylphosphonates, but, even when reactive halides are utilized, extended reaction times ranging from 48 to 78 h are required (Scheme 7.97).[232,559–565] "Normal" alkyl halides fail to react with sodium enolates under the same conditions, suggesting the weak nucleophilic character and hence highly stabilized nature of the β-oxophosphonate anions. The more reactive potassium enolates give a mixture of mono- and dialkylated products with reactive halides,[566] and only elimination is observed when the potassium enolate is treated with "normal" alkyl halides. Lithium enolates in DMF give only a modest yield of alkylated products,[567] although satisfactory yield is obtained with phenylselenyl chloride.[568]

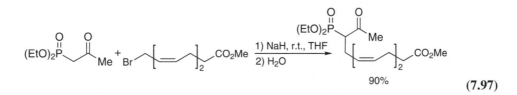

(7.97)

Ion-pair extractive phase-transfer catalysis is employed for the alkylation of cyclic β-ketophosphonates.[569] Cyclic β-ketophosphonates are converted in good yields into a mixture of *C*- and *O*-alkylated products utilizing activated halides (benzyl or allyl bromide) in the presence of TBAB as the phase-transfer reagent in CH$_2$Cl$_2$. It has been found that heating the reaction mixture of cyclic β-ketophosphonate, alkyl halide, and tetrabutylammonium hydroxide at 56°C in toluene instead of CH$_2$Cl$_2$ gives better product conversion.[569] The alkylation of acyclic β-ketophosphonates was investigated with the same technique. Mostly *C*-alkylated products are formed, along with a small amount of *O*-alkylated and dialkylated products. Thus, because acyclic β-ketophosphonates are less sterically hindered than the analogous cyclic β-ketophosphonates, the steric nature of the phosphonate-stabilized carbanion plays an important role in determining the amount of *O*-alkylation product.[569]

7.2.3.2.2 γ-Alkylation

It has been shown that treatment of dialkyl 2-oxopropylphosphonates with NaH in THF produces a monoanion, and subsequent metallation with *n*-BuLi generates the β-ketophosphonate 1,3-dianion.[570,571] When β-ketophosphonate 1,3-dianions are treated with a variety of alkylating agents, γ-monoalkylated products are isolated in good yields (65–75%, Scheme 7.98).[237,268,570–574] The procedure has been extended to the alkylation of γ-substituted β-ketophosphonates. Thus, starting with dimethyl 2-oxopropylphosphonate, it is possible to generate the 1,3-dianion, to alkylate with R[1]X, to generate the dianion of this alkylated product with an additional equivalent of n-BuLi, and to add a second alkylating agent R[2]X to yield γ-disubstituted β-ketophosphonates.[570]

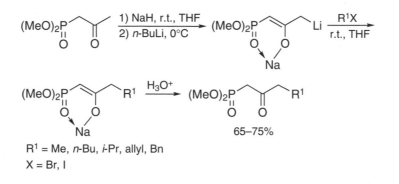

$$R^1 = \text{Me}, \text{n-Bu}, \text{i-Pr}, \text{allyl}, \text{Bn}$$
$$X = \text{Br}, \text{I}$$

(7.98)

These two procedures have been combined for the preparation of α,α′-disubstituted disymmetric ketones (see Section 7.2.3.13).[575–577]

7.2.3.2.3 δ-Alkylation

The generation of organolithium reagents from organotin compounds by tin/lithium exchange has been exploited for the preparation of β-ketophosphonate 1,4-dianions. Thus, the diisoprpyl 4-(tri-n-butylstannyl)-2-oxobutylphosphonate, on treatment with NaH in THF at room temperature followed by addition of n-BuLi at low temperature, generates the 1,4-dianion.[578] The success of tin/lithium exchange results from the stabilization of the organolithium reagent via the intramolecular chelation of lithium by the enolate oxygen. The 1,4-dianion reacts with electrophiles at the δ-carbon, giving rise to terminally substituted β-ketophosphonates (Scheme 7.99).[578]

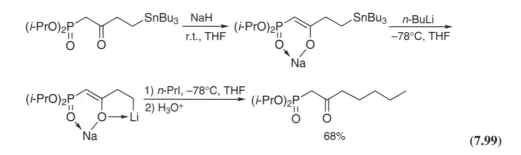

(7.99)

7.2.3.3 Halogenation

Treatment of the sodium enolate of diethyl 2-oxoheptylphosphonate with perchloryl fluoride (FClO₃) in toluene at −35°C gives diethyl 1-fluoro-2-oxoheptylphosphonate in 29% yield (Scheme 7.100)[579,580]

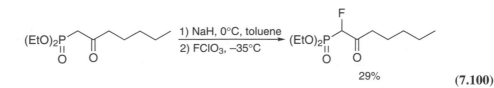

(7.100)

Similarly, the α-monochlorinated and α-monobrominated dimethyl 2-oxoalkylphosphonates are prepared in satisfactory yields (50%) by reaction of the sodium enolate of dimethyl 2-oxoalkyl-phosphonates with NCS or NBS in DME at room temperature[581,582] or with bromine in Et$_2$O (53–71%)[583] or THF (60%).[584] Synthesis of diethyl 1-chloro-2-oxopropylphosphonate has also been reported through reduction of the 1,1-dichloro derivative[585] with sodium sulfite.[584] Iodination in the γ-position of β-ketophosphonate 1,3-dianions prepared with K$_2$CO$_3$ in MeOH has been reported. These dianions react with benzaldehyde in a Horner–Wadsworth–Emmons–Darzens reaction sequence to produce the α,β-unsaturated α',β'-epoxyketones in 60–79% yields (Scheme 7.101).[586]

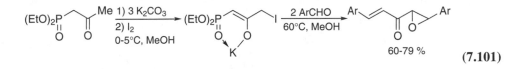

$$(7.101)$$

7.2.3.4 Alkoxycarbonylation

A general and convenient procedure for the synthesis of diethyl 2-oxo-3-(ethoxycarbonyl)alkyl-phosphonates involves the alkoxycarbonylation at the γ position of the α,γ-dimetallated complex generated from the readily available diethyl 2-oxoalkylphosphonates.[587,588] Deprotonation of diethyl 2-oxoalkylphosphonates with NaH (1 eq) in THF at room temperature followed by LDA (2 eq) at 0°C leads to the quantitative generation of the intermediate stable dianion. The latter undergoes smooth nucleophilic acylation with ethyl chloroformate, yielding a *bis*-chelated enolate. In view of the acidity of these products, it is necessary to use an excess of LDA (2 eq) to secure the quantitative generation of the *bis*-chelated enolate. The overall yields of diethyl 2-oxo-3-(ethoxy-carbonyl)alkylphosphonates obtained by this procedure are in the range 72–88% (Scheme 7.102).[587,588] The advantages of this route are more clearly evidenced by comparison with the Michaelis–Becker reaction. For example, formation of diethyl 2-oxo-3-(ethoxycarbonyl)propylphosphonate has been reported in only 50% yield by the Michaelis–Becker reaction between sodium diethyl phosphite and sodium γ-bromoacetoacetate.[155,589–595]

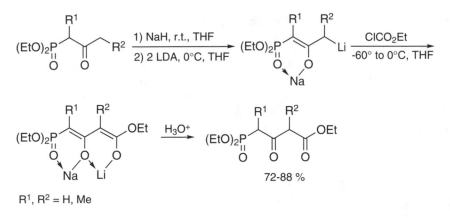

R^1, R^2 = H, Me

$$(7.102)$$

7.2.3.5 Reduction

Chiral β-hydroxy and α-amino β-hydroxyphosphonic acid analogues of carboxylic acids have received increasing interest in the last decade because they serve as surrogates for the corresponding amino acids. One versatile strategy for effecting Ru(II)-catalyzed asymmetric hydrogenation at the carbonyl center of β-ketophosphonates features a general and practical route to ruthenium com-plexes of a wide range of diphosphines (Scheme 7.103).[264,596–598] The chiral Ru(II) catalysts are prepared from the commercially available $(COD)Ru(\eta^3\text{-}C_4H_7)_2$ and diphosphines including *P*-chiral species. Very recently, the Ru-catalysts have been synthesized *in situ* under mild conditions in a one-step reaction from $(COD)Ru(\eta^3\text{-}C_4H_7)_2$.[599,600] The *in situ* prepared (P*P)RuBr$_2$ catalyzes the enantioselective hydrogenation of β-ketophosphonates in MeOH to give the corresponding β-hydroxyphosphonates with excellent enantiofacial discrimination either at atmospheric pressure or 100 bars of H$_2$ with temperature varying from room temperature to 50°C.[597] Similarly, the asymmetric hydrogenation of dimethyl 2-oxopropylphosphonate using (P*P)RuBr$_2$ [(P*P) = 1,2-bis(*tert*-butylmethylphosphino)ethane] has been reported, but the results (yields and ees) are not so high.[601]

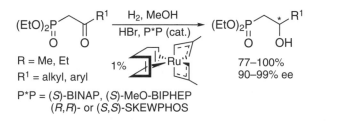

$$ (7.103) $$

Syntheses of α-amino β-hydroxyphosphonic acids based on the BINAP-Ru(II)-catalyzed hydrogenation of racemic α-acetamido β-ketophosphonates via dynamic kinetic resolution has been described.[130,602] The combination of the configurational lability of the α-substituted β-ketophosphonate, the electronegative nature of the α-amido group, and the chiral discriminating properties of (*R*)-BINAP as well as the appropriate reaction conditions leads to the formation of (*R,R*)-α-acetamido β-hydroxyphosphonates of high enantiomeric and high diastereomeric purity in excellent yield.[130,602]

Enantioselective reduction of diisopropyl 2-phenyl-2-oxoethylphosphonate to diisopropyl 2-hydroxy-2-phenylethylphosphonate has been achieved using catecholborane as well as BH$_3$·Me$_2$S as reducing agents via *B-n*-butyloxazaborolidine catalysis. The reactions conducted with catecholborane, a much less reactive reducing agent than BH$_3$·Me$_2$S, give an excellent enantiomeric excess (91%). By contrast, the introduction of more reactive BH$_3$·Me$_2$S gives lower ee's.[545,546]

7.2.3.6 Addition of Organometallic Nucleophiles

A few cases of nucleophilic additions to β-ketophosphonates have been reported. When the α-carbon atom to phosphorus is fully substituted, it appears unlikely that the reaction at phosphorus is competitive with the addition to carbonyl group. Thus, treatment of diethyl 1-fluoro-1-ethoxy-carbonyl-2-ethoxycarbonyl-2-oxoethylphosphonate with Grignard reagents at low temperature in THF gives an (*E*)/(*Z*) mixture of α-fluoro-α,β-unsaturated diesters in 49–68% yields. The initial step is the nucleophilic attack of Grignard reagent at the carbonyl group, followed by intramolecular elimination of diethyl phosphate (Scheme 7.104).[603,604]

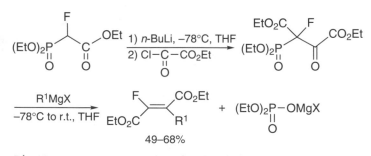

R^1 = Me, Et, *n*-Pr, *i*-Pr, *t*-Bu, CH$_2$=CH, Cy, Ph, C$_6$F$_5$

$$ (7.104) $$

In a related reaction, treatment of the lithium salt of diethyl (1-phenylsulfonyl)ethylphosphonate with trifluoroacetic anhydride affords trifluoroacetylated phosphonate, which, without isolation, reacts with Grignard reagents in THF at reflux to give trifluoromethylated vinylsulfones in 42–55% yields, with the formation of (E)-isomer exclusively or predominantly.[284] Treatment of the lithium salt of diethyl 1-fluoro-1-(ethoxycarbonyl)methylphosphonate with trimethylsilyl trifluoroacetate leads to the enolate of ethyl (trifluoroacetyl)fluoroacetate, which can further undergo protonation, alkylation, or allylation.[605]

Addition to the carbonyl group of β-ketophosphonates is successful even in the presence of a relatively acidic hydrogen.[606,607] A number of cyclic and acyclic β-ketophosphonates have been subjected to reaction with allylmagnesium chloride at 0°C in THF for several hours. The results, which vary dramatically depending on the β-ketophosphonates utilized, are greatly improved in the presence of BF$_3$·Et$_2$O. Nucleophilic additions of other Grignard reagents (PhMgBr, MeMgBr, CH$_2$=CH$_2$MgBr) are not as successful under the same conditions (Scheme 7.105).

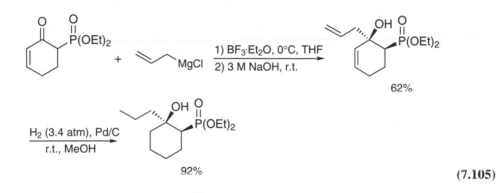

(7.105)

By contrast, the allylzinc reagents appear to be advantageous relative to allylmagnesium chloride–BF$_3$·Et$_2$O. Addition of allylzinc to β-ketophosphonates gives the corresponding β-hydroxyphosphonates in yields that are generally greater than those obtained using Grignard reagents. Addition of the zinc reagents derived from crotyl and prenyl halides, though, proceeds with complete allylic transposition.[606,607]

7.2.3.7 Conversion into 3-Oxoalkylphosphonates

Room-temperature treatment of simple β-ketoalkylphosphonates with the Furukawa-modified Simmons–Smith reagent provides rapid and efficient preparation of γ-ketoalkylphosphonates.[397] The chain extension of simple, unfunctionalized β-ketophosphonates requires an excess of both Et$_2$Zn and CH$_2$I$_2$ (6 eq). The presence of α-substituents on the β-ketophosphonate does not diminish the efficiency of the reaction (Scheme 7.106).

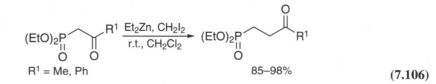

(7.106)

Exposure of diethyl 2-oxocyclopentylphosphonate to ethyl(iodomethyl)zinc results in efficient ring expansion, providing diethyl 3-oxocyclohexylphosphonate in 83% yield (Scheme 7.107).[397]

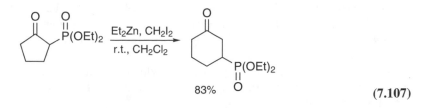

$$(7.107)$$

7.2.3.8 Ring Expansion Reactions of Cyclic 2-Oxoalkylphosphonates

The conjugate Michael addition of cyclic β-ketophosphonate anions with dimethyl acetylenedicarboxylate has been reported. For example, sodium enolates derived from cyclic β-ketophosphonate react readily as Michael donors with dimethyl acetylenedicarboxylate to afford [n + 2] ring-expanded products in reasonable yields (Scheme 7.108).[608,609] The reaction proceeds via a tandem Michael–aldol-fragmentation mechanism to give the ring-enlarged products. Investigation into the scope of the reaction revealed the existence of an alternate reaction pathway, the abnormal Michael reaction, in which the nucleophile generated from the initial Michael addition attacked the phosphorus, leading to a net 1,3-migration of the phosphoryl group.[608,609]

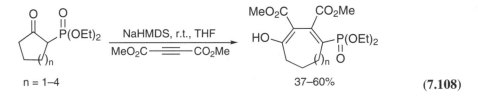

$$(7.108)$$

7.2.3.9 Decomposition of 1-Diazo-2-oxoalkylphosphonates

α-Diazo-β-ketophosphonates are prepared by a diazo-transfer reaction between α-metallated dialkyl 2-oxoalkylphosphonates and tosyl azide.[610,611] It has been reported that sodium enolates prepared from NaH in C_6H_6/THF solution[195] give superior results (80–96%, Scheme 7.109) to t-BuOK in C_6H_6/THF (20–50%), Et_3N in MeCN,[191] or K_2CO_3 in MeCN.[612–617]

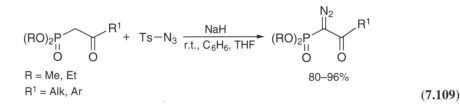

$$(7.109)$$

The Rh(II)-catalyzed decomposition of dialkyl α-diazo-β-ketophosphonates leads to the intermediate rhodium carbenoids, which can undergo a variety of reactions such as insertion into O–H, S–H, N–H, or unactivated C–H bonds and Wolff rearrangement generating a ketene. In this area, the intramolecular carbenoid cyclization of α-diazocarbonyl compounds is of considerable importance for the formation of carbocyclic rings. For example, addition of dialkyl 1-diazo-2-oxoalkylphosphonates to a refluxing suspension of $Rh_2(OAc)_4$ in CH_2Cl_2 leads to the formation of dialkyl 2-oxocycloalkylphosphonates in fair to good yields,[231,618] as illustrated by the conversion of diethyl 1-diazo-2-oxo-5-methylhexylphosphonate into diethyl 2-oxo-5,5-dimethylcyclopentylphosphonate in 70% yield (Scheme 7.110).[618] Related reaction sequences have been described for the synthesis of a 3-phosphonocarbapenem by rhodium-catalyzed insertion into a N–H bond[191] and for the synthesis of 1-alkoxy-2-oxoalkylphosphonates by insertion into the O–H bond of a mono-*tert*-butyldimethylsilyl protected diol.[619,620]

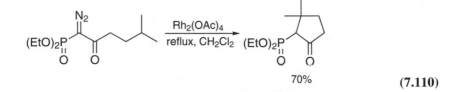

$$(7.110)$$

The phosphoryl group, which is less electron-withdrawing than the carbonyl group, renders the carbenoid intermediate less electrophilic, allowing the competitive Wolff rearrangement to take place. Thus, diethyl 1-ethoxycarbonyl-1-butylmethylphosphonate is obtained in 30% yield from diethyl 1-diazo-2-oxohexylphosphonate by a Wolff rearrangement (Scheme 7.111).[618]

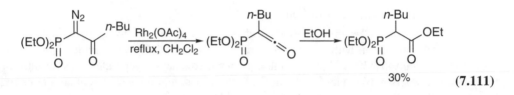

$$(7.111)$$

The Wolff rearrangement has also been explored to prepare several phosphonylated heterocycles such as 2-oxospiropiperidino- or 2-oxospiromorpholinoindolium enolates,[612,616] 1*H*-2-benzopyranes,[612,617] fused isoxazoles,[613] and hydroxynaphthalenephosphonates.[614,615]

7.2.3.10 Preparation of Vinylphosphonates

At room temperature, diethyl 1-phenyl-2-oxoalkylphosphonates undergo attack at the carbonyl by sodium dialkyl phosphites in THF to give transient hydroxyphosphonates, which eliminate sodium diethyl phosphate according to a Horner–Wadsworth–Emmons-type reaction to provide stable dialkyl (*E*)-1-phenylalkenylphosphonates in moderate to good yields (20–94%) (Scheme 7.112). The elimination is favored by the presence of the phenyl group.[621]

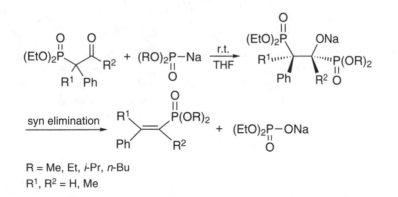

$$(7.112)$$

7.2.3.11 Preparation of 1-Alkynylphosphonates

Sodium enolates of diethyl 2-oxoalkylphosphonates react with diethyl chlorophosphate to produce enol phosphates. Low-temperature treatment of enol phosphates with *tert*-BuOK induces the β-elimination of potassium diethyl phosphate and formation of diethyl 1-alkynylphosphonates in good yields (43–95%, Scheme 7.113). However, the reaction is prone to prototropic isomerization and diethyl 2-oxobutylphosphonate leads to a mixture of diethyl 1-butynylphosphonate (43%) and diethyl 2-butynylphosphonate (51%).[622]

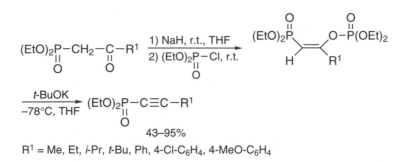

$$R^1 = Me, Et, \textit{i}\text{-Pr}, \textit{t}\text{-Bu}, Ph, 4\text{-Cl-C}_6H_4, 4\text{-MeO-C}_6H_4$$

(7.113)

7.2.3.12   Preparation of Heterocycles

7.2.3.12.1   From Enaminophosphonates

Phosphonylated *N*-benzyl enamine, formed from benzylamine and diethyl 2-oxopropylphosphonate, undergoes aza-annulation with acryloyl chloride in refluxing THF to provide a phosphonylated unsaturated lactam in 72% yield.[623,624] Subsequent hydrogenation generates a 78:22 ratio of diastereomeric products in 67% yield (Scheme 7.114). These compounds represent an interesting class of rotationally constrained β-aminophosphonic acid analogues.

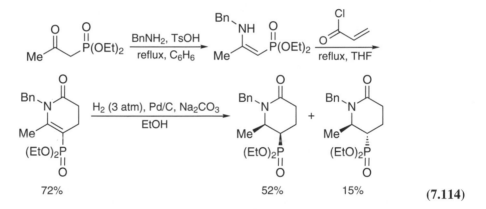

(7.114)

3-Phosphonylated aminoquinolines are prepared from phosphonylated arylamines obtained as above in refluxing toluene from diethyl 2-phenyl-2-oxoethylphosphonate and arylamines [mixture of the β-imino- and (*Z*)-β-enamino derivatives]. Metallation of β-imino and β-enaminophosohonates with *n*-BuLi in THF followed by sequential treatment with isocyanates, triphenylphosphine, hexachloroethane, Et₃N, and aqueous workup produces, after deprotection, 3-phosphonylated-4-aminoquinolines (Scheme 7.115).[625]

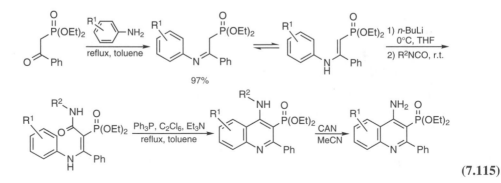

(7.115)

Phosphonylated enamines bearing a functional group (COMe, COPh) in α-position are effective precursors of phosphonylated heterocycles. They have been shown to undergo nucleophilic cyclization reactions with compounds containing mobile hydrogen atoms, providing the basis of a new synthesis of heterocycles. The dimethylamino group of phosphonylated enamines can be selectively and totally transaminated by treatment with symmetric or disymmetric hydrazines, guanidine, acetamidine, and methylisothiourea in refluxing EtOH (Scheme 7.116). The corresponding phosphonylated pyrazolones (R^1 = OH, R^2 = H), pyrazoles (Scheme 7.116a,b), and pyrimidines (Scheme 7.116c,d) are isolated in excellent yield (90–95%).[626,627]

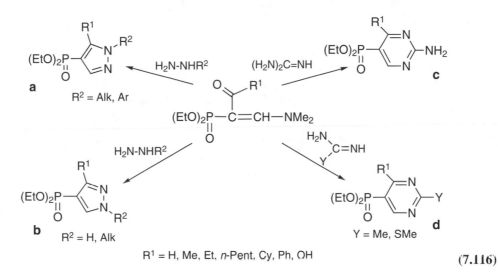

R^1 = H, Me, Et, *n*-Pent, Cy, Ph, OH

(7.116)

7.2.3.12.2 From ω-Amino-2-Oxopropylphosphonates

The diethyl 3-bromo-2-oxopropylphosphonate undergoes nucleophilic attack by arylamines in Et_2O at room temperature to give a series of diethyl 3-arylamino-2-oxopropylphosphonates in satisfactory yields (48–76%). These compounds, by merely heating in refluxing toluene for 30 min in the presence of $ZnCl_2$, produce diethyl 3-indolylmethylphosphonates in high yields (60–91%, Scheme 7.117).[584]

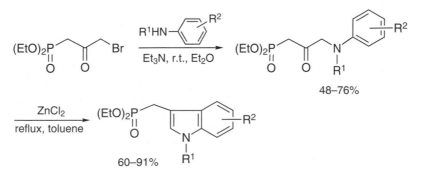

(7.117)

The reaction of the sodium salt of *N*-phenylhydroxylamine with diethyl allenylphosphonate in THF in the presence of CF_3CO_2Li gives, by Michael addition and Cope rearrangement, γ-(*ortho*-aminoaryl)-β-ketophosphonates, which are easily converted into diethyl 2-indolylmethylphosphonates in high yields (Scheme 7.118).[628,629]

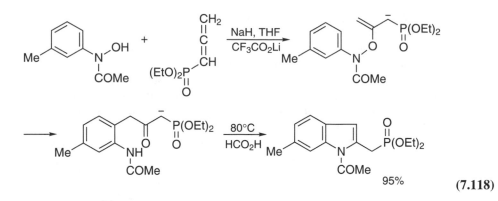

(7.118)

Similarly, diethyl 3-aryloxy-2-oxopropylphosphonates, prepared from diethyl 3-bromo-2-diethylaminopropenylphosphonate and substituted sodium phenoxides, undergo cyclodehydration in refluxing toluene in the presence of polyphosphoric acid to produce 3-benzofurylmethylphosphonates.[630]

7.2.3.12.3   With Thioaroylketene S,N-Acetals

The formation of thienylphosphonates has seldom been reported, and the following one represents an efficient route to 2-thienylphosphonate. Thus, room-temperature treatment of thioarylketene S,N-acetals with diethyl 2-oxopropylphosphonate in CH_2Cl_2 in the presence of Hg(II) acetate affords 2-phosphonylated-3-amino-5-arylthiophenes in good yields (Scheme 7.119).[631,632]

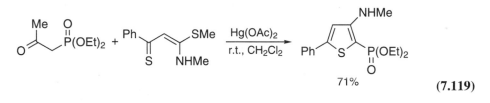

(7.119)

7.2.3.12.4   With Vinylic Acetates

Diethyl 3-furylphosphonates are prepared in fair to good yields (55–90%) by a two-step procedure based on ceric ammonium nitrate (CAN)-promoted oxidative addition of diethyl 2-oxoalkyl- or 2-oxo-2-phenylphosphonates to vinylic acetates followed by an acid-induced Pall–Knorr cyclization reaction (Scheme 7.120).[633]

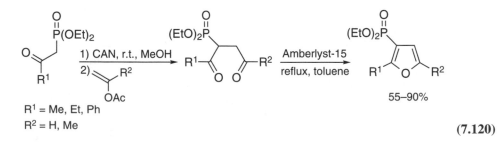

(7.120)

7.2.3.13   Dephosphonylation: Preparation of Ketones

In contrast to the ready reductive cleavage of α-sulfonyl ketones, reductive dephosphonylation of β-ketophosphonates is surprisingly difficult.[175] All attempts to produce methyl cyclohexyl ketone from di-*tert*-butyl 2-cyclohexyl-2-oxoethylphosphonate by various reducing agents failed

[Zn/AcOH, Li/naphthalene, Li/n-PrNH$_2$, Li/(CH$_2$NH$_2$)$_2$]. The only successful cleavage of the P–C bond occurred on pyrolysis of the phosphonic acid (200°C, 1 atm) to afford methyl cyclohexyl ketone in 78% yield. However, the harsh conditions of this method are incompatible with the presence of the other functional group in the molecule.[175] It was early observed that Li(t-BuO)$_3$AlH hydrogenolysed the P–C bond in diethyl 1-ethoxycarbonyl-2-oxoalkylphosphonates with conversion into the corresponding β-keto esters.[286]

The first complete dephosphonylation of β-ketophosphonates was observed when the sodium enolates of β-ketophosphonates are treated with LiAlH$_4$ in THF at room temperature (Scheme 7.121). Further attempts to find mild and effective reducing agents [LiBEt$_3$H, DIBAL-H, Li(t-BuO)$_3$AlH] for compatibility with a variety of functional groups failed under the same conditions. The reaction is rather general, and the ketones are isolated in high yields (57–98%).[575,576,634] Although the mechanism is not completely clear, results of experiments using LiAlD$_4$/H$_2$SO$_4$ and LiAlH$_4$/D$_2$SO$_4$ suggest that the cleavage of the P–C bond is not caused by the direct attack of hydride on the α- or the γ-carbon atoms. The acid might also play a significant role in dephosphonylation.[575]

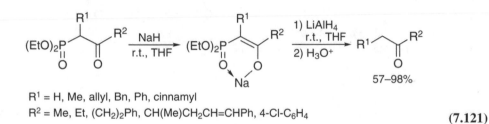

R^1 = H, Me, allyl, Bn, Ph, cinnamyl
R^2 = Me, Et, (CH$_2$)$_2$Ph, CH(Me)CH$_2$CH=CHPh, 4-Cl-C$_6$H$_4$

(7.121)

One versatile strategy for the synthetic approach to α,α′-disubstituted asymmetric ketones has been elaborated (Scheme 7.122). For example, metallation of diethyl alkylphosphonates and acylation of the copper reagents with acyl halides followed by two consecutive alkylations, at the α-carbon and then at the γ-carbon, dephosphonylation, and hydrolysis produces highly substituted ketones in 43–73% yields.[577]

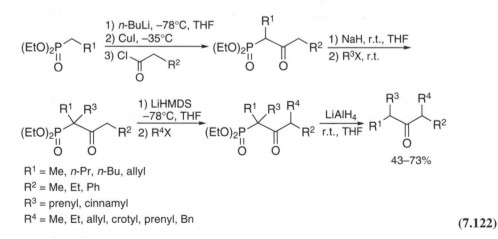

R^1 = Me, n-Pr, n-Bu, allyl
R^2 = Me, Et, Ph
R^3 = prenyl, cinnamyl
R^4 = Me, Et, allyl, crotyl, prenyl, Bn

(7.122)

7.2.3.14 Preparation of 1,2-Diketones and α-Hydroxyketones

Preparations of 1,2-diketones and α-hydroxyketones have been obtained by the reaction of masked benzoyl anion equivalent with acyl chlorides followed by successive deprotection of the carbonyl groups.[276,279] Thus, metallation of diethyl 1-phenyl-1-(trimethylsilyloxy)methylphosphonate at low temperature with LDA generates the corresponding lithium salt, which, on treatment with acyl chlorides, gives the acylated products in yields subjected to steric influence of acyl chlorides. The alkaline treatment (1 M NaOH/EtOH) gives a mixture of 1,2-diketones and α-hydroxyketones.[276] By contrast, under acid conditions, the diketones are isolated in 23–83% yields (Scheme 7.123).[279]

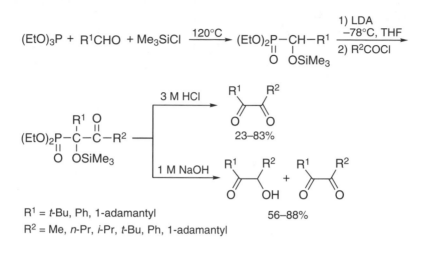

R^1 = t-Bu, Ph, 1-adamantyl
R^2 = Me, n-Pr, i-Pr, t-Bu, Ph, 1-adamantyl

(7.123)

7.2.3.15 Ohira's Reagent

One ingenious and elegant improvement in the standard synthetic procedure for the preparation of alkynes was introduced in 1989 with the use of dimethyl 1-diazo-2-oxopropylphosphonate.[635,636] This valuable and efficient reagent is prepared in a single step by reaction of the α-metallated derivative of dimethyl 2-oxopropylphosphonate with tosyl azide. It was observed that the sodium enolate prepared from NaH in C_6H_6/THF solution gives superior results (80–96%) to t-BuOK in C_6H_6/THF (20–50%) or Et_3N/MeCN (Scheme 7.124).[195]

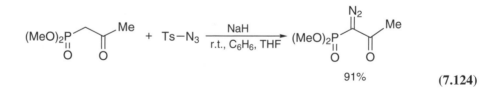

91%

(7.124)

The one-carbon homologation of aldehydes to alkynes is achieved in high yield using Ohira's reagent in the presence of K_2CO_3 in MeOH at room temperature. This reagent is successfully used in the synthesis of ethynylglycine derivatives,[637–640] ethynyl dioxolane,[641] glyco-1-ynitol derivatives,[642] multitopic bidentate ligands,[643] enyne epoxides,[644] alkynylated phenothiazines,[645] trispiro alkynes,[646] and natural compounds (Scheme 7.125).[647–665]

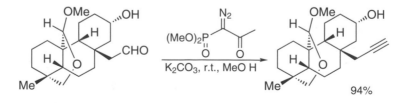

94%

(7.125)

7.2.3.16 Reductive Amination

An efficient procedure for the conversion of dialkyl 2-oxoalkyl-, aryl-, or heteroarylphosphonates to the corresponding dialkyl 2-aminoalkyl-, aryl-, or heteroarylphosphonates has been developed using the reductive amination of the carbonyl in the presence of $NaBH_3CN$ (Scheme 7.126).[251,400] The method appears to be fairly general and can be used for the preparation of diethyl 2-aminoalkylphosphonates containing primary, secondary, or tertiary amino groups in mild conditions by simply treating the carbonyl compound at room temperature with $AcONH_4$ and primary or secondary aliphatic amines[251,400] and protected diamines.[666] It appears that the reaction of 2-oxoalkylphosphonates is subjected to steric influence. The reaction of α- or γ-substituted phosphonates and long-chain phosphonates is often sluggish and requires lengthy reaction times. The reaction generally proceeds well with more nucleophilic primary and secondary amines, and the results of conversion of diethyl 2-oxopropylphosphonate to 2-aminoalkylphosphonates are quite satisfying. Hydrolysis of diethyl aminoalkylphosphonates with 8 M HCl followed by purification using ion-exchange resins (Amberlite IRA 410) provides pure aminoalkylphosphonic acids in moderate to good yields (34–81%, Scheme 7.126).

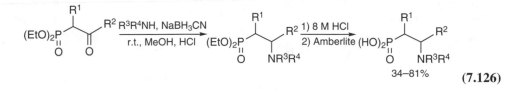

$$(7.126)$$

Reductive amination has been explored with β-ketophosphonates containing functional groups. For example, the reductive amination of 3-bromo-2-oxopropylphosphonate in the same conditions as above is accompanied by reduction of the halogen, whereas 3,4-alkenyl-2-oxoalkylphosphonates undergo simultaneous reduction of the conjugated double bond and reductive amination of the carbonyl group (Scheme 7.127).[251,400]

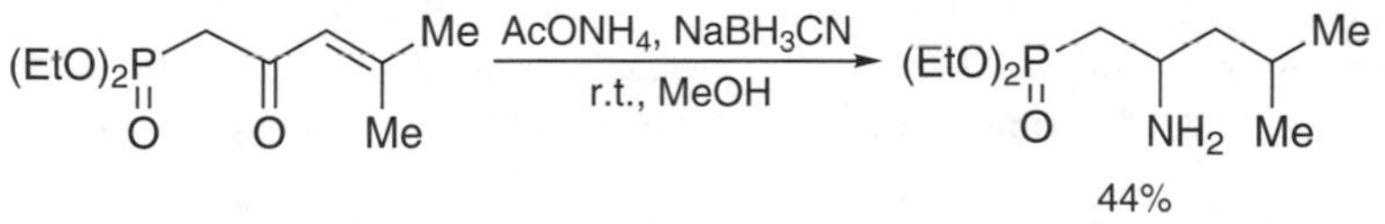

$$(7.127)$$

Reductive amination of 2-aryl-2-oxoethylphosphonates with $AcONH_4$ generally proceeds well in satisfactory yields, but because of the presence of aryl groups, the carbonyl is partially inactivated, and the reaction becomes sluggish.

7.2.3.17 Strecker and Related Reactions

In analogy with formylphosphonates, ω-oxoalkylphosphonates can be converted by Strecker and Bucherer–Berg reactions into ω-amino-ω-(hydroxycarbonyl)alkylphosphonic acids.[403,667,668] For example, for amino phosphonocyclopentane- and phosphonocyclohexanecarboxylic acids, conformationally restricted L-2-amino-4-phosphonobutanoic acid (APB) analogues are prepared from the racemic diethyl 3-oxocyclopentyl and 3-oxocyclohexylphosphonates using NH_4Cl and NaCN in concentrated NH_4OH to give the aminonitriles in good yields as a mixture of *cis-* and *trans-*isomers (Scheme 7.128). The diastereomeric aminonitriles are separated by column chromatography. Acid hydrolysis with 6 M HCl provides *cis-* and *trans-*3-amino-3-(hydroxycarbonyl)cyclopentyl- or cyclohexylphosphonic acids purified by ion-exchange chromatography.[668]

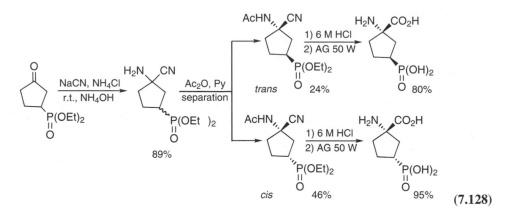

(7.128)

Similarly, the diethyl 2-(3-oxocyclopentyl)- and 2-(4-oxocyclohexyl)ethylphosphonates have been subjected to the standard Bucherer–Berg reaction conditions [KCN, $(NH_4)_2CO_3$, NH_4Cl in EtOH–H_2O at 60°C] to give spirohydantoins, which are isolated as a mixture of *cis*- and *trans*-isomers. Hydrolysis of hydantoins needs heating at 150°C in a sealed tube with concentrated HCl to provide the amino acid hydrochlorides.[403]

7.2.4 REACTIONS OF THE PHOSPHONYL GROUP: DEALKYLATION

Reaction of dialkyl 1- or 2-oxoalkylphosphonates with Me_3SiCl (2 eq) in the presence of anhydrous NaI or KI (2 eq) in dry MeCN,[669,670] Me_3SiI (2 eq) in $CHCl_3$ or CCl_4,[671,672] or Me_3SiBr (2–3 eq)[673–676] occurs rapidly and smoothly at room temperature to give the corresponding *bis*(trimethylsilyl)phosphonates quantitatively (Scheme 7.129). More recently, it has been reported that Me_3SiBr used in excess in dioxane at 60°C chemoselectively cleaves isopropyl groups of diisopropyl phosphonates.[677] In all cases, the reaction proceeds to completion with total selectivity for P–O dealkylation. However, Me_3SiI has been found to effect transesterification of methyl and ethyl esters of phosphonic acids under the mildest conditions. It is superior in use to the combination Me_3SiCl/NaI, both with respect to the facility of the operations involved and with regard to its greater specificity.[671,672,678] The corresponding acids are readily prepared by treatment with a small excess of MeOH in dry acetone and isolated by transformation into anisidium, cyclohexylammonium, or lithium salts.

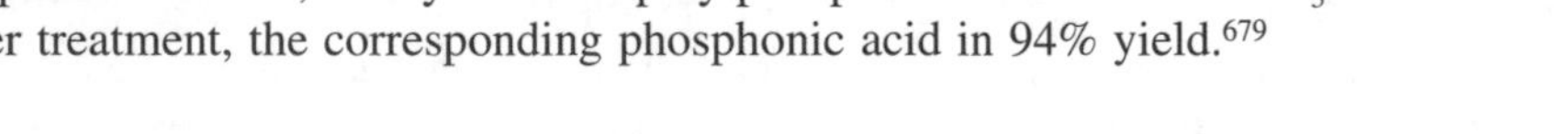

$$(EtO)_2\underset{O}{\overset{O}{P}}-\underset{O}{\overset{O}{C}}-R^1 \xrightarrow[\text{MeCN, r.t.}]{2\ Me_3SiCl,\ KI} (Me_2SiO)_2\underset{O}{\overset{O}{P}}-\underset{O}{\overset{O}{C}}-R^1 \xrightarrow{MeOH} (HO)_2\underset{O}{\overset{O}{P}}-\underset{O}{\overset{O}{C}}-R^1$$

(7.129)

Lithium bromide in dry MeCN or NaI in acetone at room temperature induces selective monodemethylation[12] or monodebenzylation[27] of the corresponding 1-oxoalkylphosphonate ester groups in good yields.

Recently it has been shown that boron tribromide is an effective reagent for dealkylation of β-ketophosphonates. Thus, diethyl 2-oxohepthylphosphonate reacts with BBr_3 in toluene at 70°C to give, after treatment, the corresponding phosphonic acid in 94% yield.[679]

REFERENCES

1. Kabachnik, M.I., and Rossiiskaya, P.A., Esters of α-keto phosphonic acids, *Izv. Akad. Nauk SSSR, Ser. Khim.*, 364, 1945; *Chem. Abstr.*, 40, 4688e, 1946.
2. Kabachnik, M.I., Rossiiskaya, P.A., and Shepeleva, E.S., Esters of α-keto phosphonic acids. Part 3. Two types of carboxylic acid derivatives, *Izv. Akad. Nauk SSSR, Ser. Khim.*, 163, 1947; *Chem. Abstr.*, 42, 4132i, 1948.

3. Zhdanov, Y.A., Uzlova, L.A., and Glebova, Z.I., α-Ketophosphonic acid esters. Synthesis, structure, and reactions, *Usp. Khim.*, 49, 1730, 1980; *Russ. Chem. Rev.*, 49, 843, 1980.

4. Breuer, E., Acylphosphonates and their derivatives, in *The Chemistry of Organophosphorus Compounds,* Vol. 4, Hartley, F.R., Ed., John Wiley & Sons, New York, 1996, p. 653.

5. McKenna, C.E., and Kashemirov, B.A., Recent progress in carbonylphosphonate chemistry, in *Topics in Current Chemistry,* Vol. 220. *New Aspects in Phosphorus Chemistry,* Vol. 1, Majoral, J.-P., Ed., Springer-Verlag, Berlin, 2002, p. 201.

6. Ackerman, B., Jordan, T.A., Eddy, C.R., and Swern, D., Phosphorus derivatives of fatty acids. Part 1. Preparation and properties of diethyl acylphosphonates, *J. Am. Chem. Soc.*, 78, 4444, 1956.

7. Okamoto, Y., Nitta, T., and Sakurai, H., The synthesis of long-chain α-hydroxycarboxylic acids from acyl phosphonates, *Kogyo Kagaku Zasshi,* 71, 187, 1968; *Chem. Abstr.*, 69, 35342, 1968.

8. Hall, L.A.R., and Stephens, C.W., Preparation and pyrolyses of some organophosphonates, *J. Am. Chem. Soc.*, 78, 2565, 1956.

9. Marmor, R.S., and Seyferth, D., The copper-catalyzed decomposition of some dimethylphosphono-substituted diazoalkanes, *J. Org. Chem.*, 36, 128, 1971.

10. Szpala, A., Tebby, J.C., and Griffiths, D.V., Reaction of phosphites with unsaturated acid chlorides. Synthesis and reactions of dimethyl but-2-enoylphosphonate, *J. Chem. Soc., Perkin Trans. 1,* 1363, 1981.

11. Theis, W., and Regitz, M., Investigations on diazo compounds and azides. Part 62. Synthesis and reactions of α-diazo phosphonates with a conjugated 1,3-diene unit, *Tetrahedron,* 41, 2625, 1985.

12. Karaman, R., Goldblum, A., Breuer, E., and Leader, H., Acylphosphonic acids and methyl hydrogen acylphosphonates. Physical and chemical properties and theoretical calculations, *J. Chem. Soc., Perkin Trans. 1,* 765, 1989.

13. Schuster, T., and Evans, S.A., Jr., Hetero-Diels–Alder reactions of dimethyl 1-oxo-(*E,E*)-2,4-hexadienephosphonate with 2-alkylidene-1,3-dithianes, *Phosphorus, Sulfur Silicon Relat. Elem.,* 103, 259, 1995.

14. Telan, L.A., Poon, C.-D., and Evans, S.A., Jr., Diastereoselectivity in the Mukaiyama–Michael reaction employing α-acylphosphonates, *J. Org. Chem.*, 61, 7455, 1996.

15. Evans, D.A., and Johnson, J.S., Catalytic enantioselective hetero Diels–Alder reactions of α,β-unsaturated acyl phosphonates with enol ethers, *J. Am. Chem. Soc.*, 120, 4895, 1998.

16. Evans, D.A., Johnson, J.S., and Olhava, E.J., Enantioselective synthesis of dihydropyrans. Catalysis of hetero Diels–Alder reactions by *bis*(oxazoline) copper(II) complexes, *J. Am. Chem. Soc.*, 122, 1635, 2000.

17. Depature, M., Diewok, J., Grimaldi, J., and Hatem, J., Tin-mediated free-radical cyclization of β-allenylbenzoyloximes, *Eur. J. Org. Chem.*, 275, 2000.

18. Berlin, K.D., Hellwerge, D.M., and Nagabhushanam, M., Dialkyl esters of acylphosphonic acids, *J. Org. Chem.*, 30, 1265, 1965.

19. Yamashita, M., Kojima, M., Yoshida, H., Ogata, T., and Inokawa, S., New synthesis and hydroboration of vinylphosphonates, *Bull. Chem. Soc. Jpn.*, 53, 1625, 1980.

20. Arbuzov, B.A., and Zolotova, M.V., Esters of α-oxoaminophosphonic acids, *Izv. Akad. Nauk SSSR, Ser. Khim.*, 1793, 1964; *Bull. Acad. Sci. USSR, Div. Chem. Sci. (Engl. Transl.),* 1701, 1964.

21. Zhdanov, Y.A., and Uzlova, L.A., Carbon–phosphorus bond in carbohydrates, *Zh. Obshch. Khim.*, 40, 2138, 1970; *J. Gen. Chem. USSR (Engl. Transl.),* 40, 2124, 1970.

22. Vishnyakova, G.M., Smirnova, T.V., and Tarakanova, L.A., Synthesis of coumarin-containing α-keto phosphonates, *Zh. Obshch. Khim.*, 55, 1205, 1985; *J. Gen. Chem. USSR (Engl. Transl.),* 55, 1076, 1985.

23. Khoukhi, N., Vaultier, M., and Carrié, R., Synthesis and reactivity of methyl γ-azidobutyrates and ethyl δ-azidovalerates and of the corresponding acid chlorides as useful reagents for the aminoalkylation, *Tetrahedron,* 43, 1811, 1987.

24. Yurchenko, R.I., Malyutina, I.V., Klepa, T.I., Povolotskii, M.I., Tuzhikov, O.I., and Pinchuk, A.M., Reduction of α-keto phosphonates to primary phosphines, *Zh. Obshch. Khim.*, 60, 697, 1990; *J. Gen. Chem. USSR (Engl. Transl.),* 60, 611, 1990.

25. Yurchenko, R.I., Klepa, T.I., Baklan, V.F., Fokina, N.A., and Kudryavtsev, A.A., Phosphorylated adamantanes. Part 14. α-Ketophosphonates of adamantanes, *Zh. Obshch. Khim.*, 62, 1760, 1992; *J. Gen. Chem. USSR (Engl. Transl.),* 62, 1447, 1992.

26. Bowman, M.P., Senet, J.P.G., Malfroot, T., and Olofson, R.A., An efficient synthesis of some substituted vinylic chloroformates. Reaction scope and limitations, *J. Org. Chem.*, 55, 5982, 1990.

27. Khomutov, R.M., Osipova, T.I., and Zhukov, Y.N., Synthesis of α-ketophosphonic acids, *Izv. Akad. Nauk SSSR, Ser. Khim.*, 1391, 1978; *Bull. Acad. Sci. USSR, Div. Chem. Sci. (Engl. Transl.)*, 1210, 1978.

28. Benech, J., Coindet, M., El Manouni, D., and Leroux, Y., Synthesis of new α-ketophosphonates, *Phosphorus, Sulfur Silicon Relat. Elem.*, 123, 377, 1997.

29. Gefter, E.L., and Kabachnik, M.I., β-Chloroethyl esters of some phosphorus acids, *Izv. Akad. Nauk SSSR, Ser. Khim.*, 194, 1957; *Chem. Abstr.*, 51, 11238d, 1957.

30. Goldstein, J.A., McKenna, C., and Westheimer, F.H., α-Diazobenzylphosphonate dianions, *J. Am. Chem. Soc.*, 98, 7327, 1976.

31. Yurchenko, R.I., Klepa, T.I., Bobrova, O.B., Yurchenko, A.G., and Pinchuk, A.M., Phosphorylated adamantanes. Part 2. Adamantyl esters of phosphorous acid in an Arbuzov reaction, *Zh. Obshch. Khim.*, 51, 786, 1981; *J. Gen. Chem. USSR (Engl. Transl.)*, 51, 647, 1981.

32. Orlov, N.F., and Kaufman, B.L., Reaction of complete triorganosilyl- and *O*-alkyl-*O*-triorganosilyl phosphites with acyl chlorides, *Zh. Obshch. Khim.*, 38, 1842, 1968; *J. Gen. Chem. USSR (Engl. Transl.)*, 38, 1791, 1968.

33. D'Yakov, V.M., Gusakova, G.S., Pokrovskii, E.I., and D'Yakova, T.L., Structure and IR spectra of *p*- and *o*-substituted *bis*(triorganosilyl) benzoylphosphonates, *Zh. Obshch. Khim.*, 41, 1035, 1971; *J. Gen. Chem. USSR (Engl. Transl.)*, 41, 1040, 1971.

34. Sekine, M., and Hata, T., Convenient synthesis of unesterified acylphosphonic acids, *J. Chem. Soc., Chem. Commun.*, 285, 1978.

35. Lutsenko, I.F., and Kraits, Z.S., Arbuzov rearrangement of vinyl esters of phosphorous and phenylphosphonous acids, *Dokl. Akad. Nauk SSSR*, 132, 612, 1960; *Chem. Abstr.*, 54, 24346d, 1960.

36. Gurudata, N., Benezra, C., and Cohen, H., Synthesis and nuclear magnetic resonance spectroscopy of some substituted aryldiazoalkanephosphonates, *Can. J. Chem.*, 51, 1142, 1973.

37. Cohen, H., and Benezra, C., Steric and electronic factors in 1,3-dipolar cycloadditions. Stereochemical course of the addition of dimethyl aryl- and alkyldiazomethylphosphonates to norbornadiene, *Can. J. Chem.*, 52, 66, 1974.

38. Griffiths, D.V., Griffiths, P.A., Whitehead, B.J., and Tebby, J.C., Novel quasiphosphonium ylides from the reaction of trialkyl phosphites with dialkyl benzoylphosphonates. Evidence for carbene intermediates in the intramolecular cyclization of 2-substituted dialkyl benzoylphosphonates, *J. Chem. Soc., Perkin Trans. 1*, 479, 1992.

39. Wrobel, J., and Dietrich, A., Preparation of L-(phosphonodifluoromethyl)phenylalanine derivatives as non-hydrolyzable mimetics of *O*-phosphotyrosine, *Tetrahedron Lett.*, 34, 3543, 1993.

40. Griffiths, D.V., Griffiths, P.A., Karim, K., and Whitehead, B.J., Reactions of carbene intermediates from the reaction of trialkyl phosphites with dialkyl benzoylphosphonates. Intramolecular cyclizations of 2-substituted dialkyl benzoylphosphonates, *J. Chem. Soc., Perkin Trans. 1*, 555, 1996.

41. Griffiths, D.V., Karim, K., and Harris, J.E., Dialkyl phosphonates and tetraalkyl bis(phosphonate)s from the decomposition of quasi-phosphonium ylidic phosphonates in aqueous conditions, *J. Chem. Soc., Perkin Trans. 1*, 2539, 1997.

42. Ishmaeva, E.A., Zimin, M.G., Galeeva, R.M., and Pudovik, A.N., Dipole moments of organophosphorus compounds. Part 2. Benzoylphosphonates, benzoylphosphinates, and benzoylphosphine oxide, *Izv. Akad. Nauk SSSR, Ser. Khim.*, 538, 1971; *Bull. Acad. Sci. USSR, Div. Chem. Sci. (Engl. Transl.)*, 473, 1971.

43. Afarinkia, K., Echenique, J., and Nyburg, S.C., Facile enolization of α-ketophosphonates, *Tetrahedron Lett.*, 38, 1663, 1997.

44. Takamizawa, A., Sato, Y., and Tanaka, S., Pyrimidine derivatives and related compounds. Part 33. Reactions of phosphonates with sodium salts of thiophenol and thiamine. Part 1, *Yakugaku Zasshi*, 85, 298, 1965; *Chem. Abstr.*, 63, 9940h, 1965.

45. Breuer, E., Karaman, R., Goldblum, A., and Leader, H., Sulfonic acid-induced fragmentation of dialkyl acylphosphonates. Formation of alkyl carboxylates and alkyl sulfonates, *J. Chem. Soc., Perkin Trans. 2*, 2029, 1988.

46. Arbuzov, B.A., and Bogonostseva, N.P., The action of phosphorus trichloride on ethyl orthopropionate and ethyl orthosilicate, *Izv. Akad. Nauk SSSR, Ser. Khim.*, 484, 1953; *Chem. Abstr.*, 48, 9905i, 1954.

47. Bligh, S.W. A., Choi, N., McGrath, C.M., McPartlin, M., and Woodroffe, T.M., Synthesis and structural properties of metal complexes of dialkyl α-hydroxyiminophosphonates, *J. Chem. Soc., Dalton Trans.*, 2587, 2000.

48. Zolotova, M.V., and Konstantinova, T.V., Reaction of trialkyl phosphites with substituted carboxylic acid chlorides, *Zh. Obshch. Khim.*, 40, 2131, 1970; *Chem. Abstr.*, 74, 42433, 1971.

49. Berlin, K.D., and Taylor, H.A., The reactions of aroyl halides with phosphites. Esters of aroylphosphonic acids, *J. Am. Chem. Soc.*, 86, 3862, 1964.

50. Meier, C., Laux, W.H.G., and Bats, J.W., Asymmetric synthesis of chiral, nonracemic dialkyl α-hydroxyarylmethyl- and α-, β-, and γ-hydroxyalkylphosphonates from keto phosphonates, *Liebigs Ann.*, 1963, 1995.

51. Pudovik, A.N., Gur'yanova, I.V., and Zimin, M.G., Reaction of the monoethyl ester of ethylthionophosphinic acid with some substituted benzoylphosphonates, *Zh. Obshch. Khim.*, 38, 1533, 1968; *J. Gen. Chem. USSR (Engl. Transl.)*, 38, 1483, 1968.

52. Hudson, H.R., Pianka, M., and Jun, W., Interaction of triethyl phosphite with 3-fluoropropanoyl chloride. The concurrent formation of 1:2 and 2:1 reaction products, *J. Chem. Soc., Chem. Commun.*, 2445, 1996.

53. Brittelli, D.R., A study of the reaction of 2-haloacyl halides with trialkyl phosphites. Synthesis of 2-substituted acyl phosphonates, *J. Org. Chem.*, 50, 1845, 1985.

54. Gazizov, M.B., and Razumov, A.I., Reaction of dialkyl chlorophosphites with carboxylic acid anhydrides, *Zh. Obshch. Khim.*, 39, 2600, 1969; *J. Gen. Chem. USSR (Engl. Transl.)*, 39, 2541, 1969.

55. Pudovik, A.N., and Gazizov, T.K., Thermal isomerization of diethyl acetyl phosphite and reactions of dialkyl acetyl phosphites with α-oxo phosphonic acid esters, *Zh. Obshch. Khim.*, 38, 140, 1968; *J. Gen. Chem. USSR (Engl. Transl.)*, 38, 139, 1968.

56. Chiusoli, G.P., Cometti, G., and Bellotti, V., Ring-forming deoxygenation of *ortho*-substituted aroylphosphonic diesters, *J. Chem. Soc., Chem. Commun.*, 216, 1977.

57. Chiusoli, G.P., Cometti, G., and Bellotti, V., Organonickel chemistry of aroylphosphonates, *Gazz. Chim. Ital.*, 107, 217, 1977.

58. Griffiths, D.V., Jamali, H.A.R., and Tebby, J.C., Reactions of phosphites with acid chlorides. Phosphite attack at the carbonyl oxygen of α-ketophosphonates, *Phosphorus Sulfur*, 11, 95, 1981.

59. Griffiths, D.V., Jamali, H.A.R., and Tebby, J.C., The reaction of phosphites with aroylphosphonates in the presence of proton donors. Reverse nucleophilic addition to carbonyl groups, *Phosphorus Sulfur*, 25, 173, 1985.

60. Griffiths, D.V., and Tebby, J.C., Carbene intermediates in the reaction of trialkyl phosphites with dialkyl aroylphosphonates. Formation of novel quasiphosphonium ylides, *J. Chem. Soc., Chem. Commun.*, 871, 1986.

61. Griffiths, D.V., Griffiths, P.A., Karim, K., and Whitehead, B.J., Reaction of 3-chloro-3-phenylphthalide with trimethyl phosphite, *Zh. Obshch. Khim.*, 63, 2245, 1993; *Russ. J. Gen. Chem. (Engl. Transl.)*, 63, 1560, 1993.

62. Griffiths, D.V., Griffiths, P.A., Karim, K., and Whitehead, B.J., The reactions of carbene intermediates from the reaction of trialkyl phosphites with dialkyl benzoylphosphonates. Intramolecular cyclisations of 2-substituted dialkyl benzoylphosphonates, *J. Chem. Res. (M)*, 901, 1996.

63. Griffiths, D.V., Griffiths, P.A., Karim, K., and Whitehead, B.J., The reactions of carbene intermediates from the reaction of trialkyl phosphites with dialkyl benzoylphosphonates. Intramolecular cyclizations of 2-substituted dialkyl benzoylphosphonates, *J. Chem. Res. (S)*, 176, 1996.

64. Griffiths, D.V., and Karim, K., Studies of the reactions of 2-substituted dimethyl benzoylphosphonates with trimethyl phosphite, *Phosphorus, Sulfur Silicon Relat. Elem.*, 111, 107, 1996.

65. Griffiths, D.V., Harris, J.E., and Whitehead, B.J., Reactions of carbene intermediates from the reaction of trialkyl phosphites with dialkyl benzoylphosphonates. Preparation and reactions of dimethyl 2-(*N*-*tert*-butyl-*N*-methylamino)benzoylphosphonate and dimethyl 2-(methylamino)benzoylphosphonate, *J. Chem. Soc., Perkin Trans. 1*, 2545, 1997.

66. D'Yakov, V.M., and Voronkov, M.G., Trialkylsilyl esters of polyfluoroacylphosphonic acids, *Izv. Akad. Nauk SSSR, Ser. Khim.*, 399, 1973; *Bull. Acad. Sci. USSR, Div. Chem. Sci. (Engl. Transl.)*, 379, 1973.

67. Gazizov, T.K., Sudarev, Y.I., and Pudovik, A.N., Reactions of dialkyl trimethylsilyl phosphite and ethyl trimethylsilyl ethylphosphonite with alkyl and acyl halides, *Zh. Obshch. Khim.*, 47, 1660, 1977; *J. Gen. Chem. USSR (Engl. Transl.)*, 47, 1977.

68. Novikova, Z.S., Mashoshina, S.N., Sapozhnikova, T.A., and Lutsenko, I.F., Reaction of trimethylsilyldiethylphosphite with carbonyl compounds, *Zh. Obshch. Khim.*, 41, 2622, 1971; *J. Gen. Chem. USSR (Engl. Transl.)*, 41, 2655, 1971.

69. Öhler, E., and Zbiral, E., Reaction of dialkyl phosphites with α-enones. Part 1. Synthesis and allylic rearrangement of dimethyl (1-hydroxy-2-alkenyl)- and (1-hydroxy-2-cycloalkenyl)phosphonates, *Chem. Ber.*, 124, 175, 1991.

70. Li, Y.-F., and Hammerschmidt, F., Enzymes in organic chemistry. Part 1. Enantioselective hydrolysis of α-(acyloxy)phosphonates by esterolytic enzymes, *Tetrahedron: Asymmetry*, 4, 109, 1993.

71. Benayoud, F., and Hammond, G.B., An expedient synthesis of (α,α-difluoroprop-2-ynyl)phosphonate esters, *J. Chem. Soc., Chem. Commun.*, 1447, 1996.

72. Tao, M., Bihovsky, R., Wells, G.J., and Mallamo, J.P., Novel peptidyl phosphorus derivatives as inhibitors of human calpain I, *J. Med. Chem.*, 41, 3912, 1998.

73. Wang, Z., Gu, Y., Zapata, A.J., and Hammond, G.B., An improved preparation of α-fluorinated propargylphosphonates and the solid phase synthesis of α-hydroxy-γ-TIPS propargylphosphonate ester, *J. Fluorine Chem.*, 107, 127, 2001.

74. Glebova, Z.I., Uzlova, L.A., and Zhdanov, Y.A., Selective oxidation of carbohydrate-containing α-hydroxy phosphonates, *Zh. Obshch. Khim.*, 55, 1435, 1985; *J. Gen. Chem. USSR (Engl. Transl.)*, 55, 1279, 1985.

75. Smyth, M.S., Ford, H., and Burke, T.R., Jr., A general method for the preparation of benzylic α,α-difluorophosphonic acids. Non-hydrolyzable mimetics of phosphotyrosine, *Tetrahedron Lett.*, 33, 4137, 1992.

76. Liao, Y., Shabany, H., and Spilling, C.D., The preparation of acyl phosphonates by the heterogeneous oxidation of 1-hydroxy phosphonates, *Tetrahedron Lett.*, 39, 8389, 1998.

77. Burke, T.R., Jr., Smyth, M.S., Nomizu, M., Otaka, A., and Roller, P.P., Preparation of fluoro- and hydroxy-4-(phosphonomethyl)-D,L-phenylalanine suitably protected for solid-phase synthesis of peptides containing hydrolytically stable analogues of *O*-phosphotyrosine, *J. Org. Chem.*, 58, 1336, 1993.

78. Ganzhorn, A.J., Hoflack, J., Pelton, P.D., Strasser, F., Chanal, M.-C., and Piettre, S.R., Inhibition of *myo*-inositol monophosphatase isoforms by aromatic phosphonates, *Bioorg. Med. Chem.*, 6, 1865, 1998.

79. Kaboudin, B., Surface-mediated solid-phase reactions. The preparation of acyl phosphonates by oxidation of 1-hydroxyphosphonates on the solid surface, *Tetrahedron Lett.*, 41, 3169, 2000.

80. Kaboudin, B., and Nazari, R., A convenient and mild procedure for the preparation of α-keto phosphonates of 1-hydroxyphosphonates under solvent-free conditions using microwave, *Synth. Commun.*, 31, 2245, 2001.

81. Firouzabadi, H., Iranpoor, N., Sobhani, S., and Sardarian, A.R., High yield preparation of α-keto-phosphonates by oxidation of α-hydroxyphosphonates with zinc dichromate trihydrate (ZnCr$_2$O$_7$·3H$_2$O) under solvent-free conditions, *Tetrahedron Lett.*, 42, 4369, 2001.

82. Burke, T.R., Jr., Smyth, M.S., Otaka, A., and Roller, P.P., Synthesis of 4-phosphono(difluoromethyl)-D,L-phenylalanine and *N*-Boc and *N*-Fmoc derivatives suitably protected for solid-phase synthesis of nonhydrolysable phosphotyrosyl peptide analogues, *Tetrahedron Lett.*, 34, 4125, 1993.

83. Ye, B., and Burke, T.R., Jr., Synthesis of a difluorophosphonomethyl-containing phosphatase inhibitor designed from the X-ray structure of a PTP1B-bound ligand, *Tetrahedron*, 52, 9963, 1996.

84. Yao, Z.-J., Ye, B., Wu, X.-W., Wang, S., Wu, L., Zhang, Z.-Y., and Burke, T.R., Jr., Structure-based design and synthesis of small molecule protein-tyrosine phosphatase 1B inhibitors, *Bioorg. Med. Chem.*, 6, 1799, 1998.

85. Iorga, B., Mouriès, V., and Savignac, P., Carbanionic displacement reactions at phosphorus. Synthesis and reactivity of 5,5-dimethyl-2-oxo-2-(1,3-dithian-2-yl)-1,3,2-dioxaphosphorinane, *Bull. Soc. Chim. Fr.*, 134, 891, 1997.

86. Coutrot, P., Grison, C., and Lecouvey, M., Preparation of the phosphonic acid analog of 3-deoxy-D-manno-2-octulosonic acid (KDO), *Tetrahedron Lett.*, 37, 1595, 1996.

87. Arbuzov, A.E., and Razumov, A.I., Tautomeric conversions of some phosphoorganic compounds, *Zh. Obshch. Khim.*, 4, 834, 1934; *Chem. Abstr.*, 28, 2145, 1935.

88. Lichtenthaler, F.W., The chemistry and properties of enol phosphates, *Chem. Rev.*, 61, 607, 1961.

89. Arbuzov, B.A., Lugovkin, B.P., and Bogonostseva, N.P., Action of α- and γ-bromoacetoacetic ester and 2-chlorocyclohexanone on triethyl phosphite and sodium diethyl phosphite, *Zh. Obshch. Khim.*, 20, 1468, 1950; *Chem. Abstr.*, 45, 1506a, 1951.

90. Dawson, N.D., and Burger, A., Some alkyl thiazolephosphonates, *J. Am. Chem. Soc.*, 74, 5312, 1952.

91. Arbuzov, B.A., and Vinogradova, V.S., Synthesis of esters of some β-keto phosphonic acids, *Dokl. Akad. Nauk SSSR*, 99, 85, 1954; *Chem. Abstr.*, 49, 13925g, 1955.

92. Arbuzov, B.A., and Vinogradova, V.S., Esters of 2-oxophosphonic acids. Part 2. Esters of aromatic and carbocyclic series, *Izv. Akad. Nauk SSSR, Ser. Khim.*, 284, 1957; *Chem. Abstr.*, 51, 14587a, 1957.

93. Arbuzov, B.A., and Vinogradova, V.S., Study of esters of β-oxo phosphonic acids by the method of titration with bromine, *Dokl. Akad. Nauk SSSR*, 106, 263, 1956; *Chem. Abstr.*, 50, 13787d, 1956.

94. Arbuzov, B.A., and Vinogradova, V.S., Esters of β-oxophosphonic acids. Part 1. Phosphonoacetic ester, phosphonoacetone, and their homologs, *Izv. Akad. Nauk SSSR, Ser. Khim.*, 54, 1957; *Chem. Abstr.*, 51, 10365c, 1957.

95. Perkow, W., Ullerich, K., and Meyer, F., New phosphoric acid ester with miotic activity, *Naturwissenschaften*, 39, 353, 1952; *Chem. Abstr.*, 47, 8248e, 1953.

96. Perkow, W., Reactions of alkyl phosphites. Part 1. Conversions with Cl_3CCHO and Br_3CCHO, *Chem. Ber.*, 87, 755, 1954.

97. Pudovik, A.N., Anomalous reaction of α-halo ketones with esters of phosphorous acid, *Dokl. Akad. Nauk SSSR*, 105, 735, 1955; *Chem. Abstr.*, 50, 11230f, 1956.

98. Pudovik, A.N., and Lebedeva, N.M., Reaction of chloro- and bromoacetone with triethyl phosphite, *Dokl. Akad. Nauk SSSR*, 101, 889, 1955; *Chem. Abstr.*, 50, 3219d, 1956.

99. Allen, J.F., and Johnson, O.H., The synthesis of monovinyl esters of phosphorus(V) acids, *J. Am. Chem. Soc.*, 77, 2871, 1955.

100. Whetstone, R.R., and Harman, D., Insecticidally active esters of phosphorus acids and preparation of the same, *Shell Oil*, U.S. Patent Appl. US 2956073, 1960; *Chem. Abstr.*, 55, 6380a, 1961.

101. Anliker, R., Beriger, E., Geiger, M., and Schmid, K., Synthesis of phosphamidon and its degradation in plants, *Helv. Chim. Acta*, 44, 1622, 1961.

102. Wadsworth, W.S., Jr., and Emmons, W.D., Bicyclic phosphites, *J. Am. Chem. Soc.*, 84, 610, 1962.

103. Chopard, P.A., Clark, V.M., Hudson, R.F., and Kirby, A.J., Mechanism of the reaction between trialkyl phosphites and α-halogenated ketones, *Tetrahedron*, 21, 1961, 1965.

104. Hudson, R.F., and Salvadori, G., Reaction between the compounds of trivalent phosphorus and α-halo-2,4,6-trimethylacetophenone, *Helv. Chim. Acta*, 49, 96, 1965.

105. Borowitz, I.J., Anschel, M., and Firstenberg, S., Oganophosphorus chemistry. Part 4. Reactions of trialkyl phosphites with α-halo ketones, *J. Org. Chem.*, 32, 1723, 1967.

106. Miller, J.A., Reactions of dimethyl phosphite with tetracyclone, *Tetrahedron Lett.*, 11, 3427, 1970.

107. Borowitz, I.J., Firstenberg, S., Borowitz, G.B., and Schuessler, D., Organophosphorus chemistry. Part 17. Kinetics and mechanism of the Perkow reaction, *J. Am. Chem. Soc.*, 94, 1623, 1972.

108. Gaydou, E.M., Freze, R., and Buono, G., Enol phosphates. Part 5. Synthesis and physical properties of substituted phosphates, *Bull. Soc. Chim. Fr.*, 2279, 1973.

109. Baboulene, M., and Sturtz, G., 1-Aminomethyl-2-benzoylcyclopropanes. Part 1. Synthesis, *Bull. Soc. Chim. Fr.*, 1585, 1974.

110. Attia, S.Y., Berry, J.P., Koshy, K.M., Leung, Y.K., Lyznicki, E.P., Jr., Nowlan, V.J., Oyama, K., and Tidwell, T.T., Acid-catalyzed hydrolysis of vinyl phosphates and vinyl acetates. The substituent effects of diethyl phosphoryloxy and acetoxy groups, *J. Am. Chem. Soc.*, 99, 3401, 1977.

111. Sekine, M., Okimoto, K., Yamada, K., and Hata, T., Silyl phosphites. Part 15. Reactions of silyl phosphites with α-halo carbonyl compounds. Elucidation of the mechanism of the Perkow reaction and related reactions with confirmed experiments, *J. Org. Chem.*, 46, 2097, 1981.

112. Sekine, M., Nakajima, M., and Hata, T., Silyl Phosphites. Part 16. Mechanism of the Perkow reaction and the Kukhtin-Ramirez reaction. Elucidation by means of a new type of phosphoryl rearrangements utilizing silyl phosphites, *J. Org. Chem.*, 46, 4030, 1981.

113. Mlotkowska, B., Majewski, P., Koziara, A., Zwierzak, A., and Sledzinski, B., Reactions of 2,4- and 2,6-dichlorophenacylidene halides with trialkylphosphites in protic solvents. Direct evidence for the "enolate anion" pathway, *Pol. J. Chem.*, 55, 631, 1981.

114. Petnehazy, I., Szakal, G., Töke, L., Hudson, H.R., Powroznyk, L., and Cooksey, C.J., Quasiphosphonium intermediates. Part 4. Isolation and identification of intermediates in the Arbuzov and Perkow reactions of neopentyl esters of phosphorus(III) acids with α-halogenoacetophenones, *Tetrahedron*, 39, 4229, 1983.

115. Taira, K., and Gorenstein, D.G., Experimental tests of the stereoelectronic effect at phosphorus. Michaelis–Arbuzov reactivity of phosphite esters, *Tetrahedron*, 40, 3215, 1984.

116. Morita, I., Tsuda, M., Kise, M., and Sugiyama, M., Reaction of cyclic phosphites with haloacetones, *Chem. Pharm. Bull.*, 35, 4711, 1987.

117. Hudson, H.R., Matthews, R.W., McPartlin, M., Pryce, M.A., and Shode, O.O., Quasiphosphonium intermediates. Part 7. The preparation of trinorborn-1-yl phosphite and its reactions with halogeno compounds. Stable intermediates of the Arbuzov and Perkow reactions and their structural character- ization by X-ray diffraction, NMR spectroscopy, and fast-atom-bombardment mass spectrometry, *J. Chem. Soc., Perkin Trans. 2*, 1433, 1993.

118. Pudovik, A.N., and Aver'yanova, V.P., Anomalous reaction of α-halo ketones with esters of phospho- rous acid. Part 2. Reaction of chloro- and bromoacetone with esters of phosphorous and phenylphosph- onous acids, *Zh. Obshch. Khim.*, 26, 1426, 1956; *Chem. Abstr.*, 50, 14512f, 1956.

119. Jacobson, H.I., Griffin, M.J., Preis, S., and Jensen, E.V., Phosphonic acids. Part 4. Preparation and reactions of β-ketophosphonate and enol phosphate esters, *J. Am. Chem. Soc.*, 79, 2608, 1957.

120. Kreutzkamp, N., and Kayser, H., Carbonyl- and cyanophosphonic esters. Part 2. The course of reaction of phosphites with bromo- and chloroacetone, *Chem. Ber.*, 89, 1614, 1956.

121. Pudovik, A.N., Anomalous reaction of α-halo ketones with triethyl ester of phosphorus acid, *Zh. Obshch. Khim.*, 35, 2173, 1955; *Chem. Abstr.*, 50, 8486i, 1956.

122. Pudovik, A.N., and Biktimirova, L.G., Anomalous reaction of α-halo ketones with esters of phospho- rous acid. Part 5. Reactions of halogen derivatives of methyl ketone and acetophenone with triethyl phosphite, *Zh. Obshch. Khim.*, 27, 1708, 1957; *Chem. Abstr.*, 52, 3714b, 1958.

123. Speziale, A.J., and Freeman, R.C., Reactions of phosphorus compounds. Part 1. Diethyl carbamoyl- methylphosphonates, *J. Org. Chem.*, 23, 1883, 1958.

124. Sturtz, G., Application of β-oxophosphonates to the synthesis of α-ethylenic ketones and alkyl cyclopropyl ketones, *Bull. Soc. Chim. Fr.*, 2349, 1964.

125. Machleidt, H., and Strehlke, G.U., Rearrangement between phosphonate and enolphosphate ester, *Angew. Chem.*, 76, 494, 1964.

126. Sasaki, K., Syntheses of isomeric methylpseudoionones and isomeric pseudoionones by the Wittig reaction, *Bull. Chem. Soc. Jpn.*, 39, 2703, 1966.

127. Aso, Y., Iyoda, M., and Nakagawa, M., A naphtho-tri-*t*-butyltrisdehydro[16]annulene, *Tetrahedron Lett.*, 20, 4217, 1979.

128. Fang, F.G., Feigelson, G.B., and Danishefsky, S.J., A total synthesis of magallanesine. DMF acetal mediated cyclodehydration of a methyl ketone thioimide, *Tetrahedron Lett.*, 30, 2743, 1989.

129. Cotton, F.A., and Schunn, R.A., Metal salts and complexes of dialkoxyphosphonylacetyl-methanide ions, *J. Am. Chem. Soc.*, 85, 2394, 1963.

130. Kitamura, M., Tokunaga, M., and Noyori, R., Asymmetric hydrogenation of β-keto phosphonates. A practical way to fosfomycin, *J. Am. Chem. Soc.*, 117, 2931, 1995.

131. Adiyaman, M., Lawson, J.A., Hwang, S.-W., Khanapure, S.P., FitzGerald, G.A., and Rokach, J., Total synthesis of a novel isoprostane iPF$_{2\alpha}$-I and its identification in biological fluids, *Tetrahedron Lett.*, 37, 4849, 1996.

132. Friesen, R.W., and Blouin, M., Preparation of γ,δ-unsaturated β-ketophosphonates from tertiary α- allenic alcohols. The synthesis of (±)-(*E*)-α-atlantone, *J. Org. Chem.*, 61, 7202, 1996.

133. Adiyaman, M., Li, H., Lawson, J.A., Hwang, S.-W., Khanapure, S.P., FitzGerald, G.A., and Rokach, J., First total synthesis of isoprostane iPF$_{2\alpha}$-III, *Tetrahedron Lett.*, 38, 3339, 1997.

134. Hwang, S.W., Adiyaman, M., Lawson, J.A., FitzGerald, G.A., and Rokach, J., Synthesis of iPF$_{2\alpha}$-V. A new route, *Tetrahedron Lett.*, 40, 6167, 1999.

135. Buono, G., and Peiffer, G., Reactivity of trivalent phosphorus derivatives toward α-allene ketones, *Tetrahedron Lett.*, 13, 149, 1972.

136. Pudovik, A.N., and Salekhova, L.G., Anomalous reaction of α-halo ketones with esters of phosphorous acid. Part 3. Reaction of α,α- and α,γ-dihalo ketones with triethyl phosphite, *Zh. Obshch. Khim.*, 26, 1431, 1956; *Chem. Abstr.*, 50, 14513c, 1956.

137. Pudovik, A.N., and Chebotareva, E.G., Anomalous reaction of phosphites with α-halo ketones. Part 8. Reaction of mixed phosphites with chloro- and dichloroacetone, *Zh. Obshch. Khim.*, 28, 2492, 1958; *Chem. Abstr.*, 53, 3117b, 1959.

138. Pudovik, A.N., Anomalous reaction of α-halo ketones with esters of phosphorous acid. Part 4. Reactions of esters of phosphorus acid with mono- and dichloroacetylacetone, phosphonoacetone, and acetoacetic ester, *Zh. Obshch. Khim.*, 26, 2238, 1956; *Chem. Abstr.*, 51, 1827a, 1957.

139. Arbuzov, B.A., Vinogradova, V.S., Polezhaeva, N.A., and Shamsutdinova, A.K., Esters of β-oxophosphonic acids. Part 12. Structure of reaction products of some α-halo ketones of the aromatic series with triethyl phosphite and diethyl sodium phosphite, *Izv. Akad. Nauk SSSR, Ser. Khim.*, 1380, 1963; *Bull. Acad. Sci. USSR, Div. Chem. Sci. (Engl. Transl.)*, 1257, 1963.

140. Arbuzov, B.A., Vinogradova, V.S., and Polezhaeva, N.A., Esters of β-oxo phosphonic acids. Part 8. Reaction of 2,6-dibromocyclohexanone with triethyl phosphite, *Izv. Akad. Nauk SSSR, Ser. Khim.*, 2013, 1961; *Chem. Abstr.*, 56, 11456i, 1962.

141. Welch, S.C., Assercq, J.M., Loh, J.P., and Glase, S.A., 3-Chloro-2-[(diethoxyphosphoryl)oxy]-1-propene. A new reagent for a one-pot cyclopentenone annelation. Synthesis of desoxyallethrolone, *cis*-jasmone, and methylenomycin B, *J. Org. Chem.*, 52, 1440, 1987.

142. Corbel, B., Medinger, L., Haelters, J.P., and Sturtz, G., An efficient synthesis of dialkyl 2-oxoalkanephosphonates and diphenyl-2-oxoalkylphosphine oxides from 1-chloroalkyl ketones, *Synthesis*, 1048, 1985.

143. Yuan, C., and Xie, R., Studies on organophosphorus compounds. Part 87. A convenient procedure for the preparation of diethyl 2-oxoalkylphosphonates, *Phosphorus, Sulfur Silicon Relat. Elem.*, 90, 47, 1994.

144. Mao, L.-J., and Chen, R.-Y., Highly selective Arbuzov reaction of α-chlorocarbonyl compounds with P(OEt)₃ and substituted amino urea, *Phosphorus, Sulfur Silicon Relat. Elem.*, 111, 167, 1996.

145. Nikolova, R., Bojilova, A., and Rodios, N.A., Reaction of 3-bromobenzyl and 3-bromoacetyl coumarin with phosphites. Synthesis of some new phosphonates and phosphates in the coumarin series, *Tetrahedron*, 54, 14407, 1998.

146. Maier, L., and Diel, P.J., Organic phosphorus compounds. Part 105. Synthesis and properties of 2-amino-2-arylethylphosphonic acids and derivatives, *Phosphorus, Sulfur Silicon Relat. Elem.*, 107, 245, 1995.

147. Maier, L., and Diel, P.J., Synthesis and properties of 2-amino-2-arylethylphosphonic acids and derivatives, *Phosphorus, Sulfur Silicon Relat. Elem.*, 109–110, 341, 1996.

148. Arbuzov, B.A., Vinogradova, V.S., and Polezhaeva, N.A., Esters of β-oxophosphonic acids. Part 3. Structure of reaction products of some halogenated ketones with triethyl phosphite and sodium diethyl phosphite, *Izv. Akad. Nauk SSSR, Ser. Khim.*, 41, 1959; *Chem. Abstr.*, 53, 15035e, 1959.

149. Arbuzov, B.A., and Movsesyan, M.E., Esters of β-oxophosphonic acids. Part 4. Infrared spectra of the reaction products of α-halo ketones with triethyl phosphite and diethyl sodiophosphite, *Izv. Akad. Nauk SSSR, Ser. Khim.*, 267, 1959; *Chem. Abstr.*, 53, 19850f, 1959.

150. Arbuzov, B.A., Vinogradova, V.S., and Polezhaeva, N.A., Esters of β-oxo phosphonic acids. Part 5. Structure of the reaction products of some α-halo ketones of the carbocyclic series with triethyl phosphite and sodium diethyl phosphite, *Izv. Akad. Nauk SSSR, Ser. Khim.*, 832, 1960; *Chem. Abstr.*, 54, 24454e, 1960.

151. Sturtz, G., The action of sodium phosphites on the ω-halogenated ketones, *Bull. Soc. Chim. Fr.*, 2333, 1964.

152. Sturtz, G., Action of sodium phosphites, sodium phosphonites, and phosphinites on halogenated ketones in the form of ketals and enol ethers, *Bull. Soc. Chim. Fr.*, 2340, 1964.

153. Boeckman, R.K., and Thomas, A.J., Methodology for the synthesis of phosphorus-activated tetramic acids. Applications to the synthesis of unsaturated 3-acyltetramic acids, *J. Org. Chem.*, 47, 2823, 1982.

154. Boeckman, R.K., Perni, R.B., Macdonald, J.E., and Thomas, A.J., 6-Diethylphosphonomethyl-2,2-dimethyl-1,3-dioxen-4-one {Phosphonic acid, [(2,2-dimethyl-4-oxo-4H-1,3-dioxin-6-yl)methyl]-, diethyl ester}, *Org. Synth. Coll.*, VIII, 192, 1993.

155. Paterson, I., Gardner, M., and Banks, B.J., Studies in marine cembranolide synthesis. A synthesis of 2,3,5-trisubstituted furan intermediates for lophotoxin and pukalide, *Tetrahedron*, 45, 5283, 1989.

156. Schlessinger, R.H., and Graves, D.D., A synthesis of the tetramic acid subunit of streptolydigin. A reactivity definition of this subunit as an Emmons reagent, *Tetrahedron Lett.*, 28, 4385, 1987.

157. Rosen, T., Fernandes, P.B., Marovich, M.A., Shen, L., Mao, J., and Pernet, A.G., Aromatic dienoyl tetramic acids. Novel antibacterial agents with activity against anaerobes and *Staphylococci*, *J. Med. Chem.*, 32, 1062, 1989.

158. Sturtz, G., Charrier, C., and Normant, H., Effect of 1-bromoacetylene derivatives on sodium dialkyl phosphites. Preparation of long-chain β-oxo phosphonates, *Bull. Soc. Chim. Fr.*, 1707, 1966.

159. Christov, V.C., Aladinova, V.M., and Prodanov, B., Hydration reactions of phosphorylated 1,3-enynes, *Phosphorus, Sulfur Silicon Relat. Elem.*, 155, 67, 1999.

160. Poss, A.J., and Belter, R.K., Diethyl 3-iodopropynylphosphonate. An alkylative β-keto phosphonate equivalent, *J. Org. Chem.*, 52, 4810, 1987.

161. Todd, R.S., Reeve, M., and Davidson, A.H., Preparation of phosphorus containing alkynyl derivatives useful as intermediates in the preparation of keto phosphonates and mevinolinic acid derivatives, *British Bio-Technology*, Int. Patent Appl. WO 9322321, 1992; *Chem. Abstr.*, 120, 245504, 1994.

162. Corey, E.J., and Virgil, S.C., Enantioselective total synthesis of a protosterol, 3β,20-dihydroxyprotost-24-ene, *J. Am. Chem. Soc.*, 112, 6429, 1990.

163. Guile, S.D., Saxton, J.E., and Thornton-Pett, M., Synthetic studies towards paspalicine. Part 2. An alternative approach to the synthesis of the C/D ring system, *J. Chem. Soc., Perkin Trans. 1,* 1763, 1992.

164. Pudovik, A.N., and Khusainova, N.G., Addition of alcohols, amines, and dialkyl hydrogen phosphites to esters of γ,γ-dimethylallenylphosphonic acid, *Zh. Obshch. Khim.*, 36, 1236, 1966; *Chem. Abstr.*, 65, 16994h, 1966.

165. Chattha, M.S., and Aguiar, A.M., Enamine phosphonates. Their use in the synthesis of α,β-ethylenic ketimines and the corresponding ketones, *Tetrahedron Lett.*, 12, 1419, 1971.

166. Chattha, M.S., and Aguiar, A.M., Organophosphorus enamines. Part 7. Synthesis and stereochemistry of enamine phosphonates, *J. Org. Chem.*, 38, 820, 1973.

167. Chattha, M.S., and Aguiar, A.M., Organophosphorus enamines. Part 8. Convenient preparation of diethyl β-ketophosphonates, *J. Org. Chem.*, 38, 2908, 1973.

168. Chattha, M.S., Synthesis and NMR spectrum of diethyl 3-buten-3-methyl-2-oxo-1-phosphonate, *Chem. Ind. (London)*, 1031, 1976.

169. Ruder, S.M., and Norwood, B.K., Cycloaddition of enamines with alkynylphosphonates. A route to functionalized medium sized rings, *Tetrahedron Lett.*, 35, 3473, 1994.

170. Sturtz, G., Preparation of α,β-unsaturated ketones from β-oxo phosphonates, *Bull. Soc. Chim. Fr.*, 2477, 1967.

171. Sturtz, G., Elimination–addition reactions of β-halo-β-ethylenic phosphonates. Obtaining of β-oxo compounds, *Bull. Soc. Chim. Fr.*, 1345, 1967.

172. Sturtz, G., and Baboulene, M., α-Aminoalcohols. γ-Amino-β-hydroxyphosphonates, *Chim. Ther.*, 4, 195, 1969.

173. Baboulene, M., Belbeoch, A., and Sturtz, G., A convenient synthesis of dialkyl 3-alkoxy-, 3-aryloxy-, and 3-arylthio-2-oxoalkanephosphonates, *Synthesis*, 240, 1977.

174. Altenbach, H.-J., and Korff, R., Phosphorus- and sulfur-substituted allenes in synthesis. Part 1. Simple synthesis of β-ketophosphonates from 1-alkyn-3-ols, *Tetrahedron Lett.*, 22, 5175, 1981.

175. Denmark, S.E., and Marlin, J.E., Carbanion-accelerated Claisen rearrangements. Part 7. Phosphine oxide and phosphonate anion stabilizing groups, *J. Org. Chem.*, 56, 1003, 1991.

176. Corey, E.J., and Kwiatkowski, G.T., The synthesis of *cis* and *trans* olefins via β-keto and β-hydroxy phosphonamides, *J. Am. Chem. Soc.*, 88, 5653, 1966.

177. Corey, E.J., Vlattas, I., Andersen, N.H., and Harding, K., A new total synthesis of prostaglandins of the E_1 and F_1 series including 11-epiprostaglandins, *J. Am. Chem. Soc.*, 90, 3247, 1968.

178. Büchi, G., and Powell, J.E., Jr., Claisen rearrangement of 3,4-dihydro-2*H*-pyranylethylenes. Synthesis of cyclohexenes, *J. Am. Chem. Soc.*, 92, 3126, 1970.

179. Dauben, W.G., Ahlgren, G., Leitereg, T.J., Schwarzel, W.C., and Yoshioko, M., Steroidal antibiotics. Total synthesis of the fusidic acid tetracyclic ring system, *J. Am. Chem. Soc.*, 94, 8593, 1972.

180. Burka, L.T., Harris, T.M., and Wilson, B.J., Synthesis of racemic ipomeamarone and *epi*ipomeamarone, *J. Org. Chem.*, 39, 2212, 1974.

181. Dauben, W.G., Beasley, G.H., Broadhurst, M.D., Muller, B., Peppard, D.J., Pesnelle, P., and Suter, C., Synthesis of (±)-cembrene, a fourteen-membered ring diterpene, *J. Am. Chem. Soc.*, 97, 4973, 1975.

182. Overman, L.E., and Jessup, P.J., Synthetic applications of *N*-acylamino-1,3-dienes. An efficient stereospecific total synthesis of *dl*-pumiliotoxin C, and a general entry to *cis*-decahydroquinoline alkaloids, *J. Am. Chem. Soc.*, 100, 5179, 1978.

183. Meyers, A.I., and Smith, R.K., A total asymmetric synthesis of (+)-*ar*-turmerone, *Tetrahedron Lett.*, 20, 2749, 1979.

184. Nakai, H., Hamanaka, N., and Kurono, M., Synthesis of 16,16-(2-fluorotrimethylene)prostaglandins and 16,16-(2,2-difluorotrimethylene)prostaglandins, *Chem. Lett.*, 63, 1979.

185. Begley, M.J., Cooper, K., and Pattenden, G., Novel dimerisation of bicyclo[3.3.0]oct-2-en-3-one, *Tetrahedron Lett.*, 22, 257, 1981.

186. Begley, M.J., Cooper, K., and Pattenden, G., Application of the intermolecular Wadsworth–Emmons reaction to bicyclo[3.3.0]oct-$\Delta^{1,2}$-en-3-ones. Synthesis and X-ray structure of a novel dimer of the parent member, *Tetrahedron*, 37, 4503, 1981.

187. Nicolaou, K.C., Pavia, M.R., and Seitz, S.P., Carbohydrates in organic synthesis. Synthesis of 16-membered-ring macrolide antibiotics. Part 5. Total synthesis of *O*-mycinosyltylonolide. Synthesis of key intermediates, *J. Am. Chem. Soc.*, 104, 2027, 1982.

188. Overman, L.E., Lesuisse, D., and Hashimoto, M., Importance of allylic interactions and stereoelectronic effects in dictating the steric course of the reaction of iminium ions with nucleophiles. An efficient total synthesis of (±)-gephyrotoxin, *J. Am. Chem. Soc.*, 105, 5373, 1983.

189. Tang, K.-C., and Coward, J.K., Synthesis of acyl phosphonate analogues of biologically important acyl phosphates. *N*-(2-Amino-10-methylpteroyl)-5-amino-2-oxopentanephosphonic acid, *J. Org. Chem.*, 48, 5001, 1983.

190. Iyoda, M., Ogura, F., Akiyama, S., and Nakagawa, M., A synthetic approach to annulenoannulenoannulene, an anthracene-like system, *Chem. Lett.*, 1883, 1983.

191. Mak, C.-P., Mayerl, C., and Fliri, H., Synthesis of carbapenem-3-phosphonic acid derivatives, *Tetrahedron Lett.*, 24, 347, 1983.

192. Aboujaoude, E.E., Collignon, N., and Savignac, P., Synthesis of β-carbonylated phosphonates. Part 1. Carbanionic route, *J. Organomet. Chem.*, 264, 9, 1984.

193. Mikolajczyk, M., and Balczewski, P., A new, total synthesis of methylenomycin B using organic sulfur and phosphorus reagents, *Synthesis*, 691, 1984.

194. Andrus, A., Christensen, B.G., and Heck, J.V., Synthesis of 3-methylphosphonyl thienamycin and related 3-phosphonyl carbapenems, *Tetrahedron Lett.*, 25, 595, 1984.

195. Callant, P., D'Haenens, L., and Vandewalle, M., An efficient preparation and the intramolecular cyclopropanation of α-diazo-β-ketophosphonates and α-diazophosphonoacetates, *Synth. Commun.*, 14, 155, 1984.

196. Callant, P., D'Haenens, L., van der Eycken, E., and Vandewalle, M., Photoinduced Wolff rearrangement of α-diazo-β-ketophosphonates. A novel entry into substituted phosphonoacetates, *Synth. Commun.*, 14, 163, 1984.

197. Nicolaou, K.C., Duggan, M.E., Hwang, C.K., and Somers, P.K., Activation of 6-*endo* over 5-*exo* epoxide openings. Ring-selective formation of tetrahydropyran systems and stereocontrolled synthesis of the ABC ring framework of brevetoxin B, *J. Chem. Soc., Chem. Commun.*, 1359, 1985.

198. Connolly, P.J., and Heathcock, C.H., An approach to the total synthesis of dendrobine, *J. Org. Chem.*, 50, 4135, 1985.

199. Guzzi, U., Ciabatti, R., Padova, G., Battaglia, F., Cellentani, M., Depaoli, A., Galliani, G., Schiatti, P., and Spina, G., Structure–activity studies of 16-methoxy-16-methyl prostaglandins, *J. Med. Chem.*, 29, 1826, 1986.

200. Ichihara, A., Kawagishi, H., Tokugawa, N., and Sakamura, S., Stereoselective total synthesis and stereochemistry of diplodiatoxin, a mycotoxin from *Diplodia maydis*, *Tetrahedron Lett.*, 27, 1347, 1986.

201. Chakravarty, P.K., Combs, P., Roth, A., and Greenlee, W.J., An efficient synthesis of γ-amino-β-ketoalkylphosphonates from α-amino acids, *Tetrahedron Lett.*, 28, 611, 1987.

202. Nicolaou, K.C., Daines, R.A., Uenishi, J., Li, W.S., Papahatjis, D.P., and Chakraborty, T.K., Stereocontrolled construction of key building blocks for the total synthesis of amphoteronolide B and amphotericin B, *J. Am. Chem. Soc.*, 109, 2205, 1987.

203. Nicolaou, K.C., Daines, R.A., and Chakraborty, T.K., Total synthesis of amphoteronolide B, *J. Am. Chem. Soc.*, 109, 2208, 1987.

204. Nicolaou, K.C., Daines, R.A., Uenishi, J., Li, W.S., Papahatjis, D.P., and Chakraborty, T.K., Total synthesis of amphoteronolide B and amphotericin B. Part 1. Strategy and stereocontrolled construction of key building blocks, *J. Am. Chem. Soc.*, 110, 4672, 1988.

205. Berkowitz, W.F., and Arafat, A.F., Preparation of a prostanoid intermediate from loganin, *J. Org. Chem.*, 53, 1100, 1988.

206. Coppola, G.M., The chemistry of 2*H*-3,1-benzoxazine-2,4(1*H*)-dione (isatoic anhydride). Part 20. Synthesis and Wittig reactions of dimethyl (4-oxo-1,4-dihydro-quinolin-2-yl)methanephosphonates, *Synthesis*, 81, 1988.

207. Just, G., and O'Connor, B., Synthesis of the 5*R*,8*R*,9*S*,11*R* dephosphorylated derivative of CI-920, a novel antitumor agent, *Tetrahedron Lett.*, 29, 753, 1988.

208. Nickson, T.E., Stereospecific synthesis of (*Z*)- and (*E*)-diethyl (3,3,3-trifluoro-1-propenyl)phosphonate, *J. Org. Chem.*, 53, 3870, 1988.

209. Chakravarty, P.K., Greenlee, W.J., Parsons, W.H., Patchett, A.A., Combs, P., Roth, A., Busch, R.D., and Mellin, T.N., (3-Amino-2-oxoalkyl)phosphonic acids and their analogues as novel inhibitors of D-alanine. D-Alanine ligase, *J. Med. Chem.*, 32, 1886, 1989.

210. Koskinen, A.M.P., and Krische, M.J., γ-Amino-β-keto phosphonates in synthesis. Synthesis of the sphingosine skeleton, *Synlett*, 665, 1990.

211. Karanewsky, D.S., Malley, M.F., and Gougoutas, J.Z., Practical synthesis of an enantiomerically pure synthon for the preparation of mevinic acid analogues, *J. Org. Chem.*, 56, 3744, 1991.

212. Plata, D.J., Leanna, M.R., and Morton, H.E., The stereospecific preparation of an hydroxyethylene isostere precursor via a novel piperidine-2,5-dione template, *Tetrahedron Lett.*, 32, 3623, 1991.

213. Kojima, K., and Saito, S., A novel synthesis of dimethyl 2-oxoalkenylphosphonate from pulegone hydrogen chloride adduct, *Synthesis*, 949, 1992.

214. Balreddy, K., Dong, L., Simpson, D.M., and Titmas, R., Synthesis of phosphonate analogs of lipid X, *Tetrahedron Lett.*, 34, 3037, 1993.

215. Roush, W.R., Warmus, J.S., and Works, A.B., Synthesis and transannular Diels–Alder reactions of (*E,E,E*)-cyclotetradeca-2,8,10-trienones, *Tetrahedron Lett.*, 34, 4427, 1993.

216. Shen, Y., and Qi, M., New synthesis of dialkyl fluoroalkynylphosphonates, *J. Chem. Soc., Perkin Trans. 1*, 2153, 1993.

217. Andriamiadanarivo, R., Pujol, B., Chantegrel, B., Deshayes, C., and Doutheau, A., Preparation of functionalized cyclobutenones and phenolic compounds from α-diazo β-ketophosphonates, *Tetrahedron Lett.*, 34, 7923, 1993.

218. Gaoni, Y., Chapman, A.G., Parvez, N., Pook, P.C.K., Jane, D.E., and Watkins, J.C., Synthesis, NMDA receptor antagonist activity, and anticonvulsant action of 1-aminocyclobutanecarboxylic acid derivatives, *J. Med. Chem.*, 37, 4288, 1994.

219. Narkunan, K., and Nagarajan, M., Preparation of sugar-derived β-keto phosphonates and their use in the synthesis of higher sugars, *J. Org. Chem.*, 59, 6386, 1994.

220. Kim, D., Shin, K.J., Kim, I.Y., and Park, S.W., A total synthesis of (−)-reiswigin A via sequential Claisen rearrangement-intramolecular ester enolate alkylation, *Tetrahedron Lett.*, 35, 7957, 1994.

221. Weibel, J.-M., and Heissler, D., A new access to *trans-syn-trans* perhydrophenanthrenic systems. Synthesis of (9β*H*)-8α-methylpodocarpan-13-one, *Tetrahedron Lett.*, 35, 473, 1994.

222. Scialdone, M.A., and Johnson, C.R., Building blocks for skipped polyols. *syn*-1,3-Acetonides by chemoenzymatic synthesis from cycloheptatriene, *Tetrahedron Lett.*, 36, 43, 1995.

223. Denmark, S.E., and Amburgey, J.S., Stereoselective synthesis of trisubstituted olefins, *Phosphorus, Sulfur Silicon Relat. Elem.*, 111, 196, 1996.

224. Berkowitz, D.B., Eggen, M., Shen, Q., and Shoemaker, R.K., Ready access to fluorinated phosphonate mimics of secondary phosphates. Synthesis of the (α,α-difluoroalkyl)phosphonate analogues of L-phosphoserine, L-phosphoallothreonine, and L-phosphothreonine, *J. Org. Chem.*, 61, 4666, 1996.

225. Ciufolini, M.A., and Roschangar, F., Total synthesis of (+)-camptothecin, *Angew. Chem. Int. Ed. Engl.*, 35, 1692, 1996.

226. Horita, K., Inoue, T., Tanaka, K., and Yonemitsu, O., Stereoselective total synthesis of lysocellin, the representative polyether antibiotic of the lysocellin family. Part 1. Synthesis of C1–C9 and C16–C23 subunits, *Tetrahedron*, 52, 531, 1996.

227. Neidlein, R., and Feistauer, H., Syntheses of 2,3-dioxoalkylphosphonates and other novel β-ketophosphonates as well as of a phosphinopyruvamide, *Helv. Chim. Acta*, 79, 895, 1996.

228. Chattopadhyay, S.K., and Pattenden, G., Towards a total synthesis of ulapualide A. A concise synthesis of the *tris*-oxazole macrolide core and entire carbon skeleton, *Synlett*, 1345, 1997.

229. Bouix, C., Bisseret, P., and Eustache, J., Stereoselective synthesis of arabinose-derived phosphonates, *Tetrahedron Lett.*, 39, 825, 1998.

230. Boger, D.L., and Hong, J., Total synthesis of nothapodytine B and (–)-mappicine, *J. Am. Chem. Soc.*, 120, 1218, 1998.

231. Mikolajczyk, M., and Zurawinski, R., Synthesis of (±)-rosaprostol, *J. Org. Chem.*, 63, 8894, 1998.

232. Angiolini, M., Belvisi, L., Poma, D., Salimbeni, A., Sciammetta, N., and Scolastico, C., Design and synthesis of nonpeptide angiotensin II receptor antagonists featuring acyclic imidazole-mimicking structural units, *Bioorg. Med. Chem.*, 6, 2013, 1998.

233. Jakeman, D.L., Ivory, A.J., Williamson, M.P., and Blackburn, G.M., Highly potent bisphosphonate ligands for phosphoglycerate kinase, *J. Med. Chem.*, 41, 4439, 1998.

234. Jarosz, S., and Ciunik, Z., Reactivity and crystal structure of 1,2:3,4:5,6-tri-*O*-isopropylidene-D-gluconolactone, *Pol. J. Chem.*, 72, 1182, 1998.

235. Jarosz, S., and Skora, S., A convenient route to enantiomerically pure highly oxygenated decalins from sugar allyltin derivatives, *Tetrahedron: Asymmetry*, 11, 1433, 2000.

236. Rocca, J.R., Tumlinson, J.H., Glancey, B.M., and Lofgren, C.S., Synthesis and stereochemistry of tetrahydro-3,5-dimethyl-6-(1-methylbutyl)-2*H*-pyran-2-one, a component of the queen recognition pheromone of *Solenopsis invicta*, *Tetrahedron Lett.*, 24, 1893, 1983.

237. Mermet-Mouttet, M.-P., Gabriel, K., and Heissler, D., A synthesis of the ring system of terpestacin, *Tetrahedron Lett.*, 40, 843, 1999.

238. Williams, D.R., Clark, M.P., and Berliner, M.A., Synthetic studies toward phorboxazole A. Stereoselective synthesis of the C3–C19 and C20–C32 subunits, *Tetrahedron Lett.*, 40, 2287, 1999.

239. Bouvet, D., and O'Hagan, D., The synthesis of 1-fluoro- and 1,1-difluoro- analogues of 1-deoxy-D-xylulose, *Tetrahedron*, 55, 10481, 1999.

240. Mikolajczyk, M., and Mikina, M., A general approach to 3-phosphorylmethyl cycloalkenones by intramolecular Horner-Wittig reaction of *bis*-β-ketophosphonates, *J. Org. Chem.*, 59, 6760, 1994.

241. Teulade, M.P., Savignac, P., Aboujaoude, E.E., and Collignon, N., Acylation of diethyl α-chloromethylphosphonates by carbanionic route. Preparation of α-clorinated β-carbonylated phosphonates, *J. Organomet. Chem.*, 287, 145, 1985.

242. Aboujaoude, E.E., Collignon, N., Teulade, M.-P., and Savignac, P., Near-quantitative preparation of β-carbonyl phosphonates using the intermediary base, LDA [(Me$_2$CH)$_2$NLi], *Phosphorus Sulfur*, 25, 57, 1985.

243. Lequeux, T.P., and Percy, J.M., Facile syntheses of α,α-difluoro-β-ketophosphonates, *J. Chem. Soc., Chem. Commun.*, 2111, 1995.

244. Blades, K., Lequeux, T.P., and Percy, J.M., A reproducible and high-yielding cerium-mediated route to α,α-difluoro-β-ketophosphonates, *Tetrahedron*, 53, 10623, 1997.

245. Phillion, D.P., and Cleary, D.G., Disodium salt of 2-[(dihydroxyphosphinyl)difluoromethyl]-propenoic acid. An isopolar and isosteric analogue of phosphoenolpyruvate, *J. Org. Chem.*, 57, 2763, 1992.

246. Savignac, P., and Mathey, F., A simple and efficient route to β-keto phosphonates, *Tetrahedron Lett.*, 17, 2829, 1976.

247. Mathey, F., and Savignac, P., A direct synthesis of diethyl 1-alkyl-2-oxoalkanephosphonates, *Synthesis*, 766, 1976.

248. Mathey, F., and Savignac, P., The α-cuprophosphonates. Preparation and application to the synthesis of β-ketophosphonates, *Actes Congr. Int. Composés Phosphorés, 1st*, 617, 1977; *Chem. Abstr.*, 89, 146979, 1978.

249. Mathey, F., and Savignac, P., α-Cuprophosphonates. Part 3. Application to the synthesis of β-ketophosphonates, *Tetrahedron*, 34, 649, 1978.

250. Coutrot, P., and Savignac, P., α-Cuprophosphonates. Part 3. Synthesis of ethyl dialkylphosphonopyruvates, *Synthesis*, 36, 1978.

251. Varlet, J.M., Collignon, N., and Savignac, P., A new route to 2-aminoalkanephosphonic acids, *Synth. Commun.*, 8, 335, 1978.

252. Varlet, J.M., Collignon, N., and Savignac, P., Synthesis and reductive amination of phosphonopyruvates. Preparation of 2-amino-2-carboxyalkylphosphonic acids (β-phosphonoalanine), *Can. J. Chem.*, 57, 3216, 1979.

253. Motoyoshiya, J., Miyajima, M., Hirakawa, K., and Kakurai, T., Dimethyl (2-oxo-4-methyl-3-pentenyl)phosphonate as a precursor of α,α'-dienones. Short syntheses of (±)-α-atlantone and (±)-*ar*-turmerone, *J. Org. Chem.*, 50, 1326, 1985.

254. Kuo, F., and Fuchs, P.L., Use of 1-penten-3-one-4-phosphonate as a kinetic ethyl vinyl ketone equivalent in the Robinson annulation reaction, *Synth. Commun.*, 16, 1745, 1986.

255. Freeman, S., Seidel, H.M., Schwalbe, C.H., and Knowles, J.R., Phosphonate biosynthesis. The stereochemical course of phosphoenolpyruvate phosphomutase, *J. Am. Chem. Soc.*, 111, 9233, 1989.

256. Schwalbe, C.H., and Freeman, S., The crystal structure of the tri(cyclohexylammonium) salt of phosphonopyruvate. Theoretical calculations on the equilibrium constant for the rearrangement of phosphonopyruvate to phosphoenolpyruvate, *J. Chem. Soc., Chem. Commun.*, 251, 1990.

257. Seidel, H.M., Freeman, S., Schwalbe, C.H., and Knowles, J.R., Phosphonate biosynthesis. The stereochemical course of phosphonoenolpyruvate mutase, *J. Am. Chem. Soc.*, 112, 8149, 1990.

258. Motoyoshiya, J., Yazaki, T., and Hayashi, S., Nazarov reaction of trisubstituted dienones. Mechanism involving Wagner–Meerwein shift, *J. Org. Chem.*, 56, 735, 1991.

259. McQueney, M.S., Lee, S.-l., Swartz, W.H., Ammon, H.L., Mariano, P.S., and Dunaway-Mariano, D., Evidence for an intramolecular, stepwise reaction pathway for PEP phosphomutase catalyzed P-C bond formation, *J. Org. Chem.*, 56, 7121, 1991.

260. Takemoto, M., Koshida, A., Miyazima, K., Suzuki, K., and Achiwa, K., Prostanoids and related compounds. Part 4. Total synthesis of clavulones, *Chem. Pharm. Bull.*, 39, 1106, 1991.

261. Freeman, S., Irwin, W.J., and Schwalbe, C.H., Synthesis and hydrolysis studies of phosphonopyruvate, *J. Chem. Soc., Perkin Trans. 2*, 263, 1991.

262. Rudisill, D.E., and Whitten, J.P., Synthesis of *(R)*-4-oxo-5-phosphononorvaline, an *N*-methyl-D-aspartic acid receptor selective β-keto phosphonate, *Synthesis*, 851, 1994.

263. Feistauer, H., and Neidlein, R., Phosphonopyruvates. Syntheses, NMR investigations, and reactions, *Helv. Chim. Acta*, 78, 1806, 1995.

264. Gautier, I., Ratovelomanana-Vidal, V., Savignac, P., and Genêt, J.-P., Asymmetric hydrogenation of β-ketophosphonates and β-ketothiophosphonates with chiral Ru(II) catalysts, *Tetrahedron Lett.*, 37, 7721, 1996.

265. Takatori, K., Tanaka, K., Matsuoka, K., Morishita, K., and Kajiwara, M., An enantioselective synthesis of (−)-nonactic acid and (+)-8-*epi*-nonactic acid using microbial reduction, *Synlett*, 159, 1997.

266. Ornstein, P.L., Bleisch, T.J., Arnold, M.B., Wright, R.A., Johnson, B.G., and Schoepp, D.D., 2-Substituted (2*SR*)-2-amino-2-(1*SR*,2*SR*)-2-carboxycycloprop-1-yl)glycines as potent and selective antagonists of group II metabotropic glutamate receptors. Part 1. Effects of alkyl, arylalkyl, and diarylalkyl substitution, *J. Med. Chem.*, 41, 346, 1998.

267. Ornstein, P.L., Bleisch, T.J., Arnold, M.B., Kennedy, J.H., Wright, R.A., Johnson, B.G., Tizzano, J.P., Helton, D.R., Kallman, M.J., and Schoepp, D.D., 2-Substituted (2*SR*)-2-amino-2-((1*SR*,2*SR*)-2-carboxycycloprop-1-yl)glycines as potent and selective antagonists of group II metabotropic glutamate receptors. Part 2. Effects of aromatic substitution, pharmacological characterization, and bioavailability, *J. Med. Chem.*, 41, 358, 1998.

268. Gil, J.M., Hah, J.H., Park, K.Y., and Oh, D.Y., One pot synthesis of mono- and spirocyclic α-phosphonato-α,β-unsaturated cycloenones, *Tetrahedron Lett.*, 39, 3205, 1998.

269. Balczewski, P., and Mikolajczyk, M., An expeditious synthesis of (±)-desepoxy-4,5-didehydromethylenomycin A methyl ester, *Org. Lett.*, 2, 1153, 2000.

270. Aucagne, V., Gueyrard, D., Tatibouët, A., Quinsac, A., and Rollin, P., Synthetic approaches to *C*-glucosinolates, *Tetrahedron*, 56, 2647, 2000.

271. Yuan, C.-Y., Wang, K., and Li, Z.-Y., Studies on organophosphorus compounds. Part 110. Enantioselective reduction of 2-keto-3-haloalkane phosphonates by baker's yeast, *Heteroat. Chem.*, 12, 551, 2001.

272. Gassama, A., d'Angelo, J., Cave, C., Mahuteau, J., and Riche, C., Regioselective annulation of 1,5-diketones. Access to functionalized Hagemann's esters, *Eur. J. Org. Chem.*, 3165, 2000.

273. Hoffmann, R.W., and Ditrich, K., Total synthesis of mycinolide V, *Liebigs Ann. Chem.*, 23, 1990.

274. Villieras, J., Reliquet, A., and Normant, J.F., Nucleophilic alkylation of carbenoids. Part 3. Copper(I)-catalysed replacement of the chlorine atoms in 1,1-dichloroalkanephosphonates by various substituents, *Synthesis*, 27, 1978.

275. Villieras, J., Perriot, P., and Normant, J.F., α-Chloroenolates. Part 6. Diethyl 1-chloro-1-lithio-2-oxoalkanephosphonates. Synthesis of α-chloro-α,β-unsaturated ketones, *Synthesis*, 29, 1978.

276. Hata, T., Hashizume, A., Nakajima, M., and Sekine, M., α-Hydroxybenzyl anion equivalent. A convenient method for the synthesis of α-hydroxy ketones utilizing α-(trimethylsilyloxy)benzylphosphonate, *Chem. Lett.*, 519, 1979.

277. Binder, J., and Zbiral, E., Diethyl (trimethylsilylethoxymethyl)phosphonates as new reagents for variable strategies for synthesis of carbonyl compounds, α-hydroxycarbonyl compounds and vinylphosphonates, *Tetrahedron Lett.*, 25, 4213, 1984.

278. Thenappan, A., and Burton, D.J., An expedient synthesis of α-fluoro-β-keto esters, *Tetrahedron Lett.*, 30, 6113, 1989.

279. Olah, G.A., and Wu, A.H., Synthetic methods and reactions. Part 159. Preparation of 1,2-diketones from nonenolizable aliphatic and aromatic acyl chlorides with diethyl 1-alkyl(aryl)-1-(trimethylsiloxy)methanephosphonates, *J. Org. Chem.*, 56, 902, 1991.

280. Thenappan, A., and Burton, D.J., Acylation of fluorocarbethoxy-substituted ylids. A simple and general route to α-fluoro β-keto esters, *J. Org. Chem.*, 56, 273, 1991.

281. Tsai, H.-J., Synthesis of phenyl substituted fluoro-olefins, *Tetrahedron Lett.*, 37, 629, 1996.

282. Shen, Y., and Ni, J., Synthesis of perfluoroalkylated 4-cyanoalka-1,4-dienes, *J. Chem. Res. (S)*, 358, 1997.

283. Shen, Y., Ni, J., Li, P., and Sun, J., One-pot stereoselective synthesis of trifluoromethylated penta-(2Z,4E)-dienenitriles via double olefination, *J. Chem. Soc., Perkin Trans. 1*, 509, 1999.

284. Shen, Y., Wang, G., and Sun, J., A novel stereoselective route to trifluoromethylated vinyl sulfones, *Synthesis*, 389, 2001.

285. Shen, Y., Wang, G., and Sun, J., Stereocontrolled synthesis of trifluoromethylated (*E*)- or (*Z*)-ynenyl sulfones via sequential transformations, *J. Chem. Soc., Perkin Trans. 1*, 519, 2001.

286. Durrant, G., and Sutherland, J.K., Wadsworth–Emmons olefination reaction, *J. Chem. Soc., Perkin Trans. 1*, 2582, 1972.

287. Kondo, K., Liu, Y., and Tunemoto, D., Preparation and reaction of sulphonium ylide stabilizes by a phosphinyl substituent, *J. Chem. Soc., Perkin Trans. 1*, 1279, 1974.

288. Hong, S., Chang, K., Ku, B., and Oh, D.Y., New synthesis of β-keto phosphonates, *Tetrahedron Lett.*, 30, 3307, 1989.

289. Baldwin, I.C., Beckett, R.P., and Williams, J.M.J., Conjugate addition of organocuprates to diethyl vinylphosphonate, *Synthesis*, 34, 1996.

290. Burton, D.J., Ishihara, T., and Maruta, M., A useful zinc reagent for the preparation of 2-oxo-1,1-difluoroalkylphosphonates, *Chem. Lett.*, 755, 1982.

291. Burton, D.J., and Sprague, L.G., Preparation of difluorophosphonoacetic acid and its derivatives, *J. Org. Chem.*, 53, 1523, 1988.

292. Lindell, S.D., and Turner, R.M., Synthesis of potential inhibitors of the enzyme aspartate transcarbamoylase, *Tetrahedron Lett.*, 31, 5381, 1990.

293. Tsai, H.-J., Preparation and synthetic application of diethyl 2-oxo-1,1-difluorophosphonates, *Phosphorus, Sulfur Silicon Relat. Elem.*, 122, 247, 1997.

294. Cahiez, G., Alexakis, A., and Normant, J.F., Organomanganese(II) reagents. Part 5. Organomanganous chlorides as a new reagent for ketone synthesis via mixed or symmetrical anhydrides, *Synth. Commun.*, 9, 639, 1979.

295. Kim, D.Y., Kong, M.S., and Rhie, D.Y., A new synthesis of 2-aryl-2-oxoalkylphosphonates from triethyl phosphonoacetate, *Synth. Commun.*, 25, 2865, 1995.

296. Corbel, B., L'Hostis-Kervella, I., and Haelters, J.-P., On a safe and practical method for the preparation of β-keto phosphonates, *Synth. Commun.*, 26, 2561, 1996.

297. Kim, D.Y., Kong, M.S., and Kim, T.H., A practical synthesis of β-keto phosphonates from triethyl phosphonoacetate, *Synth. Commun.*, 26, 2487, 1996.

298. Kim, D.Y., Kong, M.S., and Lee, K., Acylation of *in situ* generated trimethylsilyl diethylphosphonoacetate using magnesium chloride-triethylamine. A practical synthesis of β-keto phosphonates, *J. Chem. Soc., Perkin Trans. 1*, 1361, 1997.

299. Corbel, B., L'Hostis-Kervella, I., and Haelters, J.-P., Acylation of diethyl phosphonoacetic acid via the MgCl$_2$/Et$_3$N system. A practical synthesis of β-keto phosphonates, *Synth. Commun.*, 30, 609, 2000.

300. Kim, D.Y., and Suh, K.H., Solid phase acylation of phosphonoacetates. Synthesis of β-keto phosphonates from polymer bound phosphonoacetate, *Synth. Commun.*, 29, 1271, 1999.

301. Kim, D.Y., and Choi, Y.J., Synthesis of α-fluoro-β-keto phosphonates from α-fluoro phosphonoacetic acid, *Synth. Commun.*, 28, 1491, 1998.

302. Kim, D.Y., Rhie, D.Y., and Oh, D.Y., Acylation of diethyl (ethoxycarbonyl)fluoromethylphosphonate using magnesium chloride-triethylamine. A facile synthesis of α-fluoro β-keto esters, *Tetrahedron Lett.*, 37, 653, 1996.

303. Kim, D.Y., Choi, J.S., and Rhie, D.Y., P–C bond cleavage of triethyl 2-fluoro-3-oxo-2-phosphonoacetates with magnesium chloride. A synthesis of α-fluoro-β-keto esters, *Synth. Commun.*, 27, 1097, 1997.

304. Kim, D.Y., and Kim, J.Y., A new synthesis of α-fluoromalonates from α-fluoro-α-phosphonyl malonates using P–C bond cleavage, *Synth. Commun.*, 28, 2483, 1998.

305. Kim, D.Y., Lee, Y.M., and Choi, Y.J., Acylation of α-fluorophosphonoacetate derivatives using magnesium chloride-triethylamine, *Tetrahedron*, 55, 12983, 1999.

306. Kim, D.Y., A facile P–C bond cleavage of 2-fluoro-2-phosphonyl-1,3-dicarbonyl compounds on silica gel, *Synth. Commun.*, 30, 1205, 2000.

307. Ezquerra, J., de Mendoza, J., Pedregal, C., and Ramirez, C., Regioselective nucleophilic attack on *N*-Boc-pyroglutamate ethyl ester, *Tetrahedron Lett.*, 33, 5589, 1992.

308. Dufour, M.-N., Jouin, P., Poncet, J., Pantaloni, A., and Castro, B., Synthesis and reduction of α-amino ketones derived from leucine, *J. Chem. Soc., Perkin Trans. 1*, 1895, 1986.

309. Theisen, P.D., and Heathcock, C.H., Improved procedure for preparation of optically active 3-hydroxyglutarate monoesters and 3-hydroxy-5-oxoalkanoic acids, *J. Org. Chem.*, 53, 2374, 1988.

310. Blackwell, C.M., Davidson, A.H., Launchbury, S.B., Lewis, C.N., Morrice, E.M., Reeve, M.M., Roffley, J.A.R., Tipping, A.S., and Todd, R.S., An improved procedure for the introduction of the δ-lactone portion of HMG-CoA reductase inhibitors, *J. Org. Chem.*, 57, 1935, 1992.

311. Ditrich, K., and Hoffmann, R.W., Simple conversion of γ- and δ-lactones into 5-(or 6)-silyloxy-3-keto-phosphonates, *Tetrahedron Lett.*, 26, 6325, 1985.

312. Hanessian, S., Roy, P.J., Petrini, M., Hodges, P.J., Fabio, R.D., and Carganico, G., Synthetic studies on the mevinic acids using the chiron approach. Total synthesis of (+)-dihydromevinolin, *J. Org. Chem.*, 55, 5766, 1990.

313. Aristoff, P.A., Practical synthesis of 6*a*-carbaprostaglandin I$_2$, *J. Org. Chem.*, 46, 1954, 1981.

314. House, H.O., Haack, J.L., McDaniel, W.C., and VanDerveer, D., Enones with strained double bonds. Part 8. The bicyclo[3.2.1]octane systems, *J. Org. Chem.*, 48, 1643, 1983.

315. Lim, M. III, and Marquez, V.E., Total synthesis of (−)-neplanocin A, *Tetrahedron Lett.*, 24, 5559, 1983.

316. Altenbach, H.-J., Holzapfel, W., Smerat, G., and Finkler, S.H., A simple, regiospecific synthesis of cycloalkenones and lactones, *Tetrahedron Lett.*, 26, 6329, 1985.

317. Lin, C.-H., Aristoff, P.A., Johnson, P.D., McGrath, J.P., Timko, J.M., and Robert, A., Benzindene prostaglandins. Synthesis of optically pure 15-deoxy-U-68,215 and its enantiomer via a modified intramolecular Wadsworth–Emmons–Wittig reaction, *J. Org. Chem.*, 52, 5594, 1987.

318. Marquez, V.E., Lim, M., III, Tseng, C.K.H., Markovac, A., Priest, M.A., Khan, M.S., and Kaskar, B., Total synthesis of (−)-neplanocin A, *J. Org. Chem.*, 53, 5709, 1988.

319. Keller, T.H., and Weiler, L., Diastereoselective reduction of 9-oxo-13-tetradecanolide and 10,10-dimethyl-9-oxo-13-tetradecanolide, *J. Am. Chem. Soc.*, 112, 450, 1990.

320. Fukase, H., and Horii, S., Synthesis of valiolamine and its *N*-substituted derivatives AO-128, validoxylamine G, and validamycin G via branched-chain inosose derivatives, *J. Org. Chem.*, 57, 3651, 1992.

321. Morgan, B.P., Holland, D.R., Matthews, B.W., and Bartlett, P.A., Structure-based design of an inhibitor of the zinc peptidase thermolysin, *J. Am. Chem. Soc.*, 116, 3251, 1994.

322. Cai, S., Stroud, M.R., Hakomori, S., and Toyokuni, T., Synthesis of carbocyclic analogues of guanosine 5′-(β-L-fucopyranosyl diphosphate) (GDP-fucose) as potential inhibitors of fucosyltransferases, *J. Org. Chem.*, 57, 6693, 1992.

323. Comin, M.J., and Rodriguez, J.B., First synthesis of (−)-neplanocin C, *Tetrahedron*, 56, 4639, 2000.

324. Snider, B.B., and Liu, T., Total synthesis of (±)-deoxypenostatin A, *J. Org. Chem.*, 64, 1088, 1999.

325. Snider, B.B., and Liu, T., Total synthesis of (±)-deoxypenostatin A. Approaches to the syntheses of penostatins A and B, *J. Org. Chem.*, 65, 8490, 2000.

326. Savignac, P., and Coutrot, P., A direct conversion of aldehydes into 2-oxoalkanephosphonates via the diethyl α-lithiochloromethanephosphonate anion, *Synthesis*, 682, 1978.

327. Coutrot, P., and Ghribi, A., A facile and general, one-pot synthesis of 2-oxoalkane phosphonates from diethylphosphonocarboxylic acid chlorides and organometallic reagents, *Synthesis*, 661, 1986.

328. Coutrot, P., and Grison, C., General synthesis of α-fluoro-β-ketophosphonate precursors of α-fluoro enones. Application to the pyrethrene series, *Tetrahedron Lett.*, 29, 2655, 1988.

329. Coutrot, P., Grison, C., Lachgar, M., and Ghribi, A., A general and efficient synthesis of β-ketophosphonates, *Bull. Soc. Chim. Fr.*, 132, 925, 1995.

330. Arnone, A., Bravo, P., Frigerio, M., Viani, F., and Zappala, C., Synthesis of 3'-arylsulfonyl-4'-[(diethoxyphosphoryl)difluoromethyl]thymidine analogs, *Synthesis*, 1511, 1998.

331. Arnone, A., Bravo, P., Frigerio, M., Mele, A., Vergani, B., and Viani, F., Synthesis of enantiomerically pure 2',3',5'-trideoxy-4'-[(diethoxyphosphoryl)difluoromethyl]thymidine analogues, *Eur. J. Org. Chem.*, 2149, 1999.

332. Snyder, C.D., Bondinell, W.E., and Rapoport, H., Synthesis of chlorobiumquinone, *J. Org. Chem.*, 36, 3951, 1971.

333. Lim, M., III, and Marquez, V.E., A synthetic approach towards neplanocin A. Preparation of the optically active cyclopentene moiety from D-ribose, *Tetrahedron Lett.*, 24, 4051, 1983.

334. Page, P., Blonski, C., and Perié, J., Synthesis of phosphono analogues of dihydroxyacetone phosphate and glyceraldehyde 3-phosphate, *Bioorg. Med. Chem.*, 7, 1403, 1999.

335. Ermolenko, M.S., and Pipelier, M., A carbohydrate-based synthetic approach to quadrone, *Tetrahedron Lett.*, 38, 5975, 1997.

336. Shimura, T., Komatsu, C., Matsumura, M., Shimada, Y., Ohta, K., and Mitsunobu, O., Preparation of the C1–C10 fragment of carbonolide B. A relay approach to carbomycin B, *Tetrahedron Lett.*, 38, 8341, 1997.

337. Akita, H., Yamada, H., Matsukura, H., Nakata, T., and Oishi, T., Total synthesis of the aglycone of venturicidins A and B. Part 1. Synthesis of C1–C14 segment, *Tetrahedron Lett.*, 31, 1731, 1990.

338. Nakajima, N., Hamada, T., Tanaka, T., Oikawa, Y., and Yonemitsu, O., Chiral synthesis of polyketide-derived natural products. Part 10. Stereoselective synthesis of pikronolide, the aglycon of the 14-membered ring macrolide pikromycin, from D-glucose. Role of MPM and DMPM protection, *J. Am. Chem. Soc.*, 108, 4645, 1986.

339. Tatsuta, K., Ishiyama, T., Tajima, S., Koguchi, Y., and Gunji, H., The total synthesis of oleandomycin, *Tetrahedron Lett.*, 31, 709, 1990.

340. Öhler, E., and Zbiral, E., Oxidative rearrangement of phosphorus containing tertiary allylic alcohols. Synthesis of (3-oxo-1-cycloalkenyl)phosphonates, -methylphosphonates, -methyldiphenylphosphine oxides and their epoxy derivatives, *Synthesis*, 357, 1991.

341. de Macedo Puyau, P., and Perié, J.J., Synthesis of substrate analogues and inhibitors for the phosphoglycerate mutase enzyme, *Phosphorus, Sulfur Silicon Relat. Elem.*, 129, 13, 1997.

342. Chorghade, M.S., and Cseke, C.T., Biorational design of herbicides. Synthesis of inhibitors of the PFP enzyme, *Heterocycles*, 40, 213, 1995.

343. Sturtz, G., and Pondaven-Raphalen, A., Synthesis of keto- and aldophosphonates by applying the Wacker method to the corresponding ethylenic compounds, *J. Chem. Res. (M)*, 2512, 1980.

344. Poss, A.J., and Smyth, M.S., Diethyl 3-bromopropenephosphonate, an alkylative β-keto phosphonate equivalent, *Synth. Commun.*, 17, 1735, 1987.

345. Hammond, G.B., Calogeropoulou, T., and Wiemer, D.F., The 1,3-migration of phosphorus from oxygen to carbon. A new synthesis of β-ketophosphonates from enol phosphates, *Tetrahedron Lett.*, 27, 4265, 1986.

346. Calogeropoulou, T., Hammond, G.B., and Wiemer, D.F., Synthesis of β-keto phosphonates from vinyl phosphates via a 1,3-phosphorus migration, *J. Org. Chem.*, 52, 4185, 1987.

347. Gloer, K.B., Calogeropoulou, T., Jackson, J.A., and Wiemer, D.F., Regiochemistry of the rearrangement of cyclohexenyl and cyclohexadienyl phosphates to β-keto phosphonates, *J. Org. Chem.*, 55, 2842, 1990.

348. Perrone, R., Berardi, F., Colabufo, N.A., Leopoldo, M., Tortorella, V., Fornaretto, M.G., Caccia, C., and McArthur, R.A., Structure–activity relationship studies on the 5-HT$_{1A}$ receptor affinity of 1-phenyl-4-[ω-(α- or β-tetralinyl)alkyl]piperazines. Part 4, *J. Med. Chem.*, 39, 4928, 1996.

349. An, Y.-Z., and Wiemer, D.F., Regiochemistry of vinyl phosphate/β-keto phosphonate rearrangements in functionalized cyclohexanones and cyclohexenones, *J. Org. Chem.*, 57, 317, 1992.

350. Jackson, J.A., Hammond, G.B., and Wiemer, D.F., Synthesis of α-phosphono lactones and esters through a vinyl phosphate-phosphonate rearrangement, *J. Org. Chem.*, 54, 4750, 1989.

351. Baker, T.J., and Wiemer, D.F., Regiospecific vinyl phosphate/β-keto phosphonate rearrangements initiated by halogen–metal exchange, *J. Org. Chem.*, 63, 2613, 1998.

352. An, Y.-Z., An, J.G., and Wiemer, D.F., Rearrangements of nonracemic vinyl phosphates to β-keto phosphonates, *J. Org. Chem.*, 59, 8197, 1994.

353. An, J., Wilson, J.M., An, Y.-Z., and Wiemer, D.F., Diastereoselective vinyl phosphate/β-keto phosphonate rearrangement, *J. Org. Chem.*, 61, 4040, 1996.

354. Lee, K., and Wiemer, D.F., Reaction of diethyl phosphorochloridite with enolates. A general method for synthesis of β-keto phosphonates and α-phosphono esters through C-P bond formation, *J. Org. Chem.*, 56, 5556, 1991.

355. Sampson, P., Hammond, G.B., and Wiemer, D.F., A new synthesis of β-keto phosphonates and β-keto silanes, *J. Org. Chem.*, 51, 4342, 1986.

356. Roussis, V., and Wiemer, D.F., Synthesis of phosphonates from α-hydroxy carbonyl compounds and dialkyl phosphorochloridites, *J. Org. Chem.*, 54, 627, 1989.

357. Arbuzov, B.A., Polozov, A.M., and Polezhaeva, N.A., Catalytic and thermal decomposition of 2-diazo-1,3-diphenyl-1,3-propanedione in dimethyl hydrogen phosphite and *O,O*-diethyl hydrogen phosphorothioite, *Zh. Obshch. Khim.*, 54, 1517, 1984; *J. Gen. Chem. USSR (Engl. Transl.)*, 54, 1351, 1984.

358. Arbuzov, B.A., Polozov, A.M., and Polezhaeva, N.A., P-H addition of carbenoids and carbenes as a method for the synthesis of 2-phospho-substituted 1,3-dicarbonyl compounds, *Dokl. Akad. Nauk SSSR, Ser. Khim.*, 287, 849, 1986; *Dokl. Chem. (Engl. Transl.)*, 287, 69, 1986.

359. Polozov, A.M., Polezhaeva, N.A., Mustaphin, A.H., Khotinen, A.V., and Arbuzov, B.A., A new one-pot synthesis of dialkyl phosphonates from diazo compounds and dialkyl hydrogen phosphites, *Synthesis*, 515, 1990.

360. Polozov, A.M., and Mustafin, A.K., P-H internal carbenes. Part 4. Formation of *O,O*-dimethyl-3-chloro-2-oxopropylphosphonate, *Zh. Obshch. Khim.*, 62, 1039, 1992; *J. Gen. Chem. USSR (Engl. Transl.)*, 62, 850, 1992.

361. Lee, K., and Oh, D.Y., Synthesis of α-hetero atom substituted β-keto and enamine phosphonates, *Synth. Commun.*, 21, 279, 1991.

362. Lee, C.-W., and Oh, D.Y., New synthetic route to 3-furylphosphonates, *Heterocycles*, 43, 1171, 1996.

363. Kouno, R., Okauchi, T., Nakamura, M., Ichikawa, J., and Minami, T., Synthesis and synthetic utilization of α-functionalized vinylphosphonates bearing β-oxy or β-thio substituents, *J. Org. Chem.*, 63, 6239, 1998.

364. Minami, T., Kouno, R., Okauchi, T., Nakamura, M., and Ichikawa, J., Synthetic utilization of α-phosphonovinyl anions, *Phosphorus, Sulfur Silicon Relat. Elem.*, 144, 689, 1999.

365. Koh, Y.J., and Oh, D.Y., A new synthesis of β-keto phosphonate from aryl epoxysulfones and dialkyl hydrogen phosphite, *Tetrahedron Lett.*, 34, 2147, 1993.

366. Baboulene, M., and Sturtz, G., Reactions of diethyl 1,2-epoxyalkanephosphonates. A new synthesis of 2-oxoalkanephosphonates, *Synthesis*, 456, 1978.

367. Öhler, E., Kang, H.-S., and Zbiral, E., A convenient synthesis of new diethyl (2,4-dioxoalkyl)phosphonates, *Synthesis*, 623, 1988.

368. Gasteiger, J., and Herzig, C., β-Ketophosphonic ester from α-chlorooxiranes, *Tetrahedron Lett.*, 21, 2687, 1980.

369. Kim, D.Y., and Kong, M., A new synthesis of cyclic β-keto phosphonates from α-nitro epoxides and a dialkyl phosphite, *J. Chem. Soc., Perkin Trans. 1*, 3359, 1994.

370. Kim, D.Y., Mang, J.Y., and Oh, D.Y., Reaction of silyl enol ethers with phosphite using hypervalent iodine compound. A new synthesis of 2-aryl-2-oxoalkylphosphonates, *Synth. Commun.*, 24, 629, 1994.

371. Holmquist, C.R., and Roskamp, E.J., Tin(II) chloride catalyzed addition of diazo sulfones, diazo phosphine oxides, and diazo phosphonates to aldehydes, *Tetrahedron Lett.*, 33, 1131, 1992.

372. Kim, D.Y., Lee, K., and Oh, D.Y., A new synthesis of 1-aryl-2-oxoalkylphosphonates from nitroalkenes and diethyl phosphite, *J. Chem. Soc., Perkin Trans. 1*, 2451, 1992.

373. Motoyoshiya, J., and Hirata, K., Cycloaddition reactions of (diethylphosphono)ketenes, *Chem. Lett.*, 211, 1988.

374. Hirai, S., Harvey, R.G., and Jensen, E.V., Phosphonic acids. Part 21. Steroidal 21*a*-phosphonate esters, *Tetrahedron*, 22, 1625, 1966.

375. Birum, G.H., and Richardson, G.A., Silicon-phosphorus compounds, *Monsanto Chemical*, U.S. Patent Appl. US 3113139, 1963; *Chem. Abstr.*, 60, 5551g, 1964.

376. Sekine, M., Yamamoto, I., Hashizume, A., and Hata, T., Silyl phosphites. Part 5. The reactions of *tris*(trimethylsilyl) phosphite with carbonyl compounds, *Chem. Lett.*, 485, 1977.

377. Evans, D.A., Hurst, K.M., Truesdale, L.K., and Takacs, J.M., The carbonyl insertion reactions of mixed tervalent phosphorus–organosilicon reagents, *Tetrahedron Lett.*, 18, 2495, 1977.

378. Evans, D.A., Hurst, K.M., and Takacs, J.M., New silicon–phosphorus reagents in organic synthesis. Carbonyl and conjugate addition reactions of silicon phosphite esters and related systems, *J. Am. Chem. Soc.*, 100, 3467, 1978.

379. Hata, T., Hashizume, A., Nakajima, M., and Sekine, M., A convenient method of ketone synthesis utilizing the reaction of diethyl trimethylsilyl phosphite with carbonyl compounds, *Tetrahedron Lett.*, 19, 363, 1978.

380. Evans, D.A., Takacs, J.M., and Hurst, K.M., Phosphonamide stabilized allylic carbanions. New homoenolate anion equivalents, *J. Am. Chem. Soc.*, 101, 371, 1979.

381. Liotta, D., Sunay, U., and Ginsberg, S., Phosphonosilylations of cyclic enones, *J. Org. Chem.*, 47, 2227, 1982.

382. Green, K., Trimethylaluminium promoted conjugate additions of dimethylphosphite to α,β-unsaturated esters and ketones, *Tetrahedron Lett.*, 30, 4807, 1989.

383. Mori, I., Kimura, Y., Nakano, T., Matsunaga, S.-I., Iwasaki, G., Ogawa, A., and Hayakawa, K., Trimethylsilyl triflate promoted 1,4-addition of silyl phosphites to cyclic enones, *Tetrahedron Lett.*, 38, 3543, 1997.

384. Simoni, D., Invidiata, F.P., Manferdini, M., Lampronti, I., Rondanin, R., Roberti, M., and Pollini, G.P., Tetramethylguanidine (TMG)-catalyzed addition of dialkyl phosphites to α,β-unsaturated carbonyl compounds, alkenenitriles, aldehydes, ketones and imines, *Tetrahedron Lett.*, 39, 7615, 1998.

385. Simoni, D., Rondanin, R., Morini, M., Baruchello, R., and Invidiata, F.P., 1,5,7-Triazabicyclo[4.4.0]dec-1-ene (TBD), 7-methyl-TBD (MTBD) and the polymer-supported TBD (P-TBD). Three efficient catalysts for the nitroaldol (Henry) reaction and for the addition of dialkyl phosphites to unsaturated systems, *Tetrahedron Lett.*, 41, 1607, 2000.

386. Appleton, D., Duguid, A.B., Lee, S.-K., Ha, Y.-J., Ha, H.-J., and Leeper, F.J., Synthesis of analogues of 5-aminolaevulinic acid and inhibition of 5-aminolaevulinic acid dehydratase, *J. Chem. Soc., Perkin Trans. 1*, 89, 1998.

387. Sekhar, B.B.V.S., and Bentrude, W.G., Photochemical SET induced 1,4-conjugate additions of silyl phosphites to cyclic enones, *Tetrahedron Lett.*, 40, 1087, 1999.

388. McClure, C.K., and Jung, K.-Y., Pentacovalent oxaphosphorane chemistry in organic synthesis. A new route to substituted phosphonates, *J. Org. Chem.*, 56, 867, 1991.

389. McClure, C.K., and Jung, K.-Y., Pentacovalent oxaphosphorane chemistry in organic synthesis. Part 2. Total syntheses of (±)-*trans*- and (±)-*cis*-neocnidilides, *J. Org. Chem.*, 56, 2326, 1991.

390. McClure, C.K., and Grote, C.W., α,β-Functionalization of enones via pentacovalent oxaphospholenes, *Tetrahedron Lett.*, 32, 5313, 1991.

391. McClure, C.K., Grote, C.W., and Rheingold, A.L., Novel and efficient synthesis of uracil phosphonate derivatives via pentacovalent oxaphospholenes, *Tetrahedron Lett.*, 34, 983, 1993.

392. McClure, C.K., Jung, K.-Y., Grote, C.W., and Hansen, K., Pentacovalent phosphorus in organic synthesis. A new route to substituted phosphonates, *Phosphorus, Sulfur Silicon Relat. Elem.*, 75, 23, 1993.

393. McClure, C.K., Mishra, P.K., and Grote, C.W., Synthetic studies toward the preparation of phosphonate analogs of sphingomyelin and ceramide 1-phosphate using pentacovalent organophospholene methodology, *J. Org. Chem.*, 62, 2437, 1997.

394. Savignac, P., Breque, A., Mathey, F., Varlet, J.M., and Collignon, N., α-Cuprophosphonates. Part 5. The reaction of 3- and 4-oxoalkanephosphonates and their use in aminoalkanephosphonic acid synthesis, *Synth. Commun.*, 9, 287, 1979.

395. Retherford, C., Chou, T.-S., Schelkun, R.M., and Knochel, P., Preparation and reactivity of β-zinc and copper phosphonates, *Tetrahedron Lett.*, 31, 1833, 1990.

396. Li, N.-S., Yu, S., and Kabalka, G.W., Synthesis of 3-oxoalkylphosphonates via the conjugate addition of acylcuprates to diethyl vinylphosphonate, *Organometallics*, 18, 1811, 1999.

397. Verbicky, C.A., and Zercher, C.K., Zinc-mediated chain extension of β-keto phosphonates, *J. Org. Chem.*, 65, 5615, 2000.

398. McClure, C.K., and Hansen, K.B., Diels–Alder reactivity of a ketovinylphosphonate with cyclopentadiene and furan, *Tetrahedron Lett.*, 37, 2149, 1996.

399. McClure, C.K., Herzog, K.J., and Bruch, M.D., Structure determination of the Diels–Alder product of a ketovinylphosphonate with *E*-1-acetoxy-1,3-butadiene, *Tetrahedron Lett.*, 37, 2153, 1996.

400. Varlet, J.M., Collignon, N., and Savignac, P., Reductive amination of β-keto phosphonates. Preparation of aminoalkylphosphonic acids, *Tetrahedron*, 37, 3713, 1981.

401. Padwa, A., and Eastman, D., The reaction of organophosphorus compounds with α- and β-diphenylacyl bromides, *J. Org. Chem.*, 35, 1173, 1970.

402. Sarin, V., Tropp, B.E., and Engel, R., Isosteres of natural phosphates. Part 7. The preparation of 5-carboxy-4-hydroxy-4-methylpentyl-1-phosphonic acid, *Tetrahedron Lett.*, 18, 351, 1977.

403. Alonso, F., Mico, I., Najera, C., Sansano, J.M., Yus, M., Èzquerra, J., Yruretagoyena, B., and Garcia, I., Synthesis of 3- and 4-substituted cyclic α-amino acids structurally related to ACPD, *Tetrahedron*, 51, 10259, 1995.

404. Rao, Y.K., and Nagarajan, M., Synthesis of (±)-silphinene, *Tetrahedron Lett.*, 29, 107, 1988.

405. Rao, Y.K., and Nagarajan, M., Formal total synthesis of (±)-silphinene via radical cyclization, *J. Org. Chem.*, 54, 5678, 1989.

406. Darling, S.D., Muralidharan, F.N., and Muralidharan, V.B., Enolate alkylations with vinylphosphonates, *Synth. Commun.*, 9, 915, 1979.

407. Goumain, S., Jubault, P., Feasson, C., and Collignon, N., A new and efficient electrosynthesis of 2-substituted 1,1-cyclopropanediyl*bis*(phosphonates), *Synthesis*, 1903, 1999.

408. Enders, D., Wahl, H., and Papadopoulos, K., Asymmetric Michael additions via SAMP/RAMP hydrazones. Enantioselective synthesis of 2-substituted 4-oxophosphonates, *Liebigs Ann. Org. Bioorg. Chem.*, 1177, 1995.

409. Enders, D., Wahl, H., and Papadopoulos, K., Diastereo- and enantioselective synthesis of 2,3- and 1,2-disubstituted 4-oxophosphonates via asymmetric Michael addition, *Tetrahedron*, 53, 12961, 1997.

410. Hu, C.-M., and Chen, J., Addition of diethyl bromodifluoromethylphosphonate to various alkenes initiated by Co(III)/Zn bimetal redox system, *J. Chem. Soc., Perkin Trans. 1*, 327, 1993.

411. Yokomatsu, T., Abe, H., Yamagishi, T., Suemune, K., and Shibuya, S., Convenient synthesis of cyclopropylalkanol derivatives possessing a difluoromethylenephosphonate group at the ring, *J. Org. Chem.*, 64, 8413, 1999.

412. Yang, Z.-Y., and Burton, D.J., A novel, general method for the preparation of α,α-difluoro functionalized phosphonates, *Tetrahedron Lett.*, 32, 1019, 1991.

413. Yang, Z.-Y., and Burton, D.J., A novel and practical preparation of α,α-difluoro functionalized phosphonates from iododifluoromethylphosphonate, *J. Org. Chem.*, 57, 4676, 1992.

414. Denieul, M.-P., Quiclet-Sire, B., and Zard, S.Z., Trifluoroacetonyl radicals. A versatile approach to trifluoromethyl ketones, *J. Chem. Soc., Chem. Commun.*, 2511, 1996.

415. Boivin, J., Ramos, L., and Zard, S.Z., An expedient radical based approach to difluorophosphonate analogues of thionucleosides, *Tetrahedron Lett.*, 39, 6877, 1998.

416. Boivin, J., Boutillier, P., and Zard, S.Z., An unusual and highly efficient access to thieno[2,3-*b*]benzothiopyran structures, *Tetrahedron Lett.*, 40, 2529, 1999.

417. Cholleton, N., Gauthier-Gillaizeau, I., Six, Y., and Zard, S.Z., A new approach to cyclohexenes and related structures, *J. Chem. Soc., Chem. Commun.*, 535, 2000.

418. Balczewski, P., and Mikolajczyk, M., Free radical reaction of α-haloalkylphosphonates with alkenes and alkynes. A new approach to modified phosphonates, *Synthesis*, 392, 1995.

419. Balczewski, P., and Pietrzykowski, W.M., A free radical approach to functionalization of phosphonates utilizing novel 2- and 3-phosphonyl radicals, *Tetrahedron*, 52, 13681, 1996.

420. Balczewski, P., and Pietrzykowski, W.M., A new, effective approach for the C-C bond formation utilizing 1-, 2- and 3-phosphonyl substituted radicals derived from iodoalkylphosphonates and *n*-Bu₃SnH/Et₃B/O₂ system, *Tetrahedron*, 53, 7291, 1997.

421. Piers, E., Abeysekera, B., and Scheffer, J.R., Preparation of, and alkylation of enolate anions with, dimethyl 3-bromo-2-ethoxypropenylphosphonate. An efficient, convergent synthesis of 2-cyclopenten-1-ones, *Tetrahedron Lett.*, 20, 3279, 1979.

422. Callant, P., Wilde, H.D., and Vandewalle, M., The synthesis of functionalized *cis*-bicyclo[3.3.0]octanes, *Tetrahedron*, 37, 2079, 1981.

423. Dewanckele, J.M., Zutterman, F., and Vandewalle, M., Sesquiterpene lactones. A total synthesis of (±)-quadrone, *Tetrahedron*, 39, 3235, 1983.

424. Castagnino, E., Corsano, S., and Strappaveccia, G.P., The preparation of a novel 3-oxo-cyclopenten-2-phosphonate derivative, useful intermediate for 2-alkyl-substituted cyclopentenones synthesis, *Tetrahedron Lett.*, 26, 93, 1985.

425. D'Onofrio, F., Piancatelli, G., and Nicolai, M., Photo-oxidation of 2-furylalkylphosphonates. Synthesis of new cyclopentenone derivatives, *Tetrahedron*, 51, 4083, 1995.

426. Raeppel, F., Weibel, J.-M., and Heissler, D., Synthesis of the *trans-syn-trans* perhydrobenz[*e*]indene moiety of the stellettins and of the stelliferins, *Tetrahedron Lett.*, 40, 6377, 1999.

427. Yokomatsu, T., Yamagishi, T., Suemune, K., Abe, H., Kihara, T., Soeda, S., Shimeno, H., and Shibuya, S., Stereoselective reduction of cyclopropylalkanones possessing a difluoromethylenephosphonate group at the ring. Application to stereoselective synthesis of novel cyclopropane nucleotide analogues, *Tetrahedron*, 56, 7099, 2000.

428. Butt, A.H., Percy, J.M., and Spencer, N.S., A rearrangement-based approach to secondary difluorophosphonates, *J. Chem. Soc., Chem. Commun.*, 1691, 2000.

429. Ryglowski, A., and Kafarski, P., Preparation of 1-aminoalkylphosphonic acids and 2-aminoalkylphosphonic acids by reductive amination of oxoalkylphosphonates, *Tetrahedron*, 52, 10685, 1996.

430. Kosolapoff, G.M., Synthesis of amino-substituted phosphonic acids. Part 1, *J. Am. Chem. Soc.*, 69, 2112, 1947.

431. Kaushik, M.P., Lal, B., Raghuveeran, C.D., and Vaidyanathaswamy, R., Stereochemistry of aroylphosphonate phenylhydrazones and their conversion to 1*H*-indazole-3-phosphonates, *J. Org. Chem.*, 47, 3503, 1982.

432. Haelters, J. P., Corbel, B., and Sturtz, G., Synthesis of indolephosphonates by Fischer cyclization of phosphonate arylhydrazones, *Phosphorus Sulfur*, 37, 41, 1988.

433. Ben Akacha, A., Barkallah, S., and Baccar, B., Reaction of *O,O*-diethyl 1-oxoalkanephosphonates with hydrazine derivatives. Synthetic procedure for α-phosphonate hydrazones and hydrazides, *Phosphorus, Sulfur Silicon Relat. Elem.*, 69, 163, 1992.

434. Yuan, C., Chen, S., Xie, R., Feng, H., and Maier, L., Studies on organophosphorus compounds. Part 94. Syntheses of 1-hydrazino- and 2-hydrazino-alkylphosphonic acids and derivatives thereof, *Phosphorus, Sulfur Silicon Relat. Elem.*, 106, 115, 1995.

435. Kudzin, Z.H., and Kotynski, A., Synthesis of *O,O*-dialkyl 1-aminoalkanephosphonates, *Synthesis*, 1028, 1980.

436. Chen, H., Qian, D.-Q., Xu, G.-X., Liu, Y.-X., Chen, X.-D., Shi, X.-D., Cao, R.-Z., and Liu, L.-Z., New strategy for the synthesis of phosphonyl pyrazoles, *Synth. Commun.*, 29, 4025, 1999.

437. Berlin, K.D., Claunch, R.T., and Gaudy, E.T., α-Aminoarylmethylphosphonic acids and diethyl α-aminoarylmethylphosphonate hydrochlorides. Aluminum-amalgam reduction of oximes of diethyl aroylphosphonates, *J. Org. Chem.*, 33, 3090, 1968.

438. Berlin, K.D., Roy, N.K., Claunch, R.T., and Bude, D., A novel route to α-aminoalkylphosphonic acids and dialkyl α-aminoalkylphosphonate hydrochlorides, *J. Am. Chem. Soc.*, 90, 4494, 1968.

439. Asano, S., Kitahara, T., Ogawa, T., and Matsui, M., Synthesis of α-amino phosphonic acids, *Agr. Biol. Chem.*, 37, 1193, 1973.

440. Zhdanov, Y.A., Uzlova, L.A., and Glebova, Z.I., The Arbuzov reaction of higher 2-deoxyaldonic acid chlorides, *Dokl. Akad. Nauk SSSR, Ser. Khim.*, 255, 870, 1980; *Chem. Abstr.*, 95, 7660, 1981.

441. Kowalik, J., Kupczyk-Subotkowska, L., and Mastalerz, P., Preparation of dialkyl 1-aminoalkane phosphonates by reduction of dialkyl 1-hydroxy imino alkane phosphonates with zinc in formic acid, *Synthesis*, 57, 1981.

442. Anderson, D.W., Campbell, M.M., Malik, M., Prashad, M., and Wightman, R.H., Phosphonopeptides and 3-aminophosphononocardicinic acid, *Tetrahedron Lett.*, 31, 1759, 1990.

443. Maier, L., and Diel, P.J., Organic phosphorus compounds. Part 97. Synthesis and properties of 1-amino-2-aryl- and 2-pyridylethylphosphonic acids and derivatives, *Phosphorus, Sulfur Silicon Relat. Elem.*, 62, 15, 1991.

444. Rogers, R.S., and Stern, M.K., An improved synthesis of the phosphonic acid analog of tryptophan, *Synlett*, 708, 1992.

445. Yuan, C., Chen, S., Zhou, H., and Maier, L., Studies on organophosphorus compounds. Part 75. A facile synthesis of 1-(hydroxyamino)alkyl(or aryl)phosphonic acids, *Synthesis*, 955, 1993.

446. Ryglowski, A., and Kafarski, P., The facile synthesis of dialkyl 1-aminoalkylphosphonates, *Synth. Commun.*, 24, 2725, 1994.

447. Kaushik, M.P., and Vaidyanathaswamy, R., Synthesis and characterization of oximino pyridoyl phosphonates, *Phosphorus, Sulfur Silicon Relat. Elem.*, 102, 45, 1995.

448. Green, D.S. C., Gruss, U., Hägele, G., Hudson, H.R., Lindblom, L., and Pianka, M., The preparation and characterization of some fluorinated α-aminoarylmethanephosphonic acids, *Phosphorus, Sulfur Silicon Relat. Elem.*, 113, 179, 1996.

449. Breuer, E., Safadi, M., Chorev, M., and Gibson, D., Novel amino acid derivatives. Preparation and properties of aminoacylphosphonates and amino hydroxyimino phosphonates, *J. Org. Chem.*, 55, 6147, 1990.

450. Breuer, E., Karaman, R., Goldblum, A., Gibson, D., Leader, H., Potter, B.V.L., and Cummins, J.H., α-Oxyiminophosphonates. Chemical and physical properties. Reactions, theoretical calculations, and X-ray crystal structures of (E)- and (Z)-dimethyl α-hydroxyiminobenzylphosphonates, *J. Chem. Soc., Perkin Trans. 1*, 3047, 1988.

451. Neidlein, R., Keller, H., and Boese, R., Mild preparation of [1-[(benzyloxy)imino]alkyl]phosphonic dichlorides. Application to the synthesis of cyclic phosphonic diesters and cyclic monoester amides, *Heterocycles*, 35, 1185, 1993.

452. Neidlein, R., and Keller, H., Syntheses of 1-benzyloxyaminoalkylphosphonates, *Heterocycles*, 36, 1925, 1993.

453. Elhaddadi, M., Jacquier, R., Gastal, F., Petrus, C., and Petrus, F., A convenient synthesis of alkyl and dialkyl 1-benzyloxyamino alkyl phosphonates and phosphinates, *Phosphorus, Sulfur Silicon Relat. Elem.*, 54, 143, 1990.

454. Green, D., Patel, G., Elgendy, S., Baban, J.A., Claeson, G., Kakkar, V.V., and Deadman, J., A facile synthesis of O,O-dialkyl 1-aminoalkanephosphonates, *Tetrahedron Lett.*, 34, 6917, 1993.

455. Zon, J., Synthesis of diisopropyl 1-nitroalkanephosphonates from diisopropyl 1-oxoalkanephosphonates, *Synthesis*, 661, 1984.

456. Boivin, J., Schiano, A.M., and Zard, S.Z., Iminyl radicals by stannane mediated cleavage of oxime esters, *Tetrahedron Lett.*, 35, 249, 1994.

457. Zard, S.Z., Iminyl radicals. A fresh look to a forgotten species (and some of its relatives), *Synlett*, 1148, 1996.

458. Öhler, E., El-Badawi, M., and Zbiral, E., Synthesis of heteroaryl- and heteroarylvinylphosphonates from 2-bromo-1-oxoalkyl- and 4-bromo-3-oxo-1-alkenylphosphonates, *Chem. Ber.*, 117, 3034, 1984.

459. Stevens, C., De Buyck, L., and De Kimpe, N., The acylphosphonate function as an activating and masking moiety for the α-chlorination of fatty acids, *Tetrahedron Lett.*, 39, 8739, 1998.

460. Stevens, C.V., and Vanderhoydonck, B., Use of acylphosphonates for the synthesis of α-chlorinated carboxylic and α,α'-dichloro dicarboxylic acids and their derivatives, *Tetrahedron*, 57, 4793, 2001.

461. Pudovik, A.N., Zimin, M.G., and Evdokimova, V.V., Reactions of α-keto phosphonates with the cyanoacetic ester and dinitrile of malonic acid, *Zh. Obshch. Khim.*, 42, 1489, 1972; *J. Gen. Chem. USSR (Engl. Transl.)*, 42, 1481, 1972.

462. Huang, W., Zhang, Y., and Yuan, C., Studies on organophosphorus compounds. Part 99. A novel stereoselective synthesis of dialkyl (Z)-1-alkyl-2-ethoxycarbonyl-2-formylaminoeth-1-enylphosphonates, *Synthesis*, 162, 1997.

463. Seyferth, D., Marmor, R.S., and Hilbert, P., Some reactions of dimethylphosphono-substituted diazoalkanes. (MeO)$_2$P(O)CR transfer to olefins and 1,3-dipolar additions of (MeO)$_2$P(O)C(N$_2$)R[1], *J. Org. Chem.*, 36, 1379, 1971.

464. Scherer, H., Hartmann, A., Regitz, M., Tunggal, B.D., and Günther, H., Carbenes. Part 5. 7-Phosphono-7-arylnorcaradienes, *Chem. Ber.*, 105, 3357, 1972.

465. Hartmann, A., Welter, W., and Regitz, M., Carbenes. Part 7. Intramolecular reactions of vinyl- and allylphosphoryl carbenes, *Tetrahedron Lett.*, 15, 1825, 1974.

466. Maas, G., and Regitz, M., Carbenes. Part 10. Substituent dependence of the norcaradiene/cycloheptatriene-equilibrium investigated in 7-phosphoryl- and 7-carbonylsubstituted systems, *Chem. Ber.*, 109, 2039, 1976.

467. Welter, W., Hartmann, A., and Regitz, M., Carbenes. Part 18. Isomerization reactions of phosphoryl-vinyl-carbenes to phosphorylated cyclopropenes, allenes, acetylenes, indenes, and 1,3-butadienes, *Chem. Ber.*, 111, 3068, 1978.

468. Theis, W., and Regitz, M., Investigations on diazo compounds and azides. Part 60. Reactions of 3H-1,2,4-triazole-3,5(4H)-diones with vinyl diazo compounds, *Chem. Ber.*, 118, 3396, 1985.

469. Regitz, M., Chemistry of phosphorylcarbenes, *Angew. Chem.*, 87, 259, 1975; *Angew. Chem. Int. Ed. Engl.*, 14, 222, 1975.

470. Rosini, G., Baccolini, G., and Cacchi, S., Reactions of 1-oxoalkanephosphonate tosylhydrazones with sodium borohydride. New synthesis of dimethyl alkanephosphonates and dimethyl 1-diazoalkanephosphonates, *Synthesis*, 44, 1975.

471. Mastryukova, T.A., Baranov, G.M., Perekalin, V.V., and Kabachnik, M.I., Organophosphorus nitro alcohols. *O,O*-Dialkyl-1-methyl-1-hydroxy-2-nitroalkylphosphonates and their derivatives, *Dokl. Akad. Nauk SSSR*, 171, 1341, 1966; *Chem. Abstr.*, 66, 54959, 1967.

472. Serdyukova, A.V., Baranov, G.M., and Perekalin, V.V., Synthesis and properties of some derivatives of 1-hydroxy-2-nitroalkylphosphonic acids, *Zh. Obshch. Khim.*, 44, 1243, 1974; *J. Gen. Chem. USSR (Engl. Transl.)*, 44, 1220, 1974.

473. Richtarski, G., and Mastalerz, P., Deamination and rearrangement of (1-phenyl-1-hydroxy-2-amino-ethyl)phosphonic acid, *Tetrahedron Lett.*, 14, 4069, 1973.

474. Yuan, C., Cui, S., Wang, G., Feng, H., Chen, D., Li, C., Ding, Y., and Maier, L., Studies on organophosphorus compounds. Part 53. A new procedure for the synthesis of 1-alkyl or 1-aryl-1-hydroxy-2-nitroethylphosphonates under phase-transfer catalysis conditions, *Synthesis*, 258, 1992.

475. Baranov, G.M., and Perekalin, V.V., Phosphorus-containing nitroalkenes, *Zh. Obshch. Khim.*, 57, 793, 1987; *J. Gen. Chem. USSR (Engl. Transl.)*, 57, 699, 1987.

476. Yuan, Q., He, P., and Yuan, C., Studies on organophosphorus compounds. Part 102. A convenient stereospecific synthesis of dialkyl 1-alkyl(aryl)-2-nitroeth-1-enylphosphonates, *Phosphorus, Sulfur Silicon Relat. Elem.*, 127, 113, 1997.

477. Mastryukova, T.A., Baranov, G.M., Perekalin, V.V., and Kabachnik, M.I., Rearrangement in the condensation of dialkyl trichloroacetylphosphonates with nitromethane, *Izv. Akad. Nauk SSSR, Ser. Khim.*, 2842, 1968; *Chem. Abstr.*, 70, 77233, 1969.

478. Pudovik, A.N., and Konovalova, I.V., Rearrangement of methyl*bis*(diethylphosphono)carbinol, *Dokl. Akad. Nauk SSSR*, 143, 875, 1962; *Chem. Abstr.*, 57, 3480a, 1962.

479. Fitch, S.J., and Moedritzer, K., Nuclear magnetic resonance study of the P-C(OH)-P to P-CO-P rearrangement. Tetraethyl 1-hydroxyalkylidenediphosphonates, *J. Am. Chem. Soc.*, 84, 1876, 1962.

480. Nicholson, D.A., and Vaughn, H., General method of preparation of tetramethyl alkyl-1-hydroxy-1,1-diphosphonates, *J. Org. Chem.*, 36, 3843, 1972.

481. Pudovik, A.N., and Konovalova, I.V., Reaction of chlorides of carboxylic acids and anhydrides of carboxylic acids with diethyl sodiophosphonate, *Zh. Obshch. Khim.*, 33, 98, 1963; *Chem. Abstr.*, 59, 656a, 1963.

482. Pudovik, A.N., Gur'yanova, I.V., and Zimin, M.G., Reaction of salts of incomplete esters of phosphinic and ethylphosphinic acids with benzoyl chlorides, *Zh. Obshch. Khim.*, 37, 2088, 1967; *J. Gen. Chem. USSR (Engl. Transl.)*, 37, 1979, 1967.

483. Pudovik, A.N., Gur'yanova, I.V., Banderova, L.V., and Romanov, G.V., Phosphonate–phosphate rearrangement of α-hydroxylakylphosphinic acids, *Zh. Obshch. Khim.*, 38, 143, 1968; *Chem. Abstr.*, 69, 96839, 1968.

484. Pudovik, A.N., Konovalova, I.V., and Buriaeva, L.A., Reactions of phosphorous acid esters with phenylglyoxal and the ethyl ester of α,β-dioxobutyric acid, *Zh. Obshch. Khim.*, 41, 2413, 1971; *J. Gen. Chem. USSR (Engl. Transl.)*, 41, 2439, 1971.

485. Pudovik, A.N., Romanov, G.V., and Pozhidaev, V.M., Synthesis and thermal properties of α-diphosphacarbinols, *Izv. Akad. Nauk SSSR, Ser. Khim.*, 452, 1979; *Chem. Abstr.*, 90, 187057, 1979.

486. Okamoto, Y., and Sakurai, H., The reactions of diethyl acyl phosphates with diethyl phosphite, *Kogyo Kagaku Zasshi*, 71, 310, 1968; *Chem. Abstr.*, 69, 86263, 1968.

487. Nguyen, L.M., Niesor, E., and Bentzen, C.L., *gem*-Diphosphonate and *gem*-phosphonate-phosphate compounds with specific high density lipoprotein inducing activity, *J. Med. Chem.*, 30, 1426, 1987.

488. Pudovik, A.N., Konovalova, I.V., and Dedova, L.V., Rearrangement of α-hydroxyphosphonic and α-hydroxyphosphonothioic esters into phosphonates and phosphorothioates, *Dokl. Akad. Nauk SSSR*, 153, 616, 1963; *Chem. Abstr.*, 60, 8060a, 1964.

489. Baker, R., and Broughton, H.B., Mechanism and inhibition of inositol monophosphatase, *Phosphorus, Sulfur Silicon Relat. Elem.*, 109–110, 337, 1996.

490. Tromelin, A., El Manouni, D., and Burgada, R., α-Ketophosphonates and cyclic esters of hydroxymethylenediphosphonates. Synthesis, structure, and hydrolysis, *Phosphorus Sulfur*, 27, 301, 1986.

491. Leroux, Y., El Manouni, D., Safsaf, A., Neuman, A., Gillier, H., and Burgada, R., Structural study of 1-hydroxyethane-1,1'-diphosphonic acid tetramethyl ester and of 1-hydroxy-1-phenylmethane-1,1'-diphosphonic acid tetramethyl ester, *Phosphorus, Sulfur Silicon Relat. Elem.*, 56, 95, 1991.

492. Griffiths, D.V., Hughes, J.M., Brown, J.W., Caesar, J.C., Swetnam, S.P., Cumming, S.A., and Kelly, J.D., The synthesis of 1-amino-2-hydroxy- and 2-amino-1-hydroxyethylene-1,1-bisphosphonic acids and their *N*-methylated derivatives, *Tetrahedron*, 53, 17815, 1997.

493. Mizrahi, D.M., Waner, T., and Segall, Y., α-Amino acid derived bisphosphonates. Synthesis and anti-resorptive activity, *Phosphorus, Sulfur Silicon Relat. Elem.*, 173, 1, 2001.

494. Turhanen, P.A., Ahlgren, M.J., Jarvinen, T., and Vepsalainen, J.J., Bisphosphonate prodrugs. Synthesis and identification of (1-hydroxyethylidene)-1,1-bisphosphonic acid tetraesters by mass spectrometry, NMR spectroscopy and X-ray crystallography, *Phosphorus, Sulfur Silicon Relat. Elem.*, 170, 115, 2001.

495. Pudovik, A.N., Batyeva, E.S., and Zamaletdinova, G.U., Reaction of a trimethylsilyl phosphite with *O,O*-diethyl acetylphosphonate, *Zh. Obshch. Khim.*, 43, 680, 1973; *J. Gen. Chem. USSR (Engl. Transl.)*, 43, 676, 1973.

496. Prishchenko, A.A., Min'ko, S.V., Livantsov, M.V., and Petrosyan, V.S., Reaction of diethyl trimeth-ylsilyl phosphite with diethyl pivaloylphosphonite and its derivatives, *Zh. Obshch. Khim.*, 62, 2393, 1992; *J. Gen. Chem. USSR (Engl. Transl.)*, 62, 1975, 1992.

497. Kabachnik, M.M., Khomutova, Y.A., and Beletskaya, I.P., Reaction of diphenyl(trimethylsilyl)phos-phine with α-oxo phosphonates, *Izv. Akad. Nauk, Ser. Khim.*, 2379, 1999; *Russ. Chem. Bull. (Engl. Transl.)*, 2352, 1999.

498. Konovalova, I.V., Burnaeva, L.A., and Pudovik, A.N., Reaction of trimethylsilyldiethylphosphite with diethyl ester of trichloroacetyl phosphonic acid, *Zh. Obshch. Khim.*, 45, 2567, 1975; *Chem. Abstr.*, 84, 44258, 1976.

499. Smyth, M.S., and Burke, T.R., Jr., Enantioselective synthesis of *N*-Boc and *N*-Fmoc protected diethyl 4-phosphono(difluoromethyl)-L-phenylalanine. Agents suitable for the solid-phase synthesis of pep-tides containing nonhydrolyzable analogues of *O*-phosphotyrosine, *Tetrahedron Lett.*, 35, 551, 1994.

500. Solas, D., Hale, R.L., and Patel, D.V., An efficient synthesis of *N*-α-Fmoc-4-(phosphonodifluorome-thyl)-L-phenylalanine, *J. Org. Chem.*, 61, 1537, 1996.

501. Liu, W.-Q., Roques, B.P., and Garbay, C., Synthesis of L-2,3,5,6-tetrafluoro-4-(phosphonomethyl)phe-nylalanine, a novel nonhydrolyzable phosphotyrosine mimetic and L-4-(phosphonodifluorome-thyl)phenylalanine, *Tetrahedron Lett.*, 38, 1389, 1997.

502. Caplan, N.A., Pogson, C.I., Hayes, D.J., and Blackburn, G.M., The synthesis of novel bisphosphonates as inhibitors of phosphoglycerate kinase (3-PGK), *J. Chem. Soc., Perkin Trans. 1*, 421, 2000.

503. Kojima, M., Yamashita, M., Yoshida, H., Ogata, T., and Inokawa, S., Useful method for the preparation of α,β-unsaturated phosphonates (1-methylenealkanephosphonates), *Synthesis*, 147, 1979.

504. Harris, R.L.N., and McFadden, H.G., Acylphosphonates as substrates for Wittig and Horner–Wittig reactions. Unusual stereoselectivity in the synthesis of β-phosphinoylacrylates, *Aust. J. Chem.*, 37, 417, 1984.

505. Shahak, I., and Peretz, J., Reactions of dimethyl benzoylphosphonate, *Isr. J. Chem.*, 9, 35, 1971.

506. Öhler, E., El-Badawi, M., and Zbiral, E., Simple route to α-substituted (*E*)-3-oxo-1-alkenylphospho-nates, *Monatsh. Chem.*, 116, 77, 1985.

507. Öhler, E., and Kanzler, S., Synthesis of phosphonic acids related to the antibiotic fosmidomycin from allylic α- and γ-hydroxyphosphonates, *Phosphorus, Sulfur Silicon Relat. Elem.*, 112, 71, 1996.

508. Öhler, E., and Zbiral, E., Valence isomerism between 2,4-dienones and 2*H*-pyrans with dialkoxyphos-phinyl substituents. First synthesis of dialkyl 2*H*-pyran-4-ylphosphonates, *Chem. Ber.*, 118, 2917, 1985.

509. Malenko, D.M., and Gololobov, Y.G., *O*-Phosphorylation of ketophosphonates, *Zh. Obshch. Khim.*, 50, 682, 1980; *Chem. Abstr.*, 93, 114632, 1980.

510. Al'fonsov, V.A., Zamaletdinova, G.U., Nizamov, I.S., Batyeva, E.S., and Pudovik, A.N., Vinylation of acetylphosphonates. Geminal-substituted vinylphosphonates, *Zh. Obshch. Khim.*, 54, 1485, 1984; *J. Gen. Chem. USSR (Engl. Transl.)*, 54, 1324, 1984.

511. Okauchi, T., Yano, T., Fukamachi, T., Ichikawa, J., and Minami, T., α-Phosphonovinyl nonaflate.Their synthesis and cross-coupling reactions, *Tetrahedron Lett.*, 40, 5337, 1999.

512. Longmire, C.F., and Evans, S.A., Jr., Stereoselective aldol reactions via enolates of α-acylphosphonate diesters, *J. Chem. Soc., Chem. Commun.*, 922, 1990.

513. Burk, M.J., Stammers, T.A., and Straub, J.A., Enantioselective synthesis of α-hydroxy and α-amino phosphonates via catalytic asymmetric hydrogenation, *Org. Lett.*, 1, 387, 1999.

514. Takaki, K., Itono, Y., Nagafuji, A., Naito, Y., Shishido, T., Takehira, K., Makioka, Y., Taniguchi, Y., and Fujiwara, Y., Three-component coupling of acylphosphonates and two carbonyl compounds promoted by low-valent samariums. One-pot synthesis of β-hydroxyphosphonates, *J. Org. Chem.*, 65, 475, 2000.

515. Evans, D.A., Johnson, J.S., Burgey, C.S., and Campos, K.R., Reversal in enantioselectivity of *tert*-butyl *versus* phenyl-substituted *bis*(oxazoline) copper(II) catalyzed hetero Diels–Alder and ene reactions. Crystallographic and mechanistic studies, *Tetrahedron Lett.*, 40, 2879, 1999.

516. Kluger, R., Pike, D.C., and Chin, J., Kinetics and mechanism of the reaction of dimethyl acetylphosphonate with water. Expulsion of a phosphonate ester from a carbonyl hydrate, *Can. J. Chem.*, 56, 1792, 1978.

517. Kluger, R., Nakaoka, K., and Tsui, W.-C., Substrate analog studies of the specificity and catalytic mechanism of D-3-hydroxybutyrate dehydrogenase, *J. Am. Chem. Soc.*, 100, 7388, 1978.

518. Kabachnik, M.I., and Rossiiskaya, P.A., Esters of α-ketophosphonic acids. Part 2. Acid splitting, *Izv. Akad. Nauk SSSR, Ser. Khim.*, 597, 1945; *Chem. Abstr.*, 41, 88d, 1947.

519. Pudovik, A.N., Gazizov, T.K., and Pashinkin, A.P., Breaking of a phosphorus–carbon bond in chloro-substituted esters of acetophosphonic acids, *Zh. Obshch. Khim.*, 38, 2812, 1968; *J. Gen. Chem. USSR (Engl. Transl.)*, 38, 2712, 1968.

520. Terauchi, K., and Sakurai, H., Photochemical studies of the esters of aroylphosphonic acids, *Bull. Chem. Soc. Jpn.*, 43, 883, 1970.

521. Pashinkin, A.P., Gazizov, T.K., and Pudovik, A.N., Breaking the phosphorus–carbon bond in α-keto phosphonates, *Zh. Obshch. Khim.*, 40, 28, 1970; *Chem. Abstr.*, 72, 100827, 1970.

522. Evdakov, V.P., Beketov, V.P., and Svergun, V.I., Reaction of amidophosphites and acetylphosphites with acetic acid, alcohols, and phenol, *Zh. Obshch. Khim.*, 43, 55, 1973; *Chem. Abstr.*, 78, 123523, 1973.

523. Sekine, M., Kume, A., and Hata, T., Acylphosphonates. Part 2. Dialkyl acylphosphonates. New acylating agents for alcohols, *Tetrahedron Lett.*, 22, 3617, 1981.

524. Soroka, M., and Mastalerz, P., Cleavage of a phosphorus–carbon bond in α-oxophosphonates, *Zh. Obshch. Khim.*, 44, 463, 1974; *Chem. Abstr.*, 80, 121058, 1974.

525. Sekine, M., Satoh, M., Yamagata, H., and Hata, T., Acylphosphonates. Phosphorus–carbon bond cleavage of dialkyl acylphosphonates by means of amines. Substituent and solvent effects for acylation of amines, *J. Org. Chem.*, 45, 4162, 1980.

526. Fujii, M., Ozaki, K., Kume, A., Sekine, M., and Hata, T., Acylphosphonates. Part 5. A new method for stereospecific generation of phosphorothioate via aroylphosphonate intermediate, *Tetrahedron Lett.*, 27, 935, 1986.

527. Fujii, M., Ozaki, K., Sekine, M., and Hata, T., Acylphosphonates. Part 7. A new method for stereospecific and stereoselective generation of dideoxyribonucleoside phosphorothioates via the acylphosphonate intermediates, *Tetrahedron*, 43, 3395, 1987.

528. Shahak, I., and Bergmann, E.D., Reaction of α-oxophosponates with organometallic compounds and sodium borohydride, *Isr. J. Chem.*, 4, 225, 1966.

529. Maeda, H., Takahashi, K., and Ohmori, H., Reactions of acyl tributylphosphonium chlorides and dialkyl acylphosphonates with Grignard and organolithium reagents, *Tetrahedron*, 54, 12233, 1998.

530. Gordon, N.J., and Evans, S.A., Jr., Acyl phosphates from acyl phosphonates. A novel Baeyer–Villiger rearrangement, *J. Org. Chem.*, 58, 4516, 1993.

531. Nakazawa, H., Matsuoka, Y., Yamaguchi, H., Kuroiwa, T., Miyoshi, K., and Yoneda, H., First example of catalytic decarbonylation and metathesis reactions of α-ketophosphonates promoted by a palladium complex, *Organometallics*, 8, 2272, 1989.

532. Nakazawa, H., Matsuoka, Y., Nakagawa, I., and Miyoshi, K., Catalytic carbon–phosphorus bond activation by palladium complexes. Decarbonylation and metathesis reactions of α-ketophosphonates and isolation of aroyl(phosphonato)palladium complexes as intermediates, *Organometallics*, 11, 1385, 1992.

533. Nakazawa, H., Matsuoka, Y., Nakagawa, I., and Miyoshi, K., Palladium catalyzed carbon–phosphorus bond activation. Decarbonylation reaction of α-ketophosphonates, *Phosphorus, Sulfur Silicon Relat. Elem.*, 77, 13, 1993.

534. Okamoto, Y., and Sakurai, H., Synthesis of fatty aldehyde through the intermediate of diethyl α-hydroxyalkylphosphonate, *Kogyo Kagaku Zasshi*, 70, 797, 1967; *Chem. Abstr.*, 68, 48979, 1968.

535. Okamoto, Y., and Sakurai, H., Synthesis and properties of α-hydroxyalkylphosphonic acids, *Kogyo Kagaku Zasshi*, 70, 1603, 1967; *Chem. Abstr.*, 68, 22018, 1968.

536. Horner, L., and Roeder, H., Organic phosphorus compounds. Part 67. Reductive conversion of carboxylic acids into aldehydes, *Chem. Ber.*, 103, 2984, 1970.

537. Öhler, E., and Kotzinger, S., Thermal rearrangement of trichloroacetimidic esters of allylic α-hydroxyphosphonates. A convenient way to (3-amino-1-alkenyl)phosphonic acids, *Synthesis*, 497, 1993.

538. Öhler, E., and Kotzinger, S., Synthesis of (3-amino-1-alkenyl)phosphonic acids from allylic α- and γ-hydroxyphosphonates. Sigmatropic rearrangement of dialkyl (1-azido-2-alkenyl)phosphonates, *Liebigs Ann. Chem.*, 269, 1993.

539. Öhler, E., and Kanzler, S., [(2*H*-Azirin-2-yl)methyl]phosphonates. Synthesis from allylic α- and γ-hydroxyphosphonates and application to diastereoselective formation of substituted [(azirin-2-yl)methyl]phosphonates, *Liebigs Ann. Chem.*, 867, 1994.

540. Ziora, Z., Maly, A., Lejczak, B., Kafarski, P., Holband, J., and Wojcik, G., Reactions of *N*-phthalyl-amino acid chlorides with trialkyl phosphites, *Heteroat. Chem.*, 11, 232, 2000.

541. Barco, A., Benetti, S., Bergamini, P., De Risi, C., Marchetti, P., Pollini, G.P., and Zanirato, V., Diastereoselective synthesis of β-amino-α-hydroxy phosphonates via oxazaborolidine catalyzed reduction of β-phthalimido-α-keto phosphonates, *Tetrahedron Lett.*, 40, 7705, 1999.

542. Chen, R., and Breuer, E., Direct approach to α-hydroxyphosphonic and α,ω-dihydroxyalkane-α,ω-bisphosphonic acids by the reduction of (*bis*)acylphosphonic acids, *J. Org. Chem.*, 63, 5107, 1998.

543. Creary, X., Geiger, C.C., and Hilton, K., Mesylate derivatives of α-hydroxy phosphonates. Formation of carbocations adjacent to the diethyl phosphonate group, *J. Am. Chem. Soc.*, 105, 2851, 1983.

544. Gajda, T., Enantioselective synthesis of diethyl 1-hydroxyalkylphosphonates via oxazaborolidine catalyzed borane reduction of diethyl α-ketophosphonates, *Tetrahedron: Asymmetry*, 5, 1965, 1994.

545. Meier, C., and Laux, W.H.G., Enantioselective synthesis of diisopropyl α-, β-, and γ-hydroxyaryl-alkylphosphonates from ketophosphonates. A study on the effect of the phosphonyl group, *Tetrahedron*, 52, 589, 1996.

546. Meier, C., and Laux, W.H.G., Asymmetric synthesis of chiral, nonracemic dialkyl α-, β-, and γ-hydroxyalkylphosphonates via a catalyzed enantioselective catecholborane reduction, *Tetrahedron: Asymmetry*, 6, 1089, 1995.

547. Corey, E.J., and Link, J.O., A general, catalytic, and enantioselective synthesis of α-amino acids, *J. Am. Chem. Soc.*, 114, 1906, 1992.

548. Khomutov, R.M., and Osipova, T.I., $^3H_\alpha$-α-Aminoisobutylphosphonic acid and its anhydride with adenosine 5′-phosphoric acid, *Izv. Akad. Nauk SSSR, Ser. Khim.*, 1110, 1979; *Chem. Abstr.*, 91, 14158, 1979.

549. Khomutov, R.M., Osipova, T.I., Zhukov, Y.N., and Gandurina, I.A., Organophosphorus analogs of biologically active substances. Part 5. Synthesis of α-aminophosphonic acids and some of their derivatives, *Izv. Akad. Nauk SSSR, Ser. Khim.*, 2118, 1979; *Bull. Acad. Sci. USSR, Div. Chem. Sci. (Engl. Transl.)*, 1949, 1979.

550. Pudovik, A.N., Yastrebova, G.E., and Nikitina, V.I., Condensation of diethyl acetonylphosphonate with benzaldehyde and cinnamaldehyde, *Zh. Obshch. Khim.*, 37, 510, 1967; *Chem. Abstr.*, 67, 32739, 1967.

551. Pudovik, A.N., Nikitina, V.I., Shakirova, A.M., and Danilov, N.A., Reaction of diethyl acetophenone phosphonates with aldehydes, *Zh. Obshch. Khim.*, 38, 1788, 1968; *J. Gen. Chem. USSR (Engl. Transl.)*, 38, 1742, 1968.

552. Pudovik, A.N., Shakirova, A.M., and Nikitina, V.I., Condensation and phosphoryl-olefination of phosphonoacetophenones, *Dokl. Akad. Nauk SSSR*, 182, 1338, 1968; *Chem. Abstr.*, 70, 28141, 1969.

553. Sakoda, R., Matsumoto, H., and Seto, K., Novel synthesis of α-acetylstyrylphosphonates, *Synthesis*, 705, 1993.

554. Al-Badri, H., Maddaluno, J., Masson, S., and Collignon, N., Hetero-Diels–Alder reactions of α-carbonylated styrylphosphonates with enol ethers. High-pressure influence on reactivity and diastereoselectivity, *J. Chem. Soc., Perkin Trans. 1*, 2255, 1999.

555. Pudovik, A.N., and Lebedeva, N.M., New method of synthesis of esters of phosphonic and thiophosphonic acids. Part 23. Addition of phosphonoacetic ester, phosphonoacetone, and its homologs to unsaturated compounds, *Zh. Obshch. Khim.*, 25, 1920, 1955; *Chem. Abstr.*, 50, 8442c, 1956.

556. Pudovik, A.N., Shchelkina, L.P., and Bashirova, L.A., Substitution reactions of phosphonoacetic ester and phosphonoacetone, *Zh. Obshch. Khim.*, 27, 2367, 1957; *Chem. Abstr.*, 52, 7134i, 1958.

557. Pudovik, A.N., Nikitina, V.I., and Kurguzova, A.M., Reactions of organophosphorus compounds containing active methylene groups with methyl β-chlorovinyl ketone and α-halo ethers, *Zh. Obshch. Khim.*, 40, 291, 1970; *J. Gen. Chem. USSR (Engl. Transl.)*, 40, 261, 1970.

558. Pudovik, A.N., and Khusainova, N.G., Oxidative condensation and addition reactions of 1-dialkoxyphosphinyl-1-acetyl-3-butynes, *Zh. Obshch. Khim.*, 40, 1419, 1970; *J. Gen. Chem. USSR (Engl. Transl.)*, 40, 1403, 1970.

559. Clark, R.D., Kozar, L.G., and Heathcock, C.H., Alkylation of diethyl 2-oxoalkanephosphonates, *Synthesis*, 635, 1975.

560. Clark, R.D., Kozar, L.G., and Heathcock, C.H., Cyclopentenones from β,ε-diketo phosphonates. Synthesis of *cis*-jasmone, *Synth. Commun.*, 5, 1, 1975.

561. Khoukhi, M., Vaultier, M., and Carrié, R., The use of ω-iodo azides as primary protected electrophilic reagents. Alkylation of some carbanions derived from active methylene compounds and *N,N*-dimethylhydrazones, *Tetrahedron Lett.*, 27, 1031, 1986.

562. Miller, D.B., Raychaudhuri, S.R., Avasthi, K., Lal, K., Levison, B., and Salomon, R.G., Levuglandin E₂. Enantiocontrolled total synthesis of a biologically active rearrangement product from the prostaglandin endoperoxide PGH₂, *J. Org. Chem.*, 55, 3164, 1990.

563. Subbanagounder, G., Salomon, R.G., Murthi, K.K., Brame, C., and Roberts, J.L., II, Total synthesis of *iso*[4]-levuglandin E₂, *J. Org. Chem.*, 62, 7658, 1997.

564. Roy, S.C., Nagarajan, L., and Salomon, R.G., Total synthesis of *iso*[7]-levuglandin D₂, *J. Org. Chem.*, 64, 1218, 1999.

565. Sha, W., and Salomon, R.G., Total synthesis of 17-isolevuglandin E₄ and the structure of C22-PGF₄ₐ, *J. Org. Chem.*, 65, 5315, 2000.

566. Mikolajczyk, M., and Zatorski, A., Methylenomycin B. New syntheses based on β- and γ-keto phosphonates and γ-keto phosphine oxides, *J. Org. Chem.*, 56, 1217, 1991.

567. Diana, G.D., Zalay, E.S., Salvador, U.J., Pancic, F., and Steinberg, B., Synthesis of some phosphonates with antiherpetic activity, *J. Med. Chem.*, 27, 691, 1984.

568. Snider, B.B., and Yang, K., A short enantiospecific synthesis of the ceroplastin nucleus, *J. Org. Chem.*, 57, 3615, 1992.

569. Ruder, S.M., and Kulkarni, V.R., Phase transfer catalysed alkylation of 2-(diethoxyphosphinyl)cyclohexanone, *Synthesis*, 945, 1993.

570. Grieco, P.A., and Pogonowski, C.S., Alkylation of the dianion of β-keto phosphonates. Versatile synthesis of dimethyl (2-oxoalkyl)phosphonates, *J. Am. Chem. Soc.*, 95, 3071, 1973.

571. Grieco, P.A., and Finkelhor, R.S., Dianions of β-ketophosphonates. Two-step synthesis of (+)-*ar*-turmerone, *J. Org. Chem.*, 38, 2909, 1973.

572. Mata, E.G., and Thomas, E.J., Development of a synthesis of lankacidins. An investigation into 17-membered ring formation, *J. Chem. Soc., Perkin Trans. 1*, 785, 1995.

573. Mikolajczyk, M., Zurawinski, R., Kielbasinski, P., Wieczorek, M.W., Blaszczyk, J., and Majzner, W.R., Total synthesis of racemic and optically active sarkomycin, *Synthesis*, 356, 1997.

574. Haney, B.P., and Curran, D.P., Round trip radical reactions from acyclic precursors to tricyclo[5.3.1.0²·⁶]undecanes. A new cascade radical cyclization approach to (±)-isogymnomitrene and (±)-gymnomitrene, *J. Org. Chem.*, 65, 2007, 2000.

575. Lee, S.Y., Hong, J.E., Jang, W.B., and Oh, D.Y., Dephosphonylation of α-fully substituted β-keto phosphonates with LiAlH₄. Regioselective alkylation of ketones employing phosphonate as a temporary activating group, *Tetrahedron Lett.*, 38, 4567, 1997.

576. Lee, S.Y., Lee, C.-W., and Oh, D.Y., Dephosphonylation of β-carbonyl phosphonates, *J. Org. Chem.*, 64, 7017, 1999.

577. Lee, S.Y., Lee, C.-W., and Oh, D.Y., Regiocontrolled synthetic approach to α,α′-disubstituted unsymmetrical ketones, *J. Org. Chem.*, 65, 245, 2000.

578. Goswami, R., A novel β-ketophosphonate 1,4-dianion. Tin/lithium exchange, *J. Am. Chem. Soc.*, 102, 5973, 1980.

579. Grieco, P.A., Yokoyama, Y., Nicolaou, K.C., Barnette, W.E., Smith, J.B., Ogletree, M., and Lefer, A.M., Total synthesis of 14-fluoroprostaglandin F₂ₐ and 14-fluoroprostacyclin, *Chem. Lett.*, 1001, 1978.

580. Grieco, P.A., Schillinger, W.J., and Yokoyama, Y., Carbon-14 fluorinated prostaglandins. Synthesis and biological evaluation of the methyl esters of (+)-14-fluoro-, (+)-15-*epi*-14-fluoro-, (+)-13(*E*)-14-fluoro-, and (+)-13(*E*)-15-*epi*-14-fluoroprostaglandin F₂ₐ, *J. Med. Chem.*, 23, 1077, 1980.

581. Iseki, K., Shinoda, M., Ishiyama, C., Hayasi, Y., Yamada, S.-I., and Shibasaki, M., Synthesis of (Z)-4,5,13,14-tetradehydro-9(0)-methano-$\Delta^{6(9\alpha)}$-PGI$_1$, *Chem. Lett.*, 559, 1986.

582. Tomiyama, T., Wakabayashi, S., and Yokota, M., Synthesis and biological activity of novel carbacyclins having bicyclic substituents on the ω-chain, *J. Med. Chem.*, 32, 1988, 1989.

583. Grinev, G.V., Chervenyuk, G.I., and Dombrovskii, A.V., Synthesis of α-bromophosphonates, *Zh. Obshch. Khim.*, 39, 1253, 1969; *Chem. Abstr.*, 71, 101932, 1969.

584. Haelters, J.P., Corbel, B., and Sturtz, G., Synthesis of [1- or 3-(arylamino)-2-oxopropyl]phosphonates. Bischler cyclization to indolylphosphonates, *Phosphorus Sulfur*, 37, 65, 1988.

585. Moskva, V.V., Guseinov, F.I., Ismailov, V.M., and Gallyamov, M.R., α,α-Dichloro-α-(diethoxyphosphoryl)acetone, *Zh. Obshch. Khim.*, 57, 234, 1987; *Chem. Abstr.*, 107, 217736, 1987.

586. Iman, M., and Chenault, J., A convenient one-pot synthesis of 1,5-diaryl-1,2-epoxy-4-penten-3-ones, *Synthesis*, 124, 1989.

587. Fouqué, D., About-Jaudet, E., Collignon, N., and Savignac, P., A convenient synthesis of diethyl 2,4-dioxoalkylphosphonates, *Synth. Commun.*, 22, 219, 1992.

588. Fouqué, D., Al-Badri, H., About-Jaudet, E., and Collignon, N., α-Pyrazolylalkylphosphonates. Part 1. Synthesis of 1-(3- or 5-pyrazolyl)-alkylphosphonates, *Bull. Soc. Chim. Fr.*, 131, 992, 1994.

589. Bodalski, R., Pietrusiewicz, K.M., Monkiewicz, J., and Koszuk, J., A new efficient synthesis of substituted Nazarov reagents. A Wittig–Horner–Emmons approach, *Tetrahedron Lett.*, 21, 2287, 1980.

590. Bodalski, R., Pietrusiewicz, K.M., Monkiewicz, J., and Koszuk, J., A simple method for [3 + 3] annulation on chalcones with ethyl-4-(diethylphosphono)-3-oxobutanoate, *Pol. J. Chem.*, 57, 315, 1983.

591. Pietrusiewicz, K.M., Monkiewicz, J., and Bodalski, R., (3-Carbethoxy-2-oxopropylidene)triphenylphosphorane. A reagent for "3 + 3" cyclohexenone annulation, *J. Org. Chem.*, 48, 788, 1983.

592. Taylor, E.C., and Davies, H.M.L., Rhodium(II) acetate-catalyzed reaction of ethyl 2-diazo-3-oxopent-4-enoates. Simple routes to 4-aryl-2-hydroxy-1-naphthoates and β,γ-unsaturated esters. The dianion of ethyl 4-(diethylphosphono)acetoacetate as a propionate homoenolate equivalent, *Tetrahedron Lett.*, 24, 5453, 1983.

593. Karagiri, N., Takashima, K., Haneda, T., and Kato, T., Synthesis of pyrazofurin and its analogues, *J. Chem. Soc., Perkin Trans. 1*, 553, 1984.

594. Moorhoff, C.M., and Schneider, D.F., Comments on the reaction of ethyl 4-(diethoxyphosphinyl)-3-oxobutanoate and related phosphonate esters with enals, *Tetrahedron Lett.*, 28, 559, 1987.

595. Ley, S.V., and Woodward, P.R., The use of β-ketothioesters for the exceptionally mild preparation of β-ketoamides, *Tetrahedron Lett.*, 28, 3019, 1987.

596. Blanc, D., Henry, J.-C., Ratovelomanana-Vidal, V., and Genêt, J.-P., Skewphos-Ru(II). An efficient catalyst for asymmetric hydrogenation of functionalized ketones, *Tetrahedron Lett.*, 38, 6603, 1997.

597. Ratovelomanana-Vidal, V., and Genêt, J.-P., Enantioselective ruthenium-mediated hydrogenation. Developments and applications, *J. Organomet. Chem.*, 567, 163, 1998.

598. Madec, J., Pfister, X., Phansavath, P., Ratovelomanana-Vidal, V., and Genêt, J.P., Asymmetric hydrogenation reactions using a practical *in situ* generation of chiral ruthenium-diphosphine catalysts from anhydrous RuCl$_3$, *Tetrahedron*, 57, 2563, 2001.

599. Genêt, J.P., Pinel, C., Ratovelomanana-Vidala, V., Mallart, S., Pfister, X., Caño De Andrade, M.C., and Laffitte, J.A., Novel, general synthesis of the chiral catalysts diphosphine-ruthenium (II) diallyl complexes and a new practical *in situ* preparation of chiral ruthenium (II) catalysts, *Tetrahedron: Asymmetry*, 5, 665, 1994.

600. Genêt, J.P., Pinel, C., Ratovelomanana-Vidala, V., Mallart, S., Pfister, X., Bischoff, L., Caño De Andrade, M.C., Darses, S., Galopin, C., and Laffitte, J.A., Enantioselective hydrogenation reactions with a full set of preformed and prepared *in situ* chiral diphosphine–ruthenium(II) catalysts, *Tetrahedron: Asymmetry*, 5, 675, 1994.

601. Yamano, T., Taya, N., Kawada, M., Huang, T., and Imamoto, T., Enantioselective hydrogenation of β-keto esters catalyzed by *P*-chiral *bis*(dialkylphosphino)ethanes-Ru(II), *Tetrahedron Lett.*, 40, 2577, 1999.

602. Kitamura, M., Tokunaga, M., Pham, T., Lubell, W.D. and Noyori, R., Asymmetric synthesis of α-amino β-hydroxy phosphonic acids via BINAP-ruthenium catalyzed hydrogenation, *Tetrahedron Lett.*, 36, 5769, 1995.

603. Tsai, H.-J., Thenappan, A., and Burton, D.J., An expedient synthesis of α-fluoro-α,β-unsaturated diesters, *Tetrahedron Lett.*, 33, 6579, 1992.

604. Tsai, H.-J., Thenappan, A., and Burton, D.J., A novel intramolecular Horner–Wadsworth–Emmons reaction. A simple and general route to α-fluoro-α,β-unsaturated diesters, *J. Org. Chem.*, 59, 7085, 1994.

605. Tsai, H.-J., Application of fluorocarbethoxy-substituted phosphonate. A facile entry to substituted 2-fluoro-3-oxoesters, *Phosphorus, Sulfur Silicon Relat. Elem.*, 126, 1, 1997.

606. Lentsch, L.M., and Wiemer, D.F., Addition of organometallic nucleophiles to β-keto phosphonates, *J. Org. Chem.*, 64, 5205, 1999.

607. Lentsch, L.M., and Wiemer, D.F., Nucleophilic additions to β-keto phosphonates, *Phosphorus, Sulfur Silicon Relat. Elem.*, 144, 573, 1999.

608. Ruder, S.M., and Kulkarni, V.R., Michael-type additions of 2-(diethoxyphosphinyl) cyclohexanone to activated alkenes and alkynes, *J. Chem. Soc., Chem. Commun.*, 2119, 1994.

609. Ruder, S.M., and Kulkarni, V.R., Ring expansion reactions of cyclic β-keto phosphonates, *J. Org. Chem.*, 60, 3084, 1995.

610. Regitz, M., Anschutz, W., Bartz, W., and Liedhegener, A., Reactions of CH-active compounds with azides. Part 22. Synthesis and some properties of α-diazophosphine oxides and α-diazophosphonates, *Tetrahedron Lett.*, 9, 3171, 1968.

611. Regitz, M., Anschütz, W., and Liedhegener, A., Reactions of CH-active compounds with azides. Part 23. Synthesis of α-diazophosphonic acid esters, *Chem. Ber.*, 101, 3734, 1968.

612. Chantegrel, B., Deshayes, C., and Faure, R., Tandem Wolff rearrangement-"*tert*-amino effect" sequence. Synthesis of 2-oxoindolinium enolate and 1*H*-2-benzopyrane derivatives, *Tetrahedron Lett.*, 36, 7859, 1995.

613. Chen, Y.P., Chantegrel, B., and Deshayes, C., A thermal Wolff rearrangement-benzannulation route to naphth[2,1-*d*]isoxazoles, [1]benzofuro[6,7-*d*]- or [5,4-*d*]isoxazoles and 1,2-benzisoxazoles, *Heterocycles*, 41, 175, 1995.

614. Collomb, D., Deshayes, C., and Doutheau, A., Synthesis of functionalized phenolic derivatives via the benzannulation of dienylketenes formed by a thermal Wolff rearrangement of α-diazo-β-keto compounds, *Tetrahedron*, 52, 6665, 1996.

615. Collomb, D., Chantegrel, B., and Deshayas, C., Chemoselectivity in the rhodium(II) acetate catalysed decomposition of α-diazo-β-keto-γ,δ-alkenyl-δ-aryl compounds. Aromatic C–H insertion reaction or Wolff rearrangement–electrocyclization, *Tetrahedron*, 52, 10455, 1996.

616. Léost, F., Chantegrel, B., and Deshayes, C., Tandem Wolff rearrangement-"*tert*-amino effect" sequence. Synthesis of 2-oxoindolinium enolate derivatives, *Tetrahedron*, 53, 7557, 1997.

617. Léost, F., Chantegrel, B., and Deshayes, C., Tandem Wolff rearrangement-"α-cyclization of tertiary amines" sequence. Synthesis of some 1*H*-2-benzopyran derivatives, *Tetrahedron*, 54, 6457, 1998.

618. Corbel, B., Hernot, D., Haelters, J.-P., and Sturtz, G., Synthesis of α-phosphorylated cyclopentanones by intramolecular carbenoid cyclizations of α-diazo β-keto alkylphosphonates and phosphine oxides, *Tetrahedron Lett.*, 28, 6605, 1987.

619. Moody, C.J., Sie, E.-R.H.B., and Kulagowski, J.J., Diazophosphonates in cyclic ether synthesis. Use of the intramolecular Wadsworth–Emmons reaction, *Tetrahedron Lett.*, 32, 6947, 1991.

620. Kim, S., Sutton, S.C., Guo, C., LaCour, T.G., and Fuchs, P.L., Synthesis of the north 1 unit of the cephalostatin family from hecogenin acetate, *J. Am. Chem. Soc.*, 121, 2056, 1999.

621. Koh, Y.J., and Oh, D.Y., A new synthesis of vinyl phosphonates from α-phenyl-β-oxo phosphonates and dialkyl phosphite, *Synth. Commun.*, 25, 2587, 1995.

622. Hong, J.E., Lee, C.-W., Kwon, Y., and Oh, D.Y., Facile synthesis of 1-alkynylphosphonates, *Synth. Commun.*, 26, 1563, 1996.

623. Paulvannan, K., and Stille, J.R., Conformationally restricted β-amino acid isosteres prepared through regioselectivity controlled aza-annulation, *Tetrahedron Lett.*, 34, 8197, 1993.

624. Paulvannan, K., and Stille, J.R., Heterocycle formation through aza-annulation. Stereochemically controlled syntheses of (±)-5-epitashiromine and (±)-tashiromine, *J. Org. Chem.*, 59, 1613, 1994.

625. Palacios, F., Ochoa de Retana, A.M., and Oyarzabal, J., A simple synthesis of 3-phosphonyl-4-aminoquinolines from β-enaminophosphonates, *Tetrahedron*, 55, 5947, 1999.

626. Aboujaoude, E.E., Collignon, N., and Savignac, P., α-Functional dialkyl formyl-1-methylphosphonates. Part 2. Thermical preparation and conversion into α-phosphonic heterocycles, *Tetrahedron*, 41, 427, 1985.

627. Aboujaoude, E.E., Collignon, N., and Savignac, P., Synthesis of α-phosphonic heterocycles. New developments, *Phosphorus Sulfur*, 31, 231, 1987.

628. Blechert, S., Hetero-Cope rearrangements. Part 2. Simple and flexible syntheses for 2-substituted indoles, *Tetrahedron Lett.*, 25, 1547, 1984.

629. Blechert, S., Hetero-Cope rearrangements. Part 4. Regio-controlled synthesis of indoles, *Helv. Chim. Acta*, 68, 1835, 1985.

630. Haelters, J.P., Corbel, B., and Sturtz, G., Synthesis of 1- or 3-aryloxy-2-oxopropylphosphonates and their cyclodehydration to benzofuran phosphonates, *Phosphorus, Sulfur Silicon Relat. Elem.*, 42, 85, 1989.

631. Kim, K., Kim, B.S., and Choi, K.S., Thioaroylketene *S,N*-acetals. Versatile intermediates for the synthesis of 3-alkylamino-5-arylthiophenes with a variety of functional group at C-2, *Phosphorus, Sulfur Silicon Relat. Elem.*, 153, 393, 1999.

632. Kim, B.S., and Kim, K., A facile and convenient synthesis of 3-alkylamino-5-arylthiophenes with a variety of substituents at C-2 and studies of reaction mechanisms, *J. Org. Chem.*, 65, 3690, 2000.

633. Ruzziconi, R., Couthon-Gourves, H., Gourves, J.-P., and Corbel, B., A facile approach to alkyl- and aryl-substituted 3-furylphosphonates based on ceric ammonium nitrate-promoted radical reactions, *Synlett*, 703, 2001.

634. Hong, J. E., Shin, W.S., Jang, W.B., and Oh, D.Y., Dephosphonylation of β-keto phosphonates with LiAlH$_4$, *J. Org. Chem.*, 61, 2199, 1996.

635. Ohira, S., Methanolysis of dimethyl (1-diazo-2-oxopropyl)phosphonate. Generation of dimethyl (diazomethyl)phosphonate and reaction with carbonyl compounds, *Synth. Commun.*, 19, 561, 1989.

636. Eymery, F., Iorga, B., and Savignac, P., The usefulness of phosphorus compounds in alkyne synthesis, *Synthesis*, 185, 2000.

637. Meffre, P., Gauzy, L., Perdigues, C., Desanges-Levecque, F., Branquet, E., Durand, P., and Le Goffic, F., En route to optically active ethynylglycine derivatives, *Tetrahedron Lett.*, 36, 877, 1995.

638. Meffre, P., Gauzy, L., Branquet, E., Durand, P., and Le Goffic, F., Synthesis of optically active β,γ-alkynylglycine derivatives, *Tetrahedron*, 52, 11215, 1996.

639. Crisp, G.T., Jiang, Y.-L., Pullman, P.J., and de Savi, C., Elaboration of the side-chain of amino acid derivatives by palladium catalysed couplings, *Tetrahedron*, 53, 17489, 1997.

640. Serrat, X., Cabarrocas, G., Rafel, S., Ventura, M., Linden, A., and Villalgordo, J.M., A highly efficient and straightforward stereoselective synthesis of novel chiral α-acetylenic ketones, *Tetrahedron: Asymmetry*, 10, 3417, 1999.

641. Pietruszka, J., and Witt, A., Enantiomerically pure cyclopropylboronic esters. Auxiliary- *versus* substrate-control, *J. Chem. Soc., Perkin Trans. 1*, 4293, 2000.

642. Thiery, J.-C., Frechou, C., and Demailly, G., Reaction of dimethyl (diazomethyl)phosphonate with reducing sugars. Synthesis of glyco-1-ynitols, *Tetrahedron Lett.*, 41, 6337, 2000.

643. Romero-Salguero, F.J., and Lehn, J.-M., Synthesis of multitopic bidentate ligands based on alternating pyridine and pyridazine rings, *Tetrahedron Lett.*, 40, 859, 1999.

644. Cao, G.-A., Wang, Z.-X., Tu, Y., and Shi, Y., Chemo- and enantioselective epoxidation of enynes, *Tetrahedron Lett.*, 39, 4425, 1998.

645. Kraemer, C.S., Zeitler, K., and Müller, T.J.J., Synthesis of functionalized ethynylphenothiazine fluorophores, *Org. Lett.*, 2, 3723, 2000.

646. Paquette, L.A., Tae, J., and Gallucci, J.C., Synthesis and crystal structure of a unique linear homoditopic ligand bifacially complexed to lithium picrate, *Org. Lett.*, 2, 143, 2000.

647. Evina, C.M., and Guillerm, G., Synthesis of uracil polyoxin C from uridine, *Tetrahedron Lett.*, 37, 163, 1996.

648. Nicolaou, K.C., and Koide, K., Synthetic studies on maduropeptin chromophore. Part 1. Construction of the aryl ether and attempted synthesis of the [7.3.0] bicyclic system, *Tetrahedron Lett.*, 38, 3667, 1997.

649. Corey, E.J., and Liu, K., Enantioselective total synthesis of the potent anti-HIV agent neotripterifordin. Reassignment of stereochemistry at C(16), *J. Am. Chem. Soc.*, 119, 9929, 1997.

650. Saito, M., Kawamura, M., Hiroya, K., and Ogasawara, K., First enantiocontrolled syntheses of (+)-uleine and (+)-dasycarpidone, *J. Chem. Soc., Chem. Commun.*, 765, 1997.

651. Taber, D.F., and Wang, Y., Synthesis of (−)-haliclonadiamine, *J. Am. Chem. Soc.*, 119, 22, 1997.

652. Ghosh, A.K., and Cappielo, J., Stereoselective synthesis of dihydroisocoumarin moiety of microbial agent AI-77-B. A Diels–Alder based strategy, *Tetrahedron Lett.*, 39, 8803, 1998.

653. Takahashi, S., and Nakata, T., Total synthesis of an anticancer agent, mucocin. Part 1. Stereoselective synthesis of the left-half segment, *Tetrahedron Lett.*, 40, 723, 1999.

654. Nicolaou, K.C., Xu, J., Murphy, F., Barluenga, S., Baudoin, O., Wei, H.-X., Gray, D.L.F., and Ohshima, T., Total synthesis of sanglifehrin A, *Angew. Chem.*, 111, 2599, 1999; *Angew. Chem. Int. Ed. Engl.*, 38, 2447, 1999.

655. Nazare, M., and Waldman, H., Synthesis of the (9S,18R)-*seco* acid of the leukocyte adhesion inhibitor cyclamenol A, *Tetrahedron Lett.*, 41, 625, 2000.

656. Cid, M.B., and Pattenden, G., Towards the synthesis of amphidinolide B. An intramolecular Stille coupling approach, *Tetrahedron Lett.*, 41, 7373, 2000.

657. Guanti, G., and Riva, R., A new diastereoselective approach to simplified dynemicin analogues, *J. Chem. Soc., Chem. Commun.*, 1171, 2000.

658. Nicolaou, K.C., Murphy, F., Barluenga, S., Ohshima, T., Wei, H., Xu, J., Gray, D.L. F., and Baudoin, O., Total synthesis of the novel immunosuppressant sanglifehrin A, *J. Am. Chem. Soc.*, 122, 3830, 2000.

659. Paterson, I., Doughty, V.A., McLeod, M.D., and Trieselmann, T., Total synthesis of (+)-concanamycin F, *Angew. Chem.*, 112, 1364, 2000; *Angew. Chem. Int. Ed. Engl.*, 39, 1308, 2000.

660. Liptak, V.P., and Wulff, W.D., The preparation and evaluation of electron poor benzylidene Fischer carbene complexes. Studies toward the total synthesis of (+)-olivin, *Tetrahedron*, 56, 10229, 2000.

661. Lee, K.L., Goh, J.B., and Martin, S.F., Novel entry to the ergot alkaloids via ring closing metathesis, *Tetrahedron Lett.*, 42, 1635, 2001.

662. Ovaska, T.V., Reisman, S.E., and Flynn, M.A., Facile entry to the tetracyclic 5–7–6–3 tigliane ring system, *Org. Lett.*, 3, 115, 2001.

663. Evans, D.A., and Burch, J.D., Asymmetric synthesis of the chlorocyclopropane-containing callipeltoside A side chain, *Org. Lett.*, 3, 503, 2001.

664. White, J.D., Carter, R.G., Sundermann, K.F., and Wartmann, M., Total synthesis of epothilone B, epothilone D, and *cis*- and *trans*-9,10-dehydroepothilone D, *J. Am. Chem. Soc.*, 123, 5407, 2001.

665. Marshall, J.A., and Adams, N.D., Total synthesis of bafilomycin V_1. A methanolysis product of the macrolide bafilomycin C_2, *J. Org. Chem.*, 67, 733, 2002.

666. Kinney, W.A., Abou-Gharbia, M., Garrison, D.T., Schmid, J., Kowal, D.M., Bramlett, D.R., Miller, T.L., Tasse, R.P., Zaleska, M.M., and Moyer, J.A., Design and synthesis of [2-(8,9-dioxo-2,6-diazabicyclo[5.2.0]non-1(7)-cn-2-yl)-ethyl]phosphonic acid (EAA-090), a potent *N*-methyl-D-aspartate antagonist, via the use of 3-cyclobutene-1,2-dione as an achiral α-amino acid bioisostere, *J. Med. Chem.*, 41, 236, 1998.

667. Gruszecka, E., Soroka, M., and Mastalerz, P., Phosphonic analogs of α-methylaspartic and α-methylglutamic acids, *Pol. J. Chem.*, 53, 2327, 1979.

668. Crooks, S.L., Robinson, M.B., Koerner, J.F., and Johnson, R.L., Cyclic analogues of 2-amino-4-phosphonobutanoic acid (APB) and their inhibition of hippocampal excitatory transmission and displacement of [³H]APB binding, *J. Med. Chem.*, 29, 1988, 1986.

669. Morita, T., Okamoto, Y., and Sakurai, H., A convenient dealkylation of dialkyl phosphonates by chlorotrimethylsilane in the presence of sodium iodide, *Tetrahedron Lett.*, 19, 2523, 1978.

670. Morita, T., Okamoto, Y., and Sakurai, H., Dealkylation reaction of acetals, phosphonate, and phosphate esters with chlorotrimethylsilane/metal halide reagent in acetonitrile, and its application to the synthesis of phosphonic acids and vinyl phosphates, *Bull. Chem. Soc. Jpn.*, 54, 267, 1981.

671. Blackburn, G.M., and Ingleson, D., Specific dealkylation of phosphonate esters using iodotrimethylsilane, *J. Chem. Soc., Chem. Commun.*, 870, 1978.

672. Blackburn, G.M., and Ingleson, D., The dealkylation of phosphate and phosphonate esters by iodotrimethylsilane. A mild and selective procedure, *J. Chem. Soc., Perkin Trans. 1*, 1150, 1980.

673. McKenna, C.E., Higa, M.T., Cheung, N.H., and McKenna, M.C., The facile dealkylation of phosphonic acid dialkyl esters by bromotrimethylsilane, *Tetrahedron Lett.*, 18, 155, 1977.

674. Morita, T., Okamoto, Y., and Sakurai, H., The preparation of phosphonic acids having labile functional groups, *Bull. Chem. Soc. Jpn.*, 51, 2169, 1978.

675. McKenna, C.E., and Schmidhauser, J., Functional selectivity in phosphonate ester dealkylation with bromotrimethylsilane, *J. Chem. Soc., Chem. Commun.*, 739, 1979.

676. Sekine, M., Futatsugi, T., Yamada, K., and Hata, T., Silyl phosphites. Part 20. A facile synthesis of phosphoenolpyruvate and its analog utilizing *in situ* generated trimethylsilyl bromide, *J. Chem. Soc., Perkin Trans. 1*, 2509, 1982.

677. Salomon, C.J., and Breuer, E., Efficient and selective dealkylation of phosphonate diisopropyl esters using Me_3SiBr, *Tetrahedron Lett.*, 36, 6759, 1995.

678. Zygmunt, J., Kafarski, P., and Mastalerz, P., Preparation of oxoalkanephosphonic acids, *Synthesis*, 609, 1978.

679. Gauvry, N., and Mortier, J., Dealkylation of dialkyl phosphonates with boron tribromide, *Synthesis*, 553, 2001.

8 The Carboxyphosphonates

The carboxyphosphonates are the oldest and the most popular functionalized phosphonates. They are precursors of several reagents often used in organic synthesis such as diethyl 1-(hydroxycarbonyl)methylphosphonate, dialkyl 1-(alkoxycarbonyl)vinylphosphonates, and dialkyl 1-(alkoxycarbonyl)-1-diazomethylphosphonates. The are known for their valuable role in organic synthesis and in the recent development of the chemistry of Foscarnet and ω-amino-ω-(hydroxycarbonyl)alkylphosphonic acids. This chapter deals with their preparations and transformations.

8.1 SYNTHESIS OF CARBOXYPHOSPHONATES

8.1.1 DIALKYL (ALKOXYCARBONYL)PHOSPHONATES

8.1.1.1 Michaelis–Arbuzov Reactions

The Michaelis–Arbuzov reaction is generally applicable to the preparation of dialkyl alkoxy- or aryloxycarbonylphosphonates from trialkyl phosphites and alkyl or aryl chloroformates. The reaction proceeds on heating when alkyl or aryl chloroformate is added to trialkyl phosphite at a rate that will maintain a steady evolution of alkyl chloride. The reaction has been performed with a great variety of trialkyl phosphites and alkyl or aryl chloroformates to provide the corresponding phosphonoformates in good to excellent yields (64–97%, Scheme 8.1).[1–9]

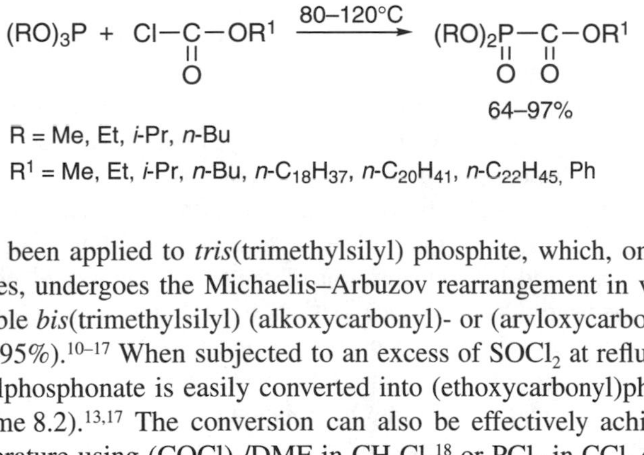

$$(RO)_3P + Cl-\underset{\underset{O}{\|}}{C}-OR^1 \xrightarrow{80-120°C} (RO)_2\underset{\underset{O}{\|}}{P}-\underset{\underset{O}{\|}}{C}-OR^1$$

64–97%

R = Me, Et, *i*-Pr, *n*-Bu

R^1 = Me, Et, *i*-Pr, *n*-Bu, *n*-C$_{18}$H$_{37}$, *n*-C$_{20}$H$_{41}$, *n*-C$_{22}$H$_{45}$, Ph

(8.1)

The reaction has been applied to *tris*(trimethylsilyl) phosphite, which, on reaction with alkyl or aryl chloroformates, undergoes the Michaelis–Arbuzov rearrangement in very mild conditions to produce the valuable *bis*(trimethylsilyl) (alkoxycarbonyl)- or (aryloxycarbonyl)phosphonates in excellent yields (80–95%).[10–17] When subjected to an excess of SOCl$_2$ at reflux, the *bis*(trimethylsilyl) ethoxycarbonylphosphonate is easily converted into (ethoxycarbonyl)phosphonic dichloride in 81% yield (Scheme 8.2).[13,17] The conversion can also be effectively achieved in comparable yields at room temperature using (COCl)$_2$/DMF in CH$_2$Cl$_2$[18] or PCl$_5$ in CCl$_4$ or CH$_2$Cl$_2$.[19]

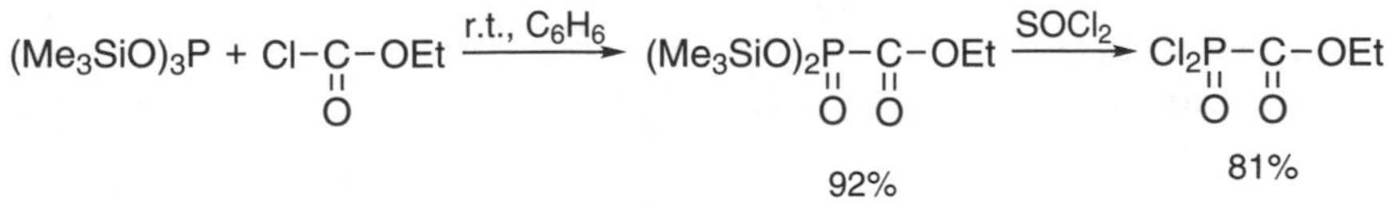

$$(Me_3SiO)_3P + Cl-\underset{\underset{O}{\|}}{C}-OEt \xrightarrow{r.t., C_6H_6} (Me_3SiO)_2\underset{\underset{O}{\|}}{P}-\underset{\underset{O}{\|}}{C}-OEt \xrightarrow{SOCl_2} Cl_2\underset{\underset{O}{\|}}{P}-\underset{\underset{O}{\|}}{C}-OEt$$

92%　　　　　　　81%

(8.2)

(Ethoxycarbonyl)phosphonic dichloride is currently used as a phosphorylating agent by sequential substitution of the two chlorides. It is coupled with primary alcohols in THF or DMF to produce phosphonochloridates with great selectivity in 60–80% yields. The intermediate acid chlorides are either hydrolyzed with excess 2 M NaOH to provide the disodium salts or subjected to aqueous workup followed by neutralization of the free acid with 1 M NaOH to form the monosodium salts.[13,17,20] Another traditional method of synthesizing diesters involves preparing the triester from phosphonic dichlorides and alcohols in the presence of a nitrogen base, followed by P-monodeesterification with NaI.[6,7]

A recent solution to this problem involves attenuating the reactivity of the alcohol by transforming it into the corresponding trimethylsilyl ether, whose reactivity toward the dichloride is lower. This means that selectivity for monosubstitution can be greatly improved.[21–23] The procedure has the additional advantage of ease of isolation of the highly reactive phosphonoformate di- or triester products because no nitrogen base is required. Thus, the first and second chlorine atoms in the phosphonic dichloride can be displaced selectively and sequentially to allow the synthesis of mixed diesters (Scheme 8.3).[21–23] Furthermore, cyclic phosphonoformate esters that cannot be made via the traditional method are accessible using trimethylsilyl ethers.

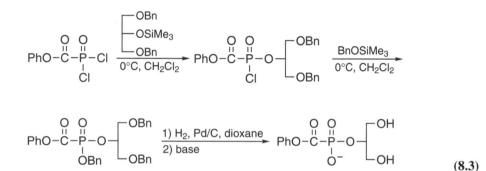

(8.3)

Two other strategies have been developed: coupling of a nucleophilic alkoxy- or aryloxyphosphonic acid silver salt with an alkyl halide[14,16,24] and coupling of a nucleophilic alcohol with a methyl (alkoxycarbonyl)phosphonochloridate.[7,25] This latter reagent is prepared by selective removal of one methyl ester group from dimethyl (alkoxycarbonyl)phosphonates by treatment with PCl$_5$ in refluxing CCl$_4$[7] or triphenylphosphine dichloride in CHCl$_3$ at room temperature.[26]

Free acylphosphonic acids are easily obtained by treatment of *bis*(trimethylsilyl) (alkoxycarbonyl)phosphonates with EtOH or by exposure to air for several hours.[10] Diethyl (alkoxycarbonyl)phosphonates are prepared by condensation of the corresponding alcohols with the diethyl (chlorocarbonyl)phosphonate in the presence of Et$_3$N or DMAP.[17]

In attempts to improve the delivery of phosphonoformate, the kinetics and products of hydrolysis of various triesters of phosphonoformic acid (PFA) in MeCN–H$_2$O mixtures at −1 < pH < 14 have been studied. Phosphonoformate triesters are hydrolyzed to give mixtures of phosphites, phosphonoformate esters, and free acids. The rates and product distribution are dependent on pH and ester leaving group abilities.[27–30]

Triethyl phosphonoformate is hydrolyzed in alkaline solution to give a crystalline trisodium salt as a hexahydrate (Scheme 8.4).[2,4,5,31–33] Treatment of *bis*(trimethylsilyl) (alkoxycarbonyl)phosphonates with NaOH has also been reported.[15] In acid solution the phosphonoformic acid decarboxylates rapidly.[4,34]

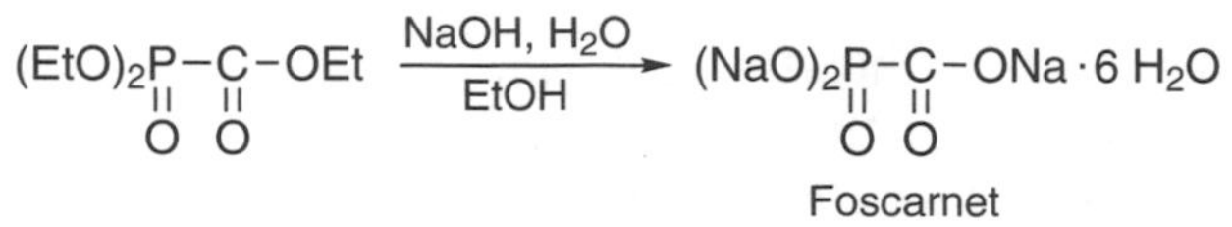

$$(EtO)_2\underset{O}{\overset{O}{P}}-\underset{O}{\overset{O}{C}}-OEt \quad \xrightarrow[EtOH]{NaOH, H_2O} \quad (NaO)_2\underset{O}{\overset{O}{P}}-\underset{O}{\overset{O}{C}}-ONa \cdot 6\,H_2O$$

Foscarnet

(8.4)

The trisodium salt of phosphonoformic acid (PFA), called Foscarnet, is an effective antiviral agent with potential application in acquired immunodeficiency syndrome (AIDS) therapy. PFA inhibits replication of HIV-1 in the laboratory and reduces viremia in patients with AIDS. It also inhibits DNA polymerase from cytomegalovirus (CMV), herpes simplex virus (HSV), and other DNA viruses and thus has been useful in the treatment of secondary viral infections in AIDS patients. Because PFA bears a structural resemblance to pyrophosphate, it can be viewed as a product analogue as opposed to a substrate analogue.

Simply passing ammonia into diethyl (methoxycarbonyl)phosphonate at 40°C results in rapid precipitation of the amide in good yield (69%).[2,3] Similarly, adding primary amines to diethyl (alkoxycarbonyl)phosphonates generates diethyl (N-alkylaminocarbonyl)phosphonates in moderate to good yields (48–80%).[3,35]

Benzonitrile isopropylide, generated from 2,2-dimethyl-3-phenyl-2H-azirine, undergoes a cycloaddition reaction to the ester carbonyl group of diethyl (ethoxycarbonyl)phosphonate to give the diethyl 5-ethoxy-2,2-dimethyl-4-phenyl-3-oxazolin-5-ylphosphonate in 96% yield.[36]

Reaction of diethyl (ethoxycarbonyl)phosphonate with Me$_3$SiCl in the presence of anhydrous NaI (2 eq) in dry MeCN[37] or Me$_3$SiBr (2–3 eq)[6,7,38,39] occurs rapidly and smoothly at room temperature to give bis(trimethylsilyl) (ethoxycarbonyl)phosphonate quantitatively. In all cases the reaction proceeds to completion with total selectivity for P–O dealkylation. The corresponding acids are readily prepared by treatment with a small excess of H$_2$O in dry acetone and isolation in vacuo (Scheme 8.5). Tetra-n-butylammonium iodide in THF[40] and NaI in DMF/THF/acetone[6,7] at room temperature induce selective monodealkylation of dialkyl (alkoxycarbonyl)phosphonates in good yields.

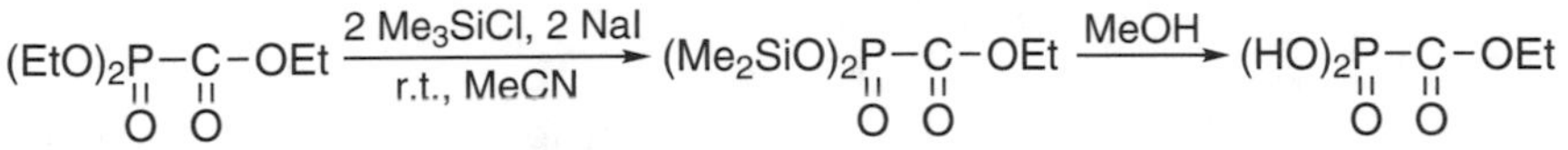

$$(EtO)_2\underset{O}{\overset{\parallel}{P}}-\underset{O}{\overset{\parallel}{C}}-OEt \xrightarrow[\text{r.t., MeCN}]{\text{2 Me}_3\text{SiCl, 2 NaI}} (Me_2SiO)_2\underset{O}{\overset{\parallel}{P}}-\underset{O}{\overset{\parallel}{C}}-OEt \xrightarrow{\text{MeOH}} (HO)_2\underset{O}{\overset{\parallel}{P}}-\underset{O}{\overset{\parallel}{C}}-OEt$$

(8.5)

8.1.2   Dialkyl 1-(Alkoxycarbonyl)alkylphosphonates

8.1.2.1   Michaelis–Arbuzov Reactions

Diethyl 1-(ethoxycarbonyl)methylphosphonate was obtained for the first time by A. E. Arbuzov and Dunin by the reaction of ethyl bromoacetate with triethyl phosphite.[1,41,42] Subsequently, the Arbuzov–Dunin method was successfully extended for the synthesis of the alkyl esters of differently substituted dialkyl phosphonoacetates (R = Me, Et, i-Pr, n-Pr, n-Bu, i-Pent, n-Pent, n-C$_6$H$_{13}$, n-C$_{10}$H$_{21}$, Ph)[43–51] and phosphonoacetates containing unsaturated substituents at either the phosphorus atom or the carboxy group.[52–57] The reaction can be used to prepare any dialkyl 1-(alkoxycarbonyl)methylphosphonates from chloro-[53,58–68] or bromoacetates[1,45,46,50,69–79] in fair to excellent yields (35–99%, Scheme 8.6, Table 8.1). Although chloro- or bromoacetates may be employed for the preparation of 1-(ethoxycarbonyl)methylphosphonates, the bromo derivatives, which give the better yields, are often preferred. The use of sensitive tris(trimethylsilyl) phosphite,[80,81] bis(trimethylsilyl) alkyl phosphite,[82] and trimethylsilyl bromoacetates[73] has been reported.

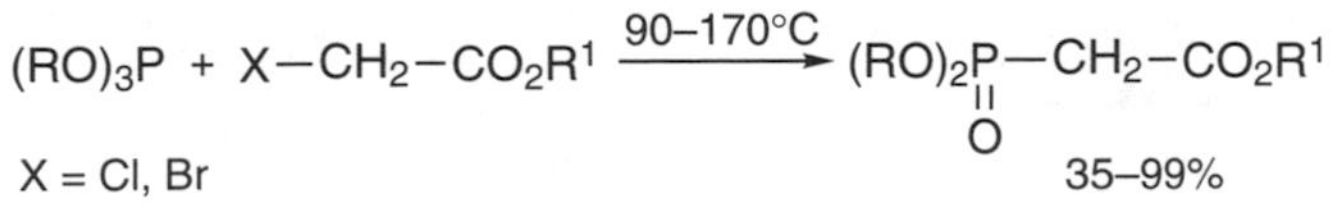

$$(RO)_3P + X-CH_2-CO_2R^1 \xrightarrow{90–170°C} (RO)_2\underset{O}{\overset{\parallel}{P}}-CH_2-CO_2R^1$$

X = Cl, Br   35–99%

(8.6)

Optically active phosphonoacetates are prepared from triethyl phosphite and the corresponding (−)-menthyl bromoacetate and (+)-bornyl bromoacetate in 87% and 86% yields, respectively.[83–85]

TABLE 8.1

	R	R^1	X	Temp. (°C)	Yield (%)[Ref.]
a	Me,[51] Et[63]	Me	Cl	160	43,[51] 95[63]
b	i-Pr	Me	Br	150	72[46]
c	Et	Et	Cl Br	128–140,[61,87] 140–152[58] 120,[1,60] 170[70,88]	67,[61] 76,[58] 85[87] —,[1] 76,[88] 80,[70] 95[60]
d	n-Pent, n-Hex, n-C$_{10}$H$_{21}$	Et	Br	185–205	37–64[50]
e	CH$_2$CO$_2$Et	Et	Br	160	63[45]
f	Et	t-Bu	Br	140	73[89]
g	CH$_2$CH=CH$_2$	Me, Et, i-Pr, n-Bu, i-Bu	Cl	110–120	—[54,90]
h	i-Pr	Me, n-Pr, (CH$_2$)$_2$Ph, (CH$_2$)$_3$Ph	Br	120–130	82–99[79]
i	Et	CH=CH$_2$	Cl,[52,54,90] Br[55]	110–120,[54,90] 157[52] 160–190[55]	—,[54,90] 54[52] 41[55]
j	CH$_2$CH=CH$_2$	CH$_2$CH=CH$_2$	Cl Br	110–120,[54,90] 160[53] reflux	—,[54,90] 51[53] 90[91]
k	Me, Et, i-Pr, n-Pr, n-Bu, i-Bu	CH$_2$CH=CH$_2$	Cl	110–120	—[54,90]
l	Me	(CH$_2$)$_8$CO$_2$Me	Br	110 (toluene)	78[75]
m	Me	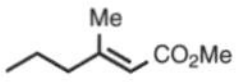 [structure: Me, CO$_2$Me]	Br	100	81[74]
n	Me	[structure: dithiane / dioxolane, OHC, Me]	Cl	90	67[66]
o	Et	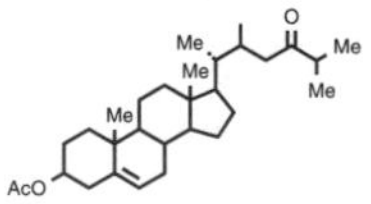 [steroid structure: AcO, Me, O, Me]	Br	130	88[72]
p	Me	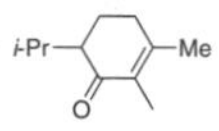 [structure: i-Pr, Me, O]	Br	110 (toluene)	68[92]
q	Me	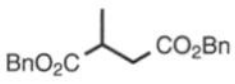 [structure: BnO$_2$C, CO$_2$Bn]	Br	reflux	60[93]
r	Me$_3$Si	Bn	Cl	165	83[82]

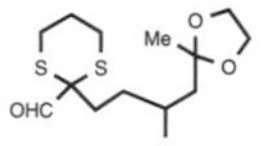

TABLE 8.1 (continued)

	R	R¹	X	Temp. (°C)	Yield (%)[Ref.]
s	Me	Bn	Br	80	91[71]
t	Me	Ph	Cl	140	58[62]
u	Et	Ph	Cl	100–150	35[59]
v	Et	4-Cl-C₆H₄	Cl	100–150	50[59]
w	Et	SiMe₃	Br	95	92[73]

A chiral cyclic phosphonate having C_2 symmetry has been prepared in 67% yield from (S)-1,1'-bi-2-naphtyl methyl phosphite and bromoacetate without any loss of optical purity.[86]

Synthesis of labeled phosphonoacetates proceeds on heating trialkyl phosphites with methyl[94] or ethyl[95] [1-^{14}C]bromoacetate, methyl [2-^{14}C]bromoacetate,[96–99] ethyl [2-^{14}C]bromoacetate,[100] ethyl [2-^{13}C]bromoacetate,[101] and methyl[102] or ethyl[103] [1-^{13}C]bromoacetate. Dialkyl 1-(methoxy-carbonyl)methyl[1,1-^{2}H$_2$]phosphonates are prepared from trialkyl phosphites and ethyl [2,2-^{2}H$_2$]bromoacetate[104] or by simple treatment of the nondeuterated phosphonoacetate with ^{2}H$_2$O in the presence of K_2CO_3.[105]

Michaelis–Arbuzov reaction of 5,5-dimethyl-2-methoxy-1,3,2-dioxaphosphorinane with chloroacetic esters at 145°C produces the expected 6-membered cyclic phosphonates in high yields.[106] In contrast, 4,5-dimethyl-2-methoxy-1,3,2-dioxaphospholane reacts sluggishly with ethyl chloroacetate and only ethyl bromoacetate leads to the desired five-membered cyclic phosphonates as mixtures of the racemic and meso isomers.[106] The 2-ethoxy-benzo[1,3,2]dioxaphosphole on reaction with chloro-, bromo-, or iodoacetate at 200–230°C leads to the 2-(ethoxycarbonyl)methyl-2-oxo-benzo[1,3,2]dioxaphosphole in variable yields (42–58%).[68]

Recent improvements in the standard synthetic procedure using haloacetates have featured the use of trialkyl phosphites and activated acetates in the presence of trimethylsilyl trifluoromethanesulfonate (TMSOTf) as activating reagent.[107–110] This reaction has proved to be of great value in constructing complex sugar phosphonates. For example, in the presence of TMSOTf, the Michaelis–Arbuzov reaction between protected ketose acetates and triethyl phosphite takes place in CH_2Cl_2 at room temperature to produce, after deprotection, the expected diethyl 1-(methoxycarbonyl)alkyl-phosphonates in high yields.[108–110]

Substituted haloacetates (R¹ and R² ≠ H) are exceptions to the rule that branched-chain halides are unreactive in the Michaelis–Arbuzov reaction. A large variety of easily available α-chloro or bromo esters[1,55,111–132] react with trialkyl phosphites at 160–190°C to produce α-substituted phosphonoacetic esters in fair to excellent yields (31–96%, Scheme 8.7, Table 8.2). Because secondary α-haloacetates are stable compounds that may be either easily prepared on laboratory scale or may be obtained commercially, the Michaelis–Arbuzov rearrangement appears as a reaction of special importance.

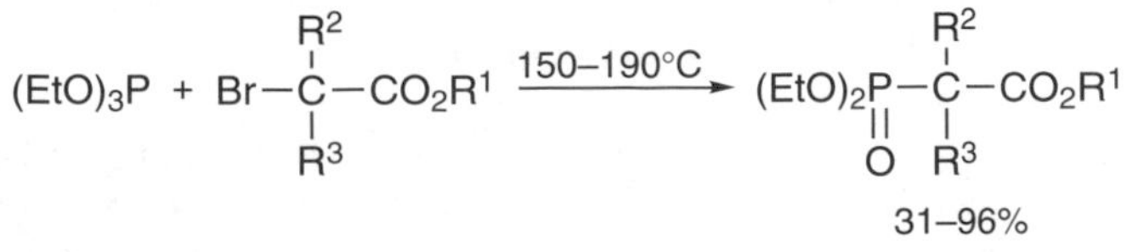

$$(EtO)_3P + Br-\underset{\underset{R^3}{|}}{\overset{\overset{R^2}{|}}{C}}-CO_2R^1 \xrightarrow{150–190°C} (EtO)_2\underset{\underset{R^3}{|}}{\overset{\overset{R^2}{|}}{\underset{O}{\overset{}{P}}}}-\overset{}{C}-CO_2R^1$$

31–96%

(8.7)

TABLE 8.2

	R	R¹	R²	R³	X	Temp. (°C)	Yield (%)[Ref.]
a	Me	Me	MeO	H	Cl	80	78[120]
b	Et	Me	n-Bu	H	Br	150	81[124]
c	i-Pr	Me	MeO	H	Cl	150	62[131]
d	Me, Et, (CH$_2$)$_2$Cl	Me	PhO, 2,4-Cl$_2$C$_6$H$_3$O	H	Br	—	55–87[125]
e	Et	Me,[122,134] Et[1,123,127]	Me	H	Br	reflux,[122,127] 160–190[1,123,134]	—,[1] 32,[134] 65,[127] 67,[123] 68[122]
f	Et	Et	Ph / Me / Et	H	Cl / I / Br	175 / 110–120 / 160–165,[1] reflux[127]	53[115] / —[1] / —,[1] 54[127]
g	i-Pr, i-Bu, i-Pent	Et	Me	H	Br	190	38–59[43,111]
h	Et,[121] i-Pr,[43,111] i-Bu[43,111]	Et	Me	Me	Br	160,[121] 210[43,111]	36–49,[43,111] 53[121]
i	i-Pr	Et	F	H	Br	145	74[129]
j	Et	Et	MeO	H	Cl	110	81[120]
k	Et	Et	EtO	H	Cl	150	83[118]
l	Et	Et	F	H	I	Δ / 30 Torr	74[135]
m	Et	Et	F, Ph, 4-Cl-C$_6$H$_4$, Et, n-Pr, i-Pr, n-Bu, n-Hex, n-C$_{10}$H$_{21}$, Ph	H	Br	145–150, reflux, 160	75–99[117,126,130] 81–83[136] 67–87[121]
n	Et	Et,[112,114] CH=CH$_2$[55]	(CH$_2$)$_n$CH$_3$ (n = 1–15)	H	Br	160–190	38–73,[55] 60–96[112,114]

Ethyl 2-bromoacetoacetate reacts with triethyl phosphite to give a mixture of diethyl 1-(ethoxycarbonyl)-2-oxopropylphosphonate (29%, Michaelis–Arbuzov product) and diethyl 1-methyl-2-(ethoxycarbonyl)vinyl phosphate (31%, Perkow product).[133] By contrast, the reaction of diethyl bromomalonate with trialkyl phosphites takes only one course, and the products formed at either temperature are the enol phosphates (Perkow product).[61]

Tetraethyl 1,4-*bis*(alkoxycarbonyl)-1,4-butylenediphosphonates have been isolated in 49–80% yields from the reaction of dialkyl 2,4-dibromoglutarates[137] or dialkyl 2,5-dibromoadipates[138,139] and triethyl phosphite at 160°C.

Synthetically useful phosphonylglycines are commonly obtained in high yields by Michaelis–Arbuzov reaction of trialkyl phosphites with α-halo-[128,140–152] or α-alkoxyglycine[153,154] esters (Scheme 8.8). Similarly, protected phosphonylsarcosines are prepared in high yields (64–87%) by a Michaelis–Arbuzov reaction from trialkyl phosphites and 4-bromo-3-methyl-2,2-*bis*(trifluoro-methyl)-1,3-oxazolidin-5-one in CH$_2$Cl$_2$ at room temperature. Unmasking of the latent aminocarboxylic group by acid hydrolysis produces the phosphonylsarcosine in 62% yield.[155]

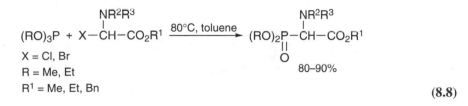

$$X = Cl, Br$$
$$R = Me, Et$$
$$R^1 = Me, Et, Bn$$

$$\text{80–90\%}$$

(8.8)

8.1.2.2 Michaelis–Becker Reactions

Nylen successfully used the Michaelis–Becker reaction for the preparation of phosphonoacetic esters.[2,156] Thus, diethyl 1-(ethoxycarbonyl)methylphosphonate can be obtained in satisfactory yield (50%) by the reaction of sodium diethyl phosphite with ethyl chloroacetate in refluxing Et$_2$O (Scheme 8.9).[2,156] The choice of the reaction solvent is crucial to the success of the reaction.[157] For example, diethyl 1-(ethoxycarbonyl)methylphosphonate is prepared in 56% yield in ligroin, 60% in Et$_2$O, and 95% in EtOH. A variety of phosphonoacetates are obtained in fair to good yields by this method.[47,158–167] Alkyl fluorochloroacetates react with sodium diethyl phosphite to give the corresponding α-fluorophosphonoacetates.[168]

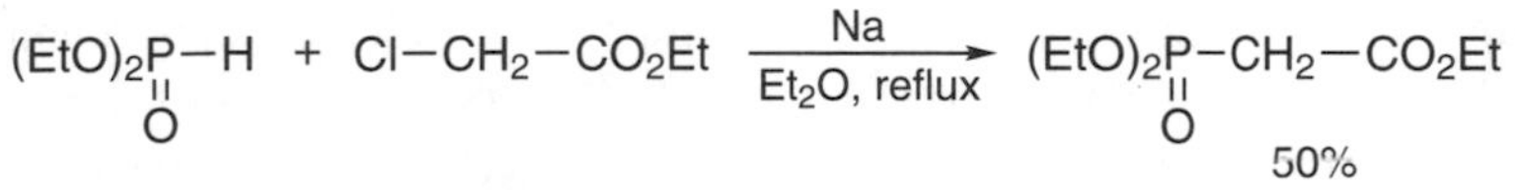

$$(EtO)_2\overset{\text{O}}{\underset{\text{||}}{P}}-H \ + \ Cl-CH_2-CO_2Et \ \xrightarrow[\text{Et}_2\text{O, reflux}]{\text{Na}} \ (EtO)_2\overset{\text{O}}{\underset{\text{||}}{P}}-CH_2-CO_2Et$$

50%

(8.9)

The Michaelis–Becker reaction has been investigated with α-chloro-, α-bromo-, and α-iodoacetates. In contrast to α-chloroacetates, on treatment with sodium diethyl phosphite, α-bromo- and α-iodoacetates undergo reduction to the corresponding acetates.[167] Low yields (18–30%) of unsubstituted and α-alkyl substituted *bis*(2,2,2-trifluoroethyl) 1-(ethoxycarbonyl)methylphosphonates, prepared from sodium *bis*(2,2,2-trifluoroethyl) phosphite and the corresponding α-bromo esters, have recently been reported.[169]

Reaction of dialkyl phosphites with chloroacetates can be conveniently and efficiently carried out under solid–liquid[170] or liquid–liquid phase transfer catalysis.[171] When CH$_2$Cl$_2$ solutions of dialkyl phosphites and chloroacetates containing a catalytic amount of triethylbenzylammonium chloride (TEBA) are treated at 0–5°C with 50% aqueous NaOH, dialkyl 1-(alkoxycarbonyl)methylphosphonates are isolated in fair to good yields (55–91%). However, changes of the reaction conditions leads to the formation of complex products and low yields of the expected products.[171] Recently, the nucleophilic substitution of ethyl bromoacetate with diphenyl phosphite has been examined in the presence of base (Et$_3$N, *i*-Pr$_2$EtN, Py, NaH).[172,173] Thus, when diphenyl phosphite is treated with ethyl bromoacetate and Et$_3$N in C$_6$H$_6$, diphenyl 1-(ethoxycarbonyl)methylphosphonate is isolated in low yield (19%). Although extension of the reaction time offered no advantage, changing the C$_6$H$_6$ to CH$_2$Cl$_2$ improves the yield (53%).[172,173]

Several innovations have significantly extended the scope and synthetic utility of the classical Michaelis–Becker phosphonoacetate preparation. For example, the coupling of the Michaelis–Becker and Horner–Wadsworth–Emmons reactions for the synthesis of α-substituted acrylic acids represents a useful modification.[174] According to Scheme 8.10, Michaelis–Becker alkylation of a dialkyl phosphite with a haloacetic acid in the presence of 3 eq of a base (one to neutralize the carboxyl group, one to form the phosphite conjugate base, and one to deprotonate the initially formed alkylation product) leads to the phosphoryl-stabilized anion directly. Treatment of the anion generated *in situ* with aromatic carbonyl substrates afford acrylic acids by a Horner–Wadsworth–Emmons reaction.[174]

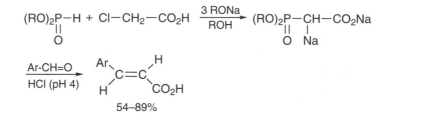

$$(8.10)$$

This simple procedure constitutes a very convenient, low-cost, and efficient synthesis of cinnamic acids. For example, reaction of a 25% MeONa in MeOH solution containing dimethyl phosphite with chloroacetic acid and then with aryl aldehydes affords (*E*)-cinnamic acids in the range of 54–89%.[174] This procedure is also applicable to the carboxyvinylation of aldehydes or ketones with 2-bromoalkanoic acids and diethyl phosphite using NaH in DME as the base/solvent system. One of the major advantages of the overall procedure is that it avoids the need to synthesize precursor reagents.

8.1.2.3 Alkoxycarbonylation of Phosphonate Carbanions

Of the numerous methods available for the synthesis of dialkyl 1-(alkoxycarbonyl)alkylphosphonates, one of the most convenient entails the nucleophilic attack of lithiated phosphonate carbanions on dialkyl carbonates[175–179] or alkyl chloroformates[178,180–188] and eventually of difluoromethylphosphonyl zinc bromide on chloroformates.[189–191] This especially useful methodology may be used for the preparation in high yields (74–87%, Scheme 8.11) of dialkyl 1-(alkoxycarbonyl)alkylphosphonates bearing a variety of α-substituents R^1 not available by the classical Michaelis–Arbuzov method. By merely treating the intermediate lithiated phosphonoenolate with carbonyl compounds, α-substituted α,β-unsaturated esters may also be obtained directly in fair to good yields (47–95%, Scheme 8.11).[177,178] In addition to this methodology, a procedure for the direct preparation of α-substituted acrylic acids exploiting BOC-F or DIBOC as alkoxycarbonylation agent has been developed.[178] These two procedures are complementary with respect to the reactivity of the lithiated enolate toward different carbonyl compounds.

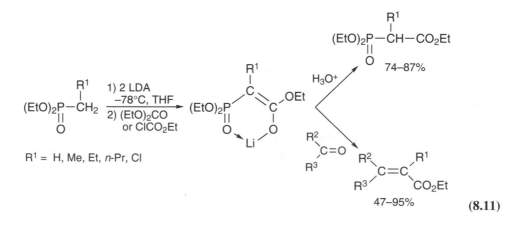

$$(8.11)$$

For example, this attractive procedure has been applied to the synthesis of a large variety of synthetically useful diethyl 1-fluoro-1-(alkoxycarbonyl)methylphosphonates from diethyl fluorodibromophosphonate and alkyl chloroformates in excellent yields (80–91%, Scheme 8.12).[192]

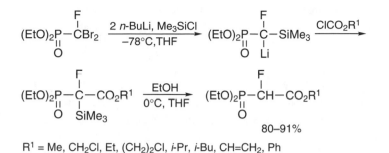

$$R^1 = Me, CH_2Cl, Et, (CH_2)_2Cl, i\text{-}Pr, i\text{-}Bu, CH=CH_2, Ph \tag{8.12}$$

8.1.2.4 Phosphonylation of α-Lithiated Carboxylic Esters

Preparation of dialkyl 1-(alkoxycarbonyl)alkylphosphonates through the nucleophilic attack of α-carbanions of α-branched and straight-chain acid esters on phosphoryl chlorides is a well-established reaction, but it is not sufficiently straightforward to be employed regularly. The first attempts were made by treating the Mg derivative of phenylacetic acid and the Ca derivative of ethyl phenylacetate with diethyl chlorophosphate in Et$_2$O at reflux.[193] In these conditions, diethyl 1-(ethoxycarbonyl)-1-phenylmethylphosphonate was isolated in 29% yield.[193] Formation of dialkyl 1-(alkoxycarbonyl)alkylphosphonates can be improved by using lithiated α-carbanions of esters prepared from LDA at low temperature in THF. A large variety of carbanions have been employed to test the general applicability of this electrophilic phosphonylation. Thus, in the presence of LDA (1 eq), methyl hexanoate and ethyl acetate fail to react with dialkyl chlorophosphates at low temperature to give the expected phosphonates,[194] whereas methyl and ethyl isobutyrates give good yields (62–80%) of the expected dialkyl 1-(alkoxycarbonyl)-1-methylethylphosphonates.[194] In the light of further experiments applying this methodology, it appears that the presence of LDA in excess (2 eq) is crucial for the success of the rearrangement of the initially formed enol phosphate to form the expected phosphonate. When additional LDA is added to the reaction mixture, the rearrangement of vinyl phosphate to phosphonate takes place in 12–41% yields.[195–197]

The impact of steric effects in the carboxylic ester has been examined with both isopropyl and *tert*-butyl propionate,[196] and it appears that there is no significant difference in yields over that observed with ethyl propionate.

Recent investigations in this standard synthetic procedure have featured the use of LDA (3 eq) or *t*-BuLi in the presence of HMPA at low temperature for promotion of the 1,2-phosphoryl migration conducive to formation of diethyl 1-(alkoxycarbonyl)methylphosphonates in satisfactory yields (58–74%, Scheme 8.13).[198]

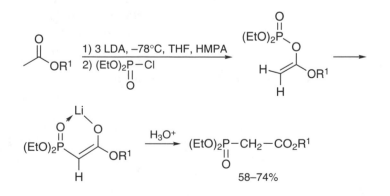

$$\tag{8.13}$$

A more recent technique is based on the property of diethyl chlorophosphite to react with ester enolates to give mainly *C*-phosphirylation instead of *O*-phosphirylation, thus avoiding the rearrangement step. Thus, the reaction of ester enolates, prepared from LDA (1 eq) in Et$_2$O at low temperature, with diethyl chlorophosphite proceeds with high regioselectivity to give mainly the phosphonites resulting from a C–P bond formation. An *in situ* oxidation of the phosphonite to phosphonate is accomplished by leaving the product open to air.[199–201] With ethyl acetate and propionate, isolated yields of the corresponding phosphonates are high (93% and 79%). With ethyl butyrate and isovalerate, the expected phosphonates are obtained in lower yields (68% and 63%), and with *tert*-butyl acetate the phosphonate is obtained in 46% yield. In the earlier cases of ethyl acetate, propionate, and butyrate, the C–P/O–P ratio is very high.[201] The same C–P versus O–P selectivity had been observed in the two-step preparation of diethyl 1,1-*bis*(ethoxycarbonyl)methylphosphonate in 60% yield from reaction of diethyl chlorophosphite with diethyl malonate followed by air oxidation.[202]

8.1.2.5 Phosphonylation by Michael Addition

Michael addition of 1-(alkoxycarbonyl)methylphosphonate anions to unsaturated compounds provides a methodology for the elaboration of new reagents and also for the preparation of phosphonylated heterocycles. Thus, in the presence of basic catalysts, diethyl 1-(ethoxycarbonyl)alkyl- and 1-(ethoxycarbonyl)methylphosphonates add to α,β-unsaturated esters and nitriles.[203–207] Addition of diethyl 1-(ethoxycarbonyl)methylphosphonates under basic conditions to methyl or ethyl acrylates, acrylonitrile, and benzalacetophenone occurs readily and gives rise to products of addition to one and two molecules of the unsaturated compound (Scheme 8.14).[203–207] Reaction of α-substituted phosphonoacetates with acrylates is less vigorous, and attempts at addition to crotonic and methacrylic esters fail.[203–206,208]

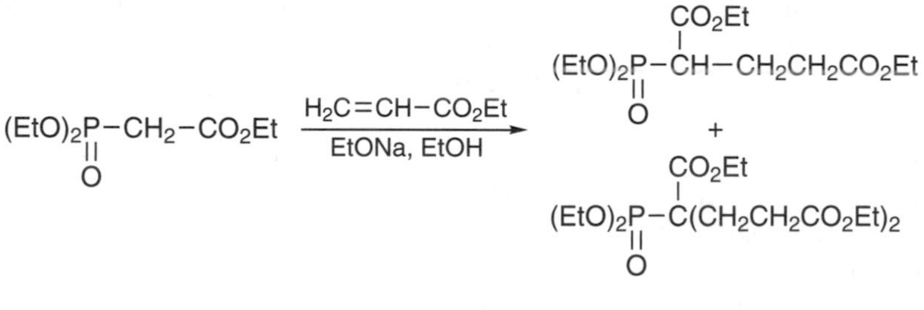

$$(EtO)_2\underset{\underset{O}{\|}}{P}-CH_2-CO_2Et \quad \xrightarrow[\text{EtONa, EtOH}]{H_2C=CH-CO_2Et} \quad \begin{array}{c} CO_2Et \\ | \\ (EtO)_2\underset{\underset{O}{\|}}{P}-CH-CH_2CH_2CO_2Et \\ + \\ CO_2Et \\ | \\ (EtO)_2\underset{\underset{O}{\|}}{P}-C(CH_2CH_2CO_2Et)_2 \end{array}$$

$$(8.14)$$

Addition of phosphonoacetates to nitroolefins is of special interest because the obtained nitroalkylphosphonates are the precursors of the corresponding aminoalkylphosphonates. For example, diethyl 1-(ethoxycarbonyl)methylphosphonate adds easily to 1-nitro-3-methyl-1-butene and to 1-nitropropene in absolute EtOH in the presence of an equimolar amount of EtONa to give the corresponding nitroalkylphosphonates in 54% and 48% yields, respectively.[209,210] The addition of phosphonoacetates to other unsaturated compounds such as ethyl vinyl sulfone,[209] divinyl sulfone,[211] *para*-*bis*(β-nitrovinyl)benzene,[211] and β-substituted aliphatic enones[212] and to esters of maleic,[213] cinnamic,[204] crotonic,[214] and propynylphosphonic[215] acids has been achieved. The esters of α,β-unsaturated carboxylic acids can be arranged in the following sequence in terms of their reactivity in reactions involving the addition of phosphonoacetic ester: acrylates > crotonates > methacrylates > cinnamates.

Similarly, when a nucleophile is added under basic conditions to a solution of dimethyl 2-(tosylmethyl)fumarate in THF at room temperature, a mixture of S$_N$2′/S$_N$2 products is obtained.[216] Thus, reaction of sodium diethyl 1-(ethoxycarbonyl)methylphosphonate with this sulfone affords exclusively the S$_N$2′ product in 58% yield. This reaction may be regarded as a Michael addition–tosyl elimination process.[216]

In the presence of a catalytic amount of MeONa, the reaction of conjugated azoalkenes with diethyl 1-methyl-1-(ethoxycarbonyl)methylphosphonate proceeds at room temperature in THF by 1,4-addition of the phosphonate carbanion to the azo–ene system of conjugated azoalkenes.[217] The adduct can be isolated and then submitted to internal ring closure in MeOH to produce 3-phosphonylated 1-aminopyrrol-2-ones. The same reaction with diethyl 1-(ethoxycarbonyl)methylphosphonate occurs easily at room temperature in THF, but all attempts to cyclize phosphonohydrazones into related pyrrole drivatives failed.[217]

8.1.2.6 α-Functionalization of 1-(Alkoxycarbonyl)methylphosphonates

8.1.2.6.1 Knoevenagel Reactions

Condensation of diethyl 1-(ethoxycarbonyl)methylphosphonate with benzaldehyde on heating at 160–170°C with Ac_2O was reported by Pudovik and Lebedev in 1953 to give a 37–40% yield of diethyl 1-ethoxycarbonyl-2-phenylvinylphosphonate.[218] In a reinvestigation of this work, much better yields (63–70%) of diethyl 1-ethoxycarbonyl-2-phenylvinylphosphonate were obtained by using milder reaction conditions (refluxing C_6H_6, piperidine, and AcOH or benzoic acid as catalysts) and longer reaction time.[166,219–223] The reaction has also been decomposed in two independent steps: condensation in refluxing MeOH containing piperidine and then dehydration in refluxing C_6H_6 containing catalytic amounts of TsOH.[224,225] Attempts were made to condense ketones, but they do not react under these conditions, and no products other than starting materials were isolated.

Several innovations have significantly extended the scope and synthetic utility of the classical Knoevenagel reaction. Thus, in the presence of $TiCl_4$ and N-methylmorpholine in THF, diethyl 1-(ethoxycarbonyl)methylphosphonate undergoes condensation with aliphatic and aromatic aldehydes and aromatic ketones to give the corresponding alkylidene or arylidene phosphonoacetates having the thermodynamically more stable (E)-configuration in 55–85% for aliphatic aldehydes and 87–96% for aromatic aldehydes (Scheme 8.15).[85,226–228]

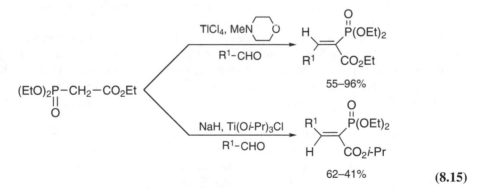

(8.15)

An interesting variant of the Knoevenagel reaction describing a kinetically controlled titanium-mediated condensation leading to the opposite stereoselectivity has been reported.[229] Thus, addition of an aldehyde to titanated diethyl 1-(ethoxycarbonyl)methylphosphonate, prepared from the sodium diethyl 1-(ethoxycarbonyl)methylphosphonate and $Ti(Oi\text{-}Pr)_3Cl$ in THF at room temperature, results in Knoevenagel condensation with preferrential formation of the thermodynamically less stable (Z)-isomer in 62–74% yield (Scheme 8.15).[229] Under these reaction conditions, chemoselective transesterification at the carbonyl function is observed when the ethoxy groups at phosphorus are not exchanged.

8.1.2.6.2 Alkylation

A. E. Arbuzov and Dunin[41,158,230] were the first to show that the reaction of alkyl halides (MeI, BnCl) with the sodium or potassium salts of diethyl 1-(ethoxycarbonyl)methylphosphonate results in the formation of diethyl 1-(ethoxycarbonyl)alkylphosphonates. By this procedure, the combinations of magnesium-1-(ethoxycarbonyl)methylphosphonate, prepared with magnesium in liquid ammonia,[231] and lithium, sodium, or potassium metal-1-(ethoxycarbonyl)methylphosphonate in Et$_2$O,[165,203,204,232–235] xylene at reflux,[236–238] or C$_6$H$_6$[162,239–241] give low to fair yields of α-substituted 1-(ethoxycarbonyl)methylphosphonates.

The best results are obtained with the use of alkali metal hydrides (NaH, KH) in THF, DME, or DMF. The reaction works well in THF or DME with activated halides such as ethyl bromoacetate,[242] *tert*-butyl bromoacetate,[243–245] ethyl 2-bromobutyrate,[246] ethyl 4-bromobutyrate,[247] (iodomethyl)trimethylstannane,[248] (iodomethyl)trimethylsilane,[249–259] benzoyl bromide,[260] benzyl bromide,[261–263] farnesyl bromide,[264] alkyl 4-bromocrotonates,[265] 1-(bromomethyl)naphtalene,[266] and *N*-bromomethylphthalimide[136] but gives poor results with primary alkyl halides.[267] Primary and secondary alkyl halides, bromides and iodides (Scheme 8.16), react satisfactorily in DMF[76,268–278] or DMSO,[279–285] although bulky electrophiles give poor results.[286] In DMSO the expected product is frequently contaminated by the dialkylation product.[287,288]

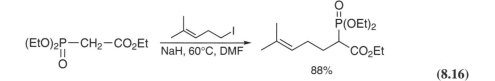

$$ (8.16) $$

The potassium salt of *bis*(2,2,2-trifluoroethyl) 1-(methoxycarbonyl)methylphosphonate is alkylated with homogeranyl iodide in THF–HMPA in the presence of 18-crown-6 (1 eq).[289] The use of Et$_2$O[290] or toluene at reflux[291] gives less satisfactory results. Treatment of diethyl 1-(ethoxycarbonyl)methylphosphonate with [^{14}C]methyl iodide[292] and allyl bromide[48] using EtONa in absolute EtOH produces diethyl 1-[^{14}C]methyl and 1-allyl-1-(ethoxycarbonyl)methylphosphonates in good yields. The tandem Michael addition–alkylation of 1-(ethoxycarbonyl)vinylphosphonate has been reported with methyl cuprate, but the alkylation is complete only in the presence of Lewis bases like HMPA or TMU.[293]

Alkylation of sodium 1-(alkoxycarbonyl)methylphosphonates proceeds equally with acetates in THF from low to room temperature[294] or in DME at reflux.[295] The asymmetric allylic alkylation of the sodium diethyl 1-(ethoxycarbonyl)methylphosphonate with 3-acetoxy-1,3-diphenyl-1-propene and cyclic allylic acetates in the presence of a chiral palladium catalyst, prepared from chiral phosphine and palladium acetate, in THF at room temperature proceeds in good yields (44–88%) and high ee's.[296–298]

When the alkylation of the sodium salt of 1-(alkoxycarbonyl)methylphosphonates is realized with tosylates in THF at room temperature, the yields are roughly comparable to those obtained from iodides in DMSO.[279] The use of phosphonium salts gives moderate to good yields (40–82%) of alkylation products.[299] A recently introduced procedure for the high-yielding (95%) alkylation of the lithium salt of dimethyl 1-(methoxycarbonyl)methylphosphonate in THF via the triflate derivative is especially attractive.[300] The reaction of diethyl 1-(ethoxycarbonyl)methylphosphonate with butadienemonoxide in the presence of Pd(PPh$_3$)$_4$ in THF gives the allylic alcohol in a high regio- and stereoselective manner in good yield (85%).[301]

Intramolecular carbometallation reaction of diethyl 1-(ethoxycarbonyl)-5-hexynyl- and -5-hexenylphosphonates has been examined under a variety of conditions. Thus, in the presence of TiCl$_4$, Et$_3$N, and I$_2$, they undergo iodocarbocyclization reactions.[302,303] With a catalytic amount of NaH or *n*-BuLi in THF at reflux, the hydrocarbocyclization reaction proceeds through a proton-transfer mechanism to give the corresponding methylenecyclopentane derivatives in satisfactory yield (67%).[304] In the presence of SnCl$_4$ and Et$_3$N, the intramolecular carbostannation proceeds in good yields to provide, after reaction of the resulting Sn intermediates with electrophiles, functionalized cyclopentane and cyclohexane derivatives.[305]

In the route to amino phosphonic acids, dialkyl 1-(alkoxycarbonyl)methylphosphonates, in the presence of NaH in THF or DME, are alkylated with activated aziridines,[306] 3-alkoxyazetidinium salts,[307,308] and cyclic sulfate derived from (S)-1,2-propanediol.[309]

With dibromoalkanes, the diethyl 1,1-cycloalkyl-1-(ethoxycarbonyl)methylphosphonates are readily obtained by two successive alkylations in fair to good yields using NaH in THF–DMSO (Scheme 8.17),[310–313] K_2CO_3 in MeCN–DMSO,[313–315] NaOH (50%)/benzyl tributylammonium bromide[316] or EtONa/EtOH, Et_2O.[317] By contrast, reaction of the potassium salt of diethyl 1-(ethoxycarbonyl)methylphosphonate with 1,3-dibromopropane leads to the formation of 1,2-oxaphosphorinane, resulting from alkylation at the methylene carbon and at phosphorus.[318]

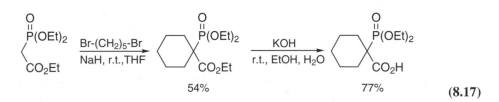

$$(8.17)$$

The ion-pair extraction method (phase-transfer catalysis) was rather disappointing; it can be used for the specific monoalkylation of diethyl 1-(ethoxycarbonyl)methylphosphonate but has low yield (30%).[319]

8.1.2.6.3 Halogenation

The strategic generation of 1-fluoro-1-(ethoxycarbonyl)methylphosphonate is often used to prepare fluoro acrylates by a Horner–Wadsworth–Emmons reaction or to enhance biologic activity in anticipation that fluoro substitution improves bioisosterism with the parent phosphates. Thus, treatment of the potassium enolate of diethyl 1,2-bis(ethoxycarbonyl)-2-oxoethylphosphonate with perchloryl fluoride ($FClO_3$) in EtOH at −10°C gives diethyl 1-fluoro-1-(ethoxycarbonyl)methylphosphonate in 77% yield.[320] Based on this example, the highly reactive and hazardous $FClO_3$ has been successfully replaced by F_2 to produce selectively the expected diethyl 1-fluoro-1-(ethoxycarbonyl)methylphosphonate in excellent yield (97%, Scheme 8.18).[321] In contrast, treatment of unprotected 1-(ethoxycarbonyl)methylphosphonate with diluted F_2 in MeCN in the presence of Cu(II) nitrate leads to a mixture of monofluoro- and difluorophosphonoacetate in low yields.[322,323] When the sodium 1-(ethoxycarbonyl)methylphosphonate is treated with F_2 in MeCN, fair yields of the desired 1-fluoro-1-(ethoxycarbonyl)methylphosphonate (35%) are obtained but, once again, are contaminated with the difluoro compound (15%).[322] It presently appears that the use of protected phosphonates may be useful for the selective preparation of monofluorophosphonates by anionic procedures.

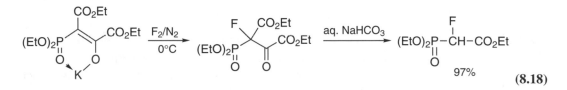

$$(8.18)$$

The N–F electrophilic fluorinating agents that are safe and easy to handle without special equipment have been introduced to replace highly reactive, corrosive, and toxic materials such as F_2 and $FClO_3$. For example, low-temperature treatment of the sodium diethyl 1-(ethoxycarbonyl)methylphosphonate with N-fluoro-o-benzenedisulfonimide (NFOBS) in THF produces selectively the diethyl 1-fluoro-1-(ethoxycarbonyl)methylphosphonate in good yield (78%). The reaction mixture is exceptionally clean because the disulfonimide byproduct, being highly water soluble, is removed during the aqueous workup.[324] Two other fluorinating agents, N-fluorobenzenesulfonimide

(NFBS) and 1-chloromethyl-4-fluoro-1,4-diazabicyclo[2.2.2]octane *bis*(tetrafluoroborate) (F-TEDA-BF$_4$, Selectfluor™), have been investigated.[325] When the sodium salt of diethyl 1-(ethoxy-carbonyl)methylphosphonate is allowed to react with F-TEDA-BF$_4$ in a THF–DMF solvent mixture at room temperature, only poor yields of the desired diethyl 1-fluoro-1-(ethoxycarbonyl)meth-ylphosphonate (17%) are obtained. One root of the problem in this procedure is the poor solubility of the F-TEDA-BF$_4$ in THF and even the use of DMF cosolvent fails to solvate the fluorinating agent. The uncharged NFBS reagent, which is more soluble in THF than F-TEDA–BF$_4$, has been introduced for monofluorination, but the amount of isolated diethyl 1-fluoro-1-(ethoxycarbo-nyl)methylphosphonate is negligible.[325] In this context the alkoxycarbonylation of diethyl 1-lithio-1-fluoro-1-(trimethylsilyl)methylphosphonate appears to be a valuable route to diethyl 1-fluoro-1-(ethoxycarbonyl)methylphosphonate.[184,192] The F-TEDA–BF$_4$ has been used in THF/DMF for flu-orination of diethyl 1-fluoro-1-(ethoxycarbonyl)methylphosphonate in good chemical yields.[326,327]

In a one-pot synthesis of α-halo-α,β-unsaturated carboxylates from dialkyl 1-(ethoxycarbo-nyl)methylphosphonate, the intermediate 1-chloro- and 1-bromo-1-(ethoxycarbonyl)methylphos-phonates were prepared by halogenation of sodium salt of dialkyl 1-(ethoxycarbonyl)methylphos-phonates with NCS or NBS in THF at room temperature.[328] Cupric chloride and cupric bromide have proved to be efficient reagents for the chlorination or bromination of sodium diethyl 1-(ethoxycarbonyl)ethylphosphonate in DMSO at room temperature in high yields (91%).[329]

Diethyl 1-(ethoxycarbonyl)methylphosphonate, on treatment with freshly prepared sodium hypochlorite or hypobromite solution at 0°C[330–333] or with chlorine under UV irradiation at 28–30°C[334] and at 50–70°C in CCl$_4$[335] or with SO$_2$Cl$_2$ at 40–70°C in CHCl$_3$,[334] is converted, respectively, into the dichloro or dibromo derivatives in excellent yields. A subsequent selective reduction converts the dichloro products into the monohalo esters. Sodium sulfite is highly selective in the reduction of diethyl 1,1-dichloro-1-(ethoxycarbonyl)methylphosphonate to diethyl 1-chloro-1-(ethoxycarbonyl)methylphosphonate in high yield (90–98%).[331,336] However, the dibromo ester is completely debrominated on treatment with sodium sulfite, so this method is not useful for the preparation of monobromo ester.[331,336] Reduction of the dibromoester with SnCl$_2$ (1 eq) in EtOH/H$_2$O gives the monobromo ester in 70–85% isolated yields.[331–333]

The diethyl 1-iodo-1-(ethoxycarbonyl)methylphosphonate is prepared by addition of iodine to the potassium salt of diethyl 1-(ethoxycarbonyl)methylphosphonate prepared from K$_2$CO$_3$ and TEBAC in refluxing C$_6$H$_6$.[337] The use of K$_2$CO$_3$ appears important because strong bases such as solid *t*-BuOK and MeONa or their DMF solutions promote undesired side reactions. Preparation of diethyl 1-iodo-1-(ethoxycarbonyl)methylphosphonate has also been reported by iodination of the sodium salt of diethyl 1-(ethoxycarbonyl)methylphosphonate with iodine in DME at 0°C[338] or 25–40°C.[339]

8.1.2.6.4 Arylation

When the reaction of iodobenzene with sodium diethyl 1-(ethoxycarbonyl)methylphosphonate is carried out in HMPA in the presence of CuI, diethyl 1-phenyl-1-(ethoxycarbonyl)methylphospho-nate is isolated in variable yields according to reaction conditions (Scheme 8.19).[340] The highest yield (84%) is obtained when 2 eq of sodium salt of diethyl 1-(ethoxycarbonyl)methylphosphonate and CuI relative to iodobenzene are used. The same reaction can be carried out in DMF instead of HMPT without lowering the yields. Bromobenzene in a similar coupling reaction provides a much lower yield of phosphonate (27%).[340]

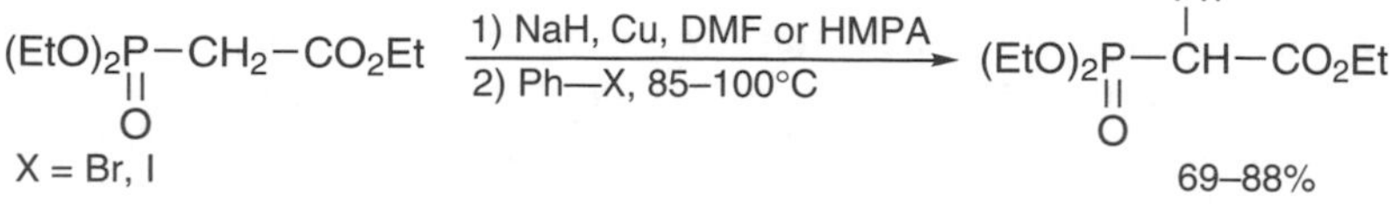

$$(EtO)_2P-CH_2-CO_2Et \quad \xrightarrow[\text{2) Ph—X, 85–100°C}]{\text{1) NaH, Cu, DMF or HMPA}} \quad (EtO)_2P-\overset{\overset{\text{Ph}}{|}}{\underset{\underset{O}{||}}{C}}H-CO_2Et$$

X = Br, I   69–88%

(8.19)

Nucleophilic aromatic substitution of perhaloarenes such as pentafluoropyridine and octafluorotoluene by anions of diethyl 1-(ethoxycarbonyl)methylphosphonate has been reported. The reaction is carried out with satisfactory yields (60–85%) in solid–liquid system using 2–3 eq CsF or NaH as bases or 3 eq K_2CO_3 in the presence of TEBAC as phase-transfer agent.[341–343]

8.1.2.6.5 Acylation

It has not proved possible to obtain C-alkylated products from diethyl 1-(ethoxycarbonyl)methylphosphonate by the action of acetyl chloride on the alkali metal salts.[344] However, when the ethoxymagnesium salt of diethyl 1-(ethoxycarbonyl)methylphosphonate is employed, diethyl 1-ethoxycarbonyl-1-acetylmethylphosphonate is isolated in a good yield (73%, Scheme 8.20).[344] Similarly, diethyl 1-ethoxycarbonyl-1-benzoylmethylphosphonate has been prepared in 56% yield from diethyl 1-(ethoxycarbonyl)methylphosphonate and benzoyl chloride.[345,346] The diethyl 1-lithio-1-fluoro-1-(ethoxycarbonyl)methylphosphonate generated with n-BuLi in THF reacts with acid chlorides,[347] fluorine-substituted acid chlorides,[348] and oxalylchloride[349] to form the corresponding acylated phosphonates. By contrast, phosphorylation of the sodium diethyl 1-(ethoxycarbonyl)methylphosphonate leads to the O-phosphorylation product.[350] The synthetically useful potassium enolate of diethyl 1,2-bis(ethoxycarbonyl)-2-oxoethylphosphonate is prepared by reaction of the potassium salt of diethyl 1-(ethoxycarbonyl)methylphosphonate with diethyl oxalate.[320]

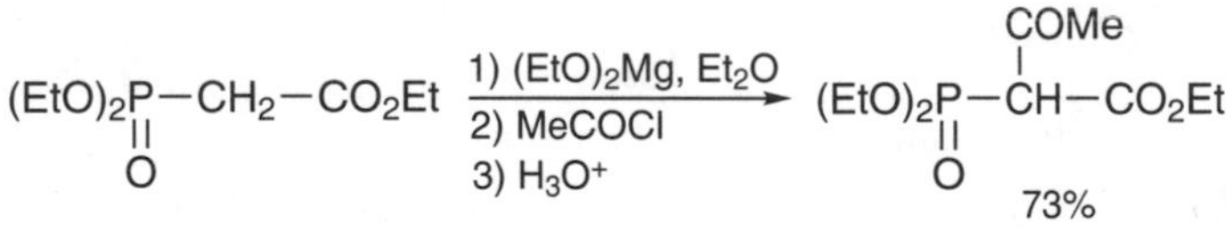

$$(EtO)_2\underset{\underset{O}{\|}}{P}-CH_2-CO_2Et \xrightarrow[\substack{2)\ MeCOCl \\ 3)\ H_3O^+}]{1)\ (EtO)_2Mg,\ Et_2O} (EtO)_2\underset{\underset{O}{\|}}{P}-\underset{\underset{}{\overset{\overset{COMe}{|}}{}}}{C}H-CO_2Et$$

$$73\%$$

$$(8.20)$$

The diethyl 1,4-bis(ethoxycarbonyl)butylphosphonate, on treatment with EtONa in refluxing C_6H_6, undergoes internal acylation to give the diethyl 1-ethoxycarbonyl-2-oxocyclopentylphosphonate in 63% yield.[351] In the same conditions the diethyl 4-(ethoxycarbonyl)butylphosphonate gives only a 6% yield of diethyl 2-oxocyclopentylphosphonate. Similarly, the methyl 3-bromopropyl 1-(methoxycarbonyl)methylphosphonate, on treatment with NaH in DME, undergoes an internal acylation to give the 3-(methoxycarbonyl)-2-methoxy-2-oxo-1,2-oxaphosphorinane in 78% yield.[352]

Several innovations have significantly extended the scope and synthetic utility of this classical acylation reaction. It has been shown that the acylated phosphonoacetates prepared from diethyl 1-(ethoxycarbonyl)methylphosphonate and aromatic or aliphatic carboxylic acid chlorides using $Mg(OEt)_2$ in THF[353] or $MgCl_2$ and Et_3N in CH_2Cl_2 or toluene[354,355] undergo an acid-catalyzed decarboxylation to give diethyl 2-oxoalkylphosphonates in good yields (60–90%) from aroyl chlorides and in fair yields (26–70%) from acyl chlorides. The solid support version of this reaction has also been reported.[356] Similarly, acylation of diethyl 1-fluoro-1-(hydroxycarbonyl)methylphosphonate with carboxylic acid chlorides in the presence of n-BuLi in THF at low temperature followed by decarboxylation appears to be an expeditious route for the synthesis of diethyl 1-fluoro-2-oxoalkylphosphonates in good yields (66–78%, Scheme 8.21).[357]

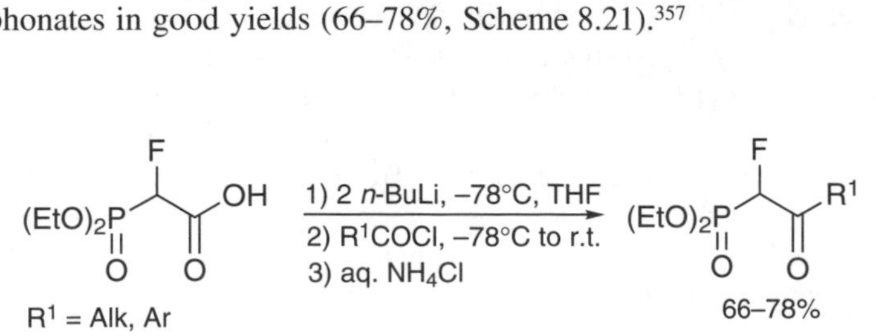

$$(EtO)_2\underset{\underset{O}{\|}}{P}\overset{\overset{F}{|}}{\underset{\underset{O}{\|}}{C}}OH \xrightarrow[\substack{2)\ R^1COCl,\ -78°C\ to\ r.t. \\ 3)\ aq.\ NH_4Cl}]{1)\ 2\ n\text{-}BuLi,\ -78°C,\ THF} (EtO)_2\underset{\underset{O}{\|}}{P}\overset{\overset{F}{|}}{\underset{\underset{O}{\|}}{C}}R^1$$

$$R^1 = Alk,\ Ar$$

$$66\text{–}78\%$$

$$(8.21)$$

The procedure has been extended to the synthesis of 2-fluoro-1,3-dicarbonyl compounds from 2-fluoro-2-phosphonyl-1,3-dicarbonyl compounds readily obtained by reaction of diethyl 1-fluoro-1-(ethoxycarbonyl)methylphosphonate with aromatic carboxylic acid chlorides or alkyl chloroformates. The reaction proceeds smoothly by the MgCl$_2$-induced P-C bond cleavage at room temperature of 1-acyl or 1-(alkoxycarbonyl)-1-fluoro-1-(ethoxycarbonyl)methylphosphonates.[358-362]

Formylation of dimethyl 1-(methoxycarbonyl)methylphosphonate with methyl formate under basic conditions using MeONa/MeOH results in the quantitative formation of the sodium enolate.[363] An alternative procedure is the preparation of masked aldehydes by reaction of *tert*-butoxy-*bis*(dimethylamino)methane[364] or *N,N*-dimethylformamide dimethyl acetal[365-367] with diethyl 1-(ethoxycarbonyl)methylphosphonate. The former reacts at 160°C for 3 h with diethyl 1-(ethoxy-carbonyl)methylphosphonate to give the corresponding enaminophosphonate in 72% yield,[364] whereas the use of the latter increases reaction yields (82–95%) and rates (2 h).[365,366] The subsequent hydrolysis of the enaminophosphonate thus produced under acid conditions (2 M HCl) in biphasic medium at room temperature leads to the diethyl 1-(ethoxycarbonyl)-1-formylmethylphosphonate in 89% yield.[366]

8.1.2.6.6   *Alkoxycarbonylation*

The sodium salt of diethyl 1-(ethoxycarbonyl)methylphosphonate reacts wth 2-chloroethyl chloro-formate in THF at room temperature to produce the diethyl 1-ethoxycarbonyl-(1,3-dioxolan-2-ylidene)methylphosphonate in 57% yield.[368] The same reaction conducted with ethyl chloroformate in C$_6$H$_6$ or THF proceeds with the formation of a complex reaction mixture.[61,369] By contrast, when the ethoxymagnesium salt is employed, the diethyl 1,1-*bis*(ethoxycarbonyl)methylphosphonate is isolated in 67% yield.[344] The diethyl 1-lithio-1-fluoro-1-(ethoxycarbonyl)methylphosphonate generated with *n*-BuLi in THF reacts with ethyl chloroformate to give only the diethyl 1-fluoro-1,1-*bis*(ethoxycarbonyl)methylphosphonate.[347]

Phosphorylglycines are easily prepared, but in limited yield, by alkoxycarbonylation of the α-lithiated Schiff base of diethyl aminomethylphosphonate in THF at low temperature.[370,371] The acylated Schiff base is converted in high yield (80%) to the corresponding β-lactam by photochemical reaction with the [(methoxy)(methyl)carbene]chromium complexes in CH$_2$Cl$_2$.[372]

8.1.2.6.7   *Amination*

There are a number of important methods worthy of merit for the synthesis of phosphorylglycines. Thus, electrophilic amination of diethyl 1-(ethoxycarbonyl)methylphosphonate proceeds in one simple operation via the sodium salt in THF or DME. A variety of aminating agents are used with variable yields, such as *O*-mesitylenesulfonylhydroxylamine (39–47%),[373] chloramine (23–84%),[374] and diphenyl (*O*-hydroxylamine)phosphine oxide (60%).[375,376] Diethyl 1-ethoxycarbonyl- or 1-(*tert*-butoxycarbonyl)methylphosphonate may also be efficiently transformed into oxime or diazo derivatives and then converted to an amine by a further reduction step. The oximes are prepared from EtONO via the sodium diethyl 1-(*tert*-butoxycarbonyl)methylphosphonate (17% or 36%),[374] from NOCl and diethyl 1-(ethoxycarbonyl)methylphosphonate in the presence of (EtO)$_2$Mg,[377] Al–Hg, or Al(O*i*-Pr)$_3$,[378,379] from NOCl and diethyl 1-(methoxycarbonyl)ethylphosphonate,[380,381] or from NOCl/ROH and diethyl 1-(chlorocarbonyl)methylphosphonate.[382]

Diethyl 1-*tert*-butoxycarbonyl-1-diazomethylphosphonate is efficiently prepared in 81% yield from the sodium salt of diethyl 1-(*tert*-butoxycarbonyl)methylphosphonate and tosyl azide in DME (Scheme 8.22).[374] In the case of sodium and potassium salts of diethyl 1-(methoxycarbonyl)ethylphosphonate, the desired azides are not formed. However, with alkoxymagnesium derivatives, the 1-azido-1-(methoxycarbonyl)ethylphosphonates are obtained in modest yields.[383] Diazotation of diethyl 1-amino-1-(ethoxycarbonyl)methylphosphonate with *n*-PrONO in AcOH gives diethyl 1-ethoxycarbonyl-1-diazomethylphosphonate in 56–74% yields.[384]

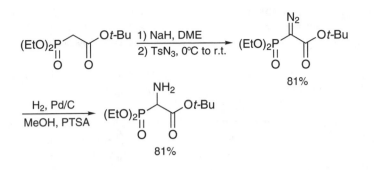

(8.22)

Other preparative methods for phosphorylglycines include the addition of sodium diethyl phosphite to the Schiff base prepared from benzylamine and ethyl or *tert*-butyl glyoxalate,[385] the reaction of Me$_3$SiBr with the triethoxyphosphorane prepared from *N*-protected alkyl oxalamides and triethyl phosphite,[386] and the Rh(II)-catalyzed N–H insertion of diethyl 1-ethoxycarbonyl-1-diazomethylphosphonate into carbamates.[148]

8.1.2.6.8 Miscellaneous

The sodium salts of dialkyl 1-(alkoxycarbonyl)methylphosphonates, in the presence of a further equivalent of NaH, undergo addition of carbon disulfide in Et$_2$O at room temperature to give alkenedithiolates, which are characterized by protonation, alkylation (MeI, EtI, BnCl), oxidation, and phosgenation. For example, reaction with methyl iodide provides the dialkyl 1-alkoxycarbonyl-2,2-*bis*(methylthio)vinylphosphonates in good yields (68–75%), whereas room-temperature treatment with acetyl or benzoyl chloride produces 1,3-dithietane-2,4-diylidene-*bis*[(alkoxycarbonyl)methylphosphonates] in moderate yields (18–39%, Scheme 8.23).[387,388] The same oxidation product is obtained by reaction with iodine.

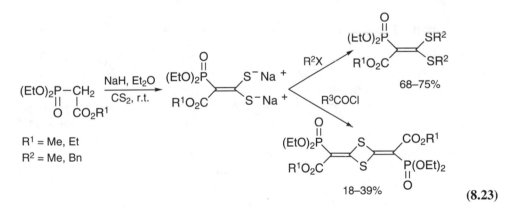

(8.23)

Alkenedithiolates, on reaction with phosgene in CH$_2$Cl$_2$ at room temperature, gives a transient 4-alkylidene-1,3-dithietane-2-one, which, on reaction with secondary amines, produces the corresponding thioamides in moderate yields (25–28%).[387] The sodium 1-substituted 1-(alkoxycarbonyl)methylphosphonates may also undergo the previous sequential treatment with CS$_2$ and methyl iodide to give the corresponding dithioesters in fair yields (26–65%).[387]

The reaction of diethyl 1-(ethoxycarbonyl)methylphosphonate with EtSH proceeds at room temperature in CH$_2$Cl$_2$ in the presence of Et$_2$AlCl, Me$_2$AlCl, or EtAlCl$_2$ to produce the phosphonoketene dithioacetals in good yields (59–83%).[389] Similarly, room-temperature treatment of phosphonosuccinate with EtSH (6–8 eq) and Et$_2$AlCl in CH$_2$Cl$_2$ results in the formation of diethyl

1-thioethoxycarbonyl-3,3-*bis*(ethylthio)-2-propenylphosphonate as a major product (61–63%) together with a small amount of expected diethyl 1,1,4,4-*tetra*(ethylthio)-1,3-butadien-2-ylphosphonate.[242] Using EtAlCl$_2$ instead of Et$_2$AlCl favors the expected phosphonate and isolates it in better yield.[242]

The readily accessible dialkyl 1-alkoxycarbonyl-2,2-*bis*(methylthio)vinylphosphonates react smoothly with lithium dimethylcuprate to give the diethyl 1-(alkoxycarbonyl)-2,2-dimethylpropyl-phosphonates in good yields (75–80%).[388]

Diisopropyl 1-(isopropoxycarbonyl)methylphosphonate undergoes mono- or disulfenylation with phenyl- or cyanosulfenyl chloride in the presence of Al(O*i*-Pr$_2$)$_3$ in toluene to give the expected phosphonate in 72–81% yields.[390] The reaction of sodium diethyl 1-(ethoxycarbonyl)methylphosphonate with phenylthioiosocyanate has also been reported.[391]

8.1.2.7 Other Reactions

Several other preparations of dialkyl 1-(ethoxycarbonyl)methylphosphonates have been reported. They include the CuSO$_4$-[392–394] or Cu(acac)-catalyzed[395,396] addition of dialkyl phosphites to ethyl diazoacetate in refluxing C$_6$H$_6$. The method using Cu(acac) gives better yields and purer products in 1-(ethoxycarbonyl)methylphosphonates. The reaction proceeds to fair to good yields (40–83%) via carbenoid insertion into the P–H bond of the phosphite.[395,396]

Heating a mixture of dimethyl phosphite and methyl glyoxalate hydrate under acidic conditions with continuous removal of water produces dimethyl 1-hydroxy-1-(methoxycarbonyl)methylphosphonate in over 90% yield.[397] The alcohol can be protected by standard methods using a variety of protecting groups (ethyl vinyl ether, dihydropyran, 2-methoxypropene, *tert*-butyldimethylsilyl chloride).[397–399] The same preparation has been realized in 97% yield by heating at 80°C diethyl phosphite and the hemiacetal, ethyl 2-ethoxy-2-hydroxyacetate.[400] Similarly, by refluxing dimethyl phosphite and *para*-nitrobenzyl glyoxalate hydrate in AcOEt, dimethyl 1-hydoxy-1-(*para*-nitrobenzyloxycarbonyl)methylphosphonate is isolated in 69% yield after workup.[401]

8.1.3 Diethyl 1-(Hydroxycarbonyl)methylphosphonate

Dialkyl 1-(hydroxycarbonyl)methylphosphonates have been the subject of numerous investigations, with the efforts largely confined to diethyl 1-(hydroxycarbonyl)methylphosphonate, whose reactivity parallels that of α-substituted derivatives. Several procedures for the synthesis of diethyl 1-(hydroxycarbonyl)methylphosphonate have been developed. They include the hydrogenolysis of the benzyloxycarbonyl ester (EtOH, Pd/C, H$_2$),[47,49,402–404] the alkaline hydolysis of alkoxycarbonyl ester (KOH or NaOH, EtOH, room temperature),[56,273,405–413] and the addition of carbon dioxide to lithium (Scheme 8.24)[405,414–425] or magnesium[426,427] phosphonate carbanions. Whereas the first two methods appear to have only limited synthetic potential, the third one (Scheme 8.24) appears to be the most general method, giving access to both dialkyl 1-(hydroxycarbonyl)methyl- or 1-(hydroxycarbonyl)alkylphosphonates from readily available alkylphosphonates.

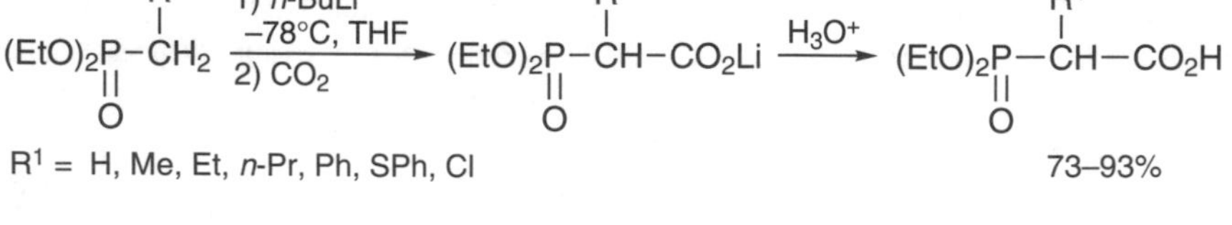

$$
\underset{\underset{O}{\|}}{(EtO)_2P}-CH_2 \quad \xrightarrow[\text{2) CO}_2]{\underset{-78°C,\ THF}{1)\ n\text{-BuLi}}} \quad \underset{\underset{O}{\|}}{(EtO)_2P}-\overset{R^1}{\underset{}{CH}}-CO_2Li \quad \xrightarrow{H_3O^+} \quad \underset{\underset{O}{\|}}{(EtO)_2P}-\overset{R^1}{\underset{}{CH}}-CO_2H
$$

R^1 = H, Me, Et, *n*-Pr, Ph, SPh, Cl   73–93%   **(8.24)**

Oxidation of dialkyl 1-formylmethylphosphonates with peracetic acid[428] or alkaline permanganate[429] has been reported to afford dialkyl 1-(hydroxycarbonyl)methylphosphonates in high yields, but the general synthetic utility of this procedure remains limited.

Diethyl 1-(hydroxycarbonyl)methylphosphonate is thermally stable up to 150°C.[402] Decomposition associated with distillation begins at 160°C to give diethyl methylphosphonate (17%), diethyl 1-(ethoxycarbonyl)methylphosphonate (34%), and 1-(ethoxycarbonyl)methylphosphonic acid (35%).[402]

Dialkyl 1-(hydroxycarbonyl)methylphosphonates are converted into the corresponding esters via the 1-(chlorocarbonyl)methylphosphonates and alcohols.[419,430–436] The acid chlorides are prepared from acids using SOCl$_2$/DMF/dioxane,[437] SOCl$_2$/CH$_2$Cl$_2$,[382,423,438–440] SOCl$_2$/HMPT,[419] SO$_2$Cl$_2$/CH$_2$Cl$_2$,[421,441] (COCl)$_2$/CH$_2$Cl$_2$/DMF,[409,430] or (COCl)$_2$/C$_6$H$_6$.[403,442] On treatment with Et$_3$N, the diethyl 1-(chlorocarbonyl)methylphosphonate is converted to phosphonoketene.[441,443] Esterification can also be induced by coupling dialkyl 1-(hydroxycarbonyl)methylphosphonates and alcohols under smooth conditions using DCC in C$_6$H$_6$ at room temperature[422,444–446] or various efficient combinations such as DCC/DMAP/CH$_2$Cl$_2$,[413,447–457] EDCI·MeI/HOBt/CH$_2$Cl$_2$,[458,459] 2-chloropyridinium iodide/Et$_3$N,[460] bis(2-pyridyl)thiocarbonate/DMAP/toluene,[461] 1-cyclohexyl-3-(2-morpholinoethyl)carbodiimide/DMAP/CH$_2$Cl$_2$,[462] and BOP/DMAP.[463,464]

Esterification of alcohols is also effected in good yields by using the mixed anhydride prepared from trifluoroacetic anhydride and dialkyl 1-(hydroxycarbonyl)methylphosphonate (CH$_2$Cl$_2$, Py).[465–467]

Several reports show that it is possible to carry out a transesterification selectively at the carboxylic ester moiety of dialkyl 1-(alkoxycarbonyl)methylphosphonates rather than at the phosphonic ester moiety with primary and secondary alcohols in the presence of NaH[468,469] or DMAP.[470–474] For example, a number of chiral phosphonates have been prepared in high yields (80–96%) by DMAP-catalyzed transesterification of dialkyl 1-(alkoxycarbonyl)methylphosphonates with 8-phenylmenthol in toluene at reflux.[454,475–478]

Preparation of diethyl 1-(cyclohexyloxycarbonyl)methylphosphonate has been described by two separate methods. One involves the reaction of 2-bromocyclohexanone with the potassium salt of diethyl 1-(hydroxycarbonyl)methylphosphonate in CH$_2$Cl$_2$ at room temperature for 4 days (91%),[479] the other the Cu(acac)-catalyzed insertion reaction at room temperature of α-diazocyclohexanone to diethyl 1-(hydroxycarbonyl)methylphosphonate (78%).[480]

Dialkyl 1-(aminocarbonyl)alkylphosphonates are prepared by coupling amines with dialkyl 1-(alkoxycarbonyl)alkylphosphonates,[111,334,481–485] with dialkyl (chlorocarbonyl)methylphosphonates,[486–488] and with dialkyl 1-(hydroxycarbonyl)alkylphosphonates in the presence of DCC,[489–494] HOBt/DCC/CH$_2$Cl$_2$,[495] 2-chloro-1-methylpyridinium iodide/Et$_3$N/CH$_2$Cl$_2$,[496] or CDI/THF.[56,412,497,498] Hydrazides are obtained in the same manner.[499–501] Diethyl 1-(hydroxycarbonyl)methylphosphonate, on treatment with chloroformate in THF in the presence of Et$_3$N, gives the mixed anhydride, which is reacted in situ with amino acids.[502]

The readily available diethyl 1-(hydroxycarbonyl)methylphosphonate can be converted into diethyl 2-oxoalkylphosphonates.[361,503] The reaction is based on the fact that α-acylated 1-(hydroxycarbonyl)methylphosphonates decompose to give β-ketophosphonates. For example, treatment of diethyl 1-(hydroxycarbonyl)methylphosphonate with Me$_3$SiCl and Et$_3$N in dry toluene at 0°C gives the trimethylsilyl ester, which, on addition of MgCl$_2$ and acyl chloride, affords diethyl 2-oxoalkylphosphonates after workup in modest to good yields (38–97%, Scheme 8.25).[503] The method has been applied to a number of aromatic and aliphatic carboxylic acid chlorides. The aromatic carboxylic acid chlorides give higher yields than the aliphatic ones. The reaction proceeds by acylation of the trimethylsilyl ester with carboxylic acid chloride in the presence of MgCl$_2$ as chelating agent followed by decarboxylation.[503] Similarly, diethyl 2-oxoalkylphosphonates have been prepared in 77–90% yields from unprotected diethyl 1-(hydroxycarbonyl)methylphosphonate and aromatic acid chlorides. However, the yields are lower (40%) with aliphatic acid chlorides.[504]

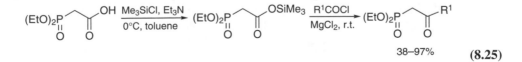

$$\text{38–97\%}\qquad\textbf{(8.25)}$$

The Mannich condensation of diethyl 1-(hydroxycarbonyl)methylphosphonate with formaldehyde and various secondary aliphatic amines in refluxing C_6H_6 leads to diethyl 1-(N,N-dialkylaminomethyl)vinylphosphonates in moderate to good yields (29–85%).[505,506] Diethyl 1-(hydroxycarbonyl)methylphosphonate reacts with aminomalononitrile *para*-toluenesulfonate in Py at room temperature in the presence of DCC to give 2-phosphonomethyl-5-amino-4-cyano-1,3-oxazole in 86% yield.[507]

8.1.4 DIMETHYL 1-(METHOXYCARBONYL)VINYLPHOSPHONATE AND DIETHYL 1-(ETHOXYCARBONYL)VINYLPHOSPHONATE

Diethyl 1-(ethoxycarbonyl)vinylphosphonate was first prepared in 50% yield by a Knoevenagel reaction from diethyl 1-(ethoxycarbonyl)methylphosphonate and formaldehyde in refluxing MeOH in the presence of piperidine.[508–510] Recently, each step in the standard synthetic procedure has been carefully monitored with the aim of making a large scale-preparation of vinylphosphonate while avoiding any purification procedure.[511] Thus, reaction of dimethyl 1-(methoxycarbonyl)methylphosphonate with excess paraformaldehyde in MeOH at reflux in the presence of piperidine gives a mixture containing mainly the dimethyl 1-methoxycarbonyl-2-methoxyethylphosphonate. To convert the methyl ether into vinylphosphonate the MeOH is replaced by toluene at reflux in the presence of TsOH. Thus, dimethyl 1-(methoxycarbonyl)vinylphosphonate is isolated clean enough to be used directly.[511]

Three laboratories have reported independently an efficient procedure for the synthesis of diethyl 1-(ethoxycarbonyl)vinylphosphonate based on the thermal elimination either of PhSOH[338,512] from diethyl 1-ethoxycarbonyl-1-(phenylsulfoxyl)ethylphosphonate or PhSeOH (82%)[513] from diethyl 1-ethoxycarbonyl-1-(phenylselenoxyl)ethylphosphonate. Very recently, the standard conditions of the latter reaction have been improved, each step proceeding with more than 90% yield (Scheme 8.26).[514] Another version of the elimination procedure using diethyl 1-ethoxycarbonyl-2-chloroethylphosphonate as precursor of diethyl 1-(ethoxycarbonyl)vinylphosphonate has been explored, but the method, having only limited utility, has not been developed.[515–519]

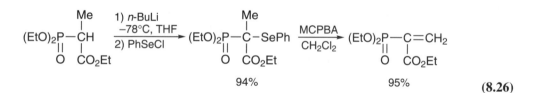

$$(8.26)$$

The use of diethyl 1-(ethoxycarbonyl)vinylphosphonate is based on Michael addition of anionic nucleophiles to the double bond to produce stabilized phosphonate anions capable of undergoing subsequent alkenylation reaction with aldehydes or ketones by a Horner–Wadsworth–Emmons process (Scheme 8.27).[512,513]

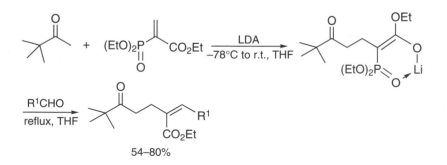

$$(8.27)$$

This sequential Michael and Horner–Wadsworth–Emmons reactions has been developed with a large variety of charged nucleophiles such as lithium carbanions of phenylethynyl, methyl[(methylsulfoxyl)methyl]sulfide and ethyl α-(methylthio)acetate[520] or lithium diethyl phosphite[514,521,522] to prepare functionalized dienes, trienes, and their analogues.

The base-catalyzed addition of an easily available 1-deoxy-1-nitroribose derivative to dimethyl 1-(methoxycarbonyl)vinylphosphonate proceeds rapidly, but the products, being unstable, decompose to a mixture containing only small amounts of the desired compound.[523] C-Glycosidic phosphonates have been synthesized by addition of pyranosidic and furanosidic glycosyl radicals, generated from acylated halogenoses by irradiation in the presence of tris(trimethylsilyl)silane (TTMS) as radical starter, to dimethyl 1-(methoxycarbonyl)vinylphosphonate, which appears as a useful substrate for the synthesis of complex sugar molecules.[524]

Under phase-transfer catalysis (NaOH/TEBAC/CH$_2$Cl$_2$) condensation of diethyl bromomalonate with diethyl 1-(methoxycarbonyl)vinylphosphonate leads to the cyclopropanation product, diethyl 2,2-bis(ethoxycarbonyl)-1-(methoxycarbonyl)cyclopropylphosphonate, in 65% yield.[525]

Oxygen and nitrogen nucleophiles add to diethyl 1-(ethoxycarbonyl)vinylphosphonate to generate the formation of various fused heterocyclic compounds. Thus, Δ^1-carbapenems are readily prepared by the Michael addition of N-lithio-4-vinylazetidin-2-one to diethyl 1-(ethoxycarbonyl)vinylphosphonate followed by ozonolysis of the adduct and subsequent treatment of the generated aldehyde by intramolecular Horner–Wadsworth–Emmons reaction (Scheme 8.28).[243]

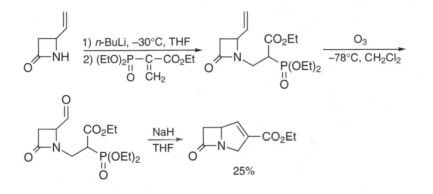

(8.28)

An imine–enamine annulation has been used in the synthesis of the indoloquinolizidine alkaloid (±)-deplancheine.[526] The annulation of dialkyl (1-alkoxycarbonyl)vinylphosphonates via a Horner–Wadsworth–Emmons reaction has been developed in the synthesis of [3.3.0] fused pyrazolidinones from monocyclic pyrazolidinones.[527,528] Treatment of diethyl 1-(ethoxycarbonyl)vinylphosphonate in excess (2 eq) with imide anions such as phthalimide, maleimide, and succinimide successfully produces the corresponding six-membered fused heterocycles.[529] Similarly, synthesis of functionalized cyclohexenylphosphonates is achieved by condensation of diethyl 1-(ethoxycarbonyl)vinylphosphonate (2 eq) with cyclopentanone enolates (Scheme 8.29).[529]

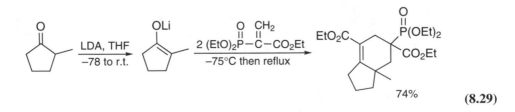

74%   (8.29)

When subjected to reaction with the sodium salt of salicylaldehyde, diethyl 1-(ethoxycarbonyl)vinylphosphonate leads to 3-(ethoxycarbonyl)-3-chromene by cycloaddition reaction.[338] The addition of ketoxime anions, generated by deprotonation with NaH, to diethyl 1-(ethoxycarbonyl)vinylphosphonate followed by a Horner–Wadsworth–Emmons reaction with aromatic and heteroaromatic aldehydes afford O-[2-(ethoxycarbonyl)-3-arylallyl]oximes in 75–86% yields.[530]

Addition of sulfur nucleophiles such as phenylthiolate to diethyl 1-(ethoxycarbonyl)vinylphosphonate has been used in the synthesis of β-acetoxy- and β-hydroxy-α-methylene-γ-butyrolactones.[531] Similarly, addition of isopropylthiolate to diethyl 1-(ethoxycarbonyl)vinylphosphonate represents a key step in the synthesis of α-methylidenebutyrolactones such as confertin (Scheme 8.30),[510,532] arteannuin B,[533] rhopaloic acid A,[534–537] and hippospongic acid A.[538] Addition of mercaptoacetaldehyde to diethyl 1-(ethoxycarbonyl)vinylphosphonate gives methyl 2,5-dihydrothiophene-3-carboxylate,[511] which thermally decomposes to methyl 1,3-butadiene-2-carboxylate.[539]

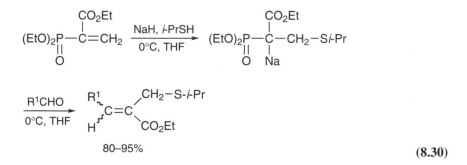

$$(8.30)$$

Dialkyl 1-(alkoxycarbonyl)vinylphosphonates undergo ene reactions with a variety of alkenes using $EtAlCl_2$ as Lewis acid catalyst.[540] They also undergo [4 + 2] cycloadditions with cyclopentadiene,[541] isoprene,[542] 6,6-diphenylfulvene,[543] and N-buta-1,3-dienylsuccinimide (Scheme 8.31).[544–546] Stereoselective $SnCl_4$-promoted [2 + 1] cycloaddition of 1-seleno-2-silylethene with dialkyl 1-(alkoxycarbonyl)vinylphosphonates leads to highly functionalized cyclopropanephosphonates in high yields.[547–549] Radicals derived from isopropylidene uronic esters of N-hydroxy-2-thiopyridone add to dimethyl 1-(methoxycarbonyl)vinylphosphonate affording derivatives of phosphonic analogues of nucleotides.[550,551]

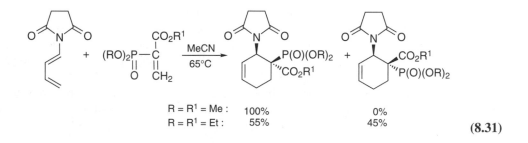

$$(8.31)$$

The dicyclohexylammonium salt of diethyl 1-(hydroxycarbonyl)methylphosphonate and formaldehyde in the presence of Et_3N undergoes a Knoevenagel reaction to give the dicyclohexylammonium salt of 1-(hydroxycarbonyl)vinylphosphonate (70%).[552] This compound is able to react readily with a variety of secondary amines,[553] 1,3-dicarbonyl and monocarbonyl nucleophiles,[552] to give the corresponding Michael adducts.

A synthesis of diethyl 1-ethoxycarbonyl-2-arylvinylphosphonates based on the Pd-catalyzed Heck reaction of aryldiazonium salts with diethyl 1-(ethoxycarbonyl)vinylphosphonate has recently been reported.[554]

8.1.5 Dialkyl 1-(Alkoxycarbonyl)-1-diazomethylphosphonates

Diethyl 1-ethoxycarbonyl-1-diazomethylphosphonate was first prepared in moderate to good yields (37–65%) by reaction of tosyl azide with the potassium salt of diethyl 1-(ethoxycarbonyl)methylphosphonate in C_6H_6 at room temperature.[555–557] Several variants have been introduced to extend the synthetic utility of the reaction and to replace tosyl azide by less dangerous reagents. Significant improvements in yields (83–99%) are observed with NaH in a THF–C_6H_6 solution[448] or cesium carbonate in THF.[558] The use of 2-naphthalenesulfonyl azide, which is a safer reagent than tosyl azide, is another useful modification, but the yields remain modest (40–61%).[559]

The azido*tris*(diethylamino)phosphonium bromide appears to be an exceptionaly safe diazo reagent stable to shock, friction, heating, and even flame.[560] It reacts smoothly with diethyl 1-(ethoxycarbonyl)methylphosphonate in Et_2O at room temperature in the presence of *t*-BuOK or another alkoxide to give diethyl 1-ethoxycarbonyl-1-diazomethylphosphonate in 70% yield (Scheme 8.32).[456,560] Mild diazo transfer has recently been achieved in excellent yields (85%) through the use of *para*-acetamidobenzenesulfonyl azide at 0°C.[57]

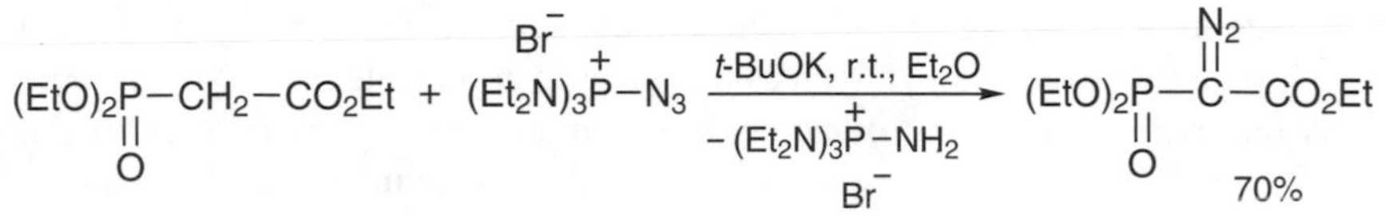

$$(EtO)_2\underset{\underset{O}{\|}}{P}-CH_2-CO_2Et \; + \; (Et_2N)_3\overset{\overset{Br^-}{+}}{P}-N_3 \; \xrightarrow[\substack{+\\-(Et_2N)_3P-NH_2\\Br^-}]{\textit{t-BuOK, r.t., Et}_2O} \; (EtO)_2\underset{\underset{O}{\|}}{P}-\overset{\overset{N_2}{\|}}{C}-CO_2Et \qquad 70\%$$

$$\text{(8.32)}$$

An attractive nonbasic variant involving the use of Al_2O_3_KF as solid base results in the formation of diethyl 1-ethoxycarbonyl-1-diazomethylphosphonate in excellent yield (96%).[561] This method has the advantages of simple workup, easy product isolation, and mild conditions.

The $Rh_2(OAc)_4$-catalyzed decomposition of dialkyl 1-alkoxycarbonyl-1-diazomethylphosphonate gives rise to a range of reactions including insertion reactions, intramolecular cyclopropanations, and Wolff rearrangements. An elegant synthesis of α-alkoxyphosphonoacetates involves elaboration of a phosphonate group by coupling to the O–H group of alcohols using the Rh(II)-catalyzed insertion of dialkyl 1-alkoxycarbonyl-1-diazomethylphosphonate. Thus, the insertion of dimethyl 1-methoxycarbonyl-1-diazomethylphosphonate into primary sugar alcohols catalyzed by $Rh_2(OAc)_4$ in refluxing C_6H_6 gives good yields (75–80%) of dimethyl 1-alkoxy-1-(methoxycarbonyl)methylphosphonates as diastereomeric mixtures (Scheme 8.33).[562,563]

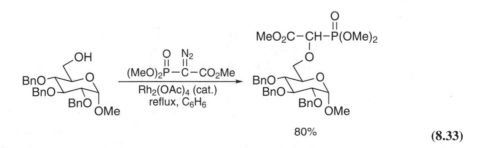

$$\text{(8.33)}$$

The reaction has been investigated with MeOH,[564] *i*-PrOH,[564] *t*-BuOH,[564] phenols,[564] cyclic allylic alcohols,[565–567] homoallylic alcohols,[568] 3,4-hydroxybenzoate,[569] and mono *t*-butyldimethylsilyl protected diols.[570,571] A subsequent Horner–Wadsworth–Emmons olefination with a range of aldehydes affords the corresponding enol ethers.[562,563,565,566,570]

Carbenoid intermediates generated by catalytic decomposition of dimethyl 1-(alkoxycarbonyl)-1-diazomethylphosphonates react with diallylsulfide in refluxing CH_2Cl_2 to give dimethyl 1-(allylthio)-1-(alkoxycarbonyl)-3-butenylphosphonates. In the presence of Grubbs' catalyst, they undergo methatesis to give cyclic α-thiophosphonates in good yields (Scheme 8.34).[572,573] Under the same conditions, the reaction of diethyl 1-ethoxycarbonyl-1-diazomethylphosphonate with 4-thiaazetidinones affords the carbene insertion product in 58% yield.[574]

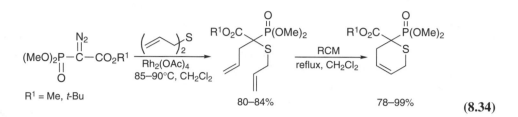

$$(8.34)$$

The N–H insertion reactions of rhodium carbenoids derived from α-diazophosphonates represents a simple route to α-aminoalkylphosphonates.[575] α-Diazophosphonates are more stable thermally and toward $Rh_2(OAc)_4$ than the corresponding diazo esters, and higher reaction temperatures are often needed. Therefore, the $Rh_2(OAc)_4$ decomposition of diethyl 1-ethoxycarbonyl-1-diazomethylphosphonate in the presence of N-protected amino acid amides carried out in boiling toluene results in the formation of phosphonylglycine derivatives in good yields. The N–H insertion reaction is completely regioselective, and no competing insertion into the carbamate N–H bond is observed (Scheme 8.35).[576,577] The reaction has also been applied to the synthesis of N-arylphosphonylglycines.[578]

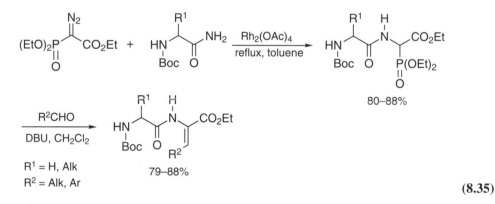

$$(8.35)$$

Thermal decomposition of diethyl 1-ethoxycarbonyl-1-diazomethylphosphonate catalyzed by $Rh_2(OAc)_4$ in the presence of propylene oxide gives diethyl 1-ethoxycarbonyl-1-oxomethylphosphonate.[579] Although stable under anhydrous conditions, this compound hydrates quantitatively in H_2O, giving the gem-diol, whereas MeOH adds quantitatively to form the hemiketal.[579]

Diethyl 1-ethoxycarbonyl-1-diazomethylphosphonate reacts with benzonitrile in refluxing $CHCl_3$ in the presence of $Rh_2(OAc)_4$ to give diethyl 5-ethoxy-2-phenyloxazole-4-phosphonate in poor yield (16%). However, the yield can be increased to 53% when $Rh_2(NHCOCF_3)_4$ is used as catalyst.[580] Moreover, it has been clearly shown that in a molecule containing two differently substituted diazo groups, the more reactive diazo system can undergo a selective Rh(II)-mediated O–H insertion reaction leaving the other intact when $Rh_2(OAc)_4$ is used as catalyst. By contrast, with $Rh_2(NHCOCF_3)_4$, a more active catalyst, both diazo groups react.[456]

The dimethyl 1-allyloxycarbonyl-1-diazomethylphosphonate and diallyl 1-tert-butoxycarbonyl-1-diazomethylphosphonate under $Rh_2(OAc)_4$ catalysis[57] or in refluxing cyclohexane in the presence of copper powder[448] undergo intramolecular cyclopropanation reaction in good yields and with moderate levels of diastereoselectivity.

In contrast to α-diazo-β-ketophosphine oxides, the α-diazo-β-ketophosphonates predominantly react on irradiation via the ketene patway. Trapping of the ketene intermediates by an alcohol provides a new route to α-substituted 1-(alkoxycarbonyl)methylphosphonates.[581,582]

8.1.6 DIALKYL 2-(ALKOXYCARBONYL)ALKYLPHOSPHONATES

8.1.6.1 Michaelis–Arbuzov Reactions

Triethyl phosphite, on heating at 120–125°C with ethyl 3-iodopropionate[583] or at 150–155°C with ethyl 3-bromopropionate,[584,585] undergoes the Michaelis–Arbuzov rearrangement leading to diethyl 2-(ethoxycarbonyl)ethylphosphonate in moderate yields (46–63%). These cannot be improved further because of the competing production of acrylates and diethyl phosphite resulting from an intramolecular β-elimination involving the quasiphosphonium salt intermediate.[584,586] Analogously, ethyl 2-acetoxy-3-chloropropionate, on reaction with triethyl phosphite at 160–170°C, gives the expected phosphonate, according to the Michaelis–Arbuzov reaction, which eliminates AcOH with production of diethyl 2-(ethoxycarbonyl)vinylphosphonate.[587] The AcOH reacts with triethyl phosphite, and the resulting diethyl phosphite undergoes addition at the α-carbon atom of 2-(ethoxycarbonyl)vinylphosphonate to produce tetraethyl 2-(ethoxycarbonyl)ethylidenediphosphonate in 70% yield (Scheme 8.36).

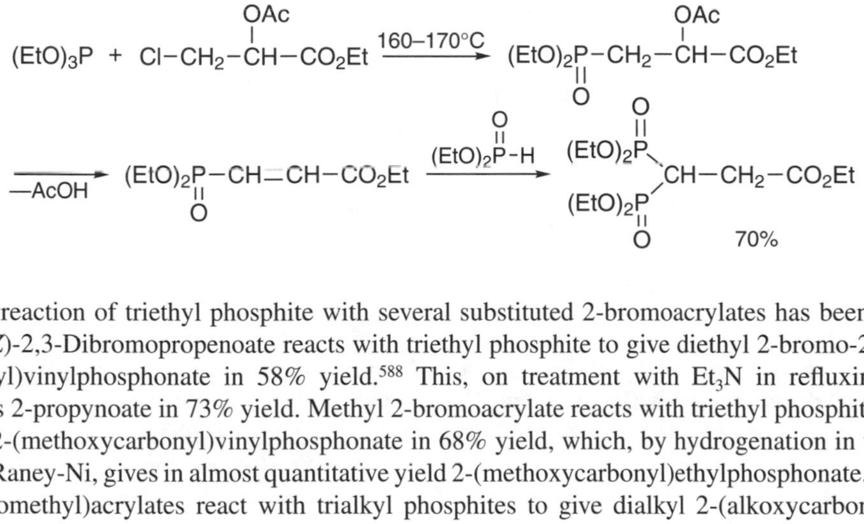

$$
\text{(EtO)}_3\text{P} + \text{Cl-CH}_2\text{-CH(OAc)-CO}_2\text{Et} \xrightarrow{160-170°C} \text{(EtO)}_2\text{P(=O)-CH}_2\text{-CH(OAc)-CO}_2\text{Et}
$$

$$
\xrightarrow[\text{-AcOH}]{} \text{(EtO)}_2\text{P(=O)-CH=CH-CO}_2\text{Et} \xrightarrow{\text{(EtO)}_2\text{P(=O)-H}} \begin{array}{c}\text{(EtO)}_2\text{P(=O)}\\\text{(EtO)}_2\text{P(=O)}\end{array}\!\!\!\!\text{CH-CH}_2\text{-CO}_2\text{Et}
$$

$$\text{70\%} \qquad (8.36)$$

The reaction of triethyl phosphite with several substituted 2-bromoacrylates has been investigated. (Z)-2,3-Dibromopropenoate reacts with triethyl phosphite to give diethyl 2-bromo-2-(ethoxycarbonyl)vinylphosphonate in 58% yield.[588] This, on treatment with Et$_3$N in refluxing Et$_2$O, produces 2-propynoate in 73% yield. Methyl 2-bromoacrylate reacts with triethyl phosphite to give diethyl 2-(methoxycarbonyl)vinylphosphonate in 68% yield, which, by hydrogenation in the presence of Raney-Ni, gives in almost quantitative yield 2-(methoxycarbonyl)ethylphosphonate.[113] Alkyl 2-(bromomethyl)acrylates react with trialkyl phosphites to give dialkyl 2-(alkoxycarbonyl)allylphosphonates in excellent yields.[589,590]

The dimethyl 2-methoxycarbonyl-2-(methoxyimino)ethylphosphonate used in the Horner–Wadsworth–Emmons reaction as an amino acid synthon is prepared by masking the carbonyl group of methyl bromopyruvate with methoxyamine hydrochloride before submission to a Michaelis–Arbuzov reaction with trimethyl phosphite (Scheme 8.37).[591–593] By contrast, unprotected 3-benzoyl-3-halopropionic esters react with trialkyl phosphites by both the Perkow and Michaelis–Arbuzov schemes.[594]

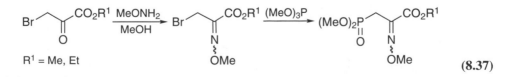

$$(8.37)$$

Diethyl 2-(ethoxycarbonyl)ethylphosphonate has also been prepared from Mannich bases. For example, triethyl phosphite reacts with ethyl 3-(diethylamino)propionate at 170°C in the presence of AcOH to give diethyl 2-(ethoxycarbonyl)ethylphosphonate in 58% yield with elimination of diethylamine and AcOEt (Scheme 8.38).[595]

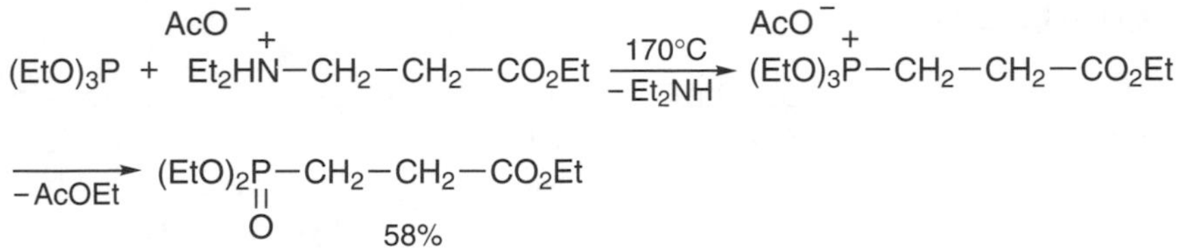

AcO⁻ +
(EtO)3P + Et2HN—CH2—CH2—CO2Et
170°C
- Et2NH
AcO⁻ +
(EtO)3P—CH2—CH2—CO2Et
- AcOEt
(EtO)2P—CH2—CH2—CO2Et
O
58%

(8.38)

A new approach to diethyl 2-methoxycarbonyl-2-alkenylphosphonates is based on the thermally induced Michaelis–Arbuzov rearrangement of diethyl allyl phosphites without added alkyl halide (Scheme 8.39).[596,597] The procedure involves treatment of the readily accessible allyl alcohols with diethyl chlorophosphite followed by heating of the crude intermediates for several hours at 70–100°C. The presence of a methoxycarbonyl group facilitates the rearrangement, the optimum temperatures being lower than those reported for unfunctionalized analogues.[596,597]

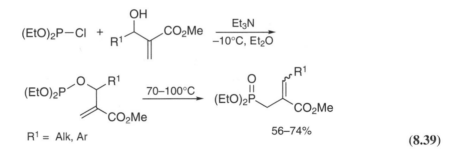

(EtO)2P—Cl +
OH
R1
CO2Me
Et3N
–10°C, Et2O
(EtO)2P
O
R1
CO2Me
70–100°C
O
(EtO)2P
R1
CO2Me
R1 = Alk, Ar
56–74%

(8.39)

An attractive procedure for the preparation of phosphonic analogues of aspartic acid by a Michaelis–Arbuzov-type reaction features the ready displacement of the acetoxy group of 4-acetoxyazetidin-2-one by triethyl phosphite at 110–120°C (Scheme 8.40).[598–600] The Michaelis–Arbuzov reaction is general, with only *tris*-(2,2,2-trichloroethyl) phosphite failing because of its reduced nucleophilicity. Acid hydrolysis of the 2-oxoazetidin-4-ylphosphonates provides a convenient route to α-aminophosphonic analogues of aspartic acid.[598–600]

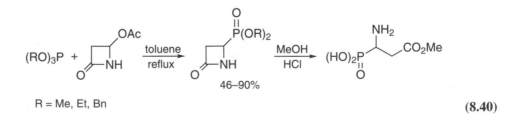

(RO)3P +
OAc
NH
O
toluene
reflux
O
P(OR)2
NH
O
46–90%
MeOH
HCl
NH2
(HO)2P
O
CO2Me
R = Me, Et, Bn

(8.40)

In the presence of copper(II) acetate, ethyl 2-iodo- and 2-bromobenzoates react with triethyl phosphite in boiling EtOH to give diethyl 2-(ethoxycarbonyl)phenylphosphonate in good yield (84%). Ease of halogen replacement follows the sequence I > Br > Cl. The iodo compound reacts in 1 h, whereas reaction of the chloro compound is incomplete after 24 h.[601]

8.1.6.2 Michaelis–Becker Reactions

The diethyl 2-(ethoxycarbonyl)ethylphosphonate is prepared in 35% yield from sodium diethyl phosphite and ethyl 3-iodopropionate.[156] Similarly, methyl 2,3-dibromopropionate reacts with sodium diethyl phosphite with formation of diethyl 2-(methoxycarbonyl)ethylphosphonate in low yield (15%).[602] Most likely, these reactions occur via an elimination–addition process promoted by the sodium diethyl phosphite. Treatment of ethyl 3-bromobutyrate with sodium dibutyl phosphite in toluene at 70°C affords dibutyl 1-methyl-2-(ethoxycarbonyl)ethylphosphonate in 62% yield,[603]

Ethyl 2-iodobenzoate reacts with potassium diethyl phosphite in liquid ammonia under irradiation in a Bunnett-type reaction to produce the coupling product in 64% yield (Scheme 8.41).[604]

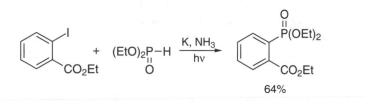

$$(8.41)$$

8.1.6.3 From β-Lactones

Triethyl phosphite attacks β-propiolactones at the β-carbon at 160°C to give predominantly diethyl 2-(ethoxycarbonyl)ethylphosphonate resulting from a C–O bond cleavage of the lactone ring.[605–612] Bases such as tertiairy amines and sodium alkoxides catalyze the reaction of lactones with phosphites.[606]

An attractive route to (R)- and (S)-2-amino-3-phosphonopropanoic acid (AP-3) involves the intermediacy of serine-derived β-lactones.[613] The nucleophilic addition of trimethyl phosphite to (R)- and (S)-N-(tert-butoxycarbonyl)-3-amino-2-oxetanone at 70°C for 42 h gives the (R)- and (S)-methyl N-(tert-butoxycarbonyl)-2-amino-3-(dimethoxyphosphinyl)propanoates in excellent yields. Exhaustive acid hydrolysis of the (S)- and (R)-isomers, followed by treatment with propylene oxide, affords the corresponding (R)- and (S)-2-amino-3-phosphonopropanoic acids as their zwitterions (D- and L-AP-3) (Scheme 8.42).[613]

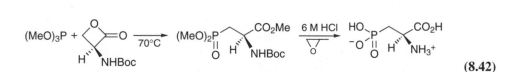

$$(8.42)$$

Recently, it has been shown that nucleophilic addition of neat dimethyl trimethylsilyl phosphite to serine-derived β-lactone at 100°C for 24 h leads specifically to the carboxylic trimethylsilyl ester by preferential transfer of the trimethylsilyl group.[614,615] A simple aqueous workup induces hydrolysis of the silyl ester to give the N- and P-protected free carboxylic acid ready for activation and coupling.

8.1.6.4 Alkylation of Phosphonate Carbanions

The ready alkylation of sodium phosphonate carbanions with chloro- or bromoacetates in THF provides a useful access to dialkyl 2-(alkoxycarbonyl)ethylphosphonates bearing a variety of alkyl, aryl, cyano, keto, or phosphoryl groups at the α-carbon (Scheme 8.43).[616–622] By using a chiral auxiliary, the alkylation of phosphorus-stabilized benzylic carbanions with bromoacetate proceeds with high diastereoselectivity to provide an easy access to optically active alkylphosphonic acids.[623]

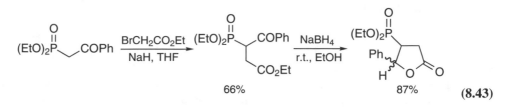

66%   87%   (8.43)

Diethyl 2-pyrrolidinoalkenylphosphonates, available from pyrrolidine and diethyl 1-formylal-kylphosphonates, on low-temperature treatment with s-BuLi in THF, undergo metallation under kinetic control in the β-position to phosphorus. Acylation of the resulting α-pyrrolidino anions with methyl or allyl chloroformate proceeds uneventfully to produce diethyl phosphonopyruvates in moderate to good yields after mild acid hydrolysis with oxalic acid on SiO_2.[624]

From the reaction of diisopropyl 1-lithio-1-fluoromethylphosphonate with ethyl pyruvate in THF at low temperature, diisopropyl 1-fluoro-2-hydroxy-2-(ethoxycarbonyl)propylphosphonate is isolated in moderate yield (36%).[625]

8.1.6.5 Phosphonylation by Michael Addition (Pudovik Reaction)

Unquestionably, one of the most attractive synthetic methods for the preparation of dialkyl 2-(alkoxycarbonyl)ethylphosphonates involves the conjugate addition of a dialkyl phosphite to the carbon–carbon double bond of α,β-unsaturated carboxylates (Pudovik reaction, Scheme 8.44). In the presence of EtONa, diethyl phosphite reacts vigorously with acrylate in EtOH to give the Michael addition product in 84% yield.[626,627] The reaction is applied with success to the synthesis of dialkyl 2-(alkoxycarbonyl)ethylphosphonates bearing a large variety of acyclic (symmetric[603,626,628–632] or disymmetric[633,634]) or cyclic (symmetric[635] or disymmetric[635]) substituents at phosphorus.

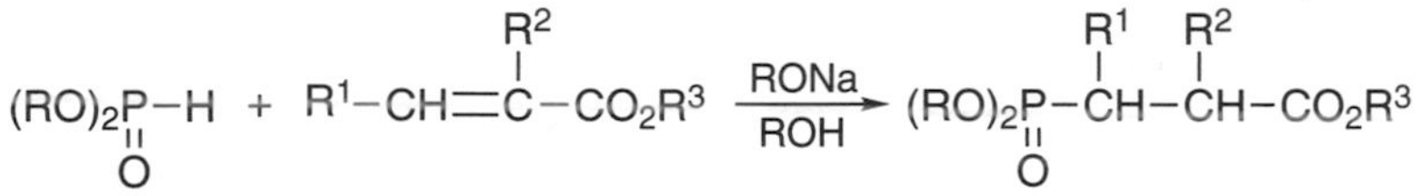

$$(RO)_2\underset{O}{\overset{\overset{\displaystyle ||}{}}{P}}-H \; + \; R^1-CH=\underset{\underset{R^2}{\overset{\displaystyle |}{}}}{C}-CO_2R^3 \xrightarrow[ROH]{RONa} (RO)_2\underset{O}{\overset{\overset{\displaystyle ||}{}}{P}}-\underset{\underset{R^1}{\overset{\displaystyle |}{}}}{CH}-\underset{\underset{R^2}{\overset{\displaystyle |}{}}}{CH}-CO_2R^3$$

R = Me, Et, i-Pr, n-Bu, i-Bu
R^1, R^2 = H, Me
R^3 = Me, Et, n-Pr

(8.44)

The reaction has been extended to methacrylates,[630,634–639] crotonates,[603,640] cinnamates,[603,641–643] ethyl 2-cyano crotonate,[644] alkyl 3-(trifluoromethyl)crotonates,[645,646] ethyl 2-(diethylamino)acry-late,[647] ethyl 2-(diethylamino)methacrylate,[647] ethyl 2-oxo-azetidine-1-acrylate,[648,649] ethylene dimethacrylate,[650] ethyl 2-ethylideneacetoacetate,[651] maleates,[638,652–654] fumarates,[628,652] diethyl ace-tamidomethylenemalonate,[655] 2-furylacrylates,[656–658] dibenzyl α-methyleneglutamate (Scheme 8.45),[659] ethyl 2-benzylacrylate,[660] ethyl 4,4,4-trichloro-2-cyano-2-butenoate,[661] and diethyl 2-(ethoxycarbo-nyl)-2-(fluorophenyl)vinylphosphonate.[662] The yields are variable (61–78%), depending on the substituents attached at phosphorus and the structure of the acrylate. On the other hand, alkyl 2-chloro- or 2-bromoacrylates, when subjected to the action of dialkyl phosphites under Michael conditions, undergo an addition–elimination reaction to provide dialkyl 2-(alkoxycarbo-nyl)vinylphosphonates.[663]

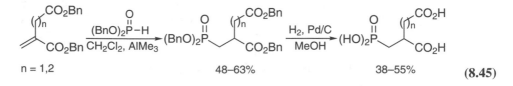

n = 1,2   48–63%   38–55%   (8.45)

Recent progress in the standard synthetic procedure has featured the use of AlMe$_3$[659,664] and tetramethylguanidine (TMG).[665,666] On addition of acrylate to a mixture of dimethyl phosphite and AlMe$_3$, high yields (89–95%) of the conjugate addition products are obtained.[664] The reaction proceeds smoothly in CH$_2$Cl$_2$ at 0°C and accommodates increasing substitution at the α- or β-carbon of the acrylate. In the presence of TMG, addition of diethyl phosphite to acrylates takes place smoothly at 0°C, giving rise to diethyl 2-(ethoxycarbonyl)ethylphosphonate in 50–70% yield after an easy workup. Recently, it has been found that 1,5,7-triazabicyclo[4.4.0]dec-5-ene (TBD) and its 7-methyl derivative (MTBD) are superior catalysts. In addition, polymer-supported TBD also proved to be an efficient promoter.[666] Dry reactions of diethyl phosphite on KF–Al$_2$O$_3$ with methyl acrylate produces good yields (80%) of diethyl 2-(methoxycarbonyl)ethylphosphonate in very simple experimental conditions.[667]

Addition of diethyl phosphite to ethyl tert-butyl 2-methyleneglutarate has been efficiently performed using solid-state K$_2$CO$_3$ as base and tetrabutylammonium hydrogen sulfate (2%) as phase-transfer reagent in the absence of any solvent. Dieckman-type cyclization is then performed in THF in the presence of NaH to give diethyl 5-(ethoxycarbonyl)-2-oxocyclopentylphosphonate, which undergoes Horner–Wadsworth–Emmons reaction with aqueous formaldehyde under heterogeneous conditions to give (±)-sarkomycin ethyl ester in good overall yields (80–100%, Scheme 8.46).[668] The addition of diethyl phosphite to acrylates can be accomplished with the K$_2$CO$_3$/EtOH catalytic system at room temperature.[669,670]

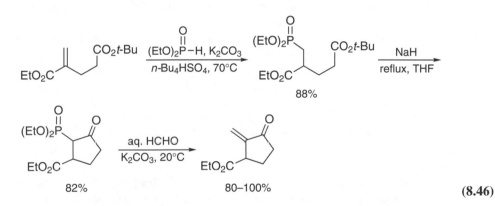

(8.46)

Michael addition has been applied to the formation of cyclopropanic systems. Thus, the addition of phosphonate carbanions generated by LDA in THF,[671] by NaH in THF,[672] by thallium(I) ethoxide in refluxing THF,[673] or by electrochemical technique[674–677] to α,β-unsaturated esters provides a preparation of substituted 2-(alkoxycarbonyl)cyclopropylphosphonates in moderate to good yields via a tandem Michael addition–cyclization sequence (Scheme 8.47). The cyclopropanation has also been achieved via oxidation with iodine in the presence of KF–Al$_2$O$_3$.[678] Nitrile ylides prepared from acyl chlorides and diethyl isocyanomethylphosphonate in the presence of Et$_3$N react with methyl acrylates by a 1,3-dipolar cycloaddition to give phosphoryl pyrrolines or pyrroles.[679]

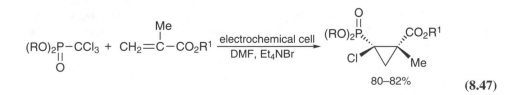

(8.47)

In Michael-type reactions, unsaturated phosphorus substrates such as dimethyl vinylphosphonate and tetraethyl ethenylidenediphosphonate undergo addition of carbanions from ethyl

2-bromoacetate[680] or methyl 2-chloropropionate[681] to give 2-(alkoxycarbonyl)cyclopropylphospho-nates in fair yields.

The reaction of trialkyl phosphites with α,β-unsaturated esters has been less thoroughly inves-tigated. By contrast with the hydrophosphonylation of acrylates with dialyl phosphites, the addition of trialkyl phosphites to α,β-unsaturated esters requires severe conditions. It involves a nucleophilic attack by the trivalent phosphorus reagent at the terminal carbon atom of the conjugated system followed by valency expansion of phosphorus in agreement with the Michaelis–Arbuzov mechanism (Scheme 8.48).

$$(EtO)_3P \ + \ CH_2\!=\!CH\!-\!CO_2Et \ \xrightarrow[90-100°C]{EtOH \ or \ PhOH} \ (EtO)_2\overset{\text{O}}{\underset{\parallel}{P}}\!-\!CH_2\!-\!CH_2\!-\!CO_2Et$$

$$50\text{--}90\%$$

$$(8.48)$$

Thus, trialkyl phosphites react with acrylates and methacrylates to afford trialkyl phosphono-propionates and phosphonoisobutyrates in low yields (20–30%). These reactions require high temperatures (140–150°C) and long reaction times in the presence of hydroquinone, so the method has only limited utility.[682] By contrast, trialkyl phosphites react smoothly with alkyl 2-bromoacryl-ates on heating to produce dialkyl 2-(alkoxycarbonyl)vinylphosphonates.[663]

A more convenient and more efficient synthetic procedure has featured the use of protic solvents (alcohol or phenol) on heating. When ethyl acrylate is treated with triethyl phosphite in these conditions, diethyl 2-(ethoxycarbonyl)ethylphosphonate is isolated in fair to good yields (50–90%, Scheme 8.48). The use of protic solvents provides a source of proton for the anionic site of the zwitterionic adduct and a nucleophile for the dealkylation step.[683] However, it has been demonstrated that superior yields of the expected phosphonates are obtained if phenol is used as the proton source–dealkylating agent rather than a simple alcohol.[683] Analogously, the reaction has been performed in the presence of an acid (AcOH, CF$_3$CO$_2$H).[684,685] For example, 2-(methoxycarbonyl)-1,4-benzoquinone reacts with trialkyl phosphites in the presence of AcOH to give dialkyl 3,6-dihydroxy-2-(methoxycarbonyl)phenylphosphonates in 30–47% yields.[684] In the presence of ammo-nium halides (NH$_4$I) as proton donors, ethyl acrylate is hydrophosphonylated in good yield (62%) by heating at 70°C for 2 h.[686,687]

Tris(trimethylsilyl) phosphite,[688] dimethyl trimethylsilyl phosphite,[412] diethyl trimethylsilyl phosphite,[689] and trialkyl phosphites in the presence of Me$_3$SiCl[690] participate in conjugate addition reactions with α,β-unsaturated esters. When ethyl crotonate is allowed to react with *tris*(trimeth-ylsilyl) phosphite in C$_6$H$_6$ at reflux, the 1,4-addition products are obtained exclusively with transfer of the silyl ester linkage to the carboxyl oxygen.[688,691] On addition of water or alcohol, the silylated products are smoothly hydrolyzed to the corresponding phosphonates. The 1,4-addition products are equally obtained by reaction of diethyl trimethylsilyl phosphite with acrylates and methacryl-ates.[689] In the presence of Me$_3$SiCl, trialkyl phosphites react with alkyl acrylates to give dialkyl 3-alkoxy-3-(trimethylsiloxy)-2-propenephosphonates in good yields.[690] Other substituted α,β-unsat-urated esters such as methyl methacrylate, methyl cinnamate, methyl crotonate, and dimethyl maleate do not bring about the reaction under the same conditions.[690]

The reactions of trialkyl phosphites with dialkyl acetylenedicarboxylates have been studied in some detail,[692–695] and the products explained as proceeding via 1:1 intermediates. Thus, the addition of a solution of dimethyl acetylenedicarboxylate to either (MeO)$_3$P or (EtO)$_3$P (2 eq) at 0°C leads to the formation of the corresponding 1,2-diphosphoranes.[694] Treatment of the diphosphorane with anhydrous HBr leads to protonation and subsequent dealkylation to give tetramethyl 1,2-*bis*(meth-oxycarbonyl)-1,2-ethylenediphosphonate in 94% yield (Scheme 8.49).[694] Trialkyl phosphites add vigorously to acetylenedicarboxylic acid in Et$_2$O to give tetraalkyl 1,2-*bis*(alkoxycarbonyl)-1,2-ethylenediphosphonates in modest yields.[696]

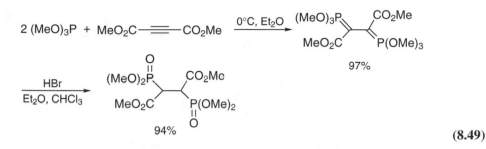

$$(8.49)$$

Similarly, the addition of diethyl phosphite to dimethyl acetylenedicarboxylate in toluene at 120°C produces tetraethyl 1,2-*bis*(methoxycarbonyl)-1,2-ethylenediphosphonate in 80% yield.[697]

The reactive 1:1 intermediate formed in the initial reaction between trialkyl phosphite and dialkyl acetylenedicarboxylates can be trapped. Thus, reaction of dialkyl acetylenedicarboxylates with trimethyl phosphite in the presence of indane-1,3-dione leads to dimethyl 2-(3-methoxy-1-oxoindenyl)-1,2-*bis*(alkoxycarbonyl)ethylphosphonate in high yields.[698] Treatment of trialkyl phosphites with C$_{60}$ and dimethyl acetylenedicarboxylate at 80°C give fullerenecyclopropyl phosphite ylides, which, on hydrolysis at ambient temperature, give the corresponding phosphonates.[699]

Addition of diethyl phosphite to ethyl propiolate or 3-substituted ethyl propiolates under basic conditions produces a mixture of mono- and bisphosphonyl esters.[700–702] When β-enaminophosphonates are allowed to react with an equimolecular amount of dialkyl acetylenedicarboxylate in THF at 50°C, substituted 2(1*H*)-pyridones containing phosphoryl and alkoxycarbonyl groups attached to the ring are obtained.[703]

The conjugate Michael addition of phosphonate-stabilized anions with dimethyl acetylenedicarboxylate has been described. For example, when the sodium salt of dimethyl 1-(methoxycarbonyl)methylphosphonate is treated with dimethyl acetylenedicarboxylate, an (*E*)/(*Z*) mixture of two isomeric β,γ-unsaturated phosphonates is isolated in modest yield (37%).[704] Addition of enolates derived from diethyl 2-oxocycloalkylphosphonates to dimethyl acetylenedicarboxylate in aprotic conditions results in [n + 2] ring-expanded products in reasonable yields (Scheme 8.50).[705] The reaction proceeds via a tandem Michael–aldol-fragmentation mechanism to give the ring enlarged products.[705]

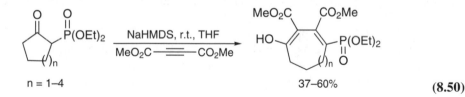

$$(8.50)$$

8.1.6.6 Phosphonomethylation

Dialkyl 1-copper(I)alkylphosphonates (1 eq), readily and quantitatively generated by transmetallation from diakyl 1-lithioalkylphosphonates, on treatment with ethyl[706,707] or 2-(trimethylsilyl)ethyl[708] oxalyl chlorides (1 eq) in THF at −30°C are converted without undesired side reactions into ethyl dialkylphosphonopyruvates in moderate to good yields (39–72%, Scheme 8.51). [^{16}O,^{17}O,^{18}O]Phosphonopyruvates of one configuration at phosphorus have also been synthesized by the same sequence.[709] Treatment of triethyl phosphonopyruvate with Me$_3$SiBr followed by hydrolysis of the *bis*(trimethylsilyl)phosphonopyruvate product gives phosphonopyruvic acid, which is isolated either by crystallization as *tris*(cyclohexylammonium) salt in 56% yield or by anion- then cation-exchange chromatography to give the free acid.[710,711]

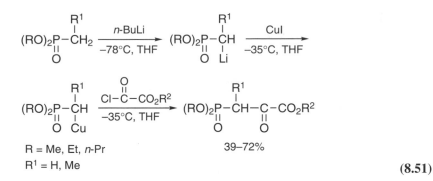

$$R = Me, Et, \textit{n}\text{-Pr}$$
$$R^1 = H, Me \qquad\qquad 39\text{–}72\%$$

(8.51)

Addition of diethyl 1-lithio-1,1-difluoromethylphosphonate to di-*tert*-butyl oxalate at low temperature in THF affords exclusively the addition product, an unusually stable *tert*-butyl hemiketal.[712] When a solution of this material is heated at reflux in C_6H_6 or chromatographed on silica gel, only a few percent is converted to the hydrate. This conversion is best effected by dissolving the hemiketal in MeCN followed by reaction with aqueous $NaHCO_3$. Subsequent azeotropic dehydration with C_6H_6 produces diethyl 1,1-difluoro-2-oxo-2-(*tert*-butoxycarbonyl)ethylphosphonate.[712]

An unusual electrophilic phosphonomethylation using diethyl phosphonomethyltriflate[713] has been reported.[714,715] A protected analogue of 3-deoxy-D-*manno*-2-octulosonic acid is metallated in the anomeric position at low temperature in THF and then reacted with diethyl phosphonomethyltriflate to give the phosphonylated β-*C*-glycosyl derivative in 50% yield.[714,715]

8.1.6.7 Other Reactions

Several other procedures have been introduced for the synthesis of phosphonopropionates. They include the oxidation at room temperature of diethyl 1-(benzyloxycarbonylamino)-3-hydroxypropylphosphonate with $KMnO_4/AcOH$ in acetone,[716] the oxidation of diethyl 2-formylethylphosphonate either with KIO_4 and ruthenium(III) chloride under biphasic conditions ($CCl_4/MeCN/H_2O$)[717] or with dimethyldioxirane in acetone,[718] the hydrogenation of diethyl 1-methyl-2-(ethoxycarbonyl)vinylphosphonate under pressure in the presence of a rhodium catalyst,[719] the thermally induced Claisen rearrangement of suitablty substituted diethoxyphosphorylallyl vinyl ethers[720] and the reaction of α-phosphoryl radical with electron-rich alkenes in toluene in the presence of *n*-$Bu_3SnH/AIBN$.[721]

Two Pd(0)-catalyzed coupling reactions are reported for the synthesis of dialkyl 2-(ethoxycarbonyl)vinylphosphonates. Thus, alkenyl triflates in DMF at room temperature and β-bromo acrylates in toluene at 80°C undergo very facile and efficient $Pd(PPh_3)_4$-catalyzed coupling with dialkyl phosphites to provide alkenyl phosphonates in high isolated yields (87–90%, Scheme 8.52).[722,723]

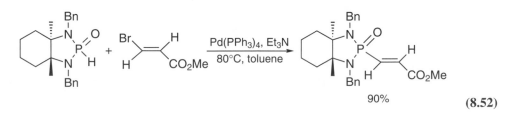

90%   (8.52)

8.1.7 Dialkyl 3-,4-,5- and 6-(Alkoxycarbonyl)alkylphosphonates

8.1.7.1 Michaelis–Arbuzov Reactions

The Michaelis–Arbuzov reaction is currently used in aliphatic, aromatic, and heteroaromatic series. In aliphatic series it is used with success for the synthesis of dialkyl 3-(ethoxycarbonyl)propyl- and

4-(ethoxycarbonyl)butylphosphonates in 55–80% yields from ethyl 4-chloro- or 4-bromobutyrate and 5-chlorovalerate and trialkyl phosphites at 180–200°C[585,724,725] and of dimethyl 5-(4-methoxy-benzyloxycarbonyl)pentylphosphonate in 62% yield from trimethyl phosphite and 4-methoxybenzyl 6-bromohexanoate at 170°C.[726]

This reaction is currently used for the preparation of synthetically useful phosphonate reagents employed in modified retinal studies. Thus, diethyl 3-alkoxycarbonyl-2-propenylphosphonates are prepared in 72–91% yields by the reaction of methyl[727] or ethyl[728,729] 4-bromocrotonates with triethyl phosphite at 150–160°C. Similarly, diethyl 3-(ethoxycarbonyl)-2-methyl-2-propenylphosphonate is prepared in 81% yield from triethyl phosphite and ethyl 3-methyl-4-chlorocrotonate by heating at 180–200°C.[260] The diethyl (E)- and (Z)-3-ethoxycarbonyl-3-fluoro-2-methyl-2-propenylphospho-nates are respectively obtained from triethyl phosphite and (E)- or (Z)-4-bromo-2-fluoro-3-methyl-2-butenoates at 140°C.[730]

The Michaelis–Arbuzov reaction has been developed in the construction of a methylene ana-logue of 3-deoxy-D-arabinoheptulosonic acid 7-phosphate (DAHP) from the C-7 bromo derivative of methyl (methyl 3-deoxy-D-arabinoheptulopyranosid)onate and triethyl phosphite in 83% yield (Scheme 8.53)[731] and of protected phosphono and phosphonomethyl derivatives of 3-dehydro-quinate (DHQ) from the corresponding halogenated compounds and trialkyl phosphites.[732,733] In the synthesis of hydroxyphosphonyl-containing mevinic acid analogues, the optically active diiso-propyl 2-(tert-butyldiphenylsilyloxy)-3-(methoxycarbonyl)proylphosphonate has been prepared at 150–160°C from triisopropyl phosphite and methyl 2-(tert-butyldiphenylsilyloxy)-1-iodobutyrate in 75% yield.[734-736]

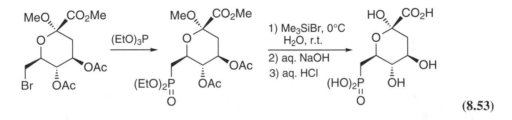

(8.53)

In the aromatic series, methyl or ethyl benzoates bearing chloro- or bromomethyl groups at the C-6 position serve essentially as electrophiles to effect the Michaelis–Arbuzov rearrangement in the presence of trialkyl phosphites.[737-742] 1-Substituted-5- or 3-bromomethyl-4-ethoxycarbonyl-3- or 5-methylpyrazoles react with triethyl phosphite at 130°C to give the corresponding phosphonates in quantitative yields.[743]

A relatively general procedure for the preparation of aromatic phosphonates containing alkox-ycarbonyl groups has been developed from triethyl phosphite and a variety of substituted haloben-zenes.[744,745] Thus, direct phosphonylation of ethyl 2-, 3-, or 4-bromobenzoates with triethyl phos-phite at 160°C, catalyzed by $NiCl_2$, affords the corresponding diethyl *ortho*-, *meta*-, or *para*-(ethoxycarbonyl)phenylphosphonates in variable yields (Scheme 8.54).

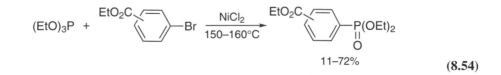

11–72%

(8.54)

8.1.7.2 Michaelis–Becker Reactions

The diethyl 5-(benzyloxycarbonyl)pentylphosphonate has been prepared from sodium diethyl phosphite and benzyl 6-bromohexanoate in DMF at room temperature.[746] Similarly, dimethyl

4-(methoxycarbonyl)benzylphosphonate is conveniently obtained in 50% yield from 4-(methoxy-carbonyl)benzyl chloride and dimethyl phosphite with MeONa in boiling C_6H_6.[747]

In the reactions of chloromethyl derivatives of furancarboxylic esters with sodium diethyl phosphite in C_6H_6 at 80°C, the corresponding phosphonates are formed according to the Michaelis–Becker scheme. By contrast, when the bromo derivatives are used, the formation of both phosphonates and reduction products resulting from nucleophilic and halophilic attack by the phosphite anion takes place (Scheme 8.55).[748–751]

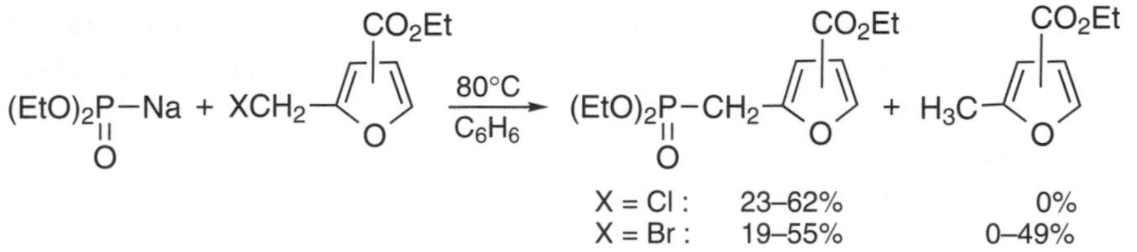

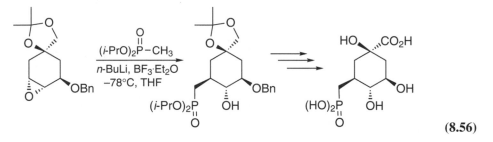

$$X = Cl : \quad 23\text{–}62\% \qquad 0\%$$
$$X = Br : \quad 19\text{–}55\% \qquad 0\text{–}49\%$$

(8.55)

8.1.7.3 Phosphonomethylation

In the synthesis of 5-phosphonomethyl-5-deoxyquinate, the phosphonomethyl moiety is introduced by the intermediacy of the ring-opening reaction of epoxides by the diisopropyl 1-lithiomethylphosphonate at −78°C in THF in the presence of $BF_3 \cdot Et_2O$ (Scheme 8.56).[752,753]

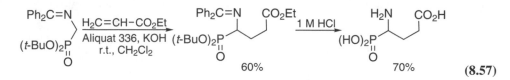

(8.56)

8.1.7.4 Michael Additions

The synthesis of dialkyl 3-(alkoxycarbonyl)propylphosphonates by Michael addition can be effected by two different routes, with the phosphorus reagent being the nucleophile or the Michael acceptor. Thus, on treatment with diethyl 1-lithioalkylphosphonates in THF at low temperature, *tert*-butyl crotonates undergo stereoselective 1,4-addition to give diethyl 1,2-dialkyl-3-(*tert*-butoxycarbonyl)propylphosphonates in good yields (69–84%). The use of methyl, ethyl, or isopropyl crotonate gives a mixture of 1,2- and 1,4-adducts.[754]

Addition of phosphonylated Schiff base to ethyl acrylate, under phase-transfer catalytic conditions in the presence of Aliquat 336 and KOH at room temperature, leads to the 1:1 Michael adduct with 60% yield (Scheme 8.57).[755,756] The *tert*-butyl and ketimine protecting groups are eliminated under mild conditions (1 M HCl, room temperature) to provide the phosphonic analogue of glutamic acid.

$$\text{(8.57)}$$

Addition of the lithium derivative of the Schiff base formed from (+)-camphor and diethyl aminomethylphosphonate to substituted acrylic esters proceeds smoothly with a high degree of stereoselectivity.[757,758] After hydrolysis of the adducts, the enantiomerically pure 3- and 4-substituted phosphonic derivatives of pyroglutamic acid are obtained. The esters are converted into diastereo-merically pure phosphonoprolines by reduction of the lactams with LiBH$_4$/BF$_3$·Et$_2$O (Scheme 8.58).[757,758] The strategy for stereocontrolled generation of 2-(ethoxycarbonyl)cyclopro-pylphosphonoglycine involves the 1,4-Michael addition of the lithium derivative of the Schiff base formed from chiral 2-hydroxypinan-3-one and diethyl aminomethylphosphonate to ethyl 4-bromo-crotonate.[759]

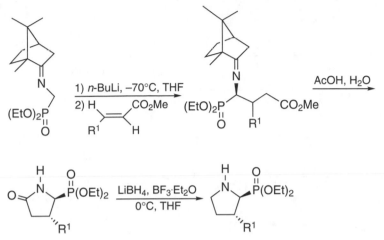

(8.58)

The method of choice for the reaction of diethyl 2-oxocyclohexylphosphonate with ethyl acrylate involves the use of a catalytic amount of EtONa in EtOH. With these conditions, good yields (69%) of Michael adduct are obtained (Scheme 8.59).[760] Use of a molar amount of base in an aprotic solvent results in recovery of only unreacted starting material. The reaction of diethyl 2-oxocyclohexylphosphonate is prone to steric factors because no Michael addition occurs when β-substituted Michael acceptors such as methyl crotonate or ethylidene malonate are used.[760]

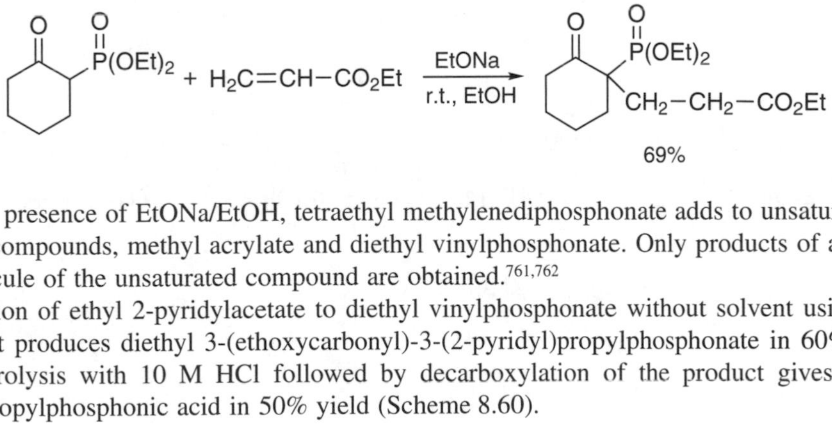

O=C—cyclohexyl—P(OEt)$_2$ + H$_2$C=CH—CO$_2$Et →[EtONa / r.t., EtOH] product

69%

(8.59)

In the presence of EtONa/EtOH, tetraethyl methylenediphosphonate adds to unsaturated elec-trophilic compounds, methyl acrylate and diethyl vinylphosphonate. Only products of addition to one molecule of the unsaturated compound are obtained.[761,762]

Addition of ethyl 2-pyridylacetate to diethyl vinylphosphonate without solvent using EtONa as catalyst produces diethyl 3-(ethoxycarbonyl)-3-(2-pyridyl)propylphosphonate in 60% yield.[763] Acid hydrolysis with 10 M HCl followed by decarboxylation of the product gives the 3-(2-pyridyl)propylphosphonic acid in 50% yield (Scheme 8.60).

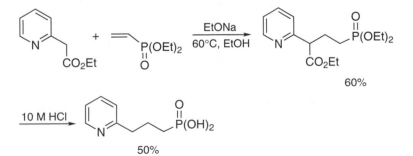

(8.60)

Addition of diethyl malonate to diethyl vinylphosphonate,[764,765] tetraethyl ethenylidenediphos-phonate,[766] and diethyl 1-ethenylidene-2-oxoalkylphosphonates[767] under basic conditions has been described. The *gem*-diphosphonate resulting from the addition of diethyl malonate to tetraethyl ethenylidenediphosphonate in the presence of EtONa/EtOH after saponification with KOH in aqueous THF and acidification is decarboxylated to give the tetraethyl 3-(hydroxycarbonyl)-1,1-propylidenediphosphonate in 65% yield (Scheme 8.61).[474,766]

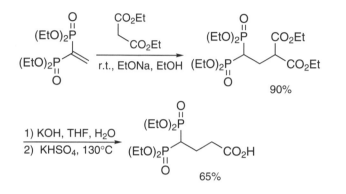

(8.61)

Addition of ethyl *N*-benzylideneglycinate to tetraethyl ethenylidenediphosphonate in EtONa/EtOH medium produces tetraethyl 3-amino-3-(ethoxycarbonyl)-1,1-propylidenediphospho-nate in excellent yield after deprotection (95%, Scheme 8.62).[766,768,769] A complementary procedure uses DBU in THF. Under these conditions, the reaction of methyl benzoate with tetraethyl ethe-nylidenediphosphonate proceeds well at 50°C to give tetraethyl 3-phenyl-3-(ethoxycarbonyl)-1,1-propylidenediphosphonate.[770]

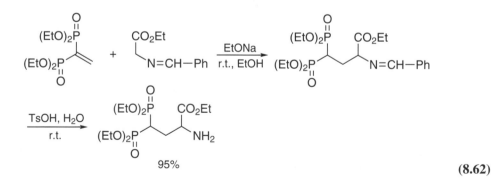

(8.62)

The lithium salt of diethyl 2-propenylphosphonate behaves as an ambident nucleophile in reactions with a variety of electrophilic reagents. In reaction with methyl methacrylate, the inter-mediate formed in the first step (conjugate addition at the γ-carbon of the phosphonate) has a carbanionic center developed at C-5. The second step involves intermolecular conjugate addition to a second molecule of methyl methacrylate, yielding an intermediate capable of the final 1,6-cyclization to produce a phosphonylated polyfunctionalized derivative of cyclohexane in good overall yield (60%, Scheme 8.63).[771,772]

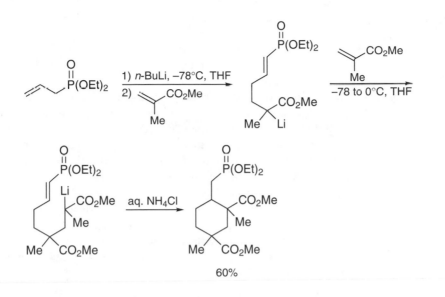

(8.63)

The addition of trihalogenomethylphosphonates to ethyl acrylate has also been reported. Thus, in the presence of the cobaloxime(III)/Zn redox system, diethyl bromodifluoromethylphosphonate adds smoothly to ethyl acrylate in EtOH at room temperature to give the 1:1 Michael adduct in good yields (60–67%).[773] Ethyl methacrylate gives satisfactory results (72%), whereas ethyl crotonate gives a lower yield (34%). This has been attributed to steric effects. Similarly, the copper(I)-catalyzed addition of diethyl trichloromethylphosphonate to ethyl acrylate gives the Michael adduct in satisfactory yield (59%).[774]

8.1.7.5 Alkylation of Phosphonyl or Carboxyl Carbanions

Room-temperature alkylation of the sodium enolate of diethyl 2-oxopropylphosphonate with ethyl 4-iodobutyrate in THF for 10 days provides the expected α-substituted β-ketophosphonate in 55% yields. Further attempts to improve the process using higher temperature, additives (HMPA), potassium instead of the sodium enolate, or excess of 4-iodobutyrate do not increase the yield.[775]

The *reverse* reaction has been reported. Thus, low-temperature treatment of sodium enolates of ethyl 2-methylacetoacetate or ethyl 2-oxocyclopentanecarboxylate with diethyl 3-iodo-1-propynylphosphonate in THF gives the corresponding alkylation products in better yields. These acetylenic phosphonates are hydrolyzed with acidic HgSO$_4$ to produce diethyl 4-(ethoxycarbonyl)-2,5-dioxoalkylphosphonates in good yields (78–80%, Scheme 8.64).[776]

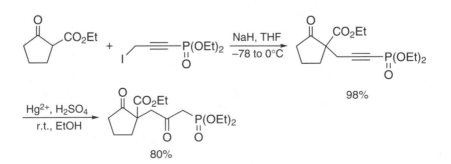

(8.64)

8.1.7.6 Horner–Wadsworth–Emmons Reactions

Preparation of phosphonate analogues of phosphates is conveniently accomplished by the Horner–Wadsworth–Emmons olefination of carbonyl compounds using the anion of tetraethyl methylenediphosphonate followed by catalytic hydrogenation. This attractive synthetic procedure, which is of general applicability, has been developed for the preparation of the methylene and fluoromethylene analogues of 3-phospho-D-glyceric acid.[777] The Horner–Wadsworth–Emmons condensation of methyl 2,3-O-isopropylidene-β-D-ribopentodialdehydo-1,4-furanoside with tetraethyl 1-lithio-1-fluoromethylenediphosphonate provides exclusively the (E)-isomer of α-fluoroalkenylphosphonate. This compound, on catalytic reduction using Pd/C, is converted to the α-fluorophosphonate as a 1:1 mixture of diastereomers. Attempts to achieve a stereoselective reduction using RhCl(PPh₃)₃ or RuHCl(PPh₃) were not successful. The corresponding condensation with the *lyxo*-isomer smoothly provides the α-fluorovinylphosphonate exclusively as the (E)-isomer, which undergoes catalytic hydrogenation with Pd/C to give the α-fluorophosphonate as an 85:15 mixture of diastereomers. This product is converted into a mixture of the α- and β-anomers of the methyl furanoside, oxidized with periodic acid, and then dealkylated to give the α-fluoromethylphosphonate analogue of 3-phospho-D-glyceric acid (Scheme 8.65).[777]

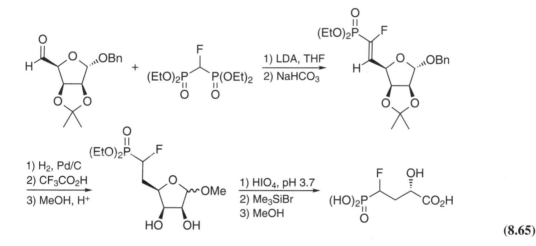

(8.65)

Similarly, the methylene analogue of 3-deoxy-D-arabinoheptulosonic acid 7-phosphate (DAHP) has been prepared from the C₇ aldehyde derived from methyl (methyl 3-deoxy-D-arabinoheptulopyranosid)onate in 60% yield.[778] The protected 3-dehydroquinate (DHQ) has also been used as starting material in the synthesis of phosphonic DHQ synthase inhibitors by the same methylenediphosphonate approach in 56% yield (Scheme 8.66).[733,779] Preparation of the phosphonate analogue of porphobilinogen is another illustration of the advantages of this synthetic procedure.[780,781]

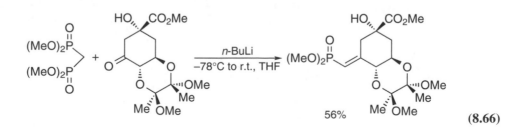

(8.66)

8.1.7.7 Addition Reactions

Addition of diethyl 1-iodo-1,1-difluoromethylphosphonate to ethyl 2-methyl-4-pentenoate catalyzed by Pd(PPh$_3$)$_4$ proceeds readily at room temperature in the absence of solvent to give a good yield of the corresponding adduct in a few minutes.[782,783] On treatment of iododifluoroacetate with diethyl allyl- or homoallylphosphonates, neat, in the presence of copper at 50–60°C, the corresponding adducts are isolated in 77% and 88% yields, respectively.[784] Similarly, the stable diethyl difluoromethylphosphonyl zinc bromide, prepared from zinc dust and diethyl bromodifluoromethylphosphonate, reacts with 2-bromoacrylic acid in THF at room temperature in the presence of catalytic amount of CuBr to give the 1,4-adduct in low yield (20–33%).[785,786] Reactions with methyl 2-(bromomethyl)acrylate and 2-(bromomethyl)acrylic acid give the corresponding 1,4-adducts in better yields (49% and 83%, respectively), resulting from an addition–elimination process.

An efficient, versatile protocol for the synthesis of highly enantioenriched α-aminophosphonate has been devised. The addition of lithium diethyl phosphite in THF at room temperature to imines prepared from methyl and methoxymethyl (MOM) ethers of (R)-(−)-2-phenylglycinol and a wide variety of aldehydes has been explored.[787,788] The reaction generates predominantly the (R,R) diastereomers whose hydrogenolysis produces α-aminophosphonates in good yields (Scheme 8.67).

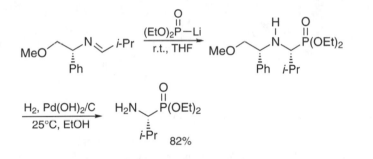

$$(8.67)$$

Zinc and copper β-metallated alkylphosphonates, prepared by the insertion of zinc into dialkyl β-bromoalkylphosphonates followed by transmetallation with CuCN·2LiCl, react with ethyl propiolate and dimethyl acetylenedicarboxylate to give stereospecifically the *syn*-adduct in 85% and 91% yields, respectively.[789]

8.1.7.8 Other Reactions

Preparations of aromatic phosphonates containing alkoxycarbonyl groups have been developed by the Pd(0)-catalyzed reaction of aryl bromides[790] or aryl triflates[791] with dialkyl phosphites. Although these two procedures employ the same catalyst, Pd(PPh$_3$)$_4$, at the same temperature in the presence of tertiary amine, the use of triflates increases reaction rates and yields (Scheme 8.68).[791]

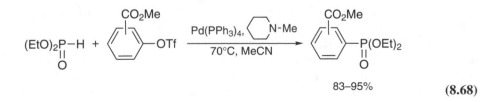

83–95%

$$(8.68)$$

Under microwave irradiation, methyl 3- and 4-iodobenzoates react with diethyl phosphite in the presence of (Ph$_3$P)$_2$PdCl$_2$ and Et$_3$SiH. The yields are comparable to those obtained in the classical conditions,[790] but the reaction rates are dramatically increased, and the workup is very simple.[792]

The palladium-catalyzed coupling reaction of dialkyl 1-copper(I)methylphosphonates with aryl iodides proceeds readily in the presence of $Pd(OAc)_2$ (4 mol%) and dppe on refluxing THF to give arylphosphonates containing alkoxycarbonyl groups in good yields (75%).[793] Similarly, when (E)-β-iodoacrylate is treated with the zinc reagent generated from diethyl 1-bromo-1,1-difluoromethylphosphonate in the presence of a stoichiometric amount of CuBr in DMF at room temperature, the desired coupling reaction proceeds to give diethyl (E)-3-(ethoxycarbonyl)-1,1-difluoro-2-propenylphosphonate in 79% isolated yield.[794] This reaction, which proceeds with complete retention of the starting geometry, has been extended to (Z)-iodoacrylates, 3-iodocrotonates, and 3-iodomethacrylates in reasonable yields.[794]

8.1.8 DIALKYL 1-OXO-ω-(ALKOXYCARBONYL)ALKYLPHOSPHONATES

The preparations and properties of diethyl 1-(ethoxycarbonyl)-1-oxomethylphosphonate (triethyl phosphonoglyoxylate) have recently been described.[795] The dialkyl 2-ethoxycarbonyl-1-oxoethylphosphonates are prepared in moderate yields by a Michaelis–Arbuzov reaction between trialkyl phosphites and ethyl 3-chloro-3-oxopropionate (Scheme 8.69).[796–798] Dialkyl 3-methoxycarbonyl-1-oxopropylphosphonate[799–801] and 5-ethoxycarbonyl-1-oxopentylphosphonate[802] are prepared similarly.

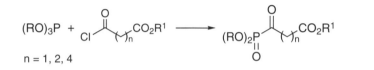

$$(8.69)$$

8.1.9 DIALKYL 2-OXO-ω-(ALKOXYCARBONYL)ALKYLPHOSPHONATES

Because dialkyl 2-oxo-ω-(alkoxycarbonyl)alkylphosphonates are valuable synthetic intermediates that may be either converted to the corresponding amino compounds[803–805] or treated with a variety of carbonyl compounds in a Horner–Wadsworth–Emmons reaction,[593,806–831] procedures that effect their preparation are of special importance. The Michaelis–Arbuzov reaction, which furnishes low yields, does not appear to be an appropriate method for the preparation of dialkyl 2-oxo-ω-(alkoxycarbonyl)alkylphosphonates.[593] One of the most attractive synthetic methods involves the chemoselective reaction of α-metallated dialkyl alkylphosphonates with acylating reagents such as carboxylic acid chlorides, cyclic anhydrides, or esters.

The major disadvantage that attends the use of α-metallated alkylphosphonates in reaction with carboxylic acid chlorides is the difficulty that may be encountered in their chemoselective coupling in the presence of a carboxylic ester moiety. For example, dimethyl 1-lithiomethylphosphonate reacts with a carboxylic acid chloride bearing a carboxylic ester function in the γ position to give substantial amounts of products of addition to both the ester and acid chloride carbonyls.[817] By contrast, dialkyl 1-copper(I)alkylphosphonates (1 eq), which is readily and quantitatively generated by transmetallation from diakyl 1-lithioalkylphosphonates, on treatment with carboxylic acid chlorides (1 eq) are converted without undesired side reactions into dialkyl 2-oxo-alkylphosphonates containing a carboxylic ester group (Scheme 8.70)[706,821,832–835] or an N-(benzyloxycarbonyl)-4-oxazolidone group[836] in good to excellent yields. The readily available dialkyl 1-bromo-1,1-difluoromethylphosphonates can also participate in the synthesis of functionalized β-ketophosphonates. Thus, dibutyl 1-bromo-1,1-difluoromethylphosphonate, on reaction with activated zinc metal, is converted into difluoromethylphosphonylzinc bromide, which is coupled with carboxylic acid chlorides in triglyme at room temperature to give dibutyl 1,1-difluoro-2-oxoalkylphosphonates containing a carboxylic ester function in the δ position.[837]

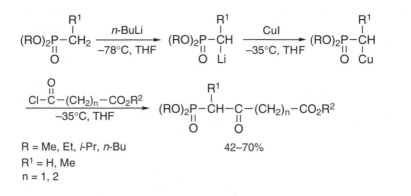

$$R = Me, Et, i\text{-}Pr, n\text{-}Bu$$
$$R^1 = H, Me$$
$$n = 1, 2$$

42–70%

(8.70)

Instead of carboxylic acid chloride, 3-acylthiazolidine-2-thiones proved to be suitable to produce 2-oxoalkylphosphonates.[838] These acylating reagents are prepared by condensation of the methyl monoesters of ω,ω′-diacids with 2-mercaptothiazoline. Addition of 3-acylthiazolidine-2-thiones to dimethyl 1-lithiomethylphosphonate (2.5 eq) leads to functionalized 2-oxoalkylphosphonates in 61–70% contaminated by a slight amount of symmetric ω,ω′-dioxoalkylphosphonate resulting from attack at the ester function.[838]

Cyclic anhydrides such as succinic, glutaric, and adipic anhydrides (1 eq) undergo nucleophilic opening with dialkyl 1-lithioalkylphosphonates (2 eq) to provide the intermediate β-ketophosphonates in the chelated enolate form bearing a lithium carboxylate salt. No further reaction can normally occur because the lithium carboxylate function is not reactive toward 1-lithioalkylphosphonates. Protonation of the lithiated enolate with oxalic acid in MeOH followed by esterification of the carboxylic acid with diazomethane in Et$_2$O gives in fair to good yields the expected functionalized β-ketophosphonates (Scheme 8.71).[806,838,839]

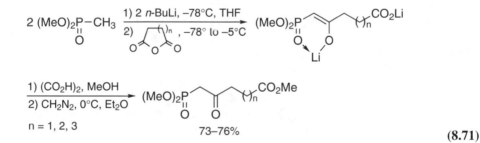

n = 1, 2, 3   73–76%

(8.71)

Another useful technique for the preparation of dialkyl 2-oxo-ω-(alkoxycarbonyl)alkylphosphonates is based on the reaction of dimethyl 1-lithiomethylphosphonate with appropriately protected and functionalized symmetric or disymmetric carboxylic esters.[803,811,840–847] For example, treatment of methyl tert-butyl glutarate with dimethyl 1-lithiomethylphosphonate (4 eq) produces dimethyl 2-oxo-5-(tert-butoxycarbonyl)pentylphosphonate in 28% yield.[803]

The same methodology has been extensively developed for the synthesis of optically active β-ketophosphonates from (3R,1′R)-methyl 1′-phenylethyl 3-hydroxypentanedioate. Low-temperature condensation of this unprotected hydroxy diester with dimethyl 1-lithiomethylphosphonate in excess in THF gives the desired optically active functionalized β-ketophosphonate in fair yield (43%). It has been observed that the methyl ester reacts faster than the 1′-phenethyl ester, and it is necessary to operate without the silyl protecting group to avoid a β-elimination reaction.[813,814,816,818–820] By an analogous route, the reaction of dimethyl 1-lithiomethylphosphonate (1.3 eq) with the corresponding protected hydroxy ester–amide (Nahm–Weinreb amide) provides the same β-ketophosphonate in better yield (62%, Scheme 8.72).[848] At low temperature, the attack of the phosphonate anion occurs exclusively at the amide, and no product resulting from the elimination of lithium *tert*-butyldimethylsilanolate is obtained.

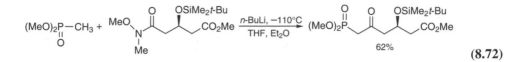

$$(8.72)$$

The addition of dimethyl 1-lithiomethylphosphonate (5.5 eq) to the protected amide–acid (Nahm–Weinreb amide) gives a clean reaction, and treatment of the crude product with CH_2N_2 affords the β-ketophosphonate in high yield (77%, Scheme 8.73).[822,823] To avoid the use of CH_2N_2, an alternative and safer route uses a reactive phosphonyl enol lactone, which is converted into the same β-ketophosphonate by treatment with MeONa in MeOH.[827,849]

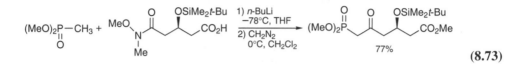

$$(8.73)$$

The synthesis of this valuable optically active β-ketophosphonate has been improved by using the protected hydroxy half-ester as substrate.[850] Thus, it has been found that low-temperature treatment of the protected 1-adamantanamine salt of chiral monomethyl ester in THF with dimethyl 1-lithiomethylphosphonate in excess followed by esterification of the crude product with CH_2N_2 affords the optically active dimethyl 2-oxo-4-*tert*-butyldimethylsilyloxy-5-(methoxycarbonyl)pentylphosphonate in 78–82% overall yield from half-ester salt.[850]

Analogously, it has been observed that treatment of dimethyl homophthalate with dimethyl 1-lithiomethylphosphonate provides only very low yields of the desired β-ketophosphonate.[851] Competitive enolization of the acidic benzylic methylene appears to be the dominant reaction under these conditions. The unwanted enolization is partially suppressed when the monomethyl ester is used as substrate.[851]

The addition of dimethyl 1-lithiomethylphosphonate to *N*-protected methyl *tert*-butyl aspartate[836,852] or to *tert*-butyl *N*-(PhF)-ω-(methoxycarbonyl)-α-aminoalkanoates[830] has been investigated in depth. In the case of *tert*-butyl methyl (2*S*)-*N*-(PhF)-α-aminoglutamate, it appears that with *n*-BuLi as base the yield of β-ketophosphonate increases with lower concentrations of dimethyl methylphosphonate in THF. Switching solvents from THF to Et_2O increases the reaction yields and selectivity.[830]

A general and convenient procedure for the synthesis of diethyl 2-oxo-3-(ethoxycarbonyl)alkylphosphonates involves the γ-acylation of 1,3-dianions generated from the readily available 2-oxoalkylphosphonates.[804,805,853] Deprotonation of diethyl 2-oxoalkylphosphonates with NaH (1 eq) in THF at room temperature followed by LDA (2 eq) at 0°C leads to the quantitative generation of the intermediate stable 1,3-dianions. These undergo smooth nucleophilic acylation with ethyl chloroformate yielding a *bis*-chelated enolate. In view of the acidity of these products, it is necessary to use an excess of LDA (2 eq) to secure the quantitative generation of the bis-chelated enolate. The overall yields of diethyl 2-oxo-3-(ethoxycarbonyl)alkylphosphonates obtained by this procedure are in the range 72–88% (Scheme 8.74).[804,805,853] The advantages of this route are more clearly evidenced by comparison with the Michaelis–Becker reaction. For example, formation of diethyl 2-oxo-3-(ethoxycarbonyl)propylphosphonate has been reported in only 50% yield by the Michaelis–Becker reaction between sodium diethyl phosphite and sodium γ-bromoacetoacetate.[807,809,810,812,815,854–856]

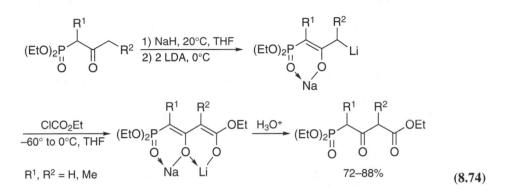

$$(8.74)$$

An attractive procedure for the preparation of 2-oxo-ω-(alkoxycarbonyl)alkylphosphonates involves the C-phosphonylation of β-keto ester dianions with dialkyl chlorophosphate (45–67%)[857] or C-phosphirylation of the monoanion of substituted dioxinones with diethyl chlorophosphite followed by oxidation.[858]

A procedure for the transformation of a formyl moiety into β-ketophosphonate has been described. It involves the selective addition of dimethyl 1-lithiomethylphosphonate to the formyl group followed by oxidation with CrO_3-Py of the intermediate β-hydroxyphosphonate.[859]

Alkylation of the lithium enolate of 2-(methoxycarbonyl)tricyclo[3.3.0.0^{2,8}]octan-3-one with dimethyl 2-ethoxy-3-iodo-1-propenylphosphonate affords reasonable yields of the enol ether, which may then be converted into the corresponding β-ketophosphonate in 87% yield by HCl-acetone hydrolysis at 0°C.[808]

The diethyl 2-oxo-4-ethoxycarbonyl-3-triphenylphosphoranylidenebutylphosphonate has been prepared in 64% yield by reaction of diethyl 1-(chlorocarbonyl)methylphosphonate with (ethoxycarbonyl)methylenetriphenylphosphorane (2 eq) in C_6H_6 at room temperature.[409,860]

8.1.10 ω-AMINO-ω-(HYDROXYCARBONYL)ALKYLPHOSPHONIC ACIDS

Excitatory amino acids (EAA) such as glutamic acid have been found to be important neurotransmitters in the mammalian central nervous system. These excitatory amino acids exert their effects on four major subtypes of receptors: the N-methyl-D-aspartic acid (NMDA), the kainic acid, the D,L-α-amino-3-hydroxy-5-methylisoxazole-4-propionic acid (AMPA), and the metabotropic receptors. Abnormal physiologic conditions, such as epilepsy, Huntington's chorea, memory disorders, and neuronal damage, which occur following an ischemic episode, have been associated with hyperactivity of these receptor complexes. The utility of selective antagonists of EAA receptors was first demonstrated by competitive NMDA-receptor antagonists such as 4-(3-phosphonopropyl)-2-piperazinecarboxylic acid, which has been shown to prevent ischemic brain damage in gerbils, to prevent NMDA-induced convulsions in mice, and to prevent 1-methyl-4-phenyl-1,2,3,6-tetrahydropyridine (MPTP)-induced Parkinsonian-like symptoms in rats. For these reasons, NMDA-receptor antagonists have been considered potential therapeutic agents for the treatment of epilepsy, stroke, and neurodegenerative disorders such as Alzheimer's and Parkinson's diseases. Several studies have shown that phosphonate analogues of glutamic acid with side chain lengths of four to six carbon atoms [(R)- and (S)-AP-4, (R)- and (S)-AP-5, and (RS)-AP-6] are activators of the NMDA receptor site, AP-5 being the most potent (Figure 8.1).[861]

Generally, ω-amino-ω-(hydroxycarbonyl)alkylphosphonic acids are made by one of two general reaction classes, which can be differentiated by the stage at which the phosphorus–carbon bond is introduced. In this particular reaction the P–C bond formation may precede (Sections 8.1.10.1–3) or follow (Sections 8.1.10.4–6) the incorporation of the aminocarboxylic functional group.

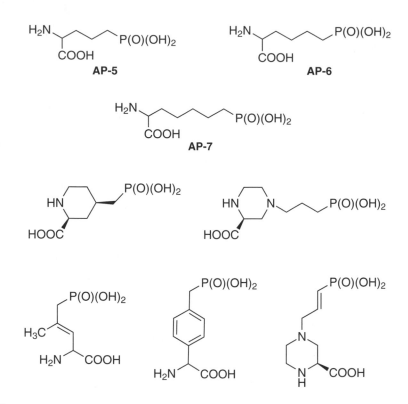

FIGURE 8.1

8.1.10.1 Strecker and Related Reactions

The Strecker and Bucherer–Berg reactions for the synthesis of ω-amino-ω-carboxyalkylphosphonic acids have been applied to a variety of phosphonylated aldehydes. Formylphosphonates are first converted to amino nitriles and then treated with concentrated HCl to provide ω-amino-ω-(hydroxycarbonyl)alkylphosphonic acids. In this method, treatment in the absence of light of freshly purified phosphonylated aldehydes at room temperature in water (or MeOH/H$_2$O or MeCN) with NaCN (or KCN) and NH$_4$Cl (or NH$_4$OH, HCO$_2$NH$_4$, or (NH$_4$)$_2$CO$_3$ at 50–60°C in the Bucherer–Berg reaction) produces selectively amino nitriles in moderate to high yields (44–91%, Scheme 8.75).[862–870] Improved yields are reported with the use of alumina/ultrasound[866] or modified Strecker reaction.[871] Hydrolysis of amino nitriles with 6–8 M HCl at reflux delivers the crude amino acids, purified by ion-exchange chromatography using acidic ion-exchange resin (Dowex 50W). The Bucherer–Berg route is not so efficient because of difficulties in the hydrolysis of hydantoin.[863]

$$(EtO)_2P(O)-CH(R^1)-(CH_2)_n-CHO \xrightarrow[H_2O]{H_2NR^2} (EtO)_2P(O)-CH(R^1)-(CH_2)_n-CH=NR^2 \xrightleftharpoons{HCN}$$

$$(EtO)_2P(O)-CH(R^1)-(CH_2)_n-CH(NHR^2)-CN \xrightarrow[\text{2) Dowex 50W}]{\text{1) 8 M HCl}} (HO)_2P(O)-CH(R^1)-(CH_2)_n-CH(NHR^2)-CO_2H$$

n = 1,2
R^1 = H, Me, Et
R^2 = H, Me, Et, n-Bu, allyl

(8.75)

In analogy with formylphosphonates, β-ketophosphonates can be converted by Strecker and Bucherer–Berg reactions into ω-amino-ω-(hydroxycarbonyl)alkylphosphonic acids.[872–875] The amino phosphonocyclopentane- and phosphonocyclohexanecarboxylic acids, the conformationally restricted L-2-amino-4-phosphonobutanoic acid (APB) analogues, are prepared from the racemic diethyl 3-oxocyclopentyl and 3-oxocyclohexylphosphonates. With NH$_4$Cl and NaCN in concentrated NH$_4$OH they are converted into aminonitriles in good yields as a mixture of *cis*- and *trans*-isomers, which are separated by column chromatography. Acid hydrolysis in 6 M HCl provides *cis*- and *trans*-3-amino-3-(hydroxycarbonyl)cyclopentyl- or cyclohexylphosphonic acids purified by ion-exchange chromatography (Scheme 8.76).[873]

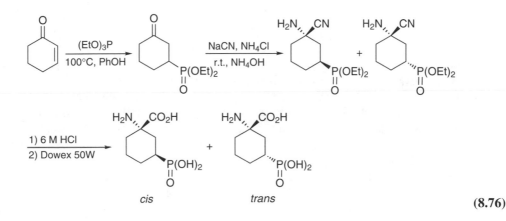

(8.76)

Similarly, diethyl 2-(3-oxocyclopentyl)- and 2-(4-oxocyclohexyl)ethylphosphonates have been subjected to the standard Bucherer–Berg reaction conditions [KCN, (NH$_4$)$_2$CO$_3$, NH$_4$Cl in EtOH/H$_2$O at 60°C] to give spirohydantoins, which are isolated as a mixture of *cis*- and *trans*-isomers. Hydrolysis of hydantoins requires heating at 150°C in a sealed tube with concentrated HCl to provide the amino acid hydrochlorides.[874]

Several attemps have been undertaken to develop synthetic protocols that would allow preparation of optically active compounds. Early reported preparations of optically active ω-amino-ω-carboxyalkylphosphonic acids describe the preparation of (*S*)-AP-3 from an optically active amino nitrile prepared by reaction of diethyl 1-formylmethylphosphonate with hydrogen cyanide and (*S*)-(−)-α-methylbenzylamine.[876] Acid hydrolysis, enrichment of the diastereomers by fractional recrystallization, and debenzylation lead to (*S*)-AP-3 with 66% yield and 86% ee.[876] Recently reported procedures that use chemoenzymatic processes offer a particularly attractive approach for the mild generation of optically active aminophosphonic acids. Enzymatic hydrolysis of amides using penicillinacylase (EC 3.5.1.11) from *Escherichia coli* provides a mild, highly enantioselective, high-yielding method of resolution.[869] The enzymatic hydrolysis of esters uses Subtilisin A[867] or Carlsberg esterase in phosphate buffer.[877] For example, hydrolysis of a diastereomeric mixture of hydantoins, epimeric at the C-5 hydantoin carbon atom, with D-hydantoinase enzyme from a strain of *Agrobacterium* in alkaline buffer results in efficient conversion into the D-carbamoyl acid.[867]

Another route to ω-amino-ω-(hydroxycarbonyl)alkylphosphonic acids involves the intermediacy of ethyl α-azidoacetate. Addition of MeONa in MeOH to a solution of dialkyl 4-formylbenzylphosphonate and ethyl α-azidoacetate in MeOH at −30°C provides the vinyl azide. Subsequent hydrogenation with 10% Pd/C in MeOH gives the aminoester, which, at reflux with 3 M HCl, delivers the free amino acid (Scheme 8.77).[878–880]

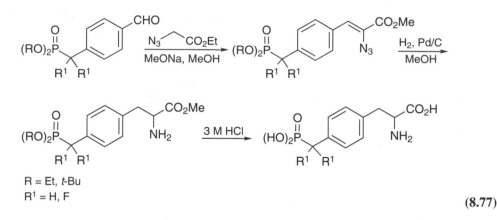

CHO
O
(RO)2P
R1 R1
N3 CO2Et
MeONa, MeOH
CO2Me
N3
H2, Pd/C
MeOH
CO2Me
NH2
3 M HCl
(HO)2P
CO2H
NH2
R = Et, t-Bu
R1 = H, F

(8.77)

An attractive procedure for the preparation of 2-aminocarboxyalkylphosphonic acids from phosphonopyruvates has been developed using reductive amination of the carbonyl group in the presence of $NaBH_3CN$.[881–883] The reaction is run at pH 7 in EtOH and is general for ammonia and primary amines. Steric hindrance represents a limiting factor because α-substituted phosphonopyruvates react sluggishly and secondary amines do not react. On hydrolysis with 6 M HCl a large variety of phosphonoalanine derivatives are isolated in fair to good yields (25–70%).[881–883]

An asymmetric synthesis of 2-amino-4-phosphonobutanoic acid has been developed from diethyl 3-oxo-4-(ethoxycarbonyl)propylphosphonate and L-*erythro*-(1*S*,2*R*)-1,2-diphenyl-2-hydroxylamine as chiral source. Formation of Schiff's base and acid-catalyzed transesterification are successfully performed in boiling *n*-BuOH. After reduction of the imino linkage with Al–Hg in moist DME, debenzylation [H_2, $Pd(OH)_2/C$] and hydrolysis with 4 M HCl, (*R*)-(−)-2-amino-4-phosphonobutanoic acid is isolated in 66% yield and 67% ee.[884]

8.1.10.2 Coupling of Glycine Synthons with ω-Halogenated Phosphonates

Metallated racemic or chiral glycine synthons such as sodium acetamidomalonate (Scheme 8.78),[885–893] potassium *N*-(diphenylmethylene)glycine esters,[890] potassium *N*-benzylideneglycine esters,[892,894] lithium *bis*lactim ethers,[895,896] sodium oxazinone derived from D-phenylglycinol,[897] lithium 1,3-imidazolidin-4-ones,[898–900] nickel(II) complex of the Schiff base derived from glycine and (*S*)- or (*R*)-*N*-(*N*-benzylprolyl)-*ortho*-aminobenzophenone,[901–904] and 1-(phenylmethyl) (+)-1,2-piperazinedicarboxylate[905] are highly versatile reagents that may be smoothly alkylated with a variety of dialkyl ω-bromoalkylphosphonates to produce with high diastereoselectivity the target intermediates.

Refluxing these intermediates with 6 M HCl results in the removal of all protecting groups to afford, after purification using a strongly acidic ion-exchange resin (Dowex 50W), the free aminocarboxyalkylphosphonic acids in good overall yields (65–70%, Scheme 8.78). However, complete hydrolysis requires temperatures equal to 140°C or higher, partial racemization is observed at 140°C, and complete racemization is found at 180°C. The intramolecular participation of the phosphoryl group might be involved in the racemization process (Scheme 8.79).[899]

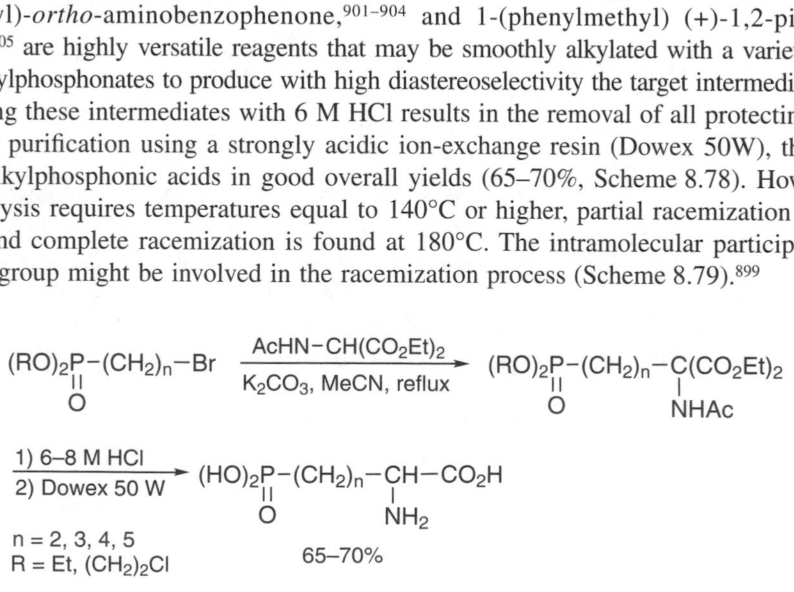

(RO)2P−(CH2)n−Br
O
AcHN−CH(CO2Et)2
K2CO3, MeCN, reflux
(RO)2P−(CH2)n−C(CO2Et)2
NHAc
1) 6–8 M HCl
2) Dowex 50 W
(HO)2P−(CH2)n−CH−CO2H
NH2
n = 2, 3, 4, 5
R = Et, (CH2)2Cl
65–70%

(8.78)

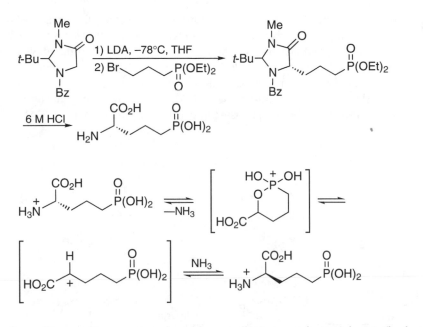

$$(8.79)$$

Starting from dibenzoyl-D- or L-tartaric acid, pure diastereomeric ω-amino-ω-(hydroxycarbo-nyl)alkylphosphonic acids are isolated by repeated crystallizations of dibenzoyl-D- or L-tartrates followed by treatment with Na_2CO_3 and hydrolysis with 6 M HCl.[892] The NMDA antagonist (R)-α-amino-6-[^{14}C]methyl-7-methyl-3-(phosphonomethyl)quinolinepropanoic acid is prepared in enantiomerically pure form by subjecting 2,6-dibromo-3-(diethoxyphosphonyl)-7-methylquinoline to a Pd-catalyzed coupling reaction with the zinc reagent prepared from protected β-iodo-(R)-alanine. The best yields are obtained using a system comprising $Pd(OAc)_2$ and $tris$(2-furyl)phos-phine. In the radiolabeling step, the zinc reagent obtained from Zn-Cu and [^{14}C]-MeI is reacted under ultrasonic activation in the presence of $NiCl_2$(dppp)$_2$ to completely replace the 6-bromo substituent.[906]

8.1.10.3   Addition of Glycine Synthons to Vinylphosphonates

Another attractive, though limited, route to fully protected ω-amino-ω-(hydroxycarbonyl)alkylphos-phonic acids involves the addition of sodium acetamidomalonate or potassium N-(diphenylmeth-ylene)glycine esters to dialkyl vinylphosphonates (Scheme 8.80).[907–911]

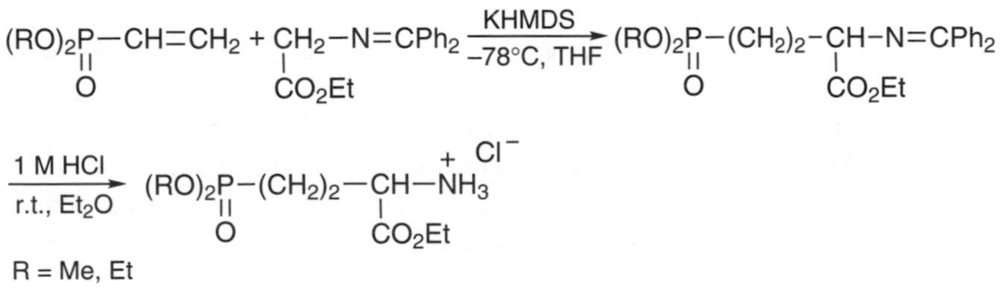

$$(8.80)$$

Among the various chiral glycine equivalents, the Schöllkopf's *bis*lactim ether derived from cyclo(L-Val-Gly) appears very attractive. Both enantiomers (derived from D- and L-valine) can be purchased or readily synthesized, making both L- and D-amino acids equally accessible. The hydrolysis of the *bis*lactim ether function to the corresponding amino ester can be performed in very mild conditions. Thus, when a solution of dialkyl 2-bromoethylphosphonates containing 10%

of dialkyl vinylphosphonates is added to the lithium *bis*lactim ether, the desired phosphonylated products are obtained in good yields with high diastereoselectivity.[912] Because the addition of the lithium *bis*lactim ether to dialkyl vinylphosphonates is much faster than the alkylation of dialkyl 2-bromoethylphosphonates, the Michael adduct is trapped by dialkyl 2-bromoethylphosphonates in a dehydrohalogenation reaction, regenerating the dialkyl vinylphosphonates. Mild acid hydrolysis of the resulting *bis*lactim ethers (0.25 M HCl, THF/MeCN, room temperature) affords dialkyl 3-amino-3-(hydroxycarbonyl)propylphosphonates in excellent yields (Scheme 8.81).[912,913]

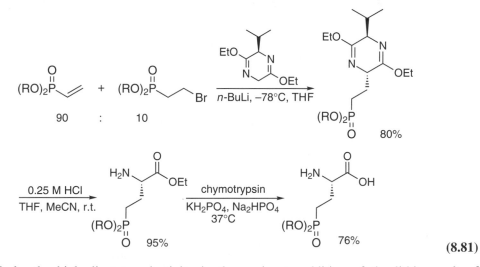

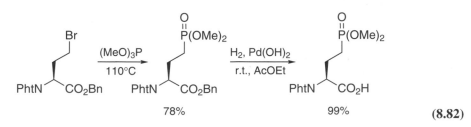

(8.81)

Similarly, the high diastereoselectivity in the conjugate addition of the lithium salt of Schöllkopf's *bis*lactim ether to (*E*)- and (*Z*)-1-propenylphosphonate has been used to achieve a sterocontrolled synthesis of all four diastereoisomers of 3-amino-3-(hydroxycarbonyl)-2-methyl-propylphosphonic acid.[914–916] Extension of the reaction to 4-substituted or 1,4- or 3,4-disubstituted (1*E*,3*E*)-butadienylphosphonates allows a direct and stereocontrolled access to semirigid AP-6 analogues, 5-amino-5-(hydroxycarbonyl)-2-pentenylphosphonic acid derivatives.[917] The conjugate addition of the lithium salt of Schöllkopf's *bis*lactim ether derived from cyclo(Ala-D-Val) to prochiral 1-alkenylphosphonates provides optically pure MAP-4 analogues, 3-amino-3-(hydroxy-carbonyl)butylphosphonic acid derivatives.[918]

8.1.10.4 Phosphonylation of ω-Halogenated Amino Acid Synthons

Another attractive route to ω-amino-ω-(hydroxycarbonyl)alkylphosphonic acids involves the use of the classical Michaelis–Arbuzov and Michaelis–Becker reactions. For example, the suitably protected 2-amino-ω-bromoalkanoates are readily converted into 3-amino-3-carboxypropylphosphonates (Scheme 8.82)[890,919–922] or substituted 6-amino-6-carboxyhexylphosphonates[923] by the Michaelis–Arbuzov rearrangement with trialkyl phosphites. However, it appears that, when enantiomerically pure bromide is used, the Michaelis–Arbuzov reaction is responsible for some racemization. It has been found that the extent of racemization is greater when the reaction is run at 150°C using triethyl phosphite than with trimethyl phosphite at 110°C for only the required amount of time.[922]

(8.82)

The use of *tris*(trimethylsilyl) phosphite for the generation of sensitive *bis*(trimethylsilyl) aminocarboxyalkylphosphonates appears to be a useful modification.[924] The Michaelis–Becker reaction of sodium diethyl phosphite with an iodoethyloxazolidinone derived from L-alanine in C_6H_6 at reflux gives the corresponding protected aminocarboxyalkylphosphonate in 83% yield.[925]

A series of substituted (phosphonoalkyl)phenylglycine and -phenylalanine derivatives as well as a series of *N*-(phosphonoalkenyl)- and *N*-(phosphonoaryl)glycine and -alanine have been prepared with combinations of several methodologies including Michaelis–Arbuzov,[926,927] Michaelis–Becker,[865,927] and coupling reactions.[865,927] For example, for the preparation of phosphonylated phenylglycine derivatives, Pd(PPh₃)₄ in toluene at 115°C in the presence of Et₃N is used to catalyze the direct attachment of diethyl phosphite to the aromatic ring. Alternatively, in a Heck reaction, (Ph₃P)₂PdCl₂ in DMF at 100°C catalyzes the coupling of aryl bromides with either diethyl vinylphosphonate or diethyl allylphosphonate to establish a two- or three-carbon spacer.[865,927,928] The (2-hydroxycarbonyl-4-piperidyl)methylphosphonic acid can be prepared in high yields from alkyl 4-(hydroxymethyl)pyridine-2-carboxylates by two different approaches. One implies a reduction–bromination–Michaelis–Arbuzov phosphonylation sequence,[929] whereas the second uses a chlorination–Michaelis–Arbuzov phosphonylation–reduction sequence.[930] Similarly, isopropyl 3-(hydroxymethyl)pyridine-2-carboxylate has been converted into diethyl (2-isopropoxycarbonyl-3-pyridyl)methylphosphonate by a bromination–Michaelis–Arbuzov phosphonylation sequence. This undergoes either direct hydrolysis with 6 M HCl at reflux to give (2-hydroxycarbonyl-3-pyridyl)methylphosphonic acid or hydrogenation followed by hydrolysis to give the corresponding piperidyl derivative.[931]

Asymmetric synthesis of (3-amino-3-hydroxycarbonyl-5-pyrrolidyl)methylphosphonic acid results from a combination of Michaelis–Arbuzov phosphonylation and Bucherer–Berg aminocarboxylation reactions.[932] Thus, starting from an easily accessible methyl *N*-(benzyloxycarbonyl)prolinate, a first reaction sequence including reduction, iodination, and Michaelis–Arbuzov rearrangement with triisopropyl phosphite produces diisopropyl [*N*-(benzyloxycarbonyl)-3-*tert*-butyldimethylsilyloxy-5-pyrrolidyl]methylphosphonate. In a second reaction sequence, it undergoes oxidation of the C-3 alcohol to a ketone and a Bucherer–Berg reaction [(NH₄)₂CO₃, KCN, EtOH–H₂O, 55°C] to give spirohydantoins, which are further hydrolyzed with 6 M HCl to give the desired free ω-phosphono-α-amino acids.[86]

Other related methods for the preparation of ω-amino-ω-(hydroxycarbonyl)alkylphosphonic acids have been reported using dialkyl trimethylsilyl phosphites and aspartaldehyde derivatives (Scheme 8.83).[933–935]

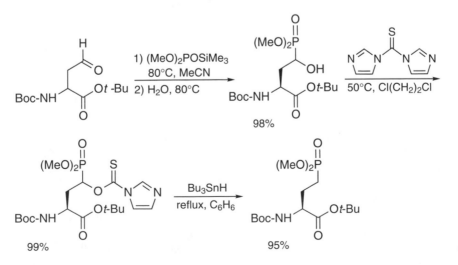

(8.83)

The 2-amino-4-phosphonobutanoic acid is available by a Michaelis–Arbuzov reaction from triethyl phosphite and ethyl 4-bromobutanoate followed by acid hydrolysis of the ester groups, bromination, and ammonolysis. After treatment with ion-exchange resin (Dowex 50W), the pure acid is obtained in 48% yield.[936]

8.1.10.5 Michael Additions

The synthesis of phosphonic analogues of aspartic acid and asparagine is accomplished by addition of diethyl phosphite to diethyl acetamidomethylenemalonate.[937,938] The strongly activated double bond readily adds diethyl phosphite in an exothermic reaction catalyzed by EtONa (Scheme 8.84). Hydrolysis of the crude addition product with concentrated HCl at reflux delivers the amino acid purified using acidic ion-exchange resin (Dowex 50W).

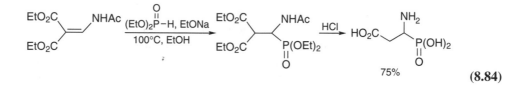

$$(8.84)$$

Regioselective addition of diisopropyl phosphite to protected derivatives of L-vinylglycine, racemic or enantiomerically pure, in xylene at 120°C catalyzed by *tert*-butyl per-2-ethylhexanoate affords the protected aminocarboxyalkylphosphonate in only moderate yield (35%, Scheme 8.85).[939] The L-2-amino-4-phosphonobutanoic acid (APB) is isolated as hydrochlorides by acidic hydrolysis with 6 M HCl.

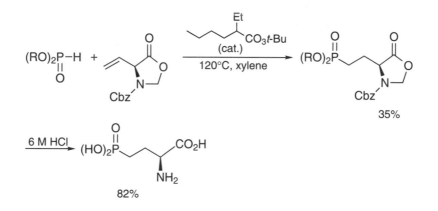

$$(8.85)$$

α,β-Unsaturated amino acids prepared from serine methyl ester are coupled with dimethyl 2-nitroethyl- or 4-nitrobutylphosphonates in moderate to good yields (37–64%). The most suitable catalyst for this reaction proved to be KF–Al$_2$O$_3$, which gives only monoaddition products. High-pressure hydrogenation with 10% Pd/C in MeOH produces the key diaminoesters in good yields (55–65%). When the purified diamino esters are treated with diethoxymethyl acetate in DMF at room temperature, the cyclization takes place smoothly and cleanly to give, after hydrolysis with 6 M HCl, tetrahydropyrimidine derivatives (Scheme 8.86) that represent 5-aza-5,6(1,6)-dehydro analogues of *cis*-4-(phosphonomethyl)-2-piperidinecarboxylic acid (CGS 19755).[940]

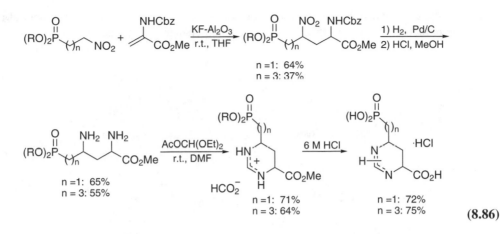

(8.86)

Michael addition of metallated camphor derivatives of diethyl aminomethylphosphonate to diethyl vinylphosphonate followed by hydrolysis with 90% aqueous AcOH leads in 73% yield and high diastereoselectivity to tetraethyl *N*-acetyl 2-amino-1,3-propylenediphosphonate, a phosphonic analogue of glutaric acid (Scheme 8.87). No racemization occurs under the hydrolysis conditions.[758] Similarly, methyl acrylate, after hydrolysis of diastereomerically pure adduct, leads to the phosphonic analogue of pyroglutamic acid further reduced with LiBH$_4$/BF$_3$·Et$_2$O to the phosphoproline diethylester in 61% yield (Scheme 8.87).[758]

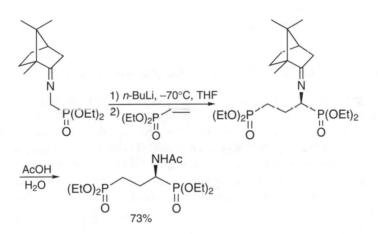

(8.87)

8.1.10.6 Horner–Wadsworth–Emmons Reactions

Introduction of the phosphonate moiety can be conveniently accomplished by the Horner–Wadsworth–Emmons olefination of carbonyl compounds using tetraethyl methylenediphosphonate. For example, a bicyclic ketone prepared from *meta*-tyrosine is used in reaction with tetraethyl methylenediphosphonate to give the vinylphosphonate, which undergoes catalytic hydrogenation in good conditions as illustrated in Scheme 8.88.[923] A significant limitation of this approach is the inability to engineer in a high degree of stereoselectivity in such hydrogenations. Attempts to control the outcome to favor one or the other diastereomer were unsuccessful. Thus, hydrogenation of the vinylphosphonate gives inseparable mixtures of the corresponding epimeric phosphorus compounds.[923] It was observed that by removing the methyl carbamate before hydrogenation, the reduction becomes stereoselective, and only one amino acid is obtained (Scheme 8.88).[923,941] Complete epimerization has been observed in reduction using NaBH$_4$ in EtOH and Mg in MeOH.[942,943]

The $[(Ph_3P)CuH]_6$, which is specific for the reduction of α,β-unsaturated esters in neutral media, gives the reduction product at 25°C with only a small change in the diastereomeric ratio.[943]

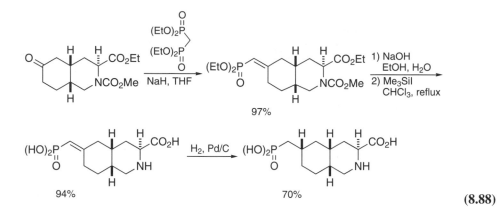

(8.88)

Synthesis of (2-hydroxycarbonyl-3-pyridyl)ethylphosphonic acid from 2-(isopropoxycarbonyl)pyridine-3-carboxaldehyde and tetramethyl methylenediphosphonate proceeds via the corresponding vinylphosphonate obtained as a mixture of (E)- and (Z)-isomers (72%). Catalytic hydrogenation of the (E)-isomer over 5% Pd/C in EtOH can be followed either by hydrolysis with 6 M HCl at reflux to give (2-hydroxycarbonyl-3-pyridyl)ethylphosphonic acid (51%) or by catalytic hydrogenation over PtO_2 to give the piperidine derivative.[931] The (N-acetyl-2-ethoxycarbonyl-4-piperidyl)acetaldehyde undergoes the same reaction sequence to give (2-hydroxycarbonyl-4-piperidyl)propylphosphonic acid.[929]

8.1.10.7 Coupling of Chiral Synthons with Phosphonate Carbanions

Phosphonate carbanions can participate to the elaboration of aminocarboxyalkylphosphonates by nucleophilic reactions. Several examples have been described, including ring opening of chiral building blocks (Boc-protected L-pyroglutamate, Boc-protected β-lactam) with diethyl 1-lithiomethylphosphonates (Scheme 8.89),[944,945] alkylation of diethyl 1-lithio-1,1-difluoromethylphosphonate with chiral oxazolidinone-derived triflates[946] or imidazolidinone-derived halides,[947] condensation of diethyl 1-lithio-1,1-difluoromethylphosphonate with D-serine-derived Garner's aldehyde[948] or ester,[949] coupling of protected D-aspartic acid chloride with diethyl 1-copper(I)alkylphosphonates,[950] and Cu(I)-promoted coupling of iodoarenes with cadmium derivative of diethyl bromodifluoromethylphosphonate.[951]

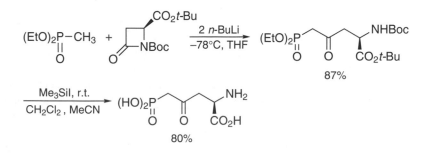

(8.89)

Alkylation of chiral imidazolidinones with 1,3-dibromopropane in the presence of LDA followed by a second condensation with the anion of the (−)-camphor imine of diethyl aminomethylphosphonate gives access to the phosphonic acid analogues of diaminopimelic acid (DAP) as pure diastereomers.[947]

Of special interest is the formation of AP-5 from (*S*)-serine-derived aldehyde and diethyl chlorophosphite via a phosphite–phosphonate rearrangement.[952–956] Thus, the alkynol prepared by reaction of lithium (trimethylsilyl)acetylene with the aldehyde derived from (*S*)-serine, on reaction with diethyl chlorophosphite gives a propargyl phosphite which undergoes a rearrangement to the enantiomerically pure allenylphosphonate. Reduction of the double bond with H_2 in the presence of PtO_2, followed by opening of the oxazolidinone and deprotection, produces AP-5 (Scheme 8.90).[957]

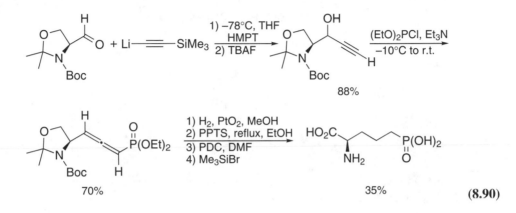

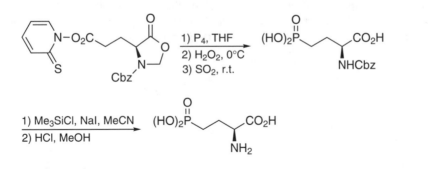

(8.90)

8.1.10.8 Other Reactions

White phosphorus in THF reacts in a long radical chain reaction with carbon radicals derived from Barton's PTOC esters. Oxidation of the adducts (H_2O_2 or SO_2) provides a convenient synthesis of phosphonic acids. For sensitive products the further transformation to phosphonic acids is best carried out with an excess of SO_2. The reaction is illustrated by the synthesis of L-2-amino-4-phosphonobutyric acid in 58% yield from L-glutamic acid using the appropriate protecting groups (Scheme 8.91).[958]

(8.91)

8.2 SYNTHETIC APPLICATIONS OF CARBOXYPHOSPHONATES

8.2.1 REACTIONS OF THE ALKOXYCARBONYL GROUP: PREPARATION OF ω-AMINOALKYLPHOSPHONIC ACIDS

The classical methods of conversion of a carboxy group into an amino group, the Curtius, Hofmann, and Schmidt reactions, have frequently been used to prepare amino-substituted organophosphorus compounds. Thus, the reaction of diethyl 1-(ethoxycarbonyl)akylphosphonates with hydrazine gives the corresponding hydrazides, which, after treatment with sodium nitrite and EtOH and hydrolysis with concentrated HCl, are converted into α-aminoalkylphosphonic acids in yields of 16–80%

(Scheme 8.92).[121,632,887,959] Although the intermediate compounds in this synthesis were not isolated, there is no doubdt that the process is a Curtius reaction. Alternatively, the synthesis of *N*-(benzyloxycarbonyl)aminoalkylphosphonates has been effected by degradation of diethyl 1-(hydroxycarbonyl)alkylphosphonates by means of diphenylphosphoryl azide in the presence of benzyl alcohol. The resulting urethanes are converted into the corresponding aminoalkylphosphonic acids.[408]

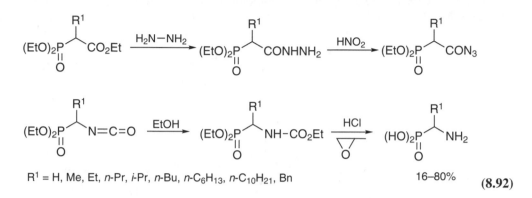

R1
(EtO)2P
CO2Et
O
H2N−NH2
CONHNH2
HNO2
CON3
N=C=O
EtOH
NH−CO2Et
HCl
(HO)2P
NH2
R1 = H, Me, Et, n-Pr, i-Pr, n-Bu, n-C6H13, n-C10H21, Bn
16–80%

(8.92)

Hydrazoic acid can also be used to convert (alkoxycarbonyl)alkylphosphonic esters into aminoalkylphosphonic acids (Schmidt rearrangement, Scheme 8.93).[162,960]

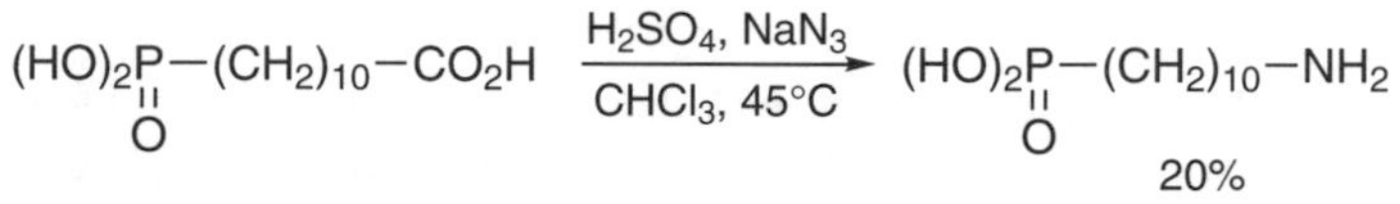

(HO)2P−(CH2)10−CO2H
H2SO4, NaN3
CHCl3, 45°C
(HO)2P−(CH2)10−NH2
O
20%

(8.93)

The Hofmann degradation is applicable to the synthesis of 2-aminoalkylphosphonic acids.[438,481,961] For example, 2-aminoethylphosphonic acid has been obtained by this method in 75% yield (Scheme 8.94).[961] To avoid the drastic conditions required by the hypobromite method, two new Hofmann-like reagents, iodosobenzene and *I*-hydroxy-*I*-*p*-toluenesulfonyloxyiodobenzene (HTIB), have been introduced. Under these conditions, the rearrangement takes place smoothly to produce the desired aminoalkylphosphonic acids in high yields (65–93%).[962]

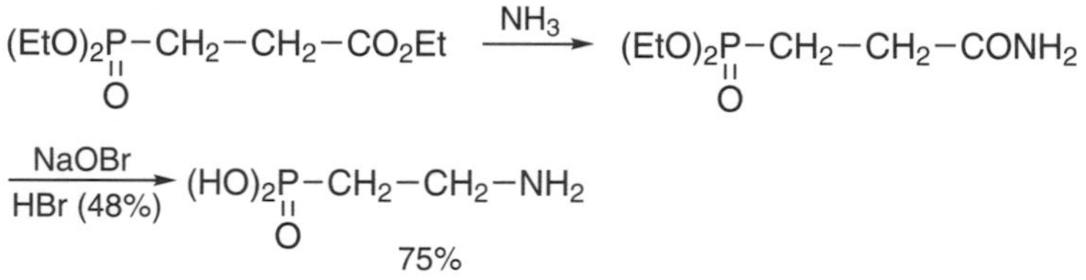

(EtO)2P−CH2−CH2−CO2Et
NH3
(EtO)2P−CH2−CH2−CONH2
O
NaOBr
HBr (48%)
(HO)2P−CH2−CH2−NH2
75%

(8.94)

The reaction of diisopropyl 2-(methoxycarbonyl)ethylphosphonate with Grignard reagents mediated by $Ti(Oi\text{-}Pr)_4$ or $MeTi(Oi\text{-}Pr)_3$ in THF at 0°C is completely diastereoselective and gives diisopropyl 2-(2-substituted-1-hydroxy-cyclopropyl)ethylphosphonates in good yields (60–71%).[963]

8.2.2 Horner–Wadsworth–Emmons Reactions of Dialkyl 1-(Alkoxycarbonyl)methylphosphonates: Preparation of α,β-Unsaturated Esters

Dialkyl 1-(alkoxycarbonyl)methylphosphonates are routine reagents readily accessible on laboratory scale and also available from a number of chemical suppliers. Since the review by Pudovik and

Yastrebova in 1970,[964] the use of dialkyl 1-(alkoxycarbonyl)methylphosphonates in Horner–Wadsworth–Emmons reactions has carefully been covered in several exhaustive and excellent reviews.[965–969] All the factors governing the reaction (nature of the carbanion and carbonyl group, reaction conditions, mechanism, and stereochemistry) have been studied in depth. We invite the reader to refer to these papers. We discuss here the synthetic applications resulting directly from the use of dialkyl 1-(alkoxycarbonyl)methylphosphonates, which is the pathway of choice for the preparation of α,β-unsaturated alcohols and related compounds via the α,β-unsaturated esters.

8.2.3 HORNER–WADSWORTH–EMMONS REACTIONS OF PHOSPHONOCROTONATES: PREPARATION OF α,β,γ,δ-UNSATURATED ESTERS

Diethyl 3-alkoxycarbonyl-2-methyl-2-propenylphosphonate,[970–978] diethyl 3-ethoxycarbonyl-3-fluoro-2-methyl-2-propenylphosphonate,[730] and 3-alkoxycarbonyl-2-propenylphosphonate[979] are used in the synthesis of modified retinals. When structural analogues of β-ionylideneacetaldehyde are allowed to react with phosphonocrotonates under Horner–Wadsworth–Emmons conditions (NaH, THF, 25°C or n-BuLi, THF, −78°C, in the presence of DMPU), *trans*-retinoates or analogues are produced. Subsequent saponification of the ester with KOH in MeOH provides the corresponding carboxylic acids in good overall yields.[974,975] The ester–aldehyde functional group interconversion is performed by DIBAL-H or AlLiH$_4$ reduction to primary alcohol followed by MnO$_2$ oxidation to give all-*trans*-retinals or analogues (Scheme 8.95).

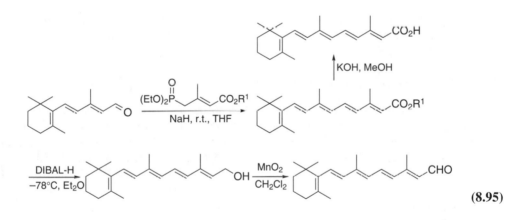

(8.95)

8.2.4 ELABORATION OF HORNER–WADSWORTH–EMMONS-PRODUCED α,β-UNSATURATED ESTERS

8.2.4.1 Into Allylic Alcohols

The most frequently used application of α,β-unsaturated esters is their conversion into allylic alcohols. The mixture of (*E*)- and (*Z*)-isomers resulting from the Horner–Wadsworth–Emmons reaction is reacted directly or after separation by column chromatography. Reduction is carried out with LiAlH$_4$ in Et$_2$O (or LiAlH$_4$, AlCl$_3$),[95,980–997] with DIBAL-H in CH$_2$Cl$_2$ (or C$_6$H$_6$ or THF) at low temperature[281,973,998–1023] or with NaAlH$_2$(OCH$_2$CH$_2$OMe)$_2$ in C$_6$H$_6$ at room temperature.[1024] The overall yields obtained by these procedures are roughly comparable (65–97%, Scheme 8.96).

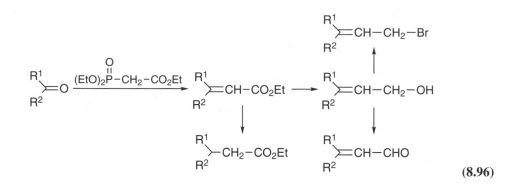

(8.96)

8.2.4.2 Into α,β-Unsaturated Aldehydes

The α,β-unsaturated alcohols are usually converted into the corresponding aldehydes by room-temperature treatment with MnO_2 in CH_2Cl_2.[982,984,994,998,1001,1019,1020]

8.2.4.3 Into Allylic Halides

The α,β-unsaturated alcohols can be converted into bromides by treatment with PBr_3–Py (1:1) in various solvents (C_6H_6, Et_2O, petroleum ether, CH_2Cl_2).[980,987,988,993,996]

8.2.4.4 Into Saturated Esters

The conversion of α,β-unsaturated esters into saturated esters proceeds by selective reduction with triethylsilane in the presence of Wilkinson catalyst[1025] or by hydrogenation using 5–10% Pd/C in EtOH[995,1026–1028] or AcOEt[1029] or using PtO_2 in dioxane.[1030]

8.2.5 REACTIONS OF THE PHOSPHONYL GROUP

The reaction of diethyl 1-(ethoxycarbonyl)methylphosphonate with Me_3SiCl/NaI or KI in MeCN occurs rapidly at room temperature to give quantitatively the corresponding *bis*(trimethylsilyl) 1-(ethoxycarbonyl)methylphosphonate (Scheme 8.97). The dealkylation proceeds to completion in a few minutes, and the rate of dealkylation decreases according to the bulkiness of the alkoxy groups attached to phosphorus.[37,1031] Another version of this reaction employs the system Me_3SiCl/LiI in CCl_4 at 50°C.[1032] Silylation of diethyl 1-(ethoxycarbonyl)methylphosphonate has also been carried out with neat Me_3SiBr at room temperature,[38,1033] and in this case the reaction is complete within 1 h. The Me_3SiBr version can be used in MeCN at room temperature[1034,1035] or in dioxane at 60°C.[1036] The dimethyl 1-methoxycarbonyl-1-diazomethylphosphonate reacts with Me_3SiBr in C_6H_6 at 50°C to give the silyl esters, which can be transformed into the *tert*-butylammonium salts.[1037] Very recently, it has been reported that BBr_3 in toluene, on heating, converts diethyl 1-(ethoxycarbonyl)methylphosphonate cleanly and quantitatively into the phosphonic acid after methanolysis.[1038]

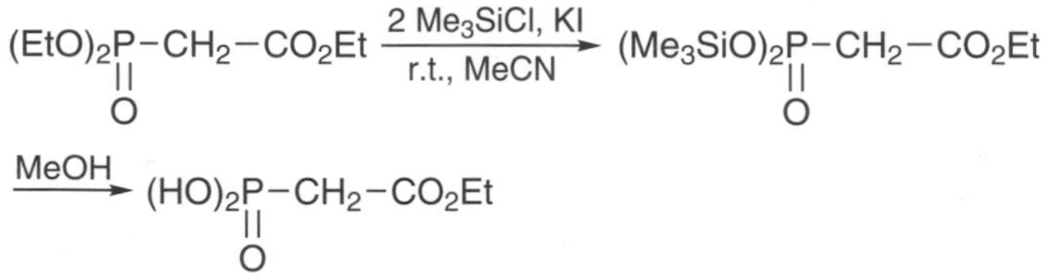

$$(EtO)_2\underset{\underset{O}{\|}}{P}-CH_2-CO_2Et \xrightarrow[\text{r.t., MeCN}]{2\ Me_3SiCl,\ KI} (Me_3SiO)_2\underset{\underset{O}{\|}}{P}-CH_2-CO_2Et$$

$$\xrightarrow{\text{MeOH}} (HO)_2\underset{\underset{O}{\|}}{P}-CH_2-CO_2Et$$

(8.97)

Room-temperature treatment of silyl esters with MeOH gives 1-(ethoxycarbonyl)methylphosphonic acid in high yields,[37,38] whereas treatment with a solution of $NaHCO_3$ in H_2O gives the disodium salt.[1039]

The 1-(hydroxycarbonyl)methylphosphonic acid is prepared in quantitative yield by heating dialkyl 1-(alkoxycarbonyl)methylphosphonate with concentrated HCl (or HBr) at 120–140°C[65,238,1040] or with catalytic H_2SO_4 and HCO_2H at 135°C.[1041] Dry HCl converts dialkyl (*tert*-butoxycarbonyl)methylphosphonate at 40–50°C into pure dialkyl 1-(hydroxycarbonyl)methylphosphonate.[1042]

Treatment of protected ω-amino-ω-(alkoxycarbonyl)alkylphosphonates with 4–8 M HCl selectively removes the ester groups to generate the free acids in 70–75% yields. Treatment of acids with acidic ion-exchange resin (Dowex 50W) or with propylene oxide followed by recrystallization from $EtOH–H_2O$ affords pure ω-amino-ω-(hydroxycarbonyl)alkylphosphonic acids.[86,865,890,919,925,927,933]

Sodium thiolates and thiophenates in EtOH at reflux[1043] and alkali halides at room temperature (for LiI, NaI, or KI)[1039] or on heating (for LiCl, LiBr, NaBr, or KBr)[44,1044] induce monodealkylation of diethyl 1-(ethoxycarbonyl)methylphosphonate.

The *bis*(trimethylsilyl) 1-(ethoxycarbonyl)methylphosphonate reacts smoothly with PCl_5 (2 eq) at room temperature in CH_2Cl_2 or CCl_4[19,431] or without solvent at 70°C[1045] to give the phosphonic dichloride in good yield (77%). Diethyl 1-(ethoxycarbonyl)methylphosphonate, on treatment with PCl_5 (2 eq) on heating[239,1046] or with a mixture of PCl_5 and $POCl_3$,[1047] leads to the phosphonic dichloride in good to excellent yields, whereas treatment with $POCl_3$ neat[352,1048] or with PCl_5 (1 eq) in CCl_4[1049] leads to the phosphonochloridate. The 1-(ethoxycarbonyl)methylphosphonic acid can be converted into phosphonic dichloride by treatment with PCl_5.[239,1050]

REFERENCES

1. Arbuzov, A.E., and Dunin, A.A., Action of halogen derivatives of aliphatic esters on alkyl phosphites, *J. Russ. Phys. Chem. Soc.*, 46, 295, 1914; *Chem. Abstr.*, 8, 2551, 1914.
2. Nylen, P., Organic phosphorus compounds, *Chem. Ber.*, 57B, 1023, 1924; *Chem. Abstr.*, 18, 3167, 1924.
3. Reetz, T., Chadwick, D.H., Hardy, E.E., and Kaufman, S., Carbamoylphosphonates, *J. Am. Chem. Soc.*, 77, 3813, 1955.
4. Warren, S., and Williams, M.R., The acid catalysed decarboxylation of phosphonoformic acid, *J. Chem. Soc. (B)*, 618, 1971.
5. Bucha, H.C., Langsdorf, P., Jr., and Jelinek, A.G., New plant growth regulants, *Du Pont de Nemours*, German Patent Appl. DE 2040367, 1971; *Chem. Abstr.*, 74, 112211, 1971.
6. Noren, J.O., Helgstrand, E., Johansson, N.G., Misiorny, A., and Stening, G., Synthesis of esters of phosphonoformic acid and their antiherpes activity, *J. Med. Chem.*, 26, 264, 1983.
7. Rosowsky, A., Fu, H., Pai, N., Mellors, J., Richman, D.D., and Hostetler, K.Y., Synthesis and *in vitro* activity of long-chain 5′-O-[(alkoxycarbonyl)phosphinyl]-3′-azido-3′-deoxythymidines against wild-type and AZT- and foscarnet-resistant strains of HIV-1, *J. Med. Chem.*, 40, 2482, 1997.
8. Maeda, H., Takahashi, K., and Ohmori, H., Reactions of acyl tributylphosphonium chlorides and dialkyl acylphosphonates with Grignard and organolithium reagents, *Tetrahedron*, 54, 12233, 1998.
9. Collins, D.C., Gagliardi, A.R., and Nickel, P., Phosphonated agents and their antiangiogenic and antitumorigenic use, *University of Kentucky Research Foundation*, U.S. Patent Appl. US 6096730, 2000; *Chem. Abstr.*, 133, 144896, 2000.
10. Sekine, M., and Hata, T., Convenient synthesis of unesterified acylphosphonic acids, *J. Chem. Soc., Chem. Commun.*, 285, 1978.
11. Sekine, M., Okimoto, K., Yamada, K., and Hata, T., Silyl phosphites. Part 15. Reactions of silyl phosphites with α-halo carbonyl compounds. Elucidation of the mechanism of the Perkow reaction and related reactions with confirmed experiments, *J. Org. Chem.*, 46, 2097, 1981.

12. Sekine, M., Mori, H., and Hata, T., Protection of phosphonate function by means of ethoxycarbonyl group. A new method for generation of reactive silyl phosphite intermediates, *Bull. Chem. Soc. Jpn.*, 55, 239, 1982.

13. Vaghefi, M.M., McKernan, P.A., and Robins, R.K., Synthesis and antiviral activity of certain nucleoside 5′-phosphonoformate derivatives, *J. Med. Chem.*, 29, 1389, 1986.

14. Iyer, R.P., Phillips, L.R., Biddle, J.A., Thakker, D.R., Egan, W., Aoki, S., and Mitsuya, H., Synthesis of acyloxyalkyl acylphosphonates as potential prodrugs of the antiviral, trisodium phosphonoformate (foscarnet sodium), *Tetrahedron Lett.*, 30, 7141, 1989.

15. Balszuweit, A., Issleib, K., Mögelin, W., and Stiebitz, B., A new approach to P-C bond formation by using trimethylsilyl phosphites, *Phosphorus, Sulfur Silicon Relat. Elem.*, 51/52, 308, 1990.

16. Iyer, R.P., Boal, J.H., Phillips, L.R., Thakker, D.R., and Egan, W., Synthesis, hydrolytic behavior, and anti-HIV activity of selected acyloxyalkyl esters of trisodium phosphonoformate (foscarnet sodium), *J. Pharm. Sci.*, 83, 1269, 1994.

17. Charvet, A.-S., Camplo, M., Faury, P., Graciet, J.-C., Mourier, N., Chermann, J.-C., and Kraus, J.-L., Inhibition of human immunodeficiency virus type 1 replication by phosphonoformate- and phosphonoacetate-2′,3′-dideoxy-3′-thiacytidine conjugates, *J. Med. Chem.*, 37, 2216, 1994.

18. Bhongle, N.N., Notter, R.H., and Turcotte, J.G., Expedient and high-yield synthesis of alkylphosphonyl dichlorides under mild, neutral conditions. Reaction of *bis*(trimethylsilyl)alkyl phosphonates with oxalyl chloride/dimethylformamide, *Synth. Commun.*, 17, 1071, 1987.

19. Morita, T., Okamoto, Y., and Sakurai, H., A mild and facile synthesis of alkyl- and arylphosphonyl dichlorides under neutral conditions. Reaction of *bis*(trimethylsilyl) phosphonates with PCl$_5$, *Chem. Lett.*, 435, 1980.

20. Biller, S.A., Forster, C., Gordon, E.M., Harrity, T., Rich, L.C., Marretta, J., and Ciosek, C.P., Jr., Isopropenyl phosphinylformates. New inhibitors of squalene synthetase, *J. Med. Chem.*, 34, 1912, 1991.

21. Gorin, B.I., Ferguson, C.G., and Thatcher, G.R.J., A novel esterification procedure applied to synthesis of biologically active esters of foscarnet, *Tetrahedron Lett.*, 38, 2791, 1997.

22. Ferguson, C.G., and Thatcher, G.R.J., A simple synthesis of phosphonoformamides, *Synlett*, 1325, 1998.

23. Ferguson, C.G., Gorin, B.I., and Thatcher, G.R.J., Design of novel derivatives of phosphonoformate (foscarnet) as prodrugs and antiviral agents, *J. Org. Chem.*, 65, 1218, 2000.

24. Mitchell, A.G., Nicholls, D., Irwin, W.J., and Freeman, S., Prodrugs of phosphonoformate. The effect of *para*-substituents on the products, kinetics and mechanism of hydrolysis of dibenzyl methoxycarbonylphosphonate, *J. Chem. Soc., Perkin Trans. 2*, 1145, 1992.

25. Griengl, H., Hayden, W., Penn, G., De Clercq, E., and Rosenwirth, B., Phosphonoformate and phosphonoacetate derivatives of 5-substituted 2′-deoxyuridines. Synthesis and antiviral activity, *J. Med. Chem.*, 31, 1831, 1988.

26. Ylagan, L., Benjamin, A., Gupta, A., and Engler, R., Organophosphorus chemistry. Ester–chloride conversion under mild conditions at phosphorus, *Synth. Commun.*, 18, 285, 1988.

27. Shiotani, S., and Kometani, T., Reaction of alkyl dialkoxyphosphinyl formate with strong base, *Chem. Pharm. Bull.*, 21, 1160, 1973.

28. Krol, E.S., Davis, J.M., and Thatcher, G.R.J., Hydrolysis of phosphonoformate esters. Product distribution and reactivity patterns, *J. Chem. Soc., Chem. Commun.*, 118, 1991.

29. Mitchell, A.G., Nicholls, D., Walker, I., Irwin, W.J., and Freeman, S., Prodrugs of phosphonoformate. Products, kinetics and mechanisms of hydrolysis of dibenzyl (methoxycarbonyl)phosphonate, *J. Chem. Soc., Perkin Trans. 2*, 1297, 1991.

30. Krol, E.S., and Thatcher, G.R.J., Hydrolysis of phosphonoformate triesters. Rate acceleration of a millionfold in nucleophilic substitution at phosphorus, *J. Chem. Soc., Perkin Trans. 2*, 793, 1993.

31. McIntosh, C.L., Phosphonates and their use as plant growth regulants, *Du Pont de Nemours*, German Patent Appl. DE 2435407, 1975; *Chem. Abstr.*, 83, 73487, 1975.

32. McIntosh, C.L., Carboxyphosphonates, *Du Pont de Nemours*, U.S. Patent Appl. US 4018854, 1977; *Chem. Abstr.*, 87, 135910, 1977.

33. Jakupovic, E., and Stenhede, J., Method for the synthesis of trisodium phosphonoformate hexahydrate, *Aktiebolaget Astra*, U.S. Patent Appl. US 5591889, 1997; *Chem. Abstr.*, 126, 118078, 1997.

34. Nylen, P., The kinetics of the hydrolysis of phosphonoformic acid, *Z. Anorg. Allg. Chem.*, 235, 33, 1937; *Chem. Abstr.*, 32, 2815, 1938.

35. Breuer, E., Reich, R., and Salomon, C., Compositions comprising oxophosphonate-based metalloproteinase inhibitors, *Yissum Research Development Company of the Hebrew University of Jerusalem*, Int. Patent Appl. WO 2001026661, 2001; *Chem. Abstr.*, 134, 311432, 2001.

36. Gakis, N., Heimgartner, H., and Schmid, H., Photoreactions. Part 34. Photochemical cycloadditions of 3-phenyl-2*H*-azirines with benzoyl-, ethoxycarbonyl-, and vinylphosphonates, *Helv. Chim. Acta*, 58, 748, 1975.

37. Morita, T., Okamoto, Y., and Sakurai, H., Dealkylation reaction of acetals, phosphonate, and phosphate esters with chlorotrimethylsilane/metal halide reagent in acetonitrile, and its application to the synthesis of phosphonic acids and vinyl phosphates, *Bull. Chem. Soc. Jpn.*, 54, 267, 1981.

38. Morita, T., Okamoto, Y., and Sakurai, H., The preparation of phosphonic acids having labile functional groups, *Bull. Chem. Soc. Jpn.*, 51, 2169, 1978.

39. McKenna, C.E., and Schmidhauser, J., Functional selectivity in phosphonate ester dealkylation with bromotrimethylsilane, *J. Chem. Soc., Chem. Commun.*, 739, 1979.

40. Briggs, A.D., Camplo, M., Freeman, S., Lundström, J., and Pring, B.G., Acyloxymethyl and 4-acyloxybenzyl diester prodrugs of phosphonoformate, *Tetrahedron*, 52, 14937, 1996.

41. Arbuzov, A.E., and Dunin, A.A., Phosphocarboxylic acids, *Chem. Ber.*, 60B, 291, 1927; *Chem. Abstr.*, 21, 1627, 1927.

42. Arbuzov, B.A., Michaelis–Arbuzov and Perkow reactions, *Pure Appl. Chem.*, 9, 307, 1964; *Chem. Abstr.*, 62, 6353a, 1965.

43. Kamai, G., and Shugurova, E.I., Alkyl esters of phosphonocarboxylic acids, *Dokl. Akad. Nauk SSSR*, 72, 301, 1950; *Chem. Abstr.*, 45, 542e, 1951.

44. Abramov, V.S., and Samoilova, O.D., Action of halide salts of alkali metals on esters of alkanephosphonic acids, *Zh. Obshch. Khim.*, 22, 914, 1952; *Chem. Abstr.*, 47, 4838d, 1953.

45. Razumov, A.I., and Korobkova, E.I., Derivatives of alkylphosphonous and phosphonic acids. Part 11. Synthesis of dicarbethoxymethyl esters of ethyl- and carbethoxymethylphosphonic acids, *Trudy Kazan. Khim. Tekhnol. Inst.*, 23, 215, 1957; *Chem. Abstr.*, 52, 9948d, 1958.

46. Balsiger, R.W., Jones, D.G., and Montgomery, J.A., Synthesis of potential anticancer agents. Part 18. Analogs of carbamoyl phosphate, *J. Org. Chem.*, 24, 434, 1959.

47. Magerlein, B.J., and Kagan, F., The conversion of alkyl halides to the next higher homologous phosphonates, *J. Am. Chem. Soc.*, 82, 593, 1960.

48. Korte, F., and Wiese, F.F., Acyllactone rearrangement. Part 30. Synthesis of α-acyl-Δ-thiollactones and Δ²-dihydrothiopyrans, *Chem. Ber.*, 97, 1963, 1964.

49. Martin, D.J., and Griffin, C.E., The determination of polar substituent constants for the dialkoxy- and diarylphosphono and trialkyl- and triarylphosphonium groups, *J. Org. Chem.*, 30, 4034, 1965.

50. Imaev, M.G., Shakirova, A.M., and Galeeva, R.A., Organophosphorus compounds with an active methylene group. Part 3. The synthesis of some esters and carboxyanilides of phosphonoacetic acid, *Zh. Obshch. Khim.*, 36, 1230, 1966; *J. Gen. Chem. USSR (Engl. Transl.)*, 36, 1245, 1966.

51. Findlay, J.A., Epoxidized hydrocarbon acid esters having juvenile hormone activity, *Findlay*, U.S. Patent Appl. US 3761495, 1973; *Chem. Abstr.*, 80, 37332, 1974.

52. Wiley, R.H., Vinyl phosphonocarboxylate esters, *Du Pont de Nemours*, U.S. Patent Appl. US 2478441, 1949; *Chem. Abstr.*, 44, 2010b, 1950.

53. Kamai, G., and Kukhtin, V.A., Some unsaturated esters of phosphonoacetic and oxo phosphonic acids and their copolymerization with methyl methacrylate, *Trudy Kazan. Khim. Tekhnol. Inst.*, 29, 1951; *Chem. Abstr.*, 51, 5720b, 1957.

54. Kamai, G., and Kukhtin, V.A., Preparation and polymerization of some unsaturated esters of phosphonic acids, *Dokl. Akad. Nauk SSSR*, 89, 309, 1953; *Chem. Abstr.*, 48, 7539i, 1954.

55. Sasin, R., Nauman, R.M., and Swern, D., Phosphorus derivatives of fatty acids. Part 5. Vinyl α-diethylphosphonates, *J. Am. Chem. Soc.*, 80, 6336, 1958.

56. Ondetti, M.A., and Petrillo, E.W., Phosphonoacyl prolines and related compounds, *E.R. Squibb and Sons*, U.S. Patent Appl. US 4151172, 1979; *Chem. Abstr.*, 91, 108230, 1979.

57. Hanson, P.R., Sprott, K.T., and Wrobleski, A.D., Intramolecular cyclopropanation reactions en route to novel *P*-heterocycles, *Tetrahedron Lett.*, 40, 1455, 1999.

58. Speziale, A.J., and Freeman, R.C., Reactions of phosphorus compounds. Part 1. Diethyl carbamoyl-methylphosphonates, *J. Org. Chem.*, 23, 1883, 1958.

59. Mel'nikov, N.N., Mandel'baum, Y.A., and Lomakina, V.I., Organic insectofungicides. Part 46. Synthesis of some derivatives of phosphonoacetic acid, *Zh. Obshch. Khim.*, 29, 3289, 1959; *Chem. Abstr.*, 54, 14215i, 1960.

60. Suga, K., Ishkawa, Y., and Watanabe, S., Synthesis of ethyl β-ionylideneacetate by the Wittig reaction, *Nippon Kagaku Zasshi*, 84, 991, 1963; *Chem. Abstr.*, 60, 10723e, 1964.

61. Thompson, J.E., Enol and acyl phosphates as intermediates in the synthesis of nonrandom triglycerides, *J. Org. Chem.*, 30, 4276, 1965.

62. Ward, L.F., Jr., Whetstone, R.R., Pollard, G.E., and Phillips, D.D., Novel 1-thiovinyl phosphates and related materials, *J. Org. Chem.*, 33, 4470, 1968.

63. Findlay, J.A., Methyl and ethyl 10,11-epoxy-7-ethyl-3,11-dimethyl-2,6-tridecadienoate, *Findlay*, German Patent Appl. DE 2000274, 1970; *Chem. Abstr.*, 73, 55959, 1970.

64. Fleck, F., and Heller, J., 4,4'-Divinylstilbene derivatives as fluorescent whiteners, *Sandoz*, German Patent Appl. DE 2602750, 1976; *Chem. Abstr.*, 85, 161886, 1976.

65. Maier, L., and Crutchfield, M.M., Organic phosphorus compounds. Part 70. Preparation and properties of new phosphorus-containing chelating agents for calcium and magnesium ions, *Phosphorus Sulfur*, 5, 45, 1978.

66. Burri, K.F., Cardone, R.A., Chen, W.Y., and Rosen, P., Preparation of macrolides via the Wittig reaction. A total synthesis of (−)-vermiculine, *J. Am. Chem. Soc.*, 100, 7069, 1978.

67. Schmidt, A., and Schwartzenbach, K., Hydroxyphenyl-substituted esters and amides as stabilizers, *Ciba-Geigy*, German Patent Appl. DE 2512895, 1975; *Chem. Abstr.*, 84, 32016, 1976.

68. Ismagilov, R.K., Razumov, A.I., Zykova, V.V., and Niyazov, N.A., Synthesis and properties of pyro-catecholphosphonoacetic esters, *Zh. Obshch. Khim.*, 53, 2000, 1983; *J. Gen. Chem. USSR (Engl. Transl.)*, 53, 1805, 1983.

69. Hainaut, D., Toromanoff, E., and Demoute, J.P., Methyl *trans,trans,cis*-3,11-dimethyl-7-ethyl-2,6,10-tridecatrienoate, *Roussel-UCLAF*, German Patent Appl. DE 2155920, 1972; *Chem. Abstr.*, 77, 87893, 1972.

70. Harding, K.E., and Tseng, C.-Y., Observations on the steric requirement of Wittig reactions with trialkylphosphonoacetates, *J. Org. Chem.*, 40, 929, 1975.

71. Bradley, J.C., and Büchi, G., A short synthesis of camptothecin, *J. Org. Chem.*, 41, 699, 1976.

72. Weihe, G.R., and McMorris, T.C., Stereoselective synthesis of 2,3-deoxyantheridiol, *J. Org. Chem.*, 43, 3942, 1978.

73. Lombardo, L., and Taylor, R.J.K., Silicon in organic chemistry. An improved preparation of α,β-unsaturated acids from carbonyl compounds, *Synthesis*, 131, 1978.

74. White, J.D., Carter, J.P., and Kezar, H.S., III, Stereoselective synthesis of the macrocycle segment of verrucarin J, *J. Org. Chem.*, 47, 929, 1982.

75. Kozikowski, A.P., and Sorgi, K.L., Use of the anomeric allylation reaction in natural products synthesis. A stereocontrolled synthesis of methyl deoxypseudomonate B, *Tetrahedron Lett.*, 25, 2085, 1984.

76. Hammond, G.B., Cox, M.B., and Wiemer, D.F., Stereocontrol in Horner–Wadsworth–Emmons condensations of a *gem*-dimethylcyclopropyl aldehyde with α-substituted phosphono acetates, *J. Org. Chem.*, 55, 128, 1990.

77. Kokin, K., Motoyoshiya, J., Hayashi, S., and Aoyama, H., Highly *cis*-selelective Horner–Wadsworth–Emmons (HWE) reaction of methyl *bis*(2,4-difluorophenyl)phosphonoacetate, *Synth. Commun.*, 27, 2387, 1997.

78. Kokin, K., Iitake, K.-I., Takaguchi, Y., Aoyama, H., Hayashi, S., and Motoyoshiya, J., A study on the Z-selective Horner–Wadsworth–Emmons (HWE) reaction of methyl diarylphosphonoacetates, *Phosphorus, Sulfur Silicon Relat. Elem.*, 133, 21, 1998.

79. Hubbard, R.D., and Miller, B.L., Lewis acid catalyzed Diels–Alder reactions of highly hindered dienophiles, *J. Org. Chem.*, 63, 4143, 1998.

80. Herrin, T.R., and Fairgrieve, J.S., Carboxylic esters of phosphonoacetic acid, *Abbott Labs*, U.S. Patent Appl. US 4052439, 1977; *Chem. Abstr.*, 88, 7055, 1978.

81. Herrin, T.R., and Fairgrieve, J.S., Triglyceride ester of phosphonoacetic acid having antiviral activity, *Abbott Labs*, U.S. Patent Appl. US 4150125, 1979; *Chem. Abstr.*, 91, 39648, 1979.

82. Herrin, T.R., Fairgrieve, J.S., Bower, R.R., Shipkowitz, N.L., and Mao, J.C.-H., Synthesis and anti-herpes simplex activity of analogs of phosphonoacetic acid, *J. Med. Chem.*, 20, 660, 1977.

83. Tömösközi, I., Mechanism of the reaction of styrene oxide with phosphonate-carbanions, *Tetrahedron*, 19, 1969, 1963.

84. Nozaki, H., Kondo, K., Nakanisi, O., and Sisido, K., Partial asymmetric synthesis of *trans*-2-phenyl-cyclopropanecarboxylic acid, *Tetrahedron*, 19, 1617, 1963.

85. Takagi, R., Hashizume, M., Nakamura, M., Begum, S., Hiraga, Y., Kojima, S., and Ohkata, K., Stereochemical considerations on the stereoselective cyclopropanation reactions of 3-aryl-2-phosphonoacrylates induced by the (−)-8-phenylmenthyl group as a chiral auxiliary, *J. Chem. Soc., Perkin Trans. 1*, 179, 2002.

86. Tanaka, K., Ohta, Y., and Fuji, K., Differentiation of enantiotopic carbonyl groups by the Horner–Wadsworth–Emmons reaction, *Tetrahedron Lett.*, 34, 4071, 1993.

87. Tullar, B.F., and Lorenz, R.R., 2-Hydroxy-3-[(3,3,5-trimethylcyclohexyl)alkyl]-1,4-naphthoquinones and their preparation, *Sterling Drug*, U.S. Patent Appl. US 3578686, 1971; *Chem. Abstr.*, 75, 20026, 1971.

88. Reimann, E., and Voss, D., Bicyclic α-amino acids. Part 2. The synthesis of 3-(1-tetralyl)- and 3-[5-(5,6,7,8-tetrahydro)quinolyl]alanine, *Arch. Pharm. (Weinheim)*, 310, 2, 1977.

89. Griffiths, G.F., Kenner, G.W., McCombie, S.W., Smith, K.M., and Sutton, M.J., Pyrroles and related compounds. Part 34. Acrylic esters in the porphyrin series, *Tetrahedron*, 32, 275, 1976.

90. Kamai, G., and Kukhtin, V.A., Polymerization of some unsaturated esters of phosphonocarboxylic acids, *Zh. Obshch. Khim.*, 24, 1855, 1954; *J. Gen. Chem. USSR (Engl. Transl.)*, 24, 1819, 1954.

91. Kennedy, J., Lane, E.S., and Robinson, B.K., Synthesis of metal-complexing polymers. Part 1. Phosphorylated polymers, *J. Appl. Chem.*, 8, 459, 1958; *Chem. Abstr.*, 53, 3032h, 1959.

92. Schneider, D.F., and Viljoen, M.S., Synthesis of annulated 4-alkylidenebutenolides from natural occurring diosphenols, *Synth. Commun.*, 27, 3349, 1997.

93. Farrington, G.K., Kumar, A., and Wedler, F.C., Design and synthesis of new transition-state analogue inhibitors of aspartate transcarbamylase, *J. Med. Chem.*, 28, 1668, 1985.

94. Suga, T., Hirata, T., Aoki, T., and Shishibori, T., Interconversion and cyclization of acyclic allylic pyrophosphates in the biosynthesis of cyclic monoterpenoids in higher plants, *Phytochemistry*, 25, 2769, 1986.

95. Dawson, R.M., Godfrey, I.M., Hogg, R.W., and Knox, J.R., Synthetic routes to some isotopically labelled intermediates for diterpenoid biosynthesis, *Aust. J. Chem.*, 42, 561, 1989.

96. Walter, W.M., Jr., Synthesis of geranylgeraniol-2-^{14}C, *J. Labelled Compd.*, 3, 54, 1967.

97. Cornforth, J.W., Mallaby, R., and Ryback, G., Synthesis of (±)-[2–^{14}C]-abscisic acid, *J. Chem. Soc. (C)*, 1565, 1968.

98. Oguni, I., and Uritani, I., The synthesis of farnesol-2-^{14}C involving a modified Wittig reaction, *Agr. Biol. Chem.*, 33, 1654, 1969.

99. Bu'Lock, J.D., Quarrie, S.A., and Taylor, D.A., Synthesis of methyl *trans*-retinoate-10-carbon-14 and *trans*-retinol-10-carbon-14, *J. Labelled Compd.*, 9, 311, 1973.

100. Rabinowitz, J.L., and Zanger, M., The synthesis of ^{14}C-*trans*-3-methyl-2-hexenoic acid labeled in various positions, *J. Labelled Compd.*, 8, 657, 1972.

101. Khachik, F., Beecher, G.R., Li, B.W., and Englert, G., Synthesis of ^{13}C-labelled (*all-E,3R,3′R*)-β,β-carotene-3,3′-diol (zeaxanthin) at C(12), C(13), C(12′), and C(13′) via *all-E*-2,7-dimethylocta-2,4,6-triene-1,8-dial-^{13}C$_4$, *J. Labelled Compd. Radiopharm.*, 36, 1157, 1995.

102. Yamamoto, H., Hoshino, T., and Uchiyama, T., Convenient preparation and quantification of 5,5′-diferulic acid, *Biosci. Biotechnol. Biochem.*, 63, 390, 1999.

103. Davidson, R.M., and Kenyon, G.L., Analogues of phosphoenol pyruvate. Part 3. New synthetic approaches to α-(dihydroxyphosphinylmethyl)acrylic acid and unequivocal assignments of the vinyl protons in its nuclear magnetic resonance spectrum, *J. Org. Chem.*, 42, 1030, 1977.

104. Goerger, M.M., and Hudson, B.S., Synthesis of *all-trans*-parinaric acid-d_8 specifically deuterated at all vinyl positions, *J. Org. Chem.*, 53, 3148, 1988.

105. Eguchi, T., Morita, M., and Kakinuma, K., Multigram synthesis of mevalonolactone-d_9 and its application to stereochemical analysis by ^{1}H NMR of the saturation reaction in the biosynthesis of the 2,3-di-*O*-phytanyl-*sn*-glycerol core of the *Archaeal* membrane lipid, *J. Am. Chem. Soc.*, 120, 5427, 1998.

106. Breuer, E., and Bannet, D.M., The preparation of some cyclic phosphonates and their use in olefin synthesis, *Tetrahedron*, 34, 997, 1978.

107. Meuwly, R., and Vasella, A., Synthesis of 1,2-*cis*-configurated glycosylphosphonates, *Helv. Chim. Acta*, 69, 25, 1986.

108. Imamura, M., and Hashimoto, H., Synthesis of novel CMP-NeuNAc analogues having a glycosyl phosphonate structure, *Tetrahedron Lett.*, 37, 1451, 1996.

109. Dondoni, A., Daninos, S., Marra, A., and Formaglio, P., Synthesis of ketosyl and ulosonyl phosphonates by Arbuzov-type glycosidation of thiazolylketol acetates, *Tetrahedron*, 54, 9859, 1998.

110. Müller, B., Martin, T.J., Schaub, C., and Schmidt, R.R., Synthesis of phosphonate analogues of CMP-Neu5Ac. Determination of $\alpha(2–6)$-sialyltransferase inhibition, *Tetrahedron Lett.*, 39, 509, 1998.

111. Kamai, G., and Bastanov, E.S., Some new esters of phosphonocarboxylic acids and their derivatives, *Zh. Obshch. Khim.*, 21, 2188, 1951; *Chem. Abstr.*, 46, 7517g, 1952.

112. Ackerman, B., Chladek, R.M., and Swern, D., Phosphorus derivatives of fatty acids. Part 3. Trialkyl α-phosphonates, *J. Am. Chem. Soc.*, 79, 6524, 1957.

113. Coover, H.W., Jr., McCall, M.A., and Dickey, J.B., Reaction of triethyl phosphite with 2-haloacrylates, *J. Am. Chem. Soc.*, 79, 1963, 1957.

114. Swern, D., Phosphonates, *U.S. Dept. of Agriculture*, U.S. Patent Appl. US 2973380, 1961; *Chem. Abstr.*, 55, 13318g, 1961.

115. Kreutzkamp, N., and Cordes, G., Carbonyl- and cyanophosphonic acid esters. Part 6. α-Substituted benzylphosphonic acid and its derivatives, *Arch. Pharm. Ber. Dtsch. Pharm. Ges.*, 295, 276, 1962.

116. Machleidt, H., and Wessendorf, R., Organic fluorine compounds. Part 4. Carbonyl fluoroolefination, *Justus Liebigs Ann. Chem.*, 674, 1, 1964.

117. Machleidt, H., Wessendorf, R., and Strehlke, G., Lower alkyl esters of β-cyclocitrylidene fluoroacetic acid and β-ionylidene fluoroacetic acid, *Olin Mathieson Chem.*, U.S. Patent Appl. US 3277147, 1966; *Chem. Abstr.*, 66, 55609, 1967.

118. Grell, W., and Machleidt, H., Synthesis of organophosphorous compounds. Part 2. PO-activated alkoxyolefination, *Liebigs Ann. Chem.*, 699, 53, 1966.

119. Tömösközi, I., Absolute configuration and partial asymmetric synthesis of 2-phenyl-1,1-dimethylcyclopropane, *Tetrahedron*, 22, 179, 1966.

120. Gross, H., Engelhardt, G., Freiberg, J., Bürger, W., and Costisella, B., α-Halo ethers. Part 31. Michaelis–Arbuzov reactions with α-chloro ethers and α-chloro amines, *Liebigs Ann. Chem.*, 707, 35, 1967.

121. Berry, J.P., Isbell, A.F., and Hunt, G.E., Amino phosphonic acids. Part 2. Aminoalkylphosphonic acids, *J. Org. Chem.*, 37, 4396, 1972.

122. Bottin-Strzalko, T., Effect of structure on the reversibility of the aldol condensation. Reaction of phosphono esters with benzaldehyde, *Tetrahedron*, 29, 4199, 1973.

123. Gallagher, G., Jr., and Webb, R.L., Tetrasubstituted acrylates. The Wittig–Horner reaction of ketones with triethyl α-phosphonopropionate, *Synthesis*, 122, 1974.

124. Harkin, S.A., Singh, O., and Thomas, E.J., Approaches to cytochalasan synthesis. Preparation and Diels–Alder reactions of 3-alkyl- and 3-acyl-Δ^3-pyrrolin-2-ones, *J. Chem. Soc., Perkin Trans. 1*, 1489, 1984.

125. Hartmann, W., Steinke, W., Glöde, J., and Gross, H., α-Substituted phosphonates. Part 48. Synthesis of α-phosphorylated derivatives of aryloxyacetic acid, *Z. Chem.*, 26, 131, 1986.

126. Elkik, E., and Imbeaux, M., A convenient synthesis of ethyl (diethoxyphosphoryl)fluoroacetate from ethyl fluoroacetate, *Synthesis*, 861, 1989.

127. Henin, F., Mortezaei, R., Muzart, J., Pete, J.-P., and Piva, O., Enantioselective photodeconjugation of conjugated esters and lactones in the presence of ephedrine, *Tetrahedron*, 45, 6171, 1989.

128. Daumas, M., Vo-Quang, L., and Le Goffic, F., An improved synthesis of methyl 2-*N*-formylamino-2-(dimethoxyphosphinyl)acetate, *Synth. Commun.*, 20, 3395, 1990.

129. Thenappan, A., and Burton, D.J., Alkylation of (fluorocarbethoxymethylene)-tri-*n*-butylphosphorane. A facile entry to α-fluoroalkanoates, *J. Org. Chem.*, 55, 2311, 1990.

130. Tsai, H.-J., Application of fluorocarbethoxy-substituted phosphonate. A facile entry to substituted 2-fluoro-3-oxoesters, *Phosphorus, Sulfur Silicon Relat. Elem.*, 126, 1, 1997.

131. Paterson, I., and McLeod, M.D., Studies in macrolide synthesis. Stereocontrolled synthesis of a C_1–C_{13} segment of concanamycin A, *Tetrahedron Lett.*, 38, 4183, 1997.

132. Tanyeli, C., and Caliskan, Z.Z., A facile synthesis of various butenolides, *Synth. Commun.*, 30, 2855, 2000.

133. Kreutzkamp, N., and Kayser, H., Interaction of triethylphosphite with halogenated β-dicarbonyl derivatives, *Liebigs Ann. Chem.*, 609, 39, 1957.

134. Ueda, K., and Matsui, M., Studies on chrysanthemic acid. Part 20. Synthesis of four geometrical isomers of (±)-pyrethric acid, *Agr. Biol. Chem.*, 34, 1119, 1970.

135. Kvicala, J., Vlaskova, R., Plocar, J., Paleta, O., and Pelter, A., Preparation of intermediates for fluorinated ligands by conjugated and tandem additions on 3-fluorofuran-2(5H)-one, *Collect. Czech. Chem. Commun.*, 65, 772, 2000.

136. Prager, R.H., and Schafer, K., Potential GABA$_B$ receptor antagonists. Part 10. The synthesis of further analogues of baclofen, phaclofen and saclofen, *Aust. J. Chem.*, 50, 813, 1997.

137. Vogel, E., Deger, H.M., Sombroek, J., Palm, J., Wagner, A., and Lex, J., Synthesis of CH$_2$-bridged [4n + 2]annulenes by a building block approach, *Angew. Chem.*, 92, 43, 1980; *Angew. Chem. Int. Ed. Engl.*, 19, 41, 1980.

138. Moreau, J.P., Chance, L.H., Boudreaux, G.J., and Drake, G.L., Jr., α,α'-*Bis*(phosphono)dicarboxylic acid derivatives, *USA Secretary of Agriculture*, U.S. Patent Appl. US 3979533, 1976; *Chem. Abstr.*, 86, 16789, 1977.

139. Vogel, E., Deger, H.M., Hebel, P., and Lex, J., Geometrically induced loss of aromaticity in bridged [14]annulenes. *syn*-1,6-Ethano-8,13-methano[14]annulene, *Angew. Chem.*, 92, 943, 1980; *Angew. Chem. Int. Ed. Engl.*, 19, 919, 1980.

140. Gross, H., and Freiberg, J., Syntheses with methyl methoxychloroacetate, *Angew. Chem. Int. Ed. Engl.*, 4, 975, 1965.

141. Schmidt, U., Lieberknecht, A., Schanbacher, U., Beuttler, T., and Wild, J., Facile preparation of N-acyl-2-(diethoxyphosphoryl)glycine esters and their use in the synthesis of dehydroamino acid esters, *Angew. Chem. Int. Ed. Engl.*, 21, 776, 1982.

142. Kober, R., and Steglich, W., Reaction of acylaminobromomalonates and acylaminobromoacetates with trialkylphosphites. A simple synthesis of ethyl 2-amino-2-(diethoxyphosphoryl)acetate, *Liebigs Ann. Chem.*, 599, 1983.

143. Schmidt, U., Lieberknecht, A., and Wild, J., Amino acids and peptides. Part 43. Dehydroamino acids. Part 18. Synthesis of dehydroamino acids and amino acids from N-acyl-2-(dialkyloxyphosphinyl)glycin esters. Part 2, *Synthesis*, 53, 1984.

144. Horenstein, B.A., and Nakanishi, K., Synthesis of unprotected (±)-tunichrome An-1, a tunicate blood pigment, *J. Am. Chem. Soc.*, 111, 6242, 1989.

145. Moran, E.J., and Armstrong, R.W., Highly convergent approach to the synthesis of the epoxy–amide fragment of the azinomycins, *Tetrahedron Lett.*, 32, 3807, 1991.

146. Narukawa, Y., Juneau, K.N., Snustad, D., Miller, D.B., and Hegedus, L.S., Synthesis of optically active β-lactams by the photolytic reaction of imines with optically active chromiumcarbene complexes. Part 2. Synthesis of 1-carbacephalothin and 3-ANA relays, *J. Org. Chem.*, 57, 5453, 1992.

147. Kunze, T., Phosphono analogues of glutathione as new inhibitors of glutathione S-transferases, *Arch. Pharm. (Weinheim)*, 329, 503, 1996.

148. Ferris, L., Haigh, D., and Moody, C.J., N–H Insertion reactions of rhodium carbenoids. Part 2. Preparation of N-substituted amino(phosphoryl)acetates (N-substituted phosphorylglycine esters), *J. Chem. Soc., Perkin Trans. 1*, 2885, 1996.

149. Humphrey, J.M., Aggen, J.B., and Chamberlin, A.R., Total synthesis of the serine-threonine phosphatase inhibitor microcystin-LA, *J. Am. Chem. Soc.*, 118, 11759, 1996.

150. Debenham, S.D., Cossrow, J., and Toone, E.J., Synthesis of α- and β-carbon-linked serine analogues of the Pk trisaccharide, *J. Org. Chem.*, 64, 9153, 1999.

151. Schiavi, B.M., and Richard, D.J., Total synthesis of isoroquefortine C, *J. Org. Chem.*, 67, 620, 2002.

152. Boaz, N.W., and Debenham, S.D., Preparation of chiral α-(2-oxo-1-azacycloalkyl)akanoates, *Eastman Chem.*, Int. Patent Appl. WO 2002026705, 2002; *Chem. Abstr.*, 136, 294725, 2002.

153. Ku, B., and Oh, D.Y., Facile synthesis of α-phosphorylated α-amino acids, *Tetrahedron Lett.*, 29, 4465, 1988.

154. Shankar, R., and Scott, A.I., A convenient synthesis of 2-(diethoxyphosphonyl)glycine and its derivatives, *Tetrahedron Lett.*, 34, 231, 1993.

155. Burger, K., Heistracher, E., Simmerl, R., and Eggersdorfer, M., Application of hexafluoroacetone as protecting and activating reagent in amino acid and peptide chemistry. Part 8. Synthesis of phosphorus-containing sarcosine derivatives via a new electrophilic sarcosine synthon, *Z. Naturforsch., Ser. B*, 47, 424, 1992.

156. Nylen, P., Organic phosphorus compounds. Part 2. β-Phosphonopropionic acid and γ-phosphono-butyric acid, *Chem. Ber.*, 59, 1119, 1926; *Chem. Abstr.*, 20, 2978, 1926.

157. Arbuzov, A.E., and Kamai, G.K., The preparation of phosphoacetic ester, *J. Russ. Phys. Chem. Soc.*, 61, 619, 1929; *Chem. Abstr.*, 23, 4443, 1929.

158. Arbuzov, A.E., and Razumov, A.I., Tautomeric conversions of some phosphoorganic compounds, *Zh. Obshch. Khim.*, 4, 834, 1934; *Chem. Abstr.*, 28, 2145, 1935.

159. Kosolapoff, G.M., Isomerization of alkyl phosphites. Part 5. The synthesis of phosphonoacetic and phosphonomalonic esters, *J. Am. Chem. Soc.*, 68, 1103, 1946.

160. Arbuzov, A.E., and Kamai, G., The reaction of the preparation of phosphonoacetic and phosphono-malonic esters, *Zh. Obshch. Khim.*, 17, 2149, 1947; *Chem. Abstr.*, 42, 4523g, 1948.

161. Chavane, V., and Rumpf, P., The abnormal reaction of α-halogenated carboxylic esters with dialkyl sodium phosphites, *C.R. Acad. Sci. (Paris)*, 225, 1322, 1947; *Chem. Abstr.*, 42, 2575g, 1948.

162. Chavane, V., Aliphatic phosphonic acids and their amino derivatives. Part 1. General, *Ann. Chim. (Paris)*, 4, 352, 1949; *Chem. Abstr.*, 45, 7516g, 1951.

163. Shugurova, E.I., and Kamai, G., Diallyl phosphite and its derivatives, *Zh. Obshch. Khim.*, 21, 658, 1951; *Chem. Abstr.*, 45, 8970a, 1951.

164. Razumov, A.I., Kukhtin, V., and Sazonova, N., Reaction of some halo derivatives of phosphorus with esters of glycolic acid, *Zh. Obshch. Khim.*, 22, 920, 1952; *Chem. Abstr.*, 47, 4836a, 1953.

165. Arbuzov, B.A., and Vinogradova, V.S., Esters of β-oxophosphonic acids. Part 1. Phosphonoacetic ester, phosphonoacetone, and their homologs, *Izv. Akad. Nauk SSSR, Ser. Khim.*, 54, 1957; *Chem. Abstr.*, 51, 10365c, 1957.

166. Patai, S., and Schwartz, A., Condensation of triethyl phosphonoacetate with aromatic aldehydes, *J. Org. Chem.*, 25, 1232, 1960.

167. Dembkowski, L., and Rachon, J., Reactivity of the acids of trivalent phosphorus and their derivatives. Part 1. Reductive debromination in the reactions of the >P-O⁻ ions with 2-bromoesters, *Phosphorus, Sulfur Silicon Relat. Elem.*, 88, 27, 1994.

168. Fokin, A.V., Zimin, V.I., Studnev, Y.N., and Rapkin, A.I., Reactions of fluorochloroacetic acid deriv-atives with nucleophilic reagents, *Zh. Org. Khim.*, 7, 249, 1971; *J. Org. Chem. USSR (Engl. Transl.)*, 7, 241, 1971.

169. Ciszewski, G.M., and Jackson, J.A., Michaelis–Becker synthesis of *bis*(2,2,2-trifluoroethyl)phosphono esters, *Org. Prep. Proced. Int.*, 31, 240, 1999.

170. Makosza, M., and Wojciechowski, K., Synthesis of phosphonic acid esters in solid–liquid catalytic two-phase system, *Bull. Pol. Acad. Sci., Chem.*, 32, 175, 1984; *Chem. Abstr.*, 102, 204043, 1985.

171. Ye, W., and Liao, X., Synthesis of dialkyl alkoxycarbonylmethanephosphonates (alkyl dialkoxyphos-phinylacetates) using phase-transfer catalysis, *Synthesis*, 986, 1985.

172. Ando, K., Convenient preparations of (diphenylphosphono)acetic acid esters and the comparison of the Z-selectivities of their Horner–Wadsworth–Emmons reaction with aldehydes depending on the ester moiety, *J. Org. Chem.*, 64, 8406, 1999.

173. Ando, K., Oishi, T., Hirama, M., Ohno, H., and Ibuka, T., Z-Selective Horner–Wadsworth–Emmons reaction of ethyl (diarylphosphono)acetates using sodium iodide and DBU, *J. Org. Chem.*, 65, 4745, 2000.

174. Brittelli, D.R., Phosphite-mediated *in situ* carboxyvinylation. A new general acrylic acid synthesis, *J. Org. Chem.*, 46, 2514, 1981.

175. Aboujaoude, E.E., Collignon, N., and Savignac, P., Synthesis of β-carbonylated phosphonates. Part 1. Carbanionic route, *J. Organomet. Chem.*, 264, 9, 1984.

176. Teulade, M.P., Savignac, P., Aboujaoude, E.E., and Collignon, N., Acylation of diethyl α-chlorome-thylphosphonates by carbanionic route. Preparation of α-clorinated β-carbonylated phosphonates, *J. Organomet. Chem.*, 287, 145, 1985.

177. Tay, M.K., About-Jaudet, E., Collignon, N., Teulade, M.P., and Savignac, P., α-Lithioalkylphospho-nates as functional group carriers. An *in situ* acrylic ester synthesis, *Synth. Commun.*, 18, 1349, 1988.

178. Teulade, M.P., Savignac, P., About-Jaudet, E., and Collignon, N., α-Lithioalkylphosphonates as functional group carriers. Part 2. A direct (E)-cinnamic and 2-alkyl-cinnamic acids synthesis, *Synth. Commun.*, 19, 71, 1989.

179. Ott, G.R., and Heathcock, C.H., A method for constructing the C44–C51 side chain of altohyrtin C, *Org. Lett.*, 1, 1475, 1999.

180. Villieras, J., Reliquet, A., and Normant, J.F., Nucleophilic alkylation of carbenoids. Part 3. Copper(I)-catalysed replacement of the chlorine atoms in 1,1-dichloroalkanephosphonates by various substituents, *Synthesis*, 27, 1978.

181. Binder, J., and Zbiral, E., Diethyl (trimethylsilylethoxymethyl)phosphonates as new reagents for variable strategies for synthesis of carbonyl compounds, α-hydroxycarbonyl compounds and vinylphosphonates, *Tetrahedron Lett.*, 25, 4213, 1984.

182. Blackburn, G.M., Brown, D., Martin, S.J., and Parratt, M.J., Studies on selected transformations of some fluoromethanephosphonate esters, *J. Chem. Soc., Perkin Trans. 1*, 181, 1987.

183. Patois, C., Savignac, P., About-Jaudet, E., and Collignon, N., A convenient method for the synthesis of *bis*-trifluoroethyl phosphonoacetates, *Synth. Commun.*, 21, 2391, 1991.

184. Patois, C., and Savignac, P., Preparation of triethyl 2-fluoro-2-phosphonoacetate, *Synth. Commun.*, 24, 1317, 1994.

185. Geirsson, J.K.F., and Njardarson, J.T., A facile synthesis of α-phosphono esters through methoxycarbonylation of α-phosphono carbanions, *Tetrahedron Lett.*, 35, 9071, 1994.

186. Oberhauser, T., and Meduna, V., On the stereochemical purity of (+)-7-aminocephalosporanic acid, *Tetrahedron*, 52, 7691, 1996.

187. Liu, X., Zhang, X.-R., and Blackburn, G.M., Synthesis of three novel supercharged β,γ-methylene analogues of adenosine triphosphate, *J. Chem. Soc., Chem. Commun.*, 87, 1997.

188. Al-Badri, H., and Collignon, N., First one-pot synthesis of mikanecic acid derivatives from allylic phosphonates, via a tandem-sequence Horner–Wadsworth–Emmons and Diels–Alder reactions, *Synthesis*, 282, 1999.

189. Burton, D.J., Sprague, L.G., Pietrzyk, D.J., and Edelmuth, S.H., A safe facile synthesis of difluorophosphonoacetic acid, *J. Org. Chem.*, 49, 3437, 1984.

190. Zhang, X., Qiu, W., and Burton, D.J., The preparation of $(EtO)_2P(O)CFHZnBr$ and $(EtO)_2P(O)CFHCu$ and their utility in the preparation of functionalized α-fluorophosponates, *Tetrahedron Lett.*, 40, 2681, 1999.

191. Tsai, H.-J., Preparation and synthetic application of diethyl 2-oxo-1,1-difluorophosphonates, *Phosphorus, Sulfur Silicon Relat. Elem.*, 122, 247, 1997.

192. Waschbüsch, R., Carran, J., and Savignac, P., New routes to diethyl 1-fluoromethylphosphonocarboxylates and diethyl 1-fluoromethylphosphonocarboxylic acid, *Tetrahedron*, 53, 6391, 1997.

193. Kirilov, M., Ivanov, D., Petrov, G., and Golemshinski, G., Phosphonation of polyfunctional organometallic reagents. Part 1. Metal derivatives of phenylacetic acid and its ethyl ester, *Bull. Soc. Chim. Fr.*, 3051, 1973.

194. Brocksom, T.J., Petragnani, N., and Rodrigues, R., Ester enolates. A new preparation of malonates, phosphonoacetates, and α-selenyl and sulfinyl esters, *J. Org. Chem.*, 39, 2114, 1974.

195. Kandil, A.A., Porter, T.M., and Slessor, K.N., One-step synthesis of stabilized phosphonates, *Synthesis*, 411, 1987.

196. Jackson, J.A., Hammond, G.B., and Wiemer, D.F., Synthesis of α-phosphono lactones and esters through a vinyl phosphate-phosphonate rearrangement, *J. Org. Chem.*, 54, 4750, 1989.

197. Ogasa, T., Saito, H., Hashimoto, Y., Sato, K., and Hirata, T., Synthesis and biological evaluation of optically active 3-H-1-carbacephem compounds, *Chem. Pharm. Bull.*, 37, 315, 1989.

198. Cabezas, J.A., and Oehlschlager, A.C., A new method for the preparation of 1-ethynyl ethers, *J. Org. Chem.*, 59, 7523, 1994.

199. Lee, K., and Wiemer, D.F., Reaction of diethyl phosphorochloridite with enolates. A general method for synthesis of β-keto phosphonates and α-phosphono esters through C–P bond formation, *J. Org. Chem.*, 56, 5556, 1991.

200. Lee, K., and Wiemer, D.F., A convenient preparation of α-phosphono esters and lactones via C–P bond formation, *Phosphorus, Sulfur Silicon Relat. Elem.*, 75, 87, 1993.

201. Lee, K., and Wiemer, D.F., Synthesis of nucleoside 3'-alkylphosphonates. Intermediates for assembly of cabon-bridged dinucleotide analogues, *J. Org. Chem.*, 58, 7808, 1993.

202. Kolodyazhnyi, O.I., Repina, L.A., and Gololobov, Y.G., C-Phosphorylation of malonic ester, *Zh. Obshch. Khim.*, 44, 1275, 1974; *J. Gen. Chem. USSR (Engl. Transl.)*, 44, 1253, 1974.

203. Pudovik, A.N., and Lebedeva, N.M., Addition of phosphonoacetic ester and its homologs to unsaturated electrophilic reagents, *Zh. Obshch. Khim.*, 22, 2128, 1952; *Chem. Abstr.*, 48, 564h, 1954.

204. Pudovik, A.N., and Lebedeva, N.M., New method of synthesis of esters of phosphonic and thiophosphonic acids. Part 23. Addition of phosphonoacetic ester, phosphonoacetone, and its homologs to unsaturated compounds, *Zh. Obshch. Khim.*, 25, 1920, 1955; *Chem. Abstr.*, 50, 8442c, 1956.

205. Fiszer, B., and Michalski, J., Synthesis of organophosphorus compounds based on phosphonoacetic ester and its analogs. Addition of phosphonoacetic ester, alkylated phosphonoacetic esters, and phosphonoacetic nitrile to α,β-unsaturated esters and nitriles, *Rocz. Chem.*, 28, 185, 1954; *Chem. Abstr.*, 49, 9493e, 1955.

206. Fiszer, B., and Michalski, J., Organophosphorus compounds with active methylene group. Part 3. Addition of phosphinylacetic esters and their analogues to α,β-unsaturated ethylenic derivatives, *Rocz. Chem.*, 34, 1461, 1960.

207. Auel, T., and Ulrich, H., (Carboxyalkyl)phosphonic acids, *Hoechst*, German Patent Appl. DE 2333151, 1975; *Chem. Abstr.*, 82, 171201, 1975.

208. Geffers, H., Radt, W., Schliebs, R., and Schulz, H., 2-Phosphonobutane-1,2,4-tricarboxylic acids, *Farbenfabriken Bayer*, German Patent Appl. DE 2061838, 1971; *Chem. Abstr.*, 77, 114565, 1972.

209. Pudovik, A.N., and Sitdikova, F.N., Addition of partial esters of phosphorus acids to nitroisoamylene and ethyl vinyl sulfone, *Dokl. Akad. Nauk SSSR*, 125, 826, 1959; *Chem. Abstr.*, 53, 19850c, 1959.

210. Mastryukova, T.A., Lazareva, M.V., and Perekalin, V.V., Synthesis of nitro- and aminoalkylphosphonates, *Izv. Akad. Nauk SSSR, Ser. Khim.*, 1164, 1972; *Bull. Acad. Sci. USSR, Div. Chem. Sci. (Engl. Transl.)*, 1114, 1972.

211. Pudovik, A.N., and Sitdykova, F.H., Addition of organophosphorus compounds with active hydrogen to divinyl sulfone and *p*-bis(β-nitrovinyl)benzene, *Zh. Obshch. Khim.*, 34, 1682, 1964; *J. Gen. Chem. USSR (Engl. Transl.)*, 34, 1692, 1964.

212. Zhdankina, G.M., Kadentsev, V.I., Kryshtal, G.V., Serebryakov, E.P., and Yanovskaya, L.A., 1,4-Addition of phosphonacetate esters to 6,10-dimethyl-3-undecen-2-one and 10-methoxy-6,10-dimethyl-3-undecen-2-one under Horner–Emmons reaction conditions in a superbasic medium, *Izv. Akad. Nauk SSSR, Ser. Khim.*, 1179, 1987; *Bull. Acad. Sci. USSR, Div. Chem. Sci. (Engl. Transl.)*, 1092, 1987.

213. Pudovik, A.N., Moshkina, T.M., and Konovalova, I.V., New method of synthesis of esters of phosphonic and thiophosphonic acids. Part 31. Addition of phosphorous and hypophosphorous acids, dialkyl hydrogen phosphites, and esters of phosphonoacetic acid to esters of maleic acid, *Zh. Obshch. Khim.*, 29, 3338, 1959; *Chem. Abstr.*, 54, 15223f, 1960.

214. Fiszer, B., and Michalski, J., Addition of phosphonoacetic esters and their analogs to α,β-unsaturated ethylenic derivatives, *Rocz. Chem.*, 26, 293, 1952.

215. Pudovik, A.N., Khusainova, N.G., and Galeeva, R.G., Addition of compounds with an active hydrogen atom in a methylene group to esters of propynylphosphonic acid, *Zh. Obshch. Khim.*, 36, 69, 1966; *J. Gen. Chem. USSR (Engl. Transl.)*, 36, 73, 1966.

216. Chinchilla, R., Galindo, N., and Najera, C., Dimethyl 2-(tosylmethyl)fumarate. An allyl sulfone as electrophilic reagent for the synthesis of itaconate ester derivatives, *Tetrahedron*, 52, 1035, 1996.

217. Attanasi, O.A., Filippone, P., Giovagnoli, D., and Mei, A., Conjugated azoalkenes. Part 18. Synthesis of functionally substituted 3-phosphono-1-aminopyrrol-2-ones and phosphonohydrazones by reaction of conjugated azoalkenes with β-phosphonocarboxylates, *Synth. Commun.*, 24, 453, 1994.

218. Pudovik, A.N., and Lebedeva, N.M., Reaction of addition and condensation of phosphonoacetone and phosphonoacetic ester, *Dokl. Akad. Nauk SSSR*, 90, 799, 1953; *Chem. Abstr.*, 50, 2429d, 1956.

219. Robinson, C.N., and Addison, J.F., Condensation of triethyl phosphonoacetate with aromatic aldehydes, *J. Org. Chem.*, 31, 4325, 1966.

220. Robinson, C.N., Li, P.K., and Addison, J.F., Reduction and hydrolysis of triethyl α-phosphonocinnamate and its derivatives, *J. Org. Chem.*, 37, 2939, 1972.

221. Robinson, C.N., and Slater, C.D., Substituent-induced chemical shifts and coupling constants in ^{31}P nuclear magnetic resonance spectra of ethyl α-diethylphosphonocinnamates and α-diethylphosphonocinnamonitriles, *J. Org. Chem.*, 52, 2011, 1987.

222. Al-Badri, H., Maddaluno, J., Masson, S., and Collignon, N., Hetero-Diels–Alder reactions of α-carbonylated styrylphosphonates with enol ethers. High-pressure influence on reactivity and diastereoselectivity, *J. Chem. Soc., Perkin Trans. 1*, 2255, 1999.

223. Janecki, T., Kus, A., Krawczyk, H., and Blaszczyk, E., A new, general approach to substituted 3-diethoxyphosphoryl-2,5-dihydro-2-furanones, *Synlett*, 611, 2000.

224. Minami, T., Yamanouchi, T., Takenaka, S., and Hirao, I., Synthesis of butadienylphosphonates containing electronegative substituents and their synthetic applications to functionalized cyclopentenylphosphonates, *Tetrahedron Lett.*, 24, 767, 1983.

225. Minami, T., Yamanouchi, T., Tokumasu, S., and Hirao, I., The reaction of butadienylphosphonates with a oxosulfonium ylide, phosphonium ylides, and ketone enolates, *Bull. Chem. Soc. Jpn.*, 57, 2127, 1984.

226. Lehnert, W., Knoevenagel condensations with titanium tetrachloride-base. Part 4. Condensation of aldehydes and ketones with phosphonoacetate and methylenediphosphonates, *Tetrahedron*, 30, 301, 1974.

227. Chen, S.F., Kumar, S.D., and Tishler, M., The synthesis of D,L-phosphotryptophan, *Tetrahedron Lett.*, 24, 5461, 1983.

228. Chiefari, J., Galanopoulos, S., Janowski, W.K., Kerr, D.I.B., and Prager, R.H., The synthesis of phosphonobaclofen, an antagonist of baclofen, *Aust. J. Chem.*, 40, 1511, 1987.

229. Reetz, M.T., Peter, R., and von Itzstein, M., Titanium-mediated stereoselective Knoevenagel condensation of ethyl (diethoxyphosphoryl)acetate with aldehydes, *Chem. Ber.*, 120, 121, 1987.

230. Arbuzov, A.E., and Razumov, A.I., Syntheses with phosphonoacetic ester, *J. Russ. Phys. Chem. Soc.*, 61, 623, 1929; *Chem. Abstr.*, 23, 4444, 1929.

231. Kirilov, M., and Petrov, G., Preparation, structure, and reactivity of magnesium diethyl phosphonoacetone. Infrared spectroscopic data on its cationotropy in solution, *Chem. Ber.*, 100, 3139, 1967.

232. Pudovik, A.N., Shchelkina, L.P., and Bashirova, L.A., Substitution reactions of phosphonoacetic ester and phosphonoacetone, *Zh. Obshch. Khim.*, 27, 2367, 1957; *Chem. Abstr.*, 52, 7134i, 1958.

233. Tolochko, A.F., Ganushchak, N.I., and Dombrovskii, A.V., Reaction of chloroarylbutenes with sodium diethyl phosphonate and sodium diethyl phosphonoethyl acetate, *Zh. Obshch. Khim.*, 38, 1112, 1968; *Chem. Abstr.*, 69, 106825, 1968.

234. Pudovik, A.N., and Khusainova, N.G., Esters and nitriles of propynylphosphonoacetic acid, *Zh. Obshch. Khim.*, 39, 2426, 1969; *J. Gen. Chem. USSR (Engl. Transl.)*, 39, 2366, 1969.

235. Pudovik, A.N., Nikitina, V.I., and Kurguzova, A.M., Reactions of organophosphorus compounds containing active methylene groups with methyl β-chlorovinyl ketone and α-halo ethers, *Zh. Obshch. Khim.*, 40, 291, 1970; *J. Gen. Chem. USSR (Engl. Transl.)*, 40, 261, 1970.

236. Kosolapoff, G.M., and Powell, J.S., Alkylation of triethyl phosphonoacetate and related esters, *J. Am. Chem. Soc.*, 72, 4198, 1950.

237. Diana, G.D., Zalay, E.S., Salvador, U.J., Pancic, F., and Steinberg, B., Synthesis of some phosphonates with antiherpetic activity, *J. Med. Chem.*, 27, 691, 1984.

238. Hollis, L.S., Miller, A.V., Amundsen, A.R., Schurig, J.E., and Stern, E.W., *cis*-Diamineplatinum(II) complexes containing phosphono carboxylate ligands as antitumor agents, *J. Med. Chem.*, 33, 105, 1990.

239. Bodnarchuk, N.D., Malovik, V.V., and Derkach, G.I., Phosphono carboxylic acid derivatives, *Zh. Obshch. Khim.*, 40, 1210, 1970; *J. Gen. Chem. USSR (Engl. Transl.)*, 40, 1201, 1970.

240. Kirilov, M., and Petrov, G., Synthesis, structure, and reactivity of some derivatives of diethyl ethoxycarbonylmethanephosphonate, *Monatsh. Chem.*, 103, 1651, 1972.

241. Ivanov, D., Vassilev, G., and Panayotov, I., Syntheses and reactions of organolithium reagents derived from weakly acidic C–H compounds, *Synthesis*, 83, 1975.

242. Okauchi, T., Fukamachi, T., Nakamura, F., Ichikawa, J., and Minami, T., 2-Phosphono-1,1,4,4-tetrathio-1,3-butadienes. Synthesis and synthetic application to highly functionalized dienes and heterocycles, *Bull. Soc. Chim. Belg.*, 106, 525, 1997.

243. Venugopalan, B., Hamlet, A.B., and Durst, T., A short synthesis of the Δ¹-carbopenem ring system, *Tetrahedron Lett.*, 22, 191, 1981.

244. Bailey, S., Billotte, S., Derrick, A.M., Fish, P.V., James, K., and Thomson, N.M., Preparation of oxazolyl- and oxadiazolyl-containing hydroxamic acids useful as procollagen *C*-proteinase inhibitors, *Pfizer*, Int. Patent Appl. WO 2001047901, 2001; *Chem. Abstr.*, 135, 92633, 2001.

245. Derrick, A.M., and Thomson, N.M., Preparation of itaconates and succinates via olefination of aldehydes with succinate phosphonates or phosphoranes followed by optional asymmetric reduction, *Pfizer*, Eur. Patent Appl. EP 1199301, 2002; *Chem. Abstr.*, 136, 325323, 2002.

246. Compagnone, R.S., and Rapoport, H., Chirospecific synthesis of (+)-pilocarpine, *J. Org. Chem.*, 51, 1713, 1986.

247. Samarat, A., Fargeas, V., Villieras, J., Lebreton, J., and Amri, H., A new synthesis of (±)-homosarko-mycin ethyl ester, *Tetrahedron Lett.*, 42, 1273, 2001.

248. Hutchinson, D.K., and Fuchs, P.L., A chemodirected, triply convergent total synthesis of *d*-(+)-carbacyclin, *J. Am. Chem. Soc.*, 109, 4755, 1987.

249. Henning, R., and Hoffmann, H.M.R., A novel approach to complex terpenoid methylenecyclohexanes, *Tetrahedron Lett.*, 23, 2305, 1982.

250. Hoffmann, H.M.R., and Henning, R., Synthesis of 2-norzizaene and 9,10-dehydro-2-norzizaene (7,7-dimethyl-6-methylidenetricyclo[6.2.1.0^{1,5}]undec-9-ene) via intramolecular allyl cation induced cycloaddition, *Helv. Chim. Acta*, 66, 828, 1983.

251. Giguere, R.J., Duncan, S.M., Bean, J.M., and Purvis, L., Intramolecular [3 + 4] allyl cation cycloaddition. Novel route to hydroazulenes, *Tetrahedron Lett.*, 29, 6071, 1988.

252. Kuroda, C., Shimizu, S., and Satoh, J.Y., A short-step synthesis of 14,15-dinoreudesmanolides using intramolecular cyclization of an allylsilane, *J. Chem. Soc., Perkin Trans. 1*, 519, 1990.

253. Yee, N.K.N., and Coates, R.M., Total synthesis of (+)-9,10-*syn*- and (+)-9,10-*anti*-copalol via epoxy trienylsilane cyclizations, *J. Org. Chem.*, 57, 4598, 1992.

254. Nishitani, K., Fukuda, H., and Yamakawa, K., Studies on the terpenoids and related alicyclic compounds. Part 42. Diastereoselective cyclization of ω-formylated allylsilanes into bicyclic α-methylene-γ-butyrolactones. A facile synthesis of *p*-menthanolides, *Heterocycles*, 33, 97, 1992.

255. Nishitani, K., Nakamura, Y., Orii, R., Arai, C., and Yamakawa, K., Stereoselective intramolecular cyclization of β-alkoxycarbonyl-ω-formylallylsilanes into bicyclic α-methylene-γ-lactones, *Chem. Pharm. Bull.*, 41, 822, 1993.

256. Kuroda, C., Shimizu, S., Haishima, T., and Satoh, J.Y., Synthesis of a stereoisomer of frullanolide utilizing the intamolecular cyclization of ω-formyl-2-alkenylsilane, *Bull. Chem. Soc. Jpn.*, 66, 2298, 1993.

257. Kuroda, C., Mitsumata, N., and Tang, C.Y., An acylative C–C single-bond cleavage and a self-cyclization of ethyl 2-(trimethylsilylmethyl)penta-2,4-dienoate of its free acid under Ritter condition, *Bull. Chem. Soc. Jpn.*, 69, 1409, 1996.

258. Kuroda, C., and Ito, K., Synthesis of cadinanolide type of a tricyclic α-methylene-γ-lactone using intramolecular cyclization of α-trimethylsilylmethyl-α,β-unsaturated ester with cyclic ketone, *Bull. Chem. Soc. Jpn.*, 69, 2298, 1996.

259. Kuroda, C., Nogami, H., Ohnishi, Y., Kimura, Y., and Satoh, J.Y., Stereochemistry of Lewis acid and fluoride promoted intramolecular cyclization of β-(alkoxycarbonyl)allylsilane with enones. Synthesis of bicyclo[4.3.0]nonanes, *Tetrahedron*, 53, 839, 1997.

260. Fujiwara, K., The reaction of phosphonates with α-halo ketones, *Nippon Kagaku Zasshi*, 84, 656, 1963; *Chem. Abstr.*, 60, 542g, 1964.

261. Rodriguez, M., Heitz, A., and Martinez, J., "Carba" peptide bond surrogates. Synthesis of BOC-L-Leu-Ψ(CH$_2$-CH$_2$)-L-Phe-OH and BOC-L-Leu-Ψ(CH$_2$-CH$_2$)-D-Phe-OH through a Horner–Emmons reaction, *Tetrahedron Lett.*, 31, 7319, 1990.

262. Maier, L., Organic phosphorus compounds. Part 91. Synthesis and properties of 1-amino-2-aryleth-ylphosphonic and -phosphinic acids as well as -phosphine oxides, *Phosphorus, Sulfur Silicon Relat. Elem.*, 53, 43, 1990.

263. Compagnone, R.S., and Suarez, A., Stereoselective synthesis of methoxy substituted 2,3-dibenzyl-γ-butyrolactones using organic phosphonates as intermediates, *Phosphorus, Sulfur Silicon Relat. Elem.*, 75, 35, 1993.

264. Holstein, S.A., Cermak, D.M., Wiemer, D.F., Lewis, K., and Hohl, R.J., Phosphonate and bisphos-phonate analogues of farnesyl pyrophosphate as potential inhibitors of farnesyl protein transferase, *Bioorg. Med. Chem.*, 6, 687, 1998.

265. Dingwall, J.G., Cook, B., and Marshall, A., Phosphonocarboxylic acid compounds, *Ciba-Geigy A.G.*, German Patent Appl. DE 2756678, 1980; *Chem. Abstr.*, 89, 129707, 1978.

266. Meyer, J.H., and Bartlett, P.A., Macrocyclic inhibitors of penicillopepsin. Part 1. Design, synthesis, and evaluation of an inhibitor bridged between P1 and P3, *J. Am. Chem. Soc.*, 120, 4600, 1998.

267. Noguchi, H., Aoyama, T., and Shioiri, T., Total synthesis of analogs of topostin B, a DNA topoisomerase I inhibitor. Part 1. Synthesis of fragments of topostin B-1 analogs, *Tetrahedron*, 51, 10531, 1995.

268. Tius, M.A., and Fauq, A.II., Total synthesis of (+)-desepoxyasperdiol, *J. Am. Chem. Soc.*, 108, 1035, 1986.

269. Kondo, A., Ochi, T., Iio, H., Tokoroyama, T., and Siro, M., A synthetic approach to furanocembranolides, *Chem. Lett.*, 1491, 1987.

270. Kormachev, V.V., Kolyamshin, O.A., and Mitrasov, Y.N., Alkylation of phosphorus-containing CII-acids by *gem*-dichlorocyclopropylmethyl halides, *Zh. Obshch. Khim.*, 62, 2391, 1992; *J. Gen. Chem. USSR (Engl. Transl.)*, 62, 1974, 1992.

271. Mandai, T., Kaihara, Y., and Tsuji, J., A new candidate for a properly substituted CD ring component of vitamin D_3 via intramolecular asymmetric olefination of a 1,3-cyclopentadienone derivative, *J. Org. Chem.*, 59, 5847, 1994.

272. Kawamura, M., and Ogasawara, K., Stereo- and enantio-controlled synthesis of (+)-juvabione and (+)-epijuvabione from (+)-norcamphor, *J. Chem. Soc., Chem. Commun.*, 2403, 1995.

273. Magnin, D.R., Biller, S.A., Dickson, J.K., Jr., Logan, J.V., Lawrence, R.M., Chen, Y., Sulsky, R.B., Ciosek, C.P., Jr., Harrity, T.W., Jolibois, K.G., Kunselman, L.K., Rich, L.C., and Slusarchyk, D.A., 1,1-Bisphosphonate squalene synthase inhibitors. Interplay between the isoprenoid subunit and the diphosphate surrogate, *J. Med. Chem.*, 38, 2596, 1995.

274. Robl, J.A., Sieber-McMaster, E., and Sulsky, R., Synthetic routes for the generation of 7,7-dialkyl azepin-2-ones, *Tetrahedron Lett.*, 37, 8985, 1996.

275. Fujiwara, H., Egawa, S., Terao, Y., Aoyama, T., and Shioiri, T., Total synthesis of analogs of topostin B, a DNA topoisomerase I inhibitor. Part 4. Synthesis of topostin B-2 analogs, *Tetrahedron*, 54, 565, 1998.

276. Li, Y., Liu, Z., Lan, J., Li, J., Peng, L., Li, W.Z., and Chan, A.S.C., Enantioselective total synthesis of natural 11,12-epoxycembrene-C, *Tetrahedron Lett.*, 41, 7465, 2000.

277. Liu, Z., Li, W.Z., Peng, L., Li, Y., and Li, Y., First enantioselective total synthesis of (natural) (+)-11,12-epoxy-11,12-dihydrocembrene-C and (−)-7,8-epoxy-7,8-dihydrocembrene-C, *J. Chem. Soc., Perkin Trans. 1*, 4250, 2000.

278. Liu, Z., Li, W.Z., and Li, Y., Enantioselective total synthesis of (+)-3,4-epoxycembrene-A, *Tetrahedron: Asymmetry*, 12, 95, 2001.

279. Marshall, J.A., DeHoff, B.S., and Cleary, D.G., Condensation of long-chain α-phosphono carboxylates with aldehydes, *J. Org. Chem.*, 51, 1735, 1986.

280. Wuts, P.G.M., Putt, S.R., and Ritter, A.R., Synthesis of the dipeptide hydroxyethylene isostere of Leu–Val, a transition state mimic for the control of enzyme function, *J. Org. Chem.*, 53, 4503, 1988.

281. Marshall, J.A., and Andersen, M.W., On the mechanism of Lewis acid promoted ene cyclizations of ω-unsaturated aldehydes, *J. Org. Chem.*, 57, 5851, 1992.

282. Poss, M.A., and Reid, J.A., Synthesis of the hydroxyethylene dipeptide isostere, (2S,4S,5S)-5-amino-6-cyclohexyl-4-hydroxy-2-isopropyl hexanoic acid *n*-butyl amide, *Tetrahedron Lett.*, 33, 1411, 1992.

283. Handa, S., Pattenden, G., and Li, W.-S., A new approach to steroid ring construction based on a novel radical cascade sequence, *J. Chem. Soc., Chem. Commun.*, 311, 1998.

284. Zhang, J., and Xu, X., Total synthesis of 6-*epi*-sarsolilide A, *Tetrahedron Lett.*, 41, 941, 2000.

285. Boehm, H.M., Handa, S., Pattenden, G., Roberts, L., Blake, A.J., and Li, W.-S., Cascade radical cyclizations leading to steroid ring constructions. Regio- and stereo-chemical studies using ester- and fluoroalkene substituted polyene acyl radical intermediates, *J. Chem. Soc., Perkin Trans. 1*, 3522, 2000.

286. Hampton, A., Sasaki, T., and Paul, B., Synthesis of 6′-cyano-6′-deoxyhomoadenosine-6′-phosphonic acid and its phosphoryl and pyrophosphoryl anhydrides and studies of their interactions with adenine nucleotide utilizing enzymes, *J. Am. Chem. Soc.*, 95, 4404, 1973.

287. Ando, K., Z-Selective Horner–Wadsworth–Emmons reaction of α-substituted ethyl (diarylphosphono)acetates with aldehydes, *J. Org. Chem.*, 63, 8411, 1998.

288. Ceruti, M., Rocco, F., Viola, F., Balliano, G., Milla, P., Arpicco, S., and Cattel, L., 29-Methylidene-2,3-oxidosqualene derivatives as stereospecific mechanism-based inhibitors of liver and yeast oxidosqualene cyclase, *J. Med. Chem.*, 41, 540, 1998.

289. Medina, J.C., Guajardo, R., and Kyler, K.S., Vinyl group rearrangement in the enzymatic cyclization of squalenoids. Synthesis of 30-oxysterols, *J. Am. Chem. Soc.*, 111, 2310, 1989.
290. Böhme, H., Lauer, R., and Matusch, R., 2-Alkoxymethyl and 2-alkylthiomethyl esters of 2-(dialkoxyphosphinyl)propionic acid, *Arch. Pharm. (Weinheim)*, 312, 60, 1979.
291. Flitsch, W., Rosche, J., and Lubisch, W., Synthetic routes to 1-substituted butadienylphosphonates, *Liebigs Ann. Chem.*, 661, 1987.
292. Moppett, C.E., and Sutherland, J.K., The biosynthesis of glauconic acid. C_9 precursors, *J. Chem. Soc., Chem. Commun.*, 772, 1966.
293. Vieth, S., Costisella, B., and Schneider, M., Tandem Michael addition alkylation of vinylphosphonates, *Tetrahedron*, 53, 9623, 1997.
294. Greengrass, C.W., and Hoople, D.W.T., Reaction of 4-acetoxy-2-azetidinone with tertiary carbanions. Preparation of 4-alkyl- and 4-alkylidene-2-azetidinones, *Tetrahedron Lett.*, 22, 1161, 1981.
295. Rezgui, F., and El Gaïed, M.M., Regiospecific reaction of stabilized carbanions with 2-(acetoxymethyl)cyclohex-2-en-1-one. Synthesis of bicyclic dienones, *Tetrahedron*, 53, 15711, 1997.
296. Okada, Y., Minami, T., Sasaki, Y., Umezu, Y., and Yamaguchi, M., The first synthesis of chiral phosphinocarboxylic acid ligands, *trans*-2-(diphenylphosphino)cycloalkanecarboxylic acids. The phosphine–palladium complexes catalyzed asymmetric allylic alkylation, *Tetrahedron Lett.*, 31, 3905, 1990.
297. Okada, Y., Minami, T., Umezu, Y., Nishikawa, S., Mori, R., and Nakayama, Y., Synthesis of a novel type of chiral phosphinocarboxylic acids. The phosphine–palladium complexes catalyzed asymmetric allylic alkylation, *Tetrahedron: Asymmetry*, 2, 667, 1991.
298. Minami, T., Okada, Y., Otaguro, T., Tawaraya, S., Furuichi, T., and Okauchi, T., Development of chiral phosphine ligands bearing a carboxyl group and their application to catalytic asymmetric reaction, *Tetrahedron: Asymmetry*, 6, 2469, 1995.
299. Zbiral, E., and Drescher, M., A novel approach to heteroarylmethyl- and heteroarylethylphosphonates, *Synthesis*, 735, 1988.
300. Stamos, D.P., Chen, S.S., and Kishi, Y., New synthetic route to the C14–C38 segment of halichondrins, *J. Org. Chem.*, 62, 7552, 1997.
301. Zucco, M., Le Bideau, F., and Malacria, M., Palladium-catalyzed intramolecular cyclization of vinyloxirane. Regioselective formation of cyclobutanol derivative, *Tetrahedron Lett.*, 36, 2487, 1995.
302. Kitagawa, O., Suzuki, T., Inoue, T., and Taguchi, T., Intramolecular carbotitanation reaction of active methine compounds having an unactivated alkyne mediated by $TiCl_4$–Et_3N, *Tetrahedron Lett.*, 39, 7357, 1998.
303. Kitagawa, O., Suzuki, T., Inoue, T., Watanabe, Y., and Taguchi, T., Carbocyclization reaction of active methine compounds with unactivated alkenyl or alkynyl groups mediated by $TiCl_4$–Et_3N, *J. Org. Chem.*, 63, 9470, 1998.
304. Kitagawa, O., Suzuki, T., Fujiwara, H., Fujita, M., and Taguchi, T., Alkaline metallic reagent-catalyzed hydrocarbocyclization reaction of various active methine compounds having an unactivated 4-alkynyl or allenyl group, *Tetrahedron Lett.*, 40, 4585, 1999.
305. Kitagawa, O., Fujiwara, H., Suzuki, T., Taguchi, T., and Shiro, M., Intramolecular carbostannation reactions of active methine compounds with an unactivated C–C π-bond mediated by $SnCl_4$–Et_3N, *J. Org. Chem.*, 65, 6819, 2000.
306. Baumann, T., Buchholz, B., and Stamm, H., Aziridines. Part 65. Acylic and cyclic γ-amidopropanephosphonic esters by amidoethylation of Horner reagents with activated aziridines, *Synthesis*, 44, 1995.
307. Helinski, J., Skrzypczynski, Z., and Michalski, J., Ring-opening of 3-benzyloxyazetidinium salts with organophosphorus nucleophiles. Application to the synthesis of polyfunctional aminophosphonic acids analogues, *Tetrahedron Lett.*, 36, 9201, 1995.
308. Bakalarz-Jeziorna, A., Helinski, J., and Krawiecka, B., Synthesis of multifunctionalized phosphonic acid esters via opening of oxiranes and azetidinium salts with phosphoryl-substituted carbanions, *J. Chem. Soc., Perkin Trans. 1*, 1086, 2001.
309. Hercouet, A., Le Corre, M., and Carboni, B., Asymmetric synthesis of a phosphonic analogue of (−)-*allo*-norcoronamic acid, *Tetrahedron Lett.*, 41, 197, 2000.
310. Oda, R., Yoshimura, T., and Shono, T., Small ring compounds. Part 12. Syntheses of cyclobutanephosphonic acids, *Kogyo Kagaku Zasshi*, 70, 215, 1967; *Chem. Abstr.*, 67, 90885, 1967.

311. Ismailov, V.M., Guliev, A.N., and Moskva, V.V., 1-(Dialkoxyphosphinyl)cyclopropanecarboxylic esters, *Zh. Obshch. Khim.*, 55, 2393, 1985; *J. Gen. Chem. USSR (Engl. Transl.)*, 55, 2127, 1985.

312. Neidlein, R., and Eichinger, T., Substituted methylphosphonates as synthons for alicyclic α-functionalized phosphonates, *Monatsh. Chem.*, 123, 1037, 1992.

313. Nasser, J., About-Jaudet, E., and Collignon, N., α-Functional cycloalkylphosphonates. Part 1. Synthesis, *Phosphorus, Sulfur Silicon Relat. Elem.*, 54, 171, 1990.

314. Kazakov, P.V., Kovalenko, L.V., Odinets, I.L., and Mastryukova, T.A., Esters of 1-phosphorylated cyclopropanecarboxylic acids, *Izv. Akad. Nauk SSSR, Ser. Khim.*, 2150, 1989; *Bull. Acad. Sci. USSR, Div. Chem. Sci. (Engl. Transl.)*, 1979, 1989.

315. Kazakov, P.V., Odinets, I.L., Laretina, A.P., Shcherbina, T.M., Petrovskii, P.V., Kovalenko, L.V., and Mastryukova, T.A., Cycloalkylation of phosphorus substituted acetic acid derivatives, *Izv. Akad. Nauk SSSR, Ser. Khim.*, 1873, 1990; *Bull. Acad. Sci. USSR, Div. Chem. Sci. (Engl. Transl.)*, 1702, 1990.

316. Hanrahan, J.R., and Hutchinson, D.W., Phase transfer catalysed synthesis of alicyclic tetraalkyl 1,1-bisphosphonates and trialkyl 1,1-phosphonocarboxylates, *Tetrahedron Lett.*, 34, 3767, 1993.

317. Zon, J., and Amrhein, N., Inhibitors of phenylalanin ammonia-lyase. 2-Aminoindan-2-phosphonic acid and related compounds, *Liebigs Ann. Chem.*, 625, 1992.

318. Ismailov, V.M., Moskva, V.V., Guliev, A.N., and Zykova, T.V., Synthesis of 1,2-oxaphosphorinane, *Zh. Obshch. Khim.*, 53, 2791, 1983; *J. Gen. Chem. USSR (Engl. Transl.)*, 53, 2516, 1983.

319. D'Incan, E., and Seyden-Penne, J., Ion pair extraction. Alkylation of phosphonates and the Wittig–Horner reaction, *Synthesis*, 516, 1975.

320. Grell, W., and Machleidt, H., Syntheses with organophosphorus compounds. Part 1. Ester olefinations, *Ann. Chem.*, 693, 134, 1966.

321. Tessier, J., Demoute, J.P., and Thuong, T.V., Process for preparing fluorinated derivatives of phosphonic acid, and products obtained using this process, *Roussel-UCLAF*, Eur. Patent Appl. EP 224417, 1987; *Chem. Abstr.*, 107, 96889, 1987.

322. Chambers, R.D., and Hutchinson, J., Elemental fluorine. Part 9. Catalysis of the direct fluorination of 2-substituted carbonyl compounds, *J. Fluorine Chem.*, 92, 45, 1998.

323. Chambers, R.D., and Hutchinson, J., Preparation of fluorinated phosphonate compounds, *F2 Chemicals*, Int. Patent Appl. WO 9905080, 1999; *Chem. Abstr.*, 130, 139451, 1999.

324. Davis, F.A., Han, W., and Murphy, C.K., Selective, electrophilic fluorinations using *N*-fluoro-*o*-benzenedisulfonimide, *J. Org. Chem.*, 60, 4730, 1995.

325. Hamilton, C.J., and Roberts, S.M., Synthesis of fluorinated phosphonoacetate derivatives of carbocyclic nucleoside monophosphonates and activity as inhibitors of HIV reverse transcriptase, *J. Chem. Soc., Perkin Trans. 1*, 1051, 1999.

326. Arnone, A., Bravo, P., Frigerio, M., Viani, F., and Zappala, C., Synthesis of 3′-arylsulfonyl-4′-[(diethoxyphosphoryl)difluoromethyl]thymidine analogs, *Synthesis*, 1511, 1998.

327. Arnone, A., Bravo, P., Frigerio, M., Mele, A., Vergani, B., and Viani, F., Synthesis of enantiomerically pure 2′,3′,5′-trideoxy-4′-[(diethoxyphosphoryl)difluoromethyl]thymidine analogues, *Eur. J. Org. Chem.*, 2149, 1999.

328. Braun, N.A., Klein, I., Spitzner, D., Vogler, B., Braun, S., Borrmann, H., and Simon, A., Cascade reactions with chiral Michael acceptors. Synthesis of enantiomerically pure tricyclo[3.2.1.0^{2,7}]- and bicyclo[3.2.1]octanes, *Liebigs Ann.*, 2165, 1995.

329. Shi, X.-X., and Dai, L.X., Mild halogenation of stabilized ester enolates by cupric halides, *J. Org. Chem.*, 58, 4596, 1993.

330. Quimby, O.T., and Prentice, J.B., Hypohalogenation of *gem*-diphosphonate esters and phosphonoacetate esters, *Procter and Gamble*, U.S. Patent Appl. US 3772412, 1973; *Chem. Abstr.*, 80, 37278, 1974.

331. McKenna, C.E., and Khawli, L.A., Synthesis of halogenated phosphonoacetate esters, *J. Org. Chem.*, 51, 5467, 1986.

332. Tago, K., and Kogen, H., *Bis*(2,2,2-trifluoroethyl) bromophosphonoacetate, a novel HWE reagent for the preparation of (*E*)-α-bromoacrylates. A general and stereoselective method for the synthesis of trisubstituted alkenes, *Org. Lett.*, 2, 1975, 2000.

333. Tago, K., and Kogen, H., A highly stereoselective synthesis of (*E*)-α-bromoacrylates, *Tetrahedron*, 56, 8825, 2000.

334. Bodnarchuk, N.D., Malovik, V.V., and Derkach, G.I., Derivatives of dialkoxyphosphonoacetic acids, *Zh. Obshch. Khim.*, 39, 1707, 1969; *J. Gen. Chem. USSR (Engl. Transl.)*, 39, 1673, 1969.

335. Guseinov, F.I., Klimentova, G.Y., Kol'tsova, O.L., Egereva, T.N., and Moskva, V.V., Intramolecular nucleophilic reactions of dialkyl 1,1-dichloro-2-hydroxyethylphosphonates, *Zh. Obshch. Khim.*, 66, 455, 1996; *Russ. J. Gen. Chem. (Engl. Transl.)*, 66, 441, 1996.

336. Nicholson, D.A., and Vaughn, H., New approaches to the preparation of halogenated methylenediphosphonates, phosphonoacetates, and malonates, *J. Org. Chem.*, 36, 1835, 1971.

337. Töke, L., Jaszay, Z.M., Petnehazy, I., Clementis, G., Vereczkey, G., Rockenbauer, A., and Kovats, K., A versatile building block for the synthesis of substituted cyclopropanephosphonic acid esters, *Tetrahedron*, 51, 9167, 1995.

338. Ide, J., Endo, R., and Muramatsu, S., Synthesis of ethyl α-(diethylphosphono)acrylate and its homologs. Versatile synthetic reagents, *Chem. Lett.*, 401, 1978.

339. Musicki, B., and Vevert, J.-P., Synthesis of conformationally restricted analogues of an angiotensin II receptor antagonist. General synthetic approach to functionalized imidazo[1,5-*a*]pyridine derivatives, *Tetrahedron Lett.*, 35, 9391, 1994.

340. Minami, T., Isonaka, T., Okada, Y., and Ichikawa, J., Copper(I) salt-mediated arylation of phosphinyl-stabilized carbanions and synthetic application to heterocyclic compounds, *J. Org. Chem.*, 58, 7009, 1993.

341. Artamkina, G.A., Tarasenko, E.A., Lukashev, N.V., and Beletskaya, I.P., Synthesis of perhaloaromatic diethyl methylphosphonates containing α-electron-withdrawing group, *Tetrahedron Lett.*, 39, 901, 1998.

342. Tarasenko, E.A., Artamkina, G.A., Voevodskaya, T.I., Lukashev, N.V., and Beletskaya, I.P., Nucleophilic substitution in perhalogenated aromatic compounds by carbanions derived from substituted dialkyl methylphosphonates, *Zh. Org. Khim.*, 34, 1523, 1998; *Russ. J. Org. Chem. (Engl. Transl.)*, 34, 1459, 1998.

343. Tarasenko, E.A., Mukhaiimana, P., Tsvetkov, A.V., Lukashev, N.V., and Beletskaya, I.P., Michael addition of phosphorylated CH acids under conditions of phase-transfer catalysis, *Zh. Org. Khim.*, 34, 64, 1998; *Russ. J. Org. Chem. (Engl. Transl.)*, 34, 52, 1998.

344. Kreutzkamp, N., Carbonyl and cyanophosphoric acid esters. Part 1. Preparation of phosphonoacetoacetic ester and -malonic ester by acylation reactions, *Chem. Ber.*, 88, 195, 1955.

345. Gough, S.T.D., and Trippett, S., A new synthesis of acetylenes. Part 1, *J. Chem. Soc.*, 2333, 1962.

346. Sakhibullina, V.G., Polezhaeva, N.A., and Arbuzov, B.A., Structure of acylated and benzoylated derivatives of phosphonoacetic trimethyl ester, *Zh. Obshch. Khim.*, 52, 1265, 1982; *J. Gen. Chem. USSR (Engl. Transl.)*, 52, 1112, 1982.

347. Thenappan, A., and Burton, D.J., Acylation of fluorocarbethoxy-substituted ylids. A simple and general route to α-fluoro β-keto esters, *J. Org. Chem.*, 56, 273, 1991.

348. Thenappan, A., and Burton, D.J., An expedient synthesis of α-fluoro-β-ketoesters, *Tetrahedron Lett.*, 30, 6113, 1989.

349. Tsai, H.-J., Isolation and characterization of intermediate in the synthesis of α-fluorodiesters, *J. Chin. Chem. Soc. (Taipei)*, 45, 543, 1998; *Chem. Abstr.*, 129, 244849, 1998.

350. Sakhibullina, V.G., Polezhaeva, N.A., and Arbuzov, B.A., Reactions of enolate anions of β-oxophosphonates and β-dicarbonyl compounds with electrophiles, *Zh. Obshch. Khim.*, 56, 2479, 1986; *J. Gen. Chem. USSR (Engl. Transl.)*, 56, 2193, 1986.

351. Henning, H.G., and Petzold, G., Dieckmann cyclization with ω-acylalkanephosphonic acid esters, *Z. Chem.*, 7, 183, 1967.

352. Tasz, M.K., Rodriguez, O.P., Cremer, S.E., Hussain, M.S., and Mazhar-ul, H., Conformational study of six-membered phostones. *Cis-* and *trans*-3-(methoxycarbonyl)-2-methoxy-2-oxo-1,2-oxaphosphorinane, *J. Chem. Soc., Perkin Trans. 2*, 2221, 1996.

353. Kim, D.Y., Kong, M.S., and Rhie, D.Y., A new synthesis of 2-aryl-2-oxoalkylphosphonates from triethyl phosphonoacetate, *Synth. Commun.*, 25, 2865, 1995.

354. Corbel, B., L'Hostis-Kervella, I., and Haelters, J.-P., On a safe and practical method for the preparation of β-keto phosphonates, *Synth. Commun.*, 26, 2561, 1996.

355. Kim, D.Y., Kong, M.S., and Kim, T.H., A practical synthesis of β-keto phosphonates from triethyl phosphonoacetate, *Synth. Commun.*, 26, 2487, 1996.

356. Kim, D.Y., and Suh, K.H., Solid phase acylation of phosphonoacetates. Synthesis of β-keto phosphonates from polymer bound phosphonoacetate, *Synth. Commun.*, 29, 1271, 1999.

357. Kim, D.Y., and Choi, Y.J., Synthesis of α-fluoro-β-keto phosphonates from α-fluoro phosphonoacetic acid, *Synth. Commun.*, 28, 1491, 1998.

358. Kim, D.Y., Rhie, D.Y., and Oh, D.Y., Acylation of diethyl (ethoxycarbonyl)fluoromethylphosphonate using magnesium chloride-triethylamine. A facile synthesis of α-fluoro β-keto esters, *Tetrahedron Lett.*, 37, 653, 1996.

359. Kim, D.Y., Choi, J.S., and Rhie, D.Y., P–C bond cleavage of triethyl 2-fluoro-3-oxo-2-phosphonoacetates with magnesium chloride. A synthesis of α-fluoro-β-keto esters, *Synth. Commun.*, 27, 1097, 1997.

360. Kim, D.Y., and Kim, J.Y., A new synthesis of α-fluoromalonates from α-fluoro-α phosphonyl malonates using P–C bond cleavage, *Synth. Commun.*, 28, 2483, 1998.

361. Kim, D.Y., Lee, Y.M., and Choi, Y.J., Acylation of α-fluorophosphonoacetate derivatives using magnesium chloride-triethylamine, *Tetrahedron*, 55, 12983, 1999.

362. Kim, D.Y., A facile P–C bond cleavage of 2-fluoro-2-phosphonyl-1,3-dicarbonyl compounds on silica gel, *Synth. Commun.*, 30, 1205, 2000.

363. Katagiri, N., Yamamoto, M., and Kaneko, C., 3-Acetoxy-2-dimethylphosphonoacrylates. New dienophiles and their use for the synthesis of carbocyclic *C*-nucleoside precursors by the aid of RRA reaction, *Chem. Lett.*, 1855, 1990.

364. Kantlehner, W., Wagner, F., and Bredereck, H., Orthoamides. Part 32. Reactions of *tert*-butoxy-*N*,*N*,*N'*,*N'*-tetramethylmethanediamine with NH- and CH-acidic compounds, *Liebigs Ann. Chem.*, 344, 1980.

365. Mel'nikov, N.N., Kozlov, V.A., Churusova, S.G., Buvashkina, N.I., Ivanchenko, V.I., Negrebetskii, V.V., and Grapov, A.F., β-Phosphorylated enamines, *Zh. Obshch. Khim.*, 53, 1689, 1983; *J. Gen. Chem. USSR (Engl. Transl.)*, 53, 1519, 1983.

366. Aboujaoude, E.E., Collignon, N., and Savignac, P., α-Functional dialkyl formyl-1-methylphosphonates. Part 2. Thermical preparation and conversion into α-phosphonic heterocycles, *Tetrahedron*, 41, 427, 1985.

367. Aboujaoude, E.E., Collignon, N., and Savignac, P., Synthesis of α-phosphonic heterocycles. New developments, *Phosphorus Sulfur*, 31, 231, 1987.

368. Neidlein, R., and Eichinger, T., (1,3-Dioxolan-2-ylidene)methylphosphonates and -phosphinates as (simple) synthons in heterocyclic synthesis, *Helv. Chim. Acta*, 75, 124, 1992.

369. Arbuzov, B.A., Sakhibullina, V.G., Polezhaeva, N.A., and Vinogradova, V.S., Tetramethyl ester phosphonomalonic acid, *Izv. Akad. Nauk SSSR, Ser. Khim.*, 2139, 1976; *Bull. Acad. Sci. USSR, Div. Chem. Sci. (Engl. Transl.)*, 2001, 1976.

370. Ratcliffe, R.W., and Christensen, B.G., Total synthesis of β-lactam antibiotics. Part 1. α-Thioformamido-diethylphosphonoacetates, *Tetrahedron Lett.*, 14, 4645, 1973.

371. Guthikonda, R.N., Cama, L.D., and Christensen, B.G., Total synthesis of β-lactam antibiotics. Part 8. Stereospecific total synthesis of (±)-1-carbacephalothin, *J. Am. Chem. Soc.*, 96, 7584, 1974.

372. Hegedus, L.S., Schultze, L.M., Toro, J., and Yijun, C., Photolytic reaction of chromium and molybdenum carbene complexes with imines. Synthesis of cepham, oxapenam, and oxacepham derivatives, *Tetrahedron*, 41, 5833, 1985.

373. Scopes, D.I.C., Kluge, A.F., and Edwards, J.A., Enolate amination with *O*-mesitylenesulfonylhydroxylamine, *J. Org. Chem.*, 42, 376, 1977.

374. Shiraki, C., Saito, H., Takahashi, K., Urakawa, C., and Hirata, T., Preparation of amino(diethoxyphosphoryl)acetic esters. Catalytic hydrogenation of diazo compounds to amines, *Synthesis*, 399, 1988.

375. Colvin, E.W., Kirby, G.W., and Wilson, A.C., *O*-(Diphenylphosphinyl)hydroxylamine. A new reagent for electrophilic *C*-amination, *Tetrahedron Lett.*, 23, 3835, 1982.

376. Stocksdale, M.G., Ramurthy, S., and Miller, M.J., Asymmetric total synthesis of an important 3-(hydroxymethyl)carbacephalosporin, *J. Org. Chem.*, 63, 1221, 1998.

377. Khokhlov, P.S., Kashemirov, B.A., and Strepikheev, Y.A., Nitrosation of the triethyl ester of phosphonacetic acid, *Zh. Obshch. Khim.*, 51, 2145, 1981; *J. Gen. Chem. USSR (Engl. Transl.)*, 51, 1847, 1981.

378. Khokhlov, P.S., Kashemirov, B.A., and Strepikheev, Y.A., Nitrosation of (dialkoxyphosphinyl)- and (alkoxyalkylphosphinyl)-acetic acid, *Zh. Obshch. Khim.*, 52, 2800, 1982; *J. Gen. Chem. USSR (Engl. Transl.)*, 52, 2468, 1982.

379. Khokhlov, P.S., Kashemirov, B.A., Mikityuk, A.D., and Strepikheev, Y.A., Preparation of phosphorylated hydantoins, *Zh. Obshch. Khim.*, 53, 2146, 1983; *J. Gen. Chem. USSR (Engl. Transl.)*, 53, 1936, 1983.

380. Kashemirov, B.A., Khokhlov, P.S., Polenov, E.A., Sokol, O.G., and Kakadii, L.I., Nitrosation of methyl 2-(diethoxyphosphinyl)propionate, *Zh. Obshch. Khim.*, 55, 460, 1985; *J. Gen. Chem. USSR (Engl. Transl.)*, 55, 407, 1985.

381. Kashemirov, B.A., Skoblikova, L.I., and Kokhlov, P.S., Reduction of nitrosophosphonates, *Zh. Obshch. Khim.*, 58, 702, 1988; *J. Gen. Chem. USSR (Engl. Transl.)*, 58, 621, 1988.

382. Kashemirov, B.A., Mikityuk, A.D., Strepikheev, Y.A., and Khokhlov, P.S., Nitrosation of (diethoxy-phosphinyl)acetyl chloride, *Zh. Obshch. Khim.*, 56, 957, 1986; *J. Gen. Chem. USSR (Engl. Transl.)*, 56, 843, 1986.

383. Kashemirov, B.A., Osipov, V.N., and Khokhlov, P.S., Azido(alkoxycarbonyl)ethylphosphonates, *Zh. Obshch. Khim.*, 62, 470, 1992; *J. Gen. Chem. USSR (Engl. Transl.)*, 62, 381, 1992.

384. Khokhlov, P.S., Kashemirov, B.A., Mikityuk, A.D., Strepikheev, Y.A., and Chimiskyan, A.L., Diazo-tization of amino(dialkoxyphosphinyl)acetic asters, *Zh. Obshch. Khim.*, 54, 2785, 1984; *J. Gen. Chem. USSR (Engl. Transl.)*, 54, 2495, 1984.

385. Hakimelahi, G.H., and Just, G., Two simple methods for the synthesis of trialkyl α-aminophospho-noacetates. Trifluoromethanesulfonyl azide as an azide-transfer agent, *Synth. Commun.*, 10, 429, 1980.

386. Seki, M., and Matsumoto, K., A facile preparation of amino(diethoxyphosphoryl)acetic esters. Trans-formation of phosphoranes to phosphonic esters, *Synthesis*, 580, 1996.

387. Schaumann, E., and Grabley, F.-F., Thioketene synthesis. Part 3. Reactions of phosphonate carbanions with carbon disulfide, *Liebigs Ann. Chem.*, 1715, 1979.

388. Schaumann, E., and Fittkau, S., Synthesis and Wittig–Horner reactions of 1-(functionally)substituted 2,2-dimethylpropanephosphonic esters (1-*tert*-butyl-substituted phosphonic esters), *Synthesis*, 449, 1983.

389. Minami, T., Okauchi, T., Matsuki, H., Nakamura, M., Ichikawa, J., and Ishida, M., Synthesis and synthetic application of phosphonoketene dithioacetals. New synthesis of dithioallenes and (α-dithio-carboxyvinyl)phosphonates, *J. Org. Chem.*, 61, 8132, 1996.

390. Kashemirov, V.A., Osipov, V.N., Emel'yanovich, A.M., and Khokhlov, P.S., Reaction of triisopropyl phosphoacetate with sulfenyl chlorides upon catalysis by aluminum triisopropoxide, *Zh. Obshch. Khim.*, 62, 1195, 1992; *J. Gen. Chem. USSR (Engl. Transl.)*, 62, 982, 1992.

391. Kozlov, V.A., Dol'nikova, T.Y., Ivanchenko, V.I., Negrebetskii, V.V., Grapov, A.F., and Mel'nikov, N.N., β-Phosphorylated *N,S*-ketenaminomercaptals, *Zh. Obshch. Khim.*, 53, 2229, 1983; *J. Gen. Chem. USSR (Engl. Transl.)*, 53, 2008, 1983.

392. Arbuzov, B.A., and Vizel, A.O., Reaction of diazoacetic ester with phosphorus acid and its esters, *Izv. Akad. Nauk SSSR, Ser. Khim.*, 749, 1963; *Chem. Abstr.*, 59, 7362g, 1963.

393. Arbuzov, B.A., Polozov, A.M., and Polezhaeva, N.A., P–H addition of carbenoids and carbenes as a method for the synthesis of 2-phospho-substituted 1,3-dicarbonyl compounds, *Dokl. Akad. Nauk SSSR, Ser. Khim.*, 287, 849, 1986; *Dokl. Chem. (Engl. Transl.)*, 287, 69, 1986.

394. Pudovik, A.N., and Gareev, R.D., Reactions of carbethoxycarbene and diazomethane with unsaturated organophosphorus compounds and dialkyl phosphonates, *Zh. Obshch. Khim.*, 34, 3942, 1964; *J. Gen. Chem. USSR (Engl. Transl.)*, 34, 4003, 1964.

395. Polozov, A.M., Polezhaeva, N.A., Mustaphin, A.H., Khotinen, A.V., and Arbuzov, B.A., A new one-pot synthesis of dialkyl phosphonates from diazo compounds and dialkyl hydrogen phosphites, *Synthesis*, 515, 1990.

396. Polozov, A.M., Mustaphin, A.H., and Khotinen, A.V., Insertion of carbenes into P–H bonds. Part 5. Synthesis of new phosphonates and phosphinates in reactions catalysed by Cu, Pd, Rh, Ni complexes, *Phosphorus, Sulfur Silicon Relat. Elem.*, 73, 153, 1992.

397. Nakamura, E., New acyl anion equivalent. A short route to the enol lactam intermediate in cytochalasin synthesis, *Tetrahedron Lett.*, 22, 663, 1981.

398. Stork, G., and Nakamura, E., A simplified total synthesis of cytochalasins via an intramolecular Diels–Alder reaction, *J. Am. Chem. Soc.*, 105, 5510, 1983.

399. Ando, K., Koike, F., Kondo, F., and Takayama, H., An improved synthesis of 24,24-difluoro-1α,25-dihydroxyvitamin D_3 from readily available vitamin D_2, *Chem. Pharm. Bull.*, 43, 189, 1995.

400. Schmidt, U., Langner, J., Kirschbaum, B., and Braun, C., Synthesis and enantioselective hydrogenation of α-acyloacrylates, *Synthesis*, 1138, 1994.

401. Jurgens, A.R., Green, K., Ruso, E.R., Jennings, M.N., Blum, D.M., and Feigelson, G.B., Process improved preparation of a versatile α-ketoester acyl anion synthon, *Synth. Commun.*, 24, 1171, 1994.

402. Fiszer, B., Organophosphorus compounds with an active methylene group. Part 5. Thermal decomposition of diethoxyphosphinylacetic acid, *Rocz. Chem.*, 37, 949, 1963.

403. Zhang, Y., Takeda, S., Kitagawa, T., and Irie, H., A synthesis of 15,16-dimethoxyerythrin-6-en-8-one, *Heterocycles*, 24, 2151, 1986.

404. Jennings, L.J., Macchia, M., and Parkin, A., Synthesis of analogues of 5-iodo-2'-deoxyuridine 5' diphosphate, *J. Chem. Soc., Perkin Trans. 1*, 2197, 1992.

405. Malevannaya, R.A., Tsvetkov, E.N., and Kabachnik, M.I., Dialkoxyphosphinylacetic acids and some of their analogs, *Zh. Obshch. Khim.*, 41, 1426, 1971; *J. Gen. Chem. USSR (Engl. Transl.)*, 41, 1432, 1971.

406. Lombardo, L., and Taylor, R.J.K., An improved procedure for the conversion of carbonyl compounds to α,β-unsaturated carboxylic acids, *Synth. Commun.*, 8, 463, 1978.

407. Clayton, J.P., Luk, K., and Rogers, N.H., The chemistry of pseudomonic acid. Part 2. The conversion of pseudomonic acid A into monic acid A and its esters, *J. Chem. Soc., Perkin Trans. 1*, 308, 1979.

408. Zon, J., Direct synthesis of diethyl N-carbobenzoxy-aminoalkanephosphonates, *Pol. J. Chem.*, 53, 541, 1979.

409. Cooke, M.P., Jr., and Biciunas, K.P., A new synthesis of unsaturated acylphosphoranes by the Wittig–Horner olefination, *Synthesis*, 283, 1981.

410. Bel'skii, V.E., Kurguzova, A.M., and Efremova, M.V., The kinetics of hydrolysis of (acetylalkyl)- and (carboxyalkyl)phosphonates, *Izv. Akad. Nauk SSSR, Ser. Khim.*, 310, 1986; *Bull. Acad. Sci. USSR, Div. Chem. Sci. (Engl. Transl.)*, 284, 1986.

411. Austin, G.N., Baird, P.D., Fleet, G.W.J., Peach, J.M., Smith, P.W., and Watkin, D.J., 3,6-Dideoxy-3,6-imino-1,2-O-isopropylidene-α-D-glucofuranose as a divergent intermediate for the synthesis of hydroxylated pyrrolidines. Synthesis of 1,4-dideoxy-1,4-imino-L-gulitol, 1,4-dideoxy-1,4-imino-D-lyxitol, 2S,3S,4R,-3,4-dihydroxyproline and (1S,2R,8S,8aR)-1,2,8-trihydroxyoctahydroindolizine [8-epi-swainsonine]. X-ray crystal structure of (1S,2R,8S,8aR)-1,2,8-trihydroxy-5-oxo-octahydroindolizine, *Tetrahedron*, 43, 3095, 1987.

412. Patel, D.V., Schmidt, R.J., Biller, S.A., Gordon, E.M., Robinson, S.S., and Manne, V., Farnesyl diphosphate-based inhibitors of ras farnesyl protein transferase, *J. Med. Chem.*, 38, 2906, 1995.

413. Roush, W.R., and Sciotti, R.J., Enantioselective total synthesis of (−)-chlorothricolide via the tandem inter- and intramolecular Diels–Alder reaction of a hexaenoate intermediate, *J. Am. Chem. Soc.*, 120, 7411, 1998.

414. Dolfini, J.E., and Breuer, H., Bactericidal potassium 6-[phenyl(diethylphosphono)acetamido]penicillanate, *Chemische Fabrik von Heyden*, German Patent Appl. DE 2208272, 1972; *Chem. Abstr.*, 77, 164719, 1972.

415. Koppel, G.A., and Kinnick, M.D., Carboxyvinylation. A one-step synthesis of α,β-unsaturated acids, *Tetrahedron Lett.*, 15, 711, 1974.

416. Koppel, G.A., α-(Dibenzylphosphono)acetic acids, *Eli Lilly*, U.S. Patent Appl. US 3897518, 1975; *Chem. Abstr.*, 83, 164374, 1975.

417. Savignac, P., Snoussi, M., and Coutrot, P., Carboxychloro olefination. A convenient synthesis of α-chloro α,β-ethylenic carboxylic acids, *Synth. Commun.*, 8, 19, 1978.

418. Coutrot, P., Snoussi, M., and Savignac, P., An improvement in the Wittig–Horner synthesis of 2-alkenoic acids, *Synthesis*, 133, 1978.

419. Perriot, P., Villieras, J., and Normant, J.F., Diethyl 1-chloro-1-lithioalkanephosphonates. A general synthesis of diethyl 1,2-epoxyalkanephosphonates and 1-alkoxycarbonyl-1-chloroalkanephosphonates, *Synthesis*, 33, 1978.

420. Blackburn, G.M., Brown, D., and Martin, S.J., A novel synthesis of fluorinated phosphonoacetic acid, *J. Chem. Res. (S)*, 92, 1985.

421. Coutrot, P., and Ghribi, A., A facile and general, one-pot synthesis of 2-oxoalkane phosphonates from diethylphosphonocarboxylic acid chlorides and organometallic reagents, *Synthesis*, 661, 1986.

422. Mikolajczyk, M., and Midura, W.H., Diethyl α-(methylthio)phosphonoacetic acid, a new bifunctional reagent. Synthesis of unsaturated five- and six-membered lactones, *Synlett*, 245, 1991.

423. Coutrot, P., Grison, C., and Charbonnier-Gérardin, C., Synthesis of modified peptides incorporating a phosphorane moiety in a terminal nitrogen or carbon, *Tetrahedron*, 48, 9841, 1992.

424. Lamothe, M., Perrin, D., Blotières, D., Leborgne, M., Gras, S., Bonnet, D., Hill, B.T., and Halazy, S., Inhibition of farnesyl protein transferase by new farnesyl phosphonate derivatives of phenylalanine, *Bioorg. Med. Chem. Lett.*, 6, 1291, 1996.

425. Kim, Y., Singer, R.A., and Carreira, E.M., Total synthesis of macrolactin A with versatile catalytic, enantioselective dienolate aldol addition reactions, *Angew. Chem.*, 110, 1321, 1998; *Angew. Chem. Int. Ed. Engl.*, 37, 1261, 1998.

426. Blicke, F.F., and Raines, S., The preparation of organic phosphorus compounds by Ivanov reactions. Part 2, *J. Org. Chem.*, 29, 2036, 1964.

427. Coutrot, P., Youssefi-Tabrizi, M., and Grison, C., Diethyl (chloromagnesio)methanephosphonate. A novel Grignard reagent and its use in organic synthesis, *J. Organomet. Chem.*, 316, 13, 1986.

428. Razumov, A.I., and Moskva, V.V., Derivatives of phosphonic and phosphinic acids. Part 26. Synthesis of phosphorylated carboxylic acids from aldehydes and acetals, *Zh. Obshch. Khim.*, 35, 1149, 1965; *J. Gen. Chem. USSR (Engl. Transl.)*, 35, 1151, 1965.

429. Isbell, A.F., Englert, L.F., and Rosenberg, H., Phosphonoacetaldehyde, *J. Org. Chem.*, 34, 755, 1969.

430. Floyd, D.M., and Fritz, A.W., Studies directed toward the preparation of polyene macrolide mimics, *Tetrahedron Lett.*, 22, 2847, 1981.

431. Burton, D.J., and Sprague, L.G., Preparation of difluorophosphonoacetic acid and its derivatives, *J. Org. Chem.*, 53, 1523, 1988.

432. Tsuda, Y., Ishiura, A., Hosoi, S., and Isobe, K., Studies toward total synthesis of non-aromatic *Erythrina* alkaloids. Part 2. A general method for synthesis of perhydro-6*H*-pyrido[2,1-*i*]indole derivatives. Synthesis of isoerythroidine skeleton, *Chem. Pharm. Bull.*, 40, 1697, 1992.

433. Ohmori, K., Suzuki, T., Nishiyama, S., and Yamamura, S., Synthetic studies on bryostatins, potent antineoplastic agents. Synthesis of the C_{17}–C_{27} fragment of C_{20} deoxybryostatins, *Tetrahedron Lett.*, 36, 6515, 1995.

434. White, J.D., Tiller, T., Ohba, Y., Porter, W.J., Jackson, R.W., Wang, S., and Hanselmann, R., Total synthesis of rutamycin B via Suzuki macrocyclization, *J. Chem. Soc., Chem. Commun.*, 79, 1998.

435. Lafontaine, J.A., Provençal, D.P., Gardelli, C., and Leahy, J.W., The enantioselective total synthesis of the antitumor macrolide natural product rhizoxin D, *Tetrahedron Lett.*, 40, 4145, 1999.

436. Keck, G.E., Wager, C.A., Wager, T.T., Savin, K.A., Covel, J.A., McLaws, M.D., Krishnamurthy, D., and Cee, V.J., Asymmetric total synthesis of rhizoxin D, *Angew. Chem. Int. Ed. Engl.*, 40, 231, 2001.

437. Schultz, R.G., and Starks, F.W., *N*-(Phosphonacetyl)-(+)-L-aspartic acid derivatives, *Starks Assoc.*, German Patent Appl. DE 2849396, 1979; *Chem. Abstr.*, 91, 108232, 1979.

438. Soroka, M., and Mastalerz, P., Hofmann degradation and bromination of amides derived from phosphonoacetic acid, *Tetrahedron Lett.*, 14, 5201, 1973.

439. Coutrot, P., and Grison, C., General synthesis of α-fluoro-β-ketophosphonate precursors of α-fluoro enones. Application to the pyrethrene series, *Tetrahedron Lett.*, 29, 2655, 1988.

440. Sturtz, G., Preparation of acrylic or methacrylic phosphonate or *gem*-bisphosphonate compounds, French Patent Appl. FR 2767829, 1999; *Chem. Abstr.*, 130, 282173, 1999.

441. Motoyoshiya, J., and Hirata, K., Cycloaddition reactions of (diethylphosphono)ketenes, *Chem. Lett.*, 211, 1988.

442. Meyers, A.I., Babiak, K.A., Campbell, A.L., Comins, D.L., Fleming, M.P., Henning, R., Heuschmann, M., Hudspeth, J.P., Kane, J.M., Reider, P.J., Roland, D.M., Shimizu, K., Tomioka, K., and Walkup, R.D., Total synthesis of (−)-maysine, *J. Am. Chem. Soc.*, 105, 5015, 1983.

443. Kolodyazhnyi, O.I., New method of phosphorylated-ketene synthesis, *Zh. Obshch. Khim.*, 49, 716, 1979; *J. Gen. Chem. USSR (Engl. Transl.)*, 49, 621, 1979.

444. Heider, J., Nickl, J., Eberlein, W., Dahms, G., and Kobinger, W., Cardenolide genins, *Dr. Karl Thomae*, German Patent Appl. DE 2015850, 1971; *Chem. Abstr.*, 76, 46404, 1972.

445. Eberlein, W., Nickl, J., Heider, J., Dahms, G., and Machleidt, H., Steroidal Cardiotonics. Part 1. Introduction of substituents into the butenolide ring of heart glycosides, *Chem. Ber.*, 105, 3686, 1972.

446. Raddatz, P., and Winterfeldt, E., Cyclopentenones. Part 4. Stereospecificity of the brefeldin A cyclization, *Angew. Chem.*, 93, 281, 1981; *Angew. Chem. Int. Ed. Engl.*, 20, 286, 1981.

447. Roush, W.R., and Blizzard, T.A., Synthesis of epoxytrichothecenes. Verrucarin J and verrucarin J isomers, *J. Org. Chem.*, 49, 1772, 1984.

448. Callant, P., D'Haenens, L., and Vandewalle, M., An efficient preparation and the intramolecular cyclopropanation of α-diazo-β-ketophosphonates and α-diazophosphonoacetates, *Synth. Commun.*, 14, 155, 1984.

449. Keck, G.E., Kachensky, D.F., and Enholm, E.J., Pseudomonic acid C from L-lyxose, *J. Org. Chem.*, 50, 4317, 1985.

450. Hori, K., Arai, M., Nomura, K., and Yoshii, E., An efficient 3(*C*)-acylation of tetramic acids involving acyl migration of 4(*O*)-acylates, *Chem. Pharm. Bull.*, 35, 4368, 1987.

451. Tamura, H., Fujita, A., Takagi, Y., Kitahara, T., and Mori, K., Simple synthesis of dehydrololiolide, *Biosci. Biotechnol. Biochem.*, 58, 1902, 1994.

452. Jung, S.H., Lee, Y.S., Park, H., and Kwon, D.-S., Transannular Diels–Alder reaction of a macrocyclic triene, (*E,E,E*)-13-trideca-2,8,10-trienolactone, *Tetrahedron Lett.*, 36, 1051, 1995.

453. Kazmaier, U., Synthesis of quaternary amino acids containing β,γ- as well as γ,δ-unsaturated side chains via chelate-enolate Claisen rearrangement, *Tetrahedron Lett.*, 37, 5351, 1996.

454. Gais, H.-J., Schmiedl, G., and Ossenkamp, R.K.L., Total synthesis of (+)-3-oxacarbocyclin. Part 1. Retrosynthesis and asymmetric olefination through Horner–Wadsworth–Emmons, Peterson and Martin reactions, *Liebigs Ann.*, 2419, 1997.

455. Vincent, S., Grenier, S., Valleix, A., Salesse, C., Lebeau, L., and Mioskowski, C., Synthesis of enzymatically stable analogues of GDP for binding studies with transducin, the G-protein of the visual photoreceptor, *J. Org. Chem.*, 63, 7244, 1998.

456. Moody, C.J., and Miller, D.J., Reactivity of differentially substituted *bis*(diazo) esters in rhodium(II) mediated O–H insertion reactions, *Tetrahedron*, 54, 2257, 1998.

457. Cermak, D.M., Wiemer, D.F., Lewis, K., and Hohl, R.J., 2-(Acyloxy)ethylphosphonate analogs of prenyl pyrophosphates. Synthesis and biological characterization, *Bioorg. Med. Chem.*, 8, 2729, 2000.

458. Forsyth, C.J., Ahmed, F., Cink, R.D., and Lee, C.S., Total synthesis of phorboxazole A, *J. Am. Chem. Soc.*, 120, 5597, 1998.

459. Davenport, R.J., and Regan, A.C., Synthesis of a C1–C9 fragment of rhizoxin, *Tetrahedron Lett.*, 41, 7619, 2000.

460. Okada, Y., Minami, T., Yamamoto, T., and Ichikawa, J., A novel type of chiral diphosphine ligand, *trans*-2,3-*bis*(diphenylphosphino)-1-methyl-1-cyclopropanecarboxylic acid and asymmetric allylic alkylation by the use of its palladium complex, *Chem. Lett.*, 547, 1992.

461. Saitoh, K., Shiina, I., and Mukaiyama, T., *O,O′*-Di(2-pyridyl) thiocarbonate as an efficient reagent for the preparation of carboxylic esters from highly hindered alcohols, *Chem. Lett.*, 679, 1998.

462. Williams, D.R., Werner, K.M., and Feng, B., Total synthesis of rhizoxin D, *Tetrahedron Lett.*, 38, 6825, 1997.

463. Rychnovsky, S.D., Griesgraber, G., and Kim, J., Rapid construction of the roflamycoin system, *J. Am. Chem. Soc.*, 116, 2621, 1994.

464. Rychnovsky, S.D., Yang, G., Hu, Y., and Khire, U.R., Prins desymmetrization of a C_2-symmetric diol. Application to the synthesis of 17-deoxyroflamycoin, *J. Org. Chem.*, 62, 3022, 1997.

465. Roush, W.R., Blizzard, T.A., and Basha, F.Z., Methodology for the synthesis of the acyclic portions of verrucarins A and J, *Tetrahedron Lett.*, 23, 2331, 1982.

466. Roush, W.R., and Blizzard, T.A., Synthesis of verrucarin J, *J. Org. Chem.*, 48, 758, 1983.

467. Lambert, R.W., Martin, J.A., Thomas, G.J., Duncan, I.B., Hall, M.J., and Heimer, E.P., Synthesis and antiviral activity of phosphonoacetic and phosphonoformic acid esters of 5-bromo-2′-deoxyuridine and related pyrimidine nucleosides and acyclonucleosides, *J. Med. Chem.*, 32, 367, 1989.

468. Brown, I.J.S., Clarkson, R., Crossley, N.S., and McLoughlin, B.J., 3-Aminoalkoxycarbonylmethylene steroids, *Imperial Chem. Ind.*, U.K. Patent Appl. GB 1175219, 1969; *Chem. Abstr.*, 72, 111712, 1970.

469. Brown, I.J.S., Clarkson, R., Crossley, N.S., and McLoughlin, B.J., Diethyl 2-(dimethylamino)ethoxycarbonylmethylphosphonate, *Imperial Chem. Ind.*, U.K. Patent Appl. GB 1175220, 1969; *Chem. Abstr.*, 72, 132950f, 1970.

470. Hatakeyama, S., Satoh, K., Sakurai, K., and Takano, S., A synthesis of (−)-pyrenophorin using 4-DMAP-catalyzed ester exchange reaction of phosphonoacetates with lactols, *Tetrahedron Lett.*, 28, 2717, 1987.

471. Duplantier, A.J., and Masamune, S., Pimaricin. Stereochemistry and synthesis of its aglycon (pimarolide) methyl ester, *J. Am. Chem. Soc.*, 112, 7079, 1990.

472. Nishioka, T., Iwabuchi, Y., Irie, H., and Hatakeyama, S., Concise enantioselective synthesis of (+)-aspicilin based on a ruthenium catalyzed olefin metathesis reaction, *Tetrahedron Lett.*, 39, 5597, 1998.

473. Williams, D.R., and Clark, M.P., The macrocyclic domain of phorboxazole A. A stereoselective synthesis of the C_1–C_{32} macrolactone, *Tetrahedron Lett.*, 40, 2291, 1999.

474. Page, P.C.B., Moore, J.P., Mansfield, I., McKenzie, M.J., Bowler, W.B., and Gallagher, J.A., Synthesis of bone-targeted oestrogenic compounds for the inhibition of bone resorption, *Tetrahedron*, 57, 1837, 2001.

475. Hatakeyama, S., Satoh, K., Sakurai, K., and Takano, S., Efficient ester exchange reaction of phosphonoacetates, *Tetrahedron Lett.*, 28, 2713, 1987.

476. Gais, H.-J., Schmiedl, G., Ball, W.A., Bund, J., Hellmann, G., and Erdelmeier, I., Synthesis of optically active 3-oxa-carbacyclin precursors featuring asymmetric Horner–Emmons reaction, *Tetrahedron Lett.*, 29, 1773, 1988.

477. Rehwinkel, H., Skupsch, J., and Vorbrüggen, H., *E-* or *Z-*selective Horner–Wittig reactions of substituted bicyclo[3.3.0]octane-3-ones with chiral phosphonoacetates, *Tetrahedron Lett.*, 29, 1775, 1988.

478. Tullis, J.S., Vares, L., Kann, N., Norrby, P.-O., and Rein, T., Reagent control of geometric selectivity and enantiotopic group preference in asymmetric Horner–Wadsworth–Emmons reactions with *meso-*dialdehydes, *J. Org. Chem.*, 63, 8284, 1998.

479. Krauser, S.F., and Watterson, A.C., Jr., New mild conditions for the synthesis of α,β-unsaturated γ-lactones. β-(2-Phthalimidoethyl)-$\Delta^{\alpha,\beta}$-butenolide, *J. Org. Chem.*, 43, 3400, 1978.

480. Shinada, T., Kawakami, T., Sakai, H., Takada, I., and Ohfune, Y., An efficient synthesis of α-acyloxyketone by Cu(acac)$_2$-catalyzed insertion reaction of α-diazoketone to carboxylic acid, *Tetrahedron Lett.*, 39, 3757, 1998.

481. Finkelstein, J., The preparation of β-aminoethanephosphonic acid, *J. Am. Chem. Soc.*, 68, 2397, 1946.

482. Kamai, G., and Shugurova, E.I., Amides of dialkylphosphonocarboxylic acids and their derivatives, *Dokl. Akad. Nauk SSSR*, 79, 605, 1951; *Chem. Abstr.*, 49, 15768c, 1955.

483. Zieloff, K., Paul, H., and Hilgetag, G., The reaction of aromatic amines and the triethyl ester of phosphonoacetic acid, *Z. Chem.*, 4, 148, 1964.

484. Tawfik, D.S., Eshhar, Z., and Green, B.S., From phosphonates to catalytic antibodies. A novel route to phosphonoester transition state analogs and haptens, *Phosphorus, Sulfur Silicon Relat. Elem.*, 76, 123, 1993.

485. Hamdouchi, C., de Blas, J., del Prado, M., Gruber, J., Heinz, B.A., and Vance, L., 2-Amino-3-substituted-6-[(*E*)-1-phenyl-2-(*N*-methylcarbamoyl)vinyl]imidazo[1,2-*a*]pyridines as a novel class of inhibitors of human rhinovirus. Stereospecific synthesis and antiviral activity, *J. Med. Chem.*, 42, 50, 1999.

486. Odinets, I.L., Kazakov, P.V., Amanov, R.U., Antipin, M.Y., Kovalenko, L.V., Struchkov, Y.T., and Mastryukova, T.A., Some properties of 1-phosphorus(IV)-substituted cycloalkanecarboxylic acids and their derivatives, *Izv. Akad. Nauk, Ser. Khim.*, 1879, 1992; *Bull. Russ. Acad. Sci., Div. Chem. Sci. (Engl. Transl.)*, 1466, 1992.

487. Chen, Y.P., Chantegrel, B., and Deshayes, C., A thermal Wolff rearrangement-benzannulation route to naphth[2,1-*d*]isoxazoles, [1]benzofuro[6,7-*d*]- or [5,4-*d*]isoxazoles and 1,2-benzisoxazoles, *Heterocycles*, 41, 175, 1995.

488. Hoffman, R.V., Reddy, M.M., and Cervantes-Lee, F., Improved methodology for the generation and trapping of α-lactams by weak nucleophiles, *J. Org. Chem.*, 65, 2591, 2000.

489. Guzman, A., Muchowski, J.M., Strosberg, A.M., and Sims, J.M., Replacement of the butenolide moiety of digitoxigenin by cyclic Michael acceptor systems, *Can. J. Chem.*, 59, 3241, 1981.

490. Lieberknecht, A., and Griesser, H., What is the structure of barettin? Novel synthesis of unsaturated diketopiperazines, *Tetrahedron Lett.*, 28, 4275, 1987.

491. Glöde, J., Gross, H., Henklein, P., Niedrich, H., Tanneberger, S., and Tschiersch, B., Synthesis of *N-*(phosphonoacetyl)-L-aspartic acid (PALA), *Pharmazie*, 43, 434, 1988.

492. Macchia, M., Jannitti, N., Gervasi, G., and Danesi, R., Geranylgeranyl diphosphate-based inhibitors of post-translational geranygeranylation of cellular proteins, *J. Med. Chem.*, 39, 1352, 1996.

493. Giambastiani, G., Pacini, B., Porcelloni, M., and Poli, G., A new palladium-catalyzed intramolecular allylation to pyrrolidin-2-ones, *J. Org. Chem.*, 63, 804, 1998.

494. Tilley, J., Kaplan, G., Fotouhi, N., Wolitzky, B., and Rowan, K., Carbacyclic peptide mimetics as VCAM-VLA-4 antagonists, *Bioorg. Med. Chem. Lett.*, 10, 1163, 2000.

495. Lindell, S.D., and Turner, R.M., Synthesis of potential inhibitors of the enzyme aspartate transcarbamoylase, *Tetrahedron Lett.*, 31, 5381, 1990.

496. Minami, T., Kamitamari, M., Utsunomiya, T., Tanaka, T., and Ichikawa, J., Synthesis of annulated γ-lactams via intramolecular 1,3-dipolar cycloadditions of functionalized *N-*allyl α-diazo amides, *Bull. Chem. Soc. Jpn.*, 66, 1496, 1993.

497. Efimtseva, E.V., Mikhailov, S.N., Jasko, M.V., Malakhov, D.V., Semizarov, D.G., Fomicheva, M.V., and Kern, E.R., Acyclic nucleoside and nucleotide analogues with amide bond, *Nucleosides Nucleotides*, 14, 373, 1995.

498. Malakhov, D.V., Semizarov, D.G., and Yas'ko, M.V., Synthesis and biochemical properties of phosphonyl acyclic analogs of 2'-deoxyadenosine nucleotides, *Bioorg. Khim.*, 21, 539, 1995; *Russ. J. Bioorg. Chem. (Engl. Transl.)*, 21, 464, 1995.

499. Kreutzkamp, N., and Schindler, H., Carbonyl and cyanophosphonic acid esters. Part 5. Conversion of cyano and carboxylic acid ester groups to substituted phosphonic acid esters, *Arch. Pharm. Ber. Dtsch. Pharm. Ges.*, 295, 28, 1962.

500. Gurevich, P.A., Razumov, A.I., Yavarova, R.L., Komina, T.V., and Kutumova, F.K., Synthesis of 2-oxo-3-phosphorylated indoles by the cyclization of arylhydrazides, *Zh. Obshch. Khim.*, 51, 1671, 1981; *J. Gen. Chem. USSR (Engl. Transl.)*, 51, 1424, 1981.

501. Griesgraber, G., Or, Y.S., Chu, D.T.W., Nilius, A.M., Johnson, P.M., Flamm, R.K., Henry, R.F., and Plattner, J.J., 3-Keto-11,12-carbazate derivatives of 6-*O*-methylerythromycin A. Synthesis and *in vitro* activity, *J. Antibiot.*, 49, 465, 1996.

502. Henklein, P., and Glöde, J., Unusual aminolysis of a carboxylic acid–carbonic acid anhydride, *Z. Chem.*, 29, 19, 1989.

503. Kim, D.Y., Kong, M.S., and Lee, K., Acylation of *in situ* generated trimethylsilyl diethylphosphonoacetate using magnesium chloride-triethylamine. A practical synthesis of β-keto phosphonates, *J. Chem. Soc., Perkin Trans. 1*, 1361, 1997.

504. Corbel, B., L'Hostis-Kervella, I., and Haelters, J.-P., Acylation of diethyl phosphonoacetic acid via the MgCl$_2$/Et$_3$N system. A practical synthesis of β-keto phosphonates, *Synth. Commun.*, 30, 609, 2000.

505. Krawczyk, H., The Mannich reaction of diethyl phosphonoacetic acid, a novel route to 1-(*N,N*-dialkylamino)methylvinylphosphonates, *Synth. Commun.*, 24, 2263, 1994.

506. Krawczyk, H., The Mannich reaction of diethylphosphonoacetic acid. Part 2. A new route to 1-(*N*-alkylamino)methylvinylphosphonates, *Phosphorus, Sulfur Silicon Relat. Elem.*, 101, 221, 1995.

507. Freeman, F., Chen, T., and van der Linden, J.B., Synthesis of highly functionalized 1,3-oxazoles, *Synthesis*, 861, 1997.

508. Pudovik, A.N., Yastrebova, G.E., and Nikitina, V.I., Condensation of diethylphosphonoacetic acid esters with formaldehyde and acetaldehyde, *Zh. Obshch. Khim.*, 37, 2790, 1967; *J. Gen. Chem. USSR (Engl. Transl.)*, 37, 2660, 1967.

509. Pudovik, A.N., Yastrebova, G.E., and Cherkasova, O.A., Condensation and addition of the diethyl ester of (carbamoylmethyl)phosphonic acid, *Zh. Obshch. Khim.*, 41, 88, 1972; *Chem. Abstr.*, 77, 19732, 1972.

510. Semmelhack, M.F., Tomesch, J.C., Czarny, M., and Boettger, S., Preparation of 2-(alkylthiomethyl)acrylates, *J. Org. Chem.*, 43, 1259, 1978.

511. Leonard, J., Hague, A.B., Jones, M.F., and Ward, R.A., A practical large-scale synthesis of 3-carbomethoxy-3-sulfolene, *Synthesis*, 507, 1999.

512. Minami, T., Suganuma, H., and Agawa, T., Synthesis and reactions of vinylphosphonates bearing electronegative substituents, *Chem. Lett.*, 285, 1978.

513. Kleschick, W.A., and Heathcock, C.H., Synthesis and chemistry of ethyl 2-diethylphosphonoacrylate, *J. Org. Chem.*, 43, 1256, 1978.

514. Crist, R.M., Reddy, P.V., and Borhan, B., Synthesis of isomeric 1,4-[^{13}C]$_2$-labelled 2-ethoxycarbonyl-1,4-diphenylbutadienes, *Tetrahedron Lett.*, 42, 619, 2001.

515. Abramov, V.S., and Il'ina, N.A., Mechanism of the Arbuzov rearrangement. Part 3. Reaction of the nitrile and methyl ester of α,β-dibromo- and α,β-dichloropropionic acids with phosphites, *Zh. Obshch. Khim.*, 26, 2014, 1956; *Chem. Abstr.*, 51, 1822c, 1957.

516. Abramov, V.S., and Bol'shakova, A.I., Mechanism of the Arbuzov rearrangement. Part 4. Reaction of α,β-dibromoethyl alkyl ethers with triisopropyl phosphite, *Zh. Obshch. Khim.*, 27, 441, 1957; *Chem. Abstr.*, 51, 15397d, 1957.

517. Yakubovich, A.Y., Soborovskii, L.Z., Muler, L.I., and Faemark, V.S., Synthesis of vinyl monomers. Part 1. α-Substituted derivatives of vinylphosphonic acid, *Zh. Obshch. Khim.*, 28, 317, 1958; *Chem. Abstr.*, 52, 13613c, 1958.

518. Boyce, C.B.C., Webb, S.B., and Phillips, L., The phosphorus trichloride–oxygen–olefin reaction. Scope and mechanism, *J. Chem. Soc., Perkin Trans. 1*, 1650, 1974.

519. Okamoto, Y., and Sakurai, H., Properties of α-substituted diethyl α-ethoxycarbonylphosphonates, *Kogyo Kagaku Zasshi*, 73, 2664, 1970; *Chem. Abstr.*, 74, 125798, 1971.

520. Minami, T., Nishimura, K., Hirao, I., Suganuma, H., and Agawa, T., Reactions of vinylphosphonates. Part 2. Synthesis of functionalized dienes, trienes, and their analogues. Synthetic applications to regioselectively functionalized benzene derivatives, *J. Org. Chem.*, 47, 2360, 1982.

521. Minami, T., Tokumasu, S., and Hirao, I., Reactions of vinylphosphonates. Part 3. One-pot synthesis of dienes and their analogs from vinylphosphonates, aldehydes and diethyl phosphonate, *Bull. Chem. Soc. Jpn.*, 58, 2139, 1985.

522. Schoen, W.R., and Parsons, W.H., Synthesis and reactions of 3-substituted-2-phosphomethyl acrylates, *Tetrahedron Lett.*, 29, 5201, 1988.

523. Mirza, S., and Vasella, A., Deoxy-nitrosugars. Part 7. Synthesis of methyl shikimate and of diethyl phosphashikimate from D-ribose, *Helv. Chim. Acta*, 67, 1562, 1984.

524. Junker, H.-D., Phung, N., and Fessner, W.-D., Diastereoselective free-radical synthesis of α-substituted *C*-glycosyl phosphonates, and their use as building blocks in the HWE reaction, *Tetrahedron Lett.*, 40, 7063, 1999.

525. McIntosh, J.M., and Khalil, H., Phase-transfer catalyzed synthesis of activated cyclopropanes, *Can. J. Chem.*, 56, 2134, 1978.

526. Calabi, L., Danieli, B., Lesma, G., and Palmisano, G., Imine-enamine annelation. Stereoselective syntheses of (±)-deplancheine, *Tetrahedron Lett.*, 23, 2139, 1982.

527. Ternansky, R.J., and Draheim, S.E., The chemistry of substituted pyrazolidinones. Applications to the synthesis of bicyclic derivatives, *Tetrahedron*, 48, 777, 1992.

528. Neel, D.A., Holmes, R.E., and Paschal, J.W., Synthesis of a 3-keto bicyclic pyrazolidinone using a Curtius rearrangement, *Tetrahedron Lett.*, 37, 4891, 1996.

529. Minami, T., Watanabe, K., and Hirakawa, K., A new synthesis of fused heterocyclic and carbocyclic compounds via the reaction of the vinylphosphonate with imide and ketone enolate anions, *Chem. Lett.*, 2027, 1986.

530. Shen, Y., and Jiang, G.-F., Sequential transformations of phosphonates. One-pot synthesis of ethoxycarbonyl allyl substituted oxime ethers, *Synthesis*, 502, 2000.

531. Corbet, J.-P., and Benezra, C., Allergenic α-methylene-γ-lactones. General method for the preparation of β-acetoxy- and β-hydroxy-α-methylene-γ-butyrolactones from sulfoxides. Application to the synthesis of a tuliposide B derivative, *J. Org. Chem.*, 46, 1141, 1981.

532. Semmelhack, M.F., Yamashita, A., Tomesch, J.C., and Hirotsu, K., Total synthesis of confertin via metal-promoted cyclization–lactonization, *J. Am. Chem. Soc.*, 100, 5565, 1978.

533. Goldberg, O., Deja, I., and Dreiding, A.S., The synthesis of stereoisomers of arteannuin B, *Helv. Chim. Acta*, 63, 2455, 1980.

534. Takagi, R., Sasaoka, A., Kojima, S., and Ohkata, K., Stereoselectivity in the formation of 2,5-disubstituted tetrahydropyrans by intramolecular hetero-Michael addition, *Heterocycles*, 45, 2313, 1997.

535. Takagi, R., Sasaoka, A., Kojima, S., and Ohkata, K., Synthesis of norsesterterpene *rac*- and *ent*-rhopaloic acid A, *J. Chem. Soc., Chem. Commun.*, 1887, 1997.

536. Takagi, R., Sasaoka, A., Nishitani, H., Kojima, S., Hiraga, Y., and Ohkata, K., Stereoselective synthesis of (+)-rhopaloic acid A and (−)-*ent*- and (±)-*rac*-rhopaloic acid A, *J. Chem. Soc., Perkin Trans. 1*, 925, 1998.

537. Nishitani, H., Sasaoka, A., Tokumasu, M., and Ohkata, K., Asymmetric synthesis of rhopaloic acid A analogs and their biological properties, *Heterocycles*, 50, 35, 1999.

538. Tokumasu, M., Ando, H., Hiraga, Y., Kojima, S., and Ohkata, K., Synthesis of *rac*-hippospongic acid A and revision of the structure, *J. Chem. Soc., Perkin Trans. 1*, 489, 1999.

539. McIntosh, J.M., and Sieler, R.A., 2-Carbomethoxy-1,3-butadiene. A convenient synthesis of a stable precursor and a survey of its Diels–Alder reactions, *J. Org. Chem.*, 43, 4431, 1978.

540. Snider, B.B., and Phillips, G.B., Ene reactions of 2-phosphonoacrylates, *J. Org. Chem.*, 48, 3685, 1983.

541. Pudovik, A.N., and Kuzovleva, R.G., Diels–Alder reactions of esters of α- and β-carbethoxyvinylphosphonic acid, *Zh. Obshch. Khim.*, 34, 1031, 1964; *J. Gen. Chem. USSR (Engl. Transl.)*, 34, 1024, 1964.

542. McIntosh, J.M., and Pillon, L.Z., Dihydrothiophenes. Part 10. The preparation and Diels–Alder reactions of some sulfur and phosphorus-substituted dienophiles and 2-aza-substituted 1,3-dienes, *Can. J. Chem.*, 62, 2089, 1984.

543. Siegel, H., Some preparatively useful [4 + 2]cycloadditions of 6,6-diphenylfulvene, *Synthesis*, 798, 1985.

544. Defacqz, N., Touillaux, R., Tinant, B., Declerq, J.-P., Peeters, D., and Marchand-Brynaert, J., Diels–Alder reactivity of trialkyl 2-phosphonoacrylates with *N*-buta-1,3-dienylsuccinimide, *J. Chem. Soc., Perkin Trans. 2*, 1965, 1997.

545. Defacqz, N., Touillaux, R., and Marchand-Brynaert, J., [4 + 2]Cycloaddition of *N*-buta-1,3-dienyl-succinimide to *gem*-substituted vinyl phosphonates, *J. Chem. Res. (S)*, 512, 1998.

546. Defacqz, N., Touillaux, R., Cordi, A., and Marchand-Brynaert, J., β-Aminophosphonic compounds derived from methyl 1-dimethoxy-phosphoryl-2-succinimidocyclohex-3-ene-1-carboxylates, *J. Chem. Soc., Perkin Trans. 1*, 2632, 2001.

547. Yamazaki, S., Imanishi, T., Moriguchi, Y., and Takada, T., Highly efficient [2 + 1] cycloaddition reactions of a 1-seleno-2-silylethene to 2-phosphonoacrylates. Synthesis of novel functionalized cyclo-propanephosphonic acid esters, *Tetrahedron Lett.*, 38, 6397, 1997.

548. Yamazaki, S., Takada, T., Imanishi, T., Moriguchi, Y., and Yamabe, S., Lewis acid-promoted [2 + 1] cycloaddition reactions of a 1-seleno-2-silylethene to 2-phosphonoacrylates. Stereoselective synthesis of a novel functionalized α-aminocyclopropanephosphonic acid, *J. Org. Chem.*, 63, 5919, 1998.

549. Yamazaki, S., Yanase, Y., and Yamamoto, K., The mechanism of [2 + 1] and [2 + 2] cycloaddition reactions of 1-phenylseleno-2-(trimethylsilyl)ethene. An isotopic labelling study, *J. Chem. Soc., Perkin Trans. 1*, 1991, 2000.

550. Barton, D.H.R., Géro, S.D., Quiclet-Sire, B., and Samadi, M., New synthesis of sugar, nucleoside and α-amino acid phosphonates, *Tetrahedron*, 48, 1627, 1992.

551. Barton, D.H.R., Géro, S.D., Quiclet-Sire, B., and Samadi, M., Radical addition to vinyl phosphonates. A new synthesis of isosteric phosphonates and phosphonate analogues of α-amino acids, *J. Chem. Soc., Chem. Commun.*, 1000, 1989.

552. Krawczyk, H., Michael addition mediated by an internal catalyst. A novel route to 2-diethylphospho-noalkanoic acids, *Synlett*, 1114, 1998.

553. Krawczyk, H., Nitrogen pronucleophiles in the self-catalytic Michael reaction, *Synth. Commun.*, 30, 1787, 2000.

554. Brunner, H., Le Cousturier de Courcy, N., and Genêt, J.-P., Heck reactions using aryldiazonium salts towards phosphonic derivatives, *Synlett*, 201, 2000.

555. Petzold, G., and Henning, H.G., α-Diazophosphoryl compounds by diazo group transfer, *Naturwis-senschaften*, 54, 469, 1967.

556. Regitz, M., Anschütz, W., and Liedhegener, A., Reactions of CH-active compounds with azides. Part 23. Synthesis of α-diazophosphonic acid esters, *Chem. Ber.*, 101, 3734, 1968.

557. Maas, G., and Regitz, M., Carbenes. Part 10. Substituent dependence of the norcaradiene/cyclohep-tatriene-equilibrium investigated in 7-phosphoryl- and 7-carbonylsubstituted systems, *Chem. Ber.*, 109, 2039, 1976.

558. Lee, J.C., and Yuk, J.Y., An improved and efficient method for diazo transfer reaction of active methylene compounds, *Synth. Commun.*, 25, 1511, 1995.

559. Khare, A.B., and McKenna, C.E., An improved synthesis of tetraalkyl diazomethylenediphosphonates and alkyl diazo(dialkoxyphosphoryl)acetates, *Synthesis*, 405, 1991.

560. McGuiness, M., and Shechter, H., Azidotris(diethylamino)phosphonium bromide. A self-catalyzing diazo transfer reagent, *Tetrahedron Lett.*, 31, 4987, 1990.

561. Alloum, A.B., and Villemin, D., Potassium fluoride on alumina. An easy preparation of diazocarbonyl compounds, *Synth. Commun.*, 19, 2567, 1989.

562. Paquet, F., and Sinaÿ, P., Intramolecular oximercuration–demercuration reaction. A new stereocon-trolled approach to sialic acid containing disaccharides, *Tetrahedron Lett.*, 25, 3071, 1984.

563. Paquet, F., and Sinaÿ, P., New stereocontrolled approach to 3-deoxy-D-manno-2-octulosonic acid containing disaccharides, *J. Am. Chem. Soc.*, 106, 8313, 1984.

564. Cox, G.G., Miller, D.J., Moody, C.J., Sie, E.-R.H.B., and Kulagowski, J.J., Rhodium-carbenoid mediated O–H insertion reactions. O–H insertion *vs.* H-abstraction and effect of catalyst, *Tetrahedron*, 50, 3195, 1994.

565. Delany, J.J., III, and Berchtold, G.A., Enzymatic and nonenzymatic reactions of 1β-[(1-carboxyethe-nyl)oxy]-4α-hydroxycyclohex-2-ene-1-carboxylate, *J. Org. Chem.*, 53, 3262, 1988.

566. Wood, H.B., Buser, H.-P., and Ganem, B., Phosphonate analogues of chorismic acid. Synthesis and evaluation as mechanism-based inactivators of chorismate mutase, *J. Org. Chem.*, 57, 178, 1992.

567. Mattia, K.M., and Ganem, B., Is there another common intermediate beyond chorismic acid in the shikimate pathway? Synthesis of *trans*-3-[(1-carboxyvinyl)oxy]-6-hydroxycyclohexa-1,4-diene-1-carboxylic acid, *J. Org. Chem.*, 59, 720, 1994.

568. Alberg, D.G., and Bartlett, P.A., Potent inhibition of 5-enolpyruvylshikimate-3-phosphate synthase by a reaction intermediate analogue, *J. Am. Chem. Soc.*, 111, 2337, 1989.

569. Shah, A., Font, J.L., Miller, M.J., Ream, J.E., Walker, M.C., and Sikorski, J.A., New aromatic inhibitors of EPSP synthase incorporating hydroxymalonates as novel 3-phosphate replacements, *Bioorg. Med. Chem.*, 5, 323, 1997.

570. Moody, C.J., Sie, E.-R.H.B., and Kulagowski, J.J., Diazophosphonates in cyclic ether synthesis. Use of the intramolecular Wadsworth–Emmons reaction, *Tetrahedron Lett.*, 32, 6947, 1991.

571. Moody, C.J., and Sie, E.-R.H.B., The use of diazophosphonates in the synthesis of cyclic ethers, *Tetrahedron*, 48, 3991, 1992.

572. Takano, S., Tomita, S.i., Takahashi, M., and Ogasawara, K., Efficient route to γ,δ-unsaturated carbonyl compounds from allyl sulfides and α-diazocarbonyls using a rhodium catalyst, *Chem. Lett.*, 1569, 1987.

573. Moore, J.D., Sprott, K.T., and Hanson, P.R., New strategies to cyclic α-thiophosphonates, *Synlett*, 605, 2001.

574. Prasad, K., Kneussel, P., Schulz, G., and Stütz, P., A new method of carbon extension at C-4 of azetidinones, *Tetrahedron Lett.*, 23, 1247, 1982.

575. Aller, E., Buck, R.T., Drysdale, M.J., Ferris, L., Haigh, D., Moody, C.J., Pearson, N.D., and Sanghera, J.B., N–H Insertion reaction of rhodium carbenoids. Part 1. Preparation of α-amino acid and α-aminophosphonic acid derivatives, *J. Chem. Soc., Perkin Trans. 1*, 2879, 1996.

576. Moody, C.J., Ferris, L., Haigh, D., and Swann, E., A new approach to peptide synthesis, *J. Chem. Soc., Chem. Commun.*, 2391, 1997.

577. Buck, R.T., Clarke, P.A., Coe, D.M., Drysdale, M.J., Ferris, L., Haigh, D., Moody, C.J., Pearson, N.D., and Swann, E., The carbenoid approach to peptide synthesis, *Chem. Eur. J.*, 6, 2160, 2000.

578. Brown, J.A., Synthesis of *N*-arylindole-2-carboxylates via an intramolecular palladium-catalyzed annulation of didehydrophenylalanine derivatives, *Tetrahedron Lett.*, 41, 1623, 2000.

579. McKenna, C.E., and Levy, J.N., α-Keto phosphonoacetates, *J. Chem. Soc., Chem. Commun.*, 246, 1989.

580. Doyle, K.J., and Moody, C.J., The rhodium carbenoid route to oxazoles. Synthesis of 4-functionalised oxazoles. Three step preparation of a *bis*-oxazole, *Tetrahedron*, 50, 3761, 1994.

581. Callant, P., D'Haenens, L., van der Eyken, E., and Vandewalle, M., Photoinduced Wolff rearrangement of α-diazo-β-ketophosphonates. A novel entry into substituted phosphonoacetates, *Synth. Commun.*, 14, 163, 1984.

582. Tomioka, H., and Hirai, K., Neighbouring phosphonate group participation in carbene chemistry, *J. Chem. Soc., Chem. Commun.*, 1611, 1990.

583. Arbuzov, A.E., Konstantinova, T., and Anzyfrova, T., Preparation of β-phosphonopropionic acid, *Izv. Akad. Nauk SSSR, Ser. Khim.*, 179, 1946; *Chem. Abstr.*, 42, 6315b, 1946.

584. Garner, A.Y., Chapin, E.C., and Scanlon, P.M., Mechanism of the Michaelis–Arbuzov reaction. Olefin formation, *J. Org. Chem.*, 24, 532, 1959.

585. Jakeman, D.L., Ivory, A.J., Williamson, M.P., and Blackburn, G.M., Highly potent bisphosphonate ligands for phosphoglycerate kinase, *J. Med. Chem.*, 41, 4439, 1998.

586. Mel'nikov, N.N., Mandel'baum, Y.A., and Lomakina, V.I., Organic insectofungicides. Part 58. Synthesis of some derivatives of dialkoxyphosphonopropionic, butyric, and toluic acids, *Zh. Obshch. Khim.*, 31, 849, 1961; *J. Gen. Chem. USSR (Engl. Transl.)*, 31, 781, 1961.

587. Romanova, I.P., Muslinkin, A.A., and Musin, R.Z., Reaction of triethyl phosphite with β-chlorolactic acid derivatives, *Zh. Obshch. Khim.*, 61, 2698, 1991; *J. Gen. Chem. USSR (Engl. Transl.)*, 61, 2504, 1991.

588. Hall, R.G., and Trippett, S., The preparation and Diels–Alder reactivity of ethyl (diethoxyphosphinyl)propynoate, *Tetrahedron Lett.*, 23, 2603, 1982.

589. Coover, H.W., Jr., and Shearer, N.H., Jr., Organophosphorus compounds derived from alkyl methacrylates, *Eastman Kodak*, U.S. Patent Appl. US 2856390, 1958; *Chem. Abstr.*, 53, 5131g, 1959.

590. Düttmann, H., and Weyerstahl, P., 1,3-Butadiene-2-carboxylic acids starting from the Wittig salt of 2-(bromomethyl)acrylic acid, *Chem. Ber.*, 112, 3480, 1979.

591. Bicknell, A.J., Burton, G., and Elder, J.S., Novel phosphorane and phosphonate synthons for vinyl glycines, *Tetrahedron Lett.*, 29, 3361, 1988.

592. Noguchi, H., Aoyama, T., and Shioiri, T., Determination of the absolute configuration and total synthesis of radiosumin, a trypsin inhibitor from a freshwater blue-green alga, *Tetrahedron Lett.*, 38, 2883, 1997.

593. Adiyaman, M., Lawson, J.A., Hwang, S.-W., Khanapure, S.P., FitzGerald, G.A., and Rokach, J., Total synthesis of a novel isoprostane $iPF_{2\alpha}$-I and its identification in biological fluids, *Tetrahedron Lett.*, 37, 4849, 1996.

594. Azizova, S.A., Mel'nikov, N.N., Vladimirova, I.L., and Negrebetskii, V.V., Synthesis of carboxylic esters containing a phosphoric or phosphonic ester grouping, *Zh. Obshch. Khim.*, 42, 816, 1972; *J. Gen. Chem. USSR (Engl. Transl.)*, 42, 808, 1972.

595. Ivanov, B.E., Zheltukhin, V.F., and Vavilova, T.G., Reactions of Mannich bases with triethyl phosphite. Part 1. Reactions in presence of acetic acid, *Izv. Akad. Nauk SSSR, Ser. Khim.*, 1285, 1967; *Bull. Acad. Sci. USSR, Div. Chem. Sci. (Engl. Transl.)*, 1239, 1967.

596. Janecki, T., and Bodalski, R., A convenient method for the synthesis of substituted 2-methoxycarbonyl- and 2-cyanoallylphosphonates. The allyl phosphite-allylphosphonate rearrangement, *Synthesis*, 799, 1990.

597. Janecki, T., A novel route to substituted trienes and tetraenes, *Synth. Commun.*, 23, 641, 1993.

598. Campbell, M.M., and Carruthers, N., Synthesis of α-aminophosphonic and α-aminophosphinic acids and derived dipeptides from 4-acetoxyazetidin-2-ones, *J. Chem. Soc., Chem. Commun.*, 730, 1980.

599. Campbell, M.M., Carruthers, N.I., and Mickel, S.J., Aminophosphonic and aminophosphinic acid analogues of aspartic acid, *Tetrahedron*, 38, 2513, 1982.

600. Chollet-Gravey, A.-M., Vo-Quang, L., Vo-Quang, Y., and Le Goffic, F., A preparative synthesis of 1-amino-3-hydroxypropylphosphonic acid (phosphonic analogue of homoserine), *Synth. Commun.*, 21, 1847, 1991.

601. Hall, N., and Price, R., The copper-promoted reaction of *o*-halogenodiarylazo-compounds with nucleo-philes. Part 1. The copper-promoted reaction of *o*-bromodiarylazo-compounds with trialkyl phosphites. A novel method for the preparation of dialkyl arylphosphonates, *J. Chem. Soc., Perkin Trans. 1*, 2634, 1979.

602. Arbuzov, B.A., Novosel'skaya, A.D., and Vinogradova, V.S., Reaction of α,β-dibromopropionitrile with sodium diethyl phosphite, *Izv. Akad. Nauk SSSR, Ser. Khim.*, 1153, 1972; *Bull. Acad. Sci. USSR, Div. Chem. Sci. (Engl. Transl.)*, 1103, 1972.

603. Bochwic, B., and Michalski, J., Organic phosphorus compounds. Part 1. Addition of dialkyl hydrogen phosphonates to ethylenic derivatives, *Rocz. Chem.*, 25, 338, 1951; *Chem. Abstr.*, 48, 12013a, 1954.

604. Issleib, K., and Vollmer, R., *o*-Substituted benzenephosphonic acid diethyl ester and *o*-amino, *o*-hydroxy, and *o*-mercaptophenyl phosphine, *Z. Chem.*, 18, 451, 1978.

605. Coover, H.W., Jr., and Dickey, J.B., Reaction of lactones with trialkyl phosphite, *Eastman Kodak*, U.S. Patent Appl. US 2652416, 1953; *Chem. Abstr.*, 48, 10053f, 1954.

606. McConnell, R.L., and Coover, H.W., Jr., New methods of preparing phosphonates, *J. Am. Chem. Soc.*, 78, 4453, 1956.

607. Cade, J.A., Preparation of diethyl 4-phosphonovalero-4-lactone, *J. Org. Chem.*, 23, 1372, 1958.

608. McConnell, R.L., and Coover, H.W., Jr., Organophosphorus derivatives of lactones, *Eastman Kodak*, U.S. Patent Appl. US 3121105, 1964; *Chem. Abstr.*, 60, 13273c, 1964.

609. Martin, D.J., Gordon, M., and Griffin, C.E., Phosphonic acids and esters. Part 18. Preparation of β,γ-unsaturated phosphonates by the Stobbe condensation with diethyl β-carbethoxyethylphosphonate. Stereochemical dependencies of homoallylic ^{31}P-^{1}H spin–spin coupling constants, *Tetrahedron*, 23, 1831, 1967.

610. Koketsu, J., Kojima, S., and Ishii, Y., Reactions between β-propiolactone and dialkyl dimethylphos-phoramidite. Ambient character of β-propiolactone and of dialkyl dimethylphosphoramidite, *Bull. Chem. Soc. Jpn.*, 43, 3232, 1970.

611. Norbeck, D.W., Kramer, J.B., and Lartey, P.A., Synthesis of an isosteric phosphonate analogue of cytidine 5′-monophospho-3-deoxy-D-manno-2-octulosonic acid, *J. Org. Chem.*, 52, 2174, 1987.

612. Tadanier, J., Lee, C.-M., Hengeveld, J., Rosenbrook, W., Jr., Whittern, D., and Wideburg, N., 3-Deoxy-2-(substituted-methyl)analogs of β-Kdop, *Carbohydr. Res.*, 201, 209, 1990.

613. Smith, E.C.R., McQuaid, L.A., Paschal, J.W., and DeHoniesto, J., An enantioselective synthesis of D-(−)- and L-(+)-2-amino-3-phosphonopropanoic acid, *J. Org. Chem.*, 55, 4472, 1990.

614. Hutchinson, J.P.E., and Parkes, K.E.B., A short synthesis of (S)-3-(dimethylphosphono)-2-((9-fluorenyl)methoxycarbamoyl)propionic acid, a protected phosphonic acid analogue of aspartic acid, *Tetrahedron Lett.*, 33, 7065, 1992.

615. Lohse, P.A., and Felber, R., Incorporation of a phosphonic acid isostere of aspartic acid into peptides using Fmoc-solid phase synthesis, *Tetrahedron Lett.*, 39, 2067, 1998.

616. Roy, C.H., Substituted methylenediphosphonic acids, *Procter and Gamble*, U.K. Patent Appl. GB 1026366, 1966; *Chem. Abstr.*, 65, 5685a, 1966.

617. Clark, R.D., Kozar, L.G., and Heathcock, C.H., Alkylation of diethyl 2-oxoalkanephosphonates, *Synthesis*, 635, 1975.

618. Porskamp, P.A.T.W., Lammerink, B.H.M., and Zwanenburg, B., Synthesis and reactions of phosphoryl-substituted sulfines, *J. Org. Chem.*, 49, 263, 1984.

619. Suarez, A., Lopez, F., and Compagnone, R.S., Stereospecific syntheses of the lignans. 2-S-(3,4-Dimethoxybenzyl)-3-R-(3,4,5-trimethoxybenzyl)butyrolactone, and its positional isomeric lactone, *Synth. Commun.*, 23, 1991, 1993.

620. Janecki, T., Bodalski, R., Wieczorek, M., and Bujacz, G., Horner–Wadsworth–Emmons olefination of nonstabilized phosphonates. A new synthetic approach to β,γ-unsaturated amides, *Tetrahedron*, 51, 1721, 1995.

621. Compagnone, R.S., Suarez, A.I., Zambrano, J.L., Pina, I.C., and Dominguez, J.N., A short and versatile synthesis of 3-substituted 2-aminoquinolines, *Synth. Commun.*, 27, 1631, 1997.

622. Balczewski, P., and Mikolajczyk, M., An expeditious synthesis of (±)-desepoxy-4,5-didehydromethylenomycin A methyl ester, *Org. Lett.*, 2, 1153, 2000.

623. Denmark, S.E., and Dorow, R.L., Stereoselective alkylations of chiral, phosphorus-stabilized benzylic carbanions, *J. Org. Chem.*, 55, 5926, 1990.

624. Boeckman, R.K., Jr., Walters, M.A., and Koyano, H., Regiocontrolled metalation of diethyl β-dialkylaminovinylphosphonates. A new synthesis of substituted β-ketophosphonates, *Tetrahedron Lett.*, 30, 4787, 1989.

625. Blackburn, G.M., and Parratt, M.J., The synthesis of α-fluoroalkylphosphonates, *J. Chem. Soc., Perkin Trans. 1*, 1425, 1986.

626. Pudovik, A.N., Addition of dialkyl phosphites to unsaturated compounds. Part 5. Addition of dialkyl phosphites to methyl acrylate, vinyl acetate, and vinyl butyl ether, *Zh. Obshch. Khim.*, 22, 473, 1952; *Chem. Abstr.*, 47, 2687h, 1953.

627. Pudovik, A.N., and Konovalova, I.V., Addition reactions of esters of phosphorus(III) acids with unsaturated systems, *Synthesis*, 81, 1979.

628. Ladd, E.C., and Harvey, M.P., Organophosphorus compounds, *U.S. Rubber*, U.S. Patent Appl. US 2971019, 1953; *Chem. Abstr.*, 55, 16427h, 1961.

629. Pudovik, A.N., New method of synthesis of esters of phosphonic and thiophosphonic acids, *Usp. Khim.*, 23, 547, 1954; *Chem. Abstr.*, 49, 8788i, 1955.

630. Pudovik, A.N., and Krupnov, G.P., New method of synthesis of esters of phosphonic and thiophosphonic acids. Part 36. Synthesis of derivatives of phosphonic acids with cyclic radicals in the ester groups, *Zh. Obshch. Khim.*, 31, 4053, 1961; *J. Gen. Chem. USSR (Engl. Transl.)*, 31, 3782, 1961.

631. Pudovik, A.N., and Sudakova, T.M., Addition of diethyl phosphorus acid to the esters and nitriles of unsaturated carboxylic acids, *Zh. Obshch. Khim.*, 39, 613, 1969; *J. Gen. Chem. USSR (Engl. Transl.)*, 39, 581, 1969.

632. Isbell, A.F., Berry, J.P., and Tansey, L.W., Amino phosphonic acids. Part 3. The synthesis and properties of 2-aminoethylphosphonic and 3-aminopropylphosphonic acids, *J. Org. Chem.*, 37, 4399, 1972.

633. Pudovik, A.N., and Krupnov, V.K., Synthesis of the ethyl ester of diethylamidophosphorous acid and its addition to unsaturated compounds, *Zh. Obshch. Khim.*, 38, 1406, 1968; *J. Gen. Chem. USSR (Engl. Transl.)*, 38, 1359, 1968.

634. Maklyaev, F.L., Kirillov, N.V., Fokin, A.V., and Rudnitskaya, L.S., Synthesis of esters of phosphonocarboxylic acids with different kinds of radicals, *Zh. Obshch. Khim.*, 40, 1014, 1970; *J. Gen. Chem. USSR (Engl. Transl.)*, 40, 999, 1970.

635. Pudovik, A.N., and Golitsyna, G.A., Addition reactions of alkylene glycol phosphites and phosphorothioites, *Zh. Obshch. Khim.*, 34, 876, 1964; *J. Gen. Chem. USSR (Engl. Transl.)*, 34, 870, 1964.

636. Pudovik, A.N., and Arbuzov, B.A., Addition of dialkyl phosphites to unsaturated ketones, nitriles, and esters, *Dokl. Akad. Nauk SSSR*, 73, 327, 1950; *Chem. Abstr.*, 45, 2853a, 1951.

637. Pudovik, A.N., and Arbuzov, B.A., Addition of dialkylphosphorous acids (dialkyl phosphites) to unsaturated compounds. Part 2. Addition of dialkylphosphorous acids to acrylonitrile and to methyl methacrylate, *Zh. Obshch. Khim.*, 21, 1837, 1951; *Chem. Abstr.*, 46, 6082e, 1952.

638. Pudovik, A.N., New method of synthesis of esters of phosphonocarboxylic acids and their derivatives, *Dokl. Akad. Nauk SSSR*, 85, 349, 1952; *Chem. Abstr.*, 47, 5351i, 1953.

639. Moshkina, T.M., and Pudovik, A.N., Synthesis of glycol diphosphates and some derivatives of esters of phosphonic acids, *Zh. Obshch. Khim.*, 32, 1671, 1962; *J. Gen. Chem. USSR (Engl. Transl.)*, 32, 1654, 1962.

640. Johnston, F., Esters of phosphono derivatives of monofunctional compounds, *Union Carbide and Carbon*, U.S. Patent Appl. US 2754320, 1956; *Chem. Abstr.*, 51, 466i, 1957.

641. Pudovik, A.N., Addition of dialkyl phosphites to unsaturated compounds. Part 6. Addition of dialkyl phosphites to ethyl esters of cinnamic and crotonic acids, *Zh. Obshch. Khim.*, 22, 1143, 1952; *Chem. Abstr.*, 47, 4836h, 1953.

642. Galkin, V.I., Kurdi, K.A., Khabibullina, A.B., and Cherkasov, R.A., Pudovik reaction in the series of α,β-unsaturated carbonyl compounds. Kinetic patterns of the addition of dialkyl phosphites to C=O and C=C bonds, *Zh. Obshch. Khim.*, 60, 92, 1990; *J. Gen. Chem. USSR (Engl. Transl.)*, 60, 81, 1990.

643. Galkin, V.I., Loginova, I.V., Konovalova, I.V., Burnaeva, L.M., and Cherkasov, R.A., Effect of fluoroalkyl substituents on the reactivities of dialkyl hydrogen phosphites in Pudovik and Abramov reactions, *Zh. Obshch. Khim.*, 63, 2467, 1993; *Russ. J. Gen. Chem. (Engl. Transl.)*, 63, 1712, 1993.

644. Shin, C.-G., Yonezawa, Y., Sekine, Y., and Yoshimura, J., α,β-Unsaturated carboxylic acid derivatives. Part 7. Reaction of ethyl α,β-unsaturated α-carboxylates with triethyl or diethyl phosphonate, *Bull. Chem. Soc. Jpn.*, 48, 1321, 1975.

645. Hägele, G., Worms, K.H., and Blum, H., Preparation and NMR spectroscopic investigation of β-trifluoromethyl-β-phosphonobutyric acid derivatives, *Phosphorus Sulfur*, 5, 277, 1975.

646. Blum, H., and Worms, K.H., β-Trifluoromethyl-β-phosphonobutyric acids and derivatives, *Henkel*, German Patent Appl. DE 2439281, 1976; *Chem. Abstr.*, 85, 5877, 1976.

647. Korshunov, M.A., and Kuzovleva, R.G., Esters of α,β-unsaturated acids with functional groups in the alkoxy radical. Part 5. Reactions of aminoalkyl acrylates and methacrylates with dialkylphosphorous acids, *Zh. Obshch. Khim.*, 38, 2548, 1968; *J. Gen. Chem. USSR (Engl. Transl.)*, 38, 2462, 1968.

648. Hakimelahi, G.H., and Jarrahpour, A.A., Synthesis of ethyl *cis*-2-[(diethoxyphosphoryl)methyl]-7-oxo-3-phenyl-6-phthalimido-1-azabicyclo[3.2.0]hept-3-ene-2-carboxylate and methyl *cis*-2-bromo-3-methyl-8-oxo-7-phthalimido-4-oxa-1-azabicyclo[4.2.0]octane-2-carboxylate, *Helv. Chim. Acta*, 72, 1501, 1989.

649. Hakimelahi, G.G., Moosavi-Movahedi, A.A., Tsay, S.-C., Tsai, F.-Y., Wright, J.D., Dudev, T., Hakimelahi, S., and Lim, C., Design, synthesis, and SAR of novel carbapenem antibiotics with high stability to *Xanthomonas maltophilia* oxyiminocephalosporinase type II, *J. Med. Chem.*, 43, 3632, 2000.

650. Pudovik, A.N., Pudovik, M.A., and Spirina, L.V., Addition reactions of dialkyl hydrogen phosphites with ethylene and propylene dimethacrylates, *Zh. Obshch. Khim.*, 37, 515, 1967; *J. Gen. Chem. USSR (Engl. Transl.)*, 37, 487, 1967.

651. Zimin, M.G., Burilov, A.R., and Pudovik, A.N., Reactions of derivatives of P(III) acids with thioketones and properties of the addition products, *Zh. Obshch. Khim.*, 54, 41, 1984; *J. Gen. Chem. USSR (Engl. Transl.)*, 54, 35, 1984.

652. Pudovik, A.N., New synthesis of esters of phosphonic and thiophosphonic acids. Part 12. Addition of dialkyl phosphites to unsaturated dibasic acids and esters, *Izv. Akad. Nauk SSSR, Ser. Khim.*, 926, 1952; *Chem. Abstr.*, 47, 10467e, 1953.

653. Johnston, F., Plastic composition comprising a vinyl halide polymer and a phosphorus-containing compound as a plasticizer, *Union Carbide and Carbon*, U.S. Patent Appl. US 2668800, 1954; *Chem. Abstr.*, 48, 8586h, 1954.

654. Johnston, F., Esters of phosphono derivatives of polyfunctional compounds, *Union Carbide and Carbon*, U.S. Patent Appl. US 2754319, 1956; *Chem. Abstr.*, 51, 466b, 1957.

655. Merrett, J.H., Spurden, W.C., Thomas, W.A., Tong, B.P., and Whitcombe, I.W.A., The synthesis and rotational isomerism of 1-amino-2-imidazol-4-ylethylphosphonic acid [phosphonohistidine, His(P)] and 1-amino-2-imidazol-2-ylethylphosphonic acid [phosphonoisohistidine, Isohis(P)], *J. Chem. Soc., Perkin Trans. 1,* 61, 1988.

656. Pevzner, L.M., Ignat'ev, V.M., and Ionin, B.I., Addition of diethyl phosphite to furylalkenes, *Zh. Obshch. Khim.,* 54, 69, 1984; *J. Gen. Chem. USSR (Engl. Transl.),* 54, 59, 1984.

657. Pevzner, L.M., Ignat'ev, V.M., and Ionin, B.I., Addition of diethyl phosphite to furylalkenes and furyl-1,3-alkadienes, *Zh. Obshch. Khim.,* 55, 2010, 1985; *J. Gen. Chem. USSR (Engl. Transl.),* 55, 1785, 1985.

658. Pevzner, L.M., Ignat'ev, V.M., and Ionin, B.I., Reaction of diethyl phosphite with 3-(2-furyl)acrylic acid derivatives, *Zh. Obshch. Khim.,* 56, 1470, 1986; *J. Gen. Chem. USSR (Engl. Transl.),* 56, 1302, 1986.

659. Jackson, P.F., Cole, D.C., Slusher, B.S., Stetz, S.L., Ross, L.E., Donzanti, B.A., and Trainor, D.A., Design, synthesis, and biological activity of a potent inhibitor of the neuropeptidase *N*-acetylated α-linked acidic dipeptidase, *J. Med. Chem.,* 39, 619, 1996.

660. Bühlmayer, P., Caselli, A., Fuhrer, W., Göschke, R., Rasetti, V., Rüeger, H., Stanton, J.L., Criscione, L., and Wood, J.M., Synthesis and biological activity of some transition-state inhibitors of human renin, *J. Med. Chem.,* 31, 1839, 1988.

661. Gaudemar-Bardone, F., Mladenova, M., and Gaudemar, M., Synthesis of 3-substituted ethyl 4,4,4-trichloro-2-cyanobutanoates via Michael addition to ethyl 4,4,4-trichloro-2-cyano-2-butenoate, *Synthesis,* 611, 1988.

662. Classen, R., and Hägele, G., A novel method for the preparation of fluoroaryl- and fluoroalkyl-substituted *bis-* and *tris*-phosphonic acids, *J. Fluorine Chem.,* 77, 71, 1996.

663. Dickey, J.B., and Coover, H.W., Jr., Acrylic acid esters containing a dialkylphosphono group and their polymers, *Eastman Kodak,* U.S. Patent Appl. US 2559854, 1951; *Chem. Abstr.,* 45, 8810e, 1951.

664. Green, K., Trimethylaluminium promoted conjugate additions of dimethylphosphite to α,β-unsaturated esters and ketones, *Tetrahedron Lett.,* 30, 4807, 1989.

665. Simoni, D., Invidiata, F.P., Manferdini, M., Lampronti, I., Rondanin, R., Roberti, M., and Pollini, G.P., Tetramethylguanidine (TMG)-catalyzed addition of dialkyl phosphites to α,β-unsaturated carbonyl compounds, alkenenitriles, aldehydes, ketones and imines, *Tetrahedron Lett.,* 39, 7615, 1998.

666. Simoni, D., Rondanin, R., Morini, M., Baruchello, R., and Invidiata, F.P., 1,5,7-Triazabicyclo[4.4.0]dec-1-ene (TBD), 7-methyl-TBD (MTBD) and the polymer-supported TBD (P-TBD). Three efficient catalysts for the nitroaldol (Henry) reaction and for the addition of dialkyl phosphites to unsaturated systems, *Tetrahedron Lett.,* 41, 1607, 2000.

667. Villemin, D., and Racha, R., Anionic activation of diethyl phosphite by potassium fluoride, *Tetrahedron Lett.,* 27, 1789, 1986.

668. Amri, H., Rambaud, M., and Villieras, J., A short large scale synthesis of (±)-sarkomycin esters, *Tetrahedron Lett.,* 30, 7381, 1989.

669. Platonov, A.Y., Sivakov, A.A., Chistokletov, V.N., and Maiorova, E.D., Transformations of electrophilic reagents in a diethyl phosphite–potassium carbonate–ethanol system, *Izv. Akad. Nauk, Ser. Khim.,* 369, 1999; *Russ. Chem. Bull. (Engl. Transl.),* 370, 1999.

670. Platonov, A.Y., Sivakov, A.A., Chistokletov, V.N., and Maiorova, E.D., Reaction of diethyl hydrogen phosphite with activated alkenes in heterophase K_2CO_3/ethanol system, *Zh. Obshch. Khim.,* 69, 514, 1999; *Russ. J. Gen. Chem. (Engl. Transl.),* 69, 493, 1999.

671. Hanessian, S., Cantin, L.-D., Roy, S., Andreotti, D., and Gomtsyan, A., The synthesis of enantiomerically pure, symmetrically substituted cyclopropane phosphonic acids. A constrained analog of the GABA antagonist phaclophen, *Tetrahedron Lett.,* 38, 1103, 1997.

672. Kondo, K., Liu, Y., and Tunemoto, D., Preparation and reaction of sulphonium ylide stabilized by a phosphinyl substituent, *J. Chem. Soc., Perkin Trans. 1,* 1279, 1974.

673. Yuan, C., Li, C., and Ding, Y., Studies on organophosphorus compounds. Part 51. A new and facile route to 2-substituted 1,1-cyclopropanediylbis(phosphonic acids), *Synthesis,* 854, 1991.

674. Jubault, P., Goumain, S., Feasson, C., and Collignon, N., A new and efficient electrosynthesis of polysubstituted cyclopropylphosphonates, using electrochemically activated magnesium, *Tetrahedron,* 54, 14767, 1998.

675. Goumain, S., Jubault, P., Feasson, C., and Quirion, J.-C., First synthesis of α-fluorinated cyclopropylphosphonates using magnesium electrochemical activation, *Tetrahedron Lett.*, 40, 8099, 1999.

676. Goumain, S., Jubault, P., Feasson, C., and Collignon, N., A new and efficient electrosynthesis of 2-substituted 1,1-cyclopropanediylbis(phosphonates), *Synthesis*, 1903, 1999.

677. Duquenne, C., Goumain, S., Jubault, P., Feasson, C., and Quirion, J.-C., Electrosynthesis of α-arylated β-substituted cyclopropylphosphonates. Synthesis of a phosphonic analogue of minalcipran, *Org. Lett.*, 2, 453, 2000.

678. Villemin, D., Thibault-Starzyk, F., and Hachemi, M., A convenient synthesis of cyclopropanes from olefins and carbon acid compounds. Synthesis of tetraethyl cyclopropanediyldiphosphonates, *Synth. Commun.*, 24, 1425, 1994.

679. Huang, W.-S., Zhang, Y.-X., and Yuan, C., Studies on organophosphorus compounds. Part 96. Nucleophilicity of the isocyano carbon atom in diethyl isocyanomethylphosphonate. First generation of a phosphorylated nitrile ylide and new syntheses of pyrrolinephosphonates and pyrrolephosphonates, *J. Chem. Soc., Perkin Trans. 1*, 1893, 1996.

680. Couthon, H., Gourvès, J.-P., Guervenou, J., Corbel, B., and Sturtz, G., Synthesis of two novel 2-aminocyclopropylidene-1,1-bisphosphonates, *Synth. Commun.*, 29, 4251, 1999.

681. Hellmuth, E.W., Kaczynski, J.A., Low, J., and McCoy, L.L., Three-membered rings. Part 7. Solvent control of the *cis-trans* isomer ratio in the preparation of a phosphonate substituted cyclopropane, *J. Org. Chem.*, 39, 3125, 1974.

682. Ginsburg, V.A., and Yakubovich, A.Y., The addition of trialkyl phosphites to acrylic systems, *Zh. Obshch. Khim.*, 30, 3987, 1960; *J. Gen Chem. USSR (Engl. Transl.)*, 30, 3944, 1960.

683. Harvey, R.G., Reactions of triethyl phosphite with activated olefins, *Tetrahedron*, 22, 2561, 1966.

684. Müller, P., Venakis, T., and Eugster, C.H., Activated quinones. *O-* versus *C*-addition of phenols. New regiospecific syntheses of xanthones, thioxanthones and *N*-methyl-9-acridones, *Helv. Chim. Acta*, 62, 2350, 1979.

685. Kolomnikova, G.D., Krylova, T.O., Chernoglazova, I.V., Petrovskii, P.V., and Gololobov, Y.G., 2-Cyanoacrylates in the conjugate addition of trifluoroacetic acid and nucleophilic reagents, *Izv. Akad. Nauk SSSR, Ser. Khim.*, 1245, 1993; *Russ. Chem. Bull. (Engl. Transl.)*, 1188, 1993.

686. Pande, K.C., Ammonium salts as proton sources in hydrophosphinylation reactions, *Chem. Ind. (London)*, 1048, 1968.

687. Pande, K.C., Hydrophosphinylation of activated alkenes or alkynes, *Dow Chemical*, U.S. Patent Appl. US 3622654, 1971; *Chem. Abstr.*, 76, 46303, 1972.

688. Sekine, M., Yamamoto, I., Hashizume, A., and Hata, T., Silyl phosphites. Part 5. The reactions of *tris*(trimethylsilyl) phosphite with carbonyl compounds, *Chem. Lett.*, 485, 1977.

689. Novikova, Z.S., Mashoshina, S.N., Sapozhnikova, T.A., and Lutsenko, I.F., Reactions of diethyl trimethylsilyl phosphite with carbonyl compounds, *Zh. Obshch. Khim.*, 41, 2622, 1971; *J. Gen. Chem. USSR (Engl. Transl.)*, 41, 2655, 1971.

690. Okamoto, Y., Azuhata, T., and Sakurai, H., Dialkyl 3-alkoxy-3-(trimethylsiloxy)-2-propenephosphonate. A one step preparation of (dialkoxyphosphinyl)methyl-substituted ketene alkyl trimethylsilyl acetal, *Chem. Lett.*, 1265, 1981.

691. Okamoto, Y., and Sakurai, H., Preparation of (dialkoxyphosphinyl)-methyl-substituted ketene alkyl trimethylsilyl acetal derivatives, *Synthesis*, 497, 1982.

692. Griffin, C.E., and Mitchell, T.D., Phosphonic acids and esters. Part 10. Oxygen to carbon methyl migration in the reaction of trimethyl phosphite with dimethyl acetylenedicarboxylate, *J. Org. Chem.*, 30, 2829, 1965.

693. Caesar, J.C., Griffiths, D.V., Tebby, J.C., and Willetts, S.E., Direct formation of λ^5-phospholes from trialkyl phosphites and dimethyl acetylenedicarboxylate. Alkylphosphonium ylides as reactive intermediates and stable products, *J. Chem. Soc., Perkin Trans. 1*, 1627, 1984.

694. Caesar, J.C., Griffiths, D.V., Griffiths, P.A., and Tebby, J.C., The preparation and reactions of the 1,2-diphosphoranes formed on reaction of trialkyl phosphites with dimethyl acetylenedicarboxylate, *Phosphorus Sulfur*, 34, 155, 1987.

695. Caesar, J.C., Griffiths, D.V., Griffiths, P.A., and Tebby, J.C., Reactions of ylides formed from trialkyl phosphites with dialkyl acetylenedicarboxylates in the presence of carbon dioxide, *J. Chem. Soc., Perkin Trans. 1*, 2329, 1990.

696. Kirillova, L.M., and Kukhtin, V.A., New forms of the Arbuzov rearrangement. Part 19. Addition of trialkyl phosphites to acetylenedicarboxylic acids, *Zh. Obshch. Khim.*, 35, 1146, 1965; *J. Gen. Chem. USSR (Engl. Transl.)*, 35, 1148, 1965.

697. Nicholson, D.A., and Campbell, D., 1,2-Dihydroxy-1,2-diphosphonosuccinic acid, its alkyl esters and alkali metal salts, *Procter and Gamble*, U.S. Patent Appl. US 3579570, 1971; *Chem. Abstr.*, 75, 36337, 1971.

698. Yavari, I., Mosslemin, M.H., and Montahaei, A.R., Stereoselective synthesis of dialkyl 2-(3-methoxy-1-oxoindan-2-yl)-3-(dimethylphosphonato)butanedioates, *J. Chem. Res. (S)*, 576, 1998.

699. Chuang, S.-C., Lee, D.-D., Santhosh, K.C., and Cheng, C.-H., Fullerene derivatives bearing a phosphite ylide, phosphonate, phosphine oxide, and phosphonic acid. Synthesis and reactivities, *J. Org. Chem.*, 64, 8868, 1999.

700. Pudovik, A.N., and Yarmukhametova, D.K., New method of synthesis of esters of phosphonic and thiophosphonic acids. Part 16. Synthesis of esters of mono- and diphosphono- and thiophosphono-carboxylic acids, *Izv. Akad. Nauk SSSR, Ser. Khim.*, 636, 1954; *Chem. Abstr.*, 49, 8789a, 1955.

701. Miller, L.A., Diphosphonate alkyl esters, *Monsanto*, U.S. Patent Appl. US 3093672, 1963; *Chem. Abstr.*, 59, 13823d, 1963.

702. Benayoud, F., deMendonca, D.J., Digits, C.A., Moniz, G.A., Sanders, T.C., and Hammond, G.B., Efficient syntheses of (α-fluoropropargyl)phosphonate esters, *J. Org. Chem.*, 61, 5159, 1996.

703. Palacios, F., Garcia, J., Ochoa de Retana, A.M., and Oyarzabal, J., Synthesis of 5-phosphonyl-2(1*H*)-pyridones from primary β-enaminophosphonate and acetylenic esters, *Heterocycles*, 41, 1915, 1995.

704. Davidson, R.M., and Kenyon, G.L., Analogues of phosphoenolpyruvate. Part 4. Syntheses of some new vinyl- and methylene-substituted phosphonate derivatives, *J. Org. Chem.*, 45, 2698, 1980.

705. Ruder, S.M., and Kulkarni, V.R., Ring expansion reactions of cyclic β-keto phosphonates, *J. Org. Chem.*, 60, 3084, 1995.

706. Coutrot, P., and Savignac, P., α-Cuprophosphonates. Part 3. Synthesis of ethyl dialkylphosphonopyruvates, *Synthesis*, 36, 1978.

707. Feistauer, H., and Neidlein, R., Phosphonopyruvates. Syntheses, NMR investigations, and reactions, *Helv. Chim. Acta*, 78, 1806, 1995.

708. McQueney, M.S., Lee, S.-l., Swartz, W.H., Ammon, H.L., Mariano, P.S., and Dunaway-Mariano, D., Evidence for an intramolecular, stepwise reaction pathway for PEP phosphomutase catalyzed P-C bond formation, *J. Org. Chem.*, 56, 7121, 1991.

709. Freeman, S., Seidel, H.M., Schwalbe, C.H., and Knowles, J.R., Phosphonate biosynthesis. The stereochemical course of phosphoenolpyruvate phosphomutase, *J. Am. Chem. Soc.*, 111, 9233, 1989.

710. Schwalbe, C.H., and Freeman, S., The crystal structure of the tri(cyclohexylammonium) salt of phosphonopyruvate. Theoretical calculations on the equilibrium constant for the rearrangement of phosphonopyruvate to phosphoenolpyruvate, *J. Chem. Soc., Chem. Commun.*, 251, 1990.

711. Freeman, S., Irwin, W.J., and Schwalbe, C.H., Synthesis and hydrolysis studies of phosphonopyruvate, *J. Chem. Soc., Perkin Trans. 2*, 263, 1991.

712. Phillion, D.P., and Cleary, D.G., Disodium salt of 2-[(dihydroxyphosphinyl)difluoromethyl] propenoic acid. An isopolar and isosteric analogue of phosphoenolpyruvate, *J. Org. Chem.*, 57, 2763, 1992.

713. Phillion, D.P., and Andrew, S.S., Synthesis and reactivity of diethyl phosphonomethyltriflate, *Tetrahedron Lett.*, 27, 1477, 1986.

714. Waglund, T., Luthman, K., and Orbe, M., Synthesis of *C*-(β-D-glycosyl) analogues of 3-deoxy-D-manno-2-octulosonic acid (Kdo) as potential inhibitors of CMP-Kdo synthetase, *Carbohydr. Res.*, 206, 269, 1990.

715. Orbe, M., Luthman, K., Waglund, T., Claesson, A., and Csoeregh, I., Conformational analysis of isopropylidene-protected *C*-glycosyl derivatives of 3-deoxy-D-manno-2-octulosonic acid (Kdo) in the solid state and in solution, *Carbohydr. Res.*, 211, 1, 1991.

716. Vasella, A., and Vöffray, R., Asymmetric synthesis of α-aminophosphonic acids by cycloaddition of *N*-glycosyl-*C*-dialkoxyphosphonoylnitrones, *Helv. Chim. Acta*, 65, 1953, 1982.

717. Chambers, R.D., Jaouhari, R., and O'Hagan, D., Fluorine in enzyme chemistry. Part 2. The preparation of difluoromethylenephosphonate analogues of glycolytic phosphates. Approaching an isosteric and isoelectronic phosphate mimic, *Tetrahedron*, 45, 5101, 1989.

718. de Macedo Puyau, P., and Perié, J.J., Selective epoxidation of polar substrates by dimethyldioxirane, *Synth. Commun.*, 28, 2679, 1998.

719. Kadyrov, R., Selke, R., Giernoth, R., and Bargon, J., Synthesis of new 3-(dialkoxyphosphoryl)butenoates and their enantioselective hydrogenation, *Synthesis*, 1056, 1999.

720. Koszuk, J.F., Preparation of new 1-alkenylphosphonates and 2-alkenylphosphonates by Claisen rearrangement, *Synth. Commun.*, 25, 2533, 1995.

721. Balczewski, P., Pietrzykowski, W.M., and Mikolajczyk, M., Studies on the free radical carbon–carbon bond formation in the reaction of α-phosphoryl sulfides and selenides with alkenes, *Tetrahedron*, 51, 7727, 1995.

722. Holt, D.A., and Erb, J.M., Palladium-catalyzed phosphorylation of alkenyl triflates, *Tetrahedron Lett.*, 30, 5393, 1989.

723. Wyatt, P.B., Villalonga-Barber, C., and Motevalli, M., Synthesis and Diels–Alder reactivity of chiral 2-(alk-1-enyl)-1,3,2-diazaphospholidine 2-oxides, *Tetrahedron Lett.*, 40, 149, 1999.

724. Sasin, R., Nauman, R.M., and Swern, D., Phosphorus derivatives of fatty acids. Part 6. ω-Dialkyl phosphonoundecanoates, *J. Am. Chem. Soc.*, 81, 4335, 1959.

725. Falbe, J., Paatz, R., and Korte, F., Synthesis of phosphorus-substituted carboxylic acid esters, *Chem. Ber.*, 98, 2312, 1965.

726. Ikeda, S., Weinhouse, M.I., Janda, K.D., Lerner, R.A., and Danishefsky, S.J., Asymmetric induction via a catalytic antibody, *J. Am. Chem. Soc.*, 113, 7763, 1991.

727. Burden, R.S., and Crombie, L., Amides of vegetable origin. Part 12. A new series of alka-2,4-dienoic tyramine-amides from *Anacyclus pyrethrum* D.C. *(compositae)*, *J. Chem. Soc. (C)*, 2477, 1969.

728. Bæckström, P., Jacobsson, U., Norin, T., and Unelius, C.R., Synthesis and characterization of all four isomers of methyl 2,4-decadienoate for an investigation of the pheromone components of *Pityogenes chalcographus*, *Tetrahedron*, 44, 2541, 1988.

729. Kryshtal, G.V., Serebryakov, E.P., Suslova, L.M., and Yanovskaya, L.A., Stereochemistry of the Horner–Emmons reaction of 3-functionalized 2-methyl-2-propenylphosphonates with aliphatic aldehydes. Part 1. Steric effect of the phosphonate ester groups, *Izv. Akad. Nauk SSSR, Ser. Khim.*, 2377, 1988; *Chem. Abstr.*, 111, 78160, 1989.

730. Wei, H.-X., and Schlosser, M., Fluorinated analogs of retinoids. Stereocontrolled synthesis employing fluoroisoprenoidal Horner synthesis, *Eur. J. Org. Chem.*, 2603, 1998.

731. Reimer, L.M., Conley, D.L., Pompliano, D.L., and Frost, J.W., Construction of an enzyme-targeted organophosphonate using immobilized enzyme and whole cell synthesis, *J. Am. Chem. Soc.*, 108, 8010, 1986.

732. Montchamp, J.-L., Peng, J., and Frost, J.W., Inversion of an asymmetric center in carbocyclic inhibitors of 3-dehydroquinate synthase. Examining and exploiting the mechanism for *syn*-elimination during substrate turnover, *J. Org. Chem.*, 59, 6999, 1994.

733. Montchamp, J.-L., and Frost, J.W., Cyclohexenyl and cyclohexylidene inhibitors of 3-dehydroquinate synthase. Active site interactions relevant to enzyme mechanism and inhibitor design, *J. Am. Chem. Soc.*, 119, 7645, 1997.

734. Karanewsky, D.S., Badia, M.C., Ciosek, C.P., Jr., Robl, J.A., Sofia, M.J., Simpkins, L.M., DeLange, B., Harrity, T.W., Biller, S.A., and Gordon, E.M., Phosphorus-containing inhibitors of HMG-CoA reductase. Part 1. 4-[(2-Arylethyl)hydroxyphosphinyl]-3-hydroxybutanoic acids. A new class of cell-selective inhibitors of cholesterol biosynthesis, *J. Med. Chem.*, 33, 2952, 1990.

735. Robl, J.A., Duncan, L.A., Pluscec, J., Karanewsky, D.S., Gordon, E.M., Ciosek, C.P., Jr., Rich, L.C., Dehmel, V.C., Slusarchyk, D.A., Harrity, T.W., and Obrien, K.A., Phosphorus-containing inhibitors of HMG-CoA reductase. Part 2. Synthesis and biological activites of a series of substituted pyridines containing a hydroxyphosphinyl moiety, *J. Med. Chem.*, 34, 2804, 1991.

736. Karanewsky, D.S., and Badia, M.C., Phosphorus-containing inhibitors of HMG-CoA reductase. Part 3. Synthesis of hydroxyphosphinyl-analogues of the mevinic acids, *Tetrahedron Lett.*, 34, 39, 1993.

737. Napolitano, E., Spinelli, G., Fiaschi, R., and Marsili, A., A simple total synthesis of the isoindolobenzazepine alkaloids lennoxamine and chilenamine, *J. Chem. Soc., Perkin Trans. 1*, 785, 1986.

738. Moody, C.J., and Warrellow, G.J., Synthesis of the isoindolobenzazepine alkaloid lennoxamine, *Tetrahedron Lett.*, 28, 6089, 1987.

739. Yamagiwa, Y., Ohashi, K., Sakamoto, Y., Hirakawa, S., Kamikawa, T., and Kubo, I., Syntheses of anacardic acids and ginkgoic acid, *Tetrahedron*, 43, 3387, 1987.

740. Moody, C.J., and Warrellow, G.J., Vinyl azides in heterocyclic synthesis. Part 10. Synthesis of the isoindolobenzazepine alkaloid lennoxamine, *J. Chem. Soc., Perkin Trans. 1*, 2929, 1990.

741. de Macedo Puyau, P., and Perié, J.J., Synthesis of substrate analogues and inhibitors for the phosphoglycerate mutase enzyme, *Phosphorus, Sulfur Silicon Relat. Elem.*, 129, 13, 1997.

742. Söderberg, B.C., Chisnell, A.C., O'Neil, S.N., and Shriver, J.A., Synthesis of indoles isolated from *Tricholoma* species, *J. Org. Chem.*, 64, 9731, 1999.

743. Deshayes, C., Chabannet, M., and Gelin, S., Synthesis of some 1-substituted 3- or 5-styrylpyrazoles, *J. Heterocycl. Chem.*, 18, 1057, 1981.

744. Tavs, P., Esters of aryl- and vinylphosphonic acids, aryl- and vinylphosphinic acids and aryl- and vinylphosphine oxides, *Shell*, U.S. Patent Appl. DE 1810431, 1970; *Chem. Abstr.*, 73, 77387, 1970.

745. Tavs, P., Reaction of aryl halides with trialkyl phosphites and dialkyl benzenephosphonites to aromatic phosphonates and phosphinates by nickel salt catalysed arylation, *Chem. Ber.*, 103, 2428, 1970.

746. Ashley, J.A., Lin, C.-H., Wirsching, P., and Janda, K.D., Monitoring chemical warfare agents. A new method for the detection of methylphosphonic acid, *Angew. Chem.*, 111, 1909, 1999; *Angew. Chem. Int. Ed. Engl.*, 38, 1793, 1999.

747. Arient, J., Symmetrical and unsymmetrical derivatives of distyrylbenzene, *Collect. Czech. Chem. Commun.*, 46, 101, 1981.

748. Pevzner, L.M., Terekhova, M.I., Ignat'ev, V.M., Petrov, E.S., and Ionin, B.I., Synthesis and CH acidities of some diethyl (furylmethyl)phosphonates, *Zh. Obshch. Khim.*, 54, 1990, 1984; *J. Gen. Chem. USSR (Engl. Transl.)*, 54, 1775, 1984.

749. Pevzner, L.M., Ignat'ev, V.M., and Ionin, B.I., Michaelis–Becker reaction in the series of halogenomethylfurans, *Zh. Obshch. Khim.*, 60, 1061, 1990; *J. Gen. Chem. USSR (Engl. Transl.)*, 60, 936, 1990.

750. Pevzner, L.M., Ignat'ev, V.M., and Ionin, B.I., (Halomethylfuran)carboxylic esters in the Michaelis–Becker reaction, *Zh. Obshch. Khim.*, 62, 797, 1992; *J. Gen. Chem. USSR (Engl. Transl.)*, 62, 658, 1992.

751. Pevzner, L.M., Ignat'ev, V.M., and Ionin, B.I., Reaction of polysubstituted halomethylfurans with sodium diethyl phosphite, *Zh. Obshch. Khim.*, 64, 1108, 1994; *Russ. J. Gen. Chem. (Engl. Transl.)*, 64, 1000, 1994.

752. Montchamp, J.-L., and Frost, J.W., Irreversible inhibition of 3-dehydroquinate synthase, *J. Am. Chem. Soc.*, 113, 6296, 1991.

753. Montchamp, J.-L., Piehler, L.T., and Frost, J.W., Diastereoselection and *in vivo* inhibition of 3-dehydroquinate synthase, *J. Am. Chem. Soc.*, 114, 4453, 1992.

754. Yamaguchi, M., Tsukamoto, Y., Hayashi, A., and Minami, T., A stereoselective Michael addition of α-lithiated phosphonates to α,β-unsaturated esters, *Tetrahedron Lett.*, 31, 2423, 1990.

755. Genêt, J.P., Uziel, J., Port, M., Touzin, A.M., Roland, S., Thorimbert, S., and Tanier, S., A practical synthesis of α-aminophosphonic acids, *Tetrahedron Lett.*, 33, 77, 1992.

756. Kim, D.Y., Suh, K.H., Huh, S.C., and Lee, K., Michael addition of N-(diphenylmethylene)aminomethylphosphonate to acrylates using phase transfer catalysis conditions. Synthesis of 1-(N-diphenylmethylene)amino-3-(alkoxycarbonyl)propylphosphonates, *Synth. Commun.*, 31, 3315, 2001.

757. Groth, U., Richter, L., and Schöllkopf, U., Asymmetric syntheses of α-aminophosphonic acids. Part 5. Synthesis of enantiomerically pure diethyl 2-pyrrolidinyl- and 5-oxo-2-pyrrolidinylphosphonates, *Liebigs Ann. Chem.*, 903, 1992.

758. Groth, U., Richter, L., and Schöllkopf, U., Asymmetric synthesis of enantiomerically pure phosphonic analogues of glutamic acid and proline, *Tetrahedron*, 48, 117, 1992.

759. Hannour, S., Roumestant, M.-L., Viallefont, P., Riche, C., Martinez, J., El Hallaoui, A., and Ouazzani, F., Asymmetric synthesis of (2S,3R,4R)-ethoxycarbonylcyclopropyl phosphonoglycine, *Tetrahedron: Asymmetry*, 9, 2329, 1998.

760. Ruder, S.M., and Kulkarni, V.R., Michael-type additions of 2-(diethoxyphosphinyl)cyclohexanone to activated alkenes and alkynes, *J. Chem. Soc., Chem. Commun.*, 2119, 1994.

761. Pudovik, A.N., Yastrebova, G.E., and Pudovik, O.A., Tetraethyl methylenediphosphonate in condensation reactions and in addition reactions with unsaturated compounds, *Zh. Obshch. Khim.*, 40, 499, 1970; *J. Gen. Chem. USSR (Engl. Transl.)*, 40, 462, 1970.

762. Blum, H., and Worms, K.H., Diphosphonoalkane carboxylic acids, *Henkel*, German Patent Appl. DE 2602030, 1977; *Chem. Abstr.*, 87, 168197, 1977.

763. Maruszewska-Wieczorkowska, E., and Michalski, J., Alkyl and alkenyl pyridines. Part 7. 3-(2′-Pyridyl) propylphosphonic acid, *Rocz. Chem.*, 37, 1315, 1963.

764. Tawney, P.O., 3-Phosphonopropane-1,1-dicarboxylic acid esters and method of preparation, *U.S. Rubber*, U.S. Patent Appl. US 2535173, 1950; *Chem. Abstr.*, 45, 3409b, 1951.

765. Pudovik, A.N., and Grishina, O.N., Synthesis and properties of vinylphosphonic esters. Part 2. Phosphonoethylation reaction. Addition of malonic, cyanoacetic, acetoacetic ester, and their homologs to vinylphosphonic ester, *Zh. Obshch. Khim.*, 23, 267, 1953; *Chem. Abstr.*, 48, 2573c, 1954.

766. Sturtz, G., and Guervenou, J., Synthesis of novel functionalized *gem*-biphosphonates, *Synthesis*, 661, 1991.

767. Minami, T., Nakayama, M., Fujimoto, K., and Matsuo, S., A new approach to cyclopentane annulated compounds via 1-(cyclopent-1-enylcarbonyl)vinylphosphonates, *J. Chem. Soc., Chem. Commun.*, 190, 1992.

768. Guervenou, J., and Sturtz, G., Synthesis of *gem*-bisphosphonic conjugates of cortisone derivatives, *Phosphorus, Sulfur Silicon Relat. Elem.*, 88, 1, 1994.

769. Fabulet, O., and Sturtz, G., Synthesis of *gem*-bisphosphonic doxorubicin conjugates, *Phosphorus, Sulfur Silicon Relat. Elem.*, 101, 225, 1995.

770. Nugent, R.A., Schlachter, S.T., Murphy, M., Dunn, C.J., Staite, N.D., Galinet, L.A., Shields, S.K., Wu, H., Aspar, D.G., and Richard, K.A., Carbonyl-containing bisphosphonate esters as novel antiin-flammatory and antiarthritic agents, *J. Med. Chem.*, 37, 4449, 1994.

771. Phillips, A.M.M.M., and Modro, T.A., Phosphonic systems. Part 3. Diethyl prop-2-enylphosphonate, a new and versatile substrate in carbon–carbon bond formation, *J. Chem. Soc., Perkin Trans. 1*, 1875, 1991.

772. Gerber, K.P., Modro, T.A., Muller, E.L., and Phillips, A.M., Chemistry of dialkyl alkenylphosphonates. Synthetic implications, *Phosphorus, Sulfur Silicon Relat. Elem.*, 75, 19, 1993.

773. Hu, C.-M., and Chen, J., Addition of diethyl bromodifluoromethylphosphonate to various alkenes initiated by Co(III)/Zn bimetal redox system, *J. Chem. Soc., Perkin Trans. 1*, 327, 1993.

774. Villemin, D., Sauvaget, F., and Hajek, M., Addition of diethyl trichloromethylphosphonate to olefins catalysed by copper complexes, *Tetrahedron Lett.*, 35, 3537, 1994.

775. Subbanagounder, G., Salomon, R.G., Murthi, K.K., Brame, C., and Roberts, J.L., II, Total synthesis of iso[4]-levuglandin E_2, *J. Org. Chem.*, 62, 7658, 1997.

776. Poss, A.J., and Belter, R.K., Diethyl 3-iodopropynylphosphonate. An alkylative β-keto phosphonate equivalent, *J. Org. Chem.*, 52, 4810, 1987.

777. Blackburn, G.M., and Rashid, A., Enantiospecific syntheses of the methylene- and α-fluoromethylene-phosphonate analogues of 3-phospho-D-glyceric acid, *J. Chem. Soc., Chem. Commun.*, 40, 1989.

778. Myrvold, S., Reimer, L.M., Pompliano, D.L., and Frost, J.W., Chemical inhibition of dehydroquinate synthase, *J. Am. Chem. Soc.*, 111, 1861, 1989.

779. Tian, F., Montchamp, J.-L., and Frost, J.W., Inhibitor ionization as a determinant of binding to 3-dehydroquinate synthase, *J. Org. Chem.*, 61, 7373, 1996.

780. Leeper, F.J., and Rock, M., Modified substrates for tetrapyrrole biosynthesis. Analogues of porpho-bilinogen showing unusual inhibition of porphobilinogen deaminase, *J. Chem. Soc., Chem. Commun.*, 242, 1992.

781. Leeper, F.J., Rock, M., and Appleton, D., Synthesis of analogues of posphobilinogen, *J. Chem. Soc., Perkin Trans. 1*, 2633, 1996.

782. Yang, Z.-Y., and Burton, D.J., A novel, general method for the preparation of α,α-difluoro function-alized phosphonates, *Tetrahedron Lett.*, 32, 1019, 1991.

783. Yang, Z.-Y., and Burton, D.J., A novel and practical preparation of α,α-difluoro functionalized phosphonates from iododifluoromethylphosphonate, *J. Org. Chem.*, 57, 4676, 1992.

784. Yang, Z.-Y., and Burton, D.J., A new approach to α,α-difluoro-functionalized esters, *J. Org. Chem.*, 56, 5125, 1991.

785. Kawamoto, A.M., and Campbell, M.M., Novel class of difluorovinylphosphonate analogues of PEP, *J. Chem. Soc., Perkin Trans. 1*, 1249, 1997.

786. Kawamoto, A.M., and Campbell, M.M., A new method for the synthesis of a phosphonic acid analogue of phosphoserine via a novel 1,1-difluorophosphonate intermediate, *J. Fluorine Chem.*, 81, 181, 1997.

787. Yager, K.M., Taylor, C.M., and Smith, A.B., III, Asymmetric synthesis of α-aminophosphonates via diastereoselective addition of lithium diethyl phosphite to chelating imines, *J. Am. Chem. Soc.*, 116, 9377, 1994.

788. Smith, A.B., III, Yager, K.M., and Taylor, C.M., Enantioselective synthesis of diverse α-amino phosphonate diesters, *J. Am. Chem. Soc.*, 117, 10879, 1995.

789. Retherford, C., Chou, T.-S., Schelkun, R.M., and Knochel, P., Preparation and reactivity of β-zinc and copper phosphonates, *Tetrahedron Lett.*, 31, 1833, 1990.

790. Hirao, T., Masunaga, T., Yamada, N., Ohshiro, Y., and Agawa, T., Palladium-catalyzed new carbon–phosphorus bond formation, *Bull. Chem. Soc. Jpn.*, 55, 909, 1982.

791. Petrakis, K.S., and Nagabhushan, T.L., Palladium-catalyzed substitutions of triflates derived from tyrosine-containing peptides and simpler hydroxyarenes forming 4-(diethoxyphosphinyl)phenylalanines and diethyl arylphosphonates, *J. Am. Chem. Soc.*, 109, 2831, 1987.

792. Villemin, D., Jaffrès, P.-A., and Siméon, F., Rapid and efficient phosphonation of aryl halides catalysed by palladium under microwaves irradiation, *Phosphorus, Sulfur Silicon Relat. Elem.*, 130, 59, 1997.

793. Tarasenko, E.A., Lukashev, N.V., and Beletskaya, I.P., Palladium-catalyzed copper(I) salt-mediated arylation of a *bis*(dimethylamino)phosphonyl-stabilized carbanion, *Tetrahedron Lett.*, 41, 1611, 2000.

794. Yokomatsu, T., Abe, H., Yamagishi, T., Suemune, K., and Shibuya, S., Convenient synthesis of cyclopropylalkanol derivatives possessing a difluoromethylenephosphonate group at the ring, *J. Org. Chem.*, 64, 8413, 1999.

795. McKenna, C.E., and Kashemirov, B.A., Recent progress in carbonylphosphonate chemistry, in *Topics in Current Chemistry,* Vol. 220. *New Aspects in Phosphorus Chemistry,* Vol. 1, Majoral, J.P., Ed., Springer-Verlag, Berlin, 2002, p. 201.

796. Maguire, M.H., Ralph, R.K., and Shaw, G., Purines, pyrimidines, and glyoxalines. Part 10. Synthesis of uracil-6-phosphonic acid, an analog of orotic acid, *J. Chem. Soc.*, 2299, 1958.

797. Polozov, A.M., Pavlov, V.A., Polezhaeva, N.A., Liorber, B.G., Tarzivolova, T.A., and Arbuzov, B.A., Migration of a phosphoryl group in the photochemical and thermal Wolff rearrangement, *Zh. Obshch. Khim.*, 56, 1217, 1986; *J. Gen. Chem. USSR (Engl. Transl.)*, 56, 1072, 1986.

798. Liorber, B.G., Tarzivolova, T.A., Pavlov, V.A., Zykova, T.V., Kisilev, V.V., Tumasheva, N.A., Slizkii, A.Y., and Shagvaleev, F.S., Phosphorylation of β-dicarbonyl compounds, *Zh. Obshch. Khim.*, 57, 534, 1987; *J. Gen. Chem. USSR (Engl. Transl.)*, 57, 465, 1987.

799. Kreutzkamp, N., and Mengel, W., Carbonyl and cyanophosphonic esters. Part 10. Preparation of γ- and α-dicarbonylphosphonates, *Arch. Pharm. Ber. Dtsch. Pharm. Ges.*, 300, 389, 1967.

800. Khomutov, R.M., Osipova, T.I., and Zhukov, Y.N., Synthesis of α-ketophosphonic acids, *Izv. Akad. Nauk SSSR, Ser. Khim.*, 1391, 1978; *Bull. Acad. Sci. USSR, Div. Chem. Sci. (Engl. Transl.)*, 27, 1210, 1978.

801. Oleksyszyn, J., Gruszecka, E., Kafarski, P., and Mastalerz, P., New phosphonic analogs of aspartic and glutamic acid by aminoalkylation of trivalent phosphorus chlorides with ethyl acetyloacetate or ethyl levulinate and benzyl carbamate, *Monatsh. Chem.*, 113, 59, 1982.

802. Diel, P.J., and Maier, L., Organic phosphorus compounds. Part 79. Preparation and properties of α-amino-ω-carboxyalkylphosphonic and -phosphinic acids, *Phosphorus Sulfur*, 29, 201, 1987.

803. Tang, K.-C., and Coward, J.K., Synthesis of acyl phosphonate analogues of biologically important acyl phosphates. *N*-(2-Amino-10-methylpteroyl)-5-amino-2-oxopentanephosphonic acid, *J. Org. Chem.*, 48, 5001, 1983.

804. Fouqué, D., About-Jaudet, E., Collignon, N., and Savignac, P., A convenient synthesis of diethyl 2,4-dioxoalkylphosphonates, *Synth. Commun.*, 22, 219, 1992.

805. Fouqué, D., Al-Badri, H., About-Jaudet, E., and Collignon, N., α-Pyrazolylalkylphosphonates. Part 1. Synthesis of 1-(3- or 5-pyrazolyl)-alkylphosphonates, *Bull. Soc. Chim. Fr.*, 131, 992, 1994.

806. Taub, D., Zelawski, Z.S., and Wendler, N.L., A stereoselective total synthesis of 7α-hydroxy-5,11-diketotetranorprostane-1,16-dioic acid, the major human urinary metabolite of PGE$_1$ and PGE$_2$, *Tetrahedron Lett.*, 3667, 1975.

807. Bodalski, R., Pietrusiewicz, K.M., Monkiewicz, J., and Koszuk, J., A new efficient synthesis of substituted Nazarov reagents. A Wittig–Horner–Emmons approach, *Tetrahedron Lett.*, 21, 2287, 1980.

808. Callant, P., De Wilde, H., and Vandewalle, M., The synthesis of functionalized *cis*-bicyclo[3.3.0]octanes, *Tetrahedron*, 37, 2079, 1981.

809. Taylor, E.C., and Davies, H.M.L., Rhodium(II) acetate-catalyzed reaction of ethyl 2-diazo-3-oxopent-4-enoates. Simple routes to 4-aryl-2-hydroxy-1-naphthoates and β,γ-unsaturated esters. The dianion of ethyl 4-(diethylphosphono)acetoacetate as a propionate homoenolate equivalent, *Tetrahedron Lett.*, 24, 5453, 1983.

810. Pietrusiewicz, K.M., Monkiewicz, J., and Bodalski, R., (3-Carbethoxy-2-oxopropylidene)triphenylphosphorane. A reagent for "3 + 3" cyclohexenone annulation, *J. Org. Chem.*, 48, 788, 1983.

811. Corey, E.J., and Shimoji, K., Total synthesis of the major human urinary metabolite of prostaglandin D_2, a key diagnostic indicator, *J. Am. Chem. Soc.*, 105, 1662, 1983.

812. Karagiri, N., Takashima, K., Haneda, T., and Kato, T., Synthesis of pyrazofurin and its analogues, *J. Chem. Soc., Perkin Trans. 1*, 553, 1984.

813. Rosen, T., and Heathcock, C.H., Total synthesis of (+)-compactin, *J. Am. Chem. Soc.*, 107, 3731, 1985.

814. Hecker, S.J., and Heathcock, C.H., Total synthesis of (+)-dihydromevinolin, *J. Am. Chem. Soc.*, 108, 4586, 1986.

815. Moorhoff, C.M., and Schneider, D.F., Comments on the reaction of ethyl 4-(diethoxyphosphinyl)-3-oxobutanoate and related phosphonate esters with enals, *Tetrahedron Lett.*, 28, 559, 1987.

816. Heathcock, C.H., Hadley, C.R., Rosen, T., Theisen, P.D., and Hecker, S.J., Total synthesis and biological evaluation of structural analogues of compactin and dihydromevinolin, *J. Med. Chem.*, 30, 1858, 1987.

817. Burke, S.D., Piscopio, A.D., and Buchanan, J.L., Total synthesis of (±)-pulo'upone, *Tetrahedron Lett.*, 29, 2757, 1988.

818. Heathcock, C.H., Davis, B.R., and Hadley, C.R., Synthesis and biological evaluation of a monocyclic, fully functional analogue of compactin, *J. Med. Chem.*, 32, 197, 1989.

819. Burke, S.D., and Deaton, D.N., Synthesis of (+)-dihydrocompactin and (+)-compactin via vinylsilane terminated cationic cyclization, *Tetrahedron Lett.*, 32, 4651, 1991.

820. Sliskovic, D.R., Picard, J.A., Roark, W.H., Roth, B.D., Ferguson, E., Krause, B.R., Newton, R.S., Sekerke, C., and Shaw, M.K., Inhibitors of cholesterol biosynthesis. Part 4. *Trans*-6-[2-(substituted-quinolinyl)ethenyl/ethyl]tetrahydro-4-hydroxy-2*H*-pyran-2-ones, a novel series of HMG-CoA reductase inhibitors, *J. Med. Chem.*, 34, 367, 1991.

821. Takemoto, M., Koshida, A., Miyazima, K., Suzuki, K., and Achiwa, K., Prostanoids and related compounds. Part 4. Total synthesis of clavulones, *Chem. Pharm. Bull.*, 39, 1106, 1991.

822. Blackwell, C.M., Davidson, A.H., Launchbury, S.B., Lewis, C.N., Morrice, E.M., Reeve, M.M., Roffley, J.A.R., Tipping, A.S., and Todd, R.S., An improved procedure for the introduction of the δ-lactone portion of HMG-CoA reductase inhibitors, *J. Org. Chem.*, 57, 1935, 1992.

823. Blackwell, C.M., Davidson, A.H., Launchbury, S.B., Lewis, C.N., Morrice, E.M., Reeve, M.M., Roffley, J.A.R., Tipping, A.S., and Todd, R.S., Total synthesis of dihydromevinolin and a series of related 3-hydroxy-3-methylglutaryl coenzyme A reductase inhibitors, *J. Org. Chem.*, 57, 5596, 1992.

824. Meese, C.O., Novel syntheses of 2,3,4,5-tetranor- and 6-oxo-prostaglandins, *Liebigs Ann. Chem.*, 1217, 1992.

825. Bone, E.A., Davidson, A.H., Lewis, C.N., and Todd, R.S., Synthesis and biological evaluation of dihydroeptastatin, a novel inhibitor of 3-hydroxy-3-methylglutaryl coenzyme A reductase, *J. Med. Chem.*, 35, 3388, 1992.

826. Takatori, K., Kajiwara, M., Sakamoto, Y., Shimayama, T., Yamada, H., and Takahashi, T., A Diels–Alder reaction approach to a homoisocarbacyclin, *Tetrahedron Lett.*, 35, 5669, 1994.

827. Araki, Y., and Konoike, T., Enantioselective total synthesis of (+)-6-*epi*-mevinolin and its analogs. Efficient construction of the hexahydronaphthalene moiety by high pressure-promoted intramolecular Diels–Alder reaction of (*R*,2*Z*,8*E*,10*E*)-1-[(*tert*-butyldimethylsilyl)oxy]-6-methyl-2,8,10-dodecatrien-4-one, *J. Org. Chem.*, 62, 5299, 1997.

828. Nokami, J., Furukawa, A., Okuda, Y., Hazato, A., and Kurozumi, S., Palladium-catalyzed coupling reactions of bromobenzaldehydes with 3,4-di(*tert*-butylmethylsilyloxy)-1-alkene to (3,4-dihydroxy-alkenyl)benzaldehydes in the synthesis of lipoxin analogues, *Tetrahedron Lett.*, 39, 1005, 1998.

829. Adiyaman, M., Lawson, J.A., FitzGerald, G.A., and Rokach, J., Synthesis and identification of the most abundant urinary type VI isoprostanes, *Tetrahedron Lett.*, 39, 7039, 1998.

830. Gosselin, F., and Lubell, W.D., An olefination entry for the synthesis of enantiopure α,ω-diaminodi-carboxylates and azabicyclo[X.Y.0]alkane amino acids, *J. Org. Chem.*, 63, 7463, 1998.

831. Gosselin, F., and Lubell, W.D., Rigid dipeptide surrogates. Syntheses of enantiopure quinolizidinone and pyrroloazepinone amino acids from a common diaminodicarboxylate precursor, *J. Org. Chem.*, 65, 2163, 2000.

832. Mathey, F., and Savignac, P., The α-cuprophosphonates. Preparation and application to the synthesis of β-ketophosphonates, *Actes Congr. Int. Composés Phosphorés, 1st*, 1977; *Chem. Abstr.*, 89, 146979, 1978.

833. Mathey, F., and Savignac, P., α.-Cuprophosphonates. Part 3. Application to the synthesis of β-ketophosphonates, *Tetrahedron*, 34, 649, 1978.

834. Ornstein, P.L., Bleisch, T.J., Arnold, M.B., Kennedy, J.H., Wright, R.A., Johnson, B.G., Tizzano, J.P., Helton, D.R., Kallman, M.J., and Schoepp, D.D., 2-Substituted (2*SR*)-2-amino-2-((1*SR*,2*SR*)-2-carboxycycloprop-1-yl)glycines as potent and selective antagonists of group II metabotropic glutamate receptors. Part 2. Effects of aromatic substitution, pharmacological characterization, and bioavailability, *J. Med. Chem.*, 41, 358, 1998.

835. Ornstein, P.L., Bleisch, T.J., Arnold, M.B., Wright, R.A., Johnson, B.G., and Schoepp, D.D., 2-Substituted (2*SR*)-2-amino-2-(1*SR*,2*SR*)-2-carboxycycloprop-1-yl)glycines as potent and selective antagonists of group II metabotropic glutamate receptors. Part 1. Effects of alkyl, arylalkyl, and diarylalkyl substitution, *J. Med. Chem.*, 41, 346, 1998.

836. Rudisill, D.E., and Whitten, J.P., Synthesis of (*R*)-4-oxo-5-phosphononorvaline, an *N*-methyl-D-aspartic acid receptor selective β-keto phosphonate, *Synthesis*, 851, 1994.

837. Burton, D.J., Ishihara, T., and Maruta, M., A useful zinc reagent for the preparation of 2-oxo-1,1-difluoroalkylphosphonates, *Chem. Lett.*, 755, 1982.

838. Delamarche, I., and Mosset, P., New syntheses of some functionalized and acetylenic β-keto phosphonates, *J. Org. Chem.*, 59, 5453, 1994.

839. Durand, T., Guy, A., Vidal, J.-P., Viala, J., and Rossi, J.-C., Total synthesis of 4-(*RS*)-F$_{4t}$-isoprostane methyl ester, *Tetrahedron Lett.*, 41, 3859, 2000.

840. Lieb, F., 2-Oxoalkanephosphonic acid dialkylesters useful in the synthesis of prostaglandin analogs, *Bayer*, German Patent Appl. DE 2711009, 1978; *Chem. Abstr.*, 89, 215553, 1978.

841. Gandolfi, C., Faustini, F., Andreoni, A., Fumagalli, A., Passarotti, C., and Ceserani, R., 5,6-Dihydro-prostacyclin derivatives and their pharmaceutical and veterinary use, *Carlo Erba*, German Patent Appl. DE 2853849, 1979; *Chem. Abstr.*, 92, 75958, 1980.

842. Nicolaou, K.C., Daines, R.A., and Chakraborty, T.K., Total synthesis of amphoteronolide B, *J. Am. Chem. Soc.*, 109, 2208, 1987.

843. Tolstikov, G.A., Akhmetvaleev, R.R., Zhurba, V.M., and Miftakhov, M.S., Prostanoids. Part 54. Alternatives to natural prostaglandins, *Zh. Org. Khim.*, 28, 1862, 1992; *J. Org. Chem. USSR (Engl. Transl.)*, 28, 1491, 1992.

844. Gaoni, Y., Chapman, A.G., Parvez, N., Pook, P.C.K., Jane, D.E., and Watkins, J.C., Synthesis, NMDA receptor antagonist activity, and anticonvulsant action of 1-aminocyclobutanecarboxylic acid derivatives, *J. Med. Chem.*, 37, 4288, 1994.

845. Miftakhov, M.S., Akhmetvaleev, R.R., Imaeva, L.R., Vostrikov, N.S., Saitova, M.Y., Zarudii, F.S., and Tolstikov, G.A., Synthesis and properties of (+)-2-decarboxy-2-methyl-20-*nor*-19-carboxyprostaglandin F$_{2α}$ and its 15β-epimer, *Khim. Farm. Zh.*, 30, 22, 1996; *Pharm. Chem. J. (Engl. Transl.)*, 30, 513, 1996.

846. Miftakhov, M.S., Imaeva, L.R., Fatykhov, A.A., and Akhmetvaleev, R.R., Prostanoids. Part 66. *A priori* postulated "9-LO prostanoids," *Zh. Org. Khim.*, 33, 47, 1997; *Russ. J. Org. Chem. (Engl. Transl.)*, 33, 47, 1997.

847. Harada, K., Matsushita, A., Sasaki, H., and Kawachi, Y., Improved preparation of optically active phosphinyl-containing oxyglutaric acid ester derivatives, *Ube Industries*, U.S. Patent Appl. US 5717124, 1998; *Chem. Abstr.*, 128, 167563, 1998.

848. Theisen, P.D., and Heathcock, C.H., Improved procedure for preparation of optically active 3-hydroxyglutarate monoesters and 3-hydroxy-5-oxoalkanoic acids, *J. Org. Chem.*, 53, 2374, 1988.

849. Konoike, T., and Araki, Y., Practical synthesis of chiral synthons for the preparation of HMG-CoA reductase inhibitors, *J. Org. Chem.*, 59, 7849, 1994.

850. Karanewsky, D.S., Malley, M.F., and Gougoutas, J.Z., Practical synthesis of an enantiomerically pure synthon for the preparation of mevinic acid analogues, *J. Org. Chem.*, 56, 3744, 1991.

851. Roush, W.R., and Murphy, M., An improved synthesis of naphthoate precursors to olivin, *J. Org. Chem.*, 57, 6622, 1992.

852. Werner, R.M., Williams, L.M., and Davis, J.T., The *C*-glycosyl analog of an *N*-linked glycoamino acid, *Tetrahedron Lett.*, 39, 9135, 1998.

853. Fouqué, D., About-Jaudet, E., and Collignon, N., Synthesis and use of new α-pyrazolylalkylphosphonates, *Phosphorus, Sulfur Silicon Relat. Elem.*, 77, 178, 1993.

854. Bodalski, R., Pietrusiewicz, K.M., Monkiewicz, J., and Koszuk, J., A simple method for [3 + 3] annulation on chalcones with ethyl-4-(diethylphosphono)-3-oxobutanoate, *Pol. J. Chem.*, 57, 315, 1983.

855. Ley, S.V., and Woodward, P.R., Preparation of *tert*-butyl 4-diethylphosphono-3-oxobutanethioate and use in the synthesis of (*E*)-4-alkenyl-3-oxo esters and macrolides, *Tetrahedron Lett.*, 28, 345, 1987.

856. Paterson, I., Gardner, M., and Banks, B.J., Studies in marine cembranolide synthesis. A synthesis of 2,3,5-trisubstituted furan intermediates for lophotoxin and pukalide, *Tetrahedron*, 45, 5283, 1989.

857. Lavallée, J.-F., Spino, C., Ruel, R., Hogan, K.T., and Deslongchamps, P., Stereoselective synthesis of *cis*-decalins via Diels–Alder and double Michael addition of substituted Nazarov reagents, *Can. J. Chem.*, 70, 1406, 1992.

858. Boeckman, R.K., Jr., Kamanecka, T.M., Nelson, S.G., Pruitt, J.R., and Barta, T.E., *C*-Phosphorylation of enolates. An alternative route to complex carbonyl-activated phosphonates, *Tetrahedron Lett.*, 32, 2581, 1991.

859. Kennedy, R.M., Abiko, A., Takemasa, T., Okumoto, H., and Masamune, S., A synthesis of 19-dehydroamphoteronolide B, *Tetrahedron Lett.*, 29, 451, 1988.

860. Cooke, M.P., Jr., and Jaw, J.Y., Cyclopropane derivatives through charge-directed conjugate addition reactions of unsaturated acylphosphoranes, *J. Org. Chem.*, 51, 758, 1986.

861. Kinney, W.A., Lee, N.E., Garrison, D.T., Podlesny, E.J., Simmonds, J.T., Bramlett, D., Notvest, R.R., Kowal, D.M., and Tasse, R.P., Bioisosteric replacement of the α-amino carboxylic acid functionality in 2-amino-5-phosphonopentanoic acid yields unique 3,4-diamino-3-cyclobutene-1,2-dione containing NMDA antagonists, *J. Med. Chem.*, 35, 4720, 1992.

862. Gruszecka, E., Mastalerz, P., and Soroka, M., New synthesis of phosphinothricin and analogues, *Rocz. Chem.*, 49, 2127, 1975.

863. Varlet, J.M., Fabre, G., Sauveur, F., Collignon, N., and Savignac, P., Preparation and conversion of ω-formylalkylphosphonates to aminocarboxyalkylphosphonic acids, *Tetrahedron*, 37, 1377, 1981.

864. Natchev, I.A., Total synthesis, enzyme-substrate interactions and herbicidal activity of plumbemicin A and B (N-1409), *Tetrahedron*, 44, 1511, 1988.

865. Bigge, C.F., Drummond, J.T., Johnson, G., Malone, T., Probert, A.W., Jr., Marcoux, F.W., Coughenour, L.L., and Brahce, L.J., Exploration of phenyl-spaced 2-amino-(5–9)-phosphonoalkanoic acids as competitive *N*-methyl-D-aspartic acid antagonists, *J. Med. Chem.*, 32, 1580, 1989.

866. Dappen, M.S., Pellicciari, R., Natalini, B., Monahan, J.B., Chiorri, C., and Cordi, A.A., Synthesis and biological evaluation of cyclopropyl analogues of 2-amino-5-phosphonopentanoic acid, *J. Med. Chem.*, 34, 161, 1991.

867. Hamilton, G.S., Huang, Z., Yang, X.-J., Patch, R.J., Narayanan, B.A., and Ferkany, J.W., Asymmetric synthesis of a potent selective competitive NMDA antagonist, *J. Org. Chem.*, 58, 7263, 1993.

868. Yokomatsu, T., Nakabayashi, N., Matsumoto, K., and Shibuya, S., Lipase-catalysed kinetic resolution of *cis*-1-diethylphosphonomethyl-2-hydroxymethylcyclohexane. Application to enantioselective synthesis of 1-diethylphosphonomethyl-2-(5′-hydantoinyl)cyclohexane, *Tetrahedron: Asymmetry*, 6, 3055, 1995.

869. Hanrahan, J.R., Taylor, P.C., and Errington, W., The synthesis of 3-phosphonocyclobutyl amino acid analogues of glutamic acid via diethyl 3-oxocyclobutylphosphonate, a versatile synthetic intermediate, *J. Chem. Soc., Perkin Trans. 1*, 493, 1997.

870. Gasparini, F., Inderbitzin, W., Francotte, E., Lecis, G., Richert, P., Dragic, Z., Kuhn, R., and Flor, P.J., (+)-(4-Phosphonophenyl)glycine (PPG). A new group-III-selective metabotropic glutamate receptor agonist, *Bioorg. Med. Chem. Lett.*, 10, 1241, 2000.

871. Crossley, R., and Curran, A.C.W., A novel synthesis of thiols from α-amino-nitriles, *J. Chem. Soc., Perkin Trans. 1*, 2327, 1974.

872. Gruszecka, E., Soroka, M., and Mastalerz, P., Phosphonic analogs of α-methylaspartic and α-methylglutamic acids, *Pol. J. Chem.*, 53, 2327, 1979.

873. Crooks, S.L., Robinson, M.B., Koerner, J.F., and Johnson, R.L., Cyclic analogues of 2-amino-4-phosphonobutanoic acid (APB) and their inhibition of hippocampal excitatory transmission and displacement of [³H]APB binding, *J. Med. Chem.*, 29, 1988, 1986.

874. Alonso, F., Mico, I., Najera, C., Sansano, J.M., Yus, M., Ezquerra, J., Yruretagoyena, B., and Garcia, I., Synthesis of 3- and 4-substituted cyclic α-amino acids structurally related to ACPD, *Tetrahedron*, 51, 10259, 1995.

875. Conway, S.J., Miller, J.C., Howson, P.A., Clark, B.P., and Jane, D.E., Synthesis of phenylglycine derivatives as potent and selective antagonists of group III metabotropic glutamate receptors, *Bioorg. Med. Chem. Lett.*, 11, 777, 2001.

876. Villanueva, J.M., Collignon, N., Guy, A., and Savignac, P., Preparation of optically active aminocar-boxyalkylphosphonic acids, *Tetrahedron*, 39, 1299, 1983.

877. Garbay-Jaureguiberry, C., McCort-Tranchepain, I., Barbe, B., Ficheux, D., and Roques, B.P., Improved synthesis of (*p*-phosphono and *p*-sulfo)methylphenylalanine. Resolution of (*p*-phosphono-, *p*-sulfo, *p*-carboxy- and *p*-*N*-hydroxycarboxamido-)methylphenylalanine, *Tetrahedron: Asymmetry*, 3, 637, 1992.

878. Burke, T.R., Jr., Russ, P., and Lim, B., Preparation of 4-[*bis*(*tert*-butoxy)phosphorylmethyl]-*N*-(fluo-ren-9-ylmethoxycarbonyl)-DL-phenylalanine. A hydrolytically stable analogue of *O*-phosphotyrosine potentially suitable for peptide synthesis, *Synthesis*, 1019, 1991.

879. Burke, T.R., Jr., Smyth, M.S., Otaka, A., and Roller, P.P., Synthesis of 4-phosphono(difluoromethyl)-D,L-phenylalanine and *N*-Boc and *N*-Fmoc derivatives suitably protected for solid-phase synthesis of nonhydrolysable phosphotyrosyl peptide analogues, *Tetrahedron Lett.*, 34, 4125, 1993.

880. Burke, T.R., Jr., Smyth, M.S., Nomizu, M., Otaka, A., and Roller, P.P., Preparation of fluoro- and hydroxy-4-(phosphonomethyl)-D,L-phenylalanine suitably protected for solid-phase synthesis of pep-tides containing hydrolytically stable analogues of *O*-phosphotyrosine, *J. Org. Chem.*, 58, 1336, 1993.

881. Varlet, J.M., Collignon, N., and Savignac, P., Synthesis and reductive amination of phosphonopyru-vates. Preparation of 2-amino-2-carboxyalkylphosphonic acids (β-phosphonoalanine), *Can. J. Chem.*, 57, 3216, 1979.

882. Varlet, J.M., Collignon, N., and Savignac, P., Reductive amination of β-keto phosphonates. Preparation of aminoalkylphosphonic acids, *Tetrahedron*, 37, 3713, 1981.

883. Savignac, P., and Collignon, N., Some aspects of aminoalkylphosphonic acids. Synthesis by the reductive amination approach, *ACS Symp. Ser. (Phosphorus Chem.)*, 171, 255, 1981; *Chem. Abstr.*, 96, 35344, 1982.

884. Jiao, X.-Y., Chen, W.-Y., and Hu, B.-F., A novel asymmetric synthesis of *S*-(+)-2-amino-4-phospho-nobutanoic acid, *Synth. Commun.*, 22, 1179, 1992.

885. Mastalerz, P., Synthesis of γ-phosphonoglutamic acid, *Acta Biochim. Polon.*, 4, 19, 1957; *Chem. Abstr.*, 53, 18879b, 1959.

886. Mastalerz, P., Synthesis of phosphonic acids related structurally to glutamic acid, *Rocz. Chem.*, 33, 985, 1959; *Chem. Abstr.*, 54, 6602b, 1960.

887. Chambers, J.R., and Isbell, A.F., A new synthesis of amino phosphonic acids, *J. Org. Chem.*, 29, 832, 1964.

888. Matoba, K., Yonemoto, H., Fukui, M., and Yamazaki, T., Structural modification of bioactive com-pounds. Part 2. Syntheses of aminophosphonic acids, *Chem. Pharm. Bull.*, 32, 3918, 1984.

889. Levenson, C.H., and Meyer, R.B., Jr., Design and synthesis of tetrahedral intermediate analogues as potential dihydroorotase inhibitors, *J. Med. Chem.*, 27, 228, 1984.

890. Aboujaoude, E.E., Collignon, N., Savignac, P., and Bensoam, J., Synthesis of ω-amino-ω-carboxy-alkylphosphonic acids, *Phosphorus Sulfur*, 34, 93, 1987.

891. Bigge, C.F., Drummond, J.T., and Johnson, G., Synthesis and NMDA receptor binding of 2-amino-7,7-difluoro-7-phosphonoheptanoic acid, *Tetrahedron Lett.*, 30, 7013, 1989.

892. Baudy, R.B., Greenblatt, L.P., Jirkovsky, I.L., Conklin, M., Russo, R.J., Bramlett, D.R., Emrey, T.A., Simmonds, J.T., Kowal, D.M., Stein, R.P., and Tasse, R.P., Potent quinoxaline-spaced phosphono α-amino acids of the AP-6 type as competitive NMDA antagonists. Synthesis and biological evaluation, *J. Med. Chem.*, 36, 331, 1993.

893. Rozhko, L.F., Ragulin, V.V., and Tsvetkov, E.N., Phosphorus-containing aminocarboxylic acids. Part 6. Improved synthesis of 2-amino-5-phosphonopentanoic acid, *Zh. Obshch. Khim.*, 66, 1093, 1996; *Russ. J. Gen. Chem. (Engl. Transl.)*, 66, 1065, 1996.

894. Saratovskikh, I.V., Kalashnikov, V.V., and Ragulin, V.V., Synthesis of α-substituted aminocarboxylic acids, *Zh. Obshch. Khim.*, 69, 1218, 1999; *Russ. J. Gen. Chem. (Engl. Transl.)*, 69, 1173, 1999.

895. Schöllkopf, U., Busse, U., Lonsky, R., and Hinrichs, R., Asymmetric syntheses via heterocyclic intermediates. Part 31. Asymmetric synthesis of various non-proteinogenic amino acid methyl esters (functionalized in the carbon chain) and amino acids by the bislactim ether method, *Liebigs Ann. Chem.*, 2150, 1986.

896. Schöllkopf, U., Pettig, D., Busse, U., Egert, E., and Dyrbusch, M., Asymmetric synthesis via hetero-cyclic intermediates. Part 30. Asymmetric synthesis of glutamic acids and derivatives thereof by the bislactim-ether method. Michael-addition of methyl 2-alkenoates to the lithiated bislactim-ether of cyclo-(L-Val-Gly), *Synthesis*, 737, 1986.

897. Ornstein, P.L., An enantioselective synthesis of D-(−)-2-amino-5-phosphonopentanoic acid, *J. Org. Chem.*, 54, 2251, 1989.

898. Müller, W., Kipfer, P., Lowe, D.A., and Urwyler, S., Syntheses of biphenyl analogues of AP7, a new class of competitive *N*-methyl-D-aspartate (NMDA) receptor antagonists, *Helv. Chim. Acta*, 78, 2026, 1995.

899. Garcia-Barradas, O., and Juaristi, E., Highly enantioselective synthesis of (*R*)- and (*S*)-2-amino-5-phosphonopentanoic acids [(*R*)- and (*S*)-AP5] via modified Seebach imidazolidinones, *Tetrahedron*, 51, 3423, 1995.

900. Garcia-Barradas, O., and Juaristi, E., Enantioselective synthesis of both enantiomers of 2-amino-6-phosphonohexanoic acid [(*R*)- and (*S*)-AP6], a potent and specific agonist of AMPA receptor subtype, *Tetrahedron: Asymmetry*, 8, 1511, 1997.

901. Soloshonok, V.A., Belokon, Y.N., Kuzmina, N.A., Maleev, V.I., Svistunova, N.Y., Solodenko, V.A., and Kukhar', V.P., Asymmetric synthesis of phosphorus analogues of dicarboxylic α-amino acids, *J. Chem. Soc., Perkin Trans. 1*, 1525, 1992.

902. Soloshonok, V.A., Svistunova, N.Y., Kukhar', V.P., Gudima, A.O., Kuz'mina, N.A., and Belokon', Y.N., Asymmetric synthesis of organoelemental analogs of natural compounds. Part 7. (2*R*,3*S*)-2-amino-3-hydroxy-5-phosphonovaleric acid, *Izv. Akad. Nauk SSSR, Ser. Khim.*, 1172, 1992; *Bull. Russ. Acad. Sci., Div. Chem. Sci. (Engl. Transl.)*, 922, 1992.

903. Soloshonok, V.A., Svistunova, N.Y., Kukhar', V.P., Solodenko, V.A., Kuz'mina, N.A., Rozhenko, A.B., Galushko, S.V., Shishkina, I.P., Gudima, A.O., and Belokon', Y.N., Asymmetric synthesis of hetero-organic analogs of natural products. Part 6. (*S*)-α-Amino-ω-phosphonocarboxylic acids, *Izv. Akad. Nauk, Ser. Khim.*, 41, 397, 1992; *Bull. Russ. Acad. Sci., Div. Chem. Sci. (Engl. Transl.)*, 41, 311, 1992.

904. Andronova, I.G., Maleev, V.I., Ragulin, V.V., Il'in, M.M., Tsvetkov, E.N., and Belokon, Y.N., Phosphorus-containing aminocarboxylic acids. Part 7. Asymmetric synthesis of ω-phosphono-α-aminocarboxylic acids, *Zh. Obshch. Khim.*, 66, 1096, 1996; *Russ. J. Gen. Chem. (Engl. Transl.)*, 66, 1068, 1996.

905. Hays, S.J., Bigge, C.F., Novak, P.M., Drummond, J.T., Bobovski, T.P., Rice, M.J., Johnson, G., Brahce, L.J., and Coughenour, L.L., New and versatile approaches to the synthesis of CPP-related competitive NMDA antagonists. Preliminary stucture–activity relationships and pharmacological evaluation, *J. Med. Chem.*, 33, 2916, 1990.

906. Swahn, B.-M., Andersson, F., Pelcman, B., Soederberg, J., and Claesson, A., Synthesis of ¹⁴C-labelled (*R*)-α-amino-6,7-dimethyl-3-(phosphonomethyl)-2-quinolinepropanoic acid, *J. Labelled Compd. Radiopharm.*, 39, 259, 1997.

907. Minowa, N., Hirayama, M., and Fukatsu, S., Asymmetric synthesis of (+)-phosphinothricin and (+)-2-amino-4-phosphonobutyric acid, *Tetrahedron Lett.*, 25, 1147, 1984.

908. Sturtz, G., and Guillamot, G., Amethopterine (methotrexate) phosphonoglutamic acid analogs. Part 1. Synthesis, *Eur. J. Med. Chem. Chim. Ther.*, 19, 267, 1984.

909. Sturtz, G., Guillamot, G., Bourdeaux, M., and Chauvet, M., Amethopterine (methotrexate) phosphonoglutamic acid analogs. Part 2. Dihydrofolate reductase inhibition, *Eur. J. Med. Chem. Chim. Ther.*, 19, 274, 1984.

910. Minowa, N., Hirayama, M., and Fukatsu, S., Asymmetric synthesis of (+)-phosphinothricin and related compounds by the Michael addition of glycine Schiff bases to vinyl compounds, *Bull. Chem. Soc. Jpn.*, 60, 1761, 1987.

911. Hamilton, R., Shute, R.E., Travers, J., Walker, B., and Walker, B.J., A convenient synthesis of phosphonate isosteres of serine phosphates, *Tetrahedron Lett.*, 35, 3597, 1994.

912. Shapiro, G., Buechler, D., Ojea, V., Pombo-Villar, E., Ruiz, M., and Weber, H.-P., Synthesis of both D- and L-Fmoc-Abu[PO(OCH₂CH=CH₂)₂]-OH for solid phase phosphonopeptide synthesis, *Tetrahedron Lett.*, 34, 6255, 1993.

913. Schick, A., Kolter, T., Giannis, A., and Sandhoff, K., Synthesis of phosphonate analogues of sphinganine-1-phosphate and sphingosine-1-phosphate, *Tetrahedron*, 51, 11207, 1995.

914. Ojea, V., Ruiz, M., Shapiro, G., and Pombo-Villar, E., Enantiospecific synthesis of 2-amino-3-methyl-4-phosphonobutanoic acids via 1,4-addition of lithiated Schöllkopf anion to prop-2-enylphosphonates, *Tetrahedron Lett.*, 35, 3273, 1994.

915. Ruiz, M., Ojea, V., Shapiro, G., Weber, H.-P., and Pombo-Willar, E., Asymmetric synthesis of a protected phosphonate isostere of phosphothreonine for solid-phase peptide synthesis, *Tetrahedron Lett.*, 35, 4551, 1994.

916. Ojea, V., Ruiz, M., Shapiro, G., and Pombo-Villar, E., Conjugate addition of 1-propenylphosphonates to metalated Schöllkopf's bis-lactim ether. Stereocontrolled access to 2-amino-3-methyl-4-phosphonobutanoic acids, *J. Org. Chem.*, 65, 1984, 2000.

917. Ojea, V., Conde, S., Ruiz, M., Fernandez, M.C., and Quintela, J.M., Conjugate addition of lithiated Schöllkopf's bislactim ether to 1*E*,3*E*-butadienylphosphonates. Stereocontrolled access to 2,3-*anti*-4*E*-2-amino-6-phosphono-4-hexenoic acid derivatives, *Tetrahedron Lett.*, 38, 4311, 1997.

918. Ruiz, M., Ojea, V., Fernandez, M.C., Conde, S., Diaz, A., and Quintela, J.M., Synthesis of 2-amino-2-methyl-4-phosphonobutanoic acids by conjugate addition of lithiated bislactim ether derived from cyclo[Ala-D-Val] to vinylphosphonates, *Synlett*, 1903, 1999.

919. Logusch, E.W., Facile synthesis of D,L-phosphinothricin from methyl 4-bromo-2-phthalimidobutyrate, *Tetrahedron Lett.*, 27, 5935, 1986.

920. Ragulin, V.V., Bofanova, M.E., and Tsvetkov, E.N., Phosphorus-containing aminocarboxylic acids. Part 1. Method of preparation of phosphonate-type compounds, *Izv. Akad. Nauk SSSR, Ser. Khim.*, 2590, 1989; *Bull. Acad. Sci. USSR, Div. Chem. Sci. (Engl. Transl.)*, 2377, 1989.

921. Malachowski, W.P., and Coward, J.K., The chemistry of phosphapeptides. Investigations on the synthesis of phosphonamidate, phosphonate, and phosphinate analogues of glutamyl-γ-glutamate, *J. Org. Chem.*, 59, 7625, 1994.

922. Nair, S.A., Lee, B., and Hangauer, D.G., Synthesis of orthogonally protected L-homocysteine and L-2-amino-4-phosphonobutanoic acid from L-homoserine, *Synthesis*, 810, 1995.

923. Ornstein, P.L., Augenstein, N.K., and Arnold, M.B., Stereoselective synthesis of 6-substituted decahydroisoquinoline-3-carboxylates. Intermediates for the preparation of conformationally constrained acidic amino acids, *J. Org. Chem.*, 59, 7862, 1994.

924. Ragulin, V.V., Bofanova, M.E., and Tsvetkov, E.N., Phosphorus-containing aminocarboxylic acids. Part 4. A convenient method of phosphonic acids synthesis, *Phosphorus, Sulfur Silicon Relat. Elem.*, 62, 237, 1991.

925. Ma, D., Ma, Z., Jiang, J., Yang, Z., and Zheng, C., Synthesis of (2*S*,1′*S*,2′*S*)-2-methyl-2-(carboxycyclopropyl)glycine and (*S*)-2-amino-2-methyl-4-phosphonobutyric acid from L-alanine, *Tetrahedron: Asymmetry*, 8, 889, 1997.

926. Dorville, A., McCort-Tranchepain, I., Vichard, D., Sather, W., Maroun, R., Ascher, P., and Roques, B.P., Preferred antagonist binding state of the NMDA receptor. Synthesis, pharmacology, and computer modeling of (phosphonomethyl)phenylalanine derivatives, *J. Med. Chem.*, 35, 2551, 1992.

927. Bigge, C.F., Johnson, G., Ortwine, D.F., Drummond, J.T., Retz, D.M., Brahce, L.J., Coughenour, L.L., Marcoux, F.W., and Probert, A.W., Jr., Exploration of *N*-phosphonoalkyl-, *N*-phosphonoalkenyl-, and *N*-(phosphonoalkyl)phenyl-spaced α-amino acids as competitive *N*-methyl-D-aspartic acid antagonists, *J. Med. Chem.*, 35, 1371, 1992.

928. Ortwine, D.F., Malone, T.C., Bigge, C.F., Drummond, J.T., Humblet, C., Johnson, G., and Pinter, G.W., Generation of *N*-methyl-D-aspartate agonist and competitive antagonist pharmacophore models. Design and synthesis of phosphonoalkyl-substituted tetrahydroisoquinolines as novel antagonists, *J. Med. Chem.*, 35, 1345, 1992.

929. Hutchison, A.J., Williams, M., Angst, C., de Jesus, R., Blanchard, L., Jackson, R.H., Wilusz, E.J., Murphy, D.E., Bernard, P.S., Schneider, J., Campbell, T., Guida, W., and Sills, M.A., 4-(Phosphonoalkyl)- and 4-(phosphonoalkenyl)-2-piperidinecarboxylic acids. Synthesis, activity at *N*-methyl-D-aspartic acid receptors, and anticonvulsant activity, *J. Med. Chem.*, 32, 2171, 1989.

930. Martin, I., Anvelt, J., Vares, L., Kühn, I., and Claesson, A., An alternative synthesis of the NMDA antagonist CGS 19755 via free radical carbamoylation of ethyl isonicotinate, *Acta Chem. Scand.*, 49, 230, 1995.

931. Ornstein, P.L., Schaus, J.M., Chambers, J.W., Huser, D.L., Leander, J.D., Wong, D.T., Paschal, J.W., Jones, N.D., and Deeter, J.B., Synthesis and pharmacology of a series of 3- and 4-(phosphonoalkyl)pyridine- and piperidine-2-carboxylic acids. Potent *N*-methyl-D-aspartate receptor antagonists, *J. Med. Chem.*, 32, 827, 1989.

932. Tanaka, K.-i., Iwabuchi, H., and Sawanishi, H., Synthesis of homochiral 4-amino-4-carboxy-2-phosphonomethylpyrrolidines via a diastereoselective Bucherer–Bergs reaction of 4-oxopyrrolidine derivative. Novel conformationally restricted AP5 analogues, *Tetrahedron: Asymmetry*, 6, 2271, 1995.

933. Valerio, R.M., Alewood, P.F., and Johns, R.B., Synthesis of optically active 2-(*tert*-butyloxycarbonylamino)-4-dialkoxyphosphorylbutanoate protected isosteres of *O*-phosphonoserine for peptide synthesis, *Synthesis*, 786, 1988.

934. Tong, G., Perich, J.W., and Johns, R.B., Synthesis Leu-Abu(P) and Glu-Abu(P)-Leu. Isosteres of Ser(P)-peptides, *Tetrahedron Lett.*, 31, 3759, 1990.

935. Tong, G., Perich, J.W., and Johns, R.B., The improved synthesis of Boc-Abu(PO$_3$Me$_2$)-OH and its use for the facile synthesis of Glu-Abu(P)-Leu, *Aust. J. Chem.*, 45, 1225, 1992.

936. Wasielewski, C., and Antczak, K., Aminophosphonic acids. Part 16. A new and facile synthesis of phosphinothricine and 2-amino-4-phosphonobutanoic acid, *Synthesis*, 540, 1981.

937. Soroka, M., and Mastalerz, P., Phosphonic and phosphinic analogues of aspartic acid and asparagine, *Rocz. Chem.*, 48, 1119, 1974.

938. Soroka, M., and Mastalerz, P., The synthesis of phosphonic and phosphinic analogs of aspartic acid and asparagine, *Rocz. Chem.*, 50, 661, 1976.

939. Zeiss, H.-J., Enantioselective synthesis of L-phosphinothricin from L-methionine and L-glutamic acid via L-vinylglycine, *Tetrahedron*, 48, 8263, 1992.

940. Bigge, C.F., Wu, J.-P., and Drummond, J.R., Synthesis of 6-phosphonoalkyl tetrahydro-4-pyrimidinecarboxylic acids as NMDA receptor antagonists, *Tetrahedron Lett.*, 32, 7659, 1991.

941. Ornstein, P.L., Arnold, M.B., Allen, N.K., and Schoepp, D.D., Synthesis and characterization of phosphonic acid-substituted amino acids as excitatory amino acid receptor antagonists, *Phosphorus, Sulfur Silicon Relat. Elem.*, 109–110, 309, 1996.

942. David, C., Bischoff, L., Meudal, H., Mothé, A., De Mota, N., DaNascimento, S., Llorens-Cortes, C., Fournié-Zaluski, M.-C., and Roques, B.P., Investigation of subsite preferences in aminopeptidase A (EC 3.4.11.7) led to the design of the first highly potent and selective inhibitors of this enzyme, *J. Med. Chem.*, 42, 5197, 1999.

943. David, C., Bischoff, L., Roques, B.P., and Fournié-Zaluski, M.-C., Straightforward and diastereoselective synthesis of tetrafunctionalized thiol synthons for the design of metallopeptidase inhibitors, *Tetrahedron*, 56, 209, 2000.

944. Ezquerra, J., de Mendoza, J., Pedregal, C., and Ramirez, C., Regioselective nucleophilic attack on *N*-Boc-pyroglutamate ethyl ester, *Tetrahedron Lett.*, 33, 5589, 1992.

945. Baldwin, J.E., Adlington, R.M., Russell, A.T., and Smith, M.L., Carbon based nucleophilic ring opening of activated monocyclic β-lactams. Synthesis and stereochemical assignment of the ACE inhibitor WF-10129, *Tetrahedron*, 51, 4733, 1995.

946. Berkowitz, D.B., Shen, Q., and Maeng, J.-H., Synthesis of the (α,α-difluoroalkyl)phosphonate analogue of phosphoserine, *Tetrahedron Lett.*, 35, 6445, 1994.

947. Song, Y., Niederer, D., Lane-Bell, P.M., Lam, L.K.P., Crawley, S., Palcic, M.M., Pickard, M.A., Pruess, D.L., and Vederas, J.C., Stereospecific synthesis of phosphonate analogues of diaminopimelic acid (DAP), their interaction with DAP enzymes, and antibacterial activity of peptide derivatives, *J. Org. Chem.*, 59, 5784, 1994.

948. Otaka, A., Miyoshi, K., Burke, T.R., Jr., Roller, P.P., Kubota, H., Tamamura, H., and Fujii, N., Synthesis and application of *N*-Boc-L-2-amino-4-(diethylphosphono)-4,4-difluorobutanoic acid for solid-phase synthesis of nonhydrolyzable phosphoserine peptide analogues, *Tetrahedron Lett.*, 36, 927, 1995.

949. Berkowitz, D.B., Eggen, M., Shen, Q., and Shoemaker, R.K., Ready access to fluorinated phosphonate mimics of secondary phosphates. Synthesis of the (α,α-difluoroalkyl)phosphonate analogues of L-phosphoserine, L-phosphoallothreonine, and L-phosphothreonine, *J. Org. Chem.*, 61, 4666, 1996.

950. Whitten, J.P., Baron, B.M., Muench, D., Miller, F., White, H.S., and McDonald, I.A., (*R*)-4-Oxo-5-phosphononorvaline. A new competitive glutamate antagonist at the NMDA receptor complex, *J. Med. Chem.*, 33, 2961, 1990.

951. Park, S.B., and Standaert, R.F., α,α-Difluorophosphonomethyl azobenzene derivatives as photoregulated phosphoamino acid analogs. Part 1. Design and synthesis, *Tetrahedron Lett.*, 40, 6557, 1999.

952. Pudovik, A.N., and Aladzheva, I.M., Acetylene–allene–acetylenic rearrangements of phosphites with a β,γ-acetylene linkage in the ester radical, *Zh. Obshch. Khim.*, 33, 707, 1963; *J. Gen. Chem. USSR (Engl. Transl.)*, 33, 700, 1963.

953. Pudovik, A.N., and Aladzheva, I.M., Thermal or "pseudo-Claisen" rearrangement of allyl and propargyl esters of phosphorous acid, *Dokl. Akad. Nauk SSSR*, 151, 1110, 1963; *Dokl. Chem. (Engl. Transl.)*, 151, 634, 1963.

954. Pudovik, A.N., Aladzheva, I.M., and Yakovenko, L.M., Synthesis and rearrangement of diethyl propargyl phosphite, *Zh. Obshch. Khim.*, 33, 3444, 1963; *J. Gen. Chem. USSR (Engl. Transl.)*, 33, 3373, 1963.

955. Boisselle, A.P., and Meinhardt, N.A., Acetylene–allene rearrangements. Reactions of trivalent phosphorus chlorides with α-acetylenic alcohols and glycols, *J. Org. Chem.*, 27, 1828, 1962.

956. Mark, V., A facile $S_N i'$ rearrangement. The formation of 1,2-alkadienylphosphonates from 2-alkynyl phosphites, *Tetrahedron Lett.*, 3, 281, 1962.

957. Muller, M., Mann, A., and Taddei, M., A new method for the preparation of (2*R*)-2-amino-5-phosphonopentanoic acid, *Tetrahedron Lett.*, 34, 3289, 1993.

958. Barton, D.H.R., and Embse, R.A.V., The invention of radical reactions. Part 39. The reaction of white phosphorus with carbon-centered radicals. An improved procedure for the synthesis of phosphonic acids and further mechanistic insights, *Tetrahedron*, 54, 12475, 1998.

959. Huber, J.W., III, and Gilmore, W.F., Synthesis and antimicrobial evaluation of *N*-D-alanyl-1-aminoethylphosphonic acid, *J. Med. Chem.*, 18, 106, 1975.

960. Oesterlin, M., Preparation of amines from carboxylic acids by means of hydronitric acid, *Angew. Chem.*, 45, 536, 1932; *Chem. Abstr.*, 26, 5550, 1932.

961. Barycki, J., Mastalerz, P., and Soroka, M., Simple synthesis of 2-aminoethylphosphonic acid and related compounds, *Tetrahedron Lett.*, 11, 3147, 1970.

962. Wasielewski, C., Topolski, M., and Dembkowski, L., Application of hypervalent iodine reagents to the synthesis of aminophosphonic acids, *J. Prakt. Chem.*, 331, 507, 1989.

963. Winsel, H., Gazizova, V., Kulinkovich, O., Pavlov, V., and de Meijere, A., Facile preparation of (phosphorylalkyl)-functionalized cyclopropanols and cyclopropylamines, *Synlett*, 1999, 1999.

964. Pudovik, A.N., and Yastrebova, G.E., Organophosphorus compounds with an active methylene group, *Usp. Khim.*, 39, 1190, 1970; *Russ. Chem. Rev.*, 39, 562, 1970.

965. Boutagy, J., and Thomas, R., Olefin synthesis with organic phosphonate carbanions, *Chem. Rev.*, 74, 87, 1974.

966. Wadsworth, W.S., Jr., Synthetic applications of phosphoryl-stabilised anions, *Org. React.*, 25, 73, 1977.

967. Walker, B.J., *Transformations via Phosphorus-Stabilised Anions. Part 2. PO-Activated Olefination,* Academic Press, New York, 1979, p. 155.

968. Maryanoff, B.E., and Reitz, A.B., The Wittig olefination reaction and modifications involving phosphoryl-stabilized carbanions. Stereochemistry, mechanism, and selected synthetic aspects, *Chem. Rev.*, 89, 863, 1989.

969. Johnson, A.W., *Ylides and Imines of Phosphorus*, John Wiley & Sons, New York, 1993.

970. Baasov, T., and Sheves, M., Model compounds for the study of spectroscopic properties of visual pigments and bacteriorhodopsin, *J. Am. Chem. Soc.*, 107, 7524, 1985.

971. Münstedt, R., and Wannagat, U., Sila-retinol, the first sila-substituted vitamin, *J. Organomet. Chem.*, 322, 11, 1987.

972. Wada, A., Sakai, M., Kinumi, T., Tsujimoto, K., Yamauchi, M., and Ito, M., Conformational study of retinochrome chromophore. Synthesis of 8,18-ethanoretinal and a new retinochrome analog, *J. Org. Chem.*, 59, 6922, 1994.

973. Zhang, H., Lerro, K.A., Takekuma, S.-i., Baek, D.-J., Moquin-Pattey, C., Boehm, M.F., and Nakanishi, K., Orientation of the retinal 9-methyl group in bacteriorhodopsin as studied by photoaffinity labeling, *J. Am. Chem. Soc.*, 116, 6823, 1994.

974. Zhang, L., Nadzan, A.M., Heyman, R.A., Love, D.L., Mais, D.E., Croston, G., Lamph, W.W., and Boehm, M.F., Discovery of novel retinoic acid receptor agonists having potent antiproliferative activity in cervical cancer cells, *J. Med. Chem.*, 39, 2659, 1996.

975. Bennani, Y.L., Marron, K.S., Mais, D.E., Flatten, K., Nadzan, A.M., and Boehm, M.F., Synthesis and characterization of a highly potent and selective isotopically labeled retinoic acid receptor ligand, ALRT1550, *J. Org. Chem.*, 63, 543, 1998.

976. Borhan, B., Kunz, R., Wang, A.Y., Nakanishi, K., Bojkova, N., and Yoshihara, K., Chemoenzymatic synthesis of 11-*cis*-retinal photoaffinity analog by use of squid retinochrome, *J. Am. Chem. Soc.*, 119, 5758, 1997.

977. Groesbeek, M., and Smith, S.O., Synthesis of 19-fluororetinal and 20-fluororetinal, *J. Org. Chem.*, 62, 3638, 1997.

978. Valla, A., Prat, V., Laurent, A., Andriamialisoa, Z., Giraud, M., Labia, R., and Potier, P., Synthesis of 9-methylene analogs of retinol, retinal, retinonitrile and retinoic acid, *Eur. J. Org. Chem.*, 1731, 2001.

979. Gärtner, W., Hopf, H., Hull, W.E., Oesterhelt, D., Scheutzow, D., and Towner, P., On the photoisomerisation of 13-desmethyl-retinal, *Tetrahedron Lett.*, 21, 347, 1980.

980. Corey, E.J., and Hamanaka, E., A new synthetic approach to medium-size carbocyclic systems, *J. Am. Chem. Soc.*, 86, 1641, 1964.

981. Kanno, S., Kato, T., and Kitahara, Y., Biogenetic-type synthesis of (±)-α-chamigrene, *J. Chem. Soc., Chem. Commun.*, 1257, 1967.

982. Bondinell, W.E., Snyder, C.D., and Rapoport, H., Synthesis of chlorobiumquinone, *J. Am. Chem. Soc.*, 91, 6889, 1969.

983. Corey, E.J., Achiwa, K., and Katzenellenbogen, J.A., Total synthesis of *dl*-sirenin, *J. Am. Chem. Soc.*, 91, 4318, 1969.

984. Snyder, C.D., Bondinell, W.E., and Rapoport, H., Synthesis of chlorobiumquinone, *J. Org. Chem.*, 36, 3951, 1971.

985. Andrewes, A.G., and Liaaen-Jensen, S., Animal carotenoids. Part 8. Synthesis of β,γ-carotene and γ,γ-carotene, *Acta Chem. Scand.*, 27, 1401, 1973.

986. Normant, J.F., Cahiez, G., Chuit, C., and Villieras, J., Reactivity of vinylcopper compounds. Application to the stereospecific synthesis of substituted allylic alcohols, *Tetrahedron Lett.*, 14, 2407, 1973.

987. Crombie, L., Kneen, G., and Pattenden, G., Synthesis of casbene, *J. Chem. Soc., Chem. Commun.*, 66, 1976.

988. Sato, K., Inoue, S., Takagi, Y., and Morii, S., A new synthesis of α-santalol, *Bull. Chem. Soc. Jpn.*, 49, 3351, 1976.

989. Schmidlin, T., and Tamm, C., Approaches to the total synthesis of cytochalasans. A convergent synthesis of the octahydroisoindolone moiety related to proxiphomin, *Helv. Chim. Acta*, 61, 2096, 1978.

990. Fraser-Reid, B., and Walker, D.L., Pillaromycin-A. Structural and synthetic studies related to the sugar moiety, pillarose, *Can. J. Chem.*, 58, 2694, 1980.

991. Dauben, W.G., Saugier, R.K., and Fleischhauer, I., Synthetic studies directed toward cembranolides. Synthesis of the basic nucleus of crassin acetate, *J. Org. Chem.*, 50, 3767, 1985.

992. Le Gall, T., Lellouche, J.-P., and Beaucourt, J.-P., An organo-iron mediated chiral synthesis of (+)-(*S*)-[6]-gingerol, *Tetrahedron Lett.*, 30, 6521, 1989.

993. Novak, L., Rohaly, J., Poppe, L., Hornyanszky, G., Kolonits, P., Zelei, I., Feher, I., Fekete, J., Szabo, E., Zahorski, U., Javor, A., and Szantay, C., Synthesis of novel HMG-CoA reductase inhibitors. Part 1. Naphthalene analogs of mevinolin, *Liebigs Ann. Chem.*, 145, 1992.

994. Kuroda, C., and Hirono, Y., Synthesis of spiro[4.5]decane ring system through allylsilane promoted spiroannulation, *Tetrahedron Lett.*, 35, 6895, 1994.

995. Tori, M., Nakashima, K., Seike, M., Asakawa, Y., Wright, A.D., König, G.M., and Sticher, O., Revised structure of a brasilane-type sesquiterpene isolated from the red alga *Laurencia implicata* and its absolute configuration, *Tetrahedron Lett.*, 35, 3105, 1994.

996. Uenishi, J.i., Kawahama, R., and Yonemitsu, O., Total synthesis of (−)-ircinianin and (+)-wistarin, *J. Org. Chem.*, 62, 1691, 1997.

997. Hashimoto, K., Yoshioka, T., Morita, C., Sakai, M., Okuno, T., and Shirahama, H., Synthesis of a photoaffinity-labeling analog of alternariolide (AM-toxin I), a host-specific phytotoxin, *Chem. Lett.*, 203, 1998.

998. Darby, N., Cresp, T.M., and Sondheimer, F., Synthesis of benzannelated *bis*dehydro[14]-, [16]-, [18]-, and -[20]annulenes, *J. Org. Chem.*, 42, 1960, 1977.

999. Finan, J.M., and Kishi, Y., Reductive ring openings of allyl-alcohol epoxides, *Tetrahedron Lett.*, 23, 2719, 1982.

1000. Takano, S., Akiyama, M., and Ogasawara, K., The total synthesis of (±)-latifine, *Chem. Lett.*, 505, 1985.

1001. Nicolaou, K.C., Chakraborty, T.K., Daines, R.A., and Simpkins, N.S., Retrosynthetic and synthetic chemistry on amphotericin B. Synthesis of C(1)–C(20) and C(21)–C(38) fragments and construction of the 38-membered macrocycle, *J. Chem. Soc., Chem. Commun.*, 413, 1986.

1002. Hashimoto, K., Konno, K., Shirahama, H., and Matsumoto, T., Synthesis of acromelic acid B, a toxic principle of *Clitocybe acromelalga*, *Chem. Lett.*, 1399, 1986.

1003. Marshall, J.A., and Trometer, J.D., Stereoselective S_N2' additions to chiral acyclic vinyloxiranes, *Tetrahedron Lett.*, 28, 4985, 1987.

1004. Schreiber, S.L., Wang, Z., and Schulte, G., Group selective reduction of acetals related to the ansa chain of the streptovaricins. Conformational and stereochemical analysis, *Tetrahedron Lett.*, 29, 4085, 1988.

1005. Keck, G.E., and Tafesh, A.M., Free-radical addition-fragmentation reactions in synthesis. A "second generation" synthesis of (+)-pseudomonic acid C, *J. Org. Chem.*, 54, 5845, 1989.

1006. Hatakeyama, S., Numata, H., Osanai, K., and Takano, S., Enantioselective synthesis of a C/D-ring synthon for the preparation of vitamin D_3 metabolites, *J. Chem. Soc., Chem. Commun.*, 1893, 1989.

1007. Nicolaou, K.C., and Ahn, K.H., Synthesis and stereochemical assignment of the C_1–C_{10} fragment of nystatin A_1, *Tetrahedron Lett.*, 30, 1217, 1989.

1008. Thijs, L., Egenberger, D.M., and Zwanenburg, B., An enantioselective total synthesis of the macrolide patulolide C, *Tetrahedron Lett.*, 30, 2153, 1989.

1009. Pak, H., Canalda, I.I., and Fraser-Reid, B., Carbohydrates to carbocycles. A synthesis of (−)-α-pipitzol, *J. Org. Chem.*, 55, 3009, 1990.

1010. Knapp, S., and Kukkola, P.J., Stereocontrolled lincomycin synthesis, *J. Org. Chem.*, 55, 1632, 1990.

1011. Sabol, J.S., and Cregge, R.J., Conformationally restricted leukotriene antagonists. Asymmetric synthesis of a *nor*-leukotriene D_4 analog. Part 2, *Tetrahedron Lett.*, 31, 27, 1990.

1012. Sakai, N., and Ohfune, Y., Efficient synthesis of the revised structure of (−)-galantinic acid, *Tetrahedron Lett.*, 31, 4151, 1990.

1013. Okamura, H., Kuroda, S., Tomita, K., Ikegami, S., Sugimoto, Y.-i., Sakaguchi, S.-i., Katsuki, T., and Yamaguchi, M., Synthesis of aplysiatoxin. Stereoselective synthesis of key fragments, *Tetrahedron Lett.*, 32, 5137, 1991.

1014. Barrett, A.G.M., Doubleday, W.W., Tustin, G.J., White, A.J.P., and Williams, D.J., Approaches to the assembly of the antifungal agent FR-900848. Studies on the asymmetric synthesis of bicyclopropanes and an X-ray crystallographic analysis of (4R,5R)-2-[(1R,3S,4S,6R)-6-phenyl-1-bicyclopropyl]-1,3-dimethyl-4,5-diphenylimidazolidine, *J. Chem. Soc., Chem. Commun.*, 1783, 1994.

1015. Kigoshi, H., Ojika, M., Suenaga, K., Mutou, T., Hirano, J., Sakakura, A., Ogawa, T., Nishiwaki, M., and Yamada, K., Synthetic studies on aplyronine A, a potent antitumor substance of marine origin. Stereocontrolled synthesis of the C21–C34 segment, *Tetrahedron Lett.*, 35, 1247, 1994.

1016. Rao, A.V.R., Gurjar, M.K., Ramana, D.V., and Chheda, A.K., Synthesis of optically active *O,O,O*-trimethylkorupensamines A and B, *Heterocycles*, 43, 1, 1996.

1017. Kiguchi, T., Yuumoto, Y., Ninomiya, I., and Naito, T., Syntheses of proposed structure of pseudodistomin A triacetate and its regioisomers on dienyl side-chain, *Heterocycles*, 42, 509, 1996.

1018. Kuehne, M.E., and Xu, F., Syntheses of strychnan- and aspidospermatan-type alkaloids. Part 10. An enantioselective synthesis of (−)-strychnine through the Wieland–Gumlich aldehyde, *J. Org. Chem.*, 63, 9427, 1998.

1019. Kuroda, C., and Koshio, H., New cyclization reaction of 2-(trimethylsilylmethyl)pentadienal. Synthesis of spiro[4,5]decane ring system, *Chem. Lett.*, 962, 2000.

1020. Suzuki, Y., Nishimaki, R., Ishikawa, M., Murata, T., Takao, K.-i., and Tadano, K.-i., Total synthesis of (−)-mniopetal E, a novel biologically intriguing drimane sesquiterpenoid, *J. Org. Chem.*, 65, 8595, 2000.

1021. Baba, Y., Saha, G., Nakao, S., Iwata, C., Tanaka, T., Ibuka, T., Ohishi, H., and Takemoto, Y., Asymmetric total synthesis of halicholactone, *J. Org. Chem.*, 66, 81, 2001.

1022. Jauch, J., Synthetic studies towards mniopetals. Part 2. A short synthesis of mniopetal E, *Synlett*, 87, 2001.

1023. Mergott, D.J., Frank, S.A., and Roush, W.R., Application of the intramolecular vinylogous Morita–Baylis–Hillman reaction toward the synthesis of the spinosyn A tricyclic nucleus, *Org. Lett.*, 4, 3157, 2002.

1024. Kotsuki, H., Kawamura, A., Ochi, M., and Tokoroyama, T., Intramolecular Diels–Alder reactions of vinylfuran derivatives. A novel approach to benzofurans, *Chem. Lett.*, 917, 1981.

1025. Arai, Y., Yamamoto, M., and Koizumi, T., Enantioselective synthesis of the functionalized bicyclo[2.2.1]heptane derivatives, key intermediates for the chiral synthesis of santalenes and santalols, *Chem. Lett.*, 1225, 1986.

1026. Kende, A.S., Gesson, J.-P., and Demuth, T.P., A regiospecific approach to 6-deoxyanthracyclinones. The structure of γ-citromycinone, *Tetrahedron Lett.*, 22, 1667, 1981.

1027. Nishida, F., Mori, Y., Rokkaku, N., Isobe, S., Furuse, T., Suzuki, M., Meevootisom, V., Flegel, T.W., Thebtaranonth, Y., and Intararuangsorn, S., Structure elucidation of glycosidic antibiotics, glykenins, from *Basidiomycetes* sp. Part 2. Absolute structures of unusual polyhydroxylated C26-fatty acids, aglycones of glykenins, *Chem. Pharm. Bull.*, 38, 2381, 1990.

1028. Paquette, L.A., Stearns, B.A., Mooney, P.A., and Tae, J., Synthesis of *meso*- and *dl*-2,2′-methylene*bis*[tetrahydro-2-furanmethanol]. Potential building blocks for the construction of ionophores housing spirotetrahydrofuranyl motifs, *Heterocycles*, 50, 27, 1999.

1029. Montforts, F.-P., Meier, A., Scheurich, G., Haake, G., and Bats, J.W., Chlorins designed for photodynamic tumor therapy and as model systems for photosynthesis, *Angew. Chem.*, 104, 1650, 1992; *Angew. Chem. Int. Ed. Engl.*, 31, 1592, 1992.

1030. Hirai, H., Ueda, K., and Matsui, M., Studies on chrysanthemic acid. Part 25. Alternative synthesis of chrysanthemic acid via tetrahydrofuran derivatives, *Agr. Biol. Chem.*, 40, 153, 1976.

1031. Morita, T., Okamoto, Y., and Sakurai, H., A convenient dealkylation of dialkyl phosphonates by chlorotrimethylsilane in the presence of sodium iodide, *Tetrahedron Lett.*, 19, 2523, 1978.

1032. Machida, Y., Nomoto, S., and Saito, I., A useful method for the dealkylation of dialkyl phosphonates, *Synth. Commun.*, 9, 97, 1979.

1033. Gross, H., Böck, C., Costisella, B., and Glöde, J., α-Substituted phosphonates. Part 30. Dealkylation of phosphonates with labile functional groups using trimethylsilyl bromide, *J. Prakt. Chem.*, 320, 344, 1978.

1034. Valerio, R.M., Perich, J.W., Alewood, P.F., Tong, G., and Johns, R.B., Synthesis of the simple peptide model Ac-Abu(PO$_3$H$_2$)-NHMe, *Aust. J. Chem.*, 45, 777, 1992.

1035. Carrick, J.M., Kashemirov, B.A., and McKenna, C.E., Indirect photo-induced phosphorylation via a photolabile troika acid *C*-ester. *o*-Nitrobenzyl (*E*)-(hydroxyimino)(dihydroxyphosphinyl)acetate, *Tetrahedron*, 56, 2391, 2000.

1036. Salomon, C.J., and Breuer, E., Efficient and selective dealkylation of phosphonate diisopropyl esters using Me$_3$SiBr, *Tetrahedron Lett.*, 36, 6759, 1995.

1037. Regitz, M., and Martin, R., Study of diazo compounds and azides. Part 49. *tert*-Butylammonium salts of α-diazophosphinic and α-diazophosphonic acids, *Tetrahedron*, 41, 819, 1985.

1038. Gauvry, N., and Mortier, J., Dealkylation of dialkyl phosphonates with boron tribromide, *Synthesis*, 553, 2001.

1039. Mitchell, A.G., Thomson, W., Nicholls, D., Irwin, W.J., and Freeman, S., Bioreversible protection for the phospho group. Bioactivation of the di(4-acyloxybenzyl) and mono(4-acyloxybenzyl) phosphoesters of methylphosphonate and phosphonoacetate, *J. Chem. Soc., Perkin Trans. 1*, 2345, 1992.

1040. Fild, M., and Rieck, H.-P., Reactions of (dichlorophosphinyl)methanesulfonyl chloride, *Chem. Ber.*, 113, 142, 1980.

1041. Auel, T., and Heymer, G., Carboxyalkylphosphonic acids and carboxyalkylphosphinic acids, *Hoechst*, German Patent Appl. DE 2518144, 1976; *Chem. Abstr.*, 86, 72871, 1977.

1042. Bodnarchuk, N.D., Malovik, V.V., Derkach, G.I., and Kirsanov, A.V., Dialkylphosphonoacetic acids and their derivatives, *Zh. Obshch. Khim.*, 41, 1464, 1971; *J. Gen. Chem. USSR (Engl. Transl.)*, 41, 1470, 1971.

1043. Savignac, P., and Lavielle, G., Monodealkylation of phosphoric and phosphonic esters by thiolates and thiophenates, *Bull. Soc. Chim. Fr.*, 1506, 1974.

1044. Krawczyk, H., A convenient route for monodealkylation of diethyl phosphonates, *Synth. Commun.*, 27, 3151, 1997.

1045. Stephenson, G.R., and Milne, K., Evaluation of phosphoramide derivatives of carboxylic acid esters for the alkylation of tricarbonyliron complexes with weak regiodirecting groups, *Aust. J. Chem.*, 47, 1605, 1994.

1046. Ando, K., Practical synthesis of Z-unsaturated esters by using a new Horner–Emmons reagent, ethyl diphenylphosphonoacetate, *Tetrahedron Lett.*, 36, 4105, 1995.

1047. Patois, C., Berté-Verrando, S., and Savignac, P., Easy preparation of alkylphosphonyl dichlorides, *Bull. Soc. Chim. Fr.*, 130, 485, 1993.

1048. Iorga, B., Carmichael, D., and Savignac, P., Phosphonate–phosphonochloridate conversion, *C.R. Acad. Sci., Ser. IIc*, 821, 2000.

1049. Hewitt, D.G., and Teese, M.W., Organophosphorus chemistry. Part 4. Synthesis and reactions of 2-ethoxy-6-oxo-1,2-azaphosphinane 2-oxide, *Aust. J. Chem.*, 37, 1631, 1984.

1050. Malevannaya, R.A., Tsvetkov, E.N., and Kabachnik, M.I., Potassium salts of tetraalkyldiamidophosphinylacetic acids, *Zh. Obshch. Khim.*, 42, 765, 1972; *J. Gen. Chem. USSR (Engl. Transl.)*, 42, 757, 1972.

Appendix

SYNTHESES OF COMPOUNDS USEFUL IN PHOSPHONATE CHEMISTRY DESCRIBED IN *ORGANIC SYNTHESES* AND *INORGANIC SYNTHESES*

Ford-Moore, A.H., and Perry, B.J., Diisopropyl methylphosphonate (Phosphonic acid, methyl-, diisopropyl ester), *Org. Synth.*, 31, 33, 1951; *Org. Synth. Coll. Vol. IV*, 325, 1963.

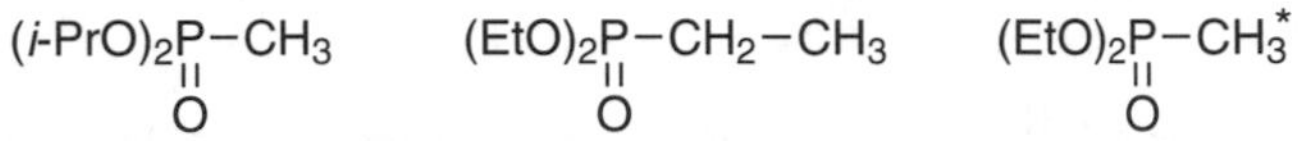

$$(i\text{-PrO})_2\overset{\text{O}}{\underset{\|}{\text{P}}}-CH_3 \qquad (EtO)_2\overset{\text{O}}{\underset{\|}{\text{P}}}-CH_2-CH_3 \qquad (EtO)_2\overset{\text{O}}{\underset{\|}{\text{P}}}-CH_3^*$$

* Mixture with diethyl ethylphosphonate. For a preparation in pure form see:
Teulade, M.-P., Savignac, P., Aboujaoude, E. E. and Collignon, N.
J. Organomet. Chem., 312, 283, 1986.

(A.1)

Ford-Moore, A.H., and Perry, B.J., Triethyl phosphite (Ethyl phosphite), *Org. Synth.*, 31, 111, 1951; *Org. Synth. Coll. Vol. IV*, 955, 1963.

$$(EtO)_3P$$

(A.2)

Malowan, J.E., Diethyl phosphite, *Inorg. Synth.*, 4, 58, 1953.

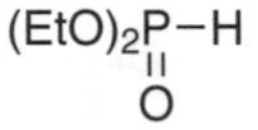

$$(EtO)_2\overset{\text{O}}{\underset{\|}{\text{P}}}-H$$

(A.3)

Malowan, J.E., Dioctyl phosphite, *Inorg. Synth.*, 4, 61, 1953.

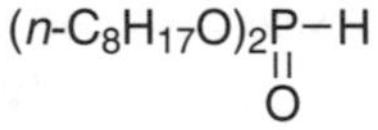

$$(n\text{-}C_8H_{17}O)_2\overset{\text{O}}{\underset{\|}{\text{P}}}-H$$

(A.4)

Malowan, J.E., Diethyl monochlorophosphate, *Inorg. Synth.*, 4, 78, 1953.

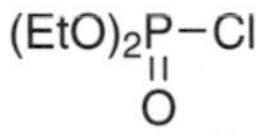

$$(EtO)_2\overset{\text{O}}{\underset{\|}{\text{P}}}-Cl$$

(A.5)

Kennard, K.C., and Hamilton, C.S., Trichloromethylphosphonyl dichloride [Phosphonic dichloride, (trichloromethyl)-], *Org. Synth.*, 37, 82, 1957; *Org. Synth. Coll. Vol. IV*, 950, 1963.

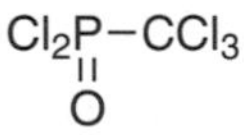

$$Cl_2\overset{\text{O}}{\underset{\|}{\text{P}}}-CCl_3$$

(A.6)

Wadsworth, W.S., Jr., and Emmons, W.D., Ethyl cyclohexylideneacetate (Δ^1-α-Cyclohexaneacetic acid, ethyl ester), *Org. Synth.*, 45, 44, 1965; *Org. Synth. Coll. Vol. V*, 547, 1973.

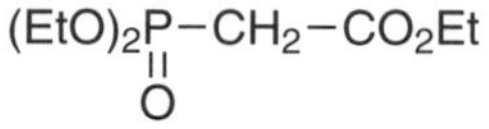

$$(EtO)_2\underset{\underset{O}{\|}}{P}-CH_2-CO_2Et$$

(A.7)

Schmutzler, R., Styrylphosphonic dichloride (Phosphonic dichloride, styryl-), *Org. Synth.*, 45, 99, 1965; *Org. Synth. Coll. Vol. V*, 1005, 1973.

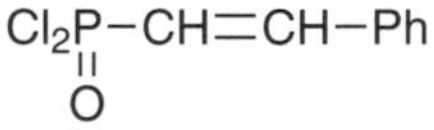

$$Cl_2\underset{\underset{O}{\|}}{P}-CH{=}CH-Ph$$

(A.8)

Jorgenson, M.J., and Thacher, A.F., 1,1-Diphenylcyclopropane (Cyclopropane, 1,1-diphenyl-), *Org. Synth.*, 48, 75, 1968; *Org. Synth. Coll. Vol. V*, 509, 1973.

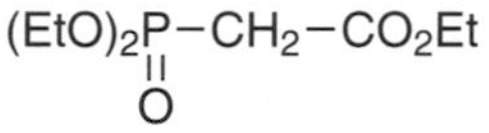

$$(EtO)_2\underset{\underset{O}{\|}}{P}-CH_2-CO_2Et$$

(A.9)

Nagata, W., Wakabayashi, T., and Hayase, Y., Diethyl 2-(cyclohexylamino)vinylphosphonate (Phosphonic acid, [2-(cyclohexylamino)ethenyl]-, diethyl ester), *Org. Synth.*, 53, 44, 1973; *Org. Synth. Coll. Vol. VI*, 448, 1988.

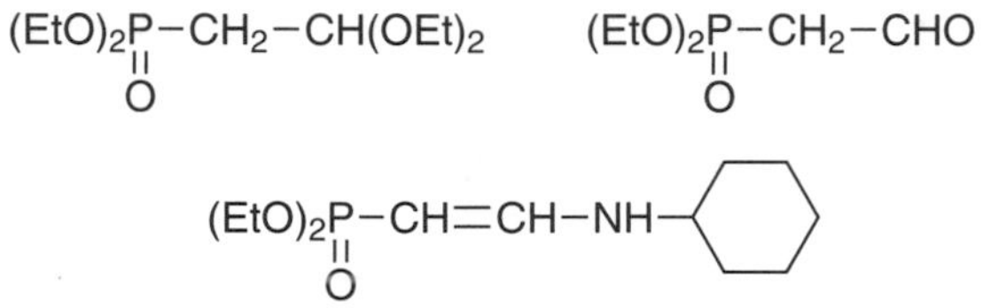

$$(EtO)_2\underset{\underset{O}{\|}}{P}-CH_2-CH(OEt)_2 \qquad (EtO)_2\underset{\underset{O}{\|}}{P}-CH_2-CHO$$

$$(EtO)_2\underset{\underset{O}{\|}}{P}-CH{=}CH-NH{-}\bigcirc$$

(A.10)

Nagata, W., Wakabayashi, T., and Hayase, Y., Preparation of α,β-unsaturated aldehydes via the Wittig reaction. Cyclohexylideneacetaldehyde (Acetaldehyde, cyclohexylidene-), *Org. Synth.*, 53, 104, 1973; *Org. Synth. Coll. Vol. VI*, 358, 1988.

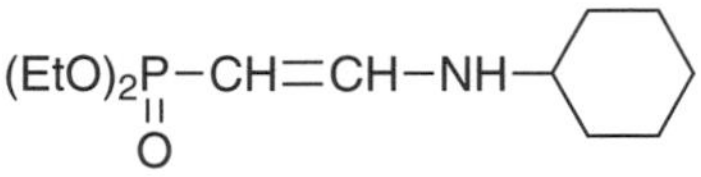

$$(EtO)_2\underset{\underset{O}{\|}}{P}-CH{=}CH-NH{-}\bigcirc$$

(A.11)

Bunnett, J.F., and Weiss, R.H., Radical anion arylation. Diethyl phenylphosphonate (phosphonic acid, phenyl-, diethyl ester), *Org. Synth.*, 58, 134, 1978; *Org. Synth. Coll. Vol. VI*, 451, 1988.

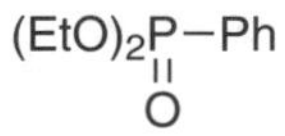

$$(EtO)_2\underset{\underset{O}{\|}}{P}-Ph$$

(A.12)

Kluge, A.F., Diethyl [(2-tetrahydropyranyloxy)methyl]phosphonate (Phosphonic acid, [(tetrahydro-2*H*-pyran-2-yl)oxy]methyl-, diethyl ester), *Org. Synth.*, 64, 80, 1986; *Org. Synth. Coll. Vol. VII*, 160, 1990.

$$(EtO)_2\underset{\underset{O}{\|}}{P}-CH_2-OH \qquad (EtO)_2\underset{\underset{O}{\|}}{P}-CH_2-O{-}\bigcirc$$

(A.13)

Bowen, S.M., and Paine, R.T., Dialkyl [(*N,N*-diethylcarbamoyl)methyl]phosphonates, *Inorg. Synth.*, 24, 101, 1986.

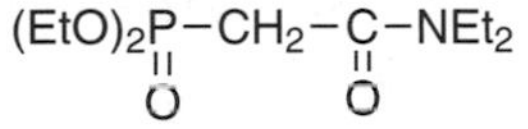

$$(EtO)_2\underset{\underset{O}{\|}}{P}-CH_2-\underset{\underset{O}{\|}}{C}-NEt_2$$

(A.14)

Davidsen, S.K., Phillips, G.W., and Martin, S.F., Geminal acylation–alkylation at a carbonyl center using diethyl *N*-benzylideneaminomethylphosphonate. Preparation of 2-methyl-2-phenyl-4-pentenal (4-Pentenal, 2-methyl-2-phenyl-), *Org. Synth.*, 65, 119, 1987; *Org. Synth. Coll. Vol. VIII*, 451, 1993.

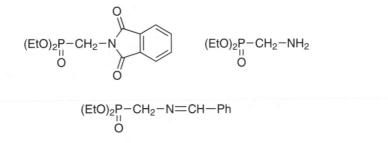

$$(EtO)_2\underset{\underset{O}{\|}}{P}-CH_2-N=CH-Ph$$

(A.15)

Boeckman, R.K., Jr., Perni, R.B., Macdonald, J.E., and Thomas, A.J., 6-Diethylphosphonomethyl-2,2-dimethyl-1,3-dioxen-4-one (Phosphonic acid, [(2,2-dimethyl-4-oxo-4*H*-1,3-dioxin-6-yl)methyl]-, diethyl ester), *Org. Synth.*, 66, 194, 1988; *Org. Synth. Coll. Vol. VIII*, 192, 1993.

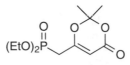

(A.16)

Villieras, J., and Rambaud, M., Ethyl α-(hydroxymethyl)acrylate ([2-Propenoic acid, 2-hydroxymethyl)-, ethyl ester], *Org. Synth.*, 66, 220, 1988; *Org. Synth. Coll. Vol. VIII*, 265, 1993.

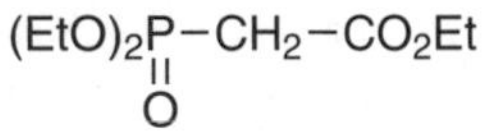

$$(EtO)_2\underset{\underset{O}{\|}}{P}-CH_2-CO_2Et$$

(A.17)

Jacques, J., and Fouquey, C., Enantiomeric (*S*)-(+) and (*R*)-(–)-1,1′-binaphthyl-2,2′-diyl hydrogen phosphate (Dinaphtho[2,1-*d*:1′,2′-*f*][1,3,2]dioxaphosphepin, 4-hydroxy-, 4-oxide), *Org. Synth.*, 67, 1, 1989; *Org. Synth. Coll. Vol. VIII*, 50, 1993.

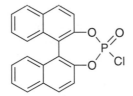

(A.18)

McCarthy, J.R., Matthews, D.P., and Paolini, J.P., Stereoselective synthesis of 2,2-disubstituted 1-fluoro-alkenes. ((*E*)-[[Fluoro(2-phenylcyclohexylidene)-methyl]sulfonyl]benzene and (*Z*)-[2-(fluoromethylene)-cyclohexyl]benzene (Benzene, [[fluoro(2-phenylcyclohexylidene)methyl]sulfonyl]-, (*E*)-(±) and Benzene, [2-(fluoromethylene)cyclohexyl]-, (*Z*)-(±)-), *Org. Synth.*, 72, 216, 1995; *Org. Synth. Coll. Vol. IX*, 442, 1998.

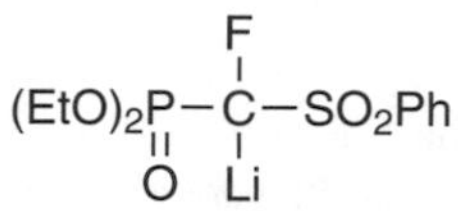

$$(EtO)_2 \overset{\overset{\displaystyle F}{|}}{\underset{\overset{\displaystyle ||}{O}}{P}} - \overset{\overset{\displaystyle F}{|}}{\underset{\overset{\displaystyle |}{Li}}{C}} - SO_2Ph$$

(A.19)

Savignac, P., and Patois, C., Diethyl 1-propyl-2-oxoethylphosphonate [Phosphonic acid (1-formylbutyl)-, diethyl ester], *Org. Synth.*, 72, 241, 1995; *Org. Synth. Coll. Vol. IX*, 239, 1998.

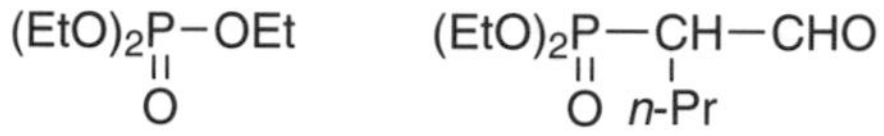

$$(EtO)_2\underset{\overset{||}{O}}{P} - OEt \qquad (EtO)_2\underset{\overset{||}{O}}{P} - \overset{|}{\underset{n\text{-}Pr}{CH}} - CHO$$

(A.20)

Patois, C., Savignac, P., About-Jaudet, E., and Collignon, N., *Bis*(trifluoroethyl) (carboethoxymethyl)phospho-nate (Acetic acid, [*bis*(2,2,2-trifluoroethoxy)phosphinyl]-, ethyl ester), *Org. Synth.*, 73, 152, 1996; *Org. Synth. Coll. Vol. IX*, 88, 1998.

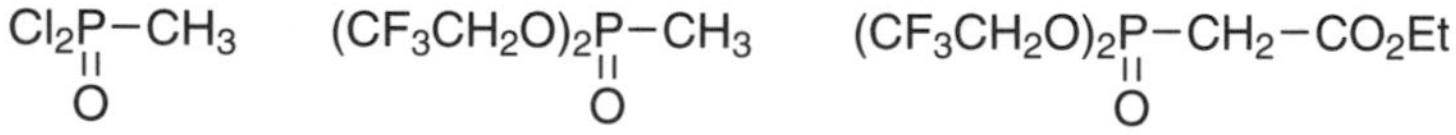

$$Cl_2\underset{\overset{||}{O}}{P} - CH_3 \qquad (CF_3CH_2O)_2\underset{\overset{||}{O}}{P} - CH_3 \qquad (CF_3CH_2O)_2\underset{\overset{||}{O}}{P} - CH_2 - CO_2Et$$

(A.21)

Steuer, B., Wehner, V., Lieberknecht, A., and Jager, V., (−)-2-*O*-Benzyl-L-glyceraldehyde and ethyl (*R,E*)-4-*O*-benzyl-4,5-dihydroxy-2-pentenoate (Propanal, 3-hydroxy-2-(phenylmethoxy)-, (*R*) and 2-Pentenoic acid, 5-hydroxy-4-(phenylmethoxy)-, ethyl ester, [*R*-(*E*)]-), *Org. Synth.*, 74, 1, 1997; *Org. Synth. Coll. Vol. IX*, 39, 1998.

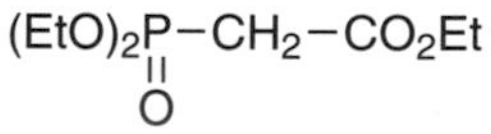

$$(EtO)_2\underset{\overset{||}{O}}{P} - CH_2 - CO_2Et$$

(A.22)

Marinetti, A., and Savignac, P., Diethyl (dichloromethyl)phosphonate. Preparation and use in the synthesis of alkynes. (4-Methoxyphenyl)ethyne (Phosphonic acid, (dichloromethyl)-, diethyl ester to prepare Benzene, 1-ethynyl-4-methoxy-), *Org. Synth.*, 74, 108, 1997; *Org. Synth. Coll. Vol. IX*, 230, 1998.

$$(EtO)_2\underset{\overset{||}{O}}{P} - CCl_3 \qquad (EtO)_2\underset{\overset{||}{O}}{P} - CHCl_2$$

(A.23)

Smith, A.B., III, Yager, K., Phillips, B.W., and Taylor, C.M., Asymmetric synthesis of diethyl (*R*)-(−)-(1-amino-3-methylbutyl)phosphonate (Phosphonic acid, (1-amino-3-methylbutyl)-, diethyl ester, (*R*)-), *Org. Synth.*, 75, 19, 1998.

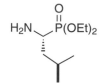

(A.24)

Fengler-Veith, M., Schwardt, O., Kautz, U., Krämer, B., and Jäger, V., (1′R)-(-)-2,4-O-Ethylidene-D-erythrose and ethyl (E)-(−)-4,6-O-ethylidene-(4S,5R,1′R)-4,5,6-trihydroxy-2-hexenoate [1,3-Dioxane-(R)-4-carboxalde-hyde, 5-hydroxy-2-methyl-, [2R-(2α,4α,5β)]] and D-*erythro*-Hex-2-enonic acid, 2,3-dideoxy-4,6-O-ethylidene, ethyl ester [2E, 4(S)]-], *Org. Synth.*, 78, 123, 2002.

$$(EtO)_2\underset{\underset{O}{\|}}{P}-CH_2-CO_2Et$$

(A.25)

Enders, D., von Berg, S., and Jandeleit, B., Diethyl [(phenylsulfonyl)methyl]phosphonate (Phosphonic acid, [(phenylsulfonyl)methyl]-, diethyl ester), *Org. Synth.*, 78, 169, 2002.

$$(EtO)_2\underset{\underset{O}{\|}}{P}-CH_2-SPh \qquad (EtO)_2\underset{\underset{O}{\|}}{P}-CH_2-SO_2Ph$$

(A.26)

Enders, D., von Berg, S., and Jandeleit, B., Synthesis of (−)-(E,S)-3-(benzyloxy)-1-butenyl phenyl sulfone via a Horner–Wadsworth–Emmons reaction of (−)-(S)-2-(benzyloxy)propanal [(Benzene, [[[1-methyl-3-(phenyl-sulfonyl)-2-propenyl]oxy]methyl]-, [S-(E)]-) from (Propanal, 2-(phenylmethoxy)-, (S)-)], *Org. Synth.*, 78, 177, 2002.

$$(EtO)_2\underset{\underset{O}{\|}}{P}-CH_2-SO_2Ph$$

(A.27)

IMPORTANT LITERATURE RESOURCES IN PHOSPHONATE CHEMISTRY

REVIEWS

1957   Freedman, L.D., and Doak, G.O., The preparation and properties of phosphonic acids, *Chem. Rev.*, 57, 479, 1957.

1961   Lichtenthaler, F.W., The chemistry and properties of enol phosphates, *Chem. Rev.*, 61, 607, 1961.

1964   Arbuzov, B.A., Michaelis–Arbuzov and Perkow reactions, *Pure Appl. Chem.*, 9, 307, 1964.

        Cox, J.R., Jr., and Ramsay, O.B., Mechanisms of nucleophilic substitution in phosphate esters, *Chem. Rev.*, 64, 317, 1964.

1970   Pudovik, A.N., and Yastrebova, G.E., Organophosphorus compounds with an active methylene group, *Usp. Khim.*, 39, 1190, 1970; *Russ. Chem. Rev.*, 39, 562, 1970.

1971   Redmore, D., Heterocyclic systems bearing phosphorus substituents. Synthesis and chemistry, *Chem. Rev.*, 71, 315, 1971.

1973   Razumov, A.I., Liorber, B.G., Moskva, V.V., and Sokolov, M.P., Phosphorylated aldehydes, *Usp. Khim.*, 42, 1199, 1973; *Russ. Chem. Rev.*, 42, 538, 1973.

1974   Boutagy, J., and Thomas, R., Olefin Synthesis with organic phosphonate carbanions, *Chem. Rev.*, 74, 87, 1974.

1977   Engel, R., Phosphonates as analogues of natural phosphates, *Chem. Rev.*, 77, 349, 1977.

1980   Zhdanov, Y.A., Uzlova, L.A., and Glebova, Z.I., α-Ketophosphonic acid esters. Synthesis, structure, and reactions, *Usp. Khim.*, 49, 1730, 1980; *Russ. Chem. Rev.*, 49, 843, 1980.

1981   Bhattacharya, A.K., and Thyagarajan, G., The Michaelis–Arbuzov rearrangement, *Chem. Rev.*, 81, 415, 1981.

1987   Kukhar, V.P., and Solodenko, V.A., The phosphorus analogues of aminocarboxylic acids, *Usp. Khim.*, 56, 1504, 1987; *Russ. Chem. Rev.*, 56, 859, 1987.

1988   Kadyrov, A.A., and Rokhlin, E.M., Fluoroalkenylphosphonates, *Usp. Khim.*, 57, 1488, 1988; *Russ. Chem. Rev.*, 57, 852, 1988.

        Töke, L., Petnehazy, I., Keglevich, G., and Szakal, G., Reaction of trialkyl phosphites and α-halo-ketones in aprotic media (Perkow–Arbuzov reaction) and in protic solvents, *Period. Polytech., Chem. Eng.*, 32, 101, 1988.

1989   Maryanoff, B.E., and Reitz, A.B., The Wittig olefination reaction and modifications involving phosphoryl-stabilized carbanions. Stereochemistry, mechanism, and selected synthetic aspects, *Chem. Rev.*, 89, 863, 1989.

        Wozniak, L., and Chojnowski, J., Silyl esters of phophorus — common intermediates in synthesis, *Tetrahedron*, 45, 2465, 1989.

1990   Gubnitskaya, E.S., Peresypkina, L.P., and Samarai, L.I., β-Aminophosphonates and β-aminophosphinates, synthesis, and properties, *Usp. Khim.*, 59, 1386, 1990; *Russ. Chem. Rev.*, 59, 807, 1990.

1991   Mastryukova, T.A., and Kabachnik, M.I., Enolisation of the phosphoryl group, *Usp. Khim.*, 60, 2167, 1991; *Russ. Chem. Rev.*, 60, 1115, 1991.

1992   Burton, D.J., and Yang, Z.-Y., Fluorinated organometallics. Perfluoroalkyl and functionalized perfluoroalkyl organometallic reagents in organic synthesis, *Tetrahedron*, 48, 189, 1992.

        Khaskin, B.A., Molodova, O.D., and Torgasheva, N.A., Reactions of phosphorus-containing compounds with diazo-compounds and carbenes, *Usp. Khim.*, 61, 564, 1992; *Russ. Chem. Rev.*, 61, 306, 1992.

        Minami, T., and Motoyoshiya, J., Vinylphosphonates in organic synthesis, *Synthesis*, 333, 1992.

1993   Furin, G.G., Phosphorus-containing nucleophiles in reactions with polyfluorinated organic compounds, *Usp. Khim.*, 62, 267, 1993; *Russ. Chem. Rev.*, 62, 243, 1993.

        Kukhar, V.P., Svistunova, N.Y., Solodenko, V.A., and Soloshonok, V.A., Asymmetric synthesis of fluorine and phosphorus-containing analogues of aminoacids, *Usp. Khim.*, 62, 284, 1993; *Russ. Chem. Rev.*, 62, 261, 1993.

        Zolotukhina, M.M., Krutikov, V.I., and Lavrent'ev, A.N., Derivatives of phosphonic acids. Synthesis and biological activity, *Usp. Khim.*, 62, 691, 1993; *Russ. Chem. Rev.*, 62, 647, 1993.

1996   Breuer, E., Acylphosphonates and their derivatives, in *The Chemistry of Organophosphorus Compounds,* Vol. 4, Hartley, F.R., Ed., John Wiley & Sons, New York, 1996, p. 653.

        Burton, D.J., Yang, Z.-Y., and Qiu, W., Fluorinated ylides and related compounds, *Chem. Rev.*, 96, 1641, 1996.

1997   Alabugin, I.V., and Brel, V.K., Phosphorylated allenes. Structure and interaction with electrophiles, *Usp. Khim.*, 66, 225, 1997; *Russ. Chem. Rev.*, 66, 205, 1997.

Mitchell, M.C., and Kee, T.P., Recent developments in phosphono-transfer chemistry, *Coord. Chem. Rev.*, 158, 359, 1997.

Waschbüsch, R., Carran, J., Marinetti, A., and Savignac, P., The synthesis of dialkyl α-halogenated methylphosphonates, *Synthesis*, 727, 1997.

Waschbüsch, R., Carran, J., Marinetti, A., and Savignac, P., Synthetic applications of dialkyl (chloromethyl)phosphonates and N,N,N',N'-tetraalkyl(chloromethyl)phosphonic diamides, *Chem. Rev.*, 97, 3401, 1997.

Wiemer, D.F., Synthesis of nonracemic phosphonates, *Tetrahedron*, 53, 16609, 1997.

Yuan, C., Li, S., Li, C., Chen, S., Huang, W., Wang, G., Pan, C., and Zhang, Y., New strategy for the synthesis of functionalized phosphonic acids, *Heteroat. Chem.*, 8, 103, 1997.

1998   Cherkasov, R.A., and Galkin, V.I., The Kabachnik–Fields reaction. Synthetic potential and the problem of the mechanism, *Usp. Khim.*, 67, 940, 1998; *Russ. Chem. Rev.*, 67, 857, 1998.

Iorga, B., Eymery, F., Mouriès, V., and Savignac, P., Phosphorylated aldehydes. Preparations and synthetic uses, *Tetrahedron*, 54, 14637, 1998.

Kolodiazhnyi, O.I., Asymmetric synthesis of organophosphorus compounds, *Tetrahedron:Asymmetry*, 9, 1279, 1998.

1999   Eymery, F., Iorga, B., and Savignac, P., Synthesis of phosphonates by nucleophilic substitution at phosphorus. The SNP(V) reaction, *Tetrahedron*, 55, 13109, 1999.

Fields, S.C., Synthesis of natural products containing a C–P bond, *Tetrahedron*, 55, 12237, 1999.

Iorga, B., Eymery, F., and Savignac, P., The syntheses and properties of 1,2-epoxyphosphonates, *Synthesis*, 207, 1999.

Seto, H., and Kuzuyama, T., Bioactive natural products with carbon-phosphorus bonds and their biosynthesis, *Nat. Prod. Rep.*, 589, 1999.

2000   Eymery, F., Iorga, B., and Savignac, P., The usefulness of phosphorus compounds in alkyne synthesis, *Synthesis*, 185, 2000.

Iorga, B., Carmichael, D., and Savignac, P., Phosphonate–phosphonochloridate conversion, *C. R. Acad. Sci., Ser. IIc*, 821, 2000.

Iorga, B., Eymery, F., Carmichael, D., and Savignac, P., Dialkyl 1-alkynylphosphonates, a range of promising reagents, *Eur. J. Org. Chem.*, 3103, 2000.

2001   Iorga, B., and Savignac, P., Controlled reactivity of phosphonates by temporary silicon connection, *Synlett*, 447, 2001.

Minami, T., Okauchi, T., and Kouno, R., α-Phosphonovinyl carbanions in organic synthesis, *Synthesis*, 349, 2001.

2002   van Staden, L.F., Gravestock, D., and Ager, D.J., New developments in the Peterson olefination reaction, *Chem. Soc. Rev.*, 31, 195, 2002.

BOOKS

1950   Kosolapoff, G.M., *Organophosphorus Compounds*, John Wiley & Sons, New York, 1950.

1963   Sasse, K., Phosphonsäuren und deren Derivate, in *Methoden der organischen Chemie* (Houben–Weyl), Band XII, Part 1, *Organische Phosphorverbindungen*, Georg Thieme Verlag, Stuttgart, 1963, pp. 338–619.

1964   Berlin, K.D., Austin, T.H., Peterson, M., and Nagabhushanam, M., Nucleophilic displacement reactions on phosphorus halides and esters by Grignard and lithium reagents, in *Topics in Phosphorus Chemistry*, Vol. 1, John Wiley & Sons, New York, 1964, pp. 17–56.

Harvey, R.G., and De Sombre, E.R., The Michaelis–Arbuzov and related reactions, in *Topics in Phosphorus Chemistry*, Vol. 1, John Wiley & Sons, New York, 1964, pp. 57–112.

1965   Hudson, R.F., *Structure and Mechanism in Organo-Phosphorus Chemistry*, Academic Press, New York, 1965.

Miller, B., Reactions between trivalent phosphorus derivatives and positive halogen sources, in *Topics in Phosphorus Chemistry*, Vol. 2, John Wiley & Sons, New York, 1965, pp. 133–200.

1967   Kirby, A.J., and Warren, S.G., *The Organic Chemistry of Phosphorus*, Elsevier, Amsterdam, 1967.

Quin, L.D., The natural occurrence of compounds with the carbon–phosphorus bond, in *Topics in Phosphorus Chemistry,* Vol. 4, John Wiley & Sons, New York, 1967, pp. 23–48.

Mark, V., Dungan, C.H., Crutchfield, M.M., and Van Wazer, J.R., Compilation of P^{31} NMR data, in *Topics in Phosphorus Chemistry,* Vol. 5, John Wiley & Sons, New York, 1967, pp. 227–458.

1969   Berlin, K.D., and Hellwege, D.M., Carbon–phosphorus heterocycles, in *Topics in Phosphorus Chemistry,* Vol. 6, John Wiley & Sons, New York, 1969, pp. 1–186.

Blackburn, G.M., and Cohen, J.S., Chemical oxidative phosphorylation, in *Topics in Phosphorus Chemistry,* Vol. 6, John Wiley & Sons, New York, 1969, pp. 187–234.

Corbridge, D.E.C., The infrared spectra of phosphorus compounds, in *Topics in Phosphorus Chemistry,* Vol. 6, John Wiley & Sons, New York, 1969, pp. 235–366

1972   Shaw, M.A., and Ward, R.S., Addition reactions of tertiary phosphorus compounds with electrophilic olefins and acetylenes, in *Topics in Phosphorus Chemistry,* Vol. 7, John Wiley & Sons, New York, 1972, pp. 1–36.

Curry, J.D., Nicholson, D.A., and Quimby, O.T., Oligophosphonates, in *Topics in Phosphorus Chemistry,* Vol. 7, John Wiley & Sons, New York, 1972, pp. 37–102.

1976   Worms, K.H., and Schmidt-Dunker, M., Phosphonic acids and derivatives, in *Organic Phosphorus Compounds,* Vol. 7, Kosolapoff, G.M., and Maier, L., Eds., John Wiley & Sons, New York, 1976.

Redmore, D., The chemistry of P–C–N systems, in *Topics in Phosphorus Chemistry,* Vol. 8, John Wiley & Sons, New York, 1976, pp. 515–586.

1979   Cadogan, J.I.G., *Organophosphorus Reagents in Organic Synthesis*, Academic Press, New York, 1979.

Edmundson, R.S., Phosphoric acid derivatives, in *Comprehensive Organic Chemistry. The Synthesis and Reactions of Organic Compounds, Vol. 2. Nitrogen Compounds, Carboxylic Acids, Phosphorus Compounds,* Sutherland, I.O., Ed., Pergamon Press, London, 1979, pp. 1257–1300.

1982   Gallenkamp, B., Hofer, W., Krüger, B.W., Maurer, F., and Pfister, T., Phosphonsäuren und ihre Derivate, in *Methoden der organischen Chemie* (Houben–Weyl), Band E(2), *Organische Phosphorverbindungen,* Regitz, M., Ed., Georg Thieme Verlag, Stuttgart, 1982, pp. 300–476.

1983   Hildebrand, R.L., *The Role of Phosphonates in Living Systems*, CRC Press, Boca Raton, 1983.

Hildebrand, R.L., Curley-Joseph, J., Lubansky, H.J., and Henderson, T.O., Biology of alkylphosphonic acids. A review of the distribution, metabolism, and structure of naturally occurring alkylphosphonic acids, in *Topics in Phosphorus Chemistry,* Vol. 11, John Wiley & Sons, New York, 1983, pp. 297–338.

Hudson, H.R., Quasi-phosphonium intermediates and compounds, in *Topics in Phosphorus Chemistry,* Vol. 11, John Wiley & Sons, New York, 1983, pp. 339–436.

1984   Hori, T., Horiguchi, M., and Hayashi, A., *Biochemistry of Natural C–P Compounds*, Japanese Association for Research on the Biochemistry of C–P Compounds, Tokyo, 1984.

1987   Verkade, J.G., and Quin, L.D., *Phosphorus-31 NMR Spectroscopy in Stereochemical Analysis*, VCH, Deerfield Beach, 1987.

1988   Engel, R., *Synthesis of Carbon–Phosphorus Bonds*, CRC Press, Boca Raton, 1988.

Edmundson, R.S., *Dictionary of Organophosphorus Compounds*, Chapman & Hall, London, 1988.

1989   Pudovik, A.N., *Chemistry of Organophosphorus Compounds*, MIR, Moscow, 1989.

1991   Tebby, J.C., *Handbook of Phosphorus-31 Nuclear Magnetic Resonance Data*, CRC Press, Boca Raton, 1991.

1992   Engel, R., *Handbook of Organophosphorus Chemistry*, Marcel Dekker, New York, 1992.

1993   Johnson, A.W., *Ylides and Imines of Phosphorus*, John Wiley & Sons, New York, 1993.

1994   Quin, L.D., and Verkade, J.G., *Phosphorus-31 NMR Spectral Properties in Compound Characterization and Structural Analysis*, VCH, New York, 1994.

1996   Hartley, F.R., *The Chemistry of Organophosphorus Compounds, Vol. 4. Ter- and Quinque-Valent Phosphorus Acids and Their Derivatives*, John Wiley & Sons, New York, 1996.

2000   Kukhar, V.P., and Hudson, H.R., *Aminophosphonic and Aminophosphinic Acids Chemistry and Biological Activity,* John Wiley & Sons, New York, 2000.

Quin, L.D., *A Guide to Organophosphorus Chemistry,* John Wiley & Sons, New York, 2000.

2002   McKenna, C.E., and Kashemirov, B.A., Recent progress in carbonylphosphonate chemistry, in *Topics in Current Chemistry,* Vol. 220. New Aspects in Phosphorus Chemistry, Vol. 1, Majoral, J.-P., Ed., Springer-Verlag, Berlin, 2002, pp. 201–238.

pK$_a$ VALUES OF DIETHYL PHOSPHONATES

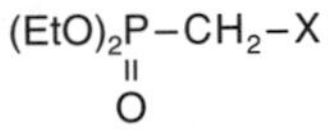

$$(EtO)_2P-CH_2-X$$
$$\underset{O}{\overset{\|}{}}$$

X	pK$_a$ calc.[1,3-9]	pK$_a$ exp.[2]
H	36.1	–
Me	33.6	–
SiMe$_3$	–	28.8
Ph	27.8	27.6
CF$_3$	26.9	–
Cl	–	26.2
(EtO)$_2$P(O)	23.0	–
C(O)NEt$_2$	22.7	–
C(O)OEt	19.3	18.6
Me$_3$N$^+$	17.9	–
CN	17.6	16.4
MeC(O)	15.8	–
PhC(O)	13.5	–
Bu$_3$P$^+$	12.1	–
Ph$_3$P$^+$	7.9	–
NO$_2$	7.2	–

1. Kabachnik, M.I., Analysis of CH acidity of organophosphorus compounds, *Zh. Obshch. Khim.*, 64, 1101, 1994; *Russ. J. Gen. Chem. (Engl. Transl.)*, 64, 993, 1994.
2. Bordwell, F.G., personal communication (http://www.chem.wisc.edu/areas/reich/pkatable/).
3. Mastryukova, T.A., Melent'eva, T.A., Lur'e, E.P., and Kabachnik, M.I., Conjugation in systems with a tetrahedral phosphorus atom. Substituted benzoyltriphenylphosphinemethylenes, *Dokl. Akad. Nauk SSSR*, 172, 611, 1966; *Dokl. Chem. (Engl. Transl.)*, 79, 1966.
4. Mesyats, S.P., Tsvetkov, E.N., Petrov, E.S., Shelganova, N.N., Shcherbina, T.M., Shatenshtein, A.I., and Kabachnik, M.I., Equilibrial CH acidity of phosphorus-containing organic compounds. Part 2. Substituted benzylphosphine oxides and several of their analogs, *Izv. Akad. Nauk SSSR, Ser. Khim.*, 2497, 1974; *Bull. Acad. Sci. USSR, Div. Chem. Sci. (Engl. Transl.)*, 2406, 1974.
5. Mesyats, S.P., Tsvetkov, E.N., Petrov, E.S., Terekhova, M.I., Shatenshtein, A.I., and Kabachnik, M.I., Equilibrial CH acidity of phosphorus-containing organic compounds. Part 1. Effect of solvent on the CH acidity of phosphine oxides and sulfides, *Izv. Akad. Nauk SSSR, Ser. Khim.*, 2489, 1974; *Bull. Acad. Sci. USSR, Div. Chem. Sci. (Engl. Transl.)*, 2399, 1974.
6. Petrov, E.S., Tsvetkov, E.N., Mesyats, S.P., and Shatenshtein, A.I., Equilibrium CH-acidity of organophosphorus compounds. Part 4. Acidifying effects of diphenylphosphinyl, carbethoxy, nitrile, and phenyl groups, *Izv. Akad. Nauk SSSR, Ser. Khim.*, 782, 1976; *Bull. Acad. Sci. USSR, Div. Chem. Sci. (Engl. Transl.)*, 762, 1976.
7. Petrov, E.S., Tsvetkov, E.N., Terekhova, M.I., Malevannaya, R.A., Shatenshtein, A.I., and Kabachnik, M.I., Equilibrium CH-acidity of organophosphorus compounds. Part 3. Esters of phosphinylacetic acids, *Izv. Akad. Nauk SSSR, Ser. Khim.*, 534, 1976; *Bull. Acad. Sci. USSR, Div. Chem. Sci. (Engl. Transl.)*, 517, 1976.
8. Kabachnik, M.I., and Mastryukova, T.A., Correlational analysis of CH acidity, *Dokl. Akad. Nauk SSSR*, 260, 893, 1981; *Chem. Abstr.*, 95, 219648, 1981.
9. Kabachnik, M.I., and Mastryukova, T.A., σρ Analysis of CH acidities of organophosphorus compounds, *Zh. Obshch. Khim.*, 54, 2161, 1984; *Russ. J. Gen. Chem. (Engl. Transl.)*, 54, 1931, 1984.

Index